Regulation of Bacterial Virulence

Regulation of Bacterial Virulence

EDITED BY

Michael L. Vasil
University of Colorado School of Medicine
Aurora, Colorado

Andrew J. Darwin
New York University School of Medicine
New York, New York

ASM PRESS WASHINGTON, DC

Library of Congress Cataloging-in-Publication Data

Regulation of bacterial virulence / edited by Michael L. Vasil and Andrew J. Darwin.
    p. ; cm.
  Includes bibliographical references and index.
  ISBN 978-1-55581-676-6 (hardcover : alk. paper) — ISBN 1-55581-676-2 (hardcover : alk. paper)
  I. Vasil, Michael Lawrence, 1945– II. Darwin, Andrew J.
  [DNLM: 1. Bacteria—pathogenicity. 2. Bacteria—genetics. 3. Bacterial Toxins. 4. Bacteriophages—physiology.
5. Virulence—physiology. 6. Virulence Factors—physiology. QW 50]
  572.8′293—dc23
                                 2012016281
  eISBN: 978-1-55581-852-4

10 9 8 7 6 5 4 3 2 1

All Rights Reserved
Printed in the United States of America

Address editorial correspondence to ASM Press, 1752 N St., N.W., Washington, DC 20036-2904, USA

Send orders to ASM Press, P.O. Box 605, Herndon, VA 20172, USA
Phone: 800-546-2416; 703-661-1593
Fax: 703-661-1501
E-mail: books@asmusa.org
Online: http://estore.asm.org

Cover image: Perle Fine, *Roaring Wind*, 1958. Oil collage on canvas with aluminum foil, 42 × 52 inches. Collection of Drs. Thomas and Marika Herskovic. From the book *American Abstract and Figurative Expressionism* (New York School Press, Franklin Lakes, NJ, 2009).
Cover design: Ed Atkeson, Berg Design.

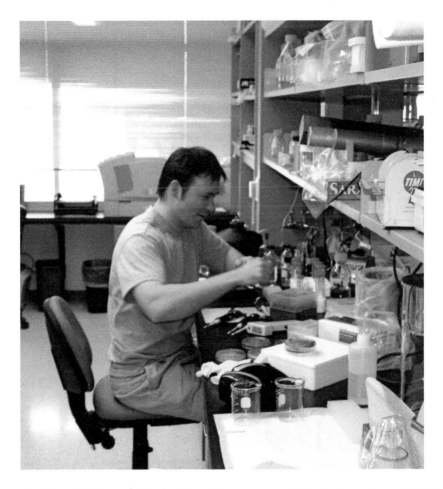

*Michael Vasil dedicates this book to the memory of Martin Stonehouse, Ph.D.,
who relished science and loved life to the fullest.
He left his loving wife, Carly, his sons, Ronan and Morgan, his family,
and all of us much too soon, 29 October 2011.*

*Andrew Darwin dedicates this book to his parents, Frank and Pauline.
They have never pushed but always supported.*

CONTENTS

CONTRIBUTORS

Meredith A. Benson
Department of Microbiology
New York University School of Medicine
New York, NY 10016

Sandra Billig
Department of Microbiology
University of Osnabrück
D-49076 Osnabrück, Germany

Alex Böhm
Institute for Molecular Infection Biology
University of Würzburg
Würzburg, Germany

Evan Bradley
Department of Molecular Biology & Microbiology
Tufts University School of Medicine
136 Harrison Avenue
Boston, MA 02111

Daniel J. Bretl
Department of Microbiology and Molecular Genetics
Center for Infectious Disease Research
Medical College of Wisconsin
Milwaukee, WI 53226

Andrew Camilli
Department of Molecular Biology & Microbiology
Tufts University School of Medicine
136 Harrison Avenue
Boston, MA 02111

Paul E. Carlson, Jr.
Department of Microbiology and Immunology
University of Michigan Medical School, Box 0620
Ann Arbor, MI 48108

Glen P. Carter
Department of Microbiology
Monash University
Clayton, Victoria 3800, Australia

Josephine R. Chandler
Department of Microbiology
University of Washington School of Medicine
1705 NE Pacific Street
Seattle, WA 98195

Yanjie Chao
Institute for Molecular Infection Biology
University of Würzburg
Würzburg, Germany

Jackie K. Cheung
Department of Microbiology
Monash University
Clayton, Victoria 3800, Australia

Peter J. Christie
Department of Microbiology and Molecular Genetics
University of Texas Medical School at Houston
Houston, TX 77030

Colin P. Corcoran
Institute for Molecular Infection Biology
University of Würzburg
Würzburg, Germany

Shipan Dai
Center for Microbial Interface Biology
Department of Microbial Infection and Immunity
The Ohio State University
Columbus, OH 43210

Jennifer L. Dale
Department of Microbiology and
Molecular Genetics
University of Texas—Houston Medical School
Houston, TX 77030

Shandee D. Dixon
Department of Microbiology and Immunology
University of Michigan Medical School, Box 0620
Ann Arbor, MI 48108

Robert K. Ernst
Department of Microbial Pathogenesis
University of Maryland, Baltimore
Baltimore, MD 21201

Allison J. Farrand
Department of Pathology
Microbiology and Immunology
Vanderbilt University Medical Center
Nashville, TN 37232

Alfonso Felipe-López
Department of Microbiology
University of Osnabrück
D-49076 Osnabrück, Germany

Nancy E. Freitag
Department of Microbiology and Immunology
University of Illinois at Chicago College
of Medicine
Chicago, IL 60612

Kathrin S. Fröhlich
Institute for Molecular Infection Biology
University of Würzburg
Würzburg, Germany

Joanna B. Goldberg
Department of Microbiology, Immunology,
and Cancer Biology
University of Virginia
Charlottesville, VA 22908

E. Peter Greenberg
Department of Microbiology
University of Washington, School of Medicine
1705 NE Pacific Street
Seattle, WA 98195-7242

Charley Gruber
Department of Microbiology
UT Southwestern Medical Center
Dallas, TX 75390

John S. Gunn
Center for Microbial Interface Biology
Department of Microbial Infection and Immunity
The Ohio State University
Columbus, OH 43210

Sanjay K. Gupta
Institute for Molecular Infection Biology
University of Würzburg
Würzburg, Germany

Philip C. Hanna
Department of Microbiology and Immunology
University of Michigan Medical School, Box 0620
Ann Arbor, MI 48108

Nadja Heidrich
Institute for Molecular Infection Biology
University of Würzburg
Würzburg, Germany

Calvin A. Henard
Department of Microbiology
University of Colorado Denver
School of Medicine
Aurora, CO 80045

Michael Hensel
Department of Microbiology
University of Osnabrück
D-49076 Osnabrück, Germany

Thomas J. Hiscox
Department of Microbiology
Monash University
Clayton, Victoria 3800, Australia

Lauren E. Hittle
Department of Microbial Pathogenesis
University of Maryland, Baltimore
Baltimore, MD 21201

Ansel Hsiao
Center for Genome Sciences & Systems Biology
Washington University
School of Medicine
St. Louis, MO 63110

Scott J. Hultgren
Department of Molecular Microbiology
Washington University School of Medicine
St. Louis, MO 63110

Michael P. Jennings
The Institute for Glycomics
Griffith University
Gold Coast Campus
Parklands Drive
Southport, QLD 4222, Australia

Barbara I. Kazmierczak
Department of Medicine
Yale University School of Medicine
333 Cedar Street, Box 208022
New Haven, CT 06520-8022

Erica N. Kintz
Department of Microbiology, Immunology,
and Cancer Biology
University of Virginia
Charlottesville, VA 22908

Theresa M. Koehler
Department of Microbiology and Molecular
Genetics
University of Texas—Houston Medical School
Houston, TX 77030

Jenny A. Laverde-Gomez
Department of Microbiology and Molecular
Genetics
University of Texas Medical School at Houston
Houston, TX 77030

Audrey Le Gouellec
TheREx, TIMC-IMAG Laboratory
UMR 5525 CNRS
Université Joseph Fourier
Grenoble, France

Lee-Yean Low
Department of Microbiology
Monash University
Clayton, Victoria 3800, Australia

Dena Lyras
Department of Microbiology
Monash University
Clayton, Victoria 3800, Australia

Kate E. Mackin
Department of Microbiology
Monash University
Clayton, Victoria 3800, Australia

Charlotte D. Majerczyk
Department of Microbiology
University of Washington School of Medicine
1705 NE Pacific Street
Seattle, WA 98195

EmilyKate McDonough
Department of Molecular Biology & Microbiology
Tufts University School of Medicine
136 Harrison Avenue
Boston, MA 02111

Kathleen A. McDonough
Wadsworth Center
New York State Department of Health
Albany, NY 12201-2002

Masatoshi Miyakoshi
Institute for Molecular Infection Biology
University of Würzburg
Würzburg, Germany

Nrusingh P. Mohapatra
Center for Microbial Interface Biology
Department of Microbial Infection and Immunity
The Ohio State University
Columbus, OH 43210

Thomas S. Murray
Department of Basic Medical Sciences
Quinnipiac University School of Medicine
275 Mt. Carmel Avenue, N1-HSC
Hamden, CT 06518-1908

Abiodun D. Ogunniyi
Research Centre for Infectious Diseases
School of Molecular and Biomedical Science
University of Adelaide
Adelaide, SA 5005, Australia

Yuta Okkotsu
Department of Microbiology
University of Colorado School of Medicine
Aurora, CO 80045

Gregory C. Palmer
Institute for Cellular and Molecular Biology
The University of Texas at Austin
Austin, TX 78712

Kai Papenfort
Department of Molecular Biology
Princeton University
Princeton, NJ 08544

Matthew R. Parsek
Department of Microbiology
University of Washington
Seattle, WA 98195

James C. Paton
Research Centre for Infectious Diseases
School of Molecular and Biomedical Science,
University of Adelaide
Adelaide, SA 5005, Australia

Ian R. Peak
The Institute for Glycomics
Griffith University
Gold Coast Campus
Parklands Drive
Southport, QLD 4222, Australia

Robert D. Perry
Department of Microbiology, Immunology, and
Molecular Genetics
University of Kentucky
Lexington, KY 40536-0298

Benoit Polack
TheREx, TIMC-IMAG Laboratory
UMR 5525 CNRS
Université Joseph Fourier
Grenoble, France

Daniel A. Powell
Department of Microbial Pathogenesis
University of Maryland, Baltimore
Baltimore, MD 21201

Christopher L. Pritchett
East Tennessee State University
Department of Health Sciences
Johnson City, TN 37614

Julian I. Rood
Department of Microbiology
Monash University
Clayton, Victoria 3800, Australia

Mayukh Sarkar
Department of Microbiology and
Molecular Genetics
University of Texas Medical School at Houston
Houston, TX 77030

Michael J. Schurr
Department of Microbiology
University of Colorado School of Medicine
Aurora, CO 80045

Drew J. Schwartz
Department of Molecular Microbiology
Washington University School of Medicine
St. Louis, MO 63110

Cynthia M. Sharma
Research Centre of Infectious Diseases
University of Würzburg
Würzburg, Germany

Dakang Shen
School of Cellular and Molecular Medicine
University of Bristol
University Walk
Bristol BS8 1TD, United Kingdom

Eric P. Skaar
Department of Pathology, Microbiology and
Immunology
Vanderbilt University Medical Center
Nashville, TN 37232

Karen Skorupski
Department of Microbiology and Immunology
Dartmouth Medical School
Hanover, NH 03755

Vanessa Sperandio
Department of Microbiology
UT Southwestern Medical Center
Dallas, TX 75390

Yogitha N. Srikhanta
Department of Microbiology and Immunology
The University of Melbourne
Royal Parade, Parkville
Melbourne, VIC 3010, Australia

Andrew M. Stern
Department of Microbiology
Perelman School of Medicine
University of Pennsylvania
Philadelphia, PA 19104

Ronald K. Taylor
Department of Microbiology and Immunology
Dartmouth Medical School
Hanover, NH 03755

Victor J. Torres
Department of Microbiology
New York University School of Medicine
New York, NY 10016

Bertrand Toussaint
TheREx, TIMC-IMAG Laboratory
UMR 5525 CNRS
Université Joseph Fourier
Grenoble, France

Boo Shan Tseng
Department of Microbiology
University of Washington
Seattle, WA 98195

Andrés Vázquez-Torres
Department of Microbiology
University of Colorado Denver School
of Medicine
Aurora, CO 80045

Jörg Vogel
Institute for Molecular Infection Biology
University of Würzburg
Würzburg, Germany

Jovanka M. Voyich
Department of Immunology and
Infectious Diseases
Montana State University
Bozeman, MT 59718

Aimee K. Wessel
Section of Molecular Genetics and Microbiology
The University of Texas at Austin
Austin, TX 78712

Marvin Whiteley
Institute for Cellular and Molecular Biology and
Section of Molecular Genetics and Microbiology
The University of Texas at Austin
Austin, TX 78712

Bobbi Xayarath
Department of Microbiology and Immunology
University of Illinois at Chicago College of Medicine
Chicago, IL 60612

Thomas C. Zahrt
Department of Microbiology and Molecular Genetics
Center for Infectious Disease Research
Medical College of Wisconsin
Milwaukee, WI 53226

Jun Zhu
Department of Microbiology
Perelman School of Medicine
University of Pennsylvania
Philadelphia, PA 19104

Wilma Ziebuhr
Institute for Molecular Infection Biology
University of Würzburg
Würzburg, Germany

PREFACE

Arguably, the theme of virulence regulation within the field of bacterial pathogenesis began as early as the 1930s, with a relatively straightforward observation about the inhibitory effect of iron on the in vitro production of diphtheria toxin by *Corynebacterium diphtheriae*. Three independent laboratories reported this important discovery (those of Pappenheimer and Johnson, Locke and Main, and Pope). It was then nearly two decades later before the next major leap of insight into the regulation of diphtheria toxin came about. In 1951, Freeman reported in the *Journal of Bacteriology* that the conversion of a nontoxigenic (i.e., avirulent) strain of *C. diphtheriae* to one that expresses diphtheria toxin required exposure of the avirulent strain to lysates containing phage B (β) but not phage A. Ultimately, these two discoveries provided an extraordinary amount of stimulating fodder to generations of other investigators. First, they established a solid foundation for the further understanding of the mechanisms of *C. diphtheriae* toxin regulation. Second, they offered novel and fascinating paradigms that were clearly worthy of further investigation in the context of the regulation of virulence in a plethora of other animal, as well as plant, bacterial pathogens.

In the time following those key discoveries, there have been thousands of publications directly relating to the topic of this book (>8,000 references found in a PubMed search from 1980, with the query "Regulation of Bacterial Virulence"). Clearly, this field is advancing at a remarkable pace. As a consequence, we felt that it would be worthwhile at this time to assemble a compendium of many of the more fascinating and contemporary insights relating to this topic, from outstanding authorities in the field, with the wish to stimulate further research efforts.

Therefore, in this book we have attempted to provide a wide range of topics that represent a balance between the newest information along more established lines of investigation (e.g., iron, chapters 5, 6, and 16), as well as information describing refreshing new paradigms that have been investigated within only the past few years (e.g., vesicle formation and host signaling, chapters 23 and 27). It is true that the book devotes significant focus toward some areas,

such as the effects of iron on bacterial virulence. Most likely this is a consequence of both its early discovery in relation to the regulation of bacterial virulence (see above) and the increasing realization that the role of environmental iron levels in virulence is magnificently complex, from the standpoint of both the pathogen and the host. That is, iron has an impact that reaches far beyond simply regulating the expression of virulence determinants. Although iron was subsequently discovered to affect the expression of other major bacterial toxins (e.g., Shiga toxin and *Pseudomonas aeruginosa* exotoxin A), environmental iron levels have also been shown to have an extraordinary impact on increasingly intricate processes relating to bacterial virulence, including biofilm formation, basic physiological processes, resistance to oxidative stress, and basic intermediary metabolism (see chapters 1, 5, 6, 9, 16, and 22).

Another example of how early observations can establish an important paradigm is provided by the requirement of a bacteriophage in the regulation of bacterial virulence, as described above with the β phage of *C. diphtheriae*. Decades later came the observations about the requirement of a different type of bacteriophage in the production of cholera toxin. In fact, cholera toxin provides an amazingly complex story about virulence gene regulation, as well as the intricate overlapping control mechanisms of different virulence factors (see chapter 12). For this reason, *Vibrio cholerae* features prominently in more than one chapter. Even so, it is clear that much still needs to be explored about the regulation of cholera toxin expression and how phage-associated genes affect the virulence of *V. cholerae*.

We have also provided chapters (see chapters 2, 27, and 28) from outstanding authors who are investigating the regulation of extremely complex behaviors of bacterial pathogens. These include descriptions of how some bacteria (e.g., *P. aeruginosa*) control gene regulation before, during, and after their transition from an acute infection to a more chronic one. Along similar lines, also included is a chapter (chapter 28) that provides new insights about the regulatory transition of *V. cholerae* from inside a human host to its more natural environments, such as estuaries, where it exists in planktonic form as well as in biofilms, and then back into a human host.

Last, but not least, we gratefully acknowledge all the other outstanding chapters we were not able to mention above, due to space constraints of this preface. The omission of any chapter in this book would most certainly diminish its value. As the editors, we offer our sincere thanks to all of the authors for their dedication and hard work toward the production of this book.

It is hoped that the exciting discoveries described by all of the wonderful authors of this book will be as inspirational to both young and more seasoned investigators, as the early observations about the regulation of diphtheria toxin were to scores of scientists for decades. We can only hope that this will most certainly be so.

Michael L. Vasil
Andrew J. Darwin

I. GLOBAL CHANGES DURING AND BETWEEN DIFFERENT STATES OF INFECTIONS

Regulation of Bacterial Virulence
Edited by Michael L. Vasil and Andrew J. Darwin
© 2013 ASM Press, Washington, DC   doi:10.1128/9781555818524.ch1

Chapter 1

Factors That Impact *Pseudomonas aeruginosa* Biofilm Structure and Function

BOO SHAN TSENG AND MATTHEW R. PARSEK

INTRODUCTION

Biofilm microbiology has been an area of recent intense research. The importance of biofilms both in the environment and in human disease has researchers trying to understand the molecular determinants involved in their formation. The hope is that by identifying such determinants, we might be able to control biofilms.

Approximately a decade ago, the hypothesis that biofilm formation is a developmental process was put forth (O'Toole et al., 2000a; Stoodley et al., 2002). The thought is that like long-studied developmental processes, such as spore formation in *Bacillus subtilis*, biofilm development is a sophisticated, genetically orchestrated series of events. Subsequently, several researchers have attempted to identify genetic elements that play a role in biofilm development. We have since learned that different regulatory and signaling systems in different species are important. We have also learned that environmental conditions are critical, with sudden changes capable of both promoting and dissolving biofilms.

Chronic infections are linked to the biofilm mode of growth. Since bacteria within a biofilm are less susceptible to antimicrobial treatment than planktonic cells, the relative antimicrobial tolerance of these communities has also been a focus of biofilm-related research. Understanding the mechanisms underpinning tolerance may be key in the therapeutic targeting of chronic infections. Currently, the antimicrobial tolerance of biofilms is thought to be multifactorial. Slow-growing subpopulations within the community, biofilm-related patterns of gene expression, and reduced penetration of antimicrobials have all been linked to tolerance. Not surprisingly, the development of biofilms has also been linked to tolerance. Different biofilm development patterns have been shown to result in different mature biofilm structures. These different structures have been shown to differ in their susceptibilities towards antimicrobial treatment.

This review focuses on the relationships among biofilm development, the environment, and antimicrobial tolerance for the paradigm organism *Pseudomonas aeruginosa*.

WHAT IS A BIOFILM?

Historically, the seemingly simple question of what a biofilm is has generated heated debate within the biofilm research community. Is a pellicle at the air-liquid interface of a standing liquid culture or a colony on a plate a biofilm? Do single cells attached to a surface constitute a biofilm? If not, how many cells are required? At the root of this controversy are the qualitative definitions traditionally used to designate a "biofilm," usually involving some type of microscopic assessment.

As researchers learn more about the molecular mechanisms that control the switch from planktonic to biofilm-associated lifestyles, some patterns are emerging that aid in defining the term biofilm. While no universal molecular definition exists to diagnose the biofilm mode of growth, elevated levels of the intracellular signal cyclic-di-GMP (c-di-GMP) appear to be a key focal point for defining biofilms among many gram-negative species (Dow et al., 2007). A high level of this signal is crucial in shutting off flagellar production and function while promoting expression of key biofilm matrix polysaccharides and extracellular proteins. There is also evidence in gram-positive species, such as *Bacillus subtilis*, that the regulation of motility and matrix production are carefully coordinated in an inverse fashion (Blair et al., 2008).

Although a universal molecular determinant indicative of the biofilm lifestyle may not exist, diagnosing the biofilm state should be achievable for

Boo Shan Tseng and Matthew R. Parsek • Department of Microbiology, University of Washington, Seattle, WA 98195.

some species. The ability to define cells as being in the biofilm lifestyle may be beneficial in studying complex environmental or clinical samples. It may also be useful in analyzing cells that have features of both biofilm and planktonic behavior, such as cells that are swarming on a surface. As researchers continue to explore the basis of biofilm formation, a more universal molecular definition of a biofilm may emerge.

BIOFILM DEVELOMENT AND STRUCTURE IN *PSEUDOMONAS AERUGINOSA*

The old, conventional view of a biofilm was of a structurally homogeneous collection of cells enclosed by a slimy matrix. The work of Lawrence and others in the 1990s revolutionized this view of biofilm structure (Lawrence et al., 1991; Stoodley et al., 1994; Massol-Deya et al., 1995). Using confocal scanning laser microscopy, they demonstrated that biofilms are morphologically complex communities. These studies have led to the suggestion that biofilm production is a highly regulated developmental process (O'Toole et al., 2000a; Stoodley et al., 2002). Investigators have discovered that biofilms grown in vitro reproducibly form specific structures that are affected by a plethora of conditions. For *P. aeruginosa*, two general biofilm shapes have been observed using the flow cell system: structured biofilms and flat biofilms. The biological significance of the different three-dimensional biofilm structures produced by the same species is still unclear. Are different cellular functions or processes used to produce these different structures? Are these structures regulated by different pathways? Is the antimicrobial tolerance of the biofilm affected by its structure? What is the relevance of the structure in the modeling of environmental or clinical biofilms? While these questions remain unresolved, many researchers have made headway in our understanding of biofilm morphology.

Structured Biofilms

A structured biofilm consists of a thin layer of cells on a substratum that is punctuated by large cellular clusters. Some of these biofilms resemble mushrooms, with a large cap-like population of cells atop a "stalk" of cells. Others of this class have more mound-like or pillar-like cellular clusters, which are composed of large cellular aggregates. These types of biofilms are all characterized by their "rough" and fairly heterogeneous nature (Costerton et al., 1995). There are at least two routes by which *P. aeruginosa* produces a structured biofilm. The first we call

"structured biofilm I" (Fig. 1). Several developmentally discrete steps have been linked to their formation. First, bacterial cells interact with a surface in a reversible manner, in which they can still detach from the surface. Some of these cells become irreversibly attached to the surface. Irreversible attachment is thought to be linked to the production of exopolysaccharide, which aids in the adherence to the surface. Cells proceed to expand clonally, producing an aggregate that is primarily derived from the initial attached cell. Ultimately, cells in the interior of the aggregates disperse, leaving a doughnut-like structure behind (Stoodley et al., 2002).

The development of the second type of structured biofilm ("structured biofilm II" [Fig. 1]) is more complex. This involves the interaction of at least two subpopulations of cells: "stalk" and "cap" cells. By growing two differentially labeled isogenic populations of *P. aeruginosa* in a flow cell system with glucose minimal medium, Klausen and colleagues showed that in an individual mushroom, the stalk cells consisted primarily of one population of labeled cells, while the caps were mixtures of the two labeled populations (Klausen et al., 2003a). This result suggests that the stalk is created through the clonal expansion of a surface-adhered, nonmotile cell and the cap is created through the migration of motile cells on top of the stalk. In agreement with this model, after photobleaching a section of a green fluorescent protein (GFP)-labeled mushroom, the cap recovered fluorescence, while the stalk did not (Haagensen et al., 2007). This result suggests that new cells can move into the cap region of the biofilm but not into the stalk region. This pattern of development, however, has only been shown under specific culturing conditions, namely, a defined medium with glucose as the primary carbon source.

Flat Biofilms

In contrast to structured biofilms, flat biofilms are more homogeneous in appearance. These biofilms are suggested to form initially in a manner similar to that of structured biofilms. By mixing two differentially labeled isogenic populations of *P. aeruginosa* in a flow cell system with a citrate-based defined medium, Klausen and colleagues showed that in early stages of biofilm development, flat biofilms start as microcolonies of one or the other labeled population, suggesting that these microcolonies form from clonal expansion, similar to the structured biofilms. Unlike the structured biofilms, however, cells move, divide, and intermix as the biofilm matures. The substratum between the microcolonies become filled with motile cells of either population, and eventually a flat biofilm

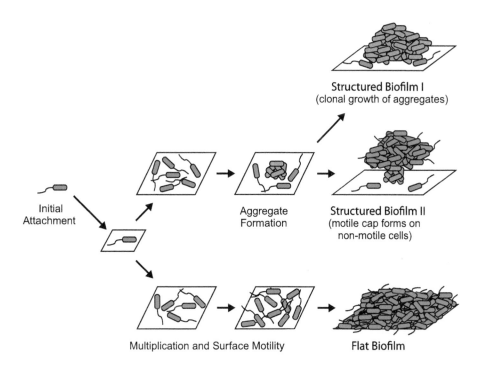

Figure 1. Models of flat versus structured biofilm formation in *P. aeruginosa*. Initial steps of biofilm formation are suggested to be the same for flat and structured biofilms. After initial steps, however, maturation of the different structures is suggested to follow different developmental pathways. There are two forms of structured biofilms (structured biofilms I and II). Aggregates of cells clonally expand to produce the structured biofilm I phenotype, while motile cells move to top nonmotile aggregates of cells to form mushroom-shaped structures in the structured biofilm II phenotype. Flat biofilms are formed through the clonal expansion of motile cells. Blue cylinders represent motile cells, and orange cylinders represent nonmotile cells. Adapted from Kirisits and Parsek, 2006, with permission from John Wiley and Sons. doi:10.1128/9781555818524.ch1f1

is produced (Klausen et al., 2003b). Data from Shrout and colleagues indicate a role for swarming motility in this process (Shrout et al., 2006).

FACTORS THAT INFLUENCE BIOFILM STRUCTURE

While a specific strain of *P. aeruginosa* consistently produces the same biofilm structure under one laboratory culturing condition, the same strain produces a very different biofilm structure under different culturing conditions. While the environmental sensing mechanisms and the regulatory pathways leading to the formation of specific biofilm structures have not been fully elucidated, it is clear that many factors are important for this process.

Motility

One of the major differences between flat and structured biofilm development is related to the motility of the cells (Shrout et al., 2006). The two biofilms are suggested to start similarly, with individual

attached cells growing clonally to create very small cellular aggregates. Cells forming flat biofilms move on the surface, grow, and fill in the space between microcolonies. On the other hand, cells that are destined to produce structured biofilms have two possible developmental paths. To create the structured biofilm I phenotype, cells statically grow, producing larger aggregates, while structured biofilm II is created by a motile population of cells migrating to the top of existing aggregates of nonmotile cells. Monte Carlo simulations suggest that modulating cellular motility affects the biofilm structure (Shrout et al., 2006). In agreement with the mathematical modeling, mutations in the biosynthesis of motility appendages (i.e., type IV pili and flagella) affect biofilm morphology (O'Toole and Kolter, 1998a; Heydorn et al., 2002; Klausen et al., 2003a; Klausen et al., 2003b; Landry et al., 2006; Barken et al., 2008; Yang et al., 2009). Furthermore, the production of rhamnolipids, which is suggested to be important for a type of surface motility called swarming, has been shown to be important for biofilm formation (Davey et al., 2003; Lequette and Greenberg, 2005; Pamp and Tolker-Nielsen, 2007; Glick et al., 2010).

Type IV pili

Twitching motility is a type IV pilus-dependent translocation of bacteria on a surface. Type IV pilus-dependent surface motility was traditionally thought to be based on a grappling-hook-type model, in which the polar type IV pilus extends, binds to the surface, and retracts, dragging the cell body behind it (Mattick, 2002). Recent data, however, suggest that type IV pilus-dependent motility is more complex (Gibiansky et al., 2010; Jin et al., 2011). Through the inactivation of the genes involved in pilus biogenesis, many studies have shown that type IV pili and twitching are important in multiple steps of biofilm development (O'Toole and Kolter, 1998a; Heydorn et al., 2002; Klausen et al., 2003a; Klausen et al., 2003b; Barken et al., 2008; Yang et al., 2009).

Attachment. Using the PA14 strain of *P. aeruginosa*, O'Toole and Kolter showed that under static conditions with a glucose-casamino acids-based minimal medium, various mutants in type IV pilus biogenesis were defective in adhering to a plastic surface (O'Toole and Kolter, 1998b). However, a PAO1 mutant with a deletion in *pilA*, which encodes the major structural subunit of the pilus, was not defective in adhering to glass under flowing conditions with a citrate minimal medium (Klausen et al., 2003b). While strain and/or surface chemistry differences could possibly explain this discrepancy (O'Toole and Kolter, 1998a; Heydorn et al., 2000), Klausen and colleagues suggested that the difference between the two results was based on the nutritional conditions (see more below on how nutrition affects biofilm morphology). Consistent with the O'Toole and Kolter study, under static conditions, the PAO1 Δ*pilA* mutant was defective for attachment in a glucose-casamino acids minimal medium but not a citrate minimal medium (Klausen et al., 2003b). Landry et al. further supported this result, showing that a PAO1 Δ*pilA* mutant was defective in attaching to a glass surface under flowing conditions with a glucose-rich medium (Landry et al., 2006). Therefore, the role of the type IV pili in the attachment of bacteria to a surface is nutrient dependent.

Flat biofilm formation. Under conditions in which *P. aeruginosa* forms flat biofilms, type IV pilus mutants form irregular rough, structured biofilms (Fig. 2). These biofilms appear as isolated microcolonies and fail to spread on the substratum (Heydorn et al., 2002; Klausen et al., 2003a; Klausen et al., 2003b). In addition, on the basis of findings obtained by mixing two differentially labeled populations of Δ*pilA* mutants, Klausen et al. suggest that these microcolonies arose from clonal growth (Fig. 2A) (Klausen et al., 2003a). Since wild-type flat biofilms start as isolated microcolonies, pilus mutants are defective in progressing to the next developmental stage where cells spread to occupy the areas between the microcolonies. This observation has led to the hypothesis that twitching-based motility is in part responsible for the flat biofilm structure (Klausen et al., 2003b). As further evidence for this hypothesis, cells that have increased twitch rates are correlated with the formation of flat biofilms (Singh et al., 2002; Landry et al., 2006).

Structured biofilm formation. Under conditions that lead to wild-type cells forming structured biofilms, Δ*pilA* mutants form biofilms with isolated microcolonies (Klausen et al., 2003a). The Δ*pilA* mutant strain was unable to form mushroom caps, as biofilms formed from a mixture of cyan fluorescent protein (CFP)-labeled Δ*pilA* mutants and YFP-labeled wild-type cells have clonal populations of CFP-labeled stalks and yellow fluorescent protein (YFP)-labeled caps. Oddly, most of the stalks are CFP labeled in these experiments, suggesting that cells defective in twitching are better at forming the stalks than those that can twitch (Klausen et al., 2003a). While this defect in cap formation of Δ*pilA* mutants may be attributed to the inability to twitch, it could also be due to the lack of pilus structures on the surface of the cells. Suggesting that the latter may be the case, Barken et al. showed that Δ*pilH* mutants, which have increased surface piliation but decreased twitching motility, can form caps and irregular mushroom-shaped biofilms (Barken et al., 2008).

Flagella

The flagellum is a highly regulated, complex structure, composed essentially of a filament and a motor that drives filament rotation (Berg, 2003). *P. aeruginosa* has a single polar flagellum. Similar to that of type IV pili, the role of the flagellum in biofilm development is nutrient dependent (O'Toole and Kolter, 1998a; Klausen et al., 2003b; Shrout et al., 2006; Barken et al., 2008). O'Toole and Kolter showed that under static conditions with a glucose-casamino acids minimal medium, a mutant in flagellar biosynthesis was defective in attachment to a plastic surface (O'Toole and Kolter, 1998a). This phenotype, however, can be rescued by using a different growth medium (O'Toole and Kolter, 1998a; Klausen et al., 2003b). In addition to attachment, mutants in flagellar biosynthesis form mature biofilms that are different from that of wild-type cells. Under conditions where wild-type cells form flat biofilms, a Δ*fliM*

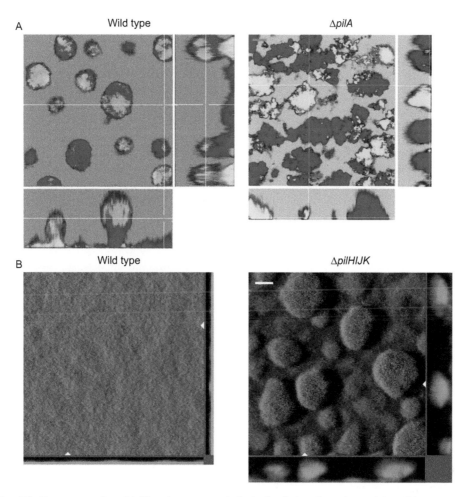

Figure 2. Type IV pilus mutants form biofilms that are morphologically distinct from those of the wild-type strain. (A) Four-day-old biofilms of wild-type and Δ*pilA* strains of *P. aeruginosa* grown in glucose-based minimal medium. A 1:1 mixture of CFP- and YFP-expressing cells of the same strain was used to initiate the biofilm. (B) Ninety-eight-hour-old biofilms of GFP-expressing wild-type and Δ*pilHIJK* strains of *P. aeruginosa* grown in citrate-based minimal medium. Crosshairs (A) and white triangles (B) indicate the positions of vertical cross sections shown on the right and bottom of each image. Scale bar, 20 μm. Modified from Klausen et al., 2003a (A), and Heydorn et al., 2000 (B), with permission from John Wiley and Sons and the Society for General Microbiology, respectively.
doi:10.1128/9781555818524.ch1f2

mutant strain produced structured biofilms that contain more biomass than wild-type biofilms, suggesting that flagellum-dependent motility is important for producing flat biofilms (Klausen et al., 2003b). In support of this hypothesis, multiple studies have shown that flagellar mutants are defective in surface-associated motility (Kohler et al., 2000; Landry et al., 2006; Shrout et al., 2006; Barken et al., 2008).

Chemotaxis

Chemotaxis is traditionally defined as the directed motion of swimming bacteria in planktonic culture. Chemotaxis, however, has also been observed on a surface and is suggested to play a role in biofilm formation (Kearns et al., 2001). A mutant in CheR1,

a methyltransferase that regulates chemotaxis, samples less of the surface than a wild-type cell and is defective in changing direction (Schmidt et al., 2011). Furthermore, mutants in chemotaxis produce mound-like biofilms, similar to flagellar mutants, under conditions where wild-type cells form flat biofilms (Li et al., 2007; Barken et al., 2008). While it is unknown what the chemotactic machinery is sensing during biofilm formation, these results provide further evidence that flagellum-mediated motility and chemotaxis are important for producing the mature biofilm structure.

Rhamnolipids

Rhamnolipids are surfactants that aid in swarming motility, and their production is quorum sensing

(QS) regulated (Kohler et al., 2000; Lequette and Greenberg, 2005). While rhamnolipids were originally suggested to be necessary for swarming (Kohler et al., 2000), the requirement for rhamnolipids in swarming on agar surfaces is nutrient dependent (Shrout et al., 2006). However, since there are no genetic approaches for manipulating swarming without directly interfering with flagellum- or pilus-dependent motility, most studies have used mutants that eliminate rhamnolipid production to investigate the role of swarming in biofilm formation. Therefore, these studies make a statement about the role of rhamnolipids in biofilm formation, while the role of swarming is less clear.

Biofilms formed by a strain containing a mutation in *rhlA*, a gene encoding a rhamnolipid biosynthetic enzyme (Soberon-Chavez et al., 2005), are structurally distinct from those of wild-type cells. Under conditions where wild-type biofilms are structured with large aggregates, a Δ*rhlA* mutant forms flat biofilms (Davey et al., 2003; Boles et al., 2005; Pamp and Tolker-Nielsen, 2007). However, if a Δ*rhlA* mutant is mixed with cells that can produce rhamnolipids, Δ*rhlA* cells can participate in structured biofilm formation, suggesting that rhamnolipids serve as a community good (Pamp and Tolker-Nielsen, 2007). Interestingly, in these mixed-population studies, the cells that produce rhamnolipids form the stalks, while the cells deficient for rhamnolipid production form the caps of the aggregates, with a developmental pattern similar to that of structured biofilm II (Pamp and Tolker-Nielsen, 2007). In agreement with this observation, Lequette and Greenberg observed that in structured wild-type biofilms, *rhlA* was expressed mainly in the stalk, with little expression in the cap portion of the mushroom (Lequette and Greenberg, 2005). Together, these results suggest that rhamnolipid production in the stalk is important for cap formation under conditions where *P. aeruginosa* forms biofilms of the structured biofilm II class.

Completely contrary to the above-described flat biofilm morphology for Δ*rhlA* mutants, under iron-limiting conditions where wild-type cells form flat biofilms, Δ*rhlA* cells formed structured biofilms (Glick et al., 2010). This phenotype is not completely surprising, as iron levels influence rhamnolipid production. Under iron-limiting conditions, wild-type *P. aeruginosa* increases expression of rhamnolipid biosynthetic genes (Glick et al., 2010). Furthermore, Glick and colleagues showed that in iron-limiting media, Δ*rhlA* cells are defective for twitching, but this deficiency can be rescued by using iron-replete media (Glick et al., 2010). Interestingly, the Δ*rhlA* cells can also form structured biofilms, like wild-type cells, under iron-replete conditions. Glick and colleagues suggest that the timing of *rhlA* expression under iron-limiting versus

replete conditions in wild-type biofilms may explain these results (Glick et al., 2010). Ultimately, these results with those described above show that the role of rhamnolipid production, and possibly swarming, in biofilm development is complicated and confounded by multiple environmental factors.

In addition to their role in biofilm morphology, rhamnolipids affect both biofilm initiation and dispersal. Exogenous addition of rhamnolipids can block biofilm initiation (Davey et al., 2003) and disperse preformed biofilms (Boles et al., 2005). This effect of excess rhamnolipids on biofilm initiation and dispersal is suggested to be due to their surfactant property, disrupting cell-cell and cell-surface interactions. Supporting this interpretation, Boles and colleagues show that another surfactant, sodium dodecyl sulfate (SDS), can also induce dispersal of a preformed biofilm in a manner similar to that of rhamnolipids (Boles et al., 2005). Since these studies are based on addition of excess rhamnolipids exogenously, they suggest that rhamnolipid production, which starts in planktonic cells at the transition into stationary phase (Lequette and Greenberg, 2005), is tightly regulated throughout biofilm development.

Nutrition

P. aeruginosa is renowned for its ability to grow in almost any environment and can use a wide variety of sources to meet its nutritional requirements. The nutritional conditions under which biofilms develop greatly affect their morphology. For instance, changing the carbon source in the media leads to drastically different biofilm structures (Wolfaardt et al., 1994; Klausen et al., 2003a; Klausen et al., 2003b; Shrout et al., 2006). Based on mathematical modeling, nutrient limitation is a major driving force in determining biofilm structure (Wimpenny and Colasanti, 1997; Picioreanu et al., 1998). In these models, rougher structured biofilms form under limiting nutrient conditions and growth was limited by nutrient transfer. In comparison, denser, flat biofilms form under nutrient-rich conditions and were limited by the growth rate of the microorganisms. These simplistic models, however, do not fully describe the complexity with which nutrients, such as carbon source and iron levels, affect biofilm morphology.

Carbon sources

From halogenated aromatics to intermediates of the tricarboxylic acid (TCA) cycle, *P. aeruginosa* can use a diverse group of organic compounds as a sole carbon source. Most defined media for laboratory culturing of *P. aeruginosa* provide either TCA cycle intermediates or carbohydrates as the primary carbon source. In the presence of TCA cycle intermediates, *P. aeruginosa* represses carbohydrate catabolic pathways and preferentially utilizes the TCA cycle intermediates as a carbon source (Wolff et al., 1991). The carbon source greatly affects the final biofilm structure (Klausen et al., 2003a; Klausen et al., 2003b; Shrout et al., 2006). For instance, biofilms grown on a glass surface with a glucose-based minimal medium under continuous flow are structured, while biofilms grown on other carbon sources (e.g., citrate, casamino acids, benzoate, glutamate, and succinate) are flat (Klausen et al., 2003b; Shrout et al., 2006). While how these different carbon sources lead to the specific morphologies is not clear, Shrout and colleagues show that cells grown on glucose are less motile on a glass surface than those grown on succinate. They proposed that relative surface motility correlates with biofilm structure, with higher degrees of surface motility producing flat biofilms. Consistent with this observation, cells grown on glucose swarmed less on agar plates than cells grown on succinate. In these assays, twitching and swimming motility were not affected by the carbon source (Shrout et al., 2006). These results suggest that the carbon source may affect biofilm formation by modulating the swarming motility of the cells. Since type IV pili and flagellar motility both contribute to swarming (Kohler et al.,

2000), this hypothesis would also explain why mutants of pili and/or flagellar motility do not form flat biofilms.

In addition to the effect on biofilm morphology, carbon source differences can also explain the variability reported in the literature as to whether a specific factor plays a role in biofilm formation. Two such examples are the role of motility appendages in surface attachment and the role of QS in biofilm development. As earlier mentioned, pili and flagella play a role in the attachment of cells to a surface in glucose-casamino acids, but not citrate, minimal medium (Klausen et al., 2003b). Similarly, biofilms of QS mutants are indistinguishable from wild-type biofilms in glucose or glutamate minimal medium. However, in succinate minimal medium, QS mutants produce drastically different biofilms compared to those of the wild type (Fig. 3) (Shrout et al., 2006). Such studies underline the importance of considering the carbon source when interpreting the effects of mutants on biofilm development and highlight that biofilm development is dynamic and not a hard-wired developmental pathway.

While mechanistically little is known about how a carbon source leads to changes in motility and biofilm development, Crc, a regulator of carbohydrate metabolism, is proposed to partially mediate this regulation (O'Toole et al., 2000b). Since *P. aeruginosa*

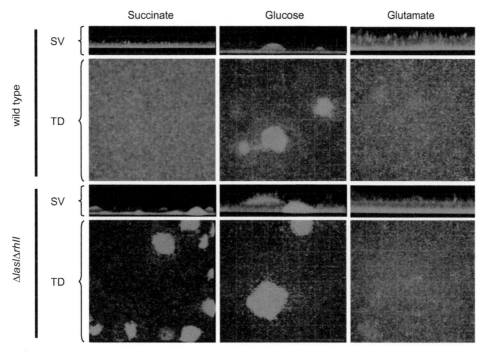

Figure 3. The carbon source affects biofilm morphology. Shown are 48-hour-old biofilms of GFP-expressing wild-type and Δ*lasI*Δ*rhlI* strains of *P. aeruginosa* grown in media with different carbon sources. SV, side view (x, z-plane); TD, top down view (x, y-plane). Reproduced from Shrout et al., 2006, with permission from John Wiley and Sons. doi:10.1128/9781555818524.ch1f3

prefers TCA intermediates to carbohydrates as carbon sources, Crc negatively regulates the utilization of carbohydrates when TCA intermediates are present (Wolff et al., 1991). O'Toole and colleagues showed that *crc* mutants are defective in microcolony formation on plastic in comparison to wild-type cells. In addition, these mutants had reduced twitching motility relative to wild-type cells (O'Toole et al., 2000b). These results suggest that metabolizing carbohydrates leads to reduced motility, thereby affecting biofilm morphology.

Iron levels

Iron is essential for growth. Iron acquisition, however, is problematic due to iron's poor solubility in aerobic environments. Under conditions of iron limitation, *P. aeruginosa* responds with multiple mechanisms, including the production of siderophores (e.g., pyoverdine and pyochelin) (Cox and Graham, 1979; Cox and Adams, 1985) and heme-binding proteins (e.g., PhuS and HasA) (Ochsner et al., 2000), to aid in iron uptake. While iron is essential, it can also be toxic to cells and produces harmful reactive oxygen species. Therefore, iron uptake is tightly regulated by the master ferric uptake regulator, Fur (Vasil and Ochsner, 1999; Cornelis, 2010). When in complex with iron, Fur represses transcription of iron uptake genes. In addition, Fur represses the expression of small RNAs (PrrF1 and PrrF2 in *P. aeruginosa*) that posttranscriptionally regulate the stability of mRNAs encoding a wide variety of proteins involved in iron regulation, QS, and cellular metabolism (Wilderman et al., 2004; Masse et al., 2007; Oglesby et al., 2008).

While the role of iron in pellicles at the air-liquid interface has not been examined, Fur, pyoverdine, and iron levels can all affect biofilm formation on a glass surface (Singh et al., 2002; Banin et al., 2005; Musk et al., 2005; Kaneko et al., 2007; Yang et al., 2007; Patriquin et al., 2008; Yang et al., 2009; Glick et al., 2010). The effect of iron on biofilm formation is concentration dependent. Under conditions where growth is not affected, cells that develop in media either limiting in or in excess of iron produce very thin, flat biofilms (Singh et al., 2002; Musk et al., 2005; Yang et al., 2009; Glick et al., 2010). Furthermore, addition of excess iron (250 µM) to a preformed biofilm causes dispersal (Musk et al., 2005). Under intermediate iron-replete conditions (50 µM), however, biofilms are structured (Glick et al., 2010). Under these conditions, cells are less motile than under iron-limiting conditions, which may explain why biofilms are flat in iron-limiting conditions (Yang et al., 2007; Patriquin et al., 2008; Glick et al., 2010). Interestingly, iron limitation induces wild-type cells

to produce a heat-stable soluble molecule that can subsequently induce increased motility in cells under iron-replete conditions (Patriquin et al., 2008).

Pyoverdine is a key siderophore in *P. aeruginosa* and is important for biofilm development. Mutants in pyoverdine biosynthesis form flat biofilms under conditions where wild-type cells produce structured biofilms (Banin et al., 2005; Kaneko, et al., 2007; Yang et al., 2009). This flat biofilm phenotype of pyoverdine mutants is similar to the biofilm observed for wild-type cells under iron-depleted conditions and can be rescued by the addition of extra iron to the medium (Patriquin et al., 2008). While the mechanistic underpinning for these biofilm phenotypes is unclear, pyoverdine is obviously important for the development of structured biofilms. In wild-type structured biofilms, the pyoverdine synthase gene, *pvdA*, is expressed in the stalk of the mushroom-shaped structures (Yang et al., 2009). Biofilms formed by *pvdA* mutants are flat. However, when *pvdA* mutants, which are motile but cannot produce pyoverdine, are mixed with Δ*pilA* mutants, which are nonmotile but can produce pyoverdine, structured biofilms form. In these biofilms, the pyoverdine-competent Δ*pilA* cells form the stalk, while the *pvdA* mutant forms the caps (Yang et al., 2009). This result suggests that cells in the stalk must be able to produce pyoverdine to support structured biofilm formation. The role of pyoverdine in this process, however, can be bypassed by providing the cells with an iron source, such as ferric citrate, that does not require pyoverdine for uptake (Yang et al., 2009), showing that the role of pyoverdine in biofilm formation is simply for iron acquisition.

Since the expression of pyoverdine biosynthesis genes is regulated by Fur (Ochsner et al., 1995), it is not surprising that Fur regulates the effect iron levels have on biofilm formation. Under iron-limiting conditions where a wild-type strain produces a flat biofilm, a strain containing a missense mutation in *fur* forms a structured biofilm (Banin et al., 2005). This mutant *fur* strain constitutively expresses siderophores, even under iron-replete conditions (Barton et al., 1996), which may explain the biofilm phenotype. Interestingly, a strain containing deletions of the PrrF1 and PrrF2 small RNAs forms biofilms that resemble that of a wild-type strain under both iron-limiting and -replete conditions (Banin et al., 2005). Together, these results suggest that Fur regulates iron-dependent biofilm structure in a PrrF1- and PrrF2-independent manner.

Las and Rhl QS Systems

Bacteria can sense and respond to population density via QS. *P. aeruginosa* has three QS modules:

Las, Rhl, and PQS. These QS modules are coregulated as well as interconnected (Williams and Camara, 2009). The *P. aeruginosa las* and *rhl* systems produce two major signals, *N*-3-oxododecanoylhomoserine lactone (3OC12-HSL) and *N*-butanoyl-homoserine lactone (C4-HSL), respectively. LasI and RhlI synthesize these molecules, while the LasR and RhlR transcriptional regulators sense these molecules (Williams and Camara, 2009). As the *las* and *rhl* systems affect many downstream targets, such as rhamnolipid production, it is not surprising that mutants of *las* and *rhl* can affect biofilm morphology (Davies et al., 1998; De Kievit et al., 2001; Lequette and Greenberg, 2005; Shrout et al., 2006; Pamp and Tolker-Nielsen, 2007; Patriquin et al., 2008). Davies and colleagues were the first to describe a role for QS in biofilm development. In their system, a *lasI* mutant formed a thin, flat uniform biofilm, while the wild type formed structured biofilms. The *lasI* mutant produced a biofilm similar to that of the wild-type cells when 3OC12-HSL was exogenously provided in the media, confirming that the difference in the biofilm structure was due to production of the QS signal (Davies et al., 1998). Heydorn and colleagues, however, failed to see an effect of a *lasI* mutation on biofilm formation. In their system, the *lasI* mutant produced a flat biofilm similar to that of the wild-type cells (Heydorn et al., 2002). One possible explanation for this discrepancy between the results is that *lasI* mutants will always form flat biofilms, and whether an observable difference between the QS mutant and wild-type cells is purely based on nutrient-dependent structure the wild-type cells make, as suggested by the results from De Kievit and colleagues (De Kievit et al., 2001). This explanation, however, does not fully account for the discrepancy in results, as QS mutants can form structured biofilms under certain conditions (Shrout et al., 2006). Shrout and colleagues show that while QS itself is not nutritionally dependent, the role of QS in biofilm morphology is dependent on the nutrient conditions (Fig. 3). The rationale behind the nutrition-dependent role of QS in biofilm development is still unclear. However, under conditions where wild-type cells form flat biofilms and the QS mutant forms structured biofilms, differences in motility are seen for the two cell types. Under these conditions, QS mutants are less motile on glass surfaces and show decreased swarming on agar surfaces relative to wild-type cells (Shrout et al., 2006). This result is congruent with the proposed models for flat versus structured biofilm development (Fig. 1).

While mutations in *las* produce obvious effects on biofilm morphology under the specific conditions, the role of the *rhl* system is less apparent. While originally reported to not influence biofilm formation (Davies et al., 1998), *rhl* mutants have subsequently been shown to produce biofilms that are different from that of wild-type cells. Under conditions where wild-type cells formed mushroom-shaped biofilms of the structured biofilm II class, an *rhlI* mutant formed a structured biofilm that resembled mushrooms without caps (Lequette and Greenberg, 2005), suggesting that *rhl* plays a role in cap formation under these conditions. This role of *rhl* in biofilm morphology may be based on its regulation of rhamnolipid production, which is needed in stalk cells for formation of the cap in mushroom-shaped structures in the structured biofilm II developmental program (Pamp and Tolker-Nielsen, 2007). In agreement with this hypothesis, a strain containing a mutation in RpoN, which regulates rhamnolipid production and *rhlI*, has a biofilm phenotype similar to that of an *rhlI* mutant (Thompson et al., 2003). In addition to this role in cap formation, *rhl* may also play a role in regulating motility during biofilm formation. An *rhl* mutant forms a biofilm with heterogeneous structure under iron-limiting conditions where a wild-type cell forms a flat biofilm (Patriquin et al., 2008). Using these conditions, Patriquin and colleagues showed that spent media from wild-type cells induced twitching motility in cells, while spent media from *rhl* mutants did not. These results suggest that during the development of a flat biofilm, the Rhl QS system is involved in creating a soluble signal that triggers increased motility in cells (Patriquin et al., 2008).

Matrix Components

The matrix provides structure to the biofilm. In *P. aeruginosa*, the matrix is composed of exopolysaccharides, proteins, and extracellular DNA (eDNA) (Stoodley et al., 2002). While these matrix components are critical for biofilm formation, overproduction of any of these components leads to changes in the biofilm structure (Hentzer et al., 2001; Nivens et al., 2001; Stapper et al., 2004; Kirisits et al., 2005; Allesen-Holm et al., 2006; Barken et al., 2008; Borlee et al., 2010; Colvin et al., 2011; Yang et al., 2011). Mutations in specific genes encoding matrix components can also severely impair biofilm development. The following matrix components have been shown to influence *P. aeruginosa* biofilm development.

Exopolysaccharides

Extracellular polysaccharides are essential for biofilm formation in many bacterial species, including *P. aeruginosa* (Ryder et al., 2007), *Escherichia coli* (Danese et al., 2000; Wang et al., 2004), *B. subtilis* (Marvasi et al., 2010), *Staphylococcus* (Gotz, 2002), and *Vibrio* (Yildiz and Visick, 2009). Three exopolysaccharides have been identified in *P. aeruginosa*: alginate, PSL, and PEL.

Alginate. Alginate is a linear copolymer of D-mannuronic acid and L-guluronic acid. Alginate is overproduced in mucoid isolates of *P. aeruginosa* (May et al., 1991). Mucoidy can arise through a mutation in *mucA*, which encodes an anti-sigma factor that negatively regulates alginate production (Martin et al., 1993). Under nutritional conditions where wild-type cells produce flat biofilms, *mucA* mutants form heterogeneous, structured biofilms (Hentzer et al., 2001; Nivens et al., 2001; Stapper et al., 2004). Early in biofilm development, *mucA* biofilms produce more microcolonies relative to an isogenic nonmucoid strain (Nivens et al., 2001). As development progresses, biofilms of *mucA* mutants are thicker and contain more biomass relative to those of wild-type cells (Hentzer et al., 2001). This increased biomass is due to the increase in the amount of polysaccharide in the matrix of mucoid biofilms. While it is unclear how the cells organize the excess alginate in the biofilm, *mucA* cells are motility defective, which at least partially explains the observed structured biofilm phenotype (Stapper et al., 2004).

PSL and PEL. PSL and PEL are two exopolysaccharides involved in biofilm formation (Friedman and Kolter, 2004a, 2004b; Jackson et al., 2004; Matsukawa and Greenberg, 2004). PSL is composed of a repeating pentasaccharide containing D-mannose, D-glucose, and L-rhamnose (Byrd et al., 2009), while the structure for PEL has not been elucidated. In surface-grown continuous-flow biofilms, PSL is the major exopolysaccharide in biofilms of the *P. aeruginosa* PAO1 strain, while PEL is dominant in the PA14 strain (Colvin et al., 2011; Yang et al., 2011). PEL is less important than PSL in PAO1, and PA14 harbors a three-gene deletion in the PSL biosynthetic operon (Friedman and Kolter, 2004b; Colvin et al., 2011; Yang et al., 2011).

Mutants unable to produce PSL are defective in surface attachment and biofilm initiation (Jackson et al., 2004; Matsukawa and Greenberg, 2004; Overhage et al., 2005; Ma et al., 2006). Furthermore, continued production of PSL in a biofilm is important for maintaining biofilm biomass, suggesting that PSL needs to be continually produced, even in a mature biofilm (Ma et al., 2006). Under conditions where wild-type cells form structured biofilms, cells that do not produce PSL form very thin, flat biofilms (Colvin et al., 2011; Yang et al., 2011). During the development of these mutant biofilms, microcolonies do not

form (Yang et al., 2011). Since microcolonies are suggested to be derived from nonmotile cell populations (Klausen et al., 2003a), these results suggest that *psl* mutants are hypermotile (Yang et al., 2011). This proposed hypermotility, however, has not been directly tested. To further examine the biofilm formation defect of *psl* mutants, Yang and colleagues mixed nonmotile PSL-producing cells with the *psl* mutants. When wild-type cells are mixed with nonmotile PSL-producing cells, wild-type cells cap the stalks produced by the nonmotile population. The *psl* mutants, however, fail to cap the stalks, suggesting that PSL production in the cap subpopulation is important for forming this structure (Yang et al., 2011). In agreement with this proposal, PSL localizes to the exterior shell of the cap in mushroom-shaped biofilms (Ma et al., 2009). In addition, PSL production also appears to be important for stalk formation. When nonmotile *psl* mutants are mixed with wild-type cells, wild-type cells form mushroom-shaped structures that exclude the nonmotile *psl* mutant cells, suggesting that the wild-type cells are unable to cap the stalks made by the other population (Yang et al., 2011). Interestingly, although PSL is secreted into the extracellular space, these results suggest that PSL is not a community good.

The importance and role of PEL in biofilm formation are prominent in the PA14 strain. In the case of this strain, *pel* mutants produce thin, flat biofilms, while wild-type cells produce structured biofilms (Colvin et al., 2011; Yang et al., 2011). While these wild-type PA14 biofilms required PEL production to continue growing, the mature biofilm does not require continued PEL production to maintain the biofilm, unlike that of PSL in PAO1 biofilms (Ma et al., 2006; Colvin et al., 2011). Furthermore, unlike PAO1 *psl* mutants, PA14 *pel* mutants do not appear to be defective in surface attachment. Cells that overproduce PEL, however, attach to glass surfaces better than wild-type cells (Colvin et al., 2011). These results suggest that PEL can play a minor role in surface attachment, as previously proposed by Vasseur and colleagues (Vasseur et al., 2005). While PA14 *pel* mutants are not defective in surface attachment, these cells are defective in microcolony formation (Colvin et al., 2011; Yang et al., 2011). Using laser-trapping assays, Colvin et al. showed that *pel* mutants fail to form stable aggregates, which may explain the microcolony formation defect (Colvin et al., 2011). Whether PEL plays a role in the formation of structured biofilms subsequent to microcolony formation has not been studied in the PA14 biofilm.

Overproducing these two exopolysaccharides also affects biofilm morphology (Fig. 4B). Mutants that overproduce both PSL and PEL form so-called rough small-colony variants (RSCVs) on agar. RSCVs

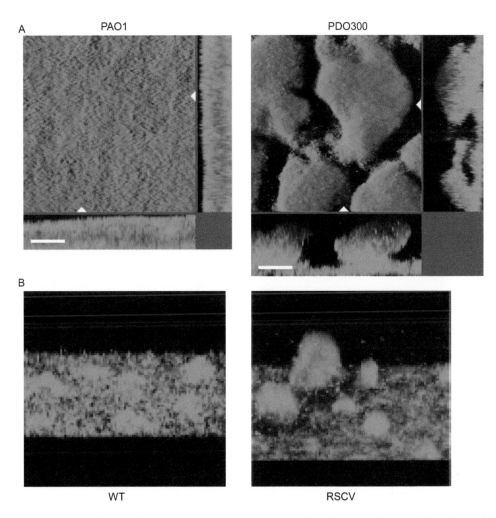

Figure 4. Overproduction of exopolysaccharides creates more structured biofilms. (A) Five-day-old biofilms of the GFP-expressing wild type (PAO1) and an isogenic, alginate-overproducing *mucA22* (PDO300) strain of *P. aeruginosa* grown in defined rich medium. White triangles indicate the position of vertical cross sections shown on the right and bottom of each image. Bar, 20 μm. (B) Three-day-old biofilms of GFP-expressing wild type (WT) and RSCV MJK8 of *P. aeruginosa* grown in defined rich medium. Modified from Hentzer et al., 2001 (A), and Kirisits et al., 2005 (B), with permission from the American Society for Microbiology.
doi:10.1128/9781555818524.ch1f4

naturally arise in biofilms of wild-type cells, suggesting that RSCVs exploit a special niche in the biofilm (Kirisits et al., 2005). Kirisits and colleagues showed that these mutants aggregate when grown in liquid, are less motile than wild-type cells, and adhere to plastic better than wild-type cells. RSCVs form highly structured biofilms under conditions where wild-type cells form flat biofilms (Kirisits et al., 2005). RSCV biofilms also contain more biomass than wild-type biofilms of a comparable age. This increased biomass is most likely due to the increased amount of exopolysaccharide in the matrix, which retains more progeny cells within the biofilm. Similar to the case with alginate in mucoid biofilms, the mechanisms underlying how the cells organize this excess polysaccharide within the biofilm structure are unclear.

Matrix structural proteins: CdrA

The protein component of biofilms has generally been less well studied than the polysaccharide component. While multiple studies have shown differences to the total proteome of planktonic versus biofilm cells in *P. aeruginosa* (Sauer et al., 2002; Vilain et al., 2004; Southey-Pillig et al., 2005; Mikkelsen et al., 2007), few structural components of the biofilm have been identified. Some of the structural biofilm proteins have been shown to function in cell-to-cell aggregation (CdrA in *P. aeruginosa* [Borlee et al., 2010], Bap in *Staphylococcus aureus* [Latasa et al., 2006], Bap1 and RbmC in *Vibrio cholerae* [Fong et al., 2006; Fong and Yildiz, 2007; Absalon et al., 2011], and TasA in *B. subtilis* [Romero et al., 2010]).

Of the proteins associated with the *P. aeruginosa* biofilm matrix, CdrA is the only structural protein identified thus far (Borlee et al., 2010). CdrA is a secreted adhesin that interacts with PSL. While little is known about this protein, CdrA clearly plays a role in biofilm formation. Under conditions where wild-type cells form structured biofilms, Δ*cdrA* cells produce flat biofilms. Overproduction of CdrA produces a structured biofilm that is much thicker than the wild-type biofilm. By using a strain background that overproduces PSL and PEL, Borlee and collaborators showed that CdrA plays a role in compacting the cells in the biofilm and in stabilizing the interaction between PSL and the cells (Borlee et al., 2010).

Extracellular DNA

Double-stranded DNA is a key component of the biofilm matrix (Matsukawa and Greenberg, 2004). This eDNA is identical to chromosomal DNA and is in part produced from autolysis of cells within the biofilm (Steinberger and Holden, 2005; Allesen-Holm et al., 2006; Mulcahy et al., 2010). The eDNA is an integral component of the biofilm that provides structural stability, as biofilms dissolve when treated with DNase I (Whitchurch et al., 2002). In structured biofilms, eDNA localizes to the outer region of the stalk and on the substratum between mushroom structures, but not in the cap portion of the mushroom (Allesen-Holm et al., 2006). The production of eDNA in the stalk portion of the structured biofilm II class has been suggested to be important for the formation of the cap (Yang et al., 2009).

While the mechanism is unclear, eDNA production is clearly a regulated process. QS, motility, and PSL production all appear to impact its production. Biofilms formed by mutants with alterations of these processes had less eDNA than wild-type biofilms (Allesen-Holm et al., 2006; Yang et al., 2007; Yang et al., 2011). While this decrease in eDNA production may be the basis for the differences seen in the biofilm morphologies of these mutants, the role of eDNA in structure formation has not been directly studied, as no genetic handle exists for manipulating eDNA without affecting other processes. More studies are needed to better understand the role of eDNA in biofilm formation.

EFFECT OF BIOFILM STRUCTURE ON ANTIBIOTIC TOLERANCE

Biofilm structure influences chemical gradients. One obvious example is the sharp oxygen gradient encountered in aerobically growing aggregates of *P. aeruginosa*. Xu and colleagues found that oxygen was quickly consumed at the periphery of aggregates (Xu et al., 1998). This decrease in oxygen levels was directly correlated with a general decrease of metabolic activity inside the aggregate. Slow-growing populations in the interior of a biofilm have been shown to be recalcitrant to antimicrobial treatment, while the metabolically active subpopulation that has access to oxygen is susceptible. Borriello et al. showed that a variety of antibiotics (i.e., carbenicillin, ceftazidime, chloramphenicol, ciprofloxacin, tetracycline, and tobramycin) from different classes was impaired for killing *P. aeruginosa* colony biofilms and that these phenotypes can largely be explained by the effects of oxygen depletion on metabolism (Borriello et al., 2004). Indeed, Alvarez-Ortega and Harwood demonstrated that three different high-affinity terminal oxidases produced by *P. aeruginosa* are required for optimal oxygen utilization and are required for production of normal, mature biofilms, highlighting the importance of oxygen in dictating biofilm structure (Alvarez-Ortega and Harwood, 2007).

The inability of individual antibiotics to target both slowly and actively growing subpopulations simultaneously is an issue in eliminating biofilms. Two groups have investigated the strategy of combining two antimicrobials, one targeting the slowly growing population and the other targeting the actively growing population. In this fashion, Kaneko and colleagues employed the transition metal analog gallium (Kaneko et al., 2007). Gallium poisons *P. aeruginosa* by substituting for iron in cellular proteins. Kaneko et al. showed that iron-starved *P. aeruginosa* cells imported gallium and were killed. In a biofilm community, cells in the interior of a biofilm were iron starved and gallium susceptible. The active cells at the exterior of the biofilm had access to iron and, hence, were gallium resistant (Kaneko et al., 2007). Tobramycin, an aminoglycoside that targets actively growing *P. aeruginosa*, produced the opposite killing pattern in the community, with the active cells at the periphery killed and the slow-growing interior cells resistant. By combining the two antimicrobials, they achieved complete killing of the biofilm (Kaneko et al., 2007). Similarly, Pamp and Tolker-Nielsen demonstrated that the antimicrobial peptide colistin effectively killed *P. aeruginosa* cells in the interior of the biofilm (Pamp and Tolker-Nielsen, 2007). This susceptibility to colistin was due to the inability of the inactive subpopulation to carry out efflux of the antibiotic. The active subpopulation was able to carry out efflux, allowing this subpopulation to adapt by changing its lipopolysaccharide structure, using the Pmr system (Pamp and Tolker-Nielsen, 2007). Like Kaneko et al. (Kaneko et al., 2007), they were able to use conventional antibiotics that targeted metabolically active cells, in combination with colistin, to kill most of the biofilm community (Pamp and Tolker-Nielsen, 2007).

Besides the important distinction between metabolically active and inactive subpopulations, the structure of the biofilm itself can influence the susceptibility of the community to antimicrobials. Landry et al. demonstrated that flat biofilms were generally more susceptible to tobramycin than structured biofilms (Landry et al., 2006). Two other studies also showed that flat biofilms were more susceptible than structured biofilms to tobramycin and SDS (Davies et al., 1998; Bjarnsholt et al., 2005). A systematic study that compares the relative susceptibilities of biofilms possessing different three-dimensional structures, however, has not been carefully conducted.

SIGNALING DETERMINANTS THAT REGULATE BIOFILM FORMATION

A focus of recent work has been on identifying the signal transduction and regulatory pathways that control biofilm formation and that integrate different environmental signals during biofilm development. While this chapter does not provide an extensive review of these regulatory pathways, a few comments are made about the regulation of biofilm formation by c-di-GMP and two-component systems (TCSs). For a more extensive discussion of molecular determinants that regulate biofilm formation, see Dow et al., 2007; Mikkelsen et al., 2011; and Karatan and Watnick, 2009.

c-di-GMP

c-di-GMP is a ubiquitous second messenger in signal transduction for many gram-negative bacterial species, which has been shown to affect virulence, motility, biofilm formation, and cell cycle progression (Hengge, 2009). c-di-GMP is created from two molecules of GTP by the enzymatic activity of a diguanylate cyclase (DGC) and is broken down into pGpG by phosphodiesterase (PDE) activity (Hengge, 2009). DGCs typically harbor a conserved GGDEF motif, while PDEs contain conserved EAL or HD-GYP motifs (Galperin, 2005). *P. aeruginosa* contains 17 GGDEF-containing proteins, 5 EAL-containing proteins, 3 HD-GYP-containing proteins, and 16 proteins that contain both GGDEF and EAL domains (Kulasakara et al., 2006). While it is unclear if all these proteins function as DGCs and/or PDEs, mutations within some of these genes impact biofilm formation (Kulasakara, et al., 2006; Ryan, et al., 2009).

Generally, increased levels of c-di-GMP are associated with increased biofilm formation. Strains with a deletion in PDEs (such as BifA/PA4367, RbdA/PA0861, PvrR/PA3949, FimX/PA4959, PA4108, and PA4781) or with hyperactivated DGCs have increased c-di-GMP levels, decreased motility, increased exopolysaccharide production, and enhanced biofilm formation, while strains with a deletion of DGCs (such as SadC/PA4332, RoeA/PA1107, and MorA/PA4601) have the opposite effect (D'Argenio et al., 2002; Hickman et al., 2005; Kazmierczak et al., 2006; Kuchma et al., 2007; Meissner et al., 2007; Merritt et al., 2007; Ryan et al., 2009; Starkey et al., 2009; An et al., 2010; Borlee et al., 2010; Merritt et al., 2010). Hickman et al. showed that in a strain containing a deletion in a negative regulator of WspR, a DGC, cellular levels of c-di-GMP are increased and biofilm formation is enhanced. While this mutant initially formed a thicker biofilm faster than the wild-type strain, both strains formed equally thick biofilms after 48 h. The mutant strain, however, formed a mature biofilm that was different from that of the wild-type strain (Hickman et al., 2005). Strains with mutations in this negative regulator have increased transcription of *cdrA* and the PSL and PEL biosynthetic genes and have decreased transcription of the flagellar motility genes, which may explain the effect of the mutation on the biofilm structure (Hickman et al., 2005; Starkey et al., 2009).

While increased levels of c-di-GMP can increase the transcript level of the PEL biosynthetic operon, c-di-GMP can also affect PEL biosynthesis at the posttranscriptional level. PelD, a PEL biosynthetic protein, contains a c-di-GMP binding motif. PelD mutants that cannot bind c-di-GMP fail to complement the biofilm phenotype of a Δ*pelD* strain, suggesting that c-di-GMP-bound PelD is needed for biofilm formation (Lee et al., 2007). These results help explain how changes in c-di-GMP levels can adjust the amount of PEL without affecting the transcription of the PEL biosynthetic machinery, as seen in certain mutants (Kuchma et al., 2007; Merritt et al., 2007; Merritt et al., 2010). In addition to this mechanism of posttranscriptional regulation, the regulation of biofilm formation by c-di-GMP may be even more complicated (Hoffman et al., 2005). Different mutations that lead to increased c-di-GMP levels have different phenotypes. For instance, strains with a deletion in either of the DGCs SadC and RoeA have similarly increased levels of c-di-GMP. The Δ*sadC* strain, however, has a strong impairment in flagellar motility and a minimal reduction in PEL production, while the Δ*roeA* strain has no obvious flagellar motility defects and a severe reduction in PEL production (Merritt et al., 2010). Merritt and colleagues show that SadC and RoeA have different localization patterns in the cell and propose that localized changes in c-di-GMP play a larger role in biofilm regulation than total changes in c-di-GMP levels (Merritt et al., 2010).

Two Component Systems

TCSs are regulatory modules that are often used by the cell to respond to environmental changes. In its simplest form, a TCS contains a sensor histidine kinase and a response regulator. The sensor consists of an input domain that is often in the periplasm and a transmitter domain in the cytoplasm, while the regulator is composed of a receiver domain and a DNA-binding output domain. A signal activates the input domain of the sensor, which causes the sensor to autophosphorylate its transmitter domain on a conserved histidine. The phospho group on the transmitter domain of the sensor is transferred to a conserved aspartate on the receiver domain of the response regulator, activating the regulator as a transcription factor (Mikkelsen et al., 2011). The *P. aeruginosa* genome contains 63 putative histidine kinases and 64 putative response regulators (Rodrigue et al., 2000). Of the many TCSs in *P. aeruginosa*, a few have been shown to regulate biofilm formation (Parkins et al., 2001; Kuchma et al., 2005; Kulasekara et al., 2005; Petrova and Sauer, 2009).

One of the TCSs that affect biofilm formation in *P. aeruginosa* is the GacS/GacA system (Parkins et al., 2001). Upon activation by GacS, GacA activates the transcription of two small RNAs (RsmY and RsmZ) (Brencic et al., 2009). These two small RNAs bind to RsmA, an RNA binding protein that modulates mRNA translation. RsmY and RsmZ essentially titrate RsmA away from its targets, thereby alleviating RsmA regulation (Mikkelsen et al., 2011). Within the large RsmA regulon are genes involved in twitching motility and PEL and PSL biosynthesis (Brencic and Lory, 2009; Irie et al., 2010). In accordance with the observation that *gacA* mutant strains are impaired in biofilm formation (Parkins et al., 2001), RsmA directly represses the translation of the *psl* operon (Irie et al., 2010).

In addition to regulating factors that affect biofilm structure, TCSs appear to regulate biofilm maturation. While mutants in the SadARS system (also known as the Roc1 system [Kulasekara et al., 2005]) are not defective in biofilm initiation in comparison to wild-type cells, these mutants develop mature biofilms that are thicker and less heterogeneous than those formed by the wild-type strain, suggesting that SadARS regulates a postinitiation step in biofilm formation (Kuchma et al., 2005). Similarly, mutants with alterations in the BfiSR, BfmSR, and MifRS TCSs also affect biofilm maturation (Petrova and Sauer, 2009). While mutants with alterations in these TCSs are similar to wild-type cells in their abilities to attach

to polystyrene, to be motile, and to produce exopolysaccharides, these mutants fail to produce biofilms that resemble that of a wild-type strain. Interestingly, under conditions where wild-type cells formed structured biofilms, these TCS mutants develop endpoint biofilms that resemble different stages of wild-type biofilm maturation. Petrova and Sauer suggest that these TCS mutants are arrested in different stages of biofilm maturation and that these TCSs regulate the transition of biofilms from one stage of maturation to the next (Petrova and Sauer, 2009). Supporting this hypothesis, Petrova and colleagues showed that the BfiSR system regulates RsmZ levels (Petrova and Sauer, 2010) and the BfmSR system regulates eDNA production (Petrova et al., 2011).

CONCLUSIONS

It is an exciting time in biofilm microbiology. Laboratory biofilms represent an experimentally tractable, spatially structured, complex system. Many microbes in the environment and in disease exist in spatially structured environments that undoubtedly impact their physiology and behavior. This is particularly true in the natural environment, where high-cell-density, multispecies communities are common. Examples include photosynthetic mats and the rhizosphere, where different species engage in important intra- and interspecies interactions. Not surprisingly, researchers in these areas have developed carefully controlled model laboratory biofilm systems to ask specific questions. The rapid progress being made in genomics, metagenomics, transcriptional, and proteomic profiling-related technologies has also propelled biofilm-related research.

The general importance of c-di-GMP signaling for biofilm development in gram-negative species has also been an exciting development. Evidence is mounting that species have several environmental signals that can modulate c-di-GMP levels, suggesting that c-di-GMP represents one of the major signaling pathways controlling the decision to attach or swim. Researchers are currently trying to determine the nature of these environmental signals, and as they do, our understanding of which environments promote or impair biofilm formation will grow. Therapeutically, a better understanding of c-di-GMP signaling may lead to the development of a "silver bullet" for disrupting biofilm infections.

As the field continues to mature, our ability to link molecular events to community-wide phenomena will improve. Our amassing knowledge may aid not only in disrupting biofilms but also in designing synthetic communities. Microbial fuel cells and wastewater treatment communities are a few examples of engineered, structured communities that could benefit from such an approach. Thus, we predict that the next decade will have many exciting advances and discoveries in store for us.

REFERENCES

Absalon, C., K. Van Dellen, and P. I. Watnick. 2011. A communal bacterial adhesin anchors biofilm and bystander cells to surfaces. *PLoS Pathog.* 7:e1002210.

Allesen-Holm, M., K. B. Barken, L. Yang, M. Klausen, J. S. Webb, S. Kjelleberg, S. Molin, M. Givskov, and T. Tolker-Nielsen. 2006. A characterization of DNA release in *Pseudomonas aeruginosa* cultures and biofilms. *Mol. Microbiol.* 59:1114–1128.

Alvarez-Ortega, C., and C. S. Harwood. 2007. Responses of *Pseudomonas aeruginosa* to low oxygen indicate that growth in the cystic fibrosis lung is by aerobic respiration. *Mol. Microbiol.* 65:153–165.

An, S., J. Wu, and L. H. Zhang. 2010. Modulation of *Pseudomonas aeruginosa* biofilm dispersal by a cyclic-di-GMP phosphodiesterase with a putative hypoxia-sensing domain. *Appl. Environ. Microbiol.* 76:8160–8173.

Banin, E., M. L. Vasil, and E. P. Greenberg. 2005. Iron and *Pseudomonas aeruginosa* biofilm formation. *Proc. Natl. Acad. Sci. USA* 102:11076–11081.

Barken, K. B., S. J. Pamp, L. Yang, M. Gjermansen, J. J. Bertrand, M. Klausen, M. Givskov, C. B. Whitchurch, J. N. Engel, and T. Tolker-Nielsen. 2008. Roles of type IV pili, flagellum-mediated motility and extracellular DNA in the formation of mature multicellular structures in *Pseudomonas aeruginosa* biofilms. *Environ. Microbiol.* 10:2331–2343.

Barton, H. A., Z. Johnson, C. D. Cox, A. I. Vasil, and M. L. Vasil. 1996. Ferric uptake regulator mutants of *Pseudomonas aeruginosa* with distinct alterations in the iron-dependent repression of exotoxin A and siderophores in aerobic and microaerobic environments. *Mol. Microbiol.* 21:1001–1017.

Berg, H. C. 2003. The rotary motor of bacterial flagella. *Annu. Rev. Biochem.* 72:19–54.

Bjarnsholt, T., P. O. Jensen, M. Burmolle, M. Hentzer, J. A. Haagensen, H. P. Hougen, H. Calum, K. G. Madsen, C. Moser, S. Molin, N. Hoiby, and M. Givskov. 2005. *Pseudomonas aeruginosa* tolerance to tobramycin, hydrogen peroxide and polymorphonuclear leukocytes is quorum-sensing dependent. *Microbiology* 151:373–383.

Blair, K. M., L. Turner, J. T. Winkelman, H. C. Berg, and D. B. Kearns. 2008. A molecular clutch disables flagella in the *Bacillus subtilis* biofilm. *Science* 320:1636–1638.

Boles, B. R., M. Thoendel, and P. K. Singh. 2005. Rhamnolipids mediate detachment of *Pseudomonas aeruginosa* from biofilms. *Mol. Microbiol.* 57:1210–1223.

Borlee, B. R., A. D. Goldman, K. Murakami, R. Samudrala, D. J. Wozniak, and M. R. Parsek. 2010. *Pseudomonas aeruginosa* uses a cyclic-di-GMP-regulated adhesin to reinforce the biofilm extracellular matrix. *Mol. Microbiol.* 75:827–842.

Borriello, G., E. Werner, F. Roe, A. M. Kim, G. D. Ehrlich, and P. S. Stewart. 2004. Oxygen limitation contributes to antibiotic tolerance of *Pseudomonas aeruginosa* in biofilms. *Antimicrob. Agents Chemother.* 48:2659–2664.

Brencic, A., and S. Lory. 2009. Determination of the regulon and identification of novel mRNA targets of *Pseudomonas aeruginosa* RsmA. *Mol. Microbiol.* 72:612–632.

Brencic, A., K. A. McFarland, H. R. McManus, S. Castang, I. Mogno, S. L. Dove, and S. Lory. 2009. The GacS/GacA signal transduction system of *Pseudomonas aeruginosa* acts exclusively through its control over the transcription of the RsmY and RsmZ regulatory small RNAs. *Mol. Microbiol.* 73:434–445.

Byrd, M. S., I. Sadovskaya, E. Vinogradov, H. Lu, A. B. Sprinkle, S. H. Richardson, L. Ma, B. Ralston, M. R. Parsek, E. M. Anderson, J. S. Lam, and D. J. Wozniak. 2009. Genetic and biochemical analyses of the *Pseudomonas aeruginosa* Psl exopolysaccharide reveal overlapping roles for polysaccharide synthesis enzymes in Psl and LPS production. *Mol. Microbiol.* 73:622–638.

Colvin, K. M., V. D. Gordon, K. Murakami, B. R. Borlee, D. J. Wozniak, G. C. Wong, and M. R. Parsek. 2011. The Pel polysaccharide can serve a structural and protective role in the biofilm matrix of *Pseudomonas aeruginosa*. *PLoS Pathog.* 7:e1001264.

Cornelis, P. 2010. Iron uptake and metabolism in pseudomonads. *Appl. Microbiol. Biotechnol.* 86:1637–1645.

Costerton, J. W., Z. Lewandowski, D. E. Caldwell, D. R. Korber, and H. M. Lappin-Scott. 1995. Microbial biofilms. *Annu. Rev. Microbiol.* 49:711–745.

Cox, C. D., and P. Adams. 1985. Siderophore activity of pyoverdin for *Pseudomonas aeruginosa*. *Infect. Immun.* 48:130–138.

Cox, C. D., and R. Graham. 1979. Isolation of an iron-binding compound from *Pseudomonas aeruginosa*. *J. Bacteriol.* 137:357–364.

Danese, P. N., L. A. Pratt, and R. Kolter. 2000. Exopolysaccharide production is required for development of *Escherichia coli* K-12 biofilm architecture. *J. Bacteriol.* 182:3593–3596.

D'Argenio, D. A., M. W. Calfee, P. B. Rainey, and E. C. Pesci. 2002. Autolysis and autoaggregation in *Pseudomonas aeruginosa* colony morphology mutants. *J. Bacteriol.* 184:6481–6489.

Davey, M. E., N. C. Caiazza, and G. A. O'Toole. 2003. Rhamnolipid surfactant production affects biofilm architecture in *Pseudomonas aeruginosa* PAO1. *J. Bacteriol.* 185:1027–1036.

Davies, D. G., M. R. Parsek, J. P. Pearson, B. H. Iglewski, J. W. Costerton, and E. P. Greenberg. 1998. The involvement of cell-to-cell signals in the development of a bacterial biofilm. *Science* 280:295–298.

De Kievit, T. R., R. Gillis, S. Marx, C. Brown, and B. H. Iglewski. 2001. Quorum-sensing genes in *Pseudomonas aeruginosa* biofilms: their role and expression patterns. *Appl. Environ. Microbiol.* 67:1865–1873.

Dow, J. M., Y. Fouhy, J. Lucey, and R. P. Ryan. 2007. Cyclic di-GMP as an intracellular signal regulating bacterial biofilm formation, p. 71–93. *In* S. Kjelleberg and M. Givskov (ed.), *The Biofilm Mode of Life: Mechanisms and Adaptations*. Horizon Bioscience, Norfolk, United Kingdom.

Fong, J. C., K. Karplus, G. K. Schoolnik, and F. H. Yildiz. 2006. Identification and characterization of RbmA, a novel protein required for the development of rugose colony morphology and biofilm structure in *Vibrio cholerae*. *J. Bacteriol.* 188:1049–1059.

Fong, J. C., and F. H. Yildiz. 2007. The *rbmBCDEF* gene cluster modulates development of rugose colony morphology and biofilm formation in *Vibrio cholerae*. *J. Bacteriol.* 189:2319–2330.

Friedman, L., and R. Kolter. 2004a. Genes involved in matrix formation in *Pseudomonas aeruginosa* PA14 biofilms. *Mol. Microbiol.* 51:675–690.

Friedman, L., and R. Kolter. 2004b. Two genetic loci produce distinct carbohydrate-rich structural components of the *Pseudomonas aeruginosa* biofilm matrix. *J. Bacteriol.* 186:4457–4465.

Galperin, M. Y. 2005. A census of membrane-bound and intracellular signal transduction proteins in bacteria: bacterial IQ, extroverts and introverts. *BMC Microbiol.* 5:35.

Gibiansky, M. L., J. C. Conrad, F. Jin, V. D. Gordon, D. A. Motto, M. A. Mathewson, W. G. Stopka, D. C. Zelasko, J. D. Shrout, and G. C. Wong. 2010. Bacteria use type IV pili to walk upright and detach from surfaces. *Science* 330:197.

Glick, R., C. Gilmour, J. Tremblay, S. Satanower, O. Avidan, E. Deziel, E. P. Greenberg, K. Poole, and E. Banin. 2010. Increase in rhamnolipid synthesis under iron-limiting conditions influences surface motility and biofilm formation in *Pseudomonas aeruginosa*. *J. Bacteriol.* **192:**2973–2980.

Gotz, F. 2002. Staphylococcus and biofilms. *Mol. Microbiol.* **43:**1367–1378.

Haagensen, J. A., M. Klausen, R. K. Ernst, S. I. Miller, A. Folkesson, T. Tolker-Nielsen, and S. Molin. 2007. Differentiation and distribution of colistin- and sodium dodecyl sulfate-tolerant cells in *Pseudomonas aeruginosa* biofilms. *J. Bacteriol.* **189:**28–37.

Hengge, R. 2009. Principles of c-di-GMP signalling in bacteria. *Nat. Rev. Microbiol.* **7:**263–273.

Hentzer, M., G. M. Teitzel, G. J. Balzer, A. Heydorn, S. Molin, M. Givskov, and M. R. Parsek. 2001. Alginate overproduction affects *Pseudomonas aeruginosa* biofilm structure and function. *J. Bacteriol.* **183:**5395–5401.

Heydorn, A., B. Ersboll, J. Kato, M. Hentzer, M. R. Parsek, T. Tolker-Nielsen, M. Givskov, and S. Molin. 2002. Statistical analysis of *Pseudomonas aeruginosa* biofilm development: impact of mutations in genes involved in twitching motility, cell-to-cell signaling, and stationary-phase sigma factor expression. *Appl. Environ. Microbiol.* **68:**2008–2017.

Heydorn, A., A. T. Nielsen, M. Hentzer, C. Sternberg, M. Givskov, B. K. Ersboll, and S. Molin. 2000. Quantification of biofilm structures by the novel computer program COMSTAT. *Microbiology* **146**(Pt. 10)**:**2395–2407.

Hickman, J. W., D. F. Tifrea, and C. S. Harwood. 2005. A chemosensory system that regulates biofilm formation through modulation of cyclic diguanylate levels. *Proc. Natl. Acad. Sci. USA* **102:**14422–14427.

Hoffman, L. R., D. A. D'Argenio, M. J. MacCoss, Z. Zhang, R. A. Jones, and S. I. Miller. 2005. Aminoglycoside antibiotics induce bacterial biofilm formation. *Nature* **436:**1171–1175.

Irie, Y., M. Starkey, A. N. Edwards, D. J. Wozniak, T. Romeo, and M. R. Parsek. 2010. *Pseudomonas aeruginosa* biofilm matrix polysaccharide Psl is regulated transcriptionally by RpoS and post-transcriptionally by RsmA. *Mol. Microbiol.* **78:**158–172.

Jackson, K. D., M. Starkey, S. Kremer, M. R. Parsek, and D. J. Wozniak. 2004. Identification of *psl*, a locus encoding a potential exopolysaccharide that is essential for *Pseudomonas aeruginosa* PAO1 biofilm formation. *J. Bacteriol.* **186:**4466–4475.

Jin, F., J. C. Conrad, M. L. Gibiansky, and G. C. Wong. 2011. Bacteria use type-IV pili to slingshot on surfaces. *Proc. Natl. Acad. Sci. USA* **108:**12617–12622.

Kaneko, Y., M. Thoendel, O. Olakanmi, B. E. Britigan, and P. K. Singh. 2007. The transition metal gallium disrupts *Pseudomonas aeruginosa* iron metabolism and has antimicrobial and antibiofilm activity. *J. Clin. Investig.* **117:**877–888.

Karatan, E., and P. Watnick. 2009. Signals, regulatory networks, and materials that build and break bacterial biofilms. *Microbiol. Mol. Biol. Rev.* **73:**310–347.

Kazmierczak, B. I., M. B. Lebron, and T. S. Murray. 2006. Analysis of FimX, a phosphodiesterase that governs twitching motility in *Pseudomonas aeruginosa*. *Mol. Microbiol.* **60:**1026–1043.

Kearns, D. B., J. Robinson, and L. J. Shimkets. 2001. *Pseudomonas aeruginosa* exhibits directed twitching motility up phosphatidylethanolamine gradients. *J. Bacteriol.* **183:**763–767.

Kirisits, M. J., and M. R. Parsek. 2006. Does *Pseudomonas aeruginosa* use intercellular signalling to build biofilm communities? *Cell. Microbiol.* **8:**1841–1849.

Kirisits, M. J., L. Prost, M. Starkey, and M. R. Parsek. 2005. Characterization of colony morphology variants isolated from *Pseudomonas aeruginosa* biofilms. *Appl. Environ. Microbiol.* **71:**4809–4821.

Klausen, M., A. Aaes-Jorgensen, S. Molin, and T. Tolker-Nielsen. 2003a. Involvement of bacterial migration in the development of complex multicellular structures in *Pseudomonas aeruginosa* biofilms. *Mol. Microbiol.* **50:**61–68.

Klausen, M., A. Heydorn, P. Ragas, L. Lambertsen, A. Aaes-Jorgensen, S. Molin, and T. Tolker-Nielsen. 2003b. Biofilm formation by *Pseudomonas aeruginosa* wild type, flagella and type IV pili mutants. *Mol. Microbiol.* **48:**1511–1524.

Kohler, T., L. K. Curty, F. Barja, C. van Delden, and J. C. Pechere. 2000. Swarming of *Pseudomonas aeruginosa* is dependent on cell-to-cell signaling and requires flagella and pili. *J. Bacteriol.* **182:**5990–5996.

Kuchma, S. L., K. M. Brothers, J. H. Merritt, N. T. Liberati, F. M. Ausubel, and G. A. O'Toole. 2007. BifA, a cyclic-Di-GMP phosphodiesterase, inversely regulates biofilm formation and swarming motility by *Pseudomonas aeruginosa* PA14. *J. Bacteriol.* **189:**8165–8178.

Kuchma, S. L., J. P. Connolly, and G. A. O'Toole. 2005. A three-component regulatory system regulates biofilm maturation and type III secretion in *Pseudomonas aeruginosa*. *J. Bacteriol.* **187:**1441–1454.

Kulasakara, H., V. Lee, A. Brencic, N. Liberati, J. Urbach, S. Miyata, D. G. Lee, A. N. Neely, M. Hyodo, Y. Hayakawa, F. M. Ausubel, and S. Lory. 2006. Analysis of *Pseudomonas aeruginosa* diguanylate cyclases and phosphodiesterases reveals a role for bis-(3'-5')-cyclic-GMP in virulence. *Proc. Natl. Acad. Sci. USA* **103:**2839–2844.

Kulasekara, H. D., I. Ventre, B. R. Kulasekara, A. Lazdunski, A. Filloux, and S. Lory. 2005. A novel two-component system controls the expression of *Pseudomonas aeruginosa* fimbrial cup genes. *Mol. Microbiol.* **55:**368–380.

Landry, R. M., D. An, J. T. Hupp, P. K. Singh, and M. R. Parsek. 2006. Mucin-*Pseudomonas aeruginosa* interactions promote biofilm formation and antibiotic resistance. *Mol. Microbiol.* **59:**142–151.

Latasa, C., C. Solano, J. R. Penades, and I. Lasa. 2006. Biofilm-associated proteins. *C. R. Biol.* **329:**849–857.

Lawrence, J. R., D. R. Korber, B. D. Hoyle, J. W. Costerton, and D. E. Caldwell. 1991. Optical sectioning of microbial biofilms. *J. Bacteriol.* **173:**6558–6567.

Lee, V. T., J. M. Matewish, J. L. Kessler, M. Hyodo, Y. Hayakawa, and S. Lory. 2007. A cyclic-di-GMP receptor required for bacterial exopolysaccharide production. *Mol. Microbiol.* **65:**1474–1484.

Lequette, Y., and E. P. Greenberg. 2005. Timing and localization of rhamnolipid synthesis gene expression in *Pseudomonas aeruginosa* biofilms. *J. Bacteriol.* **187:**37–44.

Li, Y., H. Xia, F. Bai, H. Xu, L. Yang, H. Yao, L. Zhang, X. Zhang, Y. Bai, P. E. Saris, T. Tolker-Nielsen, and M. Qiao. 2007. Identification of a new gene PA5017 involved in flagella-mediated motility, chemotaxis and biofilm formation in *Pseudomonas aeruginosa*. *FEMS Microbiol. Lett.* **272:**188–195.

Ma, L., M. Conover, H. Lu, M. R. Parsek, K. Bayles, and D. J. Wozniak. 2009. Assembly and development of the *Pseudomonas aeruginosa* biofilm matrix. *PLoS Pathog.* **5:**e1000354.

Ma, L., K. D. Jackson, R. M. Landry, M. R. Parsek, and D. J. Wozniak. 2006. Analysis of *Pseudomonas aeruginosa* conditional Psl variants reveals roles for the Psl polysaccharide in adhesion and maintaining biofilm structure postattachment. *J. Bacteriol.* **188:**8213–8221.

Martin, D. W., M. J. Schurr, M. H. Mudd, J. R. Govan, B. W. Holloway, and V. Deretic. 1993. Mechanism of conversion to mucoidy in *Pseudomonas aeruginosa* infecting cystic fibrosis patients. *Proc. Natl. Acad. Sci. USA* **90:**8377–8381.

Marvasi, M., P. T. Visscher, and L. Casillas Martinez. 2010. Exopolymeric substances (EPS) from *Bacillus subtilis*: polymers and genes encoding their synthesis. *FEMS Microbiol. Lett.* **313:**1–9.

Masse, E., H. Salvail, G. Desnoyers, and M. Arguin. 2007. Small RNAs controlling iron metabolism. *Curr. Opin. Microbiol.* **10:**140–145.

Massol-Deya, A. A., J. Whallon, R. F. Hickey, and J. M. Tiedje. 1995. Channel structures in aerobic biofilms of fixed-film reactors treating contaminated groundwater. *Appl. Environ. Microbiol.* **61:**769–777.

Matsukawa, M., and E. P. Greenberg. 2004. Putative exopolysaccharide synthesis genes influence *Pseudomonas aeruginosa* biofilm development. *J. Bacteriol.* **186:**4449–4456.

Mattick, J. S. 2002. Type IV pili and twitching motility. *Annu. Rev. Microbiol.* **56:**289–314.

May, T. B., D. Shinabarger, R. Maharaj, J. Kato, L. Chu, J. D. DeVault, S. Roychoudhury, N. A. Zielinski, A. Berry, R. K. Rothmel, et al. 1991. Alginate synthesis by *Pseudomonas aeruginosa*: a key pathogenic factor in chronic pulmonary infections of cystic fibrosis patients. *Clin. Microbiol. Rev.* **4:**191–206.

Meissner, A., V. Wild, R. Simm, M. Rohde, C. Erck, F. Bredenbruch, M. Morr, U. Romling, and S. Haussler. 2007. *Pseudomonas aeruginosa* cupA-encoded fimbriae expression is regulated by a GGDEF and EAL domain-dependent modulation of the intracellular level of cyclic diguanylate. *Environ. Microbiol.* **9:**2475–2485.

Merritt, J. H., K. M. Brothers, S. L. Kuchma, and G. A. O'Toole. 2007. SadC reciprocally influences biofilm formation and swarming motility via modulation of exopolysaccharide production and flagellar function. *J. Bacteriol.* **189:**8154–8164.

Merritt, J. H., D. G. Ha, K. N. Cowles, W. Lu, D. K. Morales, J. Rabinowitz, Z. Gitai, and G. A. O'Toole. 2010. Specific control of *Pseudomonas aeruginosa* surface-associated behaviors by two c-di-GMP diguanylate cyclases. *mBio* **1:**e00183-10.

Mikkelsen, H., Z. Duck, K. S. Lilley, and M. Welch. 2007. Interrelationships between colonies, biofilms, and planktonic cells of *Pseudomonas aeruginosa*. *J. Bacteriol.* **189:**2411–2416.

Mikkelsen, H., M. Sivaneson, and A. Filloux. 2011. Key two-component regulatory systems that control biofilm formation in *Pseudomonas aeruginosa*. *Environ. Microbiol.* **13:**1666–1681.

Mulcahy, H., L. Charron-Mazenod, and S. Lewenza. 2010. *Pseudomonas aeruginosa* produces an extracellular deoxyribonuclease that is required for utilization of DNA as a nutrient source. *Environ. Microbiol.* **12:**1621–1629.

Musk, D. J., D. A. Banko, and P. J. Hergenrother. 2005. Iron salts perturb biofilm formation and disrupt existing biofilms of *Pseudomonas aeruginosa*. *Chem. Biol.* **12:**789–796.

Nivens, D. E., D. E. Ohman, J. Williams, and M. J. Franklin. 2001. Role of alginate and its O acetylation in formation of *Pseudomonas aeruginosa* microcolonies and biofilms. *J. Bacteriol.* **183:**1047–1057.

Ochsner, U. A., Z. Johnson, and M. L. Vasil. 2000. Genetics and regulation of two distinct haem-uptake systems, *phu* and *has*, in *Pseudomonas aeruginosa*. *Microbiology* **146**(Pt. 1):185–198.

Ochsner, U. A., A. I. Vasil, and M. L. Vasil. 1995. Role of the ferric uptake regulator of *Pseudomonas aeruginosa* in the regulation of siderophores and exotoxin A expression: purification and activity on iron-regulated promoters. *J. Bacteriol.* **177:**7194–7201.

Oglesby, A. G., J. M. Farrow III, J. H. Lee, A. P. Tomaras, E. P. Greenberg, E. C. Pesci, and M. L. Vasil. 2008. The influence of iron on *Pseudomonas aeruginosa* physiology: a regulatory link between iron and quorum sensing. *J. Biol. Chem.* **283:**15558–15567.

O'Toole, G., H. B. Kaplan, and R. Kolter. 2000a. Biofilm formation as microbial development. *Annu. Rev. Microbiol.* **54:**49–79.

O'Toole, G. A., K. A. Gibbs, P. W. Hager, P. V. Phibbs, Jr., and R. Kolter. 2000b. The global carbon metabolism regulator Crc

is a component of a signal transduction pathway required for biofilm development by *Pseudomonas aeruginosa*. *J. Bacteriol.* **182:**425–431.

O'Toole, G. A., and R. Kolter. 1998a. Flagellar and twitching motility are necessary for *Pseudomonas aeruginosa* biofilm development. *Mol. Microbiol.* **30:**295–304.

O'Toole, G. A., and R. Kolter. 1998b. Initiation of biofilm formation in *Pseudomonas fluorescens* WCS365 proceeds via multiple, convergent signalling pathways: a genetic analysis. *Mol. Microbiol.* **28:**449–461.

Overhage, J., M. Schemionek, J. S. Webb, and B. H. Rehm. 2005. Expression of the *psl* operon in *Pseudomonas aeruginosa* PAO1 biofilms: PslA performs an essential function in biofilm formation. *Appl. Environ. Microbiol.* **71:**4407–4413.

Pamp, S. J., and T. Tolker-Nielsen. 2007. Multiple roles of biosurfactants in structural biofilm development by *Pseudomonas aeruginosa*. *J. Bacteriol.* **189:**2531–2539.

Parkins, M. D., H. Ceri, and D. G. Storey. 2001. *Pseudomonas aeruginosa* GacA, a factor in multihost virulence, is also essential for biofilm formation. *Mol. Microbiol.* **40:**1215–1226.

Patriquin, G. M., E. Banin, C. Gilmour, R. Tuchman, E. P. Greenberg, and K. Poole. 2008. Influence of quorum sensing and iron on twitching motility and biofilm formation in *Pseudomonas aeruginosa*. *J. Bacteriol.* **190:**662–671.

Petrova, O. E., and K. Sauer. 2009. A novel signaling network essential for regulating *Pseudomonas aeruginosa* biofilm development. *PLoS Pathog.* **5:**e1000668.

Petrova, O. E., and K. Sauer. 2010. The novel two-component regulatory system BfiSR regulates biofilm development by controlling the small RNA *rsmZ* through CafA. *J. Bacteriol.* **192:**5275–5288.

Petrova, O. E., J. R. Schurr, M. J. Schurr, and K. Sauer. 2011. The novel *Pseudomonas aeruginosa* two-component regulator BfmR controls bacteriophage-mediated lysis and DNA release during biofilm development through PhdA. *Mol. Microbiol.* **81:**767–783.

Picioreanu, C., M. C. van Loosdrecht, and J. J. Heijnen. 1998. Mathematical modeling of biofilm structure with a hybrid differential-discrete cellular automaton approach. *Biotechnol. Bioeng.* **58:**101–116.

Rodrigue, A., Y. Quentin, A. Lazdunski, V. Mejean, and M. Foglino. 2000. Two-component systems in *Pseudomonas aeruginosa*: why so many? *Trends Microbiol.* **8:**498–504.

Romero, D., C. Aguilar, R. Losick, and R. Kolter. 2010. Amyloid fibers provide structural integrity to *Bacillus subtilis* biofilms. *Proc. Natl. Acad. Sci. USA* **107:**2230–2234.

Ryan, R. P., J. Lucey, K. O'Donovan, Y. McCarthy, L. Yang, T. Tolker-Nielsen, and J. M. Dow. 2009. HD-GYP domain proteins regulate biofilm formation and virulence in *Pseudomonas aeruginosa*. *Environ. Microbiol.* **11:**1126–1136.

Ryder, C., M. Byrd, and D. J. Wozniak. 2007. Role of polysaccharides in *Pseudomonas aeruginosa* biofilm development. *Curr. Opin. Microbiol.* **10:**644–648.

Sauer, K., A. K. Camper, G. D. Ehrlich, J. W. Costerton, and D. G. Davies. 2002. *Pseudomonas aeruginosa* displays multiple phenotypes during development as a biofilm. *J. Bacteriol.* **184:**1140–1154.

Schmidt, J., M. Musken, T. Becker, Z. Magnowska, D. Bertinetti, S. Moller, B. Zimmermann, F. W. Herberg, L. Jansch, and S. Haussler. 2011. The *Pseudomonas aeruginosa* chemotaxis methyltransferase CheR1 impacts on bacterial surface sampling. *PLoS One* **6:**e18184.

Shrout, J. D., D. L. Chopp, C. L. Just, M. Hentzer, M. Givskov, and M. R. Parsek. 2006. The impact of quorum sensing and swarming motility on *Pseudomonas aeruginosa* biofilm formation is nutritionally conditional. *Mol. Microbiol.* **62:**1264–1277.

Singh, P. K., M. R. Parsek, E. P. Greenberg, and M. J. Welsh. 2002. A component of innate immunity prevents bacterial biofilm development. *Nature* **417:**552–555.

Soberon-Chavez, G., F. Lepine, and E. Deziel. 2005. Production of rhamnolipids by *Pseudomonas aeruginosa. Appl. Microbiol. Biotechnol.* **68:**718–725.

Southey-Pillig, C. J., D. G. Davies, and K. Sauer. 2005. Characterization of temporal protein production in *Pseudomonas aeruginosa* biofilms. *J. Bacteriol.* **187:**8114–8126.

Stapper, A. P., G. Narasimhan, D. E. Ohman, J. Barakat, M. Hentzer, S. Molin, A. Kharazmi, N. Hoiby, and K. Mathee. 2004. Alginate production affects *Pseudomonas aeruginosa* biofilm development and architecture, but is not essential for biofilm formation. *J. Med. Microbiol.* **53:**679–690.

Starkey, M., J. H. Hickman, L. Ma, N. Zhang, S. De Long, A. Hinz, S. Palacios, C. Manoil, M. J. Kirisits, T. D. Starner, D. J. Wozniak, C. S. Harwood, and M. R. Parsek. 2009. *Pseudomonas aeruginosa* rugose small-colony variants have adaptations that likely promote persistence in the cystic fibrosis lung. *J. Bacteriol.* **191:**3492–3503.

Steinberger, R. E., and P. A. Holden. 2005. Extracellular DNA in single- and multiple-species unsaturated biofilms. *Appl. Environ. Microbiol.* **71:**5404–5410.

Stoodley, P., D. Debeer, and Z. Lewandowski. 1994. Liquid flow in biofilm systems. *Appl. Environ. Microbiol.* **60:**2711–2716.

Stoodley, P., K. Sauer, D. G. Davies, and J. W. Costerton. 2002. Biofilms as complex differentiated communities. *Annu. Rev. Microbiol.* **56:**187–209.

Thompson, L. S., J. S. Webb, S. A. Rice, and S. Kjelleberg. 2003. The alternative sigma factor RpoN regulates the quorum sensing gene *rhlI* in *Pseudomonas aeruginosa. FEMS Microbiol. Lett.* **220:**187–195.

Vasil, M. L., and U. A. Ochsner. 1999. The response of *Pseudomonas aeruginosa* to iron: genetics, biochemistry and virulence. *Mol. Microbiol.* **34:**399–413.

Vasseur, P., I. Vallet-Gely, C. Soscia, S. Genin, and A. Filloux. 2005. The *pel* genes of the *Pseudomonas aeruginosa* PAK strain are involved at early and late stages of biofilm formation. *Microbiology* **151:**985–997.

Vilain, S., P. Cosette, M. Hubert, C. Lange, G. A. Junter, and T. Jouenne. 2004. Comparative proteomic analysis of planktonic and immobilized *Pseudomonas aeruginosa* cells: a multivariate statistical approach. *Anal. Biochem.* **329:**120–130.

Wang, X., J. F. Preston III, and T. Romeo. 2004. The *pgaABCD* locus of *Escherichia coli* promotes the synthesis of a polysaccharide adhesin required for biofilm formation. *J. Bacteriol.* **186:**2724–2734.

Whitchurch, C. B., T. Tolker-Nielsen, P. C. Ragas, and J. S. Mattick. 2002. Extracellular DNA required for bacterial biofilm formation. *Science* **295:**1487.

Wilderman, P. J., N. A. Sowa, D. J. FitzGerald, P. C. FitzGerald, S. Gottesman, U. A. Ochsner, and M. L. Vasil. 2004. Identification of tandem duplicate regulatory small RNAs in *Pseudomonas aeruginosa* involved in iron homeostasis. *Proc. Natl. Acad. Sci. USA* **101:**9792–9797.

Williams, P., and M. Camara. 2009. Quorum sensing and environmental adaptation in *Pseudomonas aeruginosa*: a tale of regulatory networks and multifunctional signal molecules. *Curr. Opin. Microbiol.* **12:**182–191.

Wimpenny, J. W. T., and R. Colasanti. 1997. A unifying hypothesis for the structure of microbial biofilms based on cellular automaton models. *FEMS Microbiol. Ecol.* **22:**1–16.

Wolfaardt, G. M., J. R. Lawrence, R. D. Robarts, S. J. Caldwell, and D. E. Caldwell. 1994. Multicellular organization in a degradative biofilm community. *Appl. Environ. Microbiol.* **60:**434–446.

Wolff, J. A., C. H. MacGregor, R. C. Eisenberg, and P. V. Phibbs, Jr. 1991. Isolation and characterization of catabolite repression control mutants of *Pseudomonas aeruginosa* PAO. *J. Bacteriol.* **173:**4700–4706.

Xu, K. D., P. S. Stewart, F. Xia, C. T. Huang, and G. A. McFeters. 1998. Spatial physiological heterogeneity in *Pseudomonas aeruginosa* biofilm is determined by oxygen availability. *Appl. Environ. Microbiol.* **64:**4035–4039.

Yang, L., K. B. Barken, M. E. Skindersoe, A. B. Christensen, M. Givskov, and T. Tolker-Nielsen. 2007. Effects of iron on DNA release and biofilm development by *Pseudomonas aeruginosa. Microbiology* **153:**1318–1328.

Yang, L., Y. Hu, Y. Liu, J. Zhang, J. Ulstrup, and S. Molin. 2011. Distinct roles of extracellular polymeric substances in *Pseudomonas aeruginosa* biofilm development. *Environ. Microbiol.* **13:**1705–1717.

Yang, L., M. Nilsson, M. Gjermansen, M. Givskov, and T. Tolker-Nielsen. 2009. Pyoverdine and PQS mediated subpopulation interactions involved in *Pseudomonas aeruginosa* biofilm formation. *Mol. Microbiol.* **74:**1380–1392.

Yildiz, F. H., and K. L. Visick. 2009. *Vibrio* biofilms: so much the same yet so different. *Trends Microbiol.* **17:**109–118.

Regulation of Bacterial Virulence
Edited by Michael L. Vasil and Andrew J. Darwin
© 2013 ASM Press, Washington, DC   doi:10.1128/9781555818524.ch2

Chapter 2

Chronic versus Acute *Pseudomonas aeruginosa* Infection States

BARBARA I. KAZMIERCZAK AND THOMAS S. MURRAY

INTRODUCTION

Pseudomonas aeruginosa is a Gram-negative bacterium easily isolated from environmental reservoirs, such as freshwater and soil. It has long been appreciated to be a causative agent of acute opportunistic infections in immunocompromised individuals or those with compromised epithelial barriers. *P. aeruginosa* is also a frequent pathogen in individuals with preexisting respiratory tract disease, particularly cystic fibrosis (CF). In CF patients, chronic infection with *P. aeruginosa* is strongly associated with declining respiratory function and increased mortality rate. Acute and chronic infections exhibit distinct clinical courses that play out over days to weeks versus months to years, prompting many investigators to examine how a single bacterium can cause such a seemingly broad range of disease. Much of this work has focused on identifying specific virulence factors and phenotypes that are associated with acute versus chronic infection and on defining the regulatory pathways that control the expression of these factors.

In this chapter, we review the observations supporting the idea that acute and chronic *P. aeruginosa* infections represent distinct modes of host-pathogen interaction. The virulence factors associated with acute and chronic infections are discussed, with a focus on data obtained from human subject-based studies, when possible. Lastly, selected global regulatory networks that control expression of these virulence factors are reviewed, with the goal of understanding whether these serve as "switches" between states that favor acute versus chronic infection.

ACUTE VERSUS CHRONIC INFECTION

P. aeruginosa rarely causes infections in healthy humans despite its frequent isolation from environmental sources. Nonetheless, it is a causative agent of acute infections in immunocompetent and immunocompromised hosts under certain circumstances. Immunocompetent individuals usually develop *P. aeruginosa* infections when local barrier functions of an epithelial or mucosal tissue are compromised, e.g., corneal ulcerations and keratitis in contact lens wearers, superinfections in burn victims, or pneumonia in mechanically ventilated patients. Acute infections present with clinical signs and symptoms of inflammation (fever, swelling, pus, erythema, and pain), which typically resolve with appropriate antibiotic therapy but may still result in significant morbidity and mortality if inadequately treated. Host innate immune defenses are particularly important in controlling *P. aeruginosa* infections, and individuals in whom these are compromised are at increased risk of *P. aeruginosa* bacteremia and sepsis (Lavoie et al., 2011). Neutrophils in particular play a critical role in *P. aeruginosa* clearance (Koh et al., 2009), and patients with reduced numbers of circulating neutrophils (e.g., those receiving cancer chemotherapy) are particularly susceptible to acute infection and sepsis (Bodey, 2009).

P. aeruginosa expresses many virulence factors that can damage host cells and which contribute to infection in both humans and animal models. Chief among these is a type 3 secretion system (T3SS) that translocates a subset of cytotoxic bacterial effector proteins across both the bacterial cell envelope and eukaryotic plasma membrane (Table 1). The four

Barbara I. Kazmierczak • Department of Medicine (Infectious Diseases) & Microbial Pathogenesis, Yale University School of Medicine, 333 Cedar St., Box 208022, New Haven, CT 06520-8022.   Thomas S. Murray • Department of Basic Medical Sciences, Quinnipiac University School of Medicine, 275 Mt. Carmel Ave., N1-HSC, Hamden, CT 06518-1908.

Table 1. Activities associated with T3SS effectors of *P. aeruginosa*[a]

Effector	Enzymatic activity	In vitro phenotypes
Exoenzyme S	ADP-ribosyltransferase Rho family GTPase-activating protein	Actin cytoskeleton disruption Inhibition of DNA synthesis Inhibition of endocytosis and vesicular trafficking Cell death (apoptosis?)
Exoenzyme T	ADP-ribosyltransferase Rho family GTPase-activating protein	Actin cytoskeleton disruption Inhibition of cell migration and proliferation Inhibition of cytokinesis Cell death (pyroptosis?)
Exoenzyme U	Phospholipase A2	Cell death (necrosis) Inhibition of caspase-1 activation Release of arachidonic acid
Exoenzyme Y	Adenylate cyclase	Altered host gene expression Increased endothelial permeability Actin cytoskeleton disruption

[a]Based upon Engel and Balachandran, 2009, and Hauser, 2009, and references therein.

identified *P. aeruginosa* exoenzymes—ExoS, ExoT, ExoU, and ExoY—target functions that are important for maintaining and repairing epithelial tissue integrity. They are also capable of disrupting signal transduction pathways that control actin cytoskeleton rearrangement, thereby compromising the ability of professional phagocytes to move toward and phagocytose bacteria. Several excellent papers have recently reviewed the extensive primary literature that details the function of these translocated effectors in vitro and in animal models of acute infection (Engel and Balachandran, 2009; Hauser, 2009). Importantly, acute human respiratory tract infections (particularly ventilator-associated pneumonia) due to clinical isolates which express the T3SS are associated with increased morbidity and mortality rates (Roy-Burman et al., 2001; Hauser et al., 2002; El Solh et al., 2008). Expression of genes encoding the T3SS effectors and secretion apparatus is regulated by poorly understood environmental cues that can be mimicked in vitro by divalent cation chelation or by contact with tissue culture cells or serum. An excellent recent review describes how T3SS expression is controlled by the transcriptional activator ExsA and a set of interacting proteins that link ExsA activity to gating of the T3SS apparatus (Diaz et al., 2011).

P. aeruginosa also expresses a Xcp type 2 secretion system (T2SS) that secretes virulence factors such as exotoxin A, multiple proteases (alkaline protease, elastase, and protease IV), lipases (LipA, LipC, and EstA) and phospholipases (PlcH, PlcN, and PlcB) across the bacterial cell envelope. A role for the T2SS in virulence in a murine acute pneumonia model has recently been demonstrated (Jyot et al., 2011; Le Berre et al., 2011).

P. aeruginosa also causes chronic infections in certain human hosts, operationally defined as the continued persistence of the bacterium for at least

6 months as documented by repeatedly positive cultures from clinical specimens (Johansen and Hoiby, 1992). Such chronic infections are most often seen in patients with pulmonary disease and bronchiectasis from CF. *P. aeruginosa* can persist for decades in the damaged airways of CF patients despite aggressive antibiotic therapy and a robust, but ultimately ineffective, host immune response characterized by the presence of large numbers of neutrophils in the airways. Chronic infections are clearly symptomatic, yet they rarely lead to the rapid clinical decompensation and sepsis that can follow acute *P. aeruginosa* respiratory tract infections. Chronic infections are thought to result from infections by environmental strains that are initially cleared by the CF host (Burns et al., 2001). A small number of strains may survive the initial host immune response and adapt to persist in the host, often as organized communities of surface-associated bacteria or biofilms (Romling et al., 1994; Bjarnsholt et al., 2009). Acute exacerbations of clinical symptoms are a feature of chronic *P. aeruginosa* infections, and some data argue that these might be caused by subpopulations of bacteria that express "planktonic" rather than biofilm-associated traits and virulence factors (VanDevanter and Van Dalfsen, 2005). A recently published review of chronic pulmonary infection in CF patients provides a comprehensive discussion of the adaptations that might promote *P. aeruginosa* persistence in this environment (Hauser et al., 2011).

Do *P. aeruginosa* Strains That Cause Persistent Infection Differ from Acute Infection Isolates?

The observation that *P. aeruginosa* can cause either acute clinical deterioration or a persistent infection that lasts for years has prompted many investigators to ask whether bacteria associated with these differing clinical presentations might be phenotypically and/or genotypically distinct. Several studies have suggested that fundamental differences do exist between *P. aeruginosa* strains isolated from acutely versus chronically infected individuals. In a *Dictyostelium discoideum* model of infection, where virulence is defined as the ability of *P. aeruginosa* to kill this soil-dwelling amoeba, a study of matched initial and late *P. aeruginosa* isolates from 16 CF patients found 15 of 16 initial isolates to be virulent, while only 8 of 16 late isolates continued to exhibit virulence (Lelong et al., 2011). This study, however, did not evaluate in vitro growth characteristics of the isolates that might have affected their ability to replicate in *D. discoideum*. Bragonzi et al. examined 25 isolates from six CF patients in a murine model of pulmonary infection where 1×10^6 to 2×10^6 *P. aeruginosa* organisms were mixed with agar beads and inoculated into the lung. Eight of these isolates were isolated within a year of initial infection (early), six 1 to 5 years after initial infection (intermdiate), and 11 after 7 to 16 years of persistent infection (late) (Bragonzi et al., 2009). Strains were deemed capable of causing acute infection if they were associated with bacteremia in the murine model, while chronic infection was defined as the persistence of >1,000 bacteria in the lung after 14 days. Early isolates from a given patient were more likely to cause bacteremia than clonally related late isolates. However, no difference was observed in the ability of isolates to establish persistence at 14 days (Bragonzi et al., 2009). A strength of this study was that the results were consistent across mice from a variety of genetic backgrounds (Bragonzi et al., 2009). However, the traits required for bacterial survival for 14 days in a mouse may not be equivalent to the adaptations that allow *P. aeruginosa* to persist in the human lung for months to years.

Many investigators have asked why late isolates recovered from patients with chronic infection are less capable of causing acute infection in model organisms than are early isolates or environmental strains. Loss of T3SS expression by late isolates is likely to be one factor. One study examined clinical isolates recovered from three small cohorts of CF patients: one group without a history of *P. aeruginosa* infection and two groups (one adult and one pediatric) defined as chronically infected, with at least 2 years of positive cultures (Jain et al., 2008). Only 12% of isolates recovered from patients with prolonged infection were T3SS positive, versus 30% of isolates recovered from patients with early infection (Jain et al., 2008). Another study used microarray analysis to compare gene expression between early and late clonal isolates recovered from three patients with chronic

P. aeruginosa infection (Rau et al., 2010). Two of these individuals showed downregulation of T3SS gene expression in the late isolates, along with downregulation of genes encoding proteins required for type IV pilus (TFP) and flagellar assembly and function. Many other studies confirm that reduced T3SS expression is common among isolates recovered from persistently colonized CF patients. However, some authors have reported that occasional patients retain subpopulations of T3SS-positive isolates even after prolonged infection, suggesting that phenotypic diversity may be maintained within the host and may allow bacteria to exploit distinct host microenvironments (Jain et al., 2008). In general, however, the T3SS appears to be dispensable for continued persistence of *P. aeruginosa* in the airways of CF patients. It may, however, be relevant to the initial establishment of infection in this population, just as it appears to be in non-CF patients (Hauser et al., 2002; El Solh et al., 2008). Such a role is suggested by the observation that children with CF whose first *P. aeruginosa* isolates were T3SS positive were 2.5 times more likely to have *P. aeruginosa* isolated from a subsequent culture than were children with T3SS-negative initial isolates (Jain et al., 2008).

Many investigators have sequenced clinical isolates from CF patients to look for mutations, often in specific virulence factors, that arise during chronic infection. A seminal study by Smith et al. compared the whole-genome sequences of two *P. aeruginosa* isolates taken from the same CF patient over 7 years apart (Smith et al., 2006). After identifying a number of genetic mutations that might favor long-term bacterial persistence, the authors examined 91 *P. aeruginosa* isolates collected from another 29 CF patients, looking for mutations in these candidate genes (Smith et al., 2006). Examples of genes with mutations identified in strains isolated from multiple patients from this and other studies are listed in Table 2. Smith et al. identified mutations in genes encoding global regulators such as the virulence factor regulator (Vfr) and the cyclic AMP (cAMP)-generating adenylate cyclase CyaB, quorum sensing receptors (LasR and, less frequently, RhlR), and proteins required for the expression of specific virulence factors, including the T3SS, flagella, and TFP (Table 2) (Smith et al., 2006). A Scandinavian study that examined the sequences of genes from the Las and Rhl quorum sensing systems (238 isolates from 152 patients) found that 111 isolates had mutations in *lasR* and 33 isolates carried mutations in *rhlR* (Bjarnsholt et al., 2010). A Danish study of several hundred isolates of the highly transmissible *P. aeruginosa* DK2 lineage recovered from patients from 1973 to 2008 found that the majority of measured phenotypic changes between these DK2

Table 2. Examples of mutations accumulated by *P. aeruginosa* during chronic pulmonary infection in CF patients

System	Gene	Function(s)	Reference(s)	Comments
Quorum sensing	*lasR* (PA1430)	Master regulator of Las quorum sensing, influences metabolism and antibiotic resistance	Smith et al., 2006; Feliziani et al., 2010	Loss-of-function mutations in signaling and DNA binding domain
	rhlR (PA3477)	Regulator of Rhl quorum sensing system	Feliziani et al., 2010	
	rhlB (PA3861)	ATP-dependent RNA helicase	Feliziani et al., 2010	
Cell surface	*mucA* (PA0763)	Anti-sigma factor binds Alg-T to inhibit alginate production, regulates T3SS	Smith et al., 2006; Feliziani et al., 2010	Some *mucA* strains lose mucoid phenotype
	wspF (PA3703)	Chemosensory transduction		
Efflux pumps/ antibiotic resistance	*mexZ* (PA2020)	Transcriptional regulator of MexXY-OprM efflux pump	Smith et al., 2006; Feliziani et al., 2010	
	mexA (PA0425)	Membrane component of drug efflux pump	Smith et al., 2006	
	mexT (PA2492)	Transcriptional regulation of drug efflux pump	Smith et al., 2006	
	ampD (PA4522)	Regulator of beta-lactamase	Smith et al., 2006	
	nalD (PA3574)	Transcriptional regulator of MexAB-OprM	Smith et al., 2006	
Hypermutator	*mutS* (PA3620)	DNA mismatch repair	Hogardt et al., 2007; Feliziani et al., 2010	
	mutL (PA4946)	DNA mismatch repair	Feliziani et al., 2010	
Global virulence regulation	*vfr* (PA0652)	cAMP-dependent global regulator of virulence factor production	Smith et al., 2006	
	cyaB (PA3217)	Adenylate cyclase	Smith et al., 2006	
	ladS (PA3974)	Hybrid sensor kinase/response regulator affecting T3SS and biofilm	Smith et al., 2006	
Type III secretion	*exsA* (PA1713)	Transcriptional regulator of T3SS	Smith et al., 2006	
Motility	*pilY1* (PA4554)	TFP biogenesis protein	Bianconi et al., 2011	
	pilB (PA4526)	ATPase required for pilus biogenesis	Smith et al., 2006	
	rpoN (PA4462)	RNA polymerase sigma factor 54	Smith et al., 2006	
	fleQ (PA1097)	Master regulator of flagellar gene transcription	Smith et al., 2006	
	flgB (PA1077)	Flagellar basal body/rod protein		
Metabolism	*accC* (PA4848)	Biotin carboxylase	Smith et al., 2006	
	mexS (PA2491)	Probable oxidoreductase	Smith et al., 2006	
	PA0366	Probable aldehyde dehydrogenase	Smith et al., 2006	
	PA0506	Acyl coenzyme A dehydrogenase	Smith et al., 2006	

isolates and wild-type *P. aeruginosa* could be explained by mutations in three genes: the sigma factor *rpoN* gene, *lasR*, and *mucA*, which encodes an anti-sigma factor that represses expression of the mucoid exopolysaccharide (EPS) alginate (Yang et al., 2011).

Chronic infection is also associated with the emergence of hypermutator strains. These strains carry mutations in genes required for DNA mismatch repair (*mutS* and *mutL*), resulting in increased mutation rates that may allow for more rapid adaptation to the host environment. A study of 38 isolates from 26 Argentinian CF patients found high frequencies of mutations in *mutS* (40%) and *mutL* (60%) (Feliziani et al., 2010). Additional genes mutated with high frequency were *mucA* (63%), *mexZ* (a regulator of an efflux pump related to antibiotic resistance) (79%), and *lasR* (39%) (Feliziani et al., 2010). Hypermutator *P. aeruginosa* isolates were also isolated and characterized in another study of CF patients. These strains, which lost traits required for cytotoxicity towards tissue culture cells and for survival in environmental habitats (e.g., tap water), became dominant in the CF lung with increasing time of colonization (Hogardt et al., 2007). Interestingly, the authors also isolated nonmutator strains from these chronically colonized patients and showed that they persisted as a minority population still capable of producing virulence-associated toxins and exoenzymes. This phenotypic (and genotypic) diversity of *P. aeruginosa* strains seems to be a common feature in the chronically colonized CF lung, even if certain phenotypes become dominant in this environment.

One challenge in interpreting such sequencing studies lies in linking genotype with phenotype. Thus, mutations in the *mexZ* regulator do not necessarily correlate with antibiotic resistance phenotypes, while mutations in *lasR* are linked to increased beta-lactamase production and increased antibiotic resistance (D'Argenio et al., 2007). Regulatory genes such as *lasR*, *mexZ*, and *mucA* directly and indirectly affect the expression of multiple phenotypes relevant to persistent infection (e.g., LasR affects both antibiotic resistance and quorum sensing; MucA impacts both alginate production and T3SS expression). When multiple phenotypes are altered by mutations in a single gene, it becomes increasingly difficult to understand which phenotypic change is under selection in the context of the persistently infected CF lung.

One limitation of many sequencing-based studies is their focus on genes known to be associated with the phenotypes that characterize isolates from persistently infected individuals, or on mutations identified in strains recovered from very small numbers of patients. This is likely to lead to an underestimation of the number and types of mutations that are selected for during chronic infection. In support of this idea, a recent study of the entire genomes of two different CF isolates, one a mucoid strain from Boston, Massachusetts, and the second the Manchester (UK) epidemic strain, found tremendous genetic diversity between the two strains (a difference of 600 predicted open reading frames) (Mathee et al., 2008). Much of this diversity was due to horizontal gene transfer and impacted metabolic pathways that promote survival in different environments (Mathee et al., 2008). Metabolic adaptation to the CF lung environment is likely to be a significant selective force for *P. aeruginosa*, as suggested by D'Argenio and colleagues in their study of *lasR* mutants (D'Argenio et al., 2007). Indeed, isolates from patients with chronic pulmonary infections display increased expression of genes encoding proteins required for growth under anaerobic and microaerophilic conditions (Hoboth et al., 2009).

Rodent chronic infection models have also been used to identify traits or patho-adaptive mutations that might favor persistent infection in the CF host. A recent example of this approach was published by Bianconi et al., who used an agar bead mouse model of chronic pulmonary infection to identify transposon mutants that were favored during persistent infection (Bianconi et al., 2011). The authors considered a mouse to be chronically infected if >1,000 CFU of bacteria were isolated 14 days after inoculation (Bianconi et al., 2011). Mutations in a number of genes appeared to favor chronic infection, including those encoding proteins required for TFP function, flagellar function, and rhamnolipid secretion (Table 2) (Bianconi et al., 2011). The authors then tested the predictive value of their screen by sequencing the genes they had identified in 14 clinical isolates. Mutations in PilY1, a TFP biogenesis gene identified in the animal screen, were identified in 7 of 14 clinical strains (Bianconi et al., 2011). This study suggests that there may be a much broader set of genes whose mutation confers a selectable advantage during chronic infection; more information is likely to emerge as whole-genome sequencing of clinical isolates becomes more common.

Summary and Future Directions

A large and still-growing body of literature has documented the genotypes of *P. aeruginosa* clinical isolates recovered from patients infected for various lengths of time. The phenotypes of these strains have been compared to those of environmental strains in vitro and in a variety of worm, insect, and mammalian infection models. Tremendous genetic and phenotypic diversity exists between clinical strains recovered from different CF patients. In particular, mutations in a small number of global regulators (*lasR*, *mucA*, *vfr*, and *rpoN*) often characterize chronic isolates and can profoundly alter bacterial phenotypes that might promote persistence in vivo. As these genes regulate many phenotypes, however, it is not always clear how their mutation might confer a selective advantage in the environment of the CF lung. As an example, mutations in *lasR*, which are frequently observed in CF isolates, alter quorum sensing, antibiotic resistance, and preferred substrates for bacterial metabolism. Each of these changes might confer a selective advantage in the human lung. Whole-genome sequencing of a few *P. aeruginosa* CF isolates has revealed tremendous genetic diversity, suggesting that our understanding of the adaptations that promote chronic colonization of the CF respiratory tract is woefully incomplete. Given the diverse microenvironments and bacterial competitors present in the lungs of patients with CF, mutations in very different genes may provide a selective advantage under specific conditions. Understanding such complex associations is likely to be an important area of future study, but limitations inherent in the studies discussed above will need to be surmounted. These include reliance on rodent models of infection that are poor mimics of human pulmonary CF disease; in vitro characterizations of clinical isolates that are unlikely to reveal relevant in vivo behavior; sampling methods that rarely capture the genotypic and phenotypic diversity of *P. aeruginosa* strains present in the infected human airway, much less that of the resident microbiota; and experimental and analytic methods that are not well suited for studying the complex and multifactorial associations present among host, pathogen, and environment.

REGULATORY "SWITCHES" FOR VIRULENCE FACTOR EXPRESSION

The identification of distinct sets of proteins associated with acute versus chronic infection suggests that these represent different physiological states for *P. aeruginosa*. It is highly unlikely that the behaviors associated with acute and chronic *P. aeruginosa* infection evolved to allow bacterial exploitation of human tissues or to cause disease; rather, they arose to allow this organism to exploit a large range of environmental niches in which other bacteria and simple eukaryotic organisms represent both predator and prey. The expression of proteins associated with acute or chronic infection is tightly controlled by multiple, often overlapping regulatory pathways; several of these are discussed below. The signals—in the environment or in the human host—that control the activity of these different regulatory networks remain largely unknown, although it is likely that they were selected to promote *P. aeruginosa* survival in whatever environment it might encounter. Exactly how *P. aeruginosa* integrates all the inputs discussed below to determine which genes are expressed and to what levels remains a key question in the field—and one that must be addressed if we are to successfully manipulate virulence traits as part of the management of *P. aeruginosa* disease.

Virulence Factor Regulator (Vfr): a Global Regulator of Secretion Systems and Motility

Vfr is a 24-kDa DNA binding protein with homology to the 3′–5′ cyclic monophosphate (cAMP)

receptor protein of *Escherichia coli*. The crystal structure of Vfr, which has recently been solved, confirms the presence of two cAMP binding sites on this protein (Cordes et al., 2011). Vfr activity has a tremendous impact on gene transcription in *P. aeruginosa*, as the expression of more than 200 genes is altered in a Δ*vfr* mutant compared to the wild type (Wolfgang et al., 2003). Δ*vfr* strains display decreased expression of genes encoding the T3SS apparatus and effectors, genes required for TFP and flagellum biogenesis, and genes associated with the T2SS (Fig. 1) (Wolfgang et al., 2003). Many of these changes in gene expression are likely to be indirect; however, a wide range of genes appear to be directly controlled by Vfr, including those for exotoxin A (*toxA*), the Las and Rhl quorum sensing regulators (*lasR* and *rhlR*), and the master regulator of flagellar gene expression (*fleQ*) (Wolfgang et al., 2003; Fuchs et al., 2010a; Croda-Garcia et al., 2011).

Vfr-dependent regulation of most virulence factors under its control requires the allosteric activator cAMP. cAMP binding by Vfr is required for the activation of the *toxA* promoter, for positive autoregulation of *vfr* transcription, and for increased transcription of *cpdA*, which encodes the phosphodiesterase that hydrolyzes cAMP (Fuchs et al., 2010a, 2010b). CpdA activity serves as a brake on Vfr-cAMP activity, as demonstrated by the phenotype of phosphodiesterase-deficient Δ*cpdA* bacteria, in which cAMP levels and Vfr expression are both increased (Fuchs et al., 2010a).

As Δ*vfr* bacteria have decreased expression of a number of gene products associated with virulence in animal models of acute infection, these bacteria might be expected to be less virulent in vivo. Indeed, Smith et al. observed decreased pulmonary recovery of Δ*vfr* bacteria compared with wild-type controls at 16 h postinfection in a BALB/c adult mouse acute pneumonia model (Smith et al., 2004). *P. aeruginosa* lacking the two adenylate cyclases encoded by this organism, *cyaA* and *cyaB*, also displayed decreased expression of acute virulence factors and reduced virulence in this murine pneumonia model, confirming that intracellular cAMP levels regulate virulence production during acute infection (Smith et al., 2004). Bacteria singly deficient in *cyaB* were more attenuated for virulence than their Δ*cyaA* counterparts, suggesting that CyaB is the primary adenylate cyclase involved in Vfr-dependent transcriptional regulation (Fig. 1). In the acute murine pneumonia model, the reduced virulence of Δ*cyaA* Δ*cyaB* bacteria was reversed by overexpressing the transcriptional activator of the T3SS, ExsA (Smith et al., 2004). This result underscores the importance of the T3SS in murine acute lung infection and confirms that Vfr-dependent

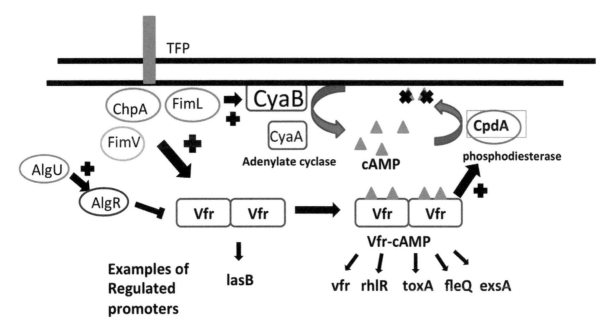

Figure 1. Vfr-cAMP dependent regulation of virulence factor expression. CyaB is the predominate adenylate cyclase that generates cAMP, which activates transcription of virulence genes upon binding to Vfr. Vfr also activates lasR transcription in a cAMP-independent fashion. The TFP biogenesis factors FimV, FimL, and ChpA all positively regulate Vfr-cAMP. FimL affects Vfr levels during growth on agar surfaces when TFP-dependent motility is high. MucA is a negative regulator of AlgU, which itself negatively regulates Vfr via AlgR. CpdA is a phosphodiesterase that breaks down cAMP. doi:10.1128/9781555818524.ch2f1

upregulation of the T3SS occurs in vivo in the lung. When clinical isolates recovered from patients with ventilator-associated pneumonia were tested in a murine acute pneumonia model, the two in vitro phenotypes that correlated with increased virulence were the secretion of T3SS effectors and the LasB elastase, both controlled by Vfr (Le Berre et al., 2011).

While most virulence factors controlled by Vfr are activated by Vfr-cAMP binding, DNA-binding assays with Vfr and the *lasR* promoter region demonstrate a cAMP-independent *lasR* activation (Fuchs et al., 2010a). This finding separates Vfr control of quorum sensing from its cAMP-dependent regulon and may allow tighter control of Vfr-dependent virulence factor expression (Fuchs et al., 2010a). Interestingly, a *vfr* mutation (ΔEQERS) that disrupts one of the cAMP binding sites results in a loss of TFP-dependent twitching motility, but not of elastase or pyocyanin production (Beatson et al., 2002). As both elastase and pyocyanin are under the control of Las quorum sensing system, this may be related to a cAMP-independent effect of Vfr on *lasR*.

Links between Vfr and TFP

Given the importance of Vfr-cAMP in determining acute virulence factor expression, several investigators have tried to identify systems that regulate

intracellular cAMP levels. A screen of a *P. aeruginosa* transposon library for mutants with decreased cAMP levels identified multiple genes involved in TFP assembly and twitching motility, many of which mapped to the Chp gene cluster (Fulcher et al., 2010). The Chp chemosensory system, previously shown to be important for twitching motility, influences cAMP levels through its control of CyaB (Fulcher et al., 2010). Analysis of isolates with deletions in the Chp system or in genes encoding proteins required for TFP assembly and function demonstrated that cAMP levels were critical for TFP assembly but that the Chp system also exhibited cAMP-independent regulation of TFP function, i.e., twitching motility (Fulcher et al., 2010). These findings again highlight the complex regulation of *P. aeruginosa* virulence factor expression. In this case, appropriate intracellular cAMP levels are required to assemble TFP on the bacterial surface, while twitching motility is regulated in a cAMP-independent manner (Fulcher et al., 2010). The molecular mechanism that couples assembly and function is not understood. ChpA is a large hybrid sensor kinase with a *cheY* domain and eight histidine phosphotransfer (Hpt) domains, raising the possibility that multiple signals that control TFP assembly and activity are integrated by ChpA itself.

A number of proteins implicated in TFP biogenesis are important for Vfr activity and influence intracellular cAMP levels. For example, *fimV*, which encodes a protein originally implicated in TFP assembly (Semmler et al., 2000), was identified in a screen of mutants with reduced type 2 secretion (Michel et al., 2011). A Δ*fimV* strain displayed reduced secretion of the LasB elastase due to decreased *lasB* transcription. Western blot analysis of Δ*fimV* cell lysates showed reduced steady-state levels of Vfr compared to the wild type that were restored after complementation with FimV (Michel et al., 2011). Δ*fimV* bacteria also exhibited decreased cAMP levels, suggesting a role for FimV in the Vfr-cAMP-dependent regulation of gene expression (Michel et al., 2011). FimV has a putative LysM peptidoglycan binding domain, so FimV might respond to changes in bacterial cell wall metabolism, but the mechanism by which FimV influences Vfr and cAMP levels remains unknown.

FimL is a protein with homology to the amino terminus of ChpA that was initially identified in a screen of genes required for TFP-dependent epithelial cell cytotoxicity (Whitchurch et al., 2005; Inclan et al., 2011). The Hpt domains of FimL are degenerate and would not be predicted to support phosphotransfer. The loss of FimL resulted in decreased expression of both TFP and T3SS proteins (Whitchurch et al., 2005). Δ*fimL* bacteria were characterized by decreased expression of Vfr, as reported for Δ*fimV*, but this phenotype was observed only in Δ*fimL* bacteria recovered from agar plates and not in those grown in liquid culture (Whitchurch et al., 2005). Importantly, *P. aeruginosa* growth on solid or semisolid surfaces results in increased numbers of TFP on the cell surface, suggesting that these bacteria recognize and respond to surface growth. How this might influence FimL effects on Vfr expression, however, is not clear.

Another link between FimL and the Vfr-cAMP regulon was established by a screen designed to identify extragenic suppressors of FimL that restored twitching motility. This screen identified the cAMP phosphodiesterase CpdA, suggesting that FimL directly or indirectly regulated intracellular levels of cAMP (Inclan et al., 2011). Since Δ*fimL* and Δ*fimL*Δ*cyaB* strains had identical intracellular cAMP levels, while decreased cAMP levels were observed in Δ*fimL* Δ*cyaA* bacteria, FimL appeared to positively regulate the CyaB-dependent generation of cAMP (Inclan et al., 2011). Δ*fimL* strains also resembled Δ*cyaB* strains with reduced virulence factor expression and cAMP levels (Fulcher et al., 2010; Inclan et al., 2011). Deletion of *fimL* did not alter CyaB levels, and both FimL and CyaB localized to the cell pole, leading Inclan et al. to hypothesize that FimL interactions with CyaB increase its adenylate cyclase activity, thereby influencing virulence factor expression (Inclan et al., 2011).

It is not clear whether other second messengers besides cAMP influence Vfr-dependent virulence factor expression. cGMP is one potential allosteric regulator of Vfr activity, predicted to bind Vfr (based on structural modeling) and hypothesized to control Vfr-dependent elastase and pyocyanin expression. The effect of cGMP on Vfr is controversial: one group of investigators reported that cGMP activated Vfr at millimolar concentrations, while another group observed that Vfr activity was inhibited by cGMP, also at high concentrations (Fulcher et al., 2010; Serate et al., 2011). Given the low levels of endogenous cGMP in *P. aeruginosa*, both groups concluded that any effects of cGMP on Vfr activity observed in vitro are unlikely to be relevant in vivo and that cAMP is the primary allosteric activator of Vfr activity (Fulcher et al., 2010; Serate et al., 2011). Cyclic di-GMP, an important regulator of surface-associated colonization and biofilm formation, has no effect on Vfr-cAMP interactions and is unlikely to influence Vfr-dependent regulation of motility and virulence factor expression (Fulcher et al., 2010).

Numerous lines of evidence suggest that the absence of Vfr function confers an adaptive advantage during growth conditions that favor chronic infection. Several studies have reported that mutations in *vfr* are found in clinical isolates recovered from CF patients with chronic infection (Smith et al., 2006) (Table 2). After static growth in rich medium for several days, spontaneous *vfr* mutations emerge that outcompete the wild-type strain (Fox et al., 2008). One of these point mutants (T193C), which changes residue 65 from Tyr to His and inactivates Vfr, disrupts the structural connection between the cAMP binding pocket, the DNA binding domain, and the dimerization domain (Cordes et al., 2011). As Vfr coordinately regulates the expression of multiple virulence factors, such observations suggest that Vfr inactivation might allow *P. aeruginosa* to downregulate expression of virulence factors important for initial infection but detrimental during prolonged colonization of the mammalian host.

Links between Vfr and regulators of alginate expression

One hallmark of prolonged chronic *P. aeruginosa* infection is the emergence of mucoid strains that overproduce large amounts of the EPS alginate. This is most often associated with mutations in *mucA*, which encodes an anti-sigma factor that interacts with the sigma factor AlgU. Inactivation of MucA allows increased expression of AlgU-dependent genes that encode proteins required for alginate biosynthesis, thereby leading to the mucoid phenotype (Fig. 1).

This adaptation to long-term survival in the host is accompanied by decreased expression of the T3SS, decreased protease production, and decreased TFP-dependent twitching motility (Wu et al., 2004; Rau et al., 2010). Recent evidence demonstrates that both Δ*mucA* and the common naturally occurring mutation *mucA22* are associated with decreased expression of Vfr and reduced expression of the T3SS effector ExoS, TFP, and elastase (Jones et al., 2010). Wild-type and *mucA22* bacteria have similar intracellular cAMP levels, suggesting that the effect of MucA on Vfr expression is direct. Ectopic expression of Vfr in *mucA22* strains restored expression of ExoS, TFP, and elastase, confirming that downregulation of Vfr is responsible for decreased virulence factor expression in mucoid strains under the conditions evaluated in this study (Jones et al., 2010). AlgU is a sigma factor required both for alginate production and for the synthesis of AlgR, which is part of the AlgR/AlgZ two-component system required for the inhibition of the T3SS in mucoid strains (Wu et al., 2004). Deletion of *algU* or *algRZ* in a *mucA22* background restored Vfr and acute virulence factor production to wild-type levels. Deletion of *algRZ* in a wild-type (nonmucoid) background led to increased transcription from the *vfr* promoter, which could be inhibited by expressing AlgR alone in Δ*algRZ*. In aggregate, these results suggest that AlgU, via AlgR, negatively regulates Vfr production and influences basal levels of acute virulence factor expression even in nonmucoid bacteria (Fig. 1) (Jones et al., 2010). Consistent with this hypothesis, Δ*algU* bacteria were more virulent than their wild-type counterparts in a murine C57BL/6 septicemia model, as measured by a 10-fold decrease in the 50% lethal dose for the AlgU-deficient strain (Yu et al., 1996).

Vfr: Summary and future questions

The role of Vfr-cAMP as a global regulator of multiple virulence factors is well established, as are the identities of the proteins (CyaB and CpdA) that control cAMP metabolism in *P. aeruginosa*. However, a number of interesting questions regarding the regulation and function of Vfr remain unanswered. The first regards the relationship between cAMP-dependent and -independent Vfr regulation of gene expression and how this impacts virulence factor production. A second set of questions revolves around the mechanisms by which the AlgU/AlgR and ChpA/FimL regulatory pathways influence Vfr levels and how their activity is integrated to control Vfr levels within the cell. The extracellular signals that control these regulatory pathways are unknown, although surface growth appears to be one signal that is sensed by this system. Lastly, while it is clear that regulators of TFP biogenesis and function (i.e., ChpA, FimL, and FimV) are intimately associated with the control of Vfr and cAMP expression, the mechanism that links twitching motility and Vfr remains to be elucidated.

Global Regulatory Networks: the Sensor Kinases GacS/RetS/LadS and Small RNA (sRNA)-Mediated Regulation of Virulence Factor Expression

The loss of T3SS and motility and the expression of alginate and EPSs are frequently observed over time in isolates from chronically infected CF patients. The notion that a "switch" might exist between these groups of virulence factors, however, gained traction with the discovery of a group of two-component system proteins that seemed to inversely control factors associated with acute versus chronic infection: the response regulator GacA, the sensor kinase GacS, and two hybrid sensor kinases, RetS and LadS.

The hybrid sensor kinase RetS (RtsM) was independently discovered by three laboratories using different screening strategies. Zolfaghar et al. isolated a PA103 *retS*::Tn5 mutant based on its altered colony morphology and diminished twitching motility and demonstrated that it had reduced cytotoxicity in vitro and in vivo (Zolfaghar et al., 2005). Laskowski et al. identified a PA103Δ*pscJ rtsM*::Tn5 mutant in a screen for noninvasive *P. aeruginosa* and demonstrated that PA103Δ*rtsM* bacteria were significantly attenuated for virulence in vitro and in vivo due to their inability to induce expression of T3SS genes (Laskowski et al., 2004). Goodman et al. systematically replaced 39 putative response regulator genes in PAK with a gentamicin resistance cassette (*aacC1*) and found that inactivation of PA4856 (RetS) resulted in accelerated biofilm formation in a static biofilm assay (Goodman et al., 2004). Goodman and colleagues then performed microarray analyses that revealed increased transcript levels for genes within the *pel* and *psl* EPS synthetic operons, as well as decreased transcripts for genes associated with T3SS, T2SS, and TFP assembly/twitching motility. A second screen looking for extragenic suppressors of the Δ*retS* hyperbiofilm phenotype yielded multiple "hits" in another sensor kinase, GacS, in its cognate response regulator, GacA, and in a gene encoding the sRNA *rsmZ*. All of these extragenic suppressors also resulted in suppression of the T3SS defect observed in the Δ*retS* parent.

A screen for loss of biofilm formation in PAK Δ*pilA* led to the identification of a third sensor kinase, LadS (PA3974), involved in reciprocal regulation of T3SS and EPS genes (Ventre et al., 2006). LadS appeared to act in an opposite manner to RetS: LadS overexpression led to hyperbiofilm formation and increased *pel* expression, while a Δ*ladS* mutant showed increased T3SS expression. Microarray analysis showed that a substantial number of genes (ca. 40) were inversely regulated by RetS versus LadS, including the operons for the T3SS, the T6SS, and *pel*. This inverse regulation was thought to result from the opposite effects of LadS and RetS on transcription of the sRNA *rsmZ*.

The only direct targets of GacA in *P. aeruginosa* are *rsmY* and *rsmZ*

rsmZ and *rsmY* are two sRNAs that interact with the RNA binding protein RsmA. RsmA is a member of the CsrA family of proteins, first described for *Escherichia coli* (Romeo et al., 1993), which inhibit translational initiation by binding to or near Shine-Dalgarno sequences of target mRNAs (Lapouge et al., 2008). Brencic et al. recently demonstrated that *rsmZ* and *rsmY* are the only two genes directly regulated by phosphorylated GacA in *P. aeruginosa* (Brencic et al., 2009). *rsmY* and *rsmZ* each contain ca. four to six binding sites for RsmA and function by sequestering free RsmA (Lapouge et al., 2008). Thus, the large changes in gene expression that accompany GacA phosphorylation are achieved in essence indirectly, through sRNA modulation of RsmA activity. Models illustrating how this system might function to control the expression of virulence factors associated with chronic infection (GacA active) versus acute

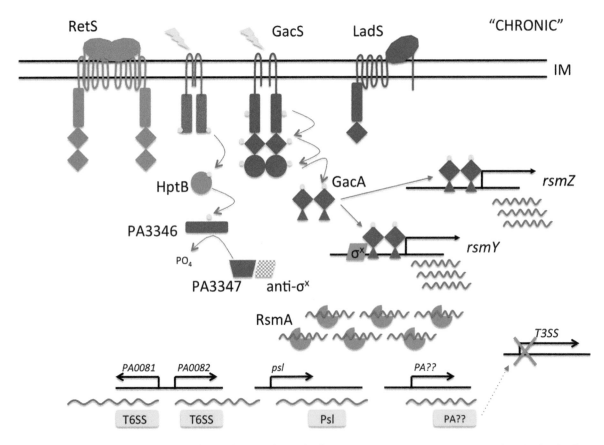

Figure 2. Expression of genes associated with chronic infection by the GacS/A two-component system. GacA activation by the GacS results in GacA phosphorylation, dimerization, and (presumably) binding to the *rsmY* and *rsmZ* promoters. The *rsmY* and *rsmZ* sRNAs bind to and sequester RsmA, thereby allowing translation of mRNAs encoding T6SS genes and enzymes involved in Psl (and Pel) EPS synthesis and secretion. Hypothetical negative regulators of the T3SS are presumably also expressed under these conditions. The grey molecule represents one of the three sensor kinases (PA1611, PA1976, and PA2824) shown to phosphorylate HptB in vitro. Phosphorylated HptB can phosphorylate and thereby activate the serine/threonine phosphatase activity of PA3346, for which PA3347 is a substrate. Dephosphorylation of PA3347 is hypothesized to favor sequestration of an anti-sigma factor, thereby allowing a putative sigma factor to bind to and activate the *rsmY* promoter. RetS is presumably inactive (dimerized?) under these conditions. Proteins whose deletion inhibits chronic gene expression are colored blue; those whose deletion favors chronic gene expression are colored red. Sensor kinase domains are schematized as follows: HK domains, rectangles; receiver domains, diamonds; and histidine phosphotransfer domains, circles. Yellow circles indicate domains with phosphoacceptor histidines or aspartates, while curved blue arrows indicate phosphotransfer reactions documented in vitro. The functions of LadS in this signaling pathway are not established. IM, inner membrane.
doi:10.1128/9781555818524.ch2f2

infection (GacA inactive) are shown in Fig. 2 and 3, respectively.

Two groups have examined the RsmA regulon using microarray approaches. Burrowes et al. compared PAO1 and PAO1Δ*rsmA* transcripts harvested from cells grown to late exponential phase (optical density, 0.8), a time when RsmA levels were reportedly maximal (Burrowes et al., 2006). Brencic and Lory carried out a similar comparison in PAK but harvested RNA from stationary-phase bacteria (optical density, 6.0), the time at which they documented expression of *rsmZ*, *rsmY*, and RsmA to be maximal (Brencic and Lory, 2009). Of the 506 genes found to be up- or downregulated in the PAO1 study, only 67 (13%) showed altered expression in the PAK study—and only 36 of these showed congruent expression changes. Nonetheless, both studies found that levels of transcripts for T3SS and TFP genes were decreased in the absence of RsmA. Interestingly, only the PAK experiment documented an increase in T6SS

transcript levels in Δ*rsmA* bacteria, despite the fact that the T6SS is strongly upregulated in PAO1 Δ*retS* bacteria, in which free RsmA levels are predicted to be low.

Brencic and Lory also attempted to determine which mRNAs could be copurified with RsmA (Brencic and Lory, 2009). This approach was relatively insensitive, in large part due to nonspecific binding by noncoding (16S and 23S) RNAs, but did identify two genes (PA0081, a T6SS gene, and PA4492, a gene of unknown function) whose transcript levels were also increased by Δ*rsmA* mutation. Brencic and Lory demonstrated that RsmA could directly bind these mRNAs, as well as that for a second T6SS gene, PA0082, but did not bind mRNAs for genes whose transcript levels decreased in Δ*rsmA* bacteria (*exoS*, *exsD*, *exsC*, and *pilM*). Transcriptional and translational *lacZ* reporter constructs were employed to demonstrate that the RetS/GacS/GacA and *rsmZ*/*rsmY*/RsmA pathway exerted

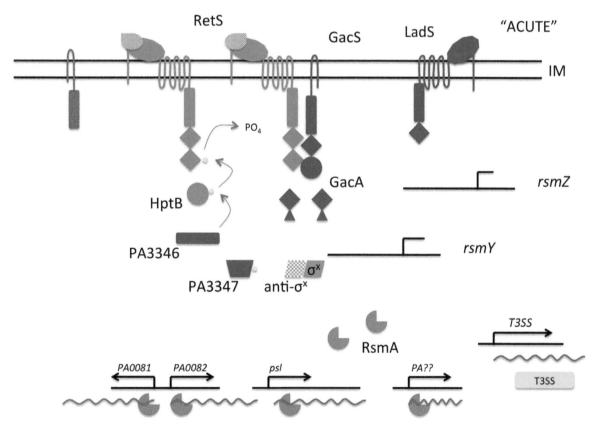

Figure 3. Expression of genes associated with acute infection by the GacS/A two-component system. Binding of an unknown ligand to the periplasmic domain of RetS is hypothesized to disrupt RetS homodimerization and to increase the likelihood of RetS-GacS interactions, which inhibit GacS autophosphorylation and phosphotransfer to GacA. Under these conditions, sRNA transcript levels are low and free RsmA can bind to target mRNAs, including those encoding T6SS genes and the *psl* operon. Blue arrows indicate phosphotransfer reactions that occur in vitro; their occurrence as an in vivo signaling pathway is speculative but based on the phenotypes of bacteria in which these gene products (RetS, HptB, PA3346, and PA3347) are not expressed. IM, inner membrane.
doi:10.1128/9781555818524.ch2f3

direct posttranscriptional control over transcripts negatively regulated by RsmA (T6SS and PA4492) (Brencic and Lory, 2009). A recent report from Irie et al. likewise showed RsmA binding to the 5′ untranslated region of *psl* mRNA, resulting in decreased mRNA translation (Irie et al., 2010). In contrast, the mechanism by which RsmA "positively" regulates gene expression is likely to be indirect. Mutations in the RetS/GacS/GacA and *rsmZ/rsmY*/RsmA signaling network result in changes in transcription, rather than translation, of T3SS reporters. Thus, RsmA's positive effects on the T3SS are likely to be the consequence of diminished expression of negative T3SS regulators, though no candidates for this role have been identified (Fig. 2).

At first glance, *rsmY* and *rsmZ* appear to be functionally redundant: overexpression of either appears to favor biofilm formation and suppress T3SS gene expression, while deletion of both *rsmY* and *rsmZ* is required to phenocopy a Δ*gacA* strain for loss of biofilm formation and increased T3SS gene expression (Brencic et al., 2009). There are some observations, however, which suggest that these sRNAs might also be subject to independent regulation and may not be functionally interchangeable (see below). For example, MvaT and MvaU, members of the H-NS family of global regulatory proteins (Castang et al., 2008), bind *rsmZ* but not *rsmY*. As these histone-like proteins are usually associated with transcription repression, this binding might account for the observation that *rsmY* is generally expressed at higher levels than *rsmZ* (Brencic et al., 2009).

How do RetS and LadS interact with GacS/A?

Microarray analyses carried out in PAK show an essentially perfect concordance between genes that are upregulated in Δ*retS* and downregulated in Δ*retS gacS*, and vice versa (Goodman et al., 2004; Brencic et al., 2009), strongly suggesting that RetS acts exclusively by controlling the activity of GacS/A. GacS is nonorthodox sensor kinase, which contains two transmembrane (TM) helices followed by HAMP, histidine kinase (HK), receiver domain (RD), and histidine phosphotransferase (Hpt) modules (Fig. 2 and 3). The cytoplasmic portion of GacS is capable of autophosphorylation in vitro and can transfer phosphate to its cognate response receiver GacA (Goodman et al., 2009). RetS has an unusual domain structure, with a large periplasmic domain followed by 7 TM helices (7TM-DISMED2) that is predicted to bind carbohydrate; this is followed by an HK domain and two tandem RDs. The mechanism by which RetS affects the GacS-GacA signaling relay has been addressed by several investigators. A structure-function

approach was taken by Laskowski and Kazmierczak, who constructed alleles of RetS in which the conserved histidine (H424) of the HK domain, predicted phosphoacceptor aspartates in RD1 (D713) and RD2 (D858), or residues implicated in phosphotransfer in RD1 (D664 D665) and RD2 (E814 D815) were mutated to alanine (Laskowski and Kazmierczak, 2006). Only mutations in RD2 residues, singly or in combination, resulted in a protein that was completely unable to restore T3SS gene expression to PA103Δ*retS* in vitro or to restore virulence in a murine acute pneumonia model. Bacteria expressing RetS(H424A) were still virulent in mice, suggesting that kinase activity was dispensable for RetS function. Indeed, the purified cytoplasmic portion of RetS shows no autophosphorylation activity in vitro (Hsu et al., 2008; Goodman et al., 2009). These authors postulated that RetS might function primarily as a phosphatase, with GacA~P as a possible target. Subsequently, Hsu et al. demonstrated that a Hpt module-containing protein, HptB, could transfer phosphate to RetS and that these two proteins interacted by bacterial two-hybrid assay (Hsu et al., 2008). Indeed, Δ*hptB* strains show increased biofilm formation (Burrowes et al., 2006; Hsu et al., 2008; Bordi et al., 2010), and increased *pelA-lacZ* transcription and decreased *exoS-lacZ* transcription, much like a Δ*retS* strain (Bordi et al., 2010).

Goodman et al. proposed a different model for RetS action. They demonstrated that the purified cytoplasmic portion of RetS could physically interact with the cytoplasmic portion of GacS in vitro and that this interaction inhibited GacS autophosphorylation (Goodman et al., 2009). Neither the conserved histidine nor the phosphoacceptor aspartates in either RD1 or RD2 were required for this inhibition. Moreover, the authors reported that a RetS protein lacking all of these residues could complement a PAKΔ*retS* strain, as assessed by the ability to suppress expression of a *rsmZ-lacZ* reporter construct. It is not clear why there is a discrepancy between this observation and the earlier finding that RetS alleles lacking the RD2 phosphoacceptor aspartate could not complement PA103 Δ*retS* (Laskowski and Kazmierczak, 2006). The two strains used in these studies, PAK and PA103, do differ significantly with regard to the RetS/GacS/LadS signaling pathway, as the *ladS* gene in PA103 has a nonsense mutation at codon 3 (M. A. Laskowski and B. I. Kazmierczak, unpublished observation). However, the use of only a single readout of RetS/GacS activity, namely, *rsmZ-lacZ* transcription, might not adequately reflect the interactions of RetS with other signaling proteins that also influence the activity of GacS/A, such as HptB.

Several unresolved issues remain regarding GacA-dependent signaling. For example, the absence of RetS results in increases in *rsmZ-lacZ* and *rsmY-lacZ* transcriptional fusion activity (Bordi et al., 2010). This is not the case for Δ*hptB*, in which only *rsmY-lacZ* is upregulated. Nonetheless, HptB control of *rsmY* appears to be mediated by GacA, as phenotypes of a Δ*hptB* mutant (increased biofilm production and decreased T3SS effector expression) are suppressed by the introduction of *gacS* or *gacA* mutations into the Δ*hptB* background (Bordi et al., 2010). It is not clear how this differential expression of the two sRNAs is achieved, though presumably additional regulatory components must play a role. HptB appears to antagonize the activity of two proteins that are encoded upstream of it in the same operon, PA3346 and PA3347 (Hsu et al., 2008; Bordi et al., 2010). PA3346 encodes a putative response regulator with a phosphatase 2C output domain, which can dephosphorylate PA3347, a putative anti-anti-sigma factor (Hsu et al., 2008). The phosphatase activity of PA3346 is increased when it itself is phosphorylated; Hsu et al. demonstrated in vitro phosphotransfer from three sensor kinases (PA1611, PA1976, and PA2824) to HptB and then to PA3346 (Hsu et al., 2008). Like HptB, PA3346 and PA3347 affect transcription of *rsmY*, but not *rsmZ*, leading Bordi et al. to propose that these proteins serve to control the activity of an unidentified sigma factor that specifically regulates *rsmY* transcription (Bordi et al., 2010). These observations are incorporated into Fig. 2 and 3. As overexpression of PA3346 or PA3347 phenocopies deletion of HptB, it seems likely that HptB primarily functions to remove, rather than donate, phosphate from PA3346.

The paper by Bordi et al. also provides the first data that *rsmY* and *rsmZ* are not functionally redundant. Increased expression of T6SS genes, whose translation is negatively regulated by RsmA, specifically requires increased expression of the *rsmZ* sRNA. Thus, Δ*hptB* bacteria, in which only *rsmY* transcription is increased, show no induction of T6SS expression (Bordi et al., 2010). Whether this reflects specific interactions between this sRNA and these mRNA targets, or instead is the consequence of qualitatively different interactions between RsmA and *rsmY* versus *rsmZ*, remains to be experimentally determined.

Many fewer data are available regarding the mechanism by which LadS interfaces with GacS/GacA in *P. aeruginosa*. A bacterial two-hybrid analysis carried out with the *Pseudomonas fluorescens* CHA0 GacS and GacA proteins, as well as the RetS and LadS homologs, demonstrated that RetS and LadS can each interact with GacS but not with GacA (Workentine et al., 2009). Whether a similar interaction occurs in *P. aeruginosa* is not yet known.

What signals regulate RetS/GacS/LadS?

The identity of the signals that activate the RetS, GacS, and LadS sensor kinases is still unknown. RetS and LadS share a similar 7TMR-DISMED2 fold at their amino termini. The periplasmic 7TMR-DISMED2 domain has similarity to carbohydrate binding domains. Two groups have crystallized the periplasmic domain of RetS (RetSp). Jing et al. observed a monomer that had adopted a beta-sandwich or jelly roll fold formed by two opposing antiparallel beta-sheets (Jing et al., 2010). A DALI (distance alignment matrix method)-guided search of the protein database (PDB) identified carbohydrate binding modules and enzymes involved in carbohydrate degradation as the closest structural homologs of RetSp. An asymmetric dimer made up the asymmetric unit of the crystal, which the authors judged unlikely to represent a biologically relevant dimer. Cross-linking studies revealed that a portion of RetSp formed a dimer in solution, suggesting a monomer-dimer equilibrium, while a fluorescence resonance energy transfer (FRET)-based strategy was used to estimate a dissociation constant (K_d) of 580 ± 50 nM for RetSp dimerization.

Vincent et al. crystallized the same portion of RetS and observed two distinct structures (Vincent et al., 2010). One, a monomer, adopted much the same fold as described by Jing et al., while the other was a domain-swapped dimer. Small-angle X-ray scattering and multiangle light scattering studies suggested that the domain-swapped dimer was not adopted by RetSp in solution but, rather, that monomer-monomer interactions might be mediated by a four-beta-strand interface.

The ligand binding domain of RetSp cannot be mapped experimentally, as the ligand for this protein is not known. Both Jing et al. and Vincent et al. compared their RetSp structures to those of carbohydrate binding modules (CBMs). The two groups concurred that a helical domain interposed between b1 and b2 occludes what would otherwise be a carbohydrate binding site based on homology with CBMs. Instead, the surface of the five-strand beta sheet b1-b3-b8-b5-b6 was proposed by both groups of authors as a likely ligand binding pocket. Vincent et al. also proposed a second hydrophobic patch protruding from the four-strand beta sheet (b2-b9-b4-b7) as a candidate carbohydrate binding surface, though the equivalent site in CBMs has not been shown to bind carbohydrate. Vincent et al. next modeled the 7TMR-DISMED2 domain of LadS and found that aromatic residues implicated in the first carbohydrate binding site of RetSp were well conserved in LadSp. This was not the case for the second putative binding site. Both Jing et al. and Vincent et al. hypothesized that ligand binding might disrupt RetS homodimerization and favor RetS-GacS interactions, resulting in diminished expression of biofilm and increased expression of T3SS. This is consistent with prior observations that a RetS allele lacking the periplasmic region could fully complement PA103 Δ*retS* for T3SS expression in vitro and virulence in vivo (Laskowski and Kazmierczak, 2006).

Summary and future questions

The GacS/GacA two-component system influences—positively and negatively—the expression of many genes associated with *P. aeruginosa* virulence. Progress has been made in understanding how RetS might influence GacS/GacA activity, although it is not yet clear how proteins such as HptB intersect with this signaling axis. Although GacA directly regulates only *rsmY* and *rsmZ* expression, other factors are also likely to influence the expression of these sRNAs. The activity of *rsmY* and *rsmZ* is primarily exerted by their interactions with the RNA binding protein RsmA, which prevent RsmA from functioning as a posttranscriptional regulator of gene expression. To date, RsmA has been shown to inhibit translation of the mRNAs with which it interacts, either by blocking the start codon or Shine-Dalgarno sequence of these mRNAs (PA0081, PA0082, and PA4492) (Brencic and Lory, 2009) or by stabilizing mRNA structures that mask these ribosome binding sites (*psl*) (Brencic and Lory, 2009; Irie et al., 2010). Although the expression of many gene products is increased by RsmA (e.g., T3SS genes), this regulation is exerted at the transcriptional level and is likely indirect, i.e., via inhibition of negative regulators of gene expression (Brencic et al., 2009). The identity of such RsmA-targeted negative regulators is still unknown.

Lastly, the signals that regulate activity of the GacS, LadS, and RetS sensor kinases have not been identified. In *P. fluorescens*, the expression of *rsmXYZ* is strongly influenced by mutations that alter the relative abundance of TCA cycle intermediates (fumarate/succinate versus pyruvate) (Takeuchi et al., 2009). Likewise, shifts in specific carbon/nitrogen substrate availability (discussed below) alter the expression of many virulence factors, such as the T3SS, biofilm, and motility, though the mechanistic basis for these changes in expression is largely not understood (Dacheux et al., 2002; Rietsch et al., 2004; Rietsch and Mekalanos, 2006). The presence of predicated carbohydrate binding domains in the periplasmic portions of the RetS and LadS sensor kinases raises the possibility that these proteins might

also be sensing nutrient availability and growth by responding to turnover of cell wall peptidoglycan.

Metabolic Signals That Influence Virulence Factor Expression

Most bacterial pathogens, *P. aeruginosa* included, use their virulence factors for the primary purpose of gaining access to nutrients rather than for causing damage to specific hosts. It is not surprising, therefore, that the expression of many virulence genes is influenced by the nutrient supply available to bacteria. In *Escherichia coli* and related proteobacteria, many virulence genes are under carbon catabolite repression (reviewed in Gorke and Stulke, 2008). In the absence of a preferred carbon source, such as glucose, the phosphorylated EIIA^{Glc} component of the phosphoenolpyruvate-carbohydrate phosphotransferase system activates adenylate cyclase. cAMP produced by this enzyme binds to the cAMP receptor protein (CRP), and the cAMP-CRP complex functions as a transcriptional activator of genes that encode S fimbriae in uropathogenic *E. coli* (Schmoll et al., 1990), the plasminogen activator gene of *Yersinia pestis* (Kim et al., 2007), and the SirA response regulator that governs expression of the SPI1 T3SS of *Salmonella enterica* (Teplitski et al., 2006).

Although pseudomonads possess a CRP homolog, Vfr, which serves as a global regulator of many genes associated with virulence (discussed above), Vfr has no clear relationship to carbon catabolite repression in *P. aeruginosa* (Suh et al., 2002). Indeed, carbon catabolite repression in pseudomonads, the subject of an excellent recent review (Rojo, 2010), is mechanistically quite distinct from this process in *E. coli* or *Bacillus subtilis*. The preferred carbon sources for pseudomonads are certain organic acids and amino acids, rather than glucose, which may reflect the availability of these compounds in a soil environment in which decomposing plant and animal tissues are present (Rojo, 2010). Many of these amino acids are present in sputum derived from CF patients and are preferentially utlized by *P. aeruginosa* grown in CF sputum (Palmer et al., 2007). The primary system governing assimilation of preferred amino acids and inhibition of transport and catabolism of nonpreferred amino acids is controlled by the global regulator Crc (Moreno et al., 2009a).

Crc is an RNA binding protein that regulates gene expression post-transcriptionally (Moreno et al., 2007; Moreno et al., 2009b), likely by inhibiting formation of a productive translational initiation complex. Crc activity appears to be regulated by an sRNA, named *crcZ*, which binds Crc in vitro with high affinity and appears to sequester the protein in cells grown under conditions that do not generate catabolite repression (Sonnleitner et al., 2009). Although Crc protein levels do not vary significantly as a function of carbon source, levels of the *crcZ* sRNA do vary and correlate with the strength of the carbon catabolite repression effect (Sonnleitner et al., 2009).

The *crcZ* sRNA is encoded downstream of the CbrA/CbrB two-component sensor-regulator system, which activates *crcZ* transcription in an RpoN (sigma 54)-dependent fashion (Sonnleitner et al., 2009; Abdou et al., 2011). Although the signals that activate CbrA/B are not known, this two-component system controls utilization of certain carbon and nitrogen sources in response to variation in the carbon/nitrogen ratio (Nishijyo et al., 2001). CbrA/B was initially implicated in control of virulence factor expression by a study that showed that T3SS expression was inhibited in a transposon mutant that exhibited excessive histidine uptake and catabolism (Rietsch et al., 2004). This inhibitory phenotype could be suppressed by overexpressing *cbrB* and deleting *cbrA* in the transposon mutant, even though the same changes in CbrB/A expression had no effect on T3SS in a wild-type PAO1 background (Rietsch et al., 2004). This result argued that CbrB was not a direct activator of T3SS but, rather, corrected the metabolic imbalance that was associated with loss of T3SS expression in the original transposon mutant (Rietsch et al., 2004).

Deletion of Crc in *P. aeruginosa* has pleiotropic effects on the expression of many virulence factors (Linares et al., 2010). Linares et al. used two-dimensional sodium dodecyl sulfate-polyacrylamide gel electrophoresis followed by matrix-assisted laser desorption ionization–time of flight mass spectrometry to identify proteins differentially expressed between wild-type PAO1 and Δ*crc* bacteria. Phenotypic assays measuring motility, biofilm formation, and protein secretion revealed that Δ*crc* bacteria showed increased clumping, EPS production, and biofilm formation, while swimming and swarming motility was decreased. Secretion of T3SS proteins was slightly diminished in the Δ*crc* mutant, and a T3SS-dependent form of virulence toward *D. discoideum* was attenuated. Although this inverse correlation between biofilm-associated behaviors and the T3SS is reminiscent of the regulatory events seen with mutations in the GacS/A-RsmA pathway, some virulence factors were regulated quite differently by Crc mutation than by RsmA mutation. These included a T6SS protein (PA0085), which was strongly downregulated in the Δ*crc* strain even though T6SS proteins are coordinately regulated with EPS production by RsmA (Brencic and Lory, 2009; Linares et al., 2010).

A second analysis of the Crc pathway was carried out by Yeung et al., who characterized PA14 mutants with alterations in Crc, CbrA, CbrB, and CrcZ (Yeung et al., 2011). These authors measured transcript levels by quantitative real-time PCR (qRT-PCR) for several T3SS genes, which decreased in a Δ*crc* strain, as did *pelD* and *pelF* transcript levels. This coordinate change in T3SS- and EPS-associated genes was confirmed phenotypically: PA14 Δ*crc* bacteria were attenuated for static biofilm formation and had decreased Congo red staining, while also showing diminished cytotoxicity toward cultured bronchial epithelial cells as measured by lactate dehydrogenase release. Mutation of *crcZ* resulted in the opposite phenotypes, namely, increased cytotoxicity toward cultured cells and increased biofilm production and Congo red staining. Deletion of CbrA and CbrB, which are thought to be positive regulators of *crcZ* expression, generally phenocopied the results

Table 3. Other regulators implicated in *P. aeruginosa* virulence factor expression

Regulator(s)	Target(s)	Signal(s)	Reference(s)
PsrA	T3SS (direct) TFP Psl operon *rpoS*	Long-chain fatty acids, antimicrobial peptides	Shen et al., 2006; Gooderham et al., 2008; Kang et al., 2009
NirS, NarGH	T3SS	Nitric oxide	Van Alst et al., 2009
SadA/R/S (RocS1/R/A1)	T3SS Biofilm Fimbriae (*cup*)	Unknown	Kuchma et al., 2005; Kulasakara et al., 2005
PhoP/Q	Motility Biofilm	Low Mg^{2+}, antimicrobial peptides	McPhee et al., 2006; Gooderham et al., 2009
Fur	Exotoxin A Siderophores	Iron limitation, oxygen tension	Vasil, 2007
GbdR PhoB	Hemolytic phospholipase C	Phosphate, osmoprotectants	Ostroff et al., 1989; Shortridge et al., 1992; Wargo et al., 2009

observed for Δ*crcZ* bacteria. The authors made the interesting observation that Δ*cbrA* bacteria showed enhanced biofilm formation when grown in media supplemented with succinate but not with glucose (Yeung et al., 2011); this analysis, however, was not extended to other mutants or to other phenotypes influenced by genetic manipulation of the Crc pathway.

Linares et al. (Linares et al., 2010) and Yeung et al. (Yeung et al., 2011) observed essentially opposite phenotypes for their Δ*crc* strains vis-à-vis biofilm formation, and the reason for this discordance is not immediately apparent. The conditions under which biofilm formation was assayed were not well described by Linares et al., leaving open the possibility that differences in assay conditions between the two studies might account for some of the variation in phenotype. It is also possible that differences between PAO1 and PA14 influence the observed phenotypes, as phenotypic and genotypic differences between wild-type strains have become increasingly apparent. Both studies, however, reported regulatory phenotypes that differ from those associated with mutations in the GacS/A-RsmA pathway, reinforcing the point that multiple regulatory pathways converge upon the same virulence-associated genes—and that these regulatory inputs result in distinct patterns of gene expression.

Other regulators of virulence-associated genes

In this chapter, we have focused on a few examples of regulatory networks that broadly impact virulence factor expression and whose function we understand at the molecular level, albeit incompletely. Some fundamental regulatory systems, like quorum sensing, are discussed elsewhere in this volume and have been only touched upon in this chapter. However, it should be made clear that many virulence regulators have been described, and these often appear to have wide-ranging effects on the expression of a range of virulence-related gene products. Table 3 summarizes some of these regulators and their apparent targets, direct and/or indirect. Many of these regulators appear to respond to metabolic cues or environmental stress. This is not particularly surprising, as these regulatory networks evolved to allow *P. aeruginosa* to exploit a very broad range of natural environments for nutrient acquisition and survival. One of the current challenges in the field of *P. aeruginosa* regulation is understanding how all of these different regulatory inputs are integrated during *P. aeruginosa* exploitation of environmental niches—and how they allow this opportunistic pathogen to colonize and persist within the environment of the mammalian lung.

Acknowledgments. Work in the Kazmierczak laboratory is supported by grants R01 AI075051 and R01 AI081825 from the National Institutes of Health and an Investigator in Pathogenesis of Infectious Diseases Award from the Burroughs Wellcome Fund (to B.I.K.).

REFERENCES

Abdou, L., H. T. Chou, D. Haas, and C. D. Lu. 2011. Promoter recognition and activation by the global response regulator CbrB in *Pseudomonas aeruginosa. J. Bacteriol.* **193:**2784–2792.

Beatson, S., C. B. Whitchurch, J. L. Sargent, R. C. Levesque, and J. S. Mattick. 2002. Differential regulation of twitching motility and elastase production by Vfr in *Pseudomonas aeruginosa. J. Bacteriol.* **184:**3605–3613.

Bianconi, I., A. Milani, C. Cigana, M. Paroni, R. C. Levesque, G. Bertoni, and A. Bragonzi. 2011. Positive signature-tagged mutagenesis in *Pseudomonas aeruginosa*: tracking patho-adaptive mutations promoting airways chronic infection. *PLoS Pathog.* **7:**e1001270.

Bjarnsholt, T., P. O. Jensen, M. J. Flandaca, J. Pedersen, C. R. Hansen, C. B. Andersen, T. Pressler, M. Givskov, and N. Hoiby. 2009. *Pseudomonas aeruginosa* biofilms in the respiratory tract of cystic fibrosis patients. *Pediatr. Pulmonol.* **44:**547–558.

Bjarnsholt, T., P. O. Jensen, T. H. Jakobsen, R. Phipps, A. K. Nielsen, M. T. Rybtke, T. Tolker-Nielsen, M. Givskov, N. Hoiby, O. Ciofu, and the Scandinavian Cystic Fibrosis Study Consortium. 2010. Quorum sensing and virulence of *Pseudomonas aeruginosa* during lung infection of cystic fibrosis patients. *PLoS One* **5:**e10115.

Bodey, G. P. 2009. Fever and neutropenia: the early years. *J. Antimicrob. Chemother.* **63**(Suppl. 1)**:**i3–i13.

Bordi, C., M.-C. Lamy, I. Ventre, E. Termine, A. Hachani, S. Fillet, B. Roche, S. Bleves, V. Mejean, A. Lazdunski, and A. Filloux. 2010. Regulatory RNAs and HptB/RetS signalling pathways fine-tune *Pseudomonas aeruginosa* pathogenesis. *Mol. Microbiol.* **76:**1427–1443.

Bragonzi, A., M. Paroni, A. Nonis, N. Cramer, S. Montanari, J. Rejman, C. Di Serio, G. Doring, and B. Tummler. 2009. *Pseudomonas aeruginosa* microevolution during cystic fibrosis lung infection establishes clones with adapted virulence. *Am. J. Respir. Crit. Care Med.* **180:**138–145.

Brencic, A., and S. Lory. 2009. Determination of the regulon and identification of novel mRNA targets of *Pseudomonas aeruginosa* RsmA. *Mol. Microbiol.* **72:**612–632.

Brencic, A., K. A. McFarland, H. R. McManus, S. Castang, I. Mogno, S. L. Dove, and S. Lory. 2009. The GacS/GacA signal transduction system of *Pseudomonas aeruginosa* acts exclusively through its control over the transcription of the RsmY and RsmZ regulatory small RNAs. *Mol. Microbiol.* **73:**434–445.

Burns, J. L., R. L. Gibson, S. McNamara, D. Yim, J. Emerson, M. Rosenfeld, P. Hiatt, K. McCoy, R. Castile, A. L. Smith, and B. W. Ramsey. 2001. Longitudinal assessment of *Pseudomonas aeruginosa* in young children with cystic fibrosis. *J. Infect.Dis.* **183:**444–452.

Burrowes, E., C. Baysse, C. Adams, and F. O'Gara. 2006. Influence of the regulatory protein RsmA on cellular functions in *Pseudomonas aeruginosa* PAO1, as revealed by transcriptome analysis. *Microbiology* **152:**405–418.

Castang, S., H. R. McManus, K. H. Turner, and S. L. Dove. 2008. H-NS family members function coordinately in an opportunistic pathogen. *Proc. Natl. Acad. Sci. USA* **105:**18947–18952.

Cordes, T. J., G. A. Worzalla, A. M. Ginster, and K. T. Forest. 2011. Crystal structure of the *Pseudomonas aeruginosa* virulence factor regulator. *J. Bacteriol.* **193:**4069–4074.

Croda-Garcia, G., V. Grosso-Becerra, A. Gonzalez-Valdez, L. Servin-Gonzalez, and G. Soberon-Chavez. 2011. Transcriptional regulation of *Pseudomonas aeruginosa rhlR*: role of the CRP orthologue Vfr (virulence factor regulator) and quorum-sensing regulators LasR and RhlR. *Microbiology* 157:2545–2555.

Dacheux, D., O. Epaulard, A. de Groot, B. Guery, R. Leberre, I. Attree, B. Polack, and B. Toussaint. 2002. Activation of the *Pseudomonas aeruginosa* type III secretion system requires an intact pyruvate dehydrogenase *aceAB* operon. *Infect. Immun.* 70:3973–3977.

D'Argenio, D. A., M. Wu, L. R. Hoffman, H. Kulasekara, E. Deziel, E. E. Smith, H. Nguyen, R. K. Ernst, T. J. L. Freeman, D. H. Spencer, M. Brittnacher, H. S. Hayden, S. Selgrade, M. Klausen, D. R. Goodlett, J. L. Burns, B. W. Ramsey, and S. I. Miller. 2007. Growth phenotypes of *Pseudomonas aeruginosa lasR* mutants adapted to the airways of cystic fibrosis patients. *Mol. Microbiol.* 64:512–533.

Diaz, M. R., J. M. King, and T. L. Yahr. 2011. Intrinsic and extrinsic regulation of type III secretion gene expression in *Pseudomonas aeruginosa*. *Front. Microbiol.* 2:89.

El Solh, A. A., M. E. Akinnusi, J. P. Wiener-Kronish, S. V. Lynch, L. A. Pineda, and K. Szarpa. 2008. Persistent infection with *Pseudomonas aeruginosa* in ventilator-associated pneumonia. *Am. J. Respir. Crit. Care Med.* 178:513–519.

Engel, J., and P. Balachandran. 2009. Role of *Pseudomonas aeruginosa* type III effectors in disease. *Curr. Opin. Microbiol.* 12:61–66.

Feliziani, S., A. M. Lujan, A. J. Moyano, C. Sola, J. L. Bocco, P. Montanaro, L. F. Canigia, C. E. Argarana, and A. M. Smania. 2010. Mucoidy, quorum sensing, mismatch repair and antibiotic resistance in *Pseudomonas aeruginosa* from cystic fibrosis chronic airways infections. *PLoS One* 5:e12669.

Fox, A., D. Haas, C. Reimmann, S. Heeb, A. Filloux, and R. Voulhoux. 2008. Emergence of secretion-defective sublines of *Pseudomonas aeruginosa* PAO1 resulting from spontaneous mutations in the *vfr* global regulatory gene. *Appl. Environ. Microbiol.* 74:1902–1908.

Fuchs, E. L., E. D. Brutinel, A. K. Jones, N. B. Fulcher, M. L. Urbanowski, T. L. Yahr, and M. C. Wolfgang. 2010a. The *Pseudomonas aeruginosa* Vfr regulator controls global virulence factor expression through cyclic AMP-dependent and -independent mechanisms. *J. Bacteriol.* 192:3553–3564.

Fuchs, E. L., E. D. Brutinel, E. R. Klem, A. R. Fehr, T. L. Yahr, and M. C. Wolfgang. 2010b. In vitro and in vivo characterization of the *Pseudomonas aeruginosa* cyclic AMP (cAMP) phosphodiesterase CpdA, required for cAMP homeostasis and virulence factor regulation. *J. Bacteriol.* 192:2779–2790.

Fulcher, N. B., P. M. Holliday, E. Klem, M. J. Cann, and M. C. Wolfgang. 2010. The *Pseudomonas aeruginosa* Chp chemosensory system regulates intracellular cAMP levels by modulating adenylate cyclase activity. *Mol. Microbiol.* 76:889–904.

Gooderham, W. J., M. Bains, J. B. McPhee, I. Wiegand, and R. E. Hancock. 2008. Induction by cationic antimicrobial peptides and involvement in intrinsic polymyxin and antimicrobial peptide resistance, biofilm formation, and swarming motility of PsrA in *Pseudomonas aeruginosa*. *J. Bacteriol.* 190:5624–5634.

Gooderham, W. J., S. L. Gellatly, F. Sanschagrin, J. B. McPhee, M. Bains, C. Cosseau, R. C. Levesque, and R. E. Hancock. 2009. The sensor kinase PhoQ mediates virulence in *Pseudomonas aeruginosa*. *Microbiology* 155:699–711.

Goodman, A. L., B. R. Kulasekara, A. Rietsch, D. Boyd, R. S. Smith, and S. Lory. 2004. A signaling network reciprocally regulates genes associated with acute infection and chronic persistence in *Pseudomonas aeruginosa*. *Dev. Cell* 7:745–754.

Goodman, A. L., M. Merighi, M. Hyodo, I. Ventre, A. Filloux, and S. Lory. 2009. Direct interaction between sensor kinase proteins mediates acute and chronic disease phenotypes in a bacterial pathogen. *Genes Dev.* 23:249–259.

Gorke, B., and J. Stulke. 2008. Carbon catabolite repression in bacteria: many ways to make the most out of nutrients. *Nat. Rev. Microbiol.* 6:613–624.

Hauser, A., E. Cobb, M. Bodi, D. Mariscal, J. Valles, J. Engel, and J. Rello. 2002. Type III protein secretion is associated with poor clinical outcomes in patients with ventilator-associated pneumonia caused by *Pseudomonas aeruginosa*. *Crit. Care Med.* 30:521–528.

Hauser, A. R. 2009. The type III secretion system of *Pseudomonas aeruginosa*: infection by injection. *Nat. Rev. Microbiol.* 7:654–665.

Hauser, A. R., M. Jain, M. Bar-Meir, and S. A. McColley. 2011. Clinical significance of microbial infection and adaptation in cystic fibrosis. *Clin. Microbiol. Rev.* 24:29–70.

Hoboth, C., R. Hoffmann, A. Eichner, C. Henke, S. Schmoldt, A. Imhof, J. Heesemann, and M. Hogardt. 2009. Dynamics of adaptive microevolution of hypermutable *Pseudomonas aeruginosa* during chronic pulmonary infection in patients with cystic fibrosis. *J. Infect. Dis.* 200:118–130.

Hogardt, M., C. Hoboth, S. Schmoldt, C. Henke, L. Bader, and J. Heesemann. 2007. Stage-specific adaptation of hypermutable *Pseudomonas aeruginosa* isolates during chronic pulmonary infection in patients with cystic fibrosis. *J. Infect. Dis.* 195:70–80.

Hsu, J.-L., H.-C. Chen, H.-L. Peng, and H.-Y. Chang. 2008. Characterization of the histidine-containing phosphotransfer protein B-mediated multistep phosphorelay system in *Pseudomonas aeruginosa* PAO1. *J. Biol. Chem.* 283:9933–9944.

Inclan, Y. F., M. J. Huseby, and J. N. Engel. 2011. FimL regulates cAMP synthesis in *Pseudomonas aeruginosa*. *PLoS One* 6:e15867.

Irie, Y., M. Starkey, A. N. Edwards, D. J. Wozniak, T. Romeo, and M. R. Parsek. 2010. *Pseudomonas aeruginosa* biofilm matrix polysaccharide Psl is regulated transcriptionally by RpoS and post-transcriptionally by RsmA. *Mol. Microbiol.* 78:158–172.

Jain, M., M. Bar-Meir, S. McColley, J. Cullina, E. Potter, C. Powers, M. Prickett, R. Seshadri, B. Jovanovic, A. Petrocheilou, J. D. King, and A. R. Hauser. 2008. Evolution of *Pseudomonas aeruginosa* type III secretion in cystic fibrosis: a paradigm of chronic infection. *Transl. Res.* 152:257–264.

Jing, X., J. Jaw, H. H. Robinson, and F. D. Schubot. 2010. Crystal structure and oligomeric state of the RetS signaling kinase sensory domain. *Proteins* 78:1631–1640.

Johansen, H. K., and N. Hoiby. 1992. Seasonal onset of initial colonisation and chronic infection with *Pseudomonas aeruginosa* in patients with cystic fibrosis in Denmark. *Thorax* 47:109–111.

Jones, A. K., N. B. Fulcher, G. J. Balzer, M. L. Urbanowski, C. L. Pritchett, M. J. Schurr, T. L. Yahr, and M. C. Wolfgang. 2010. Activation of the *Pseudomonas aeruginosa* AlgU regulon through *mucA* mutation inhibits cyclic AMP/Vfr signaling. *J. Bacteriol.* 192:5709–5717.

Jyot, J., V. Balloy, G. Jouvion, A. Verma, L. Touqui, M. Huerre, M. Chignard, and R. Ramphal. 2011. Type II secretion system of *Pseudomonas aeruginosa*: in vivo evidence of a significant role in death due to lung infection. *J. Infect. Dis.* 203:1369–1377.

Kang, Y., V. V. Lunin, T. Skarina, A. Savchenko, M. J. Schurr, and T. T. Hoang. 2009. The long-chain fatty acid sensor, PsrA, modulates the expression of *rpoS* and the type III secretion *exsCEBA* operon in *Pseudomonas aeruginosa*. *Mol. Microbiol.* 73:120–136.

Kim, T. J., S. Chauhan, V. L. Motin, E. B. Goh, M. M. Igo, and G. M. Young. 2007. Direct transcriptional control of the plasminogen activator gene of *Yersinia pestis* by the cyclic AMP receptor protein. *J. Bacteriol.* 189:8890–8900.

Koh, A. Y., G. R. Priebe, C. Ray, N. van Rooijen, and G. B. Pier. 2009. Inescapable need for neutrophils as mediators of cellular innate immunity to acute *Pseudomonas aeruginosa* pneumonia. *Infect. Immun.* **77:**5300–5310.

Kuchma, S. L., J. P. Connolly, and G. A. O'Toole. 2005. A three-component regulatory system regulates biofilm maturation and type III secretion in *Pseudomonas aeruginosa. J. Bacteriol.* **187:**1441–1454.

Kulasakara, H. D., I. Ventre, B. R. Kulasekara, A. Lazdunski, A. Filloux, and S. Lory. 2005. A novel two-component system controls the expression of *Pseudomonas aeruginosa* fimbrial *cup* genes. *Mol. Microbiol.* **55:**368–380.

Lapouge, K., M. Schubert, F. H.-T. Allain, and D. Haas. 2008. Gac/Rsm signal transduction pathway of g-proteobacteria: from RNA recognition to regulation of social behaviour. *Mol. Microbiol.* **67:**241–253.

Laskowski, M. A., and B. I. Kazmierczak. 2006. Mutational analysis of RetS, an unusual sensor kinase-response regulator hybrid required for *Pseudomonas aeruginosa* virulence. *Infect. Immun.* **74:**4462–4473.

Laskowski, M. A., E. Osborn, and B. I. Kazmierczak. 2004. A novel sensor kinase-response regulator hybrid regulates type III secretion and is required for virulence in *Pseudomonas aeruginosa. Mol. Microbiol.* **54:**1090–1103.

Lavoie, E. G., T. Wangdi, and B. I. Kazmierczak. 2011. Innate immune responses to *Pseudomonas aeruginosa* infection. *Microbes Infect.* **13:**1133–1145.

Le Berre, R., S. Nguyen, E. Nowak, E. Kipnis, M. Pierre, L. Quenee, F. Ader, S. Lancel, R. Courcol, B. P. Guery, and K. Faure. 2011. Relative contribution of three main virulence factors in *Pseudomonas aeruginosa* pneumonia. *Crit. Care Med.* **39:**2113–2120.

Lelong, E., A. Marchetti, M. Simon, J. L. Burns, C. van Delden, T. Kohler, and P. Cosson. 2011. Evolution of *Pseudomonas aeruginosa* virulence in infected patients revealed in a *Dictyostelium discoideum* host model. *Clin. Microbiol. Infect.* **17:**1415–1420.

Linares, J. F., R. Moreno, A. Fajardo, L. Martinez-Solano, R. Escalante, F. Rojo, and J. L. Martinez. 2010. The global regulator Crc modulates metabolism, susceptibility to antibiotics and virulence in *Pseudomonas aeruginosa. Environ. Microbiol.* **12:**3196–3212.

Mathee, K., G. Narasimhan, C. Valdes, X. Qiu, J. M. Matewish, M. Koehrsen, A. Rokas, C. N. Yandava, R. Engels, E. Zeng, R. Olavarietta, M. Doud, R. S. Smith, P. Montgomery, J. R. White, P. A. Godfrey, C. Kodira, B. Birren, J. E. Galagan, and S. Lory. 2008. Dynamics of *Pseudomonas aeruginosa* genome evolution. *Proc. Natl. Acad. Sci. USA* **105:**3100–3105.

McPhee, J. B., M. Bains, G. Winsor, S. Lewenza, A. Kwasnicka, M. D. Brazas, F. S. Brinkman, and R. E. Hancock. 2006. Contribution of the PhoP-PhoQ and PmrA-PmrB two-component regulatory systems to Mg^{2+}-induced gene regulation in *Pseudomonas aeruginosa. J. Bacteriol.* **188:**3995–4006.

Michel, G. P., A. Aguzzi, G. Ball, C. Soscia, S. Bleves, and R. Voulhoux. 2011. Role of *fimV* in type II secretion system-dependent protein secretion of *Pseudomonas aeruginosa* on solid medium. *Microbiology* **157:**1945–1954.

Moreno, R., M. Martinez-Gomariz, L. Yuste, C. Gil, and F. Rojo. 2009a. The *Pseudomonas putida* Crc global regulator controls the hierarchical assimilation of amino acids in a complete medium: evidence from proteomic and genomic analyses. *Proteomics* **9:**2910–2928.

Moreno, R., S. Marzi, P. Romby, and F. Rojo. 2009b. The Crc global regulator binds to an unpaired A-rich motif at the *Pseudomonas putida alkS* mRNA coding sequence and inhibits translation initiation. *Nucleic Acids Res.* **37:**7678–7690.

Moreno, R., A. Ruiz-Manzano, L. Yuste, and F. Rojo. 2007. The *Pseudomonas putida* Crc global regulator is an RNA binding protein that inhibits translation of the AlkS transcriptional regulator. *Mol. Microbiol.* **64:**665–675.

Nishijyo, T., D. Haas, and Y. Itoh. 2001. The CbrA-CbrB two-component regulatory system controls the utilization of multiple carbon and nitrogen sources in *Pseudomonas aeruginosa. Mol. Microbiol.* **40:**917–931.

Ostroff, R. M., B. Wretlind, and M. L. Vasil. 1989. Mutations in the hemolytic-phospholipase C operon result in decreased virulence of *Pseudomonas aeruginosa* PAO1 grown under phosphate-limiting conditions. *Infect. Immun.* **57:**1369–1373.

Palmer, K. L., L. M. Aye, and M. Whiteley. 2007. Nutritional cues control *Pseudomonas aeruginosa* multicellular behavior in cystic fibrosis sputum. *J. Bacteriol.* **189:**8079–8087.

Rau, M. H., S. K. Hansen, H. K. Johansen, L. E. Thomsen, C. T. Workman, K. F. Nielsen, L. Jelsbak, N. Hoiby, L. Yang, and S. Molin. 2010. Early adaptive developments of *Pseudomonas aeruginosa* after the transition from life in the environment to persistent colonization in the airways of human cystic fibrosis hosts. *Environ. Microbiol.* **12:**1643–1658.

Rietsch, A., and J. J. Mekalanos. 2006. Metabolic regulation of type III secretion gene expression in *Pseudomonas aeruginosa. Mol. Microbiol.* **59:**807–820.

Rietsch, A., M. C. Wolfgang, and J. J. Mekalanos. 2004. Effect of metabolic imbalance on expression of type III secretion genes in *Pseudomonas aeruginosa. Infect. Immun.* **72:**1383–1890.

Rojo, F. 2010. Carbon catabolite repression in *Pseudomonas*: optimizing metabolic versatility and interactions with the environment. *FEMS Microbiol. Rev.* **34:**658–684.

Romeo, T., G. M, M. Y. Liu, and A. M. Brun-Zinkernagel. 1993. Identification and molecular characterization of *csrA*, a pleiotropic gene from *Escherichia coli* that affects glycogen biosynthesis, gluconeogenesis, cell size, and surface properties. *J. Bacteriol.* **175:**4744–4755.

Romling, U., B. Fiedler, J. Bosshammer, D. Grothues, J. Greipel, H. von der Hardt, and B. Tummler. 1994. Epidemiology of chronic *Pseudomonas aeruginosa* infections in cystic fibrosis. *J. Infect. Dis.* **170:**1616–1621.

Roy-Burman, A., R. H. Savel, S. Racine, B. L. Swanson, N. S. Revadigar, J. Fujimoto, T. Sawa, D. W. Frank, and J. P. Wiener-Kronish. 2001. Type III protein secretion is associated with death in lower respiratory and systemic *Pseudomonas aeruginosa* infections. *J. Infect. Dis.* **183:**1767–1774.

Schmoll, T., M. Ott, B. Oudega, and J. Hacker. 1990. Use of a wild-type gene fusion to determine the influence of environmental conditions on expression of the S fimbrial adhesin in an *Escherichia coli* pathogen. *J. Bacteriol.* **172:**5103–5111.

Semmler, A. B., C. B. Whitchurch, A. J. Leech, and J. S. Mattick. 2000. Identification of a novel gene, *fimV*, involved in twitching motility in *Pseudomonas aeruginosa. Microbiology* **146:**1321–1332.

Serate, J., G. P. Roberts, O. Berg, and H. Youn. 2011. Ligand responses of Vfr, the virulence factor regulator from *Pseudomonas aeruginosa. J. Bacteriol.* **193:**4859–4868.

Shen, D. K., D. Filopon, L. Kuhn, B. Polack, and B. Toussaint. 2006. PsrA is a positive transcriptional regulator of the type III secretion system in *Pseudomonas aeruginosa. Infect. Immun.* **74:**1121–1129.

Shortridge, V. D., A. Lazdunski, and M. L. Vasil. 1992. Osmoprotectants and phosphate regulate expression of phospholipase C in *Pseudomonas aeruginosa. Mol. Microbiol.* **6:**863–871.

Smith, E. E., D. G. Buckley, Z. Wu, C. Saenphimmachak, L. R. Hoffman, D. A. D'Argenio, S. I. Miller, B. W. Ramsey, D. P. Speert, S. M. Moskowitz, J. L. Burns, R. Kaul, and M. V. Olson.

2006. Genetic adaptation by *Pseudomonas aeruginosa* to the airways of cystic fibrosis patients. *Proc. Natl. Acad. Sci. USA* 103:8487–8492.

Smith, R. S., M. C. Wolfgang, and S. Lory. 2004. An adenylate cyclase-controlled signaling network regulates *Pseudomonas aeruginosa* virulence in a mouse model of acute pneumonia. *Infect. Immun.* 72:1677–1684.

Sonnleitner, E., L. Abdou, and D. Haas. 2009. Small RNA as global regulator of carbon catabolite repression in *Pseudomonas aeruginosa*. *Proc. Natl. Acad. Sci. USA* 106:21866–21871.

Suh, S. J., L. J. Runyen-Janecky, T. C. Maleniak, P. Hager, C. H. MacGregor, N. A. Zielinski-Mozny, P. V. Phibbs, Jr., and S. E. H. West. 2002. Effect of *vfr* mutation on global gene expression and catabolite repression control of *Pseudomonas aeruginosa*. *Microbiology* 148:1561–1569.

Takeuchi, K., P. Kiefer, C. Reimmann, C. Keel, C. Dubuis, J. Rolli, J. A. Vorholt, and D. Haas. 2009. Small RNA-dependent expression of secondary metabolism is controlled by Krebs cycle function in *Pseudomonas fluorescens*. *J. Biol. Chem.* 284:34976–34985.

Teplitski, M., R. I. Goodier, and B. M. M. Ahmer. 2006. Catabolite repression of the SirA regulatory cascade in *Salmonella enterica*. *Int. J. Med. Microbiol.* 296:449–466.

Van Alst, N. E., M. Wellington, V. L. Clark, C. G. Haidaris, and B. H. Iglewski. 2009. Nitrite reductase NirS is required for type III secretion system expression and virulence in the human monocyte cell line THP-1 by *Pseudomonas aeruginosa*. *Infect. Immun.* 77:4446–4454.

Van Devanter, D. R., and J. M. Van Dalfsen. 2005. How much do *Pseudomonas* biofilms contribute to symptoms of pulmonary exacerbation in cystic fibrosis? *Pediatr. Pulmonol.* 39:504–506.

Vasil, M. L. 2007. How we learnt about iron acquisition in *Pseudomonas aeruginosa*: a series of very fortunate events. *Biometals* 20:587–601.

Ventre, I., A. L. Goodman, I. Vallet-Gely, P. Vasseur, C. Soscia, S. Molin, S. Bleves, A. Lazdunski, S. Lory, and A. Filloux. 2006. Multiple sensors control reciprocal expression of *Pseudomonas aeruginosa* regulatory RNA and virulence genes. *Proc. Natl. Acad. Sci. USA* 103:171–176.

Vincent, F., A. Round, A. Reynaud, C. Bordi, A. Filloux, and Y. Bourne. 2010. Distinct oligomeric forms of the *Pseudomonas aeruginosa* RetS sensor domain modulate accessibility to the ligand binding site. *Environ. Microbiol.* 12:1775–1786.

Wargo, M. J., T. C. Ho, M. J. Gross, L. A. Whittaker, and D. A. Hogan. 2009. GbdR regulates *Pseudomonas aeruginosa plcH* and *pchP* transcription in response to choline catabolites. *Infect. Immun.* 77:1103–1111.

Whitchurch, C. B., S. A. Beatson, J. C. Comolli, T. Jakobsen, J. L. Sargent, J. J. Bertrand, J. West, M. Klausen, L. L. Waite, P. J. Kang, T. Tolker-Nielsen, J. S. Mattick, and J. N. Engel. 2005. *Pseudomonas aeruginosa fimL* regulates multiple virulence functions by intersecting with Vfr-modulated pathways. *Mol. Microbiol.* 55:1357–1378.

Wolfgang, M. C., V. T. Lee, M. E. Gilmore, and S. Lory. 2003. Coordinate regulation of bacterial genes by a novel adenylate cyclase signaling pathway. *Dev. Cell* 4:253–263.

Workentine, M. L., L. Chang, H. Ceri, and R. J. Turner. 2009. The GacS-GacA two-component regulatory system of *Pseudomonas fluorescens*: a bacterial two-hybrid analysis. *FEMS Microbiol. Lett.* 292:50–56.

Wu, W., H. Badrane, S. Arora, H. V. Baker, and S. Jin. 2004. MucA-mediated coordination of type III secretion and alginate synthesis in *Pseudomonas aeruginosa*. *J. Bacteriol.* 186:7575–7585.

Yang, L., L. Jelsbak, R. L. Marvig, S. Damkiaer, C. T. Workman, M. H. Rau, S. K. Hansen, A. Folkesson, H. K. Johansen, O. Ciofu, N. Hoiby, M. O. Sommer, and S. Molin. 2011. Evolutionary dynamics of bacteria in a human host environment. *Proc. Natl. Acad. Sci. USA* 108:7481–7486.

Yeung, A. T., M. Bains, and R. E. Hancock. 2011. The sensor kinase CbrA is a global regulator that modulates metabolism, virulence, and antibiotic resistance in *Pseudomonas aeruginosa*. *J. Bacteriol.* 193:918–931.

Yu, H., J. C. Boucher, N. S. Hibler, and V. Deretic. 1996. Virulence properties of *Pseudomonas aeruginosa* lacking the extreme-stress sigma factor AlgU (s^E). *Infect. Immun.* 64:2774–2781.

Zolfaghar, I., A. A. Angus, P. J. Kang, A. To, D. J. Evans, and S. M. J. Fleiszig. 2005. Mutation of *retS*, encoding a putative hybrid two-component regulatory protein in *Pseudomonas aeruginosa*, attenuates multiple virulence mechanisms. *Microbes Infect.* 7:1305–1316.

Regulation of Bacterial Virulence
Edited by Michael L. Vasil and Andrew J. Darwin
© 2013 ASM Press, Washington, DC   doi:10.1128/9781555818524.ch3

Chapter 3

Quorum Sensing in *Burkholderia*

CHARLOTTE D. MAJERCZYK, E. PETER GREENBERG, AND JOSEPHINE R. CHANDLER

INTRODUCTION

Bacteria, like higher life forms, are social creatures that exhibit a range of extraordinary behaviors in groups (Costerton et al., 1999; West et al., 2006). Many of these behaviors are involved in virulence and for this reason have gained much attention in recent years, but they can also be beneficial in other environments (Diggle et al., 2007a). Among these social traits is the ability to communicate with one another by using different chemical signals. The signals of gram-positive bacteria are typically short oligopeptides, and *Proteobacteria* commonly communicate with enzymatically produced small molecules (Waters and Bassler, 2005). Bacterial communication can modulate a wide range of microbial processes, including pathogenesis in many species (Whitehead et al., 2001). Most communication systems become activated when a bacterial population reaches a critical density. For this reason, these types of systems are deemed "quorum sensing" systems (Fuqua et al., 1994). When a quorum is achieved, individuals can coordinate behavioral changes on a population-wide scale. Thus, bacteria, like more complex organisms, are thought to use communication to direct group activities (Diggle et al., 2007a; West et al., 2006).

One of the best-studied quorum sensing systems in *Proteobacteria* uses small signal molecules that are acylated homoserine lactones (AHLs) (some examples are shown in Fig. 1). AHLs are synthesized by a dedicated enzyme and specifically interact with a cytoplasmic DNA-binding receptor protein. AHL-mediated quorum sensing was first described to occur in a marine bacterium, *Vibrio fischeri*, in which it controls light production during symbiosis with marine animals such as the Hawaiian bobtail squid (for reviews, see Nealson and Hastings, 1979, and Fuqua et al., 2001). In *V. fischeri*, persistence in the host is impaired by mutations that abolish light production or quorum sensing (Visick et al., 2000).

At first, AHL signaling was thought to be unique to bioluminescent marine bacteria, but similar systems were later described for a number of other *Proteobacteria* (for reviews, see Whitehead et al., 2001; Fuqua et al., 2001; Fuqua and Greenberg, 2002; Fuqua and Winans, 1994; and Fuqua et al., 1994). Many of these bacterial species are opportunistic pathogens of plants and animals and use AHLs to regulate virulence factors (see Table 1). For this reason, AHL-dependent quorum sensing is thought to be important for the transition from an environmental to a host-associated lifestyle (Parsek and Greenberg, 2000). AHL synthases and receptors are also encoded in the genomes of many bacteria that are strictly free-living or host associated (Table 1), suggesting that quorum sensing may have alternative functions. In this chapter, we first summarize the general features of quorum sensing and highlight its importance in two model organisms, *V. fischeri* and an opportunistic pathogen, *Pseudomonas aeruginosa*. This will reveal as-yet-unanswered questions, which we believe can be addressed by studying quorum sensing in *Burkholderia*. Then we will turn our attention to three related *Burkholderia* species that adapted to fill disparate pathogenic and nonpathogenic ecological niches: *Burkholderia pseudomallei*, *Burkholderia thailandensis*, and *Burkholderia mallei* (the Bptm group). Through studying the Bptm group, we hope to gain a new perspective on the evolution of quorum sensing and its use in different environments.

OVERVIEW OF AHL SIGNALING

AHL Synthesis and Response

AHL synthesis is typically carried out by members of a family of proteins that are homologs of the *V. fischeri* LuxI signal synthase enzyme (Fuqua et al., 1994). Most described LuxI homologs use

Charlotte D. Majerczyk, E. Peter Greenberg, and Josephine R. Chandler • Department of Microbiology, School of Medicine, University of Washington, Seattle, WA 98195.

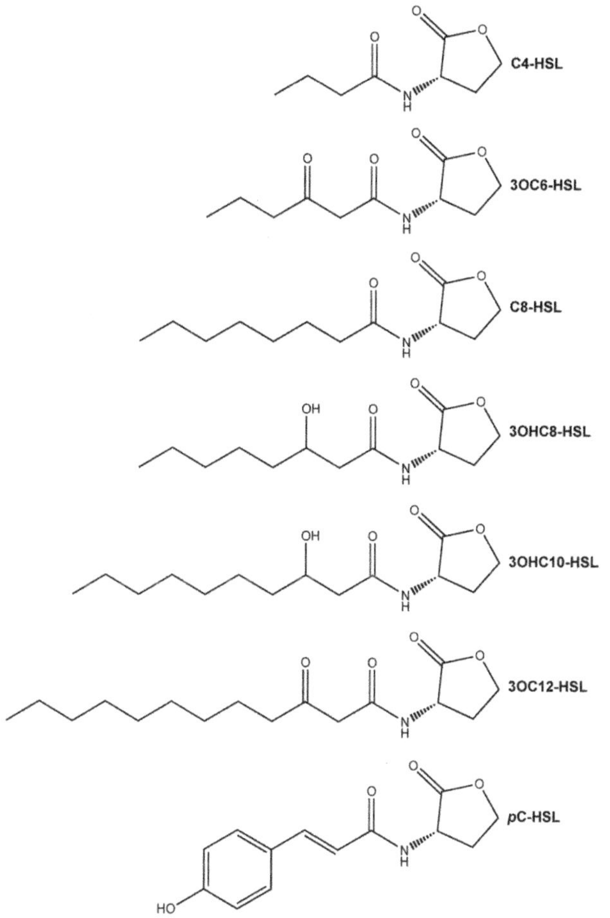

Figure 1. Some examples of AHL quorum sensing signals. The AHL structures and corresponding names are shown, organized by chain length or complexity. The signal made by RhlI of *P. aeruginosa* is C4-HSL. LuxI of *V. fisheri* makes 3OC6-HSL. The Bptm signals of QS-1, QS-2, and QS-3 are C8-HSL, 3OHC10-HSL, and 3OHC8-HSL, respectively. LasI of *P. aeruginosa* produces 3OC12-HSL, and RpaI of *R. palustris* synthesizes *para*-coumaroyl-HSL (*p*C-HSL).
doi:10.1128/9781555818524.ch3f1

S-adenosylmethionine as a substrate for the homoserine lactone ring and an acyl-acyl carrier protein derived from fatty acid biosynthesis for the acyl side group (Schaefer et al., 1996; More et al., 1996; Parsek et al., 1999). LuxI proteins create system specificity by incorporating different fatty-acyl side groups into the AHL signal (see Fig. 1 for examples of AHL structures). Fatty-acyl side chains can vary in length from 4 to 18 carbons and can have different degrees of saturation or substitution on the third carbon position. Recently, two noteworthy LuxI homologs were identified in *Rhodopseudomonas palustris* and a species of *Bradyrhizobium* (Schaefer et al., 2008; Ahlgren et al., 2011), which use an aromatic acid instead of a fatty acid intermediate in AHL synthesis. This results in an AHL that has an aromatic ring

instead of a fatty acid intermediate for the side chain (the *R. palustris* AHL is shown in Fig. 1). These AHL synthases expand the range of possible substrates for AHL synthesis.

AHL receptors share similarity with the LuxR protein from *V. fischeri*. LuxR homologs contain a specific N-terminal AHL-binding domain and a C-terminal helix-turn-helix DNA binding domain (Hanzelka and Greenberg, 1995; Stevens et al., 1994; Zhang et al., 2002). LuxI and LuxR proteins are commonly found as a cognate pair that synthesize and detect a specific AHL with high but not perfect fidelity. The genes encoding a cognate LuxI and LuxR pair are often genetically linked, but the orientation with respect to each other can vary between systems and species. Upon binding its cognate AHL, LuxR homologs change conformation; they dimerize and can then bind target DNA (for a review of LuxI and LuxR homologs, see Fuqua et al., 2001). DNA binding typically occurs upstream of the target gene in the promoter region. The *V. fischeri* LuxR binds to a 20-bp inverted repeat sequence called the *lux* box (Devine et al., 1989). LuxR homologs often bind to similar inverted repeats (Fuqua et al., 2001), although there is not always a discernible *lux* box-like element in promoters of directly regulated genes (Schuster et al., 2004). Most known AHL-bound LuxR family proteins activate transcription of target genes; however, several act as repressors (Andersson et al., 2000; Minogue et al., 2002; von Bodman et al., 1998). The repressors bind promoter DNA and block transcription in the absence of their cognate AHL. AHL binding results in the release of DNA and allows transcription to occur (Nasser et al., 1998). All known AHL-responsive repressors are encoded by species in the *Enterobacteriales* group of *Proteobacteria* (Tsai and Winans, 2010).

AHL-dependent activation by LuxR homologs is correlated with population density. At low cell density, each bacterium produces a basal level of AHLs that diffuse through the cell membrane and are rapidly diluted outside of the cell (Kaplan and Greenberg, 1985; Pearson et al., 1994). When the population density increases, the local concentration of AHLs accumulates until it reaches a sufficient concentration to bind and activate the LuxR homolog (Engebrecht and Silverman, 1984; Engebrecht et al., 1983). AHL-bound LuxR homologs often activate expression of their cognate *luxI*. This creates a positive-feedback loop that further elevates the concentration of AHL upon initial induction (Engebrecht et al., 1983).

In some cases, a gene coding for a LuxR homolog is not adjacent to a gene coding for an AHL synthase. In such situations, either the encoded LuxR homologs outnumber the LuxI homologs or they are

Table 1. Some AHL quorum sensing-regulated processes in diverse *Proteobacteria*

Organism	Host/habitat	Quorum sensing-regulated processes	References
Opportunistic pathogens			
Erwinia carotovora	Plants	Virulence, antimicrobials (carbapenem), exoenzymes (protease, cellulase)	Jones et al., 1993; Pirhonen et al., 1993; Bainton et al., 1992a; Molina et al., 2003
Agrobacterium tumefaciens	Plants	Conjugation of a virulence-associated plasmid	Piper et al., 1993; Fuqua and Winans, 1994; Zhang et al., 1993
Pseudomonas aeruginosa	Plants, animals, invertebrates	Virulence, toxins/antimicrobials (pyocyanin, hydrogen cyanide), biofilm formation (EPS), central metabolism, exoproducts (siderophores, surfactant, proteases), oxidative stress response, motility	Ochsner et al., 1994; Pearson et al., 1994; Gambello and Iglewski, 1991; Jones et al., 1993; Stintzi et al., 1998; Hassett et al, 1999; Kohler et al., 2000; Meyer et al., 1996; Smith and Iglewski, 2003; Chugani et al., 2001
Burkholderia cenocepacia (Bcc)	Plants, humans	Virulence, exoproducts (proteases, chitinases, polygalacturonase)	Aguilar et al., 2003; Huber et al., 2001; Lewenza and Sokol, 2001; Lewenza et al., 1999; Kothe et al., 2003; Sokol et al., 2003; Sousa et al., 2007
Burkholderia pseudomallei	Plants, animals	Virulence, exoproducts (PLC, MprA protease, siderophores), oxidative stress response, biofilm formation	Ulrich et al., 2004a; Valade et al., 2004; Song et al., 2005; Lumjiaktase et al., 2006; Gamage et al., 2011
Free-living nonpathogens			
Rhodobacter sphaeroides	Water	Biofilm dispersal (putative EPS)	Puskas et al., 1997
Rhodopseudomonas palustris	Soil and water	Unknown	Schaefer et al., 2008
Burkholderia thailandensis	Soil and water	Antimicrobials (bactobolin), aggregation (EPS), motility	Duerkop et al., 2009; Chandler et al., 2009; Ulrich et al., 2004c
Methylobacterium	Soil	EPS and unknown	Penalver et al., 2006; Poonguzhali et al., 2007
Acidithiobacillus ferrooxidans	Soil	Metal resistance	Wenbin et al., 2011; Rivas et al., 2005
Host-adapted pathogens			
Burkholderia mallei	Horses, donkeys	Virulence	Ulrich et al., 2004b
Brucella melitensis	Sheep, goats, camels, humans	Virulence	Delrue et al., 2005; Rambow-Larsen et al., 2008

in a genome with no LuxI homolog. When LuxR homologs do not have a cognate LuxI homolog, they are called orphans or solos (Fuqua, 2006). Orphan LuxRs are found in a number of species (for a review of orphan LuxR homologs, see Patankar and Gonzalez, 2009). Some respond to AHLs produced by a LuxI encoded elsewhere in the genome. There is also evidence that some orphans act differently and have alternative functions. For example, *Salmonella enterica* encodes only one LuxR homolog, the orphan receptor SdiR, and does not have an AHL synthase gene. It is thought that SdiR responds to AHLs from other species in the guts of certain animals where *S. enterica* is typically found (Smith et al., 2008; Dyszel et al., 2010; Ahmer et al., 1998; Kanamaru et al., 2000; Michael et al., 2001). Additionally, orphan LuxRs in two species of *Xanthomonas* do not detect AHLs but respond to unidentified molecules made by plants (Ferluga et al., 2007; Zhang et al., 2007). Orphan LuxR receptors are still not well understood,

but their study may provide insight about the evolution of AHL quorum sensing.

AHL Signaling in *Vibrio fischeri*

V. fischeri colonizes the light organ of the squid *Euprymna scolopes*, where it produces light and other factors (Antunes et al., 2007; Callahan and Dunlap, 2000; Graf and Ruby, 1998; Ruby and McFall-Ngai, 1992; McFall-Ngai and Ruby, 1991; Nealson and Hastings, 1979; Ruby, 1996). Light production by *V. fischeri* is thought to provide a counterillumination mechanism that hides the silhouette of this nocturnal invertebrate (reviewed in Visick and McFall-Ngai, 2000). During colonization, *V. fischeri* can acquire amino acids and other nutrients from the squid host and grow to the high cell densities required for quorum sensing induction. In the surrounding nutrient-poor seawater, high bacterial densities cannot be reached and a quorum cannot be achieved (Graf and

Ruby, 1998). Quorum sensing and AHL-dependent light production promote *V. fischeri* persistence in the squid light organ (Visick et al., 2000), although it is not known how luminescence is advantageous to the bacteria. It has been proposed that the light-producing luciferase enzyme protects bacteria from reactive oxygen species in the squid by consuming oxygen during light production (Ruby and McFall-Ngai, 1999; Bose et al., 2007; Visick and McFall-Ngai, 2000).

In *V. fischeri*, a single LuxIR pair controls bio-luminescence (Eberhard et al., 1981; Engebrecht and Silverman, 1984; Engebrecht et al., 1983). The AHL synthase LuxI produces *N*-3-oxo-hexanoyl homoserine lactone (3OC6-HSL, shown in Fig. 1) (Eberhard et al., 1981). The AHL receptor LuxR specifically binds this signal (Adar and Ulitzur, 1993; Hanzelka and Greenberg, 1995) and activates transcription of the *lux* operon, containing the *luxI* signal synthase and the *luxCDABEG* genes required for light production (Engebrecht and Silverman, 1984; Engebrecht et al., 1983) (see Fig. 2A for an overview). Of course, the system is considerably more complex, but this over-simplification is appropriate for the purposes of this chapter (Gilson et al., 1995; Lupp et al., 2003).

In addition to the *lux* operon, LuxR-bound 3OC6-HSL regulates about a dozen other genes that are distributed around the genome (Antunes et al., 2007; Callahan and Dunlap, 2000). Four of these are predicted to code for peptidases and proteases, which may play a role in utilization of amino acids supplied by the host (Graf and Ruby, 1998). Two other AHL-dependent factors have known or putative functions associated with colonization of animals (Luo et al., 2000; Peterson and Mekalanos, 1988; Antunes et al., 2007; Callahan and Dunlap, 2000). Thus, a number of AHL-dependent processes in *V. fischeri* appear to be important for symbiosis.

AHL Signaling in *P. aeruginosa*

Pseudomonas aeruginosa is another bacterium in which quorum sensing has been extensively studied. *P. aeruginosa* is an opportunistic pathogen that is widely distributed in diverse soil and water habitats worldwide (Green et al., 1974; Hardalo and Edberg, 1997) and also has the potential to cause severe or fatal infections in humans, other animals, plants, and several invertebrates. In immunocompromised humans, it causes acute or chronic infections (Lyczak et al., 2000). The lungs of patients with the genetic disease cystic fibrosis are often chronically colonized by *P. aeruginosa*, which can cause progressive lung deterioration and eventual death (Lyczak et al., 2002).

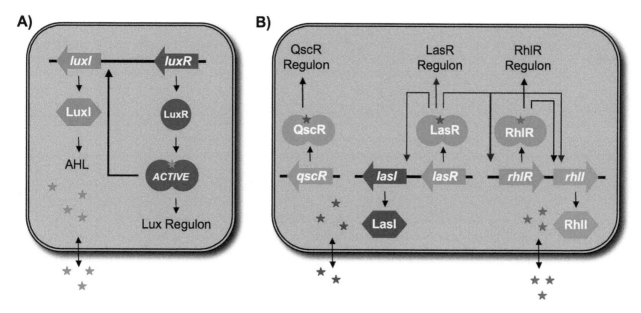

Figure 2. AHL signaling in *V. fischeri* (A) and *P. aeruginosa* (B). AHL signals (see Fig. 1) are made by members of the LuxI family of signal synthases and specifically interact with LuxR family transcription factors. At high cell density, AHLs accumulate and interact with LuxR homologs. AHL interaction causes the LuxR protein to change conformation and become active, which induces target gene regulation. (A) In *V. fischeri*, LuxI and LuxR produce and respond to 3OC6-HSL (red stars), respectively. (B) In *P. aeruginosa*, the LasIR system produces and responds to 3OC12-HSL (purple stars), and the RhlR system produces and responds to C4-HSL (green stars). QscR is an orphan LuxR receptor that is not genetically linked to a *luxI* synthase gene. QscR responds to 3OC12-HSL produced by LasI. Each quorum sensing regulon is shown as a distinct entity in the figure, but in reality there exists some overlapping regulation among the controlled genes. doi:10.1128/9781555818524.ch3f2

P. aeruginosa virulence relies on the production of a variety of secreted virulence factors. Many of these are under AHL-mediated quorum sensing control (for a more detailed review, see Smith and Iglewski, 2003). AHL quorum sensing contributes to the pathogenesis of *P. aeruginosa* in several mouse models of infection (Rumbaugh et al., 1999; Tang et al., 1996), in plants (Rahme et al., 1995), and in invertebrates (Tan et al., 1999a; Mahajan-Miklos et al., 1999; Tan et al., 1999b).

P. aeruginosa encodes two LuxIR pairs, LasIR and RhlIR, and one orphan LuxR receptor, QscR (Fig. 2B). LasR, RhlR, and QscR each regulate distinct and overlapping sets of genes (Lequette et al., 2006; Schuster et al., 2003). The LasIR system produces and responds to *N*-3-oxo-dodecanoyl homoserine lactone (3OC12-HSL, shown in Fig. 1) (Pearson et al., 1994; Passador et al., 1993), and the RhlIR system produces and responds to *N*-butanoyl homoserine lactone (C4-HSL, shown in Fig. 1) (Pearson et al., 1995; Ochsner and Reiser, 1995; Ochsner et al., 1994). Like the LuxIR system in *V. fischeri*, each of the *P. aeruginosa* systems is autoregulated (Seed et al., 1995; Latifi et al., 1996). LasR-bound 3OC12-HSL also activates the RhlIR system by directly inducing transcription of *rhlR* and *rhlI* (Pearson et al., 1997; Latifi et al., 1996). The orphan receptor QscR responds to the signal produced by LasI, 3OC12-HSL (Lee et al., 2006). Thus, there is cross-regulation between the quorum sensing systems in *P. aeruginosa*, and LasR is required for activity of both RhlIR and QscR.

AHL signaling in *P. aeruginosa* regulates a substantially larger gene set than that controlled by quorum sensing in *V. fischeri*. Recently, investigators have taken global experimental approaches using a combination of genetics, biochemistry, microarray, and proteomics analyses to show that between 100 and 300 genes are controlled by quorum sensing in *P. aeruginosa* (Hentzer et al., 2003; Wagner et al., 2003; Schuster et al., 2003; Whiteley et al., 1999; Lequette et al., 2006; Gilbert et al., 2009). Many of these are likely indirectly regulated by other AHL-independent factors (Gilbert et al., 2009). The category of genes most overrepresented in the quorum sensing regulon is those predicted to encode secreted factors or factors involved in their production. Some of these secreted factors include the toxins pyocyanin and hydrogen cyanide, iron-scavenging siderophores, and proteases (Table 1). Many of these factors are important for virulence, supporting the idea that *P. aeruginosa* uses quorum sensing to discriminate its host-associated lifestyle from its free-living lifestyle (for reviews, see Parsek and Greenberg, 2000, and Smith and Iglewski, 2003). Quorum sensing also regulates a few general

metabolic processes (Schuster and Greenberg, 2006). For these, the role in virulence is less clear. This suggests that quorum sensing may also institute broad physiological changes in the cell that prepare it for stress in a range of environments encountered at high cell density.

Although *P. aeruginosa* associates with mammalian hosts, we believe that it is primarily found in mixed-species communities in soil. There is evidence supporting the idea that *P. aeruginosa* uses quorum sensing communication to compete with other species (Hibbing et al., 2010; An et al., 2006). This suggests that quorum sensing is important to *P. aeruginosa* in the environment as well as during infection. Intriguingly, though quorum sensing promotes virulence in multiple acute infection models, many *P. aeruginosa* isolates from chronic human infections are *lasR* mutants (Heurlier et al., 2006; Fothergill et al., 2007; D'Argenio et al., 2007; Wilder et al., 2009; Smith et al., 2006; Hoffman et al., 2009; Tingpej et al., 2007). This suggests that *lasR* mutants have an advantage during the chronic disease state (D'Argenio et al., 2007; Heurlier et al., 2005). Alternatively, *lasR* mutants may arise because they can exploit quorum sensing-intact individuals with which they coexist. This would allow them the advantage of utilizing shared quorum sensing-controlled products without the metabolic burden of producing them. For this reason, these naturally occurring quorum-sensing mutants might be social cheaters (Diggle et al., 2007b; Sandoz et al., 2007; Rumbaugh et al., 2009). It is unclear if social cheaters occur in the environment; however, a small survey of *P. aeruginosa* environmental isolates showed that many retained their quorum sensing systems (Cabrol et al., 2003). The importance of quorum sensing and social cheating in infection is particularly important in light of current interest in developing novel anti-quorum sensing therapeutics to treat *P. aeruginosa* infections (Kohler et al., 2010). More research is needed to better understand how quorum sensing benefits *P. aeruginosa* in different environments and disease states.

Common Themes in Quorum Sensing

In addition to *V. fischeri* and *P. aeruginosa*, AHL-mediated quorum sensing has been described to occur in dozens of other *Proteobacteria*. In many instances, quorum sensing acts as a global regulatory circuit that controls broad physiological functions. These functions can be species specific to promote survival in certain ecological habitats, such as quorum sensing control of bioluminescence in *V. fischeri* (Table 1). AHL-dependent factors can also fall into generalized groups that are conserved. Many AHL-dependent factors are secreted. Some of the most common types are toxins (e.g., virulence factors and antimicrobials), exoenzymes (e.g., proteases), and biofilm components (e.g., extracellular polysaccharides [EPS]) (see Table 1 for specific occurrences). The overlap of AHL-regulated targets across diverse species and environments suggests that these factors provide a general benefit to bacteria in a community structure.

What is the specific benefit of quorum sensing? Many quorum sensing-controlled processes may be most valuable when carried out by populations of a sufficient cell density. For example, biofilm formation is a community behavior that is likely most beneficial to bacteria in groups rather than to isolated individuals. Bacteria in biofilm communities secrete EPS and other biofilm matrix components that surround and protect the group, but this may be a wasteful process for an isolated individual. In this way, quorum sensing provides an efficient method to regulate metabolically expensive products. An alternative hypothesis that has been proposed is that AHLs serve as a proxy for environmental diffusion potential and prevent production of secreted goods under conditions of high diffusion (Redfield, 2002). According to this hypothesis, quorum sensing also acts to efficiently regulate the production of expensive products by acting as an environmental sensor. Because many quorum sensing-controlled factors are involved in virulence, it has also been suggested that quorum sensing allows low-density populations to evade immune responses by not displaying immunogens until the population has reached a sufficient density to mount an effective attack on the host (Parsek and Greenberg, 2000). This strategy may also be useful outside of the host because delayed production of antimicrobial factors may deprive the competitor of the ability to acclimate to subinhibitory antimicrobial concentrations (Hibbing et al., 2010).

Apparently, AHL quorum sensing was established in an ancient species of *Proteobacteria* (Lerat and Moran, 2004) and has likely evolved to benefit bacteria in many different environments. However, we do not have a clear understanding of how quorum sensing specifically benefits bacteria in different host and nonhost environments. Many of the well-characterized quorum sensing systems are in species that alternate between free-living and host-associated lifestyles. In these species, it is difficult to tease apart the specific or overlapping functions of quorum sensing in each niche. For the remainder of this review, we focus on quorum sensing in *Burkholderia*, specifically the Bptm group. This group has highly conserved quorum sensing systems, yet each species occupies strikingly different environments. These two features allow the opportunity to ask how homologous

regulatory systems evolve to coordinate a spectrum of functions important in nonpathogenic environments, pathogenic environments, or the switch between free-living and host-associated environments.

QUORUM SENSING IN *BURKHOLDERIA*

The genus *Burkholderia* is a large and metabolically diverse group comprised of many species (Caballero-Mellado et al., 2007). Phylogenetic analysis shows that the members of this group are divided into two major clusters (Martinez-Aguilar et al., 2008; Caballero-Mellado et al., 2007). The first includes saprophytic nonpathogens that in most described cases are associated with plants (Caballero-Mellado et al., 2007). The other cluster is comprised of mainly pathogenic species. This includes the plant pathogen *Burkholderia glumae*, as well as a dozen species that make up the *B. cepacia* complex (Bcc), some of which can cause severe infections in immunocompromised humans (Mahenthiralingam et al., 2005). The Bptm group also populates this cluster and includes the pathogens *B. pseudomallei* and *B. mallei* and the saprophyte *B. thailandensis*.

Quorum sensing systems have been found in all of the *Burkholderia* species studied to date (Vial et al., 2007). With the exception of *B. thailandensis*, the nonpathogenic *Burkholderia* species possess a system similar to the *P. aeruginosa* LasIR system, which produces and detects 3OC12-HSL (Suarez-Moreno et al., 2008). All of the other *Burkholderia* species possess at least one system that produces and detects *N*-octanoyl homoserine lactone (C8-HSL) (Lutter et al., 2001; Lewenza et al., 1999; Kim et al., 2004). Because of its common occurrence in human disease, quorum sensing in the Bcc has been intensively studied and reviewed elsewhere (Venturi et al., 2004; Drevinek and Mahenthiralingam, 2010; Loutet and Valvano, 2010). In the related *Burkholderia* clade containing *B. pseudomallei*, *B. thailandensis*, and *B. mallei*, quorum sensing is complex. All three species contain multiple systems that likely have unique roles in each species.

The Bptm Group

Prior to discussing the quorum sensing components of the Bptm group, it is important to understand the evolutionary history and lifestyle of each species. Evidence indicates that approximately 47 million years ago, a common ancestor diverged to form the two related species *B. thailandensis* and *B. pseudomallei* (Yu et al., 2006). *B. mallei* then evolved as a host-restricted pathogen from a single clone of

B. pseudomallei. The genomes of *B. thailandensis* and *B. pseudomallei* are highly conserved: approximately 85% of their genes are shared. Species-specific genes occur in genomic islands or in homologous coding sequences that contain simple sequence repeats resulting in the formation of pseudogenes (Holden et al., 2004; Yu et al., 2006). Other genomic differences that account for the speciation of *B. thailandensis* and *B. pseudomallei* include the acquisition of virulence determinants in *B. pseudomallei*, four large-scale (>10-kb) genomic inversions, and factors conferring divergent metabolic capacities in each species (Kim et al., 2005; Yu et al., 2006).

The third member of this group, *B. mallei*, is thought to have evolved from an ancestral *B. pseudomallei* isolate following animal infection. The expansion of genomic insertion sequences then facilitated reductive evolution (Godoy et al., 2003; Nierman et al., 2004). *B. mallei* has lost over 1,000 genes (~1.4 Mb) relative to *B. pseudomallei* (Nierman et al., 2004). The majority of the lost genetic content is believed to be associated with environmental survival, as *B. mallei* is isolated only from humans or other animal hosts such as horses and donkeys. In comparison with *B. pseudomallei* and *B. thailandensis*, there are few *B. mallei* species-specific genes, suggesting that the host environment offers little opportunity for gene acquisition (Losada et al., 2010). Despite this, the *B. mallei* genome remains highly plastic. The insertion sequence elements that mediated the genome reduction of *B. mallei* also promote ongoing rearrangement by providing sites for intragenomic homologous recombination (Losada et al., 2010; Nierman et al., 2004; Song et al., 2010). Furthermore, like those of *B. thailandensis* and *B. pseudomallei*, the *B. mallei* genome encodes many simple sequence repeats that allow for the formation of pseudogenes (Nierman et al., 2004).

B. thailandensis and *B. pseudomallei* are saprophytic bacteria found in the soil and water in tropical regions common to Southeast Asia, northern Australia, South America, the Middle East, and some regions in Africa (Brett et al., 1998; Cheng and Currie, 2005). *B. pseudomallei* is also an opportunistic pathogen causing melioidosis in humans and other mammals (Cheng and Currie, 2005). Human melioidosis appears to be an emerging disease in Southeast Asia. Diabetes is a high risk factor for developing melioidosis (Currie et al., 2010; Suputtamongkol et al., 1999). The disease most commonly manifests as abscess formation on internal organs; however, pneumonia is reported for nearly half of subacute or acute cases (Leelarasamee, 2004). Humans and other mammals contract melioidosis by inhalation, by ingestion, or through breaks in the skin with contaminated environmental reservoirs. Melioidosis can cause severe acute infection or persist for months to years in a latent form (Lazar Adler et al., 2009). Asymptomatic and localized forms of melioidosis have mortality rates below 10%. However, the mortality rate of disseminated septicemic melioidosis (involving more than one organ) is 90% (Leelarasamee, 2004). Diagnosis and treatment of melioidosis are challenging because the disease presents with various symptoms and *B. pseudomallei* is intrinsically multidrug resistant. Effective treatment involves a multiweek oral and intravenous antibiotic regimen (Leelarasamee, 2004). In regions where it is endemic, melioidosis incidence is rising. Specifically, melioidosis is now the third most common cause of death from an infectious disease, after AIDS and tuberculosis in northeast Thailand (Limmathurotsakul et al., 2010).

Though *B. mallei* can be found in regions where *B. pseudomallei* and *B. thailandensis* are endemic, *B. mallei* has a markedly different lifestyle than its close relatives (Gregory and Waag, 2007). *B. mallei* is a host-restricted pathogen that has been isolated only from animal infections, including humans and its natural equine hosts. *B. mallei* causes glanders in horses and donkeys, in which it most commonly presents in a chronic disease state. Though less frequent, *B. mallei* also infects humans who have come in contact with diseased animals. Glanders in horses can be traced back to the time of Aristotle (330 BC) and is noted through history as a reemerging disease prior to eradication programs established in the 19th and 20th centuries (Gregory and Waag, 2007). In humans, glanders is predominantly an acute disease and has a high mortality rate if untreated. Acute disease traits of *B. mallei* infection resemble melioidosis, showing similar organ tropism for the lungs, spleen, and liver in laboratory models of infection (Fritz et al., 2000; Lever et al., 2003; Fritz et al., 1999; Gauthier et al., 2001; Hoppe et al., 1999; West et al., 2008).

B. mallei and *B. pseudomallei* are classified as category B select agents of concern for bioterrorism by the Centers for Disease Control and Prevention (CDC) because they are highly infectious, cause severe human disease, and have no vaccine options (Gilad et al., 2007). It is suspected that *B. mallei* has been used as a bioweapon on numerous occasions. The best-documented examples include use by Germany in World War I and use by Japan when it initiated glanders biowarfare studies on horses, civilians, and prisoners of war in the mid-20th century (Gregory and Waag, 2007). Because *B. thailandensis* is highly related to the pathogenic *Burkholderia*, it serves as a model organism for the study of *B. pseudomallei* and

B. mallei, which require confinement in a biosafety level 3 laboratory.

Quorum Sensing Networks in *B. pseudomallei*, *B. thailandensis*, and *B. mallei*

Quorum sensing was first described to occur in the Bptm group within the past decade. The quorum sensing circuits in these bacteria are among the most complex AHL systems described. *B. thailandensis* and *B. pseudomallei* each possess three complete LuxIR-dependent systems termed QS-1, QS-2, and QS-3 (Fig. 3). *B. mallei* also contains homologous systems, but it lost QS-2 during its reductive evolution from *B. pseudomallei*. Additionally, each of these species harbors two orphan LuxR homologs (R4 and R5) (Ulrich et al., 2004a; Ulrich et al., 2004c). The protein sequences of the quorum sensing systems and orphan LuxRs share an exceptionally high degree of conservation, ranging from 95 to 100% amino acid identity. It is intriguing to wonder how these related bacteria use essentially the same systems in their divergent lifestyles.

A summary of the quorum sensing components and AHL signals for each species is shown in Fig. 3. The genes for *B. thailandensis*, *B. pseudomallei*, and *B. mallei* are designated *bta*, *bps*, and *bma*, respectively. Given the high degree of identity among the quorum sensing circuits in these bacteria, it is not surprising that the corresponding LuxI homologs from all three species produce the same signals. The QS-1 LuxI homologs (BpsI1, BtaI1, and BmaI1) produce C8-HSL and are related to the C8-HSL synthases in other pathogenic *Burkholderia* species (Chandler et al., 2009; Duerkop et al., 2007; Gamage et al., 2011; Lumjiaktase et al., 2006; Song et al., 2005). The QS-2 LuxI homologs (BpsI2 and BtaI2) produce *N*-3-hydroxy-decanoyl homoserine lactone (3OHC10-HSL) (Duerkop et al., 2009; Gamage et al., 2011). Because *B. mallei* does not contain the QS-2 system, it does not produce 3OHC10-HSL (Song et al., 2010; Ulrich et al., 2004b). The QS-3 LuxI homologs (BpsI3, BtaI3, and BmaI3) produce *N*-3-hydroxy-octanoyl homoserine lactone (3OHC8-HSL) (Chandler et al., 2009; Duerkop et al., 2008; Gamage et al., 2011). The structures of these signals are shown in Fig. 1.

Early studies suggested that other signals, including *N*-hexanoyl homoserine lactone (C6-HSL), *N*-decanoyl homoserine lactone (C10-HSL), and *N*-oxo-tetradecanoyl homoserine lactone (3OC14-HSL), were associated with the different systems (Ulrich et al., 2004a; Ulrich et al., 2004b; Ulrich et al., 2004c). However, later, more quantitative work has led to our current understanding (Chandler et al., 2009; Duerkop et al., 2008; Duerkop et al., 2007; Duerkop et al., 2009; Gamage et al., 2011; Lumjiaktase et al., 2006; Song et al., 2005). The relevant AHLs may have been misidentified for a combination of reasons. Different groups used varied identification and detection methods. Some of these methods have biases towards the detection of certain AHLs (Schaefer et al., 2000). Additionally, minor products may have been identified without confirmation of physiological significance. Furthermore, several studies used *Burkholderia* strains with efflux pump mutations. Although there are conflicting reports in the literature on *B. pseudomallei*

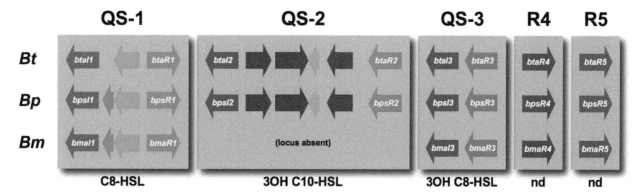

Figure 3. Quorum sensing circuits of *B. thailandensis*, *B. pseudomallei*, and *B. mallei*. Shown are the genetic context of the homologous quorum sensing circuits (QS-1, QS-2, and QS-3) in *B. thailandensis* (*Bt*), *B. pseudomallei* (*Bp*), and *B. mallei* (*Bm*). The cognate signal of each system is shown below the QS designation, and each structure can be found in Fig. 1. The signals that bind the orphan LuxR homologs have not been determined (nd). The genes for the QS-1 LuxIR homologs are separated by a small region that contains one or two open reading frames of unknown function. The genes coding for the QS-2 LuxIR homologs are found within the bactobolin biosynthetic gene cluster and are separated by three open reading frames predicted to contribute to bactobolin synthesis. The genes coding for the LuxIR homologs of the QS-3 system are separated by a small intergenic region that does not contain additional open reading frames.
doi:10.1128/9781555818524.ch3f3

regarding efflux pumps and quorum sensing (Chan et al., 2007; Chan and Chua, 2005; Mima and Schweizer, 2010), studies in *P. aeruginosa* suggest that AHL accumulation and signaling are mildly altered by efflux pumps (Aendekerk et al., 2002; Kohler et al., 2001). For this reason, care should be taken when interpreting quorum sensing studies done in strains with efflux pump mutations.

Currently, little is known about the interplay between the three complete quorum sensing systems and orphan LuxR homologs in the Bptm group. Some studies have investigated the specificity of the LuxR homologs by using *Escherichia coli* as a heterologous host. The QS-1 and QS-2 LuxR homologs are most sensitive to their cognate signals, but they can also be activated by other AHLs (Duerkop et al., 2007; Duerkop et al., 2009; Gamage et al., 2011). The QS-1 and QS-2 systems also exhibit autoregulation where the AHL-bound LuxR activates expression of its cognate *luxI* synthase gene in all of the species tested (Gamage et al., 2011; Duerkop et al., 2009; Kiratisin and Sanmee, 2008; Duerkop et al., 2007). In contrast, the QS-3 LuxR homolog does not activate the promoter of its cognate *luxI* synthase gene in a heterologous host, but it can weakly activate the QS-1 *luxI* gene (Gamage et al., 2011; Duerkop et al., 2008). This suggests that QS-3 may activate QS-1. This idea is supported by a study in *B. pseudomallei* showing that *bpsI3* is required for production of its cognate signal (3OHC8-HSL) as well as the QS-1 signal (C8-HSL) (Gamage et al., 2011). If such regulation occurs, a hierarchy of activation may exist among the quorum sensing systems of these *Burkholderia* species similar to that of *P. aeruginosa*. Further work needs to be done to verify this.

Quorum Sensing in *Burkholderia pseudomallei*

Quorum sensing in *B. pseudomallei* has many parallels with quorum sensing in *P. aeruginosa*. Both bacteria are saprophytes and opportunistic pathogens of humans and other animals. They both contain multiple LuxIR pairs and orphan LuxR homologs. In *P. aeruginosa*, quorum sensing regulates factors important for host colonization during disease and may also be used for interspecies competition. This section describes evidence that like for *P. aeruginosa*, quorum sensing in *B. pseudomallei* is used during infection and may also contribute to saprophytic survival.

The QS-1 system of *B. pseudomallei* is the best studied. Multiple groups have demonstrated that genetic disruption of QS-1 in three different *B. pseudomallei* isolates reduced bacterial virulence. These studies collectively used five models of infection, including *Caenorhabditis elegans* feeding on bacterial

lawns, acute murine pneumonic infection (by aerosol exposure or intranasal inoculation), acute murine infection (by subcutaneous or intraperitoneal injection), and acute pneumonic hamster infection (by aerosol exposure) (Ulrich et al., 2004a; Valade et al., 2004; Song et al., 2005). The precise role of QS-2 and QS-3 in *B. pseudomallei* virulence is not clear. However, early studies by Ulrich et al. suggested that QS-3 and, to a lesser extent, QS-2 contribute to virulence (Ulrich et al., 2004a). These mutant strains also carried efflux pump mutations, and thus, these studies may need to be revisited.

Several QS-1-controlled phenotypes have been described and shed light on which QS-regulated processes may contribute to virulence. QS-1 was reported to control the production of the MprA protease (Valade et al., 2004), siderophores, and phospholipase C (PLC) activity (Song et al., 2005). However, an *mprA* mutant was not attenuated, and therefore, MprA is not predicted to significantly contribute to the virulence defect of the QS-1 mutant in the model examined in that study (Valade et al., 2004). Furthermore, quorum sensing control of protease was not observed in a study using a different strain (Song et al., 2005), suggesting that there may be strain-to-strain variation and that this may not be a conserved QS-1-regulated process. Though we do not know if quorum sensing directly regulates PLC activity or siderophore production, these factors have been associated with virulence. PLCs are generally thought to interact with phospholipids in the eukaryotic cell membrane. In *B. pseudomallei*, they are cytotoxic and contribute to plaque formation (indicative of intercellular bacterial spread) on eukaryotic cells (Korbsrisate et al., 2007). Additionally, genetic disruption of a putative *B. pseudomallei* PLC precursor caused attenuation in a hamster model of infection (Tuanyok et al., 2006), and antibodies in pooled sera from melioidosis patients recognized another *B. pseudomallei* PLC (Korbsrisate et al., 1999). Siderophore production has not been linked directly to *B. pseudomallei* virulence, but it is a virulence factor in other opportunistic pathogens, including *P. aeruginosa*.

Additionally, the QS-1 system in *B. pseudomallei* directly activates the *dpsA* gene (Lumjiaktase et al., 2006). DpsA protects DNA from damage by acid and oxidative stress (Loprasert et al., 2004). This may serve to protect bacterial DNA from environmental stressors or phagocytic killing in the host (Loprasert et al., 2004; Lumjiaktase et al., 2006). Quorum sensing control of the oxidative stress response is another instance in which we observe parallels between *B. pseudomallei* and *P. aeruginosa* (Table 1).

Bacterial adherence, aggregation into microcolonies, and biofilm formation are important survival factors during the saprophytic and host-associated lifestyle of many opportunistic pathogens (Costerton et al., 1995). Biofilm formation often involves a secreted EPS, which can also cause colony wrinkling and promote surface attachment (Costerton et al., 1999). EPS production has not been thoroughly explored in *B. pseudomallei*, but this bacterium is known to form wrinkled colonies and can attach to surfaces (Chantratita et al., 2007; Galyov et al., 2010; Sawasdidoln et al., 2010; Taweechaisupapong et al., 2005). Recently, Gamage et al. investigated the role of quorum sensing in static adherence assays. These studies showed that in *B. pseudomallei*, QS-1 promotes surface attachment. The QS-3 AHL synthase *bpsI3* was also important for adherence, but to a lesser degree. This may be mediated through the QS-1 system, as the *bpsI3* mutant was unable to make detectable amounts of C8-HSL (the QS-1 signal) (Gamage et al., 2011). High- and low-biofilm-producing *B. pseudomallei* isolates show no difference in virulence (Taweechaisupapong et al., 2005); however, a biofilm-deficient strain with defined mutations has not been tested.

B. pseudomallei also contains two orphan LuxR homologs, BpsR4 and BpsR5. The precise role of these proteins is unknown; however, genetic disruption of *bpsR5* in the efflux pump mutant background caused a mild reduction in murine virulence and a notable attenuation in the hamster model of infection (Ulrich et al., 2004a). It is not known if any AHLs interact with BpsR5. In fact, it has been suggested that this orphan LuxR exhibits intrinsic activity independent of AHL interaction and may activate the *bpsI1* promoter when expressed exogenously in *E. coli* (Kiratisin and Sanmee, 2008). Future work needs to be done to understand the role of the orphan LuxRs during virulence and to characterize their association with the other quorum sensing systems in *Burkholderia*.

There is still much to learn about quorum sensing in *B. pseudomallei*. Like with many opportunistic pathogens, the ability to quorum sense provides *B. pseudomallei* an advantage during infection. *B. pseudomallei* may also use quorum sensing during saprophytic growth, as many components present in its genome have also been retained in the nonpathogenic *B. thailandensis*.

Quorum Sensing in *Burkholderia thailandensis*

The quorum sensing-controlled phenotypes observed in *B. thailandensis* are consistent with the idea that this bacterium uses quorum sensing during its saprophytic lifestyle. *B. thailandensis* has a low potential to cause human disease (Lertpatanasuwan et al., 1999; Glass et al., 2006), but at sufficiently high doses it can cause lethal infection in mice (West et al., 2008; Wiersinga et al., 2008). Under such conditions, quorum sensing is not a virulence factor (Chandler et al., 2009). This is unlike the case with *B. pseudomallei*, which requires quorum sensing for full virulence in a similar infection model (see previous section). Such an observation is not surprising, as *B. thailandensis* is essentially a nonpathogen, but it does highlight the fact that homologous quorum sensing systems may regulate species-specific factors. One can speculate that *B. thailandensis* and *B. pseudomallei* both use quorum sensing for saprophytic survival, but *B. pseudomallei* has extended its quorum sensing-controlled regulon to also promote infection.

In *B. thailandensis*, QS-1 controls cell aggregation under certain conditions (Chandler et al., 2009). This phenotype appears to be linked to the production of an EPS related to the cepacian EPS, which is produced by members of the Bcc and contributes to the virulence of several Bcc species (Conway et al., 2004; Cunha et al., 2004; Richau et al., 2000; Sousa et al., 2007). Homologs of Bcc genes coding for cepacian biosynthesis are found in *B. thailandensis* (Ferreira et al., 2010) and are quorum sensing regulated (Chandler et al., 2009). Interestingly, *B. pseudomallei* also contains homologs of the cepacian biosynthesis genes (Ferreira et al., 2010), but it is not known if these genes contribute to biofilm formation or virulence or are quorum sensing controlled. The common occurrence of EPS in *B. thailandensis* and other saprophytes supports the idea that EPS may provide resilience in the environment. Further work is required to characterize biofilm formation in *B. thailandensis* and *B. pseudomallei* and to understand how this factor may have multiple functions in different habitats.

The *B. thailandensis* genome contains a number of gene clusters predicted to be involved in the synthesis of secondary metabolites with putative antimicrobial activity (Brett et al., 1998; Knappe et al., 2008; Nguyen et al., 2008; Yu et al., 2006). We have discovered one of these clusters and showed that it is required for the production of a family of broad-spectrum antibiotics, bactobolins A to H, that are under QS-2 control (Carr et al., 2011; Duerkop et al., 2009; Seyedsayamdost et al., 2010). This cluster encompasses approximately 3.3 kb of DNA and encodes a large polyketide synthase/nonribosomal peptide synthase hybrid biosynthetic pathway. The QS-2 LuxI and LuxR homologs (*btaI2* and *btaR2*) are also encoded within this region (Duerkop et al., 2009; Seyedsayamdost et al., 2010). Bactobolin synthesis requires *btaR2* and 3OHC10-HSL. 3OHC8-HSL can serve as a poor substitute for 3OHC10-HSL

(Duerkop et al., 2009). Bactobolins A to C had been previously identified, but the genes responsible for their biosynthesis and their dependence on quorum sensing were not known until recently. In fact, prior to the characterization of QS-2, it was not known that *B. thailandensis* produced these molecules (Adachi and Nishimura, 2003; Hori et al., 1981; Kondo et al., 1979; Seyedsayamdost et al., 2010). Bactobolins A and C are the most potent bactobolins in terms of antibiotic activity (Carr et al., 2011; Seyedsayamdost et al., 2010). They exhibit broad-spectrum activity against a variety of *Proteobacteria* and gram-positive bacteria. The list includes methicillin-resistant *Staphylococcus aureus*, vancomycin-resistant *Enterococcus faecalis*, *E. coli*, and the nonpathogenic saprophyte *Bacillus subtilis* (Seyedsayamdost et al., 2010).

B. thailandensis and *B. pseudomallei*, but not *B. mallei*, contain QS-2 and flanking DNA encoding the bactobolin biosynthetic genes. The absence of this genetic content in *B. mallei* supports the idea that QS-2 and bactobolin are important for saprophytic growth outside of the host. Many other saprophytic *Proteobacteria* also use quorum sensing to control the production of antimicrobials (Bainton et al., 1992b; McGowan et al., 1995; Pierson et al., 1994; Park et al., 2001), suggesting that quorum sensing-controlled antibiotic production may be commonly used to compete in mixed-species environments.

Other phenotypes have been associated with quorum sensing in *B. thailandensis*. An early study by Ulrich et al. showed that *B. thailandensis* quorum sensing controlled lipase production, siderophore production, hemolytic activity, twitching motility, swarming, carbon source utilization, and colony morphology (Ulrich, 2004; Ulrich et al., 2004c). Protease production was not affected by quorum sensing (Ulrich et al., 2004c). Chandler et al. later confirmed that protease is not quorum sensing controlled and that QS-1 represses motility and alters colony morphology, but the observation that quorum sensing controls hemolytic activity and carbon source utilization could not be confirmed (Chandler et al., 2009). The studies by Chandler et al. were done in a wild-type *B. thailandensis* background, compared to the efflux pump mutant used by Ulrich et al., and suggested that hemolysis and carbon source utilization may not be conserved quorum sensing-regulated processes in *B. thailandensis*.

Quorum Sensing in *Burkholderia mallei*:
Perspectives

B. mallei contains the QS-1 and QS-3 systems and the orphan LuxR homologs BmaR4 and BmaR5 (Ulrich et al., 2004b). As quorum sensing has traditionally been associated with environmental shifts common to bacteria that need to discriminate between a free-living and a host-associated state, it is interesting to consider how *B. mallei* may use quorum sensing in its strictly host-adapted lifestyle. Furthermore, *B. mallei* is the only documented example of an obligate animal pathogen with a complete LuxI and LuxR quorum sensing system. One other obligate pathogen, *Brucella melitensis*, has been reported to produce an AHL (Taminiau et al., 2002). However, the AHL *N*-dodecanoyl homoserine lactone (C12-HSL) was made in exceptionally small amounts, and there is no *luxI* homolog in the *B. melitensis* genome (Taminiau et al., 2002). In *B. melitensis*, two proteins with similarity to LuxR family regulators have been identified, but there is not yet evidence that these proteins directly interact with AHLs (Rambow-Larsen et al., 2008; Delrue et al., 2005). Transcriptomic analyses showed that addition of C12-HSL to wild-type *B. melitensis* was able to induce changes in transcription, which partially overlapped with similar studies identifying the transcriptome of one of the putative LuxR proteins, VjbR (Weeks et al., 2010). Inactivation of both of these LuxR homologs led to reduced virulence (Delrue et al., 2005, Rambow-Larsen et al., 2008). Further work needs to be done to elucidate how quorum sensing functions and contributes to pathogenesis in *B. melitensis*.

An early study showed that QS-1, QS-3, and the orphan BmaR5 promote *B. mallei* survival in murine and hamster models of infection (Ulrich et al., 2004b). The strains used in this study were destroyed by direction from the CDC because they were made with an antibiotic resistance marker that is no longer approved for laboratory experiments. Unpublished observations from our group showed that independently reconstructed quorum-sensing *B. mallei* mutants are not attenuated in rodent models of infection. Despite the controversial role for quorum sensing in *B. mallei* acute infection, we believe that this bacterium requires quorum sensing for its host-associated lifestyle. This is strongly supported by the observation that the QS-1, QS-3, R4, and R5 regulatory genes have been retained in all sequenced *B. mallei* genomes, with the exception of the genome of the attenuated strain SAVPI (which does not contain the QS-3 system). The precise role of *B. mallei* quorum sensing in host infections is currently under investigation in our laboratory.

The *B. mallei* genome contains a nonfunctional general secretory cluster found in *B. pseudomallei* that is required for protease and PLC activity (Nierman et al., 2004; DeShazer et al., 1999). Thus, *B. mallei* is nonproteolytic (Gauthier et al., 2000) and is believed to have deficiencies in the production of many secreted products that are made by *B. pseudomallei* (Galyov et al., 2010). The absence of secreted factors is thought to help *B. mallei* evade immune detection during chronic host colonization. Interestingly, secreted factors comprise a large portion of described quorum sensing-controlled processes in other bacteria (Table 1). To our knowledge, other than *bmaI1* itself, there have not been any quorum sensing-controlled factors identified in *B. mallei*. It will be informative to determine if quorum sensing in this bacterium controls unique processes that promote persistence in the host. Ongoing studies in our laboratory are aimed at characterizing such factors in *B. mallei* and evaluating the role of quorum sensing during acute and chronic animal infections. We believe that understanding the role of quorum sensing in *B. mallei* will expand our knowledge of quorum sensing in pathogenic bacteria and the evolution of AHL quorum sensing systems.

FUTURE CHALLENGES

AHL signaling is common to many *Proteobacteria* with diverse habitats, but little is known about the specific contributions of quorum sensing in diverse environments. Studying quorum sensing in *B. thailandensis*, *B. pseudomallei*, and *B. mallei* provides a unique opportunity to learn about the evolution of AHL signaling. As quorum sensing is a form of bacterial communication that enables cooperation, adaptation, and often survival, it is interesting to think about how homologous signaling systems can achieve different benefits in this group. We predict that *B. thailandensis* and *B. pseudomallei* use quorum sensing for survival in mixed microbial communities common to the saprophytic lifestyle and that *B. pseudomallei* has also evolved to use quorum sensing for adaptation and survival in the mammalian host. *B. mallei* represents an entirely different situation because it is a host-restricted bacterium without any known environmental reservoir. Future work should be directed towards characterizing and comparing quorum sensing-controlled processes in each organism to understand the relevance of these factors in each of these lifestyles.

REFERENCES

Adachi, H., and Y. Nishimura. 2003. Synthesis and biological activity of bactobolin glucosides. *Nat. Prod. Res.* 17:253–257.

Adar, Y. Y., and S. Ulitzur. 1993. GroESL proteins facilitate binding of externally added inducer by LuxR protein-containing *E. coli* cells. *J. Biolumin. Chemilumin.* 8:261–266.

Aendekerk, S., B. Ghysels, P. Cornelis, and C. Baysse. 2002. Characterization of a new efflux pump, MexGHI-OpmD, from *Pseudomonas aeruginosa* that confers resistance to vanadium. *Microbiology* 148:2371–2381.

Aguilar, C., A. Friscina, G. Devescovi, M. Kojic, and V. Venturi. 2003. Identification of quorum-sensing-regulated genes of *Burkholderia cepacia*. *J. Bacteriol.* 185:6456–6462.

Ahlgren, N. A., C. S. Harwood, A. L. Schaefer, E. Giraud, and E. P. Greenberg. 2011. Aryl-homoserine lactone quorum sensing in stem-nodulating photosynthetic bradyrhizobia. *Proc. Natl. Acad. Sci. USA* 108:7183–7188.

Ahmer, B. M., J. van Reeuwijk, C. D. Timmers, P. J. Valentine, and F. Heffron. 1998. *Salmonella typhimurium* encodes an SdiA homolog, a putative quorum sensor of the LuxR family, that regulates genes on the virulence plasmid. *J. Bacteriol.* 180:1185–1193.

An, D., T. Danhorn, C. Fuqua, and M. R. Parsek. 2006. Quorum sensing and motility mediate interactions between *Pseudomonas aeruginosa* and *Agrobacterium tumefaciens* in biofilm cocultures. *Proc. Natl. Acad. Sci. USA* 103:3828–3833.

Andersson, R. A., A. R. Eriksson, R. Heikinheimo, A. Mae, M. Pirhonen, V. Koiv, H. Hyytiainen, A. Tuikkala, and E. T. Palva. 2000. Quorum sensing in the plant pathogen *Erwinia carotovora* subsp. *carotovora*: the role of $expR_{Ecc}$. *Mol. Plant-Microbe Interact.* 13:384–393.

Antunes, L. C., A. L. Schaefer, R. B. Ferreira, N. Qin, A. M. Stevens, E. G. Ruby, and E. P. Greenberg. 2007. Transcriptome analysis of the *Vibrio fischeri* LuxR-LuxI regulon. *J. Bacteriol.* 189:8387–8391.

Bainton, N. J., B. W. Bycroft, S. R. Chhabra, P. Stead, L. Gledhill, P. J. Hill, C. E. Rees, M. K. Winson, G. P. Salmond, G. S. Stewart, et al. 1992a. A general role for the *lux* autoinducer in bacterial cell signalling: control of antibiotic biosynthesis in *Erwinia*. *Gene* 116:87–91.

Bainton, N. J., P. Stead, S. R. Chhabra, B. W. Bycroft, G. P. Salmond, G. S. Stewart, and P. Williams. 1992b. N-(3-Oxohexanoyl)-L-homoserine lactone regulates carbapenem antibiotic production in *Erwinia carotovora*. *Biochem. J.* 288(Pt. 3):997–1004.

Bose, J. L., U. Kim, W. Bartkowski, R. P. Gunsalus, A. M. Overley, N. L. Lyell, K. L. Visick, and E. V. Stabb. 2007. Bioluminescence in *Vibrio fischeri* is controlled by the redox-responsive regulator ArcA. *Mol. Microbiol.* 65:538–553.

Brett, P. J., D. DeShazer, and D. E. Woods. 1998. *Burkholderia thailandensis* sp. nov., a *Burkholderia pseudomallei*-like species. *Int. J. Syst. Bacteriol.* 48(Pt. 1):317–320.

Caballero-Mellado, J., J. Onofre-Lemus, P. Estrada-de Los Santos, and L. Martinez-Aguilar. 2007. The tomato rhizosphere, an environment rich in nitrogen-fixing *Burkholderia* species with capabilities of interest for agriculture and bioremediation. *Appl. Environ. Microbiol.* 73:5308–5319.

Cabrol, S., A. Olliver, G. B. Pier, A. Andremont, and R. Ruimy. 2003. Transcription of quorum-sensing system genes in clinical and environmental isolates of *Pseudomonas aeruginosa*. *J. Bacteriol.* 185:7222–7230.

Callahan, S. M., and P. V. Dunlap. 2000. LuxR- and acyl-homoserine-lactone-controlled non-*lux* genes define a quorum-sensing regulon in *Vibrio fischeri*. *J. Bacteriol.* 182:2811–2822.

Carr, G., M. R. Seyedsayamdost, J. R. Chandler, E. P. Greenberg, and J. Clardy. 2011. Sources of diversity in bactobolin biosynthesis by *Burkholderia thailandensis* E264. *Org. Lett.* 13:3048–3051.

Chan, Y. Y., H. S. Bian, T. M. Tan, M. E. Mattmann, G. D. Geske, J. Igarashi, T. Hatano, H. Suga, H. E. Blackwell, and K. L. Chua. 2007. Control of quorum sensing by a *Burkholderia pseudomallei* multidrug efflux pump. *J. Bacteriol.* 189:4320–4324.

Chan, Y. Y., and K. L. Chua. 2005. The *Burkholderia pseudomallei* BpeAB-OprB efflux pump: expression and impact on quorum sensing and virulence. *J. Bacteriol.* 187:4707–4719.

Chandler, J. R., B. A. Duerkop, A. Hinz, T. E. West, J. P. Herman, M. E. Churchill, S. J. Skerrett, and E. P. Greenberg. 2009. Mutational analysis of *Burkholderia thailandensis* quorum sensing and self-aggregation. *J. Bacteriol.* **191**:5901–5909.

Chantratita, N., V. Wuthiekanun, K. Boonbumrung, R. Tiyawisutsri, M. Vesaratchavest, D. Limmathurotsakul, W. Chierakul, S. Wongratanacheewin, S. Pukritiyakamee, N. J. White, N. P. Day, and S. J. Peacock. 2007. Biological relevance of colony morphology and phenotypic switching by *Burkholderia pseudomallei*. *J. Bacteriol.* **189**:807–817.

Cheng, A. C., and B. J. Currie. 2005. Melioidosis: epidemiology, pathophysiology, and management. *Clin. Microbiol. Rev.* **18**:383–416.

Chugani, S. A., M. Whiteley, K. M. Lee, D. D'Argenio, C. Manoil, and E. P. Greenberg. 2001. QscR, a modulator of quorum-sensing signal synthesis and virulence in *Pseudomonas aeruginosa*. *Proc. Natl. Acad. Sci. USA* **98**:2752–2757.

Conway, B. A., K. K. Chu, J. Bylund, E. Altman, and D. P. Speert. 2004. Production of exopolysaccharide by *Burkholderia cenocepacia* results in altered cell-surface interactions and altered bacterial clearance in mice. *J. Infect. Dis.* **190**:957–966.

Costerton, J. W., Z. Lewandowski, D. E. Caldwell, D. R. Korber, and H. M. Lappin-Scott. 1995. Microbial biofilms. *Annu. Rev. Microbiol.* **49**:711–745.

Costerton, J. W., P. S. Stewart, and E. P. Greenberg. 1999. Bacterial biofilms: a common cause of persistent infections. *Science* **284**:1318–1322.

Cunha, M. V., S. A. Sousa, J. H. Leitao, L. M. Moreira, P. A. Videira, and I. Sa-Correia. 2004. Studies on the involvement of the exopolysaccharide produced by cystic fibrosis-associated isolates of the *Burkholderia cepacia* complex in biofilm formation and in persistence of respiratory infections. *J. Clin. Microbiol.* **42**:3052–3058.

Currie, B. J., L. Ward, and A. C. Cheng. 2010. The epidemiology and clinical spectrum of melioidosis: 540 cases from the 20 year Darwin prospective study. *PLoS Negl. Trop. Dis.* **4**:e900.

D'Argenio, D. A., M. Wu, L. R. Hoffman, H. D. Kulasekara, E. Deziel, E. E. Smith, H. Nguyen, R. K. Ernst, T. J. Larson Freeman, D. H. Spencer, M. Brittnacher, H. S. Hayden, S. Selgrade, M. Klausen, D. R. Goodlett, J. L. Burns, B. W. Ramsey, and S. I. Miller. 2007. Growth phenotypes of *Pseudomonas aeruginosa lasR* mutants adapted to the airways of cystic fibrosis patients. *Mol. Microbiol.* **64**:512–533.

Delrue, R. M., C. Deschamps, S. Leonard, C. Nijskens, I. Danese, J. M. Schaus, S. Bonnot, J. Ferooz, A. Tibor, X. De Bolle, and J. J. Letesson. 2005. A quorum-sensing regulator controls expression of both the type IV secretion system and the flagellar apparatus of *Brucella melitensis*. *Cell. Microbiol.* **7**:1151–1161.

DeShazer, D., P. J. Brett, M. N. Burtnick, and D. E. Woods. 1999. Molecular characterization of genetic loci required for secretion of exoproducts in *Burkholderia pseudomallei*. *J. Bacteriol.* **181**:4661–4664.

Devine, J. H., G. S. Shadel, and T. O. Baldwin. 1989. Identification of the operator of the *lux* regulon from the *Vibrio fischeri* strain ATCC7744. *Proc. Natl. Acad. Sci. USA* **86**:5688–5692.

Diggle, S. P., A. Gardner, S. A. West, and A. S. Griffin. 2007a. Evolutionary theory of bacterial quorum sensing: when is a signal not a signal? *Philos. Trans. R. Soc. Lond. B* **362**:1241–1249.

Diggle, S. P., A. S. Griffin, G. S. Campbell, and S. A. West. 2007b. Cooperation and conflict in quorum-sensing bacterial populations. *Nature* **450**:411–414.

Drevinek, P., and E. Mahenthiralingam. 2010. *Burkholderia cenocepacia* in cystic fibrosis: epidemiology and molecular mechanisms of virulence. *Clin. Microbiol. Infect.* **16**:821–830.

Duerkop, B. A., J. P. Herman, R. L. Ulrich, M. E. Churchill, and E. P. Greenberg. 2008. The *Burkholderia mallei* BmaR3-BmaI3 quorum-sensing system produces and responds to *N*-3-hydroxy-octanoyl homoserine lactone. *J. Bacteriol.* **190**:5137–5141.

Duerkop, B. A., R. L. Ulrich, and E. P. Greenberg. 2007. Octanoyl-homoserine lactone is the cognate signal for *Burkholderia mallei* BmaR1-BmaI1 quorum sensing. *J. Bacteriol.* **189**:5034–5040.

Duerkop, B. A., J. Varga, J. R. Chandler, S. B. Peterson, J. P. Herman, M. E. Churchill, M. R. Parsek, W. C. Nierman, and E. P. Greenberg. 2009. Quorum-sensing control of antibiotic synthesis in *Burkholderia thailandensis*. *J. Bacteriol.* **191**:3909–3918.

Dyszel, J. L., J. N. Smith, D. E. Lucas, J. A. Soares, M. C. Swearingen, M. A. Vross, G. M. Young, and B. M. Ahmer. 2010. *Salmonella enterica* serovar Typhimurium can detect acyl homoserine lactone production by *Yersinia enterocolitica* in mice. *J. Bacteriol.* **192**:29–37.

Eberhard, A., A. L. Burlingame, C. Eberhard, G. L. Kenyon, K. H. Nealson, and N. J. Oppenheimer. 1981. Structural identification of autoinducer of *Photobacterium fischeri* luciferase. *Biochemistry* **20**:2444–2449.

Engebrecht, J., K. Nealson, and M. Silverman. 1983. Bacterial bioluminescence: isolation and genetic analysis of functions from *Vibrio fischeri*. *Cell* **32**:773–781.

Engebrecht, J., and M. Silverman. 1984. Identification of genes and gene products necessary for bacterial bioluminescence. *Proc. Natl. Acad. Sci. USA* **81**:4154–4158.

Ferluga, S., J. Bigirimana, M. Hofte, and V. Venturi. 2007. A LuxR homologue of *Xanthomonas oryzae* pv. *oryzae* is required for optimal rice virulence. *Mol. Plant Pathol.* **8**:529–538.

Ferreira, A. S., J. H. Leitao, I. N. Silva, P. F. Pinheiro, S. A. Sousa, C. G. Ramos, and L. M. Moreira. 2010. Distribution of cepacian biosynthesis genes among environmental and clinical *Burkholderia* strains and role of cepacian exopolysaccharide in resistance to stress conditions. *Appl. Environ. Microbiol.* **76**:441–450.

Fothergill, J. L., S. Panagea, C. A. Hart, M. J. Walshaw, T. L. Pitt, and C. Winstanley. 2007. Widespread pyocyanin overproduction among isolates of a cystic fibrosis epidemic strain. *BMC Microbiol.* **7**:45.

Fritz, D. L., P. Vogel, D. R. Brown, D. Deshazer, and D. M. Waag. 2000. Mouse model of sublethal and lethal intraperitoneal glanders (*Burkholderia mallei*). *Vet. Pathol.* **37**:626–636.

Fritz, D. L., P. Vogel, D. R. Brown, and D. M. Waag. 1999. The hamster model of intraperitoneal *Burkholderia mallei* (glanders). *Vet. Pathol.* **36**:276–291.

Fuqua, C. 2006. The QscR quorum-sensing regulon of *Pseudomonas aeruginosa*: an orphan claims its identity. *J. Bacteriol.* **188**:3169–3171.

Fuqua, C., and E. P. Greenberg. 2002. Listening in on bacteria: acyl-homoserine lactone signalling. *Nat. Rev. Mol. Cell Biol.* **3**:685–695.

Fuqua, C., M. R. Parsek, and E. P. Greenberg. 2001. Regulation of gene expression by cell-to-cell communication: acyl-homoserine lactone quorum sensing. *Annu. Rev. Genet.* **35**:439–468.

Fuqua, W. C., and S. C. Winans. 1994. A LuxR-LuxI type regulatory system activates *Agrobacterium* Ti plasmid conjugal transfer in the presence of a plant tumor metabolite. *J. Bacteriol.* **176**:2796–2806.

Fuqua, W. C., S. C. Winans, and E. P. Greenberg. 1994. Quorum sensing in bacteria: the LuxR-LuxI family of cell density-responsive transcriptional regulators. *J. Bacteriol.* **176**:269–275.

Galyov, E. E., P. J. Brett, and D. DeShazer. 2010. Molecular insights into *Burkholderia pseudomallei* and *Burkholderia mallei* pathogenesis. *Annu. Rev. Microbiol.* **64**:495–517.

Gamage, A. M., G. Shui, M. R. Wenk, and K. L. Chua. 2011. *N*-Octanoyl homoserine lactone signaling mediated by the BpsI-BpsR quorum sensing system plays a major role in biofilm formation of *Burkholderia pseudomallei*. *Microbiology* **157**:1176–1186.

Gambello, M. J., and B. H. Iglewski. 1991. Cloning and characterization of the *Pseudomonas aeruginosa lasR* gene, a transcriptional activator of elastase expression. *J. Bacteriol.* **173**:3000–3009.

Gauthier, Y. P., R. M. Hagen, G. S. Brochier, H. Neubauer, W. D. Splettstoesser, E. J. Finke, and D. R. Vidal. 2001. Study on the pathophysiology of experimental *Burkholderia pseudomallei* infection in mice. *FEMS Immunol. Med. Microbiol.* **30**:53–63.

Gauthier, Y. P., F. M. Thibault, J. C. Paucod, and D. R. Vidal. 2000. Protease production by *Burkholderia pseudomallei* and virulence in mice. *Acta Trop.* **74**:215–220.

Gilad, J., I. Harary, T. Dushnitsky, D. Schwartz, and Y. Amsalem. 2007. *Burkholderia mallei* and *Burkholderia pseudomallei* as bioterrorism agents: national aspects of emergency preparedness. *Isr. Med. Assoc. J.* **9**:499–503.

Gilbert, K. B., T. H. Kim, R. Gupta, E. P. Greenberg, and M. Schuster. 2009. Global position analysis of the *Pseudomonas aeruginosa* quorum-sensing transcription factor LasR. *Mol. Microbiol.* **73**:1072–1085.

Gilson, L., A. Kuo, and P. V. Dunlap. 1995. AinS and a new family of autoinducer synthesis proteins. *J. Bacteriol.* **177**:6946–6951.

Glass, M. B., J. E. Gee, A. G. Steigerwalt, D. Cavuoti, T. Barton, R. D. Hardy, D. Godoy, B. G. Spratt, T. A. Clark, and P. P. Wilkins. 2006. Pneumonia and septicemia caused by *Burkholderia thailandensis* in the United States. *J. Clin. Microbiol.* **44**:4601–4604.

Godoy, D., G. Randle, A. J. Simpson, D. M. Aanensen, T. L. Pitt, R. Kinoshita, and B. G. Spratt. 2003. Multilocus sequence typing and evolutionary relationships among the causative agents of melioidosis and glanders, *Burkholderia pseudomallei* and *Burkholderia mallei*. *J. Clin. Microbiol.* **41**:2068–2079.

Graf, J., and E. G. Ruby. 1998. Host-derived amino acids support the proliferation of symbiotic bacteria. *Proc. Natl. Acad. Sci. USA* **95**:1818–1822.

Green, S. K., M. N. Schroth, J. J. Cho, S. K. Kominos, and V. B. Vitanza-jack. 1974. Agricultural plants and soil as a reservoir for *Pseudomonas aeruginosa*. *Appl. Microbiol.* **28**:987–991.

Gregory, B. C., and D. M. Waag. 2007. Glanders, p. 121–146. *In* Z. F. Dembek (ed.), *Medical Aspects of Biological Warfare*. Office of the Surgeon General, Washington, DC.

Hanzelka, B. L., and E. P. Greenberg. 1995. Evidence that the N-terminal region of the *Vibrio fischeri* LuxR protein constitutes an autoinducer-binding domain. *J. Bacteriol.* **177**:815–817.

Hardalo, C., and S. C. Edberg. 1997. *Pseudomonas aeruginosa*: assessment of risk from drinking water. *Crit. Rev. Microbiol.* **23**:47–75.

Hassett, D. J., J. F. Ma, J. G. Elkins, T. R. McDermott, U. A. Ochsner, S. E. West, C. T. Huang, J. Fredericks, S. Burnett, P. S. Stewart, G. McFeters, L. Passador, and B. H. Iglewski. 1999. Quorum sensing in *Pseudomonas aeruginosa* controls expression of catalase and superoxide dismutase genes and mediates biofilm susceptibility to hydrogen peroxide. *Mol. Microbiol.* **34**:1082–1093.

Hentzer, M., H. Wu, J. B. Andersen, K. Riedel, T. B. Rasmussen, N. Bagge, N. Kumar, M. A. Schembri, Z. Song, P. Kristoffersen, M. Manefield, J. W. Costerton, S. Molin, L. Eberl, P. Steinberg, S. Kjelleberg, N. Hoiby, and M. Givskov. 2003. Attenuation of *Pseudomonas aeruginosa* virulence by quorum sensing inhibitors. *EMBO J.* **22**:3803–3815.

Heurlier, K., V. Denervaud, and D. Haas. 2006. Impact of quorum sensing on fitness of *Pseudomonas aeruginosa*. *Int. J. Med. Microbiol.* **296**:93–102.

Heurlier, K., V. Denervaud, M. Haenni, L. Guy, V. Krishnapillai, and D. Haas. 2005. Quorum-sensing-negative (*lasR*) mutants of *Pseudomonas aeruginosa* avoid cell lysis and death. *J. Bacteriol.* **187**:4875–4883.

Hibbing, M. E., C. Fuqua, M. R. Parsek, and S. B. Peterson. 2010. Bacterial competition: surviving and thriving in the microbial jungle. *Nat. Rev. Microbiol.* **8**:15–25.

Hoffman, L. R., H. D. Kulasekara, J. Emerson, L. S. Houston, J. L. Burns, B. W. Ramsey, and S. I. Miller. 2009. *Pseudomonas aeruginosa lasR* mutants are associated with cystic fibrosis lung disease progression. *J. Cyst. Fibros.* **8**:66–70.

Holden, M. T., R. W. Titball, S. J. Peacock, A. M. Cerdeno-Tarraga, T. Atkins, L. C. Crossman, T. Pitt, C. Churcher, K. Mungall, S. D. Bentley, M. Sebaihia, N. R. Thomson, N. Bason, I. R. Beacham, K. Brooks, K. A. Brown, N. F. Brown, G. L. Challis, I. Cherevach, T. Chillingworth, A. Cronin, B. Crossett, P. Davis, D. DeShazer, T. Feltwell, A. Fraser, Z. Hance, H. Hauser, S. Holroyd, K. Jagels, K. E. Keith, M. Maddison, S. Moule, C. Price, M. A. Quail, E. Rabbinowitsch, K. Rutherford, M. Sanders, M. Simmonds, S. Songsivilai, K. Stevens, S. Tumapa, M. Vesaratchavest, S. Whitehead, C. Yeats, B. G. Barrell, P. C. Oyston, and J. Parkhill. 2004. Genomic plasticity of the causative agent of melioidosis, *Burkholderia pseudomallei*. *Proc. Natl. Acad. Sci. USA* **101**:14240–14245.

Hoppe, I., B. Brenneke, M. Rohde, A. Kreft, S. Haussler, A. Reganzerowski, and I. Steinmetz. 1999. Characterization of a murine model of melioidosis: comparison of different strains of mice. *Infect. Immun.* **67**:2891–2900.

Hori, M., K. Suzukake, C. Ishikawa, H. Asakura, and H. Umezawa. 1981. Biochemical studies on bactobolin in relation to actinobolin. *J. Antibiot.* (Tokyo) **34**:465–468.

Huber, B., K. Riedel, M. Hentzer, A. Heydorn, A. Gotschlich, M. Givskov, S. Molin, and L. Eberl. 2001. The *cep* quorum-sensing system of *Burkholderia cepacia* H111 controls biofilm formation and swarming motility. *Microbiology* **147**:2517–2528.

Jones, S., B. Yu, N. J. Bainton, M. Birdsall, B. W. Bycroft, S. R. Chhabra, A. J. Cox, P. Golby, P. J. Reeves, S. Stephens, et al. 1993. The *lux* autoinducer regulates the production of exoenzyme virulence determinants in *Erwinia carotovora* and *Pseudomonas aeruginosa*. *EMBO J.* **12**:2477–2482.

Kanamaru, K., I. Tatsuno, T. Tobe, and C. Sasakawa. 2000. SdiA, an *Escherichia coli* homologue of quorum-sensing regulators, controls the expression of virulence factors in enterohaemorrhagic *Escherichia coli* O157:H7. *Mol. Microbiol.* **38**:805–816.

Kaplan, H. B., and E. P. Greenberg. 1985. Diffusion of autoinducer is involved in regulation of the *Vibrio fischeri* luminescence system. *J. Bacteriol.* **163**:1210–1214.

Kim, H. S., M. A. Schell, Y. Yu, R. L. Ulrich, S. H. Sarria, W. C. Nierman, and D. DeShazer. 2005. Bacterial genome adaptation to niches: divergence of the potential virulence genes in three *Burkholderia* species of different survival strategies. *BMC Genomics* **6**:174.

Kim, J., J. G. Kim, Y. Kang, J. Y. Jang, G. J. Jog, J. Y. Lim, S. Kim, H. Suga, T. Nagamatsu, and I. Hwang. 2004. Quorum sensing and the LysR-type transcriptional activator ToxR regulate toxoflavin biosynthesis and transport in *Burkholderia glumae*. *Mol. Microbiol.* **54**:921–934.

Kiratisin, P., and S. Sanmee. 2008. Roles and interactions of *Burkholderia pseudomallei* BpsIR quorum-sensing system determinants. *J. Bacteriol.* **190**:7291–7297.

Knappe, T. A., U. Linne, S. Zirah, S. Rebuffat, X. Xie, and M. A. Marahiel. 2008. Isolation and structural characterization of capistruin, a lasso peptide predicted from the genome sequence of *Burkholderia thailandensis* E264. *J. Am. Chem. Soc.* **130**:11446–11454.

Kohler, T., L. K. Curty, F. Barja, C. van Delden, and J. C. Pechere. 2000. Swarming of *Pseudomonas aeruginosa* is dependent on cell-to-cell signaling and requires flagella and pili. *J. Bacteriol.* **182**:5990–5996.

Kohler, T., C. van Delden, L. K. Curty, M. M. Hamzehpour, and J. C. Pechere. 2001. Overexpression of the MexEF-OprN multidrug efflux system affects cell-to-cell signaling in *Pseudomonas aeruginosa. J. Bacteriol.* **183**:5213–5222.

Kohler, T., G. G. Perron, A. Buckling, and C. van Deldon. 2010. Quorum sensing inhibition selects for virulence and cooperation in *Pseudomonas aeruginosa. PLoS Pathog.* **6**:e1000883.

Kondo, S., Y. Horiuchi, M. Hamada, T. Takeuchi, and H. Umezawa. 1979. A new antitumor antibiotic, bactobolin produced by *Pseudomonas. J. Antibiot.* (Tokyo) **32**:1069–1071.

Korbsrisate, S., N. Suwanasai, A. Leelaporn, T. Ezaki, Y. Kawamura, and S. Sarasombath. 1999. Cloning and characterization of a nonhemolytic phospholipase C gene from *Burkholderia pseudomallei. J. Clin. Microbiol.* **37**:3742–3745.

Korbsrisate, S., A. Tomaras, S. Damin, J. Ckumdee, V. Srinon, I. Lengwehasatit, M. Vasil, and S. Suparak. 2007. Characterization of two distinct phospholipase C enzymes from *Burkholderia pseudomallei. Microbiology* **153**:1907–1915.

Kothe, M., M. Antl, B. Huber, K. Stoecker, D. Ebrecht, I. Steinmetz, and L. Eberl. 2003. Killing of *Caenorhabditis elegans* by *Burkholderia cepacia* is controlled by the *cep* quorum-sensing system. *Cell. Microbiol.* **5**:343–351.

Latifi, A., M. Foglino, K. Tanaka, P. Williams, and A. Lazdunski. 1996. A hierarchical quorum-sensing cascade in *Pseudomonas aeruginosa* links the transcriptional activators LasR and RhIR (VsmR) to expression of the stationary-phase sigma factor RpoS. *Mol. Microbiol.* **21**:1137–1146.

Lazar Adler, N. R., B. Govan, M. Cullinane, M. Harper, B. Adler, and J. D. Boyce. 2009. The molecular and cellular basis of pathogenesis in melioidosis: how does *Burkholderia pseudomallei* cause disease? *FEMS Microbiol. Rev.* **33**:1079–1099.

Lee, J. H., Y. Lequette, and E. P. Greenberg. 2006. Activity of purified QscR, a *Pseudomonas aeruginosa* orphan quorum-sensing transcription factor. *Mol. Microbiol.* **59**:602–609.

Leelarasamee, A. 2004. Recent development in melioidosis. *Curr. Opin. Infect. Dis.* **17**:131–136.

Lequette, Y., J. H. Lee, F. Ledgham, A. Lazdunski, and E. P. Greenberg. 2006. A distinct QscR regulon in the *Pseudomonas aeruginosa* quorum-sensing circuit. *J. Bacteriol.* **188**:3365–3370.

Lerat, E., and N. A. Moran. 2004. The evolutionary history of quorum-sensing systems in bacteria. *Mol. Biol. Evol.* **21**:903–913.

Lertpatanasuwan, N., K. Sermsri, A. Petkaseam, S. Trakulsomboon, V. Thamlikitkul, and Y. Suputtamongkol. 1999. Arabinose-positive *Burkholderia pseudomallei* infection in humans: case report. *Clin. Infect. Dis.* **28**:927–928.

Lever, M. S., M. Nelson, P. I. Ireland, A. J. Stagg, R. J. Beedham, G. A. Hall, G. Knight, and R. W. Titball. 2003. Experimental aerogenic *Burkholderia mallei* (glanders) infection in the BALB/c mouse. *J. Med. Microbiol.* **52**:1109–1115.

Lewenza, S., B. Conway, E. P. Greenberg, and P. A. Sokol. 1999. Quorum sensing in *Burkholderia cepacia*: identification of the LuxRI homologs CepRI. *J. Bacteriol.* **181**:748–756.

Lewenza, S., and P. A. Sokol. 2001. Regulation of ornibactin biosynthesis and N-acyl-L-homoserine lactone production by CepR in *Burkholderia cepacia. J. Bacteriol.* **183**:2212–2218.

Limmathurotsakul, D., S. Wongratanacheewin, N. Teerawattanasook, G. Wongsuvan, S. Chaisuksant, P. Chetchotisakd, W. Chaowagul, N. P. Day, and S. J. Peacock. 2010. Increasing incidence of human melioidosis in Northeast Thailand. *Am. J. Trop. Med. Hyg.* **82**:1113–1117.

Loprasert, S., W. Whangsuk, R. Sallabhan, and S. Mongkolsuk. 2004. DpsA protects the human pathogen *Burkholderia pseudomallei* against organic hydroperoxide. *Arch. Microbiol.* **182**:96–101.

Losada, L., C. M. Ronning, D. DeShazer, D. Woods, N. Fedorova, H. S. Kim, S. A. Shabalina, T. R. Pearson, L. Brinkac, P. Tan, T.

Nandi, J. Crabtree, J. Badger, S. Beckstrom-Sternberg, M. Saqib, S. E. Schutzer, P. Keim, and W. C. Nierman. 2010. Continuing evolution of *Burkholderia mallei* through genome reduction and large-scale rearrangements. *Genome Biol. Evol.* **2**:102–116.

Loutet, S. A., and M. A. Valvano. 2010. A decade of *Burkholderia cenocepacia* virulence determinant research. *Infect. Immun.* **78**:4088–4100.

Lumjiaktase, P., S. P. Diggle, S. Loprasert, S. Tungpradabkul, M. Daykin, M. Camara, P. Williams, and M. Kunakorn. 2006. Quorum sensing regulates *dpsA* and the oxidative stress response in *Burkholderia pseudomallei. Microbiology* **152**:3651–3659.

Luo, Y., E. A. Frey, R. A. Pfuetzner, A. L. Creagh, D. G. Knoechel, C. A. Haynes, B. B. Finlay, and N. C. Strynadka. 2000. Crystal structure of enteropathogenic *Escherichia coli* intimin-receptor complex. *Nature* **405**:1073–1077.

Lupp, C., M. Urbanowski, E. P. Greenberg, and E. G. Ruby. 2003. The *Vibrio fischeri* quorum-sensing systems *ain* and *lux* sequentially induce luminescence gene expression and are important for persistence in the squid host. *Mol. Microbiol.* **50**: 319–331.

Lutter, E., S. Lewenza, J. J. Dennis, M. B. Visser, and P. A. Sokol. 2001. Distribution of quorum-sensing genes in the *Burkholderia cepacia* complex. *Infect. Immun.* **69**:4661–4666.

Lyczak, J. B., C. L. Cannon, and G. B. Pier. 2000. Establishment of *Pseudomonas aeruginosa* infection: lessons from a versatile opportunist. *Microbes Infect.* **2**:1051–1060.

Lyczak, J. B., C. L. Cannon, and G. B. Pier. 2002. Lung infections associated with cystic fibrosis. *Clin. Microbiol. Rev.* **15**:194–222.

Mahajan-Miklos, S., M. W. Tan, L. G. Rahme, and F. M. Ausubel. 1999. Molecular mechanisms of bacterial virulence elucidated using a *Pseudomonas aeruginosa-Caenorhabditis elegans* pathogenesis model. *Cell* **96**:47–56.

Mahenthiralingam, E., T. A. Urban, and J. B. Goldberg. 2005. The multifarious, multireplicon *Burkholderia cepacia* complex. *Nat. Rev. Microbiol.* **3**:144–156.

Martinez-Aguilar, L., R. Diaz, J. J. Pena-Cabriales, P. Estrada-de Los Santos, M. F. Dunn, and J. Caballero-Mellado. 2008. Multichromosomal genome structure and confirmation of diazotrophy in novel plant-associated *Burkholderia* species. *Appl. Environ. Microbiol.* **74**:4574–4579.

McFall-Ngai, M. J., and E. G. Ruby. 1991. Symbiont recognition and subsequent morphogenesis as early events in an animal-bacterial mutualism. *Science* **254**:1491–1494.

McGowan, S., M. Sebaihia, S. Jones, B. Yu, N. Bainton, P. F. Chan, B. Bycroft, G. S. Stewart, P. Williams, and G. P. Salmond. 1995. Carbapenem antibiotic production in *Erwinia carotovora* is regulated by CarR, a homologue of the LuxR transcriptional activator. *Microbiology* **141**(Pt. 3):541–550.

Meyer, J. M., A. Neely, A. Stintzi, C. Georges, and I. A. Holder. 1996. Pyoverdin is essential for virulence of *Pseudomonas aeruginosa. Infect. Immun.* **64**:518–523.

Michael, B., J. N. Smith, S. Swift, F. Heffron, and B. M. Ahmer. 2001. SdiA of *Salmonella enterica* is a LuxR homolog that detects mixed microbial communities. *J. Bacteriol.* **183**:5733–5742.

Mima, T., and H. P. Schweizer. 2010. The BpeAB-OprB efflux pump of *Burkholderia pseudomallei* 1026b does not play a role in quorum sensing, virulence factor production, or extrusion of aminoglycosides but is a broad-spectrum drug efflux system. *Antimicrob. Agents Chemother.* **54**:3113–3120.

Minogue, T. D., M. Wehland-von Trebra, F. Bernhard, and S. B. von Bodman. 2002. The autoregulatory role of EsaR, a quorum-sensing regulator in *Pantoea stewartii* ssp. *stewartii*: evidence for a repressor function. *Mol. Microbiol.* **44**:1625–1635.

Molina, L., F. Constantinescu, L. Michel, C. Reimmann, B. Duffy, and G. Defago. 2003. Degradation of pathogen quorum-sensing molecules by soil bacteria: a preventive and curative biological control mechanism. *FEMS Microbiol. Ecol.* **45:**71–81.

More, M. I., L. D. Finger, J. L. Stryker, C. Fuqua, A. Eberhard, and S. C. Winans. 1996. Enzymatic synthesis of a quorum-sensing autoinducer through use of defined substrates. *Science* **272:**1655–1658.

Nasser, W., M. L. Bouillant, G. Salmond, and S. Reverchon. 1998. Characterization of the *Erwinia chrysanthemi expI-expR* locus directing the synthesis of two N-acyl-homoserine lactone signal molecules. *Mol. Microbiol.* **29:**1391–1405.

Nealson, K. H., and J. W. Hastings. 1979. Bacterial bioluminescence: its control and ecological significance. *Microbiol. Rev.* **43:**496–518.

Nguyen, T., K. Ishida, H. Jenke-Kodama, E. Dittmann, C. Gurgui, T. Hochmuth, S. Taudien, M. Platzer, C. Hertweck, and J. Piel. 2008. Exploiting the mosaic structure of trans-acyltransferase polyketide synthases for natural product discovery and pathway dissection. *Nat. Biotechnol.* **26:**225–233.

Nierman, W. C., D. DeShazer, H. S. Kim, H. Tettelin, K. E. Nelson, T. Feldblyum, R. L. Ulrich, C. M. Ronning, L. M. Brinkac, S. C. Daugherty, T. D. Davidsen, R. T. Deboy, G. Dimitrov, R. J. Dodson, A. S. Durkin, M. L. Gwinn, D. H. Haft, H. Khouri, J. F. Kolonay, R. Madupu, Y. Mohammoud, W. C. Nelson, D. Radune, C. M. Romero, S. Sarria, J. Selengut, C. Shamblin, S. A. Sullivan, O. White, Y. Yu, N. Zafar, L. Zhou, and C. M. Fraser. 2004. Structural flexibility in the *Burkholderia mallei* genome. *Proc. Natl. Acad. Sci. USA* **101:**14246–14251.

Ochsner, U. A., A. K. Koch, A. Fiechter, and J. Reiser. 1994. Isolation and characterization of a regulatory gene affecting rhamnolipid biosurfactant synthesis in *Pseudomonas aeruginosa. J. Bacteriol.* **176:**2044–2054.

Ochsner, U. A., and J. Reiser. 1995. Autoinducer-mediated regulation of rhamnolipid biosurfactant synthesis in *Pseudomonas aeruginosa. Proc. Natl. Acad. Sci. USA* **92:**6424–6428.

Park, J. H., J. Hwang, J. W. Kim, S. O. Lee, B. A. Conway, E. P. Greenberg, and K. Lee. 2001. Characterization of quorum-sensing signaling molecules produced by *Burkholderia cepacia* G4. *Microbiol. Biotechnol.* **11:**804–811.

Parsek, M. R., and E. P. Greenberg. 2000. Acyl-homoserine lactone quorum sensing in gram-negative bacteria: a signaling mechanism involved in associations with higher organisms. *Proc. Natl. Acad. Sci. USA* **97:**8789–8793.

Parsek, M. R., D. L. Val, B. L. Hanzelka, J. E. Cronan, Jr., and E. P. Greenberg. 1999. Acyl homoserine-lactone quorum-sensing signal generation. *Proc. Natl. Acad. Sci. USA* **96:**4360–4365.

Passador, L., J. M. Cook, M. J. Gambello, L. Rust, and B. H. Iglewski. 1993. Expression of *Pseudomonas aeruginosa* virulence genes requires cell-to-cell communication. *Science* **260:**1127–1130.

Patankar, A. V., and J. E. Gonzalez. 2009. Orphan LuxR regulators of quorum sensing. *FEMS Microbiol. Rev.* **33:**739–756.

Pearson, J. P., K. M. Gray, L. Passador, K. D. Tucker, A. Eberhard, B. H. Iglewski, and E. P. Greenberg. 1994. Structure of the autoinducer required for expression of *Pseudomonas aeruginosa* virulence genes. *Proc. Natl. Acad. Sci. USA* **91:**197–201.

Pearson, J. P., L. Passador, B. H. Iglewski, and E. P. Greenberg. 1995. A second N-acylhomoserine lactone signal produced by *Pseudomonas aeruginosa. Proc. Natl. Acad. Sci. USA* **92:**1490–1494.

Pearson, J. P., E. C. Pesci, and B. H. Iglewski. 1997. Roles of *Pseudomonas aeruginosa las* and *rhl* quorum-sensing systems in control of elastase and rhamnolipid biosynthesis genes. *J. Bacteriol.* **179:**5756–5767.

Penalver, C. G., F. Cantet, D. Morin, D. Haras, and J. A. Vorholt. 2006. A plasmid-borne truncated *luxI* homolog controls quorum-sensing systems and extracellular carbohydrate production in *Methylobacterium extorquens* AM1. *J. Bacteriol.* **188:**7321–7324.

Peterson, K. M., and J. J. Mekalanos. 1988. Characterization of the *Vibrio cholerae* ToxR regulon: identification of novel genes involved in intestinal colonization. *Infect. Immun.* **56:**2822–2829.

Pierson, L. S., III, V. D. Keppenne, and D. W. Wood. 1994. Phenazine antibiotic biosynthesis in *Pseudomonas aureofaciens* 30-84 is regulated by PhzR in response to cell density. *J. Bacteriol.* **176:**3966–3974.

Piper, K. R., S. Beck von Bodman, and S. K. Farrand. 1993. Conjugation factor of *Agrobacterium tumefaciens* regulates Ti plasmid transfer by autoinduction. *Nature* **362:**448–450.

Pirhonen, M., D. Flego, R. Heikinheimo, and E. T. Palva. 1993. A small diffusible signal molecule is responsible for the global control of virulence and exoenzyme production in the plant pathogen *Erwinia carotovora. EMBO J.* **12:**2467–2476.

Poonguzhali, S., M. Madhaiyan, and T. Sa. 2007. Production of acyl-homoserine lactone quorum-sensing signals is wide-spread in Gram-negative *Methylobacterium. J. Microbiol. Biotechnol.* **17:**226–233.

Puskas, A., E. P. Greenberg, S. Kaplan, and A. L. Schaefer. 1997. A quorum-sensing system in the free-living photosynthetic bacterium *Rhodobacter sphaeroides. J. Bacteriol.* **179:** 7530–7537.

Rahme, L. G., E. J. Stevens, S. F. Wolfort, J. Shao, R. G. Tompkins, and F. M. Ausubel. 1995. Common virulence factors for bacterial pathogenicity in plants and animals. *Science* **268:**1899–1902.

Rambow-Larsen, A. A., G. Rajashekara, E. Petersen, and G. Splitter. 2008. Putative quorum-sensing regulator BlxR of *Brucella melitensis* regulates virulence factors including the type IV secretion system and flagella. *J. Bacteriol.* **190:**3274–3282.

Redfield, R. J. 2002. Is quorum sensing a side effect of diffusion sensing? *Trends Microbiol.* **10:**365–370.

Richau, J. A., J. H. Leitao, M. Correia, L. Lito, M. J. Salgado, C. Barreto, P. Cescutti, and I. Sa-Correia. 2000. Molecular typing and exopolysaccharide biosynthesis of *Burkholderia cepacia* isolates from a Portuguese cystic fibrosis center. *J. Clin. Microbiol.* **38:**1651–1655.

Rivas, M., M. Seeger, D. S. Holmes, and E. Jedlicki. 2005. A Lux-like quorum sensing system in the extreme acidophile *Acidithiobacillus ferrooxidans. Biol. Res.* **38:**283–297.

Ruby, E. G. 1996. Lessons from a cooperative, bacterial-animal association: the *Vibrio fischeri-Euprymna scolopes* light organ symbiosis. *Annu. Rev. Microbiol.* **50:**591–624.

Ruby, E. G., and M. J. McFall-Ngai. 1992. A squid that glows in the night: development of an animal-bacterial mutualism. *J. Bacteriol.* **174:**4865–4870.

Ruby, E. G., and M. J. McFall-Ngai. 1999. Oxygen-utilizing reactions and symbiotic colonization of the squid light organ by *Vibrio fischeri. Trends Microbiol.* **7:**414–420.

Rumbaugh, K. P., S. P. Diggle, C. M. Watters, A. Ross-Gillespie, A. S. Griffin, and S. A. West. 2009. Quorum sensing and the social evolution of bacterial virulence. *Curr. Biol.* **19:**341–345.

Rumbaugh, K. P., J. A. Griswold, and A. N. Hamood. 1999. Contribution of the regulatory gene *lasR* to the pathogenesis of *Pseudomonas aeruginosa* infection of burned mice. *J. Burn Care Rehabil.* **20:**42–49.

Sandoz, K. M., S. M. Mitzimberg, and M. Schuster. 2007. Social cheating in *Pseudomonas aeruginosa* quorum sensing. *Proc. Natl. Acad. Sci. USA* **104:**15876–15881.

Sawasdidoln, C., S. Taweechaisupapong, R. W. Sermswan, U. Tattawasart, S. Tungpradabkul, and S. Wongratanacheewin. 2010. Growing *Burkholderia pseudomallei* in biofilm stimulating conditions significantly induces antimicrobial resistance. *PLoS One* 5:e9196.

Schaefer, A. L., E. P. Greenberg, C. M. Oliver, Y. Oda, J. J. Huang, G. Bittan-Banin, C. M. Peres, S. Schmidt, K. Juhaszova, J. R. Sufrin, and C. S. Harwood. 2008. A new class of homoserine lactone quorum-sensing signals. *Nature* 454:595–599.

Schaefer, A. L., B. L. Hanzelka, M. R. Parsek, and E. P. Greenberg. 2000. Detection, purification, and structural elucidation of the acylhomoserine lactone inducer of *Vibrio fischeri* luminescence and other related molecules. *Methods Enzymol.* 305:288–301.

Schaefer, A. L., D. L. Val, B. L. Hanzelka, J. E. Cronan, Jr., and E. P. Greenberg. 1996. Generation of cell-to-cell signals in quorum sensing: acyl homoserine lactone synthase activity of a purified *Vibrio fischeri* LuxI protein. *Proc. Natl. Acad. Sci. USA* 93:9505–9509.

Schuster, M., and E. P. Greenberg. 2006. A network of networks: quorum-sensing gene regulation in *Pseudomonas aeruginosa*. *Int. J. Med. Microbiol.* 296:73–81.

Schuster, M., C. P. Lostroh, T. Ogi, and E. P. Greenberg. 2003. Identification, timing, and signal specificity of *Pseudomonas aeruginosa* quorum-controlled genes: a transcriptome analysis. *J. Bacteriol.* 185:2066–2079.

Schuster, M., M. L. Urbanowski, and E. P. Greenberg. 2004. Promoter specificity in *Pseudomonas aeruginosa* quorum sensing revealed by DNA binding of purified LasR. *Proc. Natl. Acad. Sci. USA* 101:15833–15839.

Seed, P. C., L. Passador, and B. H. Iglewski. 1995. Activation of the *Pseudomonas aeruginosa lasI* gene by LasR and the *Pseudomonas* autoinducer PAI: an autoinduction regulatory hierarchy. *J. Bacteriol.* 177:654–659.

Seyedsayamdost, M. R., J. R. Chandler, J. A. Blodgett, P. S. Lima, B. A. Duerkop, K. Oinuma, E. P. Greenberg, and J. Clardy. 2010. Quorum-sensing-regulated bactobolin production by *Burkholderia thailandensis* E264. *Org. Lett.* 12:716–719.

Smith, E. E., D. G. Buckley, Z. Wu, C. Saenphimmachak, L. R. Hoffman, D. A. D'Argenio, S. I. Miller, B. W. Ramsey, D. P. Speert, S. M. Moskowitz, J. L. Burns, R. Kaul, and M. V. Olson. 2006. Genetic adaptation by *Pseudomonas aeruginosa* to the airways of cystic fibrosis patients. *Proc. Natl. Acad. Sci. USA* 103:8487–8492.

Smith, J. N., J. L. Dyszel, J. A. Soares, C. D. Ellermeier, C. Altier, S. D. Lawhon, L. G. Adams, V. Konjufca, R. Curtiss III, J. M. Slauch, and B. M. Ahmer. 2008. SdiA, an *N*-acylhomoserine lactone receptor, becomes active during the transit of *Salmonella enterica* through the gastrointestinal tract of turtles. *PLoS One* 3:e2826.

Smith, R. S., and B. H. Iglewski. 2003. *P. aeruginosa* quorum-sensing systems and virulence. *Curr. Opin. Microbiol.* 6:56–60.

Sokol, P. A., U. Sajjan, M. B. Visser, S. Gingues, J. Forstner, and C. Kooi. 2003. The CepIR quorum-sensing system contributes to the virulence of *Burkholderia cenocepacia* respiratory infections. *Microbiology* 149:3649–3658.

Song, H., J. Hwang, H. Yi, R. L. Ulrich, Y. Yu, W. C. Nierman, and H. S. Kim. 2010. The early stage of bacterial genome-reductive evolution in the host. *PLoS Pathog.* 6:e1000922.

Song, Y., C. Xie, Y. M. Ong, Y. H. Gan, and K. L. Chua. 2005. The BpsIR quorum-sensing system of *Burkholderia pseudomallei*. *J. Bacteriol.* 187:785–790.

Sousa, S. A., M. Ulrich, A. Bragonzi, M. Burke, D. Worlitzsch, J. H. Leitao, C. Meisner, L. Eberl, I. Sa-Correia, and G. Doring. 2007. Virulence of *Burkholderia cepacia* complex strains in gp-91phox-/- mice. *Cell. Microbiol.* 9:2817–2825.

Stevens, A. M., K. M. Dolan, and E. P. Greenberg. 1994. Synergistic binding of the *Vibrio fischeri* LuxR transcriptional activator domain and RNA polymerase to the *lux* promoter region. *Proc. Natl. Acad. Sci. USA* 91:12619–12623.

Stintzi, A., K. Evans, J. M. Meyer, and K. Poole. 1998. Quorum-sensing and siderophore biosynthesis in *Pseudomonas aeruginosa*: *lasR/lasI* mutants exhibit reduced pyoverdine biosynthesis. *FEMS Microbiol. Lett.* 166:341–345.

Suarez-Moreno, Z. R., J. Caballero-Mellado, and V. Venturi. 2008. The new group of non-pathogenic plant-associated nitrogen-fixing *Burkholderia spp.* shares a conserved quorum-sensing system, which is tightly regulated by the RsaL repressor. *Microbiology* 154:2048–2059.

Suputtamongkol, Y., W. Chaowagul, P. Chetchotisakd, N. Lertpatanasuwun, S. Intaranongpai, T. Ruchutrakool, D. Budhsarawong, P. Mootsikapun, V. Wuthiekanun, N. Teerawatasook, and A. Lulitanond. 1999. Risk factors for melioidosis and bacteremic melioidosis. *Clin. Infect. Dis.* 29:408–413.

Taminiau, B., M. Daykin, S. Swift, M. L. Boschiroli, A. Tibor, P. Lestrate, X. De Bolle, D. O'Callaghan, P. Williams, and J. J. Letesson. 2002. Identification of a quorum-sensing signal molecule in the facultative intracellular pathogen *Brucella melitensis*. *Infect. Immun.* 70:3004–3011.

Tan, M. W., S. Mahajan-Miklos, and F. M. Ausubel. 1999a. Killing of *Caenorhabditis elegans* by *Pseudomonas aeruginosa* used to model mammalian bacterial pathogenesis. *Proc. Natl. Acad. Sci. USA* 96:715–720.

Tan, M. W., L. G. Rahme, J. A. Sternberg, R. G. Tompkins, and F. M. Ausubel. 1999b. *Pseudomonas aeruginosa* killing of *Caenorhabditis elegans* used to identify *P. aeruginosa* virulence factors. *Proc. Natl. Acad. Sci. USA* 96:2408–2413.

Tang, H. B., E. DiMango, R. Bryan, M. Gambello, B. H. Iglewski, J. B. Goldberg, and A. Prince. 1996. Contribution of specific *Pseudomonas aeruginosa* virulence factors to pathogenesis of pneumonia in a neonatal mouse model of infection. *Infect. Immun.* 64:37–43.

Taweechaisupapong, S., C. Kaewpa, C. Arunyanart, P. Kanla, P. Homchampa, S. Sirisinha, T. Proungvitaya, and S. Wongratanacheewin. 2005. Virulence of *Burkholderia pseudomallei* does not correlate with biofilm formation. *Microb. Pathog.* 39:77–85.

Tingpej, P., L. Smith, B. Rose, H. Zhu, T. Conibear, K. Al Nassafi, J. Manos, M. Elkins, P. Bye, M. Willcox, S. Bell, C. Wainwright, and C. Harbour. 2007. Phenotypic characterization of clonal and nonclonal *Pseudomonas aeruginosa* strains isolated from lungs of adults with cystic fibrosis. *J. Clin. Microbiol.* 45:1697–1704.

Tsai, C. S., and S. C. Winans. 2010. LuxR-type quorum-sensing regulators that are detached from common scents. *Mol. Microbiol.* 77:1072–1082.

Tuanyok, A., M. Tom, J. Dunbar, and D. E. Woods. 2006. Genome-wide expression analysis of *Burkholderia pseudomallei* infection in a hamster model of acute melioidosis. *Infect. Immun.* 74:5465–5476.

Ulrich, R. L. 2004. Quorum quenching: enzymatic disruption of *N*-acylhomoserine lactone-mediated bacterial communication in *Burkholderia thailandensis*. *Appl. Environ. Microbiol.* 70:6173–6180.

Ulrich, R. L., D. Deshazer, E. E. Brueggemann, H. B. Hines, P. C. Oyston, and J. A. Jeddeloh. 2004a. Role of quorum sensing in the pathogenicity of *Burkholderia pseudomallei*. *J. Med. Microbiol.* 53:1053–1064.

Ulrich, R. L., D. Deshazer, H. B. Hines, and J. A. Jeddeloh. 2004b. Quorum sensing: a transcriptional regulatory system involved in the pathogenicity of *Burkholderia mallei*. *Infect. Immun.* 72:6589–6596.

Ulrich, R. L., H. B. Hines, N. Parthasarathy, and J. A. Jeddeloh. 2004c. Mutational analysis and biochemical characterization of the *Burkholderia thailandensis* DW503 quorum-sensing network. *J. Bacteriol.* 186:4350–4360.

Valade, E., F. M. Thibault, Y. P. Gauthier, M. Palencia, M. Y. Popoff, and D. R. Vidal. 2004. The PmlI-PmlR quorum-sensing system in *Burkholderia pseudomallei* plays a key role in virulence and modulates production of the MprA protease. *J. Bacteriol.* **186:**2288–2294.

Venturi, V., A. Friscina, I. Bertani, G. Devescovi, and C. Aguilar. 2004. Quorum sensing in the *Burkholderia cepacia* complex. *Res. Microbiol.* **155:**238–244.

Vial, L., M. C. Groleau, V. Dekimpe, and E. Deziel. 2007. *Burkholderia* diversity and versatility: an inventory of the extracellular products. *J. Microbiol. Biotechnol.* **17:** 1407–1429.

Visick, K. L., J. Foster, J. Doino, M. McFall-Ngai, and E. G. Ruby. 2000. *Vibrio fischeri lux* genes play an important role in colonization and development of the host light organ. *J. Bacteriol.* **182:**4578–4586.

Visick, K. L., and M. J. McFall-Ngai. 2000. An exclusive contract: specificity in the *Vibrio fischeri-Euprymna scolopes* partnership. *J. Bacteriol.* **182:**1779–1787.

von Bodman, S. B., D. R. Majerczak, and D. L. Coplin. 1998. A negative regulator mediates quorum-sensing control of exopolysaccharide production in *Pantoea stewartii* subsp. *stewartii*. *Proc. Natl. Acad. Sci. USA* **95:**7687–7692.

Wagner, V. E., D. Bushnell, L. Passador, A. I. Brooks, and B. H. Iglewski. 2003. Microarray analysis of *Pseudomonas aeruginosa* quorum-sensing regulons: effects of growth phase and environment. *J. Bacteriol.* **185:**2080–2095.

Waters, C. M., and B. L. Bassler. 2005. Quorum sensing: cell-to-cell communication in bacteria. *Annu. Rev. Cell Dev. Biol.* **21:**319–346.

Weeks, J. N., C. L. Galindo, K. L. Drake, G. L. Adams, H. R. Garner, and T. A. Ficht. 2010. *Brucella melitensis* VjbR and C12-HSL regulons: contributions of the N-dodecanoyl homoserine lactone signaling molecule and LuxR homologue VjbR to gene expression. *BMC Microbiol.* **10:**167.

Wenbin, N., Z. Dejuan, L. Feifan, Y. Lei, C. Peng, Y. Xiaoxuan, and L. Hongyu. 2011. Quorum-sensing system in *Acidithiobacillus ferrooxidans* involved in its resistance to Cu²⁺. *Lett. Appl. Microbiol.* **53:**84–91.

West, S. A., A. S. Griffin, A. Gardner, and S. P. Diggle. 2006. Social evolution theory for microorganisms. *Nat. Rev. Microbiol.* **4:**597–607.

West, T. E., C. W. Frevert, H. D. Liggitt, and S. J. Skerrett. 2008. Inhalation of *Burkholderia thailandensis* results in lethal necrotizing pneumonia in mice: a surrogate model for pneumonic melioidosis. *Trans. R. Soc. Trop. Med. Hyg.* **102**(Suppl. 1):S119–S126.

Whitehead, N. A., A. M. Barnard, H. Slater, N. J. Simpson, and G. P. Salmond. 2001. Quorum-sensing in Gram-negative bacteria. *FEMS Microbiol. Rev.* **25:**365–404.

Whiteley, M., K. M. Lee, and E. P. Greenberg. 1999. Identification of genes controlled by quorum sensing in *Pseudomonas aeruginosa*. *Proc. Natl. Acad. Sci. USA* **96:**13904–13909.

Wiersinga, W. J., A. F. de Vos, R. de Beer, C. W. Wieland, J. J. Roelofs, D. E. Woods, and T. van der Poll. 2008. Inflammation patterns induced by different *Burkholderia* species in mice. *Cell. Microbiol.* **10:**81–87.

Wilder, C. N., G. Allada, and M. Schuster. 2009. Instantaneous within-patient diversity of *Pseudomonas aeruginosa* quorum-sensing populations from cystic fibrosis lung infections. *Infect. Immun.* **77:**5631–5639.

Yu, Y., H. S. Kim, H. H. Chua, C. H. Lin, S. H. Sim, D. Lin, A. Derr, R. Engels, D. DeShazer, B. Birren, W. C. Nierman, and P. Tan. 2006. Genomic patterns of pathogen evolution revealed by comparison of *Burkholderia pseudomallei*, the causative agent of melioidosis, to avirulent *Burkholderia thailandensis*. *BMC Microbiol.* **6:**46.

Zhang, L., Y. Jia, L. Wang, and R. Fang. 2007. A proline iminopeptidase gene upregulated in planta by a LuxR homologue is essential for pathogenicity of *Xanthomonas campestris* pv. *campestris*. *Mol. Microbiol.* **65:**121–136.

Zhang, L., P. J. Murphy, A. Kerr, and M. E. Tate. 1993. *Agrobacterium* conjugation and gene regulation by N-acyl-L-homoserine lactones. *Nature* **362:**446–448.

Zhang, R. G., T. Pappas, J. L. Brace, P. C. Miller, T. Oulmassov, J. M. Molyneaux, J. C. Anderson, J. K. Bashkin, S. C. Winans, and A. Joachimiak. 2002. Structure of a bacterial quorum-sensing transcription factor complexed with pheromone and DNA. *Nature* **417:**971–974.

Regulation of Bacterial Virulence
Edited by Michael L. Vasil and Andrew J. Darwin
© 2013 ASM Press, Washington, DC   doi:10.1128/9781555818524.ch4

Chapter 4

Staphylococcus aureus Pathogenesis and Virulence Factor Regulation

VICTOR J. TORRES, MEREDITH A. BENSON, AND JOVANKA M. VOYICH

INTRODUCTION TO *STAPHYLOCOCCUS AUREUS*

Staphylococcus aureus is a gram-positive bacterium that is a component of the commensal flora of the skin and the nares. It is estimated that approximately 32% of the population (Kuehnert et al., 2006) is colonized by this bacterium. Although colonization is asymptomatic, upon breaching of colonization sites due to a cut or direct inoculation into the bloodstream (e.g., intravenous drug users), *S. aureus* can unveil its pathogenic traits and become one of the most fastidious human pathogens. *S. aureus* is responsible for causing significant morbidity and mortality, resulting in nearly 300,000 hospitalizations and 19,000 deaths in the United States annually (Klevens et al., 2007). Thus, the deaths from *S. aureus* in the United States exceed the deaths by AIDS and tuberculosis combined. This versatile pathogen is one of the most common causes of skin and soft tissue infections, accounting for an estimated 12 million to 13 million outpatient visits per year (Moran et al., 2006). Remarkably, a single strain of *S. aureus* can cause diverse infections, including skin infections, bacteremia, endocarditis, sepsis, necrotizing fasciitis, pneumonia, and toxic shock syndrome (Miller et al., 2005; Lowy, 1998; Klevens et al., 2007; Kazakova et al., 2005; Fridkin et al., 2005; Adem et al., 2005).

Not only is *S. aureus* capable of causing severe disease but also the pathogen has an impressive ability to acquire resistance to antibiotics (Deleo et al., 2010). The emergence of antibiotic resistance in *S. aureus* follows the introduction of antibiotics designed to treat it. Thus, shortly after the introduction of penicillin, *S. aureus* developed resistance via acquisition of a plasmid-encoded penicillinase, causing hospitals to be plagued with penicillin-resistant *S. aureus* (Laurell and Wallmark, 1953). Within 10 years, penicillin-resistant *S. aureus* was common in the community but declined due to the introduction of methicillin

(Jevons and Parker, 1964). The same trend followed the introduction of methicillin, which was introduced to treat penicillin-resistant *S. aureus*. Within 1 year of its introduction, *S. aureus* methicillin resistance was identified. Methicillin-resistant *S. aureus* (MRSA) strains occur upon horizontal transfer of the *mecA* gene, carried within the staphylococcal cassette chromosome (scc*mec*) (Katayama et al., 2000; reviewed in Lowy, 2003). *mecA* codes for an altered penicillin binding protein, PBP 2a (also called PBP 2′). The result is broad-spectrum antibiotic resistance that includes resistance to the entire class of β-lactam antibiotics, including penicillins, carbapenems, and cephalosporins (Lowy, 2003). The first significant MRSA outbreaks occurred in the United States in the late 1970s and were restricted to hospital settings. By the mid-1980s, MRSA was endemic in hospitals in the United States, and it is now globally widespread in hospitals and long-term care facilities. MRSA strains not only are often resistant to β-lactam antibiotics but also demonstrate resistance to quinolones (Lowy, 2003). Due to the pressure to use vancomycin to treat MRSA, the pathogen has developed intermediate and full resistance to vancomycin (emergence of VISA and VRSA strains, respectively) (Centers for Disease Control and Prevention, 2002a, 2002b, 2004).

Even more alarming than the resistance *S. aureus* developed in hospital settings was the emergence of MRSA in the community. The first reports of MRSA distinct from hospital-associated strains originating in the community in the United States occurred in 1982 in Detroit, Michigan, but all these patients had predisposing risk factors for development of infection, including intravenous drug abuse (Saravolatz et al., 1982). In the late 1990s, the first reports of community-associated MRSA (CA-MRSA) in healthy children with no underlying conditions described devastating illnesses resulting in death (Centers for Disease Control and Prevention, 1999). Thus, not only

Victor J. Torres and Meredith A. Benson • Department of Microbiology, New York University School of Medicine, New York, NY 10016.
Jovanka M. Voyich • Department of Immunology and Infectious Diseases, Montana State University, Bozeman, MT 59718.

were the strains carrying drug resistance in the community but, for reasons unknown, these clones were also more virulent and could cause disease not typical of community-associated *S. aureus*, including sepsis, necrotizing pneumonia, and necrotizing fasciitis (Sifri et al., 2007; Miller et al., 2005; Lowy, 2007; Kravitz et al., 2005; Gillet et al., 2002; Adem et al., 2005). The mechanisms behind the increased virulence associated with community-associated *S. aureus* strains compared to their hospital counterparts are still unknown. However, the importance of gene expression in the evolution of *S. aureus* virulence is under intense study, and it is becoming increasingly evident that differential regulation of core genome-encoded virulence factors is associated with the pathogenesis of clinically relevant strains (Wang et al., 2007; Voyich et al., 2005; Li et al., 2009)

Staphylococcal Virulence Factors

In order for *S. aureus* to infect the mammalian host, the bacterium must produce factors that facilitate bacterial growth by inhibiting the host's potent inflammatory response, which is triggered by the intruding organism. These virulence factors, collectively known as the "virulon," contribute to *S. aureus* pathogenesis by facilitating adherence to host tissues, inhibiting the innate and adaptive immune responses, and destroying tissues to promote dissemination and to release nutrients required for bacterial growth (Foster, 2005; Nizet, 2007). Although many factors that contribute to pathogenesis could be considered virulence factors, virulence factors are defined here as accessory molecules that facilitate infection without altering the growth of the microorganism during in vitro culture. *S. aureus* is armed with a repertoire of virulence factors, which include both cell-associated factors and factors secreted into the extracellular milieu (Fig. 1), which facilitate the promiscuous lifestyle of this microorganism.

Surface virulence factors

Surface-associated factors enable *S. aureus* to bind to biotic surfaces, abiotic surfaces, and itself. The ability to adhere to varied surfaces likely represents the first step during the infection process, in which the bacterium attaches and grows on different tissues. Thus, *S. aureus* produces a collection of cell wall-anchored and cell surface-associated proteinaceous and nonproteinaceous molecules. Among these, the microbial surface components recognizing adhesive matrix molecules (MSCRAMMs) bind to fibrinogen (e.g., ClfA, ClfB, IsdA, and SdrE), fibronectin (e.g., IsdA, FnBPA, and FnBPB), vitronectin (e.g., Emp),

bone sialoprotein (e.g., Bbp), elastin (e.g., FnBPA and FnBPB), collagen (e.g., Can, Emp, and Eap), and von Willebrand factor (e.g., SpA), which facilitate bacterial adhesion to host tissue and, in some instances, bacterial aggregation (for an in-depth review, see Heilmann, 2011). In addition, *S. aureus* binding to fibrinogen results in the aggregation of platelets, a process that is thought to contribute to the establishment of endocarditis. Importantly, MSCRAMMs can overlap and bind to similar host factors, resulting in significant redundancy.

Cell surface-associated proteins are also involved in binding plasma proteins like hemoglobin (e.g., IsdB and IsdA), haptoglobin (e.g., IsdH), and immunoglobulins (e.g., SpA). In addition to extracellular matrix and serum binding factors, the surface of *S. aureus* also contains molecules involved in immunomodulation. For example, the surface of *S. aureus* is decorated with a polysaccharide capsule and the antibody binding protein SpA, both contributing to the resistance of this bacterium to opsonophagocytosis by preventing the deposition of antibodies on the surface of the bacterium. Moreover, *S. aureus* produces the cell wall-anchored protein AdsA, an enzyme that converts AMP to adenosine (Thammavongsa et al., 2009), which also protects *S. aureus* from phagocytic clearance primarily due to the ability of adenosine to inhibit the superoxide burst of neutrophils (Cronstein, 1994; Nemeth et al., 2006). Several of the *S. aureus* cell wall-anchored and cell surface-associated molecules are also involved in binding to human cells, including nasal epithelial cells (e.g., wall teichoic acid, IsdA, SasG, SdrC, Pls, and ClfB), suggesting that they contribute to nasal colonization and thus to the commensal trait of this organism (Corrigan et al., 2009; Weidenmaier et al., 2004). For example, *S. aureus* strains deficient in cell wall teichoic acid, nonproteinaceous, highly charged polymers linked to the peptidoglycan, are attenuated in their ability to bind human nasal epithelial cells in vitro and show reduced colonization of nasal tissue in vivo in a rodent model of nasal colonization (Weidenmaier et al., 2004). Moreover, although *S. aureus* is considered an extracellular pathogen, recent in vitro and in vivo evidence strongly supports the notion that this bacterium can also adhere to and invade nonprofessional phagocytic cells (epithelial and endothelial cells, fibroblasts, osteoblasts, etc.) (Agerer et al., 2003; Kintarak et al., 2004; Hirschhausen et al., 2010; Oviedo-Boyso et al., 2011). This process is facilitated by FnBP-A and FnBP-B, which decorate *S. aureus* with fibronectin. Bacteria coated with fibronectin interact with $\alpha_5\beta_1$ integrin, inducing the uptake of the bacteria by host cells (Sinha et al., 1999). This ability to invade and survive within cells provides a

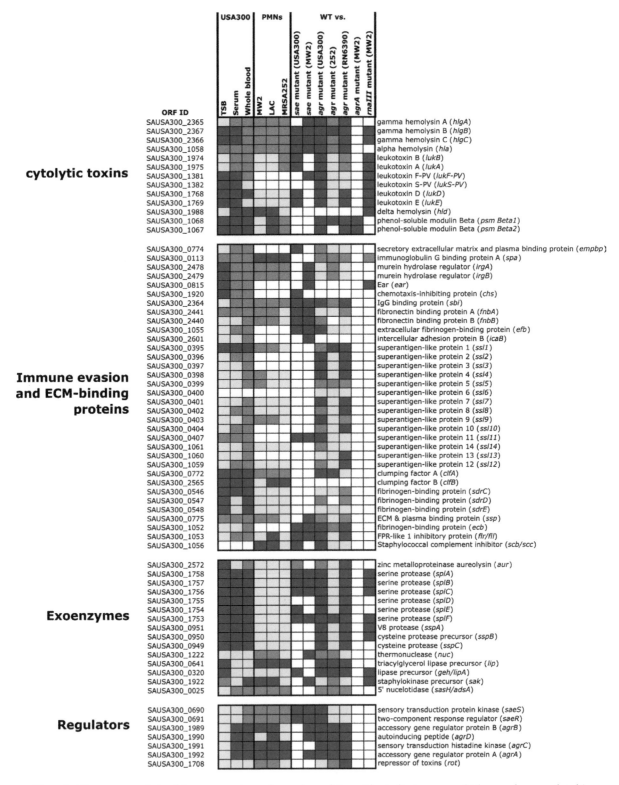

Figure 1. Microarray analysis of *S. aureus* strains and growth conditions. Blue indicates genes which were downregulated in indicated condition or mutant, red indicates upregulated genes, and yellow indicates no change. White denotes that the gene was not included in the particular microarray analysis. Transcription profiles of strains LAC and MW2 following neutrophil (PMN) phagocytosis were performed by Malachowa et al. (Malachowa et al., 2011). Transcription profiles of strains LAC and MW2 were performed by Voyich et al. (Voyich et al., 2005). Comparison of the transcription profiles between USA300 CA-MRSA wild-type (WT) strain LAC and a Δ*agr* or a Δ*sae* isogenic mutant strain were performed by Cheung et al. (Cheung et al., 2011) and Nygaard et al. (Nygaard et al., 2010), respectively. Experiments comparing the transcription profile of the USA400 CA-MRSA WT strain MW2 to that of a Δ*agrA* or Δ*rnaIII* or an Δ*sae* isogenic mutant strain were performed by Queck et al. (Queck et al., 2008) and Voyich et al. (Voyich et al., 2009), respectively.
doi:10.1128/9781555818524.ch4f1

"safe haven" for the bacteria by protecting the organism from the innate and adaptive immune responses. In addition, cell invasion is also likely to contribute to persistence of *S. aureus* infections by reducing the efficacy of antibiotic therapies.

An additional function of *S. aureus* cell surface-associated molecules is the formation of bacterial communities called biofilms that facilitate attachment to biotic and abiotic surfaces. The initial attachment to surfaces is followed by aggregates of bacteria growing in an extracellular matrix that is composed of polysaccharides, extracellular DNA, and proteins. *S. aureus* biofilms are a serious problem and are associated with chronic infections like endocarditis, osteomyelitis, and indwelling medical device infections. As with cell invasion, biofilms are highly refractory to antibiotic treatment and protect the bacterium from the immune system.

Secreted virulence factors

In addition to cell surface-associated virulence factors, *S. aureus* produces a plethora of proteins that are secreted into the extracellular milieu. The exoproteome of *S. aureus* under in vitro growth conditions consists of anywhere from 90 to 120 different proteins, depending on the growth condition and strain analyzed (Burlak et al., 2007; Torres et al., 2010). These exoproteins can be divided into four main categories: (i) cytotoxins, (ii) immunomodulators, (iii) exoenzymes, and (iv) uncharacterized proteins. *S. aureus* is notorious for the production of a large number of cytotoxins that target and kill mammalian cells from different origins. These cytotoxins include hemolysins (alpha, beta, and delta), bicomponent leukotoxins (HlgACB, LukSF-PV, LukED, and LukAB/GH), and phenol-soluble modulins (PSMα and PSMβ) (Fig. 1). Among these, LukSF-PV (also known as PVL), alpha-hemolysin, LukAB/GH, and the PSMs have all been shown to contribute to CA-MRSA pathogenesis (Bubeck Wardenburg et al., 2007a; Wang et al., 2007; Brown et al., 2009; Diep et al., 2010; Dumont et al., 2011; Ventura et al., 2010). The preferred targets of a majority of these toxins are neutrophils, innate immune effector cells absolutely required for controlling *S. aureus* infection. Alpha-hemolysin and the bicomponent toxins are members of the β-barrel pore-forming toxin family, which target the plasma membrane of host cells to form small pores that disrupt the plasma membrane potential, ultimately resulting in cell death (Menestrina et al., 2003). In addition, the ability of these toxins to lyse red blood cells is thought to facilitate the acquisition of nutrients (e.g., iron) required for bacterial growth (Torres et al., 2010).

S. aureus also secretes a large repertoire of hydrolases, which include proteases, lipases, and nucleases. These molecules are involved in tissue destruction and bacterial dissemination. In addition, *S. aureus* proteases and nucleases play a critical role in promoting the dispersal of biofilms, which is believed to contribute to the hematogenous spread of the organism (Rice et al. 2007; Boles and Horswill, 2008; Mann et al., 2009).

Finally, *S. aureus* produces proteins that protect the bacterium by modulating the immune system. These immunomodulatory proteins include superantigens, superantigen-like proteins, complement inhibitors, and chemotaxis inhibitors, among others (Fig. 1) (Foster, 2005; Fraser and Proft, 2008; Rooijakkers et al., 2005). Of interest, the majority of superantigens and chemotaxis inhibitors are carried by accessory genetic elements that include plasmids, transposons, prophages, and pathogenicity islands, which suggests that these factors were horizontally acquired by *S. aureus*.

Regulation of the *S. aureus* Virulon

The ability of *S. aureus* to infect many tissues in the mammalian body suggests that this bacterium is extremely efficient at adapting to different environments. Accordingly, *S. aureus* is able to respond to a variety of stresses under in vitro growth conditions that include, but are not limited to, nutrient limitation, pH, temperature, human serum, human blood, oxidative stress, neutrophils, oxygen tension, and neutrophil components (Fig. 1). *S. aureus* responds to these stresses by altering gene expression, including the expression of genes that code for virulence factors. The adaptation of *S. aureus* to these environments is carried out by a collection of regulatory circuits that involve two-component systems (TCS), small regulatory RNAs, alternative sigma factors, and a collection of winged helix-turn-helix DNA binding proteins (Cheung et al., 2004; Novick and Geisinger, 2008; Felden et al., 2011). Although extensive research on the contribution of these regulatory systems to the expression of the staphylococcal virulon has been performed, very little is known about the function of *S. aureus* regulatory systems in the complex environment of the human host. The ability to evade the human immune system depends on the expression of staphylococcal virulence factors at key times during infection. Failure to produce the proper set of factors in response to specific challenges during infection results in the breakdown of *S. aureus* pathogenesis. Thus, timely expression of many factors contributes to *S. aureus* pathogenesis. *S. aureus* is endowed with 16 different TCS that can survey the extracellular milieu for cues from environments encountered within the host. TCS are composed of a transmembrane sensor, usually a histidine kinase, which senses extracellular stimuli and undergoes autophosphorylation-mediated activation (Rodrigue et al., 2000; Novick, 1991; Hoch, 2000; Bronner et al., 2004). Upon activation, the receptor transfers a phosphoryl group to an aspartyl residue of the cognate response regulator, which increases the affinity of these transcription factors for their target promoters. In *S. aureus*, only a few of these TCS have been extensively studied at the molecular level. Among these, the accessory gene regulator (Agr) and the *S. aureus* exoprotein expression (Sae) are involved in the ability of *S. aureus* to regulate the expression of its virulon in response to sensing of itself (Agr) and the host (Sae).

Growth phase-dependent expression of virulence factors

In order to monitor the density of the population and prevent starvation and/or clearance by the host, most bacteria secrete small molecules into the extracellular milieu that accumulate in response to an increase in the number of bacteria in a defined space. These so-called autoinducers are sensed by the entire population, triggering a signaling cascade that informs the community that quorum has been attained. Activation of quorum sensing systems results in changes in the transcriptional response that leads to altered expression of genes encoding metabolic factors and, in the case of pathogenic bacteria, altered expression of genes that encode virulence factors.

S. aureus is armed with an impressive arsenal of virulence factors that contribute to different stages during infection, suggesting that this bacterium must tightly coordinate the expression of these virulence factors. Accordingly, the regulation of virulence factors in *S. aureus* is complex and involves checkpoints at the transcription, translation, and protein abundance level. Under laboratory growth conditions (i.e., in a flask with rich medium and aeration), *S. aureus* orchestrates a growth phase-dependent expression of its virulon (Fig. 1). This coordinated elaboration of virulence factors is thought to be crucial to the infection process. For example, the increased production of surface proteins during exponential phase (low quorum) is likely to aid the bacterium in vivo by enhancing binding to host tissues, thus initiating the infection process. Once the bacteria proliferate and reach postexponential growth, quorum is reached, resulting in a dramatic change in the expression profile of virulence factors, a behavior that is likely a response to nutrient limitation, which signals the bacterium to

disseminate. It is at this point that *S. aureus* upregulates the expression of secreted proteins and represses the expression of surface protein-encoding genes. The increased production of exoproteins, which include exoenzymes, cytotoxins, and immunomodulators, is believed to aid the bacterium in vivo by facilitating dislodging from the initial infection site, facilitating immune evasion, and promoting hematogenous spread.

The Agr system

Genetic experiments identified the *agr* locus as the regulatory system responsible for the growth phase-dependent alteration of virulence factor expression in *S. aureus* (Fig. 2) (Brown and Pattee, 1980; Janzon et al., 1989; Recsei et al., 1986). The *agr* locus contains two divergently transcribed RNA molecules regulated by the P2 and P3 promoters (Fig. 3) (Novick and Geisinger, 2008). The P2 promoter activates the transcription of RNAII, which serves as the mRNA for *agrA*, *agrC*, *agrD*, and *agrB* (Novick et al., 1995). The Agr system is activated in a quorum-dependent manner by a posttranslationally modified autoinducing peptide (AIP) (Fig. 3) (Ji et al., 1995) that is derived from AgrD, which is processed, modified, and secreted by AgrB (Novick and Geisinger, 2008; Zhang et al., 2002; Thoendel and Horswill, 2009) with help by type I signal peptidase SpsB (Kavanaugh et al., 2007). AIP is composed of a 5-amino-acid thiolactone ring with an N terminus of anywhere from 2 to 4 amino acids (Fig. 4). AIP activates the Agr system by binding to AgrC, the receptor histidine kinase of the Agr two-component module, resulting in AgrC autophosphorylation and subsequent phosphotransfer to AgrA, the cognate response regulator of the Agr system (Lina et al., 1998; Morfeldt et al., 1996). Activation of AgrA leads to a positive-feedback loop where AgrA oligomerizes and binds to direct repeats found within the P2 promoter (Koenig et al., 2004; Reynolds and Wigneshweraraj, 2011), resulting in the increased transcription of RNAII, which, in turn, results in the production of AgrBDCA.

Under laboratory growth conditions, transcriptional analyses using whole-genome arrays have revealed that the Agr regulon consists of 262 to 360 genes depending on the growth condition and the *S. aureus* strain analyzed (Fig. 1) (Dunman et al., 2001; Cassat et al., 2006; Queck et al., 2008; Cheung et al., 2011). The importance of the Agr system is further highlighted by the observation that the *agr* locus is found across all the staphylococcal species and an *agr*-like locus is found in different pathogenic gram-positive bacteria, including *Enterococcus faecalis*, *Listeria monocytogenes*, *Clostridium perfringens*,

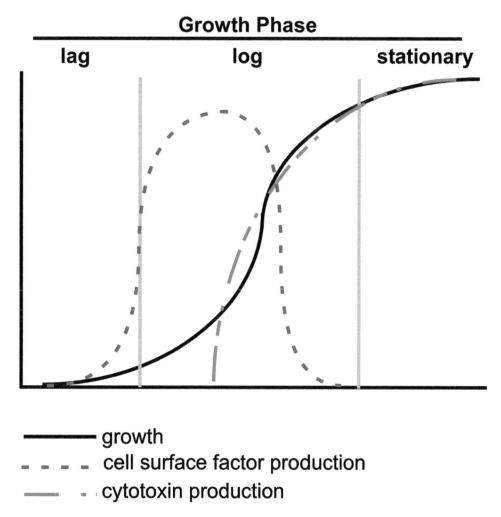

Growth Phase

lag   log   stationary

—— growth

- - - - cell surface factor production

—— • cytotoxin production

Figure 2. *S. aureus* regulates the production of virulence factors in a growth phase-dependent manner. Shown is a schematic representation of the association of *S. aureus* growth in vitro (black line) and the production of surface virulence factors (green line) and cytotoxins (red line). Early in log phase, *S. aureus* produces high levels of surface proteins and low levels of cytotoxins. In contrast, at the transition between late log and stationary phase, *S. aureus* concomitantly downregulates the production of surface proteins and upregulates the production of cytotoxins. doi:10.1128/9781555818524.ch4f2

Clostridium botulinum, *Clostridium difficile*, and several *Bacillus* species, including *Bacillus cereus* (Wuster and Babu, 2008).

The main effector of the Agr TCS is a regulatory RNA

In contrast to most TCS, where response regulators dictate the effect of the activated system, the main effector molecule of the Agr system is a regulatory RNA molecule known as RNAIII (Novick et al., 1993). In addition to activating the P2 promoter, phosphorylated AgrA binds and activates the P3 promoter, resulting in the transcription of RNAIII (Morfeldt et al., 1996; Reynolds and Wigneshweraraj, 2011). RNAIII is a 514-nucleotide stable RNA that assembles into a structure containing 14 loops that are rich

in cytosines, which regulates the vast majority of the Agr targets (Fig. 5A) (Novick et al., 1995; Novick et al., 1993; Benito et al., 2000). Encoded within RNAIII is delta-hemolysin, a 26-amino-acid cytolytic peptide. A recent study investigating the contribution of Agr to the regulation of virulence factors revealed that RNAIII is responsible for the expression of about 66% of all the Agr-regulated genes (Queck et al., 2008). RNAIII primarily functions as an antisense RNA molecule that anneals to the 5′ untranslated region (UTR) of target mRNAs masking the Shine-Dalgarno sequence, which prevents the binding of the ribosomes, inhibiting the translation of the target mRNAs (Fig. 5B). Recognition of target mRNAs by RNAIII also leads to degradation of the mRNA by RNase III, an endoribonuclease that recognizes double-stranded RNAs. Thus, RNAIII carries out its regulatory effects

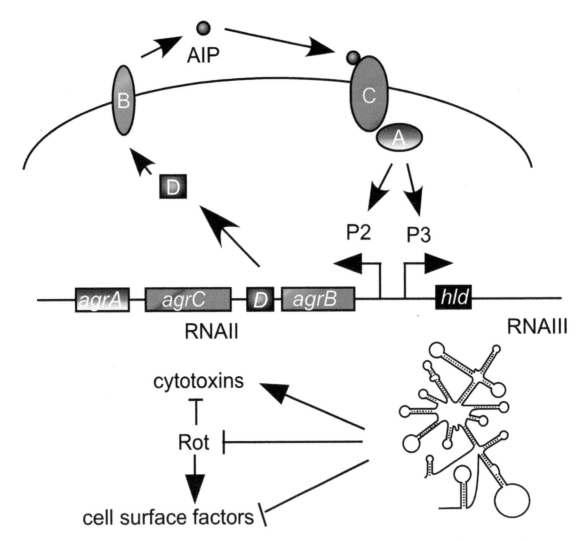

Figure 3. The *S. aureus* Agr system. The diagram depicts the Agr quorum sensing TCS. Upon reaching quorum, the Agr TCS is activated by binding of AIP to AgrC, resulting in the autophosphorylation of AgrC. AgrC then phosphorylates ArgA, which subsequently binds and activates the P2 and P3 promoters. Activation of the P2 promoter results in the expression of the multicistronic RNAII transcript and increased production of AgrB, AgrD, AgrC, and AgrA, and thus continuous activation of the system. Activation of the P3 promoter results in the expression of RNAIII transcript, which is itself a regulatory RNA and also codes for delta-hemolysin. Production of RNAIII regulates the expression and production of virulence factors directly via RNAIII-target mRNA interactions and indirectly via RNAIII-mediated inhibition of Rot synthesis. Rot (see p. 66) represses the expression of cytotoxin-encoding genes and activates the expression of genes coding for cell surface proteins. doi:10.1128/9781555818524.ch4f3

not only by inhibiting translation but also by promoting the degradation of target mRNAs. Two key targets of RNAIII are *spA* and *hla* mRNAs. *spA* encodes protein A (SpA), a multifactorial protein that contributes to *S. aureus* pathogenesis by preventing antibody-mediated opsonization by binding to the Fc region of antibodies (Peterson et al., 1977; Lindmark et al., 1983), exerting superantigen activity towards B cells (Sasso et al., 1989), and activating inflammatory signaling pathways (Gomez et al., 2006; Gomez et al., 2007; Martin et al., 2009). *spA* is the classic example of a gene that undergoes growth phase-dependent regulation in *S. aureus* (Novick et al., 1993). *spA* mRNA

is highly expressed during exponential growth, but it is downregulated during the transition from postexponential to stationary growth (Novick et al., 1993). The decreased levels of *spA* mRNA coincide with the activation of the Agr system and the production of high levels of RNAIII (Novick et al., 1993; Huntzinger et al., 2005). RNAIII binds to *spA* mRNA (Fig. 5B), resulting in translation inhibition and the degradation of the *spA* transcript as described above (Huntzinger et al., 2005).

A unique feature of RNAIII is that it can also promote the translation of target mRNAs. An example of this activity is RNAIII-mediated regulation

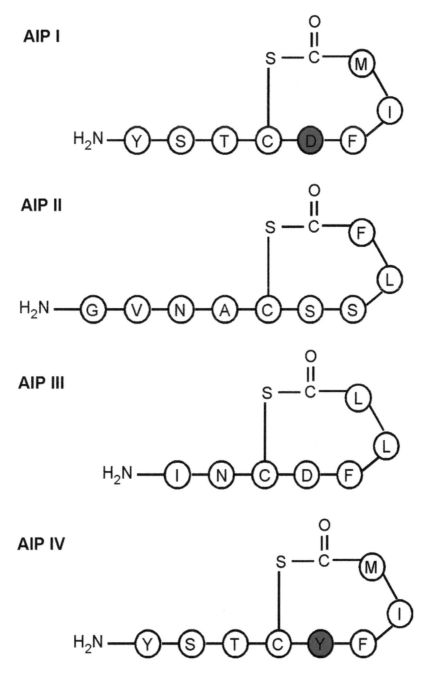

Figure 4. The Agr system AIPs. Shown is a schematic representation of the four different *S. aureus* AIP molecules (AIP I to IV). The sequences of AIP I and AIP IV differ only by a single amino acid in the thiolactone ring. doi:10.1128/9781555818524.ch4f4

of *hla* mRNA, which codes for alpha-hemolysin, also known as alpha-toxin (Novick et al., 1993; Morfeldt et al., 1995). *S. aureus* strains lacking RNAIII produce *hla* mRNA but no toxin product (Novick et al., 1993; Morfeldt et al., 1995). Analysis of *hla* mRNA revealed that this molecule assembles into a hairpin loop at the 5′ UTR masking the Shine-Dalgarno sequence and preventing translation. Sequence alignment of the 5′ UTR of the *hla* mRNA and the 5′ end of RNAIII revealed 75% complementarity, which is responsible for RNAIII binding to the 5′ UTR region of the *hla* mRNA, which results in the disassembly of the hairpin loop making the *hla* mRNA accessible for translation (Fig. 5C) (Novick et al., 1993; Morfeldt et al., 1995). Thus, RNAIII interaction with target mRNAs facilitates both positive and negative post-transcriptional regulation of key virulence factors in *S. aureus*.

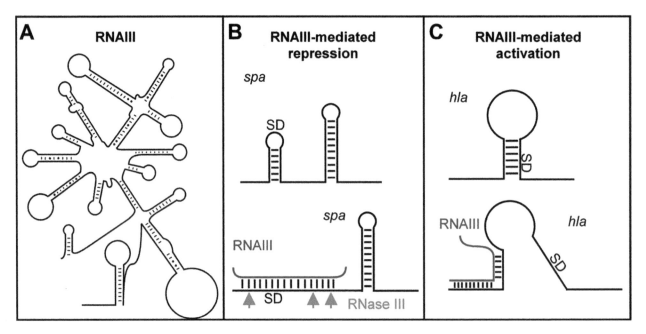

Figure 5. Structure and mechanism of action of RNAIII. The diagrams depict the secondary structure of *S. aureus* RNAIII (A) and the 5′ UTR sequences of *spA* (B) and *hla* (C). Binding of RNAIII to the *spA* 5′ UTR sequence results in the formation of a double-stranded RNA molecule that is recognized and degraded by RNase III. In contrast, binding of RNAIII to the *hla* 5′ UTR sequence enables translation of the *hla* mRNA.
doi:10.1128/9781555818524.ch4f5

RNAIII targets the transcription factor Rot to amplify the Agr signal

RNAIII directly influences the translation of several target mRNAs as described above. However, the vast majority of the RNAIII-regulated genes are altered in an indirect manner via the effects of RNAIII on global regulators, in particular on a protein known as repressor of toxins, or Rot. As with *spA* mRNA, RNAIII binds to the 5′ UTR of *rot* mRNA, halting translation and promoting the degradation of the *rot* transcript by RNase III (Geisinger et al., 2006; Boisset et al., 2007). Thus, during stationary phase, RNAIII levels increase, resulting in the reduction of Rot synthesis, which, in turn, results in the increased expression of Rot-repressed genes. Rot was initially identified in a screen for mutations that rescued the production of toxins in a strain lacking *agr* (McNamara et al., 2000). The *rot* gene codes for an ~18-kDa protein that contains a winged helix-turn-helix DNA-binding domain similar to that found on the GntR family of transcriptional regulators. Transcriptional profiles using microarrays have provided insight into the Rot regulon, revealing that Rot is involved in the negative regulation of key virulence factors, including cytotoxins, exoenzymes, and enterotoxin-encoding genes (Said-Salim et al., 2003). Importantly, Rot also acts as a positive regulator of several genes that encode cell surface molecules like

SpA and a subset of genes whose products are involved in metabolism, transport, and energy production (Said-Salim et al., 2003). The precise mechanism by which Rot orchestrates the differential expression of genes remains to be elucidated but is thought to be via both direct and indirect regulation through Rot's ability to alter the expression of additional transcription factors (Said-Salim et al., 2003).

AgrA regulates gene expression independently of RNAIII

Until recently, RNAIII was thought to be the sole effector molecule of the Agr system. Investigation into AgrA-dependent, RNAIII-independent regulation of virulence factors uncovered that AgrA also controls the expression of a subset of genes (Queck et al., 2008). Specifically, AgrA is responsible for the direct and positive regulation of the *psm* loci, which code for a collection of short secreted peptides (~20 to 40 amino acids) dubbed phenol-soluble modulins (PSMα and PSMβ) (Fig. 1) (Queck et al., 2008). The *psmα* and *psmβ* loci code for PSMα 1 to 4 and PSMβ 1 and 2, respectively, which are involved in the recruitment, activation, and lysis of the host's neutrophils (Wang et al., 2007), thus debilitating the innate immune response to *S. aureus* infection.

Surprisingly, the large majority of the genes regulated by AgrA independently of RNAIII and Rot are repressed by AgrA (Queck et al., 2008). These genes encode proteins primarily involved in metabolism and transport. Whether AgrA directly inhibits the expression of these genes remains to be elucidated, but AgrA influences the expression of additional transcriptional regulators, which could aid in the regulation of target genes. Thus, the Agr system in *S. aureus* is responsible for a regulation program that results in the coordinated differential production of virulence factors (cytotoxins versus adhesins) and modulation of the metabolic state of the bacterium.

agr types and Agr interference

Sequencing of the *agr* locus from diverse *S. aureus* strains has identified a polymorphic region within the locus that encompasses *agrB*, *agrD*, and *agrC* (Dufour et al., 2002). Based on these allelic variations, *S. aureus* strains are classified into at least four *agr* groups (*agr I* to *agrIV*) (Ji et al., 1997; Dufour et al., 2002; Jarraud et al., 2002). Variation of the protein sequence in AgrB, AgrD, and AgrC results in four different types of Agr systems that respond primarily to their cognate AIP molecule (Fig. 4) (Ji et al., 1997; Wright et al., 2005).

Experiments performed to study the specificity of different AIPs have uncovered the phenomenon of "*agr* interference," which is defined by the ability of AIPs to cross-inhibit Agr during heterologous pairing of *S. aureus* strains harboring different *agr* types (Ji et al., 1997; Wright et al., 2005). The interference of heterologous AIPs is due to competitive binding of the heterologous AIP to the AgrC receptor, which is facilitated by the high affinity of the different AIP types to their cognate AgrC receptor. The significance of *agr* interference during infections is not completely clear, but this interference suggests that strains producing high levels of AIP will be able to outcompete low-AIP-producing strains. Consistent with this supposition, *agr* interference has been observed in vivo in an insect model of infection (Fleming et al., 2006). Later in this chapter, we discuss the benefits and limitations of Agr interference as a novel therapeutic model to treat *S. aureus* infections.

Role of additional regulators and environmental stresses on the expression of the *agr* locus

As explained above, the *agr* locus is activated in a quorum-dependent manner upon phosphorylation of AgrA, which mediates the activation of the P2 and P3 promoters. Research into the regulation of the *agr* locus has identified a series of additional regulatory loci involved in the activation of the *agr* P2 promoter. Among these, the best-characterized loci are *sarA*

and *sigB* (Novick, 2003; Cheung et al., 2004). *sarA* codes for a helix-turn-helix transcription factor that binds to the P2 promoter and assists AgrA to fully activate the expression of RNAII and thus to propagate the feedback loop of the Agr system (Cheung et al., 1992; Reyes et al., 2011). *sigB* encodes the alternative sigma factor B (SigB) (Hecker et al., 2007), which is involved in the stress response and is critical for the expression of a variety of virulence factors, including surface proteins, and for the repression of cytotoxin- and exoenzyme-encoding genes (Bischoff et al., 2004). Deletion of the *sigB* locus results in the increased expression of RNAIII, suggesting that this locus inhibits *agr* expression, but the molecular mechanism of how SigB modulates RNAIII expression remains to be fully elucidated (Horsburgh et al., 2002; Lauderdale et al., 2009). In addition, the expression of *agr* is altered in response to different stimuli (Fig. 1). For example, acidic and basic pHs inhibit *agr* expression, although the mechanism by which this happens remains to be understood (Weinrick et al., 2004). Similarly, exposure to the antimicrobial peptide gramicidin, which induces a membrane stress response in *S. aureus*, also results in the inhibition of Agr activation (Benson et al., 2011). In contrast, at subinhibitory concentrations, antibiotics that inhibit protein synthesis, like tetracycline and clindamycin, have been shown to increase the expression of RNAIII and genes regulated by the Agr system (Joo et al., 2010). These observations suggest that *S. aureus* infections could be exacerbated by these antibiotics when at subinhibitory concentrations and warrant further investigation due to their clinical relevance.

Studies monitoring the transcriptional response of *S. aureus* growth in the presence of human whole blood, serum, and polymorphonuclear leukocytes (PMNs) revealed that these environments repress *agr* expression (Fig. 1) (Voyich et al., 2005; Palazzolo-Ballance et al., 2008; Malachowa et al., 2011). The reduced expression of the *agr* locus during growth with serum and in whole blood could be explained by the observation that apolipoprotein B (ApoB), a critical component of the low-density lipoprotein particle, which is in high concentration in the blood, binds AIP and prevents the activation of AgrC (Peterson et al., 2008). Similarly, hemoglobin, the oxygen-transport metalloprotein, also inhibits Agr activation (Schlievert et al., 2007; Pynnonen et al., 2011), but in contrast to ApoB, hemoglobin seems to act on the bacterial cells, most likely affecting AgrC-AIP interaction (Schlievert et al., 2007; Pynnonen et al., 2011). Finally, the expression of *agr* type I, and most likely type IV, can also be inhibited by phagocytes via oxidation of the methionine found in the AIP-I/IV molecules (Fig. 4)

(Rothfork et al., 2004). As illustrated, the regulation of *agr* expression is extremely complex and is currently an area of active investigation.

Agr and *S. aureus* pathogenesis

The importance of the Agr system to *S. aureus* pathogenesis in animal models is unequivocal, as deletion of *agr* in a variety of backgrounds results in attenuation of *S. aureus* in animal models of infection that include sepsis, endocarditis, skin infection, pneumonia, and endophthalmitis (Booth et al., 1995; Cheung et al., 1994; McNamara and Bayer, 2005; Peterson et al., 2008, Rothfork et al., 2004; Bubeck Wardenburg et al., 2007b; Wright et al., 2005; Montgomery et al., 2010). Consistent with these observations, *agr* has been demonstrated to be rapidly activated in vivo in a murine model of skin infection (Wright et al., 2005). In addition, studies comparing the pathological effects of exoproteins produced by the wild type compared to that of an isogenic *agr* mutant strain revealed that Agr-regulated virulence factors are responsible for the lesions observed in vivo when animals are infected subcutaneously with *S. aureus* (Wright et al., 2005). Collectively, these observations strongly suggest that Agr significantly contributes to human infection. This supposition is supported by the observation that the majority of strains isolated from human infections harbor a wild-type *agr* locus (Shopsin et al., 2008; Shopsin et al., 2010).

Interestingly, surveys of clinical isolates have revealed that patients can also be infected with *agr*-defective strains or with a heterogeneous population consisting of a wild-type parent and a naturally occurring isogenic *agr*-defective strain (Fowler et al., 2004; Vuong et al., 2004; Shopsin et al., 2008; Traber et al., 2008; Shopsin et al., 2010). The majority of the *agr*-inactivating mutations found in vivo have been mapped to *agrA* and *argC*, the two-component module of the Agr system (Traber et al., 2008; Shopsin et al., 2010). In addition, *agr*-defective mutants have been found to colonize individuals and have been associated with persistent bacteremia (Fowler et al., 2004; Schweizer et al., 2011). The observation that *agr*-defective strains can be isolated in vivo is paradoxical, primarily because these strains produce negligible amounts of cytotoxins, virulence factors identified to be important for *S. aureus* pathogenesis in animal models of infection. These findings suggest that under certain conditions, a fully functional Agr is detrimental to *S. aureus* pathogenesis, which could be due to the production of exoproteins that can exacerbate inflammation and potentially allow the host to clear the infection or, with too much inflammation induced, become lethal to the host. Importantly, strains lacking *agr* are not completely avirulent and are able to infect and disseminate in animal models of infection (Cheung et al., 1994; Gillaspy et al., 1995; McNamara and Bayer, 2005; Kielian et al., 2001; Benson et al., 2011), albeit to a lesser extent than their wild-type counterpart.

The observations described above highlight the fact that *S. aureus* pathogenesis is complex and depends not only on the production of exoproteins, which requires a functional Agr system, but also on the precise control of the production of specific exoproteins. Hence, increased production of Rot in *agr*-defective strains can also provide an advantage in certain scenarios in vivo. For example, a recent study investigating the expression profile of virulence factors and master regulators in persistent carriers demonstrated that *agr* was not expressed in vivo in the nares of the tested individuals (Burian et al., 2010b). Similar results were also found when a rodent model of nasal colonization was used (Burian et al., 2010a), suggesting that *agr* activation is dispensable for nasal colonization and survival of *S. aureus* in this niche. The decrease in *agr* expression in the nose could be explained by the recent observation that nasal secretions contain high levels of hemoglobin, which prevents *agr* activation (Pynnonen et al., 2011). Inhibition of *agr* results in the expression of surface factors that promote binding to tissues and to extracellular matrixes. In addition, *agr*-defective strains can thrive in vivo by overproducing SpA, which protects the pathogen from opsonophagocytic killing (Peterson et al., 1977; Lindmark et al., 1983) and promotes invasion across the epithelium (Soong et al., 2011). Moreover, inhibition of the Agr system also results in the overproduction of a subset of secreted immunomodulatory proteins known as the staphylococcal superantigen-like exoproteins (Ssls) (Pantrangi et al., 2010; Benson et al., 2011). Genetic and biochemical analyses revealed that Agr represses the expression of the *ssl* locus by controlling the synthesis of Rot (Benson et al., 2011). Rot interacts directly with the *ssl* promoters and plays a critical role for the full activation of these promoters. Thus, in contrast to the role of Rot as a repressor of cytotoxin-encoding genes, Rot acts as an activator for the *ssl* locus. An interesting trait of the Ssl proteins is that they exhibit sequence similarity to superantigens, but unlike superantigens, they are not mitogenic to lymphocytes and do not engage major histocompatibility complex class II molecules on antigen-presenting cells (Fraser and Proft, 2008). Ssls target and inhibit components of the complement system, bind to antibodies to block opsonization, and bind to the surface of neutrophils, inhibiting neutrophil rolling. Based on the function of these molecules, it is hypothesized that they play an important role in protecting *agr*-defective *S. aureus* in vivo. Experiments using a systemic model of infection revealed that indeed, deletion of the *ssl* locus in a strain lacking *agr* resulted in decreased pathogenesis of the *agr* mutant strain (Benson et al., 2011). Thus, *S. aureus* is able to partially compensate for the inactivation and/or inhibition of the Agr system by upregulating surface virulence factors and by coordinating an immunomodulatory response composed of at least SpA and the Ssl proteins.

Expression of Virulence Factors In Response to Phagocytes

Interaction of *S. aureus* with neutrophils

The emergence of community-associated *S. aureus* requires the ability of the pathogen to survive a noncompromised human innate immune system. Therefore, the ability of *S. aureus* to survive after neutrophil phagocytosis is thought to contribute significantly to the virulence of this pathogen (Wang et al., 2007; Voyich et al., 2009; Voyich et al., 2006; Voyich et al., 2005; Dumont et al., 2011). This is exemplified by the observed increase in susceptibility to *S. aureus* infections in individuals suffering from defects that alter normal neutrophil function, such as chronic granulomatous disease, leukocyte adhesion deficiency, and neutropenia (Pincus et al., 1976; Lekstrom-Himes and Gallin, 2000; Dale et al., 1979; Bodey et al., 1966). Phagocytosis of bacteria triggers production of reactive oxygen species, including superoxide, hydrogen peroxide, and hypochlorous acid. Neutrophil phagocytosis also causes the release of cytotoxic granule components, including cell wall-damaging agents such as cationic peptides, into pathogen-containing phagocytic vacuoles. Furthermore, neutrophil phagocytosis accelerates neutrophil apoptosis and downregulates their inflammatory capacity, thereby limiting potential uncontrolled inflammation caused by these potent immune effector cells (Kobayashi et al., 2003c; Kobayashi et al., 2003b; Kobayashi et al., 2003a). Collectively, neutrophil microbicidal systems are very efficient at killing ingested bacteria and limiting inflammation. Therefore, the ability to rapidly recognize neutrophils and/or neutrophil components is a virulence strategy used by pathogenic microorganisms. For example, group A *Streptococcus* uses the Ihk-Irr TCS to recognize neutrophil components, including

reactive oxygen species and primary granule proteins (Voyich et al., 2004). *Salmonella enterica* serovar Typhimurium uses a similar system to recognize host cationic antimicrobial peptides, including C18G, LL-37, polymyxin B, and protegrin, through the PhoP/PhoQ TCS (Bader et al., 2005; Bader et al., 2003). *Staphylococcus epidermidis* has a regulatory system (ApsR/S) that responds very specifically to cationic antimicrobial peptides (Li et al., 2007). Thus, recognizing host factors through sensor and/or regulatory systems is a strategy maintained and used by a broad range of bacterial pathogens to detect and respond to innate immunity.

SaeR/S and *S. aureus* neutrophil interaction

Like other pathogenic bacteria, *S. aureus* has mechanisms to sense and respond to neutrophils and neutrophil components. Comprehensive transcriptome analysis of clinically relevant *S. aureus* strains after neutrophil phagocytosis revealed that the pathogen undergoes abrupt changes in gene expression profiles (Voyich et al., 2005). Additionally, differences in gene expression following neutrophil phagocytosis revealed not only fluctuations in the mRNA levels of specific virulence factors but also subtle variations in gene regulation between strains that may explain the increased survival rate of community-associated *S. aureus* strains following phagocytosis (Fig. 1) (Voyich et al., 2005).

Recent studies have identified SaeR/S as a critical TCS essential for *S. aureus* survival following neutrophil phagocytosis (Voyich et al., 2009). This global regulatory system was designated *sae* for *S. aureus* exoprotein expression, due to altered exoprotein production observed in *sae* mutant strains (Giraudo et al., 1996; Giraudo et al., 1999). The *sae* locus consists of four open reading frames (*saePQRS*) (Fig. 6). *saeS* and *saeR* code for the two-component module of the SaeR/S-TCS, where SaeS is the histidine kinase and SaeR is the cognate response regulator. *saeQ* codes for a membrane protein that stabilizes SaeS in the membrane, and *saeP* codes for a hypothetical lipoprotein with an unknown function (Adhikari and Novick, 2008; Giraudo et al., 1999). Expression of the *sae* locus is primarily driven by two promoters, designated P1 and P3. The P1 promoter activates the expression of all four open reading frames (*saePQRS*), while the P3 promoter, which is found within *saeQ*, activates the expression of *saeRS* (Fig. 6). The P1 promoter is strongly autoregulated, while the P3 promoter exhibits constitutive activity.

The SaeR/S TCS directly and indirectly influences ~1.2 to 8% of the *S. aureus* genome depending

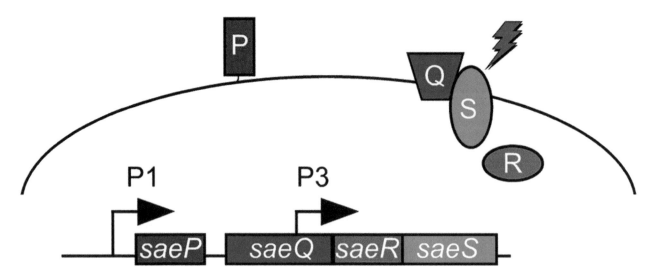

Figure 6. The *S. aureus* SaeRS TCS. The diagram depicts the organization of the *sae* locus and the predicted localization of the Sae components. The P3 promoter of the *sae* is constitutively active, resulting in basal level expression of *saeR* and *saeS*. Upon exposure to a signal (lightning bolt), SaeS is autophosphorylated, followed by phosphotransfer from SaeS to SaeR. Phosphorylation of SaeR enhances the DNA binding activity of this transcriptional regulator, resulting in binding and activation of the P1 promoter and thus autoinduction. In addition, activated SaeR binds to target promoters containing the SaeR binding sequence, resulting in their increased expression of these genes.
doi:10.1128/9781555818524.ch4f6

on the growth condition and strain analyzed (Voyich et al., 2009; Rogasch et al., 2006; Nygaard et al., 2010; Liang et al., 2006). Specifically, SaeR/S directly activates the expression of genes encoding many virulence factors, including serine proteases, nucleases, alpha- and beta-hemolysins (*hla* and *hlb*), and two-component leukotoxins (*hlgBC*, *lukED*, and *lukAB/GH)* (Fig. 1) (Rogasch et al., 2006; Liang et al., 2006; Giraudo et al., 1997), and adhesins, such as coagulase, fibronectin-binding protein A, *Eap*, and *Emp* (Rogasch et al., 2006; Novick and Jiang, 2003; Liang et al., 2006; Harraghy et al., 2005; Giraudo et al., 1994). The SaeR recognition sequence has been identified and confirmed within virulence gene promoters that are transcriptionally regulated by SaeR/S (Nygaard et al., 2010; Sun et al., 2010). These findings demonstrate that SaeR/S plays a major role during *S. aureus* pathogenesis by directly binding the SaeR recognition sequence within virulence gene promoters to upregulate transcription of factors that further bacterial survival.

SaeR/S TCS and *S. aureus* virulence

SaeR/S plays an important role in conferring *S. aureus* virulence. Whole-genome screens identified *saeR/S* as being essential for full virulence in a nematode model of *S. aureus* pathogenesis (Bae et al., 2004) and in murine models of systemic infection (Voyich et al., 2009; Benton et al., 2004). In addition, inactivation of *saeR/S* reduced *S. aureus*

invasion of human endothelial cells (Steinhuber et al., 2003). SaeR/S is also essential for full pathogenesis during sepsis (Voyich et al., 2009; Nygaard et al., 2010; Watkins et al., 2011) and during necrotizing pneumonia (Montgomery et al., 2010). The *saeR/S* phenotype during invasive staphylococcal infection can be partially explained by the induction of proinflammatory cytokines, including gamma interferon, tumor necrosis factor alpha, and interleukin 6, by SaeR/S-regulated factors (Watkins et al., 2011). Finally, SaeR/S is essential for evasion of neutrophil killing and important for neutrophil lysis (Voyich et al., 2009). Although the exact mechanisms of SaeR/S-dependent survival following neutrophil phagocytosis are not defined, among the genes positively regulated by this TCS are several virulence factors that have been shown to be important in *S. aureus* survival after neutrophil phagocytosis, including *hla*, *lukAB/GH*, *sbi*, *ear*, *sec4*, *lrgA*, *lrgB*, and *hlb* (Fig. 1) (Voyich et al., 2006; Voyich et al., 2005).

Although SaeR/S is essential for full virulence in mouse models of invasive disease, it is expendable for skin infection caused by USA400 strain MW2 but necessary for skin infection caused by USA300 strain LAC (Voyich et al., 2009; Nygaard et al., 2010). One possible explanation for the differences observed in the influence of SaeR/S on *S. aureus* skin infection caused by USA400 versus USA300 is potentially the ability to control expression of *saeR/S* and other regulatory systems. USA300 produces more *saeR/S* than USA400 (Montgomery et al., 2008), suggesting that expression of *S. aureus* regulatory systems may account for differences in virulence observed between these two CA-MRSA strains during skin infections. Further supporting this idea are the differences in fold change in alpha-toxin (*hla*) expression observed in Δ*saeR/S* strains from CA-MRSA strains MW2 (USA400) and LAC (USA300). Alpha-toxin expression has been linked to severity during CA-MRSA skin infection in mice (Kennedy et al., 2010). USA300 is a prominent cause of CA-MRSA skin and soft tissue infection, accounting for the majority of these infections (Nygaard et al., 2008). Regulation of *hla* is complex and is controlled by both the *saeR/S* and *agr* regulatory systems (Xiong et al., 2006; Montgomery et al., 2010). Transcription of *hla* in USA300 Δ*saeR/S* showed a 33-fold reduction in *hla* transcription, compared with a 4-fold reduction in USA400 Δ*saeR/S* (compared to their respective parental wild-type strains) (Voyich et al., 2009; Nygaard et al., 2010). These data demonstrate that the influence of *saeR/S* on toxin abundance can vary from strain to strain. Further supporting this concept is the well-studied Newman strain (Duthie and Lorenz, 1952; Baba et al., 2008). The Newman strain is characterized by increased toxin production and high virulence in animal models compared to other strains. The increased virulence and toxin production are attributed to high expression of the SaeR/S system as a result of a naturally occurring point mutation in *saeS* that changes the amino acid leucine to proline within the first membrane-spanning domain of SaeS (Rogasch et al., 2006; Schafer et al., 2009). This point mutation results in constitutive activation of SaeS and consequently the SaeR/S TCS and the SaeR-regulated genes (Rogasch et al., 2006; Schafer et al., 2009; Mainiero et al., 2010). Thus, changes in the activation status and/or abundance of these master regulatory circuits have major implications in the pathogenesis of *S. aureus*.

As mentioned above, very little is understood of *S. aureus* gene regulation in vivo. In device-related infections in the guinea pig model, *saeR/S* was shown to positively influence transcription of *hla* and *clfA* (Goerke et al., 2001). *saeR/S* was shown to play a key role in *hla* regulation in vivo in a rabbit model of infective endocarditis, and this function appears to be independent of *sarA*- and *agr*-mediated *hla* regulation (Xiong et al., 2006). This study also demonstrated a difference in the effect of *saeR/S* on *hla* regulation in vitro versus in vivo. For example, in vitro, alpha-hemolysin was undetectable in *sae* mutant strains; in vivo, however, alpha-hemolysin was detected in *sae* mutants, albeit at significantly lower levels (Xiong et al., 2006). Collectively, this study demonstrates the complexity of staphylococcal virulence regulation and highlights differences between in vitro and in vivo regulation of virulence.

Cross talk between transcription factor regulators to ensure proper regulation of the staphylococcal virulon

The ability to recognize the host is likely a virulence strategy used by *S. aureus*. So what triggers the activation of SaeR/S? As with most *S. aureus* TCS, the exact ligand for SaeS is not known. *saeR/S* does respond to changes in pH (Novick and Jiang, 2003) and to subinhibitory concentrations of antibiotics (Novick and Jiang, 2003; Kuroda et al., 2003; Kuroda et al., 2007), and it is upregulated in response to key microbicides of human neutrophils, including hypochlorous acid, hydrogen peroxide, and azurophilic granule components (Palazzolo-Ballance et al., 2008). Also, promoter activity of *sae* was shown to be induced in response to subinhibitory concentrations of alpha-defensins (Geiger et al., 2008), and *saeR/S* is upregulated after PMN phagocytosis in diverse strains of *S. aureus* (Voyich et al., 2005). In addition, components of whole blood upregulate transcription of *saeR/S*, but serum components are not a biological trigger (Malachowa et al., 2011). These data strongly suggest that the trigger of SaeS is a specific host product and most likely a neutrophil component.

The relationship of the SaeR/S TCS in regard to other *S. aureus* regulatory systems is not well defined and is currently an active area of investigation. The interaction of *agr* and *sae* is complex. Both regulatory systems influence exoprotein production, but this influence is distinct for each system (Novick and Jiang, 2003; Novick, 2003). As described above, Agr is critical for the inhibition of Rot synthesis and the derepression of the *S. aureus* virulon, activation of which is carried out primarily by the SaeRS TCS (Steinhuber et al., 2003; Giraudo et al., 1997; Nygaard et al., 2010). Transcription analyses have indicated that the SaeR/S TCS positively influences *agr* expression in the CA-MRSA strain MW2 (USA400). That is, *agrA*, *agrC*, *agrD*, *agrB*, and *hld* showed fold decreases in a MW2Δ*saeR/S* strain compared to MW2 (Voyich et al., 2009). However, this was not seen in a Δ*saeR/S* strain from the CA-MRSA strain LAC (USA300) (Nygaard et al., 2010). These results highlight the complexity of regulation of virulence factors in *S. aureus*. The current model suggests that the SaeR/S-Agr relationship coordinates the effects of environmental signals (detected by SaeS) with quorum sensing by Agr. Consistent with this model, comparison of the Agr and Sae regulon revealed that most of the SaeR positively regulated genes are negatively regulated by Rot (Fig. 1). Thus, an important function of Agr is to control the synthesis of Rot, while the SaeR/S TCS functions as the primary master activator of the promoters driving the expression of genes critical for staphylococcal pathogenesis. In addition, the SaeR conserved binding sequence has been identified within the promoter of additional regulators, including *rot*, and direct binding of recombinant SaeR to the *rot* promoter has been demonstrated (Nygaard et al., 2010; Sun et al., 2010). Lastly, further supporting SaeR/S and Rot cross talk is the observation that both systems are required for the activation of promoters of the genes encoding staphylococcal superantigen-like proteins (Ssls) (Pantrangi et al., 2010; Benson et al., 2011). Thus, additional understanding of the molecular details of SaeR/S interaction with Agr, Rot, and additional regulatory systems will lend significant insight into virulence gene regulation in *S. aureus*.

Prospect of developing inhibitors to target regulatory networks as a strategy to treat *S. aureus* infections

As discussed above, *S. aureus* has rapidly evolved mechanisms to reduce efficacy of our current antibiotics. For antimicrobial drugs to be effective, they must kill or prevent growth of the microbe without killing or damaging host cells. Microbial components that are different enough from human host cells have been the targets of pharmaceutical companies for years. Thus, the obvious targets for combating microbial pathogens have already been tested, without long-term success. In recent years, alternative approaches to the traditional antibiotics and vaccines have been investigated and have shown therapeutic efficacy in animal models of staphylococcal disease. Among these, vaccination with virulence factors like coagulase, SpA, and IsdB, as well as cytotoxins has shown promise in controlling *S. aureus* infection in animal models (Menzies and Kernodle, 1996; Bubeck Wardenburg and Schneewind, 2008; Ragle and Bubeck Wardenburg, 2009; Brown et al., 2009; Kennedy et al., 2010; Kim et al., 2010; Cheng et al., 2010).

Another novel approach to combat *S. aureus* infection is the development of compounds to interfere with virulence gene regulation. Ideally, compounds (antibodies, small molecules, and complementary and alternative medicines) would impede the ability of *S. aureus* to sense its environment. Without this ability, the pathogen would not properly regulate virulence factors and could be cleared by the innate immune system. One advantage of this approach is the targeting of multiple virulence factors, which is likely the most feasible approach to a successful therapeutic since *S. aureus* pathogenesis depends on multiple factors. This idea has been tested by targeting the Agr system with antagonistic AIP molecules (Ji et al., 1997; Wright et al., 2005). The in vivo application

of these molecules has been demonstrated using skin infection models where a single injection of the modified AIP is sufficient to reduce the pathology exhibited by infection with *S. aureus* (Ji et al., 1997; Wright et al., 2005). Whether or not these modified AIPs could be used as prophylactic therapeutics to treat staphylococcal infections in humans remains to be tested. Additionally, antibodies specific to an *S. aureus* AIP have also been shown to inhibit Agr signaling in vitro and to exhibit therapeutic efficacy in murine models of infection (Park et al., 2007). Together, these results highlight the potential therapy of inhibiting master regulatory systems as a strategy to control the expression of virulence factors, which, in turn, will facilitate the natural clearance of the organism by the host immune system. It is tempting to speculate that inhibition of both the Agr and Sae TCS should provide greater protection from *S. aureus* infections, since these two master regulators are required for the expression of most of the staphylococcal virulon. However, as described above, *agr*-defective strains are able to persist in vivo, suggesting that inhibition of this master regulatory system may not be sufficient to completely eradicate *S. aureus* infections. Potentially, combinatory therapies that include antibiotics together with inhibitors for Agr and Sae could attenuate *S. aureus*, facilitating the natural clearance of the organism by the host immune system. Whether or not any of the aforementioned vaccine strategies will work in humans remains to be determined, but it is clear that innovative approaches are urgently necessary to thwart the current *S. aureus* epidemic.

CONCLUDING REMARKS

S. aureus is one of the most successful pathogens on Earth. Its ability to cause a wide range of diseases is indicative of the fact that it can adapt to almost any environment within the mammalian host. In order to ensure survival, *S. aureus* produces a wide variety of virulence factors, some with seemingly redundant targets or mechanisms of action. Additionally, the complex network of regulatory factors that precisely regulate these virulence factors adds to the complexity of this pathogen. We have explored here the complex topic of the regulation of *S. aureus* virulence factors. Numerous regulatory networks work with or against one another to carefully coordinate the precise expression and production of a large collection of virulence factors that play different roles in *S. aureus* infection. A critical layer of complexity to this topic is the tremendous strain-to-strain variability seen in clinical isolates. This variability often influences the expression of virulence factors, which directly alters the pathogenic trait of clinical isolates. Due to the presence of numerous virulence factors and complex regulatory schemes and to its highly adaptive nature, *S. aureus* is an extremely complex pathogen that, despite years of research, is likely to remain a major global health problem for years to come.

REFERENCES

Adem, P. V., C. P. Montgomery, A. N. Husain, T. K. Koogler, V. Arangelovich, M. Humilier, S. Boyle-Vavra, and R. S. Daum. 2005. *Staphylococcus aureus* sepsis and the Waterhouse-Friderichsen syndrome in children. *N. Engl. J. Med.* 353:1245–1251.

Adhikari, R. P., and R. P. Novick. 2008. Regulatory organization of the staphylococcal *sae* locus. *Microbiology* 154:949–959.

Agerer, F., A. Michel, K. Ohlsen, and C. R. Hauck. 2003. Integrin-mediated invasion of *Staphylococcus aureus* into human cells requires Src family protein-tyrosine kinases. *J. Biol. Chem.* 278: 42524–42531.

Baba, T., T. Bae, O. Schneewind, F. Takeuchi, and K. Hiramatsu. 2008. Genome sequence of *Staphylococcus aureus* strain Newman and comparative analysis of staphylococcal genomes: polymorphism and evolution of two major pathogenicity islands. *J. Bacteriol.* 190:300–310.

Bader, M. W., W. W. Navarre, W. Shiau, H. Nikaido, J. G. Frye, M. McClelland, F. C. Fang, and S. I. Miller. 2003. Regulation of *Salmonella typhimurium* virulence gene expression by cationic antimicrobial peptides. *Mol. Microbiol.* 50:219–230.

Bader, M. W., S. Sanowar, M. E. Daley, A. R. Schneider, U. Cho, W. Xu, R. E. Klevit, H. Le Moual, and S. I. Miller. 2005. Recognition of antimicrobial peptides by a bacterial sensor kinase. *Cell* 122:461–472.

Bae, T., A. K. Banger, A. Wallace, E. M. Glass, F. Aslund, O. Schneewind, and D. M. Missiakas. 2004. *Staphylococcus aureus* virulence genes identified by *bursa aurealis* mutagenesis and nematode killing. *Proc. Natl. Acad. Sci. USA* 101:12312–12317.

Benito, Y., F. A. Kolb, P. Romby, G. Lina, J. Etienne, and F. Vandenesch. 2000. Probing the structure of RNAIII, the *Staphylococcus aureus agr* regulatory RNA, and identification of the RNA domain involved in repression of protein A expression. *RNA* 6:668–679.

Benson, M. A., S. Lilo, G. A. Wasserman, M. Thoendel, A. Smith, A. R. Horswill, J. Fraser, R. P. Novick, B. Shopsin, and V. J. Torres. 2011. *Staphylococcus aureus* regulates the expression and production of the staphylococcal superantigen-like secreted proteins in a Rot-dependent manner. *Mol. Microbiol.* 81:659–675.

Benton, B. M., J. P. Zhang, S. Bond, C. Pope, T. Christian, L. Lee, K. M. Winterberg, M. B. Schmid, and J. M. Buysse. 2004. Large-scale identification of genes required for full virulence of *Staphylococcus aureus*. *J. Bacteriol.* 186:8478–8489.

Bischoff, M., P. Dunman, J. Kormanec, D. Macapagal, E. Murphy, W. Mounts, B. Berger-Bachi, and S. Projan. 2004. Microarray-based analysis of the *Staphylococcus aureus* σ^B regulon. *J. Bacteriol.* 186:4085–4099.

Bodey, G. P., M. Buckley, Y. S. Sathe, and E. J. Freireich. 1966. Quantitative relationships between circulating leukocytes and infection in patients with acute leukemia. *Ann. Intern. Med.* 64:328–340.

Boisset, S., T. Geissmann, E. Huntzinger, P. Fechter, N. Bendridi, M. Possedko, C. Chevalier, A. C. Helfer, Y. Benito, A. Jacquier, C. Gaspin, F. Vandenesch, and P. Romby. 2007. *Staphylococcus aureus* RNAIII coordinately represses the synthesis of virulence

factors and the transcription regulator Rot by an antisense mechanism. *Genes Dev.* 21:1353–1366.

Boles, B. R., and A. R. Horswill. 2008. Agr-mediated dispersal of *Staphylococcus aureus* biofilms. *PLoS Pathog.* 4:e1000052.

Booth, M. C., R. V. Atkuri, S. K. Nanda, J. J. Iandolo, and M. S. Gilmore. 1995. Accessory gene regulator controls *Staphylococcus aureus* virulence in endophthalmitis. *Investig. Ophthalmol. Vis. Sci.* 36:1828–1836.

Bronner, S., H. Monteil, and G. Prevost. 2004. Regulation of virulence determinants in *Staphylococcus aureus*: complexity and applications. *FEMS Microbiol. Rev.* 28:183–200.

Brown, D. R., and P. A. Pattee. 1980. Identification of a chromosomal determinant of alpha-toxin production in *Staphylococcus aureus*. *Infect. Immun.* 30:36–42.

Brown, E. L., O. Dumitrescu, D. Thomas, C. Badiou, E. M. Koers, P. Choudhury, V. Vazquez, J. Etienne, G. Lina, F. Vandenesch, and M. G. Bowden. 2009. The Panton-Valentine leukocidin vaccine protects mice against lung and skin infections caused by *Staphylococcus aureus* USA300. *Clin. Microbiol. Infect.* 15:156–164.

Bubeck Wardenburg, J., T. Bae, M. Otto, F. R. Deleo, and O. Schneewind. 2007a. Poring over pores: alpha-hemolysin and Panton-Valentine leukocidin in *Staphylococcus aureus* pneumonia. *Nat. Med.* 13:1405–1406.

Bubeck Wardenburg, J., R. J. Patel, and O. Schneewind. 2007b. Surface proteins and exotoxins are required for the pathogenesis of *Staphylococcus aureus* pneumonia. *Infect. Immun.* 75:1040–1044.

Bubeck Wardenburg, J., and O. Schneewind. 2008. Vaccine protection against *Staphylococcus aureus* pneumonia. *J. Exp. Med.* 205:287–294.

Burian, M., M. Rautenberg, T. Kohler, M. Fritz, B. Krismer, C. Unger, W. H. Hoffmann, A. Peschel, C. Wolz, and C. Goerke. 2010a. Temporal expression of adhesion factors and activity of global regulators during establishment of *Staphylococcus aureus* nasal colonization. *J. Infect. Dis.* 201:1414–1421.

Burian, M., C. Wolz, and C. Goerke. 2010b. Regulatory adaptation of *Staphylococcus aureus* during nasal colonization of humans. *PLoS One* 5:e10040.

Burlak, C., C. H. Hammer, M. A. Robinson, A. R. Whitney, M. J. McGavin, B. N. Kreiswirth, and F. R. Deleo. 2007. Global analysis of community-associated methicillin-resistant *Staphylococcus aureus* exoproteins reveals molecules produced *in vitro* and during infection. *Cell. Microbiol.* 9:1172–1190.

Cassat, J., P. M. Dunman, E. Murphy, S. J. Projan, K. E. Beenken, K. J. Palm, S. J. Yang, K. C. Rice, K. W. Bayles, and M. S. Smeltzer. 2006. Transcriptional profiling of a *Staphylococcus aureus* clinical isolate and its isogenic *agr* and *sarA* mutants reveals global differences in comparison to the laboratory strain RN6390. *Microbiology* 152:3075–3090.

Centers for Disease Control and Prevention. 1999. Four pediatric deaths from community-acquired methicillin-resistant *Staphylococcus aureus*—Minnesota and North Dakota, 1997–1999. *JAMA* 282:1123–1125.

Centers for Disease Control and Prevention. 2002a. *Staphylococcus aureus* resistant to vancomycin—United States, 2002. *MMWR Morb. Mortal. Wkly. Rep.* 51:565–567.

Centers for Disease Control and Prevention. 2002b. Vancomycin-resistant *Staphylococcus aureus*—Pennsylvania, 2002. *MMWR Morb. Mortal. Wkly. Rep.* 51:902.

Centers for Disease Control and Prevention. 2004. Vancomycin-resistant *Staphylococcus aureus*—New York, 2004. *MMWR Morb. Mortal. Wkly. Rep.* 53:322–323.

Cheng, A. G., M. McAdow, H. K. Kim, T. Bae, D. M. Missiakas, and O. Schneewind. 2010. Contribution of coagulases towards

Staphylococcus aureus disease and protective immunity. *PLoS Pathog.* 6:e1001036.

Cheung, A. L., A. S. Bayer, G. Zhang, H. Gresham, and Y. Q. Xiong. 2004. Regulation of virulence determinants *in vitro* and *in vivo* in *Staphylococcus aureus*. *FEMS Immunol. Med. Microbiol.* 40:1–9.

Cheung, A. L., K. J. Eberhardt, E. Chung, M. R. Yeaman, P. M. Sullam, M. Ramos, and A. S. Bayer. 1994. Diminished virulence of a *sar-/agr-* mutant of *Staphylococcus aureus* in the rabbit model of endocarditis. *J. Clin. Investig.* 94:1815–1822.

Cheung, A. L., J. M. Koomey, C. A. Butler, S. J. Projan, and V. A. Fischetti. 1992. Regulation of exoprotein expression in *Staphylococcus aureus* by a locus (*sar*) distinct from *agr*. *Proc. Natl. Acad. Sci. USA* 89:6462–6466.

Cheung, G. Y., R. Wang, B. A. Khan, D. E. Sturdevant, and M. Otto. 2011. Role of the accessory gene regulator *agr* in community-associated methicillin-resistant *Staphylococcus aureus* pathogenesis. *Infect. Immun.* 79:1927–1935.

Corrigan, R. M., H. Miajlovic, and T. J. Foster. 2009. Surface proteins that promote adherence of *Staphylococcus aureus* to human desquamated nasal epithelial cells. *BMC Microbiol.* 9:22.

Cronstein, B. N. 1994. Adenosine, an endogenous anti-inflammatory agent. *J. Appl. Physiol.* 76:5–13.

Dale, D. C., D. T. Guerry, J. R. Wewerka, J. M. Bull, and M. J. Chusid. 1979. Chronic neutropenia. *Medicine* (Baltimore) 58:128–144.

Deleo, F. R., M. Otto, B. N. Kreiswirth, and H. F. Chambers. 2010. Community-associated methicillin-resistant *Staphylococcus aureus*. *Lancet* 375:1557–1568.

Diep, B. A., L. Chan, P. Tattevin, O. Kajikawa, T. R. Martin, L. Basuino, T. T. Mai, H. Marbach, K. R. Braughton, A. R. Whitney, D. J. Gardner, X. Fan, C. W. Tseng, G. Y. Liu, C. Badiou, J. Etienne, G. Lina, M. A. Matthay, F. R. DeLeo, and H. F. Chambers. 2010. Polymorphonuclear leukocytes mediate *Staphylococcus aureus* Panton-Valentine leukocidin-induced lung inflammation and injury. *Proc. Natl. Acad. Sci. USA* 107:5587–5592.

Dufour, P., S. Jarraud, F. Vandenesch, T. Greenland, R. P. Novick, M. Bes, J. Etienne, and G. Lina. 2002. High genetic variability of the *agr* locus in *Staphylococcus* species. *J. Bacteriol.* 184:1180–1186.

Dumont, A. L., T. K. Nygaard, R. L. Watkins, A. Smith, L. Kozhaya, B. N. Kreiswirth, B. Shopsin, D. Unutmaz, J. M. Voyich, and V. J. Torres. 2011. Characterization of a new cytotoxin that contributes to *Staphylococcus aureus* pathogenesis. *Mol. Microbiol.* 79:814–825.

Dunman, P. M., E. Murphy, S. Haney, D. Palacios, G. Tucker-Kellogg, S. Wu, E. L. Brown, R. J. Zagursky, D. Shlaes, and S. J. Projan. 2001. Transcription profiling-based identification of *Staphylococcus aureus* genes regulated by the *agr* and/or *sarA* loci. *J. Bacteriol.* 183:7341–7353.

Duthie, E. S., and L. L. Lorenz. 1952. Staphylococcal coagulase; mode of action and antigenicity. *J. Gen. Microbiol.* 6:95–107.

Felden, B., F. Vandenesch, P. Bouloc, and P. Romby. 2011. The *Staphylococcus aureus* RNome and its commitment to virulence. *PLoS Pathog.* 7:e1002006.

Fleming, V., E. Feil, A. K. Sewell, N. Day, A. Buckling, and R. C. Massey. 2006. Agr interference between clinical *Staphylococcus aureus* strains in an insect model of virulence. *J. Bacteriol.* 188:7686–7688.

Foster, T. J. 2005. Immune evasion by staphylococci. *Nat. Rev. Microbiol.* 3:948–958.

Fowler, V. G., Jr., G. Sakoulas, L. M. McIntyre, V. G. Meka, R. D. Arbeit, C. H. Cabell, M. E. Stryjewski, G. M. Eliopoulos, L. B. Reller, G. R. Corey, T. Jones, N. Lucindo, M. R. Yeaman, and A. S. Bayer. 2004. Persistent bacteremia due to methicillin-resistant *Staphylococcus aureus* infection is associated with *agr* dysfunction and low-level in vitro resistance to thrombin-induced platelet microbicidal protein. *J. Infect. Dis.* 190:1140–1149.

Fraser, J. D., and T. Proft. 2008. The bacterial superantigen and superantigen-like proteins. *Immunol. Rev.* 225:226–243.

Fridkin, S. K., J. C. Hageman, M. Morrison, L. T. Sanza, K. Como-Sabetti, J. A. Jernigan, K. Harriman, L. H. Harrison, R. Lynfield, and M. M. Farley. 2005. Methicillin-resistant *Staphylococcus aureus* disease in three communities. *N. Engl. J. Med.* 352:1436–1444.

Geiger, T., C. Goerke, M. Mainiero, D. Kraus, and C. Wolz. 2008. The virulence regulator Sae of *Staphylococcus aureus*: promoter activities and response to phagocytosis-related signals. *J. Bacteriol.* 190:3419–3428.

Geisinger, E., R. P. Adhikari, R. Jin, H. F. Ross, and R. P. Novick. 2006. Inhibition of *rot* translation by RNAIII, a key feature of *agr* function. *Mol. Microbiol.* 61:1038–1048.

Gillaspy, A. F., S. G. Hickmon, R. A. Skinner, J. R. Thomas, C. L. Nelson, and M. S. Smeltzer. 1995. Role of the accessory gene regulator (*agr*) in pathogenesis of staphylococcal osteomyelitis. *Infect. Immun.* 63:3373–3380.

Gillet, Y., B. Issartel, P. Vanhems, J. C. Fournet, G. Lina, M. Bes, F. Vandenesch, Y. Piemont, N. Brousse, D. Floret, and J. Etienne. 2002. Association between *Staphylococcus aureus* strains carrying gene for Panton-Valentine leukocidin and highly lethal necrotising pneumonia in young immunocompetent patients. *Lancet* 359:753–759.

Giraudo, A. T., A. Calzolari, A. A. Cataldi, C. Bogni, and R. Nagel. 1999. The *sae* locus of *Staphylococcus aureus* encodes a two-component regulatory system. *FEMS Microbiol. Lett.* 177:15–22.

Giraudo, A. T., A. L. Cheung, and R. Nagel. 1997. The *sae* locus of *Staphylococcus aureus* controls exoprotein synthesis at the transcriptional level. *Arch. Microbiol.* 168:53–58.

Giraudo, A. T., H. Rampone, A. Calzolari, and R. Nagel. 1996. Phenotypic characterization and virulence of a *sae- agr-* mutant of *Staphylococcus aureus*. *Can. J. Microbiol.* 42:120–123.

Giraudo, A. T., C. G. Raspanti, A. Calzolari, and R. Nagel. 1994. Characterization of a Tn551-mutant of *Staphylococcus aureus* defective in the production of several exoproteins. *Can. J. Microbiol.* 40:677–681.

Goerke, C., U. Fluckiger, A. Steinhuber, W. Zimmerli, and C. Wolz. 2001. Impact of the regulatory loci *agr*, *sarA* and *sae* of *Staphylococcus aureus* on the induction of alpha-toxin during device-related infection resolved by direct quantitative transcript analysis. *Mol. Microbiol.* 40:1439–1447.

Gomez, M. I., M. O'Seaghdha, M. Magargee, T. J. Foster, and A. S. Prince. 2006. *Staphylococcus aureus* protein A activates TNFR1 signaling through conserved IgG binding domains. *J. Biol. Chem.* 281:20190–20196.

Gomez, M. I., M. O. Seaghdha, and A. S. Prince. 2007. *Staphylococcus aureus* protein A activates TACE through EGFR-dependent signaling. *EMBO J.* 26:701–709.

Harraghy, N., J. Kormanec, C. Wolz, D. Homerova, C. Goerke, K. Ohlsen, S. Qazi, P. Hill, and M. Herrmann. 2005. sae is essential for expression of the staphylococcal adhesins Eap and Emp. *Microbiology* 151:1789–1800.

Hecker, M., J. Pane-Farre, and U. Volker. 2007. SigB-dependent general stress response in *Bacillus subtilis* and related gram-positive bacteria. *Annu. Rev. Microbiol.* 61:215–236.

Heilmann, C. 2011. Adhesion mechanisms of staphylococci. *Adv. Exp. Med. Biol.* 715:105–123.

Hirschhausen, N., T. Schlesier, M. A. Schmidt, F. Gotz, G. Peters, and C. Heilmann. 2010. A novel staphylococcal internalization mechanism involves the major autolysin Atl and heat shock cognate protein Hsc70 as host cell receptor. *Cell. Microbiol.* **12:**1746–1764.

Hoch, J. A. 2000. Two-component and phosphorelay signal transduction. *Curr. Opin. Microbiol.* **3:**165–170.

Horsburgh, M. J., J. L. Aish, I. J. White, L. Shaw, J. K. Lithgow, and S. J. Foster. 2002. σ^B modulates virulence determinant expression and stress resistance: characterization of a functional *rsbU* strain derived from *Staphylococcus aureus* 8325-4. *J. Bacteriol.* **184:**5457–5467.

Huntzinger, E., S. Boisset, C. Saveanu, Y. Benito, T. Geissmann, A. Namane, G. Lina, J. Etienne, B. Ehresmann, C. Ehresmann, A. Jacquier, F. Vandenesch, and P. Romby. 2005. *Staphylococcus aureus* RNAIII and the endoribonuclease III coordinately regulate *spa* gene expression. *EMBO J.* **24:**824–835.

Janzon, L., S. Lofdahl, and S. Arvidson. 1989. Identification and nucleotide sequence of the delta-lysin gene, *hld*, adjacent to the accessory gene regulator (*agr*) of *Staphylococcus aureus*. *Mol. Gen. Genet.* **219:**480–485.

Jarraud, S., C. Mougel, J. Thioulouse, G. Lina, H. Meugnier, F. Forey, X. Nesme, J. Etienne, and F. Vandenesch. 2002. Relationships between *Staphylococcus aureus* genetic background, virulence factors, *agr* groups (alleles), and human disease. *Infect. Immun.* **70:**631–641.

Jevons, M. P., and M. T. Parker. 1964. The evolution of new hospital strains of *Staphylococcus aureus*. *J. Clin. Pathol.* **17:**243–250.

Ji, G., R. Beavis, and R. P. Novick. 1997. Bacterial interference caused by autoinducing peptide variants. *Science* **276:** 2027–2030.

Ji, G., R. C. Beavis, and R. P. Novick. 1995. Cell density control of staphylococcal virulence mediated by an octapeptide pheromone. *Proc. Natl. Acad. Sci. USA* **92:**12055–12059.

Joo, H. S., J. L. Chan, G. Y. Cheung, and M. Otto. 2010. Subinhibitory concentrations of protein synthesis-inhibiting antibiotics promote increased expression of the *agr* virulence regulator and production of phenol-soluble modulin cytolysins in community-associated methicillin-resistant *Staphylococcus aureus*. *Antimicrob. Agents Chemother.* **54:**4942–4944.

Katayama, Y., T. Ito, and K. Hiramatsu. 2000. A new class of genetic element, staphylococcus cassette chromosome *mec*, encodes methicillin resistance in *Staphylococcus aureus*. *Antimicrob. Agents Chemother.* **44:**1549–1555.

Kavanaugh, J. S., M. Thoendel, and A. R. Horswill. 2007. A role for type I signal peptidase in *Staphylococcus aureus* quorum sensing. *Mol. Microbiol.* **65:**780–798.

Kazakova, S. V., J. C. Hageman, M. Matava, A. Srinivasan, L. Phelan, B. Garfinkel, T. Boo, S. McAllister, J. Anderson, B. Jensen, D. Dodson, D. Lonsway, L. K. McDougal, M. Arduino, V. J. Fraser, G. Killgore, F. C. Tenover, S. Cody, and D. B. Jernigan. 2005. A clone of methicillin-resistant *Staphylococcus aureus* among professional football players. *N. Engl. J. Med.* **352:**468–475.

Kennedy, A. D., J. Bubeck Wardenburg, D. J. Gardner, D. Long, A. R. Whitney, K. R. Braughton, O. Schneewind, and F. R. DeLeo. 2010. Targeting of alpha-hemolysin by active or passive immunization decreases severity of USA300 skin infection in a mouse model. *J. Infect. Dis.* **202:**1050–1058.

Kielian, T., A. Cheung, and W. F. Hickey. 2001. Diminished virulence of an alpha-toxin mutant of *Staphylococcus aureus* in experimental brain abscesses. *Infect. Immun.* **69:**6902–6911.

Kim, H. K., A. G. Cheng, H. Y. Kim, D. M. Missiakas, and O. Schneewind. 2010. Nontoxigenic protein A vaccine for methicillin-resistant *Staphylococcus aureus* infections in mice. *J. Exp. Med.* **207:**1863–1870.

Kintarak, S., S. A. Whawell, P. M. Speight, S. Packer, and S. P. Nair. 2004. Internalization of *Staphylococcus aureus* by human keratinocytes. *Infect. Immun.* **72:**5668–5675.

Klevens, R. M., M. A. Morrison, J. Nadle, S. Petit, K. Gershman, S. Ray, L. H. Harrison, R. Lynfield, G. Dumyati, J. M. Townes, A. S. Craig, E. R. Zell, G. E. Fosheim, L. K. McDougal, R. B. Carey, and S. K. Fridkin. 2007. Invasive methicillin-resistant *Staphylococcus aureus* infections in the United States. *JAMA* **298:**1763–1771.

Kobayashi, S. D., J. M. Voyich, K. R. Braughton, and F. R. DeLeo. 2003a. Down-regulation of proinflammatory capacity during apoptosis in human polymorphonuclear leukocytes. *J. Immunol.* **170:**3357–3368.

Kobayashi, S. D., J. M. Voyich, and F. R. DeLeo. 2003b. Regulation of the neutrophil-mediated inflammatory response to infection. *Microbes Infect.* **5:**1337–1344.

Kobayashi, S. D., J. M. Voyich, G. A. Somerville, K. R. Braughton, H. L. Malech, J. M. Musser, and F. R. DeLeo. 2003c. An apoptosis-differentiation program in human polymorphonuclear leukocytes facilitates resolution of inflammation. *J. Leukoc. Biol.* **73:**315–322.

Koenig, R. L., J. L. Ray, S. J. Maleki, M. S. Smeltzer, and B. K. Hurlburt. 2004. *Staphylococcus aureus* AgrA binding to the RNAIII-*agr* regulatory region. *J. Bacteriol.* **186:**7549–7555.

Kravitz, G. R., D. J. Dries, M. L. Peterson, and P. M. Schlievert. 2005. Purpura fulminans due to *Staphylococcus aureus*. *Clin. Infect. Dis.* **40:**941–947.

Kuehnert, M. J., D. Kruszon-Moran, H. A. Hill, G. McQuillan, S. K. McAllister, G. Fosheim, L. K. McDougal, J. Chaitram, B. Jensen, S. K. Fridkin, G. Killgore, and F. C. Tenover. 2006. Prevalence of *Staphylococcus aureus* nasal colonization in the United States, 2001-2002. *J. Infect. Dis.* **193:**172–179.

Kuroda, H., M. Kuroda, L. Cui, and K. Hiramatsu. 2007. Subinhibitory concentrations of beta-lactam induce haemolytic activity in *Staphylococcus aureus* through the SaeRS two-component system. *FEMS Microbiol. Lett.* **268:**98–105.

Kuroda, M., H. Kuroda, T. Oshima, F. Takeuchi, H. Mori, and K. Hiramatsu. 2003. Two-component system VraSR positively modulates the regulation of cell-wall biosynthesis pathway in *Staphylococcus aureus*. *Mol. Microbiol.* **49:**807–821.

Lauderdale, K. J., B. R. Boles, A. L. Cheung, and A. R. Horswill. 2009. Interconnections between sigma B, *agr*, and proteolytic activity in *Staphylococcus aureus* biofilm maturation. *Infect. Immun.* **77:**1623–1635.

Laurell, G., and G. Wallmark. 1953. Studies on *Staphylococcus aureus pyogenes* in a children's hospital. III. Results of phage-typing and tests for penicillin resistance of 2,474 strains isolated from patients and staff. *Acta Pathol. Microbiol. Scand.* **32:**438–447.

Lekstrom-Himes, J. A., and J. I. Gallin. 2000. Immunodeficiency diseases caused by defects in phagocytes. *N. Engl. J. Med.* **343:**1703–1714.

Li, M., B. A. Diep, A. E. Villaruz, K. R. Braughton, X. Jiang, F. R. DeLeo, H. F. Chambers, Y. Lu, and M. Otto. 2009. Evolution of virulence in epidemic community-associated methicillin-resistant *Staphylococcus aureus*. *Proc. Natl. Acad. Sci. USA* **106:**5883–5888.

Li, M., Y. Lai, A. E. Villaruz, D. J. Cha, D. E. Sturdevant, and M. Otto. 2007. Gram-positive three-component antimicrobial peptide-sensing system. *Proc. Natl. Acad. Sci. USA* **104:**9469–9474.

Liang, X., C. Yu, J. Sun, H. Liu, C. Landwehr, D. Holmes, and Y. Ji. 2006. Inactivation of a two-component signal transduction system, SaeRS, eliminates adherence and attenuates virulence of *Staphylococcus aureus*. *Infect. Immun.* **74:**4655–4665.

Lina, G., S. Jarraud, G. Ji, T. Greenland, A. Pedraza, J. Etienne, R. P. Novick, and F. Vandenesch. 1998. Transmembrane topology and histidine protein kinase activity of AgrC, the *agr* signal receptor in *Staphylococcus aureus*. *Mol. Microbiol.* **28:**655–662.

Lindmark, R., K. Thoren-Tolling, and J. Sjoquist. 1983 Binding of immunoglobulins to protein A and immunoglobulin levels in mammalian sera. *J. Immunol. Methods* **62:**1–13.

Lowy, F. D. 1998. *Staphylococcus aureus* infections. *N. Engl. J. Med.* **339:**520–532.

Lowy, F. D. 2003. Antimicrobial resistance: the example of *Staphylococcus aureus*. *J. Clin. Investig.* **111**:1265–1273.

Lowy, F. D. 2007. Secrets of a superbug. *Nat. Med.* **13**:1418–1420.

Mainiero, M., C. Goerke, T. Geiger, C. Gonser, S. Herbert, and C. Wolz. 2010. Differential target gene activation by the *Staphylococcus aureus* two-component system *saeRS*. *J. Bacteriol.* **192**:613–623.

Malachowa, N., A. R. Whitney, S. D. Kobayashi, D. E. Sturdevant, A. D. Kennedy, K. R. Braughton, D. W. Shabb, B. A. Diep, H. F. Chambers, M. Otto, and F. R. Deleo. 2011. Global changes in *Staphylococcus aureus* gene expression in human blood. *PLoS One* **6**:e18617.

Mann, E. E., K. C. Rice, B. R. Boles, J. L. Endres, D. Ranjit, L. Chandramohan, L. H. Tsang, M. S. Smeltzer, A. R. Horswill, and K. W. Bayles. 2009. Modulation of eDNA release and degradation affects *Staphylococcus aureus* biofilm maturation. *PLoS One* **4**:e5822.

Martin, F. J., M. I. Gomez, D. M. Wetzel, G. Memmi, M. O'Seaghdha, G. Soong, C. Schindler, and A. Prince. 2009. *Staphylococcus aureus* activates type I IFN signaling in mice and humans through the Xr repeated sequences of protein A. *J. Clin. Investig.* **119**:1931–1939.

McNamara, P. J., and A. S. Bayer. 2005. A *rot* mutation restores parental virulence to an *agr*-null *Staphylococcus aureus* strain in a rabbit model of endocarditis. *Infect. Immun.* **73**:3806–3809.

McNamara, P. J., K. C. Milligan-Monroe, S. Khalili, and R. A. Proctor. 2000. Identification, cloning, and initial characterization of *rot*, a locus encoding a regulator of virulence factor expression in *Staphylococcus aureus*. *J. Bacteriol.* **182**:3197–3203.

Menestrina, G., M. Dalla Serra, M. Comai, M. Coraiola, G. Viero, S. Werner, D. A. Colin, H. Monteil, and G. Prevost. 2003. Ion channels and bacterial infection: the case of beta-barrel pore-forming protein toxins of *Staphylococcus aureus*. *FEBS Lett.* **552**:54–60.

Menzies, B. E., and D. S. Kernodle. 1996. Passive immunization with antiserum to a nontoxic alpha-toxin mutant from *Staphylococcus aureus* is protective in a murine model. *Infect. Immun.* **64**:1839–1841.

Miller, L. G., F. Perdreau-Remington, G. Rieg, S. Mehdi, J. Perlroth, A. S. Bayer, A. W. Tang, T. O. Phung, and B. Spellberg. 2005. Necrotizing fasciitis caused by community-associated methicillin-resistant *Staphylococcus aureus* in Los Angeles. *N. Engl. J. Med.* **352**:1445–1453.

Montgomery, C. P., S. Boyle-Vavra, P. V. Adem, J. C. Lee, A. N. Husain, J. Clasen, and R. S. Daum. 2008. Comparison of virulence in community-associated methicillin-resistant *Staphylococcus aureus* pulsotypes USA300 and USA400 in a rat model of pneumonia. *J. Infect. Dis.* **198**:561–570.

Montgomery, C. P., S. Boyle-Vavra, and R. S. Daum. 2010. Importance of the global regulators Agr and SaeRS in the pathogenesis of CA-MRSA USA300 infection. *PLoS One* **5**:e15177.

Moran, G. J., A. Krishnadasan, R. J. Gorwitz, G. E. Fosheim, L. K. McDougal, R. B. Carey, and D. A. Talan. 2006. Methicillin-resistant *S. aureus* infections among patients in the emergency department. *N. Engl. J. Med.* **355**:666–674.

Morfeldt, E., D. Taylor, A. von Gabain, and S. Arvidson. 1995. Activation of alpha-toxin translation in *Staphylococcus aureus* by the trans-encoded antisense RNA, RNAIII. *EMBO J.* **14**:4569–4577.

Morfeldt, E., K. Tegmark, and S. Arvidson. 1996. Transcriptional control of the *agr*-dependent virulence gene regulator, RNAIII, in *Staphylococcus aureus*. *Mol. Microbiol.* **21**:1227–1237.

Nemeth, Z. H., B. Csoka, J. Wilmanski, D. Xu, Q. Lu, C. Ledent, E. A. Deitch, P. Pacher, Z. Spolarics, and G. Hasko. 2006. Adenosine A2A receptor inactivation increases survival in polymicrobial sepsis. *J. Immunol.* **176**:5616–5626.

Nizet, V. 2007. Understanding how leading bacterial pathogens subvert innate immunity to reveal novel therapeutic targets. *J. Allergy Clin. Immunol.* **120**:13–22.

Novick, R. P. 1991. Genetic systems in staphylococci. *Methods Enzymol.* **204**:587–636.

Novick, R. P. 2003. Autoinduction and signal transduction in the regulation of staphylococcal virulence. *Mol. Microbiol.* **48**:1429–1449.

Novick, R. P., and E. Geisinger. 2008. Quorum sensing in staphylococci. *Annu. Rev. Genet.* **42**:541–564.

Novick, R. P., and D. Jiang. 2003. The staphylococcal *saeRS* system coordinates environmental signals with *agr* quorum sensing. *Microbiology* **149**:2709–2717.

Novick, R. P., S. Projan, J. Kornblum, H. Ross, B. Kreiswirth, and S. Moghazeh. 1995. The *agr* P-2 operon: an autocatalytic sensory transduction system in *Staphylococcus aureus*. *Mol. Gen. Genet.* **248**:446–458.

Novick, R. P., H. F. Ross, S. J. Projan, J. Kornblum, B. Kreiswirth, and S. Moghazeh. 1993. Synthesis of staphylococcal virulence factors is controlled by a regulatory RNA molecule. *EMBO J.* **12**:3967–3975.

Nygaard, T. K., F. R. DeLeo, and J. M. Voyich. 2008. Community-associated methicillin-resistant *Staphylococcus aureus* skin infections: advances toward identifying the key virulence factors. *Curr. Opin. Infect. Dis.* **21**:147–152.

Nygaard, T. K., K. B. Pallister, P. Ruzevich, S. Griffith, C. Vuong, and J. M. Voyich. 2010. SaeR binds a consensus sequence within virulence gene promoters to advance USA300 pathogenesis. *J. Infect. Dis.* **201**:241–254.

Oviedo-Boyso, J., R. Cortes-Vieyra, A. Huante-Mendoza, H. B. Yu, J. J. Valdez-Alarcon, A. Bravo-Patino, M. Cajero-Juarez, B. B. Finlay, and V. M. Baizabal-Aguirre. 2011. The PI3K-Akt signaling pathway is important for *Staphylococcus aureus* internalization by endothelial cells. *Infect. Immun.* **79**:4569–4577.

Palazzolo-Ballance, A. M., M. L. Reniere, K. R. Braughton, D. E. Sturdevant, M. Otto, B. N. Kreiswirth, E. P. Skaar, and F. R. DeLeo. 2008. Neutrophil microbicides induce a pathogen survival response in community-associated methicillin-resistant *Staphylococcus aureus*. *J. Immunol.* **180**:500–509.

Pantrangi, M., V. K. Singh, C. Wolz, and S. K. Shukla. 2010. Staphylococcal superantigen-like genes, *ssl5* and *ssl8*, are positively regulated by Sae and negatively by Agr in the Newman strain. *FEMS Microbiol. Lett.* **308**:175–184.

Park, J., R. Jagasia, G. F. Kaufmann, J. C. Mathison, D. I. Ruiz, J. A. Moss, M. M. Meijler, R. J. Ulevitch, and K. D. Janda. 2007. Infection control by antibody disruption of bacterial quorum sensing signaling. *Chem. Biol.* **14**:1119–1127.

Peterson, M. M., J. L. Mack, P. R. Hall, A. A. Alsup, S. M. Alexander, E. K. Sully, Y. S. Sawires, A. L. Cheung, M. Otto, and H. D. Gresham. 2008. Apolipoprotein B is an innate barrier against invasive *Staphylococcus aureus* infection. *Cell Host Microbe* **4**:555–566.

Peterson, P. K., J. Verhoef, L. D. Sabath, and P. G. Quie. 1977. Effect of protein A on staphylococcal opsonization. *Infect. Immun.* **15**:760–764.

Pincus, S. H., L. A. Boxer, and T. P. Stossel. 1976. Chronic neutropenia in childhood. Analysis of 16 cases and a review of the literature. *Am. J. Med.* **61**:849–861.

Pynnonen, M., R. E. Stephenson, K. Schwartz, M. Hernandez, and B. R. Boles. 2011. Hemoglobin promotes *Staphylococcus aureus* nasal colonization. *PLoS Pathog.* **7**:e1002104.

Queck, S. Y., M. Jameson-Lee, A. E. Villaruz, T. H. Bach, B. A. Khan, D. E. Sturdevant, S. M. Ricklefs, M. Li, and M. Otto. 2008. RNAIII-independent target gene control by the *agr* quorum-sensing system: insight into the evolution of virulence regulation in *Staphylococcus aureus*. *Mol. Cell* **32**:150–158.

Ragle, B. E., and J. Bubeck Wardenburg. 2009. Anti-alpha-hemolysin monoclonal antibodies mediate protection against *Staphylococcus aureus* pneumonia. *Infect. Immun.* **77:**2712–2718.

Recsei, P., B. Kreiswirth, M. O'Reilly, P. Schlievert, A. Gruss, and R. P. Novick. 1986. Regulation of exoprotein gene expression in *Staphylococcus aureus* by agr. *Mol. Gen. Genet.* **202:**58–61.

Reyes, D., D. O. Andrey, A. Monod, W. L. Kelley, G. Zhang, and A. L. Cheung. 2011. Coordinated regulation by AgrA, SarA, and SarR to control agr expression in *Staphylococcus aureus*. *J. Bacteriol.* **193:**6020–6031.

Reynolds, J., and S. Wigneshweraraj. 2011. Molecular insights into the control of transcription initiation at the *Staphylococcus aureus agr* operon. *J. Mol. Biol.* **412:**862–881.

Rice, K. C., E. E. Mann, J. L. Endres, E. C. Weiss, J. E. Cassat, M. S. Smeltzer, and K. W. Bayles. 2007. The *cidA* murein hydrolase regulator contributes to DNA release and biofilm development in *Staphylococcus aureus*. *Proc. Natl. Acad. Sci. USA* **104:**8113–8118.

Rodrigue, A., Y. Quentin, A. Lazdunski, V. Mejean, and M. Foglino. 2000. Two-component systems in *Pseudomonas aeruginosa*: why so many? *Trends Microbiol.* **8:**498–504.

Rogasch, K., V. Ruhmling, J. Pane-Farre, D. Hoper, C. Weinberg, S. Fuchs, M. Schmudde, B. M. Broker, C. Wolz, M. Hecker, and S. Engelmann. 2006. Influence of the two-component system SaeRS on global gene expression in two different *Staphylococcus aureus* strains. *J. Bacteriol.* **188:**7742–7758.

Rooijakkers, S. H., K. P. van Kessel, and J. A. van Strijp. 2005. Staphylococcal innate immune evasion. *Trends Microbiol.* **13:**596–601.

Rothfork, J. M., G. S. Timmins, M. N. Harris, X. Chen, A. J. Lusis, M. Otto, A. L. Cheung, and H. D. Gresham. 2004. Inactivation of a bacterial virulence pheromone by phagocyte-derived oxidants: new role for the NADPH oxidase in host defense. *Proc. Natl. Acad. Sci. USA* **101:**13867–13872.

Said-Salim, B., P. M. Dunman, F. M. McAleese, D. Macapagal, E. Murphy, P. J. McNamara, S. Arvidson, T. J. Foster, S. J. Projan, and B. N. Kreiswirth. 2003. Global regulation of *Staphylococcus aureus* genes by Rot. *J. Bacteriol.* **185:**610–619.

Saravolatz, L. D., D. J. Pohlod, and L. M. Arking. 1982. Community-acquired methicillin-resistant *Staphylococcus aureus* infections: a new source for nosocomial outbreaks. *Ann. Intern. Med.* **97:**325–329.

Sasso, E. H., G. J. Silverman, and M. Mannik. 1989. Human IgM molecules that bind staphylococcal protein A contain VHIII H chains. *J. Immunol.* **142:**2778–2783.

Schafer, D., T. T. Lam, T. Geiger, M. Mainiero, S. Engelmann, M. Hussain, A. Bosserhoff, M. Frosch, M. Bischoff, C. Wolz, J. Reidl, and B. Sinha. 2009. A point mutation in the sensor histidine kinase SaeS of *Staphylococcus aureus* strain Newman alters the response to biocide exposure. *J. Bacteriol.* **191:**7306–7314.

Schlievert, P. M., L. C. Case, K. A. Nemeth, C. C. Davis, Y. Sun, W. Qin, F. Wang, A. J. Brosnahan, J. A. Mleziva, M. L. Peterson, and B. E. Jones. 2007. Alpha and beta chains of hemoglobin inhibit production of *Staphylococcus aureus* exotoxins. *Biochemistry* **46:**14349–14358.

Schweizer, M. L., J. P. Furuno, G. Sakoulas, J. K. Johnson, A. D. Harris, M. D. Shardell, J. C. McGregor, K. A. Thom, and E. N. Perencevich. 2011. Increased mortality with accessory gene regulator (agr) dysfunction in *Staphylococcus aureus* among bacteremic patients. *Antimicrob. Agents Chemother.* **55:**1082–1087.

Shopsin, B., A. Drlica-Wagner, B. Mathema, R. P. Adhikari, B. N. Kreiswirth, and R. P. Novick. 2008. Prevalence of agr dysfunction among colonizing *Staphylococcus aureus* strains. *J. Infect. Dis.* **198:**1171–1174.

Shopsin, B., C. Eaton, G. A. Wasserman, B. Mathema, R. P. Adhikari, S. Agolory, D. R. Altman, R. S. Holzman, B. N. Kreiswirth, and R. P. Novick. 2010. Mutations in agr do not persist in natural populations of methicillin-resistant *Staphylococcus aureus*. *J. Infect. Dis.* **202:**1593–1599.

Sifri, C. D., J. Park, G. A. Helm, M. E. Stemper, and S. K. Shukla. 2007. Fatal brain abscess due to community-associated methicillin-resistant *Staphylococcus aureus* strain USA300. *Clin. Infect. Dis.* **45:**e113–e117.

Sinha, B., P. P. Francois, O. Nusse, M. Foti, O. M. Hartford, P. Vaudaux, T. J. Foster, D. P. Lew, M. Herrmann, and K. H. Krause. 1999. Fibronectin-binding protein acts as *Staphylococcus aureus* invasin via fibronectin bridging to integrin alpha5beta1. *Cell. Microbiol.* **1:**101–117.

Soong, G., F. J. Martin, J. Chun, T. S. Cohen, D. S. Ahn, and A. Prince. 2011. *Staphylococcus aureus* protein A mediates invasion across airway epithelial cells through activation of RhoA signaling and proteolytic activity. *J. Biol. Chem.* **286:**35891–35898.

Steinhuber, A., C. Goerke, M. G. Bayer, G. Doring, and C. Wolz. 2003. Molecular architecture of the regulatory locus *sae* of *Staphylococcus aureus* and its impact on expression of virulence factors. *J. Bacteriol.* **185:**6278–6286.

Sun, F., C. Li, D. Jeong, C. Sohn, C. He, and T. Bae. 2010. In the *Staphylococcus aureus* two-component system *sae*, the response regulator SaeR binds to a direct repeat sequence and DNA binding requires phosphorylation by the sensor kinase SaeS. *J. Bacteriol.* **192:**2111–2127.

Thammavongsa, V., J. W. Kern, D. M. Missiakas, and O. Schneewind. 2009. *Staphylococcus aureus* synthesizes adenosine to escape host immune responses. *J. Exp. Med.* **206:**2417–2427.

Thoendel, M., and A. R. Horswill. 2009. Identification of *Staphylococcus aureus* AgrD residues required for autoinducing peptide biosynthesis. *J. Biol. Chem.* **284:**21828–21838.

Torres, V. J., A. S. Attia, W. J. Mason, M. I. Hood, B. D. Corbin, F. C. Beasley, K. L. Anderson, D. L. Stauff, W. H. McDonald, L. J. Zimmerman, D. B. Friedman, D. E. Heinrichs, P. M. Dunman, and E. P. Skaar. 2010. *Staphylococcus aureus fur* regulates the expression of virulence factors that contribute to the pathogenesis of pneumonia. *Infect. Immun.* **78:**1618–1628.

Traber, K. E., E. Lee, S. Benson, R. Corrigan, M. Cantera, B. Shopsin, and R. P. Novick. 2008. agr function in clinical *Staphylococcus aureus* isolates. *Microbiology* **154:**2265–2274.

Ventura, C. L., N. Malachowa, C. H. Hammer, G. A. Nardone, M. A. Robinson, S. D. Kobayashi, and F. R. DeLeo. 2010. Identification of a novel *Staphylococcus aureus* two-component leukotoxin using cell surface proteomics. *PLoS One* **5:**e11634.

Voyich, J. M., K. R. Braughton, D. E. Sturdevant, C. Vuong, S. D. Kobayashi, S. F. Porcella, M. Otto, J. M. Musser, and F. R. DeLeo. 2004. Engagement of the pathogen survival response used by group A *Streptococcus* to avert destruction by innate host defense. *J. Immunol.* **173:**1194–1201.

Voyich, J. M., K. R. Braughton, D. E. Sturdevant, A. R. Whitney, B. Said-Salim, S. F. Porcella, R. D. Long, D. W. Dorward, D. J. Gardner, B. N. Kreiswirth, J. M. Musser, and F. R. Deleo. 2005. Insights into mechanisms used by *Staphylococcus aureus* to avoid destruction by human neutrophils. *J. Immunol.* **175:**3907–3919.

Voyich, J. M., M. Otto, B. Mathema, K. R. Braughton, A. R. Whitney, D. Welty, R. D. Long, D. W. Dorward, D. J. Gardner, G. Lina, B. N. Kreiswirth, and F. R. DeLeo. 2006. Is Panton-Valentine leukocidin the major virulence determinant in community-associated methicillin-resistant *Staphylococcus aureus* disease? *J. Infect. Dis.* **194:**1761–1770.

Voyich, J. M., C. Vuong, M. DeWald, T. K. Nygaard, S. Kocianova, S. Griffith, J. Jones, C. Iverson, D. E. Sturdevant, K. R. Braughton, A. R. Whitney, M. Otto, and F. R. DeLeo. 2009. The SaeR/S gene regulatory system is essential for innate immune evasion by *Staphylococcus aureus*. *J. Infect. Dis.* **199:**1698–1706.

Vuong, C., S. Kocianova, Y. Yao, A. B. Carmody, and M. Otto. 2004. Increased colonization of indwelling medical devices by quorum-sensing mutants of *Staphylococcus epidermidis* in vivo. *J. Infect. Dis.* **190:**1498–1505.

Wang, R., K. R. Braughton, D. Kretschmer, T. H. Bach, S. Y. Queck, M. Li, A. D. Kennedy, D. W. Dorward, S. J. Klebanoff, A. Peschel, F. R. DeLeo, and M. Otto. 2007. Identification of novel cytolytic peptides as key virulence determinants for community-associated MRSA. *Nat. Med.* **13:**1510–1514.

Watkins, R. L., K. B. Pallister, and J. M. Voyich. 2011. The SaeR/S gene regulatory system induces a pro-inflammatory cytokine response during *Staphylococcus aureus* infection. *PLoS One* **6:**e19939.

Weidenmaier, C., J. F. Kokai-Kun, S. A. Kristian, T. Chanturiya, H. Kalbacher, M. Gross, G. Nicholson, B. Neumeister, J. J. Mond, and A. Peschel. 2004. Role of teichoic acids in *Staphylococcus aureus* nasal colonization, a major risk factor in nosocomial infections. *Nat. Med.* **10:**243–245.

Weinrick, B., P. M. Dunman, F. McAleese, E. Murphy, S. J. Projan, Y. Fang, and R. P. Novick. 2004. Effect of mild acid on gene expression in *Staphylococcus aureus*. *J. Bacteriol.* **186:**8407–8423.

Wright, J. S., III, R. Jin, and R. P. Novick. 2005. Transient interference with staphylococcal quorum sensing blocks abscess formation. *Proc. Natl. Acad. Sci. USA* **102:**1691–1696.

Wuster, A., and M. M. Babu. 2008. Conservation and evolutionary dynamics of the *agr* cell-to-cell communication system across firmicutes. *J. Bacteriol.* **190:**743–746.

Xiong, Y. Q., J. Willard, M. R. Yeaman, A. L. Cheung, and A. S. Bayer. 2006. Regulation of *Staphylococcus aureus* alpha-toxin gene (*hla*) expression by *agr*, *sarA*, and *sae* in vitro and in experimental infective endocarditis. *J. Infect. Dis.* **194:**1267–1275.

Zhang, L., L. Gray, R. P. Novick, and G. Ji. 2002. Transmembrane topology of AgrB, the protein involved in the post-translational modification of AgrD in *Staphylococcus aureus*. *J. Biol. Chem.* **277:**34736–34742.

Regulation of Bacterial Virulence
Edited by Michael L. Vasil and Andrew J. Darwin
© 2013 ASM Press, Washington, DC   doi:10.1128/9781555818524.ch5

Chapter 5

Regulation of Virulence by Iron in Gram-Positive Bacteria

ALLISON J. FARRAND AND ERIC P. SKAAR

INTRODUCTION

Iron is an abundant element on Earth and functions as a crucial cofactor for biological processes in nearly every living organism (Morgan and Anders, 1980). Nonetheless, iron is also a precious commodity in the context of the host-pathogen interaction and is a prize that is won through significant battle. Patients with elevated iron levels may experience exacerbated infectious disease, whereas iron deficiency often enhances resistance to infection (Gangaidzo et al., 2001; Schaible et al., 2002). Host-mediated sequestration of iron from pathogens causes a dramatic shift in bacterial protein expression and drastically affects bacterial virulence. The primary result of these changes is the upregulation of systems dedicated to increasing intracellular iron levels to support bacterial growth. However, a concomitant response to iron deprivation is the expression of factors that promote iron acquisition, virulence, and the progression of infection. In this chapter, we discuss the mechanisms by which vertebrates sequester iron from invading pathogens and the response of pathogens to this sequestration. We also provide examples of iron-regulated virulence determinants in several clinically important gram-positive bacteria (Fig. 1 and Table 1).

Availability of Iron in the Environment

Iron is a major constituent of Earth's composition, as it is the fourth most abundant element in Earth's crust (Hammond, 2004; Morgan and Anders, 1980). In addition to its ubiquity in the environment, iron is a vital component of life on this planet. In living systems, iron can exist in a variety of oxidation states but is typically found in the ferrous (Fe^{2+}) or ferric (Fe^{3+}) forms. The potential reactivity of this metal makes it a valuable element in many chemical reactions. As a consequence, iron functions as an essential cofactor in many cellular processes that are vital to the physiology of both bacteria and eukaryotes, including DNA synthesis, respiration, and antioxidant defense. Iron is crucial to the activity of ribonucleotide reductase, nitrogenase, peroxidase, catalase, and succinic dehydrogenase, and it is therefore required for the vital functions of respiration and several metabolic pathways (Crosa, 1997; Litwin and Calderwood, 1993). Iron bound in the tetrapyrrole ring of heme acts as a cofactor in numerous proteins and enzymes, providing reactivity for their function. Heme plays important roles in oxygen storage and transport in vertebrates when bound to hemoglobin and myoglobin. Additionally, heme is required for essential cellular processes such as electron transfer through cytochromes, signal transduction, and protection against reactive oxygen species (Poulos, 2007).

Although iron is important for cellular function, it also has the potential to severely damage the cell. Iron catalyzes the generation of reactive oxygen species through participation in the Fenton reaction, in which hydrogen peroxide reacts with iron to form hydroxyl radicals (Fig. 2) (Imlay et al., 1988). Reactive oxygen species contribute to cellular damage by targeting DNA, proteins, and cell membrane-associated lipids (Henle and Linn, 1997; Luo et al., 1994). Thus, a delicate balance of intracellular iron levels must be maintained to promote cellular growth and protect against the deleterious effects of excess iron. As discussed below, regulating iron levels is vital for both eukaryotic and bacterial systems and becomes an important determinant in the outcome of the host-pathogen interaction. During infection, pathogens must rely on their host as the sole source of nutrient iron. Consequently, vertebrate hosts have developed several mechanisms to sequester iron from invading organisms to prevent bacterial replication.

Allison J. Farrand and Eric P. Skaar • Department of Pathology, Microbiology and Immunology, Vanderbilt University Medical Center, Nashville, TN 37232.

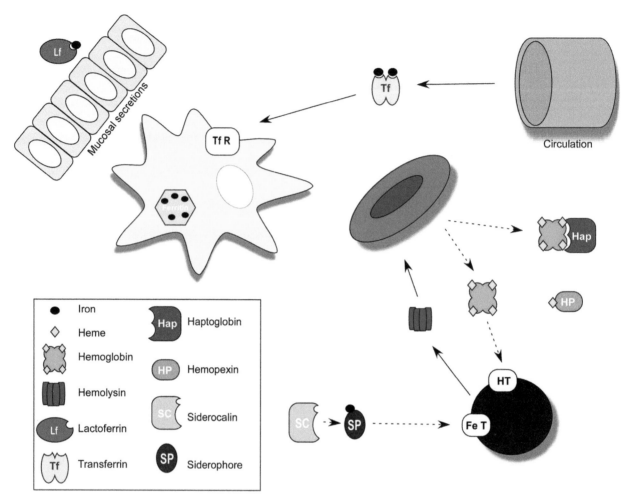

Figure 1. Nutritional immunity: host-mediated iron sequestration and theft of iron by pathogens. Vertebrate hosts employ several mechanisms to withhold nutrient iron from invading pathogens during infection. Most iron is located intracellularly, complexed to hemoglobin in erythrocytes or stored within ferritin inside host cells. Lactoferrin (Lf) and transferrin (Tf) transport iron to cells throughout the body and are internalized through the transferrin receptor (Tf R) on host cells. Invading pathogens respond by producing dedicated systems to steal host iron. In some pathogens, hemoglobin is released from erythrocytes following lysis of the cells by bacterial hemolysins and can be imported through the bacterial membrane via heme transporters (HT) or bound by host haptoglobin (Hap) or hemopexin (HP). Host iron can be obtained by the pathogen through direct import via iron transporters (Fe T) or through iron-chelating siderophores (SP), which are secreted from the pathogen to scavenge available extracellular iron. In response, the host produces siderocalin (SC), which can bind some siderophores and prevent them from being utilized by the pathogen.
doi:10.1128/9781555818524.ch5f1

Nutritional Immunity: Limitation of Iron Availability in the Host Environment

The importance of iron to cellular functions of both host and pathogen, combined with the potential damaging effect of the metal, necessitates dedicated regulatory mechanisms to control the amount of available iron within mammalian tissues. Under normal circumstances, dietary iron is absorbed by intestinal epithelial cells and released into circulation. Free iron is then bound by the iron transport molecule transferrin and delivered to cells throughout the body. Uptake of iron-bound transferrin occurs through an interaction with transferrin receptors on the cell surface, whereby the iron-bound transferrin is internalized and iron is removed and either used immediately (e.g., enzyme function) or stored intracellularly by ferritin.

Alterations in mammalian iron homeostasis can significantly affect the outcome of infection. Diseases in iron metabolism impact susceptibility to infection, exemplified by an increased frequency of infections caused by *Salmonella*, *Mycobacterium*, and *Plasmodium* in patients with high iron levels (Gangaidzo et al., 2001; Magnus et al., 1999; Moyo et al., 1997). In contrast, abnormally low iron levels

Table 1. Virulence determinants regulated by iron-dependent metalloregulators in gram-positive bacteria

Virulence determinant	Organism	Regulator	Iron-dependent effect
Bacterial surface remodeling	*Corynebacterium diphtheriae*	DtxR	↑ in low Fe
	Streptococcus pyogenes (GAS)	Mga	↓ in low Fe
	Streptococcus suis	Fur	↑ in low Fe
	Listeria monocytogenes	Fur	↑ in low Fe
	Mycobacterium tuberculosis	IdeR	↑ in low Fe
Biofilm formation	*Staphylococcus aureus*	Fur/Sae	↑ in low Fe
	Staphylococcus epidermidis	Unknown	↑ in low Fe
	Streptococcus pneumoniae	Unknown	↑ in Hb,[a] heme or ferritin
	Streptococcus mutans	SloR	↑ in low Fe
	Actinomyces naeslundii	AmdR	↑ in low Fe
Host cell attachment	*Staphylococcus aureus*	Fur	↑ in low Fe
	Corynebacterium diphtheriae	DtxR	↑ in low Fe
	Bacillus cereus	Fur	↑ in low Fe
	Streptococcus mutans	SloR	↑ in low Fe
	Listeria monocytogenes	Unknown	↑ in low Fe
Host cell invasion	*Staphylococcus aureus*	Fur	↑ in low Fe
	Bacillus anthracis	Unknown	↑ in low Fe
	Bacillus cereus	Fur	↑ in low Fe
	Listeria monocytogenes	Unknown	↑ in low Fe
Exotoxin production	*Corynebacterium diphtheriae*	DtxR	↑ in low Fe
	Staphylococcus aureus	Fur	↑ in low Fe
	Bacillus cereus	Fur	↑ in low Fe
	Streptococcus pneumoniae	Unknown	↑ in Hb, heme or ferritin
	Streptococcus pyogenes (GAS)	Unknown	↑ in low Fe
	Listeria monocytogenes	Unknown	↑ in low Fe
	Mycobacterium tuberculosis	IdeR	↑ in low Fe

[a]Hb, hemoglobin.

can result in a measure of resistance to microbial infection (Murray et al., 1978). Consistent with this observation, infection and inflammation result in a concerted reduction in the amount of iron available to invading pathogens (Wang and Cherayil, 2009). The concentration of free iron within the mammalian host is extremely low, below what is required to support bacterial growth. Such a low level of available iron is achieved through a coordinated series of sequestration mechanisms collectively termed *nutritional immunity*. Iron limitation within vertebrate tissues is due to several factors, including (i) the low solubility of iron in the presence of oxygen and at physiological pH, (ii) the intracellular location of the majority of iron, and (iii) the fact that most iron in extracellular fluids is complexed within iron-binding proteins (Fig. 1).

$$Fe^{2+} + H_2O_2 \longrightarrow Fe^{3+} + OH^\bullet + OH^-$$

Figure 2. The Fenton reaction. Ferrous iron catalyzes the reaction, in which hydrogen peroxide is broken down into a hydroxyl radical and hydroxyl anion and the metal is oxidized to ferric iron. Hydroxyl radicals are a form of oxidative stress that damages DNA, proteins, and lipids within a cell.
doi:10.1128/9781555818524.ch5f2

During infection, pathogens encounter an environment where most iron is found intracellularly (Drabkin, 1951). Approximately 80% of intracellular iron is bound to hemoglobin, a tetrameric molecule that binds the iron-containing protoporphyrin ring of heme. The heme cofactor allows hemoglobin to bind oxygen and transport it to sites throughout the body via circulating erythrocytes. The remaining portion of intracellular iron is stored in ferritin, a large molecule composed of 24 subunits that can hold up to 4,000 iron atoms (Wang et al., 2006). Ferritin stores excess iron for future use while protecting the cell from its damaging effects.

Most of the remaining iron in the body is bound by glycoproteins circulating in the extracellular milieu. The primary iron-binding protein in serum is transferrin, a monomeric protein comprised of two high-affinity iron-binding domains (Gomme et al., 2005). Transferrins deliver iron to host cells through an interaction with transferrin receptors that leads to subsequent internalization into endocytic vesicles (He et al., 2000). An additional iron-binding glycoprotein, lactoferrin, functions similarly to transferrin but is found in milk and nasal secretions as well as neutrophil granules.

As discussed in the next section, one major response of bacterial pathogens to iron deprivation is the secretion of iron-scavenging molecules called siderophores, which steal iron from host proteins. To counteract the function of some bacterial siderophores, mammalian hosts produce siderocalins, a family of carrier proteins typically involved in the transport of hydrophobic molecules such as steroids and lipids (Nelson et al., 2005). Siderocalins also function in innate immunity to bind catecholate siderophores, including enterobactin from *Escherichia coli*, preventing the iron-bound siderophore from being utilized by the invading pathogen (Flo et al., 2004). Expression of siderocalin in the respiratory mucosa increases 65-fold in response to colonization of murine nasal passages with *Streptococcus pneumoniae*, a prominent gram-positive respiratory pathogen (Nelson et al., 2005). This effect appears to be a general antimicrobial response, as a similar phenomenon was observed in the presence of the gram-negative organism *Haemophilus influenzae*. Interestingly, neither of these pathogens is known to produce siderophores, suggesting that they have evolved to evade siderocalin-mediated iron sequestration and obtain host iron through other means. Production of siderocalins, therefore, appears to represent both a counterattack against invading pathogens as they attempt to acquire iron sequestered by the host and a mechanism by which commensal and asymptomatically colonizing bacteria are controlled.

Humans also produce hepcidin, an antimicrobial peptide that is upregulated under conditions of iron overload or gamma interferon-induced inflammation. Hepcidin blocks iron export from duodenal enterocytes and macrophages by inducing the internalization and degradation of the iron export protein ferroportin 1 (De Domenico et al., 2007). The hepcidin-mediated reduction in iron levels affects invading pathogens, as hepicidin localizes to *Mycobacterium tuberculosis*-containing phagosomes in vitro and inhibits bacterial growth (Sow et al., 2007). Hence, hepcidin is believed to be a component of the host's arsenal to defend against iron-dependent propagation of intracellular pathogens.

Free heme and hemoglobin are not well tolerated by the body and are prime iron sources for many microbes. Consequently, vertebrates employ two different mechanisms to remove these molecules from circulation. First, liberated hemoglobin is bound by haptoglobin through a high-affinity interaction and removed from the bloodstream via CD163 receptor-mediated endocytosis (Kristiansen et al., 2001). Second, free heme is bound by the glycoprotein hemopexin and recycled in the liver (Tolosano and Altruda, 2002). Nevertheless, some pathogens have adapted to circumvent these scavenging mechanisms and acquire heme-iron from the host through secreted hemophores and membrane transport systems that directly import heme-iron.

Response of Bacterial Pathogens to Nutritional Immunity

As invading pathogens enter the iron-limiting environment of the host, they alter their transcriptome in order to survive and successfully colonize the host. The causative agent of Lyme disease, *Borrelia burgdorferi*, however, is an exception to this rule. *B. burgdorferi* has adapted to the low-iron environment of the host such that genes encoding iron-dependent metalloproteins either are eliminated altogether or utilize manganese as a cofactor instead of iron (Posey and Gherardini, 2000). Iron starvation is overcome in other pathogens by upregulating expression of systems dedicated to obtaining extracellular iron to promote growth. In conjunction, many virulence factors critical for successful infection are expressed in response to the environmental cue of iron limitation (Litwin et al., 1992).

Bacterial pathogens acquire iron from their host via two primary mechanisms: (i) the secretion of siderophores and (ii) the expression of surface-localized systems that directly import iron or recognize iron-containing host proteins. Siderophores are small molecules that chelate iron with a very high affinity, above 10^{30} M (Wandersman and Delepelaire, 2004). Three types of siderophores have been identified in bacteria: hydroxamate, catecholate, and hydroxyl acid (Drechsel and Jung, 1998). Iron-bound siderophores are imported into the cell by siderophore-specific transport systems. Following transport into the cytoplasm, iron is removed from the siderophore and utilized in various cellular functions. Iron can be released from siderophores via two mechanisms: reduction of siderophore-bound iron by ferrisiderophore reductases or by hydrolysis of the siderophore backbone (Drechsel and Jung, 1998). The second method of iron assimilation occurs via specific cell surface receptors and membrane transport systems which bind iron-containing host proteins (e.g., lactoferrin), extract the metal, and transport it into the bacterial cytosol. ATP-binding cassette (ABC) transporters are the primary transport systems used to import iron and heme-iron across the bacterial membrane. These transporters consist of a ligand-specific lipoprotein on the cell surface, an ATP-binding protein, and two membrane transporter proteins (Nikaido and Hall, 1998). Both siderophore synthesis and membrane transport systems are strictly regulated in iron-rich environments to prevent excess iron from collecting intracellularly and damaging the cell.

The paucity of free iron within vertebrates also serves as an environmental cue for pathogens, signifying entrance into a host. Consequently, the regulation of the expression of many virulence factors is intimately linked to that of iron acquisition systems, and in many pathogens both are controlled by iron-dependent transcriptional regulators. The assimilation of iron within the host is required for bacterial survival and pathogenesis and is widely considered a virulence determinant in its own right (Aranda et al., 2010; Bates et al., 2005; Brown et al., 2001; Dale et al., 2004; Gold et al., 2001; Harvie et al., 2005; Horsburgh et al., 2001b; Janulczyk et al., 2003; Montanez et al., 2005; Rodriguez and Smith, 2006; Skaar et al., 2004a; Torres et al., 2006). Discussion of iron-responsive virulence factors in several important pathogens can be found in subsequent sections of this chapter as well as in Chapter 6.

MECHANISMS OF IRON-DEPENDENT GENE REGULATION IN GRAM-POSITIVE BACTERIA

Transcriptional regulation of bacterial genes in response to iron occurs through the activity of metal-dependent regulators. The importance of metalloregulators to the pathogenicity of gram-positive bacteria is underscored by the significant decrease in virulence that occurs upon genetic inactivation of these regulators (Horsburgh et al., 2001b; Manabe et al., 1999; Rea et al., 2004; Torres et al., 2010). Iron-bound metalloregulators negatively regulate transcription of iron uptake genes. When intracellular iron levels are high, iron interacts with these regulators and activates them to bind operator sequences of genes within their regulon. Transcription of these genes is repressed, thereby preventing continual uptake that would result in iron overload (Fig. 3). Conversely, when iron levels are low, the regulators are not iron bound and disengage from DNA, allowing expression of downstream genes. This global regulatory cascade enables the bacteria to access iron from their environment. Not surprisingly, disruption of these regulators often leads to enhanced and uncontrolled iron uptake, increasing the potential for generation of reactive oxygen species and resulting in mutants that are more sensitive to oxidative stress (Dussurget et al., 1996; Horsburgh et al., 2001a). The two families of metalloregulators in gram-positive bacteria are briefly discussed below. More detailed descriptions of their roles in the regulation of virulence determinants in several important bacterial pathogens can be found in the following sections.

The Fur Family

The ferric uptake regulator (Fur) was initially described for the gram-negative pathogen *Salmonella*

enterica serovar Typhimurium, and orthologs have been identified in low-GC-content gram-positive pathogens in recent years (Ernst et al., 1978). In the presence of iron, Fur represses iron uptake genes as well as a variety of virulence factors. Regulation of iron acquisition and pathogenicity by Fur is critical to the progression of infection, as *fur* knockout strains are more sensitive to neutrophil killing and are significantly less virulent in infection models (Horsburgh et al., 2001b; Torres et al., 2010). Similar to some gram-negative pathogens, *Bacillus subtilis* Fur mediates iron-dependent regulation through a small RNA (FsrA) and three small, basic proteins (FbpA, -B, and -C) believed to act as RNA chaperones (Gaballa et al., 2008). Under iron-limiting conditions, the small RNA and chaperones inhibit the translation of many iron-containing proteins to prioritize the use of available iron.

Many gram-positive bacteria, including the human pathogen *Staphylococcus aureus*, express more than one Fur paralog. The Fur-like proteins PerR (peroxide response regulator) and Zur (zinc uptake regulator) typically regulate the expression of genes in response to oxidative stress and zinc, respectively (Horsburgh et al., 2001a; Lindsay and Foster, 2001). The regulation of iron acquisition and storage is intimately linked with the response to and protection from oxidative stress, as excess iron can catalyze the Fenton reaction and generate hydroxyl radicals. To overcome this problem, regulation by PerR occurs in response to hydrogen peroxide levels but remains dependent on manganese and iron (Herbig and Helmann, 2001). Expression of *fur* in *B. subtilis* is not dependent on iron levels but instead is regulated by PerR (Fuangthong et al., 2002). Fur-mediated iron regulation appears to be a complex process involving several regulators to maintain the delicate balance between iron deprivation and overload.

The DtxR Family

Diphtheria toxin regulator (DtxR) orthologs are found in gram-positive bacteria with high GC content, including *Corynebacterium* and *Mycobacterium*, and repress genes within their regulon in iron- or manganese-rich environments. As the name implies, the founding member of this family was identified as the molecule responsible for the iron-dependent regulation of diphtheria toxin, an important virulence factor produced by *Corynebacterium diphtheriae* (Fourel et al., 1989). The iron-dependent control of the DtxR regulon is a sensitive, tightly controlled process, as evidenced by the drastic change in diphtheria toxin production in response to as little as a threefold change in DtxR levels (Oram et al., 2002). Similar to Fur, DtxR

Fur Family Regulators

Pathogen	Regulator (metal)
Staphylococcus aureus	Fur (Fe) PerR (Fe/Mn) Zur (Zn)
Bacillus subtilis	Fur (Fe) PerR (Fe/Mn) Zur (Zn)
Bacillus anthracis	Fur (Fe) PerR (Fe/Mn) Zur (Zn)
Bacillus cereus	Fur (Fe) PerR (Fe/Mn) Zur (Zn)
Streptococcus pyogenes	PerR (Fe/Mn)
Streptococcus suis	Fur (Fe)
Listeria monocytogenes	Fur (Fe)
Mycobacterium tuberculosis	FurA (Fe/Mn) FurB (Zn)

DtxR Family Regulators

Pathogen	Regulator (metal)
Corynebacterium diphtheriae	DtxR (Fe)
Corynebacterium pseudotuberculosis	DtxR (Fe)
Staphylococcus aureus	MntR (Mn)
Staphylococcus epidermidis	SirR (Fe)
Bacillus subtilis	MntR (Mn)
Streptococcus pyogenes	MtsR (Fe/Mn)
Streptococcus agalactiae	MtsR (Fe/Mn)
Streptococcus mutans	SloR (Fe)
Actinomyces naeslundii	AmdR (Fe)
Mycobacterium tuberculosis	IdeR (Fe) SirR (Mn)

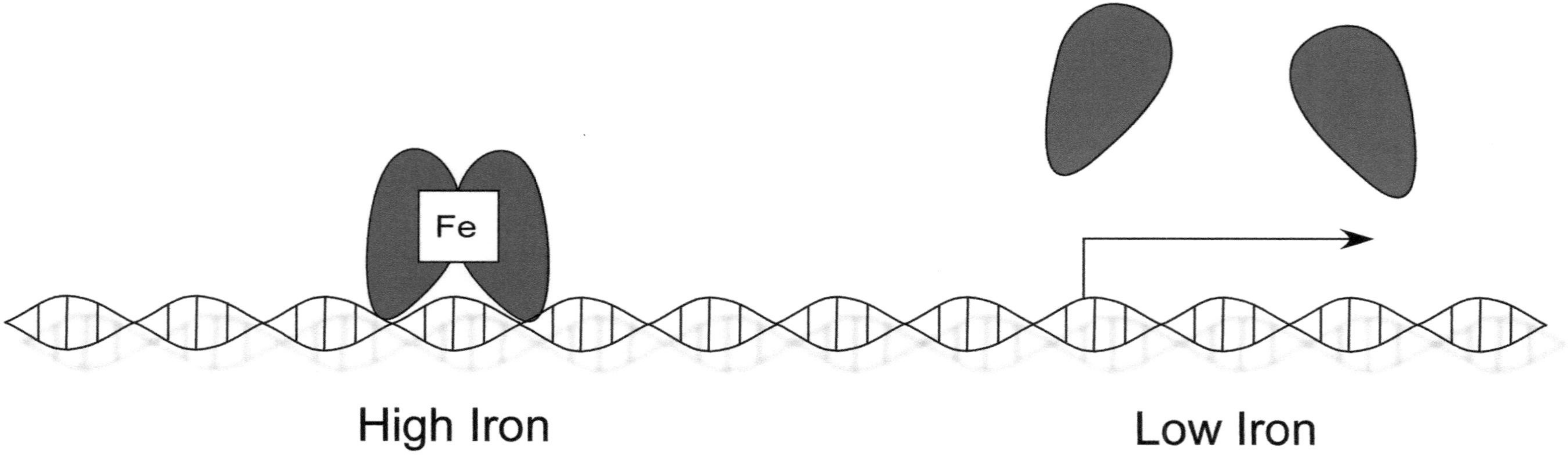

High Iron

Low Iron

Figure 3. Mechanism of iron-dependent regulation in gram-positive bacteria. The metalloregulators Fur and DtxR regulate expression of a subset of genes in response to the availability of iron. Under iron-rich conditions, the metal binds to the regulator and activates it, allowing the protein to dimerize and bind to iron boxes in the promoter regions of genes within its regulon. Dimers bind to both strands of the double helix and prevent transcription. When iron levels are low, the regulators are not iron bound and no longer remain bound to the consensus sequence, and transcription is allowed to proceed. doi:10.1128/9781555818524.ch5f3

regulates the expression of genes encoding iron acquisition systems as well as toxins. Inactivation of DtxR-type regulators often leads to increased sensitivity to hydrogen peroxide, as seen with mutation of the DtxR ortholog *ideR* of *Mycobacterium smegmatis* and *M. tuberculosis* and *mtsR* of *Streptococcus pyogenes* (Bates et al., 2005; Dussurget et al., 1996). This point highlights the direct interplay between the maintenance of intracellular iron levels and protection against oxidative stress. While DtxR primarily controls expression of its regulon in response to iron levels, it can interact in vitro with other divalent ions, including manganese, cadmium, cobalt, nickel, and zinc (Schmitt and Holmes, 1993; Tao et al., 1992).

Metalloregulator Protein Structure

Fur and DtxR have very little similarity at the amino acid level (25%) but exhibit some tertiary structural resemblance (Fig. 4). Both regulators are comprised of an N-terminal winged-helix-turn-helix motif that participates in DNA binding following recognition of a consensus sequence or an iron box. The second domain is responsible for dimerization of the regulators and contains two high-affinity metal binding sites, termed sites I and II (Stojiljkovic and Hantke, 1995). Metalloregulators also bind a single zinc atom per monomer, which is important for maintaining the structure of the complex (Bsat and Helmann, 1999). The structure of DtxR-like

regulators is unique from the Fur family in that they contain an extra C-terminal *src* homology 3 (SH3)-like domain attached to the dimerization domain by a flexible linker. The DtxR SH3 domain is believed to affect the metal-dependent activation of the regulator by stabilizing the N-terminal DNA-binding domain but does not interact directly with DNA (Wylie et al., 2004).

Mechanism of Transcriptional Regulation

Metal binding to site I in the dimerization domain alters the conformation of the metalloregulator, allowing it to adopt a structure capable of binding its consensus sequence. Metal binding to site II is believed to be important for maintaining structure of the complex (Schiering et al., 1995). Fur and DtxR recognize and bind their cognate operator sequence as homodimers in the presence of Fe^{2+}, which prevents the interaction of RNA polymerase with DNA and effectively blocks transcription of target genes (Fig. 3). When intracellular iron levels are high, the regulator binds the iron box and represses transcription. When intracellular iron levels are limiting, the repressor is no longer metal bound and dissociates from the promoter region, allowing expression of genes within its regulon.

The Fur paralog PerR is unique in mechanism in that it primarily responds to intracellular hydrogen peroxide levels, but this response is dependent

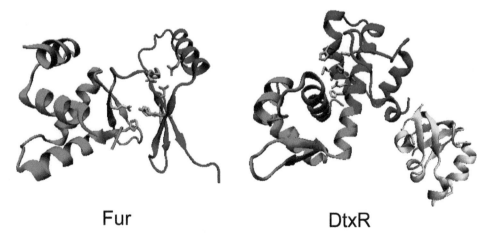

Fur   DtxR

Figure 4. Structural characteristics of gram-positive metalloregulators. The crystal structures of the two primary iron-dependent metalloregulators in gram-positive bacteria are shown in ribbon diagram (Pohl et al., 1998; Pohl et al., 2003). The ferric uptake regulator, Fur, and diphtheria toxin regulator, DtxR, share very little sequence homology but contain similar tertiary structures. The DNA binding domain of both regulators, shown in red, interacts with the consensus iron box sequence in the promoter of Fur- or DtxR-regulated genes. The dimerization domain, shown in blue, allows for two regulator subunits to interact with each other and the DNA strand. The dimerization domain also contains two metal binding sites, shown in green and orange, which bind iron or manganese to activate the regulator and promote DNA binding. DtxR also contains a unique SH3-like domain, shown in yellow, which is attached to the dimerization domain by a flexible linker and is believed to stabilize the molecule while it is bound to DNA.
doi:10.1128/9781555818524.ch5f4

on levels of both manganese and iron. PerR in *B. subtilis* can bind either manganese or iron and can therefore exist in two forms: PerR:Mn and PerR:Fe. The differentially coordinated forms of PerR respond to oxidative stress with differing levels of sensitivity, allowing for a tightly coordinated response; PerR:Fe is activated by low hydrogen peroxide levels, while PerR:Mn requires higher levels (Herbig and Helmann, 2001). Oxidative stress is sensed by PerR when histidine residues that coordinate iron are oxidized by hydrogen peroxide. Oxidation of these residues liberates iron, releasing PerR from its operator and allowing for downstream genes to be transcribed (Lee and Helmann, 2006).

Fur-like regulators themselves exhibit an interdependence and regulation by metal ions. The *B. subtilis* PerR regulates its own expression as well as that of *fur*, and both are repressed in the presence of manganese but not iron. Alteration of intracellular iron levels by mutation of *fur* leads to greater iron-dependent repression of PerR-regulated genes and a decreased repressive response to manganese (Fuangthong et al., 2002). This is consistent with the facts that (i) inactivation of *fur* leads to increased and uncontrolled iron uptake and (ii) PerR-dependent transcriptional regulation differs based on its two metallated forms.

Consensus DNA Binding Sites

Both Fur- and DtxR-like metalloregulators bind to specific iron box sequences located in the promoter regions of target genes (Table 2). Currently, three theories exist regarding the mechanism of iron box recognition by iron-dependent regulators in gram-positive bacteria. The first theory hypothesizes that Fur recognizes a fairly well-conserved 19-bp inverted repeat sequence (de Lorenzo et al., 1987). Likewise, DtxR dimers cover a 27-bp sequence in DNA

footprinting studies, and the consensus sequence recognized by each dimer was determined to be a 19-bp sequence that bears some similarities to the Fur box sequence (Boyd and Murphy, 1988). This theory is supported by studies demonstrating binding of Fur to synthetic oligonucleotides containing the 19-bp sequence. Furthermore, Fur recognizes and binds the exact 19-bp sequence within the promoter of the siderophore bacillibactin from *B. subtilis* (Bsat and Helmann, 1999; Calderwood and Mekalanos, 1988). However, this hypothesis did not explain why Fur interacts with such a large region of DNA, as other DNA-binding proteins containing helix-turn-helix motifs typically bind to a span of DNA only 12 bp in length (Harrison and Aggarwal, 1990). This apparent inconsistency led to the development of two additional hypotheses. The second theory suggests that Fur interacts with the hexamer sequence GATAAT, which is presented twice in a head-to-tail arrangement within the canonical 19-bp sequence (Escolar et al., 1998). Recognition of the shorter embedded sequence would allow Fur dimers to interact with the AT-AT core, a mechanism reminiscent of eukaryotic DNA-binding zinc finger proteins (Escolar et al., 1999). The final theory posits that two sets of Fur dimers interact with the consensus sequence by binding the two overlapping inverted repeat sequences consisting of a 7-1-7 heptamer motif within the 19-bp region, such that each dimer interacts with a sequence on opposite sides of the DNA strand (Baichoo and Helmann, 2002). Consistent with the final hypothesis, the crystal structure of DtxR bound to the *C. diphtheriae tox* promoter sequence demonstrates binding of the repressor as dimers on opposing faces of the double helix (White et al., 1998). Iron boxes recognized by most gram-positive metalloregulators exhibit sequence similarities, while some are quite divergent (Table 2). Most genes under the control of iron-dependent regulators contain an associated iron box, but this is not the case for every gene affected by Fur (Torres et al., 2010). This suggests that while the traditional mode of Fur-mediated repression appears to be the dominant mechanism of regulation, alternate routes of Fur-dependent control likely exist.

Table 2. Consensus sequences recognized by iron-dependent regulators in gram-positive bacteria

Regulator	Consensus sequence[a]
Fur	GATAATGATAATCATTATC
PerR	TTATAATNATTATAA
DtxR	TTAGGTTAGCCTAACCTAA
IdeR	TTAGGTTAGGCTAACCTAA
SirR	TTAGGTTAACCTAAACTTT
RitR	(A/T)NATTAN(A/T)(A/T)(A/T)R(A/T)YRR
MtsR	ATTAAGTTNANTTAAT
SloR	CTAATATAAAAATTAACTTGACTTAATTTTT ATATTAG
FurA	AGTCTTGACTGATTCCAGAAAAG

[a]Sequences located in the promoter regions of regulated genes are recognized and bound by the cognate metalloregulator under high-iron conditions. These iron boxes generally consist of repeats of AT-rich regions. N, any base; R, purine; Y, pyrimidine.

IRON-DEPENDENT VIRULENCE REGULATION IN *CORYNEBACTERIUM*

Corynebacterium diphtheriae

C. diphtheriae is the causative agent of diphtheria, a contagious upper respiratory tract infection that has been largely eradicated in the last century due to worldwide utilization of the diphtheria vaccines.

Diphtheria is a local infection primarily contained within the nasopharynx by the formation of a pseudomembrane. Both host and pathogen contribute to pseudomembrane formation through C. *diphtheriae*-induced necrosis of epithelial cells and an influx of the host's blood in response to the tissue damage. The pseudomembrane protects the pathogen while promoting colonization and contributes to respiratory stress characteristic of this disease. While C. *diphtheriae* does not invade other parts of the body, it produces virulence factors that enter circulation and exert a systemic effect. The principal virulence factor of C. *diphtheriae* is diphtheria toxin, which acts on host cells by inhibiting protein synthesis through ADP ribosylation of elongation factor 2. Diphtheria toxin is responsible for the pathogenesis of this disease and causes serious systemic complications, including myocarditis and polyneuritis.

Two critical determinants of C. *diphtheriae* pathogenesis are the establishment of colonization in the nasopharynx and the production of toxin. Iron limitation in the host plays an important role in both of these processes. First, the low iron availability signals association with the host and triggers dramatic changes in the gene expression profile. In response, bacterial factors that promote colonization and tissue distruction are upregulated. Remodeling of the bacterial surface, including upregulation of the lipoprotein Lrp1 and alteration of surface carbohydrates, occurs in response to iron deprivation (Moreira et al., 2003). Modification of the cell surface combined with increased biofilm production promotes adhesion of C. *diphtheriae* to host respiratory epithelial cells and erythrocytes, allowing for the establishment of colonization in the throat of the host. Regulation of iron-dependent factors required for C. *diphtheriae* colonization likely occurs through the diphtheria toxin regulator, DtxR. DtxR is primarily known to regulate expression of diphtheria toxin, the main virulence factor of C. *diphtheriae*. While the gene for diphtheria toxin, *tox*, is carried by the β family of corynebacteriophage, regulation of *tox* occurs through the chromosome-encoded metalloregulator (Fourel et al., 1989; Uchida et al., 1971). Under conditions of high iron, expression of *tox* is repressed, while in iron-limited environments such as the vertebrate host, transcription proceeds. Thus, toxin production is regulated such that it only occurs in the host.

In addition to regulating colonization and virulence factors in C. *diphtheriae*, DtxR controls expression of genes involved in iron acquisition. DtxR regulates the synthesis of corynebactin, a siderophore that removes iron from host transferrin (Russell et al., 1984; Tai et al., 1990). DtxR also controls expression of transport systems required to import both free iron and corynebactin-iron complexes (Kunkle and Schmitt, 2003; Qian et al., 2002). C. *diphtheriae* can utilize host heme as an iron source as well as free and siderophore-bound iron. To this end, a heme uptake system, encoded by *hmuTUV*, and a cytosolic heme oxygenase, *hmuO*, are regulated by the two-component signal transduction system *chrA-chrS* but remain under the iron-dependent control of DtxR (Chu et al., 1999; Drazek et al., 2000; Schmitt, 1997; Schmitt, 1999). Two-component systems typically consist of a membrane-bound histidine kinase that senses a signal and activates a response regulator, which then induces transcription of target genes to generate a response to the signal. Thus, DtxR is a crucial factor in the pathogenesis of C. *diphtheriae* that impacts nutrient acquisition as well as the establishment and progression of infection.

Corynebacterium pseudotuberculosis

C. *pseudotuberculosis* is an animal pathogen that causes infections of the lymphatic system in goats and sheep as well as cutaneous abscesses in the lower limbs of horses and cattle (Batey, 1986). While C. *pseudotuberculosis* primarily infects animals, it can also cause disease in humans, though this occurs rarely. Three forms of the disease exist, including cutaneous infection, ulcerative lymphangitis, and internal abscessation. During the course of infection, C. *pseudotuberculosis* can multiply in the extracellular milieu or in the intracellular environment of macrophages. C. *pseudotuberculosis,* therefore, is a versatile pathogen that can travel throughout the body and cause serious invasive disease (Hard, 1972; Tashjian and Campbell, 1983).

Iron-dependent regulation of C. *pseudotuberculosis* genes controls iron acquisition during infection and may impact expression of a major virulence factor. Host iron is obtained via a membrane transport system encoded by four genes termed *fagA, -B, -C,* and *-D* (for Fe acquisition genes) which exhibit homology to other gram-positive, iron-dependent ABC transporter genes (Billington et al., 2002). Interestingly, these genes are located downstream of the gene for phospholipase D, a major virulence factor of C. *pseudotuberculosis* that damages host cells and is leukotoxic (McNamara et al., 1994). As with iron uptake genes in other pathogens, the *fag* genes are expressed during iron starvation. A sequence with similarity to the DtxR consensus sequence is located upstream of the *fag* genes, suggesting a potential role for DtxR in directing the iron-dependent expression of this system. Though relatively little is known about iron-dependent regulation in C. *pseudotuberculosis*, a *fag* knockout is less virulent in a goat model of caseous lymphadenitis, demonstrating that iron acquisition is required for pathogenicity of *C. pseudotuberculosis* in vivo (Billington et al., 2002).

IRON-DEPENDENT VIRULENCE REGULATION IN STAPHYLOCOCCI

Staphylococcus aureus

S. aureus is a major human pathogen, causing significant morbidity and mortality worldwide. Infections caused by *S. aureus* range from relatively mild skin and soft tissue infections to life-threatening diseases such as toxic shock syndrome, meningitis, pneumonia, and septicemia (Weems, 2001). The prolific nature of *S. aureus* allows it to infect nearly every organ in the body and is responsible for the wide range of diseases caused by this pathogen.

Iron is central to the ability of *S. aureus* to cause disease. Accordingly, *S. aureus* has evolved redundant mechanisms for acquiring iron and has linked the expression of many genes to the availability of this metal. To regulate the expression of metal-dependent genes, *S. aureus* expresses three Fur paralogs, Fur, PerR, and Zur, and a DtxR ortholog, MntR, which is regulated by manganese and is not discussed in this chapter (Horsburgh et al., 2002). Fur is the major iron-dependent repressor of *S. aureus* that mediates the coordinated expression of multiple genes involved in host iron acquisition. These mechanisms include the production and secretion of siderophores and cell surface receptors specific for mammalian hemoproteins. Fur-dependent transcriptional regulation is key to the pathogenesis of *S. aureus* infections, as inactivation of *fur* significantly decreases virulence in murine infection models (Horsburgh et al., 2001b; Torres et al., 2010).

S. aureus produces two known siderophores that remove iron from host transferrin: staphyloferrin A, encoded by the *sfnABCD* operon, and staphyloferrin B, encoded by the *sbnABCDEFGHI* operon (Beasley et al., 2009; Cheung et al., 2009; Cotton et al., 2009). Deletion of siderophore synthesis genes greatly affects virulence of *S. aureus*, suggesting that siderophore production is essential for pathogenesis (Dale et al., 2004). *S. aureus* can also utilize siderophores from other bacterial species, including enterobactin, ferrichrome, and aerobactin (Maskell, 1980; Sebulsky and Heinrichs, 2001). Siderophore-iron complexes are imported into the bacterial cell by energy-dependent transporter systems. Staphyloferrin A is imported by the HtsABC transport system, which also imports heme-iron into the cell, while staphyloferrin B is transported by the SirABC system (Beasley

et al., 2009; Dale et al., 2004). A component of the ferric hydroxymate uptake system (*fhuCBG*), FhuC, is believed to be the ATPase that supplies energy to transport both staphyloferrin A and staphyloferrin B into the cell, since neither the *htsABC* nor the *sirABC* operon encodes an ATPase (Speziali et al., 2006). Furthermore, FhuCBG appears to be an indiscriminant transporter, as it is also responsible for the import of xenosiderophores, or siderophores produced by other bacterial species (Cabrera et al., 2001; Sebulsky et al., 2000; Speziali et al., 2006). *S. aureus* expresses an additional iron-regulated transport system, encoded by *sstABCD*, which imports catecholamine iron (Beasley et al., 2011). Expression of siderophore transport genes is induced by Fur-mediated derepression in response to a low-iron environment and repressed by Fur when sufficient iron levels are achieved (Heinrichs et al., 1999; Horsburgh et al., 2001b; Morrissey et al., 2000).

The iron-containing tetrapyrrole heme is the preferentially bound iron source of *S. aureus* (Skaar et al., 2004b). The majority of heme in the body is bound by hemoglobin, the tetrameric oxygen-carrying molecule found abundantly in circulating red blood cells (Fermi et al., 1984). *S. aureus* directs the acquisition of iron from host hemoglobin through a concerted series of events using secreted cytotoxins and the iron-regulated surface determinant (isd) system, all of which are upregulated under iron-limiting conditions (Skaar and Schneewind, 2004). *S. aureus* responds to the low-iron environment of the bloodstream and produces the membrane-damaging hemolysin α-toxin to lyse erythrocytes and liberate host hemoglobin (Torres et al., 2010). Free hemoglobin or hemoglobin complexed to haptoglobin is then bound by the staphylococcal hemoglobin receptor, IsdB, or the hemoglobin/haptoglobin receptor, IsdH (Torres et al., 2006). By an as-yet-unknown mechanism, heme is removed from hemoglobin and transferred to either IsdA or IsdC. IsdABCH contact hemoglobin and/or heme through near-iron transport (NEAT) domains, which are named based on the location of these genes with respect to iron transport systems in gram-positive bacteria (Andrade et al., 2002). Heme is then transported through the cell wall by IsdC to the IsdDEF membrane transporter, which pumps the molecule into the cytoplasm, where it is degraded by two heme oxygenases, IsdG and IsdI (Skaar et al., 2004b). In addition to functioning in heme acquisition, IsdA is an adhesin that binds fibrinogen and fibronectin, demonstrating that IsdA is a multifunctional staphylococcal virulence factor (Clarke et al., 2004).

Iron-dependent virulence gene expression in *S. aureus* involves a complex regulatory network comprised of Fur and the two-component systems Agr and Sae, which regulate quorum sensing and secreted virulence factors, respectively (Johnson et al., 2011; Voyich et al., 2005; Voyich et al., 2009). Intriguingly, there is evidence for a complex regulatory loop whereby Fur induces *sae* and *agr* transcription during iron deprivation, while Agr positively regulates *sae* transcription and represses *fur* transcription when it reaches sufficient levels. Under iron-limiting conditions, Fur and Sae positively regulate the expression of two cell surface proteins Eap and Emp, which are required for biofilm formation (Johnson et al., 2005). In addition to their roles in biofilms, these proteins are critical to the host-pathogen interaction due to their involvement in adhesion, inhibition of wound healing, cellular invasion, and promotion of invasive disease (Chavakis et al., 2002; Chavakis et al., 2005; Haggar et al., 2003; Hussain et al., 2001; Palma et al., 1999). Furthermore, Sae induces expression of some Isd genes, while Fur represses them in an iron-dependent manner. Fur also directly positively regulates Rot, a transcriptional regulator of toxin expression. The Fur-mediated regulation of *rot* indirectly affects transcription of the genes within its regulon, including hemolysins and other secreted effector proteins. Secretion of at least 58 exoproteins is affected by mutation of *fur*, illustrating the complexity of this regulon (Torres et al., 2010). Inactivation of *fur* causes several proteins to decrease in abundance, including proteins that interact directly with the host immune system, such as protein A, staphylococcal immunoglobulin G-binding protein A, superantigen-like proteins, and coagulase (Torres et al., 2010). In contrast, the *fur* mutant produces higher levels of several cytotoxins, hydrolases, and a number of immune clearance-inhibiting proteins. Moreover, many iron-dependent, Fur-regulated genes are expressed in vivo, including the *sir*, *sbn*, and *isd* operons, which are required for importing iron or heme (Allard et al., 2006). Fur, therefore, is an integral component of the response of *S. aureus* to the challenges faced in the host environment, with respect to both the acquisition of iron and the expression of virulence determinants required to promote disease and evade the host's immune system.

PerR, the peroxide response regulator, is an iron- and manganese-dependent repressor of genes involved in oxidative stress resistance. In many bacteria, PerR-mediated repression is lifted upon hydrogen peroxide exposure to counteract the damaging effects of oxidative stress. It has been observed, however, that increased iron levels induce expression of the PerR regulon in *S. aureus* (Horsburgh et al., 2001a). The *S. aureus* PerR regulon includes genes that protect against hydroxyl radicals and sequester iron, including catalase (KatA), alkyl hydroperoxide reductase

(AhpCF), ferritin (Ftn), bacterioferritin comigratory protein (Bcp), thioredoxin reductase (TrxB), and the Dps ortholog MrgA (Horsburgh et al., 2001a). Catalase expression is also regulated by Fur in *S. aureus*, though in this case transcription is activated instead of repressed in low iron (Horsburgh et al., 2001a; Xiong et al., 2000). Inactivation of *perR* in *S. aureus* renders the bacterium hyperresistant to oxidative stress and significantly decreases virulence in a murine skin abscess model of infection. These data are consistent with the role of PerR as a transcriptional repressor of oxidative stress response genes (Horsburgh et al., 2001a; Horsburgh et al., 2001b). *S. aureus perR* not only is autoregulatory but also affects expression of *fur* (Horsburgh et al., 2001b). Thus, PerR is a crucial mediator of both iron homeostasis and oxidative stress response in *S. aureus*.

Staphylococcus epidermidis

S. epidermidis is a member of the human skin microbiome but can become pathogenic when associated with indwelling medical devices, especially in immunocompromised patients (von Eiff et al., 2001). When *S. epidermidis* enters the bloodstream, it is capable of causing endocarditis or septicemia (Winston et al., 1983). One of the major virulence determinants of this organism is its ability to form biofilms. Biofilms consist of a slimy matrix made of carbohydrate polymers that are secreted by many strains of coagulase-negative staphylococci (Davenport et al., 1986; Peters et al., 1982). Formation of this matrix encases the bacteria and provides a niche to protect the organisms from antibiotics. *S. epidermidis* senses the low-iron environment of the host and responds by increasing slime production to aid in biofilm formation and promote infection (Deighton and Borland, 1993).

Much of what is known about iron-dependent transcriptional regulation in *S. epidermidis* involves the metalloregulator SirR, which is an iron-regulated, DtxR-like repressor (Hill et al., 1998). Like many other DtxR orthologs, SirR can be activated by iron or manganese in vitro (Hill et al., 1998). SirR controls iron acquisition by regulating the expression of the iron uptake system, SitABC (Hill et al., 1998). Moreover, it appears that iron acquisition and expression of virulence determinants may be linked in *S. epidermidis* and controlled by SirR. The *sitABC* operon displays homology to ABC operons found in *S. pneumoniae* that are required for cell adherence, genetic competence, and virulence (Dintilhac et al., 1997). Currently, little is known about the contribution of the SitABC system to *S. epidermidis* pathogenesis; however, it is hypothesized to be critical for the virulence of this opportunistic pathogen.

IRON-DEPENDENT REGULATION OF VIRULENCE IN BACILLI

Bacillus subtilis

While *B. subtilis* is not a human pathogen, work in this bacterium has laid the foundation for an understanding of metalloregulators in gram-positive pathogens. Similar to *S. aureus*, *B. subtilis* encodes four metalloregulators: three Fur-like regulators (Fur, PerR, and Zur) and MntR, a member of the DtxR regulator family (Bsat et al., 1996; Gaballa and Helmann, 1998; Que and Helmann, 2000). Fur is the primary iron-dependent regulator in *B. subtilis*, orchestrating expression of more than 20 operons (Baichoo et al., 2002). In some bacteria, Fur can act as a repressor or activator of transcription. However, *B. subtilis* Fur functions solely as a transcriptional repressor, as Fur-dependent activation of gene expression has not been demonstrated in this organism (Baichoo et al., 2002). The major systems regulated by Fur in *B. subtilis* include siderophore synthesis, iron acquisition systems, and several metabolic pathways (Baichoo et al., 2002). *B. subtilis* encodes several siderophores, including the catecholate siderophore bacillibactin and three hydroxamate siderophores, ferrichrome, ferrioxamine, and schizokinin (Schneider and Hantke, 1993).

PerR in *B. subtilis* can interact with manganese or iron and controls expression of oxidative stress resistance genes as well as heme biosynthesis (*hemAXCDBL*) and a zinc uptake system (*zosA*) (Bsat et al., 1996; Chen and Helmann, 1995; Chen et al., 1993). The ability of PerR to bind either manganese or iron allows for a competition between the two metals, resulting in two metal-coordinated forms of PerR, PerR:Fe and PerR:Mn. Manganese- and iron-bound PerR proteins exhibit different sensitivities to hydrogen peroxide (Herbig and Helmann, 2001, 2002). Expression of some members of the PerR regulon is affected by iron levels, including *mrgA* and *katA*, which are repressed in high iron, and *fur*, which is expressed at slightly higher levels of iron (Fuangthong et al., 2002). By this mechanism, *B. subtilis* is able to control responses to oxidative stress using a fine-tuned system driven by intracellular concentrations of manganese and iron.

Bacillus anthracis

B. anthracis is a highly virulent pathogen and the causative agent of anthrax, an easily spread and often fatal disease. *B. anthracis* is primarily a pathogen of domestic cattle, sheep, and goats, but it can infect humans that come in contact with infected animals or animal products. *B. anthracis* has gained much attention in recent years due to its potential for use as a biological weapon. Contact with *B. anthracis* spores can result in three different forms of anthrax disease depending upon the route of entry into the host. Cutaneous anthrax is the most common disease caused by *B. anthracis* and occurs through infection of a superficial wound by contaminated soil or animal products. The infection quickly forms a necrotic ulcer at the lesion site and can become systemic. Inhalational anthrax occurs when an individual inhales spores which then enter the respiratory tract, resulting in a very rapid spread to systemic illness that has a very high mortality rate. Finally, gastrointestinal anthrax occurs upon contact with spores following ingestion of undercooked, infected meat and is frequently fatal. In all three routes of infection, spores are phagocytosed by macrophages and travel to lymph nodes, where they germinate, divide rapidly, and spread throughout the body (Lincoln et al., 1965; Smith et al., 1954).

Genes believed to encode three metalloregulators, Fur, PerR, and Zur, have been identified in the *B. anthracis* genome, and studies are now beginning to elucidate the roles of these regulators in iron acquisition and infection (Cendrowski et al., 2004; Read et al., 2003). Analysis of genes affected by iron deprivation in the *B. anthracis* strain Sterne, a nonencapsulated, nonpathogenic laboratory strain, revealed immense alterations in the transcriptome that affect the global processes of metabolism, uptake systems, and transcription factors (Carlson et al., 2009). The pathogenic Ames strain genome encodes several iron acquisition genes, including 15 ABC transport systems involved in uptake of iron and iron-bound siderophores (Read et al., 2003). Anthrabactin, encoded by *bac*, is homologous to the *B. subtilis* catechol siderophore bacillibactin. Petrobactin, also called anthrachelin, is the siderophore encoded by *asbAB* and is a homolog of the *E. coli* hydroxamate siderophore aerobactin (Cendrowski et al., 2004; Pfleger et al., 2008; Read et al., 2003). Both siderophore systems are believed to be regulated by Fur in an iron-dependent fashion, as operators of these genes contain putative Fur boxes (Cendrowski et al., 2004). Anthrabactin and petrobactin are important for iron acquisition, but only the latter is required for growth in macrophages and in a mouse model of anthrax (Cendrowski et al., 2004).

B. anthracis also encodes machinery to import and utilize host heme as an iron source. Comparative genomic analyses have revealed the presence of three operons in the *B. anthracis* genome that exhibited homology to the *S. aureus* Isd system described above (Skaar et al., 2006). The largest *B. anthracis isd* operon encodes two NEAT domain-containing proteins designated IsdX1 and IsdX2. Also contained in this operon are *isdCEF*, while *isdG* and *srtB* are expressed in separate transcriptional units. Most *B. anthracis isd* genes resemble their staphylococcal counterparts; however, IsdX1 and IsdX2 are not homologous to any *S. aureus* Isd protein (Gat et al., 2008; Maresso et al., 2008). Unlike members of the *S. aureus* Isd system, which are cytoplasmic or bound to the membrane or cell wall, *B. anthracis* IsdX1 and IsdX2 are secreted hemophores that bind host hemoglobin and remove the heme moiety (Maresso et al., 2008). These proteins, therefore, represent the first identified secreted hemophores in gram-positive bacteria (Maresso et al., 2008). Exogenous heme can also be acquired through the *B. anthracis* S-layer protein K (BslK), an additional NEAT domain-containing protein that is found on the bacterial surface (Tarlovsky et al., 2010). Heme acquired by IsdX1, IsdX2, or BslK is passed to cell wall-anchored IsdC and internalized through the membrane transport system, IsdEF (Fabian et al., 2009). The *B. anthracis isd* operons appear to be regulated at least in part by Fur, as canonical 19-bp Fur boxes have been identified in the promoter regions (Gat et al., 2008). Regulation of the Isd system may be more complex, however, as these operons are repressed by high levels of carbon dioxide as well as external iron concentration (Gat et al., 2008). The full impact of the Isd system on the pathogenesis of anthrax is not well understood. Protective antisera specific for IsdC, X1, and X2 are produced during infection with *B. anthracis*, indicating that these proteins are expressed in vivo and recognized by the immune system. However, the Isd system does not appear to be required for full virulence of *B. anthracis* in murine or guinea pig anthrax infection models, though these experiments were performed using the attenuated strain Sterne (Gat et al., 2008; Skaar et al., 2006).

Aside from nutrient acquisition, *B. anthracis* also regulates many factors required for its life cycle in response to iron levels. For example, the expression of two putative internalins, annotated as GBAA0052 and GBAA1346, is upregulated in iron limitation (Carlson et al., 2009). Internalins are expressed by several pathogens and are most well studied in *Listeria monocytogenes*. These proteins are important for adhesion and invasion of host cells (Bierne et al., 2007). The internalin genes in bacilli are found exclusively in pathogenic members of the genus, suggesting that they may be important for pathogenesis (Yu, 2009). Indeed, inactivation of GBAA0052 or GBAA11346 results in significantly decreased virulence in a murine model of anthrax, highlighting the importance

of iron-regulated genes in infection (Carlson et al., 2009). These studies suggest that *B. anthracis* alters its transcriptome in response to the iron-limiting environment of the host to promote the intracellular phase of its infection cycle.

Bacillus cereus

B. cereus, a bacterium closely related to *B. anthracis*, is a human pathogen that is a common cause of food poisoning and one of the leading causes of endophthalmitis (Kotiranta et al., 2000; Kramer and Gilbert, 1989). An isolate of *B. cereus* has been identified that has acquired two *B. anthracis* plasmids and causes anthrax-like disease in humans (Hoffmaster et al., 2004). The major regulator of virulence determinants in *B. cereus* is the phospholipase C regulator, PlcR, though Fur, PerR, and Zur orthologs have been identified (Harvie et al., 2005; Lereclus et al., 2000). As with other metalloregulators, *B. cereus* Fur regulates iron uptake and storage proteins. Genes encoding several virulence factors contain Fur boxes in their promoter regions, including two adhesin-like cell surface proteins and two putative internalins, which are likely to be involved in host cell adhesion and invasion, respectively (Harvie et al., 2005). *B. cereus* Fur may also regulate expression of hemolysin (HlyII), one of the many membrane-damaging toxins produced by this pathogen (Kotiranta et al., 2000). Though relatively little is known about iron-dependent regulation in *B. cereus*, it is clear that Fur-dependent iron acquisition plays an important role in pathogenesis, as a strain inactivated for *fur* exhibits reduced virulence in an insect model of infection (Harvie et al., 2005).

IRON-DEPENDENT VIRULENCE REGULATION IN STREPTOCOCCI

Streptococcus pneumoniae

S. pneumoniae is a major cause of pneumonia, septicemia, and meningitis (Lim et al., 2001). Pneumococci and other respiratory pathogens likely encounter unique nutrient challenges in the in vivo environments of the lung and peritoneal cavity, which are hypothesized to represent iron-limiting conditions. Three genes have been identified that are activated during in vitro iron-limiting conditions as well as in an in vivo mouse infection model (Marra et al., 2002). These genes encode a polypeptide deformylase (AE007449), a hypothetical protein resembling streptococcal carbonic anhydrases (AE007320), and a hypothetical protein with unknown function (ISIP456)

(Marra et al., 2002). Expression of genes under both conditions suggests that they may be activated in vivo due to a lack of available iron. While the iron-responsive nature and role of these proteins during infection are not known, inactivation of these genes either attenuates virulence or is essential for pneumococcal growth (Marra et al., 2002).

S. pneumoniae can utilize free ferrous and ferric iron as well as heme or heme-containing proteins as an iron source, but it is unable to access the metal from host transferrin, lactoferrin, or ferritin (Brown et al., 2001; Tai et al., 1993). As it appears that *S. pneumoniae* does not express iron-scavenging siderophores, this pathogen is believed to rely on other mechanisms of iron acquisition in vivo (Tai et al., 1993; Tettelin et al., 2001). The pneumococcal surface protein PspA binds human lactoferrin, but it does not appear to obtain iron for growth from this interaction (Hammerschmidt et al., 1999; Tai et al., 1993). *S. pneumoniae* expresses three ABC transporter operons to import host iron, termed *piuBCDA* (pneumococcal iron uptake), *piaABCD* (pneumococcal iron acquisition), and *pitADBC* (pneumococcal iron transporter) (Brown et al., 2001; Brown et al., 2002). Mutation of both the *piu* and *pia* operons results in an inability to use hemoglobin as an iron source, suggesting that iron acquisition occurs via removal of heme-iron from host hemoglobin (Brown et al., 2001). Ultimately, the Pia transport system appears to be the principal iron transporter in *S. pneumoniae*, as inactivation of this operon results in the greatest loss of iron uptake and virulence in mouse pneumonia and septicemia models of infection (Brown et al., 2001; Brown et al., 2002). An orphan response regulator, RitR (repressor of iron transport), interacts with operator sites within the *piu* and potentially *pia* operons and is believed to control transcription of these genes, though it is not known if repression is iron dependent (Ulijasz et al., 2004). Consistent with its role in pneumococcal iron homeostasis, inactivation of *ritR* leads to loss of virulence in a murine pneumonia model (Throup et al., 2000; Ulijasz et al., 2004).

Iron restriction also affects pneumococcal morphology, since cells incubated in iron-limited media form long chains instead of diplococci (Gupta et al., 2009). Growth in the presence of hemoglobin induces the formation of cellular aggregates that form a biofilm-like extracellular matrix. Exposure of pneumococci to ferritin, heme, or hemoglobin results in altered expression of several virulence factors. Expression of genes encoding capsule biosynthesis, an extracellular proteinase, and several toxins is upregulated under these conditions (Gupta et al., 2009). PspA, which binds complement and lactoferrin, is downregulated in hemoglobin but upregulated

in other iron sources (Gupta et al., 2009). Moreover, several of these genes are also upregulated during nasopharyngeal colonization and bacteremia in a mouse model of pneumococcal infection (Gupta et al., 2009). These data illustrate the vast array of transcriptional changes that occur in response to the lack of iron or, conversely, the interaction with host-derived iron sources.

Streptococcus pyogenes (GAS)

Streptococcus pyogenes (group A streptococcus [GAS]) is an important human pathogen, causing a wide array of diseases ranging from the relatively harmless pharyngitis and impetigo to invasive conditions, including necrotizing fasciitis and toxic shock syndrome (Cunningham, 2000). During invasive disease, GAS encounters components of the bloodstream and utilizes numerous iron sources for growth, including heme, hemoglobin, ferritin, and myoglobin. Though GAS does not produce siderophores, its genome contains three known iron and heme acquisition systems. Expression of acquisition systems is controlled by two GAS metalloregulators, the Fur paralog PerR and the DtxR family member MtsR (multimetal transport system regulator) (Hanks et al., 2005).

GAS PerR controls expression of genes encoding a variety of functions. Inactivation of *perR* decreases expression of the cold shock protein Csp, the iron homeostasis protein MtsA, and PerR-regulated metal transporter A (PmtA) (Brenot et al., 2005; Brenot et al., 2007; Ricci et al., 2002). PerR responds to oxidative stress by inducing expression of hydrogen peroxide resistance genes, including a Dps (DNA-binding protein from starved cells)-like peroxide resistance protein, MrgA, and SodA, the only superoxide dismutase gene found in the GAS genome (Ricci et al., 2002). Regulation of these genes allows PerR to protect the cell from oxidative damage by (i) downregulating iron uptake systems to limit the amount of iron available to participate in the Fenton reaction and (ii) enhancing the peroxide resistance response. While GAS PerR appears to respond primarily to hydrogen peroxide, its regulation in response to metal ion levels in the cell is not known (Ricci et al., 2002). Nevertheless, PerR is critical to GAS pathogenesis, since inactivation results in a 10-log reduction of virulence in a murine skin air sac infection model (Ricci et al., 2002).

The iron- and manganese-dependent metalloregulator MtsR controls transcription of iron uptake genes in GAS. The streptococcal iron acquisition (*sia*) operon (also called *htsABC*) encodes a system that removes iron from heme-containing proteins.

This system consists of Shr, a receptor specific for hemoproteins, Shp, which interacts with the heme molecule, and an ABC transporter that is believed to pass heme through the membrane (Bates et al., 2003). MtsR represses transcription of the *sia* operon in the presence of high levels of iron and manganese (Bates et al., 2005; Hanks et al., 2006). The *sia* operon is important for GAS iron acquisition, since hemoglobin binding and iron uptake are both significantly decreased upon deletion of this genetic region (Bates et al., 2003). Additional iron uptake genes have been identified in GAS, including the *mts* operon, which imports iron, zinc, and manganese, and the *siu* operon (streptococcal iron uptake), which encodes an additional heme transport system (Bates et al., 2005; Janulczyk et al., 1999; Janulczyk et al., 2003; Montanez et al., 2005). A member of the Mts transport system, MtsA, has been reported to be affected by PerR as mentioned previously, but it is also repressed by MtsR in response to manganese or iron (Hanks et al., 2006). Inactivation of *sia*, *siu*, or *mtsR* leads to loss of virulence in infection models, suggesting that iron uptake and regulation are important for GAS pathogenesis (Hanks et al., 2006; Montanez et al., 2005).

Perhaps one of the most well-known virulence factors of GAS is the M protein, an important cell surface protein involved in evasion of phagocytosis and establishment of disease (Fischetti, 1989; Fischetti et al., 1988; Lancefield, 1962). Transcription at the *emm* locus, which encodes the M protein, is controlled in response to carbon dioxide by the two-component response regulator Mga (Caparon et al., 1992). However, both *mga* and *emm* genes are positively regulated by MtsR (Toukoki et al., 2010). Interestingly, M protein expression negatively correlates with iron deprivation, suggesting that production of M protein could be downregulated in the iron-limited environment of the host (McIver et al., 1995). Taken together, these data highlight the importance of iron-dependent regulation to the pathogenesis of GAS infections (Janulczyk et al., 2003).

Sequestration of iron by human transferrin and lactoferrin alters the expression of the streptococcal pyrogenic exotoxins (Spe), important GAS virulence factors that contribute to the development of invasive disease (Cunningham, 2000; Kansal et al., 2005). SpeA is a superantigen, while SpeB is a cysteine protease that appears to be important for modifying both bacterial and host proteins (Ashbaugh and Wessels, 2001; Kotb, 1995). The relative levels of *speA* and *speB* correlate with virulence of invasive M1T1 GAS isolates, as low levels of *speB* expression are associated with more severe disease (Kansal et al., 2005). The iron-binding proteins are responsible for inducing an in vivo shift in expression of the Spe proteins from the original *speB*^{high} *speA*^{low} phenotype to *speB*^{low} *speA*^{high} by day 7 of a murine infection (Kazmi et al., 2001). Importantly, only the metal-coordinated state of the transferrin or lactoferrin is required to elicit this response, as only apotransferrin (stripped of iron) and not the iron-bound holotransferrin induces expression of *speA*. Furthermore, *S. pyogenes* is not known to utilize either transferrin or lactoferrin as an iron source for growth. This suggests that transferrin-mediated iron sequestration is responsible for the alterations in *speA* and *speB* expression. The downshift of *speB* expression has a profound impact on bacterial pathogenesis, preventing phagocytosis and promoting colonization (Kansal et al., 2005). Therefore, in response to iron limitation by the host, GAS upregulates specific virulence factors that promote survival of the bacteria in vivo (Kansal et al., 2005; Welcher et al., 2002).

Streptococcus agalactiae (GBS)

Streptococcus agalactiae (group B streptococcus [GBS]) is occasionally found in the human intestinal microflora and can also become part of the vaginal microbiota of up to 40% of women. GBS is a leading cause of invasive infections in neonates following transfer of the pathogen from the mother to the infant. Infants that contract GBS can develop often fatal cases of pneumonia, meningitis, or septicemia. GBS can also cause serious disease in adults, though this occurs less frequently than in infants (Johri et al., 2006).

While iron-dependent transcriptional regulation has not been studied as extensively in GBS as in other gram-positive pathogens, some orthologous systems have been identified. Similar to GAS, GBS expresses the *mtsABC* operon, which encodes a metal transport system (Janulczyk et al., 2003). Like other ABC transport systems, the GBS MtsABC system includes a ligand-binding lipoprotein, membrane permease, and ATP-binding proteins that are required for the import of both manganese and iron (Bray et al., 2009). A potential regulatory gene, *mtsR*, is located adjacent to the *mtsABC* operon and is believed to control transcription of this locus in a mechanism similar to that of its GAS DtxR-like ortholog (Bates et al., 2005). In an effort to control the levels of intracellular manganese and iron, the expression of *mtsABC* is repressed by high levels of both metals, and presumably this occurs through MtsR (Bray et al., 2009). While the contribution of the MtsABC system to GBS pathogenesis has not been defined, this system is hypothesized to be important for virulence, similar to its homologous ABC transporter in GAS.

Streptococcus mutans

S. mutans is a common oral pathogen involved in dental caries, but it has the capacity to cause fatal cases of endocarditis if it enters the bloodstream (Dall and Herndon, 1990). The principal iron-dependent regulator of *S. mutans* is SloR, a metalloregulator that is an ortholog of the DtxR-like GAS MtsR (Bates et al., 2005). SloR represses transcription of *sloC*, an adhesin and metal transporter, in high iron or manganese (Kitten et al., 2000). SloC also contributes to biofilm formation and is an important virulence determinant in a rat model of endocarditis (Rolerson et al., 2006). While SloR negatively regulates iron acquisition, it also activates several virulence genes, including the competence regulator genes *comCDE*; the critical stress response/biofilm architecture protein gene *ropA*; the superoxide dismutase gene *sod*; *spaP*, which encodes a sucrose-independent adherence factor involved in plaque formation; and genes for two sucrose-dependent adhesion factors, *gtfB* and *gdpB* (Rolerson et al., 2006). Therefore, SloR of *S. mutans* affects expression of iron import and virulence genes and is likely important for the establishment of oral biofilms as well as the transition to invasive disease.

Streptococcus suis

S. suis causes meningitis and septicemia in swine and is therefore a zoonotic pathogen with the ability to cause meningitis in humans (Arends and Zanen, 1988; Clifton-Hadley, 1983). The role of iron-dependent regulation in *S. suis* infection is not entirely known, but recent studies have begun to uncover the transcriptional alterations in response to iron depletion. Iron starvation affects 23 genes involved in metabolism and 8 involved in cell wall construction (Li et al., 2009). In addition, three putative *S. suis* virulence factors are upregulated under iron-limiting conditions: *relA*, *arcA*, and *cpdB* (Li et al., 2009). RelA is a GTP pyrophosphokinase involved in the stringent stress response and is important for amino acid synthesis (Jain et al., 2006). RelA contributes to stress adaptation within the host in several other pathogens, including *M. tuberculosis*, *Vibrio cholerae*, and *L. monocytogenes* (Haralalka et al., 2003; Primm et al., 2000; Taylor et al., 2002). ArcA, an arginine deaminase, generates energy by converting arginine to ornithine, ammonia, and carbon dioxide (Li et al., 2009). Other pathogens that can orchestrate such a reaction, including *S. pyogenes*, can better withstand oxidative stress and acidic environments (Casiano-Colón and Marquis, 1988; Degnan et al., 2000). CpdB (2'3'-cyclic nucleotide 2'-phosphodiesterase) is a surface protein that likely interacts with integrins

on host cells during infection (Osaki et al., 2002). Approximately one-quarter of the genes affected by iron deprivation in this study appeared to have Fur boxes in their promoter regions, and the mechanism by which the remaining genes are regulated in response to iron is not known (Li et al., 2009). It is likely, however, that these proteins act as iron-dependent virulence factors and impact *S. suis* infection in vivo. Though the entire Fur regulon in *S. suis* is not known, inactivation of *fur* leads to significantly decreased virulence in a mouse model of infection, suggesting that iron-dependent Fur regulation is critical to the pathogenesis of this organism (Aranda et al., 2010).

IRON-DEPENDENT VIRULENCE REGULATION IN *LISTERIA*

Listeria monocytogenes

L. monocytogenes is the causative agent of listeriosis, a severe and often life-threatening foodborne infection (Farber and Peterkin, 1991). Several manifestations of listeriosis can occur, including pneumonia, meningitis, encephalitis, septicemia, and spontaneous abortion. *L. monocytogenes* is an intracellular pathogen that responds to low-iron environments by altering transcription of iron and heme-iron uptake genes, iron storage molecules, and several important virulence factors.

While *L. monocytogenes* does not produce siderophores, it is capable of using exogenous siderophores from other bacteria. *L. monocytogenes* chiefly acquires host iron through the use of two ABC transporter systems: the ferric hydroxamate uptake (Fhu) system and the hemin/hemoglobin uptake (Hup) system (Jin et al., 2006). *Listeria* can therefore utilize host hemoglobin, transferrin, or ferritin as an iron source, as well as xenosiderophores (Simon et al., 1995). Acquisition of iron from host hemoglobin is crucial for pathogenesis, as inactivation of the *hup* operon results in a 50-fold decrease in virulence in a mouse model of listeriosis (Jin et al., 2006).

Aside from the Hup uptake system, *L. monocytogenes* may be able to acquire host heme through another mechanism. The *svpA-srtB* operon encodes two putative heme-binding proteins, Lmo2186 and SvpA, a ferric hydroxymate ABC transport system, and sortase B, which anchors proteins to the peptidoglycan layer (Newton et al., 2005). The limited sequence similarity of Lmo2186 (also called heme/hemoglobin binding protein, Hbp1) and SvpA (Hbp2) to IsdC of *S. aureus* as well as their shared SrtB-dependent cell wall attachment suggests that both Lmo2186 and SvpA may be involved in heme binding. SrtB attaches SvpA to the cell wall under iron-limiting conditions, while SvpA is secreted in iron-replete environments, suggesting that it functions as a hemophore that scavenges exogenous heme (Newton et al., 2005; Xiao et al., 2011). Moreover, the *svpA-srtB* operon in *L. monocytogenes* is regulated by Fur in an iron-dependent manner. However, while SvpA has been shown to interact with heme in solution, it only appears to be required for heme uptake when extracellular heme levels are low, whereas the Hup transport system is the principal heme acquisition mechanism in high heme (Xiao et al., 2011). Nevertheless, heme acquisition via SvpA contributes to pathogenesis of *L. monocytogenes* in the iron- and heme-restricted environment of the host, as inactivation of *svpA* increases the 50% lethal dose by 2 logs (Borezee et al., 2000; Newton et al., 2005).

L. monocytogenes must protect itself from the potentially damaging effects of iron that it obtains during infection. To this end, upregulation of *fri*, a gene encoding a ferritin-like Dps protein, is induced to store free iron in the cell. Fri expression is induced in limited iron and is controlled by PerR and Fur, which repress transcription in the presence of iron (Fiorini et al., 2008; Olsen et al., 2005; Polidoro et al., 2002). Fri promotes growth under iron-limiting conditions and defends against oxidative stress (Olsen et al., 2005). Inactivation of *fri* results in a 10-fold reduction in intracellular survival in murine macrophage-like J774.A1 cells and significantly reduces growth of *L. monocytogenes* in an in vivo mouse model of infection (Olsen et al., 2005).

Alteration of host iron levels significantly impacts the infection cycle of *L. monocytogenes*. Adherence and invasion of *L. monocytogenes* into Caco-2 intestinal epithelial cells are increased in response to high iron levels and likely occur due to increased transcription of the internalin (*inlAB*) operon (Conte et al., 1996). Conversely, iron-limited conditions significantly inhibit internalization of *L. monocytogenes* into Caco-2 cells within the first 3 h of incubation (Conte et al., 1996). However, bacteria that are able to enter the host cells upregulate expression of the listerial surface protein ActA in response to iron deprivation. Actin polymerization by ActA on the bacterial surface drives intra- and intercellular movement and promotes pathogenicity of *L. monocytogenes* (Conte et al., 2000; Cossart, 1995; Smith and Portnoy, 1997; Tilney and Portnoy, 1989). Indeed, *actA* expression is increased over 10-fold in iron-limited environments compared to iron-rich conditions (Conte et al., 2000). In addition, expression of the major listerial virulence factor listeriolysin O, which is required for intracellular survival, is affected by iron levels. Listeriolysin

O is a cholesterol-dependent cytolysin critical for the phagosomal escape of *L. monocytogenes*, and its expression is upregulated fivefold in response to iron limitation (Cowart and Foster, 1981). These observations demonstrate that *L. monocytogenes* significantly alters its transcriptome in response to changing iron levels to obtain iron and promote intracellular survival and progression of infection.

IRON-DEPENDENT VIRULENCE REGULATION IN *CLOSTRIDIUM*

Some members of the genus *Clostridium* are pathogens that cause severe and often fatal anaerobic infections. Little information is available regarding iron acquisition and regulation of virulence in *Clostridium* in response to iron limitation, though some studies are beginning to approach the subject. Heme oxygenases have been identified in *Clostridium perfringens*, *C. tetani*, and *C. novyi*, suggesting that these pathogens utilize heme-iron for growth (Brüggemann et al., 2004; Hassan et al., 2010). The heme oxygenase in *C. perfringens*, encoded by *hemO*, is expressed in response to increasing heme concentrations and repressed by the addition of iron (Hassan et al., 2010). Transcription of *hemO* is positively regulated by the principal *C. perfringens* virulence gene regulator VirR/VirS-VR-RNA, though the dependence on iron for regulation of *hemO* has not been defined (Hassan et al., 2010). Clostridial heme oxygenases are thought to scavenge available oxygen to promote an anoxic environment suitable for the progression of infection. The presence of heme oxygenases in clostridia suggests that these pathogens also have mechanisms to import host heme. Indeed, genes encoding a putative Isd system similar to that expressed by *S. aureus* have been identified in *C. tetani* (Skaar et al., 2006). The confirmed function and role in pathogenesis of the *C. tetani* Isd system have not been determined.

IRON-DEPENDENT VIRULENCE REGULATION IN *ACTINOMYCES*

Actinomyces naeslundii

Similar to *S. mutans*, *A. naeslundii* is a common oral commensal bacterium that causes dental caries and periodontitis (Ellen et al., 1985; Schüpbach et al., 1995). Iron uptake via expression of the siderophore synthesis gene *sid* is regulated by the DtxR-like metalloregulator AmdR (*A*ctinomyces *m*etal-*d*ependent *r*epressor) (Moelling et al., 2007).

AmdR also regulates the production of biofilm slime, which is increased under low-iron conditions. Biofilm formation is a crucial virulence determinant that promotes aggregation and adherence of *A. naeslundii* to the host's dentition. Transcriptional regulation by AmdR, therefore, is thought to contribute to pathogenesis of *A. naeslundii*.

IRON-DEPENDENT VIRULENCE REGULATION IN MYCOBACTERIA

Due to the unique characteristics of their cell walls, mycobacteria are not classified as gram positive or gram negative. Mycobacteria are acid-fast because, upon staining with phenolic-containing dyes (e.g., carbol fuchsin), they resist decolorization by dilute acid and ethanol-based solutions. The mycobacterial cell wall consists of peptidoglycan and various lipids (the most abundant of which is mycolic acid), which contribute to its waxy, impermeable nature. Nevertheless, a discussion of iron-dependent regulation of virulence determinants in mycobacteria is included in this chapter since these pathogens most closely resemble gram-positive organisms.

Mycobacterium tuberculosis

M. tuberculosis is the causative agent of tuberculosis, a widespread respiratory disease responsible for significant morbidity and mortality, especially in AIDS patients and those who are otherwise immunocompromised. *M. tuberculosis* spreads through aerosolized droplets that enter the lungs of the host and are taken up by alveolar macrophages. *M. tuberculosis* prevents macrophage killing by inhibiting phagosome-lysosome fusion and multiplies inside the host cell. While the host immune system attempts to control the infection, *M. tuberculosis* can establish a latent, asymptomatic infection that can reactivate at a later date (reviewed in Glickman and Jacobs, 2001). Similar to *L. monocytogenes*, *M. tuberculosis* is an intracellular pathogen that must acquire iron from within host cells. Mycobacterial siderophore production is greatly enhanced inside macrophages, highlighting the lack of iron in the phagosome and the importance of iron acquisition to the intracellular survival of the bacterium (De Voss et al., 2000). In an effort to overcome this limitation, *M. tuberculosis* hijacks the host iron acquisition pathways during fusion of mycobacterium-containing phagosomes with vesicles containing iron-loaded transferrin (Clemens and Horwitz, 1996; Sturgill-Koszycki et al., 1996).

An analysis of the *M. tuberculosis* transcriptome revealed 155 open reading frames that are affected by changes in iron concentration (Rodriguez et al., 2002). Under iron-rich conditions, iron storage genes *bfrA* and *bfrB*, which encode bacterioferritin and ferritin, respectively, and *katG*, a catalase-peroxidase, are expressed at higher levels, illustrating the need to protect the cell from free iron (Rodriguez et al., 2002). In contrast, iron deprivation leads to upregulation of genes encoding a putative metal transport system consisting of IrpA and MtaA (Calder and Horwitz, 1998). Iron availability in mycobacteria has the potential to drastically alter the physiological state of the cell beyond the effort to import and store iron. Two sigma factors, sigma E and sigma B, are more highly expressed under iron-limiting conditions (Rodriguez et al., 2002). In addition, the *mtrA* gene, encoding a two-component response regulator, is expressed in response to iron and is essential to *M. tuberculosis* (Rodriguez et al., 2002; Zahrt and Deretic, 2000). The consequences of altered expression of sigma E, sigma B, and *mtrA* in response to iron levels have not yet been elucidated.

M. tuberculosis encodes four known metalloregulators: IdeR (iron-dependent regulator), SirR (staphylococcal iron regulatory repressor), FurA, and FurB (Cole et al., 1998). IdeR and SirR are members of the DtxR family of metal-dependent regulators, while FurA and FurB belong to the family of Fur orthologs. Since the role of SirR in *M. tuberculosis* is not entirely known and is likely involved in manganese-dependent regulation, and FurB is a zinc-dependent regulator, neither protein is further discussed in this chapter (Canneva et al., 2005; Rodriguez, 2006).

IdeR is a DtxR ortholog unique to mycobacteria and is essential for growth. IdeR can be activated in vitro by iron, manganese, zinc, cobalt, nickel, or manganese, though iron is the preferred metal activator (Rodriguez and Smith, 2003). IdeR controls several important virulence factors of *M. tuberculosis*, including genes involved in synthesis of the mycobacterial siderophore mycobactin (*mbtA*, *mbtB*, and *mbtI*) and exochelin in *M. smegmatis* (*fxbA*), bacterioferritin (*bfd*, *bfrA*; long-term iron storage), and a histidine synthesis gene (*hisE*) (Dussurget et al., 1999; Gold et al., 2001; Rodriguez et al., 2002; Rodriguez et al., 1999). Two types of mycobactins are produced by pathogenic mycobacteria: cell wall-bound mycobactin and secreted, soluble carboxymycobactin (Ratledge, 1999). It is not fully understood why mycobacteria produce both secreted and membrane-bound siderophores, but it has been demonstrated that extracellular iron scavenged by carboxymycobactin can be passed to surface-bound mycobactins, suggestive of a cooperative system of iron acquisition (Gobin and Horwitz, 1996). Mycobactins are believed to be important for iron acquisition during the intracellular life cycle of mycobacteria. This point is highlighted by the fact that mycobactin synthesis genes *mbtJ* and *mbtI* are expressed under iron-limited conditions inside bone marrow-derived macrophages (Rachman et al., 2006). Iron that is scavenged by mycobactins is transported into the cytoplasm by ABC transporters, including IrtAB, which primarily transports iron-bound carboxymycobactins (Rodriguez and Smith, 2006). IrtAB is critical for iron acquisition in mycobacteria, as inactivation of the transport system results in significantly decreased intracellular survival and virulence in a mouse model of infection (Rodriguez and Smith, 2006). *M. tuberculosis* may also acquire another form of extracellular iron, ferric dicitrate, through putative transporters homologous to *E. coli* FecB, called FecB and FecB2 (Rodriguez, 2006). Salicylic acid, a low-affinity, nonsiderophore iron-binding molecule, is produced by mycobacteria in an iron-dependent fashion, though it is not believed to be a major contributor to iron acquisition in *M. tuberculosis* (Ratledge and Winder, 1962).

Positive and negative regulation by IdeR can occur and depend on the location of the iron box; genes repressed by IdeR, such as *mbt* genes, contain a single iron box in the promoter region, whereas genes activated by IdeR, such as *bfrA* and *bfrB*, contain two iron boxes located further upstream of the gene (Gold et al., 2001). It is hypothesized that the location of the iron box in relation to the transcriptional start site of IdeR-regulated genes affects the ability of RNA polymerase to engage the promoter region and initiate transcription (Rodriguez, 2006). That is, IdeR bound to an iron box in the promoter region blocks RNA polymerase binding, while IdeR bound to iron boxes further upstream allow RNA polymerase to contact the promoter and transcribe downstream genes.

Intriguingly, several IdeR-regulated genes have been identified that do not appear to be directly linked to iron, including *acpP* and *murB*, two genes required for lipid and membrane synthesis (Gold et al., 2001). These results suggest that *M. tuberculosis* may undergo membrane remodeling in response to the low-iron environment of the host. In support of this hypothesis, iron limitation induces alterations in lipid biosynthesis and metabolism. For example, MmpL4, an important lipid transporter required for cell surface remodeling, is regulated by IdeR in response to iron (Bacon et al., 2007; Camacho et al., 1999; Rodriguez et al., 2002). The waxy mycobacterial cell envelope consists of peptidoglycan, arabinogalactan, and mycolic acid as well as a variety of lipids, and it is an important contributor to pathogenesis (Cox et al., 1999; Minnikin et al., 2002; Reed et al., 2004). Therefore, iron-dependent alterations

at the cell surface could have a significant impact on pathogenesis of *M. tuberculosis*.

Some mycobacterial secretion systems are also affected by iron deprivation. A type VII secretion system in *M. tuberculosis*, Esx-3, is expressed under the iron-dependent regulation of IdeR (Rodriguez et al., 2002). Esx-3 appears to be essential in *M. tuberculosis*, as attempts to delete this system have been unsuccessful (Siegrist et al., 2009). However, a conditional knockout strain has been generated in *M. bovis* BCG, and the *esx-3* gene was not essential in the nonpathogenic *M. smegmatis*. Using these strains, it was determined that Esx-3-dependent secretion of EsxG and EsxH, two small secreted proteins encoded by *esx-3*, occurs in response to iron limitation. Loss of *esx-3* does not affect expression or secretion of mycobacterial siderophores but instead results in decreased growth under iron-limiting conditions that cannot be rescued by the addition of iron-loaded mycobactin. This result suggests that Esx-3 is important for import of iron-bound mycobactins. Iron-dependent expression and activity of Esx-3 are required for pathogenesis, as inactivation of *esx-3* significantly reduces intracellular survival in J774 macrophages and decreases bacterial recovery from the spleens and lungs of infected mice (Siegrist et al., 2009).

FurA-dependent transcriptional regulation is intimately tied to oxidative stress. The gene encoding FurA is located upstream of *katG*, which encodes a catalase-peroxidase that is important for the oxidative stress response and pathogenesis of *M. tuberculosis* (Pym et al., 2001). Because of this, FurA is postulated to behave as a peroxide-sensing PerR ortholog in *M. tuberculosis* (Lucarelli et al., 2008). FurA controls expression of itself and *katG* through negative regulation (Sala et al., 2003). KatG has already been exploited as a target for tuberculosis treatment. The antimycobacterial drug isoniazid (isonicotinic acid hydrazide) is converted by KatG into intermediates that prevent bacterial growth by inhibiting mycolic acid synthesis (Timmins and Deretic, 2006).

CONCLUDING REMARKS

The importance of iron acquisition to the growth and propagation of pathogens during disease is highlighted by two general themes. First, systems involved in iron assimilation are typically redundant, suggesting that bacteria have developed multiple iron acquisition mechanisms to prevent reliance on any single pathway. This is also suggestive of a coevolution of pathogens and their hosts centered on obtaining iron during infection, as both organisms continually develop mechanisms to counteract iron acquisition/sequestration strategies of the other. Alternatively, iron acquisition systems may be beneficial for specific environmental niches. Therefore, the evolution of these systems could represent an adaptation to survival within a particular environment or host. Second, disruption of sequestration systems in the host and acquisition systems in the pathogen has a distinct effect on the outcome of disease. As discussed in the introduction, alteration of iron levels in the host, either by enhancing sequestration mechanisms or by increasing iron load, drastically impacts the level of sensitivity of the host to bacterial infection. Conversely, disruption of bacterial systems or regulators designed to obtain host iron during infection significantly reduces the virulence of nearly every pathogen discussed in this chapter.

Differences in Iron-Regulated Genes between Intracellular and Extracellular Pathogens

Primarily extracellular pathogens, such as *S. aureus*, tend to utilize host iron-binding proteins found in serum and tissues, such as hemoglobin or transferrin (Table 3). While hemoglobin is predominantly found within circulating erythrocytes, many pathogens have evolved exquisite mechanisms to liberate the hemoprotein from its intracellular environment for use as an iron source. In contrast, intracellular pathogens, such as *M. tuberculosis*, have developed mechanisms to subvert iron limitation while inside a host cell by hijacking the cell's own iron stores. Not surprisingly, intracellular pathogens also upregulate virulence determinants that promote invasion of host cells in response to iron deprivation.

Therapeutic Targets

The bacteria discussed in this chapter and the next represent some of the world's most virulent

Table 3. Preferred iron sources of extracellular and intracellular pathogens

Category of pathogen	Utilized host iron source	Example
Primarily intracellular life cycle	Ferritin Endocytosed transferrin Free heme	*Mycobacterium tuberculosis* *Bacillus anthracis* *Listeria monocytogenes*
Primarily extracellular life cycle	Transferrin Hemoglobin Myoglobin Free heme	*Corynebacterium diphtheriae* *Staphylococcus aureus* *Streptococcus pneumoniae* GAS *Clostridium*

pathogens that cause millions of deaths worldwide. Though these pathogens cause very different diseases affecting various systems of the body, one central theme unites them: the need to obtain iron during infection. As mentioned previously, disruption of iron acquisition in these bacteria results in decreased virulence in in vivo infection models, highlighting the importance of these systems to pathogenesis. Furthermore, many proteins involved in iron uptake in gram-positive bacteria are surface exposed and available for immune recognition, and they therefore represent potential vaccine candidates. Finally, orthologs of bacterial iron-dependent regulators do not appear to be expressed by eukaryotes (Pohl et al., 2003). It is believed, therefore, that bacterial iron-dependent metalloregulators and iron uptake systems represent excellent targets for antibacterial therapeutics.

Several drugs have already been developed to exploit the siderophore-mediated iron acquisition systems to treat bacterial infections. For example, one of the first therapeutics for tuberculosis, *p*-aminosalicylic acid, and the small molecule 5-*O*-(*N*-salicylsulfomyl) adenosine interfere with the synthesis and function of the mycobacterial siderophore, mycobactin (Adilakshmi et al., 2000; Ferreras et al., 2005). Additionally, many studies have taken advantage of the ability of siderophores to deliver iron to the bacterial organism and have developed siderophore-drug conjugates called *sideromycins* to treat microbial infection (Miller and Malouin, 1993; Roosenberg et al., 2000). The naturally occurring sideromycins albomycin and salmycin, as well as siderophores engineered with antibiotics bound to the active site, have been used against both gram-positive and gram-negative organisms (Braun et al., 2009). The siderophore-antibiotic complex is recognized by the bacterium and imported into the cell by siderophore transport systems where the antibiotic can exert its antimicrobial function. Salmycin has been utilized against the gram-positive pathogens *S. aureus* and *S. pneumoniae* and disrupts cellular function by inhibiting protein synthesis (Vértesy et al., 1995). Utilization of siderophore-mediated delivery of antibiotics directly to pathogens has a dramatic effect on the efficacy of these drugs, reducing the MICs more than 100-fold (Braun, 1999). These data support the notion that iron acquisition pathways of bacterial pathogens are excellent candidates for the development of novel therapeutics.

REFERENCES

Adilakshmi, T., P. D. Ayling, and C. Ratledge. 2000. Mutational analysis of a role for salicylic acid in iron metabolism of *Mycobacterium smegmatis*. J. Bacteriol. **182:**264–271.

Allard, M., H. Moisan, E. Brouillette, A. L. Gervais, M. Jacques, P. Lacasse, M. S. Diarra, and F. Malouin. 2006. Transcriptional modulation of some *Staphylococcus aureus* iron-regulated genes during growth *in vitro* and in a tissue cage model *in vivo*. *Microbes Infect.* 8:1679–1690.

Andrade, M. A., F. D. Ciccarelli, C. Perez-Iratxeta, and P. Bork. 2002. NEAT: a domain duplicated in genes near the components of a putative Fe^{3+} siderophore transporter from Gram-positive pathogenic bacteria. *Genome Biol.* 3:RESEARCH0047.

Aranda, J., M. E. Garrido, N. Fittipaldi, P. Cortés, M. Llagostera, M. Gottschalk, and J. Barbé. 2010. The cation-uptake regulators AdcR and Fur are necessary for full virulence of *Streptococcus suis*. *Vet. Microbiol.* 144:246–249.

Arends, J. P., and H. C. Zanen. 1988. Meningitis caused by *Streptococcus suis* in humans. *Rev. Infect. Dis.* 10:131–137.

Ashbaugh, C. D., and M. R. Wessels. 2001. Absence of a cysteine protease effect on bacterial virulence in two murine models of human invasive group A streptococcal infection. *Infect. Immun.* 69:6683–6688.

Bacon, J., L. G. Dover, K. A. Hatch, Y. Zhang, J. M. Gomes, S. Kendall, L. Wernisch, N. G. Stoker, P. D. Butcher, G. S. Besra, et al. 2007. Lipid composition and transcriptional response of *Mycobacterium tuberculosis* grown under iron-limitation in continuous culture: identification of a novel wax ester. *Microbiology* 153:1435–1444.

Baichoo, N., and J. D. Helmann. 2002. Recognition of DNA by Fur: a reinterpretation of the Fur box consensus sequence. *J. Bacteriol.* 184:5826–5832.

Baichoo, N., T. Wang, R. Ye, and J. D. Helmann. 2002. Global analysis of the *Bacillus subtilis* Fur regulon and the iron starvation stimulon. *Mol. Microbiol.* 45:1613–1629.

Bates, C., G. Montanez, C. Woods, R. Vincent, and Z. Eichenbaum. 2003. Identification and characterization of a *Streptococcus pyogenes* operon involved in binding of hemoproteins and acquisition of iron. *Infect. Immun.* 71:1042–1055.

Bates, C. S., C. Toukoki, M. N. Neely, and Z. Eichenbaum. 2005. Characterization of MtsR, a new metal regulator in group A streptococcus, involved in iron acquisition and virulence. *Infect. Immun.* 73:5743-5753.

Batey, R. G. 1986. Pathogenesis of caseous lymphadenitis in sheep and goats. *Aust. Vet. J.* 63:269–272.

Beasley, F. C., C. L. Marolda, J. Cheung, S. Buac, and D. E. Heinrichs. 2011. *Staphylococcus aureus* transporters Hts, Sir, and Sst capture iron liberated from human transferrin by staphyloferrin A, staphyloferrin B, and catecholamine stress hormones, respectively, and contribute to virulence. *Infect. Immun.* 79:2345–2355.

Beasley, F. C., E. D. Vinés, J. C. Grigg, Q. Zheng, S. Liu, G. A. Lajoie, M. E. P. Murphy, and D. E. Heinrichs. 2009. Characterization of staphyloferrin A biosynthetic and transport mutants in *Staphylococcus aureus*. *Mol. Microbiol.* 72:947–963.

Bierne, H., C. Sabet, N. Personnic, and P. Cossart. 2007. Internalins: a complex family of leucine-rich repeat-containing proteins in *Listeria monocytogenes*. *Microbes Infect.* 9:1156–1166.

Billington, S. J., P. A. Esmay, J. G. Songer, and B. H. Jost. 2002. Identification and role in virulence of putative iron acquisition genes from *Corynebacterium pseudotuberculosis*. *FEMS Microbiol. Lett.* 208:41–45.

Borezee, E., E. Pellegrini, and P. Berche. 2000. OppA of *Listeria monocytogenes*, an oligopeptide-binding protein required for bacterial growth at low temperature and involved in intracellular survival. *Infect. Immun.* 68:7069–7077.

Boyd, J., and J. R. Murphy. 1988. Analysis of the diphtheria *tox* promoter by site-directed mutagenesis. *J. Bacteriol.* 170:5949–5952.

Braun, V. 1999. Active transport of siderophore-mimicking antibacterials across the outer membrane. *Drug Resist. Updates* 2:363–369.

Braun, V., A. Pramanik, T. Gwinner, M. Köberle, and E. Bohn. 2009. Sideromycins: tools and antibiotics. *BioMetals* 22:3–13.

Bray, B., I. Sutcliffe, and D. Harrington. 2009. Expression of the MtsA lipoprotein of *Streptococcus agalactiae* A909 is regulated by manganese and iron. *Antonie van Leeuwenhoek* 95:101–109.

Brenot, A., K. Y. King, and M. G. Caparon. 2005. The PerR regulon in peroxide resistance and virulence of *Streptococcus pyogenes*. *Mol. Microbiol.* 55:221–234.

Brenot, A., B. F. Weston, and M. G. Caparon. 2007. A PerR-regulated metal transporter (PmtA) is an interface between oxidative stress and metal homeostasis in *Streptococcus pyogenes*. *Mol. Microbiol.* 63:1185–1196.

Brown, J. S., S. M. Gilliland, and D. W. Holden. 2001. A *Streptococcus pneumoniae* pathogenicity island encoding an ABC transporter involved in iron uptake and virulence. *Mol. Microbiol.* 40:572–585.

Brown, J. S., S. M. Gilliland, J. Ruiz-Albert, and D. W. Holden. 2002. Characterization of Pit, a *Streptococcus pneumoniae* iron uptake ABC transporter. *Infect. Immun.* 70:4389–4398.

Brüggemann, H., R. Bauer, S. Raffestin, and G. Gottschalk. 2004. Characterization of a heme oxygenase of *Clostridium tetani* and its possible role in oxygen tolerance. *Arch. Microbiol.* 182:259–263.

Bsat, N., L. Chen, and J. Helmann. 1996. Mutation of the *Bacillus subtilis* alkyl hydroperoxide reductase (*ahpCF*) operon reveals compensatory interactions among hydrogen peroxide stress genes. *J. Bacteriol.* 178:6579–6586.

Bsat, N., and J. D. Helmann. 1999. Interaction of *Bacillus subtilis* Fur (ferric uptake repressor) with the *dhb* operator in vitro and in vivo. *J. Bacteriol.* 181: 4299–4307.

Cabrera, G., A. Xiong, V. Uebel, V. K. Singh, and R. K. Jayaswal. 2001. Molecular characterization of the iron-hydroxamate uptake system in *Staphylococcus aureus*. *Appl. Environ. Microbiol.* 67:1001–1003.

Calder, K. M., and M. A. Horwitz. 1998. Identification of iron-regulated proteins of *Mycobacterium tuberculosis* and cloning of tandem genes encoding a low iron-induced protein and a metal transporting ATPase with similarities to two-component metal transport systems. *Microb. Pathog.* 24:133–143.

Calderwood, S. B., and J. J. Mekalanos. 1988. Confirmation of the Fur operator site by insertion of a synthetic oligonucleotide into an operon fusion plasmid. *J. Bacteriol.* 170:1015–1017.

Camacho, L. R., D. Ensergueix, E. Perez, B. Gicquel, and C. Guilhot. 1999. Identification of a virulence gene cluster of *Mycobacterium tuberculosis* by signature-tagged transposon mutagenesis. *Mol. Microbiol.* 34:257–267.

Canneva, F., M. Branzoni, G. Riccardi, R. Provvedi, and A. Milano. 2005. Rv2358 and FurB: two transcriptional regulators from *Mycobacterium tuberculosis* which respond to zinc. *J. Bacteriol.* 187:5837–5840.

Caparon, M. G., R. T. Geist, J. Perez-Casal, and J. R. Scott. 1992. Environmental regulation of virulence in group A streptococci: transcription of the gene encoding M protein is stimulated by carbon dioxide. *J. Bacteriol.* 174:5693–5701.

Carlson, P. E., Jr., K. A. Carr, B. K. Janes, E. C. Anderson, and P. C. Hanna. 2009. Transcriptional profiling of *Bacillus anthracis* Sterne (34F2) during iron starvation. *PLoS One* 4: e6988.

Casiano-Colón, A., and R. E. Marquis. 1988. Role of the arginine deiminase system in protecting oral bacteria and an enzymatic basis for acid tolerance. *Appl. Environ. Microbiol.* 54:1318–1324.

Cendrowski, S., W. MacArthur, and P. Hanna. 2004. *Bacillus anthracis* requires siderophore biosynthesis for growth in macrophages and mouse virulence. *Mol. Microbiol.* 51:407–417.

Chavakis, T., M. Hussain, S. M. Kanse, G. Peters, R. G. Bretzel, J.-I. Flock, M. Herrmann, and K. T. Preissner. 2002. *Staphylococcus aureus* extracellular adherence protein serves as anti-inflammatory factor by inhibiting the recruitment of host leukocytes. *Nat. Med.* **8:**687–693.

Chavakis, T., K. Wiechmann, K. T. Preissner, and M. Herrmann. 2005. *Staphylococcus aureus* interactions with the endothelium: the role of bacterial "secretable expanded repertoire adhesive molecules" (SERAM) in disturbing host defense systems. *Thromb. Haemost.* **94:**278–285.

Chen, L., and J. D. Helmann. 1995. *Bacillus subtilis* MrgA is a Dps(PexB) homologue: evidence for metalloregulation of an oxidative-stress gene. *Mol. Microbiol.* **18:**295-300.

Chen, L., L. P. James, and J. D. Helmann. 1993. Metalloregulation in *Bacillus subtilis*: isolation and characterization of two genes differentially repressed by metal ions. *J. Bacteriol.* **175:**5428–5437.

Cheung, J., F. C. Beasley, S. Liu, G. A. Lajoie, and D. E. Heinrichs. 2009. Molecular characterization of staphyloferrin B biosynthesis in *Staphylococcus aureus*. *Mol. Microbiol.* **74:**594–608.

Chu, G. C., K. Katakura, X. Zhang, T. Yoshida, and M. Ikeda-Saito. 1999. Heme degradation as catalyzed by a recombinant bacterial heme oxygenase (Hmu O) from *Corynebacterium diphtheriae*. *J. Biol. Chem.* **274:**21319–21325.

Clarke, S. R., M. D. Wiltshire, and S. J. Foster. 2004. IsdA of *Staphylococcus aureus* is a broad spectrum, iron-regulated adhesin. *Mol. Microbiol.* **51:**1509–1519.

Clemens, D. L., and M. A. Horwitz. 1996. The *Mycobacterium tuberculosis* phagosome interacts with early endosomes and is accessible to exogenously administered transferrin. *J. Exp. Med.* **184:**1349–1355.

Clifton-Hadley, F. A. 1983. *Streptococcus suis* type 2 infections. *Br. Vet. J.* **139:**1–5.

Cole, S. T., R. Brosch, J. Parkhill, T. Garnier, C. Churcher, D. Harris, S. V. Gordon, K. Eiglmeier, S. Gas, C. E. Barry, et al. 1998. Deciphering the biology of *Mycobacterium tuberculosis* from the complete genome sequence. *Nature* **393:**537–544.

Conte, M., C. Longhi, M. Polidoro, G. Petrone, V. Buonfiglio, S. Di Santo, E. Papi, L. Seganti, P. Visca, and P. Valenti. 1996. Iron availability affects entry of *Listeria monocytogenes* into the enterocytelike cell line Caco-2. *Infect. Immun.* **64:**3925–3929.

Conte, M. P., C. Longhi, G. Petrone, M. Polidoro, P. Valenti, and L. Seganti. 2000. Modulation of *actA* gene expression in *Listeria monocytogenes* by iron. *J. Med. Microbiol.* **49:**681–683.

Cossart, P. 1995. Actin-based bacterial motility. *Curr. Opin. Cell Biol.* **7:**94–101.

Cotton, J. L., J. Tao, and C. J. Balibar. 2009. Identification and characterization of the *Staphylococcus aureus* gene cluster coding for staphyloferrin A. *Biochemistry* **48:**1025–1035.

Cowart, R. E., and B. G. Foster. 1981. The role of iron in the production of haemolysin by *Listeria monocytogenes*. *Curr. Microbiol.* **6:**287–290.

Cox, J. S., B. Chen, M. McNeil, and W. R. Jacobs. 1999. Complex lipid determines tissue-specific replication of *Mycobacterium tuberculosis* in mice. *Nature* **402:**79–83.

Crosa, J. 1997. Signal transduction and transcriptional and posttranscriptional control of iron-regulated genes in bacteria. *Microbiol. Mol. Biol. Rev.* **61:**319–336.

Cunningham, M. 2000. Pathogenesis of group A streptococcal infections. *Clin. Microbiol. Rev.* **13:**470 –511.

Dale, S. E., A. Doherty-Kirby, G. Lajoie, and D. E. Heinrichs. 2004. Role of siderophore biosynthesis in virulence of *Staphylococcus aureus*: identification and characterization of genes involved in production of a siderophore. *Infect. Immun.* **72:**29–37.

Dall, L. H., and B. L. Herndon. 1990. Association of cell-adherent glycocalyx and endocarditis production by viridans group streptococci. *J. Clin. Microbiol.* **28:**1698–1700.

Davenport, D. S., R. M. Massanari, M. A. Pfaller, M. J. Bale, S. A. Streed, and W. J. Hierholzer. 1986. Usefulness of a test for slime production as a marker for clinically significant infections with coagulase-negative staphylococci. *J. Infect. Dis.* **153:**332–339.

De Domenico, I., D. M. Ward, C. Langelier, M. B. Vaughn, E. Nemeth, W. I. Sundquist, T. Ganz, G. Musci, and J. Kaplan. 2007. The molecular mechanism of hepcidin-mediated ferroportin down-regulation. *Mol. Biol. Cell* **18:** 2569–2578.

Degnan, B. A., M. C. Fontaine, A. H.Doebereiner, J. J. Lee, P. Mastroeni, G. Dougan, J. A. Goodacre, and M. A. Kehoe. 2000. Characterization of an isogenic mutant of *Streptococcus pyogenes* Manfredo lacking the ability to make streptococcal acid glycoprotein. *Infect. Immun.* **68:**2441–2448.

Deighton, M., and R. Borland. 1993. Regulation of slime production in *Staphylococcus epidermidis* by iron limitation. *Infect. Immun.* **61:**4473–4479.

de Lorenzo, V., S. Wee, M. Herrero, and J. B. Neilands. 1987. Operator sequences of the aerobactin operon of plasmid ColV-K30 binding the ferric uptake regulation (*fur*) repressor. *J. Bacteriol.* **169:**2624–2630.

De Voss, J. J., K. Rutter, B. G. Schroeder, H. Su, Y. Zhu, and C. E. Barry. 2000. The salicylate-derived mycobactin siderophores of *Mycobacterium tuberculosis* are essential for growth in macrophages. *Proc. Natl. Acad. Sci. USA* **97:**1252–1257.

Dintilhac, A., G. Alloing, C. Granadel, and J.-P. Claverys. 1997. Competence and virulence of *Streptococcus pneumoniae*: Adc and PsaA mutants exhibit a requirement for Zn and Mn resulting from inactivation of putative ABC metal permeases. *Mol. Microbiol.* **25:**727–739.

Drabkin, D. L. 1951. Metabolism of the hemin chromoproteins. *Physiol. Rev.* **31:**345–431.

Drazek, E. S., C. A. Hammack, Sr, and M. P. Schmitt. 2000. *Corynebacterium diphtheriae* genes required for acquisition of iron from haemin and haemoglobin are homologous to ABC haemin transporters. *Mol. Microbiol.* **36:**68–84.

Drechsel, H., and G. Jung. 1998. Peptide siderophores. *J. Pept. Sci.* **4:**147–181.

Dussurget, O., M. Rodriguez, and I. Smith. 1996. An ideR mutant of *Mycobacterium smegmatis* has derepressed siderophore production and an altered oxidative-stress response. *Mol. Microbiol.* **22:**535–544.

Dussurget, O., J. Timm, M. Gomez, B. Gold, S. Yu, S. Z. Sabol, R. K. Holmes, W. R. Jacobs, Jr., and I. Smith. 1999. Transcriptional control of the iron-responsive *fxbA* gene by the mycobacterial regulator IdeR. *J. Bacteriol.* **181:**3402–3408.

Ellen, R. P., D. W. Banting, and E. D. Fillery. 1985. CLINICAL SCIENCE longitudinal microbiological investigation of a hospitalized population of older adults with a high root surface caries risk. *J. Dent. Res.* **64:**1377–1381.

Ernst, J. F., R. L. Bennett, and L. I. Rothfield. 1978. Constitutive expression of the iron-enterochelin and ferrichrome uptake systems in a mutant strain of *Salmonella typhimurium*. *J. Bacteriol.* **135:**928–934.

Escolar, L., J. Pérez-Martín, and V. de Lorenzo. 1999. Opening the iron box: transcriptional metalloregulation by the Fur protein. *J. Bacteriol.* **181:**6223–6229.

Escolar, L., J. Pérez-Martín, and V. de Lorenzo. 1998. Binding of the Fur (ferric uptake regulator) repressor of *Escherichia coli* to arrays of the GATAAT sequence. *J. Mol. Biol.* **283:**537–547.

Fabian, M., E. Solomaha, J. S. Olson, and A. W. Maresso. 2009. Heme transfer to the bacterial cell envelope occurs via a secreted hemophore in the Gram-positive pathogen *Bacillus anthracis*. *J. Biol. Chem.* **284:**32138–32146.

Farber, J. M., and P. I. Peterkin. 1991. *Listeria monocytogenes*, a food-borne pathogen. *Microbiol. Rev.* **55:**476–511.

Fermi, G., M. F. Perutz, B. Shaanan, and R. Fourme. 1984. The crystal structure of human deoxyhaemoglobin at 1.74 Å resolution. *J. Mol. Biol.* **175**:159–174.

Ferreras, J. A., J.-S. Ryu, F. Di Lello, D. S. Tan, and L. E. N. Quadri. 2005. Small-molecule inhibition of siderophore biosynthesis in *Mycobacterium tuberculosis* and *Yersinia pestis*. *Nat. Chem. Biol.* **1**:29–32.

Fiorini, F., S. Stefanini, P. Valenti, E. Chiancone, and D. De Biase. 2008. Transcription of the *Listeria monocytogenes fri* gene is growth-phase dependent and is repressed directly by Fur, the ferric uptake regulator. *Gene* **410**:113–121.

Fischetti, V. A. 1989. Streptococcal M protein: molecular design and biological behavior. *Clin. Microbiol. Rev.* **2**:285–314.

Fischetti, V. A., D. A. Parry, B. L. Trus, S. K. Hollingshead, J. R. Scott, and B. N. Manjula. 1988. Conformational characteristics of the complete sequence of group A streptococcal M6 protein. *Proteins* **3**:60–69.

Flo, T. H., K. D. Smith, S. Sato, D. J. Rodriguez, M. A. Holmes, R. K. Strong, S. Akira, and A. Aderem. 2004. Lipocalin 2 mediates an innate immune response to bacterial infection by sequestrating iron. *Nature* **432**:917–921.

Fourel, G., A. Phalipon, and M. Kaczorek. 1989. Evidence for direct regulation of diphtheria toxin gene transcription by an Fe^{2+}-dependent DNA-binding repressor, DtoxR, in *Corynebacterium diphtheriae*. *Infect. Immun.* **57**:3221–3225.

Fuangthong, M., A. F. Herbig, N. Bsat, and J. D. Helmann. 2002. Regulation of the *Bacillus subtilis fur* and *perR* genes by PerR: not all members of the PerR regulon are peroxide inducible. *J. Bacteriol.* **184**:3276–3286.

Gaballa, A., H. Antelmann, C. Aguilar, S. K. Khakh, K.-B. Song, G. T. Smaldone, and J. D. Helmann. 2008. The *Bacillus subtilis* iron-sparing response is mediated by a Fur-regulated small RNA and three small, basic proteins. *Proc. Natl. Acad. Sci. USA* **105**:11927–11932.

Gaballa, A., and J. D. Helmann. 1998. Identification of a zinc-specific metalloregulatory protein, Zur, controlling zinc transport operons in *Bacillus subtilis*. *J. Bacteriol.* **180**:5815–5821.

Gangaidzo, I. T., V. M. Moyo, E. Mvundura, G. Aggrey, N. L. Murphree, H. Khumalo, T. Saungweme, I. Kasvosve, Z. A. R. Gomo, T. Rouault, et al. 2001. Association of pulmonary tuberculosis with increased dietary iron. *J. Infect. Dis.* **184**:936–939.

Gat, O., G. Zaide, I. Inbar, H. Grosfeld, T. Chitlaru, H. Levy, and A. Shafferman. 2008. Characterization of *Bacillus anthracis* iron-regulated surface determinant (Isd) proteins containing NEAT domains. *Mol. Microbiol.* **70**:983–999.

Glickman, M. S., and W. R. Jacobs. 2001. Microbial pathogenesis of *Mycobacterium tuberculosis*: dawn of a discipline. *Cell* **104**:477–485.

Gobin, J., and M. A. Horwitz. 1996. Exochelins of *Mycobacterium tuberculosis* remove iron from human iron-binding proteins and donate iron to mycobactins in the *M. tuberculosis* cell wall. *J. Exp. Med.* **183**:1527–1532.

Gold, B., G. M. Rodriguez, S. A. E. Marras, M. Pentecost, and I. Smith. 2001. The *Mycobacterium tuberculosis* IdeR is a dual functional regulator that controls transcription of genes involved in iron acquisition, iron storage and survival in macrophages. *Mol. Microbiol.* **42**:851–865.

Gomme, P. T., K. B. McCann, and J. Bertolini. 2005. Transferrin: structure, function and potential therapeutic actions. *Drug Disc. Today* **10**:267–273.

Gupta, R., P. Shah, P., and E. Swiatlo. 2009. Differential gene expression in *Streptococcus pneumoniae* in response to various iron sources. *Microb. Pathog.* **47**:101–109.

Haggar, A., M. Hussain, H. Lonnies, M. Herrmann, A. Norrby-Teglund, and J.-I. Flock. 2003. Extracellular adherence protein from *Staphylococcus aureus* enhances internalization into eukaryotic cells. *Infect. Immun.* **71**:2310–2317.

Hammerschmidt, S., G. Bethe, P. H. Remane, and G. S. Chhatwal. 1999. Identification of pneumococcal surface protein A as a lactoferrin-binding protein of *Streptococcus pneumoniae*. *Infect. Immun.* **67**:1683–1687.

Hammond, C. R. 2004. The elements, p. 4–32. *In* D. R. Lide (ed.), *CRC Handbook of Chemistry and Physics*, 85th ed. CRC Press, Boca Raton, FL.

Hanks, T., M. Liu, M. McClure, and B. Lei. 2005. ABC transporter FtsABCD of *Streptococcus pyogenes* mediates uptake of ferric ferrichrome. *BMC Microbiol.* **5**:62.

Hanks, T. S., M. Liu, M. J. McClure, M. Fukumura, A. Duffy, and B. Lei. 2006. Differential regulation of iron- and manganese-specific MtsABC and heme-specific HtsABC transporters by the metalloregulator MtsR of group A streptococcus. *Infect. Immun.* **74**:5132–5139.

Haralalka, S., S. Nandi, and R. K. Bhadra. 2003. Mutation in the *relA* gene of *Vibrio cholerae* affects in vitro and in vivo expression of virulence factors. *J. Bacteriol.* **185**:4672–4682.

Hard, G. C. 1972. Examination by electron microscopy of the interaction between peritoneal phagocytes and *Corynebacterium ovis*. *J. Med. Microbiol.* **5**:483–491.

Harrison, S. C., and A. K. Aggarwal. 1990. DNA recognition by proteins with the helix-turn-helix motif. *Annu. Rev. Biochem.* **59**:933–969.

Harvie, D. R., S. Vílchez, J. R. Steggles, and D. J. Ellar. 2005. *Bacillus cereus* Fur regulates iron metabolism and is required for full virulence. *Microbiology* **151**:569–577.

Hassan, S., K. Ohtani, R. Wang, Y. Yuan, Y. Wang, Y. Yamaguchi, and T. Shimizu. 2010. Transcriptional regulation of *hemO* encoding heme oxygenase in *Clostridium perfringens*. *J. Microbiol.* **48**:96–101.

He, Q. Y., A. B. Mason, V. Nguyen, R. T. MacGillivray, and R. C. Woodworth. 2000. The chloride effect is related to anion binding in determining the rate of iron release from the human transferrin N-lobe. *Biochem J.* **350**:909–915.

Heinrichs, J. H., L. E. Gatlin, C. Kunsch, G. H. Choi, and M. S. Hanson. 1999. Identification and characterization of SirA, an iron-regulated protein from *Staphylococcus aureus*. *J. Bacteriol.* **181**:1436–1443.

Henle, E. S., and S. Linn. 1997. Formation, prevention, and repair of DNA damage by iron/hydrogen peroxide. *J. Biol. Chem.* **272**:19095–19098.

Herbig, A. F., and J. D. Helmann. 2001. Roles of metal ions and hydrogen peroxide in modulating the interaction of the *Bacillus subtilis* PerR peroxide regulon repressor with operator DNA. *Mol. Microbiol.* **41**:849–859.

Herbig, A. F., and J. D. Helmann. 2002. Metal ion uptake and oxidative stress, p. 405–414. *In* A. L. Sonenshein, J. A. Hoch, and R. Losick (ed.), Bacillus subtilis *and Its Closest Relatives*. ASM Press, Washington, DC.

Hill, P. J., A. Cockayne, P. Landers, J. A. Morrissey, C. M. Sims, and P. Williams. 1998. SirR, a novel iron-dependent repressor in *Staphylococcus epidermidis*. *Infect. Immun.* **66**:4123–4129.

Hoffmaster, A. R., J. Ravel, D. A. Rasko, G. D. Chapman, M. D. Chute, C. K. Marston, B. K. De, C. T. Sacchi, C. Fitzgerald, L. W. Mayer, et al. 2004. Identification of anthrax toxin genes in a *Bacillus cereus* associated with an illness resembling inhalation anthrax. *Proc. Natl. Acad. Sci. USA* **101**:8449–8454.

Horsburgh, M. J., M. O. Clements, H. Crossley, E. Ingham, and S. J. Foster. 2001a. PerR controls oxidative stress resistance and iron storage proteins and is required for virulence in *Staphylococcus aureus*. *Infect. Immun.* **69**:3744–3754.

Horsburgh, M. J., E. Ingham, and S. J. Foster. 2001b. In *Staphylococcus aureus*, Fur is an interactive regulator with PerR, contributes to virulence, and is necessary for oxidative stress resistance through positive regulation of catalase and iron homeostasis. *J. Bacteriol.* **183:**468–475.

Horsburgh, M. J., S. J. Wharton, A. G. Cox, E. Ingham, S. Peacock, and S. J. Foster. 2002. MntR modulates expression of the PerR regulon and superoxide resistance in *Staphylococcus aureus* through control of manganese uptake. *Mol. Microbiol.* **44:**1269–1286.

Hussain, M., K. Becker, C. von Eiff, J. Schrenzel, G. Peters, and M. Herrmann. 2001. Identification and characterization of a novel 38.5-kilodalton cell surface protein of *Staphylococcus aureus* with extended-spectrum binding activity for extracellular matrix and plasma proteins. *J. Bacteriol.* **183:**6778–6786.

Imlay, J. A., S. M. Chin, and S. Linn. 1988. Toxic DNA damage by hydrogen peroxide through the Fenton reaction *in vivo* and *in vitro*. *Science* **240:**640–642.

Jain, V., R. Saleem-Batcha, A. China, and D. Chatterji. 2006. Molecular dissection of the mycobacterial stringent response protein Rel. *Protein Sci.* **15:**1449–1464.

Janulczyk, R., J. Pallon, and L. Bjorck. 1999. Identification and characterization of a *Streptococcus pyogenes* ABC transporter with multiple specificity for metal cations. *Mol. Microbiol.* **34:**596–606.

Janulczyk, R., S. Ricci, and L. Bjorck. 2003. MtsABC is important for manganese and iron transport, oxidative stress resistance, and virulence of *Streptococcus pyogenes*. *Infect. Immun.* **71:**2656–2664.

Jin, B., S. M. C. Newton, Y. Shao, X. Jiang, A. Charbit, and P. E. Klebba. 2006. Iron acquisition systems for ferric hydroxamates, haemin and haemoglobin in *Listeria monocytogenes*. *Mol. Microbiol.* **59:**1185–1198.

Johnson, M., A. Cockayne, P. H. Williams, and J. A. Morrissey. 2005. Iron-responsive regulation of biofilm formation in *Staphylococcus aureus* involves Fur-dependent and Fur-independent mechanisms. *J. Bacteriol.* **187:**8211–8215.

Johnson, M., M. Sengupta, J. Purves, E. Tarrant, P. H. Williams, A. Cockayne, A. Muthaiyan, R. Stephenson, N. Ledala, B. J. Wilkinson, et al. 2011. Fur is required for the activation of virulence gene expression through the induction of the sae regulatory system in *Staphylococcus aureus*. *Int. J. Med. Microbiol.* **301:**44–52.

Johri, A. K., L. C. Paoletti, P. Glaser, M. Dua, P. K. Sharma, G. Grandi, and R. Rappuoli. 2006. Group B streptococcus: global incidence and vaccine development. *Nat. Rev. Microbiol.* **4:**932–942.

Kansal, R. G., R. K. Aziz, and M. Kotb. 2005. Modulation of expression of superantigens by human transferrin and lactoferrin: a novel mechanism in host-streptococcus interactions. *J. Infect. Dis.* **191:**2121–2129.

Kazmi, S. U., R. Kansal, R. K. Aziz, M. Hooshdaran, A. Norrby-Teglund, D. E. Low, A.-B. Halim, and M. Kotb. 2001. Reciprocal, temporal expression of SpeA and SpeB by invasive M1T1 group A streptococcal isolates in vivo. *Infect. Immun.* **69:**4988–4995.

Kitten, T., C. L. Munro, S. M. Michalek, and F. L. Macrina. 2000. Genetic characterization of a *Streptococcus mutans* LraI family operon and role in virulence. *Infect. Immun.* **68:**4441–4451.

Kotb, M. 1995. Bacterial pyrogenic exotoxins as superantigens. *Clin. Microbiol. Rev.* **8:**411–426.

Kotiranta, A., K. Lounatmaa, and M. Haapasalo. 2000. Epidemiology and pathogenesis of *Bacillus cereus* infections. *Microbes Infect.* **2:**189–198.

Kramer, J. M., and R. J. Gilbert. 1989. *Bacillus cereus* and other *Bacillus* species, p. 21–70. *In* M. P. Doyle (ed.), *Foodborne Bacterial Pathogens*. Marcel Dekker, New York, NY.

Kristiansen, M., J. H. Graversen, C. Jacobsen, O. Sonne, H.-J. Hoffman, S. K. A. Law, and S. K. Moestrup. 2001. Identification of the haemoglobin scavenger receptor. *Nature* **409:**198–201.

Kunkle, C. A., and M. P. Schmitt. 2003. Analysis of the *Corynebacterium diphtheriae* DtxR regulon: identification of a putative siderophore synthesis and transport system that is similar to the *Yersinia* high-pathogenicity island-encoded yersiniabactin synthesis and uptake system. *J. Bacteriol.* **185:**6826–6840.

Lancefield, R. C. 1962. Current knowledge of type-specific M antigens of group A streptococci. *J. Immunol.* **89:**307–313.

Lee, J.-W., and J. D. Helmann. 2006. The PerR transcription factor senses H_2O_2 by metal-catalysed histidine oxidation. *Nature* **440:**363–367.

Lereclus, D., H. Agaisse, C. Grandvalet, S. Salamitou, and M. Gominet. 2000. Regulation of toxin and virulence gene transcription in *Bacillus thuringiensis*. *Int. J. Med. Microbiol.* **290:**295–299.

Li, W., L. Liu, H. Chen, and R. Zhou. 2009. Identification of *Streptococcus suis* genes preferentially expressed under iron starvation by selective capture of transcribed sequences. *FEMS Microbiol. Lett.* **292:**123–133.

Lim, W. S., J. T. Macfarlane, T. C. J. Boswell, T. G. Harrison, D. Rose, M. Leinonen, and P. Saikku. 2001. Study of community acquired pneumonia aetiology (SCAPA) in adults admitted to hospital: implications for management guidelines. *Thorax* **56:**296–301.

Lincoln, R. E., D. R. Hodges, F. Klein, B. G. Mahlandt, W. I. Jones, B. W. Haines, M. A. Rhian, and J. S. Walker. 1965. Role of the lymphatics in the pathogenesis of anthrax. *J. Infect. Dis.* **115:**481–494.

Lindsay, J. A., and S. J. Foster. 2001. zur: a Zn^{2+}-responsive regulatory element of *Staphylococcus aureus*. *Microbiology* **147:**1259–1266.

Litwin, C. M., S. A. Boyko, and S. B. Calderwood. 1992. Cloning, sequencing, and transcriptional regulation of the *Vibrio cholerae* fur gene. *J. Bacteriol.* **174:**1897–1903.

Litwin, C. M., and S. B. Calderwood. 1993. Role of iron in regulation of virulence genes. *Clin. Microbiol. Rev.* **6:**137–149.

Lucarelli, D., M. L. Vasil, W. Meyer-Klaucke, and E. Pohl. 2008. The metal-dependent regulators FurA and FurB from *Mycobacterium tuberculosis*. *Int. J. Mol. Sci.* **9:**1548–1560.

Luo, Y., Z. Han, S. M. Chin, and S. Linn. 1994. Three chemically distinct types of oxidants formed by iron-mediated Fenton reactions in the presence of DNA. *Proc. Natl. Acad. Sci. USA* **91:**12438–12442.

Magnus, S. A., I. R. Hambleton, F. Moosdeen, and G. R. Serjeant. 1999. Recurrent infections in homozygous sickle cell disease. *Arch. Dis. Child.* **80:**537–541.

Manabe, Y. C., B. J. Saviola, L. Sun, J. R. Murphy, and W. R. Bishai. 1999. Attenuation of virulence in *Mycobacterium tuberculosis* expressing a constitutively active iron repressor. *Proc. Natl. Acad. Sci. USA* **96:**12844–12848.

Maresso, A. W., G. Garufi, and O. Schneewind. 2008. *Bacillus anthracis* secretes proteins that mediate heme acquisition from hemoglobin. *PLoS Pathog.* **4:**e1000132.

Marra, A., J. Asundi, M. Bartilson, S. Lawson, F. Fang, J. Christine, C. Wiesner, D. Brigham, W. P. Schneider, and A. E. Hromockyj. 2002. Differential fluorescence induction analysis of *Streptococcus pneumoniae* identifies genes involved in pathogenesis. *Infect. Immun.* **70:**1422–1433.

Maskell, J. P. 1980. The functional interchangeability of enterobacterial and staphylococcal iron chelators. *Antonie van Leeuwenhoek* **46:**343–351.

McIver, K., A. Heath, and J. Scott. 1995. Regulation of virulence by environmental signals in group A streptococci: influence of osmolarity, temperature, gas exchange, and iron limitation on *emm* transcription. *Infect. Immun.* **63**:4540–4542.

McNamara, P. J., G. A. Bradley, and J. G. Songer. 1994. Targeted mutagenesis of the phospholipase D gene results in decreased virulence of *Corynebacterium pseudotuberculosis*. *Mol. Microbiol.* **12**:921–930.

Miller, M. J., and F. Malouin. 1993. Microbial iron chelators as drug delivery agents: the rational design and synthesis of siderophore-drug conjugates. *Accounts Chem. Res.* **26**:241–249.

Minnikin, D. E., L. Kremer, L. G. Dover, and G. S. Besra. 2002. The methyl-branched fortifications of *Mycobacterium tuberculosis*. *Chem. Biol.* **9**:545–553.

Moelling, C., R. Oberschlacke, P. Ward, J. Karijolich, K. Borisova, N. Bjelos, and L. Bergeron. 2007. Metal-dependent repression of siderophore and biofilm formation in *Actinomyces naeslundii*. *FEMS Microbiol. Lett.* **275**:214–220.

Montanez, G. E., M. N. Neely, and Z. Eichenbaum. 2005. The streptococcal iron uptake (Siu) transporter is required for iron uptake and virulence in a zebrafish infection model. *Microbiology* **151**:3749–3757.

Moreira, L. D. O., A. F. B. Andrade, M. D. Vale, S. M. S. Souza, R. Hirata, Jr., L. M. O. B. Asad, N. R. Asad, L. H. Monteiro-Leal, J. O. Previato, and A. L. Mattos-Guaraldi. 2003. Effects of iron limitation on adherence and cell surface carbohydrates of *Corynebacterium diphtheriae* strains. *Appl. Environ. Microbiol.* **69**:5907–5913.

Morgan, J. W., and E. Anders. 1980. Chemical composition of Earth, Venus, and Mercury. *Proc. Natl. Acad. Sci. USA* **77**:6973–6977.

Morrissey, J. A., A. Cockayne, P. J. Hill, and P. Williams. 2000. Molecular cloning and analysis of a putative siderophore ABC transporter from *Staphylococcus aureus*. *Infect. Immun.* **68**:6281–6288.

Moyo, V. M., I. T. Gangaidzo, V. R. Gordeuk, C. F. Kiire, and A. P. Macphail. 1997. Tuberculosis and iron overload in Africa: a review. *Cent. Afr. J. Med.* **43**:334–339.

Murray, M. J., A. B. Murray, M. B. Murray, and C. J. Murray. 1978. The adverse effect of iron repletion on the course of certain infections. *Br. Med. J.* **2**:1113–1115.

Nelson, A. L., J. M. Barasch, R. M. Bunte, and J. N. Weiser. 2005. Bacterial colonization of nasal mucosa induces expression of siderocalin, an iron-sequestering component of innate immunity. *Cell. Microbiol.* **7**:1404–1417.

Newton, S. M. C., P. E. Klebba, C. Raynaud, Y. Shao, X. Jiang, I. Dubail, C. Archer, C. Frehel, and A. Charbit. 2005. The *svpA-srtB* locus of *Listeria monocytogenes*: Fur-mediated iron regulation and effect on virulence. *Mol. Microbiol.* **55**:927–940.

Nikaido, H., and J. A. Hall. 1998. Overview of bacterial ABC transporters. *Methods Enzymol.* **292**:3–20.

Olsen, K. N., M. H. Larsen, C. G. M. Gahan, B. Kallipolitis, X. A. Wolf, R. Rea, C. Hill, and H. Ingmer. 2005. The Dps-like protein Fri of *Listeria monocytogenes* promotes stress tolerance and intracellular multiplication in macrophage-like cells. *Microbiology* **151**:925–933.

Oram, D. M., A. Avdalovic, and R. K. Holmes. 2002. Construction and characterization of transposon insertion mutations in *Corynebacterium diphtheriae* that affect expression of the diphtheria toxin repressor (DtxR). *J. Bacteriol.* **184**:5723–5732.

Osaki, M., D. Takamatsu, Y. Shimoji, and T. Sekizaki. 2002. Characterization of *Streptococcus suis* genes encoding proteins homologous to sortase of gram-positive bacteria. *J. Bacteriol.* **184**:971–982.

Palma, M., A. Haggar, and J. I. Flock. 1999. Adherence of *Staphylococcus aureus* is enhanced by an endogenous secreted protein with broad binding activity. *J. Bacteriol.* **181**:2840–2845.

Peters, G., R. Locci, and G. Pulverer. 1982. Adherence and growth of coagulase-negative staphylococci on surfaces of intravenous catheters. *J. Infect. Dis.* **146**:479–482.

Pfleger, B. F., Y. Kim, T. D. Nusca, N. Maltseva, J. Y. Lee, C. M. Rath, J. B. Scaglione, B. K. Janes, E. C. Anderson, N. H. Bergman, et al. 2008. Structural and functional analysis of AsbF: origin of the stealth 3,4-dihydroxybenzoic acid subunit for petrobactin biosynthesis. *Proc. Natl. Acad. Sci. USA* **105**:17133–17138.

Pohl, E., J. C. Haller, A. Mijovilovich, W. Meyer-Klaucke, E. Garman, and M. L. Vasil. 2003. Architecture of a protein central to iron homeostasis: crystal structure and spectroscopic analysis of the ferric uptake regulator. *Mol. Microbiol.* **47**:903–915.

Pohl, E., R. K. Holmes, and W. G. Hol. 1998. Motion of the DNA-binding domain with respect to the core of the diphtheria toxin repressor (DtxR) revealed in the crystal structure of apo- and holo-DtxR. *J. Biol. Chem.* **273**:22420–22427.

Polidoro, M., D. De Biase, B. Montagnini, L. Guarrera, S. Cavallo, P. Valenti, S. Stefanini, and E. Chiancone. 2002. The expression of the dodecameric ferritin in *Listeria* spp. is induced by iron limitation and stationary growth phase. *Gene* **296**:121–128.

Posey, J. E., and F. C. Gherardini. 2000. Lack of a role for iron in the Lyme disease pathogen. *Science* **288**:1651–1653.

Poulos, T. L. 2007. The Janus nature of heme. *Nat. Prod. Rep.* **24**:504–510.

Primm, T. P., S. J. Andersen, V. Mizrahi, D. Avarbock, H. Rubin, and C. E. Barry III. 2000. The stringent response of *Mycobacterium tuberculosis* is required for long-term survival. *J. Bacteriol.* **182**:4889–4898.

Pym, A. S., P. Domenech, N. Honoré, J. Song, V. Deretic, and S. T. Cole. 2001. Regulation of catalase-peroxidase (KatG) expression, isoniazid sensitivity and virulence by *furA* of *Mycobacterium tuberculosis*. *Mol. Microbiol.* **40**:879–889.

Qian, Y., J. H. Lee, and R. K. Holmes. 2002. Identification of a DtxR-regulated operon that is essential for siderophore-dependent iron uptake in *Corynebacterium diphtheriae*. *J. Bacteriol.* **184**:4846–4856.

Que, Q., and J. D. Helmann. 2000. Manganese homeostasis in *Bacillus subtilis* is regulated by MntR, a bifunctional regulator related to the diphtheria toxin repressor family of proteins. *Mol. Microbiol.* **35**:1454–1468.

Rachman, H., M. Strong, U. Schaible, J. Schuchhardt, K. Hagens, H. Mollenkopf, D. Eisenberg, and S. H. E. Kaufmann. 2006. *Mycobacterium tuberculosis* gene expression profiling within the context of protein networks. *Microbes Infect.* **8**:747–757.

Ratledge, C. 1999. Iron metabolism, p. 260–286. *In* C. Ratledge and J. Dale (ed.), *Mycobacteria: Molecular Biology and Virulence*. Blackwell Science, London, United Kingdom.

Ratledge, C., and F. G. Winder. 1962. The accumulation of salicylic acid by mycobacteria during growth on an iron-deficient medium. *Biochem. J.* **84**:501–506.

Rea, R. B., C. G. M. Gahan, and C. Hill. 2004. Disruption of putative regulatory loci in *Listeria monocytogenes* demonstrates a significant role for Fur and PerR in virulence. *Infect. Immun.* **72**:717–727.

Read, T. D., S. N. Peterson, N. Tourasse, L. W. Baillie, I. T. Paulsen, K. E. Nelson, H. Tettelin, D. E. Fouts, J. A. Eisen, S. R. Gill, et al. 2003. The genome sequence of *Bacillus anthracis* Ames and comparison to closely related bacteria. *Nature* **423**:81–86.

Reed, M. B., P. Domenech, C. Manca, H. Su, A. K. Barczak, B. N. Kreiswirth, G. Kaplan, and C. E. Barry. 2004. A glycolipid of hypervirulent tuberculosis strains that inhibits the innate immune response. *Nature* **431**:84–87.

Ricci, S., R. Janulczyk, and L. Bjorck. 2002. The regulator PerR is involved in oxidative stress response and iron homeostasis and is necessary for full virulence of *Streptococcus pyogenes*. *Infect. Immun.* **70**:4968–4976.

Rodriguez, G. M. 2006. Control of iron metabolism in *Mycobacterium tuberculosis*. *Trends Microbiol.* **14**:320–327.

Rodriguez, G. M., B. Gold, M. Gomez, O. Dussurget, and I. Smith. 1999. Identification and characterization of two divergently transcribed iron regulated genes in *Mycobacterium tuberculosis*. *Tuber. Lung Dis.* **79**:287–298.

Rodriguez, G. M., and I. Smith. 2003. Mechanisms of iron regulation in mycobacteria: role in physiology and virulence. *Mol. Microbiol.* **47**:1485–1494.

Rodriguez, G. M., and I. Smith. 2006. Identification of an ABC transporter required for iron acquisition and virulence in *Mycobacterium tuberculosis*. *J. Bacteriol.* **188**:424–430.

Rodriguez, G. M., M. I. Voskuil, B. Gold, G. K. Schoolnik, and I. Smith. 2002. *ideR*, an essential gene in *Mycobacterium tuberculosis*: role of IdeR in iron-dependent gene expression, iron metabolism, and oxidative stress response. *Infect. Immun.* **70**:3371–3381.

Rolerson, E., A. Swick, L. Newlon, C. Palmer, Y. Pan, B. Keeshan, and G. Spatafora. 2006. The SloR/Dlg metalloregulator modulates *Streptococcus mutans* virulence gene expression. *J. Bacteriol.* **188**:5033–5044.

Roosenberg, J. M. N., Y. M. Lin, Y. Lu, and M. J. Miller. 2000. Studies and syntheses of siderophores, microbial iron chelators, and analogs as potential drug delivery agents. *Curr. Med. Chem.* **7**:159–197.

Russell, L. M., S. J. Cryz, Jr., and R. K. Holmes. 1984. Genetic and biochemical evidence for a siderophore-dependent iron transport system in *Corynebacterium diphtheriae*. *Infect. Immun.* **45**:143–149.

Sala, C., F. Forti, E. Di Florio, F. Canneva, A. Milano, G. Riccardi, and D. Ghisotti. 2003. *Mycobacterium tuberculosis* FurA autoregulates its own expression. *J. Bacteriol.* **185**:5357–5362.

Schaible, U. E., H. L. Collins, F. Priem, and S. H. E. Kaufmann. 2002. Correction of the iron overload defect in beta-2-microglobulin knockout mice by lactoferrin abolishes their increased susceptibility to tuberculosis. *J. Exp. Med.* **196**:1507–1513.

Schiering, N., X. Tao, H. Zeng, J. R. Murphy, G. A. Petsko, and D. Ringe. 1995. Structures of the apo- and the metal ion-activated forms of the diphtheria tox repressor from *Corynebacterium diphtheriae*. *Proc. Natl. Acad. Sci. USA* **92**:9843–9850.

Schmitt, M. 1997. Transcription of the *Corynebacterium diphtheriae hmuO* gene is regulated by iron and heme. *Infect. Immun.* **65**:4634–4641.

Schmitt, M. P. 1999. Identification of a two-component signal transduction system from *Corynebacterium diphtheriae* that activates gene expression in response to the presence of heme and hemoglobin. *J. Bacteriol.* **181**:5330–5340.

Schmitt, M. P., and R. K. Holmes. 1993. Analysis of diphtheria toxin repressor-operator interactions and characterization of a mutant repressor with decreased binding activity for divalent metals. *Mol. Microbiol.* **9**:173–181.

Schneider, R., and K. Hantke. 1993. Iron-hydroxamate uptake systems in *Bacillus subtilis*: identification of a lipoprotein as part of a binding protein-dependent transport system. *Mol. Microbiol.* **8**:111–121.

Schüpbach, P., V. Osterwalder, and B. Guggenheim. 1995. Human root caries: microbiota in plaque covering sound, carious and arrested carious root surfaces. *Caries Res.* **29**:382–395.

Sebulsky, M. T., and D. E. Heinrichs. 2001. Identification and characterization of *fhuD1* and *fhuD2*, two genes involved in iron-hydroxamate uptake in *Staphylococcus aureus*. *J. Bacteriol.* **183**:4994–5000.

Sebulsky, M. T., D. Hohnstein, M. D. Hunter, and D. E. Heinrichs. 2000. Identification and characterization of a membrane permease involved in iron-hydroxamate transport in *Staphylococcus aureus*. *J. Bacteriol.* **182**: 4394–4400.

Siegrist, M. S., M. Unnikrishnan, M. J. McConnell, M. Borowsky, T. Y. Cheng, N. Siddiqi, S. M. Fortune, D. B. Moody, and E. J. Rubin.

2009. Mycobacterial Esx-3 is required for mycobactin-mediated iron acquisition. *Proc. Natl. Acad. Sci. USA* **106**:18792–18797.

Simon, N., V. Coulanges, P. Andre, and D. J. Vidon. 1995. Utilization of exogenous siderophores and natural catechols by *Listeria monocytogenes*. *Appl. Environ. Microbiol.* **61**:1643–1645.

Skaar, E. P., A. H. Gaspar, and O. Schneewind. 2004b. IsdG and IsdI, heme-degrading enzymes in the cytoplasm of *Staphylococcus aureus*. *J. Biol. Chem.* **279**:436–443.

Skaar, E. P., A. H. Gaspar, and O. Schneewind. 2006. *Bacillus anthracis* IsdG, a heme-degrading monooxygenase. *J. Bacteriol.* **188**:1071–1080.

Skaar, E. P., M. Humayun, T. Bae, K. L. DeBord, and O. Schneewind. 2004a. Iron-source preference of *Staphylococcus aureus* infections. *Science* **305**:1626–1628.

Skaar, E. P., and O. Schneewind. 2004. Iron-regulated surface determinants (Isd) of *Staphylococcus aureus*: stealing iron from heme. *Microbes Infect.* **6**:390–397.

Smith, G. A., and D. A. Portnoy. 1997. How the *Listeria monocytogenes* ActA protein converts actin polymerization into a motile force. *Trends Microbiol.* **5**:272–276.

Smith, H., J. Keppie, and J. L. Stanley. 1954. Observations on the cause of death in experimental anthrax. *Lancet* **267**:474–476.

Sow, F. B., W. C. Florence, A. R. Satoskar, L. S. Schlesinger, B. S. Zwilling, and W. P. Lafuse. 2007. Expression and localization of hepcidin in macrophages: a role in host defense against tuberculosis. *J. Leukoc. Biol.* **82**:934–945.

Speziali, C. D., S. E. Dale, J. A. Henderson, E. D. Vines, and D. E. Heinrichs. 2006. Requirement of *Staphylococcus aureus* ATP-binding cassette-ATPase FhuC for iron-restricted growth and evidence that it functions with more than one iron transporter. *J. Bacteriol.* **188**:2048–2055.

Stojiljkovic, I., and K. Hantke. 1995. Functional domains of the *Escherichia coli* ferric uptake regulator protein (Fur). *Mol. Gen. Genet.* **247**:199–205.

Sturgill-Koszycki, S., U. E. Schaible, and D. G. Russell. 1996. Mycobacterium-containing phagosomes are accessible to early endosomes and reflect a transitional state in normal phagosome biogenesis. *EMBO J.* **15**:6960–6968.

Tai, S.-P. S., A. E. Krafft, P. Nootheti, and R. K. Holmes. 1990. Coordinate regulation of siderophore and diphtheria toxin production by iron in *Corynebacterium diphtheriae*. *Microb. Pathog.* **9**:267–273.

Tai, S. S., C. J. Lee, and R. E. Winter. 1993. Hemin utilization is related to virulence of *Streptococcus pneumoniae*. *Infect. Immun.* **61**:5401–5405.

Tao, X., J. Boyd, and J. R. Murphy. 1992. Specific binding of the diphtheria *tox* regulatory element DtxR to the *tox* operator requires divalent heavy metal ions and a 9-base-pair interrupted palindromic sequence. *Proc. Natl. Acad. Sci. USA* **89**:5897–5901.

Tarlovsky, Y., M. Fabian, E. Solomaha, E. Honsa, J. S. Olson, and A. W. Maresso. 2010. A *Bacillus anthracis* S-layer homology protein that binds heme and mediates heme delivery to IsdC. *J. Bacteriol.* **192**:3503–3511.

Tashjian, J. J., and S. G. Campbell. 1983. Interaction between caprine macrophages and *Corynebacterium pseudotuberculosis*: an electron microscopic study. *Am. J. Vet. Res.* **44**:690–693.

Taylor, C. M., M. Beresford, H. A. S. Epton, D. C. Sigee, G. Shama, P. W. Andrew, and I. S. Roberts. 2002. *Listeria monocytogenes relA* and *hpt* mutants are impaired in surface-attached growth and virulence. *J. Bacteriol.* **184**:621–628.

Tettelin, H., K. E. Nelson, I. T. Paulsen, J. A. Eisen, T. D. Read, S. Peterson, J. Heidelberg, R. T. DeBoy, D. H. Haft, R. J. Dodson, et al. 2001. Complete genome sequence of a virulent isolate of *Streptococcus pneumoniae*. *Science* **293**:498–506.

Throup, J. P., K. K. Koretke, A. P. Bryant, K. A. Ingraham, A. F. Chalker, Y. Ge, A. Marra, N. G. Wallis, J. R. Brown, D. J. Holmes, et al. 2000. A genomic analysis of two-component signal transduction in *Streptococcus pneumoniae. Mol. Microbiol.* **35:**566–576.

Tilney, L. G., and D. A. Portnoy. 1989. Actin filaments and the growth, movement, and spread of the intracellular bacterial parasite, *Listeria monocytogenes. J. Cell Biol.* **109:**1597–1608.

Timmins, G. S., and V. Deretic. 2006. Mechanisms of action of isoniazid. *Mol. Microbiol.* **62:**1220–1227.

Tolosano, E., and F. Altruda. 2002. Hemopexin: structure, function, and regulation. *DNA Cell Biol.* **21:**297–306.

Torres, V. J., A. S. Attia, W. J. Mason, M. I. Hood, B. D. Corbin, F. C. Beasley, K. L. Anderson, D. L. Stauff, W. H. McDonald, L. J. Zimmerman, et al. 2010. *Staphylococcus aureus* Fur regulates the expression of virulence factors that contribute to the pathogenesis of pneumonia. *Infect. Immun.* **78:**1618–1628.

Torres, V. J., G. Pishchany, M. Humayun, O. Schneewind, and E. P. Skaar. 2006. *Staphylococcus aureus* IsdB is a hemoglobin receptor required for heme iron utilization. *J. Bacteriol.* **188:**8421–8429.

Toukoki, C., K. M. Gold, K. S. McIver, and Z. Eichenbaum. 2010. MtsR is a dual regulator that controls virulence genes and metabolic functions in addition to metal homeostasis in the group A streptococcus. *Mol. Microbiol.* **76:**971–989.

Uchida, T., D. M. Gill, and A. M. Pappenheimer, Jr. 1971. Mutation in the structural gene for diphtheria toxin carried by temperate phage. *Nat. New Biol.* **233:** 8–11.

Ulijasz, A. T., D. R. Andes, J. D. Glasner, and B. Weisblum. 2004. Regulation of iron transport in *Streptococcus pneumoniae* by RitR, an orphan response regulator. *J. Bacteriol.* **186:**8123–8136.

Vértesy, L., W. Aretz, H.-W. Fehlhaber, and H. Kogler. 1995. Salmycin A–D, Antibiotika aus *Streptomyces violaceus*, DSM 8286, mit Siderophor-Aminoglycosid-Struktur. *Helv. Chim. Acta* **78:** 46–60.

von Eiff, C., R. A. Proctor, and G. Peters. 2001. Coagulase-negative staphylococci. Pathogens have major role in nosocomial infections. *Postgrad. Med.* **110:**63–64, 69–70, 73–76.

Voyich, J. M., K. R. Braughton, D. E. Sturdevant, A. R. Whitney, B. Saïd-Salim, S. F. Porcella, R. D. Long, D. W. Dorward, D. J. Gardner, B. N. Kreiswirth, et al. 2005. Insights into mechanisms used by *Staphylococcus aureus* to avoid destruction by human neutrophils. *J. Immunol.* **175:**3907–3919.

Voyich, J. M., C. Vuong, M. DeWald, T. K. Nygaard, S. Kocianova, S. Griffith, J. Jones, C. Iverson, D. E. Sturdevant, K. R. Braughton, et al. 2009. The SaeR/S gene regulatory system is essential for innate immune evasion by *Staphylococcus aureus. J. Infect. Dis.* **199:**1698–1706.

Wandersman, C., and P. Delepelaire. 2004. Bacterial iron sources: from siderophores to hemophores. *Annu. Rev. Microbiol.* **58:**611–647.

Wang, L., and B. J. Cherayil. 2009. Ironing out the wrinkles in host defense: interactions between iron homeostasis and innate immunity. *J. Innate Immun.* **1:**455–464.

Wang, Z., C. Li, M. Ellenburg, E. Soistman, J. Ruble, B. Wright, J. X. Ho, and D. C. Carter. 2006. Structure of human ferritin L chain. *Acta Crystallogr. D* **62:**800–806.

Weems, J. J., Jr. 2001. The many faces of *Staphylococcus aureus* infection. Recognizing and managing its life-threatening manifestations. *Postgrad. Med.* 110, 24–26, 29–31, 35–36.

Welcher, B. C., J. H. Carra, L. DaSilva, J. Hanson, C. S. David, M. J. Aman, and S. Bavari. 2002. Lethal shock induced by streptococcal pyrogenic exotoxin A in mice transgenic for human leukocyte antigen and human CD4 receptors: implications for development of vaccines and therapeutics. *J. Infect. Dis.* **186:**501–510.

White, A., X. Ding, J. C. vanderSpek, J. R. Murphy, and D. Ringe. 1998. Structure of the metal-ion-activated diphtheria toxin repressor/*tox* operator complex. *Nature* **394:**502–506.

Winston, D. J., D. V. Dudnick, M. Chapin, W. G. Ho, R. P. Gale, and W. J. Martin. 1983. Coagulase-negative staphylococcal bacteremia in patients receiving immunosuppressive therapy. *Arch. Intern. Med.* **143:**32–36.

Wylie, G. P., V. Rangachari, E. A. Bienkiewicz, V. Marin, N. Bhattacharya, J. F. Love, J. R. Murphy, and T. M. Logan. 2004. Prolylpeptide binding by the prokaryotic SH3-like domain of the diphtheria toxin repressor: a regulatory switch. *Biochemistry* **44:**40–51.

Xiao, Q., X. Jiang, K. J. Moore, Y. Shao, H. Pi, I. Dubail, A. Charbit, S. M. Newton, and P. E. Klebba. 2011. Sortase independent and dependent systems for acquisition of haem and haemoglobin in *Listeria monocytogenes. Mol. Microbiol.* **80:**1581–1597.

Xiong, A., V. K. Singh, G. Cabrera, and R. K. Jayaswal. 2000. Molecular characterization of the ferric-uptake regulator, Fur, from *Staphylococcus aureus. Microbiology* **146:**659–668.

Yu, G. 2009. Pathogenic *Bacillus anthracis* in the progressive gene losses and gains in adaptive evolution. *BMC Bioinformatics* **10:**S3.

Zahrt, T. C., and V. Deretic. 2000. An essential two-component signal transduction system in *Mycobacterium tuberculosis. J. Bacteriol.* **182:**3832–3838.

Regulation of Bacterial Virulence
Edited by Michael L. Vasil and Andrew J. Darwin
© 2013 ASM Press, Washington, DC   doi:10.1128/9781555818524.ch6

Chapter 6

Iron Regulation and Virulence in Gram-Negative Bacterial Pathogens with *Yersinia pestis* as a Paradigm

ROBERT D. PERRY AND KATHLEEN A. McDONOUGH

INTRODUCTION

The regulation of gene expression in prokaryotes has been a fundamental theme in the study of bacterial infections of animals, as well as plants, for decades. Additionally, the examination of mechanisms of iron-regulated gene expression, not merely associated with the expression of iron acquisition systems but also associated with the expression of key virulence determinants (e.g., toxins and adherence factors), constitutes a primary subtheme in bacterial pathogenesis. The chapter on iron regulation of virulence in gram-positive organisms (Chapter 5) summarizes the difficulty in acquiring iron, which is essential for nearly all bacteria, and the necessity of regulating transport systems and metabolism in response to available iron. While difficulty in obtaining sufficient iron is common, the level of iron within bacterial cells must also be carefully regulated due to iron toxicity as described in Chapter 6. In this chapter, we cover the role of iron regulation in the virulence of gram-negative bacterial pathogens, where possible, using *Yersinia pestis* as a paradigm.

Y. pestis was among the first group of pathogens in which the relationship between iron acquisition from the host and disease outcome was noted (Jackson and Burrows, 1956). It is a member of the *Enterobacteriaceae* family and causes bubonic, septicemic, and pneumonic plague. Although responsible for three pandemics, plague is a zoonotic disease primarily affecting rodents and their associated fleas. After ingestion of an infected blood meal, the organism grows in the flea midgut. Transmission to mammals occurs by early-phase (biofilm-independent) and biofilm-dependent (or blocked-flea) mechanisms. From the site of the fleabite, bacterial cells spread via the lymphatics to a regional lymph node. Multiplication there causes the characteristic swelling (buboes) from which bubonic plague derives its name. The infection is spread through the bloodstream to internal organs (e.g., liver and spleen), again leading to overwhelming proliferation and subsequent septicemia. Sustained high bacterial numbers in the blood are required for infection of a naïve flea to complete the life cycle. In humans, there is a 2- to 8-day incubation period before the onset of relatively nonspecific symptoms—fever, weakness, chills, and regional acutely swollen lymph nodes (buboes)—with an untreated fatality rate of ~50%. Proper therapy reduces this to ~14%. Septicemic plague appears to result from direct inoculation (natural or artificial) into the bloodstream, bypassing the early lymphatic stage of the bubonic disease. In humans and nonhuman primates, infection of the lungs (secondary pneumonic plague) can lead to respiratory droplet spread (primary pneumonic plague). Onset of symptoms for primary pneumonic plague includes relatively nonspecific bronchial to multilobar pneumonia symptoms, including fever, lymphadenopathy, and coughing up of infected sputum and blood, as well as nausea, vomiting, abdominal pain, and diarrhea. Death (100% if untreated) occurs from pneumonia, although *Y. pestis* cells have spread to other internal organs via the bloodstream. Twenty-four hours after the onset of symptoms, antibiotic treatment is often too late (Perry and Fetherston, 1997; Inglesby et al., 2000; Gage and Kosoy, 2005; Eisen et al., 2006; Vetter et al., 2010).

Early observations on the effects of bacterial growth during iron deprivation included reduced growth rates, altered metabolism, and increased levels of certain outer membrane (OM) proteins, as well as increased synthesis of toxins and siderophore-dependent iron transport systems (Lankford, 1973). Numerous microarray and/or proteomic studies in a variety of bacteria have demonstrated a global response to iron limitation. Andrews et al. concisely summarized five bacterial strategies for iron homeostasis and prevention of iron toxicity: (i) high-affinity iron uptake systems to acquire iron during scarcity, (ii) intracellular iron storage to avoid toxicity during

Robert D. Perry • Department of Microbiology, Immunology, and Molecular Genetics, University of Kentucky, Lexington, KY 40536-0298.
Kathleen A. McDonough • Wadsworth Center, New York State Department of Health, Albany, NY 12201-2002.

times of iron surplus and for use during deprivation, (iii) degradation mechanisms for iron-induced reactive oxygen species, (iv) altered metabolism employing Fe-independent enzymes during iron starvation, and (v) iron-responsive regulatory systems that coordinate use of the above four mechanisms (Andrews et al., 2003). In most gram-negative bacteria, including *Y. pestis*, the alterations in mRNA and protein levels in response to iron availability noted above are due to inherent iron requirements and iron homeostasis mechanisms controlled by two primary regulatory systems: the iron (Fe) uptake regulation protein Fur and the small RNA (sRNA) RyhB or its analogs (e.g., PrrF and NrrF). This chapter focuses on these two regulators, their regulons, and their roles in virulence. While Fur and RyhB regulons are often described separately, they are in fact interrelated regulatory networks. Fe-Fur represses *ryhB* transcription, and the base-pairing regulatory sRNA RyhB in some cases indirectly represses translation of *fur* mRNA. Thus, assessment of Fur versus RyhB regulation, even with *fur* or *ryhB* mutations, is often complex. In many cases the mechanism of iron regulation of virulence factors and metabolism has not yet been elucidated. Conclusive evidence of *direct* regulation by Fur requires demonstrating interaction of the Fur protein with the promoter region of the iron-regulated gene/locus, whereas proof of direct RyhB regulation requires demonstrating RNA-RNA interaction. Evidence of such interactions is still relatively uncommon. More often, iron regulation, deregulation in a *fur* mutant, and the presence of a Fur box are used to conclude that the gene/locus is directly Fur regulated. However, this does not conclusively distinguish between direct and indirect regulation by Fur (e.g., by RyhB).

While we highlight key discoveries and advances in understanding iron regulation and virulence in gram-negative pathogens, we will give examples of the types of virulence factors regulated by iron but not attempt a comprehensive listing of the pathogens using these mechanisms. We do not cover the alternative iron-responsive regulators RirA and Irr found in the alphaproteobacteria, which have been reviewed recently (Johnston et al., 2007). Expression of some systems under iron-deficient conditions is also enhanced by siderophore signals via AraC family regulators, extracytoplasmic sigma factors (ECFs), or two-component regulators in a number of bacteria. The mechanisms for indirect iron regulation in these systems are not discussed here, but examples of these mechanisms have been covered fairly recently in reviews on yersiniabactin, regulation of ferric citrate uptake, and iron regulation in *Pseudomonas* and *Bordetella* (Braun et al., 2006; Vasil, 2007; Brickman

and Armstrong, 2009; Cornelis et al., 2009; Perry and Fetherston, 2011). Finally, we do not discuss in detail the PmrAB two-component regulatory system which has been reviewed recently. In *Salmonella*, the PmrB sensor responds to high levels of Fe as well as other signals and, with its response regulator, PmrA, regulates >20 confirmed genes and possibly up to 100. Although the level of iron required to activate PmrAB (0.1 to 1 mM) is unlikely to be encountered in mammalian hosts, PmrA activates expression of genes in a number of pathogens whose products modify lipopolysaccharide, thus increasing resistance to antimicrobial peptides (Gunn, 2008).

FUR PROPERTIES AND REGULATORY MECHANISMS

Discovery as well as Early and General Observations

The discovery and naming of Fur began with a *Salmonella enterica* serovar Typhimurium mutant that constitutively expressed a group of OM proteins, some involved in enterobactin and ferrichrome siderophore uptake, normally repressed by iron (Bennett and Rothfield, 1976; Ernst et al., 1978). Shortly thereafter, an *Escherichia coli fur* mutant was selected, the mutation mapped, and the *fur* gene cloned and used to complement the mutation and sequence the gene (Hantke, 1981, 1982, 1984; Bagg and Neilands, 1985; Schäffer et al., 1985). Fur regulation was then found in *Bordetella bronchiseptica*, *Pseudomonas aeruginosa*, *Vibrio cholerae*, *Shigella flexneri*, and *Y. pestis* (Schmitt and Payne, 1988; Stoebner and Payne, 1988; Prince et al., 1991; Staggs and Perry, 1991; Staggs et al., 1994; Brickman and Armstrong, 1995), followed by a host of other gram-negative bacteria. Indeed, Fur is now recognized as a member of a superfamily of transcriptional regulators responsive to iron (Fur), zinc (Zn, Zur), manganese (Mn, Mur), peroxides via Fe or Mn (PerR), and nickel (Ni, Nur) (Hantke, 2005; Lee and Helmann, 2007). Studies, primarily in *E. coli*, developed the now-well-accepted model of iron-responsive transcriptional repression by Fur (Fig. 1A). While *fur* genes from a variety of different bacteria complement an *E. coli fur* mutant, *fur* mutations fall into two distinct groups. In *E. coli*, *Helicobacter pylori*, *S. flexneri*, *V. cholerae*, and *Y. pestis*, *fur* deletion or disruption mutations have been constructed (Hantke, 1984; Schmitt and Payne, 1988; Stoebner and Payne, 1988; Staggs et al., 1994; Bereswill et al., 2000). However, *fur* appears to be essential in *Burkholderia pseudomallei*, *Haemophilus ducreyi*, *Legionella pneumophila*, *Neisseria*, *Pseudomonas*, and

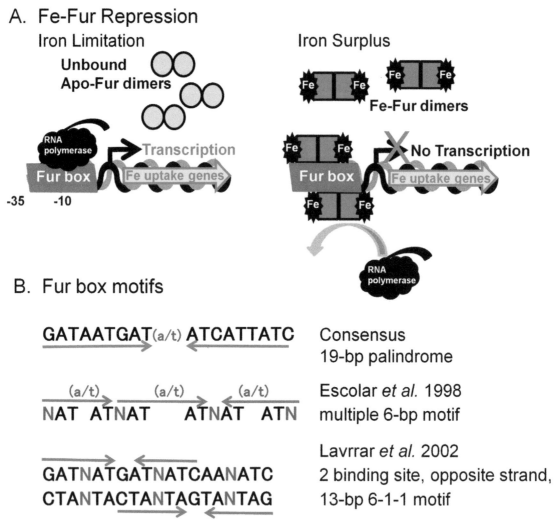

Figure 1. Model of Fe-Fur transcriptional repression and Fur box motifs. (A) Fur boxes for Fe-Fur binding generally overlap the –10 region of regulated promoters. During iron limitation, cytoplasmic Fe levels are sufficiently low so that Fur is primarily in its apo form. apo-Fur dimers have a lower affinity for binding to the Fur box, allowing RNA polymerase access, and transcription proceeds. Under iron surplus conditions, Fe^{2+} in the bacterial cytoplasm is bound by Fur. Fe-Fur dimers bind to Fur boxes, preventing access of RNA polymerase to the promoter region, thereby preventing transcription. (B) Three alternative interpretations of Fur box motifs are shown. Arrows indicate the direct or indirect repeated sequences. The Lavrrar et al. 2002 model and data provided from the crystallographic structure of Fur_{Pa} support a model in which two Fur dimers bind on opposite sides of the DNA helix of a single Fur box, in a Fur-dependent promoter, as shown in panel A (Lavrrar et al., 2002; Pohl et al., 2003). This view is also supported by the consistent observation that Fur from *E. coli* and *P. aeruginosa* protects no less than 27 to 30 bp in DNase I assays (Escolar et al., 1998; Lavrrar et al., 2002; Pohl et al., 2003). doi:10.1128/9781555818524.ch6f1

Vibrio anguillarum; at least, attempts to construct null mutations have been unsuccessful. Mutations in *fur* that affect iron regulation in these bacteria are due to amino acid changes in Fur (Berish et al., 1993; Prince et al., 1993; Thomas and Sparling, 1994; Tolmasky et al., 1994; Venturi et al., 1995; Carson et al., 1996; Hickey and Cianciotto, 1997; Loprasert et al., 2000). This suggests that Fur regulates some distinct genes in these two groups of pathogens. Perhaps Fur plays a key or unique role in regulating responses to oxidative damage in bacteria in which it appears to be essential. This speculation is supported by the fact that an *E. coli fur recA* double mutant is nonviable

under aerobic conditions yet is still able to grow anaerobically via fermentation (Touati et al., 1995).

Fur Properties

The *E. coli* 17-kDa Fur protein forms a homodimer even in the absence of Fe^{2+}. The C-terminal portion of Fur is involved in dimerization, while the N-terminal portion has a winged-helix motif that is involved in DNA binding. Unlike many other transcriptional regulators, the Fur protein level is high: 5,000 or 10,000 copies in *E. coli* and 2,500 or 7,500 copies in *V. cholerae* (exponential versus stationary phases for

```
        {   α1    }           {    α2      }  |  {    α3     }    {      α4
Ec   TDNNTALKKAGLKVTLPRLKILEVLQEPDNHHVSAEDLYKRLIDMGEEIGLATVYRVLNQ  60
Yp   TDNNKALKNAGLKVTLPRLKILEVLQNPACHHVSAEDLYKILIDIGEEIGLATVYRVLNQ  60
Vc   SDNNQALKDAGLKVTLPRLKILEVLQQPECQHISAEELYKKLIDLGEEIGLATVYRVLNQ  60
Pa   VENS-ELRKAGLKVTLPRVKILQMLDSAEQRHMSAEDVYKALMEAGEDVGLATVYRVLTQ  59
Con   :*.    *:.*********:***::*..  :*:.***::** *::  **:::*******.*

                                        |
        }  [ β1]      [   β2   ]  [β3  ]     [β4] {      α3        }   [
Ec   FDDAGIVTRHNFEGGKSVFELTQQHHHDHLICLDCGKVIEFSDDSIEARQREIAAKHGIR  120
Yp   FDDAGIVTRHNFEGGKSVFELTQQHHHDHLICLDCGKVIEFSNESIESLQREIAKQHGIK  120
Vc   FDDAGIVTRHHFEGGKSVFELSTQHHHDHLVCLDCGEVIEFSDDVIEQRQKEIAAKYNVQ  120
Pa   FEAAGLVVRHNFDGGHAVFELADSGHHDHMVCVDTGEVIEFMDAEIEKRQKEIVRERGFE  119
Con   *:  **:*.**:*:**::****:  .  ****::*:* *:****  :   **  *:**.  :  ...

          B5      ]
Ec   LTNHSLYLYGHC-AEGDCREDEHAHEGK-  147
Yp   LTNHSLYLYGHC-ETGNCREDESAHSKR-  147
Vc   LTNHSLYLYGKCGSDGSCKDNPNAHKPKK  149
Pa   LVDHNLVLY-------------VRKKK-  133
Con   *.:*.* **          .:. :
```

Figure 2. Alignment of Fur amino acid sequences from *E. coli* (Ec), *Y. pestis* (Yp), *V. cholerae* (Vc), and *P. aeruginosa* (Pa). The four sequences were aligned using ClustalW2 (Larkin et al., 2007) and the EMBL-EBI website. α-Helical regions (bold, purple text and [α] labels) and β-strands (bold, green text and [β] labels) are shown for Fur$_{Ec}$ based on X-ray data (Pecqueur et al., 2006). A consensus sequence (Con) based on these four sequences is shown. Asterisks indicate identical residues, while colons and periods indicate conservative and semiconservative changes, respectively. Residues involved in Zn binding at the Zn1 structural binding site (underlined, bold, blue text) are shown for all Fur proteins except Fur$_{Yp}$. Residues for the regulatory site for Fe^{2+} binding (site 2) are shown as underlined, bold, red text for Fur$_{Vc}$ and Fur$_{Pa}$. While these residues are conserved in Fur$_{Ec}$, four different Cys residues (also in underlined, bold, blue text) have been implicated in Zn binding and dimerization. However, Fur$_{Pa}$ has only a single Cys residue, which functions only in dimerization. Zn1, site 1, and the Cys residues are conserved in Fur$_{Yp}$ although no structure/function analyses have been performed on the *Y. pestis* protein. doi:10.1128/9781555818524.ch6f2

both bacteria). It has been speculated that the reason for the large number of Fur proteins is (i) the large number of promoters regulated, (ii) the polymerization of Fur at some promoters, and/or (iii) that Fur has a second function to sequester free iron in the cytoplasm (Coy and Neilands, 1991; Stojiljkovic and Hantke, 1995; Watnick et al., 1997; Zheng et al., 1999; Andrews et al., 2003).

Although not present in all Fur proteins, *E. coli* Fur has at least one nonregulatory or structural zinc ion per dimer. Fur binds one Fe^{2+} atom per monomer but binds other metals—all with dissociation constants ranging from 20 to 60 nM (Zn^{2+} > Co^{2+} > Fe^{2+} > Mn^{2+} > Fe^{3+}). An electrophoretic mobility shift assay (EMSA) showed Fe^{2+}-Fur binding to a synthetic oligonucleotide containing a Fur box. For convenience, Mn at a high concentration (often 100 µM) is routinely used in vitro, instead of Fe^{2+}, to demonstrate interaction of Fur with specific promoter region sequences. Although numerous articles state that the free concentrations of these other cations in the bacterial cell are too low to be physiologically relevant, several studies (see below) have demonstrated in vivo repression by Mn-Fur. Thus, we have been careful to indicate where Mn-Fur was used in in vitro

studies examining interaction of Fur with promoters. It should be noted that a number of *fur* mutations, such as those in *E. coli*, *Klebsiella*, *P. aeruginosa*, and *Serratia*, were selected by resistance to high levels of Mn (e.g., 10 mM) and that Mn is mutagenic at these high levels, affecting the fidelity of DNA polymerase (Hantke, 1987; Prince et al., 1993; Smith et al., 1996; Escolar et al., 1999; Andrews et al., 2003; Mills and Marletta, 2005).

Since the discovery of Fur, numerous groups have contributed important information on the structure and function of Fur proteins from various bacteria. A major advance was publication of the crystal structure of *P. aeruginosa* Fur (Fur$_{Pa}$). Now, crystal structures of *V. cholerae* Fur and the N-terminal domain (residues 1 to 82) of *E. coli* Fur have been obtained and analyzed. β-Strands and α-helical structures based on Fur$_{Ec}$ and Zn-binding sites are shown in Fig. 2. Two metal-binding sites (Zn1 and site 2) per monomer were identified in Fur$_{Pa}$ and Fur$_{Vc}$. During isolation of Fur proteins, these sites are often occupied by Zn. A variety of studies now favor site 2 as the regulatory Fe-binding site, with Zn1 serving as the structural Zn-binding site. For Fur$_{Vc}$, Mn or Fe^{2+} could be bound in vitro to both sites. Combinations of these three metals bound at both sites or one Zn bound to Fur$_{Vc}$ were able to interact with *fur* promoter DNA. For Fur$_{Ec}$, there are many studies that indicate that four Cys residues (Cys92, Cys95, Cys132, and Cys137; underlined, bold, blue text in Fig. 2) are involved in Zn binding at Zn1 (the structural site) and dimerization. Mutation of Cys92 or Cys95 causes a loss of Fur activity. In contrast, Fur$_{Pa}$ has only one of these two Cys residues, and this Cys plays a role in Fur$_{Pa}$ dimerization, not metal binding. Fur$_{Pa}$ also lacks the equivalent Fur$_{Ec}$ residues Cys132 and Cys137, since Fur$_{Pa}$ terminates in three Lys residues from 132 to 135 (Fig. 2). In addition, Fur$_{Vc}$ has both Cys residues, but they are not involved in metal binding; instead, Cys93 and Cys133 (equivalent to Cys92 and Cys132 in Fur$_{Ec}$) formed a disulfide linkage in the Fur$_{Vc}$ crystal (Coy et al., 1994; Pohl et al., 2003; Pecqueur et al., 2006; Sheikh and Taylor, 2009). For both Fur$_{Vc}$ and Fur$_{Pa}$, there is evidence that an Asp, Glu, and two His residues are involved in Zn binding to Zn1 (underlined, bold, blue text in Fig. 2). This discrepancy in the Fur$_{Ec}$ and Fur$_{Vc}$/Fur$_{Pa}$ Zn1 sites is unresolved. The residues involved in metal binding for regulatory site 2 are similar in Fur$_{Vc}$ and Fur$_{Pa}$ (underlined, bold, red text in Fig. 2). Fur$_{Ec}$ has these residues as well, but they have not been clearly linked to metal binding. Finally, the experimental results of Pecqueur et al. indicate that the N-terminal α-helix is present in the activated form of Fur$_{Ec}$ but absent in the unactivated form; this supports the conclusion by Pohl et al. that this N-terminal α-helix in Fur$_{Pa}$ is required for effective DNA binding (Pohl et al., 2003; Pecqueur et al., 2006).

fur Loci and Their Regulation

In *E. coli*, the *fur* locus and its regulation have been characterized (Fig. 3). An essential flavodoxin (FldA), involved in redox reactions, is encoded upstream of *fur*. An *fldA-fur* transcript is initiated from the *fldA* promoter and is induced ~10-fold by SoxS in response to redox stress from superoxide. Downstream of *fldA*, a strong *fur* promoter is activated ~10-fold by OxyR in response to H$_2$O$_2$-mediated redox stress, although Fur protein levels increase only 2-fold due to this stress. Although CRP (cyclic AMP receptor protein) regulated a *fur::lacZ* fusion, likely through interaction with a CAP box upstream of a weak *fur* promoter, this needs to be reassessed since the entire *fur* promoter region was not used. Finally, the weak *fur* promoter is autoregulated by Fur, repressing transcription under excess iron conditions (de Lorenzo et al., 1988; Zheng et al., 1999; Andrews et al., 2003; de Lorenzo et al., 2004). More recently, indirect downregulation of *fur* translation by the *E. coli* sRNA RyhB has been demonstrated (see section on RyhB below) (Večerek et al., 2007). In addition to *E. coli*, the apparent *fldA-fur* bicistronic operon is present in *Klebsiella pneumoniae*, *H. ducreyi*, *H. influenzae*, *S. flexneri*, and *Y. pestis*. However, in *B. bronchiseptica*, *H. pylori*, *L. pneumophila*, *Neisseria gonorrhoeae*, *Neisseria meningitidis*, *P. aeruginosa*, and *V. cholerae*, *fur* is not

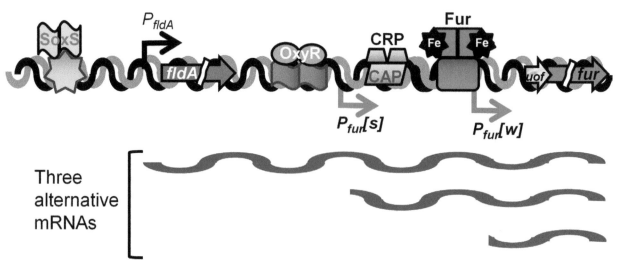

Figure 3. Genetic organization of the *E. coli fur* locus. Designated arrows represent *fur*, *fldA*, and *uof* (upstream of *fur*) genes and show their direction of transcription. Smaller arrows indicate promoters. Transcriptional regulators CRP, Fur, OxyR, and SoxS are shown "bound" to their binding sites (similarly colored figures on the double DNA strand) within promoter regions. Indirect regulation of *fur* translation by RyhB through translation of *uof* is not shown here (see Fig. 6C). The three alternative mRNAs are shown. Genes, transcriptional regulators, and promoter elements are not drawn to scale. doi:10.1128/9781555818524.ch6f3

located downstream of *fldA* (Zheng et al., 1999; NCBI Entrez Genome Project).

Autoregulation of Fur appears to be the most widespread regulatory mechanism for Fur expression, with putative Fur boxes in the *fur* promoter region of numerous bacteria. However, Fur and/or iron regulation of *fur* has been demonstrated in relatively few organisms: *E. coli*, *Edwardsiella tarda*, *H. pylori*, *N. gonorrhoeae*, and *Vibrio vulnificus* (de Lorenzo et al., 1988; Delany et al., 2002; Sebastian et al., 2002; Delany et al., 2003; Lee et al., 2003; Lee et al., 2007; Wang et al., 2008). Microarray and proteomic analyses in *Campylobacter jejuni* did not demonstrate iron regulation of *fur* expression (Holmes et al., 2005). Less common regulatory mechanisms controlling Fur expression are RpoS in *V. vulnificus* and NikR in *H. pylori* (Lee et al., 2003; Delany et al., 2005).

Regulation of *fur* in *Y. pestis* has not been investigated. *Y. pestis* KIM10+ has putative OxyR, CAP, and Fur boxes as well as a potential promoter region overlapping the Fur box between *fldA* and *fur*, as noted in *E. coli* (Fig. 1B and 2) (Deng et al., 2002; unpublished observations). Microarray analysis of *Y. pestis* strain 201 did not detect iron- or Fur-dependent regulation of *fur* (Zhou et al., 2006; Gao et al., 2008). If an *fldA-fur* transcript is made in *Y. pestis*, it is not regulated by SoxS, as there is no *soxS* gene in the *Y. pestis* genome (Deng et al., 2002).

Transcriptional Repression by Fur

In the model of iron-responsive transcriptional repression by Fur, growth with surplus iron supplies sufficient iron in the bacterial cytoplasm for Fur to bind Fe^{2+}, altering its structure so that Fe-Fur recognizes and binds to Fur boxes (also called iron boxes or Fur binding sites) in promoter regions, thus preventing transcription. During iron limitation, apo-Fur has a much lower affinity for its regulated promoters, allowing RNA polymerase access to initiate transcription (Fig. 1A). A 19-bp palindromic consensus sequence of the Fur box (Fig. 1B) was identified and became the standard for recognizing Fur boxes in gram-negative promoters, although the exact sequence is not found even in the *E. coli* genome. Indeed, some Fur boxes confer iron regulation with only 10 or 11 of 19 base pair matches (de Lorenzo et al., 1987; Ochsner and Vasil, 1996; Andrews et al., 2003; de Lorenzo et al., 2004). In A/T-rich *H. pylori*, the consensus Fur box (NNNNNAATAATNNTNANN) is clearly less conserved and is different from the *E. coli* consensus (Fig. 1B) (Merrell et al., 2003). Promoters with two or more overlapping or adjacent Fur boxes likely bind several Fur dimers. Fur binds to

the promoter for the aerobactin biosynthetic operon (*iucABCD*) at a high-affinity site, followed by polymerization of Fur dimers into regions with no apparent Fur box. The multiple Fur dimers wrap around the DNA helix. These observations led to a reinterpretation of the Fur box sequence as three repeats of a 6-bp motif (NAT[a/t]AT) (Fig. 1B), with increasing numbers of adjacent hexamer repeats allowing cooperative binding and polymerization of Fe-Fur. However, Fur residues required for this polymerization have not been identified, and the repeated hexamer motif does not entirely explain the observed corkscrew pattern of binding of Fur to some regulated promoters. Consequently, Lavrrar et al. envisioned the Fur box as overlapping 13-bp motifs with a 6-1-6 configuration (Fig. 1B). In this interpretation, two Fur dimers would bind to each Fur box on opposite sides of the DNA helix (shown in Fig. 1A) that would match the corkscrew binding of Fur (Coy et al., 1994; Escolar et al., 1998, 2000; Lavrrar et al., 2002; Andrews et al., 2003; Pohl et al., 2003; Pecqueur et al., 2006; Sheikh and Taylor, 2009).

Direct Activation by Fur

In some early studies, it was suggested not only that Fur repressed expression of some genes in the presence of iron but also that it activated gene expression. A mutation in *fur* lowered expression of these genes under iron surplus conditions. It was later found that many of these genes are directly regulated by RyhB and its analogs. Fur repression of this negative regulator appeared as activation in the *fur* mutant (Massé and Gottesman, 2002). However, some clear examples of direct Fur activation or at least Fur activation in the absence of RyhB homologs remain. The expression of *E. coli ftnA* has been shown to be directly activated by Fur independent of RyhB (Fig. 4). Fur binds to a region with five tandem Fur boxes upstream of the −35 region and competes with H-NS for binding to this region. It was proposed that H-NS binding at this site facilitates H-NS binding to additional binding sites within the promoter region. Thus, Fur binding activates transcription by displacing upstream H-NS binding, preventing H-NS-mediated DNA looping that limits RNA polymerase access to the promoter. This model of Fur activation (Fig. 4) is the first described mechanism for Fur activation and may turn out to be widespread (Nandal et al., 2010).

In *N. meningitidis*, DNase I footprinting in the promoter region of *norB* (encoding nitric oxide reductase) shows protection by Mn-Fur around a Fur box ~40 nucleotides (nt) upstream of the −35 region. In vitro transcription of *norB* required Mn-Fur and

A. H-NS Repression of *ftnA* transcription

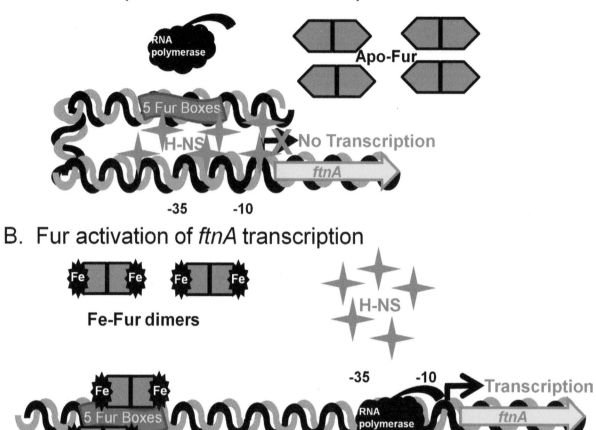

B. Fur activation of *ftnA* transcription

Figure 4. Model of direct transcriptional activation of *E. coli ftnA* by Fur. During iron limitation (A), *ftnA* transcription is repressed since apo-Fur does not bind to the five upstream Fur boxes. This allows H-NS to bind at these sites and other sites in the promoter region. Interaction among H-NS proteins bends the DNA, preventing access of RNA polymerase to the promoter region. Under iron surplus conditions (B), Fe-Fur binds to the upstream Fur boxes, preventing H-NS binding and interaction. This prevents occlusion of the promoter region, allowing RNA polymerase to bind and initiate transcription. doi:10.1128/9781555818524.ch6f4

the Fur box. Although a number of genes appear to be activated by Fe-Fur in *N. meningitidis*, direct interaction of Fur has been demonstrated only with promoters for NMB1436-1438 (required for resistance to H₂O₂ and survival in the bloodstream of mice), *norB*, *nuoA*, and *pan1/aniA*. *N. gonorrhoeae* genes *sodB* and *secY* are directly activated by Fur, exhibiting higher expression under iron surplus conditions (Sebastian et al., 2002; Delany et al., 2004; Grifantini et al., 2004). In *H. pylori*, microarray analysis using a *fur* mutant showed increased expression of 29 genes. Of these, four promoter regions have been shown to bind Mn-Fur and are activated under iron surplus conditions in a Fur-dependent manner: *oorD* (encoding a ferredoxin-like protein), *nifS*, *pdxJ* (encoding a predicted iron-sulfur protein), and *HP1432* (encoding a putative metal-binding protein) (Alamuri et al.,

2006; Gancz et al., 2006). *Y. pestis katA* and *napFDABC* are overexpressed in a *fur* mutant, and binding of Mn-Fur to these two promoter regions has been demonstrated by EMSAs (Gao et al., 2008).

Repression by Mn-Fur

Although it is generally stated that in the bacterial cell, Mn levels are insufficient to compete with Fe²⁺ for binding by Fur, one of the early observations was that the *E. coli iucABCD* promoter was repressed by growth with 5 μM Mn through Fur. Mn-Fur repression of iron/Fur-regulated promoters is not universal: *E. coli sodA* transcription is repressed by growth with excess Fe but not Mn. Despite this, Mn-Fur is routinely used for convenience in vitro to demonstrate that Fur binds to specific promoter region sequences. Escolar et al. used DNase I protection studies with Mn-Fur to define the long operator region of the *iucABCD* promoter and oligomerization of Fur along this DNA sequence (Bagg and Neilands, 1987; Privalle and Fridovich, 1993; Escolar et al., 2000). It would be interesting to determine whether this protection pattern is altered by using Fe^{2+}-Fur. In this regard, it is perhaps worthwhile noting that Fur from different organisms (i.e., Fur_{Ec} versus Fur_{Pa}) associated with some divalent cations (e.g., Zn) may demonstrate distinct footprinting patterns with the same DNA fragment, perhaps reflecting the subtle distinctions in the metal affinity and sequences of these Fur orthologs (Ochsner et al., 1995).

Reporters (*yfeA::lacZ* and *yfe::phoA*) for the *Y. pestis yfeABCD* promoter show transcriptional repression during growth with 10 µM Mn (~2-fold) or 10 µM Fe (~7-fold) that is Fur dependent. The promoter for the only other known Mn transporter in *Y. pestis*, MntH, also exhibits Fur-dependent transcriptional repression by Mn (~2- to 3-fold) and Fe (~10-fold) (Bearden et al., 1998; Perry et al., 2003, 2012). Mn transcriptional repression of the *sit* (similar to *yfe*) and *mntH* operons in *E. coli*, *Shigella*, and/or *Salmonella* has been demonstrated. This repression is due primarily to the Mn-responsive regulator, MntR, although Fur also caused some Mn-associated repression of these systems (Patzer and Hantke, 2001; Ikeda et al., 2005; Runyen-Janecky et al., 2006). It is possible that all Mn repression in *Y. pestis* occurs via Fur, because *Y. pestis* lacks the MntR transcriptional regulator. However, Mn repression of Fur-regulated promoters is not typical in *Y. pestis*. Promoter fusions to *lacZ* for the *efeUOB*, *fetMP* (both encoding putative or proven Fe^{2+} transporters), *hmuP RSTUV* (encoding a heme transporter), *ybtPQXS* (encoding yersiniabactin siderophore biosynthetic and transport components), *iucABCD* (defective for aerobactin biosynthesis), and *yiuABC* (encoding an iron/siderophore transporter) all demonstrate Fur-dependent repression by Fe but not by Mn (Fetherston et al., 1999; Gong et al., 2001; Kirillina et al., 2006; Forman et al., 2007; Perry, et al., in press; Perry et al., 2012).

The *Y. pestis mntH* and *yfeABCD* promoter regions have Fur boxes with upstream sequences with similarities to each other. A hybrid *lacZ* reporter with the *hmuP RSTUV* promoter in which 15 nucleotides upstream of the Fur box were replaced with nucleotides from a similar position in the Mn-responsive *yfeABCD* promoter converted this *hmu* hybrid promoter to one repressed (~2-fold) by Mn. Thus, at least in *Y. pestis*, sequences upstream of the Fur box

of *yfeABCD* are likely involved in Mn regulation via Fur (Perry et al., 2012).

apo-Fur Regulation

Transcriptional activation and repression by iron-free Fur (apo-Fur) are relatively recent observations in which apo-Fur binds to the promoter region or an upstream region of the affected promoter. This mechanism is still widely debated since there is a question of whether apo-Fur can be prepared in vitro for DNase I footprinting and EMSAs. However, metal chelators have been used in these assays to demonstrate Mn-dependent Fur interactions with DNA, and it is clear that the expression of some *H. pylori* and *C. jejuni* genes as well as *V. vulnificus fur* is affected by *fur* mutations in a manner consistent with apo-Fur regulation. For *V. vulnificus fur*, a region well upstream of the −35/−10 region was protected from DNase I degradation only in the presence of Fur_{Vv} and a metal chelator. Mutations in this Fur binding region abolished Fur binding in vitro along with in vivo regulation by Fur and iron availability. Unfortunately, the DNase I protection reaction buffer contained 100 µM $MnCl_2$ and an unspecified concentration of an "iron" chelator, likely 2,2′-dipyridyl (Lee et al., 2007). A reaction buffer without added $MnCl_2$ would have provided more definitive evidence of apo-Fur binding. DNase I footprinting of the *H. pylori pfr* promoter shows three distinct Fur operator regions (one overlapping and two upstream of the promoter region). Fur protects these regions in the presence of 2,2′-dipyridyl and shows a lower affinity for these sites in the presence of $MnCl_2$ or freshly prepared $FeSO_4$. This corresponds to the observation of Fur-dependent repression of *pfr* (encoding a nonheme ferritin) during iron-deficient growth of *H. pylori*. *H. pylori sodB* shows a similar pattern of Fur and iron regulation. EMSA analysis showed that binding of Fur was inhibited in the presence of Mn but occurred in the presence of the metal chelator EDTA. In contrast to *pfr*, DNase I footprinting indicated that Fur bound only to the −10/−35 promoter region. The DNA binding region of Fur_{Hp} (~50 N-terminal residues) is considerably more basic (pI, >9.0) than the corresponding more acidic regions (pIs, ~5.5) in Fur proteins of other gram-negative bacteria (e.g., *E. coli* and *P. aeruginosa*). Perhaps this basic region of Fur_{Hp} plays a role in binding DNA in the absence of a metal cofactor (Delany et al., 2001; Ernst et al., 2005; Carpenter et al., 2010). Thus, there is significant evidence for regulation by apo-Fur, but this phenomenon is currently limited to three genera.

RyhB PROPERTIES AND REGULATORY MECHANISMS

Discovery of RyhB in *E. coli* and Other Gram-Negative Bacteria

As indicated earlier in this chapter, RyhB is a base-pairing regulatory sRNA whose expression is negatively regulated by Fur at the transcriptional level. This breakthrough was made through the discovery, based on microarray data, of more than 20 distinct new sRNAs in the genome of *E. coli*. The expression of one of these previously unknown sRNAs (i.e., RyhB) was found to be iron and Fur dependent. Yet, it also repressed the expression of a set of Fur-dependent iron-responsive genes (Massé and Gottesman, 2002). The absence of Fur-mediated repression of RyhB in a Δ*fur* mutant results in constitutive expression of RyhB,

which constitutively downregulates the genes in question. Figure 5 shows *sodB* as an example of the interplay between Fur and RyhB regulation. Six genes, encoding iron-dependent tricarboxylic acid (TCA) cycle enzymes (*sdh*, *fumA*, and *acnB*), iron-dependent superoxide dismutase (SOD) (*sodB*), and iron storage proteins (*bfr* and *ftnA*), were initially identified as being regulated by RyhB (Massé and Gottesman, 2002). Although regulation of *ftnA* has been recently shown to be RyhB independent (Nandal et al., 2010) (see Fur activation section above and Fig. 4), another study has expanded the RyhB regulon to include an additional 18 mRNAs that encode a total of 56 mostly nonessential proteins involved in iron metabolism (Massé et al., 2005). While many of these new putative targets still require experimental validation, RyhB has been well established as an important global regulator in *E. coli* and other bacteria in which its primary function is to maintain iron homeostasis

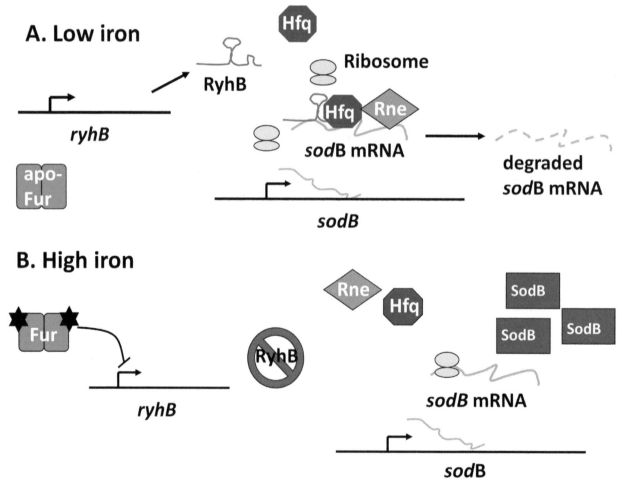

Figure 5. Model of indirect regulation of *sodB* by Fur through RyhB. RyhB is expressed under low-iron conditions (A) because apo-Fur has a low affinity for binding to the Fur box in the *ryhB* promoter and consequently does not repress transcription. RyhB sRNA base-pairs with *sodB* mRNA, causing downregulation of *sodB* translation in an Hfq- and RNase E-dependent manner. Under iron-replete conditions (B), Fe-Fur binds to the *ryhB* promoter and prevents its transcription. SodB is expressed in the absence of RyhB-mediated repression.
doi:10.1128/9781555818524.ch6f5

by downregulating expression of nonessential iron-containing proteins when iron is limited.

RyhB orthologs have also been identified in *Salmonella*, *Shigella*, *Yersinia*, *Photorhabdus*, *Klebsiella*, and *Vibrio* species by their similarity to the *E. coli* RyhB sequence (Gottesman et al., 2006). Fur regulation has been predicted or confirmed for all of these genes, although many variations exist among RyhB genes between species. For example, the single RyhB in *Vibrio* species is approximately twice the length of *E. coli* RyhB, while many other bacteria, including *Salmonella*, *Klebsiella*, *Photorhabdus*, and *Yersinia*, encode two RyhB copies. One *ryhB* gene in *Salmonella*, which is also called *isrE*, was acquired horizontally as part of a pathogenicity island (Ellermeier and Slauch, 2008; Padalon-Brauch et al., 2008). Despite their different origins, both RyhB and IsrE regulate expression of *sodB* in *S. enterica* serovar Typhimurium. Bioinformatic and functional approaches were used to identify two RyhB analogs, named PrrF1 (*P*seudomonas *r*egulatory *R*NA involving Fe) and PrrF2, in *Pseudomonas* species. The PrrF sRNAs share no sequence similarity with RyhB, but their expression is Fur controlled and they share several, but not all, targets (*sodB*, *sdhCDAB*, and *acnA*) with RhyB (Wilderman et al., 2004). Thus, RyhB and its orthologs or analogs appear to be widespread in bacteria.

RyhB Regulatory Mechanisms

The range of molecular mechanisms used by RyhB to regulate gene expression is impressive, even when only *E. coli* is considered. This diversity is likely to increase as the RyhB orthologs and analogs that have been identified in other bacteria are further studied. The two main effects of sRNA-mRNA base-pairing are regulation of translation and RNA stability (as discussed in Chapter 25 of this book). sRNAs regulate translation and stability of mRNAs in bacteria through imperfect base-pairing of a 10- to 50- nt region, often in the 5′ untranslated region of the mRNA (Gottesman et al., 2006). Base-pairing with the ribosome binding site (RBS) can block translation, although base-pairing at other sites can promote translation by relieving inhibitory RNA structures. RyhB was the first sRNA shown to induce mRNA degradation (Massé and Gottesman, 2002; Massé et al., 2003) and has the most known regulatory targets to date (Massé et al., 2005). RyhB binding has been experimentally verified for several targets (Frohlich and Vogel, 2009), including *sodB*, *uof-fur* (Večerek et al., 2007), *iscS* (Desnoyers et al., 2009), *cysE* (Salvail et al., 2010), and *shiA* (Prevost et al., 2007), and the regulatory mechanism differs somewhat with each, as described below.

Hfq, an RNA chaperone, is critical for the regulatory functions of many of the RNAs to which it binds, including RyhB, but the precise mechanisms by which it acts have not been established. Hfq was first defined as the host factor required for the replication of RNA phage QB. Like the eukaryotic SM and SM-like proteins that are involved in mRNA degradation and splicing, Hfq is a ring structure comprised of six identical subunits (Valentin-Hansen et al., 2004; Brennan and Link, 2007). One key role of Hfq is to stimulate RNA binding, which is necessary for downstream sRNA regulatory effects (Maki et al., 2008). While Hfq binding can stabilize RNA, it also promotes RNA degradation through its interactions with the major endoribonuclease of *E. coli*, RNase E, a critical component of the RNA degradosome, which also includes a polynucleotide phosphorylase, RNA helicase, and enolase (Carpousis, 2007). The RppH pyrophosphatase activity of the RNA degradosome generates 5′ monophosphorylated mRNAs in a process that is analogous to uncapping of mRNAs in eukaryotic cells (Celesnik et al., 2007; Deana et al., 2008). These monophosphorylated mRNAs stimulate RNase E's endonuclease activity, which initiates mRNA degradation. The RNA fragments generated by RNase E are then susceptible to further degradation by 3′, 5′ exonucleases (Carpousis, 2007; Belasco, 2010).

sodB, a paradigm for direct negative regulation by RyhB

sodB encodes an iron-dependent SOD whose expression is repressed by RyhB when iron is limiting. The RyhB-*sodB* regulatory interaction also provides a model for the best-studied sRNA regulatory paradigm, in which complementary binding of the sRNA to an mRNA sequence overlapping the RBS and start codon blocks translation and leads to degradation of both RNAs by RNase E and, to a lesser extent, RNase III (Fig. 5 and 6A). Both Hfq and RNase E are required for RyhB's destabilization of *sodB* mRNA, which is degraded within 2 to 3 min after RyhB binding. Hfq facilitates the interaction between RyhB and *sodB* mRNA, may stabilize unpaired RyhB by blocking an endonuclease cleavage site, and also binds to a scaffold domain in the C terminus of RNase E, tethering it near the RyhB-*sodB* RNA base-pairing region (Massé et al., 2003; Moll et al., 2003; Massé et al., 2005; Mitarai et al., 2009; Ikeda et al., 2011). Despite this close proximity, RNase E cleaves the mRNA at sites as much as 400 nt distal to the region of RhyB:*sodB* base-pairing (Prevost et al., 2007; Morita and Aiba, 2011). RNA sequence rather than position determines the site of cleavage, as this can be altered by moving an AU-rich RNA region that is essential for cleavage. Although this AU-rich region and adjacent sequences have roles in determining susceptibility to RNase E cleavage in *E. coli*, a specific RNase E recognition motif has not been identified (Morita et al., 2005).

Despite the rapid degradation of *sodB* mRNA, translational inhibition may be the most critical step in repression of SodB expression by RyhB, as this alone is sufficient to block SodB production. The subsequent degradation of *sodB* mRNA may simply serve to tighten and amplify the control, providing an ultrasensitive regulatory response system (Mitarai et al., 2009). Translational blockage may also facilitate mRNA degradation by clearing the message of ribosomes that might protect RNase E cleavage sites. However, a recent study demonstrated that the degradation of *sodB* mRNA resulting from base-pairing with RyhB is an active process and not simply a downstream effect of translational blockage. In contrast, translation blockage of *fumA*, another RyhB target, results in complete degradation of *fumA* mRNA, even when the RNase E cleavage site is deleted. These different results indicate that regulatory diversity exists even among RyhB targets which appear to fall within the same regulatory class (Prevost et al., 2007).

cysE: partial inhibition of translation by RyhB

A variation on RyhB's translational repression model occurs with *cysE*, which encodes a serine acetyltransferase that converts serine to cysteine. RyhB downregulates translation of the *cysE* mRNA by base-pairing to sequences that overlap the start codon (Fig. 6B). This allows diversion of available serine from cysteine biosynthesis to synthesis of the siderophore enterobactin under low-iron conditions, thereby increasing iron acquisition. The effect of RyhB on *cysE* is mild compared to its other targets, causing only partial repression. Since some CysE is still required for cysteine synthesis, *cysE* mRNA may not be actively degraded. This interesting case provides an example by which RyhB is able to cause subtle changes in metabolism by tempering the strength of its regulatory response (Salvail et al., 2010).

fur: indirect inhibition of translation by RyhB

In *E. coli*, Fur expression requires translation of a short overlapping reading frame upstream of *fur* (*uof*) (Fig. 3 and 6C). Translation of the *uof* mRNA is repressed by RyhB in an Hfq-dependent manner by RyhB base-pairing with an extended

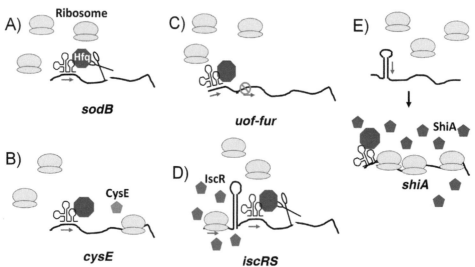

Figure 6. Models showing various RyhB regulatory mechanisms. (A) RyhB sRNA base-pairs with a region of the *sodB* mRNA that contains translation initiation signals (green arrow), blocking ribosome access (light blue oblongs). Hfq (purple octagon) binds both RyhB and RNase E (scissors), promoting cleavage of the *sodB* mRNA by the RNA degradosome (not shown). (B) RyhB base-pairing with the cysE mRNA reduces, but does not eliminate, ribosome binding and translational readthrough to produce low levels of CysE protein (red pentagon). (C) RyhB base-pairing with the *uof* open reading frame upstream of *fur* in some organisms prevents translational readthrough into *fur*, indirectly downregulating *fur* translation. (D) RyhB's interaction with the *iscRSUA* mRNA is similar to that of *sodB* mRNA, except that it base-pairs with a sequence at the start of an internal cistron rather than the 5' end. A strong hairpin structure upstream of this binding prevents the first cistron from degradation, leaving only the *iscR* portion of the mRNA intact. Blue pentagons represent IscR protein that is translated from this truncated message. (E) RyhB base-pairing with *shiA* mRNA alters the mRNA folding to make the translational start site available for ribosome binding, increasing translation of *shiA*. Blue pentagons represent ShiA protein.
doi:10.1128/9781555818524.ch6f6

region that includes the *uof* translational initiation signals (Fig. 6C). Thus, RyhB indirectly inhibits *fur* translation through direct inhibition of *uof* translation. Sequence analyses indicate that this molecular mechanism is likely to be conserved in several other enteric bacteria, including *S. flexneri*, *S. enterica*, and the pathogenic *E. coli* O157:H7. In contrast, RyhB orthologs or analogs may directly inhibit translation of *fur* mRNA in *V. cholerae* and *P. aeruginosa*, as there is sequence complementarity between these *fur* genes and their respective RhyB orthologs or analogs. RyhB regulation of *fur* is one of multiple ways in which Fur expression is controlled (Fig. 3 and 6C; discussed above) and may be designed to maintain Fur expression at a relatively constant level under iron-poor conditions in which metallo-Fur is not available for autoregulation. RyhB-mediated regulation of *fur* may only come into play during rapid shifts in iron availability to prevent excess Fur expression immediately following iron depletion with chelators such as with 2′, 2-dipyridyl and not during continued growth under low-iron conditions (Večerek et al., 2007; Salvail et al., 2010).

iscRSUA: selective degradation of an mRNA

Fe-S complexes are critical for the function of numerous enzymes and are primarily generated by one of two enzyme complexes. The ISC (iron-sulfur complex), which is encoded by the *iscSUA* genes, functions as the main housekeeping Fe-S biogenesis machine under iron-replete conditions, while the SUF (sulfur mobilization) system performs this function under iron-poor conditions. In a variation on the mRNA degradation theme described above for *sodB*, RyhB selectively promotes degradation of only the 3′ portion of the *iscRSUA* mRNA, while preserving expression of IscR, a transcription factor encoded by the first gene of the operon (Fig. 6D). As with *sodB* regulation, this RyhB-mediated mRNA degradation is dependent on the RNA degradosome as well as Hfq, which facilitates the base-pairing of RyhB with a sequence overlapping the RBS and start codon of *iscS*. This blocks translation and triggers RNase E cleavage at a sequence within the *iscS* open reading frame. Only the *iscR* sequence is protected from subsequent 3′, 5′ exonuclease degradation by a strong secondary structure that forms within the 111-nt intergenic region between *iscR* and *iscS* that has sequence similarity to repetitive extragenic palindromic sequences. This differential degradation downregulates expression of the Fe-S cluster biogenesis enzymes encoded by the *iscSUA* genes with continued expression of IscR. Apo-IscR is a positive regulator of *suf* expression; this and the differential degradation of the *iscRSUA* mRNA may play an important role in mediating an iron starvation-dependent transition between these systems. In addition, IscR regulates numerous genes involved in anaerobic respiration and oxidative stress responses. Thus, under low-iron conditions, IscR controls important functions in addition to shifting Fe-S cluster production from the ISC machinery to the SUF complex (Giel et al., 2006; Desnoyers et al., 2009).

shiA: direct positive regulation by RyhB

While *ompX* and *ygdQ* may also be upregulated by RyhB, *shiA* provides one of the few examples in which RyhB serves directly as a positive regulator of gene expression. The *shiA* gene encodes ShiA, an inner membrane permease that transports shikimate into the cell. Shikimate is involved in the biosynthesis of the enterobactin siderophore, aromatic amino acids, folic acid, and ubiquinone. In the absence of RyhB, the *shiA* mRNA forms a secondary structure that blocks access to the RBS and translational start codon, inhibiting translation initiation. In part due to the lack of protective ribosomes, *shiA* mRNA is also extremely unstable in this inhibitory conformation (Fig. 6E). This mechanism has some resemblance to the positive regulation of *rpoS* translation by DsrA. RyhB base-pairing with an extensive region in the 5′ untranslated region of *shiA* frees up the Shine-Dalgarno sequence and AUG for ribosome binding and covers 38 of 51 nt that include a core of 12 consecutive base pairs. Both translational readthrough and stability of *shiA* mRNA are significantly enhanced in the presence of RyhB. While ribosomes contribute to the increased stability of *shiA* mRNA in the presence of RyhB, additional mechanisms are also likely involved. Hfq has a surprisingly complex role: facilitating the binding of RyhB with *shiA* and serving as a direct negative regulator of *shiA* in the absence of RyhB. The mechanism underlying this negative regulation is not clear, but it most likely promotes formation of the intrinsic inhibitory structure of *shiA* mRNA (Herrmann and Weaver, 1999; Lease and Belfort, 2000a, 2000b; Massé et al., 2005; Prevost et al., 2007).

IRON REGULATION AND VIRULENCE

Iron regulation of virulence (and other) factors is often observed by comparing (i) expression of specific genes or proteins, (ii) microarray analysis, or (iii) proteomics. These studies typically compare growth under different iron availability conditions and/or compare expression in a *fur* or *ryhB* mutant with that in its parental strain. Microarray studies often expose bacterial cells to iron deprivation by adding a metal chelator for a short period. This has the advantage of avoiding indirect effects of prolonged iron starvation (slower growth, etc.); however, it misses genes that are regulated only under more severe iron stress conditions or genes affected by the adaptation of the bacterial cell to growth during iron starvation. In addition, not all genes identified as iron regulated by microarray analysis have been confirmed by other methods. In many cases, the mechanism of iron regulation of virulence factors is not established. Where there is conclusive evidence of *direct* regulation by Fur or RyhB, it will be indicated.

Fur and RyhB

Relatively few virulence studies have been performed with *fur* mutants. This is possibly due to early assumptions that Fur largely derepressed virulence factors important in the host and that mutants already expressing these systems would be fully virulent. However, the majority of *fur* mutants tested to date show, to various degrees, decreased virulence. Depending upon the strain background, an S. Typhimurium *fur* mutant showed a 10- to 1,000-fold decrease in virulence in orally infected mice (Wilmes-Riesenberg et al., 1996; Troxell et al., 2011). Competitive index analysis with a *V. cholerae* wild-type strain demonstrated lower colonization of the infant mouse gut by the *fur* mutant (Mey et al., 2005b). A *C. jejuni fur* mutant exhibited reduced colonization of the chick gastrointestinal tract. Since the enterobactin iron uptake system is needed for this colonization, the gastrointestinal environment appears to be iron restricted, suggesting that loss of colonization is not due to iron overload toxicity (Palyada et al., 2004). A *fur* point mutation in *P. aeruginosa* results in several prototypical phenotypes of Δ*fur* mutants in other bacteria; this mutant has a significantly increased 50% infective dose compared to that of the parental strain in a corneal infection model in mice, indicating a significant loss of virulence (Preston et al., 1995). An *H. pylori fur* mutant has lower gastric colonization in mice and Mongolian gerbils than does its parent strain. This may be directly related to the requirement of Fur for growth under acidic conditions and/or for ferritin expression, which is essential for colonization of the gastric mucosa of gerbils (Bijlsma et al., 2002; Waidner et al., 2002; Bury-Moné et al., 2004; Gancz et al., 2006). Finally, an *Actinobacillus pleuropneumoniae fur* mutant showed decreased clinical scores and lung lesions in a pig aerosol infection model (Jacobsen et al., 2005). Collectively, these studies indicate that (i) constitutive derepression of genes in these *fur* mutants is detrimental in the host and/or (ii) Fur directly or indirectly activates expression of genes important for survival and growth in the host. In contrast to these studies, we have recently shown that a *Y. pestis fur* mutant retains full virulence in mouse models of bubonic and pneumonic plague, at least as measured by 50% lethal dose analysis (G. Bai, E. Smith, J. D. Fetherston, A. D. Perry, and K. A. McDonough, in preparation).

Despite its typical role in regulating iron homeostasis as well as motility and biofilm production (see "Iron Regulation of Metabolism" below), RyhB is not required for *V. cholerae* virulence in an infant mouse model. This has been interpreted to mean that iron is not limiting in the mouse intestine, an idea that is supported by the lack of RyhB expression in this environment (Davis et al., 2005; Wyckoff et al., 2007). In contrast, the *V. vulnificus* RyhB ortholog is required for virulence in an iron overload murine model of infection. *V. vulnificus* is an opportunistic pathogen that causes a fatal septicemia in susceptible individuals, such as those with the iron overload diseases hemachromatosis and β-thalassemia (Alice et al., 2008). This leads to the opposite suggestion that iron limitation genes are important for infection even in an iron-overloaded host. Our analysis of a *Y. pestis ryhB1 ryhB2* double mutant failed to show a significant difference in the 50% lethal doses in mouse models of bubonic and pneumonic plague (G. Bai, E. Smith, J. D. Fetherston, A. D. Perry, and K. A. McDonough, in preparation).

Regulators

The iron status of the bacterial cell appears to regulate transcription of a diverse number of regulators in a variety of bacteria. The secondary effect of these regulators greatly expands the iron modulon, making Fur and RyhB truly global regulators in many bacteria. Many of the examples are from microarray studies comparing *fur* mutants versus the Fur+ parent and/or growth under iron-deficient versus iron-sufficient conditions.

In *V. vulnificus*, microarray analysis identified 12 putative transcriptional regulators that were iron repressible (Alice et al., 2008). In *P. aeruginosa*, Fur is predicted or proven to regulate 14 transcriptional regulators and 11 ECFs. Many, but not all, of these regulators control expression of genes involved in iron acquisition. PvdS is one example of an ECF that regulates not only synthesis and transport of a siderophore-dependent (pyoverdine) uptake system but also expression of other regulators, exotoxin A (ETA), the PrpL protease, and several extracellular proteins. A mutation in *pvdS* also shows the effect of oxygen tension on iron-regulated virulence. Two different outcomes are seen with this mutant in left-sided versus right-sided rabbit endocarditis models. An infection established on the left side of the heart showed significantly decreased colonization of the heart valves and decreased dissemination to other organs. In contrast, the mutant and the parent strain showed similar degrees of valve colonization and dissemination from infection of the right side of the heart. Oxygen tension differs between the two chambers by ~50 Hg, with the left-sided chamber being comparable to in vitro aerobic conditions and the right-sided chamber in vitro microaerobic conditions (Xiong et al., 2000; Vasil, 2007; Cornelis et al., 2009).

The literature on the roles of quorum sensing (QS) systems in virulence and biofilm development is extensive. Iron is one of the signals known to affect QS in *P. aeruginosa* during infection. Fur and the RyhB analogs PrrF1 and PrrF2 are involved in the iron-responsive regulation of one of three QS systems in this opportunistic pathogen (Oglesby et al., 2008). The 2-heptyl-3-hydroxly-4-quinolone *Pseudomonas* quinolone signal (PQS) is thought to be important for virulence because it is produced at high levels and secreted by *P. aeruginosa* into the lungs of cystic fibrosis patients during infection (Guina et al., 2003). PQS autoregulates its own expression by coinducing the transcription factor PqsR, which activates the *pqsABCDE* genes, which code for the PQS synthesis machinery. A derivative of tryptophan metabolism, anthranilate, which can be degraded by the products of the *antABC* and *catBCA* operons, is a direct precursor of PQS biosynthesis in *P. aeruginosa*. Microarray and mutational studies showed that the PrrF sRNAs repress the expression of the anthranilate-degrading enzymes encoded by these operons, under iron-limiting conditions, thereby ensuring the availability of this key PQS precursor. In contrast, under iron-replete conditions, the degradation of anthranilate ultimately provides increased availability of TCA cycle intermediates (e.g., succinate) from tryptophan or anthranilate catabolism. Direct repression of *antA* expression through base-pairing between PrrF and *antA* mRNA is proposed, although this mechanism has not yet been confirmed (Oglesby et al., 2008).

In *Shigella* species, *sodB* and *virB* are direct targets of RyhB. VirB is a transcription factor that positively regulates expression of virulence genes, including those that encode the type three secretion system and its secreted effector Ipa proteins. Both the type three secretion system and the Ipa proteins are required for invasion of epithelial cells. As with acid resistance (see below), *fur* knockout strains of *Shigella dysenteriae* are defective for *ipa* expression and epithelial cell invasion unless *ryhB* is also inactivated (Payne et al., 2006). Surprisingly, *sodB* and *virB* mRNAs are regulated by different regions of the RyhB sRNA in *S. dysenteriae* (Murphy and Payne, 2007). This suggests a mechanism by which RyhB may accommodate regulation of multiple diverse mRNA targets that requires further investigation.

Iron or Fur regulation of transcriptional regulators has been documented for a number of other pathogens. *S.* Typhimurium Fur negatively regulates expression of HilA, a regulator of the SPI-1 pathogenicity island that is involved in invasion of epithelial cells of the small intestine (Troxell et al., 2011). PerR and known PerR-regulated promoters (for expression of gene products to alleviate oxidative stress) are derepressed by a *fur* mutation in *C. jejuni*, indicating regulation in response to iron and peroxide stress (Holmes et al., 2005). In *Yersinia pseudotuberculosis*, resistance to antimicrobial peptides is enhanced during iron starvation. This is mediated by a LysR-type regulator (YPTB0333) which also enhances expression of YfeA, the periplasmic binding protein for an ABC transporter of Fe and Mn. In *Y. pestis*, microarray analysis did not detect changes in expression of the YPTB0333 homolog, YPO0276, due to a *fur* mutation or by iron deprivation (Han et al., 2007). *H. pylori* Fur binds to the operator site of the *nikR* promoter and thus likely controls expression of this Ni-responsive regulator. Since Fur and NikR compete for binding to some promoters, the levels of Fe-Fur may affect not only NikR levels but also the ability of NikR to bind to some Fur- and NikR-regulated promoters (Delany et al., 2005).

Iron Transport Systems

Entire books have been written on iron uptake mechanisms and their roles in virulence; indeed, studies showing a loss of virulence due to mutations in one or more iron/heme uptake system(s) in a wide variety of bacterial pathogens are too numerous to cite. In the many instances where iron regulation has been tested, iron uptake systems are repressed during growth with excess iron and are constitutively overexpressed in *fur* mutants. Consequently, we focus on the iron transport systems of *Y. pestis* in this section. The sequenced *Y. pestis* KIM10+ strain, a derivative of the strain isolated from a human plague case, encodes 14 putative or proven iron/hemin uptake systems. There are four siderophore-dependent systems with genes encoding biosynthetic enzymes, five ABC transporters with predicted specificities for iron or siderophores, four ferrous uptake systems, two hemin uptake systems, and one ferrous efflux system (Table 1). Of the four siderophore-dependent systems, only the yersiniabactin (Ybt) system appears to be fully functional in *Y. pestis* KIM; the aerobactin, Ynp, and Ysu systems all have frameshift mutations in genes encoding siderophore biosynthetic functions. However, the Fhu ABC transporter is capable of transporting exogenously supplied aerobactin and ferrichrome siderophores. The Yfu and Yiu ABC transporters are functional for iron uptake, while the Fit and Fiu systems have not been tested. Of the four ferrous iron uptake systems, Yfe, Feo, and Fet are functional; Efe has not been tested. The *Y. pestis* Hmu ABC transporter allows the use of hemin, hemoglobin, and these compounds complexed with albumin, hemopexin, or haptoglobin. Despite

a concerted effort, the Has hemophore system has not been demonstrated to have the ability to acquire hemin or hemoglobin (Table 1) (Bearden and Perry 1999; Thompson et al., 1999; Gong et al., 2001; Rossi et al., 2001; Deng et al., 2002; Kirillina et al., 2006; Forman et al., 2007; Perry et al., 2007; Forman et al., 2010; Perry and Fetherston, 2011; J. D. Fetherston, I. Mier, Jr., H. Truszczynska, and R. D. Perry, submitted for publication).

Table 1. Iron transport systems of *Y. pestis*[a]

| System | Iron regulation | | Functionality | Role in virulence |
	Transcriptional reporter	Microarray RT-PCR[b]		
Siderophore dependent				
Yersiniabactin (Ybt)	Yes	Yes	Functional	Yes (both)[d]
Aerobactin biosynthesis	Yes	Yes	Nonfunctional	Not tested
Yersinia nonribosomal peptide (Ynp)	Not tested	Yes	Nonfunctional	Not tested
Yersinia siderophore uptake	Not tested	Yes	Nonfunctional	Not tested
ABC Fe³⁺/siderophore transporters				
Yfu	Yes	Yes	Functional	No (bubonic)[d]
Fhu	Yes	Yes	Functional	Not tested
Yiu	Yes	Yes	Functional	No (bubonic)[d]
Fit	Not tested	Yes	Not tested	Not tested
Fiu	Not tested	No	Not tested[c]	Not tested
Fe²⁺ uptake systems				
Yfe	Yes	Yes	Functional	Yes (bubonic)[d]
Feo	Yes	Yes	Functional	Yes (bubonic)[d]
Fet	Yes	No	Functional	Not tested
Efe	Yes	No	Not tested	Not tested
Hemin uptake systems				
Hmu	Yes	Yes	Functional	No (both)[d]
Has	Yes	Yes	Nonfunctional	No (both)[d]
Fe²⁺ efflux				
FieF	Not tested	No	Not tested	Not tested
TonB system				
TonB	Not tested	Yes	Not tested	Not tested
ExbB/ExbC	Not tested	Yes	Not tested	Not tested

[a]Studies were performed in epidemic KIM strains (transcriptional reporter studies and studies on functionality and virulence) or endemic strain 201 (microarray and RT-PCR). Endemic strains cause bubonic plague in some rodents but not guinea pigs, nonhuman primates, or humans.
[b]Genes/loci positive for iron regulation by microarray were confirmed by RT-PCR. Loci without detected iron regulation were not tested by RT-PCR.
[c]*fiuB* appears to be a pseudogene due to a frameshift mutation; its functionality has not been tested experimentally. No evidence for siderophore production by Ynp or Ysu in the KIM6+ strain has been found.
[d]Results for both mouse models of bubonic and pneumonic plague. Yfu and Yiu were tested only in the bubonic plague model. A *yfe feo* double mutant is less virulent in the bubonic plague model but fully virulent in the pneumonic plague model.

The Ybt system is essential to the virulence of *Y. pestis* in a mouse model of bubonic plague, with a >10^6-fold loss of virulence compared to its Ybt$^+$ parent. This system is important but not essential for pneumonic plague: a 33- to 790-fold loss of virulence occurs, depending upon the type of mutant (unable to transport Ybt versus unable to synthesize Ybt). There is also indirect evidence that the Ybt siderophore plays a role in pathogenesis in the lung independent of its iron acquisition function. For Fe^{2+} uptake, a *yfe feo* double mutant exhibits an ~90-fold reduction in virulence via subcutaneous injection (bubonic plague) but is fully virulent by intranasal instillation (pneumonic plague), a finding that supports the hypothesis that different iron transport systems are used or important in different host organ systems. While an *hmu has* double mutant is fully virulent in both of these forms of plague, *hmu* and *hmu has* mutants displayed a moderate growth defect in J774A.1 cells compared to their Hmu$^+$ Has$^+$ parent strain. Finally, mutations in the Yfu and/or Yiu system did not cause a loss of virulence in the bubonic plague model but have not been tested in the pneumonic plague model (Gong et al., 2001; Rossi et al., 2001; Kirillina et al., 2006; Lee-Lewis and Anderson, 2010; Fetherston et al., 2010; Forman et al., 2010; Fetherston et al., submitted). While the roles of Efe, Fet, and Fit in pathogenesis remain to be tested, it is clear that only a subset of *Y. pestis* functional iron uptake systems are important in disease progression in mammals.

Except for the apparently nonfunctional Fiu ABC transporter, Fur- and iron-dependent repression has been demonstrated in vitro using transcriptional reporters and/or microarray analysis for all the other putative and proven Fe uptake systems, including TonB/ExbB/ExbD, and the MntH Mn transporter. Genes exhibiting iron regulation by microarray were confirmed by reverse transcription-PCR (RT-PCR), and direct interaction of Mn-Fur with the promoter regions of most of these systems was demonstrated by EMSAs. However, in some cases the concentration of Fur required was relatively high (0.5 to 1 μM), raising the possibility of nonspecific interactions. In addition, the in vitro microarray analysis, performed with endemic strain 201, failed to detect iron regulation of the *fetMP* and *efeUOB* loci, although transcriptional reporters for these systems in epidemic strain KIM6+ (from which the sequenced strain KIM10+ was derived) clearly show repression by iron (Table 1). This difference is likely not due to use of an endemic strain (which causes disease in some rodents but not primates) versus an epidemic strain (which causes disease in rodents and primates). Rather, the transcriptional reporter studies depleted intracellular iron stores prior to assessing transcriptional activity, while the microarray studies added an iron chelator 30 min prior to RNA isolation. The role of *E. coli* RyhB in increasing shikimate levels has been described. Shikimate is a precursor of salicylate which is incorporated into the Ybt siderophore. Thus, it is likely that RyhB contributes to Ybt synthesis in *Y. pestis*. An intriguing difference between the microarray/RT-PCR and transcriptional reporter analyses occurred. In our hands, iron- and Fur-dependent repression of transcription from *feoA::lacZ* and *feoB::lacZ* reporters occurred during static (microaerobic) growth but not during aerobic growth, while Gao et al. found repression during aerobic growth (Gao et al., 2008; Lathem et al., 2005; Chauvaux et al., 2007; Sebbane et al., 2006).

In addition to in vitro regulation, high levels of transcription of many of the genes of these systems have been detected during growth in human plasma or in the lungs of infected mice and lymph nodes of infected rats (Thompson et al., 1999; Gong et al., 2001; Rossi et al., 2001; Perry et al., 2003; Lathem et al., 2005; Kirillina et al., 2006; Chauvaux et al., 2007; Forman et al., 2007; Gao et al., 2008; Fetherston et al., submitted; Perry et al., 2012; Sebbane et al., 2006).

Toxin Production

In addition to toxin production by gram-positive pathogens (see Chapter 6), reduced Shiga toxin production by *S. dysenteriae* grown with excess iron was another early observation of iron regulation (Dubos and Geiger, 1946). Iron regulation of toxins is thought to be a mechanism for increasing expression in the iron-limited environments of the host. It has also been speculated that increased expression of toxins has a secondary benefit of providing iron to invading pathogens through the lysis of host cells. In *E. coli*, Fur mediates repression of hemolysin (proven by direct interaction) and Shiga-like toxins SltA and SltB (Calderwood and Mekalanos, 1987; Fréchon and Le Cam, 1994). Fur also negatively regulates expression of hemolysins in *V. cholerae*, *Plesiomonas shigelloides*, and *Yersinia ruckeri* (a fish pathogen) (Stoebner and Payne, 1988; Santos et al., 1999). Fur has also been implicated in the regulation of cytolysins (VacA in *H. pylori* and ClyA in *Salmonella enterica* serovar Typhi), Shiga toxin 1 (but not Stx2 in *S. dysenteriae*), an enterotoxin (Act in *Aeromonas hydrophila*), and a CHO cell elongation factor (*P. shigelloides*). Regulation of VacA in *H. pylori* appears to be indirect (Payne, 1988; Gardner et al., 1990; Sha et al., 2001; Gancz et al., 2006; Cui et al., 2009). In *P. aeruginosa*, Fur indirectly controls expression of ETA (*toxA*), the PrpL extracellular endoprotease (*prpL*), and the extracellular lipase LipA. For ETA, it has been demonstrated that Fur negatively regulates expression of the ECF PvdS, which positively regulates expression of the transcriptional regulators RegA and PtxR, which are both required for maximal expression of ETA. Signaling by the *P. aeruginosa* siderophore pyoverdine also influences expression of the virulence factors ETA and PrpL protease. Under iron-limiting conditions, binding of the siderophore to its OM receptor (FpvA) initiates a signal, and that results in increased transcription of both genes (Vasil and Ochsner, 1999; Lamont et al., 2002). Thus, iron regulation of toxin production is widespread in gram-negative pathogens.

Adherence and Invasion

An important component of pathogenesis is adherence, whether it is to remain in a preferred host niche or as the first step in host cell invasion. Here again, Fur/iron regulation appears to use the iron-deficient environment of the host as a signal to enhance expression of adhesins and invasins in a variety of pathogens. The lower colonization of the infant mouse gut by a *V. cholerae fur* mutant correlates with lower expression of the toxin-coregulated pilus in that mutant (Mey et al., 2005b). In *S.* Typhimurium, Fur negatively regulates HilA, a regulator of SPI-1 pathogenicity island genes involved in invasion of epithelial cells of the small intestine (Troxell et al., 2011). Purified *Neisseria gonorrhoeae* Fur bound to the promoter regions of 11 *opa* genes. The OM Opa proteins are involved in adhesion to and invasion of host cells (Sebastian et al., 2002). A Fur titration assay in *Aeromonas salmonicida* identified genes encoding putative pilin proteins as Fur regulated (Najimi et al., 2009). In studies using iron deprivation, increased expression of genes encoding one of three Tad (tight adherence) systems by *V. vulnificus*, a long polar fimbria by Shiga-toxingenic *E. coli*, Iha (an IrgA homolog) by Shiga-toxigenic and uropathogenic *E. coli*, CFA/I fimbriae by enterotoxigenic *E. coli*, Stg fimbriae by an avian pathogenic *E. coli* strain, OprQ (a porin) in *P. aeruginosa*, cable pili by *Burkholderia cenocepacia*, and type IV pili by *Moraxella catarrhalis* was detected. For Iha (IrgA) and Stg fimbriae, Fur is required for iron-dependent repression. It should be noted that in addition to promoting binding to human fibronectin, OprQ appears to function as a porin and negatively affects production of the virulence factor pyocyanin. Although *E. coli* Iha has been implicated in adhesion, Iha/IrgA is a TonB-dependent receptor involved in catechol siderophore uptake in *V. cholerae* and uropathogenic *E. coli*. In *V.*

cholerae, an *irgA* mutation did not affect enterobactin use (due to a redundant receptor) or virulence. Thus, a virulence role for Iha/IrgA as an adhesion factor needs to be further explored (Karjalainen et al., 1991; Mey et al., 2002; Luke et al., 2004; Tomich and Mohr, 2004; Lymberopoulos et al., 2006; Rashid et al., 2006; Torres et al., 2007; Alice et al., 2008; Herold et al., 2009; Arhin and Boucher, 2010). As with toxin production, iron regulation of some adherence mechanism is a common theme in bacterial pathogens.

Oxidative Stress and Intracellular Survival

Although reactive oxygen species are produced during aerobic metabolism, pathogens must also withstand host defenses based on superoxide, hydrogen peroxide, and nitric oxide. These oxidative stresses are countered by bacterial SODs, catalases, and peroxidases. The Mn-SOD (SodA) is directly repressed by Fe-Fur in *E. coli* and *P. aeruginosa*. In a number of other gram-negative pathogens, *sodA* transcription is repressed by Fe in a Fur-dependent manner (Niederhoffer et al., 1990; Tardat and Touati, 1993; Hassett et al., 1997). In addition, *sodB* genes from bacterial pathogens such as *Vibrio vulnificus*, *Y. pestis*, and *Y. pseudotuberculosis* are directly controlled by RyhB. In *B. pseudomallei*, expression of SodB and a peroxidase activity was repressed by a *fur* mutation (Loprasert et al., 2000; Alice et al., 2008; Bai et al., in preparation). While detoxifying enzymes such as catalases and SODs deactivate H_2O_2 and its derivatives, bacterioferritin (Bfr) and Dps also reduce H_2O_2 levels by using H_2O_2 to oxidize the iron they take up. Although Bfr uses both O_2 and H_2O_2, Dps is ~100-fold more efficient at using H_2O_2 than O_2 for iron oxidation prior to storage. The strong peroxide clearance capacity of Dps is particularly helpful for bacterial pathogens that face an oxidative burst within macrophages, as evidenced by the protection Dps provides for *S.* Typhimurium during infection of macrophages and mice (Halsey et al., 2004). The role of Dps in protecting against the damaging effects of Fenton-mediated reactions during antibiotic treatment further emphasizes the critical role of this ferritin-like protein in *Salmonella* bacteria subjected to stressful conditions (Calhoun and Kwon, 2011a). Iron and/or Fur regulation of oxidative stress defenses has been demonstrated in *C. jejuni* and *N. meningitidis* group B (Grifantini et al., 2004; Holmes et al., 2005).

Proteomic and microarray analyses both showed that expression of *Francisella tularensis iglA* and *pdpB* is repressed under iron surplus conditions. The products of both of these genes have been implicated in the intracellular survival of *F. tularensis* (Lenčo et al., 2007).

Acid Tolerance Response

The acid tolerance response (ATR) is a two-step adaptation to increased survival at low pH. Brief exposure to a moderately low pH enhances survival upon subsequent exposure to more acidic conditions. In *S.* Typhimurium, Fur regulates an ATR at pH 5.8 that is required for survival at pH 3.3. A *fur* mutant no longer induces several ATR genes and has a lower survival rate at pH 3.3. Fur regulates genes that primarily protect against organic acids. This Fur regulation is independent of iron: the bacterial iron status does not affect the ATR, and an iron-blind *fur* mutant has a normal ATR (Hall and Foster, 1996; Bearson et al., 1998). *H. pylori*, an avian septicemic *E. coli* strain, and *Shewanella oneidensis* (a fish and opportunistic human pathogen) also have ATRs regulated by Fur but not iron. In all three organisms, a *fur* mutant shows decreased survival at acidic pHs compared to their *fur⁺* parent. Microarray analysis of *H. pylori* identified 12 acid-regulated genes whose expression was affected by a *fur* mutation (Bijlsma et al., 2002; Zhu et al., 2002; Gancz et al., 2006; Yang et al., 2008; Valenzuela et al., 2011). In contrast, a *V. vulnificus fur* mutant has increased acid tolerance due to derepression of MnSOD expression (Kim et al., 2005). Most of these pathogens have an oral infection route and thus must survive transit through or residence in the acidic stomach environment, with Fur participating in control of the ATR.

Fur and RyhB affect acid resistance in *Shigella* as well. A *fur* mutant is unable to survive acidic conditions, unless *ryhB* is also deleted. Likewise, overexpression of RyhB makes *S. dysenteriae* acid sensitive. RyhB increases acid sensitivity by indirectly downregulating expression of *ydeP*, which encodes a putative oxidoreductase that is needed for acid resistance in *Shigella*. The direct target of RyhB in this pathway is likely to be *evgA*, which encodes an activator of *ydeP* expression (Oglesby et al., 2005).

Biofilm Development

It is thought that biofilms are involved in greater than 50% of bacterial infections (Costerton et al., 1999). There are a number of observations that biofilm development is inhibited during iron-limited growth. In some cases, it was not determined whether this was due simply to lower cell mass resulting from iron starvation. However, there are clear examples where iron has a regulatory effect on biofilm development.

In a few examples, biofilm development is inhibited under iron surplus growth conditions. This is the case for *Legionella pneumophila* and *Aggregatibacter* (formerly *Actinobacillus*) *actinomycetemcomitans*. Iron chelation increased exopolysaccharide levels and transcription (and corresponding translation) of tight adherence fimbriae (*tad*), an OM protein involved in expolysaccharide export (*pgaA*), and lipopolysaccharide (*rmlB*) in *A. actinomycetemcomitans* (Hindré et al., 2008; Amarasinghe et al., 2009).

In contrast, a *V. cholerae* Δ*ryhB* mutant had reduced biofilm formation under low-iron conditions. This defect in biofilm development was alleviated by adding iron or succinate. Although the mechanism of this effect has not been established, the gene expression profile suggests that the *V. cholerae ryhB* mutant is under a general iron stress. Note that *V. cholerae* RyhB regulates motility and chemotaxis as well, abilities that likely factor into biofilm regulation and development (Mey et al., 2005a).

In *E. coli* K-12, iron limitation also decreased biofilm formation. A mutation in *iscR*, which encodes the Fe-S cluster-sensitive transcriptional regulator IscR, increased biofilm formation. It was concluded that apo-IscR decreases expression of type I fimbriae by activating transcription of *fimE*. FimE is a recombinase that downregulates the *fimAICDFGH* operon by switching its promoter to the off state (Wu and Outten, 2009). Biofilm development in a uropathogenic *E. coli* was also found to be dependent upon sufficient levels of cellular iron, with genes of the Ybt iron uptake system and some genes of the enterobactin and ferric citrate systems more highly expressed by biofilm cells than by planktonic cells growing in human urine. A mutation in the OM Ybt receptor (FyuA; Psn in *Y. pestis*) greatly reduced biofilm formation without significantly affecting planktonic growth in human urine. Thus, the Ybt system seems to be critical in supplying the necessary iron for biofilm development in this model (Hancock et al., 2008).

Intracellular iron levels serve as a signal for biofilm development in *P. aeruginosa* as well; this intracellular signaling is mediated through Fur, not by PrrF1 and PrrF2 (RyhB analogs). While low iron levels, such as those that were present in the siderophore mutants used in this study, are still sufficient for full planktonic growth, biofilm formation was shown to be reduced, most likely a consequence of the inability of *P. aeruginosa* to gather sufficient levels of the increased amounts of iron required for robust biofilm formation. An additional interpretation of those data suggests that under iron-limiting conditions twitching motility is increased, thereby leading to poorer biofilm formation. In any case, it is clear that the effect of iron on any one of the several types of biofilms (e.g., Psl, Pel, or alginate) that *P. aeruginosa* can exhibit is extremely intricate and will require considerable further investigation to be fully understood (Singh, 2004; Banin et al., 2005; Musk et al., 2005; Glick et al., 2010)

In *Y. pestis*, biofilm development is temperature dependent: produced at 26 to 34°C and inhibited at 37°C. Although this temperature regulation is primarily achieved through cyclic-di-GMP levels and proteolytic degradation of proteins essential for biofilm development, a *fur* mutant exhibited high biofilm levels at 37°C. The mechanism behind this temperature-independent biofilm phenotype has not been resolved, but it does not seem to involve iron-dependent expression of three proteins essential for biofilm development. Note that in *Y. pestis*, biofilm formation is involved in one form of plague transmission from fleas to mammals. Biofilm formation is not involved in the progression of bubonic or pneumonic disease in mice; indeed, a mutation causing biofilm expression at 37°C in vitro is detrimental to lethality of bubonic plague in mice (Staggs et al., 1994; Perry et al., 2004; Eisen et al., 2006; Bobrov et al., 2011).

IRON REGULATION OF METABOLISM

As indicated above, adaptation to changing iron availability is essential in bacteria. In addition to the various virulence factors described above, metabolic changes are important for survival and growth during both iron deprivation and iron excess. A microarray analysis comparing *P. aeruginosa* cells adapted to iron-deficient and iron-sufficient growth showed a shift to metabolic enzymes without iron cofactors by iron-starved cells (Ochsner et al., 2002; Oglesby et al., 2008). Extensive proteomic profiling has recently identified three major physiological impacts of iron deprivation on *Y. pestis*. In this analysis, bacterial cells were adapted to either iron-deficient or iron-sufficient conditions; iron-starved cells exhibited a lower growth rate and lower final yield in batch cultures than did iron-replete cells. Consequently, the iron-deficient proteome reflects both direct and indirect effects but likely represents adaptation to continued growth under iron-deficient conditions.

Y. pestis cells grown under iron-limited conditions expressed proteins from all five verified iron/heme acquisition systems at significantly higher levels than they were expressed under iron-replete conditions. Similarly, levels of nonessential iron-containing proteins such as TCA cycle enzymes AcnA, AcnB, SdhA, and FumA, as well as AceA and AceB, which support the glyoxylate bypass within the TCA cycle, are reduced under iron-poor conditions. Respiration in *Y. pestis* is also affected by iron levels: iron-containing proteins such as the formate-pyruvate lyase and the NuoCD NADH:ubiquinone oxidoreductase complex are decreased under low-iron conditions. In contrast, amounts of iron-independent enzymes like PoxB (pyruvate oxidase), which converts pyruvate to acetate in a flavin-dependent reaction that generates ubiquinol, are significantly increased during iron deprivation. Together, these data suggest the use of an alternative iron-independent pyruvate degradation pathway in lieu of the TCA cycle when iron is in short supply. Use of PoxB would maintain energy metabolism while generating reducing equivalents in an iron-independent fashion to facilitate growth of *Y. pestis* under iron-poor conditions. Oxidative stress proteins in *Y. pestis* show a similar shift from iron-dependent to iron-independent variants as iron levels are reduced. In this case, iron-containing catalases (KatE and KatY) and SodB are replaced by alternative enzymes such as SodA, SodC, AhpC, Tpx, TrxB, and Gst, which inactivate oxidative compounds and/or maintain redox homeostasis without the use of iron cofactors. Proteome data also indicate a shift in iron-sulfur cluster assembly from the Isc to the SUF machinery in iron-deprived *Y. pestis*, which also occurs in *E. coli* (Outten et al., 2004; Desnoyers et al., 2009; Pieper et al., 2010). Proteins in the SUF sulfur mobilization pathway are present at higher levels in iron-starved *Y. pestis* cells, as are TauD and ErpA (essential respiratory protein A). In *E. coli*, the SUF system is a primary generator of iron-sulfur clusters under iron-poor conditions, while TauD mobilizes sulfite from taurine and ErpA transfers iron-sulfur clusters to an isoprenoid-synthesizing enzyme involved in making the electron carrier ubiquinone. All of these changes are consistent with the "iron-sparing" response that has been described for *E. coli* and *P. aeruginosa*, which is largely controlled by Fur-regulated sRNAs (e.g., RyhB and PrrF) (Massé et al., 2005; Oglesby et al., 2008; Pieper et al., 2010).

A primary role of RyhB is to maintain iron homeostasis by downregulating expression of nonessential iron-containing proteins when iron is limited. This process effectively increases the free iron levels within the cell cytoplasm (Jacques et al., 2006). RyhB exerts these effects by downregulating expression of iron-containing proteins involved in the TCA cycle, glycolysis, respiration, oxidative stress, and iron-sulfur cluster formation (Massé et al., 2005; Bollinger and Kallio, 2007), so any available iron can be redirected for use in essential iron-containing enzymes. An important outcome of this iron sparing process is a change in the use of metabolic pathways, as described above for *Y. pestis*. *E. coli* also shifts its iron-sulfur cluster machinery from the ISC to the SUF system as iron levels are depleted. RyhB facilitates this shift through selective intracistronic regulation of the *iscRSUA* operon that downregulates ISC production while promoting SUF expression (Desnoyers et al., 2009) as described above.

The two RyhB analogs in *P. aeruginosa*, PrrF1 and PrrF2, are repressed by Fur under iron-replete conditions. PrrF1 and PrrF2 downregulate expression of genes encoding iron-containing proteins such as SodB (*sodB*), aconitase A (*acnA*), and succinate dehydrogenase (*sdhCDAB*) under iron-poor conditions, as RhyB does in *E. coli* (Wilderman et al., 2004). Both PrrF1 and PrrF2 sRNAs are involved in this gene regulation, as mutation of both genes is required to induce significant regulatory effects. Microarray studies have shown that PrrF also plays an extensive role during oxidative stress and in regulating aerobic and anaerobic metabolism (Oglesby et al., 2008).

In *E. coli* K-12, RyhB also promotes iron acquisition through its positive effects on siderophore production, which occur at multiple levels. As described in the section on RyhB regulatory mechanisms, RyhB regulation increases the pools of both shikimate and serine, while indirectly contributing to expression of the enterobactin biosynthesis genes. Shikimate is a precursor for 2,3-dihydroxybenzoic acid, which is joined with serine to form the siderophore enterobactin. RyhB positively controls expression of *shiA*, which encodes the shikimate permease responsible for uptake of shikimate under iron-poor conditions. A lack of RyhB results in decreased expression of the shikimate permease, and the consequent deficiency in enterobactin causes a growth lag under iron-deficient conditions (Prevost et al., 2007). The defective growth response of an *E. coli ryhB* null mutant under low-iron conditions can be rescued by shikimate, demonstrating the direct connection between enterobactin synthesis and RyhB-mediated regulation of *shiA*. RyhB further contributes to normal expression of the genes involved in enterobactin synthesis, likely through its iron-sparing effects. However, another important way in which RyhB contributes to enterobactin production is by reducing the expression of CysE, which is a serine acetyltransferase involved in the conversion of serine to cysteine. RyhB's downregulation of CysE increases the pool of serine available for enterobactin production (Salvail et al., 2010). *Y. pestis*, various pathogenic *E. coli* strains, and some other enteric pathogens produce the Ybt siderophore, which has salicylate, three methyl groups, a malonyl linker, and three cysteines. In some strains that produce both enterobactin and Ybt, the Ybt system is more important for pathogenesis. Thus, the RyhB regulatory scheme for CysE would limit the availability of cysteine for production of the Ybt siderophore. It will be interesting to see if this CysE regulatory mechanism is absent or modified in strains producing Ybt.

In contrast, under iron surplus conditions, bacterial cells repress expression of iron transport systems, sparing resources and reducing the amount of excess iron in the cytoplasm that can generate H_2O_2 via the Fenton reaction, which generates reactive hydroxyl radicals that damage DNA, proteins, and lipids (Andrews et al., 2003). Fur regulates many iron transporters and *ftnA*, while RyhB directly regulates bacterioferritin (Bfr). Bfr and Ftn proteins have important roles in iron detoxification and protection against oxidative stress, in addition to their roles as iron storage centers for later use under iron-deprived conditions (Carrondo, 2003; Calhoun and Kwon, 2011b). Finally, metabolic pathways with iron-containing enzymes are used under iron-replete conditions (Ochsner et al., 2002; Oglesby et al., 2008; Pieper et al., 2010).

CONCLUDING REMARKS

It is clear that low iron is used as a signal by numerous bacterial pathogens. Thus, the iron-sequestered environment of the mammalian host is used against it to turn on expression of not only iron acquisition systems but also a variety of virulence factors needed for survival and growth. The literature on iron, Fur, and/or RyhB regulation of virulence factors is voluminous and still growing. Perhaps less well appreciated is the metabolic shift to iron-sparing pathways. Although clearly not virulence factors, these "survival factors" likely play a role as important as that of iron-regulated virulence factors in promoting growth in mammals.

The discovery of Fur led to a model of transcriptional repression by Fe-Fur. As with most early models, this proved somewhat simplistic, as indicated by the title of a recent minireview: "This is not your mother's repressor: the complex role of Fur in pathogenesis" (Carpenter et al., 2009). Indeed, the complexity of iron regulation occurs not only from additional Fur regulatory mechanisms but also due to the varied regulatory mechanisms of the sRNA RyhB and its orthologs and analogs. While *fur* and *ryhB* mutants have been extremely valuable in defining their regulons, the interpretation of results with these mutants is not always straightforward for a number of reasons. Fur regulates expression of RyhB and vice versa. This and the regulation of other regulators by Fur and RyhB make assigning direct effects difficult. Expression of a gene is often regulated by a number of environmental factors and regulators. In particular, the overlapping regulatory and functional responses to iron levels, oxidative stress, and Fe-S cluster assembly (e.g., IscR) make dissection of the responses to each of these signals complex. In addition, a *fur* mutation may have subtle to moderate artifactual effects. The absence of Fur may allow increased binding of other regulators to promoters where Fur competes with these transcriptional regulators (e.g., NikR in *H. pylori*). Thus, a *fur* mutant phenotype is a complex interplay of reduced competition by transcriptional regulators, effects on other regulators, overlapping regulatory responses, and loss of iron-independent Fur regulation in addition to direct iron repression by Fur.

Site-directed mutagenesis of *fur* and structural studies of the protein, including analysis of crystals, have greatly expanded our understanding of the regulatory properties of Fur. Despite this, there is much that remains unresolved. The nucleotides required for Fe-Fur binding have not been entirely resolved. Crystal structures for Fe-Fur and Fe-Fur or Mn-Fur bound to DNA have not been obtained in the nearly 10 years since the Zn-bound form of Fur$_{Pa}$ was crystallized. The mechanisms of Fur activation, Mn-Fur regulation, and apo-Fur regulation as well as the different regulatory mechanisms of RyhB are being addressed. However, the complexity of iron regulation mechanisms and their role in microbial pathogenesis will likely provide interesting questions to resolve for decades to come.

REFERENCES

Alamuri, P., N. Mehta, A. Burk, and R. J. Maier. 2006. Regulation of the *Helicobacter pylori* Fe-S cluster synthesis protein NifS by iron, oxidative stress conditions, and Fur. *J. Bacteriol.* **188**:5325–5330.

Alice, A. F., H. Naka, and J. H. Crosa. 2008. Global gene expression as a function of the iron status of the bacterial cell: influence of differentially expressed genes in the virulence of the human pathogen *Vibrio vulnificus*. *Infect. Immun.* **76**:4019–4037.

Amarasinghe, J. J., F. A. Scannapieco, and E. M. Haase. 2009. Transcriptional and translational analysis of biofilm determinants of *Aggregatibacter actinomycetemcomitans* in response to environmental perturbation. *Infect. Immun.* **77**:2896–2907.

Andrews, S. C., A. K. Robinson, and F. Rodríguez-Quiñones. 2003. Bacterial iron homeostasis. *FEMS Microbiol. Rev.* **27**:215–237.

Arhin, A., and C. Boucher. 2010. The outer membrane protein OprQ and adherence of *Pseudomonas aeruginosa* to human fibronectin. *Microbiology* **156**:1415–1423.

Bagg, A., and J. B. Neilands. 1987. Ferric uptake regulation protein acts as a repressor, employing iron(II) as a cofactor to bind the operator of an iron transport operon in *Escherichia coli*. *Biochemistry* **26**:5471–5477.

Bagg, A., and J. B. Neilands. 1985. Mapping of a mutation affecting regulation of iron uptake systems in *Escherichia coli* K-12. *J. Bacteriol.* **161**:450–453.

Banin, E., M. L. Vasil, and E. P. Greenberg. 2005. Iron and *Pseudomonas aeruginosa* biofilm formation. *Proc. Natl. Acad. Sci. USA* **102**:11076–11081.

Bearden, S. W., and R. D. Perry. 1999. The Yfe system of *Yersinia pestis* transports iron and manganese and is required for full virulence of plague. *Mol. Microbiol.* **32**:403–414.

Bearden, S. W., T. M. Staggs, and R. D. Perry. 1998. An ABC transporter system of *Yersinia pestis* allows utilization of chelated iron by *Escherichia coli* SAB11. *J. Bacteriol.* **180**:1135–1147.

Bearson, B. L., L. Wilson, and J. W. Foster. 1998. A low pH-inducible, PhoPQ-dependent acid tolerance response protects *Salmonella typhimurium* against inorganic acid stress. *J. Bacteriol.* **180**:2409–2417.

Belasco, J. G. 2010. All things must pass: contrasts and commonalities in eukaryotic and bacterial mRNA decay. *Nat. Rev. Mol. Cell Biol.* **11**:467–478.

Bennett, R. L., and L. I. Rothfield. 1976. Genetic and physiological regulation of intrinsic proteins of the outer membrane of *Salmonella typhimurium*. *J. Bacteriol.* **127**:498–504.

Bereswill, S., S. Greiner, A. H. M. van Vliet, B. Waidner, F. Fassbinder, E. Schiltz, J. G. Kusters, and M. Kist. 2000. Regulation of ferritin-mediated cytoplasmic iron storage by the ferric uptake regulator homolog (Fur) of *Helicobacter pylori*. *J. Bacteriol.* **182**:5948–5953.

Berish, S. A., S. Subbarao, C.-Y. Chen, D. L. Trees, and S. A. Morse. 1993. Identification and cloning of a *fur* homolog from *Neisseria gonorrhoeae*. *Infect. Immun.* **61**:4599–4606.

Bijlsma, J. J. E., B. Waidner, A. H. M. van Vliet, N. J. Hughes, S. Häg, S. Bereswill, D. J. Kelly, C. M. J. E. Vandenbroucke-Grauls, M. Kist, and J. G. Kusters. 2002. The *Helicobacter pylori* homologue of the ferric uptake regulator is involved in acid resistance. *Infect. Immun.* **70**:606–611.

Bobrov, A. G., O. Kirillina, D. A. Ryjenkov, C. M. Waters, P. A. Price, J. D. Fetherston, D. Mack, W. E. Goldman, M. Gomelsky, and R. D. Perry. 2011. Systematic analysis of cyclic di-GMP signalling enzymes and their role in biofilm formation and virulence in *Yersinia pestis*. *Mol. Microbiol.* **79**:533–551.

Bollinger, C. J., and P. T. Kallio. 2007. Impact of the small RNA RyhB on growth, physiology and heterologous protein expression in *Escherichia coli*. *FEMS Microbiol. Lett.* **275**:221–228.

Braun, V., S. Mahren, and A. Sauter. 2006. Gene regulation by transmembrane signaling. *BioMetals* **19**:103–113.

Brennan, R. G., and T. M. Link. 2007. Hfq structure, function and ligand binding. *Curr. Opin. Microbiol.* **10**:125–133.

Brickman, T., and S. Armstrong. 2009. Temporal signaling and differential expression of *Bordetella* iron transport systems: the role of ferrimones and positive regulators. *BioMetals* **22**:33–41.

Brickman, T. J., and S. K. Armstrong. 1995. *Bordetella pertussis fur* gene restores iron repressibility of siderophore and protein expression to deregulated *Bordetella bronchiseptica* mutants. *J. Bacteriol.* **177**:268–270.

Bury-Moné, S., J.-M. Thiberge, M. Contreras, A. Maitournam, A. Labigne, and H. De Reuse. 2004. Responsiveness to acidity via metal ion regulators mediates virulence in the gastric pathogen *Helicobacter pylori*. *Mol. Microbiol.* **53**:623–638.

Calderwood, S. B., and J. J. Mekalanos. 1987. Iron regulation of shiga-like toxin expression in *Escherichia coli* is mediated by the *fur* locus. *J. Bacteriol.* **169**:4759–4764.

Calhoun, L. N., and Y. M. Kwon. 2011a. The ferritin-like protein Dps protects *Salmonella enterica* serotype Enteritidis from the Fenton-mediated killing mechanism of bactericidal antibiotics. *Int. J. Antimicrob. Agents* **37**:261–265.

Calhoun, L. N., and Y. M. Kwon. 2011b. Structure, function and regulation of the DNA-binding protein Dps and its role in acid and oxidative stress resistance in *Escherichia coli*: a review. *J. Appl. Microbiol.* **110**:375–386.

Carpenter, B. M., H. Gancz, S. L. Benoit, S. Evans, C. H. Olsen, S. L. J. Michel, R. J. Maier, and D. S. Merrell. 2010. Mutagenesis of conserved amino acids of *Helicobacter pylori* Fur reveals residues important for function. *J. Bacteriol.* **192**:5037–5052.

Carpenter, B. M., J. M. Whitmire, and D. S. Merrell. 2009. This is not your mother's repressor: the complex role of Fur in pathogenesis. *Infect. Immun.* **77**:2590–2601.

Carpousis, A. J. 2007. The RNA degradosome of *Escherichia coli*: an mRNA-degrading machine assembled on RNase E. *Annu. Rev. Microbiol.* **61**:71–87.

Carrondo, M. A. 2003. Ferritins, iron uptake and storage from the bacterioferritin viewpoint. *EMBO J.* **22**:1959–1968.

Carson, S. D. B., C. E. Thomas, and C. Elkins. 1996. Cloning and sequencing of a *Haemophilus ducreyi fur* homolog. *Gene* **176**:125–129.

Celesnik, H., A. Deana, and J. G. Belasco. 2007. Initiation of RNA decay in *Escherichia coli* by 5 pyrophosphate removal. *Mol. Cell* **27**:79–90.

Chauvaux, S., M.-L. Rosso, L. Frangeul, C. Lacroix, L. Labarre, A. Schiavo, M. Marceau, M.-A. Dillies, J. Foulon, J.-Y. Coppée, C. Médigue, M. Simonet, and E. Carniel. 2007. Transcriptome analysis of *Yersinia pestis* in human plasma: an approach for discovering bacterial genes involved in septicaemic plague. *Microbiology* **153**:3112–3124.

Cornelis, P., S. Matthijs, and L. Van Oeffelen. 2009. Iron uptake regulation in *Pseudomonas aeruginosa*. *BioMetals* **22**:15–22.

Costerton, J. W., P. S. Stewart, and E. P. Greenberg. 1999. Bacterial biofilms: a common cause of persistent infections. *Science* **284**:1318–1322.

Coy, M., C. Doyle, J. Besser, and J. B. Neilands. 1994. Site-directed mutagenesis of the ferric uptake regulation gene of *Escherichia coli*. *BioMetals* **7**:292–298.

Coy, M., and J. B. Neilands. 1991. Structural dynamics and functional domains of the Fur protein. *Biochemistry* **30**:8201–8210.

Cui, J., H. Piao, S. Jin, H. S. Na, Y. Hong, H. E. Choy, and P. Y. Ryu. 2009. Effect of iron on cytolysin A expression in *Salmonella enterica* serovar Typhi. *J. Microbiol.* **47**:479–485.

Davis, B. M., M. Quinones, J. Pratt, Y. Ding, and M. K. Waldor. 2005. Characterization of the small untranslated RNA RyhB and its regulon in *Vibrio cholerae*. *J. Bacteriol.* **187**:4005–4014.

Deana, A., H. Celesnik, and J. G. Belasco. 2008. The bacterial enzyme RppH triggers messenger RNA degradation by 5 pyrophosphate removal. *Nature* **451**:355–358.

Delany, I., R. Ieva, C. Alaimo, R. Rappuoli, and V. Scarlato. 2003. The iron-responsive regulator Fur is transcriptionally autoregulated and not essential in *Neisseria meningitidis*. *J. Bacteriol.* **185**:6032–6041.

Delany, I., R. Ieva, A. Soragni, M. Hilleringmann, R. Rappuoli, and V. Scarlato. 2005. In vitro analysis of protein-operator interactions of the NikR and Fur metal-responsive regulators of coregulated genes in *Helicobacter pylori*. *J. Bacteriol.* **187**:7703–7715.

Delany, I., R. Rappuoli, and V. Scarlato. 2004. Fur functions as an activator and as a repressor of putative virulence genes in *Neisseria meningitidis*. *Mol. Microbiol.* **52**:1081–1090.

Delany, I., G. Spohn, A.-B. F. Pacheco, R. Ieva, C. Alaimo, R. Rappuoli, and V. Scarlato. 2002. Autoregulation of *Helicobacter pylori* Fur revealed by functional analysis of the iron-binding site. *Mol. Microbiol.* **46**:1107–1122.

Delany, I., G. Spohn, R. Rappuoli, and V. Scarlato. 2001. The Fur repressor controls transcription of iron-activated and -repressed genes in *Helicobacter pylori*. *Mol. Microbiol.* **42**:1297–1309.

de Lorenzo, V., M. Herrero, F. Giovannini, and J. B. Neilands. 1988. Fur (ferric uptake regulation) protein and CAP (catabolite-activator

protein) modulate transcription of *fur* gene in *Escherichia coli*. *Eur. J. Biochem.* **173**:537–546.

de Lorenzo, V., J. Perez-Martín, L. Escolar, G. Pesole, and G. Bertoni. 2004. Mode of binding of the Fur protein to target DNA: negative regulation of iron-controlled gene expression, p. 185–196. *In* J. H. Crosa, A. R. Mey, and S. M. Payne (ed.), *Iron Transport in Bacteria*. ASM Press, Washington, DC.

de Lorenzo, V., S. Wee, M. Herrero, and J. B. Neilands. 1987. Operator sequences of the aerobactin operon of plasmid ColV-K30 binding the ferric uptake regulation (*fur*) repressor. *J. Bacteriol.* **169**:2624–2630.

Deng, W., V. Burland, G. Plunkett III, A. Boutin, G. F. Mayhew, P. Liss, N. T. Perna, D. J. Rose, B. Mau, S. Zhou, D. C. Schwartz, J. D. Fetherston, L. E. Lindler, R. R. Brubaker, G. V. Plano, S. C. Straley, K. A. McDonough, M. L. Nilles, J. S. Matson, F. R. Blattner, and R. D. Perry. 2002. Genome sequence of *Yersinia pestis* KIM. *J. Bacteriol.* **184**:4601–4611.

Desnoyers, G., A. Morissette, K. Prevost, and E. Massé. 2009. Small RNA-induced differential degradation of the polycistronic mRNA iscRSUA. *EMBO J.* **28**:1551–1561.

Dubos, R. J., and J. W. Geiger. 1946. Preparation and properties of Shiga toxin and toxoid. *J. Exp. Med.* **84**:143–156.

Eisen, R. J., S. W. Bearden, A. P. Wilder, J. A. Montenieri, M. F. Antolin, and K. L. Gage. 2006. Early-phase transmission of *Yersinia pestis* by unblocked fleas as a mechanism explaining rapidly spreading plague epizootics. *Proc. Natl. Acad. Sci. USA* **103**:15380–15385.

Ellermeier, J. R., and J. M. Slauch. 2008. Fur regulates expression of the *Salmonella* pathogenicity island 1 type III secretion system through HilD. *J. Bacteriol.* **190**:476–486.

Ernst, F. D., G. Homuth, J. Stoof, U. Mäder, B. Waidner, E. J. Kuipers, M. Kist, J. G. Kusters, S. Bereswill, and A. H. M. van Vliet. 2005. Iron-responsive regulation of the *Helicobacter pylori* iron-cofactored superoxide dismutase SodB is mediated by Fur. *J. Bacteriol.* **187**:3687–3692.

Ernst, J. F., R. L. Bennett, and L. I. Rothfield. 1978. Constitutive expression of the iron-enterochelin and ferrichrome uptake systems in a mutant strain of *Salmonella typhimurium*. *J. Bacteriol.* **135**:928–934.

Escolar, L., J. Pérez-Martín, and V. de Lorenzo. 1998. Binding of the Fur (ferric uptake regulator) repressor of *Escherichia coli* to arrays of the GATAAT sequence. *J. Mol. Biol.* **283**:537–547.

Escolar, L., J. Pérez-Martín, and V. de Lorenzo. 1999. Opening the iron box: transcriptional metalloregulation by the Fur protein. *J. Bacteriol.* **181**:6223–6229.

Escolar, L., J. Pérez-Martín, and V. de Lorenzo. 2000. Evidence of an unusually long operator for the Fur repressor in the aerobactin promoter of *Escherichia coli*. *J. Biol. Chem.* **275**:24709–24714.

Fetherston, J. D., V. J. Bertolino, and R. D. Perry. 1999. YbtP and YbtQ: two ABC transporters required for iron uptake in *Yersinia pestis*. *Mol. Microbiol.* **32**:289–299.

Fetherston, J. D., O. Kirillina, A. G. Bobrov, J. T. Paulley, and R. D. Perry. 2010. The yersiniabactin transport system is critical for the pathogenesis of bubonic and pneumonic plague. *Infect. Immun.* **78**:2045–2052.

Forman, S., M. J. Nagiec, J. Abney, R. D. Perry, and J. D. Fetherston. 2007. Analysis of the aerobactin and ferric hydroxamate uptake systems of *Yersinia pestis*. *Microbiology* **153**:2332–2341.

Forman, S., J. Paulley, J. Fetherston, Y.-Q. Cheng, and R. Perry. 2010. *Yersinia* ironomics: comparison of iron transporters among *Yersinia pestis* biotypes and its nearest neighbor, *Yersinia pseudotuberculosis*. *BioMetals* **23**:275–294.

Fréchon, D., and E. Le Cam. 1994. Fur (ferric uptake regulation) protein interaction with target DNA: comparison of gel retardation, footprinting and elecron microscopy analyses. *Biochem. Biophys. Res. Commun.* **201:**346–355.

Frohlich, K. S., and J. Vogel. 2009. Activation of gene expression by small RNA. *Curr. Opin. Microbiol.* **12:**674–682.

Gage, K. L., and M. Y. Kosoy. 2005. Natural history of plague: perspectives from more than a century of research. *Annu. Rev. Entomol.* **50:**505–528.

Gancz, H., S. Censini, and D. S. Merrell. 2006. Iron and pH homeostasis intersect at the level of Fur regulation in the gastric pathogen *Helicobacter pylori*. *Infect. Immun.* **74:**602–614.

Gao, H., D. Zhou, Y. Li, Z. Guo, Y. Han, Y. Song, J. Zhai, Z. Du, X. Wang, J. Lu, and R. Yang. 2008. The iron-responsive Fur regulon in *Yersinia pestis*. *J. Bacteriol.* **190:**3063–3075.

Gardner, S. E., S. E. Fowlston, and W. L. George. 1990. Effect of iron on production of a possible virulence factor by *Plesiomonas shigelloides*. *J. Clin. Microbiol.* **28:**811–813.

Giel, J. L., D. Rodionov, M. Liu, F. R. Blattner, and P. J. Kiley. 2006. IscR-dependent gene expression links iron-sulphur cluster assembly to the control of O_2-regulated genes in *Escherichia coli*. *Mol. Microbiol.* **60:**1058–1075.

Glick, R., C. Gilmour, J. Tremblay, S. Satanower, O. Avidan, E. Déziel, E. P. Greenberg, K. Poole, and E. Banin. 2010. Increase in rhamnolipid synthesis under iron-limiting conditions influences surface motility and biofilm formation in *Pseudomonas aeruginosa*. *J. Bacteriol.* **192:**2973–2980.

Gong, S., S. W. Bearden, V. A. Geoffroy, J. D. Fetherston, and R. D. Perry. 2001. Characterization of the *Yersinia pestis* Yfu ABC iron transport system. *Infect. Immun.* **67:**2829–2837.

Gottesman, S., C. A. McCullen, M. Guillier, C. K. Vanderpool, N. Majdalani, J. Benhammou, K. M. Thompson, P. C. FitzGerald, N. A. Sowa, and D. J. FitzGerald. 2006. Small RNA regulators and the bacterial response to stress. *Cold Spring Harbor Symp. Quant. Biol.* **71:**1–11.

Grifantini, R., E. Frigimelica, I. Delany, E. Bartolini, S. Giovinazzi, S. Balloni, S. Agarwal, G. Galli, C. Genco, and G. Grandi. 2004. Characterization of a novel *Neisseria meningitidis* Fur and iron-regulated operon required for protection from oxidative stress: utility of DNA microarray in the assignment of the biological role of hypothetical genes. *Mol. Microbiol.* **54:**962–979.

Guina, T., S. O. Purvine, E. C. Yi, J. Eng, D. R. Goodlett, R. Aebersold, and S. I. Miller. 2003. Quantitative proteomic analysis indicates increased synthesis of a quinolone by *Pseudomonas aeruginosa* isolates from cystic fibrosis airways. *Proc. Natl. Acad. Sci. USA* **100:**2771–2776.

Gunn, J. S. 2008. The *Salmonella* PmrAB regulon: lipopolysaccharide modifications, antimicrobial peptide resistance and more. *Trends Microbiol.* **16:**284–290.

Hall, H. K., and J. W. Foster. 1996. The role of Fur in the acid tolerance response of *Salmonella typhimurium* is physiologically and genetically separable from its role in iron acquisition. *J. Bacteriol.* **178:**5683–5691.

Halsey, T. A., A. Vazquez-Torres, D. J. Gravdahl, F. C. Fang, and S. J. Libby. 2004. The ferritin-like Dps protein is required for *Salmonella enterica* serovar Typhimurium oxidative stress resistance and virulence. *Infect. Immun.* **72:**1155–1158.

Han, Y., J. Qiu, Z. Guo, H. Gao, Y. Song, D. Zhou, and R. Yang. 2007. Comparative transcriptomics in *Yersinia pestis*: a global view of environmental modulation of gene expression. *BMC Microbiol.* **7:**96.

Hancock, V., L. Ferrières, and P. Klemm. 2008. The ferric yersiniabactin uptake receptor FyuA is required for efficient biofilm formation by urinary tract infectious *Escherichia coli* in human urine. *Microbiology* **154:**167–175.

Hantke, K. 2005. Bacterial zinc uptake and regulators. *Curr. Opin. Microbiol.* **8:**196–202.

Hantke, K. 1984. Cloning of the repressor protein gene of iron-regulated systems in *Escherichia coli* K12. *Mol. Gen. Genet.* **197:**337–341.

Hantke, K. 1982. Negative control of iron uptake system in *Escherichia coli*. *FEMS Microbiol. Lett.* **15:**83–86.

Hantke, K. 1981. Regulation of ferric iron transport in *Escherichia coli* K12: isolation of a constitutive mutant. *Mol. Gen. Genet.* **182:**288–292.

Hantke, K. 1987. Selection procedure for deregulated iron transport mutants (*fur*) in *Escherichia coli* K 12: *fur* not only affects iron metabolism. *Mol. Gen. Genet.* **210:**135–139.

Hassett, D. J., M. L. Howell, U. A. Ochsner, M. L. Vasil, Z. Johnson, and G. E. Dean. 1997. An operon containing *fumC* and *sodA* encoding fumarase C and manganese superoxide dismutase is controlled by the ferric uptake regulator in *Pseudomonas aeruginosa*: *fur* mutants produce elevated alginate levels. *J. Bacteriol.* **179:**1452–1459.

Herold, S., J. C. Paton, P. Srimanote, and A. W. Paton. 2009. Differential effects of short-chain fatty acids and iron on expression of *iha* in Shiga-toxigenic *Escherichia coli*. *Microbiology* **155:**3554–3563.

Herrmann, K. M., and L. M. Weaver. 1999. The shikimate pathway. *Annu. Rev. Plant Physiol. Plant Mol. Biol.* **50:**473–503.

Hickey, E. K., and N. P. Cianciotto. 1997. An iron- and Fur-repressed *Legionella pneumophila* gene that promotes intracellular infection and encodes a protein with similarity to the *Escherichia coli* aerobactin synthetases. *Infect. Immun.* **65:**133–143.

Hindré, T., H. Brüggemann, C. Buchrieser, and Y. Héchard. 2008. Transcriptional profiling of *Legionella pneumophila* biofilm cells and the influence of iron on biofilm formation. *Microbiology* **154:**30–41.

Holmes, K., F. Mulholland, B. M. Pearson, C. Pin, J. McNicholl-Kennedy, J. M. Ketley, and J. M. Wells. 2005. *Campylobacter jejuni* gene expression in response to iron limitation and the role of Fur. *Microbiology* **151:**243–257.

Ikeda, J. S., A. Janakiraman, D. G. Kehres, M. E. Maguire, and J. M. Slauch. 2005. Transcriptional regulation of *sitABCD* of *Salmonella enterica* serovar Typhimurium by MntR and Fur. *J. Bacteriol.* **187:**912–922.

Ikeda, Y., M. Yagi, T. Morita, and H. Aiba. 2011. Hfq binding at RhlB-recognition region of RNase E is crucial for the rapid degradation of target mRNAs mediated by sRNAs in *Escherichia coli*. *Mol. Microbiol.* **79:**419–432.

Inglesby, T. V., D. T. Dennis, D. A. Henderson, J. G. Bartlett, M. S. Ascher, E. Eitzen, A. D. Fine, A. M. Friedlander, J. Hauer, J. F. Koerner, M. Layton, J. McDade, M. T. Osterholm, T. O'Toole, G. Parker, T. M. Perl, P. K. Russell, M. Schoch-Spana, and K. Tonat. 2000. Plague as a biological weapon: medical and public health management. *JAMA* **283:**2281–2290.

Jackson, S., and T. W. Burrows. 1956. The virulence-enhancing effect of iron on non-pigmented mutants of virulent strains of *Pasteurella pestis*. *Br. J. Exp. Pathol.* **37:**577–583.

Jacobsen, I., J. Gerstenberger, A. D. Gruber, J. T. Bossé, P. R. Langford, I. Hennig-Pauka, J. Meens, and G.-F. Gerlach. 2005. Deletion of the ferric uptake regulator Fur impairs the in vitro growth and virulence of *Actinobacillus pleuropneumoniae*. *Infect. Immun.* **73:**3740–3744.

Jacques, J. F., S. Jang, K. Prevost, G. Desnoyers, M. Desmarais, J. Imlay, and E. Massé. 2006. RyhB small RNA modulates the free intracellular iron pool and is essential for normal growth during iron limitation in *Escherichia coli*. *Mol. Microbiol.* **62:**1181–1190.

Johnston, A., J. Todd, A. Curson, S. Lei, N. Nikolaidou-Katsaridou, M. Gelfand, and D. Rodionov. 2007. Living without Fur: the subtlety and complexity of iron-responsive gene regulation in the symbiotic bacterium *Rhizobium* and other α-proteobacteria. *BioMetals* **20:**501–511.

Karjalainen, T. K., D. G. Evans, D. J. Evans, Jr., D. Y. Graham, and C.-H. Lee. 1991. Iron represses the expression of CFA/I fimbriae of enterotoxigenic *E. coli. Microb. Pathog.* **11:**317–323.

Kim, J.-S., M.-H. Sung, D.-H. Kho, and J. K. Lee. 2005. Induction of manganese-containing superoxide dismutase is required for acid tolerance in *Vibrio vulnificus. J. Bacteriol.* **187:**5984–5995.

Kirillina, O., A. G. Bobrov, J. D. Fetherston, and R. D. Perry. 2006. A hierarchy of iron uptake systems: Yfu and Yiu are functional in *Yersinia pestis. Infect. Immun.* **74:**6171–6178.

Lamont, I. L., P. A. Beare, U. Ochsner, A. I. Vasil, and M. L. Vasil. 2002. Siderophore-mediated signaling regulates virulence factor production in *Pseudomonas aeruginosa. Proc. Natl. Acad. Sci. USA* **99:**7072–7077.

Lankford, C. E. 1973. Bacterial assimilation of iron. *CRC Crit. Rev. Microbiol.* **2:**273–331.

Larkin, M. A., G. Blackshields, N. P. Brown, R. Chenna, P. A. McGettigan, H. McWilliam, F. Valentin, I. M. Wallace, A. Wilm, R. Lopez, J. D. Thompson, T. J. Gibson, and D. G. Higgins. 2007. Clustal W and Clustal X version 2.0. *Bioinformatics* **23:**2947–2948.

Lathem, W. W., S. D. Crosby, V. L. Miller, and W. E. Goldman. 2005. Progression of primary pneumonic plague: a mouse model of infection, pathology, and bacterial transcriptional activity. *Proc. Natl. Acad. Sci. USA* **102:**17786–17791.

Lavrrar, J. L., C. A. Christoffersen, and M. A. McIntosh. 2002. Fur-DNA interactions at the bidirectional *fepDGC-entS* promoter region in *Escherichia coli. J. Mol. Biol.* **322:**983–995.

Lease, R. A., and M. Belfort. 2000a. Riboregulation by DsrA RNA: trans-actions for global economy. *Mol. Microbiol.* **38:**667–672.

Lease, R. A., and M. Belfort. 2000b. A trans-acting RNA as a control switch in *Escherichia coli*: DsrA modulates function by forming alternative structures. *Proc. Natl. Acad. Sci. USA* **97:**9919–9924.

Lee, H.-J., S. H. Bang, K.-H. Lee, and S.-J. Park. 2007. Positive regulation of *fur* gene expression via direct interaction of Fur in a pathogenic bacterium, *Vibrio vulnificus. J. Bacteriol.* **189:**2629–2636.

Lee, H.-J., K.-J. Park, A. Y. Lee, S. G. Park, B. C. Park, K.-H. Lee, and S.-J. Park. 2003. Regulation of *fur* expression by RpoS and Fur in *Vibrio vulnificus. J. Bacteriol.* **185:**5891–5896.

Lee, J.-W., and J. Helmann. 2007. Functional specialization within the Fur family of metalloregulators. *BioMetals* **20:**485–499.

Lee-Lewis, H., and D. M. Anderson. 2010. Absence of inflammation and pneumonia during infection with nonpigmented *Yersinia pestis* reveals a new role for the *pgm* locus in pathogenesis. *Infect. Immun.* **78:**220–230.

Lenčo, J., M. Hubálek, P. Larsson, A. Fučíková, M. Brychta, A. Macela, and J. Stulík. 2007. Proteomics analysis of the *Francisella tularensis* LVS response to iron restriction: induction of the *F. tularensis* pathogenicity island proteins IglABC. *FEMS Microbiol. Lett.* **269:**11–21.

Loprasert, S., R. Sallabhan, W. Whangsuk, and S. Mongkolsuk. 2000. Characterization and mutagenesis of *fur* gene from *Burkholderia pseudomallei. Gene* **254:**129–137.

Luke, N. R., A. J. Howlett, J. Shao, and A. A. Campagnari. 2004. Expression of type IV pili by *Moraxella catarrhalis* is essential for natural competence and is affected by iron limitation. *Infect. Immun.* **72:**6262–6270.

Lymberopoulos, M. H., S. Houle, F. Daigle, S. Léveillé, A. Brée, M. Moulin-Schouleur, J. R. Johnson, and C. M. Dozois. 2006. Characterization of Stg fimbriae from an avian pathogenic *Escherichia coli* O78:K80 strain and assessment of their contribution to colonization of the chicken respiratory tract. *J. Bacteriol.* **188:**6449–6459.

Maki, K., K. Uno, T. Morita, and H. Aiba. 2008. RNA, but not protein partners, is directly responsible for translational silencing by a bacterial Hfq-binding small RNA. *Proc. Natl. Acad. Sci. USA* **105:**10332–10337.

Massé, E., F. E. Escorcia, and S. Gottesman. 2003. Coupled degradation of a small regulatory RNA and its mRNA targets in *Escherichia coli. Genes Dev.* **17:**2374–2383.

Massé, E., and S. Gottesman. 2002. A small RNA regulates the expression of genes involved in iron metabolism in *Escherichia coli. Proc. Natl. Acad. Sci. USA* **99:**4620–4625.

Massé, E., C. K. Vanderpool, and S. Gottesman. 2005. Effect of RyhB small RNA on global iron use in *Escherichia coli. J. Bacteriol.* **187:**6962–6971.

Merrell, D. S., L. J. Thompson, C. C. Kim, H. Mitchell, L. S. Tompkins, A. Lee, and S. Falkow. 2003. Growth phase-dependent response of *Helicobacter pylori* to iron starvation. *Infect. Immun.* **71:**6510–6525.

Mey, A. R., S. A. Craig, and S. M. Payne. 2005a. Characterization of *Vibrio cholerae* RyhB: the RyhB regulon and role of *ryhB* in biofilm formation. *Infect. Immun.* **73:**5706–5719.

Mey, A. R., E. E. Wyckoff, V. Kanukurthy, C. R. Fisher, and S. M. Payne. 2005b. Iron and Fur regulation in *Vibrio cholerae* and the role of Fur in virulence. *Infect. Immun.* **73:**8167–8178.

Mey, A. R., E. E. Wyckoff, A. G. Oglesby, E. Rab, R. K. Taylor, and S. M. Payne. 2002. Identification of the *Vibrio cholerae* enterobactin receptors VctA and IrgA: IrgA is not required for virulence. *Infect. Immun.* **70:**3419–3426.

Mills, S. A., and M. A. Marletta. 2005. Metal binding characteristics and role of iron oxidation in the ferric uptake regulator from *Escherichia coli. Biochemistry* **44:**13553–13559.

Mitarai, N., J. A. Benjamin, S. Krishna, S. Semsey, Z. Csiszovszki, E. Massé, and K. Sneppen. 2009. Dynamic features of gene expression control by small regulatory RNAs. *Proc. Natl. Acad. Sci. USA* **106:**10655–10659.

Moll, I., T. Afonyushkin, O. Vytvytska, V. R. Kaberdin, and U. Blasi. 2003. Coincident Hfq binding and RNase E cleavage sites on mRNA and small regulatory RNAs. *RNA* **9:**1308–1314.

Morita, T., and H. Aiba. 2011. RNase E action at a distance: degradation of target mRNAs mediated by an Hfq-binding small RNA in bacteria. *Genes Dev.* **25:**294–298.

Morita, T., K. Maki, and H. Aiba. 2005. RNase E-based ribonucleoprotein complexes: mechanical basis of mRNA destabilization mediated by bacterial noncoding RNAs. *Genes Dev.* **19:**2176–2186.

Murphy, E. R., and S. M. Payne. 2007. RyhB, an iron-responsive small RNA molecule, regulates *Shigella dysenteriae* virulence. *Infect. Immun.* **75:**3470–3477.

Musk, D. J., D. A. Banko, and P. J. Hergenrother. 2005. Iron salts perturb biofilm formation and disrupt existing biofilms of *Pseudomonas aeruginosa. Chem. Biol.* **12:**789–796.

Najimi, M., M. L. Lemos, and C. R. Osorio. 2009. Identification of iron regulated genes in the fish pathogen *Aeromonas salmonicida* subsp. *salmonicida*: genetic diversity and evidence of conserved iron uptake systems. *Vet. Microbiol.* **133:**377–382.

Nandal, A., C. C. O. Huggins, M. R. Woodall, J. McHugh, F. Rodríguez-Quiñones, M. A. Quail, J. R. Guest, and S. C. Andrews. 2010. Induction of the ferritin gene (*ftnA*) of *Escherichia coli* by Fe^{2+}-Fur is mediated by reversal of H-NS silencing and is RyhB independent. *Mol. Microbiol.* **75:**637–657.

Niederhoffer, E. C., C. M. Naranjo, K. L. Bradley, and J. A. Fee. 1990. Control of *Escherichia coli* superoxide dismutase (*sodA* and *sodB*) genes by the ferric uptake regulation (*fur*) locus. *J. Bacteriol.* **172:**1930–1938.

Ochsner, U. A., A. I. Vasil, and M. L. Vasil. 1995. Role of the ferric uptake regulator of *Pseudomonas aeruginosa* in the regulation of siderophores and exotoxin A expression: purification and activity on iron-regulated promoters. *J. Bacteriol.* **177:**7194–7201.

Ochsner, U. A., and M. L. Vasil. 1996. Gene repression by the ferric uptake regulator in *Pseudomonas aeruginosa*: cycle selection of iron-regulated genes. *Proc. Natl. Acad. Sci. USA* **93:**4409–4414.

Ochsner, U. A., P. J. Wilderman, A. I. Vasil, and M. L. Vasil. 2002. GeneChip® expression analysis of the iron starvation response in *Pseudomonas aeruginosa*: identification of novel pyoverdine biosynthesis genes. *Mol. Microbiol.* **45:**1277–1287.

Oglesby, A. G., J. M. Farrow III, J.-H. Lee, A. P. Tomaras, E. P. Greenberg, E. C. Pesci, and M. L. Vasil. 2008. The influence of iron on *Pseudomonas aeruginosa* physiology: a regulatory link between iron and quorum sensing. *J. Biol. Chem.* **283:**15558–15567.

Oglesby, A. G., E. R. Murphy, V. R. Iyer, and S. M. Payne. 2005. Fur regulates acid resistance in *Shigella flexneri* via RyhB and *ydeP*. *Mol. Microbiol.* **58:**1354–1367.

Outten, F. W., O. Djaman, and G. Storz. 2004. A *suf* operon requirement for Fe-S cluster assembly during iron starvation in *Escherichia coli*. *Mol. Microbiol.* **52:**861–872.

Padalon-Brauch, G., R. Hershberg, M. Elgrably-Weiss, K. Baruch, I. Rosenshine, H. Margalit, and S. Altuvia. 2008. Small RNAs encoded within genetic islands of *Salmonella typhimurium* show host-induced expression and role in virulence. *Nucleic Acids Res.* **36:**1913–1927.

Palyada, K., D. Threadgill, and A. Stintzi. 2004. Iron acquisition and regulation in *Campylobacter jejuni*. *J. Bacteriol.* **186:**4714–4729.

Patzer, S. I., and K. Hantke. 2001. Dual repression by Fe²⁺-Fur and Mn²⁺-MntR of the *mntH* gene, encoding an NRAMP-Like Mn²⁺ transporter in *Escherichia coli*. *J. Bacteriol.* **183:**4806–4813.

Payne, S. M. 1988. Iron and virulence in the family *Enterobacteriaceae*. *CRC Crit. Rev. Microbiol.* **16:**81–111.

Payne, S. M., E. E. Wyckoff, E. R. Murphy, A. G. Oglesby, M. L. Boulette, and N. M. Davies. 2006. Iron and pathogenesis of *Shigella*: iron acquisition in the intracellular environment. *BioMetals* **19:**173–180.

Pecqueur, L., B. D'Autréaux, J. Dupuy, Y. Nicolet, L. Jacquamet, B. Brutscher, I. Michaud-Soret, and B. Bersch. 2006. Structural changes of *Escherichia coli* ferric uptake regulator during metal-dependent dimerization and activation explored by NMR and X-ray crystallography. *J. Biol. Chem.* **281:**21286–21295.

Perry, R. D., J. Abney, I. Mier, Jr., Y. Lee, S. W. Bearden, and J. D. Fetherston. 2003. Regulation of the *Yersinia pestis* Yfe and Ybt iron transport systems. *Adv. Exp. Med. Biol.* **529:**275–283.

Perry, R. D., A. G. Bobrov, O. Kirillina, and J. D. Fetherston. *Yersinia pestis* transition metal divalent cation transporters. *Adv. Exp. Med. Biol.*, in press.

Perry, R. D., A. G. Bobrov, O. Kirillina, H. A. Jones, L. L. Pedersen, J. Abney, and J. D. Fetherston. 2004. Temperature regulation of the hemin storage (Hms⁺) phenotype of *Yersinia pestis* is post-transcriptional. *J. Bacteriol.* **186:**1638–1647.

Perry R. D., S. K. Craig, J. Abney, A. G. Bobrov, O. Kirillina, I. Mier, Jr., H. Truszczynska, and J. D. Fetherston. 2012. Manganese transporters Yfe and Mnth are Fur-regulated and important for the virulence of *Yersinia pestis*. *Microbiology* **158:**804–815.

Perry, R. D., and J. D. Fetherston. 1997. *Yersinia pestis*—etiologic agent of plague. *Clin. Microbiol. Rev.* **10:**35–66.

Perry, R. D., and J. D. Fetherston. 2011. Yersiniabactin iron uptake: mechanisms and role in *Yersinia pestis* pathogenesis. *Microbes Infect.* **13:**808–817.

Perry, R. D., I. Mier, Jr., and J. D. Fetherston. 2007. Roles of the Yfe and Feo transporters of *Yersinia pestis* in iron uptake and intracellular growth. *BioMetals* **20:**699–703.

Pieper, R., S.-T. Huang, P. P. Parmar, D. J. Clark, H. Alami, R. D. Fleischmann, R. D. Perry, and S. N. Peterson. 2010. Proteomic analysis of iron acquisition, metabolic and regulatory responses of *Yersinia pestis* to iron starvation. *BMC Microbiol.* **10:**30.

Pohl, E., J. C. Haller, A. Mijovilovich, W. Meyer-Klaucke, E. Garman, and M. L. Vasil. 2003. Architecture of a protein central to iron homeostasis: crystal structure and spectroscopic analysis of the ferric uptake regulator. *Mol. Microbiol.* **47:**903–915.

Preston, M. J., S. M. Fleiszig, T. S. Zaidi, J. B. Goldberg, V. D. Shortridge, M. L. Vasil, and G. B. Pier. 1995. Rapid and sensitive method for evaluating *Pseudomonas aeruginosa* virulence factors during corneal infections in mice. *Infect. Immun.* **63:**3497–3501.

Prevost, K., H. Salvail, G. Desnoyers, J. F. Jacques, E. Phaneuf, and E. Massé. 2007. The small RNA RyhB activates the translation of *shiA* mRNA encoding a permease of shikimate, a compound involved in siderophore synthesis. *Mol. Microbiol.* **64:**1260–1273.

Prince, R. W., C. D. Cox, and M. L. Vasil. 1993. Coordinate regulation of siderophore and exotoxin A production: molecular cloning and sequencing of the *Pseudomonas aeruginosa fur* gene. *J. Bacteriol.* **175:**2589–2598.

Prince, R. W., D. G. Storey, A. I. Vasil, and M. L. Vasil. 1991. Regulation of *toxA* and *regA* by the *Escherichia coli fur* gene and identification of a Fur homologue in *Pseudomonas aeruginosa* PA103 and PA01. *Mol. Microbiol.* **5:**2823–2831.

Privalle, C. T., and I. Fridovich. 1993. Iron specificity of the Fur-dependent regulation of the biosynthesis of the manganese-containing superoxide dismutase in *Escherichia coli*. *J. Biol. Chem.* **268:**5178–5181.

Rashid, R. A., P. I. Tarr, and S. L. Moseley. 2006. Expression of the *Escherichia coli* IrgA homolog adhesin is regulated by the ferric uptake regulation protein. *Microb. Pathog.* **41:**207–217.

Rossi, M.-S., J. D. Fetherston, S. Létoffé, E. Carniel, R. D. Perry, and J.-M. Ghigo. 2001. Identification and characterization of the hemophore-dependent heme acquisition system of *Yersinia pestis*. *Infect. Immun.* **69:**6707–6717.

Runyen-Janecky, L., E. Dazenski, S. Hawkins, and L. Warner. 2006. Role and regulation of the *Shigella flexneri* Sit and MntH systems. *Infect. Immun.* **74:**4666–4672.

Salvail, H., P. Lanthier-Bourbonnais, J. M. Sobota, M. Caza, J. A. Benjamin, M. E. Mendieta, F. Lepine, C. M. Dozois, J. Imlay, and E. Massé. 2010. A small RNA promotes siderophore production through transcriptional and metabolic remodeling. *Proc. Natl. Acad. Sci. USA* **107:**15223–15228.

Santos, J. A., C. J. González, T. M. López, A. Otero, and M. L. García-López. 1999. Hemolytic and elastolytic activities influenced by iron in *Plesiomonas shigelloides*. *J. Food Prot.* **62:**1475–1477.

Schäffer, S., K. Hantke, and V. Braun. 1985. Nucleotide sequence of the iron regulatory gene *fur*. *Mol. Gen. Genet.* **200:**110–113.

Schmitt, M. P., and S. M. Payne. 1988. Genetics and regulation of enterobactin genes in *Shigella flexneri*. *J. Bacteriol.* **170:**5579–5587.

Sebastian, S., S. Agarwal, J. R. Murphy, and C. A. Genco. 2002. The gonococcal Fur regulon: identification of additional genes involved in major catabolic, recombination, and secretory pathways. *J. Bacteriol.* **184:**3965–3974.

Sebbane, F., N. Lemaître, D. E. Sturdevant, R. Rebeil, K. Virtaneva, S. F. Porcella, and B. J. Hinnebusch. 2006. Adaptive response of *Yersinia pestis* to extracellular effectors of innate immunity during bubonic plague. *Proc. Natl. Acad. Sci. USA* **103:**11766–11771.

Sha, J., M. Lu, and A. K. Chopra. 2001. Regulation of the cytotoxic enterotoxin gene in *Aeromonas hydrophila*: characterization of an iron uptake regulator. *Infect. Immun.* **69:**6370–6381.

Sheikh, M. A., and G. L. Taylor. 2009. Crystal structure of the *Vibrio cholerae* ferric uptake regulator (Fur) reveals insights into metal co-ordination. *Mol. Microbiol.* **72:**1208–1220.

Singh, P. K. 2004. Iron sequestration by human lactoferrin stimulates *P. aeruginosa* surface motility and blocks biofilm formation. *BioMetals* **17:**267–270.

Smith, A., N. I. Hooper, N. Shipulina, and W. T. Morgan. 1996. Heme binding by a bacterial repressor protein, the gene product of the ferric uptake regulation (*fur*) gene of *Escherichia coli*. *J. Protein Chem.* **15**:575–583.

Staggs, T. M., J. D. Fetherston, and R. D. Perry. 1994. Pleiotropic effects of a *Yersinia pestis fur* mutation. *J. Bacteriol.* **176**:7614–7624.

Staggs, T. M., and R. D. Perry. 1991. Identification and cloning of a *fur* regulatory gene in *Yersinia pestis*. *J. Bacteriol.* **173**:417–425.

Stoebner, J. A., and S. M. Payne. 1988. Iron-regulated hemolysin production and utilization of heme and hemoglobin by *Vibrio cholerae*. *Infect. Immun.* **56**:2891–2895.

Stojiljkovic, I., and K. Hantke. 1995. Functional domains of the *Escherichia coli* ferric uptake regulator protein (Fur). *Mol. Gen. Genet.* **247**:199–205.

Tardat, B., and D. Touati. 1993. Iron and oxygen regulation of *Escherichia coli* MnSOD expression: competition between the global regulators Fur and ArcA for binding to DNA. *Mol. Microbiol.* **9**:53–63.

Thomas, C. E., and P. F. Sparling. 1994. Identification and cloning of *fur* homologue from *Neisseria meningitidis*. *Mol. Microbiol.* **11**:725–737.

Thompson, J. M., H. A. Jones, and R. D. Perry. 1999. Molecular characterization of the hemin uptake locus (*hmu*) from *Yersinia pestis* and analysis of *hmu* mutants for hemin and hemoprotein utilization. *Infect. Immun.* **67**:3879–3892.

Tolmasky, M. E., A. M. Wertheimer, L. A. Actis, and J. H. Crosa. 1994. Characterization of the *Vibrio anguillarum fur* gene: role in regulation of expression of the FatA outer membrane protein and catechols. *J. Bacteriol.* **176**:213–220.

Tomich, M., and C. D. Mohr. 2004. Transcriptional and posttranscriptional control of cable pilus gene expression in *Burkholderia cenocepacia*. *J. Bacteriol.* **186**:1009–1020.

Torres, A. G., L. Milflores-Flores, J. G. Garcia-Gallegos, S. D. Patel, A. Best, R. M. La Ragione, Y. Martinez-Laguna, and M. J. Woodward. 2007. Environmental regulation and colonization attributes of the long polar fimbriae (LPF) of *Escherichia coli* O157:H7. *Int. J. Med. Microbiol.* **297**:177–185.

Touati, D., M. Jacques, B. Tardat, L. Bouchard, and S. Despied. 1995. Lethal oxidative damage and mutagenesis are generated by iron in Δ*fur* mutants of *Escherichia coli*: protective role of superoxide dismutase. *J. Bacteriol.* **177**:2305–2314.

Troxell, B., M. L. Sikes, R. C. Fink, A. Vazquez-Torres, J. Jones-Carson, and H. M. Hassan. 2011. Fur negatively regulates *hns* and is required for the expression of HilA and virulence in *Salmonella enterica* serovar Typhimurium. *J. Bacteriol.* **193**:497–505.

Valentin-Hansen, P., M. Eriksen, and C. Udesen. 2004. The bacterial Sm-like protein Hfq: a key player in RNA transactions. *Mol. Microbiol.* **51**:1525–1533.

Valenzuela, M., J. P. Albar, A. Paradela, and H. Toledo. 2011. *Helicobacter pylori* exhibits a Fur-dependent acid tolerance response. *Helicobacter* **16**:189–199.

Vasil, M. L. 2007. How we learnt about iron acquisition in *Pseudomonas aeruginosa*: a series of very fortunate events. *BioMetals* **20**:587–601.

Vasil, M. L., and U. A. Ochsner. 1999. The response of *Pseudomonas aeruginosa* to iron: genetics, biochemistry and virulence. *Mol. Microbiol.* **34**:399–413.

Večerek, B., I. Moll, and U. Bläsi. 2007. Control of Fur synthesis by the non-coding RNA RyhB and iron-responsive decoding. *EMBO J.* **26**:965–975.

Venturi, V., C. Ottevanger, M. Bracke, and P. Weisbeek. 1995. Iron regulation of siderophore biosynthesis and transport in *Pseudomonas putida* WCS358: involvement of a transcriptional activator and of the Fur protein. *Mol. Microbiol.* **15**:1081–1093.

Vetter, S. M., R. J. Eisen, A. M. Schotthoefer, J. A. Montenieri, J. L. Holmes, A. G. Bobrov, S. W. Bearden, R. D. Perry, and K. L. Gage. 2010. Biofilm formation is not required for early-phase transmission of *Yersinia pestis*. *Microbiology* **156**:2216–2225.

Waidner, B., S. Greiner, S. Odenbreit, H. Kavermann, J. Velayudhan, F. Stähler, J. Guhl, E. Bissé, A. H. M. van Vliet, S. C. Andrews, J. G. Kusters, D. J. Kelly, R. Haas, M. Kist, and S. Bereswill. 2002. Essential role of ferritin Pfr in *Helicobacter pylori* iron metabolism and gastric colonization. *Infect. Immun.* **70**:3923–3929.

Wang, F., S. Cheng, K. Sun, and L. Sun. 2008. Molecular analysis of the *fur* (ferric uptake regulator) gene of a pathogenic *Edwardsiella tarda* strain. *J. Microbiol.* **46**:350–355.

Watnick, P. I., T. Eto, H. Takahashi, and S. B. Calderwood. 1997. Purification of *Vibrio cholerae* Fur and estimation of its intracellular abundance by antibody sandwich enzyme-linked immunosorbent assay. *J. Bacteriol.* **179**:243–247.

Wilderman, P. J., N. A. Sowa, D. J. FitzGerald, P. C. FitzGerald, S. Gottesman, U. A. Ochsner, and M. L. Vasil. 2004. Identification of tandem duplicate regulatory small RNAs in *Pseudomonas aeruginosa* involved in iron homeostasis. *Proc. Natl. Acad. Sci. USA* **101**:9792–9797.

Wilmes-Riesenberg, M. R., B. Bearson, J. W. Foster, and R. Curtis III. 1996. Role of the acid tolerance response in virulence of *Salmonella typhimurium*. *Infect. Immun.* **64**:1085–1092.

Wu, Y., and F. W. Outten. 2009. IscR controls iron-dependent biofilm formation in *Escherichia coli* by regulating type I fimbria expression. *J. Bacteriol.* **191**:1248–1257.

Wyckoff, E. E., A. R. Mey, and S. M. Payne. 2007. Iron acquisition in *Vibrio cholerae*. *BioMetals* **20**:405–416.

Xiong, Y. Q., M. L. Vasil, Z. Johnson, U. A. Ochsner, and A. S. Bayer. 2000. The oxygen- and iron-dependent sigma factor *pvdS* of *Pseudomonas aeruginosa* is an important virulence factor in experimental infective endocarditis. *J. Infect. Dis.* **181**:1020–1026.

Yang, Y., D. P. Harris, F. Luo, L. Wu, A. B. Parsons, A. V. Palumbo, and J. Zhou. 2008. Characterization of the *Shewanella oneidensis* Fur gene: roles in iron and acid tolerance response. *BMC Genomics* **9**:S11.

Zheng, M., B. Doan, T. D. Schneider, and G. Storz. 1999. OxyR and SoxRS regulation of *fur*. *J. Bacteriol.* **181**:4639–4643.

Zhou, D., L. Qin, Y. Han, J. Qiu, Z. Chen, B. Li, Y. Song, J. Wang, Z. Guo, J. Zhai, Z. Du, X. Wang, and R. Yang. 2006. Global analysis of iron assimilation and fur regulation in *Yersinia pestis*. *FEMS Microbiol. Lett.* **258**:9–17.

Zhu, C., M. Ngeleka, A. A. Potter, and B. J. Allan. 2002. Effect of *fur* mutation on acid-tolerance response and in vivo virulence of avian septicemic *Escherichia coli*. *Can. J. Microbiol.* **48**:458–462.

II. ADHERENCE, COLONIZATION, AND SURFACE FACTORS

Regulation of Bacterial Virulence
Edited by Michael L. Vasil and Andrew J. Darwin
© 2013 ASM Press, Washington, DC   doi:10.1128/9781555818524.ch7

Chapter 7

Uropathogenic *Escherichia coli* Virulence and Gene Regulation

DREW J. SCHWARTZ AND SCOTT J. HULTGREN

INTRODUCTION

Bacterial virulence entails an ability to persist within a host and cause disease in the face of myriad host defense mechanisms in more than one niche or under more than one immune condition. The array of virulence factors that a pathogen possesses and its ability to regulate their expression in response to environmental signals greatly contribute to the variety and specificity of niches a pathogen can colonize and impact the spectrum of diseases that occur. Colonization factors may confer a fitness advantage in a specific niche, being required in some environments and unnecessary or deleterious in others, or they may contribute globally to persistence and disease. Thus, coordinated gene expression is critical to ensure that appropriate genes are expressed in specific environments and turned off in others. For example, when a pathogen encounters a susceptible host, the ability to attach to a specific tissue or niche in a timely manner is vitally important to establishing infection in a dynamically changing local environment. A host-pathogen interaction is frequently the consequence of an adhesin expressed by the pathogen recognizing a receptor on host cells with exquisite stereochemical specificity (Finlay and Falkow, 1989; Waksman and Hultgren, 2009). Several families of bacterial adhesins exist to mediate specific interactions to various hosts, tissues, organisms, or surfaces: outer membrane proteins, secreted soluble, extracellular proteins that create an adhesive matrix, type 4 pili, autotransporters and two-partner secreted adhesins, and chaperone-usher pathway (CUP) assembled pili (see also Wright and Hultgren, 2006, and Waksman and Hultgren, 2009, for reviews). Myriad gram-negative and gram-positive organisms utilize adhesins to attach to epithelial surfaces to persist within a host. Gram-negative organisms such as *Porphyromonas gingivalis*, *Salmonella enterica*, *Yersinia enterocolitica*, and *Neisseria gonorrhoeae* utilize surface-exposed adhesins to colonize host mucosal surfaces (reviewed in Kline et al., 2009). The gram-positive organisms *Enterococcus faecalis* and *Clostridium perfringens* also encode pili or pilus-like adhesins that mediate attachment to surface structures. In this chapter, we describe the virulence and regulation of CUP pili, focusing on type 1 and P pili in uropathogenic *Escherichia coli* (UPEC), the most common cause of urinary tract infections (UTIs).

URINARY TRACT INFECTIONS

E. coli organisms are normally commensal colonizers of the gut. Several pathotypes of *E. coli* exist, however, that can cause disease in a variety of organ systems in both immunocompetent and immunocompromised individuals dependent on the arsenal of virulence factors and adhesins they express. *E. coli* organisms are generally classified into three major groups: commensals, intestinal pathogens, and extraintestinal pathogens (ExPEC) (Russo and Johnson, 2000). Commensal strains peaceably coexist with numerous other bacterial species in the guts of healthy humans and many other animals. The majority of these strains causes no disease in healthy individuals but may colonize foreign bodies if present. Additionally, these commensals lack many of the virulence genes that other *E. coli* organisms possess. Intestinal pathotypes of *E. coli* are rarely found in the asymptomatic host but, instead, are obligate pathogens of the gastrointestinal tract. Among the intestinal *E. coli* pathogens, there are several classifications based on virulence genes expressed: enterotoxigenic (ETEC), enteropathogenic (EPEC), Shiga toxin-producing/enterohemorrhagic (STEC/EHEC), enteroinvasive (EIEC), and diffusely adherent (DAEC) (reviewed in Nataro and Kaper, 1998). ETEC adheres via fimbriae to the small intestine and produces one or both of labile toxin and stable toxin, which lead to secretory

Drew J. Schwartz and Scott J. Hultgren • Department of Molecular Microbiology, Center for Women's Infectious Disease Research, Washington University School of Medicine, St. Louis, MO 63110.

diarrhea. EPEC organisms cause attaching and effacing lesions via adherence with bundle-forming pili and export of type III secretion effectors, resulting in loss of absorptive microvilli and diarrhea. EHEC produces Shiga toxin, leading to bloody diarrhea, hemolytic-uremic syndrome, and possibly death. Attachment of EHEC to the tissue has been speculated to occur via multiple different adhesins, and many of these may contribute to colonization of more than one pathotype, but no clear adherence profile exists (Bardiau et al., 2010). EIEC invades colonocytes and secretes enterotoxins, very similar to the pathogenesis of shigellae. DAEC organisms diffusely adhere to the epithelial surface and autoaggregate, yet they do not produce either of the toxins common to ETEC. Infection with DAEC results in edematous villi and mucoid diarrhea. Extraintestinal infections caused by *E. coli* (ExPEC) include meningitis, pneumonia, soft tissue infections, and UTI. ExPEC is characterized by its lack of gastrointestinal pathogenesis; however, it can stably colonize the gut (Russo and Johnson, 2000). In order to cause disease, ExPEC organisms must access the site of infection: brain, lungs, medical device, or urinary tract. ExPEC generally acquires stretches of chromosomal DNA called pathogenicity islands conferring the virulence factors, often including operons encoding adhesins necessary to colonize a particular niche. UPEC is one such member of ExPEC, primarily responsible for UTI by virtue of the adhesins and other virulence genes it expresses. The array of virulence factors and adhesins expressed by the various *E. coli* pathotypes defines their pathogenic niche and virulence profile.

UPEC causes >80% of community-acquired UTIs and ~50% of nosocomial UTIs (Johnson and Stamm, 1989; Kucheria et al., 2005) and has evolved elaborate mechanisms that allow it to survive within and exploit multiple niches in the urinary tract (Virkola 1987; Mulvey et al., 1998; Schwartz et al., 2011). Although UTIs afflict both men and women, females are disproportionately affected, a trend that begins as early as when they are 8 years old (Montini et al., 2011). Approximately 50% of women will have at least one symptomatic UTI in their lifetime, with 20 to 40% of these experiencing a recurrence within 6 months of the first infection (Stamm and Hooton, 1993; Engel and Schaeffer, 1998). Clinically, UTIs are divided into categories based on symptomatology. One category is infections of the bladder (cystitis), which are generally low-mortality infections involving bacteriuria, pain or burning during urination, urgency, and frequency. The other category includes infections of the kidneys (pyelonephritis), which in addition to the symptoms of cystitis include chills, fever, flank pain, and tenderness. Antibiotics are generally successful at clearing bacteriuria and many other symptoms. However, many women suffer frequent recurrent infections with increased morbidity, necessitating long-term daily, prophylactic antibiotics. Furthermore, studies have shown that frequently, the recurrent infection is caused by the strain that caused the original symptomatic UTI (Ikaheimo et al., 1996), emphasizing the presence of a bacterial reservoir within the host that cannot be effectively eradicated by current treatments. Additionally, certain patients experience asymptomatic bacteriuria or chronic cystitis marked by persistent bacteriuria with accompanying cystitis symptoms (Mabeck, 1972; Ferry et al., 2004). The balance between bacterial virulence factors and host genetics dictates symptoms and disease state. Certain polymorphisms and reduced expression of the Toll-like receptor 4 (TLR4) gene are associated with less severe UTI and asymptomatic bacteriuria (Ragnarsdóttir et al., 2007; Ragnarsdóttir et al., 2010). Conversely, reduced expression of the interleukin 8 (IL-8) receptor, CXCR-1, is correlated with severe pyelonephritis (Ragnarsdóttir et al., 2007). Other human genetic factors that impact acute, chronic, and recurrent UTI have not yet been elucidated, but studies to determine polymorphisms to address these common syndromes are currently being conducted. Thus, depending on the host factors and bacterial virulence attributes, UPEC causes a variety of diseases in the susceptible host.

The increasingly accelerated rate of global trafficking of antibiotic-resistant uropathogens is necessitating the development of better therapeutic strategies. In addition to the rising resistance to first-line therapies like trimethoprim-sulfamethoxazole (TMP-SMX), UPEC strains are becoming increasingly resistant to fluoroquinolones, which are being used more as a result of TMP-SMX resistance (Gupta et al., 2001; Hooton, 2003; Zhanel et al., 2006). Furthermore, strains of the clonal group ST131 that are resistant to the majority of antibiotics have globally spread, causing UTIs and bloodstream infections in North America, Europe, Asia, and Australia (Nicolas-Chanoine et al., 2008). The spread of these multidrug-resistant strains combined with the increased resistance among UPEC is poised to create a public health crisis. Thus, it is essential to thoroughly reexamine the molecular basis of UPEC infection, determining the role and regulation of essential virulence factors to create novel therapeutics that target these virulence factors while reducing the negative side effects accompanying antibiotic treatment.

CUP OVERVIEW

UPEC and *E. coli* in general heavily rely on CUP pili to mediate attachment to biotic and abiotic surfaces (Pratt and Kolter, 1998; Anderson et al., 2003; Pinkner et al., 2006; Beloin et al., 2008). The CUP is a nearly ubiquitous system among *Enterobacteriaceae* (Clegg and Gerlach, 1987) used for the assembly of surface-exposed pili (see Waksman and Hultgren, 2009, and Sauer et al., 2004, for thorough reviews). CUPs are genetically organized as operons encoding a pilus fiber, a periplasmic chaperone, and an outer membrane usher. Whole-genome sequencing has revealed the presence of many canonical or hypothetical CUP operons in UPEC genomes (Welch et al., 2002; Brzuszkiewicz et al., 2006; Chen et al., 2006) (Table 1). These operons are *fim* (type 1), *pap* (P), F17-like, *sfa* (S), *yad*, *auf*, *yfc*, *ygi*, *yeh*, *fml*, *foc* (F1C), *yde* (F9), *dra/afa* (Dr family), *fso* (F7), *fst*, and *pix* (reviewed in Korea et al., 2011) (Table 1). In fact, it was recently shown that UPEC possesses, on average, twice as many fimbrial types as commensal *E. coli* (Spurbeck et al., 2011), presumably because of the increased array of environments these organisms colonize. We focus on type 1 and P pili, given that they have been the most thoroughly characterized CUP pilus systems via mechanistic studies that have revealed their biochemical, structural, and regulatory properties as well as their contribution to UPEC pathogenesis.

Type 1 pili are encoded by the *fim* operon. They are composite fibers with an ~2-nm-wide fibrillar tip joined to a 0.3- to 1.5-μm-length rod consisting of repeating subunits arranged in a 7-nm-wide right-handed helical rod (Brinton, 1965; Jones et al., 1995; Hahn et al., 2002). Upon translation, pilus subunits are translocated across the inner membrane via the Sec translocation machinery (Fig. 1) (Dodd et al., 1984). Once in the periplasm, they are bound by the chaperone FimC, which is critical for subunit stability and serves to facilitate folding and cap interactive surfaces to prevent nonproductive aggregation (Kuehn et al., 1991). Pilus subunits have an incomplete immunoglobulin (Ig)-like fold, missing the C-terminal seventh β-strand (Choudhury et al., 1999; Sauer et al., 1999). The periplasmic chaperones attain a "boomerang" shape with two Ig-like domains connected by a linker (Holmgren and Branden, 1989; Kuehn et al., 1993). Upon binding to the chaperone, the incomplete subunit folds are templated to fold by a process termed donor strand complementation, in which four residues from the chaperone's G1 strand provide the steric information to complete the fold of the bound subunit (Choudhury et al., 1999; Sauer et al., 1999; Sauer et al., 2002). Chaperone subunit complexes are targeted to the outer membrane usher, the protein through which the growing pilus is extruded across the outer membrane (Fig. 1). The outer membrane usher is a large β-barrel protein consisting of a translocation domain, an ~120-residue N-terminal domain (NTD), and an ~170-residue C-terminal domain (CTD) that bind chaperone-subunit complexes (Phan et al., 2011). The current model suggests that incoming chaperone-subunit complexes are targeted to the usher NTD and transferred to the CTD, where donor strand exchange (DSE) occurs (Fig. 1B) (Henderson et al., 2011; Phan et al., 2011). In DSE, the chaperone is displaced by a 10- to 20-residue N-terminal extension (Nte) present on the next subunit to be incorporated into the growing pilus. This process completes the subunit Ig fold in a canonical fashion, resulting in the addition of a single subunit to the growing pilus rod. The order of pilus assembly corresponds to the affinity of the chaperone-subunit complex for the usher and the efficacy and specificity each Nte has for DSE (Dodson et al., 1993; Saulino et al., 1998; Nishiyama et al., 2003; Lee et al., 2007). Subunits are added to the base of the growing pilus, with the FimH adhesin first incorporated. Accordingly, the usher, FimD, has the highest affinity for FimC-FimH complexes (Saulino et al., 1998). After incorporation of FimH, a single subunit each of FimG and FimF is added to complete the tip fibrillum. FimC-FimA complexes are then added sequentially to yield a fully formed pilus rod containing ~1,000 FimA subunits arranged in a right-handed helical cylinder with approximately 3.2 subunits per turn (Fig. 1) (Bullitt et al., 1996; Hahn et al., 2002). The pilus is constructed from tip to base through targeting of chaperone-subunit complexes to

Table 1. Distribution of CUP pili in sequenced UPEC genomes[a]

CUP operon(s)	K-12 MG1655	UPEC		
		UTI89	CFT073	536
fim	+	+	+	+
pap	−	+	+*	+
F17-like	−	+	−	+
sfa	−	+	−	+
yad	+	+	+	+
auf	−	±	+	+
yfc	+	+	+	+
ygi	+	+	+	−
yeh	+	+	+	+
fml	±	±	+	−
foc	−	−	+	−
yde	+	−	+	−
Dr family[b]	−	−	−	−
fso[b]	−	−	−	−
fst[b]	−	−	−	−
pix	−	−	−	+

[a]+, presence; −, absence; ±, operon is present but likely nonfunctional due to mutation or deletion of genes. An asterisk indicates that two copies of the operon are present. Modified with permission from the National Academy of Sciences from Chen et al., 2006.
[b]Operon is not present in this UPEC strain but is present in many clinical isolates.

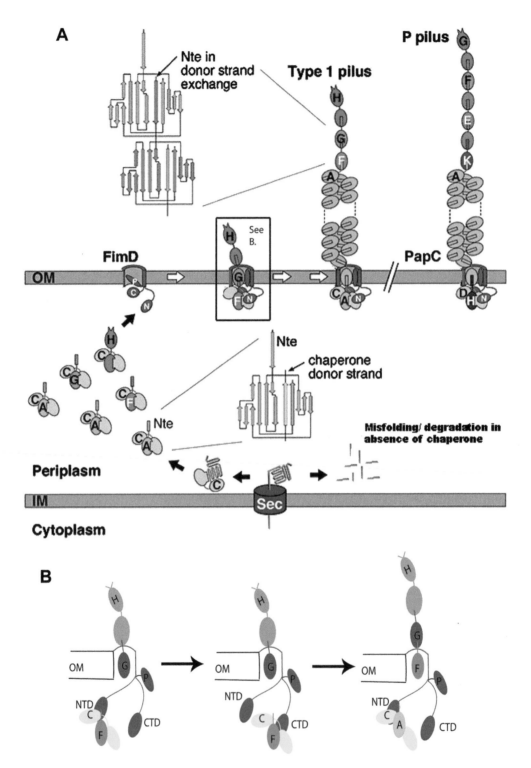

Figure 1. Model of type 1 pilus biogenesis. (A) Subunits are secreted across the Sec apparatus and immediately bound by the cognate periplasmic chaperone, the absence of which results in subunit misfolding and degradation. The chaperone donates its G1 β-strand to the subunit to complete its incomplete Ig-like fold in donor strand complementation. (B) The chaperone then delivers the subunit to the NTD of the membrane usher, FimD. The subunit is then transferred to the CTD, where DSE occurs with the previously added subunit. P pilus biogenesis occurs in a similar fashion. IM, inner membrane; OM, outer membrane. Panel A was modified with permission from John Wiley and Sons from Henderson et al., 2011, in *Molecular Microbiology*. doi:10.1128/9781555818524.ch7f1

the usher and DSE to complete the incomplete Ig-like fold of each subunit.

DSE occurs via a concerted "zip-in, zip-out" mechanism at a groove of the previously added subunit containing five crucial residues called P1 to P5 (Remaut et al., 2006). The Nte of an incoming subunit binds to the open P5 pocket of the previous subunit. The incoming subunit then zips in, replacing the hydrophobic interactions between the previous subunit and the chaperone from P5 to P1 and displacing the chaperone. The P pilus rod is terminated by the addition of a single PapH subunit. PapH contains a loop partially occluding this P5 pocket; therefore, the chaperone bound to PapH cannot be displaced and pilus rod formation is terminated (Verger et al., 2006). How type 1 pili are terminated is unclear, as no PapH equivalent has been discovered.

Based on sequence identity and genomic organization, it is very likely that other CUP systems are assembled in a manner that is similar to type 1 and P pilus biogenesis (Waksman and Hultgren, 2009). The presence of multiple CUP pili within *E. coli* genomes is thought to be required for tuning adhesive properties specific for different environmental niches (Uhlin et al., 1985; Morschhauser et al., 1993; Mulvey et al., 1998; Buts et al., 2003; Kline et al., 2010). For example, the type 1 pilus adhesin FimH and the P pilus adhesin PapG bind mannosylated and digalactoside receptors, respectively, thereby mediating UPEC colonization of bladders and kidneys (Uhlin et al., 1985; Mulvey, et al., 1998; Dodson et al., 2001; Hung et al., 2002). Transcriptional profiling and genetic studies have revealed that other UPEC CUP systems are expressed and may contribute to virulence (Beloin et al., 2008; Hadjifrangiskou et al., 2011; Spurbeck et al., 2011). However, how expression of multiple CUP operons is coordinated in response to environmental signals in different niches during infection is essentially unknown.

UPEC VIRULENCE

The pathogenesis of community-acquired UPEC UTI is thought to begin with UPEC colonization of the periurethral area from the fecal flora. Transmission of UPEC into the bladder can then occur via urethral manipulation (Bran et al., 1972; Foxman, 2010), sexual intercourse (Buckley et al., 1978), or possibly direct ascension, although the periurethral presence of UPEC does not necessarily lead to infection (Schlager et al., 1993). During infection, UPEC is capable of colonizing the urine, the bladder epithelium (both extracellularly and intracellularly), and the kidneys, with significant bacterial flux between these niches (Schwartz et al., 2011). Depending on the specific niche UPEC inhabits,

it encounters different elements of the immune response as well as different environmental pressures, necessitating precise gene regulation.

Although multiple studies have correlated particular genes with virulence (Hughes et al., 1983; Connell et al., 1996; Bokranz et al., 2005; Smith et al., 2008), to date there is no common virulence profile among cystitis isolates. However, the vast majority (>95%) encodes type 1 pili, and expression of type 1 pili is highly correlated with cystitis (Chen et al., 2006; Garofalo et al., 2007; Norinder et al., 2011). Indeed, the most consistent virulence factor expressed among cystitis isolates is type 1 pili (Norinder et al., 2011). UPEC isolated from women suffering recurrent UTI expressed type 1 pili when grown in broth cultures (Hultgren et al., 1986). Additional studies demonstrated that 75% of UPEC strains isolated from the urine of 41 adult patients had type 1 pili on their surface and were often found in association with or attached to exfoliated epithelial cells and leukocytes (Hultgren et al., 1985; Kisielius et al., 1989). Several studies have indicated that planktonic cells within the urine usually do not express type 1 pili, implying genetic regulation that is niche specific (Gunther et al., 2001; Hagan et al., 2010). However, other studies have shown that expression patterns of planktonic UPEC in the urine are not necessarily indicative of tissue-associated bacteria (Hultgren et al., 1985; Gunther et al., 2001; Reigstad et al., 2007; Hagan et al., 2010). Planktonic UPEC inhabiting the bladder lumen have limited ability to maintain residence in the urinary tract without the ability to adhere to the epithelial surface due to the flushing action of micturition except under conditions that lead to incomplete voiding or reflux.

Because of the myriad outcomes that result in human UTI, it is imperative to utilize an interdisciplinary approach in multiple model systems to completely capture the complexities of UTI pathogenesis that may be differentially exhibited spatially and temporally. Several murine models of UPEC infection have been developed (Hopkins et al., 1998; Hung et al., 2009). Each model takes advantage of different available, genetically defined, inbred mouse strains that exhibit varied immunological responses to UPEC infection to dissect several aspects of UTI progression. Studies performed in C3H/HeN, C3H/HeJ, and C57BL/6J backgrounds revealed stages of acute, subacute, and chronic infection that bear striking parallels to human UTI (Hopkins et al., 1998; Schilling et al., 2002; Mysorekar and Hultgren, 2006; Hannan et al., 2010), indicating that outcomes from these mouse models strongly parallel human disease. These and other models, utilizing CBA/J, C3H/HeOUJ, and BALB/c mouse strains, provided insights into immune checkpoints that may predispose humans to chronic UTI (Hopkins et al., 1998; Mysorekar and Hultgren, 2006; Hannan et al., 2010). Using these mouse models, significant strides have been made in understanding the onset and progression of UPEC through the urinary tract.

COMPLEX POPULATION DYNAMICS GOVERN MURINE UPEC UTI

In order to determine the population dynamics and overall niche distribution of UPEC during murine UTI in C3H/HeN mice, a library of 40 isogenic strains of UTI89, each with a unique genetic bar code, was utilized (Schwartz et al., 2011). These strains were mixed in equal numbers, introduced into the bladders of C3H/HeN mice, and tracked over time in the following niches of the urinary tract: kidneys, urine, and bladder lumen and within the gentamicin-protected compartment of the bladder. Gentamicin is an antibiotic that does not penetrate the eukaryotic epithelium, thus allowing the detection of bacteria in a protected intracellular niche within eukaryotic cells. During the acute phase of UTI in C3H/HeN mice, the majority of tagged strains were present in every niche assessed (Fig. 2). By 24 h postinfection (hpi), there was a dramatic reduction in tags in both the whole bladder and kidneys, revealing an acute population bottleneck (Fig. 2B to E); however, between 6 and 24 hpi, bacteria from the kidneys repopulated the bladder with unique UPEC genotypes (Fig. 2E). Nevertheless, the severe reduction in tags demonstrates the effect of the acute immune system and micturition in clearing the majority of tagged bacteria. The ability of UPEC to transcend this acute population bottleneck is described in detail below. It was also found that in mice that resolved bacteriuria, the bacterial strains present in the bladder were not the same as those observed in the kidneys. These complex infection dynamics involving colonization and migration between niches within a single organ, like the bladder or kidney, likely involve changes in the repertoire of genes needed for colonization and survival. Regulation of the appropriate adhesins in distinct niches is likely essential for appropriate colonization, while the ability to rapidly interchange the expressed pili is necessary for dispersal and subsequent colonization of a new niche.

TYPE 1 PILUS-DEPENDENT BLADDER INVASION AND INTRACELLULAR BACTERIAL COMMUNITY (IBC) FORMATION

The FimH adhesin located at the tip of the type 1 pilus is instrumental in mediating UPEC interactions with the bladder epithelium. FimH has been shown to bind mannose and its derivatives, thereby interacting with mannosylated moieties on abiotic and biotic surfaces. UPEC organisms expressing type 1 pili were shown *in vitro* to bind to human and mouse uroplakin Ia (UPIa) and UPIb, integral membrane glycoproteins that create a hexagonal array on the surface of the superficial facet cells of the bladder that create an impermeable barrier to toxic compounds in urine (Wu et al., 1996). Uroplakins are highly conserved across mammals both structurally and via sequence homology (Wu et al., 1994). UPIa and UPIb are members of the tetraspanin family of molecules that modulate immune signaling via leukocyte differentiation. Using high-resolution, freeze-fracture, deep-etch electron microscopy, Mulvey et al. showed that the type 1 pilus tip fibrillum interacts with the hexagonal array of uroplakins on the terminally differentiated superficial facet cell layer of C57BL/6 mice (Fig. 2A) (Mulvey et al., 1998), validating previous *in vitro* data (Wu et al., 1996). This interaction, combined with the high tensile strength of the pilus rod, enables UPEC to withstand the strong shear forces applied by urine flow and persist in the bladder (Yakovenko et al., 2008; Aprikian et al., 2011). The crystal structure of FimH with α-D-mannose bound in its recognition pocket, combined with mutational analysis, revealed the structural details of the precise specificity of FimH for mannose (Hung et al., 2002). Subsequently, analysis of more than 200 sequenced UPEC strains revealed that the FimH binding pocket is invariant, further demonstrating the importance of this domain for pathogenesis (Chen et al., 2009). FimH binding to UPIa results in global conformational changes of the uroplakin plaques, which lead to cytoskeletal rearrangement and the alignment of the cytoplasmic tails of UPIa, UPIb, UPII, and UPIIIa, initiating second messenger signaling (Thumbikat et al., 2009b; Wang et al., 2009). The resultant increase in intracellular Ca^{2+} mediates host cell apoptosis and IL-6 secretion, in an effort to control the infection and alert the innate immune system to the bacterial presence in the urinary tract (Song et al., 2007; Thumbikat et al., 2009b). Through the precise interaction between FimH on type 1 pili and uroplakins on the bladder epithelial surface, UPEC binds to the tissue, initiating downstream signaling events within the host cell.

UPEC BINDING TRIGGERS HOST AND BACTERIAL RESPONSES

The binding of type 1 pili to a surface enacts bacterial transcriptional changes in addition to host responses. Binding of UPEC to mannosylated yeast enacts transcriptional change in an *E. coli* strain

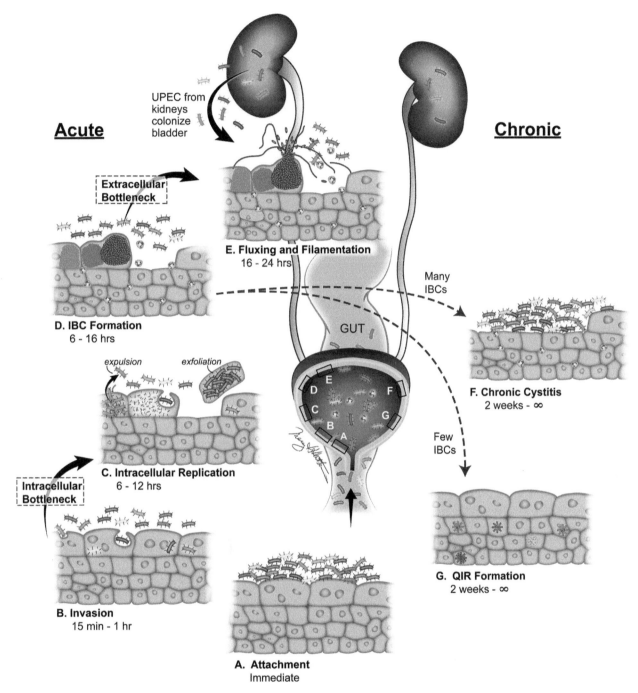

Figure 2. Pathogenesis of UPEC UTI. (A) A population of UPEC from the gut is introduced into the bladder, where the bacteria attach to the epithelial surface with the FimH adhesin at the tip of type 1 pili. (B) UPEC invades the superficial facet cells of the bladder within the first hour of infection. (C) UPEC replicates within the facet cells in a type 1 pilus-dependent manner unless the bacteria are expelled via a TLR4-dependent process or infected epithelial cells are jettisoned by an apoptotic mechanism. These processes constitute an intracellular bottleneck whereby unique genetic diversity, depicted here with colored bacteria, is decreased. (D) UPEC organisms form clonal IBCs within the cytoplasm of superficial facet cells of the bladder. (E) Micturition and neutrophils eliminate the majority of luminal bacterial clones constituting an extracellular population bottleneck. Between 16 and 24 hpi, UPEC organisms flux out of the IBC, with some bacteria becoming filamentous to become the dominant population of the bladder. Additional UPEC clones descend from the infected kidneys, adding unique diversity to the bladder population. (F) If a high number of IBCs is formed early (D), chronic cystitis is likely to develop, marked by bacterial replication in the lumen of the bladder and adherence to the tissue via type 1 pili. The genetic diversity of this bacterial population is higher than if the infection resolves with formation of QIRs (G).
doi:10.1128/9781555818524.ch7f2

(CSH50) (Bhomkar et al., 2010). In addition to up-regulation of genes involved in general metabolism, several genes classified as being involved in removal of reactive oxygen species and hydrophobic compounds were also upregulated. The latter genes are likely involved in defense against antimicrobial peptides in the urine and potentially antibiotics as well as attack from neutrophils and other inflammatory cells. Once bound, UPEC is capable of invading the superficial layer of the transitional epithelial surface in a FimH-dependent fashion (Fig. 2B) (Mulvey et al., 1998). Interaction of UPEC with α3β1 integrin in cell culture leads to actin rearrangement and bacterial engulfment in an active process requiring Rho family GTPases (Martinez et al., 2000; Martinez and Hultgren, 2002; Eto et al., 2007). The invasion process occurs via a cholesterol- and dynamin-dependent process modulated by calcium levels and clathrin (Eto et al., 2008). By an unknown mechanism, UPEC organisms escape the endocytic vesicle, where they replicate in the cytoplasm to form biofilm-like IBCs of 10^4 to 10^5 bacteria that protrude into the luminal surface of the bladder (Fig. 2D) (Anderson et al., 2003). While type 1 piliated UPEC binding and invasion into the superficial facet cells provide a niche for rapid replication, studies in C57BL/6J mice demonstrated that it also leads to a robust apoptotic response and exfoliation of the facet cells within a few hpi (Mulvey, et al., 1998), jettisoning infected cells (Fig. 2C). In addition to facet cell exfoliation, host signaling mechanisms lead to a massive influx of immune cells, mainly polymorphonuclear leukocytes, and upregulation of other immune effectors. In spite of this robust response designed to thwart infection, UPEC evolved mechanisms to subvert the elements of the innate immune response and persist within the urinary tract. Upon IBC maturation, UPEC organisms at the outer edges of an IBC become motile and disperse into the extracellular compartment of the lumen. Concomitant with this fluxing event, some UPEC organisms adopt a filamentous morphology (Justice et al., 2004) (Fig. 2E), a finding that along with IBCs has been observed in urine from women suffering recurrent UTI (Rosen et al., 2007). Once extracellular, the immune cell infiltrate attacks luminal UPEC. Filamentous UPEC organisms are resistant to phagocyte killing, providing a mechanism whereby UPEC emerging from an IBC into an inflamed environment can survive and reinitiate the intracellular cascade (Horvath et al., 2011). Deletion of the cell division inhibitor-encoding gene *sulA* prevented filamentation and led to virulence attenuation at 24 hpi onward in C3H/HeN mice (Justice et al., 2006a). Thus, the Δ*sulA* strain formed first-generation IBCs but was incapable of filamentation and second-generation IBCs in C3H/HeN mice.

Bacterial filamentation was not observed during acute infection in C3H/HeJ mice, which lack the ability to signal through TLR4 (Justice et al., 2004). Infection of C3H/HeJ mice with the Δ*sulA* strain restored the virulence of this strain (Justice et al., 2006a). Thus, filamentation and possibly second-round IBC formation are dependent, in part, on TLR4 responses in the bladder leading to regulatory responses by UPEC.

GENETIC REGULATION WITHIN IBCs

As a result of the rapid and dramatic shift in local environment, robust transcriptional changes occur in UPEC inhabiting an IBC. At 6 hpi, each C3H/HeN mouse bladder contains between 1 and 700 IBCs (Fig. 2D) (Justice et al., 2006b; Wright et al., 2007; Schwartz et al., 2011), with each IBC resulting from the replication of a single bacterial invader (Schwartz et al., 2011). In the 2 h subsequent to invasion, wild-type (WT) bacteria coalesce into loose collections 2 to 4 μm beneath the surface of superficial facet cells (Justice et al., 2004). UPEC replicate every 30 to 35 min in this environment and retain bacillary morphology. Invasion alone, however, is not sufficient to lead to IBC formation. Using a *tet*-inducible copy of FimH, it was shown that the development of IBCs requires the expression of FimH and type 1 pili (Wright et al., 2007). Additionally, deletion of the gene encoding the outer membrane protein OmpA or the capsule synthesis genes disrupted IBC formation despite an invasion efficiency equivalent to that of WT UPEC (Nicholson et al., 2009; Anderson et al., 2010). Host and bacterial transcriptional changes that accompany intracellular replication within IBCs in C3H/HeJ mice have been deduced (Reigstad et al., 2007). UPEC organisms within IBCs upregulate the siderophores enterobactin and salmochelin to scavenge available iron. Concomitantly, host upregulation of the transferrin receptor and lipocalin-2 within the IBC may restrict iron availability (Reigstad et al., 2007). The battle for iron within the host is an important event during the host-pathogen interaction for many pathogens (see chapters 5, 6, and 16). Indeed, in women suffering from recurrent UTI, urinary isolates were more likely to express the siderophores salmochelin and yersiniabactin than were rectal isolates from the same patients (Henderson et al., 2009). Thus, regulatory changes, including the upregulation of siderophores and continued expression of type 1 pili, accompany the transition from the luminal niche of the bladder to the intracellular niche within the bladder epithelium.

We discovered a substantial bottleneck that exists during acute infection that prevents the majority of clones from progressing to later stages of infection

(Fig. 2) (Schwartz et al., 2011). Using a set of 40 isogenic UTI89 isolates, each with a unique genetic tag, we demonstrated that, on average, only 25% of these tags were present at 24 hpi in the bladder. An additional population bottleneck prevented the majority of tissue invaders from progressing to mature IBCs (Fig. 2B and C). Utilizing confocal microscopy of mouse bladders infected with an equal inoculum of green fluorescent protein-expressing (GFP+) and GFP- UTI89, we demonstrated that each IBC is the result of clonal expansion from a single bacterial invader (Schwartz et al., 2011). At 1 hpi, 10^3 to 10^4 UPEC organisms invaded the tissue; however, by 6 hpi the number of IBCs varied from 3 to 700 (Fig. 2B and C). Thus, host epithelial mechanisms exist to reduce the number of successful bacterial founders of an IBC (Fig. 2C). TLR4-mediated expulsion of UPEC may play a role in limiting the numbers of UPEC that successfully form IBCs (Song et al., 2009). Epithelial exfoliation, as previously described, may also contribute to the reduction of invasion events that progress to IBCs.

In addition to these epithelial mechanisms that restrict IBC formation, the specific allele of FimH also impacts the ability of UPEC to form IBCs. The mannose binding pocket of FimH is comprised of invariant residues among all clinical UPEC strains (Chen et al., 2009). Variation outside of this pocket occurs frequently among UPEC organisms. Sokurenko et al. used a variety of methods to examine the natural variation and have shown that FimH is under positive selection in UPEC (Sokurenko et al., 1998; Weissman et al., 2007; Ronald et al., 2008). We performed in silico analysis of FimH gene sequences from 279 diverse *E. coli* isolates, which identified several amino acid residues outside of the mannose binding pocket under positive selection (Chen et al., 2009). Our structure/function analyses of these positively selected residues indicated that although they are not part of the mannose binding pocket, they can impact mannose binding, IBC formation, and virulence (Chen et al., 2009). It was previously shown that the majority of *E. coli* isolates, including the pyelonephritis isolate CFT073 and the K-12 strain MG1655, have a serine at position 62 (S62) (Johnson et al., 2001). An alanine at position 62 (A62) was associated with collagen and monomannose binding (Pouttu et al., 1999) and increased virulence in the cystitis isolate NU14 (Johnson et al., 2001). Similarly, UTI89 FimH possesses an A62 allele. We found that an A62S *fimH* mutation in the chromosome of UTI89 significantly attenuated invasion, IBC formation, and virulence and affected phase variable type 1 pilus regulation. In addition, we found that a double mutation of the positively selected residues at positions 27 and 163 (UTI89 *fimH*::A27V/V163A) had no effect on pilus assembly or mannose binding *in vitro* but

that the mutant exhibited a 10,000-fold reduction in mouse bladder colonization at 24 hpi and was unable to form IBCs even though it bound normally to mannosylated receptors in the urothelium (Chen et al., 2009). Thus, the A27V/V163A double mutation identified a function of FimH that is required, in addition to mannose binding, for IBC formation and *in vivo* fitness, suggesting that IBC formation is critical for successful UTI. Thus, many factors contribute to the ability of invaded UPEC to form IBCs, including TLR4-mediated expulsion, bladder cell exfoliation, and the *fimH* allele, which may alter regulation or the ability of UPEC to aggregate intracellularly.

RESERVOIRS FORM IN THE UNDERLYING BLADDER TISSUE

The intracellular pathogenic cascade that UPEC undergoes has been shown to be critical for several of the outcomes of infection. If the immune response effectively eradicates luminal bacteria as evidenced by sterile urine, mice may still have bacterial CFU in the bladder (Hultgren et al., 1985; Hopkins et al., 1998) (Fig. 2G). As mentioned earlier, during the acute infection cascade, infected and uninfected cells of the superficial epithelium are exfoliated in a caspase-dependent manner (Mulvey et al., 1998; Klumpp et al., 2006; Thumbikat et al., 2009a) (Fig. 2C). While this defense mechanism jettisons bacteria attached to and replicating within the epithelium, it exposes cells of the underlying, transitional cell layer, which UPEC can invade. UPEC establishes reservoirs consisting of 8 to 12 bacteria within Lamp1+ rosettes in the bladder tissue called quiescent intracellular reservoirs (QIRs) (Fig. 2G) (Mysorekar and Hultgren, 2006). The mechanism of QIR formation is unknown. UPEC does not actively replicate within these reservoirs but instead persists in a dormant state. Bacteria in the QIR can reactivate (by an unknown mechanism) to release bacteria into the lumen to begin the IBC cascade anew. Reactivation of QIRs has been triggered pharmacologically, resulting in a recurrent infection with high titers of bacteria in the urine of mice (Schilling et al., 2002; Mysorekar and Hultgren, 2006). Reservoirs can thus form in the mouse bladder that can reactivate, leading to a recurrent UTI.

AN EXUBERANT IMMUNE RESPONSE DURING ACUTE INFECTION PREDISPOSES TO CHRONIC CYSTITIS

Different murine models of UTI recapitulate the range of clinical outcomes of UTI (Hopkins et al.,

1998). WT inbred C57BL/6J mice resolve bacteriuria rapidly but are susceptible to recurrent UTI (Schilling et al., 2002), possibly as a result of the large amount of QIRs formed in the bladder tissue (Mysorekar and Hultgren, 2006) (Fig. 2G). C3H/HeOUJ mice are exquisitely susceptible to persistent bacterial replication of an inflamed bladder throughout the lifetime of the mouse, a phenomenon referred to as chronic cystitis (Hannan et al., 2010) (Fig. 2F). Elevated levels of the cytokines IL-5, IL-6, keratinocyte cytokine (KC), and granulocyte colony-stimulating factor in the serum of mice at 24 hpi predict the development of persistent bacteriuria and chronic cystitis at 4 weeks postinfection (Hannan et al., 2010). C3H/HeN mice exhibited a bimodal distribution of these two outcomes, with persistent bacteriuria and chronic cystitis occurring in 20 to 40% when infected with 10^7 CFU of UTI89, a prototypical virulent UPEC isolate. Infection of mice with a dose of 10^8 CFU increased the proportion of C3H/HeN mice that developed chronic cystitis to 50%. It was recently shown that high levels of IBC formation at 6 hpi also correlated strongly with the development of persistent bacteriuria and chronic cystitis (Fig. 2D and F) (Schwartz et al., 2011). Therefore, high levels of invading UPEC may lead to the exuberant immune response. In the bladders of mice that experienced chronic cystitis, the superficial facet cells become completely denuded, and IBC formation ceases to occur after 24 to 48 hpi. Thus, during chronic cystitis, bacteria proliferate in association with the underlying tissue but do not invade to an appreciable degree (Fig. 2F). Furthermore, treatment of these mice with an antibiotic was shown to completely sterilize the bladder (Hannan et al., 2010), in contrast to the tolerance observed for bacteria in QIRs (Schilling et al., 2002). Thus, these outcomes are likely mutually exclusive. An overexuberant immune response during acute infection to invasive UPEC leads to cytokine secretion and rampant bacterial replication.

DISSEMINATION TO THE KIDNEYS IS ACCOMPANIED BY REGULATORY CHANGES

The dynamics of UTI are complex, especially when bacteria seed the kidneys. The mouse models of UTI differ in their propensities for kidney infection (Hopkins et al., 1998). C3H/HeJ mice, which lack the ability to respond to lipopolysaccharide via TLR4 sustain persistent kidney infection (Goluszko et al., 1997). Kidney infection in C3H/HeN mice generally mirrors the outcome in the bladder; however, bacterial levels in the kidneys are generally lower than in the bladders of the same animals (Hannan et al., 2010; Schwartz et al., 2011). The ability to ascend

from the bladder to the kidneys depends on many factors, including vesicoureteral reflux and bacterial motility (Fig. 2). Bacterial inoculation into the bladder leads to reflux into the kidneys to various degrees depending on the volume and speed with which the inoculum is introduced as well as the mouse strain (Schaeffer et al., 1987; Hopkins et al., 1995; Schwartz et al., 2011). A minor but significant role for flagella has been documented: contributing to ascension to the upper urinary tract. Expression of flagella coincided with ascension to the upper urinary tract (Lane et al., 2007), and lacking flagella decreased the ability of UPEC to colonize murine kidneys (Wright et al., 2005; Lane et al., 2007). Expression levels of the flagellum promoter flhDC varied within IBCs, perhaps corresponding to their maturity and likelihood of releasing bacteria into the bladder lumen. Robust and rapid gene expression, including activation of motility, occurs prior to a transition from a bladder niche, allowing transition and colonization of the upper urinary tract.

The ability to transition from the bladders to the kidneys likely entails complex genetic changes, including the downregulation of type 1 pili with compensatory upregulation of flagella and P pili. Three different classes of the P pilus adhesin exist called PapGI, -II, and -III, each of which has different affinities for a series of Galα4Gal-containing glycolipid receptors (Stromberg et al., 1990). These glycolipids are expressed to various degrees in human and other mammalian kidneys and ureters (Lanne et al., 1995). The class I and II Pap adhesins bind similarly to globotriaosyl ceramide and globoside, both present in human kidneys. The class III adhesin binds strongest to Forssman glycolipid, which is absent in human kidneys (Breimer and Karlsson, 1983; Breimer et al., 1985; Haslam and Baenziger, 1996). The terminal sugar orientation at the cell surface governs which Pap adhesin binds with the highest affinity (Stromberg et al., 1991). Accordingly, in human clinical isolates, the PapGII class dominated in patients with pyelonephritis (Otto et al., 1993), whereas in patients with acute cystitis, PapGIII predominated (Johnson et al., 1999). Thus, certain UPEC strains are more suited for colonization of one host niche based on the adhesin and other virulence genes that they express. Of the sequenced UPEC organisms we have discussed, the pyelonephritis isolate CFT073 contains two pap clusters, each of which encodes PapGII (Welch et al., 2002), whereas the cystitis isolates UTI89 and UPEC 536 encode PapGIII (Middendorf et al., 2001; Chen et al., 2006), perhaps explaining the clinical picture of the patients from which these strains were isolated.

Bacterial strains that colonize the upper urinary tract are more likely to contain the pap operon than those strains isolated from patients suffering lower urinary tract infection only (Vejborg et al., 2011). It has been shown that several UPEC CUP pili can bind receptors in the human kidney (Virkola et al., 1988). Type 1 pili were shown to bind tissue from the proximal tubules and the vessel walls, whereas P and S fimbriae bound Bowman's capsule, the glomerulus, the renal tubules, and the vessels. F1C pili bound to endothelial cells and the collecting duct and distal tubules. The role of P pili has been clearly demonstrated in cynomolgus monkeys, where the receptor specificity more closely matches that of humans (Lanne et al., 1995). The PapG-containing strain DS17 caused significantly more pathology in the monkey kidneys as assessed histologically, as well as resulted in a greater loss of renal function than did an isogenic knockout (Roberts et al., 1994). Furthermore, immunization against PapG effectively reduced renal histopathology in monkeys (Roberts et al., 2004). The role of P pili has been difficult to assess in mouse models because the receptor for PapG is not present in the mouse kidney (Adams and Gray, 1968; Lanne et al., 1995). However, Pap-expressing UPEC organisms were isolated in higher numbers from the kidneys of CBA/J mice, a commonly used murine UTI model strain, than those lacking P pili in coinfection experiments (Hagberg et al., 1983). Knockout analysis of the *pap* genes in CFT073 was utilized to determine that P pili play only a minor role in kidney colonization in the CBA/J model (Mobley et al., 1993). It was recently shown that Ygi fimbriae bound kidney epithelial cells *in vitro* (Spurbeck et al., 2011). Therefore, P pili appear to be an essential virulence determinant in causing human UTI; however, other CUP pili may also contribute to kidney tropism in humans and animal models.

UPEC may also occupy disparate niches in the kidneys, as illustrated by the ability of different adhesins to mediate attachment to different receptors within the kidney (Virkola et al., 1988). The infection dynamics of UPEC infecting the kidneys were recently demonstrated utilizing intravital multiphoton microscopy in rats (Melican et al., 2011). Direct microinfusion of bacteria into the nephron revealed a synergistic role of P and type 1 pili. Interestingly, knockouts of FimH and P pili in distinct UPEC strains colonized the rat renal tubule, but the kinetics and dynamics of the infection differed. In the absence of P pili, attachment was temporally retarded but spatially indistinguishable from the WT. Inversely, a FimH knockout rapidly colonized the rat renal tubules but only areas adjacent to the epithelium. Likely, in the absence of FimH, UPEC was incapable of coalescing into a biofilm occupying the lumen of the renal tubule (Melican et al., 2011). FimH-mediated bacterial

aggregation may play a role in biofilm formation in the kidneys and IBC formation in the bladder (Chen et al., 2009). Further work to investigate this function of FimH and type 1 pili is ongoing. Because of the variety of niches UPEC can occupy in the bladder and kidneys, the dynamics of UPEC UTI are complex, with genetic regulatory networks likely governing the appropriate confluence of virulence factor expression for each given niche.

REGULATION OF TYPE 1 PILI

The most well-associated virulence factor with UPEC strains causing cystitis is type 1 pili, present in greater than 95% of cystitis isolates (Norinder et al., 2011; Vejborg et al., 2011). Type 1 piliated UPEC is more effective at colonizing the mouse bladder than nonpiliated UPEC (Hultgren et al., 1985; Connell et al., 1996; Totsika et al., 2011). Although type 1 pili are essential for bladder colonization, their expression is dispensable and perhaps disadvantageous in other niches such as the kidneys, colon, or environment. In fact, loss of type 1 piliation does not decrease intestinal colonization of individual animals; however, it does significantly reduce communicability (Bloch et al., 1992), further supporting the argument that *E. coli* expressing type 1 pili is better suited for life in the urinary tract than the gastrointestinal tract (Chen et al., 2009). Accordingly, type 1 pilus expression is reduced in kidney infections of mice (Schaeffer et al., 1987). Bacteria attached to epithelial cells sloughed in mouse urine were significantly more likely to express type 1 pili than planktonic bacteria in the same urine sample (Gunther et al., 2001). This niche-specific difference in type 1 pilus expression provides strong evidence for the spatial and temporal regulation of this CUP operon and its necessity in causing cystitis. Accordingly, a vaccine against FimH was shown to dramatically decrease infection in both infected mice and monkeys (Langermann et al., 1997; Langermann et al., 2000).

Type 1 pili are phase variable, transcriptionally regulated by the inversion of the 314-bp *fim* promoter, *fimS*, that is located between flanking invertible repeats which are recognized by the recombinases FimB and FimE (Fig. 3A). FimB and FimE, which are 48% identical, are encoded by two regulatory genes, *fimB* and *fimE*, 5′ of the *fimS* region (Eisenstein, 1981; Abraham et al., 1985; Klemm, 1986). FimB recombines the invertible repeats, showing no preference in promoter orientation, while FimE inverts the promoter to the OFF orientation only (Klemm, 1986; McClain et al., 1991). FimB appears to show no bias in turning the switch ON or OFF; however, the rate

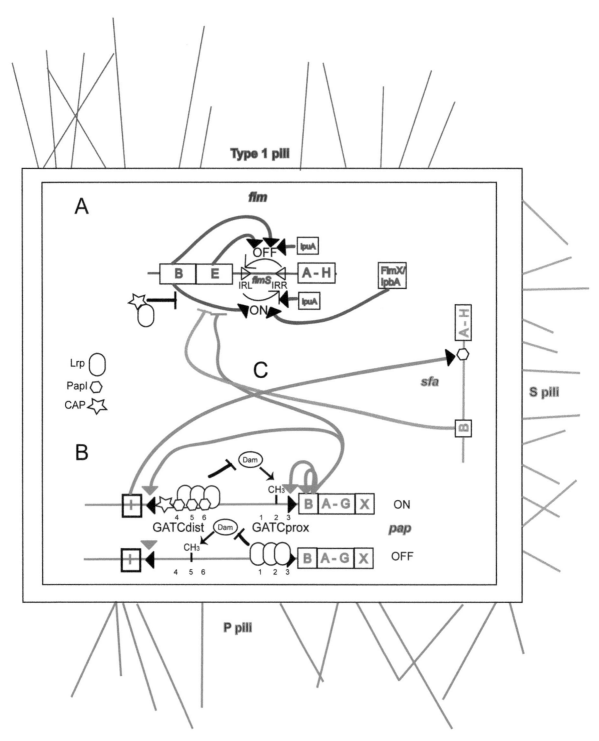

Figure 3. Type 1, P, and S pilus regulation. (A) Type 1 pili are transcriptionally regulated by recombinases that invert the promoter, *fimS*. FimE binds the inverted repeats, turning the promoter OFF. FimB can invert the promoter bidirectionally; however, its major function is turning the promoter ON. The functionally redundant recombinases FimX and IpbA can also turn the promoter ON, whereas IpuA functions similarly to FimB. Lrp and CAP inhibit FimB-mediated OFF-to-ON inversion. (B) The *pap* operon is regulated via differential Dam methylation. In the ON orientation, PapI-Lrp complexes bind the distal methylation sites, allowing RNA polymerase to transcribe the *pap* genes. Binding of Lrp to proximal GATC sites prevents Dam methylation and RNA polymerase binding, shutting down P pilus transcription. (C) The cross-regulation between pilus operons serves to ensure appropriate adhesin expression for in vivo niche. PapI binds the *sfa* operon and enhances its transcription. PapB and SfaB inhibit FimB-mediated OFF-to-ON inversion of the *fim* promoter, preventing type 1 pilus transcription. CAP, catabolite activator protein.

doi:10.1128/9781555818524.ch7f3

with which FimE turns the switch OFF is an order of magnitude greater than for FimB (Kuwahara et al., 2010). Thus, the abundance and activity of FimE and FimB appear to be the determining factors of the orientation of the phase switch.

A number of factors have been identified that in turn control expression or activation of FimB and FimE, thus contributing to modulation of type 1 pilus expression in response to environmental changes. For example, in broth cultures at 37°C, FimB is produced in excess, leading to the population distribution skewed to more piliated bacteria (Schwan et al., 1992; Gally et al., 1993). Recently, in silico analysis has demonstrated stringent control of the type 1 fimbriation switch, such that at temperatures lower than body temperature, FimE dominates, inverting the promoter to the OFF orientation (Kuwahara et al., 2010). At physiological or elevated host temperatures, during febrile episodes, for example, the ON-to-OFF switching rates are lowered dramatically via increasing activity of FimB, locking UPEC into the more fimbriate state. Additionally, mutation of one of the invertible repeats of the fim promoter such that the promoter was irreversibly "locked" in the OFF orientation resulted in significantly fewer CFUs retrieved from urine and bladders at 24 hpi than with the WT strain, CFT073 (Gunther et al., 2002). As expected, when the promoter was locked in the ON orientation, the resulting strain slightly outcompeted CFT073 in the bladder at 4 and 24 hpi. In a recently conducted study with globally spreading, multidrug-resistant UPEC strains of the clonal group ST131, Totsika et al. found that growth under static conditions, known to enrich for fim expression (Old and Duguid, 1971), increased bladder colonization without affecting urine titers at 18 hpi (Totsika et al., 2011). In contrast, strains from the same collection that lacked the ability to turn fim ON were unable to colonize the bladder (Totsika et al., 2011). Affecting the orientation of the promoter via genetic manipulation or through natural environmental changes alters the degree of piliation and fitness in the bladder.

Multiple studies have demonstrated that UPEC lacking both FimB and FimE rapidly inverted the fim promoter to the ON orientation when inoculated into mice (Bryan et al., 2006; Hannan et al., 2008). These experiments revealed that additional recombinases influence phase variation. Genome analysis of the pyelonephritis isolate CFT073 revealed genes for three such recombinases, ipuA, ipuB, and ibpA, with 65 to 70% sequence similarity to FimE and FimB (Fig. 3A) (Bryan et al., 2006). ipuA and ibpA were able to invert the fim switch in the absence of FimE and FimB in vitro and when inoculated into the urinary tract of mice. Further analysis revealed that IpuA acts similarly

to FimB, inverting the switch bidirectionally, whereas IpbA can only invert the promoter from OFF to ON. The cystitis isolate UTI89 contains FimX, an IpbA homolog with 49.1% amino acid similarity to FimB (Hannan et al., 2008), but does not contain IpuA or IpuB. A phase OFF triple fimBEX deletion mutant was completely deficient in bladder colonization at 6 hpi, whereas a phase OFF fimBE mutant colonized at WT levels, indicating that fim phase inversion was due to the activity of FimX. Indeed, complementation of the fimBEX mutant with fimX alone was sufficient to restore WT levels of bladder colonization at 6 hpi, suggesting that UPEC has several functionally redundant recombinases that enable colonization of the bladder by mediating fim promoter inversion to the ON orientation (Hannan et al., 2008). In addition, these functionally redundant recombinases may serve to fine-tune the phase status of the population in certain niches, or they may represent redundant regulators in the event of a fimB or fimE mutation, which occurs in certain UPEC strains, such as the multidrug-resistant clones of the ST131 lineage (Totsika et al., 2011). The majority (59 to 71%) of strains from this clonal group were shown to have an insertional mutation in fimB, yet 87% of tested isolates were still able to express type 1 pili as assessed by yeast agglutination and by adherence to and invasion into bladder epithelial cells (Totsika et al., 2011). One representative member of this group, EC958, contained ipbA, which may account for the observed fim promoter recombination. By virtue of its importance in bladder colonization, type 1 pili are regulated by many functionally redundant recombinases.

DNA binding proteins that impact structure of the chromosome and global regulators also influence the directionality of the fim promoter. Integration host factor (IHF) plays a dual role in governing phase status of UPEC (Dorman and Higgins, 1987). Mutations in IHF locked the phase switch in the OFF or ON orientation, thereby preventing inversion of the promoter (Eisenstein et al., 1987). Additionally, when the fim switch was locked in the ON orientation, mutation of IHF led to a sevenfold reduction in LacZ expression of a FimA-LacZ fusion (Dorman and Higgins, 1987). Mutations in the histone-like protein H-NS were shown to dramatically increase the rate of fim promoter inversion (Kawula and Orndorff, 1991). Conversely, mutations in the master regulator leucine-responsive regulatory protein (Lrp) decreased the rate of fim promoter inversion. Lrp increased the transcription of fimB and decreased fimE expression (Blomfield et al., 1993). Additionally, Lrp binds directly to the fim invertible repeat region, thereby sterically hindering FimB/FimE binding or RNA polymerase complex recruitment (Gally et al., 1994).

Binding of the catabolite repression system through cAMP-CRP (cyclic AMP [cAMP] receptor protein or catabolite activator protein) also influenced the *fim* promoter (Fig. 3A) (Müller et al., 2009). Deletion of the *crp* gene led to higher expression of *fimA* transcript while decreasing P pili and flagella. The effects of cAMP-CRP on the *fim* promoter are complex, likely because this regulator imparts pleiotropic effects in response to the nutritional status of the cell and growth phase. Additionally, in log phase of growth, cAMP-CRP repressed fimbriation, whereas in stationary phase, it appeared to have little effect. The contribution of multiple regulatory networks and proteins like the ones discussed here are likely to be critical in ensuring appropriate expression of type 1 pili during transitions between niches with varying temperatures and nutritional availabilities.

DIFFERENTIAL METHYLATION REGULATES P PILI

Unlike the DNA inversion-based phase variation of type 1 pili, phase variation of the *pap* operon is mediated by differential methylation of two DNA adenine methyltransferase (Dam) sites in the promoter (Fig. 3B) (Blyn et al., 1990), while additional regulation is accomplished by the *cis*-encoded regulators PapB (Forsman et al., 1989) and PapI (Baga et al., 1985) as well as several DNA binding proteins, including H-NS, Lrp, and cAMP-CRP. *papB* and *papI* are encoded upstream of the *pap* operon. *papI* is transcribed from its own promoter in the opposite direction to the *papBA* transcript (Baga et al., 1985). Two GATC methylation regions are present upstream of the *papBA* promoter (Blyn et al., 1990). Methylation of $GATC_{1130}$ ($GATC^{prox}$) results in phase ON cells that can be actively transcribed, whereas methylation of $GATC_{1028}$ ($GATC^{dist}$) results in phase OFF bacteria (Fig. 3B). Because methylation is the main mechanism for P pilus phase variation, P piliation is linked to cell division, during which these epigenetic changes can be enacted (Peterson and Reich, 2008). Furthermore, *pap* genes were not expressed when Dam was present in too-high or too-low quantities within the bacterial cell, implying that appropriate methylation is essential for proper regulation. The regulator PapI binds to DNA upstream of the *papBA* promoter at distal sites. With Dam absent, the proper methylation of $GATC_{1130}$ does not occur, and transcription is abrogated. Conversely, overexpression of Dam prevented phase OFF bacteria from turning ON, presumably as a result of aberrant methylation of GATC sites (Blyn et al., 1990). As for type 1 pili, Lrp also affects P piliation. Methylation of $GATC_{1028}$ (distal) and

cooperative Lrp binding at proximal sites prevented RNA polymerase from transcription at the *papBA* promoter (Peterson and Reich, 2008).

Autoregulation of the *pap* operon by PapB is exquisitely tuned to the amount of the regulator present. PapB binds to three locations within the operon: 200 bp upstream of the *papI* promoter adjacent to the binding site of CRP, adjacent to the –10 area of the *papBA* promoter, and within the PapB coding sequence (Fig. 3B) (Forsman et al., 1989; Xia et al., 1998). PapB binds upstream and activates the transcription of *papI*, perpetuating a positive-feedback loop, maintaining bacteria in the P piliated state (Hernday et al., 2002). However, overexpression of PapB led to repression of P piliation by blocking the binding of RNA polymerase at the –10 region of the *papBA* promoter (Forsman et al., 1989; Hernday et al., 2002). cAMP-CRP is essential for the transcription of the *pap* operon (Goransson et al., 1989). Binding of cAMP-CRP 215 bp upstream of the *papBA* promoter led to increased levels of transcript. Binding of Lrp 140 bp upstream of the promoter was essential for CRP-mediated *papBA* transcription (Weyand et al., 2001). Addition of PapI in *trans* dramatically increased OFF-to-ON switching rates and subsequent *papBA* transcription. The precise control of P pilus regulation by metabolite binding proteins and *cis*-encoded regulators serves to assess local environmental conditions and respond accordingly with the timely expression of the appropriate pilus system.

PILUS CROSS-REGULATION

In order to accomplish accurate and rapid exchange of expressed CUPs, regulators of the P, type 1, and S pilus systems interact with other operons, regulating their expression (Fig. 3C). These regulatory networks are likely responsible for ensuring that UPEC expresses the appropriate pili in a timely manner to colonize a specific niche. Most *E. coli* organisms express only one pilus type at a time (Nowicki et al., 1984), with cross-regulatory networks likely responsible for switching between expressed pilus systems (Holden and Gally, 2004).

Deletion of a pathogenicity operon encoding P-related fimbriae in the UPEC strain 536 diminished expression of S pili encoded by the *sfa* operon (Knapp et al., 1986; Morschhauser et al., 1994). The regulators PrfB and PrfI encoded by the P-related fimbria operon are 76 and 87% homologous to the S pilus regulators SfaB and SfaI, respectively. PrfB and PrfI act in *trans* on S fimbriae, promoting their expression, and complemented *sfaB* and *sfaI* knockouts (Morschhauser et al., 1994). Coordinate expression of these two pilus operons and other interactions not yet known may serve to enhance the probability of attaching to a host surface. It is possible that attachment feedback through master regulators selects for the expression of appropriate adhesin systems for that environment. This theory is consistent with data suggesting that attachment altered complex genetic regulatory networks (Zhang and Normark, 1996; Otto and Silhavy, 2002; Bhomkar et al., 2010). Further, negative cross-regulatory interactions between pili may serve to divert resources to the conditions most effective for persistence or transit throughout hosts. Transcription of both the P and S pilus operons repressed type 1 pilus expression (Fig. 3C) (Xia et al., 2000; Holden et al., 2001). PapB binds to several regions of the *fim* operon, enhances the expression of *fimE*, and also prevents the ability of FimB to invert *fimS* (Xia et al., 2000) (Fig. 3). SfaB represses FimB-mediated recombination without affecting FimE (Holden et al., 2001) (Fig. 3C). When expressed from its native promoter in *E. coli* K-12 and CFT073, PapB was shown to inhibit type 1 pili. Furthermore, the expression of P pili in clinical isolates from patients with various clinical UTI syndromes repressed type 1 pili (Holden et al., 2006). These data present a model of coordinate regulation of virulence factors in which during cystitis, type 1 pilus expression is dominant, whereas other CUPs such as P or S pilus expression are expressed in the kidneys during pyelonephritis, repressing type 1 pilus expression.

Altering the balance of the two-component regulatory system QseBC, which is present in many pathogenic bacteria (Rasko et al., 2008), affects many UPEC virulence factors, CUP systems chief among them (Kostakioti et al., 2009; Hadjifrangiskou et al., 2011). Aberrant and uncontrolled phosphorylation of QseB via deletion of the sensor kinase QseC in UTI89 led to decreased type 1 pilus expression, decreased flagellum expression, and increased expression of the *sfa* operon (encoding S pili), the F-17 operon, and the *fml* operon. The expression of the CUP operons *yqi*, *yeh*, and *auf* was decreased. The alteration of conserved metabolic processes likely accounts in part for these pleiotropic effects (Hadjifrangiskou et al., 2011). It is at present unclear whether these changes are direct effects of QseB phosphorylation or whether the decrease in *fim* expression leads to the concomitant changes in other CUP expression.

ANTIVIRULENCE COMPOUNDS TO TREAT UTI

UPEC colonization of the diverse urinary tract niches largely depends on its ability to adhere to different receptors that are niche specific. Therefore, inhibiting the adhesive organelles that mediate attachment poses an attractive strategy for treating or preventing UTIs. The mannose-binding pocket of FimH is invariant among sequenced UPEC organisms, and a mutation in this region rendered FimH defective in mannose binding, as well as bladder cell invasion and IBC formation (Hung et al., 2002; Chen et al., 2009). FimH is known to bind human uroplakins that coat the luminal surface of the bladder (Wu et al., 1996). Recently, low-molecular-weight orally available compounds called mannosides that bind the FimH binding pocket with low nanomolar affinities were developed using rational structure-directed design (Han et al., 2010; Klein et al., 2010; Cusumano et al., 2011; Schwardt et al., 2011). The addition of mannosides to an oral regimen of the antibiotics TMP-SMX was additive in reducing bacterial burden by the sensitive strain UTI89 (Cusumano et al., 2011). By utilizing a strain resistant to TMP-SMX, PBC-1, it was shown that mannoside addition potentiated the effects of this antibiotic combination. Mannoside prevented PBC-1 from accessing the intracellular niche of the bladder, thus partitioning organisms to the urine and bladder lumen, where antibiotic levels were above the MIC for the clinically resistant PBC-1 strain. Mannoside was also effective at reducing bacterial titers by greater than 4 logs in as little as 6 h after its oral administration to mice experiencing chronic cystitis at 2 weeks postinfection (Cusumano et al., 2011). Because mannosides do not need to access the bacterial cytosol, they are not subject to efflux pumps, degradation, or changes in outer membrane permeability, implying that resistance development is unlikely.

While type 1 pilus inhibition would likely be effective for preventing and treating cystitis, a more general approach targeting multiple CUPs could incapacitate multiple pilus systems, limiting colonization of multiple niches. Accordingly, compounds have been designed to block pilus biogenesis by taking advantage of the structural similarity between CUPs (Pinkner et al., 2006). These bicyclic 2-pyridinones, known as pilicides, reduced type 1 and P pilus-based hemagglutination and biofilm formation. Structural analysis revealed that these compounds bind to the surface of the chaperone that interacts with the usher (Pinkner et al., 2006). Therefore, pilicides function by entering the bacterial periplasm and binding to the chaperone to prevent pilus assembly through the usher. Antivirulence compounds such as pilicides and mannosides represent novel strategies to translate basic knowledge from the investigation of pilus structure and function into new therapeutics that may have efficacy in treating UTIs by affecting CUP expression and function.

CONCLUSIONS AND PERSPECTIVES

Surface-expressed organelles with terminal adhesins mediate the first interaction between host and pathogen. In colonizing a host or the environment, bacteria encounter many niches with disparate surfaces on which to attach. Accordingly, many pathogens have the capacity to express different adhesins assembled into different surface organelles to mediate attachment to the various niches and surfaces they encounter. Understanding the regulatory interplay between pilus function and regulation and the cross talk between operons has led to the development of novel antivirulence therapeutics such as mannosides and pilicides, which target the FimH adhesin expressed at the distal tip of type 1 pili or multiple chaperone-usher pili, respectively (Pinkner et al., 2006; Cusumano and Hultgren, 2009; Han, et al., 2010; Cusumano et al., 2011). Furthermore, understanding the complex population dynamics, niche occupation, and bacterial fluxing between urinary tract niches in time and space is crucial to dissecting the importance of virulence factors. Virulence gene expression is precisely regulated over time and space by the local environmental conditions, including the receptors for attachment, immunological response, and other bacterial populations nearby.

REFERENCES

Abraham, J. M., C. S. Freitag, J. R. Clements, and B. I. Eisenstein. 1985. An invertible element of DNA controls phase variation of type 1 fimbriae of *Escherichia coli*. *Proc. Natl. Acad. Sci. USA* **82:**5724–5727.

Adams, E. P., and G. M. Gray. 1968. The carbohydrate structures of the neutral ceramide glycolipids in kidneys of different mouse strains with special reference to the ceramide dihexosides. *Chem. Phys. Lipids* **2:**147–155.

Anderson, G. G., C. C. Goller, S. Justice, S. J. Hultgren, and P. C. Seed. 2010. Polysaccharide capsule and sialic acid-mediated regulation promote biofilm-like intracellular bacterial communities during cystitis. *Infect. Immun.* **78:**963–975.

Anderson, G. G., J. J. Palermo, J. D. Schilling, R. Roth, J. Heuser, and S. J. Hultgren. 2003. Intracellular bacterial biofilm-like pods in urinary tract infections. *Science* **301:**105–107.

Aprikian, P., G. Interlandi, B. A. Kidd, I. Le Trong, V. Tchesnokova, O. Yakovenko, M. J. Whitfield, E. Bullitt, R. E. Stenkamp, W. E. Thomas, and E. V. Sokurenko. 2011. The bacterial fimbrial tip acts as a mechanical force sensor. *PLoS Biol.* **9:**e1000617.

Baga, M., M. Goransson, S. Normark, and B. E. Uhlin. 1985. Transcriptional activation of a Pap pilus virulence operon from uropathogenic *Escherichia coli*. *EMBO J.* **4:**3887–3893.

Bardiau, M., M. Szalo, and J. G. Mainil. 2010. Initial adherence of EPEC, EHEC and VTEC to host cells. *Vet. Res.* **41:**57.

Beloin, C., A. Roux, and J. M. Ghigo. 2008. *Escherichia coli* biofilms. *Curr. Top. Microbiol. Immunol.* **322:**249–289.

Bhomkar, P., W. Materi, V. Semenchenko, and D. S. Wishart. 2010. Transcriptional response of *E. coli* upon FimH-mediated fimbrial adhesion. *Gene Regul. Syst. Biol.* **4:**1–17.

Bloch, C. A., B. A. Stocker, and P. E. Orndorff. 1992. A key role for type 1 pili in enterobacterial communicability. *Mol. Microbiol.* **6:**697–701.

Blomfield, I. C., P. J. Calie, K. J. Eberhardt, M. S. McClain, and B. I. Eisenstein. 1993. Lrp stimulates phase variation of type 1 fimbriation in *Escherichia coli* K-12. *J. Bacteriol.* **175:**27–36.

Blyn, L. B., B. A. Braaten, and D. A. Low. 1990. Regulation of *pap* pilin phase variation by a mechanism involving differential Dam methylation states. *EMBO J.* **9:**4045–4054.

Bokranz, W., X. Wang, H. Tschape, and U. Romling. 2005. Expression of cellulose and curli fimbriae by *Escherichia coli* isolated from the gastrointestinal tract. *J. Med. Microbiol.* **54:**1171–1182.

Bran, J. L., M. E. Levison, and D. Kaye. 1972. Entrance of bacteria into the female urinary bladder. *N. Engl. J. Med.* **286:**626–629.

Breimer, M. E., G. C. Hansson, and H. Leffler. 1985. The specific glycosphingolipid composition of human ureteral epithelial cells. *J. Biochem.* **98:**1169–1180.

Breimer, M. E., and K. A. Karlsson. 1983. Chemical and immunological identification of glycolipid-based blood group ABH and Lewis antigens in human kidney. *Biochim. Biophys. Acta* **755:**170–177.

Brinton, C. C., Jr. 1965. The structure, function, synthesis, and genetic control of bacterial pili and a model for DNA and RNA transport in gram negative bacteria. *Trans. N. Y. Acad. Sci.* **27:**1003–1054.

Bryan, A., P. Roesch, L. Davis, R. Moritz, S. Pellett, and R. A. Welch. 2006. Regulation of type 1 fimbriae by unlinked FimB- and FimE-like recombinases in uropathogenic *Escherichia coli* strain CFT073. *Infect. Immun.* **74:**1072–1083.

Brzuszkiewicz, E., H. Brüggemann, H. Liesegang, M. Emmerth, T. Olschläger, G. Nagy, K. Albermann, C. Wagner, C. Buchrieser, L. Emody, G. Gottschalk, J. Hacker, and U. Dobrindt. 2006. How to become a uropathogen: comparative genomic analysis of extraintestinal pathogenic *Escherichia coli* strains. *Proc. Natl. Acad. Sci. USA* **103:**12879–12884.

Buckley, R. M., Jr., M. McGuckin, and R. R. MacGregor. 1978. Urine bacterial counts after sexual intercourse. *N. Engl. J. Med.* **298:**321–324.

Bullitt, E., C. H. Jones, R. Striker, G. Soto, F. Jacob-Dubuisson, J. Pinkner, M. J. Wick, L. Makowski, and S. J. Hultgren. 1996. Development of pilus organelle subassemblies in vitro depends on chaperone uncapping of a beta zipper. *Proc. Natl. Acad. Sci. USA* **93:**12890–12895.

Buts, L., J. Bouckaert, E. De Genst, R. Loris, S. Oscarson, M. Lahmann, J. Messens, E. Brosens, L. Wyns, and H. De Greve. 2003. The fimbrial adhesin F17-G of enterotoxigenic *Escherichia coli* has an immunoglobulin-like lectin domain that binds N-acetylglucosamine. *Mol. Microbiol.* **49:**705–715.

Chen, S. L., C. S. Hung, J. S. Pinkner, J. N. Walker, C. K. Cusumano, Z. Li, J. Bouckaert, J. I. Gordon, and S. J. Hultgren. 2009. Positive selection identifies an in vivo role for FimH during urinary tract infection in addition to mannose binding. *Proc. Natl. Acad. Sci. USA* **106:**22439–22444.

Chen, S. L., C. S. Hung, J. Xu, C. S. Reigstad, V. Magrini, A. Sabo, D. Blasiar, T. Bieri, R. R. Meyer, P. Ozersky, J. R. Armstrong, R. S. Fulton, J. P. Latreille, J. Spieth, T. M. Hooton, E. R. Mardis, S. J. Hultgren, and J. I. Gordon. 2006. Identification of genes subject to positive selection in uropathogenic strains of *Escherichia coli*: a comparative genomics approach. *Proc. Natl. Acad. Sci. USA* **103:**5977–5982.

Choudhury, D., A. Thompson, V. Stojanoff, S. Langermann, J. Pinkner, S. J. Hultgren, and S. D. Knight. 1999. X-ray structure of the FimC-FimH chaperone-adhesin complex from uropathogenic *Escherichia coli*. *Science* **285:**1061–1066.

Clegg, S., and G. F. Gerlach. 1987. Enterobacterial fimbriae. *J. Bacteriol.* **169**:934–938.

Connell, H., W. Agace, P. Klemm, M. Schembri, S. Marild, and C. Svanborg. 1996. Type 1 fimbrial expression enhances *Escherichia coli* virulence for the urinary tract. *Proc. Natl. Acad. Sci. USA* **93**:9827–9832.

Cusumano, C. K., and S. J. Hultgren. 2009. Bacterial adhesion—a source of alternate antibiotic targets. *IDrugs* **12**:699–705.

Cusumano, C. K., J. S. Pinkner, Z. Han, S. E. Greene, B. A. Ford, J. R. Crowley, J. P. Henderson, J. W. Janetka, and S. J. Hultgren. 2011. Treatment and prevention of urinary tract infection with orally active FimH inhibitors. *Sci. Transl. Med.* **3**:109ra115.

Dodd, D. C., P. J. Bassford, Jr., and B. I. Eisenstein. 1984. Dependence of secretion and assembly of type 1 fimbrial subunits of *Escherichia coli* on normal protein transport. *J. Bacteriol.* **159**:1077–1079.

Dodson, K. W., F. Jacob-Dubuisson, R. T. Striker, and S. J. Hultgren. 1993. Outer-membrane PapC molecular usher discriminately recognizes periplasmic chaperone-pilus subunit complexes. *Proc. Natl. Acad. Sci. USA* **90**:3670–3674.

Dodson, K. W., J. S. Pinkner, T. Rose, G. Magnusson, S. J. Hultgren, and G. Waksman. 2001. Structural basis of the interaction of the pyelonephritic *E. coli* adhesin to its human kidney receptor. *Cell* **105**:733–743.

Dorman, C. J., and C. F. Higgins. 1987. Fimbrial phase variation in *Escherichia coli*: dependence on integration host factor and homologies with other site-specific recombinases. *J. Bacteriol.* **169**:3840–3843.

Eisenstein, B. I. 1981. Phase variation of type 1 fimbriae in *Escherichia coli* is under transcriptional control. *Science* **214**:337–339.

Eisenstein, B. I., D. S. Sweet, V. Vaughn, and D. I. Friedman. 1987. Integration host factor is required for the DNA inversion that controls phase variation in *Escherichia coli*. *Proc. Natl. Acad. Sci. USA* **84**:6506–6510.

Engel, J. D., and A. J. Schaeffer. 1998. Evaluation of and antimicrobial therapy for recurrent urinary tract infections in women. *Urol. Clin. North Am.* **25**:685–701.

Eto, D. S., H. B. Gordon, B. K. Dhakal, T. A. Jones, and M. A. Mulvey. 2008. Clathrin, AP-2, and the NPXY-binding subset of alternate endocytic adaptors facilitate FimH-mediated bacterial invasion of host cells. *Cell. Microbiol.* **10**:2553–2567.

Eto, D. S., T. A. Jones, J. L. Sundsbak, and M. A. Mulvey. 2007. Integrin-mediated host cell invasion by type 1-piliated uropathogenic *Escherichia coli*. *PLoS Pathog.* **3**:e100.

Ferry, S., S. Holm, H. Stenlund, R. Lundholm, and T. Monsen. 2004. The natural course of uncomplicated lower urinary tract infection in women illustrated by a randomized placebo controlled study. *Scand. J. Infect. Dis.* **36**:296–301.

Finlay, B. B., and S. Falkow. 1989. Common themes in microbial pathogenicity. *Microbiol. Rev.* **53**:210–230.

Forsman, K., M. Göransson, and B. E. Uhlin. 1989. Autoregulation and multiple DNA interactions by a transcriptional regulatory protein in *E. coli* pili biogenesis. *EMBO J.* **8**:1271–1277.

Foxman, B. 2010. The epidemiology of urinary tract infection. *Nat. Rev. Urol.* **7**:653–660.

Gally, D. L., J. A. Bogan, B. I. Eisenstein, and I. C. Blomfield. 1993. Environmental regulation of the fim switch controlling type 1 fimbrial phase variation in *Escherichia coli* K-12: effects of temperature and media. *J. Bacteriol.* **175**:6186–6193.

Gally, D. L., T. J. Rucker, and I. C. Blomfield. 1994. The leucineresponsive regulatory protein binds to the *fim* switch to control phase variation of type 1 fimbrial expression in *Escherichia coli* K-12. *J. Bacteriol.* **176**:5665–5672.

Garofalo, C. K., T. M. Hooton, S. M. Martin, W. E. Stamm, J. J. Palermo, J. I. Gordon, and S. J. Hultgren. 2007. *Escherichia*

coli from urine of female patients with urinary tract infections is competent for intracellular bacterial community formation. *Infect. Immun.* **75**:52–60.

Goluszko, P., S. L. Moseley, L. D. Truong, A. Kaul, J. R. Williford, R. Selvarangan, S. Nowicki, and B. Nowicki. 1997. Development of experimental model of chronic pyelonephritis with *Escherichia coli* O75:K5:H-bearing Dr fimbriae: mutation in the *dra* region prevented tubulointerstitial nephritis. *J. Clin. Investig.* **99**:1662–1672.

Goransson, M., K. Forsman, P. Nilsson, and B. E. Uhlin. 1989. Upstream activating sequences that are shared by two divergently transcribed operons mediate cAMP-CRP regulation of pilus-adhesin in *Escherichia coli*. *Mol. Microbiol.* **3**:1557–1565.

Gunther, N. W., V. Lockatell, D. E. Johnson, and H. L. Mobley. 2001. In vivo dynamics of type 1 fimbria regulation in uropathogenic *Escherichia coli* during experimental urinary tract infection. *Infect. Immun.* **69**:2838–2846.

Gunther, N. W., J. A. Snyder, V. Lockatell, I. Blomfield, D. E. Johnson, and H. L. Mobley. 2002. Assessment of virulence of uropathogenic *Escherichia coli* type 1 fimbrial mutants in which the invertible element is phase-locked on or off. *Infect. Immun.* **70**:3344–3354.

Gupta, K., T. M. Hooton, and W. E. Stamm. 2001. Increasing antimicrobial resistance and the management of uncomplicated community-acquired urinary tract infections. *Ann. Intern. Med.* **135**:41–50.

Hadjifrangiskou, M., M. Kostakioti, S. L. Chen, J. P. Henderson, S. E. Greene, and S. J. Hultgren. 2011. A central metabolic circuit controlled by QseC in pathogenic *Escherichia coli*. *Mol. Microbiol.* **80**:1516–1529.

Hagan, E. C., A. L. Lloyd, D. A. Rasko, G. J. Faerber, and H. L. Mobley. 2010. *Escherichia coli* global gene expression in urine from women with urinary tract infection. *PLoS Pathog.* **6**:e1001187.

Hagberg, L., R. Hull, S. Hull, S. Falkow, R. Freter, and C. Svanborg Eden. 1983. Contribution of adhesion to bacterial persistence in the mouse urinary tract. *Infect. Immun.* **40**:265–272.

Hahn, E., P. Wild, U. Hermanns, P. Sebbel, R. Glockshuber, M. Häner, N. Taschner, P. Burkhard, U. Aebi, and S. A. Müller. 2002. Exploring the 3D molecular architecture of *Escherichia coli* type 1 pili. *J. Mol. Biol.* **323**:845–857.

Han, Z., J. S. Pinkner, B. Ford, R. Obermann, W. Nolan, S. A. Wildman, D. Hobbs, T. Ellenberger, C. K. Cusumano, S. J. Hultgren, and J. W. Janetka. 2010. Structure-based drug design and optimization of mannoside bacterial FimH antagonists. *J. Med. Chem.* **53**:4779–4792.

Hannan, T. J., I. U. Mysorekar, S. L. Chen, J. N. Walker, J. M. Jones, J. S. Pinkner, S. J. Hultgren, and P. C. Seed. 2008. LeuX tRNA-dependent and -independent mechanisms of *Escherichia coli* pathogenesis in acute cystitis. *Mol. Microbiol.* **67**:116–128.

Hannan, T. J., I. U. Mysorekar, C. S. Hung, M. L. Isaacson-Schmid, and S. J. Hultgren. 2010. Early severe inflammatory responses to uropathogenic *E. coli* predispose to chronic and recurrent urinary tract infection. *PLoS Pathog.* **6**:e1001042.

Haslam, D. B., and J. U. Baenziger. 1996. Expression cloning of Forssman glycolipid synthetase: a novel member of the histoblood group ABO gene family. *Proc. Natl. Acad. Sci. USA* **93**:10697–10702.

Henderson, J. P., J. R. Crowley, J. S. Pinkner, J. N. Walker, P. Tsukayama, W. E. Stamm, T. M. Hooton, and S. J. Hultgren. 2009. Quantitative metabolomics reveals an epigenetic blueprint for iron acquisition in uropathogenic *Escherichia coli*. *PLoS Pathog.* **5**:e1000305.

Henderson, N. S., T. W. Ng, I. Talukder, and D. G. Thanassi. 2011. Function of the usher N-terminus in catalysing pilus assembly. *Mol. Microbiol.* **79**:954–967.

Hernday, A., M. Krabbe, B. Braaten, and D. Low. 2002. Self-perpetuating epigenetic pili switches in bacteria. *Proc. Natl. Acad. Sci. USA* 99(Suppl. 4):16470–16476.

Holden, N. J., and D. L. Gally. 2004. Switches, cross-talk and memory in *Escherichia coli* adherence. *J. Med. Microbiol.* 53:585–593.

Holden, N. J., M. Totsika, E. Mahler, A. J. Roe, K. Catherwood, K. Lindner, U. Dobrindt, and D. L. Gally. 2006. Demonstration of regulatory cross-talk between P fimbriae and type 1 fimbriae in uropathogenic *Escherichia coli*. *Microbiology* (Reading) 152:1143–1153.

Holden, N. J., B. E. Uhlin, and D. L. Gally. 2001. PapB paralogues and their effect on the phase variation of type 1 fimbriae in *Escherichia coli*. *Mol. Microbiol.* 42:319–330.

Holmgren, A., and C. I. Branden. 1989. Crystal structure of the chaperone protein PapD reveals an immunoglobulin fold. *Nature* 342:248–251.

Hooton, T. M. 2003. Fluoroquinolones and resistance in the treatment of uncomplicated urinary tract infection. *Int. J. Antimicrob. Agents* 22(Suppl. 2):65–72.

Hopkins, W. J., A. Gendron-Fitzpatrick, E. Balish, and D. T. Uehling. 1998. Time course and host responses to *Escherichia coli* urinary tract infection in genetically distinct mouse strains. *Infect. Immun.* 66:2798–2802.

Hopkins, W. J., J. A. Hall, B. P. Conway, and D. T. Uehling. 1995. Induction of urinary tract infection by intraurethral inoculation with *Escherichia coli*: refining the murine model. *J. Infect. Dis.* 171:462–465.

Horvath, D. J., B. Li, T. Casper, S. Partida-Sanchez, D. A. Hunstad, S. J. Hultgren, and S. S. Justice. 2011. Morphological plasticity promotes resistance to phagocyte killing of uropathogenic *Escherichia coli*. *Microbes Infect.* 13:426–437.

Hughes, C., J. Hacker, A. Roberts, and W. Goebel. 1983. Hemolysin production as a virulence marker in symptomatic and asymptomatic urinary tract infections caused by *Escherichia coli*. *Infect. Immun.* 39:546–551.

Hultgren, S. J., T. N. Porter, A. J. Schaeffer, and J. L. Duncan. 1985. Role of type 1 pili and effects of phase variation on lower urinary tract infections produced by *Escherichia coli*. *Infect. Immun.* 50:370–377.

Hultgren, S. J., W. R. Schwan, A. J. Schaeffer, and J. L. Duncan. 1986. Regulation of production of type 1 pili among urinary tract isolates of *Escherichia coli*. *Infect. Immun.* 54:613–620.

Hung, C. S., J. Bouckaert, D. Hung, J. Pinkner, C. Widberg, A. DeFusco, C. G. Auguste, R. Strouse, S. Langermann, G. Waksman, and S. J. Hultgren. 2002. Structural basis of tropism of *Escherichia coli* to the bladder during urinary tract infection. *Mol. Microbiol.* 44:903–915.

Hung, C. S., K. W. Dodson, and S. J. Hultgren. 2009. A murine model of urinary tract infection. *Nat. Protocols* 4:1230–1243.

Ikaheimo, R., A. Siitonen, T. Heiskanen, U. Karkkainen, P. Kuosmanen, P. Lipponen, and P. H. Makela. 1996. Recurrence of urinary tract infection in a primary care setting: analysis of a 1-year follow-up of 179 women. *Clin. Infect. Dis.* 22:91–99.

Johnson, J. R., C. E. Johnson, and J. N. Maslow. 1999. Clinical and bacteriologic correlates of the papG alleles among *Escherichia coli* strains from children with acute cystitis. *Pediatr. Infect. Dis. J.* 18:446–451.

Johnson, J. R., and W. E. Stamm. 1989. Urinary tract infections in women: diagnosis and treatment. *Ann. Intern. Med.* 111:906–917.

Johnson, J. R., S. J. Weissman, A. L. Stell, E. Trintchina, D. E. Dykhuizen, E. V. Sokurenko. 2001. Clonal and pathotypic analysis of archetypal *Escherichia coli* cystitis isolate NU14. *J. Infect. Dis.* 184:1556–1565.

Jones, C. H., J. S. Pinkner, R. Roth, J. Heuser, A. V. Nicholes, S. N. Abraham, and S. J. Hultgren. 1995. FimH adhesin of type 1 pili is assembled into a fibrillar tip structure in the *Enterobacteriaceae*. *Proc. Natl. Acad. Sci. USA* 92:2081–2085.

Justice, S. S., C. Hung, J. A. Theriot, D. A. Fletcher, G. G. Anderson, M. J. Footer, and S. J. Hultgren. 2004. Differentiation and developmental pathways of uropathogenic *Escherichia coli* in urinary tract pathogenesis. *Proc. Natl. Acad. Sci. USA* 101:1333–1338.

Justice, S. S., D. A. Hunstad, P. C. Seed, and S. J. Hultgren. 2006a. Filamentation by *Escherichia coli* subverts innate defenses during urinary tract infection. *Proc. Natl. Acad. Sci. USA* 103:19884–19889.

Justice, S. S., S. R. Lauer, S. J. Hultgren, and D. A. Hunstad. 2006b. Maturation of intracellular *Escherichia coli* communities requires SurA. *Infect. Immun.* 74:4793–4800.

Kawula, T. H., and P. E. Orndorff. 1991. Rapid site-specific DNA inversion in *Escherichia coli* mutants lacking the histonelike protein H-NS. *J. Bacteriol.* 173:4116–4123.

Kisielius, P. V., W. R. Schwan, S. K. Amundsen, J. L. Duncan, and A. J. Schaeffer. 1989. In vivo expression and variation of *Escherichia coli* type 1 and P pili in the urine of adults with acute urinary tract infections. *Infect. Immun.* 57:1656–1662.

Klein, T., D. Abgottspon, M. Wittwer, S. Rabbani, J. Herold, X. Jiang, S. Kleeb, C. Luthi, M. Scharenberg, J. Bezencon, E. Gubler, L. Pang, M. Smiesko, B. Cutting, O. Schwardt, and B. Ernst. 2010. FimH antagonists for the oral treatment of urinary tract infections: from design and synthesis to *in vitro* and *in vivo* evaluation. *J. Med. Chem.* 53:8627–8641.

Klemm, P. 1986. Two regulatory *fim* genes, *fimB* and *fimE*, control the phase variation of type 1 fimbriae in *Escherichia coli*. *EMBO J.* 5:1389–1393.

Kline, K. A., K. W. Dodson, M. G. Caparon, and S. J. Hultgren. 2010. A tale of two pili: assembly and function of pili in bacteria. *Trends Microbiol.* 18:224–232.

Kline, K. A., S. Fälker, S. Dahlberg, S. Normark, and B. Henriques-Normark. 2009. Bacterial adhesins in host-microbe interactions. *Cell Host Microbe* 5:580–592.

Klumpp, D. J., M. T. Rycyk, M. C. Chen, P. Thumbikat, S. Sengupta, and A. J. Schaeffer. 2006. Uropathogenic *Escherichia coli* induces extrinsic and intrinsic cascades to initiate urothelial apoptosis. *Infect. Immun.* 74:5106–5113.

Knapp, S., J. Hacker, T. Jarchau, and W. Goebel. 1986. Large, unstable inserts in the chromosome affect virulence properties of uropathogenic *Escherichia coli* O6 strain 536. *J. Bacteriol.* 168:22–30.

Korea, C. G., J. M. Ghigo, and C. Beloin. 2011. The sweet connection: solving the riddle of multiple sugar-binding fimbrial adhesins in *Escherichia coli*. Multiple *E. coli* fimbriae form a versatile arsenal of sugar-binding lectins potentially involved in surface-colonisation and tissue tropism. *Bioessays* 33:300–311.

Kostakioti, M., M. Hadjifrangiskou, J. S. Pinkner, and S. J. Hultgren. 2009. QseC-mediated dephosphorylation of QseB is required for expression of genes associated with virulence in uropathogenic *Escherichia coli*. *Mol. Microbiol.* 73:1020–1031.

Kucheria, R., P. Dasgupta, S. H. Sacks, M. S. Khan, and N. S. Sheerin. 2005. Urinary tract infections: new insights into a common problem. *Postgrad. Med. J.* 81:83–86.

Kuehn, M. J., S. Normark, and S. J. Hultgren. 1991. Immunoglobulin-like PapD chaperone caps and uncaps interactive surfaces of nascently translocated pilus subunits. *Proc. Natl. Acad. Sci. USA* 88:10586–10590.

Kuehn, M. J., D. J. Ogg, J. Kihlberg, L. N. Slonim, K. Flemmer, T. Bergfors, and S. J. Hultgren. 1993. Structural basis of pilus subunit recognition by the PapD chaperone. *Science* 262:1234–1241.

Kuwahara, H., C. Myers, and M. Samoilov. 2010. Temperature control of fimbriation circuit switch in uropathogenic *Escherichia coli*: quantitative analysis via automated model abstraction. *PLoS Comput. Biol.* 6:e1000723.

Lane, M. C., C. J. Alteri, S. N. Smith, and H. L. Mobley. 2007. Expression of flagella is coincident with uropathogenic *Escherichia coli* ascension to the upper urinary tract. *Proc. Natl. Acad. Sci. USA* 104:16669–16674.

Langermann, S., R. Mollby, J. E. Burlein, S. R. Palaszynski, C. G. Auguste, A. DeFusco, R. Strouse, M. A. Schenerman, S. J. Hultgren, J. S. Pinkner, J. Winberg, L. Guldevall, M. Soderhall, K. Ishikawa, S. Normark, and S. Koenig. 2000. Vaccination with FimH adhesin protects cynomolgus monkeys from colonization and infection by uropathogenic *Escherichia coli*. *J. Infect. Dis.* 181:774–778.

Langermann, S., S. Palaszynski, M. Barnhart, G. Auguste, J. S. Pinkner, J. Burlein, P. Barren, S. Koenig, S. Leath, C. H. Jones, and S. J. Hultgren. 1997. Prevention of mucosal *Escherichia coli* infection by FimH-adhesin-based systemic vaccination. *Science* 276:607–611.

Lanne, B., B. M. Olsson, P. A. Jovall, J. Angström, H. Linder, B. I. Marklund, J. Bergström, and K. A. Karlsson. 1995. Glycoconjugate receptors for P-fimbriated *Escherichia coli* in the mouse. An animal model of urinary tract infection. *J. Biol. Chem.* 270:9017–9025.

Lee, Y. M., K. W. Dodson, and S. J. Hultgren. 2007. Adaptor function of PapF depends on donor strand exchange in P-pilus biogenesis of *Escherichia coli*. *J. Bacteriol.* 189:5276–5283.

Mabeck, C. E. 1972. Treatment of uncomplicated urinary tract infection in non-pregnant women. *Postgrad. Med. J.* 48:69–75.

Martinez, J. J., and S. J. Hultgren. 2002. Requirement of Rho-family GTPases in the invasion of type 1-piliated uropathogenic *Escherichia coli*. *Cell. Microbiol.* 4:19–28.

Martinez, J. J., M. A. Mulvey, J. D. Schilling, J. S. Pinkner, and S. J. Hultgren. 2000. Type 1 pilus-mediated bacterial invasion of bladder epithelial cells. *EMBO J.* 19:2803–2812.

McClain, M. S., I. C. Blomfield, and B. I. Eisenstein. 1991. Roles of *fimB* and *fimE* in site-specific DNA inversion associated with phase variation of type 1 fimbriae in *Escherichia coli*. *J. Bacteriol.* 173:5308–5314.

Melican, K., R. Sandoval, A. Kader, L. Josefsson, G. Tanner, B. Molitoris, and A. Richter-Dahlfors. 2011. Uropathogenic *Escherichia coli* P and type 1 fimbriae act in synergy in a living host to facilitate renal colonization leading to nephron obstruction. *PLoS Pathog.* 7:e1001298.

Middendorf, B., G. Blum-Oehler, U. Dobrindt, I. Muhldorfer, S. Salge, and J. Hacker. 2001. The pathogenicity islands (PAIs) of the uropathogenic *Escherichia coli* strain 536: island probing of PAI II536. *J. Infect. Dis.* 183:S17–S20.

Mobley, H. L., K. G. Jarvis, J. P. Elwood, D. I. Whittle, C. V. Lockatell, R. G. Russell, D. E. Johnson, M. S. Donnenberg, and J. W. Warren. 1993. Isogenic P-fimbrial deletion mutants of pyelonephritogenic *Escherichia coli*: the role of alpha Gal(1–4) beta Gal binding in virulence of a wild-type strain. *Mol. Microbiol.* 10:143–155.

Montini, G., K. Tullus, and I. Hewitt. 2011. Febrile urinary tract infections in children. *N. Engl. J. Med.* 365:239–250.

Morschhauser, J., V. Vetter, L. Emody, and J. Hacker. 1994. Adhesin regulatory genes within large, unstable DNA regions of pathogenic *Escherichia coli*: cross-talk between different adhesin gene clusters. *Mol. Microbiol.* 11:555–566.

Morschhauser, J., V. Vetter, T. Korhonen, B. E. Uhlin, and J. Hacker. 1993. Regulation and binding properties of S fimbriae cloned from *E. coli* strains causing urinary tract infection and meningitis. *Zentralbl. Bakteriol.* 278:165–176.

Müller, C., A. Åberg, J. Straseviçiene, L. Emödy, B. E. Uhlin, and C. Balsalobre. 2009. Type 1 fimbriae, a colonization factor of uropathogenic *Escherichia coli*, are controlled by the metabolic sensor CRP-cAMP. *PLoS Pathog.* 5:e1000303.

Mulvey, M. A., Y. S. Lopez-Boado, C. L. Wilson, R. Roth, W. C. Parks, J. Heuser, and S. J. Hultgren. 1998. Induction and evasion of host defenses by type 1-piliated uropathogenic *Escherichia coli*. *Science* 282:1494–1497.

Mysorekar, I. U., and S. J. Hultgren. 2006. Mechanisms of uropathogenic *Escherichia coli* persistence and eradication from the urinary tract. *Proc. Natl. Acad. Sci. USA* 103:14170–14175.

Nataro, J. P., and J. B. Kaper. 1998. Diarrheagenic *Escherichia coli*. *Clin. Microbiol. Rev.* 11:142–201.

Nicholson, T. F., K. M. Watts, and D. A. Hunstad. 2009. OmpA of uropathogenic *Escherichia coli* promotes postinvasion pathogenesis of cystitis. *Infect. Immun.* 77:5245–5251.

Nicolas-Chanoine, M. H., J. Blanco, V. Leflon-Guibout, R. Demarty, M. P. Alonso, M. M. Canica, Y. J. Park, J. P. Lavigne, J. Pitout, and J. R. Johnson. 2008. Intercontinental emergence of *Escherichia coli* clone O25:H4-ST131 producing CTX-M-15. *J. Antimicrob. Chemother.* 61:273–281.

Nishiyama, M., M. Vetsch, C. Puorger, I. Jelesarov, and R. Glockshuber. 2003. Identification and characterization of the chaperone-subunit complex-binding domain from the type 1 pilus assembly platform FimD. *J. Mol. Biol.* 330:513–525.

Norinder, B. S., B. Köves, M. Yadav, A. Brauner, and C. Svanborg. 2011. Do *Escherichia coli* strains causing acute cystitis have a distinct virulence repertoire? *Microb. Pathog.* doi:10.1016/j.micpath.2011.08.005.

Nowicki, B., M. Rhen, V. Vaisanen-Rhen, A. Pere, and T. K. Korhonen. 1984. Immunofluorescence study of fimbrial phase variation in *Escherichia coli* KS71. *J. Bacteriol.* 160:691–695.

Old, D. C., and J. P. Duguid. 1971. Selection of fimbriate transductants of *Salmonella typhimurium* dependent on motility. *J. Bacteriol.* 107:655–658.

Otto, G., T. Sandberg, B. I. Marklund, P. Ulleryd, and C. Svanborg. 1993. Virulence factors and *pap* genotype in *Escherichia coli* isolates from women with acute pyelonephritis, with or without bacteremia. *Clin. Infect. Dis.* 17:448–456.

Otto, K., and T. J. Silhavy. 2002. Surface sensing and adhesion of *Escherichia coli* controlled by the Cpx-signaling pathway. *Proc. Natl. Acad. Sci. USA* 99:2287–2292.

Peterson, S. N., and N. O. Reich. 2008. Competitive Lrp and Dam assembly at the *pap* regulatory region: implications for mechanisms of epigenetic regulation. *J. Mol. Biol.* 383:92–105.

Phan, G., H. Remaut, T. Wang, W. J. Allen, K. F. Pirker, A. Lebedev, N. S. Henderson, S. Geibel, E. Volkan, J. Yan, M. B. Kunze, J. S. Pinkner, B. Ford, C. W. Kay, H. Li, S. J. Hultgren, D. G. Thanassi, and G. Waksman. 2011. Crystal structure of the FimD usher bound to its cognate FimC-FimH substrate. *Nature* 474:49–53.

Pinkner, J. S., H. Remaut, F. Buelens, E. Miller, V. Aberg, N. Pemberton, M. Hedenstrom, A. Larsson, P. Seed, G. Waksman, S. J. Hultgren, and F. Almqvist. 2006. Rationally designed small compounds inhibit pilus biogenesis in uropathogenic bacteria. *Proc. Natl. Acad. Sci. USA* 103:17897–17902.

Pouttu, R., T. Puustinen, R. Virkola, J. Hacker, P. Klemm, and T. K. Korhonen. 1999. Amino acid residue Ala-62 in the FimH fimbrial adhesin is critical for the adhesiveness of meningitis-associated *Escherichia coli* to collagens. *Mol. Microbiol.* 31:1747–1757.

Pratt, L. A., and R. Kolter. 1998. Genetic analysis of *Escherichia coli* biofilm formation: roles of flagella, motility, chemotaxis and type I pili. *Mol. Microbiol.* 30:285–293.

Ragnarsdóttir, B., K. Jönsson, A. Urbano, J. Grönberg-Hernandez, N. Lutay, M. Tammi, M. Gustafsson, A. C. Lundstedt, I. Leijonhufvud, D. Karpman, B. Wullt, L. Truedsson, U. Jodal,

B. Andersson, and C. Svanborg. 2010. Toll-like receptor 4 promoter polymorphisms: common TLR4 variants may protect against severe urinary tract infection. *PLoS One* **5:**e10734.

Ragnarsdóttir, B., M. Samuelsson, M. C. Gustafsson, I. Leijonhufvud, D. Karpman, and C. Svanborg. 2007. Reduced Toll-like receptor 4 expression in children with asymptomatic bacteriuria. *J. Infect. Dis.* **196:**475–484.

Rasko, D. A., C. G. Moreira, D. R. Li, N. C. Reading, J. M. Ritchie, M. K. Waldor, N. Williams, R. Taussig, S. Wei, M. Roth, D. T. Hughes, J. F. Huntley, M. W. Fina, J. R. Falck, and V. Sperandio. 2008. Targeting QseC signaling and virulence for antibiotic development. *Science* **321:**1078–1080.

Reigstad, C. S., S. J. Hultgren, and J. I. Gordon. 2007. Functional genomic studies of uropathogenic *Escherichia coli* and host urothelial cells when intracellular bacterial communities are assembled. *J. Biol. Chem.* **282:**21259–21267.

Remaut, H., R. J. Rose, T. J. Hannan, S. J. Hultgren, S. E. Radford, A. E. Ashcroft, and G. Waksman. 2006. Donor-strand exchange in chaperone-assisted pilus assembly proceeds through a concerted beta strand displacement mechanism. *Mol. Cell* **22:**831–842.

Roberts, J. A., M. B. Kaack, G. Baskin, M. R. Chapman, D. A. Hunstad, J. S. Pinkner, and S. J. Hultgren. 2004. Antibody responses and protection from pyelonephritis following vaccination with purified *Escherichia coli* PapDG protein. *J. Urol.* **171:**1682–1685.

Roberts, J. A., B.-I. Marklund, D. Ilver, D. Haslam, M. B. Kaack, G. Baskin, M. Louis, R. Mollby, J. Winberg, and S. Normark. 1994. The Gal (α1-4)Gal-specific tip adhesin of *Escherichia coli* P-fimbriae is needed for pyelonephritis to occur in the normal urinary tract. *Proc. Natl. Acad. Sci. USA* **91:**11889–11893.

Ronald, L. S., O. Yakovenko, N. Yazvenko, S. Chattopadhyay, P. Aprikian, W. E. Thomas, and E. V. Sokurenko. 2008. Adaptive mutations in the signal peptide of the type 1 fimbrial adhesin of uropathogenic *Escherichia coli*. *Proc. Natl. Acad. Sci. USA* **105:**10937–10942.

Rosen, D. A., T. M. Hooton, W. E. Stamm, P. A. Humphrey, and S. J. Hultgren. 2007. Detection of intracellular bacterial communities in human urinary tract infection. *PLoS Med.* **4:**1949–1958.

Russo, T. A., and J. R. Johnson. 2000. Proposal for a new inclusive designation for extraintestinal pathogenic isolates of *Escherichia coli*: ExPEC. *J. Infect. Dis.* **181:**1753–1754.

Sauer, F. G., K. Futterer, J. S. Pinkner, K. W. Dodson, S. J. Hultgren, and G. Waksman. 1999. Structural basis of chaperone function and pilus biogenesis. *Science* **285:**1058–1061.

Sauer, F. G., J. S. Pinkner, G. Waksman, and S. J. Hultgren. 2002. Chaperone priming of pilus subunits facilitates a topological transition that drives fiber formation. *Cell* **111:**543–551.

Sauer, F. G., H. Remaut, S. J. Hultgren, and G. Waksman. 2004. Fiber assembly by the chaperone-usher pathway. *Biochim. Biophys. Acta* **1694:**259–267.

Saulino, E. T., D. G. Thanassi, J. S. Pinkner, and S. J. Hultgren. 1998. Ramifications of kinetic partitioning on usher-mediated pilus biogenesis. *EMBO J.* **17:**2177–2185.

Schaeffer, A. J., W. R. Schwan, S. J. Hultgren, and J. L. Duncan. 1987. Relationship of type 1 pilus expression in *Escherichia coli* to ascending urinary tract infections in mice. *Infect. Immun.* **55:**373–380.

Schilling, J. D., R. G. Lorenz, and S. J. Hultgren. 2002. Effect of trimethoprim-sulfamethoxazole on recurrent bacteriuria and bacterial persistence in mice infected with uropathogenic *Escherichia coli*. *Infect. Immun.* **70:**7042–7049.

Schlager, T. A., J. O. Hendley, J. A. Lohr, and T. S. Whittam. 1993. Effect of periurethral colonization on the risk of urinary tract infection in healthy girls after their first urinary tract infection. *Pediatr. Infect. Dis. J.* **12:**988–993.

Schwan, W. R., H. S. Seifert, and J. L. Duncan. 1992. Growth conditions mediate differential transcription of *fim* genes involved in phase variation of type 1 pili. *J. Bacteriol.* **174:**2367–2375.

Schwardt, O., S. Rabbani, M. Hartmann, D. Abgottspon, M. Wittwer, S. Kleeb, A. Zalewski, M. Smiesko, B. Cutting, and B. Ernst. 2011. Design, synthesis and biological evaluation of mannosyl triazoles as FimH antagonists. *Bioorg. Med. Chem.* **19:**6454–6473.

Schwartz, D. J., S. L. Chen, S. J. Hultgren, and P. C. Seed. 2011. Population dynamics and niche distribution of uropathogenic *Escherichia coli* during acute and chronic urinary tract infection. *Infect. Immun.* **79:**4250–4259.

Smith, Y. C., S. B. Rasmussen, K. K. Grande, R. M. Conran, and A. D. O'Brien. 2008. Hemolysin of uropathogenic *Escherichia coli* evokes extensive shedding of the uroepithelium and hemorrhage in bladder tissue within the first 24 hours after intraurethral inoculation of mice. *Infect. Immun.* **76:**2978–2990.

Sokurenko, E. V., V. Chesnokova, D. E. Dykhuizen, I. Ofek, X. R. Wu, K. A. Krogfelt, C. Struve, M. A. Schembri, and D. L. Hasty. 1998. Pathogenic adaptation of *Escherichia coli* by natural variation of the FimH adhesin. *Proc. Natl. Acad. Sci. USA* **95:**8922–8926.

Song, J., B. Bishop, G. Li, R. Grady, A. Stapleton, and S. Abraham. 2009. TLR4-mediated expulsion of bacteria from infected bladder epithelial cells. *Proc. Natl. Acad. Sci. USA* **106:**14966–14971.

Song, J., M. J. Duncan, G. Li, C. Chan, R. Grady, A. Stapleton, and S. Abraham. 2007. A novel TLR4-mediated signaling pathway leading to IL-6 responses in human bladder epithelial cells. *PLoS Pathog.* **3:**e60.

Spurbeck, R. R., A. E. Stapleton, J. R. Johnson, S. T. Walk, T. M. Hooton, and H. L. Mobley. 2011. Fimbrial profiles predict virulence of uropathogenic *Escherichia coli* strains: contribution of Ygi and Yad fimbriae. *Infect. Immun.* **79:**4753–4763.

Stamm, W. E., and T. M. Hooton. 1993. Management of urinary tract infections in adults. *N. Engl. J. Med.* **329:**1328–1334.

Stromberg, N., B. I. Marklund, B. Lund, D. Ilver, A. Hamers, W. Gaastra, K. A. Karlsson, and S. Normark. 1990. Host-specificity of uropathogenic *Escherichia coli* depends on differences in binding specificity to Gal alpha 1-4Gal-containing isoreceptors. *EMBO J.* **9:**2001–2010.

Stromberg, N., P.-G. Nyholm, I. Pascher, and S. Normark. 1991. Saccharide orientation at the cell surface affects glycolipid receptor function. *Proc. Natl. Acad. Sci. USA* **88:**9340–9344.

Thumbikat, P., R. E. Berry, A. J. Schaeffer, and D. J. Klumpp. 2009a. Differentiation-induced uroplakin III expression promotes urothelial cell death in response to uropathogenic *E. coli*. *Microbes Infect.* **11:**57–65.

Thumbikat, P., R. E. Berry, G. Zhou, B. Billips, R. Yaggie, T. Zaichuk, T. Sun, A. J. Schaeffer, and D. J. Klumpp. 2009b. Bacteria-induced uroplakin signaling mediates bladder response to infection. *PLoS Pathog.* **5:**e1000415.

Totsika, M., S. A. Beatson, S. Sarkar, M. Phan, N. Petty, N. Bachmann, M. Szubert, H. Sidjabat, D. Paterson, M. Upton, and M. A. Schembri. 2011. Insights into a multidrug resistant *Escherichia coli* pathogen of the globally disseminated ST131 lineage: genome analysis and virulence mechanisms. *PLoS One* **6:**e26578.

Uhlin, B. E., M. Nogren, M. Baga, and S. Normark. 1985. Adhesion to human cells by *Escherichia coli* lacking the major subunit of a digalactoside-specific pilus. *Proc. Natl. Acad. Sci. USA* **82:**1800–1804.

Vejborg, R. M., V. Hancock, M. A. Schembri, and P. Klemm. 2011. Comparative genomics of *Escherichia coli* strains causing urinary tract infections. *Appl. Environ. Microbiol.* **77:**3268–3278.

Verger, D., E. Miller, H. Remaut, G. Waksman, and S. Hultgren. 2006. Molecular mechanism of P pilus termination in uropathogenic *Escherichia coli*. *EMBO Rep.* **7:**1228–1232.

Virkola, R. 1987. Binding characteristics of *Escherichia coli* type 1 fimbriae in the human kidney. *FEMS Microbiol. Lett.* **40:**257–262.

Virkola, R., B. Westerlund, H. Holthöfer, J. Parkkinen, M. Kekomäki, and T. K. Korhonen. 1988. Binding characteristics of *Escherichia coli* adhesins in human urinary bladder. *Infect. Immun.* 56:2615–2622.

Waksman, G., and S. J. Hultgren. 2009. Structural biology of the chaperone-usher pathway of pilus biogenesis. *Nat. Rev. Microbiol.* 7:765–774.

Wang, H., G. Min, R. Glockshuber, T. Sun, and X. P. Kong. 2009. Uropathogenic *E. coli* adhesin-induced host cell receptor conformational changes: implications in transmembrane signaling transduction. *J. Mol. Biol.* 392:352–361.

Weissman, S. J., V. Beskhlebnaya, V. Chesnokova, S. Chattopadhyay, W. E. Stamm, T. M. Hooton, and E. V. Sokurenko. 2007. Differential stability and trade-off effects of pathoadaptive mutations in the *Escherichia coli* FimH adhesin. *Infect. Immun.* 75:3548–3555.

Welch, R. A., V. Burland, G. Plunkett III, P. Redford, P. Roesch, D. Rasko, E. L. Buckles, S. R. Liou, A. Boutin, J. Hackett, D. Stroud, G. F. Mayhew, D. J. Rose, S. Zhou, D. C. Schwartz, N. T. Perna, H. L. Mobley, M. S. Donnenberg, and F. R. Blattner. 2002. Extensive mosaic structure revealed by the complete genome sequence of uropathogenic *Escherichia coli. Proc. Natl. Acad. Sci. USA* 99:17020–17024.

Weyand, N. J., B. A. Braaten, M. van der Woude, J. Tucker, and D. A. Low. 2001. The essential role of the promoter-proximal subunit of CAP in *pap* phase variation: Lrp- and helical phase-dependent activation of *papBA* transcription by CAP from −215. *Mol. Microbiol.* 39:1504–1522.

Wright, K. J., and S. J. Hultgren. 2006. Sticky fibers and uropathogenesis: bacterial adhesins in the urinary tract. *Future Microbiol.* 1:75–87.

Wright, K. J., P. C. Seed, and S. J. Hultgren. 2007. Development of intracellular bacterial communities of uropathogenic *Escherichia coli* depends on type 1 pili. *Cell. Microbiol.* 9:2230–2241.

Wright, K. J., P. C. Seed, and S. J. Hultgren. 2005. Uropathogenic *Escherichia coli* flagella aid in efficient urinary tract colonization. *Infect. Immun.* 73:7657–7668.

Wu, X. R., J. H. Lin, T. Walz, M. Haner, J. Yu, U. Aebi, and T. T. Sun. 1994. Mammalian uroplakins. A group of highly conserved urothelial differentiation-related membrane proteins. *J. Biol. Chem.* 269:13716–13724.

Wu, X. R., T. T. Sun, and J. J. Medina. 1996. In vitro binding of type 1-fimbriated *Escherichia coli* to uroplakins Ia and Ib: relation to urinary tract infections. *Proc. Natl. Acad. Sci. USA* 93:9630–9635.

Xia, Y., K. Forsman, J. Jass, and B. E. Uhlin. 1998. Oligomeric interaction of the PapB transcriptional regulator with the upstream activating region of pili adhesin gene promoters in *Escherichia coli. Mol. Microbiol.* 30:513–523.

Xia, Y., D. Gally, K. Forsman-Semb, and B. E. Uhlin. 2000. Regulatory cross-talk between adhesin operons in *Escherichia coli*: inhibition of type 1 fimbriae expression by the PapB protein. *EMBO J.* 19:1450–1457.

Yakovenko, O., S. Sharma, M. Forero, V. Tchesnokova, P. Aprikian, B. Kidd, A. Mach, V. Vogel, E. Sokurenko, and W. E. Thomas. 2008. FimH forms catch bonds that are enhanced by mechanical force due to allosteric regulation. *J. Biol. Chem.* 283:11596–11605.

Zhanel, G. G., T. L. Hisanaga, N. M. Laing, M. R. DeCorby, K. A. Nichol, B. Weshnoweski, J. Johnson, A. Noreddin, D. E. Low, J. A. Karlowsky, and D. J. Hoban. 2006. Antibiotic resistance in *Escherichia coli* outpatient urinary isolates: final results from the North American Urinary Tract Infection Collaborative Alliance (NAUTICA). *Int. J. Antimicrob. Agents* 27:468–475.

Zhang, J. P., and S. Normark. 1996. Induction of gene expression in *Escherichia coli* after pilus-mediated adherence. *Science* 273:1234–1236.

Regulation of Bacterial Virulence
Edited by Michael L. Vasil and Andrew J. Darwin
© 2013 ASM Press, Washington, DC   doi:10.1128/9781555818524.ch8

Chapter 8

Phasevarions: an Emerging Paradigm in Epigenetic Gene Regulation in Host-Adapted Mucosal Pathogens

Yogitha N. Srikhanta, Ian R. Peak, and Michael P. Jennings

INTRODUCTION

The ability of host-adapted bacteria to successfully colonize and survive within a new host depends on how rapidly they can adapt to hostile and constantly changing host environments as well as effectively evade the host immune system (Moxon and Thaler, 1997; van der Woude and Baumler, 2004). To address these challenges, some bacterial pathogens such as *Haemophilus influenzae*, *Neisseria gonorrhoeae*, *Neisseria meningitidis*, and *Helicobacter pylori* have evolved molecular mechanisms for rapidly generating genetic variation in highly mutable individual genes or "contingency" loci (Moxon et al., 1994). This strategy is called phase variation, the high-frequency reversible ON/OFF switching of gene expression, which is a common feature of many virulence determinants or factors associated with colonization in bacterial pathogens (Weiser et al., 1990; van Ham et al., 1993; Hallet 2001). The independent, random switching of these contingency loci results in phenotypically diverse bacterial populations, increasing the overall fitness of the population, as a subpopulation will express a different subset of the contingency genes. This subpopulation provides the variants that will enable rapid adaptation to changing host microenvironments or counteract host immune responses (Moxon et al., 1994; Bayliss et al., 2001). These changes are heritable and reversible between generations, and the frequency of switching is characteristic for the gene, the bacterial species, and the regulatory mechanism (van der Woude and Baumler, 2004).

Phase variation is associated with a number of different systems and can be generated by changes in DNA sequence and/or DNA structure (Dybvig, 1993; Moxon et al., 1994; Moxon and Thaler, 1997; Henderson et al., 1999). The first phase-variable systems to be understood in molecular detail involve invertible DNA

elements controlling the phase-variable expression of flagella in *Salmonella* (Kutsukake and Iino, 1980) and later elucidated in other organisms (Dybvig, 1993). This was followed by well-documented studies in epigenetic gene regulation where phase-variable gene expression of several *Escherichia coli* adhesins, such as Ag43 (Henderson and Owen, 1999), is controlled through the methylation of specific DNA sequences by DNA adenine methyltransferases (Dam) (Casadesus and Low, 2006; Wion and Casadesus, 2006). Several host-adapted bacterial mucosal pathogens have also evolved molecular mechanisms for rapidly generating genetic variation through high mutation rates in contingency loci, which are controlled by simple sequence DNA repeats, including repetitive single nucleotides or short tandem motifs (Moxon et al., 1994). The first report of repeat-mediated phase variation in a pathogen was from *Bordetella pertussis*, where changes in a homopolymeric tract of C residues in a two-component regulator gene mediate flagellar switching (Stibitz et al., 1989). The independent, random switching of genes which typically encode surface structures results in a large number of alternate combinations of surface components. Interestingly, *H. influenzae*, *N. meningitidis*, *N. gonorrhoeae*, and *H. pylori* have relatively small genome sizes but possess an abundance of loci containing tandem DNA repeats (Hood et al., 1996; Saunders et al., 2000). Most phase-variable genes encode surface-exposed proteins or synthesis of surface glycans. However, phase variation via simple tandem repeats has been identified in methyltransferase genes associated with restriction-modification (R-M) systems, which are not associated with the biosynthesis of surface antigens.

It was recently shown that in the human pathogens *H. influenzae*, *N. gonorrhoeae*, *N. meningitidis*, and *H. pylori* (Srikhanta et al., 2005; Srikhanta et al., 2009; Srikhanta et al., 2010; Srikhanta et al., 2011), phase

Yogitha N. Srikhanta • Department of Microbiology and Immunology, The University of Melbourne, Royal Parade, Parkville, Melbourne, VIC 3010, Australia.   **Ian R. Peak and Michael P. Jennings** • The Institute for Glycomics, Griffith University, Gold Coast Campus, Parklands Drive, Southport, QLD 4222, Australia.

variation of a type III DNA methyltransferase, encoded by the *mod* gene, regulates the expression of multiple genes. This novel genetic system has been termed the "phasevarion" (phase-variable regulon). The wide distribution of phase variable *mod* genes indicates that this may be a common strategy used by host-adapted bacteria to randomly switch between distinct cell types.

This chapter discusses the molecular mechanisms of phase variation and the possible roles of phase variable R-M systems in bacterial pathogens and reveals how a number of phase-variable type III R-M systems have evolved to have a new and distinct function in gene regulation that results in generation of a diverse bacterial population. We also detail the phase variable regulon, or phasevarion, of genes controlled by phase-variable type III R-M systems in four pathogens: *H. influenzae*, *N. meningitidis*, *N. gonorrhoeae*, and *H. pylori*.

MECHANISMS OF PHASE VARIATION

There are a number of mechanisms capable of mediating phase variation. Refer to Table 1 for a summary of phase variation mechanisms identified to date, highlighting the advantage this molecular mechanism confers to pathogenic bacteria (Robertson and Meyer, 1992; Hallet, 2001; van der Woude and Baumler, 2004; van der Woude, 2006). Phase variation via simple tandem repeats is by far the most common mechanism of phase variation.

PHASE VARIATION VIA SIMPLE TANDEM REPEATS

Phase variation can be directed by simple tandem repeats that are located in the open reading frame (ORF) or promoter region of genes (Levinson and Gutman, 1987; Chandler and Fayet, 1993; van Belkum et al., 1998) (Fig. 1A). The repeat tracts range from short homopolymeric tracts of seven nucleotides to di-, tetra-, and pentanucleotide repeats of several hundred base pairs. Multiple contiguous repeats of units of DNA sequence can undergo expansion and contraction in the number of repeats. The gain or loss of repeat units in these homo- or heteropolymeric tracts is thought to involve a mechanism known as slipped-strand mispairing. Slipped-strand mispairing can occur during chromosomal replication or during DNA repair and recombination, processes that require DNA synthesis, and is independent of RecA (Hallet, 2001). The result is altered expression of a gene and ultimately its product if the location of the repeats is such that transcription or translation is affected (Levinson and Gutman, 1987; Chandler and Fayet, 1993; van Belkum et al., 1998). Tracts situated within promoters can influence the degree of gene expression by regulating transcription, modifying the relative positioning of the RNA polymerase-binding sites within the promoter. For example, in *N. meningitidis* strain MC58, a change in the polycytosine repeat number in the sequence of the –10 region of the promoter of the *opc* gene, which encodes an outer membrane protein, affects the transcription of this gene (Fig. 1B). Repeat tracts located upstream or downstream of the promoter can also alter gene expression. For example, *nadA* contains a tract of TAA repeats upstream of the –35 region. Phase variation of *nadA* results from altered interactions of the transcription factor IHF (integration host factor) and its DNA binding site (Martin et al., 2005). Changes in repeat length alter the relationship between the regulators bound to their target DNA and the *nadA* promoter region. It is thought that changes in repeat lengths outside of the promoter area may affect the

Table 1. Phase variation mechanisms

Mechanism	Description	Example
Site-specific DNA rearrangements	Site-specific inversion of DNA segments results in changes in gene expression by altering the spatial relationship of promoters or regulatory elements with the genes they affect.	*hsd1* and *hsd2* type I R-M systems (Dybvig et al., 1998; Henderson et al., 1999; Hallet, 2001)
Dam methylation controlled	The presence of methylation sites within a promoter regulates the binding of regulatory proteins, altering transcription of the gene.	Antigen 43-OxyR (van der Woude et al., 1996; Heithoff et al., 1999; Haagmans and van der Woude, 2000; Low et al., 2001)
Homologous recombination	DNA recombination that results in complete or partial replacement of one expressed recipient gene with variable DNA segments from a silent copy located in a different part of the genome	*pilE* and *pilS* (Seifert, 1996; Mehr and Seifert, 1998; Hallet, 2001)
Insertional sequences	Insertion sequence elements which can site specifically reversibly insert into ORFs, disrupting expression	*siaA* (Hammerschmidt et al., 1996)
Simple tandem repeats	Changes in gene expression mediated by simple tandem repeats, located in the ORF or promoter region of genes	*pglA* (Jennings et al., 1998; Hallet, 2001) (Fig. 1)

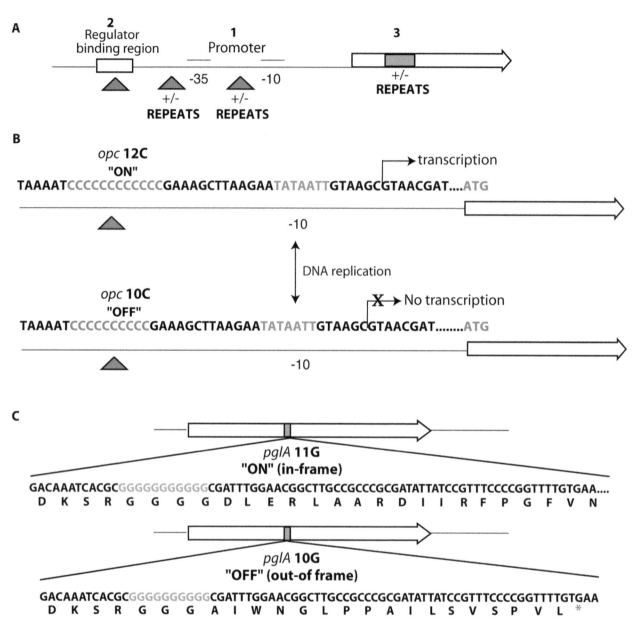

Figure 1. Phase variation as a result of slipped-strand mispairing in simple tandem repeats and its effects on gene transcription and translation. (A) Repeat sequences in the promoter region (regions 1 and 2) or within a gene (region 3) can lead to phase variation by effecting transcription initiation and translation. (B) The presence of an ON number of homopolymeric tract repeats [poly(C) tract] in the promoter region of the *opc* gene of *N. meningitidis* enables transcription to proceed. A loss of repeat units modifies the spacing between the –35 promoter and –10 promoter sequence preventing transcription initiation (Sarkari et al., 1994). (C) Effect on the translation product of a one-unit deletion due to slipped-strand mispairing in the homopolymeric tract repeat sequence [pol(G) tract] in the coding sequence of the *pglA* gene of *N. meningitidis*. A deletion changes the reading frame, which results in a premature stop codon (asterisk), leading to the expression of a truncated form of the protein (Jennings et al., 1998). Adapted from van der Woude and Baumler, 2004.
doi:10.1128/9781555818524.ch8f1

expression or stability of the transcript, by promoting the formation of specific secondary structures in RNA or DNA or by interfering with the binding of putative regulators. Tracts within coding regions result in ON/OFF switching according to whether the downstream sequences are moved in or out of frame

for expression, resulting in a fully functional protein or a truncated, nonfunctional protein (e.g., *pglA* [Fig. 1C]) (Moxon et al., 1994; van Belkum et al., 1998, Henderson et al., 1999). These random changes in repeats, then, mediate independent phase variation of individual genes (Fig. 2A).

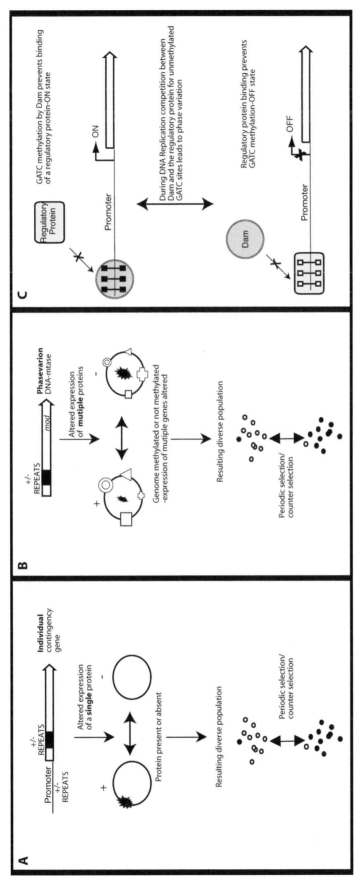

Figure 2. Schematic representation of differences in the phase variation properties of individual contingency genes, the phasevarion and Dam methylation. (A) Phase variation of an individual gene. Phase variation via changes in repeat length affects translation (repeats within the gene) or transcription initiation (repeats within the promoter region), leading to reversible, altered expression of a single protein and resulting in the presence or absence of that protein. Random switching of many individual genes leads to a large number of alternate combinations of surface components, resulting in diverse populations. (B) Phase variation via Dam methylation. During DNA replication, competition between Dam and a DNA binding regulatory protein forms DNA methylation patterns that control gene expression at a target site. The target site's methylation state affects the DNA binding of a regulatory protein, which directly regulates transcription. (C) Phasevarion (mod)-mediated control of multiple genes. Phase variation via changes in repeat length within the mod gene results in altered expression of genes that contain a specific sequence recognized by mod, affecting their transcriptional control. Thus, multiple genes can be under the control of the phasevarion, depending on the methylation state of the genome, resulting in diverse populations. Methylated sites are indicated by black squares and unmethylated sites by white squares. Black shapes represent increase gene expression and white shapes represent decreased gene expression. doi:10.1128/9781555818524.ch8f2

DNA METHYLATION AND GENE REGULATION IN BACTERIAL PATHOGENS

Phase variation mediated by DNA methylation is different from the mechanisms described above. While these mechanisms result from changes in the genome, DNA methylation is epigenetic, meaning that while the phenotype differs the DNA sequence remains unaltered (Casadesus and Low, 2006; Wion and Casadesus, 2006).

The DNA-methylating enzyme deoxyadenosine methyltransferase (Dam) is involved in phase variation of specific virulence genes in *E. coli* and *Salmonella* (Blyn et al., 1990; Haagmans and van der Woude, 2000; Nicholson and Low, 2000). Dam specifically methylates the adenine residue of GATC sequences. Competition between the Dam methylase and a particular DNA-binding regulatory protein forms DNA methylation patterns (see Fig. 2C) that control gene expression at a single target site (Low et al., 2001; van der Woude and Baumler, 2004). Examples include Ag43, an outer membrane protein involved in autoaggregation and the formation of biofilms in *E. coli* (Henderson and Owen, 1999; Danese et al., 2000; Haagmans and van der Woude, 2000) and the pyelonephritis-associated pilus (*pap*) operon in uropathogenic *E. coli* (Blomfield, 2001; Blyn et al., 1990; Nou et al., 1995; Weyand and Low, 2000). Phase variation of Ag43 is regulated at the transcriptional level by Dam and by the regulator OxyR, which activates genes involved in oxidative stress. OxyR is a repressor of *agn43* transcription, while Dam methylation activates *agn43* transcription. OxyR prevents methylation by binding to three unmethylated Dam target (GATC) sequences found in the OxyR binding site in the *agn43* regulatory region (Henderson and Owen, 1999; Henderson et al., 1999; Haagmans and van der Woude, 2000). This prevents Dam from accessing the target site resulting in the OFF state. However, when the GATC sequences are methylated, OxyR is prevented from binding and expression is in the ON state. Switching from the ON to the OFF state requires OxyR binding to hemimethylated sites following DNA replication, which competes with the hemimethylated GATC-binding protein SeqA (Correnti et al., 2002). SeqA binding is transient and enables GATC methylation, allowing the ON state. OFF-to-ON switching occurs when OxyR leaves the OxyR binding site, enabling GATC methylation. Besides preventing OxyR binding, Dam methylation also methylates the GATC sites, increasing transcription initiation (Wallecha et al., 2002). In this example the methylation pattern modulates expression of a single gene.

PHASE-VARIABLE R-M SYSTEMS

The identification of repeats within genes encoding methyltransferases suggested that coordinate methylation of multiple sites within a bacterial genome could be phase variable. Most phase-variable genes encode surface-exposed proteins or proteins involved in production of surface glycans, and such phase-variable genes may allow immune evasion, or the stochastic adaptation to changing adverse host environments (Moxon et al., 2006). The presence of simple tandem repeats in methyltransferase genes associated with R-M systems suggests that these restriction systems are also phase variable. That is, phase variation also occurs in genes not obviously involved in production of surface exposed structures. R-M systems are ubiquitous in bacteria and are mechanisms to identify and cleave foreign DNA, classically represented as protection against viral (phage) infection (Bickle et al., 1978). R-M systems are classified into three groups—types I, II, and III—on the basis of subunit composition, cleavage position, sequence specificity, and cofactor requirements (Boyer, 1971). Type III R-M systems are composed of a methyltransferase gene (modification, *mod*) and an endonuclease gene (restriction, *res*), whose products form a two-subunit enzyme: Mod and Res (Bourniquel and Bickle, 2002). Res must form a complex with Mod to be functional (Meisel et al., 1995); however, Mod can function independently of Res (Bachi et al., 1979). The Mod subunit contains several conserved motifs in the N- and C-terminal regions and central region that contains the DNA recognition domain that dictates sequence specificity (Humbelin et al., 1988).

The identification of phase-variable type III R-M systems led to discussion of their significance in the biology of bacterial pathogens, with several functions proposed, including roles in either DNA restriction activity or gene regulation.

A Barrier to Bacteriophage or Genetic Exchange via Natural Transformation

The traditional role of R-M systems is to protect bacterial cells against foreign DNA. As many organisms that contain phase-variable type III R-M systems are naturally competent, a phase-variable R-M system might allow potentially beneficial sequences to be utilized by provisional removal of this barrier (Ando et al., 2000; Donahue et al., 2000). In the case of *N. meningitidis*, *N. gonorrhoeae*, and *H. influenzae*, DNA uptake is predominantly, but not exclusively, restricted to DNA containing genus-specific uptake sequences and is the primary route of genetic exchange. The mosaic structure of many genes in these bacteria indicates a history of genetic exchange and recombination, possibly aided by phase-variable R-M systems. For example, *N. gonorrhoeae* utilizes inter- and intragenomic exchange as a mechanism to generate antigenic variation of the pilin subunit of type 4 fimbriae (Gibbs et al., 1989). Altering methylation patterns may regulate the efficiency and frequency of lateral DNA transfer within and between bacterial populations (Hallet, 2001), but this is yet to be experimentally proven for phase-variable type III R-M systems. There is a suggestion that the variable presence of restriction systems (including phase-variable type III R-M systems) is, in part, responsible for the population structure of *N. meningitidis* (Budroni et al., 2011).

Autolytic Self-DNA Degradation for Uptake by Cells

It has also been suggested that phase variation of methyltransferases may lead to autolytic self-DNA degradation by the cognate restriction enzyme, leading to cell death (Dybvig et al., 1998; Saunders et al., 1998; Hamilton and Dillard, 2006). For bacterial species possessing natural transformation systems, "bacterial suicide" by a proportion of the population would release DNA into the environment for uptake by other cells and the selective advantage of the population (Dybvig et al., 1998; Saunders et al., 1998; Hamilton and Dillard, 2006).

However, Res subunits of type I and III R-M systems are inactive without Mod. For type I R-M systems, a complex of HsdR2M2S is required for endonuclease activity (Buhler and Yuan, 1978; Willcock et al., 1994; Doronina and Murray, 2001). In addition, some type I systems can control the endonuclease activity posttranslationally (Dryden et al., 2001) through ClpXP-mediated degradation of the endonuclease (HsdR) during DNA translocation, preventing restriction from proceeding to completion (Makovets et al., 1999). This leaves a complex of HsdMS that retains only methylation activity (Murray, 2000), which provides rapid and effective protection of self-DNA should it acquire unmodified target sequences (Dryden et al., 2001). In the case of type III enzymes, Mod can form a stable homodimer that can function as a methyltransferase (Hadi et al., 1983; Ahmad and Rao, 1994) independent of the number and orientation of recognition sites. However, Res is functional only when complexed with Mod (Dryden et al., 2001) and is degraded when expressed alone (Redaschi and Bickle, 1996). Suicidal restriction of self-DNA is prevented if unmodified recognition sites are in the same orientation or paired with methylated sites (Meisel et al., 1992). In type III R-M systems, methylation is favored over restriction.

Arber et al. demonstrated that modification by EcoP1I was detectable within a few minutes after P1 bacteriophage infection, as opposed to restriction, which was evident after ~1 h (Arber et al., 1975).

For organisms with phase-variable R-M systems, the implications of *mod* switching with respect to restriction function are unclear. If a phase-variable *mod* gene associated with an active type III restriction system switches from OFF to ON, after several rounds of bacterial replication, the result may be autorestriction and death of the bacterium, as unmethylated targets in the bacterial chromosome would be vulnerable to cleavage. However, in many phase-variable R-M systems Res has been found to be inactive (see below and Srikhanta et al., 2010). Our observations with *Neisseria gonorrhoeae* indicate that ON/OFF switching of the *mod* gene is not lethal to the cell (Srikhanta et al., 2009), but the possibility remains that such a situation may be deleterious in another background. Non-phase-variable type III R-M systems are regulated in various ways, possibly providing alternate routes to prevent autorestriction. In *E. coli*, the EcoP1I and EcoP15I genes are freely transferable between cells by phage infection, transformation, and conjugation, indicating that they have no deleterious effect on the cell. In these systems Res is controlled both at the translational level and by Mod (Redaschi and Bickle, 1996). Bourniquel and Bickle have suggested that dimerization follows DNA translocation in EcoP15I and that cooperation between EcoP15I proteins, rather than DNA translocation, may be important for cleavage (Bourniquel and Bickle, 2002). Conversely, horizontal transfer of the chromosomally located *Salmonella enterica serovar* Typhimurium StyLTI system results in death of the recipient cell (Arber and Dussoix, 1962; Arber, 1974; De Backer and Colson, 1991). The PstII type III system of *Providencia stuartii* is also chromosomally encoded and not as freely transferable as EcoPI and EcoPI5I, despite similarity in their enzyme activities (Sears et al., 2005). It is proposed PstII R-M activity is controlled through subunit assembly and specific protein-protein contacts are needed to activate endonuclease activity (Sears and Szczelkun, 2005). It is not known if the molecular mechanisms that control expression of phase-variable type III R-M systems are similar to those for the non-phase-variable systems described above.

Gene Regulation

We proposed a role for phase-variable type III R-M systems in gene regulation via differential methylation of the genome (Seib et al., 2002), as it is well established that Dam-mediated methylation can affect gene expression (Blyn et al., 1990; Haagmans and van der Woude, 2000; Nicholson and Low, 2000). Furthermore, a phase-variable methyltransferase could be involved in pathogenicity by randomizing virulence factor expression via global changes in methylation.

Epigenetic gene regulation through the methylation of specific DNA sequences by methyltransferases is a common phenomenon (Casadesus and Low, 2006; Wion and Casadesus, 2006; Marinus and Casadesus, 2009). The best-studied example is methylation by the Dam methyltransferase, which is important for bacterial virulence in *Salmonella* species. Mutations in *dam* lead to attenuation, and strains that carry these mutations have been proposed as live vaccine candidates (Garcia-Del Portillo et al., 1999; Heithoff et al., 1999; Heithoff et al., 2001; Pucciarelli et al., 2002; Balbontin et al., 2006; Prieto et al., 2007; Chatti et al., 2008; Jakomin et al., 2008). In addition, mutations in Dam attenuate the virulence of several other pathogens (Julio et al., 2001; Julio et al., 2002; Chen et al., 2003; Taylor et al., 2005; Campellone et al., 2007; Falker et al., 2007; Mehling et al., 2007; Kim et al., 2008; Murphy et al., 2008). However, the Dam phenotypes used in these studies are presumed to be a consequence of missense (e.g., *dam-3*), insertion, or deletion mutations, and in contrast to the phase-variable type III methyltransferases, the Dam methyltransferase itself does not phase vary and is not known to be regulated.

As mentioned above, Dam methylation is also involved in phase variation of specific virulence genes in *E. coli* and *Salmonella* (Blyn et al., 1990; Haagmans and van der Woude, 2000; Nicholson and Low, 2000). The fundamental characteristic of these DNA methylation-dependent phase-variable systems is that the methylation state of the target site affects the DNA binding of a regulatory protein, which directly regulates transcription. As stated above, Dam does not phase vary, nor are there any examples of Dam itself being regulated by phase variation.

Our recent work supports a role for phase-variable type III systems in gene regulation. Alteration of expression of a type III DNA methyltransferase in *H. influenzae* strain Rd changes expression of multiple genes (Srikhanta et al., 2005). This novel genetic system, termed the phasevarion, regulates gene expression in at least four major bacterial pathogens: *H. influenzae*, *N. meningitidis*, *N. gonorrhoeae*, and *H. pylori*.

THE PHASEVARION

Haemophilus influenzae

In our initial study, microarray expression analysis comparing a wild-type *H. influenzae* strain Rd expressing *modA1* with a *modA1* knockout mutant strain revealed changes in expression of 15 genes (Srikhanta et al., 2005). Differential methylation of the genome in the *modA* ON and OFF states leads to changes in expression of multiple genes, effectively differentiating the bacteria into two distinct cell types, creating two phenotypically distinct populations (Fig. 2B). In previous examples of phase variation, phase-variable expression mediated by simple tandem repeats has been limited to the particular gene associated with the repeats (Fig. 2A).

Of the 15 genes influenced by the *modA1* switching in *H. influenzae* strain Rd, 7 genes were upregulated in the *modA1* mutant relative to the wild type. These included two genes encoding outer membrane proteins. One of these is *opa*, encoding an outer membrane protein orthologous to Opa adhesins of *Neisseria*. An *opa::lacZ* fusion demonstrated a direct link between ON/OFF status of the *modA1* gene and thereby the methylation state of the genome, and gene expression (Fig. 3). Genes encoding heat shock proteins HtpG, DnaK, DnaJ, GroES, and GroEL, which have important roles in cell physiology as molecular chaperones, were also within the regulon. Transient induction of heat shock proteins is a vital protective mechanism to cope with various sources of physiological and environmental stress at the cellular level. Evidence suggests that one of these chaperones, DnaK, may play a direct role in the pathogenesis of *H. influenzae* by altering the glycolipid binding specificity of *H. influenzae* following heat shock (Hartmann and Lingwood, 1997; Hartmann et al., 2001). Therefore, in this example, phase variation of *mod* can enhance or ameliorate the heat shock response. It is interesting to note that a previous study using signature-tagged mutagenesis on *H. influenzae* strain Rd b+ identified HI1056 (*modA1*) as one of 25 genes potentially essential for invasive disease (Herbert et al., 2002).

Neisseria meningitidis and Neisseria gonorrhoeae

Studies in the pathogenic *Neisseria* revealed that each strain contained a *modA* gene, like *H. influenzae*; however, the DNA recognition domain dictating the target site for methylation was distinct. Three main *modA* alleles were found: *modA11*, *-12*, and *-13* (Fig. 4). Investigating the *modA11* phasevarion of *N. meningitidis* revealed altered expression of 80 genes, of which 5 encode surface-exposed proteins, including lactoferrin binding proteins A and B (LbpA and LpbB) (Srikhanta et al., 2009), which allow acquisition of iron from lactoferrin-containing compounds. Lactoferrin is an important source of iron for some pathogenic bacteria (Pettersson et al., 1993; Pettersson et al., 1998). This is an important

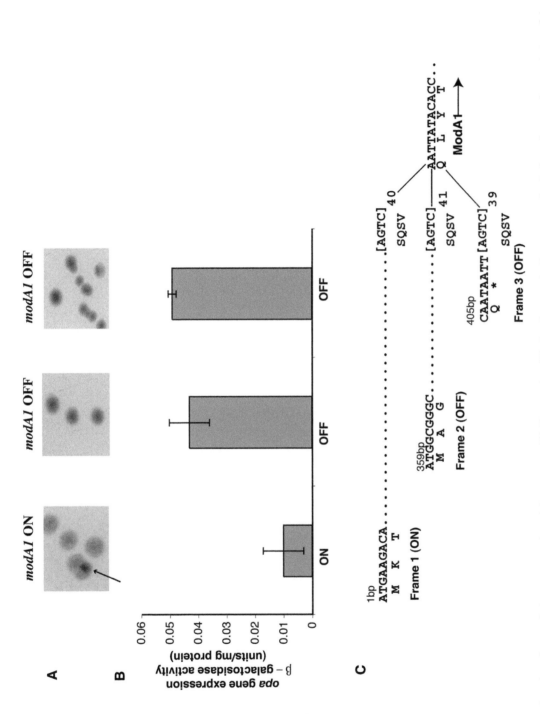

Figure 3. Effect of *modA1* phase variation on expression of the *opa* gene. (A) Phenotypic validation that *opa::lacZ* gene expression is dependent on phase variation of the *modA1* gene. R*dopa::lacZ* colonies with the *mod* 5'-AGTC-3' repeat tract in frame with the ON ATG (resulting in active Mod) were white, indicating low *opa::lacZ* expression. Colonies which phase varied to a blue phenotype (example indicated with an arrow) were observed and picked, and the *modA1* repeat region was sequenced to determine if change in *opa::lacZ* expression correlated with *modA1* phase variation. All blue colonies were found to have switched from ON (40 repeats) to be in frame with either the OFF with 41 repeats or OFF with 39 repeats. All colonies that switched back from blue to white were found to be in frame (40 repeats). (B) Beta-galactosidase assays showing quantitative differences in the level of *opa::lacZ* gene expression resulting from mod repeat tract changes (ON, OFF, or OFF). A fivefold difference in expression was observed between ON and OFF. (C) Schematic diagram showing that translation of the *modA1* gene is initiated from one of three frames (ON [40], OFF [39 or 41]) depending on the number of 5'-AGTC-3' repeats. doi:10.1128/9781555581824.ch8f3

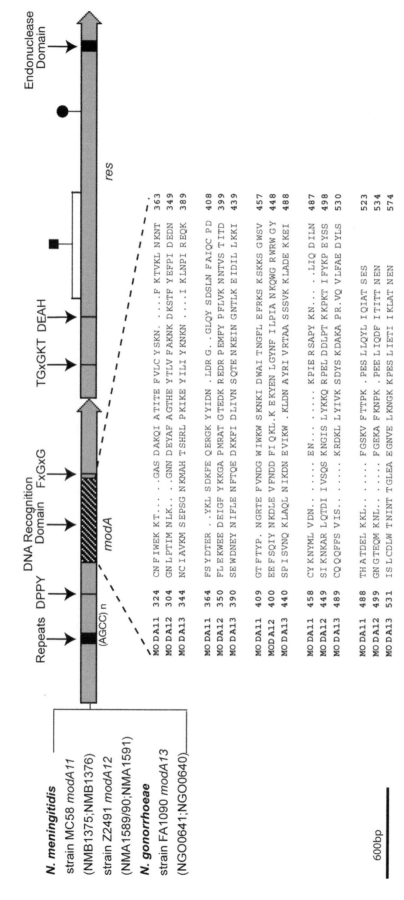

Figure 4. *modA* genes of *N. meningitidis* and *N. gonorrhoeae.* The methylase (*mod*) genes, restriction endonuclease (*res*) genes, and repeat regions that mediate phase variation are indicated. Also shown are the conserved, characteristic motifs found within type III R-M systems, which include in *mod* the catalytic region (DPPY) and the AdoMet (methyl donor) binding pocket (FXGXG) (Ahmad and Rao, 1994; Malone et al., 1995), and in *res* the ATP binding motif (TGxGKT), the motif linked to ATP hydrolysis (DEAH), and the endonuclease domain (Gorbalenya and Koonin, 1991; Pingoud and Jeltsch, 1997; Saha et al., 1998). The *mod* and *res* genes are colored to indicate differences in homology between both *mod* genes and both *res* genes, respectively. A variable region within *mod* (highlighted in stripes) contains the DNA recognition domain (Humbelin et al., 1988). Strains and accession numbers that define the *mod* alleles are shown to the left. n, indicates the number of repeats. A black circle on a line and black square on a line indicate the positions of a frameshift mutation and large deletion, respectively. doi:10.1128/9781555818524.ch8f4

observation, as LbpAB are key surface antigens and potential vaccine candidates that had not previously been known to have any capacity for ON/OFF expression. The other genes have various functions, including DNA repair, electron transport, amino acid transport, and growth. Some of these genes have previously been shown to be regulated by Fur, an iron-responsive transcription factor (Srikhanta et al., 2009).

The *modA13* gene of *N. gonorrhoeae* strain FA1090 controls a phasevarion of 17 genes. Five of these genes have obvious roles in virulence: four function in oxidative stress and one in antimicrobial susceptibility (Srikhanta et al., 2009). Functional studies confirmed that a *modA13* mutant has distinct phenotypes in antimicrobial resistance and in a primary human cervical epithelial cell model of infection. In addition, the *modA13* mutant formed biofilms more robustly than the wild type. These observations are consistent with the random generation of two cell types with distinct niche specialization.

Helicobacter pylori

A small number of genes were found to be under the control of the *modH5* phasevarion associated with *H. pylori* strain P12 (Srikhanta et al., 2011). These include two genes encoding proteins that have important roles in motility, FlaA and FliK. Motility is an essential factor for the colonization and persistence of *H. pylori* in the human stomach (Eaton et al., 1996), and consequently, flagella have an important role in virulence. In addition, *H. pylori* FlaA has a low intrinsic capacity to activate innate immunity via Toll-like receptor 5 (Lee et al., 2003; Gewirtz et al., 2004; Andersen-Nissen et al., 2005). Therefore, altered expression of flagella may be advantageous for the adaptation of *H. pylori* to alternate host environments and in evading the host immune response. The gene encoding the essential outer membrane protein for colonization, HopG (Kavermann et al., 2003), is also regulated by the *modH5* phasevarion. Bacterial adherence mediated by HopG and outer membrane proteins is thought to play an important role in the colonization of the gastric epithelium by *H. pylori* (Bernarde et al., 2010), making HopG an attractive vaccine target (Peck et al., 2001). Hence, phasevarion-mediated phase variation of *hopG* has the potential to mediate escape from the host immune response.

COMMON FEATURES

These examples from several host-adapted bacteria that colonize mucosal surfaces of humans, and have the capacity to cause disease, highlight the significant capacity that phase variation of methylases has to mediate coordinated expression of multiple genes, allowing a stochastic switch from one cellular phenotype to another. It has been noted that different *mod* alleles exist for some organisms. In each case the *mod* alleles differ in their putative DNA recognition domains, suggesting that each phase-variable *mod* allele regulates expression of a different phasevarion. Alleles may vary between strains, so that the phenotypic diversity of each strain may be constrained by the *mod* allele that it contains. Importantly, a distinct *mod* is associated with a hypervirulent clonal lineage of meningococci, and its phasevarion includes genes suggested to be virulence factors (Seib et al., 2011). As well as interstrain differences in *mod*, there are several examples of multiple phase-variable *mod* alleles within a single strain (Fig. 4) (Srikhanta et al., 2009; Srikhanta et al., 2010; Srikhanta et al., 2011). The presence of multiple phase-variable *mod* alleles suggests the possibility of distinct phasevarions existing within each strain, each regulating a different set of genes. Thus, in organisms with multiple phasevarions switching independently, multiple phenotypically different cell types can be generated.

Further evidence that type III Res-Mod systems with phase-variable *mod* do not have a traditional restriction function can be inferred from the fact that many of them have inactivating mutations in the *res* gene. To date, the majority of studies on the evolution of phase-variable type III R-M systems have focused on restriction function and have assumed that these are functional DNA restriction systems (Bayliss et al., 2006). However, restriction activity of the type III R-M system has been lost or inactivated in several strains of *H. influenzae*, *N. meningitidis*, and *N. gonorrhoeae* (Fox et al., 2007; Srikhanta et al., 2009). In *H. influenzae*, 20% of strains that contain a phase-variable *mod* gene have a frameshift mutation within the *res* gene that results in premature truncation of Res and loss of the endonuclease domain (Fox et al., 2007). No such inactivating mutations were identified in 15 strains in which the *mod* gene is non-phase variable (Fox et al., 2007). Inactivating frameshift mutations have been identified in 70% of the *res* genes associated with the *modA11* allele of *N. meningitidis* (Fig. 4). In *N. gonorrhoeae*, a 250-codon deletion that potentially inactivates the restriction function has been observed in Res associated with the *modA13* and *modA12* alleles (Fox et al., 2007; Srikhanta et al., 2009). Frameshift mutations in *res* have also been found in *modH* alleles of *H. pylori* as well as *H. pylori* AG1 (HP1315/16), while in *Mycoplasma hyopneumoniae* J (MHJ0423), *mod* is not associated with a *res* gene. Along with these obvious mutations, many other possible silencing mechanisms may exist.

It is clear that inactivation of phase-variable type III restriction function is common in many bacterial pathogens. This could be because type III Res function is redundant due to the presence of other R-M systems in the cell, thereby providing adequate protection from invasion by foreign DNA. Alternatively, Res activity might be incompatible or harmful to the cell in the context of a randomly switching *mod* gene due to autolytic self-DNA degradation when the R-M system is switched from OFF to ON. Importantly, the fact that Res function has been lost or inactivated in many strains indicates that in these cases, DNA restriction is clearly not the function of the phase-variable type III R-M system.

MECHANISM OF ACTION

Although the cognate DNA sequence of *mod* has been identified for some examples, exactly how methylation mediates changes in gene expression for any one phasevarion has not been demonstrated. For example, the genome of the *N. gonorrhoeae* strain FA1090 contains 5,135 copies of the ModA13 recognition site sequence (5'-AGAAA-3') (Srikhanta et al., 2009). Four of the genes and operons (*recN*, *metF*, *metE*, and *mtrF*) that were regulated by *modA13* phase variation contain ModA13 recognition sites within their upstream intergenic regions, although their role in Mod-mediated regulation of these genes has not yet been tested and remains an area requiring further research.

GENE REGULATION BY OTHER PHASE-VARIABLE R-M SYSTEMS

The *mod* gene, associated with a number of type II R-M systems in *H. pylori*, also contains repeats, indicating that this system may also phase vary (Lin et al., 2001). Recently, methylation by a putative phase-variable *mod* gene associated with a type II R-M system in *H. pylori* (M.HpyAIV) has been shown to influence gene expression of *katA* (Skoglund et al., 2007). The M.HpyAIV gene has been associated with the induction of a more robust host response in mice, suggesting an involvement in gene regulation (Bjorkholm et al., 2002). Bioinformatic analysis to identify genes with HpyAIV recognition sites in *H. pylori* strains J99 and 26695 identified 13 genes, including 9 genes in the cag pathogenicity island and the catalase gene *katA* (Skoglund et al., 2007). However, only *katA* was found to be differentially regulated, which may be a reflection of the culture conditions (Skoglund et al., 2007). It was previously demonstrated that in strain J99, the endonuclease is

inactive (Kong et al., 2000); however, in strain 26695, the endonuclease is active (Lin et al., 2001). In strain HPAG1, the *res* gene is truncated, supporting a dedicated role for this system in gene regulation in this strain. A survey of the type II R-M system *mod* gene of clinical strains found changes in a tract of adenine residues; loss of one adenine residue correlated with the loss of methylation activity (Skoglund et al., 2007). In addition, *H. pylori* contains a number of active methyltransferases without an associated endonuclease (Kong et al., 2000; Lin et al., 2001; Vitkute et al., 2001), suggesting that these methyltransferases may have other functions, including gene regulation (Skoglund et al., 2007). Similarly, *Mycoplasma* species contain a large number of ORFs that encode type III methyltransferases that also contain repeats but lack an associated endonuclease (Brocchi et al., 2007). Additionally, in *Mycoplasma* species phase variation has been observed in some genes of type I and type III R-M systems (Dybvig et al., 1998; Rocha and Blanchard, 2002). The presence of repeats in these systems allows phase variation and suggest a role for R-M systems in gene regulation and expression, enabling mycoplasmas to adapt to different environmental conditions (Brocchi et al., 2007). A possible role in virulence for a phase-variable type I R-M system has been proposed for *Mycoplasma pulmonis*. This R-M system may contribute to the fitness of the mycoplasmas within the host, as studies in rats found that different environments within the animal affected R-M activity (Gumulak-Smith et al., 2001).

CONCLUDING REMARKS

The list of phase-variable R-M systems is expanding, with examples in a range of gram-negative pathogens and one gram-positive pathogen. We propose that the phase-variable methylation has arisen due to the selective advantage conferred by the phasevarion enabling random switching of an organism between two distinct cell types. In organisms with multiple phasevarions switching independently, multiple differentiated cell types can be generated. The widespread distribution of phase-variable R-M systems in host-adapted pathogenic bacteria suggests that this novel mechanism may be a common strategy. The target site for the vast majority of phase-variable R-M systems is unknown, which limits the investigation of the molecular mechanism by which differential methylation influences gene expression at individual promoters. Further assessment of the roles these systems play in gene regulation, including defining the mechanisms and structural biology of these ill-defined systems, is required to advance this area.

REFERENCES

Ahmad, I., and D. N. Rao. 1994. Interaction of EcoP15I DNA methyltransferase with oligonucleotides containing the asymmetric sequence 5'-CAGCAG-3'. *J. Mol. Biol.* **242:**378–388.

Andersen-Nissen, E., K. D. Smith, K. L. Strobe, S. L. Barrett, B. T. Cookson, S. M. Logan, and A. Aderem. 2005. Evasion of Toll-like receptor 5 by flagellated bacteria. *Proc. Natl. Acad. Sci. USA* **102:**9247–9252.

Ando, T., Q. Xu, M. Torres, K. Kusugami, D. A. Israel, and M. J. Blaser. 2000. Restriction-modification system differences in *Helicobacter pylori* are a barrier to interstrain plasmid transfer. *Mol. Microbiol.* **37:**1052–1065.

Arber, W. 1974. DNA modification and restriction. *Prog. Nucleic Acid Res. Mol. Biol.* **14:**1–37.

Arber, W., and D. Dussoix. 1962. Host specificity of DNA produced by *Escherichia coli*. I. Host controlled modification of bacteriophage lambda. *J. Mol. Biol.* **5:**18–36.

Arber, W., R. Yuan, and T. A. Bickle. 1975. Strain-specific modification and restriction of DNA in bacteria. *FEBS Proc. Symp.* **9:**3–22.

Bachi, B., J. Reiser, and V. Pirrotta. 1979. Methylation and cleavage sequences of the EcoP1 restriction-modification enzyme. *J. Mol. Biol.* **128:**143–163.

Balbontin, R., G. Rowley, M. G. Pucciarelli, J. Lopez-Garrido, Y. Wormstone, S. Lucchini, F. Garcia-Del Portillo, J. C. Hinton, and J. Casadesus. 2006. DNA adenine methylation regulates virulence gene expression in *Salmonella enterica* serovar Typhimurium. *J. Bacteriol.* **188:**8160–8168.

Bayliss, C. D., M. J. Callaghan, and E. R. Moxon. 2006. High allelic diversity in the methyltransferase gene of a phase variable type III restriction-modification system has implications for the fitness of *Haemophilus influenzae*. *Nucleic Acids Res.* **34:**4046–4059.

Bayliss, C. D., D. Field, and E. R. Moxon. 2001. The simple sequence contingency loci of *Haemophilus influenzae* and *Neisseria meningitidis*. *J. Clin. Investig.* **107:**657–662.

Bernarde, C., P. Lehours, J. P. Lasserre, M. Castroviejo, M. Bonneu, F. Megraud, and A. Menard. 2010. Complexomics study of two *Helicobacter pylori* strains of two pathological origins: potential targets for vaccine development and new insight in bacteria metabolism. *Mol. Cell. Proteomics* **9:**2796–2826.

Bickle, T. A., C. Brack, and R. Yuan. 1978. ATP-induced conformational changes in the restriction endonuclease from *Escherichia coli* K-12. *Proc. Natl. Acad. Sci. USA* **75:**3099–3103.

Bjorkholm, B. M., J. L. Guruge, J. D. Oh, A. J. Syder, N. Salama, K. Guillemin, S. Falkow, C. Nilsson, P. G. Falk, L. Engstrand, and J. I. Gordon. 2002. Colonization of germ-free transgenic mice with genotyped *Helicobacter pylori* strains from a case-control study of gastric cancer reveals a correlation between host responses and HsdS components of type I restriction-modification systems. *J. Biol. Chem.* **277:**34191–34197.

Blomfield, I. C. 2001. The regulation of pap and type 1 fimbriation in *Escherichia coli*. *Adv. Microb. Physiol.* **45:**1–49.

Blyn, L. B., B. A. Braaten, and D. A. Low. 1990. Regulation of *pap* pilin phase variation by a mechanism involving differential Dam methylation states. *EMBO J.* **9:**4045–4054.

Bourniquel, A. A., and T. A. Bickle. 2002. Complex restriction enzymes: NTP-driven molecular motors. *Biochimie* **84:**1047–1059.

Boyer, H. W. 1971. DNA restrictions and modification mechanisms in bacteria. *Annu. Rev. Microbiol.* **25:**153–176.

Brocchi, M., A. Vasconcelos, and A. Zaha. 2007. Restriction-modification systems in *Mycoplasma* spp. *Genet. Mol. Biol.* **30:**236–244.

Budroni, S., E. Siena, J. C. Hotopp, K. L. Seib, D. Serruto, C. Nofroni, M. Comanducci, D. R. Riley, S. C. Daugherty, S. V.

Angiuoli, A. Covacci, M. Pizza, R. Rappuoli, E. R. Moxon, H. Tettelin, and D. Medini. 2011. *Neisseria meningitidis* is structured in clades associated with restriction modification systems that modulate homologous recombination. *Proc. Natl. Acad. Sci. USA* **108:**4494–4499.

Buhler, R., and R. Yuan. 1978. Characterization of a restriction enzyme from *Escherichia coli* K carrying a mutation in the modification subunit. *J. Biol. Chem.* **253:**6756–6760.

Campellone, K. G., A. J. Roe, A. Lobner-Olesen, K. C. Murphy, L. Magoun, M. J. Brady, A. Donohue-Rolfe, S. Tzipori, D. L. Gally, J. M. Leong, and M. G. Marinus. 2007. Increased adherence and actin pedestal formation by *dam*-deficient enterohaemorrhagic *Escherichia coli* O157:H7. *Mol. Microbiol.* **63:**1468–1481.

Casadesus, J., and D. Low. 2006. Epigenetic gene regulation in the bacterial world. *Microbiol. Mol. Biol. Rev.* **70:**830–856.

Chandler, M., and O. Fayet. 1993. Translational frameshifting in the control of transposition in bacteria. *Mol. Microbiol.* **7:**497–503.

Chatti, A., D. Daghfous, and A. Landoulsi. 2008. Effect of repeated in vivo passage (in mice) on *Salmonella typhimurium dam* mutant virulence and fitness. *Pathol. Biol.* (Paris) **56:**121–124.

Chen, L., D. B. Paulsen, D. W. Scruggs, M. M. Banes, B. Y. Reeks, and M. L. Lawrence. 2003. Alteration of DNA adenine methylase (Dam) activity in *Pasteurella multocida* causes increased spontaneous mutation frequency and attenuation in mice. *Microbiology* **149:**2283–2290.

Correnti, J., V. Munster, T. Chan, and M. Woude. 2002. Dam-dependent phase variation of Ag43 in *Escherichia coli* is altered in a seqA mutant. *Mol. Microbiol.* **44:**521–532.

Danese, P. N., L. A. Pratt, S. L. Dove, and R. Kolter. 2000. The outer membrane protein, antigen 43, mediates cell-to-cell interactions within *Escherichia coli* biofilms. *Mol. Microbiol.* **37:**424–432.

De Backer, O., and C. Colson. 1991. Transfer of the genes for the StyLTI restriction-modification system of *Salmonella typhimurium* to strains lacking modification ability results in death of the recipient cells and degradation of their DNA. *J. Bacteriol.* **173:**1328–1330.

Donahue, J. P., D. A. Israel, R. M. Peek, M. J. Blaser, and G. G. Miller. 2000. Overcoming the restriction barrier to plasmid transformation of *Helicobacter pylori*. *Mol. Microbiol.* **37:**1066–1074.

Doronina, V. A., and N. E. Murray. 2001. The proteolytic control of restriction activity in *Escherichia coli* K-12. *Mol. Microbiol.* **39:**416–428.

Dryden, D. T., N. E. Murray, and D. N. Rao. 2001. Nucleoside triphosphate-dependent restriction enzymes. *Nucleic Acids Res.* **29:**3728–3741.

Dybvig, K. 1993. DNA rearrangements and phenotypic switching in prokaryotes. *Mol. Microbiol.* **10:**465–471.

Dybvig, K., R. Sitaraman, and C. T. French. 1998. A family of phase-variable restriction enzymes with differing specificities generated by high-frequency gene rearrangements. *Proc. Natl. Acad. Sci. USA* **95:**13923–13928.

Eaton, K. A., S. Suerbaum, C. Josenhans, and S. Krakowka. 1996. Colonization of gnotobiotic piglets by *Helicobacter pylori* deficient in two flagellin genes. *Infect. Immun.* **64:**2445–2448.

Falker, S., J. Schilling, M. A. Schmidt, and G. Heusipp. 2007. Overproduction of DNA adenine methyltransferase alters motility, invasion, and the lipopolysaccharide O-antigen composition of *Yersinia enterocolitica*. *Infect. Immun.* **75:**4990–4997.

Fox, K. L., S. J. Dowideit, A. L. Erwin, Y. N. Srikhanta, A. L. Smith, and M. P. Jennings. 2007. *Haemophilus influenzae* phasevarions have evolved from type III DNA restriction systems into epigenetic regulators of gene expression. *Nucleic Acids Res.* **35:**5242–5252.

Garcia-Del Portillo, F., M. G. Pucciarelli, and J. Casadesus. 1999. DNA adenine methylase mutants of *Salmonella typhimurium* show defects in protein secretion, cell invasion, and M cell cytotoxicity. *Proc. Natl. Acad. Sci. USA* **96:**11578–11583.

Gewirtz, A. T., Y. Yu, U. S. Krishna, D. A. Israel, S. L. Lyons, and R. M. Peek, Jr. 2004. *Helicobacter pylori* flagellin evades toll-like receptor 5-mediated innate immunity. *J. Infect. Dis.* **189:**1914–1920.

Gibbs, C. P., B. Y. Reimann, E. Schultz, A. Kaufmann, R. Haas, and T. F. Meyer. 1989. Reassortment of pilin genes in *Neisseria gonorrhoeae* occurs by two distinct mechanisms. *Nature* **338:**651–652.

Gorbalenya, A. E., and E. V. Koonin. 1991. Endonuclease (R) subunits of type-I and type-III restriction-modification enzymes contain a helicase-like domain. *FEBS Lett.* **291:**277–281.

Gumulak-Smith, J., A. Teachman, A. H. Tu, J. W. Simecka, J. R. Lindsey, and K. Dybvig. 2001. Variations in the surface proteins and restriction enzyme systems of *Mycoplasma pulmonis* in the respiratory tract of infected rats. *Mol. Microbiol.* **40:**1037–1044.

Haagmans, W., and M. van der Woude. 2000. Phase variation of Ag43 in *Escherichia coli*: Dam-dependent methylation abrogates OxyR binding and OxyR-mediated repression of transcription. *Mol. Microbiol.* **35:**877–887.

Hadi, S. M., B. Bachi, S. Iida, and T. A. Bickle. 1983. DNA restriction-modification enzymes of phage P1 and plasmid p15B. Subunit functions and structural homologies. *J. Mol. Biol.* **165:**19–34.

Hallet, B. 2001. Playing Dr Jekyll and Mr Hyde: combined mechanisms of phase variation in bacteria. *Curr. Opin. Microbiol.* **4:**570–581.

Hamilton, H. L., and J. P. Dillard. 2006. Natural transformation of *Neisseria gonorrhoeae*: from DNA donation to homologous recombination. *Mol. Microbiol.* **59:**376–385.

Hammerschmidt, S., R. Hilse, J. P. van Putten, R. Gerardy-Schahn, A. Unkmeir, and M. Frosch. 1996. Modulation of cell surface sialic acid expression in *Neisseria meningitidis* via a transposable genetic element. *EMBO J.* **15:**192–198.

Hartmann, E., and C. Lingwood. 1997. Brief heat shock treatment induces a long-lasting alteration in the glycolipid receptor binding specificity and growth rate of *Haemophilus influenzae*. *Infect. Immun.* **65:**1729–1733.

Hartmann, E., C. A. Lingwood, and J. Reidl. 2001. Heat-inducible surface stress protein (Hsp70) mediates sulfatide recognition of the respiratory pathogen *Haemophilus influenzae*. *Infect. Immun.* **69:**3438–3441.

Heithoff, D. M., E. Y. Enioutina, R. A. Daynes, R. L. Sinsheimer, D. A. Low, and M. J. Mahan. 2001. *Salmonella* DNA adenine methylase mutants confer cross-protective immunity. *Infect. Immun.* **69:**6725–6730.

Heithoff, D. M., R. L. Sinsheimer, D. A. Low, and M. J. Mahan. 1999. An essential role for DNA adenine methylation in bacterial virulence. *Science* **284:**967–970.

Henderson, I. R., and P. Owen. 1999. The major phase-variable outer membrane protein of *Escherichia coli* structurally resembles the immunoglobulin A1 protease class of exported protein and is regulated by a novel mechanism involving Dam and OxyR. *J. Bacteriol.* **181:**2132–2141.

Henderson, I. R., P. Owen, and J. P. Nataro. 1999. Molecular switches—the ON and OFF of bacterial phase variation. *Mol. Microbiol.* **33:**919–932.

Herbert, M. A., S. Hayes, M. E. Deadman, C. M. Tang, D. W. Hood, and E. R. Moxon. 2002. Signature tagged mutagenesis of *Haemophilus influenzae* identifies genes required for in vivo survival. *Microb. Pathog.* **33:**211–223.

Hood, D. W., M. E. Deadman, M. P. Jennings, M. Bisercic, R. D. Fleischmann, J. C. Venter, and E. R. Moxon. 1996. DNA repeats

identify novel virulence genes in *Haemophilus influenzae*. *Proc. Natl. Acad. Sci. USA* **93:**11121–11125.

Humbelin, M., B. Suri, D. N. Rao, D. P. Hornby, H. Eberle, T. Pripfl, S. Kenel, and T. A. Bickle. 1988. Type III DNA restriction and modification systems EcoP1 and EcoP15. Nucleotide sequence of the EcoP1 operon, the EcoP15 mod gene and some EcoP1 mod mutants. *J. Mol. Biol.* **200:**23–29.

Jakomin, M., D. Chessa, A. J. Baumler, and J. Casadesus. 2008. Regulation of the *Salmonella enterica std* fimbrial operon by DNA adenine methylation, SeqA, and HdfR. *J. Bacteriol.* **190:**7406–7413.

Jennings, M. P., M. Virji, D. Evans, V. Foster, Y. N. Srikhanta, L. Steeghs, P. van der Ley, and E. R. Moxon. 1998. Identification of a novel gene involved in pilin glycosylation in *Neisseria meningitidis*. *Mol. Microbiol.* **29:**975–984.

Julio, S. M., D. M. Heithoff, D. Provenzano, K. E. Klose, R. L. Sinsheimer, D. A. Low, and M. J. Mahan. 2001. DNA adenine methylase is essential for viability and plays a role in the pathogenesis of *Yersinia pseudotuberculosis* and *Vibrio cholerae*. *Infect. Immun.* **69:**7610–7615.

Julio, S. M., D. M. Heithoff, R. L. Sinsheimer, D. A. Low, and M. J. Mahan. 2002. DNA adenine methylase overproduction in *Yersinia pseudotuberculosis* alters YopE expression and secretion and host immune responses to infection. *Infect. Immun.* **70:**1006–1009.

Kavermann, H., B. P. Burns, K. Angermuller, S. Odenbreit, W. Fischer, K. Melchers, and R. Haas. 2003. Identification and characterization of *Helicobacter pylori* genes essential for gastric colonization. *J. Exp. Med.* **197:**813–822.

Kim, J. S., J. Li, I. H. Barnes, D. A. Baltzegar, M. Pajaniappan, T. W. Cullen, M. S. Trent, C. M. Burns, and S. A. Thompson. 2008. Role of the *Campylobacter jejuni* Cj1461 DNA methyltransferase in regulating virulence characteristics. *J. Bacteriol.* **190:**6524–6529.

Kong, H., L. F. Lin, N. Porter, S. Stickel, D. Byrd, J. Posfai, and R. J. Roberts. 2000. Functional analysis of putative restriction-modification system genes in the *Helicobacter pylori* J99 genome. *Nucleic Acids Res.* **28:**3216–3223.

Kutsukake, K., and T. Iino. 1980. Inversions of specific DNA segments in flagellar phase variation of *Salmonella* and inversion systems of bacteriophages P1 and Mu. *Proc. Natl. Acad. Sci. USA* **77:**7338–7341.

Lee, S. K., A. Stack, E. Katzowitsch, S. I. Aizawa, S. Suerbaum, and C. Josenhans. 2003. *Helicobacter pylori* flagellins have very low intrinsic activity to stimulate human gastric epithelial cells via TLR5. *Microbes Infect.* **5:**1345–1356.

Levinson, G., and G. A. Gutman. 1987. Slipped-strand mispairing: a major mechanism for DNA sequence evolution. *Mol. Biol. Evol.* **4:**203–221.

Lin, L. F., J. Posfai, R. J. Roberts, and H. Kong. 2001. Comparative genomics of the restriction-modification systems in *Helicobacter pylori*. *Proc. Natl. Acad.Sci. USA* **98:**2740–2745.

Low, D. A., N. J. Weyand, and M. J. Mahan. 2001. Roles of DNA adenine methylation in regulating bacterial gene expression and virulence. *Infect. Immun.* **69:**7197–7204.

Makovets, S., V. A. Doronina, and N. E. Murray. 1999. Regulation of endonuclease activity by proteolysis prevents breakage of unmodified bacterial chromosomes by type I restriction enzymes. *Proc. Natl. Acad. Sci. USA* **96:**9757–9762.

Malone, T., R. M. Blumenthal, and X. Cheng. 1995. Structure-guided analysis reveals nine sequence motifs conserved among DNA amino-methyltransferases, and suggests a catalytic mechanism for these enzymes. *J. Mol. Biol.* **253:**618–632.

Marinus, M. G., and J. Casadesus. 2009. Roles of DNA adenine methylation in host-pathogen interactions: mismatch repair, transcriptional regulation, and more. *FEMS Microbiol. Rev.* 33:488–503.

Martin, P., K. Makepeace, S. A. Hill, D. W. Hood, and E. R. Moxon. 2005. Microsatellite instability regulates transcription factor binding and gene expression. *Proc. Natl. Acad. Sci. USA* 102:3800–3804.

Mehling, J. S., H. Lavender, and S. Clegg. 2007. A Dam methylation mutant of *Klebsiella pneumoniae* is partially attenuated. *FEMS Microbiol. Lett.* 268:187–193.

Mehr, I. J., and H. S. Seifert. 1998. Differential roles of homologous recombination pathways in *Neisseria gonorrhoeae* pilin antigenic variation, DNA transformation and DNA repair. *Mol. Microbiol.* 30:697–710.

Meisel, A., T. A. Bickle, D. H. Kruger, and C. Schroeder. 1992. Type III restriction enzymes need two inversely oriented recognition sites for DNA cleavage. *Nature* 355:467–469.

Meisel, A., P. Mackeldanz, T. A. Bickle, D. H. Kruger, and C. Schroeder. 1995. Type III restriction endonucleases translocate DNA in a reaction driven by recognition site-specific ATP hydrolysis. *EMBO J.* 14:2958–2966.

Moxon, E. R., P. B. Rainey, M. A. Nowak, and R. E. Lenski. 1994. Adaptive evolution of highly mutable loci in pathogenic bacteria. *Curr. Biol.* 4:24–33.

Moxon, E. R., and D. S. Thaler. 1997. Microbial genetics. The tinkerer's evolving tool-box. *Nature* 387:659, 661–662.

Moxon, R., C. Bayliss, and D. Hood. 2006. Bacterial contingency loci: the role of simple sequence DNA repeats in bacterial adaptation. *Annu. Rev. Genet.* 40:307–333.

Murphy, K. C., J. M. Ritchie, M. K. Waldor, A. Lobner-Olesen, and M. G. Marinus. 2008. Dam methyltransferase is required for stable lysogeny of the Shiga toxin (Stx2)-encoding bacteriophage 933W of enterohemorrhagic *Escherichia coli* O157:H7. *J. Bacteriol.* 190:438–441.

Murray, N. E. 2000. Type I restriction systems: sophisticated molecular machines (a legacy of Bertani and Weigle). *Microbiol. Mol. Biol. Rev.* 64:412–434.

Nicholson, B., and D. Low. 2000. DNA methylation-dependent regulation of pef expression in *Salmonella typhimurium*. *Mol. Microbiol.* 35:728–742.

Nou, X., B. Braaten, L. Kaltenbach, and D. A. Low. 1995. Differential binding of Lrp to two sets of pap DNA binding sites mediated by Pap I regulates Pap phase variation in *Escherichia coli*. *EMBO J.* 14:5785–5797.

Peck, B., M. Ortkamp, U. Nau, M. Niederweis, E. Hundt, and B. Knapp. 2001. Characterization of four members of a multigene family encoding outer membrane proteins of *Helicobacter pylori* and their potential for vaccination. *Microbes Infect.* 3:171–179.

Pettersson, A., T. Prinz, A. Umar, J. van der Biezen, and J. Tommassen. 1998. Molecular characterization of LbpB, the second lactoferrin-binding protein of *Neisseria meningitidis*. *Mol. Microbiol.* 27:599–610.

Pettersson, A., P. van der Ley, J. T. Poolman, and J. Tommassen. 1993. Molecular characterization of the 98-kilodalton iron-regulated outer membrane protein of *Neisseria meningitidis*. *Infect. Immun.* 61:4724–4733.

Pingoud, A., and A. Jeltsch. 1997. Recognition and cleavage of DNA by type-II restriction endonucleases. *Eur. J. Biochem.* 246:1–22.

Prieto, A. I., M. Jakomin, I. Segura, M. G. Pucciarelli, F. Ramos-Morales, F. Garcia-Del Portillo, and J. Casadesus. 2007. The GATC-binding protein SeqA is required for bile resistance and virulence in *Salmonella enterica* serovar Typhimurium. *J. Bacteriol.* 189:8496–8502.

Pucciarelli, M. G., A. I. Prieto, J. Casadesus, and F. Garcia-del Portillo. 2002. Envelope instability in DNA adenine methylase mutants of *Salmonella enterica*. *Microbiology* 148:1171–1182.

Redaschi, N., and T. A. Bickle. 1996. Posttranscriptional regulation of EcoP1I and EcoP15I restriction activity. *J. Mol. Biol.* 257:790–803.

Robertson, B. D., and T. F. Meyer. 1992. Genetic variation in pathogenic bacteria. *Trends Genet.* 8:422–427.

Rocha, E. P., and A. Blanchard. 2002. Genomic repeats, genome plasticity and the dynamics of *Mycoplasma* evolution. *Nucleic Acids Res.* 30:2031–2042.

Saha, S., I. Ahmad, Y. V. Reddy, V. Krishnamurthy, and D. N. Rao. 1998. Functional analysis of conserved motifs in type III restriction-modification enzymes. *Biol. Chem.* 379:511–517.

Sarkari, J., N. Pandit, E. R. Moxon, and M. Achtman. 1994. Variable expression of the Opc outer membrane protein in *Neisseria meningitidis* is caused by size variation of a promoter containing poly-cytidine. *Mol. Microbiol.* 13:207–217.

Saunders, N. J., A. C. Jeffries, J. F. Peden, D. W. Hood, H. Tettelin, R. Rappuoli, and E. R. Moxon. 2000. Repeat-associated phase variable genes in the complete genome sequence of *Neisseria meningitidis* strain MC58. *Mol. Microbiol.* 37:207–215.

Saunders, N. J., J. F. Peden, D. W. Hood, and E. R. Moxon. 1998. Simple sequence repeats in the *Helicobacter pylori* genome. *Mol. Microbiol.* 27:1091–1098.

Sears, A., L. J. Peakman, G. G. Wilson, and M. D. Szczelkun. 2005. Characterization of the type III restriction endonuclease PstII from *Providencia stuartii*. *Nucleic Acids Res.* 33:4775–4787.

Sears, A., and M. D. Szczelkun. 2005. Subunit assembly modulates the activities of the type III restriction-modification enzyme PstII in vitro. *Nucleic Acids Res.* 33:4788–4796.

Seib, K. L., I. R. Peak, and M. P. Jennings. 2002. Phase variable restriction-modification systems in *Moraxella catarrhalis*. *FEMS Immunol. Med. Microbiol.* 32:159–165.

Seib, K. L., E. Pigozzi, A. Muzzi, J. A. Gawthorne, I. Delany, M. P. Jennings, and R. Rappuoli. 2011. A novel epigenetic regulator associated with the hypervirulent *Neisseria meningitidis* clonal complex 41/44. *FASEB J.* 25:3622–3633.

Seifert, H. S. 1996. Questions about gonococcal pilus phase- and antigenic variation. *Mol. Microbiol.* 21:433–440.

Skoglund, A., B. Bjorkholm, C. Nilsson, A. Andersson, C. Jernberg, K. Schirwitz, C. Enroth, M. Krabbe, and L. Engstrand. 2007. Functional analysis of the M.HpyAIV DNA methyltransferase of *Helicobacter pylori*. *J. Bacteriol.* 189:8914–8921.

Srikhanta, Y. N., S. J. Dowideit, J. L. Edwards, M. L. Falsetta, H. J. Wu, O. B. Harrison, K. L. Fox, K. L. Seib, T. L. Maguire, A. H. Wang, M. C. Maiden, S. M. Grimmond, M. A. Apicella, and M. P. Jennings. 2009. Phasevarions mediate random switching of gene expression in pathogenic *Neisseria*. *PLoS Pathog.* 5:e1000400.

Srikhanta, Y. N., K. L. Fox, and M. P. Jennings. 2010. The phasevarion: phase variation of type III DNA methyltransferases controls coordinated switching in multiple genes. *Nat. Rev. Microbiol.* 8:196–206.

Srikhanta, Y. N., R. G. Gorrell, J. A. Steen, J. A. Gawthorne, T. Kwok, S. M. Grimmond, R. M. Robins-Browne, and M. P. Jennings. 2011. Phasevarion mediated epigenetic gene regulation in *Helicobacter pylori*. *PLoS One* 6:e27569.

Srikhanta, Y. N., T. L. Maguire, K. J. Stacey, S. M. Grimmond, and M. P. Jennings. 2005. The phasevarion: a genetic system controlling coordinated, random switching of expression of multiple genes. *Proc. Natl. Acad. Sci. USA* 102:5547–5551.

Stibitz, S., W. Aaronson, D. Monack, and S. Falkow. 1989. Phase variation in *Bordetella pertussis* by frameshift mutation in a gene for a novel two-component system. *Nature* 338:266–269.

Taylor, V. L., R. W. Titball, and P. C. Oyston. 2005. Oral immunization with a *dam* mutant of *Yersinia pseudotuberculosis* protects against plague. *Microbiology* **151:**1919–1926.

van Belkum, A., S. Scherer, L. van Alphen, and H. Verbrugh. 1998. Short-sequence DNA repeats in prokaryotic genomes. *Microbiol. Mol. Biol. Rev.* **62:**275–293.

van der Woude, M., B. Braaten, and D. Low. 1996. Epigenetic phase variation of the pap operon in *Escherichia coli. Trends Microbiol.* **4:**5–9.

van der Woude, M. W. 2006. Re-examining the role and random nature of phase variation. *FEMS Microbiol. Lett.* **254:**190–197.

van der Woude, M. W., and A. J. Baumler. 2004. Phase and antigenic variation in bacteria. *Clin. Microbiol. Rev.* **17:**581–611.

van Ham, S. M., L. van Alphen, F. R. Mooi, and J. P. van Putten. 1993. Phase variation of *Haemophilus influenzae* fimbriae: transcriptional control of two divergent genes through a variable combined promoter region. *Cell* **73:**1187–1196.

Vitkute, J., K. Stankevicius, G. Tamulaitiene, Z. Maneliene, A. Timinskas, D. E. Berg, and A. Janulaitis. 2001. Specificities of eleven different DNA methyltransferases of *Helicobacter pylori* strain 26695. *J. Bacteriol.* **183:**443–450.

Wallecha, A., V. Munster, J. Correnti, T. Chan, and M. van der Woude. 2002. Dam- and OxyR-dependent phase variation of *agn43*: essential elements and evidence for a new role of DNA methylation. *J. Bacteriol.* **184:**3338–3347.

Weiser, J. N., A. Williams, and E. R. Moxon. 1990. Phase-variable lipopolysaccharide structures enhance the invasive capacity of *Haemophilus influenzae. Infect. Immun.* **58:**3455–3457.

Weyand, N. J., and D. A. Low. 2000. Regulation of Pap phase variation. Lrp is sufficient for the establishment of the phase off *pap* DNA methylation pattern and repression of *pap* transcription in vitro. *J. Biol. Chem.* **275:**3192–3200.

Willcock, D. F., D. T. Dryden, and N. E. Murray. 1994. A mutational analysis of the two motifs common to adenine methyltransferases. *EMBO J.* **13:**3902–3908.

Wion, D., and J. Casadesus. 2006. N6-methyl-adenine: an epigenetic signal for DNA-protein interactions. *Nat. Rev. Microbiol.* **4:**183–192.

Regulation of Bacterial Virulence
Edited by Michael L. Vasil and Andrew J. Darwin
© 2013 ASM Press, Washington, DC   doi:10.1128/9781555818524.ch9

Chapter 9

Regulation of Exopolysaccharide Biosynthesis in *Pseudomonas aeruginosa*

YUTA OKKOTSU, CHRISTOPHER L. PRITCHETT, AND MICHAEL J. SCHURR

INTRODUCTION

Pseudomonas aeruginosa is a gram-negative, ubiquitous opportunistic pathogen that is a major causative agent of severe infections in nosocomial settings. The type of infection depends upon the immune status of the patient; common types of pathogen-associated infections include acute and chronic pneumonia, folliculitis, otitis externa, osteomyelitis, and septicemia. The innate and acquired resistance against antimicrobials makes treatment extremely challenging. The organism is capable of infecting almost any surface of the human body, but it is most prevalent in acute and chronic pneumonia. *P. aeruginosa* also produces an arsenal of cell-associated and extracellular virulence factors that enable its proliferation in various niches. Examples of these include two ADP-ribosylating toxins (ExoS and ExoT), five phospholipases (PlcB, PlcH, Pld, PlcN, and ExoU), hemolysins, flagella, type IV pili, a type III secretion system, pyocyanin, pyoverdine, and various types of exopolysaccharides that perform multiple functions for the organism (Vasil et al., 1991; Mattick, 2002; Lau et al., 2004; Vasil, 2007; Engel and Balachandran, 2009).

There are three well-characterized exopolysaccharides produced by *P. aeruginosa*, including: (i) alginate, synthesized from enzymes encoded by PA3540 to PA3551 on the PAO1 genome; (ii) a glucose-rich exopolysaccharide synthesized by enzymes encoded by the *pel* gene cluster (i.e., PA3058 to PA3064 in PAO1); and (iii) a mannose-, glucose-, and rhamnose-rich exopolysaccharide produced by proteins encoded in the *psl* gene cluster (i.e., PA2231 to PA2245 in PAO1) (Govan and Deretic, 1996; Friedman and Kolter, 2004a, 2004b). Each of these polysaccharides is associated with some stage of *P. aeruginosa* biofilm development, and mounting evidence suggests that Pel, Psl, and alginate are involved in different stages of the infectious process. Alginate was the first

P. aeruginosa exopolysaccharide discovered, when it was first associated with the mucoid phenotype of *P. aeruginosa* strains isolated from cystic fibrosis (CF) patients (Doggett, 1969). Pel, on the other hand, was initially described in terms of *P. aeruginosa* growing at an air-liquid interface in static broth-grown cultures (Friedman and Kolter, 2004a). The third exopolysaccharide, Psl, was identified from studies in which strains that were not capable of producing alginate were still able to form biofilm structures on solid glass surfaces (Wozniak et al., 2003b; Friedman and Kolter, 2004b; Jackson et al., 2004; Matsukawa and Greenberg, 2004).

Alginate is arguably the best-characterized exopolysaccharide produced by *P. aeruginosa*, and several excellent reviews have been written on the molecular biology of its production and clinical ramifications (Govan and Deretic, 1996; Mathee et al., 2002; Muhammadi and Ahmed, 2007; Pritt et al., 2007; Hay et al., 2010; Franklin et al., 2011). This chapter reviews the transcriptional and posttranscriptional factors involved in controlling and inducing alginate production. Alginate is an anionic heteropolymer consisting of β-1,4 D-mannuronic and D-guluronic acid, which is frequently produced by clinical isolates from later-stage CF patients. Clinical isolates and lab-derived strains that overproduce alginate are phenotypically referred to as mucoid. Mucoid *P. aeruginosa* has been observed in up to 40% of *Pseudomonas*-positive sputa from non-CF patients with chronic bronchitis and other obstructive lung diseases (Govan, 1988) and less frequently (up to 10%) from non-CF patients with chronic urinary tract infections (McAvoy et al., 1989) and, more rarely, from other sites in non-CF patients (Doggett, 1969; Anastassiou et al., 1987).

The production of alginate enables *P. aeruginosa* to be recalcitrant against host defenses by evasion of the innate immune system, quenching reactive oxygen intermediates, and interfering with antibody

Michael J. Schurr and Yuta Okkotsu • Department of Microbiology, University of Colorado School of Medicine, Aurora, CO 80045.
Christopher L. Pritchett • East Tennessee State University, Department of Health Sciences, Johnson City, TN 37614.

recognition by macrophages and neutrophils (Oliver and Weir, 1983, 1985; Krieg et al., 1988a, 1988b; Simpson et al., 1988, 1989, 1993; Mai et al., 1993; Pasquier et al., 1997; Mathee et al., 1999). Therefore, alginate production is an important virulence determinant for *P. aeruginosa*, particularly in chronic infections of the CF lung.

The biosynthesis of alginate has been studied for several decades, and a few of the biosynthetic enzymes have been crystallized (Regni et al., 2002; Snook et al., 2003; Keiski et al., 2010; Whitney et al., 2011). Alginate biosynthetic and excretion proteins are encoded by 13 genes on the PAO1 chromosome: a 12-gene operon containing the genes *algD*, *alg8*, *alg44*, *algK*, *algE*, *algG*, *algX*, *algL*, *algI*, *algJ*, *algF*, and *algA* (PA3540 to PA3551), and *algC* (PA5322), which is critical for alginate production but not located within the aforementioned operon. The *algC* (PA5322) gene encodes phosphomannomutase and is important for rhamnolipid, lipopolysaccharide (LPS), and Psl polysaccharide synthesis (Zielinski et al., 1991; Coyne et al., 1994; Olvera et al., 1999; Byrd et al., 2009). Some of the proteins encoded by the alginate genes are proposed to be assembled into a membrane-spanning configuration that allows excretion of alginate from the cytoplasm out of the cell (Fig. 1).

The first substrate required for alginate biosynthesis is fructose-6-phosphate, which can be created from numerous sugars that feed through the Entner-Doudoroff pathway, or other carbon sources that feed into the tricarboxylic acid (TCA) cycle (Fig. 1) (Hay et al., 2010). Fructose-6-phosphate is then converted to mannose-6-phosphate by AlgA, an enzyme that also catalyzes the conversion of mannose-1-phosphate to GDP-mannose. AlgC is required to convert mannose-6-phosphate to mannose-1-phosphate. AlgD then converts GDP-mannose to GDP-mannuronate. GDP-mannuronate synthesized in the cytoplasm likely interacts with inner membrane proteins Alg8 and Alg44

to be transported to the periplasm. Periplasmic enzymes AlgF, AlgI, and AlgJ are in a complex involved in O acetylation of the alginate polymer. AlgG, a periplasmic epimerase, converts D-mannuronic acid to L-guluronic acid in a polymer. AlgX does not have a known function but is likely associated with the AlgI/AlgF/AlgJ complex, since deletion of *algX* results in the secretion of unpolymerized alginate (Robles-Price et al., 2004). The AlgL enzyme degrades the alginate polymer and *algL* deletions have been difficult to obtain in mucoid strains, indicating that this gene might be essential. However, when a LacI^q trc promoter was used to control the alginate biosynthetic cluster, an *algL* mutant was obtained when the alginate genes were not expressed. Induction of the alginate biosynthetic pathway in the *algL* background resulted in cell lysis, and transmission electron microscopy showed an increase of a polymer in the periplasmic space (Jain and Ohman, 2005). AlgK is a putative scaffold protein located in the periplasm and is anchored to the outer membrane by a lipid moiety (Keiski et al., 2010). AlgK is thought to help guide the alginate polymer through the periplasmic space, as deletion of *algK* resulted in the secretion of short alginate degradation products (Jain and Ohman, 1998). Recently, AlgK, AlgX, and MucD have been purified as a multiprotein complex in the periplasmic space, thereby strongly suggesting a direct link between the biosynthesis and the posttranscriptional regulation of alginate production (see below and Hay et al., 2012). AlgE is an outer membrane protein that studies from the Howell laboratory have shown is the porin through which alginate is released into the extracellular milieu (Whitney et al., 2011).

One of the first studies to indicate that *P. aeruginosa* may produce exopolysaccharides other than alginate in a biofilm showed that the nonmucoid strains PAO1 and PA14 biofilm matrices were composed of rhamnose, glucose, and mannose, not mannuronate or guluronate (Wozniak et al., 2003b). Three independent

Figure 1. Biochemical pathways that leads to Pel, alginate, and Psl exopolysaccharide production. Alginate production requires fructose-6-phosphate as the precursor for GDP-mannuronic acid. AlgA, AlgC, and AlgD perform this conversion. Since *P. aeruginosa* lacks enzymes required for glycolysis, carbon sources must be converted into tricarboxylic acid (TCA) cycle intermediates before being processed to fructose-6-phosphate via gluconeogenesis. GDP-mannuronic acid is polymerized (Alg8, Alg44, AlgX, and AlgK) and modified (AlgI, AlgJ, AlgF, AlgG, and AlgL) before being exported (AlgE) out to the extracellular space. Several enzymes required for alginate production overlap with Psl and Pel production. The precursors for Psl are sugar nucleotides, including GDP-mannose, UDP-glucose, and dTDP-L-rhamnose, which are derived from fructose-6-phosphate and glucose-6-phosphate from the activity of AlgC, a Psl-specific enzyme PslB, the AlgA homolog WbpW, and an enzyme involved with rhamnose production, RmlC. These precursors are polymerized (PslA, PslE, PslF, PslC, PslH, and PslI), modified (PslG), and exported (PslD). Pel is thought to consist of linear chains of sugar moieties. Sugar nucleotides produced by the metabolic pathway are, again, polymerized (PelF), modified (PelA), and exported (PelB). Currently, the roles of PelG, PelE, and PelC have not been elucidated. c-di-GMP also activates alginate and Pel production. MucR contains a GGDEF domain to synthesize c-di-GMP. c-di-GMP, in turn, binds to the PilZ domain of Alg44 to enhance alginate polymerization. PelD also contains a PilZ domain. Acetyl-CoA, acetyl coenzyme A; IM, inner membrane; OM, outer membrane; PDG, peptidoglycan layer.
doi:10.1128/9781555818524.ch9f1

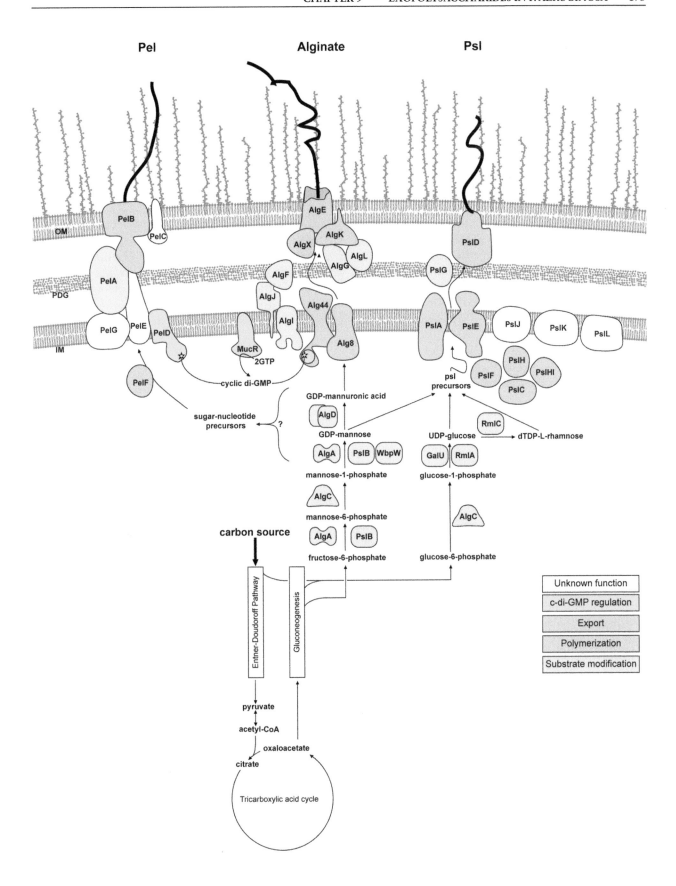

groups identified the genetic loci that produce the Psl polysaccharide, consisting of 15 cotranscribed genes in PAO1 (*pslABCDEFGHIJKLMNO*, PA2231 to PA2245 on the PAO1 chromosome) and 11 genes in *P. aeruginosa* ZK2870 (Friedman and Kolter, 2004a; Jackson et al., 2004; Matsukawa and Greenberg, 2004). Friedman and Kolter identified the first 11 genes through transposon mutagenesis and sequence analyses, whereas transcriptional analyses performed by the Wozniak and Greenberg groups extended the operon by 4 genes (Friedman and Kolter, 2004a; Jackson et al., 2004; Matsukawa and Greenberg, 2004). Friedman and Kolter also showed that PA14 contains a deletion of *pslA* to *pslD*, thereby explaining differences between their biofilm phenotypes with ZK2870 versus PA14. Analyses of the carbohydrate composition of pellicles from *psl* and *pel* mutants of ZK2879 showed that mannose was greatly reduced in the *psl* mutant pellicle, whereas glucose amounts were reduced in the *pel* mutant pellicle (Friedman and Kolter, 2004b). The pellicle is a term used to describe the biomass that forms at the air-liquid interface of bacterial cells grown under static conditions (Friedman and Kolter, 2004). The report from the Greenberg laboratory also showed decreased amounts of mannose in the PAO1 *pslA* mutant; however, they state that rhamnose levels were also low (Matsukawa and Greenberg, 2004). Deletions of *pslA* or *pslB* in PAO1 severely attenuated biofilm initiation, as assessed by a microtiter plate and flowthrough biofilm systems (Jackson et al., 2004). Further characterization of the carbohydrate composition by chemical analyses, mannose and galactose-lectin staining, confocal microscopy, and electron microscopy showed that Psl is a galactose- and mannose-rich exopolysaccharide (Ma et al., 2007). Psl can also contribute to adherence to mucin-coated surfaces and airway epithelial cells (Ma et al., 2006); however, its role is indirect in the activation of NF-κB in A549 cells (Byrd et al., 2010). In an evaluation of three biofilm mediators, bis-(3'-5')-cyclic dimeric-GMP (c-di-GMP), flagella, and quorum sensing, it was determined that Psl (a c-di-GMP-dependent polysaccharide) was associated with nonviable cells in a chinchilla ear bulla infection model (Byrd et al., 2011). Altogether, these studies provide ample evidence that Psl is required for the initial stages of biofilm formation and is important in a rugose small-colony variant (RSCV) that has been isolated from CF sputa (see below) and in other relevant biofilm infection models, such as the chinchilla ear bulla model.

The *pel* gene cluster was discovered through a random transposon mutagenesis screen looking for genes that interfered with pellicle biofilm formation in strain PA14. Further characterization of the *pel* operon in PAK showed that Pel was required for early- and late-stage biofilm formation. The early biofilm stage effects of Pel were observed only when PAK was not piliated (Vasseur et al., 2005). The *pel* and *psl* gene clusters are also induced in some CF isolates that exhibit an RSCV phenotype that autoaggregates (Starkey et al., 2009). The structure of the Pel polysaccharide has not been fully characterized, but there is evidence that it is a glucose-rich exopolysaccharide (Friedman and Kolter, 2004b). There are seven biosynthetic genes arranged into what appears to be a single operon termed *pelA–G* (PA3064 to PA3058 on the PAO1 chromosome). Since *pel* was discovered relatively recently, the functions and locations for most of the proteins have been predicted based on homologies to other proteins or enzymes involved in exopolysaccharide biosynthesis. The reader is referred to Franklin et al., who presented a cogent model for how these proteins may interact based upon structural predictions and homologies (Fig. 1) (Franklin et al., 2011).

The roles of alginate, Psl, and Pel have been examined extensively in biofilms individually (Cochran et al., 2000; Hentzer et al., 2001; Mathee et al., 2002; Friedman and Kolter, 2004a, 2004b; Jackson et al., 2004; Matsukawa and Greenberg, 2004). Most recently, mutants deficient in one or more polysaccharides and their effects on biofilm architecture were examined. Alginate-deficient mutants developed biofilms with a decreased proportion of viable cells and had more extracellular DNA containing surface structures. Deletion of *pslA* and *alg8* resulted in cells that overproduced Pel, whereas deletion of either Psl or alginate genes abrogated the characteristic mushroom-like structure observed in later stage biofilms. *P. aeruginosa* containing deletion of the genes encoding Psl and Pel lost their ability to form biofilms altogether. Deletion of Psl enhanced Pel production, and the absence of Pel enhanced alginate production. Altogether, these data show that alginate, Psl, Pel, and extracellular DNA interactively contribute to some aspect of *P. aeruginosa* biofilm architecture (Ghafoor et al., 2001).

Although alginate, Psl, and Pel are the most-studied exopolysaccharides, other exopolysaccharides important to *P. aeruginosa* include rhamnolipids and various other sugars. A recent characterization of the exopolysaccharides produced by strain PA14 in a biofilm revealed that this strain produced three extracellular carbohydrates: an LPS-like material consisting of a trisugar moiety, an extracellular O polysaccharide, and cyclic β(1,3)-glucans (Coulon et al., 2010). These results indicate that we are just beginning to characterize the exopolysaccharides produced by *P. aeruginosa*. Despite our current level of understanding regarding the exopolysaccharides produced by *P. aeruginosa*, there are multiple connections at the biosynthetic and regulatory levels for the control and production of alginate, Psl, and Pel. The similarities among the biosynthetic pathways of alginate, Pel, and Psl were recently presented in a review (Franklin et al., 2011). The regulatory mechanisms and connections among alginate, Psl, and Pel are presented in the rest of this chapter.

ALGINATE REGULATION

Two-Component Regulatory Systems That Control Alginate Production

AlgR and AlgZ

Studies examining the regulation of alginate production were initiated when AlgR was identified as a regulator of alginate production (Deretic et al., 1989). AlgR is a 27.6-kDa LytR-type transcriptional regulator that likely contains two domains, an amino-terminal CheY-like domain and a carboxyl-terminal AgrA-like DNA binding domain (Whitchurch et al., 2002; Sidote et al., 2008). AlgR interacts with the *algD* and *algC* promoters (P*algD* and P*algC*, respectively) by binding to three semiconserved sites (Mohr et al., 1990; Kato and Chakrabarty, 1991; Mohr et al., 1991; Zielinski et al., 1991; Mohr et al., 1992). Initiation of transcription from P*algD* requires AlgR, as well as the alternative sigma factor AlgU (see below and Martin et al., 1993). All three AlgR binding sites are required for maximal activation of P*algD*. The three AlgR binding sites within P*algD* (Kimbara and Chakrabarty, 1989; Mohr et al., 1990; Mohr et al., 1991; Mohr et al., 1992) have been proposed to interact when AlgR, integration host factor (IHF), and AlgP cause a looping of the P*algD* region such that the three sites come in contact with RNA polymerase (RNAP) holoenzyme and initiate transcription (Fig. 2C) (Kimbara and Chakrabarty, 1989; Mohr et al., 1992; Fujiwara et al., 1993). AlgR also regulates expression of the *algC* gene, encoding phosphomannomutase, the enzyme required for the second step of the alginate biosynthetic pathway. The promoter for the *algC* gene also contains three AlgR binding sites, termed ABSs, similar to *algD* (Zielinski et al., 1992). However, unlike for P*algD*, all three of the AlgR binding sites of P*algC* are situated in the direction of transcription, while ABS3 is located 8 bp into the *algC* open reading frame (ORF) (Zielinski et al., 1991; Zielinski et al., 1992; Fujiwara et al., 1993). In contrast to the *algD* promoter, the *algC* promoter appears to be controlled by the sigma factor RpoN,

as its expression decreased by 50% in a *rpoN* mutant (Zielinski et al., 1992). Surprisingly, the expression of *algC* has been examined, almost exclusively, in a single clinical CF isolate background, a mucoid strain (8821), its nonmucoid revertant (8822), and a stable, mucoid, ethane methyl sulfonate (EMS)-generated derivative of 8821, 8830 (Zielinski et al., 1992; Davies et al., 1993; Fujiwara and Chakrabarty, 1994; Davies and Geesey, 1995; Tavares et al., 1999). It is possible that *algC* expression is controlled by AlgU and RpoN, as has been found with the *algD* promoter, but further studies are needed in different isogenic backgrounds (Boucher et al., 2000). However, if *algC* expression is dependent upon RpoN alone, this would indicate that AlgR is able to modulate gene expression through two different alternative sigma factors in the same biosynthetic pathway. Although the AlgR binding site orientations and positions on the promoter differ between *algD* and *algC*, both promoters are activated by AlgR and are stimulated by treatment with 0.3 M NaCl (Berry et al., 1989; Zielinski et al., 1992). Other environmental signals for these promoters include starvation for carbon and nitrogen, heat shock, and oxidative stress (Krieg et al., 1986; Krieg et al., 1988; Terry et al., 1991, 1992; Schurr et al., 1995; Yu et al., 1995; Mathee et al., 1999).

The role of AlgR in *P. aeruginosa* gene expression has expanded beyond alginate regulation (Whitchurch et al., 1996; Lizewski et al., 2002; Carterson et al., 2004; Lizewski et al., 2004; Morici et al., 2007). Initially, it was determined that AlgR controls twitching motility (Whitchurch et al., 1996). Further studies with AlgR demonstrated that its expression and overexpression had a significant impact on virulence in a murine septicemia model of infection (Lizewski et al., 2002).

In microarray studies that determined which genes are AlgR dependent, it was discovered that this regulator represses many genes, but it also activates the expression of *algD* and *algC*. The microarray analyses showed that the *fimUVWXY1Y2pilE* operon was not expressed in *algR* mutants. In fact, expression of this operon from a heterologous *tac* promoter restored twitching motility in an *algR* deletion strain (Lizewski et al., 2004). These analyses also showed that AlgR was associated with nitrate, nitrite, or anaerobic metabolism, as AlgR-controlled genes overlapped with those transcriptionally activated by ANR (anaerobic regulation of arginine deiminase and nitrate reduction) (Lizewski et al., 2004). Examples of ANR- and AlgR-regulated genes include *hcnABC* (hydrogen cyanide synthase), *dnr* (dissimilatory nitrate respiration regulator), *PA1557* (*ccoN2*, cytochrome *c* oxidase), *arcDABC* (anaerobic

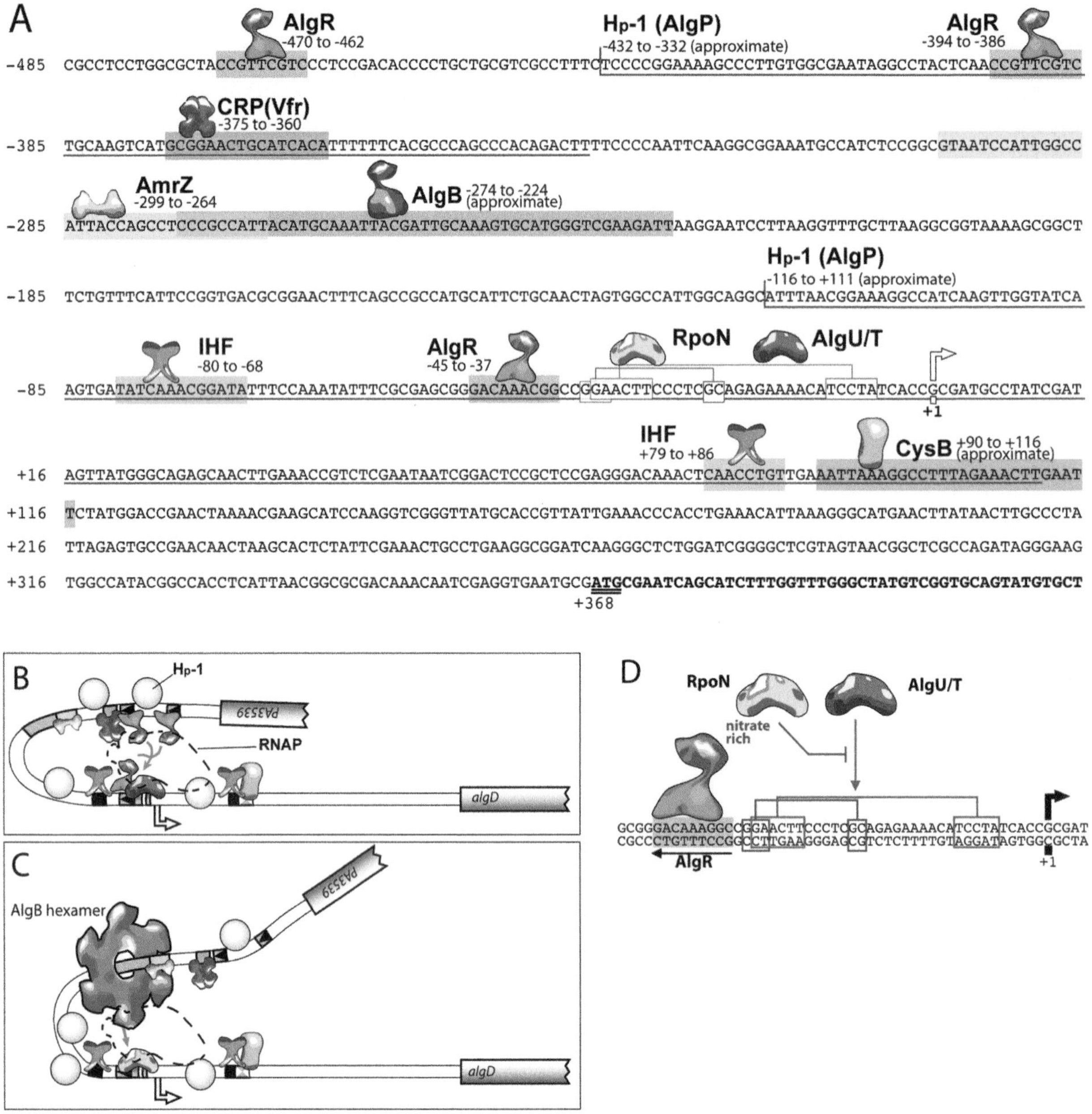

Figure 2. Transcriptional regulation of alginate biosynthetic genes. The transcription of the major operon (*algD*, *alg8*, *alg44*, *algK*, *algE*, *algG*, *algX*, *algL*, *algI*, *algJ*, *algF*, and *algA*) encoding the biosynthetic enzymes and membrane-associated polymerization, modification, and export proteins is regulated by the promoter region of *algD* (A). Transcriptional regulators (AlgR, AlgB, AmrZ, CysB, and CRP [from *E. coli*]), histone-like proteins (H$_p$-1 and IHF), and two sigma factors (AlgU/T and RpoN) associate with the DNA and are involved with P*algD* transcription. Numbers underneath the regulator name indicate the regions of the DNA (relative to the transcriptional start site) that have been found to bind the regulator through experimental evidence. Binding sites of the regulators are shaded with their respective colors. H$_p$-1 binding regions are underlined. Sigma factor consensus sequences are boxed. (B and C) Models showing AlgU/T- and RpoN-dependent transcriptional activation. DNA bending is thought to occur with the aid of IHF and H$_p$-1. Transcriptional activators in the far upstream region, such as AlgR (B) or AlgB (C), are then able to activate transcription near the +1 site by interaction with the AlgU-RNAP or RpoN-RNAP complexes, respectively. P*algD* transcription is thought to be regulated by sigma factor competition (D). RpoN and AlgU binding sites overlap, and alginate production is activated depending on the type of stress encountered by the cell (nitrogen-related stress versus cell wall stress).
doi:10.1128/9781555818524.ch9f2

arginine catabolism), and *hemN* (oxygen-independent coproporphyrinogen III oxidase) (Arai et al., 1997; Hasegawa et al., 1998; Rompf et al., 1998; Blumer and Haas, 2000; Lizewski et al., 2002; Carterson et al., 2004; Comolli and Donohue, 2004; Lizewski et al., 2004; Cody et al., 2009).

Interestingly, an ANR binding site is present in the *algD* promoter and *P. aeruginosa* is capable of alginate production under anaerobic growth conditions, indicating a putative role for ANR in *algD* transcription (Galimand et al., 1991; Hassett, 1996). However, no further studies have directly tested these observations.

The putative cognate histidine kinase, AlgZ/FimS (encoded by PA5262 on the PAO1 chromosome), was discovered to regulate twitching motility and alginate production in *P. aeruginosa* by two independent research groups, at approximately the same time (Whitchurch et al., 1996; Yu et al., 1997). Whitchurch et al. described *fimS*, a gene encoding a putative sensor kinase directly upstream of *algR*, and demonstrated that inactivation of either *fimS* or *algR* resulted in abrogation of the type IV pilus phenotype, referred to as twitching motility (Whitchurch et al., 1996). Yu et al. identified the same ORF upstream of *algR*, termed it *algZ*, and found that inactivation of *algZ* in a *mucA* mutant background (i.e. *mucA* encodes an anti-sigma factor) resulted in increased alginate production. The literature is further complicated by another AlgZ protein, now called AmzR (PA3385 in PAO1), which was found to bind the *algD* promoter (see below and Baynham and Wozniak, 1996; Baynham et al., 1999; Wozniak et al., 2003a; Ramsey et al., 2005; and Baynham et al., 2006). Interestingly, AlgZ does not contain the conserved nucleotide-binding region typical of most histidine kinases. These data suggest that AlgZ may be a phosphatase in mucoid *P. aeruginosa* because it acted as a negative regulator of alginate production. Taken together, AlgZ is an activator of twitching motility in PAO1 but a negative regulator of alginate production in a mucoid background. There is additional genetic evidence that AlgZ and AlgR interact to control cyanide production (Carterson et al., 2004; Cody, et al., 2009); however, AlgZ has not yet been purified to demonstrate any phosphotransfer activity upon AlgR.

KinB and AlgB

The *algB* gene was identified as a regulator of alginate production, by transposon Tn*501* mutagenesis and complementation in the mucoid clinical strain, FRD1. Interestingly, although *algB*::Tn*501* insertional and *algB* deletion mutants displayed a nonmucoid phenotype (indicating an alginate deficiency), alginate production was not completely abrogated (in contrast with an *algR* deletion) in most growth media. These observations suggested that the *algB* gene product is required for maximal production of alginate but that it is not directly involved in the pathway leading to its biosynthesis (Goldberg and Ohman, 1987). Further studies revealed that *algB* encodes a 50-kDa protein with similarity to the NtrC subfamily of response regulators. Accordingly, AlgB was the first ortholog of the NtrC subfamily to be described for *P. aeruginosa*. The amino-terminal and central domains of AlgB have primary sequences which are highly conserved with those in the CheY-like receiver domains of response regulators and an NtrC subfamily of transcriptional activators, respectively. The central domain contains a potential nucleotide-binding site, which has been implicated in open complex formation of σ^{54} (RpoN)-RNAP-dependent promoters. Furthermore, the carboxyl terminus of AlgB has a helix-turn-helix motif and it can bind the *algD* promoter, thereby suggesting that its principal regulatory mechanism is through DNA-protein interaction (Leech et al., 2008).

Not surprisingly, studies on P*algD* activity showed that AlgB is required for full activation of that promoter. A plasmid-borne *algD*::*cat* transcriptional fusion (encoding chloramphenicol acetyltransferase [CAT]) was tested in two isogenic *P. aeruginosa* FRD derivative strains containing Tn*501* insertions in the *algB* gene, and CAT activity in these strains was reduced at least 20-fold (Wozniak and Ohman, 1991; Goldberg and Dahnke, 1992).

In order to determine if AlgB directly, or indirectly, controls alginate biosynthesis, microarray analyses were used to examine the transcriptional profiles of PAO1 *mucA algB*, PAO1 *mucA kinB*, and PAO1 treated with D-cycloserine (an antibiotic known to increase *algD* expression). Comparison of these three transcriptional profiles showed that AlgB controlled only the *algD* operon and another gene not involved in alginate production. These data suggest that AlgB activates alginate production by binding to P*algD*. Chromosome immunoprecipitation experiments revealed that AlgB bound in vivo to P*algD* but did not bind when AlgB had an amino acid substitution that disrupted its DNA binding domain. AlgB also bound to P*algD* fragments in an electrophoretic mobility shift assay at pH 4.5 but not at pH 8.0. Furthermore, a direct approach using systematic evolution of ligands by exponential enrichment showed AlgB binding to a 50-bp fragment located at bp −224 to −274 relative to the *algD* transcriptional start (Fig. 2) (Leech et al., 2008).

The transcriptional regulation of *algB* is indirectly controlled by the alternative stress sigma factor AlgU (AlgT, σ^{22}, σ^{E}). Evidence for this includes an approximately 50% reduction in *algB::cat* activity in two different genetic backgrounds that had little or no AlgU expression. The authors of this study also mapped the transcriptional initiation site for *algB* and noted that its −10 and −35 regions are homologous to σ⁷⁰ responsive regions of genes in *Escherichia coli*. Transcriptional-mapping experiments also showed that *algB* expression was unaffected in backgrounds that have no AlgU expression. Together these data indicate that AlgU indirectly controls *algB* transcription, but further studies are needed to understand how this occurs (Wozniak and Ohman, 1993).

The gene encoding the cognate histidine kinase for the response regulator AlgB is *kinB* (PA5484 on the PAO1 chromosome), located directly downstream of *algB* on the PAO1 chromosome. Localization studies using Western blot analysis, and *lacZ*/*phoA* fusions, revealed that KinB is localized to the inner membrane and has an amino-terminal domain located in the periplasm. The purified KinB carboxyl terminus was observed to undergo progressive autophosphorylation in vitro in the presence of [γ-³²P]ATP and to transfer the labeled phosphoryl group to purified AlgB. These results provide evidence that *kinB* encodes the AlgB cognate histidine protein kinase (Ma et al., 1997).

The two alginate response regulators, AlgB and AlgR, do not require phosphorylation to activate alginate production, which calls into question the importance of the histidine kinases (KinB and AlgZ) in alginate production. CheY-like receiver domains of bacterial two-component system response regulators have a single phosphorylatable residue (canonically, an aspartate). Typically, phosphorylation alters the conformation of the response regulator to modulate its activity. Current evidence suggests that phosphorylation of the two alginate response regulators, AlgB and AlgR, is not required to activate alginate production. Phosphodeficient mutants were constructed by replacing the predicted phosphorylation sites of AlgB (D59) and AlgR(D54) with an asparagine (N). These mutations were then transferred to the chromosome of *P. aeruginosa* FRD1 *algB* or *algR* mutants in order to examine their effects on alginate biosynthesis. There were no observed differences in alginate production among the FRD1 wild-type parent, FRD *algBD59N*, and FRD *algRD54N*, thereby indicating that phosphorylation of AlgR or AlgB is not required for the regulation of alginate production. Indeed, the FRD mucoid strain expressing either recombinant AlgB.D59N or AlgBΔ1–145 (in which the first 145 amino acids containing the receiver domain were deleted) remained mucoid, indicating that neither AlgB phosphorylation nor its receiver domain

was required for alginate production (Ma et al., 1998). Consistent with these results, a null mutation of *kinB* in the mucoid background failed to abrogate alginate production as well. Additionally, deletions of the *algZ* gene, encoding the putative histidine kinase for AlgR, actually increased alginate production in the *mucA22* strain PAO568 (Yu et al., 1997).

Interestingly, even though AlgB phosphorylation did not play a role in FRD1 alginate production, a *kinB* deletion in the nonmucoid PAO1 and PAK background strains resulted in a mucoid phenotype (Ma et al., 1997; Ma et al., 1998; Damron et al., 2009; Chand et al., 2011). Since KinB is the histidine kinase for AlgB, and *kinB* deletion resulted in mucoid PAO1 and PAK, these phenotypes indicate that phosphorylated AlgB somehow represses the mucoid phenotype in these strains (Damron et al., 2009; Chand et al., 2011). It is difficult to resolve how a *kinB* deletion in nonmucoid strains led to increased alginate biosynthesis, yet the same deletion in a mucoid background failed to affect alginate levels (Ma et al., 1997; Ma et al., 1998; Damron et al., 2009; Chand et al., 2011). Alternatively, these seemingly contrasting results may be due to the different backgrounds in which the experiments were conducted, or there is another facet of AlgB phosphorylation not currently understood.

ADDITIONAL DNA-BINDING PROTEINS THAT REGULATE ALGINATE PRODUCTION: IHF, AlgP, AmrZ, CysB, AND Vfr

Since two of the three identified AlgR DNA binding sites for the *algD* promoter are located at −479 to −457 and −400 to −380 bp from the transcriptional start site, it has been suggested that DNA looping is responsible for bringing AlgR from these far upstream sites into contact with RNAP, in order to activate *algD* transcription. AlgR by itself does not bend DNA in vitro, indicating that other transcriptional regulators likely play this role in *algD* expression (Mohr and Deretic, 1992). Two different histone-like proteins have been shown to affect the transcription of the *algD* promoter: IHF and AlgP (also known as AlgR3 and Hₚ-1) (Deretic and Konyecsni, 1990; Kato et al., 1990; Konyecsni and Deretic, 1990; Deretic et al., 1992; Mohr and Deretic, 1992; Wozniak, 1994). IHF is a histone-like protein that bends DNA in order to bring transcriptional regulators into close proximity of RNA polymerase. Three independent studies identified two IHF binding sites within the *algD* promoter: one IHF binding site is located −50 to −85 relative to the transcriptional start site, and another is located 3′ to the transcriptional start site, covering bases 77 to 104 (Fig. 2A and B) (Mohr and

Deretic, 1992; Wozniak, 1994; Delic-Attree et al., 1997). Both studies were conducted with purified *E. coli* IHF. Purified *P. aeruginosa* IHF also bound to the *algD* promoter in vitro (Delic-Attree et al., 1997). Additionally, site-directed mutagenesis and deletion of the IHF binding sites within the *algD* promoter reduced plasmid-borne *algD::cat* and *algD::xylE* transcription 4- and 10-fold, respectively, supporting the idea that *P. aeruginosa* IHF binds to the *algD* promoter and influences its expression (Konyecsni and Deretic, 1988; Wozniak, 1994). Additionally, deletion of *P. aeruginosa ihf* from a mucoid strain resulted in 50% less alginate production and decreased *algD* expression (Delic-Attree et al., 1996). A subsequent study showed that in vitro relocation of the second IHF binding site to several different positions on a 175-bp DNA fragment induced looping-out of DNA (Delic-Attree et al., 1997). The apparent affinity of IHF for *ihf2* was 10-fold higher than for *ihf1*, indicating that IHF could facilitate the interaction between RNA polymerase and a factor bound to the 3′ region of the *algD* promoter (see discussion of CysB below) (Wozniak, 1994; Delic-Attree et al., 1996; Delic-Attree et al., 1997).

AlgP (AlgR3, H_p-1; PA5253 on the PAO1 chromosome) is another histone-like protein (34.5 kDa), identified independently through complementation of an alginate-deficient allele (*alg52*) of strain 8821 (Deretic and Konyecsni, 1990; Kato et al., 1990). Deletion of the H_p-1 ORF decreased plasmid-borne complementation of the *alg52* allele fivefold (Konyecsni and Deretic, 1990). The nucleic acid sequence of the carboxyl-terminal portion of H_p-1 contains multiple direct repeats organized in six highly conserved, tandemly arranged 75-bp units, which encode a KPAA motif or single amino acid substitution variations of this motif. Purified H_p-1 has not been demonstrated to directly interact with the *algD* promoter. However, a synthetic 53-amino-acid peptide, termed P616, with a sequence identical to a region of AlgP bound to the far upstream sequence in the *algD* promoter (−432 to −332) and to a region overlapping the transcriptional start site of *algD* (−116 to 111) (Fig. 2) (Deretic and Konyecsni, 1990). Importantly, the DNA binding of H_p-1 was not specific to the *algD* promoter, thereby indicating that this protein acts as a histone (Deretic and Konyecsni, 1990). Monoclonal and polyclonal antibodies raised against P616 recognized two proteins (46.4 and 41.6 kDa) in cell-free extracts of *P. aeruginosa* PAO1, indicating that H_p-1 may be posttranslationally processed or that there are other H1-like histone elements in *P. aeruginosa*. Cellular localization studies using immunogold-labeled α-H_p-1 antibody and transmission electron microscopy showed that H_p-1 is localized mainly to the ribosomal region within the cytoplasm, coinciding with the hypothesis that this protein is a histone-like element that influences transcription of many *P. aeruginosa* genes (Deretic et al., 1992).

AmrZ is a proposed member of the ribbon-helix-helix family of DNA-binding proteins originally identified by electrophoretic mobility shift assay using cell extracts derived from the mucoid strain FRD1. The authors initially termed this protein AlgZ, as the first report was published at approximately the same time that *algZ*, the putative histidine kinase for AlgR, was characterized. The AmrZ binding site on P*algD* was mapped to −282 (Fig. 2) by copper-phenanthroline footprinting, as well as by mutagenesis of the binding site, which abrogated in vitro binding and led to a 28-fold reduction in *algD::cat* transcriptional activity. Additionally, cell extracts from seven mucoid clinical isolates bound to an AmrZ-specific DNA fragment, whereas seven extracts from nonmucoid isolates failed to bind this fragment. Based upon these results, the authors concluded that AmrZ is an activator of the *algD* promoter (Baynham and Wozniak, 1996). Further characterization of the AmrZ protein, through purification and reverse genetics, identified the gene encoding AmrZ and showed that the monomeric form of this protein is estimated to be between 6 and 15 kDa. The genomic sequence of PAO1 predicts that *amrZ* (PA3385) encodes a 12.3-kDa protein, in agreement with these authors. These investigators further showed that deletion of *amrZ* in FRD1 resulted in no *algD* transcriptional activity and a nonmucoid phenotype (Baynham et al., 1999). Characterization of the *amrZ* transcriptional start site showed that AlgU is required for *amrZ* transcription, while the −35 and −10 regions upstream contain the AlgU consensus sequence. Mutation of the AlgU −10 and −35 sequences within the *amrZ* promoter reduced *amrZ* transcription over 85% (Wozniak et al., 2003a). AmrZ is also required for type IV pilus function and repressed the flagellar motility in mucoid *P. aeruginosa* through control of the *fleQ* promoter (Baynham et al., 2006; Tart et al., 2006). Recently, residues 18, 20, and 22 of AmrZ, which are associated with its β-sheet structure, were shown to be critical for its DNA binding. Moreover, AmrZ was shown to be required for full virulence of *P. aeruginosa* in an acute murine pneumonia infection model (Waligora et al., 2010).

While examining the IHF binding sites of the *algD* promoter in a CF clinical mucoid *P. aeruginosa* isolate, Deli-Attree et al. discovered that CysB, a LysR family transcriptional regulator, bound to the 3′ untranslated region of the *algD* promoter (Fig. 2A to C). Deletion of *P. aeruginosa cysB* resulted in cysteine auxotrophy and decreased P*algD* activity by 90%

in minimal media supplemented with 0.3 M NaCl (Delic-Attree et al., 1997). The authors concluded from these data that CysB acts as an activator of *algD* expression. Due to the CysB and IHF binding site locations, the authors suggest that these regulators may interact to activate *algD* transcription. They further hypothesized that *P. aeruginosa* CysB plays a role in maintaining the 3′ region of the *algD* promoter in a supercoiled state, based upon evidence that *E. coli cysB* is involved in resistance to DNA gyrase (Delic-Attree et al., 1997).

The *algD* promoter is sensitive to glucose repression both in *E. coli* and *P. aeruginosa*, indicating that catabolite repression may operate (or play a role) in the expression of alginate biosynthesis in *P. aeruginosa*. One of the mechanisms for control of catabolite repression in *E. coli* is through the DNA-binding protein cyclic AMP (cAMP) receptor protein (CRP). The *P. aeruginosa* ortholog of the *E. coli* CRP is Vfr (virulence factor regulator), which regulates the transcription of *lasR*, the type III secretion system, flagellar genes, and twitching motility (Albus et al., 1997; Beatson et al., 2002; Dasgupta et al., 2002; Wolfgang et al., 2003). In order to determine if CRP controlled the *algD* promoter, a putative consensus CRP binding sequence upstream of the *algD* promoter was deleted to render the promoter nonresponsive to glucose repression. The involvement of cAMP-CRP complex in the activation of the *algD* promoter in *E. coli* was also demonstrated directly through binding of a 255-bp DNA fragment containing the putative consensus CRP binding sequence (DeVault et al., 1991). These data indicate that Vfr and cAMP may directly control *algD* expression, but no studies have been conducted to examine this in any mucoid *P. aeruginosa* strain. Another observation that links Vfr to the alginate system is that Vfr binds to the *algZ* promoter in vitro (Kanack et al., 2006). Transcriptional fusion data, obtained using a single chromosomal copy of *algZ::lacZ* in different backgrounds, show that Vfr is required for *algZ* transcription (C. L. Pritchett and M. J. Schurr, unpublished data). Thus far, nine DNA-binding proteins have been identified to directly interact with P*algD*, making it challenging to come up with a single model that incorporates all of these proteins (Fig. 2A). There are at least three DNA-bending proteins involved in *algD* expression: IHF, AmrZ, and H$_p$-1. These three proteins may or may not bind at the same time to P*algD* to control its expression. Furthermore, creation of double and triple deletion mutants will result in assorted, difficult-to-interpret, pleiotropic effects. It is likely that these three proteins do interact with P*algD*, either individually or in some combination, to induce looping of the promoter region under certain physiological conditions that have

yet to be determined. There are also two sigma factors involved in controlling *algD* transcription, AlgU and RpoN. Consequently, it has been proposed that antagonism is a mechanism by which both are able to initiate *algD* transcription (Boucher et al., 2000) (Fig. 2D). The best-studied sigma factor for alginate production, AlgU, probably activates transcription with AlgR, AmrZ, and some combination of the other two DNA-bending proteins (Fig. 2B). Evidence for this includes the fact that AlgU directly transcribes *algR* and *amrZ*, as well as *algD*, and both AlgU and AlgR are required for alginate production in multiple mucoid strains (Martin et al., 1994; Wozniak et al., 2003a). It appears from the literature that there may be other strains besides *muc-23* that are RpoN dependent, but the frequency of these strains among clinical isolates is not known. Since AlgB is an NtrC-like transcriptional regulator, it is likely that AlgB and RpoN work in concert to transcribe P*algD* under nitrogen-limiting, or other physiological, conditions distinct from membrane perturbation (Fig. 2C). The position of the AlgB binding site is significantly removed from where a typical NtrC-like activating binding site should be located. However, all of the other DNA-bending proteins may compensate for this apparent dislocation.

POSTTRANSCRIPTIONAL REGULATORS OF ALGINATE PRODUCTION

AlgU (AlgT, σ^{22}, RpoE), MucABCD, AlgW, ClpXP, MucP (YaeL), and Prc

P. aeruginosa has numerous genes and phenotypes that contribute to the persistence in the CF lung. The phenotype known as mucoidy is caused by the overproduction of the exopolysaccharide alginate. Initially, nonmucoid environmental strains colonize the lungs of CF patients. Isolation of stable mucoid strains from sputum samples signals the onset of chronic infection of the CF lung (Govan and Deretic, 1996). One of the master regulators of alginate production is the sigma factor σ^{22}, also known as AlgU or AlgT and the *E. coli* homolog of RpoE (Flynn and Ohman, 1988; Martin et al., 1993; Hershberger et al., 1995; Yu et al., 1995; Mathee et al., 1997; Keith and Bender, 1999). AlgU activates alginate overproduction through regulating the expression of transcription factors, like *algR* and *amrZ* (Deretic et al., 1989; Martin et al., 1994; Baynham et al., 1999), leading to transcriptional activation of the *algD* biosynthetic operon (Martin et al., 1993; DeVries and Ohman, 1994; Hershberger et al., 1995).

The central paradigm for the conversion to the mucoid phenotype is centered on the operon containing the *algUmucABCD* genes (PA0762 to PA0766 on the PAO1 chromosome). Constitutively mucoid CF isolates frequently harbor mutations in *mucA* (Martin et al., 1993), but mutations in the other negative regulators of AlgU, such as *mucB* and *mucD*, have been identified in the mucoid CF isolates (Ciofu et al., 2008). When this chromosomal region was first attributed to mucoidy, it was observed to undergo recombination events analogous to phase variation in a mucoid CF isolate, FRD1. The authors referred to this "genetic switch" as *algT* and *algS* (Flynn and Ohman, 1988). Subsequent studies by another group identified this region independently through suppression of several clinical and laboratory-generated mucoid strains using a cosmid containing strain PAO1 chromosomal DNA. These investigators identified an ORF independently named *algU* that was later found to be identical to the previously characterized *algT* gene.

The *algU* gene encodes a sigma factor (AlgU), homologous to RpoE of *E. coli* (Martin et al., 1993; Hershberger et al., 1995; Schurr et al., 1995; Yu et al., 1995), that was shown to be required for in vitro transcription from the *E. coli rpoH P3* promoter (Hershberger et al., 1995). Moreover, *E. coli* RpoE complements a *P. aeruginosa* Alg⁻ insertion mutant (i.e., *algU*::Tcʳ) back to its mucoid (Alg⁺) phenotype (Yu et al., 1995). Accordingly, AlgU should be considered to be a functional ortholog of the alternative extracytoplasmic sigma factor *E. coli* RpoE (Yu et al., 1995; Keith and Bender, 1999; Wood et al., 2006). Directly 3′ to *algU*, and potentially in the same operon, are four other genes: *mucA* (Martin et al., 1993), *mucB* (*algN*) (Goldberg et al., 1993; Martin et al., 1993), *mucC* (Martinez-Salazar et al., 1996; Boucher et al., 1997), and *mucD* (Boucher et al., 1996).

MucA is a 21-kDa, membrane-spanning, anti-sigma factor that normally sequesters AlgU to the inner cytoplasmic membrane of *P. aeruginosa* (Fig. 3) (Rowen and Deretic, 2000). Mutations in *mucA* are frequently present in chronic mucoid isolates from CF patients, which can then lead to constitutive expression of mucoidy, since AlgU in these strains is then free to increase expression of *algD*, *algR*, or any other AlgU-regulated gene. MucA contains multiple protease cleavage sites and releases AlgU under certain membrane stress conditions such as heat, oxidative stress, and membrane-perturbing antibiotics (Schurr and Deretic, 1997; Mathee et al., 1999; Edwards and Saunders, 2001; Wood et al., 2006; Damron et al., 2011). MucA can be degraded by three different proteases: AlgW (*E. coli* DegP homolog), MucP (*E. coli* RseP, YaeL homolog), and ClpXP, which are all involved in sensing membrane stress in *P. aeruginosa*

(Qiu et al., 2007). The AlgW protease can be activated by accumulation of envelope proteins that have a WVF amino acid motif that activates AlgW proteolytic activity (Fig. 3) (Qiu et al., 2007; Cezairliyan and Sauer, 2009). Once an unfolded, or partially degraded, envelope protein binds to and activates AlgW, cleavage of MucA occurs at several specific residues in the periplasmic C terminus of MucA (Cezairliyan and Sauer, 2009). It is hypothesized that after AlgW cleaves MucA, MucP will then further digest MucA (Qiu et al., 2007), resulting in activated AlgU. However, ClpXP is required to digest the remaining cytoplasmic portion of MucA that is released from the inner membrane (Fig. 3) (Qiu et al., 2008). An exception to this scenario has been proposed where it appears MucP cleaves MucA, independent of AlgW (Damron and Yu, 2011).

The *mucB* gene was initially characterized as a negative regulator of alginate production encoding a 35-kDa protein located in the periplasm (Martin et al., 1993). This ORF was also termed AlgN by an independent group and shown to have the same function in different mucoid *P. aeruginosa* isolates (Goldberg et al., 1993). Recently, it has been shown that MucB dimerizes and protects the C terminus of MucA from proteolytic degradation (Cezairliyan and Sauer, 2009).

The fourth gene of the *algUmucABCD* operon encodes a 15.9-kDa protein, MucC, predicted to be an inner membrane-associated protein. Unlike insertional inactivation of *mucA*, *mucB*, or *mucD*, inactivation of *mucC* in the wild-type *P. aeruginosa* strain PAO1 did not result in alginate production. However, it affected growth of *P. aeruginosa* under conditions of combined elevated temperature and increased ionic strength or osmolarity. Inactivation of *mucC* in *mucA* or *mucB* mutant backgrounds resulted in a mucoid phenotype when the cells were grown under the same combined stresses. Interestingly, the combined stress factors were not sufficient to cause enhancement of alginate production in *mucA* or *mucB* mutants unless *mucC* was also inactivated. These findings support a negative regulatory role of *mucC* in alginate production by *P. aeruginosa* (Boucher et al., 1997). Currently, there are transposon mutants of *mucC* within the University of Washington *Pseudomonas* transposon library (http://www.gs.washington.edu/labs/manoil/libraryindex.htm) that are mucoid, in contrast to the published data (Boucher et al., 1997) (M. L. Vasil and M. J. Schurr, unpublished observations). Insertional inactivation of *mucC* in *Azotobacter vinelandii*, another bacterium capable of alginate production, also resulted in increased alginate production (Nunez et al., 2000). Taken together, these data indicate that the role of *P. aeruginosa* MucC in alginate production needs further study.

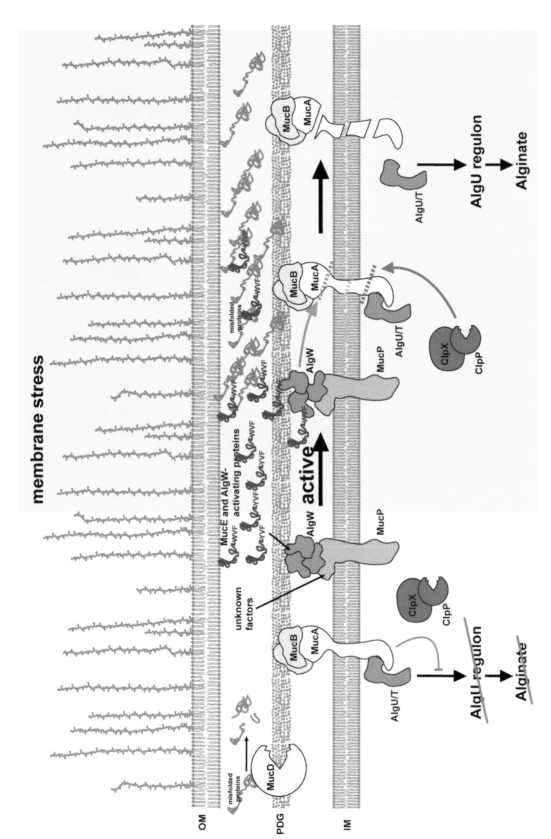

Figure 3. Posttranslational regulatory system for alginate production. The cell wall stress-sensing mechanism that is intimately linked with alginate is encoded by the genes *algUmucABCD*. MucA is a membrane-spanning anti-sigma factor that represses the AlgU sigma-factor regulon by sequestering AlgU to the membrane. MucB protects MucA from degradation. MucD is a general protease that scans the periplasm for misfolded or excess proteins to be degraded. Increase in membrane stress results in MucE or other proteins to be degraded in a way to reveal a WVF or YVF triple-residue motif, which is recognized by the protease AlgW. An unidentified signal activates the protease AlgW. AlgW, MucP, and ClpP cleave MucA at either the cytoplasmic domain, the inner membrane, or the periplasmic domain, which results in AlgU/T release and activation of *PalgD*. IM, inner membrane; OM, outer membrane; PDG, peptidoglycan layer.

doi:10.1128/9781555818524.ch9f3

The *mucD* gene was identified in conjunction with *algW*, and the predicted gene products of *mucD* and *algW* showed similarities to *E. coli* HtrA. Inactivation of *mucD* on the PAO1 chromosome resulted in conversion to the mucoid phenotype. This mutation in *mucD* also caused increased sensitivity to H_2O_2 and heat killing (Boucher et al., 1996). MucD is a serine HtrA-like chaperone protease which also participates in modulating AlgU activity through regulating protein quality in the periplasm and MucA degradation (Yorgey et al., 2001; Wood and Ohman, 2006; Damron and Yu, 2011).

Another protease that affects the mucoid phenotype is encoded by *prc* (PA3257 on the PAO1 chromosome). This periplasmic protease was identified by complementation of spontaneous revertants of PAO578 (*mucA22*), from the mucoid phenotype to the nonmucoid phenotype, on LB medium. The authors refer to these mutations as *som* (suppressor of mucoidy). Introduction of plasmid-borne *prc* restored the mucoid phenotype to two different *som* revertants. These nonmucoid *som* revertants contained mutations in *prc*. Additionally, insertional inactivation of *prc* in the mucoid strain PAO578 reduced alginate levels to those observed in *som* revertants. Further characterization of *prc* in clinical CF isolates showed that Prc affects alginate production in these strains when *prc* is insertionally inactivated. However, mutation of *prc* in PAO1 had no effect on alginate production. Taken together, these findings led the authors to speculate that Prc acts to promote the degradation of MucA by cleaving off the C terminus of truncated forms but not wild-type forms of MucA (Reiling et al., 2005).

THE ROLE OF c-di-GMP IN ALGINATE, Pel, AND Psl REGULATION

Recently, the intracellular second messenger c-di-GMP has been discovered to play a role in the production of alginate, Pel, and Psl. Mounting evidence suggests that c-di-GMP positively regulates the production of biofilm matrix components at the transcriptional and allosteric levels in *P. aeruginosa* and other gram-negative species. The PilZ domain has been recognized as a c-di-GMP binding motif and was identified at the amino-terminal region of Alg44 through bioinformatics by Amikam and Galperin (Amikam and Galperin, 2006). Subsequent studies showed that this domain binds c-di-GMP and that elevated c-di-GMP levels, through overexpression of diguanylate cyclases, increased alginate production. Conversely, decreasing c-di-GMP pools by overexpression of phosphodiesterases reduced alginate

levels. Additionally, substitution of conserved residues in the PilZ domain required for c-di-GMP binding abrogated alginate production (Merighi et al., 2007). These authors localized Alg44 to either the inner or outer membrane fraction, whereas a previous study localized Alg44 to the periplasmic space (Remminghorst and Rehm, 2006). Localization of Alg44 to the inner membrane would be consistent with a model where c-di-GMP binds to an amino-terminal cytoplasmic PilZ domain and affects the periplasmic NolF domain. Localization of Alg44 to the periplasmic space would require that c-di-GMP would have to be pumped into the periplasm by some mechanism, in order to bind Alg44. However, whether Alg44 is a single-pass transmembrane protein or whether it is merely associated with the membrane remains to be determined.

MucR is a membrane-bound diguanylate cyclase and a positive regulator of alginate production, as deletion of *mucR* abrogated alginate synthesis in a mucoid *mucA22* derivative of PAO1, PDO300 (Mathee et al., 1999). Deletion of *mucR* also affected other known c-di-GMP processes in *P. aeruginosa*, including swarming motility, biofilm formation, and alginate production. This study additionally showed, through the use of *lacZ* and *phoA* fusions, that the predicted amino-terminal MHYT domain of MucR resides on the inner membrane. MHYT domains are proposed to bind O_2, NO, or CO. Moreover, O_2 and NO are known to stimulate alginate production (Galperin et al., 2001; Worlitzsch et al., 2002; Bragonzi et al., 2005; Wood et al., 2006). The authors of this study propose a model for c-di-GMP regulation of alginate production whereby the guanylate cyclase of MucR is stimulated by a yet-to-be-identified signal, which binds to a predicted MHYT domain in the amino terminus of MucR. Binding to the MucR MHYT domain stimulates the production of c-di-GMP locally, whereby this increased local concentration is sensed by the cytoplasmic PilZ domain of Alg44, resulting in increased alginate production (Fig. 1) (Hay et al., 2009).

Starkey et al. showed that levels of c-di-GMP also control the Psl and Pel extracellular polysaccharides by transcriptional profiling (Starkey et al., 2009). One of the genes induced by increased levels of c-di-GMP is a secreted adhesin termed CdrA (PA4625 on PAO1 chromosome). CdrA is expressed as a 220-kDa precursor and then processed to 150 kDa before its extracellular secretion as a rod-shaped protein containing a β-helical motif domain. CdrA is encoded in a two-gene operon, i.e., *cdrAB*, with *cdrB* potentially encoding an outer membrane protein. Evidence that CdrA is involved in the extracellular biofilm matrix includes the following: (i) mutation of *cdrA* decreased biofilm mass, and (ii) CdrA coimmunoprecipitated with anti-Psl antibodies (Borlee et al., 2010). More recently, c-di-GMP was reported to bind and allosterically regulate the activity of PelD, thereby activating Pel synthesis and secretion (Fig. 1) (Lee et al., 2007).

Pel and Psl Regulation

In contrast to the regulation of alginate production, the conditions that control the expression of the *pel* and *psl* operons are just emerging from the literature. The second messenger c-di-GMP is one molecule that induces Pel and Psl production (see above). Quorum sensing has been suggested to positively regulate *pel* and *psl* expression and to regulate biofilm formation (Davies et al., 1998; Sakuragi and Kolter, 2007; Gilbert et al., 2009). Transcription of the *pel* operon was greatly reduced in *P. aeruginosa* PA14 *lasI* and *rhlI* mutants, and addition of 3-oxo-dodecanoyl homoserine lactone to the *lasI* mutant restored both *pel* transcription to the wild-type level and normal biofilm formation (Sakuragi and Kolter, 2007). These data indicate that the Rhl and Las quorum sensing systems either indirectly or directly control *pel* expression. Gilbert et al. provided direct evidence that the Psl system is regulated by the Las quorum sensing system. These authors identified a *psl*-dependent promoter region that is bound by LasR in vivo by employing chromatin immunoprecipitation in conjunction with microarray analysis. These analyses identified the promoter region of *pslA* as being regulated by LasR (Gilbert et al., 2009). Irie et al. recently determined that Psl expression is regulated positively by RpoS and negatively by RsmA. The authors show that *psl* mRNA has an extensive 5′ untranslated region to which the post-transcriptional regulator RsmA binds, repressing *psl* translation. The authors propose that upon binding RsmA, the *pslA* ribosome binding site region of the mRNA folds into a secondary stem-loop structure that blocks the Shine-Dalgarno sequence, preventing ribosome access and protein translation. The authors further suggest that this constitutes a novel mechanism for translational repression by this family of regulators. Interestingly, RpoS also regulates the expression of alginate production in the mucoid clinical isolate FRD1 (Suh et al., 1999). Recent data from our laboratory suggest that *rpoS* deletion abrogates *algR* expression, which would explain the loss of alginate production observed in the FRD1 *rpoS* mutant (Pritchett and Schurr, unpublished).

There are many regulatory points that overlap among the biosyntheses of Psl, Pel, and alginate polysaccharides, including c-di-GMP, RpoS, quorum sensing, and metabolic precursors used for the building blocks of each polysaccharide (Fig. 1). The current state of knowledge indicates that Psl and Pel are likely involved with the initial stages of biofilm development, whereas alginate is the stress response exopolysaccharide. RpoS regulation of Psl expression may indicate that it also is produced in response to certain stress conditions. It may be that there is an order to the expression of these exopolysaccharides depending upon which stresses are encountered first by *P. aeruginosa*. One model that could be proposed based upon the current literature is that Pel is induced initially by the Las quorum sensing system, followed by Psl as the organisms encounter starvation or other stresses. Finally, when the environment becomes so hostile that membrane damage is induced, the organism responds by making alginate. There are likely still more regulatory mechanisms yet to be discovered. The levels of alginate produced by newly generated MucA, MucB, MucC, or MucD mutants of *P. aeruginosa* PAO1 and other nonmucoid clinical isolates (i.e., non-CF isolates) was found to be inversely related to biologically relevant concentrations (e.g., <5 to 100 µM) of iron present in the media used in this study (J. R. Wiens, Y. Okkotsu, M. J. Schurr and M. L. Vasil, unpublished results). It will be exciting to follow the developments that emerge from future studies, and hopefully, a therapeutic targeted to one or all of these systems will emerge from that knowledge.

REFERENCES

Albus, A. M., E. C. Pesci, L. J. Runyen-Janecky, S. E. West, and B. H. Iglewski. 1997. Vfr controls quorum sensing in *Pseudomonas aeruginosa*. *J. Bacteriol.* 179:3928–3935.

Amikam, D., and M. Y. Galperin. 2006. PilZ domain is part of the bacterial c-di-GMP binding protein. *Bioinformatics* 22:3–6.

Anastassiou, E. D., A. C. Mintzas, C. Kounavis, and G. Dimitracopoulos. 1987. Alginate production by clinical nonmucoid *Pseudomonas aeruginosa*. *J. Clin. Microbiol.* 25:656–659.

Arai, H., T. Kodama, and Y. Igarashi. 1997. Cascade regulation of the two CRP/FNR-related transcriptional regulators (ANR and DNR) and the denitrification enzymes in *Pseudomonas aeruginosa*. *Mol. Microbiol.* 25:1141–1148.

Baynham, P. J., A. L. Brown, L. L. Hall, and D. J. Wozniak. 1999. *Pseudomonas aeruginosa* AlgZ, a ribbon-helix-helix DNA-binding protein, is essential for alginate synthesis and algD transcriptional activation. *Mol. Microbiol.* 33:1069–1080.

Baynham, P. J., D. M. Ramsey, B. V. Gvozdyev, E. M. Cordonnier, and D. J. Wozniak. 2006. The *Pseudomonas aeruginosa* ribbon-helix-helix DNA-binding protein AlgZ (AmrZ) controls twitching motility and biogenesis of type IV pili. *J. Bacteriol.* 188:132–140.

Baynham, P. J., and D. J. Wozniak. 1996. Identification and characterization of AlgZ, an AlgT-dependent DNA-binding protein required for *Pseudomonas aeruginosa* algD transcription. *Mol. Microbiol.* 22:97–108.

Beatson, S. A., C. B. Whitchurch, J. L. Sargent, R. C. Levesque, and J. S. Mattick. 2002. Differential regulation of twitching motility and elastase production by Vfr in *Pseudomonas aeruginosa*. *J. Bacteriol.* **184:**3605–3613.

Berry, A., J. D. DeVault, and A. M. Chakrabarty. 1989. High osmolarity is a signal for enhanced *algD* transcription in mucoid and nonmucoid *Pseudomonas aeruginosa* strains. *J. Bacteriol.* **171:**2312–2317.

Blumer, C., and D. Haas. 2000. Iron regulation of the *hcnABC* genes encoding hydrogen cyanide synthase depends on the anaerobic regulator ANR rather than on the global activator GacA in *Pseudomonas fluorescens* CHA0. *Microbiology* **146**(Pt. 10):2417–2424.

Borlee, B. R., A. D. Goldman, K. Murakami, R. Samudrala, D. J. Wozniak, and M. R. Parsek. 2010. *Pseudomonas aeruginosa* uses a cyclic-di-GMP-regulated adhesin to reinforce the biofilm extracellular matrix. *Mol. Microbiol.* **75:**827–842.

Boucher, J. C., J. Martinez-Salazar, M. J. Schurr, M. H. Mudd, H. Yu, and V. Deretic. 1996. Two distinct loci affecting conversion to mucoidy in *Pseudomonas aeruginosa* in cystic fibrosis encode homologs of the serine protease HtrA. *J. Bacteriol.* **178:**511–523.

Boucher, J. C., M. J. Schurr, and V. Deretic. 2000. Dual regulation of mucoidy in *Pseudomonas aeruginosa* and sigma factor antagonism. *Mol. Microbiol.* **36:**341–351.

Boucher, J. C., M. J. Schurr, H. Yu, D. W. Rowen, and V. Deretic. 1997. *Pseudomonas aeruginosa* in cystic fibrosis: role of *mucC* in the regulation of alginate production and stress sensitivity. *Microbiology* **143:**3473–3480.

Bragonzi, A., D. Worlitzsch, G. B. Pier, P. Timpert, M. Ulrich, M. Hentzer, J. B. Andersen, M. Givskov, M. Conese, and G. Doring. 2005. Nonmucoid *Pseudomonas aeruginosa* expresses alginate in the lungs of patients with cystic fibrosis and in a mouse model. *J. Infect. Dis.* **192:**410–419.

Byrd, M. S., B. Pang, W. Hong, E. A. Waligora, R. A. Juneau, C. E. Armbruster, K. E. Weimer, K. Murrah, E. E. Mann, H. Lu, A. Sprinkle, M. R. Parsek, N. D. Kock, D. J. Wozniak, and W. E. Swords. 2011. Direct evaluation of *Pseudomonas aeruginosa* biofilm mediators in a chronic infection model. *Infect. Immun.* **79:**3087–3095.

Byrd, M. S., B. Pang, M. Mishra, W. E. Swords, and D. J. Wozniak. 2010. The *Pseudomonas aeruginosa* exopolysaccharide Psl facilitates surface adherence and NF-κB activation in A549 cells. *mBio* **1:**e00140-10.

Byrd, M. S., I. Sadovskaya, E. Vinogradov, H. Lu, A. B. Sprinkle, S. H. Richardson, L. Ma, B. Ralston, M. R. Parsek, E. M. Anderson, J. S. Lam, and D. J. Wozniak. 2009. Genetic and biochemical analyses of the *Pseudomonas aeruginosa* Psl exopolysaccharide reveal overlapping roles for polysaccharide synthesis enzymes in Psl and LPS production. *Mol. Microbiol.* **73:**622–638.

Carterson, A. J., L. A. Morici, D. W. Jackson, A. Frisk, S. E. Lizewski, R. Jupiter, K. Simpson, D. A. Kunz, S. H. Davis, J. R. Schurr, D. J. Hassett, and M. J. Schurr. 2004. The transcriptional regulator AlgR controls cyanide production in *Pseudomonas aeruginosa*. *J. Bacteriol.* **186:**6837–6844.

Cezairliyan, B. O., and R. T. Sauer. 2009. Control of *Pseudomonas aeruginosa* AlgW protease cleavage of MucA by peptide signals and MucB. *Mol. Microbiol.* **72:**368–379.

Chand, N. S., J. S. Lee, A. E. Clatworthy, A. J. Golas, R. S. Smith, and D. T. Hung. 2011. The sensor kinase KinB regulates virulence in acute *Pseudomonas aeruginosa* infection. *J. Bacteriol.* **193:**2989–2999.

Ciofu, O., B. Lee, M. Johannesson, N. O. Hermansen, P. Meyer, and N. Hoiby. 2008. Investigation of the *algT* operon sequence in mucoid and non-mucoid *Pseudomonas aeruginosa* isolates from 115 Scandinavian patients with cystic fibrosis and in 88 in vitro non-mucoid revertants. *Microbiology* **154:**103–113.

Cochran, W. L., S. J. Suh, G. A. McFeters, and P. S. Stewart. 2000. Role of RpoS and AlgT in *Pseudomonas aeruginosa* biofilm

resistance to hydrogen peroxide and monochloramine. *J. Appl. Microbiol.* **88:**546–553.

Cody, W. L., C. L. Pritchett, A. K. Jones, A. J. Carterson, D. Jackson, A. Frisk, M. C. Wolfgang, and M. J. Schurr. 2009. *Pseudomonas aeruginosa* AlgR controls cyanide production in an AlgZ-dependent manner. *J. Bacteriol.* **191:**2993–3002.

Comolli, J. C., and T. J. Donohue. 2004. Differences in two *Pseudomonas aeruginosa* cbb_3 cytochrome oxidases. *Mol. Microbiol.* **51:**1193–1203.

Coulon, C., E. Vinogradov, A. Filloux, and I. Sadovskaya. 2010. Chemical analysis of cellular and extracellular carbohydrates of a biofilm-forming strain *Pseudomonas aeruginosa* PA14. *PLoS One* **5:**e14220.

Coyne, M. J., K. S. Russell, C. L. Coyle, and J. B. Goldberg. 1994. The *Pseudomonas aeruginosa algC* gene encodes phosphoglucomutase, required for the synthesis of a complete lipopolysaccharide core. *J. Bacteriol.* **176:**3500–3507.

Damron, F. H., M. R. Davis, Jr., T. R. Withers, R. K. Ernst, J. B. Goldberg, G. Yu, and H. D. Yu. 2011. Vanadate and triclosan synergistically induce alginate production by *Pseudomonas aeruginosa* strain PAO1. *Mol. Microbiol.* **81:**554–570.

Damron, F. H., D. Qiu, and H. D. Yu. 2009. The *Pseudomonas aeruginosa* sensor kinase KinB negatively controls alginate production through AlgW-dependent MucA proteolysis. *J. Bacteriol.* **191:**2285–2295.

Damron, F. H., and H. D. Yu. 2011. *Pseudomonas aeruginosa* MucD regulates the alginate pathway through activation of MucA degradation via MucP proteolytic activity. *J. Bacteriol.* **193:**286–291.

Dasgupta, N., E. P. Ferrell, K. J. Kanack, S. E. West, and R. Ramphal. 2002. *fleQ*, the gene encoding the major flagellar regulator of *Pseudomonas aeruginosa*, is sigma70 dependent and is downregulated by Vfr, a homolog of *Escherichia coli* cyclic AMP receptor protein. *J. Bacteriol.* **184:**5240–5250.

Davies, D. G., A. M. Chakrabarty, and G. G. Geesey. 1993. Exopolysaccharide production in biofilms: substratum activation of alginate gene expression by *Pseudomonas aeruginosa*. *Appl. Environ. Microbiol.* **59:**1181–1186.

Davies, D. G., and G. G. Geesey. 1995. Regulation of the alginate biosynthesis gene *algC* in *Pseudomonas aeruginosa* during biofilm development in continuous culture. *Appl. Environ. Microbiol.* **61:**860–867.

Davies, D. G., M. R. Parsek, J. P. Pearson, B. H. Iglewski, J. W. Costerton, and E. P. Greenberg. 1998. The involvement of cell-to-cell signals in the development of a bacterial biofilm. *Science* **280:**295–298.

Delic-Attree, I., B. Toussaint, A. Froger, J. C. Willison, and P. M. Vignais. 1996. Isolation of an IHF-deficient mutant of a *Pseudomonas aeruginosa* mucoid isolate and evaluation of the role of IHF in *algD* gene expression. *Microbiology* **142:**2785–2793.

Delic-Attree, I., B. Toussaint, J. Garin, and P. M. Vignais. 1997. Cloning, sequence and mutagenesis of the structural gene of *Pseudomonas aeruginosa* CysB, which can activate *algD* transcription. *Mol. Microbiol.* **24:**1275–1284.

Deretic, V., R. Dikshit, W. M. Konyecsni, A. M. Chakrabarty, and T. K. Misra. 1989. The *algR* gene, which regulates mucoidy in *Pseudomonas aeruginosa*, belongs to a class of environmentally responsive genes. *J. Bacteriol.* **171:**1278–1283.

Deretic, V., N. S. Hibler, and S. C. Holt. 1992. Immunocytochemical analysis of AlgP (Hp1), a histonelike element participating in control of mucoidy in *Pseudomonas aeruginosa*. *J. Bacteriol.* **174:**824–831.

Deretic, V., and W. M. Konyecsni. 1990. A procaryotic regulatory factor with a histone H1-like carboxy-terminal domain:

clonal variation of repeats within *algP*, a gene involved in regulation of mucoidy in *Pseudomonas aeruginosa*. *J. Bacteriol.* **172:**5544–5554.

DeVault, J. D., W. Hendrickson, J. Kato, and A. M. Chakrabarty. 1991. Environmentally regulated *algD* promoter is responsive to the cAMP receptor protein in *Escherichia coli. Mol. Microbiol.* **5:**2503–2509.

DeVries, C. A., and D. E. Ohman. 1994. Mucoid-to-nonmucoid conversion in alginate-producing *Pseudomonas aeruginosa* often results from spontaneous mutations in *algT*, encoding a putative alternate sigma factor, and shows evidence for autoregulation. *J. Bacteriol.* **176:**6677–6687.

Doggett, R. G. 1969. Incidence of mucoid *Pseudomonas aeruginosa* from clinical sources. *Appl. Microbiol.* **18:**936–937.

Edwards, K. J., and N. A. Saunders. 2001. Real-time PCR used to measure stress-induced changes in the expression of the genes of the alginate pathway of *Pseudomonas aeruginosa. J. Appl. Microbiol.* **91:**29–37.

Engel, J., and P. Balachandran. 2009. Role of *Pseudomonas aeruginosa* type III effectors in disease. *Curr. Opin. Microbiol.* **12:**61–66.

Flynn, J. L., and D. E. Ohman. 1988a. Cloning of genes from mucoid *Pseudomonas aeruginosa* which control spontaneous conversion to the alginate production phenotype. *J. Bacteriol.* **170:**1452–1460.

Flynn, J. L., and D. E. Ohman. 1988b. Use of a gene replacement cosmid vector for cloning alginate conversion genes from mucoid and nonmucoid *Pseudomonas aeruginosa* strains: *algS* controls expression of *algT. J. Bacteriol.* **170:**3228–3236.

Franklin, M. J., D. E. Nivens, J. T. Weadge, and L. P. Howell. 2011. Biosynthesis of the *Pseudomonas aeruginosa* extracellular polysaccharides alginate, Pel and Psl. *Front. Cell. Infect. Microbiol.* **2:**167.

Friedman, L., and R. Kolter. 2004a. Genes involved in matrix formation in *Pseudomonas aeruginosa* PA14 biofilms. *Mol. Microbiol.* **51:**675–690.

Friedman, L., and R. Kolter. 2004b. Two genetic loci produce distinct carbohydrate-rich structural components of the *Pseudomonas aeruginosa* biofilm matrix. *J. Bacteriol.* **186:**4457–4465.

Fujiwara, S., and A. M. Chakrabarty. 1994. Post-transcriptional regulation of the *Pseudomonas aeruginosa algC* gene. *Gene* **146:**1–5.

Fujiwara, S., N. A. Zielinski, and A. M. Chakrabarty. 1993. Enhancer-like activity of AlgR1-binding site in alginate gene activation: positional, orientational, and sequence specificity. *J. Bacteriol.* **175:**5452–5459.

Galimand, M., M. Gamper, A. Zimmermann, and D. Haas. 1991. Positive FNR-like control of anaerobic arginine degradation and nitrate respiration in *Pseudomonas aeruginosa. J. Bacteriol.* **173:**1598–1606.

Galperin, M. Y., T. A. Gaidenko, A. Y. Mulkidjanian, M. Nakano, and C. W. Price. 2001. MHYT, a new integral membrane sensor domain. *FEMS Microbiol. Lett.* **205:**17–23.

Ghafoor, A., I. D. Hay, and B. H. M. Rehm. 2011. Role of exopolysaccharides in *Pseudomonas aeruginosa* biofilm formation and architecture. *Appl. Environ. Microbiol.* **77:**5238–5246.

Gilbert, K. B., T. H. Kim, R. Gupta, E. P. Greenberg, and M. Schuster. 2009. Global position analysis of the *Pseudomonas aeruginosa* quorum-sensing transcription factor LasR. *Mol. Microbiol.* **73:**1072–1085.

Goldberg, J. B., and T. Dahnke. 1992. *Pseudomonas aeruginosa* AlgB, which modulates the expression of alginate, is a member of the NtrC subclass of prokaryotic regulators. *Mol. Microbiol.* **6:**59–66.

Goldberg, J. B., W. L. Gorman, J. Flynn, and D. E. Ohman. 1993. A mutation in *algN* permits *trans* activation of alginate production by *algT* in *Pseudomonas* species. *J. Bacteriol.* **175:**1303–1308.

Goldberg, J. B., and D. E. Ohman. 1987. Construction and characterization of *Pseudomonas aeruginosa algB* mutants: role of *algB* in high-level production of alginate. *J. Bacteriol.* **169:**1593–1602.

Govan, J. R., and V. Deretic. 1996. Microbial pathogenesis in cystic fibrosis: mucoid *Pseudomonas aeruginosa* and *Burkholderia cepacia. Microbiol. Rev.* **60:**539–574.

Govan, J. R. W. 1988. Alginate biosynthesis and other unusual characteristics associated with the pathogenesis of *Pseudomonas aeruginosa* in cystic fibrosis, p. 67–96. *In* E. Griffiths, W. E. Donachie, and J. Stephen (ed.), *Bacterial Infections of Respiratory and Gastrointestinal Mucosae.* IRL Press, Oxford, United Kingdom.

Hasegawa, N., H. Arai, and Y. Igarashi. 1998. Activation of a consensus FNR-dependent promoter by DNR of *Pseudomonas aeruginosa* in response to nitrite. *FEMS Microbiol. Lett.* **166:**213–217.

Hassett, D. J. 1996. Anaerobic production of alginate by *Pseudomonas aeruginosa*: alginate restricts diffusion of oxygen. *J. Bacteriol.* **178:**7322–7325.

Hay, I. D., Z. U. Rehman, A. Ghafoor, and B. H. Rehm. 2010. Bacterial biosynthsis of alginates. *J. Chem. Technol. Biotechnol.* **85:**752–759.

Hay, I. D., U. Remminghorst, and B. H. Rehm. 2009. MucR, a novel membrane-associated regulator of alginate biosynthesis in *Pseudomonas aeruginosa. Appl. Environ. Microbiol.* **75:**1110–1120.

Hay, I. D., O. Schmidt, J. Filitcheva, and B. H. Rehm. 2012. Identification of a periplasmic AlgK-AlgX-MucD multiprotein complex in *Pseudomonas aeruginosa* involved in biosynthesis and regulation of alginate. *Appl. Microbiol. Biotechnol.* **93:**215–227.

Hentzer, M., G. M. Teitzel, G. J. Balzer, A. Heydorn, S. Molin, M. Givskov, and M. R. Parsek. 2001. Alginate overproduction affects *Pseudomonas aeruginosa* biofilm structure and function. *J. Bacteriol.* **183:**5395–5401.

Hershberger, C. D., R. W. Ye, M. R. Parsek, Z.-D. Xie, and A. M. Chakrabarty. 1995. The *algT* (*algU*) gene of *Pseudomonas aeruginosa*, a key regulator involved in alginate biosynthesis, encodes an alternative σ factor (σ^E). *Proc. Natl. Acad. Sci. USA* **92:**7941–7945.

Irie, Y., M. Starkey, A. N. Edwards, D. J. Wozniak, T. Romeo, and M. R. Parsek. 2010. *Pseudomonas aeruginosa* biofilm matrix polysaccharide Psl is regulated transcriptionally by RpoS and post-transcriptionally by RsmA. *Mol. Microbiol.* **78:**158–172.

Jackson, K. D., M. Starkey, S. Kremer, M. R. Parsek, and D. J. Wozniak. 2004. Identification of *psl*, a locus encoding a potential exopolysaccharide that is essential for *Pseudomonas aeruginosa* PAO1 biofilm formation. *J. Bacteriol.* **186:**4466–4475.

Jain, S., and D. E. Ohman. 1998. Deletion of *algK* in mucoid *Pseudomonas aeruginosa* blocks alginate polymer formation and results in uronic acid secretion. *J. Bacteriol.* **180:**634–641.

Jain, S., and D. E. Ohman. 2005. Role of an alginate lyase for alginate transport in mucoid *Pseudomonas aeruginosa. Infect. Immun.* **73:**6429–6436.

Kanack, K. J., L. J. Runyen-Janecky, E. P. Ferrell, S. J. Suh, and S. E. West. 2006. Characterization of DNA-binding specificity and analysis of binding sites of the *Pseudomonas aeruginosa* global regulator, Vfr, a homologue of the *Escherichia coli* cAMP receptor protein. *Microbiology* **152:**3485–3496.

Kato, J., and A. M. Chakrabarty. 1991. Purification of the regulatory protein AlgR1 and its binding in the far upstream region of the *algD* promoter in *Pseudomonas aeruginosa. Proc. Natl. Acad. Sci. USA* **88:**1760–1764.

Kato, J., T. K. Misra, and A. M. Chakrabarty. 1990. AlgR3, a protein resembling eukaryotic histone H1, regulates alginate synthesis in *Pseudomonas aeruginosa. Proc. Natl. Acad. Sci. USA* **87**:2887–2891.

Keiski, C. L., M. Harwich, S. Jain, A. M. Neculai, P. Yip, H. Robinson, J. C. Whitney, L. Riley, L. L. Burrows, D. E. Ohman, and P. L. Howell. 2010. AlgK is a TPR-containing protein and the periplasmic component of a novel exopolysaccharide secretin. *Structure* **18**:265–273.

Keith, L. M., and C. L. Bender. 1999. AlgT (sigma22) controls alginate production and tolerance to environmental stress in *Pseudomonas syringae. J. Bacteriol.* **181**:7176–7184.

Kimbara, K., and A. M. Chakrabarty. 1989. Control of alginate synthesis in *Pseudomonas aeruginosa*: regulation of the *algR1* gene. *Biochem. Biophys. Res. Commun.* **164**:601–608.

Konyecsni, W. M., and V. Deretic. 1988. Broad-host-range plasmid and M13 bacteriophage-derived vectors for promoter analysis in *Escherichia coli* and *Pseudomonas aeruginosa. Gene* **74**:375–386.

Konyecsni, W. M., and V. Deretic. 1990. DNA sequence and expression analysis of *algP* and *algQ*, components of the multigene system transcriptionally regulating mucoidy in *Pseudomonas aeruginosa*: *algP* contains multiple direct repeats. *J. Bacteriol.* **172**:2511–2520.

Krieg, D. P., J. A. Bass, and S. J. Mattingly. 1986. Aeration selects for mucoid phenotype of *Pseudomonas aeruginosa. J. Clin. Microbiol.* **24**:986–990.

Krieg, D. P., J. A. Bass, and S. J. Mattingly. 1988a. Phosphorylcholine stimulates capsule formation of phosphate-limited mucoid *Pseudomonas aeruginosa. Infect. Immun.* **56**:864–873.

Krieg, D. P., R. J. Helmke, V. F. German, and J. A. Mangos. 1988b. Resistance of mucoid *Pseudomonas aeruginosa* to nonopsonic phagocytosis by alveolar macrophages in vitro. *Infect. Immun.* **56**:3173–3179.

Lau, G. W., D. J. Hassett, H. Ran, and F. Kong. 2004. The role of pyocyanin in *Pseudomonas aeruginosa* infection. *Trends Mol. Med.* **10**:599–606.

Lee, V. T., J. M. Matewish, J. L. Kessler, M. Hyodo, Y. Hayakawa, and S. Lory. 2007. A cyclic-di-GMP receptor required for bacterial exopolysaccharide production. *Mol. Microbiol.* **65**:1474–1484.

Leech, A. J., A. Sprinkle, L. Wood, D. J. Wozniak, and D. E. Ohman. 2008. The NtrC family regulator AlgB, which controls alginate biosynthesis in mucoid *Pseudomonas aeruginosa*, binds directly to the *algD* promoter. *J. Bacteriol.* **190**:581–589.

Lizewski, S. E., D. S. Lundberg, and M. J. Schurr. 2002. The transcriptional regulator AlgR is essential for *Pseudomonas aeruginosa* pathogenesis. *Infect. Immun.* **70**:6083–6093.

Lizewski, S. E., J. R. Schurr, D. W. Jackson, A. Frisk, A. J. Carterson, and M. J. Schurr. 2004. Identification of AlgR-regulated genes in *Pseudomonas aeruginosa* by use of microarray analysis. *J. Bacteriol.* **186**:5672–5684.

Ma, L., K. D. Jackson, R. M. Landry, M. R. Parsek, and D. J. Wozniak. 2006. Analysis of *Pseudomonas aeruginosa* conditional Psl variants reveals roles for the Psl polysaccharide in adhesion and maintaining biofilm structure postattachment. *J. Bacteriol.* **188**:8213–8221.

Ma, L., H. Lu, A. Sprinkle, M. R. Parsek, and D. J. Wozniak. 2007. *Pseudomonas aeruginosa* Psl is a galactose- and mannose-rich exopolysaccharide. *J. Bacteriol.* **189**:8353–8356.

Ma, S., U. Selvaraj, D. E. Ohman, R. Quarless, D. J. Hassett, and D. J. Wozniak. 1998. Phosphorylation-independent activity of the response regulators AlgB and AlgR in promoting alginate biosynthesis in mucoid *Pseudomonas aeruginosa. J. Bacteriol.* **180**:956–968.

Ma, S., D. J. Wozniak, and D. E. Ohman. 1997. Identification of the histidine protein kinase KinB in *Pseudomonas aeruginosa* and its phosphorylation of the alginate regulator *algB. J. Biol. Chem.* **272**:17952–17960.

Mai, G. T., W. K. Seow, G. B. Pier, J. G. McCormack, and Y. H. Thong. 1993. Suppression of lymphocyte and neutrophil functions by *Pseudomonas aeruginosa* mucoid exopolysaccharide (alginate): reversal by physicochemical, alginase, and specific monoclonal antibody treatments. *Infect. Immun.* **61**:559–564.

Martin, D. W., B. W. Holloway, and V. Deretic. 1993a. Characterization of a locus determining the mucoid status of *Pseudomonas aeruginosa*: AlgU shows sequence similarities with a *Bacillus* sigma factor. *J. Bacteriol.* **175**:1153–1164.

Martin, D. W., M. J. Schurr, M. H. Mudd, and V. Deretic. 1993b. Differentiation of *Pseudomonas aeruginosa* into the alginate-producing form: inactivation of *mucB* causes conversion to mucoidy. *Mol. Microbiol.* **9**:497–506.

Martin, D. W., M. J. Schurr, M. H. Mudd, J. R. W. Govan, B. W. Holloway, and V. Deretic. 1993c. Mechanism of conversion to mucoidy in *Pseudomonas aeruginosa* infecting cystic fibrosis patients. *Proc. Natl. Acad. Sci. USA* **90**:8377–8381.

Martin, D. W., M. J. Schurr, H. Yu, and V. Deretic. 1994. Analysis of promoters controlled by the putative sigma factor AlgU regulating conversion to mucoidy in *Pseudomonas aeruginosa*: relationship to sigma E and stress response. *J. Bacteriol.* **176**:6688–6696.

Martinez-Salazar, J. M., S. Moreno, R. Najera, J. C. Boucher, G. Espin, G. Soberon-Chavez, and V. Deretic. 1996. Characterization of the genes coding for the putative sigma factor AlgU and its regulators MucA, MucB, MucC, and MucD in *Azotobacter vinelandii* and evaluation of their roles in alginate biosynthesis. *J. Bacteriol.* **178**:1800–1808.

Mathee, K., O. Ciofu, C. Sternberg, P. W. Lindum, J. I. Campbell, P. Jensen, A. H. Johnsen, M. Givskov, D. E. Ohman, S. Molin, N. Hoiby, and A. Kharazmi. 1999. Mucoid conversion of *Pseudomonas aeruginosa* by hydrogen peroxide: a mechanism for virulence activation in the cystic fibrosis lung. *Microbiology* **145**(Pt. 6):1349–1357.

Mathee, K., A. Kharazmi, and N. Hoiby. 2002. Role of exopolysaccharide in biofilm matrix formation: the alginate paradigm, p. 1–34. *In* R. J. C. McLean and A. W. Decho (ed.), *Molecular Ecology of Biofilms*. Horizon Scientific Press, Wymondham, United Kingdom.

Mathee, K., C. J. McPherson, and D. E. Ohman. 1997. Posttranslational control of the AlgT (AlgU)-encoded sigma22 for expression of the alginate regulon in *Pseudomonas aeruginosa* and localization of its antagonist proteins MucA and MucB (AlgN). *J. Bacteriol.* **179**:3711–3720.

Matsukawa, M., and E. P. Greenberg. 2004. Putative exopolysaccharide synthesis genes influence *Pseudomonas aeruginosa* biofilm development. *J. Bacteriol.* **186**:4449–4456.

Mattick, J. S. 2002. Type IV pili and twitching motility. *Annu. Rev. Microbiol.* **56**:289–314.

McAvoy, M. J., V. Newton, A. Paull, J. Morgan, P. Gacesa, and N. J. Russell. 1989. Isolation of mucoid strains of *Pseudomonas aeruginosa* from non-cystic fibrosis patients and characterisation of the structure of their secreted alginate. *J. Med. Microbiol.* **28**:183–189.

Merighi, M., V. T. Lee, M. Hyodo, Y. Hayakawa, and S. Lory. 2007. The second messenger bis-(3'-5')-cyclic-GMP and its PilZ domain-containing receptor Alg44 are required for alginate biosynthesis in *Pseudomonas aeruginosa. Mol. Microbiol.* **65**:876–895.

Mohr, C. D., and V. Deretic. 1992. *In vitro* interactions of the histone-like protein IHF with the *algD* promoter, a critical site for control of mucoidy in *Pseudomonas aeruginosa. Biochem. Biophys. Res. Commun.* **189**:837–844.

Mohr, C. D., N. S. Hibler, and V. Deretic. 1991. AlgR, a response regulator controlling mucoidy in *Pseudomonas aeruginosa*, binds to the FUS sites of the *algD* promoter located unusually far upstream from the mRNA start site. *J. Bacteriol.* **173**:5136–5143.

Mohr, C. D., J. H. Leveau, D. P. Krieg, N. S. Hibler, and V. Deretic. 1992. AlgR-binding sites within the *algD* promoter make up a set of inverted repeats separated by a large intervening segment of DNA. *J. Bacteriol.* **174**:6624–6633.

Mohr, C. D., D. W. Martin, W. M. Konyecsni, J. R. Govan, S. Lory, and V. Deretic. 1990. Role of the far-upstream sites of the *algD* promoter and the *algR* and *rpoN* genes in environmental modulation of mucoidy in *Pseudomonas aeruginosa*. *J. Bacteriol.* **172**:6576–6580.

Morici, L. A., A. J. Carterson, V. E. Wagner, A. Frisk, J. R. Schurr, K. H. Zu Bentrup, D. J. Hassett, B. H. Iglewski, K. Sauer, and M. J. Schurr. 2007. *Pseudomonas aeruginosa* AlgR represses the Rhl quorum-sensing system in a biofilm-specific manner. *J. Bacteriol.* **189**:7752–7764.

Muhammadi, and N. Ahmed. 2007. Genetics of bacterial alginate: alginate genes distribution, organization and biosynthesis in bacteria. *Curr. Genomics* **8**:191–202.

Nunez, C., R. Leon, J. Guzman, G. Espin, and G. Soberon-Chavez. 2000. Role of *Azotobacter vinelandii mucA* and *mucC* gene products in alginate production. *J. Bacteriol.* **182**:6550–6556.

Oliver, A. M., and D. M. Weir. 1985. The effect of *Pseudomonas* alginate on rat alveolar macrophage phagocytosis and bacterial opsonization. *Clin. Exp. Immunol.* **59**:190–196.

Oliver, A. M., and D. M. Weir. 1983. Inhibition of bacterial binding to mouse macrophages by *Pseudomonas* alginate. *J. Clin. Lab. Immunol.* **10**:221–224.

Olvera, C., J. B. Goldberg, R. Sanchez, and G. Soberon-Chavez. 1999. The *Pseudomonas aeruginosaalgC* gene product participates in rhamnolipid biosynthesis. *FEMS Microbiol. Lett.* **179**:85–90.

Pasquier, C., N. Marty, J. L. Dournes, G. Chabanon, and B. Pipy. 1997. Implication of neutral polysaccharides associated to alginate in inhibition of murine macrophage response to *Pseudomonas aeruginosa*. *FEMS Microbiol. Lett.* **147**:195–202.

Pritt, B., L. O'Brien, and W. Winn. 2007. Mucoid *Pseudomonas* in cystic fibrosis. *Am. J. Clin. Pathol.* **128**:32–34.

Qiu, D., V. M. Eisinger, N. E. Head, G. B. Pier, and H. D. Yu. 2008. ClpXP proteases positively regulate alginate overexpression and mucoid conversion in *Pseudomonas aeruginosa*. *Microbiology* **154**:2119–2130.

Qiu, D., V. M. Eisinger, D. W. Rowen, and H. D. Yu. 2007. Regulated proteolysis controls mucoid conversion in *Pseudomonas aeruginosa*. *Proc. Natl. Acad. Sci. USA* **104**:8107–8112.

Ramsey, D. M., P. J. Baynham, and D. J. Wozniak. 2005. Binding of *Pseudomonas aeruginosa* AlgZ to sites upstream of the *algZ* promoter leads to repression of transcription. *J. Bacteriol.* **187**:4430–4443.

Regni, C., P. A. Tipton, and L. J. Beamer. 2002. Crystal structure of PMM/PGM: an enzyme in the biosynthetic pathway of *Pseudomonas aeruginosa* virulence factors. *Structure* **10**:269–279.

Reiling, S. A., J. A. Jansen, B. J. Henley, S. Singh, C. Chattin, M. Chandler, and D. W. Rowen. 2005. Prc protease promotes mucoidy in *mucA* mutants of *Pseudomonas aeruginosa*. *Microbiology* **151**:2251–2261.

Remminghorst, U., and B. H. Rehm. 2006. Alg44, a unique protein required for alginate biosynthesis in *Pseudomonas aeruginosa*. *FEBS Lett.* **580**:3883–3888.

Robles-Price, A., T. Y. Wong, H. Sletta, S. Valla, and N. L. Schiller. 2004. AlgX is a periplasmic protein required for alginate biosynthesis in *Pseudomonas aeruginosa*. *J. Bacteriol.* **186**:7369–7377.

Rompf, A., C. Hungerer, T. Hoffmann, M. Lindenmeyer, U. Romling, U. Gross, M. O. Doss, H. Arai, Y. Igarashi, and D. Jahn. 1998. Regulation of *Pseudomonas aeruginosa hemF* and *hemN*

by the dual action of the redox response regulators Anr and Dnr. *Mol. Microbiol.* **29**:985–997.

Rowen, D. W., and V. Deretic. 2000. Membrane-to-cytosol redistribution of ECF sigma factor AlgU and conversion to mucoidy in *Pseudomonas aeruginosa* isolates from cystic fibrosis patients. *Mol. Microbiol.* **36**:314–327.

Sakuragi, Y., and R. Kolter. 2007. Quorum-sensing regulation of the biofilm matrix genes (*pel*) of *Pseudomonas aeruginosa*. *J. Bacteriol.* **189**:5383–5386.

Schurr, M. J., and V. Deretic. 1997. Microbial pathogenesis in cystic fibrosis: co-ordinate regulation of heat-shock response and conversion to mucoidy in *Pseudomonas aeruginosa*. *Mol. Microbiol.* **24**:411–420.

Schurr, M. J., H. Yu, J. C. Boucher, N. S. Hibler, and V. Deretic. 1995a. Multiple promoters and induction by heat shock of the gene encoding the alternative sigma factor AlgU (sigma E) which controls mucoidy in cystic fibrosis isolates of *Pseudomonas aeruginosa*. *J. Bacteriol.* **177**:5670–5679.

Schurr, M. J., H. Yu, J. M. Martinez-Salazar, N. S. Hibler, and V. Deretic. 1995b. Biochemical characterization and posttranslational modification of AlgU, a regulator of stress response in *Pseudomonas aeruginosa*. *Biochem. Biophys. Res. Commun.* **216**:874–880.

Sidote, D. J., C. M. Barbieri, T. Wu, and A. M. Stock. 2008. Structure of the *Staphylococcus aureus* AgrA LytTR domain bound to DNA reveals a beta fold with an unusual mode of binding. *Structure* **16**:727–735.

Simpson, J. A., S. E. Smith, and R. T. Dean. 1988. Alginate inhibition of the uptake of *Pseudomonas aeruginosa* by macrophages. *J. Gen. Microbiol.* **134**:29–36.

Simpson, J. A., S. E. Smith, and R. T. Dean. 1993. Alginate may accumulate in cystic fibrosis lung because the enzymatic and free radical capacities of phagocytic cells are inadequate for its degradation. *Biochem. Mol. Biol. Int.* **30**:1021–1034.

Simpson, J. A., S. E. Smith, and R. T. Dean. 1989. Scavenging by alginate of free radicals released by macrophages. *Free Radic. Biol. Med.* **6**:347–353.

Snook, C. F., P. A. Tipton, and L. J. Beamer. 2003. Crystal structure of GDP-mannose dehydrogenase: a key enzyme of alginate biosynthesis in *Pseudomonas aeruginosa*. *Biochemistry* **42**:4658–4668.

Starkey, M., J. H. Hickman, L. Ma, N. Zhang, S. De Long, A. Hinz, S. Palacios, C. Manoil, M. J. Kirisits, T. D. Starner, D. J. Wozniak, C. S. Harwood, and M. R. Parsek. 2009. *Pseudomonas aeruginosa* rugose small-colony variants have adaptations that likely promote persistence in the cystic fibrosis lung. *J. Bacteriol.* **191**:3492–3503.

Suh, S. J., L. Silo-Suh, D. E. Woods, D. J. Hassett, S. E. West, and D. E. Ohman. 1999. Effect of *rpoS* mutation on the stress response and expression of virulence factors in *Pseudomonas aeruginosa*. *J. Bacteriol.* **181**:3890–3897.

Tart, A. H., M. J. Blanks, and D. J. Wozniak. 2006. The AlgT-dependent transcriptional regulator AmrZ (AlgZ) inhibits flagellum biosynthesis in mucoid, nonmotile *Pseudomonas aeruginosa* cystic fibrosis isolates. *J. Bacteriol.* **188**:6483–6489.

Tavares, I. M., J. H. Leitao, A. M. Fialho, and I. Sa-Correia. 1999. Pattern of changes in the activity of enzymes of GDP-D-mannuronic acid synthesis and in the level of transcription of *algA*, *algC* and *algD* genes accompanying the loss and emergence of mucoidy in *Pseudomonas aeruginosa*. *Res. Microbiol.* **150**:105–116.

Terry, J. M., S. E. Pina, and S. J. Mattingly. 1991. Environmental conditions which influence mucoid conversion in *Pseudomonas aeruginosa* PAO1. *Infect. Immun.* **59**:471–477.

Terry, J. M., S. E. Pina, and S. J. Mattingly. 1992. Role of energy metabolism in conversion of nonmucoid *Pseudomonas aeruginosa* to the mucoid phenotype. *Infect. Immun.* **60**:1329–1335.

Vasil, M. L. 2007. How we learnt about iron acquisition in *Pseudomonas aeruginosa*: a series of very fortunate events. *Biometals* 20:587–601.

Vasil, M. L., L. M. Graham, R. M. Ostroff, V. D. Shortridge, and A. I. Vasil. 1991. Phospholipase C: molecular biology and contribution to the pathogenesis of *Pseudomonas aeruginosa*. *Antibiot. Chemother.* 44:34–47.

Vasseur, P., I. Vallet-Gely, C. Soscia, S. Genin, and A. Filloux. 2005. The *pel* genes of the *Pseudomonas aeruginosa* PAK strain are involved at early and late stages of biofilm formation. *Microbiology* 151:985–997.

Waligora, E. A., D. M. Ramsey, E. E. Pryor, Jr., H. Lu, T. Hollis, G. P. Sloan, R. Deora, and D. J. Wozniak. 2010. AmrZ beta-sheet residues are essential for DNA binding and transcriptional control of *Pseudomonas aeruginosa* virulence genes. *J. Bacteriol.* 192:5390–5401.

Whitchurch, C. B., R. A. Alm, and J. S. Mattick. 1996. The alginate regulator AlgR and an associated sensor FimS are required for twitching motility in *Pseudomonas aeruginosa*. *Proc. Natl. Acad. Sci. USA* 93:9839–9843.

Whitchurch, C. B., T. E. Erova, J. A. Emery, J. L. Sargent, J. M. Harris, A. B. Semmler, M. D. Young, J. S. Mattick, and D. J. Wozniak. 2002. Phosphorylation of the *Pseudomonas aeruginosa* response regulator AlgR is essential for type IV fimbria-mediated twitching motility. *J. Bacteriol.* 184:4544–4554.

Whitney, J. C., I. D. Hay, C. Li, P. D. Eckford, H. Robinson, M. F. Amaya, L. F. Wood, D. E. Ohman, C. E. Bear, B. H. Rehm, and P. Lynne Howell. 2011. Structural basis for alginate secretion across the bacterial outer membrane. *Proc. Natl. Acad. Sci. USA* 108:13083–13088.

Wolfgang, M. C., V. T. Lee, M. E. Gilmore, and S. Lory. 2003. Coordinate regulation of bacterial virulence genes by a novel adenylate cyclase-dependent signaling pathway. *Dev. Cell* 4:253–263.

Wood, L. F., A. J. Leech, and D. E. Ohman. 2006. Cell wall-inhibitory antibiotics activate the alginate biosynthesis operon in *Pseudomonas aeruginosa*: roles of σ^{22} (AlgT) and the AlgW and Prc proteases. *Mol. Microbiol.* 62:412–426.

Wood, L. F., and D. E. Ohman. 2006. Independent regulation of MucD, an HtrA-like protease in *Pseudomonas aeruginosa*, and the role of its proteolytic motif in alginate gene regulation. *J. Bacteriol.* 188:3134–3137.

Wood, S. R., A. M. Firoved, W. Ornatowski, T. Mai, V. Deretic, and G. S. Timmins. 2007. Nitrosative stress inhibits production of the virulence factor alginate in mucoid *Pseudomonas aeruginosa*. *Free Radic. Res.* 41:208–215.

Worlitzsch, D., R. Tarran, M. Ulrich, U. Schwab, A. Cekici, K. C. Meyer, P. Birrer, G. Bellon, J. Berger, T. Weiss, K. Botzenhart, J. R. Yankaskas, S. Randell, R. C. Boucher, and G. Doring. 2002. Effects of reduced mucus oxygen concentration in airway *Pseudomonas* infections of cystic fibrosis patients. *J. Clin. Investig.* 109:317–325.

Wozniak, D. J. 1994. Integration host factor and sequences downstream of the *Pseudomonas aeruginosa algD* transcription start site are required for expression. *J. Bacteriol.* 176:5068–5076.

Wozniak, D. J., and D. E. Ohman. 1993. Involvement of the alginate *algT* gene and integration host factor in the regulation of the *Pseudomonas aeruginosa algB* gene. *J. Bacteriol.* 175:4145–4153.

Wozniak, D. J., and D. E. Ohman. 1991. *Pseudomonas aeruginosa* AlgB, a two-component response regulator of the NtrC family, is required for *algD* transcription. *J. Bacteriol.* 173:1406–1413.

Wozniak, D. J., A. B. Sprinkle, and P. J. Baynham. 2003a. Control of *Pseudomonas aeruginosa algZ* expression by the alternative sigma factor AlgT. *J. Bacteriol.* 185:7297–7300.

Wozniak, D. J., T. J. Wyckoff, M. Starkey, R. Keyser, P. Azadi, G. A. O'Toole, and M. R. Parsek. 2003b. Alginate is not a significant component of the extracellular polysaccharide matrix of PA14 and PAO1 *Pseudomonas aeruginosa* biofilms. *Proc. Natl. Acad. Sci. USA* 100:7907–7912.

Yorgey, P., L. G. Rahme, M. W. Tan, and F. M. Ausubel. 2001. The roles of *mucD* and alginate in the virulence of *Pseudomonas aeruginosa* in plants, nematodes and mice. *Mol. Microbiol.* 41:1063–1076.

Yu, H., M. Mudd, J. C. Boucher, M. J. Schurr, and V. Deretic. 1997. Identification of the *algZ* gene upstream of the response regulator AlgR and its participation in control of alginate production in *Pseudomonas aeruginosa*. *J. Bacteriol.* 179:187–193.

Yu, H., M. J. Schurr, and V. Deretic. 1995. Functional equivalence of *Escherichia coli* σ^E and *Pseudomonas aeruginosa* AlgU: *E. coli rpoE* restores mucoidy and reduces sensitivity to reactive oxygen intermediates in *algU* mutants of *P. aeruginosa*. *J. Bacteriol.* 177:3259–3268.

Zielinski, N. A., A. M. Chakrabarty, and A. Berry. 1991. Characterization and regulation of the *Pseudomonas aeruginosa algC* gene encoding phosphomannomutase. *J. Biol. Chem.* 266:9754–9763.

Zielinski, N. A., R. Maharaj, S. Roychoudhury, C. E. Danganan, W. Hendrickson, and A. M. Chakrabarty. 1992. Alginate synthesis in *Pseudomonas aeruginosa*: environmental regulation of the *algC* promoter. *J. Bacteriol.* 174:7680–7688.

Regulation of Bacterial Virulence
Edited by Michael L. Vasil and Andrew J. Darwin
© 2013 ASM Press, Washington, DC   doi:10.1128/9781555818524.ch10

Chapter 10

Regulation of Pneumococcal Surface Proteins and Capsule

Abiodun D. Ogunniyi and James C. Paton

INTRODUCTION

Streptococcus pneumoniae (the pneumococcus) is a formidable human pathogen, responsible for massive global morbidity and mortality. It causes a broad spectrum of diseases, including pneumonia, meningitis, bacteremia, and otitis media, and accounts for more deaths worldwide than any other single pathogen (O'Brien et al., 2009). The problem is being further exacerbated by the prevalence of multidrug-resistant pneumococci and the rapid global spread of antibiotic-resistant clones (Reinert, 2009a, 2009b). Vaccination represents the best prospect for managing pneumococcal disease in the 21st century. However, currently available vaccines targeted at the type-specific capsular polysaccharide (CPS) have major shortcomings. Polyvalent purified CPS vaccines introduced in the 1980s confer strictly serotype-specific protection and are poorly immunogenic in young children (Douglas et al., 1983). Immunogenicity of CPS antigens has been greatly improved by conjugating the CPS to protein carriers, and 7-, 10-, and 13-valent pneumococcal conjugate vaccines are now in use (Spratt and Greenwood, 2000; Paton, 2004; Chuck et al., 2010). However, pneumococcal conjugate vaccines are prohibitively expensive in developing countries, where the need is greatest. Furthermore, their clinical benefit is being offset by significant increases in both carriage and disease caused by nonvaccine serotypes, which have occupied the niche vacated by the vaccine types (Brueggemann et al., 2007; Hicks et al., 2007). Consequently, current global efforts are now focused on accelerating the development of alternative pneumococcal vaccines based on proteins that contribute to pathogenesis and are common to all serotypes (Briles et al., 2003; Brown et al., 2001b; Ogunniyi et al., 2000; Ogunniyi et al., 2001; Ogunniyi et al., 2007a; Paton, 1998, 2004).

The continuing problems associated with management of pneumococcal disease can be at least partially attributed to our incomplete understanding of the pathogenic process. Thus, it is important that future decisions regarding vaccination strategies are based on a deep knowledge of the complex and dynamic interactions between the pneumococcus and its host. Asymptomatic colonization of the upper respiratory tract (carriage) almost invariably precedes pneumococcal disease. Carriers are the major source for transmission of pneumococci, and communities with high rates of carriage also have high disease attack rates. Carriage rates vary markedly with both age and socioeconomic status, with the highest rates found in young children in developing countries. Individuals may be colonized by multiple strains or serotypes of *S. pneumoniae*, and these presumably compete for occupation of the nasopharyngeal niche (Lipsitch et al., 2000). The precise mechanism(s) of adherence to the mucosa is not fully understood, but adherence has been demonstrated to be mediated by interaction of pneumococcal surface proteins such as choline-binding protein A (CbpA) and/or phosphorylcholine (ChoP) moieties of the cell wall teichoic acid with receptors on the surface of human nasopharyngeal epithelial cells (Tuomanen and Masure, 1997; Weiser, 2004). Pneumococci may thus colonize the nasopharynx and be carried asymptomatically for many weeks. However, in a proportion of carriers, the pneumococcus is able to penetrate host defenses and cause invasive disease, either by translocation to sites such as the lungs or middle ear cavity or by direct invasion of the blood. The precise events associated with progression from asymptomatic carriage to invasive disease are poorly understood, although it is clearly a "watershed" in the host-pathogen interaction. This transition involves a major alteration in the microenvironment(s) to which the pneumococcus is exposed, which undoubtedly requires alteration in the levels of expression of complex sets of bacterial genes.

Examination of pneumococcal gene expression patterns in distinct niches was initially hampered by

Abiodun D. Ogunniyi and James C. Paton • Research Centre for Infectious Diseases, School of Molecular and Biomedical Science, University of Adelaide, Adelaide, SA 5005, Australia.

technical difficulties associated with isolating sufficient quantities of pure and intact bacterial RNA from infected host tissues to perform accurate, quantitative mRNA analyses. However, recent advances, particularly in bacterial RNA extraction, enrichment, and linear amplification methodologies, have enabled systematic examination of transcription patterns of key pneumococcal virulence genes in distinct host niches, including sites such as the nasopharynx, where pneumococci exist in very low numbers. Quantitative real-time reverse transcription-PCR studies showed firstly that key pneumococcal virulence genes were upregulated in host tissues relative to in vitro-grown cells (Ogunniyi et al., 2002). Differential virulence gene expression was also observed for pneumococci isolated from distinct in vivo niches (LeMessurier et al., 2006; Mahdi et al., 2008), and simultaneous gene expression patterns of selected host immunomodulatory molecules were also different between host niches (Mahdi et al., 2008). Further improvements in the yields of in vivo-derived RNA are now facilitating use of genome-wide transcriptomic techniques, including microarrays and next-generation sequencing, to comprehensively analyze pneumococcal gene expression patterns and innate host responses during pathogenesis of invasive disease. From the bacterial perspective, this is providing insights into the differential requirements for given virulence factors, as well as metabolic and stress response networks in one host niche versus another.

PNEUMOCOCCAL REGULATORY SYSTEMS

The pneumococcus employs diverse regulatory systems to adapt its gene expression patterns to meet the challenges of specific host microenvironments, and in many cases, more than one such system impacts expression of a given virulence-associated trait. An initial overview of these regulatory systems is provided below, and this is followed by an examination of regulation of specific surface components that are known to be important in pathogenesis.

Quorum Sensing

Quorum sensing is a common bacterial regulatory mechanism that modulates gene expression in a population-dependent manner. The underlying mechanism involves the production and extracellular accumulation of a signaling molecule during growth, which is sensed by a receptor that triggers signaling pathways once a critical threshold concentration is reached. Many mechanistic insights have been gained in recent years by study of quorum sensing in luminescent marine gram-negative bacteria (Bassler, 1999). However, the phenomenon was actually first recognized in *S. pneumoniae* nearly half a century ago by Tomasz, who demonstrated that cell density-dependent accumulation of a hormone-like activator substance, now known as the competence-stimulating peptide (CSP), regulated development of competence for genetic transformation (Tomasz, 1965, 1966). In the last decade or so, a more complete understanding of the mechanism of competence development has emerged, as recently reviewed (Johnsborg and Håvarstein, 2009). Briefly, the CSP precursor, which is encoded by the *comC* gene, is processed and secreted by the dedicated ComAB transporter, resulting in extracellular accumulation of mature CSP. Basal transcription of the *comCDE* operon is subjected to regulation by global regulators such as the serine/threonine protein kinase StkP and the CiaRH two-component system. Binding of CSP to its ComD receptor is believed to result in autophosphorylation of ComD and subsequent transfer of the phosphoryl group to the ComE response regulator. ComE then binds to and activates transcription from the various early competence gene promoters. ComE binding sets off increased transcription of the *comCDE* operon, leading to a boost in the production of CSP, ComD, and phosphorylated ComE. This autoinduction loop accelerates accumulation of the alternative σ factor ComX, ComW, and the ComM fratricide immunity protein (see below). ComW protects ComX from proteolytic cleavage and stimulates the latter protein to activate transcription of the late competence genes encoding the DNA uptake and recombination machinery. Interestingly, ComX also stimulates production of a bacteriocin, CibAB, and its cognate immunity protein CibC, as well as CbpD, a murein hydrolase belonging to the choline-binding protein (CBP) family. Competent cells are protected from CbpD by the product of the early gene *comM*. However, noncompetent pneumococci in the immediate vicinity are susceptible to killing by CibAB produced by their competent counterparts, leading to CbpD-mediated release of their DNA for uptake by the latter. This process of competence-induced fratricide is an effective means of facilitating genetic exchange between closely related bacteria (Johnsborg and Håvarstein, 2009).

Pneumococci also employ a second quorum-sensing regulatory system, dependent on the *luxS* gene, which enables them to respond to signals produced by other bacterial species, not just *S. pneumoniae* as in the case of CSP. Homologues of *luxS* are ubiquitous in bacteria and encode S-ribosylhomocysteine lyase, an enzyme that produces the non-species-specific quorum-sensing molecule autoinducer-2 (AI-2) as a by-product of detoxification of *S*-adenosyl-L-homocysteine as part of the central activated methyl cycle. In *S. pneumoniae*, LuxS has been shown to regulate biosynthetic and virulence gene (e.g., pneumolysin) expression (Joyce et al., 2004); it also plays a role in persistence of pneumococci in the nasopharynx and impacts systemic virulence in mice (Joyce et al., 2004; Stroeher et al., 2003). During infection, pneumococci exist mainly in sessile biofilms rather than in planktonic form, except during sepsis, and the capacity to form biofilms is believed to be important for nasopharyngeal colonization as well as disease pathogenesis. Recent studies show that pneumococcal biofilm formation requires a functional *luxS* gene and that overexpression of *luxS* leads to a hyper-biofilm-forming phenotype (Trappetti et al., 2011b). Iron, in the form of Fe^{3+}, also stimulated biofilm formation, as well as genetic competence. However, these effects required *luxS*, and supplementation of media with Fe^{3+} significantly increased expression of *luxS* in wild-type *S. pneumoniae*. LuxS also increases expression of the major pneumococcal iron uptake protein PiuA, thereby boosting cellular iron levels and providing a positive amplification loop. In addition, LuxS directly induces early and late competence gene expression, along with competence-mediated fratricide. Significantly, CbpD-mediated DNA release is critical for biofilm formation, providing an important component of the extracellular matrix that cements the pneumococci to the substratum and to each other (Trappetti et al., 2011b). Thus, competence, fratricide, and biofilm formation are closely linked in pneumococci, and *luxS* is a central regulator of these critical processes.

Two-Component Systems

As with many other bacteria (Dziejman and Mekalanos, 1995), environmental adaptation in *S. pneumoniae* may be mediated, at least in part, by two-component signal transduction systems (TCSTSs). The typical TCSTS consists of a histidine kinase (HK) and a response regulator (RR). The HK spans the cell membrane and autophosphorylates a conserved histidine residue in response to specific environmental stimuli. Subsequent transfer of this phosphate group from the HK to the RR leads to conformational changes enabling the RR to alter expression of the cognate gene(s) by binding to the promoter region and acting as a transcription factor. In some systems the HK also has phosphatase activity, enabling the HK to limit the duration of the activated state of the RR. This activity is believed to be favored when the HK is not phosphorylated (i.e., in the absence of the specific signal responsible for turning the system on).

The *S. pneumoniae* genome contains 13 TCSTSs as well as one orphan RR (RitR) lacking a cognate HK (Lange et al., 1999; Throup et al., 2000; Paterson et al., 2006). Mutagenesis of either the RR or both components of each locus demonstrated that one TCSTS was essential for growth, identifying it as a possible antimicrobial target. Interestingly, Lange et al. (1999) reported that all of the other 12 RR mutants were fully virulent in mice, as judged by survival time after intraperitoneal challenge. In contrast, Throup et al. (2000) reported that deletions in eight of the HK/RR loci significantly attenuated virulence, as determined by numbers of organisms in the lungs of mice 48 h after intranasal challenge. These seemingly discordant findings, which were obtained with different animal models, *S. pneumoniae* strains, and mutagenesis techniques, may simply reflect real differences in the importance of these regulatory systems to pathogenesis of systemic versus respiratory infections. However, studies performed later using the same animal models have detected strain-to-strain variation in the effect of otherwise identical TCSTS mutations on virulence phenotype (Blue and Mitchell, 2003; McCluskey et al., 2004; Hendriksen et al., 2007). To date, only one TCSTS (CiaRH) has been directly shown to be essential for nasopharyngeal colonization by *S. pneumoniae* (in an infant rat model), and this was attributed to effects on expression of the *htrA* gene, which encodes a putative serine protease (Sebert et al., 2002; Ibrahim et al., 2004). CiaRH has also been shown to be critical for biofilm formation and adherence to respiratory epithelial cells in vitro, as well as nasopharyngeal colonization and translocation to the lungs in a murine pneumonia model (Trappetti et al., 2011a). Fifteen pneumococcal promoters were shown to be directly controlled by the response regulator CiaR (14 are activated, and 1 is repressed) (Halfmann et al., 2007). The genes that are transcribed from these promoters include *ciaRH* itself, as well as loci that are predicted to be involved in the modification of teichoic acids (*lic*) and in sugar metabolism (*mal* and *man*), stress response (*htrA*), chromosome segregation (*parB*), protease maturation (*ppmA*), and unknown functions. Remarkably, the five strongest promoters of the CiaR regulon drive expression of small noncoding RNAs between 87 and 151 nucleotides in size, and deletion mutations in two of these affected stationary-phase autolysis (Halfmann et al., 2007). Studies emanating from our laboratory (Standish et al., 2005, 2007) showed that one of the other pneumococcal TCSTSs (designated RR/HK06) directly regulates expression of CbpA and another CBP, pneumococcal surface protein A (PspA), via two distinct mechanisms reliant on phosphorylation or nonphosphorylation of RR06.

Table 1. Pneumococcal TCSTSs and their roles in virulence

TCSTS[a]	Alternative designation(s)[b]	TCSTS family[c]	Experimental role in virulence (reference[s])
TCSTS01	480	Pho	Role in lung infection (Throup et al., 2000; Hava and Camilli, 2002)
TCSTS02	492; YycFG,VicRK, MicAB	Pho	Role in blood infection (Wagner et al., 2002; Kadioglu et al., 2003)
TCSTS03	474	Nar	No role in lung infection (Throup et al., 2000)
TCSTS04	481; PnpSR	Pho	Role in lung and blood infection (Throup et al., 2000; McCluskey et al., 2004)
TCSTS05	494; CiaRH	Pho	Role in colonization and lung and blood infection (Throup et al., 2000; Marra et al., 2002; Sebert et al., 2002; Ibrahim et al., 2004)
TCSTS06	478; RR/HK06	Pho	Role in colonization and lung and blood infection (Throup et al., 2000; Standish et al., 2005, 2007)
TCSTS07	539	Lyt	Role in lung infection (Throup et al., 2000; Hava and Camilli, 2002)
TCSTS08	484	Pho	Role in lung infection (Throup et al., 2000)
TCSTS09	488; ZmpSR	Lyt	Strain-dependent role in virulence (Lau et al., 2001; Throup et al., 2000; Hava and Camilli, 2002; Blue and Mitchell, 2003)
TCSTS10	491; VncSR	Pho	Role in lung infection (Throup et al., 2000)
TCSTS11	479	Nar	No role in lung infection (Throup et al., 2000)
TCSTS12	498; ComDE	Agr	Role in lung and blood infection (Lau et al., 2001; Bartilson et al., 2001; Hava and Camilli, 2002)
TCSTS13	486; BlpHR	Agr	Role in lung infection (Throup et al., 2000)
Orphan RR	489; RitR	Pho	Role in lung infection (Throup et al., 2000; Ulijasz et al., 2004) Experimental data show role in blood infection (unpublished data)

[a]Designations according to Lange et al., 1999.
[b]Numeric designations according to Throup et al., 2000; alternative designations by other workers.
[c]Family designations according to Throup et al., 2000.

Another gene (*regR*) encoding a LacI/GalR regulator involved in the adaptive response of the pneumococcus also appears to modulate expression of virulence genes and the TCSTS CiaRH (Chapuy-Regaud et al., 2003). Known or putative triggers, targets, and phenotypic impacts of all the pneumococcal TCSTSs are summarized in Table 1.

Metal Ion-Dependent Gene Regulation

Metal ions are essential micronutrients for growth and survival of bacteria in diverse environmental niches, including sites within higher organisms. One strategy employed by host organisms is to limit bacterial growth by restricting metal ion availability by sequestration with high-affinity metal binding proteins (Corbin et al., 2008). To overcome these innate defenses, bacteria have evolved both metal ion chelating mechanisms and high-affinity transport systems, enabling them to scavenge and assimilate essential metals in vivo. Although the best-studied systems are involved in the acquisition of Fe^{3+}, Mn^{2+} is increasingly being recognized as a critical micronutrient, particularly for pathogenic bacteria. Pneumococcal surface antigen A (PsaA) (Russell et al., 1990; Lawrence et al., 1998) is the Mn^{2+}-specific, solute-binding component of an ATP-binding cassette (ABC) cation permease encoded by the *psaBCA* locus, and it is an important virulence factor for *S. pneumoniae* (Paton, 1998; Rajam et al., 2008). The Psa permease complex belongs to the cluster A to I

(formerly cluster IX) (Dintilhac et al., 1997) family of bacterial transporters of the essential metal ions Mn^{2+}, Zn^{2+}, and Fe^{2+} (Berntsson et al., 2010). Mutation of *psaA* has been shown to result in massively reduced virulence in systemic, respiratory tract, and otitis media murine models of infection (Berry and Paton, 1996; Marra et al., 2002a; McAllister et al., 2004).

Expression of *psaA* (and other virulence genes such as *pcpA* and *prtA*) has been shown to be Mn^{2+} dependent through regulation by PsaR, a member of the DtxR family of transcriptional regulators (Johnston et al., 2006; Ogunniyi et al., 2010). Indeed, the PsaR-binding sequence was identified in the promoter regions of *psaBCA*, *pcpA*, and *prtA* but not at any other locations in the pneumococcal genome (Kloosterman et al., 2008). Interestingly, PsaR was also shown to mediate Zn^{2+}-dependent derepression of *psaA* and these other genes (Kloosterman et al., 2008). Thus, PsaR-mediated regulation of *psaA* and other genes appears to be controlled by Zn^{2+}/Mn^{2+} ratios. This could be of consequence in vivo, as our recent inductively coupled plasma mass spectrometry measurements of metal content of various tissues of mice intranasally infected with *S. pneumoniae* D39 show significantly higher Zn^{2+}/Mn^{2+} ratios in the nasopharynx and blood serum than in uninfected tissues (McDevitt et al., 2011). This scenario favors increased transcription of the *psa* operon in those niches, as reported previously (Ogunniyi et al., 2002; LeMessurier et al., 2006; Mahdi et al., 2008). The findings are also consistent with the transcriptional profiles of regulated genes in recent in vitro models of Zn^{2+}/Mn^{2+} competition (Jacobsen et al., 2011; McDevitt et al., 2011). The impact of PsaR mutation on global gene expression and pneumococcal virulence also appears to be strain specific. An earlier study showed a significant attenuation of a type 19F (EF3030) *psaR* mutant in a murine pneumonia model (Johnston et al., 2006). However, a later study using D39 and TIGR4 *psaR* mutants showed conflicting results in both pneumonia and bacteremia models (Hendriksen et al., 2009). These seemingly discordant results further highlight the potential impact of the plasticity of the pneumococcal genome and the crucial need to investigate the phenotypic effects of a given mutation in multiple strains representing diverse genetic backgrounds.

Further to the above, one microarray study identified a two-component system (RR04/HK04) that controls the expression of PsaA and regulates virulence and resistance to oxidative stress, albeit in a serotype-specific manner (McCluskey et al., 2004). It would appear that this strain-specific regulation of PsaA by RR04/HK04 is independent of the PsaR-PsaA interaction, because *psaR* transcription was not affected in the RR04 mutant (McCluskey et al., 2004). In addition, other PsaR-regulated genes, such as *pcpA* or the *rlrA* regulon, were not similarly affected (Johnston et al., 2006).

Another metal-dependent transcriptional regulator in *S. pneumoniae* and other streptococci that has been studied extensively in recent years is AdcR, a Zn^{2+}-dependent repressor belonging to the MarR family (Dintilhac and Claverys, 1997; Claverys, 2001; Panina et al., 2003). AdcR has been shown to regulate the ABC-type *adc* and *adcA*II Zn^{2+} permeases in streptococci, as well as the *pht* genes and a Zn^{2+}-containing alcohol dehydrogenase of *S. pneumoniae* (Dintilhac et al., 1997; Panina et al., 2003; Brenot et al., 2007; Loisel et al., 2008; Ogunniyi et al., 2009; Reyes-Caballero et al., 2010; Rioux et al., 2011; Shafeeq et al., 2011). Consistent with these findings, consensus 12-bp palindromic AdcR recognition sequences (TTAACYRGTTAA) have been located upstream of these genes (Panina et al., 2003; Loisel et al., 2008), and direct binding of AdcR to these promoter regions has been demonstrated (Ogunniyi et al., 2009; Reyes-Caballero et al., 2010). Although direct evidence for a role for AdcR in pneumococcal virulence is lacking, a role for its ortholog in *Streptococcus suis* virulence and resistance to oxidative stress has been demonstrated (Aranda et al., 2010). Furthermore, regulation of *phtABDE* expression by AdcR has been shown to be Zn^{2+} dependent, although it is not entirely clear whether Zn^{2+} stimulates or represses *pht* expression in vivo (Ogunniyi et al., 2009; Rioux et al., 2011; Shafeeq et al., 2011).

In addition to PsaR and AdcR, a transcriptional regulator of Fe^{2+} transport in *S. pneumoniae*, RitR (repressor of iron transport), has been characterized (Ulijasz et al., 2004). RitR was originally described as an orphan two-component signal transduction response regulator (RR489) that negatively regulates transcription of Fe^{2+} uptake genes, including *piuA* and *piuB* (Brown et al., 2001a), as well as two-component signal transduction elements and sugar transport genes. A direct link between RitR and Fe^{2+} uptake regulation was established by DNA footprinting experiments demonstrating that RitR binds to three sites in the promoter region of the *piu* operon (Ulijasz et al., 2004). More recently, in vitro phosphorylation/dephosphorylation of the DNA-binding domain of RitR by a Ser-Thr phosphokinase (StkP) and its putative cognate phosphatase (PhpP) has been demonstrated, and both proteins appear to be necessary for Piu transporter expression (Ulijasz et al., 2009). It is suggested that the primary role of RitR is to maintain iron homeostasis in *S. pneumoniae*.

Nutrient and Metabolic State Regulation

Access to genome sequences in the last decade or so has enabled deployment of a variety of genome-wide molecular approaches in an attempt to identify the full complement of virulence-associated genes in *S. pneumoniae* and other pathogenic bacteria. A common feature of these studies has been the identification of metabolic enzymes and nutrient uptake systems as being essential for survival in vivo but not for growth in vitro in rich media (Polissi et al., 1998; Hava and Camilli, 2002; Marra et al., 2002b; Orihuela et al., 2004). For many of these genes, differences in expression levels have also been observed between pneumococci isolated from distinct host niches. Clearly, nutrient availability, metal ion concentration, oxidative stress, etc., pose challenges that bacteria must meet in order to survive, proliferate, and cause disease. One regulatory pathway by which *S. pneumoniae* modulates gene expression in response to environmental nutrient concentrations in various niches is catabolite repression via carbon catabolite control protein A (CcpA; also referred to as RegM) (Rosenow et al., 1999; Giammarinaro and Paton, 2002). Catabolite repression is essentially a mechanism involving preferential upregulation of genes for utilization of a carbon source that can be rapidly metabolized in any given environment. CcpA is a member of the LacI-GalR family, which binds to catabolite-responsive elements (*cre*) located within or near promoters (Weickert and Chambliss, 1990; Weickert and Adhya, 1992). CcpA has been shown to directly affect transcription of CPS biosynthesis genes (Giammarinaro and Paton, 2002) and regulates expression of β-galactosidase (BgaA) and NanA (neuraminidase A) (Kaufman, 2007; Kaufman and Yother, 2007) as well as expression of many proteins that are involved in central and intermediary metabolism (Iyer et al., 2005). CcpA has also been demonstrated to contribute to virulence in mouse models of colonization, pneumonia, and sepsis (Giammarinaro and Paton, 2002; Iyer et al., 2005; Kaufman and Yother, 2007). Bioinformatic analysis suggests that CcpA can potentially regulate expression of other pneumococcal surface proteins, including StrH (an N-acetylglucosaminidase), GlpO (alpha-glycerophosphate oxidase), and MalX (a maltose/maltodextrin ABC transporter). Our laboratory has experimental evidence for the regulation of GlpO and MalX expression by CcpA (unpublished data), while direct molecular evidence for regulation of StrH expression is currently under investigation.

Another global regulator, CodY, has been characterized recently in *S. pneumoniae* (Kloosterman et al., 2006; Hendriksen et al., 2008; Caymaris et al., 2010). The CodY regulon consists mainly of genes involved in amino acid metabolism (Kloosterman et al., 2006), biosynthesis, and several other cellular processes, including carbon metabolism and iron uptake. Binding of CodY to its target promoters requires a 15-bp recognition site and is enhanced by branched-chain amino acids, the latter suggesting that it can regulate its own expression (Hendriksen et al., 2008). Interestingly, CodY is required for optimal levels of in vitro adherence and colonization of the murine nasopharynx, likely due to the activation of expression of *pcpA* (Hendriksen et al., 2008), strongly indicating a direct link between nutritional regulation and adherence/colonization.

Phase Variation

Colony opacity phase variation is a well-studied but poorly understood strategy employed by *S. pneumoniae* to adapt to the various microenvironments it encounters during transition from carriage to invasive disease. When pneumococcal colonies growing on agar plates are observed under oblique transmitted light, two distinct morphologies, described as "opaque" and "transparent," are observed (Weiser et al., 1994). When opaque colonies are subcultured, a small proportion of them spontaneously change to the transparent form, and vice versa, and the frequency of switching differs from strain to strain (10^{-3} to 10^{-6} per generation) (Weiser et al., 1994; Kim and Weiser, 1998). Interestingly, the transparent forms have an enhanced capacity to colonize the nasopharynx relative to opaque variants of the same strain,

which correlates with increased in vitro adherence to epithelial cells, while the opaque form is associated with substantially increased virulence in animal models of systemic disease (Weiser et al., 1994; Kim and Weiser, 1998; Briles et al., 2005). Variation in the levels of expression of CPS and certain surface proteins between the two forms has also been reported (Saluja and Weiser, 1995; Weiser et al., 1996; Kim and Weiser, 1998; Weiser, 1998; Overweg et al., 2000). The relevance of phase variation to pathogenesis of human disease is supported by the finding that when apparently identical *S. pneumoniae* strains are isolated simultaneously from the nasopharynx and blood of patients with invasive disease, those from the former niche are largely in the transparent phase, while the latter are almost all opaque (Weiser et al., 2001). Unpublished observations in our laboratory also confirm that opaque variants are significantly more virulent than their transparent counterparts in mouse intranasal and intraperitoneal models of infection. Thus, random, low-level bidirectional switching permits selection for, and clonal expansion of, whichever phenotype is best suited to the microenvironment in which the pneumococcus is located at a given time.

Studies in other pathogenic bacteria such as *Neisseria meningitidis*, *Neisseria gonorrhoeae*, and *Haemophilus influenzae* have shown that phase variation often relies on the presence of homopolymeric tracts, or tandem repeated nontrimeric sequences within the open reading frame of key genes. Such genes are highly susceptible to slipped-strand mispairing during replication, which results in frameshift mutations leading to premature termination of translation of the respective gene product (Belland et al., 1989; Weiser et al., 1989; Peak et al., 1996; Yang and Gotschlich, 1996; Jennings et al., 1999). Examination of the *S. pneumoniae* TIGR4 and R6 genome sequences (Hoskins et al., 2001; Tettelin et al., 2001) revealed the presence of a large number of potentially phase-variable genes containing such elements either within the open reading frame itself or in potential promoter regions upstream. In order to investigate if a similar mechanism operates in *S. pneumoniae*, we selected 16 such genes, either genes encoding synthesis of cell envelope components (teichoic acid, CPS, or surface proteins) or potential regulatory genes, for further study. Those selected had the longest repeat elements and hence the highest probability of undergoing slipped-strand mutation. PCR primers flanking each gene were designed on the basis of the genome sequence, and the respective regions were amplified using template DNA from opaque and transparent variants of two distinct *S. pneumoniae* strains. However, sequence analysis of the PCR products did not detect variation in the number of repeat elements between opaque and transparent variants of either strain at any of these loci (McKessar, 2003). This was quite surprising, as variation in the number of bases in homopolymeric tracts and other deletions resulting in frameshift mutations has been described for some pneumococcal genes. However, these did not correlate with differences in colony opacity phenotype (Pericone et al., 2002), and at present, the molecular mechanism(s) underlying phase variation in pneumococci remains largely unknown.

REGULATION OF CPS BIOSYNTHESIS

The CPS is often considered to be a sine qua non of pneumococcal virulence (Austrian, 1981). Although nonencapsulated strains have been associated with superficial infections such as conjunctivitis (Martin et al., 2003; Crum et al., 2004), fresh clinical isolates from other sterile sites are encapsulated, and spontaneous nonencapsulated derivatives of such strains are largely avirulent. The CPS is known to have strong antiphagocytic properties in nonimmune hosts. The majority of CPS serotypes are highly charged at physiological pH, and electrostatic repulsion may directly interfere with interactions with phagocytes (Lee et al., 1991). The capsule also forms an inert shield, which appears to prevent either the Fc region of immunoglobulin G or iC3b fixed to deeper cell surface structures (e.g., teichoic acid and surface proteins) from interacting with receptors on phagocytic cells (Winkelstein, 1981; Musher, 1992). It may also reduce the total amount of complement deposited on the bacterial surface (Abeyta et al., 2003). Clearly, maximal expression of CPS is critical for systemic virulence because of resistance to opsonophagocytic clearance. However, invasive disease is invariably preceded by colonization of the nasopharynx, and the thickness of the capsule influences the degree of exposure of other important pneumococcal surface structures, such as the adhesins, which are required during this initial colonization phase. Nonencapsulated pneumococci exhibit higher adherence to human respiratory epithelial (A549) cells in vitro than otherwise isogenic derivatives expressing either type 3 or type 19F capsules (Talbot et al., 1996), and previously encapsulated pneumococci appear to express very little CPS when in intimate contact with respiratory epithelial cells in vitro or in vivo (Hammerschmidt et al., 2005). Nevertheless, at least a basal level of CPS expression is still advantageous in the nasopharyngeal niche (Magee and Yother, 2001), as CPS promotes evasion of mucus-mediated clearance mechanisms (Nelson et al., 2007b) and reduces

entanglement of pneumococci in neutrophil extracellular traps (Wartha et al., 2007).

All but 2 of the more than 90 pneumococcal CPS serotypes are synthesized in a Wzy-polymerase-dependent fashion, whereby oligosaccharide repeat units are assembled on a lipid carrier on the inner face of the cytoplasmic membrane and then flipped to the outer face, where they are polymerized (Whitfield and Roberts, 1999). All the genes required for synthesis of a given CPS serotype are clustered on the chromosome. These *cps* loci comprise up to 20 or more cotranscribed genes, depending on the complexity of the CPS repeat unit structure, including four conserved genes at the 5′ end (*cpsABCD*) associated with modulation of CPS synthesis, as described below. The remaining two serotypes (types 3 and 37) have very simple CPS structures and are synthesized by processive transferases.

As mentioned above, alteration in the total amount of CPS produced is one of the key features of phase variation between opaque and transparent phenotypes (Kim and Weiser, 1998). Although the mechanism of phase variation is not understood, there is evidence that the level of *cps* locus gene products, such as CpsD (see below), differs in transparent and opaque variants (Weiser et al., 2001). However, differences in transcription of the *cps* genes themselves have not been reported, suggesting that the regulation is either indirect or at the posttranscriptional level. A distinct phase variation phenomenon has also been reported to occur in type 3, 8, and 37 pneumococci grown on biofilms. These strains exhibited switching from encapsulated to nonencapsulated phenotypes as a result of spontaneous tandem duplications of up to 239 bp in the *cap3A*, *cap8E*, and *tts* genes, respectively, which exhibited low-frequency reversion (Waite et al., 2001; Waite et al., 2003). However, the in vivo relevance of this phenomenon is uncertain.

Clearly, the capacity to regulate CPS production, at either the transcriptional, translational, or posttranslational level, is important for the survival of the pneumococcus in different host environments. Quantitative reverse transcription PCR has been used to show that the level of *cps* mRNA, relative to 16S rRNA, in pneumococci isolated from the blood of infected mice was approximately fourfold higher than that in pneumococci grown in vitro (Ogunniyi et al., 2002). However, in a later study, significant differences in *cps* mRNA levels between pneumococci isolated from the nasopharynx, blood, and lungs of infected mice could not be detected (LeMessurier et al., 2006). Bioinformatic analysis of sequences upstream of the pneumococcal *cps* locus promoter shows that only about 250 nucleotides are shared among all strains/serotypes, owing largely to variable presence of insertion sequence elements. Nevertheless, this region does include sequences with similarity to binding sites for known transcriptional regulators, including ComX1, CopY, MalR, GlnR, CcpA, and RitR (Moscoso and Garcia, 2009). To date, however, experimental confirmation that *cps* transcription is directly impacted by these factors is scant, except for CcpA, which upregulates *cps* expression, as described above (Giammarinaro and Paton, 2002). This may provide a mechanism whereby exposure of pneumococci to high glucose concentrations in the blood might increase CPS production, relative to levels in the nasopharynx, where glucose is low (Moscoso and Garcia, 2009). Oxygen availability may also serve as an environmental cue for modulation of CPS expression, as growth under anaerobic, rather than atmospheric, conditions significantly increased CpsD protein and total CPS levels, particularly in opaque-phase variants (Weiser et al., 2001). Yamaguchi et al. (2009) also reported that mutagenesis of the *nrc* gene, which encodes a protein with partial similarity to pneumolysin, resulted in substantial increases in transcription of several D39 *cps* locus genes and higher CPS production. Curiously, however, resistance to opsonophagocytosis was diminished, and there were also effects on growth kinetics of the mutant.

The first four genes of the *cps* locus, *cpsABCD*, are common to all pneumococcal serotypes except types 3 and 37, and the products encoded by these genes, CpsA to -D, have been shown to affect the level of CPS expression (Morona et al., 2000a). CpsA is believed to be a LytR-type transcriptional activator, and a homologue appears to function as a transcriptional enhancer in group B streptococci (Cieslewicz et al., 2001). However, to date, *cpsA* has not been shown to affect *cps* transcription in *S. pneumoniae*. Type 19F and type 2 mutants in which the *cpsA* gene had been deleted produced reduced levels of CPS (approximately two-thirds that of the respective wild-type strains) (Morona et al., 2000a; Morona et al., 2003; Morona et al., 2004). However, there were no obvious differences in the amounts of CpsB and CpsD proteins detected by Western blotting in any of the strains (Morona et al., 2000a; Bender et al., 2003).

CpsB, CpsC, and CpsD have all been implicated in regulation of CPS production at a posttranslational level. CpsC and CpsD belong to the PCP2b family of polysaccharide copolymerases and are predicted to function together in polymerization and export of CPS in a fashion similar to PCP2a proteins, such as Wzc and ExoP, in production of CPS and EPS in gram-negative bacteria (Glucksmann et al., 1993; Guidolin et al., 1994; Morona et al., 2000b; Whitfield and Paiment, 2003). CpsD has similarities

to the C-terminal domain of PCP2a proteins and is an autophosphorylating protein-tyrosine kinase (Morona et al., 2000a; Morona et al., 2000b). These kinases contain Walker A and B ATP-binding motifs and a tyrosine-rich region at the C terminus which becomes phosphorylated at multiple tyrosine residues (Morona et al., 2000a; Morona et al., 2003; Bender and Yother, 2001). In *S. pneumoniae*, the tyrosine-rich region of CpsD is arranged as an ordered YGX motif varying from two to four repeats in different serotypes. CpsC is a membrane protein with two membrane-spanning hydrophobic domains; the N and C termini are located in the cytoplasm, while the central portion is exposed on the external side of the cell membrane. It has similarities to PCP1/Wzz proteins associated with polymerization of O antigen in gram-negative bacteria and to the N-terminal domain of PCP2a proteins (Morona et al., 2000b). CpsC is required for initial CpsD tyrosine autophosphorylation but is not required for transphosphorylation between CpsD proteins (Morona et al., 2000a; Bender and Yother, 2001). CpsB has been identified as a manganese-dependent phosphotyrosine-protein phosphatase belonging to the PHP (polymerase and histidinol phosphatase) family of phosphoesterases and is required to dephosphorylate CpsD (Morona et al., 2002). CpsB is also thought to be able to bind CpsD and prevent transphosphorylation between CpsD proteins (Bender and Yother, 2001).

In a derivative of the avirulent *S. pneumoniae* strain Rx1 expressing type 19F CPS, *cpsD* point mutations affecting the ATP-binding domain (Walker A motif) or deletion of the C-terminal tyrosine-rich domain eliminated CPS production, indicating that both of these regions are important for function of CpsD (Morona et al., 2000a; Morona et al., 2003). CpsD required at least two intact YGX repeats to produce wild-type levels of CPS (Morona et al., 2003). Strains in which *cpsB* had been either deleted or mutated to eliminate the phosphotyrosine-protein phosphatase activity of CpsB had high levels of phosphorylated CpsD. These strains produced very little CPS, suggesting that the phosphorylated form of CpsD was inactive. Additionally, the loss of CpsB activity did not affect CPS production in strains which contained a CpsD protein that could no longer be phosphorylated due to mutation of all four tyrosines to phenylalanine in the $[YGX]_4$ repeat domain. Thus, autophosphorylation of CpsD at tyrosine attenuates its activity and negatively regulates CPS production in Rx1-19F (Morona et al., 2000a; Morona et al., 2003).

Morona et al. constructed D39 *cpsB* deletion mutants by transformation with an overlap-extension PCR product in which an erythromycin resistance gene replaced *cpsB*. Interestingly, whereas most mutants exhibited a small-colony phenotype, approximately 1% exhibited a mucoid phenotype (Morona et al., 2006). The small-colony phenotype produced reduced amounts of CPS, virtually all of which was attached to the cell wall, consistent with attenuation of CPS production by phosphorylated CpsD. The mucoid phenotype was caused by apparently random secondary mutations in the 5′ half of *cpsC*, which resulted in a loss of requirement for CpsB and, therefore, a loss of regulation of CPS biosynthesis via CpsD tyrosine phosphorylation. Although total CPS production was unaffected, the mucoid mutants attached only half as much CPS to the cell wall as did wild-type D39. Similar findings have recently been obtained by targeted mutagenesis of CpsC (Byrne et al., 2011). Thus, the phosphorylated form of CpsD promotes attachment of CPS to the cell wall. Interestingly, the mucoid mutants could colonize the nasopharynx and lungs of mice after intranasal challenge but were unable to translocate from the lungs to the blood, demonstrating a requirement for attachment of CPS to the cell wall for optimal systemic virulence of pneumococci (Morona et al., 2006). Derivatives of D39 with mutations in all three YGX motifs of CpsD, such that it could not be phosphorylated, also exhibit a mucoid phenotype and are defective in cell wall attachment and translocation from the lungs to the blood of mice (Morona et al., 2004). Thus, CpsB, CpsC, and CpsD are proposed to function together to generate a phosphorylation cycle across CpsD, and through this cycle CPS polymer biosynthesis and ligation are predicted to be controlled.

Total CPS production can also be influenced posttranslationally by changes in sugar metabolism that impact availability of nucleotide sugar precursors. Two key enzymes are PGM (phosphoglucomutase), which catalyzes the conversion of Glc-6-P to Glc-1-P, and GalU (Glc-1-P-uridylyltransferase), which catalyzes the formation of UDP-Glc from Glc-1-P. Pneumococcal mutants in which either the *galU* or the *pgm* genes were disrupted produced almost no CPS and also exhibited growth defects (Mollerach et al., 1998; Hardy et al., 2001). Additionally, strains in which the *pgm* gene had defined point mutations which significantly reduced, but did not eliminate, enzymatic activity still produced reduced amounts of CPS even though they no longer exhibited growth defects (Hardy et al., 2001). UDP-Glc is a precursor for the biosynthesis of all 90+ pneumococcal CPS types, as well as for other cellular structures such as teichoic acid. Thus, limiting the supply of this precursor would be expected to impact heavily upon CPS production in the pneumococcus. Indirect modulation of CPS production by controlled availability of precursors or cofactors may be one of the regulatory

mechanisms used by the pneumococcus. Substrate availability is certainly a critical factor in regulation of type 3 CPS, which is synthesized by a processive transferase (polysaccharide synthase) from UDP-Glc and UDP-GlcA precursors. UDP-GlcA is synthesized from the ubiquitous UDP-Glc by UDP-Glc dehydrogenase, which is the only enzyme required in addition to the transferase/synthase for type 3 CPS production. High UDP-GlcA concentrations favor rapid polymerization, but low concentrations favor chain termination and ejection of the nascent CPS from the polymerase (Ventura et al., 2006).

REGULATION OF PNEUMOCOCCAL SURFACE PROTEINS

The prospect of developing vaccines targeted at pneumococcal surface proteins increases the importance of understanding their role in pathogenesis, their relative expression levels in various host compartments, and the mechanism(s) whereby their expression in vivo is regulated. Current knowledge on regulatory mechanisms operating on various classes of pneumococcal surface proteins is provided below and is summarized in Table 2.

Lipoproteins

The pneumococcal genome includes over 30 putative lipoproteins, with prolipoprotein signal peptidase recognition sequences (Tettelin et al., 2001). The so-called lipobox motif directs covalent attachment of a diacyl glycerol moiety to the N-terminal Cys residue of the mature protein, anchoring it to the outer face of the plasma membrane. Thus, these lipoproteins are located beneath the cell wall and the capsule in *S. pneumoniae*. These lipoproteins have diverse functions, the commonest being substrate binding components of ABC transport systems, and many are important for growth and survival of the pneumococcus in vitro and in vivo. Indeed, the most abundant class of putative virulence genes detected in *S. pneumoniae* using signature-tagged mutagenesis appeared to encode transporters, the majority of which were ABC permeases (Lau et al., 2001).

As indicated earlier, PsaA is the solute-binding component of an ABC cation permease encoded by the *psaBCA* locus of *S. pneumoniae* (Dintilhac et al., 1997). Available physiological data for PsaA indicate that it is involved in the transport of Mn^{2+} (Dintilhac et al., 1997; Marra et al., 2002b; McAllister et al., 2004; Johnston et al., 2006), and its expression has been shown to be Mn^{2+} dependent through regulation by PsaR (Johnston et al., 2006; Ogunniyi et

Table 2. Pneumococcal surface proteins and their known or putative regulators

Gene identification (TIGR4)[a]	Protein identification	Known (or putative) regulator[b]	Evidence (reference[s])[c]
SP_0117	PspA	TCSTS06; VicR	Experimental (Standish et al., 2005, 2007; Ng et al., 2005)
SP_2190	PspC (CbpA)	TCSTS06; (CcpA)	Experimental (Standish et al., 2005, 2007); putative (van Opijnen et al., 2009)
SP_2136	PcpA	PsaR; (CcpA)	Experimental (Johnston et al., 2006; Kloosterman et al., 2008; Ogunniyi et al., 2010); putative (van Opijnen et al., 2009)
SP_2201	CbpD	Competence-stimulating peptide; LuxS	Experimental (Eldholm et al., 2010; Johnsborg and Håvarstein, 2009; Trappetti et al., 2011b)
SP_1937	LytA	(CcpA)	Putative (unpublished data)
SP_1650	PsaA	PsaR; TCSTS04; (AdcR; MerR)	Experimental (McCluskey et al., 2004; Johnston et al., 2006; Kloosterman et al., 2008; Ogunniyi et al., 2010); putative (unpublished data)
SP_1872	PiuA	RitR; CodY; (PsaR; MerR)	Experimental (Brown et al., 2001; Ulijasz et al., 2004; 2009; Caymaris et al., 2010); putative (unpublished data)
SP_1032	PiaA	PsaR, CodY, RitR, MerR	Putative
SP_1175	PhtA	AdcR	Experimental (Ogunniyi et al., 2009; Reyes-Caballero et al., 2010; Rioux et al., 2011; Shafeeq et al., 2011)
SP_1174	PhtB	AdcR	Experimental (Ogunniyi et al., 2009; Reyes-Caballero et al., 2010; Rioux et al., 2011; Shafeeq et al., 2011)
SP_1003	PhtD	AdcR	Experimental (Ogunniyi et al., 2009; Reyes-Caballero et al., 2010; Rioux et al., 2011; Shafeeq et al., 2011)
SP_1004	PhtE	AdcR	Experimental (Ogunniyi et al., 2009; Reyes-Caballero et al., 2010; Rioux et al., 2011; Shafeeq et al., 2011)
SP_0314	Hyl	RegR	Experimental (Chapuy-Regaud et al., 2003)
SP_1693	NanA	CcpA	Experimental (Kaufman, 2007; Kaufman and Yother, 2007)
SP_0648	BgaA	CcpA	Experimental (Iyer et al., 2005; Kaufman, 2007; Kaufman and Yother, 2007)
SP_0057	StrH	CcpA	Putative
SP_0462-SP_0464	Pili (RrgA, RrgB and RrgC)	PsaR, RlrA, MgrA, TCSTS08	Experimental (Hava et al., 2003; Hemsley et al., 2003; Johnston et al., 2006; Song et al., 2009
SP_2108	MalX	CcpA	Experimental (unpublished data)
SP_2185	GlpO	CcpA	Experimental (unpublished data)
SP_0366	AliA	CodY, RitR	Putative

[a]Gene identifications were obtained from the *S. pneumoniae* TIGR4 (serotype 4) genome as deposited in the Kyoto Encyclopedia of Genes and Genomes (KEGG) database.
[b]Putative regulators predicted by bioinformatic analysis are indicated in parentheses.
[c]Putative evidence is predicted by bioinformatic analysis, but there is no experimental evidence for regulation.

al., 2010) (Table 2). However, structural data for PsaA show that the metal-binding site is occupied by Zn^{2+} (Lawrence et al., 1998), and PsaR was subsequently shown to be responsible for the Zn^{2+}-dependent derepression of PsaA (Kloosterman et al., 2008). In addition, unpublished observations in our laboratory suggest possible regulation of PsaA expression by AdcR and also possibly by the MerR family transcriptional regulator (SP_1856). Furthermore, serotype-specific regulation of PsaA by a two-component system (RR04/HK04), distinct from the PsaR-PsaA interaction, has been described previously (McCluskey et al., 2004). Hence, PsaA expression appears to be controlled by multiple regulatory mechanisms. A representation of the multilevel regulation is depicted schematically in Fig. 1.

PiuA (pneumococcal iron uptake A) and PiaA (pneumococcal iron acquisition A) are lipoprotein components of two separate iron uptake ABC transporter operons, *piu* and *pia* (Brown et al., 2001a). Originally termed Pit1A (pneumococcal iron transport 1A) and Pit2A, respectively, these proteins were first characterized by following identification during a signature-tagged mutagenesis screen (Lau et al., 2001). Mutagenesis studies demonstrate that PiuA and PiaA are required for bacterial growth and full virulence in both systemic and pulmonary models of infection (Brown et al., 2001a). Moreover, the mRNA of *piaA* is upregulated in the nasopharynx and blood of mice 72 h after intranasal infection with serotype 2 strain D39 (LeMessurier et al., 2006). Growth and binding studies suggest that the substrate for these

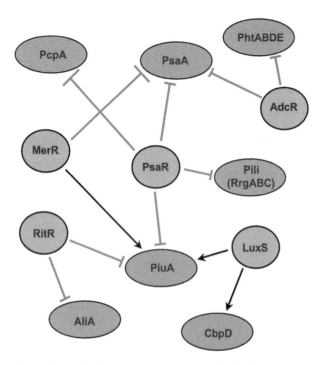

Figure 1. Simple schematic representation of multiple regulatory circuits impacting expression of surface proteins of *S. pneumoniae*. Selected surface proteins are shown in purple, and their respective or putative regulators are depicted in green. Positive regulation is shown by black arrows, while negative regulation (repression) is indicated in red.
doi:10.1128/9781555818524.ch10f1

iron uptake ABC transporters may be hemoglobin or other heme-containing proteins (Brown et al., 2001a; Tai et al., 1993; Tai et al., 1997). Available data have shown that PiuA is negatively regulated by RitR (Brown et al., 2001a; Ulijasz et al., 2004; Ulijasz et al., 2009) and CodY (Caymaris et al., 2010) (Table 2) and possibly by PsaR and MerR (our unpublished studies). Given the functional link between PiuA and PiaA, we postulate that PiaA is regulated in a similar manner. Interestingly, recent studies show that the AI-2 quorum-sensing molecule produced by LuxS increases expression of PiuA and boosts cellular iron levels (Trappetti et al., 2011b). The expression of the *luxS* gene itself is also increased by supplementation of media with Fe^{3+}, thereby providing a positive amplification loop (Trappetti et al., 2011b). The multiple regulatory circuits impacting PiuA expression suggest that a degree of redundancy facilitates maintenance of iron homeostasis in pneumococci.

Another distinct ABC transporter operon encoding proteins with similarity to iron uptake transporters is *pit* (pneumococcal iron transport) (Brown et al., 2002). However, molecular and genetic studies using the lipoprotein component, PitA, demonstrated that while it is required for acquisition of intracellular iron by *S. pneumoniae*, its contribution to growth

and virulence is relatively minor compared to that of PiuA or PiaA (Brown et al., 2002). To date, it is not known how PitA is regulated.

Pht Proteins

Polyhistidine triad (Pht) proteins are a distinct class of surface proteins found only in streptococci (Panina et al., 2003), characterized by the presence of five or six repeated histidine triad (HxxHxH) motifs in their sequences; in *S. pneumoniae* there are four members, PhtA, PhtB, PhtD, and PhtE (Adamou et al., 2001; Wizemann et al., 2001). These proteins share very high sequence homology in the N-terminal region (87% for amino acids 1 to 240), and each possesses an atypical signal peptidase II (LxxC) lipoprotein motif at the N terminus. However, they are not lipoproteins, and their mode of attachment to the cell surface is not understood (Adamou et al., 2001; Hamel et al., 2004). Signature-tagged mutagenesis studies support a role for PhtA, PhtB, and PhtD in the progression to pulmonary disease (Hava and Camilli, 2002). A role in immune evasion was also suggested by the finding that a 20-kDa fragment of PhtB (previously termed PhpA) cleaved human complement component C3 (Hostetter, 1999). Direct evidence for the involvement of Pht proteins in pneumococcal virulence was obtained using a large set of defined, nonpolar mutants of the respective genes in *S. pneumoniae* D39. In murine intraperitoneal (sepsis) and intranasal (pneumonia) challenge models, mutagenesis of all four *pht* genes was required to abrogate virulence relative to the wild type (Ogunniyi et al., 2009), suggesting significant functional redundancy among Pht proteins. Pht proteins appear to collectively interfere with deposition of complement on the surface of *S. pneumoniae* by binding the complement regulatory factor H (Ogunniyi et al., 2009), a property shared by a Pht homologue from group B streptococci (Maruvada et al., 2009). The role of pneumococcal Pht proteins in complement inhibition/factor H binding is controversial, and a subsequent study produced conflicting results (Melin et al., 2010). However, this could have been due to the use of normal human serum, which contains high (potentially inhibitory) levels of anti-Pht antibodies, as a source of factor H.

Histidines possess affinity for small cations, and the crystal structure of a histidine triad motif from PhtA revealed complexed Zn^{2+} (Riboldi-Tunnicliffe et al., 2005). Moreover, Zn^{2+} binding has been shown to confer increased thermal stability on PhtD (Loisel et al., 2011). This has led to the suggestion that Pht proteins might function in metal ion homeostasis, perhaps by scavenging and sequestering Zn^{2+}

on the cell surface (Rioux et al., 2011; Loisel et al., 2011), although direct evidence for this is lacking. Nevertheless, Zn^{2+} has a clear role in *pht* regulation. AdcR, a Zn^{2+}-dependent MarR family transcriptional regulator, modulates expression of all four *pht* genes of *S. pneumoniae* (Ogunniyi et al., 2009), as shown using an unmarked, in-frame *adcR* deletion mutant and by comparing expression of the various *pht* genes with that in the otherwise isogenic wild-type strain (D39). In vitro binding of purified recombinant AdcR to the promoter regions of each of the *pht* genes was also demonstrated by electrophoretic mobility shift assays (Ogunniyi et al., 2009). Interestingly, AdcR also regulates the *adc* operon, which encodes the ABC-type permease responsible for cellular uptake of Zn^{2+} (Dintillac et al., 1997), along with *lmb/adcAII* (which encodes a putative secondary Zn^{2+}-binding lipoprotein), which is cotranscribed with *phtD* (Reyes-Caballero et al., 2010; Rioux et al., 2011). Expression of all components of AdcR regulon is upregulated in Zn^{2+}-deficient media (Shafeeq et al., 2011). However, the in vivo relevance of this is uncertain, as during pneumococcal infection, Zn^{2+} levels in the nasopharynx, lungs, blood, and brain are in the 100 to 600 μM range, as measured by inductively coupled plasma mass spectrometry (McDevitt et al., 2011), and yet there is substantial expression of all four *pht* genes in these niches (Ogunniyi et al., 2009).

Analysis of the levels of the various *pht* transcripts in the lungs, blood, and nasopharynx of *S. pneumoniae*-infected mice revealed marked upregulation of expression of all four genes in all of the in vivo niches, relative to that in bacteria grown in vitro in semisynthetic medium with 100 μM Zn^{2+} (Ogunniyi et al., 2009). Levels of expression were also relatively higher in the nasopharynx and lungs than in the blood for all of the genes. These differences may reflect variations in the levels of Zn^{2+} in the various niches (Zn^{2+} levels are highest in the blood), although a role for additional regulatory circuits cannot be excluded. Interestingly, increasing $[Zn^{2+}]$ results in oligomerization of factor H (Nan et al., 2008), and this raises the possibility that if Pht proteins are indeed capable of sequestering Zn^{2+} on the pneumococcal surface, this might directly contribute to factor H binding. Thus, studies of the role of Zn^{2+} in the Pht-factor H interaction, as well as cellular Zn^{2+} levels in *pht* mutants, might provide further clues as to the function of Pht proteins in vivo.

Choline-Binding Proteins

The CBPs of *S. pneumoniae* are a family of proteins that are attached to ChoP moieties on cell wall teichoic acid and membrane-bound lipoteichoic acid

(Rosenow et al., 1997; Garcia et al., 1998; Gosink et al., 2000). CBPs share a common choline binding region (CBR) consisting of between 2 and 10 highly conserved 20 amino acid repeats, which bind noncovalently to ChoP (Giffard and Jacques, 1994). To date, at least 10 CBPs, exhibiting diverse functions, have been identified and characterized. Among these, pneumococcal surface protein A (PspA) (Yother and Briles, 1992), pneumococcal surface protein C (PspC; also known as choline-binding protein A [CbpA]) (Rosenow et al., 1997), choline-binding protein D (CbpD; a murein hydrolase) (Kausmally et al., 2005), the major autolysin (LytA) (Garcia et al., 1986), and pneumococcal surface protein PcpA (Sanchez-Beato et al., 1998) have been extensively studied. The various CBPs of *S. pneumoniae* play diverse roles in animal models of colonization, respiratory tract infection, and systemic disease (Briles et al., 1988; Rosenow et al., 1997; Paton, 1998; Gosink et al., 2000; Ogunniyi et al., 2007b).

Expression of CbpA and PspA has been shown to be directly regulated by the pneumococcal RR/HK06 TCSTS via two distinct mechanisms: *cbpA* is upregulated by phosphorylated RR06, while *pspA* is upregulated by the nonphosphorylated form (Standish et al., 2005, 2007). The expression of these two proteins is also oppositely regulated in opaque-versus transparent-phase variants (Kim and Weiser, 1998). Furthermore, the expression of PspA has been shown to be positively regulated by VicR of the VicRK (YycFG) TCSTS (Ng et al., 2005). Expression of PcpA has been shown to be Zn^{2+} and Mn^{2+} dependent through regulation by PsaR (Johnston et al., 2006; Kloosterman et al., 2008; Ogunniyi et al., 2010). Interestingly, expression of CbpD, which is critical for biofilm formation, is regulated in parallel with late competence genes by both the CSP and the LuxS/AI-2 quorum sensing systems (Eldholm et al., 2010; Johnsborg and Håvarstein, 2009; Trappetti et al., 20011b). Thus, within a single family of surface proteins, nearly all of the known regulatory systems deployed by the pneumococcus (TCSTS, quorum sensing, phase variation, metal-dependent repressors, etc.) are operational (Fig. 1).

Sortase-Dependent Surface Proteins

Sortase-dependent surface proteins are unique to gram-positive bacteria and are characterized by a C-terminal anchoring motif, comprising a conserved LPXTG sequence followed by a hydrophobic domain and a positively charged tail. Sortase, a membrane-localized cysteine protease, cleaves the LPXTG motif and catalyzes covalent linkage of the mature protein to the cell wall peptidoglycan (Marraffini et al.,

2006). A number of sortase-anchored proteins have been described for *S. pneumoniae*, including proven and putative virulence factors such as NanA (Kharat and Tomasz, 2003; Camara et al., 1994), Hyl (hyaluronidate lyase) (Paton et al., 1997; Berry and Paton, 2000; Jedrzejas, 2001), Iga (immunoglobulin A1 protease) (Paton et al., 1993; Wani et al., 1996; Bender and Weiser, 2006), StrH (King et al., 2006; Dalia et al., 2010), BgaA (Zähner and Hakenbeck, 2000; Kharat and Tomasz, 2003), and pili (Barocchi et al., 2006; LeMieux et al., 2006; Nelson et al., 2007a; Rosch et al., 2008).

It was found that the exoglycosidases NanA, BgaA, and StrH sequentially deglycosylate N-linked glycans on host defense molecules, thereby altering the clearance function of these molecules and allowing pneumococci to persist in the airway (King et al., 2006). Later, it was postulated that the exoglycosidase-dependent liberation of monosaccharides from these glycoconjugates in close proximity to the pneumococcal surface serves as a source of fermentable carbohydrate in vivo (Burnaugh et al., 2008). Consistent with this hypothesis, NanA and BgaA expression has been shown to be regulated by CcpA (Kaufman, 2007; Kaufman and Yother, 2007). Interestingly, a previous study has shown that transparent-phase variants have increased levels of NanA, BgaA, and StrH but decreased capsule production, while the opposite is the case for the opaque variants (King et al., 2004). We recently found that *ccpA* mRNA levels are upregulated in transparent-phase pneumococci grown in vitro (unpublished studies). This suggests that CcpA is part of a downstream effector mechanism for phase variation-induced changes in gene expression that allow *S. pneumoniae* to adapt to changing environments. Hyl expression has also been shown to be under the control of the global regulator RegR (Chapuy-Regaud et al., 2003). However, to date, evidence for regulation of immunoglobulin A1 protease expression is lacking.

Pili are present on the surface of a proportion of *S. pneumoniae* strains (Paterson and Mitchell, 2006) and comprise protofilaments consisting of three structural subunits, RrgA, RrgB, and RrgC, arranged in a coiled-coil superstructure, and encoded by the *rlrA* pathogenicity islet (Barocchi et al., 2006; LeMieux et al., 2006; Hilleringmann et al., 2008; Hilleringmann et al., 2009). These pili contain sorting signal motifs that deviate from the typical LPXTG motif (YPRTG, IPQTG, and VPDTG, respectively) (Marraffini et al., 2006). The proteins were demonstrated to be important for colonization and virulence (Hava and Camilli, 2002; Barocchi et al., 2006; LeMieux et al., 2006; Nelson et al., 2007a). The transcription factor RlrA positively regulates all

seven genes in the *rlrA* pathogenicity islet, including *rlrA* itself, as well as *srtBCD* (encoding three sortase homologues required for pilus assembly and attachment) and *rrgABC* (Hava et al., 2003). However, another transcriptional regulator, MgrA, has also been identified by microarray analysis as a repressor of the *rlrA* pathogenicity islet (Hemsley et al., 2003). In addition, expression of RrgA, RrgB, and RrgC is indirectly controlled by Mn^{2+}-dependent regulation of RlrA expression by PsaR (Johnston et al., 2006). Moreover, late-logarithmic-phase-dependent regulation of the genes in the *rlrA* pathogenicity islet by TCSTS08 was demonstrated by transcriptional profile analysis, which showed upregulation of expression of the *rlrA* islet genes in an *rr08* mutant (Song et al., 2009). Thus, multilevel regulation of expression appears to be a common theme among pneumococcal surface antigens (Fig. 1).

CONCLUDING REMARKS

S. pneumoniae is a highly successful, human-adapted pathogen, responsible for more than a million deaths each year. It has a relatively small (approximately 2.3 Mb), highly plastic genome, which nevertheless encodes sufficient products to enable survival in diverse host microenvironments. Its capacity to survive in these discrete host niches in the face of robust innate and adaptive immune responses, as well as fierce competition from other microorganisms in niches such as the nasopharynx, underscores the ability of the pneumococcus to modulate expression of its genome to optimize adaptation to each distinct microenvironment. Clearly, surface structures such as the polysaccharide capsule and surface proteins are principally responsible for mediating interactions between *S. pneumoniae* and its immediate environment. Thus, the capacity to adapt solute uptake and downstream metabolic pathways in accordance with nutrient availability, and to optimize deployment of primary virulence determinants that interact with host surface molecules, or engage innate host defenses, in discrete niches will have a major impact on disease pathogenesis. As can be seen from the examples presented herein, regulation of pneumococcal surface proteins and the capsule is very complex and multifactorial, involving overlapping regulatory mechanisms. By way of example, PiuA expression is regulated by RitR, CodY, LuxS, PsaR, and AdcR. Each of these regulators, in turn, also has multiple targets for regulation (e.g., PsaR regulates *psaA*, *piuA*, *prtA*, *pcpA*, and the *rlrA* islet). The complexity of these regulatory networks makes the task of identifying the principal determinants of virulence gene expression a challenging one. Nevertheless, a thorough dissection of the critical regulatory pathways employed by *S. pneumoniae* in discrete in vivo niches will undoubtedly provide an improved understanding of pneumococcal pathogenesis and possibly identify novel targets for intervention.

REFERENCES

Abeyta, M., G. G. Hardy, and J. Yother. 2003. Genetic alteration of capsule type but not PspA type affects accessibility of surface-bound complement and surface antigens of *Streptococcus pneumoniae*. *Infect. Immun.* **71:**218–225.

Adamou, J. E., J. H. Heinrichs, A. L. Erwin, W. Walsh, T. Gayle, M. Dormitzer, R. Dagan, Y. A. Brewah, P. Barren, R. Lathigra, S. Langermann, S. Koenig, and S. Johnson. 2001. Identification and characterization of a novel family of pneumococcal proteins that are protective against sepsis. *Infect. Immun.* **69:**949–958.

Aranda, J., M. E. Garrido, N. Fittipaldi, P. Cortes, M. Llagostera, M. Gottschalk, and J. Barbe. 2010. The cation-uptake regulators AdcR and Fur are necessary for full virulence of *Streptococcus suis*. *Vet. Microbiol.* **144:**246–249.

Austrian, R. 1981. Some observations on the pneumococcus and on the current status of pneumococcal disease and its prevention. *Rev. Infect. Dis.* **3**(Suppl.):S1–S17.

Barocchi, M. A., J. Ries, X. Zogaj, C. Hemsley, B. Albiger, A. Kanth, S. Dahlberg, J. Fernebro, M. Moschioni, V. Masignani, K. Hultenby, A. R. Taddei, K. Beiter, F. Wartha, A. von Euler, A. Covacci, D. W. Holden, S. Normark, R. Rappuoli, and B. Henriques-Normark. 2006. A pneumococcal pilus influences virulence and host inflammatory responses. *Proc. Natl. Acad. Sci. USA* **103:**2857–2862.

Bartilson, M., A. Marra, J. Christine, J. S. Asundi, W. P Schneider, and A. E. Hromockyj. 2001. Differential fluorescence induction reveals *Streptococcus pneumoniae* loci regulated by competence stimulatory peptide. *Mol. Microbiol.* **39:**126–135.

Bassler, B. L. 1999. How bacteria talk to each other: regulation of gene expression by quorum sensing. *Curr. Opin. Microbiol.* **2:**582–587.

Belland, R. J., S. G. Morrison, P. van der Ley, and J. Swanson. 1989. Expression and phase variation of gonococcal P.II genes in *Escherichia coli* involves ribosomal frameshifting and slipped-strand mispairing. *Mol. Microbiol.* **3:**777–786.

Bender, M. H., R. T. Cartee, and J. Yother. 2003. Positive correlation between tyrosine phosphorylation of CpsD and capsular polysaccharide production in *Streptococcus pneumoniae*. *J. Bacteriol.* **185:**6057–6066.

Bender, M. H., and J. N. Weiser. 2006. The atypical amino-terminal LPNTG-containing domain of the pneumococcal human IgA1-specific protease is required for proper enzyme localization and function. *Mol. Microbiol.* **61:**526–543.

Bender, M. H., and J. Yother. 2001. CpsB is a modulator of capsule-associated tyrosine kinase activity in *Streptococcus pneumoniae*. *J. Biol. Chem.* **276:**47966–47974.

Berntsson, R. P., S. H. Smits, L. Schmitt, D. J. Slotboom, and B. Poolman. 2010. A structural classification of substrate-binding proteins. *FEBS Lett.* **584:**2606–2617.

Berry, A. M., and J. C. Paton. 2000. Additive attenuation of virulence of *Streptococcus pneumoniae* by mutation of the genes encoding pneumolysin and other putative pneumococcal virulence proteins. *Infect. Immun.* **68:**133–140.

Berry, A. M., and J. C. Paton. 1996. Sequence heterogeneity of PsaA, a 37-kilodalton putative adhesin essential for virulence of *Streptococcus pneumoniae*. *Infect. Immun.* **64:**5255–5262.

Blue, C. E., and T. J. Mitchell. 2003. Contribution of a response regulator to the virulence of *Streptococcus pneumoniae* is strain dependent. *Infect. Immun.* **71:**4405–4413.

Brenot, A., B. F. Weston, and M. G. Caparon. 2007. A PerR-regulated metal transporter (PmtA) is an interface between oxidative stress and metal homeostasis in *Streptococcus pyogenes*. *Mol. Microbiol.* **63:**1185–1196.

Briles, D. E., S. K. Hollingshead, J. C. Paton, E. W. Ades, L. Novak, F. W. van Ginkel, and W. H. Benjamin, Jr. 2003. Immunizations with pneumococcal surface protein A and pneumolysin are protective against pneumonia in a murine model of pulmonary infection with *Streptococcus pneumoniae*. *J. Infect. Dis.* **188:**339–348.

Briles, D. E., L. Novak, M. Hotomi, F. W. van Ginkel, and J. King. 2005. Nasal colonization with *Streptococcus pneumoniae* includes subpopulations of surface and invasive pneumococci. *Infect. Immun.* **73:**6945–6951.

Briles, D. E., J. Yother, and L. S. McDaniel. 1988. Role of pneumococcal surface protein A in the virulence of *Streptococcus pneumoniae*. *Rev. Infect. Dis.* **10**(Suppl. 2):S372–S374.

Brown, J. S., S. M. Gilliland, and D. W. Holden. 2001a. A *Streptococcus pneumoniae* pathogenicity island encoding an ABC transporter involved in iron uptake and virulence. *Mol. Microbiol.* **40:**572–585.

Brown, J. S., A. D. Ogunniyi, M. C. Woodrow, D. W. Holden, and J. C. Paton. 2001b. Immunization with components of two iron uptake ABC transporters protects mice against systemic *Streptococcus pneumoniae* infection. *Infect. Immun.* **69:**6702–6706.

Brown, J. S., S. M. Gilliland, J. Ruiz-Albert, and D. W. Holden. 2002. Characterization of Pit, a *Streptococcus pneumoniae* iron uptake ABC transporter. *Infect. Immun.* **70:**4389–4398.

Brueggemann, A. B., R. Pai, D. W. Crook, and B. Beall. 2007. Vaccine escape recombinants emerge after pneumococcal vaccination in the United States. *PLoS Pathog.* **3:**e168.

Burnaugh, A. M., L. J. Frantz, and S. J. King. 2008. Growth of *Streptococcus pneumoniae* on human glycoconjugates is dependent upon the sequential activity of bacterial exoglycosidases. *J. Bacteriol.* **190:**221–230.

Byrne, J. P., J. K. Morona, J. C. Paton, and R. Morona. 2011. Identification of *Streptococcus pneumoniae* Cps2C residues that affect capsular polysaccharide polymerization, cell wall ligation, and Cps2D phosphorylation. *J. Bacteriol.* **193:**2341–2346.

Camara, M., G. J. Boulnois, P. W. Andrew, and T. J. Mitchell. 1994. A neuraminidase from *Streptococcus pneumoniae* has the features of a surface protein. *Infect. Immun.* **62:**3688–3695.

Caymaris, S., H. J. Bootsma, B. Martin, P. W. Hermans, M. Prudhomme, and J. P. Claverys. 2010. The global nutritional regulator CodY is an essential protein in the human pathogen *Streptococcus pneumoniae*. *Mol. Microbiol.* **78:**344–360.

Chapuy-Regaud, S., A. D. Ogunniyi, N. Diallo, Y. Huet, J. F. Desnottes, J. C. Paton, S. Escaich, and M. C. Trombe. 2003. RegR, a global LacI/GalR family regulator, modulates virulence and competence in *Streptococcus pneumoniae*. *Infect. Immun.* **71:**2615–2625.

Chuck, A. W., P. Jacobs, G. Tyrrell, and J. D. Kellner. 2010. Pharmacoeconomic evaluation of 10- and 13-valent pneumococcal conjugate vaccines. *Vaccine* **28:**5485–5490.

Cieslewicz, M. J., D. L. Kasper, Y. Wang, and M. R. Wessels. 2001. Functional analysis in type Ia group B *Streptococcus* of a cluster of genes involved in extracellular polysaccharide production by diverse species of streptococci. *J. Biol. Chem.* **276:**139–146.

Claverys, J. P. 2001. A new family of high-affinity ABC manganese and zinc permeases. *Res. Microbiol.* **152:**231–243.

Corbin, B. D., E. H. Seeley, A. Raab, J. Feldmann, M. R. Miller, V. J. Torres, K. L. Anderson, B. M. Dattilo, P. M. Dunman, R. Gerads, R. M. Caprioli, W. Nacken, W. J. Chazin, and E. P. Skaar. 2008. Metal chelation and inhibition of bacterial growth in tissue abscesses. *Science* **319:**962–965.

Crum, N. F., C. P. Barrozo, F. A. Chapman, M. A. Ryan, and K. L. Russell. 2004. An outbreak of conjunctivitis due to a novel unencapsulated *Streptococcus pneumoniae* among military trainees. *Clin. Infect. Dis.* **39:**1148–1154.

Dalia, A. B., A. J. Standish, and J. N. Weiser. 2010. Three surface exoglycosidases from *Streptococcus pneumoniae*, NanA, BgaA, and StrH, promote resistance to opsonophagocytic killing by human neutrophils. *Infect. Immun.* **78:**2108–2116.

Dintilhac, A., G. Alloing, C. Granadel, and J. P. Claverys. 1997. Competence and virulence of *Streptococcus pneumoniae*: Adc and PsaA mutants exhibit a requirement for Zn and Mn resulting from inactivation of putative ABC metal permeases. *Mol. Microbiol.* **25:**727–739.

Dintilhac, A., and J. P. Claverys. 1997. The *adc* locus, which affects competence for genetic transformation in *Streptococcus pneumoniae*, encodes an ABC transporter with a putative lipoprotein homologous to a family of streptococcal adhesins. *Res. Microbiol.* **148:**119–131.

Douglas, R. M., J. C. Paton, S. J. Duncan, and D. J. Hansman. 1983. Antibody response to pneumococcal vaccination in children younger than five years of age. *J. Infect. Dis.* **148:**131–137.

Dziejman, M., and J. J. Mekalanos. 1995. Two-component signal transduction and its role in the expression of bacterial virulence factors, p. 305–317. *In* J. A. Hoch and T. J. Silhavy (ed.), *Two-Component Signal Transduction*. ASM Press, Washington, DC.

Eldholm, V., O. Johnsborg, D. Straume, H. S. Ohnstad, K. H. Berg, J. A. Hermoso, and L. S. Havarstein. 2010. Pneumococcal CbpD is a murein hydrolase that requires a dual cell envelope binding specificity to kill target cells during fratricide. *Mol. Microbiol.* **76:**905–917.

Garcia, J. L., A. R. Sanchez-Beato, F. J. Medrano, and R. Lopez. 1998. Versatility of choline-binding domain. *Microb. Drug Resist.* **4:**25–36.

Garcia, P., J. L. Garcia, E. Garcia, and R. Lopez. 1986. Nucleotide sequence and expression of the pneumococcal autolysin gene from its own promoter in *Escherichia coli*. *Gene* **43:**265–272.

Giammarinaro, P., and J. C. Paton. 2002. Role of RegM, a homologue of the catabolite repressor protein CcpA, in the virulence of *Streptococcus pneumoniae*. *Infect. Immun.* **70:**5454–5461.

Giffard, P. M., and N. A. Jacques. 1994. Definition of a fundamental repeating unit in streptococcal glucosyltransferase glucan-binding regions and related sequences. *J. Dent. Res.* **73:**1133–1141.

Glucksmann, M. A., T. L. Reuber, and G. C. Walker. 1993. Genes needed for the modification, polymerization, export, and processing of succinoglycan by *Rhizobium meliloti*: a model for succinoglycan biosynthesis. *J. Bacteriol.* **175:**7045–7055.

Gosink, K. K., E. R. Mann, C. Guglielmo, E. I. Tuomanen, and H. R. Masure. 2000. Role of novel choline binding proteins in virulence of *Streptococcus pneumoniae*. *Infect. Immun.* **68:**5690–5695.

Guidolin, A., J. K. Morona, R. Morona, D. Hansman, and J. C. Paton. 1994. Nucleotide sequence analysis of genes essential for capsular polysaccharide biosynthesis in *Streptococcus pneumoniae* type 19F. *Infect. Immun.* **62:**5384–5396.

Halfmann, A., M. Kovacs, R. Hakenbeck, and R. Bruckner. 2007. Identification of the genes directly controlled by the response regulator CiaR in *Streptococcus pneumoniae*: five out of 15 promoters drive expression of small non-coding RNAs. *Mol. Microbiol.* **66:**110–126.

Hamel, J., N. Charland, I. Pineau, C. Ouellet, S. Rioux, D. Martin, and B. R. Brodeur. 2004. Prevention of pneumococcal disease in mice immunized with conserved surface-accessible proteins. *Infect. Immun.* **72:**2659–2670.

Hammerschmidt, S., S. Wolff, A. Hocke, S. Rosseau, E. Muller, and M. Rohde. 2005. Illustration of pneumococcal polysaccharide capsule during adherence and invasion of epithelial cells. *Infect. Immun.* **73:**4653–4667.

Hardy, G. G., A. D. Magee, C. L. Ventura, M. J. Caimano, and J. Yother. 2001. Essential role for cellular phosphoglucomutase in virulence of type 3 *Streptococcus pneumoniae*. *Infect. Immun.* **69:**2309–2317.

Hava, D. L., and A. Camilli. 2002. Large-scale identification of serotype 4 *Streptococcus pneumoniae* virulence factors. *Mol. Microbiol.* **45:**1389–1406.

Hava, D. L., C. J. Hemsley, and A. Camilli. 2003. Transcriptional regulation in the *Streptococcus pneumoniae* rlrA pathogenicity islet by RlrA. *J. Bacteriol.* **185:**413–421.

Hemsley, C., E. Joyce, D. L. Hava, A. Kawale, and A. Camilli. 2003. MgrA, an orthologue of Mga, Acts as a transcriptional repressor of the genes within the *rlrA* pathogenicity islet in *Streptococcus pneumoniae*. *J. Bacteriol.* **185:**6640–6647.

Hendriksen, W. T., H. J. Bootsma, A. van Diepen, S. Estevao, O. P. Kuipers, R. de Groot, and P. W. Hermans. 2009. Strain-specific impact of PsaR of *Streptococcus pneumoniae* on global gene expression and virulence. *Microbiology* **155:**1569–1579.

Hendriksen, W. T., T. G. Kloosterman, H. J. Bootsma, S. Estevao, R. de Groot, O. P. Kuipers, and P. W. Hermans. 2008. Site-specific contributions of glutamine-dependent regulator GlnR and GlnR-regulated genes to virulence of *Streptococcus pneumoniae*. *Infect. Immun.* **76:**1230–1238.

Hendriksen, W. T., N. Silva, H. J. Bootsma, C. E. Blue, G. K. Paterson, A. R. Kerr, A. de Jong, O. P. Kuipers, P. W. Hermans, and T. J. Mitchell. 2007. Regulation of gene expression in *Streptococcus pneumoniae* by response regulator 09 is strain dependent. *J. Bacteriol.* **189:**1382–1389.

Hicks, L. A., L. H. Harrison, B. Flannery, J. L. Hadler, W. Schaffner, A. S. Craig, D. Jackson, A. Thomas, B. Beall, R. Lynfield, A. Reingold, M. M. Farley, and C. G. Whitney. 2007. Incidence of pneumococcal disease due to non-pneumococcal conjugate vaccine (PCV7) serotypes in the United States during the era of widespread PCV7 vaccination, 1998–2004. *J. Infect. Dis.* **196:**1346–1354.

Hilleringmann, M., F. Giusti, B. C. Baudner, V. Masignani, A. Covacci, R. Rappuoli, M. A. Barocchi, and I. Ferlenghi. 2008. Pneumococcal pili are composed of protofilaments exposing adhesive clusters of Rrg A. *PLoS Pathog.* **4:**e1000026.

Hilleringmann, M., P. Ringler, S. A. Muller, G. De Angelis, R. Rappuoli, I. Ferlenghi, and A. Engel. 2009. Molecular architecture of *Streptococcus pneumoniae* TIGR4 pili. *EMBO J.* **28:**3921–3930.

Hoskins, J., W. E. Alborn, Jr., J. Arnold, L. C. Blaszczak, S. Burgett, B. S. DeHoff, S. T. Estrem, L. Fritz, D. J. Fu, W. Fuller, C. Geringer, R. Gilmour, J. S. Glass, H. Khoja, A. R. Kraft, R. E. Lagace, D. J. LeBlanc, L. N. Lee, E. J. Lefkowitz, J. Lu, P. Matsushima, S. M. McAhren, M. McHenney, K. McLeaster, C. W. Mundy, T. I. Nicas, F. H. Norris, M. O'Gara, R. B. Peery, G. T. Robertson, P. Rockey, P. M. Sun, M. E. Winkler, Y. Yang, M. Young-Bellido, G. Zhao, C. A. Zook, R. H. Baltz, S. R. Jaskunas, P. R. Rosteck, Jr., P. L. Skatrud, and J. I. Glass. 2001. Genome of the bacterium *Streptococcus pneumoniae* strain R6. *J. Bacteriol.* **183:**5709–5717.

Hostetter, M. K. 1999. Opsonic and nonopsonic interactions of C3 with *Streptococcus pneumoniae*. *Microb. Drug Resist.* **5:**85–89.

Ibrahim, Y. M., A. R. Kerr, J. McCluskey, and T. J. Mitchell. 2004. Control of virulence by the two-component system CiaR/H is mediated via HtrA, a major virulence factor of *Streptococcus pneumoniae*. *J. Bacteriol.* **186:**5258–5266.

Iyer, R., N. S. Baliga, and A. Camilli. 2005. Catabolite control protein A (CcpA) contributes to virulence and regulation of sugar metabolism in *Streptococcus pneumoniae*. *J. Bacteriol.* **187:**8340–8349.

Jacobsen, F. E., K. M. Kazmierczak, J. P. Lisher, M. E. Winkler, and D. P. Giedroc. 2011. Interplay between manganese and zinc homeostasis in the human pathogen *Streptococcus pneumoniae*. *Metallomics* **3:**38–41.

Jedrzejas, M. J. 2001. Pneumococcal virulence factors: structure and function. *Microbiol. Mol. Biol. Rev.* **65:**187–207.

Jennings, M. P., Y. N. Srikhanta, E. R. Moxon, M. Kramer, J. T. Poolman, B. Kuipers, and P. van der Ley. 1999. The genetic basis of the phase variation repertoire of lipopolysaccharide immunotypes in *Neisseria meningitidis*. *Microbiology* **145**(Pt. 11):3013–3021.

Johnsborg, O., and L. S. Håvarstein. 2009. Regulation of natural genetic transformation and acquisition of transforming DNA in *Streptococcus pneumoniae*. *FEMS Microbiol. Rev.* **33:**627–642.

Johnston, J. W., D. E. Briles, L. E. Myers, and S. K. Hollingshead. 2006. Mn²⁺-dependent regulation of multiple genes in *Streptococcus pneumoniae* through PsaR and the resultant impact on virulence. *Infect. Immun.* **74:**1171–1180.

Joyce, E. A., A. Kawale, S. Censini, C. C. Kim, A. Covacci, and S. Falkow. 2004. LuxS is required for persistent pneumococcal carriage and expression of virulence and biosynthesis genes. *Infect. Immun.* **72:**2964–2975.

Kadioglu, A., J. Echenique, S. Manco, M. C. Trombe, and P. W. Andrew. 2003. The MicAB two-component signaling system is involved in virulence of *Streptococcus pneumoniae*. *Infect. Immun.* **71:**6676–6679.

Kaufman, G. E. 2007. Characterization of a global regulatory pathway in *Streptococcus pneumoniae*. Ph.D. thesis. The University of Alabama at Birmingham.

Kaufman, G. E., and J. Yother. 2007. CcpA-dependent and -independent control of beta-galactosidase expression in *Streptococcus pneumoniae* occurs via regulation of an upstream phosphotransferase system-encoding operon. *J. Bacteriol.* **189:**5183–5192.

Kausmally, L., O. Johnsborg, M. Lunde, E. Knutsen, and L. S. Havarstein. 2005. Choline-binding protein D (CbpD) in *Streptococcus pneumoniae* is essential for competence-induced cell lysis. *J. Bacteriol.* **187:**4338–4345.

Kharat, A. S., and A. Tomasz. 2003. Inactivation of the *srtA* gene affects localization of surface proteins and decreases adhesion of *Streptococcus pneumoniae* to human pharyngeal cells in vitro. *Infect. Immun.* **71:**2758–2765.

Kim, J. O., and J. N. Weiser. 1998. Association of intrastrain phase variation in quantity of capsular polysaccharide and teichoic acid with the virulence of *Streptococcus pneumoniae*. *J. Infect. Dis.* **177:**368–377.

King, S. J., K. R. Hippe, J. M. Gould, D. Bae, S. Peterson, R. T. Cline, C. Fasching, E. N. Janoff, and J. N. Weiser. 2004. Phase variable desialylation of host proteins that bind to *Streptococcus pneumoniae in vivo* and protect the airway. *Mol. Microbiol.* **54:**159–171.

King, S. J., K. R. Hippe, and J. N. Weiser. 2006. Deglycosylation of human glycoconjugates by the sequential activities of exoglycosidases expressed by *Streptococcus pneumoniae*. *Mol. Microbiol.* **59:**961–974.

Kloosterman, T. G., W. T. Hendriksen, J. J. Bijlsma, H. J. Bootsma, S. A. van Hijum, J. Kok, P. W. Hermans, and O. P. Kuipers. 2006. Regulation of glutamine and glutamate metabolism by GlnR and GlnA in *Streptococcus pneumoniae*. *J. Biol. Chem.* **281:**25097–25109.

Kloosterman, T. G., R. M. Witwicki, M. M. van der Kooi-Pol, J. J. Bijlsma, and O. P. Kuipers. 2008. Opposite effects of Mn²⁺ and Zn²⁺ on PsaR-mediated expression of the virulence genes *pcpA*, *prtA*, and *psaBCA* of *Streptococcus pneumoniae*. *J. Bacteriol.* **190:**5382–5393.

Lange, R., C. Wagner, A. de Saizieu, N. Flint, J. Molnos, M. Stieger, P. Caspers, M. Kamber, W. Keck, and K. E. Amrein. 1999. Domain organization and molecular characterization of 13 two-component systems identified by genome sequencing of *Streptococcus pneumoniae*. *Gene* **237:**223–234.

Lau, G. W., S. Haataja, M. Lonetto, S. E. Kensit, A. Marra, A. P. Bryant, D. McDevitt, D. A. Morrison, and D. W. Holden. 2001. A functional genomic analysis of type 3 *Streptococcus pneumoniae* virulence. *Mol. Microbiol.* **40:**555–571.

Lawrence, M. C., P. A. Pilling, V. C. Epa, A. M. Berry, A. D. Ogunniyi, and J. C. Paton. 1998. The crystal structure of pneumococcal surface antigen PsaA reveals a metal-binding site and a novel structure for a putative ABC-type binding protein. *Structure* **6:**1553–1561.

Lee, C. J., S. D. Banks, and J. P. Li. 1991. Virulence, immunity, and vaccine related to *Streptococcus pneumoniae*. *Crit. Rev. Microbiol.* **18:**89–114.

LeMessurier, K. S., A. D. Ogunniyi, and J. C. Paton. 2006. Differential expression of key pneumococcal virulence genes *in vivo*. *Microbiology* **152:**305–311.

LeMieux, J., D. L. Hava, A. Basset, and A. Camilli. 2006. RrgA and RrgB are components of a multisubunit pilus encoded by the *Streptococcus pneumoniae rlrA* pathogenicity islet. *Infect. Immun.* **74:**2453–2456.

Lipsitch, M., J. K. Dykes, S. E. Johnson, E. W. Ades, J. King, D. E. Briles, and G. M. Carlone. 2000. Competition among *Streptococcus pneumoniae* for intranasal colonization in a mouse model. *Vaccine* **18:**2895–2901.

Loisel, E., S. Chimalapati, C. Bougault, A. Imberty, B. Gallet, A. M. Di Guilmi, J. Brown, T. Vernet, and C. Durmort. 2011. Biochemical characterization of the histidine triad protein PhtD as a cell surface zinc-binding protein of pneumococcus. *Biochemistry* **50:**3551–3558.

Loisel, E., L. Jacquamet, L. Serre, C. Bauvois, J. L. Ferrer, T. Vernet, A. M. Di Guilmi, and C. Durmort. 2008. AdcAII, a new pneumococcal Zn-binding protein homologous with ABC transporters: biochemical and structural analysis. *J. Mol. Biol.* **381:**594–606.

Magee, A. D., and J. Yother. 2001. Requirement for capsule in colonization by *Streptococcus pneumoniae*. *Infect. Immun.* **69:**3755–3761.

Mahdi, L. K., A. D. Ogunniyi, K. S. LeMessurier, and J. C. Paton. 2008. Pneumococcal virulence gene expression and host cytokine profiles during pathogenesis of invasive disease. *Infect. Immun.* **76:**646–657.

Marra, A., S. Lawson, J. S. Asundi, D. Brigham, and A. E. Hromockyj. 2002a. *In vivo* characterization of the *psa* genes from *Streptococcus pneumoniae* in multiple models of infection. *Microbiology* **148:**1483–1491.

Marra, A., J. Asundi, M. Bartilson, S. Lawson, F. Fang, J. Christine, C. Wiesner, D. Brigham, W. P. Schneider, and A. E. Hromockyj. 2002b. Differential fluorescence induction analysis of *Streptococcus pneumoniae* identifies genes involved in pathogenesis. *Infect. Immun.* **70:**1422–1433.

Marraffini, L. A., A. C. Dedent, and O. Schneewind. 2006. Sortases and the art of anchoring proteins to the envelopes of gram-positive bacteria. *Microbiol. Mol. Biol. Rev.* **70:**192–221.

Martin, M., J. H. Turco, M. E. Zegans, R. R. Facklam, S. Sodha, J. A. Elliott, J. H. Pryor, B. Beall, D. D. Erdman, Y. Y. Baumgartner, P. A. Sanchez, J. D. Schwartzman, J. Montero, A. Schuchat, and C. G. Whitney. 2003. An outbreak of conjunctivitis due to atypical *Streptococcus pneumoniae*. *N. Engl. J. Med.* **348:**1112–1121.

Maruvada, R., N. V. Prasadarao, and C. E. Rubens. 2009. Acquisition of factor H by a novel surface protein on group B *Streptococcus* promotes complement degradation. *FASEB J.* **23:**3967–3977.

McAllister, L. J., H. J. Tseng, A. D. Ogunniyi, M. P. Jennings, A. G. McEwan, and J. C. Paton. 2004. Molecular analysis of the *psa* permease complex of *Streptococcus pneumoniae*. *Mol. Microbiol.* **53:**889–901.

McCluskey, J., J. Hinds, S. Husain, A. Witney, and T. J. Mitchell. 2004. A two-component system that controls the expression of pneumococcal surface antigen A (PsaA) and regulates virulence and resistance to oxidative stress in *Streptococcus pneumoniae*. *Mol. Microbiol.* **51:**1661–1675.

McDevitt, C.A., A. D. Ogunniyi, E. Valkov, M. C. Lawrence, B. Kobe, A. G. McEwan, and J. C. Paton. 2011. A molecular mechanism for bacterial susceptibility to zinc. *PLoS Pathog.* 7:e1002357.

McKessar, S. 2003. The characterisation of phase variation and a novel fimbrial protein in *Streptococcus pneumoniae*. Ph.D. thesis. The University of Adelaide, Adelaide, Australia.

Melin, M., E. Di Paolo, L. Tikkanen, H. Jarva, C. Neyt, H. Kayhty, S. Meri, J. Poolman, and M. Vakevainen. 2010. Interaction of pneumococcal histidine triad proteins with human complement. *Infect. Immun.* 78:2089–2098.

Mollerach, M., R. Lopez, and E. Garcia. 1998. Characterization of the *galU* gene of *Streptococcus pneumoniae* encoding a uridine diphosphoglucose pyrophosphorylase: a gene essential for capsular polysaccharide biosynthesis. *J. Exp. Med.* 188:2047–2056.

Morona, J. K., D. C. Miller, R. Morona, and J. C. Paton. 2004. The effect that mutations in the conserved capsular polysaccharide biosynthesis genes *cpsA*, *cpsB* and *cpsD* have on virulence of *Streptococcus pneumoniae*. *J. Infect. Dis.* 189:1905–1913.

Morona, J. K., R. Morona, D. C. Miller, and J. C. Paton. 2002. *Streptococcus pneumoniae* capsule biosynthesis protein CpsB is a novel manganese-dependent phosphotyrosine-protein phosphatase. *J. Bacteriol.* 184:577–583.

Morona, J. K., R. Morona, D. C. Miller, and J. C. Paton. 2003. Mutational analysis of the carboxy-terminal (YGX)$_4$ repeat domain of CpsD, an autophosphorylating tyrosine kinase required for capsule biosynthesis in *Streptococcus pneumoniae*. *J. Bacteriol.* 185:3009–3019.

Morona, J. K., R. Morona, and J. C. Paton. 2006. Attachment of capsular polysaccharide to the cell wall of *Streptococcus pneumoniae* type 2 is required for invasive disease. *Proc. Natl. Acad. Sci. USA* 103:8505–8510.

Morona, J. K., J. C. Paton, D. C. Miller, and R. Morona. 2000a. Tyrosine phosphorylation of CpsD negatively regulates capsular polysaccharide biosynthesis in *Streptococcus pneumoniae*. *Mol. Microbiol.* 35:1431–1442.

Morona, R., L. Van Den Bosch, and C. Daniels. 2000b. Evaluation of Wzz/MPA1/MPA2 proteins based on the presence of coiled-coil regions. *Microbiology* 146:1–4.

Moscoso, M., and E. Garcia. 2009. Transcriptional regulation of the capsular polysaccharide biosynthesis locus of *Streptococcus pneumoniae*: a bioinformatic analysis. *DNA Res.* 16:177–186.

Musher, D. M. 1992. Infections caused by *Streptococcus pneumoniae*: clinical spectrum, pathogenesis, immunity, and treatment. *Clin. Infect. Dis.* 14:801–807.

Nan, R., J. Gor, I. Lengyel, and S. J. Perkins. 2008. Uncontrolled zinc- and copper-induced oligomerisation of the human complement regulator factor H and its possible implications for function and disease. *J. Mol. Biol.* 384:1341–1352.

Nelson, A. L., J. Ries, F. Bagnoli, S. Dahlberg, S. Falker, S. Rounioja, J. Tschop, E. Morfeldt, I. Ferlenghi, M. Hilleringmann, D. W. Holden, R. Rappuoli, S. Normark, M. A. Barocchi, and B. Henriques-Normark. 2007a. RrgA is a pilus-associated adhesin in *Streptococcus pneumoniae*. *Mol. Microbiol.* 66:329–340.

Nelson, A. L., A. M. Roche, J. M. Gould, K. Chim, A. J. Ratner, and J. N. Weiser. 2007b. Capsule enhances pneumococcal colonization by limiting mucus-mediated clearance. *Infect. Immun.* 75:83–90.

Ng, W. L., H. C. Tsui, and M. E. Winkler. 2005. Regulation of the *pspA* virulence factor and essential *pcsB* murein biosynthetic genes by the phosphorylated VicR (YycF) response regulator in *Streptococcus pneumoniae*. *J. Bacteriol.* 187:7444–7459.

O'Brien, K. L., L. J. Wolfson, J. P. Watt, E. Henkle, M. Deloria-Knoll, N. McCall, E. Lee, K. Mulholland, O. S. Levine, and T. Cherian. 2009. Burden of disease caused by *Streptococcus pneumoniae* in children younger than 5 years: global estimates. *Lancet* 374:893–902.

Ogunniyi, A. D., R. L. Folland, D. E. Briles, S. K. Hollingshead, and J. C. Paton. 2000. Immunization of mice with combinations of pneumococcal virulence proteins elicits enhanced protection against challenge with *Streptococcus pneumoniae*. *Infect. Immun.* 68:3028–3033.

Ogunniyi, A. D., P. Giammarinaro, and J. C. Paton. 2002. The genes encoding virulence-associated proteins and the capsule of *Streptococcus pneumoniae* are upregulated and differentially expressed *in vivo*. *Microbiology* 148:2045–2053.

Ogunniyi, A. D., M. Grabowicz, D. E. Briles, J. Cook, and J. C. Paton. 2007a. Development of a vaccine against invasive pneumococcal disease based on combinations of virulence proteins of *Streptococcus pneumoniae*. *Infect. Immun.* 75:350–357.

Ogunniyi, A. D., K. S. LeMessurier, R. M. Graham, J. M. Watt, D. E. Briles, U. H. Stroeher, and J. C. Paton. 2007b. Contributions of pneumolysin, pneumococcal surface protein A (PspA), and PspC to pathogenicity of *Streptococcus pneumoniae* D39 in a mouse model. *Infect. Immun.* 75:1843–1851.

Ogunniyi, A. D., M. Grabowicz, L. K. Mahdi, J. Cook, D. L. Gordon, T. A. Sadlon, and J. C. Paton. 2009. Pneumococcal histidine triad proteins are regulated by the Zn^{2+}-dependent repressor AdcR and inhibit complement deposition through the recruitment of complement factor H. *FASEB J.* 23:731–738.

Ogunniyi, A. D., L. K. Mahdi, M. P. Jennings, A. G. McEwan, C. A. McDevitt, M. B. Van der Hoek, C. J. Bagley, P. Hoffmann, K. A. Gould, and J. C. Paton. 2010. Central role of manganese in regulation of stress responses, physiology, and metabolism in *Streptococcus pneumoniae*. *J. Bacteriol.* 192:4489–4497.

Ogunniyi, A. D., M. C. Woodrow, J. T. Poolman, and J. C. Paton. 2001. Protection against *Streptococcus pneumoniae* elicited by immunization with pneumolysin and CbpA. *Infect. Immun.* 69:5997–6003.

Orihuela, C. J., J. N. Radin, J. E. Sublett, G. Gao, D. Kaushal, and E. I. Tuomanen. 2004. Microarray analysis of pneumococcal gene expression during invasive disease. *Infect. Immun.* 72:5582–5596.

Overweg, K., C. D. Pericone, G. G. Verhoef, J. N. Weiser, H. D. Meiring, A. P. De Jong, R. De Groot, and P. W. Hermans. 2000. Differential protein expression in phenotypic variants of *Streptococcus pneumoniae*. *Infect. Immun.* 68:4604–4610.

Panina, E. M., A. A. Mironov, and M. S. Gelfand. 2003. Comparative genomics of bacterial zinc regulons: enhanced ion transport, pathogenesis, and rearrangement of ribosomal proteins. *Proc. Natl. Acad. Sci. USA* 100:9912–9917.

Paterson, G. K., C. E. Blue, and T. J. Mitchell. 2006. Role of two-component systems in the virulence of *Streptococcus pneumoniae*. *J. Med. Microbiol.* 55:355–363.

Paterson, G. K., and T. J. Mitchell. 2006. The role of *Streptococcus pneumoniae* sortase A in colonisation and pathogenesis. *Microbes Infect.* 8:145–153.

Paton, J. C. 2004. New pneumococcal vaccines: basic science developments, p. 382–402. *In* E. I. Tuomanen, T. J. Mitchell, D. A. Morrison, and B. G. Spratt (ed.), *The Pneumococcus*. ASM Press, Washington, DC.

Paton, J. C. 1998. Novel pneumococcal surface proteins: role in virulence and vaccine potential. *Trends Microbiol.* 6:85–87; discussion, 87–88.

Paton, J. C., P. W. Andrew, G. J. Boulnois, and T. J. Mitchell. 1993. Molecular analysis of the pathogenicity of *Streptococcus pneumoniae*: the role of pneumococcal proteins. *Annu. Rev. Microbiol.* 47:89–115.

Paton, J. C., A. M. Berry, and R. A. Lock. 1997. Molecular analysis of putative pneumococcal virulence proteins. *Microb. Drug Resist.* 3:1–10.

Peak, I. R., M. P. Jennings, D. W. Hood, M. Bisercic, and E. R. Moxon. 1996. Tetrameric repeat units associated with virulence factor phase variation in *Haemophilus* also occur in *Neisseria* spp. and *Moraxella catarrhalis*. *FEMS Microbiol. Lett.* **137:**109–114.

Pericone, C. D., D. Bae, M. Shchepetov, T. McCool, and J. N. Weiser. 2002. Short-sequence tandem and nontandem DNA repeats and endogenous hydrogen peroxide production contribute to genetic instability of *Streptococcus pneumoniae*. *J. Bacteriol.* **184:**4392–4399.

Polissi, A., A. Pontiggia, G. Feger, M. Altieri, H. Mottl, L. Ferrari, and D. Simon. 1998. Large-scale identification of virulence genes from *Streptococcus pneumoniae*. *Infect. Immun.* **66:**5620–5629.

Rajam, G., J. M. Anderton, G. M. Carlone, J. S. Sampson, and E. W. Ades. 2008. Pneumococcal surface adhesin A (PsaA): a review. *Crit. Rev. Microbiol.* **34:**131–142.

Reinert, R. R. 2009a. The public health ramifications of pneumococcal resistance. *Clin. Microbiol. Infect.* **15**(Suppl. 3):1–3.

Reinert, R. R. 2009b. The antimicrobial resistance profile of *Streptococcus pneumoniae*. *Clin. Microbiol. Infect.* **15**(Suppl. 3):7–11.

Reyes-Caballero, H., A. J. Guerra, F. E. Jacobsen, K. M. Kazmierczak, D. Cowart, U. M. Koppolu, R. A. Scott, M. E. Winkler, and D. P. Giedroc. 2010. The metalloregulatory zinc site in *Streptococcus pneumoniae* AdcR, a zinc-activated MarR family repressor. *J. Mol. Biol.* **403:**197–216.

Riboldi-Tunnicliffe, A., N. W. Isaacs, and T. J. Mitchell. 2005. 1.2 Å crystal structure of the *S. pneumoniae* PhtA histidine triad domain a novel zinc binding fold. *FEBS Lett.* **579:**5353–5360.

Rioux, S., C. Neyt, E. Di Paolo, L. Turpin, N. Charland, S. Labbe, M. C. Mortier, T. J. Mitchell, C. Feron, D. Martin, and J. T. Poolman. 2011. Transcriptional regulation, occurrence and putative role of the Pht family of *Streptococcus pneumoniae*. *Microbiology* **157:**336–348.

Rosch, J. W., B. Mann, J. Thornton, J. Sublett, and E. Tuomanen. 2008. Convergence of regulatory networks on the pilus locus of *Streptococcus pneumoniae*. *Infect. Immun.* **76:**3187–3196.

Rosenow, C., M. Maniar, and J. Trias. 1999. Regulation of the alpha-galactosidase activity in *Streptococcus pneumoniae*: characterization of the raffinose utilization system. *Genome Res.* **9:**1189–1197.

Rosenow, C., P. Ryan, J. N. Weiser, S. Johnson, P. Fontan, A. Ortqvist, and H. R. Masure. 1997. Contribution of novel choline-binding proteins to adherence, colonization and immunogenicity of *Streptococcus pneumoniae*. *Mol. Microbiol.* **25:**819–829.

Russell, H., J. A. Tharpe, D. E. Wells, E. H. White, and J. E. Johnson. 1990. Monoclonal antibody recognizing a species-specific protein from *Streptococcus pneumoniae*. *J. Clin. Microbiol.* **28:**2191–2195.

Saluja, S. K., and J. N. Weiser. 1995. The genetic basis of colony opacity in *Streptococcus pneumoniae*: evidence for the effect of box elements on the frequency of phenotypic variation. *Mol. Microbiol.* **16:**215–227.

Sanchez-Beato, A. R., R. Lopez, and J. L. Garcia. 1998. Molecular characterization of PcpA: a novel choline-binding protein of *Streptococcus pneumoniae*. *FEMS Microbiol. Lett.* **164:**207–214.

Sebert, M. E., L. M. Palmer, M. Rosenberg, and J. N. Weiser. 2002. Microarray-based identification of *htrA*, a *Streptococcus pneumoniae* gene that is regulated by the CiaRH two-component system and contributes to nasopharyngeal colonization. *Infect. Immun.* **70:**4059–4067.

Shafeeq, S., T. G. Kloosterman, and O. P. Kuipers. 2011. Transcriptional response of *Streptococcus pneumoniae* to Zn^{2+} limitation and the repressor/activator function of AdcR. *Metallomics* **3:**609–618.

Song, X. M., W. Connor, K. Hokamp, L. A. Babiuk, and A. A. Potter. 2009. The growth phase-dependent regulation of the pilus locus genes by two-component system TCS08 in *Streptococcus pneumoniae*. *Microb. Pathog.* **46:**28–35.

Spratt, B. G., and B. M. Greenwood. 2000. Prevention of pneumococcal disease by vaccination: does serotype replacement matter? *Lancet* **356:**1210–1211.

Standish, A. J., U. H. Stroeher, and J. C. Paton. 2007. The pneumococcal two-component signal transduction system RR/HK06 regulates CbpA and PspA by two distinct mechanisms. *J. Bacteriol.* **189:**5591–5600.

Standish, A. J., U. H. Stroeher, and J. C. Paton. 2005. The two-component signal transduction system RR06/HK06 regulates expression of *cbpA* in *Streptococcus pneumoniae*. *Proc. Natl. Acad. Sci. USA* **102:**7701–7706.

Stroeher, U. H., A. W. Paton, A. D. Ogunniyi, and J. C. Paton. 2003. Mutation of *luxS* of *Streptococcus pneumoniae* affects virulence in a mouse model. *Infect. Immun.* **71:**3206–3212.

Tai, S. S., C. J. Lee, and R. E. Winter. 1993. Hemin utilization is related to virulence of *Streptococcus pneumoniae*. *Infect. Immun.* **61:**5401–5405.

Tai, S. S., T. R. Wang, and C. J. Lee. 1997. Characterization of hemin binding activity of *Streptococcus pneumoniae*. *Infect. Immun.* **65:**1083–1087.

Talbot, U. M., A. W. Paton, and J. C. Paton. 1996. Uptake of *Streptococcus pneumoniae* by respiratory epithelial cells. *Infect. Immun.* **64:**3772–3777.

Tettelin, H., K. E. Nelson, I. T. Paulsen, J. A. Eisen, T. D. Read, S. Peterson, J. Heidelberg, R. T. DeBoy, D. H. Haft, R. J. Dodson, A. S. Durkin, M. Gwinn, J. F. Kolonay, W. C. Nelson, J. D. Peterson, L. A. Umayam, O. White, S. L. Salzberg, M. R. Lewis, D. Radune, E. Holtzapple, H. Khouri, A. M. Wolf, T. R. Utterback, C. L. Hansen, L. A. McDonald, T. V. Feldblyum, S. Angiuoli, T. Dickinson, E. K. Hickey, I. E. Holt, B. J. Loftus, F. Yang, H. O. Smith, J. C. Venter, B. A. Dougherty, D. A. Morrison, S. K. Hollingshead, and C. M. Fraser. 2001. Complete genome sequence of a virulent isolate of *Streptococcus pneumoniae*. *Science* **293:**498–506.

Throup, J. P., K. K. Koretke, A. P. Bryant, K. A. Ingraham, A. F. Chalker, Y. Ge, A. Marra, N. G. Wallis, J. R. Brown, D. J. Holmes, M. Rosenberg, and M. K. Burnham. 2000. A genomic analysis of two-component signal transduction in *Streptococcus pneumoniae*. *Mol. Microbiol.* **35:**566–576.

Tomasz, A. 1965. Control of the competent state in *Pneumococcus* by a hormone-like cell product: an example for a new type of regulatory mechanism in bacteria. *Nature* **208:**155–159.

Tomasz, A. 1966. Model for the mechanism controlling the expression of competent state in *Pneumococcus* cultures. *J. Bacteriol.* **91:**1050–1061.

Trappetti, C., A. D. Ogunniyi, M. R. Oggioni, and J. C. Paton. 2011a. Extracellular matrix formation enhances the ability of *Streptococcus pneumoniae* to cause invasive disease. *PLoS One* **6:**e19844.

Trappetti, C., A. J. Potter, A. W. Paton, M. R. Oggioni, and J. C. Paton. 2011b. LuxS mediates iron-dependent biofilm formation, competence, and fratricide in *Streptococcus pneumoniae*. *Infect. Immun.* **79:**4550–4558.

Tuomanen, E. I., and H. R. Masure. 1997. Molecular and cellular biology of pneumococcal infection. *Microb. Drug Resist.* **3:**297–308.

Ulijasz, A. T., D. R. Andes, J. D. Glasner, and B. Weisblum. 2004. Regulation of iron transport in *Streptococcus pneumoniae* by RitR, an orphan response regulator. *J. Bacteriol.* **186:**8123–8136.

Ulijasz, A. T., S. P. Falk, and B. Weisblum. 2009. Phosphorylation of the RitR DNA-binding domain by a Ser-Thr phosphokinase:

implications for global gene regulation in the streptococci. *Mol. Microbiol.* **71:**382–390.

van Opijnen, T., K. L. Bodi, and A. Camilli. 2009. Tn-seq: high-throughput parallel sequencing for fitness and genetic interaction studies in microorganisms. *Nat. Methods* **6:**767–772.

Ventura, C. L., R. T. Cartee, W. T. Forsee, and J. Yother. 2006. Control of capsular polysaccharide chain length by UDP-sugar substrate concentrations in *Streptococcus pneumoniae*. *Mol. Microbiol.* **61:**723–733.

Wagner, C., A. de Saizieu, H.-J. Schönfeld, M. Kamber, R. Lange, C. J. Thompson, and M. G. Page. 2002. Genetic analysis and functional characterization of the *Streptococcus pneumoniae vic* operon. *Infect. Immun.* **70:**6121–6128.

Waite, R. D., D. W. Penfold, J. K. Struthers, and C. G. Dowson. 2003. Spontaneous sequence duplications within capsule genes *cap8E* and *tts* control phase variation in *Streptococcus pneumoniae* serotypes 8 and 37. *Microbiology* **149:**497–504.

Waite, R. D., J. K. Struthers, and C. G. Dowson. 2001. Spontaneous sequence duplication within an open reading frame of the pneumococcal type 3 capsule locus causes high-frequency phase variation. *Mol. Microbiol.* **42:**1223–1232.

Wani, J. H., J. V. Gilbert, A. G. Plaut, and J. N. Weiser. 1996. Identification, cloning, and sequencing of the immunoglobulin A1 protease gene of *Streptococcus pneumoniae*. *Infect. Immun.* **64:**3967–3974.

Wartha, F., K. Beiter, B. Albiger, J. Fernebro, A. Zychlinsky, S. Normark, and B. Henriques-Normark. 2007. Capsule and D-alanylated lipoteichoic acids protect *Streptococcus pneumoniae* against neutrophil extracellular traps. *Cell. Microbiol.* **9:**1162–1171.

Weickert, M. J., and S. Adhya. 1992. A family of bacterial regulators homologous to Gal and Lac repressors. *J. Biol. Chem.* **267:**15869–15874.

Weickert, M. J., and G. H. Chambliss. 1990. Site-directed mutagenesis of a catabolite repression operator sequence in *Bacillus subtilis*. *Proc. Natl. Acad. Sci. USA* **87:**6238–6242.

Weiser, J. N. 1998. Phase variation in colony opacity by *Streptococcus pneumoniae*. *Microb. Drug Resist.* **4:**129–135.

Weiser, J. N. 2004. Mechanisms of carriage, p. 169–182. *In* E. I. Tuomanen, T. J. Mitchell, D. A. Morrison, and B. G. Spratt (ed.), *The Pneumococcus*. ASM Press, Washington, DC.

Weiser, J. N., R. Austrian, P. K. Sreenivasan, and H. R. Masure. 1994. Phase variation in pneumococcal opacity: relationship between colonial morphology and nasopharyngeal colonization. *Infect. Immun.* **62:**2582–2589.

Weiser, J. N., D. Bae, H. Epino, S. B. Gordon, M. Kapoor, L. A. Zenewicz, and M. Shchepetov. 2001. Changes in availability of oxygen accentuate differences in capsular polysaccharide expression by phenotypic variants and clinical isolates of *Streptococcus pneumoniae*. *Infect. Immun.* **69:**5430–5439.

Weiser, J. N., J. M. Love, and E. R. Moxon. 1989. The molecular mechanism of phase variation of *H. influenzae* lipopolysaccharide. *Cell* **59:**657–665.

Weiser, J. N., Z. Markiewicz, E. I. Tuomanen, and J. H. Wani. 1996. Relationship between phase variation in colony morphology, intrastrain variation in cell wall physiology, and nasopharyngeal colonization by *Streptococcus pneumoniae*. *Infect. Immun.* **64:**2240–2245.

Whitfield, C., and A. Paiment. 2003. Biosynthesis and assembly of group 1 capsular polysaccharides in *Escherichia coli* and related extracellular polysaccharides in other bacteria. *Carbohydr. Res.* **338:**2491–2502.

Whitfield, C., and I. S. Roberts. 1999. Structure, assembly and regulation of expression of capsules in *Escherichia coli*. *Mol. Microbiol.* **31:**1307–1319.

Winkelstein, J. A. 1981. The role of complement in the host's defense against *Streptococcus pneumoniae*. *Rev. Infect. Dis.* **3:**289–298.

Wizemann, T. M., J. H. Heinrichs, J. E. Adamou, A. L. Erwin, C. Kunsch, G. H. Choi, S. C. Barash, C. A. Rosen, H. R. Masure, E. Tuomanen, A. Gayle, Y. A. Brewah, W. Walsh, P. Barren, R. Lathigra, M. Hanson, S. Langermann, S. Johnson, and S. Koenig. 2001. Use of a whole genome approach to identify vaccine molecules affording protection against *Streptococcus pneumoniae* infection. *Infect. Immun.* **69:**1593–1598.

Yamaguchi, M., Y. Minamide, Y. Terao, R. Isoda, T. Ogawa, S. Yokota, S. Hamada, and S. Kawabata. 2009. Nrc of *Streptococcus pneumoniae* suppresses capsule expression and enhances anti-phagocytosis. *Biochem. Biophys. Res. Commun.* **390:** 155–160.

Yang, Q. L., and E. C. Gotschlich. 1996. Variation of gonococcal lipooligosaccharide structure is due to alterations in poly-G tracts in *lgt* genes encoding glycosyl transferases. *J. Exp. Med.* **183:**323–327.

Yother, J., and D. E. Briles. 1992. Structural properties and evolutionary relationships of PspA, a surface protein of *Streptococcus pneumoniae*, as revealed by sequence analysis. *J. Bacteriol.* **174:**601–609.

Zähner, D., and R. Hakenbeck. 2000. The *Streptococcus pneumoniae* beta-galactosidase is a surface protein. *J. Bacteriol.* **182:**5919–5921.

Regulation of Bacterial Virulence
Edited by Michael L. Vasil and Andrew J. Darwin
© 2013 ASM Press, Washington, DC   doi:10.1128/9781555818524.ch11

Chapter 11

Regulation of Lipopolysaccharide Modifications and Antimicrobial Peptide Resistance[†]

Erica N. Kintz, Daniel A. Powell, Lauren E. Hittle, Joanna B. Goldberg,
and Robert K. Ernst

INTRODUCTION

Lipopolysaccharide (LPS) is the major component of the outer membrane of gram-negative bacteria and consists of three distinct structural domains: lipid A, a nonrepeating "core" oligosaccharide, and a distal repeated O-antigen polysaccharide. Lipid A, or endotoxin, is the hydrophobic anchor of LPS in the outer leaflet of the outer membrane in gram-negative bacteria, whereas the core and O antigen extend out from the surface of the membrane. In this chapter, we discuss the structure of these three regions, in order, as they extend out from the outer membrane, focusing on regulated alterations, modifications, and/or substitutions.

LIPID A BIOSYNTHESIS

Lipid A, or endotoxin, consists of a diglucosamine backbone substituted with up to eight acyl chains attached directly to the backbone sugars or acyl-oxy-acyl additions to these primary fatty acids. The acyl chains located at the 2 and 2′ positions are N linked, whereas the chains at positions 3 and 3′ are O linked. Biosynthesis of lipid A is conserved up to the stage of lipid IV_A, which includes a diglucosamine backbone, two phosphate residues, and four fatty acids (see Fig. 1). After this structure is synthesized on the inner membrane of the bacteria, a wide variety of modifications, including heterogeneity in the number of attached fatty acids, length of the fatty acids, and decorations to the terminal phosphate moieties, occur at both the inner and outer membranes (Table 1). In most cases, synthesis of lipid A is essential to bacterial growth and survival.

Modification of the base lipid IV_A structure is species specific and has profound implications for disease, particularly in humans. Lipid A modifications, both constitutive and regulated, have the potential to aid bacterial pathogens by evasion of the host innate immune system recognition and host killing mechanisms. Recognition of lipid A by the host innate immune system occurs via Toll-like receptor 4 (TLR4). Stimulation of TLR4 by lipid A leads to activation of inflammatory mediators, such as tumor necrosis factor alpha and interleukin 1β (IL-1β). While this inflammation can at times lead to bacterial clearance, it can also lead to septic shock and death. More complete reviews of the TLR4 pathway and its recognition and downstream effects have been undertaken by a number of authors (Beutler, 2005; Munford, 2008; O'Neill, 2008a, 2008b). Modifications of lipid A not only alter inflammation and pathogenesis but also can alter profiles of resistance to host antimicrobial peptides (AMPs). A majority of host AMPs are positively charged (cationic) and target the surface of the bacterial cell through electrostatic interactions. Upon binding to the bacterial outer membrane, AMPs are thought to generate a pore and bind to components in the inner membrane that will eventually lead to cell death (Koprivnjak and Peschel, 2011). Bacteria are able to resist AMPs by changing the charge or fluidity of their membranes, specifically through lipid A modifications, to deter AMPs from traversing the membrane.

Robert K. Ernst, Daniel A. Powell, and Lauren E. Hittle • Department of Microbial Pathogenesis, University of Maryland, Baltimore, 650 West Baltimore Street—8 South, Baltimore, MD 21201.   **Joanna B. Goldberg and Erica N. Kintz** • Department of Microbiology, Immunology, and Cancer Biology, University of Virginia Health System, 7230 Jordan Hall, 1300 Jefferson Park Avenue, Charlottesville, VA 22908.

[†] This chapter is dedicated to the memory of Dr. Christian R. H. Raetz from Duke University. Dr. Raetz's passion for uncovering the biochemical nature and pathway for synthesis of the gram-negative membrane served to drive the field. His intellectual curiosity and generosity to the field of lipopolysaccharide biochemistry will be sorely missed.

Lipid A Biosynthetic Enzymes

While the early stages of synthesis of the lipid A molecule are conserved and play an important role in pathogenesis, modifications beyond the base lipid IV$_A$ structure require a wide variety of biosynthetic enzymes, including acyltransferases, deacylases, phosphatases, glycosyltransferases, and hydroxylases (Table 1). Here we discuss the enzymes involved in the pathways after synthesis of lipid IV$_A$ (for a complete review of lipid IV$_A$ synthesis, see the works of Raetz and colleagues [Raetz et al., 2007; Raetz and Whitfield, 2002]). As this chapter details, these biosynthetic enzymes are either constitutively active or regulated by a variety of two-component regulatory systems, including PhoR/PhoB, PmrA/PmrB and/or PhoP/PhoQ.

Constitutively active lipid A biosynthetic enzymes

Constitutive lipid A biosynthetic enzymes work either to change membrane permeability by the addition of fatty acids or to lessen the negative charge of the membranes by the removal of phosphate groups. These enzymes work by making modifications to lipid IV$_A$ independently of environmental signals.

One of the first enzymes acting on lipid IV$_A$ is LpxL, also known as HtrB. LpxL is a constitutively active acyltransferase that adds a laurate, C$_{12:0}$ fatty acid, acyl-oxy-acyl at the 2'-position of lipid IV$_A$, thus generating a penta-acylated lipid A structure. LpxL was first identified in a screen for mutants in *Escherichia coli* that failed to grow at temperatures over 33°C, leading to its original name, *htrB*, for high-temperature regulation. However, additional analysis demonstrated that it was not temperature regulated at the transcription level and suggested a role for proper acylation patterns in membrane function (Brozek and Raetz, 1990; Clementz et al., 1996; Karow et al., 1991a; Karow and Georgopoulos, 1991; Karow et al., 1991b). LpxL acts upon lipid IV$_A$ on the cytoplasmic side of the inner membrane prior to transport by MsbA, an essential ATP-binding cassette (ABC) protein responsible for the transport of lipid A from the inner to outer leaflet of the inner membrane. For a number of bacterial species, including *E. coli*, *Haemophilus*, *Bordetella*, and *Vibrio cholerae*, LpxL activity was found to require phosphorylation of 3-deoxy-D-*manno*-oct-2-ulsonic acid (Kdo), an eight-carbon sugar that is required for the addition of core (Hankins and Trent, 2009; Mohan and Raetz, 1994; Stead et al., 2008). Alternatively, in *Helicobacter pylori* and *Pseudomonas aeruginosa*, LpxL functions independently of phosphate on Kdo. Deletion of *lpxL* had little to no effect on AMP sensitivity in *E. coli* (Vorachek-Warren et al., 2002).

Following the addition of a laurate fatty acid by LpxL/HtrB, the lipid A molecule can undergo the addition of a second acyl-oxy-acyl fatty acid, myristate or C$_{14:0}$, at the 3' fatty acid of lipid A by LpxM, also known as MsbB. This addition generates a hexa-acylated molecule that is then transported by the LPS transporter MsbA (Somerville et al., 1996). Though constitutive in most bacterial backgrounds, *msbB* appears to be active at lower temperatures (~21°C) in *Yersinia* (Perez-Gutierrez et al., 2010). Recently, deletion of *msbB* in a variety of bacterial backgrounds has been used to generate avirulent strains for possible vaccine development (Kim et al., 2009; Lee et al., 2009b; Ranallo et al., 2010). LPS isolated from an *lpxM* mutant in *E. coli* acts as an agonist for human TLR4 by saturation binding to the TLR4 coreceptor myeloid differentiation factor-2 and thus alters interactions with the host innate immune system (Coats et al., 2007) and increases sensitivity to AMPs (Somerville et al., 1999; Vaara and Nurminen, 1999).

Subsequent to the addition of myristate by LpxM/MsbB, this fatty acid can be modified by the addition of a hydroxyl group (OH) at the second carbon to generate a 2-hydroxymyristate-modified lipid A. Hydroxylation of myristate requires a membrane-bound hydroxylase, LpxO, that directly modifies lipid A containing two Kdo residues (Kdo$_2$-lipid A) on the inner leaflet of the inner membrane. LpxO enzymatic activity is present in a limited number of bacterial species, such as *Salmonella*, *Klebsiella*, *Pseudomonas*, *Bordetella*, and *Legionella*, and is constitutively active in each (Doerrler et al., 2004; Ernst et al., 1999; Gibbons et al., 2005; Gibbons et al., 2000). Additionally, the activity of LpxO is dependent on molecular oxygen during growth, as growth under anaerobic or oxygen-limited conditions yields a lipid A structure lacking 2-OH hydroxylation. Loss of LpxO activity in *Salmonella* and *V. cholerae* leads to increased resistance to AMP, suggesting a potential decrease in hydrogen bonding between individual lipid A molecules altering outer membrane permeability (Hankins et al., 2011; Hankins and Trent, 2009; Murata et al., 2007; Nikaido, 2003).

Following transport of the lipid A molecule to the periplasmic face of the inner membrane, two constitutively active phosphatases—LpxE and LpxF—remove the phosphate moieties present at the 1 and 4' positions, respectively (Fig. 1). Initially identified in *Francisella tularensis* subsp. *novicida*, phosphatase activity has been observed in *Rhizobium*, *H. pylori*, and *Porphyromonas gingivalis* (Darveau et al., 2004; Karbarz et al., 2003; Tran et al., 2006; Wang et al., 2004). When expressed in *E. coli*, LpxE from *Francisella* is dependent on MsbA, while LpxF is independent. In contrast, LpxF and LpxE enzyme

Table 1. Enzymes responsible for modification of lipid A

Enzyme	Modification	Active site	Distribution	Regulation
LpxE	Removal of 1 position phosphate	Outer leaflet of inner membrane	*Francisella, Helicobacter, Porphyromonas*[a]	Constitutive
LpxF	Removal of 4' position phosphate	Outer or inner leaflet of inner membrane	*Francisella, Helicobacter, Porphyromonas,*[a] *Leptospira*[a]	Constitutive
LpxL/ MsbB	Addition of acyloxyacyl laurate	Inner leaflet of inner membrane	*E. coli, Salmonella, Shigella, Yersinia, Bordetella, Legionella, Pseudomonas, Francisella*	Constitutive
LpxM/ HtrB	Addition of acyloxyacyl myristate	Inner leaflet of inner membrane	*E. coli, Salmonella, Shigella, Yersinia, Bordetella, Legionella, Pseudomonas*	Constitutive
LpxO	Hydroxylation of acyl chains	Inner leaflet of inner membrane	*Salmonella, Klebsiella, Pseudomonas, Bordetella, Legionella*	Constitutive oxygen dependent
LpxT	Addition of diphosphate to lipid A	Outer leaflet of inner membrane	*E. coli, Salmonella, Yersinia*[a]	PhoR/PhoB
LpxR	Removal of 3' acyl chain	Outer leaflet of outer membrane (*Salmonella*) or inner leaflet of inner membrane	*Salmonella, E. coli* O157:H7,[b] *Helicobacter, Vibrio,*[b] *Yersinia,*[b] *Francisella*[a]	PhoP/PhoQ and Ca^{2+} in *Salmonella*
PagL	Removal of 3 acyl chain	Outer leaflet of outer membrane	*Salmonella, Pseudomonas, Bordetella*	PhoP/PhoQ in *Salmonella*
PagP	Addition of palmitate	Outer leaflet of outer membrane	*E. coli, Salmonella, Shigella, Yersinia, Bordetella, Legionella, Pseudomonas*[a]	PhoP/PhoQ in *Salmonella*
ArnT/PmrK	Aminoarabinose addition to lipid A	Outer leaflet of inner membrane	*E. coli, Salmonella, Shigella, Pseudomonas, Yersinia, Francisella*[c]	PmrA
EptA (PmrC, LptA, YjdB)	Phosphoethanolamine addition to lipid A	Outer leaflet of inner membrane	*E. coli, Salmonella, Neisseria, Campylobacter, Helicobacter, Vibrio*	PmrA

[a]Enzyme presence is conferred by structure though no homologue is identified as of yet.
[b]Homologue is present in genome but function is not seen in structure as of yet.
[c]*Francisella* adds galactosamine and mannose as opposed to aminoarabinose.

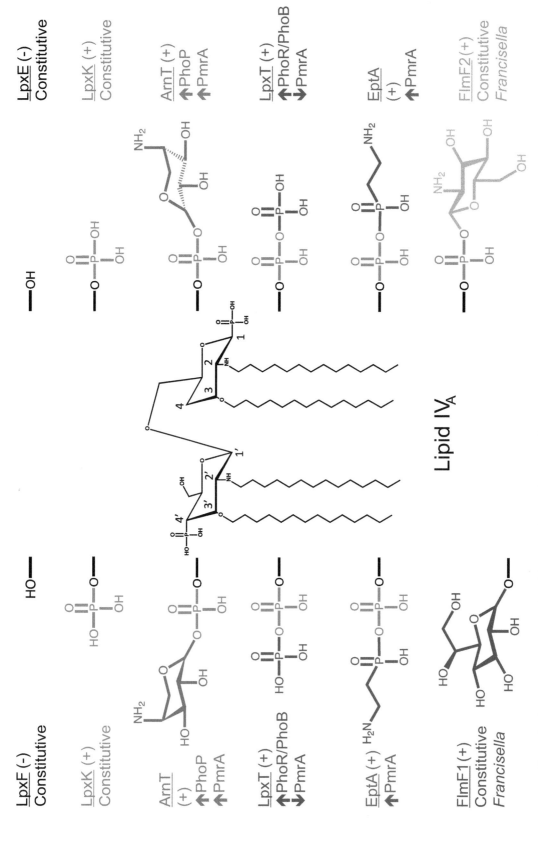

Figure 1. Schematic representation of modifications of terminal residues of lipid A. The diagram shows the possible modifications outside of the terminal residues of lipid A. Chemical groups are color coded by the enzyme responsible for their action; a plus sign indicates an addition to the base structure, and a minus sign indicates a removal. Under each enzyme is the regulatory system that controls each its action; an upward-pointing arrow indicates a positive regulator, and a downward-pointing arrow indicates a negative regulator. doi:10.1128/9781555818524.ch11f1

activity in *Francisella* and *H. pylori* is MsbA independent, indicating that both may function before lipid A is "flipped" to the periplasmic side of the membrane. Deletion of either phosphatase results in an increased sensitivity to positively charged AMPs and/or altered virulence in animal models. In *Helicobacter* and *Rhizobium*, deletion of *lpxE* results in an increased sensitivity to the AMPs, polymyxin B, and colistin (Ingram et al., 2010; McGee et al., 2011; Tran et al., 2006; Wang et al., 2004), whereas deletion of *lpxF* in *Rhizobium* leads to increased sensitivity only to polymyxin B (Ingram et al., 2010), most likely due to the increase in the overall negative charge of this bacterium's cell surface. Finally, when the *lpxF* mutant of *Francisella* is injected into mice, it is avirulent and elicits a protective immune response to lethal wildtype challenge (D. Kanistanon and R. K. Ernst, unpublished data).

Regulated lipid A biosynthetic enzymes

In addition to the constitutively expressed lipid A biosynthetic enzymes, a wide variety of enzymes are regulated in response to specific environmental cues and growth conditions. Regulated structural changes in lipid A are shown in Fig. 1 and include the addition and removal of fatty acids, the addition of amino-containing compounds such as aminoarabinose and phosphoethanolamine (pEtN), and the addition or removal of phosphate groups. These structural modifications are regulated via a series of two-component regulatory systems, which are comprised of a sensor kinase, PhoB, PhoQ, and PmrB, and its response regulator, PhoR, PhoP, and PmrA, respectively. Each cognate pair controls the expression of a specific set of genes. The sensor kinase spans the inner membrane with the sensing domain in the periplasm, while the response regulator is found in the cytoplasm. Upon sensing of a specific environmental or host niche, the sensor protein phosphorylates a conserved residue on the cytoplasmic DNA-binding domain of the response regulator that subsequently interacts with target gene(s) promoters. More detailed reviews of two-component regulatory systems can be found elsewhere (Koretke et al., 2000; Loomis et al., 1998; West and Stock, 2001).

PhoR/PhoB-regulated lipid A biosynthetic enzymes. PhoR/PhoB is a recently identified two-component system found to play a role in lipid A modification. The PhoR/PhoB system is activated under low-phosphate growth conditions, which, in turn, leads to the activation of LpxT, a phosphotransferase which specifically transfers an additional phosphate group to lipid A. LpxT, formerly YeiU, is an enzyme

originally identified in *E. coli* (Touze et al., 2008) that gave rise to a subset of lipid A, ~30% of the total, in the outer membrane that contained a substituted diphosphate at position 1 of lipid A. This second phosphate moiety, added in the periplasmic space, is cleaved from the essential carrier lipid undecaprenyl phosphate (C_{55}-P; also referred to as Und-P below) by LpxT in an MsbA-dependent manner. While *lpxT* was originally identified for its ability to remove phosphate from C_{55}-P (El Ghachi et al., 2005), purified protein analysis showed that it has both the ability to cleave the phosphate as well as add the phosphate to the lipid A (Valvano, 2008). Preliminary work on the regulation of *lpxT* found that it is downregulated in mutants of the phosphate-specific transport (Pst) system. The Pst system is a periplasmic protein-dependent transporter that operates as a primary transport mechanism for phosphate under stress conditions (Lamarche et al., 2008b). Additional analysis of the promoter region of the *lpxT* operon showed that *lpxT* was regulated by an additional mediator under the control of PhoR/PhoB (Lamarche et al., 2008a) though not under conditions that induce PmrA (see discussion of PmrA/PmrB below), suggesting a possible posttranscriptional level of control of *lpxT* (Herrera et al., 2010). Further work showed that *E. coli* grown in phosphate-limited conditions produced lipid A with smaller amounts of diphosphate at position 1 and showed an increased sensitivity to AMPs (Lamarche et al., 2008a). Finally, deletion of *lpxT* in *E. coli* resulted in an increase in sensitivity to the cationic AMP polymyxin B, although experiments elucidating roles for *lpxT* in overall pathogenesis have yet to be undertaken.

PhoP/PhoQ-regulated lipid A biosynthetic enzymes. The second two-component system discussed is the PhoP/PhoQ system. Originally described for *Salmonella typhimurium* (Groisman et al., 1989; Miller et al., 1989) (now known as *Salmonella enterica* serovar Typhimurium), PhoP/PhoQ regulates a series of genes including those involved in Mg^{2+} transport (*mgtABC*) (Soncini et al., 1996), modification of LPS (*lpxR, pagL, pagP*) (Belden and Miller, 1994; Guo et al., 1998), and activation of a third two-component regulatory system, PmrA/PmrB (discussed below) (Gunn and Miller, 1996). A variety of conditions that activate the PhoQ sensor region have been described: low concentrations of divalent cations (Garcia Vescovi et al., 1996), low pH (Alpuche Aranda et al., 1992), anaerobic growth (R. K. Ernst, unpublished data), and AMPs (Bader et al., 2005). While primarily investigated in *Salmonella*, PhoQ homologues have been found in a variety of pathogens, including *P. aeruginosa*, *Shigella flexneri*, *Yersinia* *pestis, Photorhabdus luminescens,* and *Erwinia chrysanthemi,* in which PhoQ is also indicated to sense low pH and AMPs and play a role in pathogenesis (Derzelle et al., 2004; Llama-Palacios et al., 2005; Moss et al., 2000; Oyston et al., 2000; Rebeil et al., 2004). Interestingly, the three major lipid A-modifying enzymes regulated by PhoP/PhoQ (LpxR, PagL, and PagP) are located in the outer membrane and modify previously transported lipid A.

LpxR, originally characterized in *Salmonella,* is an outer-membrane 12-stranded β-barrel protein that catalyzes the removal of the 3′ acyl chains from lipid A (Bishop, 2008; Reynolds et al., 2006; Rutten et al., 2009). Transcription of *lpxR* in *Salmonella* is regulated by the transcription factor *slyA,* under the control of the PhoP/PhoQ system. Lipid A extracted from *Salmonella* is normally acylated at the 3′ position, indicating that the enzyme is not active under normal cellular growth conditions. LpxR activity was only demonstrated using *Salmonella* membrane preparations after incubation in high levels of Ca^{2+} (Reynolds et al., 2006). However, deletion of *lpxR* in *S. typhimurium* led to decreased replication in macrophages and increased inducible nitric oxide synthetase, suggesting a role in overall pathogenesis of the organism (Kawano et al., 2010). Orthologues of *lpxR* are found in the genomes of *E. coli* O157:H7, *Yersinia enterocolitica, Yersinia pseudotuberculosis, V. cholerae,* and *H. pylori.* Interestingly, only *Y. enterocolitica, Y. pseudotuberculosis,* and *H. pylori* synthesize lipid A species that are deacylated at the 3′ position (Moran et al., 1997; Oertelt et al., 2001; Rebeil et al., 2004). Also of note, *F. tularensis* and *P. gingivalis* synthesize 3′ deacylated lipid A species, though orthologues of *lpxR* from *Salmonella* are not present in the genomes of these organisms. In these species the deacylation likely occurs before the lipid A transports to the outer membrane, indicating that these deacylases are different from LpxR in *Salmonella* (Darveau et al., 2004; Vinogradov et al., 2002).

A second deacylase, PagL, is an eight-stranded b-barrel outer-membrane enzyme that removes the acyl chain from the 3 position of lipid A (Bishop et al., 2000; Rutten et al., 2009; Trent et al., 2001a). PagL is present among a wide variety of gram-negative bacteria (Asensio et al., 2011; Kawasaki et al., 2007; Kawasaki et al., 2004a; Macarthur et al., 2011; Trent et al., 2001a), with the crystal structure elucidated from *P. aeruginosa* (Rutten et al., 2009), in which the enzyme may play an important role in adaptation to the airways of patients with cystic fibrosis through its effect on IL-8 production (Ernst et al., 1999). In contrast, removal of the 3 position fatty acid leads to decreased proinflammatory responses via TLR4 recognition and increased sensitivity to polymyxin

B in *Salmonella* (Kawasaki et al., 2007; Kawasaki et al., 2004a; Kawasaki et al., 2004b). Interestingly, PagL activity is inhibited by the addition of aminoarabinose to the terminal phosphate residue(s) to lipid A, suggesting that aminoarabinose-containing outer membranes directly inactivate PagL enzymatic activity or lipid A modified with aminoarabinose inhibits the physical interaction of LPS with PagL (Kawasaki et al., 2005). Recent reports have also shown that *pagL* from *Bordetella bronchiseptica* is constitutively active and produces a lipid A structure that both is deacylated at the 3 position and contains aminoarabinose, suggesting that PagL latency induced by aminoarabinose in *Bordetella* is species dependent (Macarthur et al., 2011).

The third PhoP/PhoQ-regulated enzyme is PagP, an eight stranded β-barrel acyltransferase located in the outer membrane of gram-negative bacteria, which transfers an acyl-oxy-acyl palmitate chain ($C_{16:0}$ fatty acid) to lipid A (Bishop et al., 2000; Hwang et al., 2002). PagP is present among a limited number of gram-negative bacteria, including the enteric organisms, as well as *Pseudomonas, Bordetella,* and *Legionella* (Pilione et al., 2004; Preston et al., 2003; Robey et al., 2001), with the crystal and nuclear magnetic resonance structure elucidated from *E. coli* (Hwang et al., 2002). The *pagP* gene was originally identified in a *Salmonella* mutant constitutively active for PhoP/PhoQ expression and was shown to play an important role in inducible AMP resistance and increased acylation of lipid A (Guo et al., 1998). Further, *pagP* transcription was subsequently confirmed to be regulated by PhoP/PhoQ (Kawasaki et al., 2004a).

Similar to the *Salmonella* PagP, addition of palmitate by *P. aeruginosa* PagP led to increased resistance to AMPs and polymyxin B under conditions that induce PhoP/PhoQ along with decreased activation of TLR4 (Ernst et al., 1999; Hajjar et al., 2002; Macfarlane et al., 2000; Macfarlane et al., 1999). *P. aeruginosa* PagP activity is induced in lab strains, environmental isolates, non-cystic fibrosis clinical isolates, and clinical specimens isolated from the airways of patients with cystic fibrosis (Ernst et al., 1999). Palmitate additions have been observed in both *Y. enterocolitica* and *Y. pseudotuberculosis* but not *Y. pestis* (Rebeil et al., 2004). Palmitoylation in *Yersinia* species is strongly induced upon shifting growth temperature from the environment to a warm-blooded host (21 to 37°C) (Rebeil et al., 2004; Wren, 2003). Addition of palmitate can also be driven in *Y. enterocolitica* under Mg^{2+}-limited conditions, presumably via PhoP/PhoQ, and results in increased resistance to AMPs (Guo et al., 1998). Divergent from *Salmonella* PagP, *Bordetella parapertussis* PagP adds the palmitate group to the 3′ position of the lipid A, as opposed to the 2 position, e.g., on the opposite side of the lipid A structure. In contrast to *Salmonella*, the *Bordetella pagP* gene is regulated not by PhoP/PhoQ but by the BvgA/BvgS two-component virulence system (Preston et al., 2003), and the deletion of this gene had no effect on resistance to AMPs but did result in increased killing by antibody-mediated complement lysis. Finally, *pagP* was required in *B. bronchiseptica* for bacterial persistence in a mouse model of infection (Pilione et al., 2004). Interestingly, in *Bordetella pertussis pagP* is inactivated by an insertion of a transposable genetic element (Preston et al., 2003).

PmrA/PmrB-regulated lipid A biosynthetic enzymes. Activation of the PmrA/PmrB system leads to activation of lipid A modifications, including the addition of aminoarabinose and pEtN additions. These modifications to the terminal phosphates of lipid A mask the negative charge affecting the electrostatic interaction of the bacterial cell surface and certain cationic AMPs (Gunn et al., 1998; Zhou et al., 2001). The genes in *Salmonella* involved in the aminoarabinose addition are *pmrE* and the *pmrH-FIJKLM* operon (also named the *arn* or *pbg* operon). All the genes of this cluster are required for lipid A modifications except for *pmrM* (Gunn et al., 2000). In *Salmonella*, PmrA/PmrB regulates greater than 20 confirmed and possibly up to 100 genes (Marchal et al., 2004; Tamayo, 2005; Tamayo et al., 2002). Activation of PmrA/PmrB can occur by both direct and indirect mechanisms. In *Salmonella*, iron (Fe^{3+}), vanadate, aluminum (Al^{3+}), and low pH are known as direct activators of PmrA/PmrB (Wosten et al., 2000; Zhou et al., 1999). PmrA/PmrB can be indirectly activated by the PhoP/PhoQ system (Gunn and Miller, 1996; Soncini et al., 1996). Activation of PhoP/PhoQ leads to the production of PmrD, which posttranscriptionally regulates PmrA. PmrD binds to and stabilizes PmrA in its phosphorylated form (Kato and Groisman, 2004; Kox et al., 2000). Conversely, PmrA represses expression of *pmrD*. The mechanism for addition of aminoarabinose to lipid A of *Salmonella* was first elucidated in mutants with constitutively active PhoP or PmrA (Guo et al., 1998; Helander et al., 1994). Activation of PhoP or PmrA, as well as high iron and low pH, led to an upregulation of *pmrK/arnT*, resulting in the addition of aminoarabinose at the 1 position of lipid IV_A; the addition at the 4′ position only happens with Kdo_2 lipid IV_A (Trent et al., 2001b).

pEtN can be added to the outer heptose of core but also to the terminal phosphates of lipid A. The original identification of this addition was in *Neisseria*, in which it is a constitutive addition (Kulshin et al., 1992; Plested et al., 1999; Tzeng et al., 2002). The enzyme responsible for the addition, LptA, was identified by homology to the heptose phosphoethanolamine transferase, Lpt3 (Cox et al., 2003; Mackinnon et al., 2002). Deletion of *lptA* leads to an increased susceptibility to complement-mediated killing and AMPs (Lewis et al., 2009; Tzeng et al., 2005). While the gene is named *lptA* in *Neisseria* for LPS phosphoethanolamine transferase for lipid A, in other species it is named *eptA*, and care should be taken to avoid confusion with the LPS transfer protein LptA. The resistance to AMPs is due to the pEtN reducing the overall negative charge and hydrophobicity of lipid A (Lee et al., 2004; Spitznagel, 1990; Tamayo et al., 2005). In *Salmonella* and *E. coli*, the PmrA/PmrB system is required for pEtN addition (Zhou et al., 2001). However, addition of pEtN is constitutive in *Helicobacter* and *Campylobacter* (Trent et al., 2006). Mutants of *Helicobacter* and *Campylobacter* that lack pEtN show increased sensitivity to AMPs (Cullen and Trent, 2010; Tran et al., 2006). In *Campylobacter*, the gene that adds pEtN to lipid A also adds pEtN to the flagellar rod protein FlgG and encourages flagellum assembly (Cullen and Trent, 2010; Golden and Acheson, 2002). This result highlights that the transferases involved in lipid A modification may very well serve in other cellular roles.

CORE BIOSYNTHESIS

Core oligosaccharide comprises the interior of typical LPS and has been shown to contribute to the permeability barrier of the outer membrane, leading to variability in defenses against antibiotics and cationic AMPs. The charge conferred by modifications of core sugars is of vital importance to proper linkage between LPS molecules through interactions with divalent cations as well as proper stabilization and incorporation of outer membrane proteins (OMPs). Experimental models where LPS fails to incorporate substitutions allowing for this negative charge through its core have been shown to allow leakage of cytoplasmic proteins as well as decreased levels of OMPs. Loss of core charge cannot be compensated for via modifications to the lipid A and/or O antigen.

In addition to interactions between proteins and other LPS molecules the core, in typical cases, attaches the lipid A component of LPS to the O antigen. In the instance where O antigen is present, the LPS is referred to as smooth LPS (S-LPS). When O antigen is not present the LPS is described as rough (R-LPS); when the core is truncated even further, lacking the outer core portion, it is referred to as deep rough LPS (Re-LPS). Instead of the archetypical LPS, some bacterial pathogens that inhabit mucosal surfaces express lipooligosaccharide (LOS), which lacks the repetitive O-antigen side chain. The LOSs of the human pathogens *Haemophilus* and *Neisseria* have been well studied and found to be variable and, in some cases, mimic host structures. Here we explore the regulation of one particular modification, that of phosphorylcholine (ChoP), on *Haemophilus* and *Neisseria* LOS.

There are two regions of core commonly discussed, inner and outer core. The actual composition of each of these components is highly variable among bacteria at the levels of genera, species, and strains, especially with regard to outer core. Generally speaking, inner core is made up of one to three Kdo groups attached to the lipid A moiety at the nonreducing glucosamine at the 6′ position. Branching from the KdoI residue is a string of heptose residues comprising the remainder of the inner core. Outer core is the most highly variable region of core. Outer core is made up of hexose residues such as D-glucose, D-mannose, and D-galactose. In addition to the main core sugars, the structure can be further modified by the addition of phosphate, galacturonic acid, ethanolamine derivatives, Kdo, rhamnose, galactose, glucosamine, N-acetylglucosamine, heptose, and D-glycero-α-D-talo-oct-2-ulopyranosonic acid (Ko) (Frirdich and Whitfield, 2005).

Since core has been thoroughly studied with regard to pathways, enzymes involved, and regulation in the enteric organisms, the focus of this section from here on is on systems primarily in *E. coli* K-12, R1, R2, R3, and R4 and *S. enterica* serovar Typhimurium unless stated otherwise. The R groups indicate variations in the outer core region and O antigen.

Biosynthesis of Inner Core

Synthesis and attachment of Kdo to lipid A have been well characterized in *E. coli* and other enteric organisms; studies in other bacteria have been steadily evolving. Generally, five enzymes are necessary for the synthesis and addition of Kdo to lipid A. The first four steps in the pathway occur in the cytoplasm, with the final enzyme attaching Kdo sugars to lipid A at the inner leaflet of the inner membrane. The first enzyme in this pathway, KdsD, mediates the conversion of D-ribulose 5-phosphate into D-arabinose 5-phosphate (Cipolla et al., 2009). In the absence of KdsD there is a redundant enzyme, GutQ, which can act in its place (Sperandeo et al., 2006). The Kdo 8-phosphate synthase, KdsA, is the second enzyme acting in this pathway, initiating the reaction of D-arabinose 5-phosphate with phosphoenolpyruvate to form Kdo 8-phosphate (Woisetschlager and Hogenauer, 1986). The phosphate is then cleaved by

the phosphatase KdsC in the third step to produce Kdo and inorganic phosphate (Biswas et al., 2009). Like KdsD, KdsC has been found to be nonessential, indicating that there may be a redundant enzyme not yet elucidated. KdsB works in the fourth step to produce the activated sugar CMP-Kdo (Ray et al., 1981). This pathway is summarized in Fig. 2. Once the active sugar is produced, it can then be attached to the lipid A molecule by the integral inner membrane protein WaaA. The incorporation of one or more Kdo molecules into the final LPS/LOS has previously been described as dependent on the specificity of the WaaA enzyme itself leading to genus-specific additions of Kdo. WaaA in *Chlamydia trachomatis* has a slightly different activity in that it is trifunctional, adding three Kdo molecules (Belunis et al., 1995); *B. pertussis* (Isobe et al., 1999) and *Haemophilus influenzae* express a monofunctional WaaA (White et al., 1997). However, new insights into the *H. pylori* Kdo pathway negate the idea that the number of Kdo moieties in the final molecule is solely dependent on WaaA. Previously, it was thought that *H. pylori* WaaA was monofunctional, as the final LPS molecule contained only one Kdo in the inner core. Instead, enzyme activity was shown to be bifunctional, with a Kdo hydrolase removing one Kdo moiety in the membrane (Stead et al., 2010). Kdo hydrolases are now being discovered in other bacteria; two such enzymes have recently been described for *F. novicida* (Zhao and Raetz, 2010), with genes for these hydrolases identified in *F. tularensis*, *H. pylori*, and *Legionella pneumophila* (Chalabaev et al., 2010).

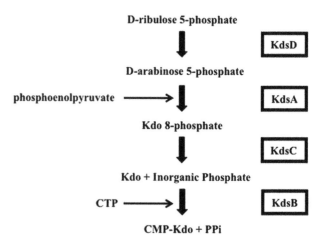

Figure 2. Pathway for the synthesis of Kdo. KdsD converts D-ribulose 5-phosphate to D-arabinose 5-phosphate. D-Arabinose 5-phosphate and phosphoenolpyruvate are combined by KdsA to form Kdo 8-phosphate. KdsC cleaves Kdo 8-phosphate to Kdo and inorganic phosphate. KdsB mediates the reaction of Kdo and CTP to form the activated sugar CMP-Kdo and PP$_i$. This active sugar is added to lipid IV$_A$ in the inner membrane. doi:10.1128/9781555818524.ch11f2

A series of integral membrane proteins catalyzes the synthesis of the inner core after Kdo$_2$-lipid IV$_A$ is established in the inner leaflet of the inner membrane. Sugars commonly incorporated are L-glycero-D-*manno*-heptopyranose (Hep). The heptosyltransferases WaaC, WaaF, and WaaQ add these Hep sugars. WaaC transfers a heptose residue from ADP-D-glycero-D-manno-heptose to Kdo$_2$-lipid IV$_A$ (Grizot et al., 2006). The heptosyl II transferase, WaaF, catalyzes the reaction of ADP-d-glycero-D-manno-heptose to HepI-Kdo$_2$-lipid IV$_A$ (Gronow et al., 2000). As previously stated, there are also nonstoichiometric substitutions, such as phosphate and pEtN added to inner core, which are necessary for core assembly (Frirdich and Whitfield, 2005). Addition of the final HepIII by the heptosyl III transferase is dependent on specific substitutions for proper enzyme activity. Two enzymes, WaaP and WaaY, mediate the phosphate/pEtN addition. WaaP makes the addition of phosphate or pEtN to HepI prior to the activities of both WaaY and WaaQ; disruption of WaaP leads to loss of phosphate addition at HepII (WaaY) and loss of the branch HepIII (WaaQ) (Yethon et al., 1998). Substitutions other than phosphate or pEtN are made to inner core. *Salmonella* and *E. coli* K-12 and R2 add a third Kdo (KdoIII) to KdoII via WaaZ (Frirdich et al., 2003), R2 adds Gal via WabA (Heinrichs et al., 1998c), and K-12 utilizes WaaS (Heinrichs et al., 1998c) to add L-Rha. In *Salmonella*, the HepI residue is modified by the addition of a pEtN by the enzyme CptA (Tamayo et al., 2005).

Biosynthesis of Outer Core

Outer core synthesis is a more diverse process, reflecting the degree of variability of outer core compared to inner core between genera and between species. The outer core pathways from *E. coli* K-12, R1, R2, R3, and R4 and *S.* Typhimurium serve as models for this section. Outer core synthesis begins with the addition of a hexose sugar to the HepI sugar; for all the above-listed *E. coli* strains and *S.* Typhimurium, this sugar is a glucose-α-1,3-heptose (GlcI) added by WaaG (Kadam et al., 1985). Subsequently, the newly added sugar is modified by an α-1,6-galactose added by the enzyme WaaB (Kadam et al., 1985; Wollin et al., 1983). *E. coli* R1, R3, and R4 do not have the *waaB* gene and therefore lack the substitution. The next addition for *E. coli* K-12, R1, R2, and R4 is the addition of α-1,3-glucose (GlcII) to GlcI by WaaO (formerly RfaI) (Creeger and Rothfield, 1979). *E. coli* R3 and *S.* Typhimurium add α-1,2-galactose residue via WaaI (Kadam et al., 1985). This HexII sugar is modified after its addition in R1 by a β-1,3-glucose (b-Glc) via WaaV (Heinrichs et al., 1998b). R3 adds

α-1,3-*N*-acetylglucosamine (GlcNAc) (Kaniuk, 2004), R4 adds β-1,4-galactose (b-Gal) via WaaX (Heinrichs et al., 1998b), while K-12, R2, and *S.* Typhimurium remain unchanged at HexII. Further additions to the straight chain by K-12 and R2 are an α-1,2-linked glucose residue at the HexIII position by WaaR and by R3 and *S.* Typhimurium by WaaJ (Kadam et al., 1985; Kaniuk et al., 2004). R1 and R4 make the subsequent straight chain additions at the HexIII of Gal I via the enzyme WaaT (Kaniuk et al., 2004). Terminal substitutions at HexIII are made through the activity of WaaK adding GlcNAc in the case of *Salmonella* and R2, WaaU adding a HepIV for *E. coli* K-12 (Heinrichs et al., 1998a), WaaW adding a Gal II for R1 and R4 (Heinrichs et al., 1998b), and WaaD adding Glc III for R3 (Kaniuk et al., 2004). Final core structures and the enzymes responsible for each addition are shown in Fig. 3.

Regulation of Core

Regulation of LPS core biosynthesis is not well understood, although some of the regulatory mechanisms for biosynthesis of Kdo and inner and outer core are emerging. As previously stated, the most active area of study with regard to regulation focuses on the pathways described above for *E. coli* and *Salmonella*. Here we look at what has been elucidated in those pathways as a model for what may emerge as other bacterial regulatory systems come to light.

Regulation of Kdo Biosynthesis

There are five major enzymes in the Kdo pathway. The genes for the first and third enzymes of the pathway, *kdsD* and *kdsC*, are found in the *yrbG-lptB* locus (Sperandeo et al., 2006). This locus is comprised of six genes: *yrbG*, a putative cation exchanger; *kdsD* (D-arabinose 5-phosphate isomerase); *kdsC* (Kdo 8-phosphate phosphatase); and *lptC*, *lptA*, and *lptB*, encoding three of the proteins in the LPS transporter. It is also important to note that directly downstream of the locus is the *rpoN* operon, which may also be transcribed from promoters in the *yrbG-lptB* locus, indicating regulation of the locus at some level by sigma factor N (σ^N), most well characterized for its activity under nitrogen limitation. Within the locus there are three promoter regions regulating three operons: *yrbG*, *kdsC*, and *lptA* (Fig. 4A). Promoters in the first two operons are described here, with the third region described in a later section. The *yrbGp* promoter can drive transcription of all six genes of the operon, *yrbG-kdsD*, or *yrbG* alone. Sequence data suggest that this promoter may be regulated by σ^D, the housekeeping sigma factor. The promoter

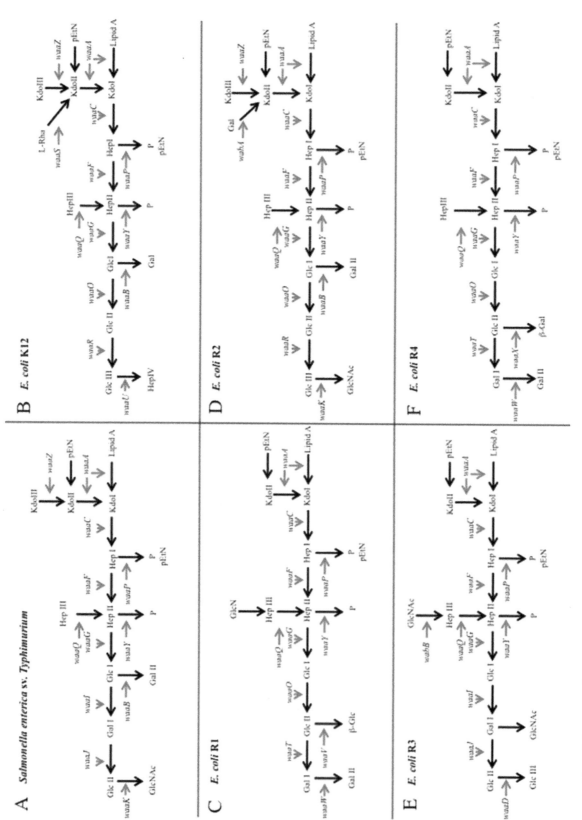

Figure 3. Structures of LPS core. Structures of known inner and outer cores for *Salmonella enterica* serovar Typhimurium (A), *E. coli* K-12 (B), and *E. coli* R1 (C), R2 (D), R3 (E), and R4 (F). All genes with known/proposed activities are indicated with gray arrows at the sites of activity. Adapted from A. Silipo and A. Molinaro, 2010. doi:10.1128/9781555818524.ch11f3

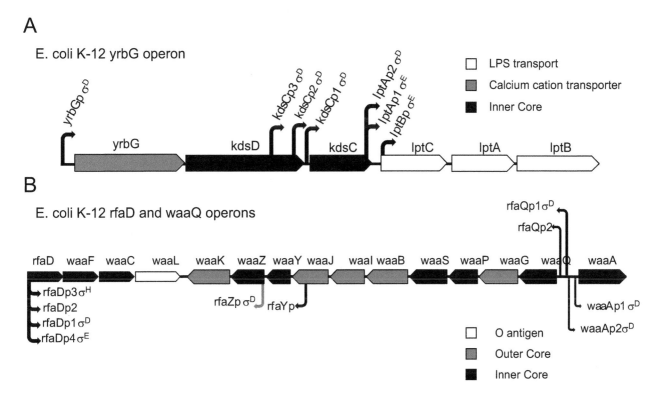

Figure 4. Genetic organization of LPS core and LPS transport genes. (A) *E. coli* K-12 *yrbG-lptB* operon and known promoters. All promoters are indicated with interacting sigma factors. Genes for LPS transport are in white; the *yrbG* gene, part of the cation antiporter family, is shown in dark blue, and inner core genes are shown in black. (B) *E. coli* K-12 *rfaD-waaL* and *waaQ-waaK* operons with the *waaA* gene and indicated promoters. Promoters are labeled with interacting sigma factors; promoters in black indicate active promoters, while blue promoters have not been proven active. Genes implicated in O-antigen synthesis are displayed in white, outer core genes are shown in blue, and inner core genes are shown in black. doi:10.1128/9781555818524.ch11f4

region containing three promoters, *kdsCp3*, *kdsCp2*, and *kdsCp1*, for the *kdsC* operon is contained within the 3′ *kdsD* gene, which may drive transcription of *kdsC-lptC* as well as *kdsC-lptB*. Although exact regulation and environmental stimuli of these promoters are not known at this time, data suggest that that σ^N, sigma factor E (σ^E) (the extracytoplasmic/extreme heat stress sigma factor), and sigma factor S (σ^S) (the starvation/stationary-phase sigma factor) are not involved (Martorana et al., 2011). The genes encoding the second and fourth enzymes in the pathway, *kdsA* and *kdsB*, are located within separate gene clusters. *kdsA* is the terminal gene in cluster of three genes that is driven primarily from a promoter located upstream of the *ychQA-kdsA* operon (Strohmaier et al., 1995). The promoter for *kdsB* drives the transcription of *ycaR* and *kdsB*. Both *kdsA* and *kdsB* are growth phase regulated at the level of transcription, with mRNA levels dropping during entry to stationary phase. Accordingly, the promoters of these genes show sequences for σ^D binding. Protein levels for KdsA are highest in late log phase, where KdsB shows expression in early phases, followed by a reemergence

in late stationary phase (Strohmaier et al., 1995). Data have yet to show what turns transcription of these genes off in stationary phase, although weak binding of σ^S may play a role. Not much is known about the regulation of *kdtA* other than its possible transcription from one of two possible promoters from the *waaQ* operon, both containing possible binding sites for σ^D (Clementz, 1992).

Regulation of Inner Core and Outer Core

The genes involved in synthesis of the heptose region of inner core and most of the genes for outer core synthesis of *E. coli* are found in one of two operons, *hldD-waaL* and *waaQ-waaK* (Fig. 4B). The *hldD* operon contains the genes *hldD*, *waaF*, *waaC*, and *waaL*, driven from three promoters. The three promoters, P1, P2, and P3, are all located upstream of *hldD* and regulated under three different conditions. The proximal promoter, P1, contains the recognition sequence for σ^D directing transcription of these genes during normal growth phases. The second promoter, P2, shows some sequence homology to the consensus sequence for σ^N, suggesting the importance of these enzymes during nitrogen-limited conditions. Lastly, the P3 promoter is regulated by the heat shock response for growth above 42°C, reaching optimal expression at 50°C (Raina and Georgopoulos, 1991). Expression studies show that at least two of the three, P1 and P3, are necessary for expression of the operon. A change of pEtN to HepI in *Salmonella* and the removal of phosphate from HepII are regulated by the two-component system PmrA/PmrB described above.

Regulation of the *waaQ* operon is RfaH dependent, which coincides with regulation of certain O antigen genes. The *waaQ* operon in *E. coli* contains a majority of the genes necessary for the remainder of inner and the outer core synthesis. The biosynthesis pathway of outer core for several *E. coli* and *Salmonella* was described above; the organization of the genes utilized in that pathway reflects the organization of the *waaQ* operon. The *waaQ* operon of *E. coli* K-12 is used here as an example; other bacteria outside of the enteric organisms show a similar operon structure based on specific outer core sugar incorporation and enzymes necessary for their synthesis. RfaH is an antiterminator interacting with RNA polymerase (RNAP), permitting read-through of termination sites for many genes involved in virulence and fertility. Antitermination by RfaH works through recognition of a 39-bp region upstream of the operon called the JUMPstart (just upstream of many polysaccharide-associated gene starts) sequence containing the *ops* (operon polarity suppressor) element on the nontemplate DNA strand downstream of the main promoter. This sequence recruits the RfaH protein while also causing RNAP to pause at the same site (Bailey et al., 2000). Binding of the N-terminal domain of RfaH causes the C-terminal domain to undergo a conformational change releasing the N-terminal domain to bind to the sigma site on RNAP (Svetlov et al., 2007). Once bound to RNAP, RfaH can act by blocking termination sites within the operon and by blocking the rebinding of sigma factors which would lead to sigma-dependent pausing (Sevostyanova et al., 2008).

Within the *waaQ* locus lie two other possible promoters, *rfaYp* and *rfaZp*, driving expression of *waaYZ* and *waaZ*, respectively. There is no strong evidence for the activity *rfaZp*, although weak activity from a fluorescent transcriptional reporter was indicated (Zaslaver et al., 2006). *rfaYp* in contrast, showed appreciable β-galactosidase activity from a *waaYp-lacZ* promoter fusion integrated into the chromosome when subjected to oxidant stress and certain antibiotics (Lee et al., 2009a). The two-component systems SoxR/SoxS and MarA/MarB were indicated as likely mechanisms of regulation of *waaYp* under

these conditions, since SoxRS is induced under oxidative stress and MarAB through antibiotic stress. This was verified since strains containing mutations in *soxR* and *marA* abolished *waaYZ* expression upon inducing conditions (Lee et al., 2009a).

Regulation of Phosphorylcholine Modification of LOS by Phase Variation

ChoP is recognized as a common surface modification of many mucosal pathogens. ChoP mimics platelet-activating factor (PAF) and binds to C-reactive protein (CRP) and PAF receptors on host cells. In addition to *H. influenzae* and *Neisseria* species, *Streptococcus pneumoniae* (Mosser and Tomasz, 1970) and *P. aeruginosa* also express this moiety on surface-associated molecules (Weiser et al., 1998).

The ChoP modification of *H. influenzae* LOS was first recognized by screening for mutants that were immunochemically distinct from the wild-type strain (Weiser et al., 1989b). These studies identified a locus, *lic1*, that was found to vary by slip-strand mispairing of a repeat sequence, 5′-CAAT-3′, within the coding region of the *licA* gene, encoding choline kinase. By varying the number of repeats, a change in the reading frame results in a high frequency (10^{-2} to 10^{-3} per generation) of spontaneous phase variation in the expression of ChoP on LOS (Weiser et al., 1989a). In the *lic1* operon, the downstream gene, *licB*, encodes a choline transporter, while the *licC* gene encodes a protein with similarity to nucleotide pyrophosphorylases, and the product of the *licD* gene is involved in the transfer of ChoP to its final location on *H. influenzae* LOS (Weiser et al., 1997). Interestingly, genes homologous to *licA* to *licD* have been identified in *S. pneumoniae* (Zhang et al., 1999), which expresses ChoP on its cell wall teichoic acid and lipoteichoic acid.

The biological role of ChoP modification of *H. influenzae* has been defined based on analysis of strains that vary in ChoP expression. ChoP-expressing strains are more sensitive to the bactericidal action of normal human serum; this is correlated with the presence of anti-ChoP antibody in the serum (Weiser et al., 1997). The phase-variable nature of the ChoP modification on LOS has been suggested to aid *H. influenzae* in evading CRP-mediated clearance. Further experiments revealed that strains that did not express ChoP were more prevalent in the blood during infection (Weiser et al., 1997), while ChoP-expressing strains are better able to persist on the mucosal surface of the respiratory tract (Weiser et al., 1998) and play a role at the early stages of infection. Other studies have confirmed that ChoP increases host cell adherence and invasion (Swords et al., 2000; Swords et al.,

2001) and resistance to host-derived antimicrobials (Lysenko et al., 2000). Additional studies have shown that ChoP is expressed in biofilms and that this modification may be disadvantageous during the planktonic phase of growth (Hong et al., 2007).

Only a few hints have emerged since these genes were originally recognized to define conditions responsible for regulating ChoP expression on *H. influenzae*. Wong and Akerley noted that ChoP was more highly expressed under conditions of low aeration and that this correlated with the expression of *licA* in *H. influenzae* (Wong and Akerley, 2005). However, increased expression of *licA* alone was not sufficient to increase ChoP modification. The authors concluded that phase variation was not the only mechanism to regulate ChoP expression. They found that a mutation in a gene encoding the global regulator CsrA showed an increase in ChoP expression under aerobic conditions compared to the wild-type strain. In *E. coli*, CsrA is a pleiotropic posttranscriptional regulator. Thus, it was concluded that the regulation of other genes involved in LOS biosynthesis are likely responsible for the modulation of ChoP expression. Such modulation could allow *H. influenzae* to adapt rapidly to changing environments to control the level of ChoP modification.

The location of ChoP on the LOS core region varies depending on the *H. influenzae* strain and is likely due to the different specificity of the *licD* gene (Lysenko et al., 2000). Other studies have linked the presence of two distinct ChoP modifications on *H. influenzae* LOS to two different copies of the *lic1* operon (Fox et al., 2008). In this case, ChoP can be added to a hexose or heptose linked to the LPS core. However, how these different *lic1* loci are regulated and, therefore, what modification is expressed under particular conditions are not currently known.

Interestingly, the ChoP modification of the LOS of *Neisseria* is found only on commensal and not pathogenic species. These commensal strains have a gene homologous to the *H. influenzae licA* gene (Serino and Virji, 2000), as well as downstream genes homologous to *licB*, *licC*, and *licD* (Serino and Virji, 2002). Again, with similarity to the *H. influenzae* system, the *Neisseria licA* gene has varying numbers of 5′-CAAT-3′ repeats within the gene. Therefore, the expression of ChoP on LOS of commensal *Neisseria* is also phase variable. The expression of ChoP on the LOS of these commensal *Neisseria* correlates with increased susceptibility to the bactericidal effects of human serum and adherence and invasion of human epithelial cells. Interestingly, and distinctly different, pathogenic *Neisseria* species, including *N. meningitidis* and *N. gonorrhoeae*, do not contain the *lic1* genes. In fact, in these species, it is the pili that are modified

with ChoP and not the LOS. ChoP is added to the surface protein, pilin, by a posttranslation mechanism (Hegge et al., 2004). It has been suggested that these modifications are also subject to phase variation due to a homopolymeric guanosine tract in the *pptA* gene, encoding a putative pilin phosphorylcholine transferase A (Snyder et al., 2001; Warren and Jennings, 2003). Similar to that found with the modification of LOS, the variable expression of ChoP on pilin likely influences the interaction of these pathogens at different stages of infection.

Regulation of LOS by PhoP/PhoQ in *Yersinia pestis*

Multiple systems are employed for the regulation of LOS, including phase variation, two-component systems, and feedback systems. Many of the regulatory mechanisms are well studied, with an emphasis on organisms such as *Neisseria*. Modification of the LOS by *Y. pestis* is important for survival in both the mammalian and flea hosts. Flea and mammalian hosts require adaptation to two temperatures, 25 and 37°C, respectively; therefore, it is not surprising that regulation of certain LOS modifications for adaptation within the host is temperature dependent. Structural studies have identified a unique 37°C structure and four glycoforms at 25°C conferring, among other modifications, a decrease in galactose incorporation into the core as temperature increases (Knirel et al., 2005). This change in galactose incorporation has been shown to be regulated by the two-component regulatory system PhoP/PhoQ, as mutants lacking *phoP* are unable to incorporate galactose, thus expressing only the terminal Hep (Hitchen et al., 2002). Although no direct link between temperature and PhoP/PhoQ activation has been elucidated, other modes of activation such as Mg^{2+} deficiency and exposure to cationic AMPs have been shown to regulate this system. A feedback loop mechanism has recently been identified for downregulation of PhoP/PhoQ, mediated by the integral membrane protein MgrB. Transcription studies found that upregulation of *mgrB* by PhoP and Δ*mgrB* strains shows increased expression of PhoP-regulated genes even under repressing conditions (Lippa and Goulian, 2009).

Influence of Core Structural Modifications on Antibiotic Resistance

Core structure is important to the overall stability and integrity of the gram-negative outer membrane; when minimal core structure is not incorporated, release of cytoplasmic enzymes into the media decreases fitness and loss of viability can be seen as a result. The importance of core structure and modification is evident when looking at survival within the host and resistance to antibiotics. In most cases AMP resistance can be equated with a loss or masking of the negative charge of the membrane and/or decrease in membrane fluidity, thus decreasing electrostatic interactions, hydrophobic interactions, and ability of AMPs to insert into the membrane. Positively charged AMPs utilize these properties to integrate into the outer membrane, making pores, which ultimately lead to bacterial lysis and death. Utilizing positively charged additions such as aminoarabinose, pEtN, and ethanolamine, bacteria can lower the affinity of AMPs for the cell surface, resulting in increased resistance. Other modifications such as the additions added to *Y. pestis* by PhoP/PhoQ confer resistance to AMPs such as polymyxin B (Hitchen et al., 2002).

Since LPS makes up the majority of the outer membrane of gram-negative bacteria, it is important to understand how different LPS molecules will interact with and affect the OMPs and membrane integrity. Some of these OMPs, such as porins, and changes in LPS affecting membrane permeability are implicated in antibiotic resistance mechanisms. Changes in OMP incorporation, stability, and composition can all potentially lead to changes in susceptibility to antibiotics. There are different schools of thought on how variation in LPS core structure can affect any of these properties; some believe that LPS core is independent of membrane protein incorporation and biogenesis, while others feel that core plays more of an active role. Recently, it was shown that an *E. coli* S17-λ (λ*pir*) mutant with a *waaQ* mutation had an OMP aggregate composition identical to that of wild-type *E. coli* K-12, thereby asserting that core does not play a role in biogenesis (Corsaro et al., 2009). A *N. meningitidis lpxA* deletion strain lacking LPS expresses a majority of normal OMP but lacks surface-exposed lipoproteins. The authors indicated that these properties are likely specific to capsule-producing *N. meningitidis* (Steeghs et al., 2001). Others have shown appreciable changes to major OMPs with LPS core truncations. An *E. coli* deep rough mutant showed inability to produce a properly folded monomer of the OMP, PhoE, compared to an isogenic rough strain; however, there was no difference between the two strains in trimerization of the monomers in the outer membrane (de Cock and Tommassen, 1996). Follow-up work from the de Cock lab utilized inner core mutants to isolate regions of inner core necessary for stable folding of the PhoE monomer and found that mutants with reduction of negative charge in the inner core reduced proper folding (de Cock et al., 1999). Some modifications change antibiotic susceptibility by decreasing membrane permeability

to lipophilic antibiotics as well as AMPs. A study in *Citrobacter rodentium* examined effects of the two-component system PmrA/PmrB, which triggers the addition of pEtN on both the lipid A by PmrC and core moiety by CptA, on susceptibility to antibiotics and AMPs. Mutants with deletions in *pmrA/pmrB*, *pmrC*, and *cptA* had increased susceptibility to antibiotics that diffuse across the membrane, increased uptake of fluorescent dyes, and leakage of periplasmic β-lactamase, all of which indicate a decrease in membrane barrier function (Viau et al., 2011).

O-ANTIGEN BIOSYNTHESIS

General Structure of the O Antigen

The core oligosaccharide of LPS is often capped with an O-antigen side chain, which is composed of various numbers of repeating sugar subunits and extends out from the surface of the cell (Fig. 5A and B). It is the composition of the O-antigen subunit that distinguishes different serotypes of bacterial species from each other. The presence of this side chain differentiates S-LPS molecules (which contain the O-antigen side chain) from R-LPS molecules (where nothing is added to the core). Some of the LPS produced in a cell can remain uncapped and rough, and the percentages can differ widely among different bacteria. For example, *E. coli* and *S. enterica* cap about 90% of their LPS molecules, while *P. aeruginosa* caps only about 10% (Wilkinson and Galbrath, 1975). The mechanism that determines how often LPS molecules will receive an O-antigen side chain remains undetermined.

Gram-negative bacteria attach O-antigen side chains of different lengths to the LPS expressed on the surface. This phenotype is evident as a laddered banding pattern when isolated LPS is run on sodium dodecyl sulfate-polyacrylamide gel electrophoresis (SDS-PAGE) and stained, where each increasing band represents an LPS molecule with one more O-antigen subunits linked together to form the side chain (Fig. 5C). Bacteria tend to have a preference for producing O-antigen side chains of specific lengths, resulting in a grouped modal banding pattern; this preference in chain length can vary depending on the species, strain, and/or serotype of the organism.

The serotype-specific LPS O antigen is generally considered an immunodominant protective antigen. *E. coli* and *Salmonella* strains with an S-LPS have been shown to interact with the terminal components of the complement cascade, yet insertion of the membrane attack complex never occurs (Joiner et al., 1982). Also, these strains are more resistant to the

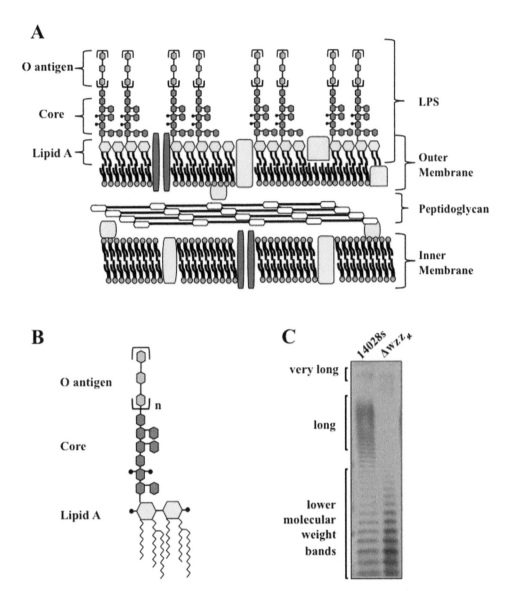

Figure 5. Structure and visualization of LPS. (A) The typical gram-negative membrane is comprised of the inner membrane, the periplasm, and the outer membrane. The inner membrane is a phospholipid bilayer with integral and peripheral membrane proteins. (B) Representation of LPS structure showing the O antigen in blue, core in purple, and lipid A in yellow. Phosphates are shown as black circles. (C) LPS from *S.* Typhimurium strain 14028s and its isogenic Δ*wzz*~st~ mutant. The lower-molecular-weight bands demonstrate the banding pattern associated with increasing O-antigen side chains. The loss of long chain lengths due to the absence of *wzz*~st~ is evident in the deletion mutant strain. Since *wzz*~fepE~ is still contained in the genome of both strains, the very long chain length is still detected at the top of the gel. LPS was detected using Salmonella O Antiserum Factor 4 from Difco Laboratories.
doi:10.1128/9781555818524.ch11f5

action of normal human serum than strains with R-LPS. The O-antigen portion of LPS has been shown to be critical for virulence, and pathogenesis has been shown for many gram-negative bacteria (Comstock and Kasper, 2006; Stenutz et al., 2006; Whitfield, 2006).

Loci Containing O-Antigen Genes

Most gram-negative bacteria contain within their genome an entire locus dedicated to the synthesis of the O-antigen subunit and assembly of the side chain. The organization and sequence of these loci differ depending on serotype, which include distinct genes encoding enzymes responsible for synthesisis of the sugars, glycosyltransferases responsible for linking them together, and other genes for assembly of the O antigen subunits (Fig. 6). Across the different serotypes of a species, these loci are generally located in the same area of the genome despite the sequence of the loci themselves being unique. In *Enterobacteriaceae*, such as *E. coli* and *Salmonella*, these gene clusters

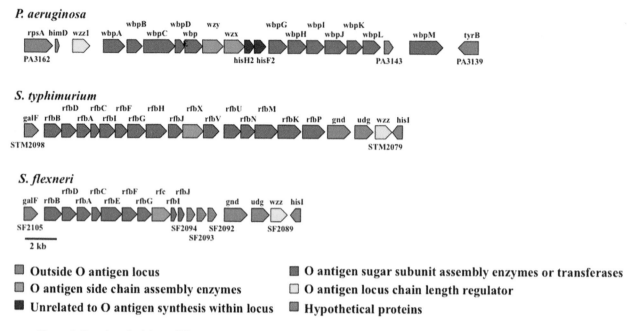

P. aeruginosa

S. typhimurium

S. flexneri

2 kb

■ **Outside O antigen locus**
■ **O antigen side chain assembly enzymes**
■ **Unrelated to O antigen synthesis within locus**

■ **O antigen sugar subunit assembly enzymes or transferases**
□ **O antigen locus chain length regulator**
■ **Hypothetical proteins**

Figure 6. O-antigen loci from different gram-negative bacteria. Organizations of loci were obtained from the Genome home of NCBI (http://www.ncbi.nlm.nih.gov/sites/genome). Strains are as follows: *P. aeruginosa* PAO1; *S.* Typhimurium LT2, and *S. flexneri* 2a str 301. Locus tag numbers are provided underneath genes at the beginning and end of the operons. For reference, some of the other O-antigen-associated genes are as follows: *P. aeruginosa, waaL* = PA4999 and *wzz2* = PA0938; *S.* Typhimurium, *rfc* (*wzy*) = STM1332 and *fepE* (second *wzz* gene) = STM0589; and *S. flexneri, waaL* = SF3666. doi:10.1128/9781555818524.ch11f6

are located next to the *his* locus (Keenleyside and Whitfield, 1999), while in *P. aeruginosa*, they are located between *himD* and *tyrB* (Raymond et al., 2002). These O-antigen loci generally have a much lower GC content than the rest of the genome, suggesting that they have been horizontally transferred and inserted into hot spots on the chromosome. In some cases, genes required for O-antigen synthesis and modification may be outside this locus. In bacteria that use the ABC transporter system for the synthesis of O antigen (see below), the genes encoding this system reside elsewhere on the chromosome.

Biosynthesis of O Antigen

To date, three different mechanisms for assembling the O-antigen side chain have been described (reviewed in Keenleyside and Whitfield, 1999). The most common, referred to as the Wzy-dependent pathway, uses specialized enzymes to generate the O-antigen side chain (Fig. 7). The second most common, the ABC transporter-dependent pathway, uses glycosyltransferases to assemble the side chain and ABC transporters to move it to the periplasm (Fig. 8). The final synthase-dependent pathway, encoded by plasmid-borne genes, has only been thus far described for *S. enterica* serovar Borreze (Keenleyside and Whitfield, 1996). This system uses one dual-activity

enzyme to both form the side chain and transport it into the periplasm. All three pathways use enzymes located within the inner membrane and end with the completed side chain on the periplasmic face of the inner membrane so it can interact with the

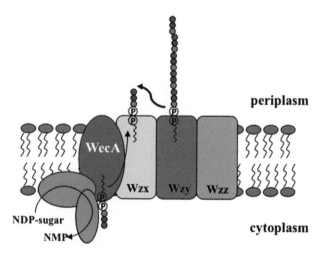

Figure 7. Wzy-dependent pathway for O-antigen synthesis and transfer across the inner membrane. The diagram shows that O antigens are synthesized on the Und-P and transferred via the Wzx, O-antigen flippase. The Wzy, O-antigen polymerase, extends the chains, the length of which is controlled by the Wzz O-antigen regulator. (Adapted from Raetz and Whitfield, 2002.) doi:10.1128/9781555818524.ch11f7

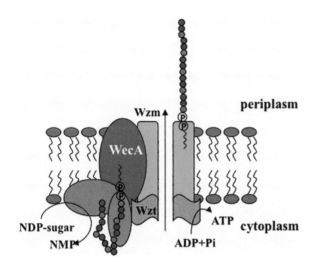

Figure 8. ABC transporter-dependent pathway for transfer across the inner membrane. The diagram shows the transfer of the Und-P-linked O antigen across the inner membrane by the ABC transporter, Wzm and Wzt. (Adapted from Raetz and Whitfield, 2002.) doi:10.1128/9781555818524.ch11f8

WaaL O-antigen ligase, linking it to lipid A plus core. Another common feature of all three pathways is the use of a carrier lipid, Und-P, to begin assembly of the side chain.

Generation of the O-antigen side chain occurs in the periplasm during Wzy-dependent assembly (Fig. 7). First, glycosyltransferases assemble the subunit onto Und-P on the cytoplasmic surface of the inner membrane. The Wzx flippase transfers the Und-P-linked O-antigen subunits to the periplasmic face of the inner membrane. Here the Wzy polymerase can link together subunits to form the side chain, requiring a new Und-P-linked subunit for each addition. The modal distribution discussed above is determined by the Wzz chain length regulator proteins (discussed in detail below). A few bacteria (*S.* Typhimurium, *S. flexneri*, and *P. aeruginosa*) produce two preferred chain lengths, leading to a bimodal distribution of lengths when LPS is visualized on gels (Daniels et al., 2002; Murray et al., 2003; Stevenson et al., 1995). Once the desired length has been reached, the entire side chain is transferred en bloc to the lipid A plus core by the WaaL ligase, and the completed LPS molecule can then be moved to the outer membrane (discussed below). Mutants that have abolished the activity of their Wzz protein produce a random distribution of side chains, where shorter side chains are produced by the Wzy polymerase acting without a regulating influence (Fig. 5C).

The ABC transporter-dependent pathway for O-side chain synthesis is typically associated with much simpler O-antigen repeats than are found with the Wzy-dependent pathway. *P. aeruginosa* uses this system to make a second O antigen, referred to as A-band or common antigen (Raymond et al., 2002). Instead of adding each O-antigen subunit to an individual Und-P molecule, as occurs in the Wzy-dependent pathway, in the ABC transport system, a "priming" sugar is added to Und-P. O-antigen side chain synthesis then occurs within the cytoplasm, anchored to the membrane by the Und-P lipid. Glycosyltransferases add sugars to the growing side chain, and thus, a dedicated Wzy polymerase is not required. Once the appropriate length is reached, the entire side chain is shuttled across the inner membrane via ABC transporters, involving the membrane-spanning protein Wzm and the ATP-binding protein Wzt. Wzz proteins are not associated with this pathway of assembly, and how these strains exhibit O-antigen modality is still under investigation. It is thought that it might be associated with size restrictions present by the opening of the transporter (Whitfield et al., 1997). Alternatively, it may be determined by the ratio of glycosyltransferases versus the ABC transporter. Bronner et al. demonstrated that overexpression of the ABC transporter led to a decrease in chain length, possibly because the increased amount of transport protein gave the glycosyltransferases a shorter time to act before the chain was moved to the periplasm. Since the ABC transporter genes are located at the beginning of the O-antigen operon, transcriptional polarity could influence protein levels and thus act as a mechanism to control chain length (Bronner et al., 1994).

In the synthase-dependent pathway, a priming sugar is also used to initiate the formation of the side chain. WbbF then adds sugars to increase the length of the side chain and also acts as the transporter that moves it to the periplasm. The side chains produced by this pathway appear largely unregulated in length; the modal distribution typically seen in other gram-negative bacteria is not evident. There is a noticeable absence of side chains containing only one or two subunits, which are common for side chains produced by the other pathways. This finding suggests that the side chain has to reach a specific length before WbbF can move it across the membrane (Keenleyside and Whitfield, 1996).

Regulation of O-Antigen Biosynthesis Genes

The O-antigen side chain, given its position on the surface, is one of the first components of the bacteria to interact with the environment. Since the side chains form a barrier that can protect the bacteria from complement and AMPs, it would seem evolutionarily advantageous that bacteria would regulate the amount of O antigen on their surface to protect themselves as they encounter changing conditions.

Several environmental conditions have been described that affect the expression of O antigen on the surface of the bacteria. These range from osmolarity to oxygen tension to Mg^{2+} concentration. One of the more extreme examples of growth conditions affecting chain length has been described for growth at different temperatures. For example, both *Aeromonas hydrophila* and *Y. enterocolitica* express more O antigen at lower temperatures (20 to 25°C) than at 37°C. For *Y. enterocolitica*, not only is more overall O antigen present at lower temperatures but also an increase in the modality conferred by Wzz is more apparent. At 37°C, the banding pattern resembles the random distribution of chain lengths of *wzz* mutant strains.

Conditions that create membrane stress also lead to changes in O-antigen chain length. In *Y. enterocolitica*, Wzz overexpression led to decreased transcription of the O-antigen locus. The altered dynamics of the inner membrane were shown to activate the CpxA/CpxR extracytoplasmic stress response; the authors hypothesized that induction of this system eventually fed back to lead to the decrease in expression of the O-antigen genes (Bengoechea et al., 2002). Similarly, extracytoplasmic stress in *E. coli* was also demonstrated to lead to changes in O antigen. Deletion of *tolA* or *pal* was associated with an upregulation of the σ^E extracytoplasmic stress response, and a decrease in the amount of the regulated O-antigen chain length was evident (Vines et al., 2005). A similar phenotype could be elicited by the addition of indole to the bacterial cultures, which is known to affect membrane stability; in this case, there was no decrease in transcription of the O-antigen locus to explain the phenotype (Vines et al., 2005). However, the authors did not investigate expression levels or protein degradation of the Wzz chain length regulator, but decreased amounts of the preferred chain length indicate that there may be a reduced amount of Wzz available during membrane stress conditions. How all these changes in conditions are recognized and how the bacteria subsequently adjust the expression of O-antigen assembly genes have not been completely characterized.

Regulation of the O-Antigen Locus

Regardless of which pathway is used to make O antigen, gram-negative bacteria contain a locus dedicated to the synthesis of the O-antigen subunit and assembly of the side chain. Despite the importance of this molecule as a major component of the bacterial surface, very little work has been done to investigate the transcriptional control and expression of O-antigen-associated genes. With a few exceptions, most of the work on O-antigen gene regulation has been performed in enteric bacteria.

One of the most common forms of regulation seen for O-antigen loci across many species is posttranscriptional control of expression through RfaH. The JUMPstart sequence, containing *ops*, has been identified in both Wzy-dependent and ABC transporter-dependent O-antigen loci (Hobbs and Reeves, 1994; Wang et al., 1998). Given the large size of the O-antigen operons, it is believed that RfaH acts as a transcriptional antiterminator to promote transcription of genes at the 3′ end of the operon. Mutants made in *rfaH* have altered O-antigen chain lengths compared to wild-type strains, and β-galactosidase activity was reduced when promoter-*lacZ* fusions using the upstream sequence to O-antigen loci were expressed in *rfaH* mutant strains (Bittner et al., 2002; Rojas et al., 2001). Similarly, it has been demonstrated that transcription of distal genes is abolished compared to genes closer to the promoter in *rfaH* mutants. In some cases, O-antigen locus promoter-*lacZ* fusions have not identified a change in expression under different conditions (Marolda and Valvano, 1998). It is believed in these cases that the O-antigen locus promoter may be constitutive and that proper expression of assembly enzymes is dependent on regulation of transcript length via RfaH. The residual expression of LPS in a *Salmonella rfaH* mutant and its modest attenuation have supported the premise that this strain might be a viable vaccine candidate (Nagy et al., 2008).

The amount of O antigen present on the surface increases as bacteria enter late exponential and stationary phases, indicating that growth phase may play a role in the regulation of the O-antigen locus (Carter et al., 2007; Rojas et al., 2001). *rfaH-lacZ* fusions in *S. enterica* serovar Typhi demonstrated that expression also increased as bacteria grew, providing an explanation for the differences in O-antigen amount (Bittner et al., 2002). Analysis of the *rfaH* promoter revealed a σ^N binding site, and mutation of *rpoN* abolished the increase in *rfaH* expression seen during stationary phase (Bittner et al., 2002). Since σ^S is associated with expression of stationary-phase genes, its role in *rfaH* regulation was also investigated (Bittner et al., 2004). Mutation of *rpoS* also reduced the expression of *rfaH* during stationary phase. However, when a double *rpoS rpoN* mutant was complemented with each individual gene, only expression of *rpoN* could rescue *rfaH* expression, leading the authors to conclude that RpoS does not directly interact with the *rfaH* promoter. Given the constitutive expression of σ^N, the authors hypothesize that σ^S regulates an as-yet-unidentified stationary-phase factor that acts with σ^N to increase expression of *rfaH*, leading to the increased amounts of O antigen seen during stationary phase.

Regulation via RfaH does not completely explain expression of the O-antigen locus or the determination of O-antigen chain length under different conditions. *rfaH-lacZ* fusions do not exhibit changes in expression when cultures are grown under different temperatures or different osmolarity conditions (Marolda and Valvano, 1998), indicating that other, unidentified factors may be regulating the transcription of the locus. Furthermore, JUMPstart sequences have not been identified upstream of the O-antigen loci in other gram-negative species, including *Y. enterocolitica* and *P. aeruginosa*, also suggesting that they may use another mechanism to ensure full transcription of the entire locus.

Regulation of *wzz* Chain Length Regulators

The Wzy-dependent pathway of O-antigen assembly relies on the Wzz chain length regulators to determine the modal chain length characteristic to each species. In a majority of the O-antigen loci whose sequences have been determined, the *wzz* gene is located near the locus, but it has been found to have its own promoter separate from the one used for the rest of the operon. This provides another level of regulation for O-antigen chain length. That is, the *wzz* genes can be expressed differently from the rest of the O-antigen assembly genes.

Transcriptional regulation of the *wzz* gene associated with the O-antigen locus has best been characterized in different *Salmonella enterica* serovars (Delgado et al., 2006). Delgado et al. investigated the role of the PmrA/PmrB and RcsC/YojN/RcsB in *wzz* expression in *S.* Typhimurium. These authors chose PmrA/PmrB to investigate, since this two-component system is involved in the regulation of other LPS modification genes (Garcia Vescovi et al., 1996; Marchal et al., 2004; Soncini et al., 1996; Tamayo et al., 2005; Tamayo et al., 2002; Wosten et al., 2000; Zhou et al., 1999). They also examined RscC/YojN/RcsB because the corresponding genes control capsular polysaccharide production, which is mechanistically similar to O-antigen synthesis. The authors demonstrated that PmrA and RcsB bind the *wzz* promoter and can independently drive expression of *wzz*. Expression of *wzz* increased under conditions that led to the activation of these two-component systems, such as growth under low-Mg^{2+} conditions or in the presence of Fe^{3+} (Belden and Miller, 1994; Gunn and Miller, 1996; Guo et al., 1998; Soncini et al., 1996).

More recently, the effect of *dam* methylation of *wzz* expression in *S. enterica* serovar Enteritidis was investigated. The *dam* mutant exhibited an altered O-antigen banding pattern compared to that of the wild type, resulting in an increased number of lower-molecular-weight bands (Sarnacki et al., 2009). *wzz* promoter-*lacZ* fusions demonstrated decreased expression of *wzz* in the *dam* mutant; this was confirmed when Western blotting also found lower levels of Wzz protein in the mutant. It was not determined if this effect was due to direct regulation of the *wzz* promoter via the Dam methyltransferase. There are a few GATC sequences upstream of, and within, the *wzz* coding sequence, but the authors noted a threefold-higher distribution of the GATC sequence within *rcsB* compared to the surrounding sequence. This suggests that the effect seen on *wzz* may be due to *dam* regulation of the *wzz* transcription factors.

Temperature regulation of the *wzz* gene has been associated with the changes in O-antigen expression seen at different growth temperatures. For example, *A. hydrophila* has decreased O antigen at 37°C compared to that at 20°C. Transcript levels of the O-antigen locus are fairly equivalent between the two temperatures, so a decrease in the levels of the rest of the O-antigen assembly enzymes does not explain the difference seen (Jimenez et al., 2008). However, Wzz levels were reduced when the bacteria were grown at 37°C. Less Wzz protein would lead to fewer of the regulated chain length being produced, leading to more lower-molecular-weight chain lengths at that temperature.

Gram-negative species that produce two different preferred chain lengths, such as some strains of *E. coli*, *S.* Typhimurium, and *S. flexneri*, also contain a second *wzz* gene that is not associated with the O-antigen locus. In *E. coli* and *S.* Typhimurium, it is found within a locus dedicated to iron transport, while *S. flexneri* stably maintains this gene on a plasmid. When investigating expression of the different O-antigen chain lengths in *S.* Typhimurium, it was noted that the amount of very long chain lengths increased when the bacteria were grown in serum compared to standard media (Murray et al., 2005). However, a *wzz$_{fepE}$* promoter-*lacZ* fusion demonstrated that expression of this gene was not upregulated under low-iron conditions, as might be expected given its location in the genome, thereby suggesting that some other condition present during growth in serum is responsible for the increased expression. More recently, PmrA, which regulates the O-antigen locus-associated *wzz* in *S.* Typhimurium, was found to also bind the promoter region of *wzz$_{fepE}$* in the same strain (Pescaretti et al., 2011). RcsB/C was not found to play a role in *wzz$_{fepE}$* expression.

The expression of both *wzzB* (associated with the O-antigen locus) and *wzz$_{pHS-2}$* (carried on the virulence plasmid pHS-2) has been investigated in *S. flexneri* using promoter-*lacZ* fusions (Carter et al., 2007). Both of the genes exhibited an increase in expression as cultures reached stationary phase, similar to that seen for genes in the O-antigen locus. However, this increased expression was not dependent on RfaH, since expression of these genes increased during stationary phase even in an *S. flexneri rfaH* mutant. This illustrates how the *wzz* genes are expressed in a different manner than the rest of the O-antigen locus despite corresponding to key proteins important for achieving proper O-antigen chain length.

Regulation of O-Antigen Modification Genes

It has been demonstrated in both *S. enterica* and *S. flexneri* that bacteriophage infection can lead to a conversion from one O-antigen serotype to another. This is due to phage-encoded glycosyltransferases that modify the sugars of the O-antigen subunit and is believed to prevent competing bacteriophages, which typically bind O antigen for invasion, from infecting the same cell. The temperate bacteriophages involved in serotype switching contain three genes (Allison and Verma, 2000). The first of these genes, two in the phage operon, are the highly conserved glucosyltransferase genes *gtrA* and *gtrB*. The third gene is unique to each phage and encodes a gene product responsible for the specific linkage of the glucosyl residues to the O antigen subunit required for seroconversion.

Similar modifications, which convert one O-antigen serotype to another, have also been seen in *P. aeruginosa*. The phage D3 has been shown to carry genes responsible for changes in the acetylation pattern and linkages between sugars within the O antigen (Kuzio and Kropinski, 1983). This phage carries an O-acetylase gene, an alternative *wzy* gene, and a gene for a *wzy* inhibitor (Newton et al., 2001). However, other than observations that the GC content of these phages genes was distinct from the rest of the *P. aeruginosa* genome (Kaluzny et al., 2007), how the genes controlling this type of modification are regulated has not been determined.

The expression of the O-antigen modification operon after insertion into the bacterial genome has been characterized for the P22 phage in *S.* Typhimurium (Broadbent et al., 2010). In this system, serotype conversion exhibited a heritable but reversible expression pattern similar to phase variation. Examination of the upstream region indicated that there were three OxyR binding sites and several GATC sequences that might be subjected to methylation via the Dam methyltransferase. Since OxyR acts as a dimer, it was determined that if OxyR bound the first two sites because the third was precluded due to methylation of the site, then OxyR acted as a positive regulator for expression of the *gtr* operon.

However, if the first binding site was methylated, leaving open the last two binding sites for OxyR, it then behaved as a negative regulator and inhibited transcription of the locus. The epigenetic regulation imposed by the methylation of the DNA would lead to the heritability of the phenotype but would also explain how during DNA replication, when the sites were not bound by transcription factors, it could occasionally switch and lead to the phase variability observed.

Regulation of O-Antigen Chain Length by Wzz Proteins

The Wzz proteins, which are members of the polysaccharide copolymerase family of proteins (Morona et al., 2009), are responsible for determining the number of O-antigen subunits that are linked together to form the side chain. In strains producing two different preferred chain lengths, two different Wzz proteins are used to regulate each chain length. While the sequence similarity among Wzz proteins from different species is relatively low, these proteins all contain two transmembrane domains at the amino- and carboxy-terminal ends that anchor them in the inner membrane. The middle portion is exposed in the periplasm, and structural analyses indicate that it is largely alpha-helical (Guo et al., 2006; Tocilj et al., 2008). Because O-antigen side chain synthesis occurs in the periplasm, this allows for the Wzz protein to interact with either the Wzy polymerase or the growing chain of O-antigen subunits to determine the number of subunits linked together before ligation to lipid A plus core.

The effect of O-antigen chain length on interaction with host cells has been investigated (Kintz et al., 2008; Murray et al., 2006). An *E. coli* strain mutated in the *wzz* gene (called *rol*, for regulator of O-antigen length) was more sensitive to the action of normal human serum than was the wild-type strain, but it was less sensitive than a rough strain completely lacking O antigen (Burns and Hull, 1998). For *Salmonella*, strains that lack the long chain length were more sensitive to the action of normal human serum and were less virulent in a mouse model of infection than either wild-type strains or those lacking the very long O-chain length (Murray et al., 2003). *P. aeruginosa wzz* mutants showed similar differential effects in in vitro and in vivo virulence models (Ivanov et al., 2011; Kintz and Goldberg, 2008).

Despite the importance of achieving proper O-antigen side chain lengths for the survival of the bacteria, how the Wzz proteins regulate this activity is not well understood. Mutational analyses have determined that single amino acid changes over the entire length of this protein could lead to shifts in preferred chain length or completely abolish the ability of Wzz to regulate any chain length (Daniels and Morona, 1999; Franco et al., 1998). Since these studies were unable to define a specific active site associated with regulating O-antigen chain length, it is now assumed that Wzz does not possess any inherent enzymatic activity for regulating chain length but, instead, acts as a structural scaffold for the other O-antigen assembling components.

Cross-linking studies indicate that Wzz proteins form homo-oligomers (Daniels and Morona, 1999), which has been confirmed via several structural analyses (Larue et al., 2009; Tang et al., 2007; Tocilj et al., 2008). This ability to homo-oligomerize is tied to length-regulating activity, because *wzz* mutants that no longer oligomerize also fail to regulate chain length (Daniels and Morona, 1999). Wzz oligomers with sizes consistent with octamers have been observed via Western blotting (Daniels et al., 2002). The crystal structures to the periplasmic region of three Wzz proteins have been solved, and all three crystallized as oligomers containing either five, eight, or nine Wzz monomers (Tocilj et al., 2008). These crystal structures suggested a general shape for Wzz oligomers, with the monomers coming together to form an open bell-shaped structure that extends about 100 Å into the perisplasmic space. Since the number of monomers forming the oligomer differed, there was a larger difference in the diameter of both the inner cavity (9 to 50 Å) and the outer portion of the oligomer closest to the membrane (73 to 95 Å). These authors noted the presence of many charged amino acids covering the outer surface of the barrel, which they suggested may be important for interacting with the growing O-antigen side chain (Tocilj et al., 2008).

Several theories have been proposed to explain Wzz activity. According to the "molecular timer" model, Wzz exists in two conformational states, with one favoring further polymerization of the side chain by Wzy, while the other terminates this process to allow for ligation to lipid A plus core (Bastin et al., 1993). Upon incubation with O antigen, Wzz is found in a more alpha-helical conformation, providing some support for this theory (Guo et al., 2006). The "molecular chaperone" theory proposes that Wzz organizes Wzy and WaaL in a complex, with the ratio of each protein affecting ligation kinetics, thus determining chain length (Morona et al., 1995). The recent in vitro study by Woodward et al. demonstrated that Wzz maintained chain length-regulating activity in the presence of only Wzy, indicating that WaaL is not a necessary component of the complex for Wzz's activity (Woodward et al., 2010). Both the molecular timer and molecular chaperone theories were introduced before evidence existed that Wzz itself homo-oligomerizes, although this information does not discount either model. More recently, Tocilj et al. proposed that the Wzz oligomer serves as a scaffold around which Wzy organizes, similar to the molecular chaperone theory. In this model, each Wzy protein completes the addition of one O-antigen subunit before passing it off to a neighboring Wzy, rather than one Wzy protein completing the entire side chain as was previously assumed (Tocilj et al., 2008). Since there is a rough correlation between the number of Wzz proteins associated together in an oligomer and the length of the O-antigen side chain produced, this "processive" model would explain how an increase in Wzy protein levels leads to an increase in chain length (Morona et al., 2009). An increased amount of Wzy assembled around a larger oligomer would allow for more rounds of polymerization to produce longer side chains. Which theory is most correct has been difficult to determine, since direct protein-protein interactions between Wzz and Wzy have yet to be detected.

More recent mutagenesis studies have indicated an important role for the conformation of the barrel when determining chain length. When a five-amino-acid linker sequence was inserted into the region constituting the base of the protein, the chain length produced by that mutated Wzz increased in length (Papadopoulos and Morona, 2010). The insertion was predicted to face the inner cavity formed by the monomer and may have been large enough to increase the diameter of the barrel. It is an intriguing possibility that O-antigen chain length may be physically measured in some way using the barrel, so that increases in barrel size would lead to increases in chain length. Similarly, several studies have found that changing amino acids along the monomer-monomer interface can affect chain length, possibly by affecting the conformation of the barrel or the stability of the oligomer complex, again indicating that the barrel of the Wzz oligomer may play a larger role in Wzz's chain length-regulating activity than was suggested by any of the previous theories of activity (Kalynych et al., 2011; Papadopoulos and Morona, 2010). However, a direct correlation between mutated versions of Wzz producing different chain lengths or ones actually having altered conformations of the barrel has not been investigated.

The amount of Wzz protein available can also have an effect on the amount of the preferred chain length produced. Overexpression of Wzz in *Y. enterocolitica* was demonstrated to produce its preferred chain length even when transcription of the rest of the O-antigen locus was decreased (Bengoechea et al., 2002). The authors hypothesize that despite the reduced amount of O-antigen assembly machinery available, the excess Wzz protein expressed from the complementation vector allowed the assembly complex to operate more efficiently, thus producing a regulated O-antigen chain length even under conditions where expression is typically repressed. Alternatively, overexpression of Wzy often leads to the production of an increased amount of shorter chain lengths, possibly because it exceeds the levels that can interact with the Wzz oligomers (Carter et al., 2009). These results demonstrate the importance of the ratio of the proteins in the hypothetical assembly complex when determining chain length.

In species producing two different chain lengths, there appears to be a competition between the two Wzz proteins for the O-antigen assembly machinery, leading to another level of regulation. Different levels of the proteins can determine which chain length is produced in a greater amount. It has been demonstrated that overexpression of *wzz* associated with the O-antigen locus can abolish the production of the chain length associated with the second Wzz

protein, despite the fact that it still being expressed (Keenleyside and Whitfield, 1999). Determining which gene is upregulated under different conditions will help explain why one chain length is expressed in a larger amount than the other as bacteria encounter different environments.

LPS TRANSPORT

After the synthesis of lipid A and the addition of the core oligosaccharide, this moiety is flipped from the inner surface of the inner membrane to the periplasmic surface of the inner membrane by the ABC transporter MsbA. The LPS can then be modified by the WaaL O-antigen ligase (or can remain rough, as is the case for *E. coli* K-12) and transported to the cell surface. The components responsible for this transport of LPS have only recently been identified. Most of the genes encoding these LPS transport (Lpt) factors have

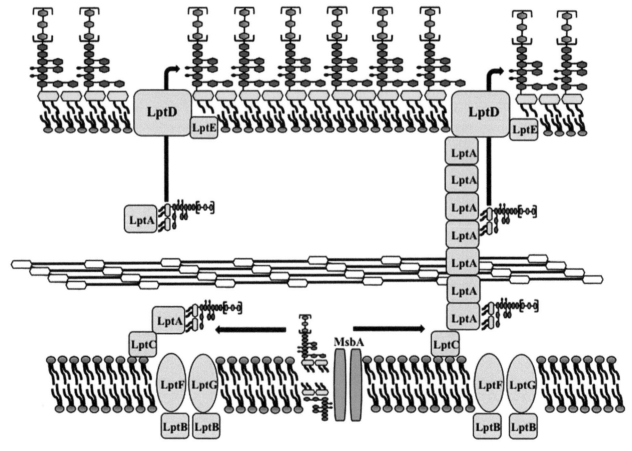

Figure 9. Diagram of the two proposed LptA transport mechanisms. In both models, MsbA transports the lipid A-core across the inner membrane. The completed Und-linked O antigen is transferred to the lipid A-core via the WaaL, O-antigen ligase (not shown). The left model shows that LptA docks with LptC and picks up the lipid A-core-O antigen and transports it across the periplasmic space, where it docks with the LptDE complex before being translocated across the outer membrane. In the right model, LptA oligomerizes to form a scaffold that allows the LPS to be transported across the periplasmic space. (Adapted from Bowyer et al., 2011.) doi:10.1128/9781555818524.ch11f9

been recognized by studies of conditional lethal mutants in *E. coli* (reviewed in Ruiz et al., 2009).

The Lpt system contains seven components. LptB appears to be a cytoplasmic protein, LptF and LptG are in the inner membrane, LptC appears to be periplasmic and anchored to the inner membrane, LptA is periplasmic, LptE (previously known as RlpB) appears periplasmic and anchored to the outer membrane, and LptD (previously known as Imp) is an OMP. There are currently two competing mechanisms proposed for how these proteins work together mainly based on studies in *E. coli* (Bowyer et al., 2011). One relies on LptA acting as a chaperone to usher the LPS through the periplasm in a similar manner to how LolA transports outer membrane lipoproteins, while another proposes a transenvelope complex of LptA that forms a bridge between the inner and outer membrane that assists the transport of the LPS (Fig. 9). The current evidence based on genetic and biophysical interaction studies of the Lpt proteins supports the transenvelope model (Bowyer et al., 2011; Chimalakonda et al., 2011; Chng et al., 2010a; Chng et al., 2010b; Sperandeo et al., 2011).

Interestingly, while these would appear to be processes that might be conserved in all gram-negative bacteria, distinctions have already been noted to exist between *E. coli* and *N. meningitidis*. Some of the Lpt proteins are found in different compartments in *N. meningitidis* compared to *E. coli*, and while mutations in the *lpt* genes affected LPS transport in *N. meningitidis*, these genes are nonessential for survival in this organism (Bos and Tommassen, 2011). The latter observation would appear to be similar to the results with the *lpxA* mutant, which does not make LPS but is nonessential in *N. meningitidis* (Steeghs et al., 1997). Whether this is due to the fact that *Neisseria* generally makes LOS rather than LPS or other currently unrecognized differences in the outer membrane is not known.

Regulation of LPS Transport

The regulation of the *lpt* genes is complex (Martorana et al., 2011). Those that have been studied in the most detail are *lptA*, *lptB*, and *lptC*. These genes are arranged in an operon starting with a nonessential gene, *yrbG*, encoding a putative cation exchange factor; the Kdo genes, *kdsD* and *kdsC*; and then *lptC*, *lptA*, and *lptB* (Fig. 4). Downstream of this operon is another operon starting with the *rpoN* gene, encoding the alternate sigma factor, σ^N. Also as mentioned, the *rpoN* operon may also be transcribed and thus regulated with these upstream genes. This has led Martorana et al. to suggest that there may be regulatory coupling between LPS biogenesis and σ^N (Martorana et al., 2011). Studies have shown that promoters for *yrbG*, *kpsC*, and *lptA* were dependent on the housekeeping σ^D and that there are multiple transcriptional start sites (Sperandeo et al., 2006). Further studies have shown that *lptA* is also regulated by the extracytoplasmic σ^E (Sperandeo et al., 2007). This regulation seems to be distinct from other σ^E-regulated genes, which suggests an added layer of regulation in the control of *lptA*. Early studies showed that *lptD* (previously named *imp*) is also part of the σ^E regulon (Dartigalongue et al., 2001).

CONCLUDING REMARKS

LPS is an integral component of the outer surface of the outer membrane of gram-negative bacteria and represents the intersection of the bacteria and the environment. The conservation of LPS structure among different species increasingly varies as it extends out from the outer membrane. Lipid A is generally well conserved, while its O-antigen structure is distinct for each serotype of a species. Therefore, it is not surprising that our understanding of the biosynthetic pathway and regulation of LPS follows a similar pattern: the pathway for the synthesis of lipid A and the environmental conditions and genes impacting this structure are well worked out. The pathway for the synthesis of the LPS core is also well conserved, while the regulation is not as well understood. Finally, the assembly of the O antigen on the Und-P and attachment to the lipid A plus core is relatively conserved across species, but the biosynthetic pathway and regulation of these functions remain elusive. Further studies into the regulation of LPS should help provide a link between signals in the environment and the resulting outer membrane composition that are likely to have the most impact on host-pathogen interactions.

Acknowledgments. We acknowledge the significant contributions of our colleagues in the LPS field. Due to the limitations of space for this chapter, valuable publications may have been unintentionally missed. The Goldberg laboratory is supported from grants from the National Institutes of Health and the Cystic Fibrosis Foundation. The Ernst Laboratory has been supported by grants from the National Institutes of Health, Army Research Office, and Cystic Fibrosis Foundation. Finally, we thank Alison Scott for critical review of the manuscript. Robert K. Ernst and Joanna B. Goldberg contributed equally to this work; Daniel A. Powell, Lauren E. Hittle, and Erica N. Kintz contributed equally to this work.

REFERENCES

Allison, G. E., and N. K. Verma. 2000. Serotype-converting bacteriophages and O-antigen modification in *Shigella flexneri*. *Trends Microbiol.* 81:17–23.

Alpuche Aranda, C. M., J. A. Swanson, W. P. Loomis, and S. I. Miller. 1992. *Salmonella typhimurium* activates virulence gene transcription within acidified macrophage phagosomes. *Proc. Natl. Acad. Sci. USA* 89:10079–10083.

Asensio, C. J., M. E. Gaillard, G. Moreno, D. Bottero, E. Zurita, M. Rumbo, P. van der Ley, A. van der Ark, and D. Hozbor. 2011. Outer membrane vesicles obtained from *Bordetella pertussis* Tohama expressing the lipid A deacylase PagL as a novel acellular vaccine candidate. *Vaccine* 29:1649–1656.

Bader, M. W., S. Sanowar, M. E. Daley, A. R. Schneider, U. Cho, W. Xu, R. E. Klevit, H. Le Moual, and S. I. Miller. 2005. Recognition of antimicrobial peptides by a bacterial sensor kinase. *Cell* 122:461–472.

Bailey, M. J., C. Hughes, and V. Koronakis. 2000. In vitro recruitment of the RfaH regulatory protein into a specialised transcription complex, directed by the nucleic acid *ops* element. *Mol. Gen. Genet.* 262:1052–1059.

Bastin, D. A., G. Stevenson, P. K. Brown, A. Haase, and P. R. Reeves. 1993. Repeat unit polysaccharides of bacteria: a model for polymerization resembling that of ribosomes and fatty acid synthetase, with a novel mechanism for determining chain length. *Mol. Microbiol.* 7:725–734.

Belden, W. J., and S. I. Miller. 1994. Further characterization of the PhoP regulon: identification of new PhoP-activated virulence loci. *Infect. Immun.* 62:5095–5101.

Belunis, C. J., T. Clementz, S. M. Carty, and C. R. Raetz. 1995. Inhibition of lipopolysaccharide biosynthesis and cell growth following inactivation of the *kdtA* gene in *Escherichia coli*. *J. Biol. Chem.* 270:27646–27652.

Bengoechea, J. A., L. Zhang, P. Toivanen, and M. Skurnik. 2002. Regulatory network of lipopolysaccharide O-antigen biosynthesis in *Yersinia enterocolitica* includes cell envelope-dependent signals. *Mol. Microbiol.* 44:1045–1062.

Beutler, B. 2005. The Toll-like receptors: analysis by forward genetic methods. *Immunogenetics* 57:385–392.

Bishop, R. E. 2008. Structural biology of membrane-intrinsic beta-barrel enzymes: sentinels of the bacterial outer membrane. *Biochim. Biophys. Acta* 1778:1881–1896.

Bishop, R. E., H. S. Gibbons, T. Guina, M. S. Trent, S. I. Miller, and C. R. Raetz. 2000. Transfer of palmitate from phospholipids to lipid A in outer membranes of gram-negative bacteria. *EMBO J.* 19:5071–5080.

Biswas, T., L. Yi, P. Aggarwal, J. Wu, J. R. Rubin, J. A. Stuckey, R. W. Woodard, and O. V. Tsodikov. 2009. The tail of KdsC: conformational changes control the activity of a haloacid dehalogenase superfamily phosphatase. *J. Biol. Chem.* 284:30594–30603.

Bittner, M., S. Saldias, F. Altamirano, M. A. Valvano, and I. Contreras. 2004. RpoS and RpoN are involved in the growth-dependent regulation of *rfaH* transcription and O antigen expression in *Salmonella enterica* serovar Typhi. *Microb. Pathog.* 36:19–24.

Bittner, M., S. Saldias, C. Estevez, M. Zaldivar, C. L. Marolda, M. A. Valvano, and I. Contreras. 2002. O-antigen expression in *Salmonella enterica* serovar Typhi is regulated by nitrogen availability through RpoN-mediated transcriptional control of the *rfaH* gene. *Microbiology* 148(Pt. 12): 3789–3799.

Bos, M. P., and J. Tommassen. 2011. The LptD chaperone LptE is not directly involved in lipopolysaccharide transport in *Neisseria meningitidis*. *J. Biol. Chem.* 286:28688–28696.

Bowyer, A., J. Baardsnes, E. Ajamian, L. Zhang, and M. Cygler. 2011. Characterization of interactions between LPS transport proteins of the Lpt system. *Biochem. Biophys. Res. Commun.* 404:1093–1098.

Broadbent, S. E., M. R. Davies, and M. W. van der Woude. 2010. Phase variation controls expression of *Salmonella* lipopolysaccharide modification genes by a DNA methylation-dependent mechanism. *Mol. Microbiol.* 77:337–353.

Bronner, D., B. R. Clarke, and C. Whitfield. 1994. Identification of an ATP-binding cassette transport system required for translocation of lipopolysaccharide O-antigen side-chains across the cytoplasmic membrane of *Klebsiella pneumoniae* serotype O1. *Mol. Microbiol.* 14:505–519.

Brozek, K. A., and C. R. Raetz. 1990. Biosynthesis of lipid A in *Escherichia coli*. Acyl carrier protein-dependent incorporation of laurate and myristate. *J. Biol. Chem.* 265:15410–15417.

Burns, S. M., and S. I. Hull. 1998. Comparison of loss of serum resistance by defined lipopolysaccharide mutants and an acapsular mutant of uropathogenic *Escherichia coli* O75:K5. *Infect. Immun.* 66:4244–4253.

Carter, J. A., C. J. Blondel, M. Zaldivar, S. A. Alvarez, C. L. Marolda, M. A. Valvano, and I. Contreras. 2007. O-antigen modal chain length in *Shigella flexneri* 2a is growth-regulated through RfaH-mediated transcriptional control of the *wzy* gene. *Microbiology* 153(Pt. 10):3499–3507.

Carter, J. A., J. C. Jimenez, M. Zaldivar, S. A. Alvarez, C. L. Marolda, M. A. Valvano, and I. Contreras. 2009. The cellular level of O-antigen polymerase Wzy determines chain length regulation by WzzB and WzzpHS-2 in *Shigella flexneri* 2a. *Microbiology* 155(Pt. 10):3260–3269.

Chalabaev, S., T. H. Kim, R. Ross, A. Derian, and D. L. Kasper. 2010. 3-Deoxy-D-manno-octulosonic acid (Kdo) hydrolase identified in *Francisella tularensis*, *Helicobacter pylori*, and *Legionella pneumophila*. *J. Biol. Chem.* 285:34330–34336.

Chimalakonda, G., N. Ruiz, S. S. Chng, R. A. Garner, D. Kahne, and T. J. Silhavy. 2011. Lipoprotein LptE is required for the assembly of LptD by the beta-barrel assembly machine in the outer membrane of *Escherichia coli*. *Proc. Natl. Acad. Sci. USA* 108:2492–2497.

Chng, S. S., L. S. Gronenberg, and D. Kahne. 2010a. Proteins required for lipopolysaccharide assembly in *Escherichia coli* form a transenvelope complex. *Biochemistry* 49:4565–4567.

Chng, S. S., N. Ruiz, G. Chimalakonda, T. J. Silhavy, and D. Kahne. 2010b. Characterization of the two-protein complex in *Escherichia coli* responsible for lipopolysaccharide assembly at the outer membrane. *Proc. Natl. Acad. Sci. USA* 107:5363–5368.

Cipolla, L., A. Polissi, C. Airoldi, P. Galliani, P. Sperandeo, and F. Nicotra. 2009. The Kdo biosynthetic pathway toward OM biogenesis as target in antibacterial drug design and development. *Curr. Drug Discov. Technol.* 6:19–33.

Clementz, T. 1992. The gene coding for 3-deoxy-manno-octulosonic acid transferase and the *rfaQ* gene are transcribed from divergently arranged promoters in *Escherichia coli*. *J. Bacteriol.* 174:7750–7756.

Clementz, T., J. J. Bednarski, and C. R. Raetz. 1996. Function of the *htrB* high temperature requirement gene of *Escherchia coli* in the acylation of lipid A: HtrB catalyzed incorporation of laurate. *J. Biol. Chem.* 271:12095–12102.

Coats, S. R., C. T. Do, L. M. Karimi-Naser, P. H. Braham, and R. P. Darveau. 2007. Antagonistic lipopolysaccharides block *E. coli* lipopolysaccharide function at human TLR4 via interaction with the human MD-2 lipopolysaccharide binding site. *Cell. Microbiol.* 9:1191–1202.

Comstock, L. E., and D. L. Kasper. 2006. Bacterial glycans: key mediators of diverse host immune responses. *Cell* 126:847–850.

Corsaro, M. M., E. Parrilli, R. Lanzetta, T. Naldi, G. Pieretti, B. Lindner, A. Carpentieri, M. Parrilli, and M. L. Tutino. 2009. The presence of OMP inclusion bodies in a *Escherichia coli* K-12 mutated strain is not related to lipopolysaccharide structure. *J. Biochem.* **146:**231–240.

Cox, A. D., J. C. Wright, J. Li, D. W. Hood, E. R. Moxon, and J. C. Richards. 2003. Phosphorylation of the lipid A region of meningococcal lipopolysaccharide: identification of a family of transferases that add phosphoethanolamine to lipopolysaccharide. *J. Bacteriol.* **185:**3270–3277.

Creeger, E. S., and L. I. Rothfield. 1979. Cloning of genes for bacterial glycosyltransferases. I. Selection of hybrid plasmids carrying genes for two glucosyltransferases. *J. Biol. Chem.* **254:**804–810.

Cullen, T. W., and M. S. Trent. 2010. A link between the assembly of flagella and lipooligosaccharide of the Gram-negative bacterium *Campylobacter jejuni*. *Proc. Natl. Acad. Sci. USA* **107:**5160–5165.

Daniels, C., C. Griffiths, B. Cowles, and J. S. Lam. 2002. *Pseudomonas aeruginosa* O-antigen chain length is determined before ligation to lipid A core. *Environ. Microbiol.* **4:**883–897.

Daniels, C., and R. Morona. 1999. Analysis of *Shigella flexneri wzz* (Rol) function by mutagenesis and cross-linking: *wzz* is able to oligomerize. *Mol. Microbiol.* **34:**181–194.

Dartigalongue, C., D. Missiakas, and S. Raina. 2001. Characterization of the *Escherichia coli* sigma E regulon. *J. Biol. Chem.* **276:**20866–20875.

Darveau, R. P., T. T. Pham, K. Lemley, R. A. Reife, B. W. Bainbridge, S. R. Coats, W. N. Howald, S. S. Way, and A. M. Hajjar. 2004. *Porphyromonas gingivalis* lipopolysaccharide contains multiple lipid A species that functionally interact with both Toll-like receptors 2 and 4. *Infect. Immun.* **72:**5041–5051.

de Cock, H., K. Brandenburg, A. Wiese, O. Holst, and U. Seydel. 1999. Non-lamellar structure and negative charges of lipopolysaccharides required for efficient folding of outer membrane protein PhoE of *Escherichia coli*. *J. Biol. Chem.* **274:**5114–5119.

de Cock, H., and J. Tommassen. 1996. Lipopolysaccharides and divalent cations are involved in the formation of an assembly-competent intermediate of outer-membrane protein PhoE of *E. coli*. *EMBO J.* **15:**5567–5573.

Delgado, M. A., C. Mouslim, and E. A. Groisman. 2006. The PmrA/PmrB and RcsC/YojN/RcsB systems control expression of the *Salmonella* O-antigen chain length determinant. *Mol. Microbiol.* **60:**39–50.

Derzelle, S., E. Turlin, E. Duchaud, S. Pages, F. Kunst, A. Givaudan, and A. Danchin. 2004. The PhoP-PhoQ two-component regulatory system of *Photorhabdus luminescens* is essential for virulence in insects. *J. Bacteriol.* **186:**1270–1279.

Doerrler, W. T., H. S. Gibbons, and C. R. Raetz. 2004. MsbA-dependent translocation of lipids across the inner membrane of *Escherichia coli*. *J. Biol. Chem.* **279:**45102–45109.

El Ghachi, M., A. Derbise, A. Bouhss, and D. Mengin-Lecreulx. 2005. Identification of multiple genes encoding membrane proteins with undecaprenyl pyrophosphate phosphatase (UppP) activity in *Escherichia coli*. *J. Biol. Chem.* **280:**18689–18695.

Ernst, R. K., E. C. Yi, L. Guo, K. B. Lim, J. L. Burns, M. Hackett, and S. I. Miller. 1999. Specific lipopolysaccharide found in cystic fibrosis airway *Pseudomonas aeruginosa*. *Science* **286:**1561–1565.

Fox, K. L., J. Li, E. K. Schweda, V. Vitiazeva, K. Makepeace, M. P. Jennings, E. R. Moxon, and D. W. Hood. 2008. Duplicate copies of *lic1* direct the addition of multiple phosphocholine residues in the lipopolysaccharide of *Haemophilus influenzae*. *Infect. Immun.* **76:**588–600.

Franco, A. V., D. Liu, and P. R. Reeves. 1998. The Wzz (Cld) protein in *Escherichia coli*: amino acid sequence variation determines O-antigen chain length specificity. *J. Bacteriol.* **180:**2670–2675.

Frirdich, E., B. Lindner, O. Holst, and C. Whitfield. 2003. Overexpression of the *waaZ* gene leads to modification of the structure of the inner core region of *Escherichia coli* lipopolysaccharide, truncation of the outer core, and reduction of the amount of O polysaccharide on the cell surface. *J. Bacteriol.* **185:**1659–1671.

Frirdich, E., and C. Whitfield. 2005. Lipopolysaccharide inner core oligosaccharide structure and outer membrane stability in human pathogens belonging to the *Enterobacteriaceae*. *J. Endotoxin Res.* **11:**133–144.

Garcia Vescovi, E., F. C. Soncini, and E. A. Groisman. 1996. Mg²⁺ as an extracellular signal: environmental regulation of *Salmonella* virulence. *Cell* **84:**165–174.

Gibbons, H. S., S. R. Kalb, R. J. Cotter, and C. R. Raetz. 2005. Role of Mg²⁺ and pH in the modification of *Salmonella* lipid A after endocytosis by macrophage tumour cells. *Mol. Microbiol.* **55:**425–440.

Gibbons, H. S., S. Lin, R. J. Cotter, and C. R. Raetz. 2000. Oxygen requirement for the biosynthesis of the S-2-hydroxymyristate moiety in *Salmonella typhimurium* lipid A. Function of LpxO, a new Fe²⁺/alpha-ketoglutarate-dependent dioxygenase homologue. *J. Biol. Chem.* **275:**32940–32949.

Golden, N. J., and D. W. Acheson. 2002. Identification of motility and autoagglutination *Campylobacter jejuni* mutants by random transposon mutagenesis. *Infect. Immun.* **70:**1761–1771.

Grizot, S., M. Salem, V. Vongsouthi, L. Durand, F. Moreau, H. Dohi, S. Vincent, S. Escaich, and A. Ducruix. 2006. Structure of the *Escherichia coli* heptosyltransferase WaaC: binary complexes with ADP and ADP-2-deoxy-2-fluoro heptose. *J. Mol. Biol.* **363:**383–394

Groisman, E. A., E. Chiao, C. J. Lipps, and F. Heffron. 1989. *Salmonella typhimurium phoP* virulence gene is a transcriptional regulator. *Proc. Natl. Acad. Sci. USA* **86:**7077–7081.

Gronow, S., W. Brabetz, and H. Brade. 2000. Comparative functional characterization in vitro of heptosyltransferase I (WaaC) and II (WaaF) from *Escherichia coli*. *Eur. J. Biochem.* **267:**6602–6611.

Gunn, J. S., K. B. Lim, J. Krueger, K. Kim, L. Guo, M. Hackett, and S. I. Miller. 1998. PmrA-PmrB-regulated genes necessary for 4-aminoarabinose lipid A modification and polymyxin resistance. *Mol. Microbiol.* **27:**1171–1182.

Gunn, J. S., and S. I. Miller. 1996. PhoP-PhoQ activates transcription of *pmrAB*, encoding a two-component regulatory system involved in *Salmonella typhimurium* antimicrobial peptide resistance. *J. Bacteriol.* **178:**6857–6864.

Gunn, J. S., S. S. Ryan, J. C. Van Velkinburgh, R. K. Ernst, and S. I. Miller. 2000. Genetic and functional analysis of a PmrA-PmrB-regulated locus necessary for lipopolysaccharide modification, antimicrobial peptide resistance, and oral virulence of *Salmonella enterica* serovar Typhimurium. *Infect. Immun.* **68:**6139–6146.

Guo, H., K. Lokko, Y. Zhang, W. Yi, Z. Wu, and P. G. Wang. 2006. Overexpression and characterization of Wzz of *Escherichia coli* O86:H2. *Protein Expr. Purif.* **48:**49–55.

Guo, L., K. B. Lim, C. M. Poduje, M. Daniel, J. S. Gunn, M. Hackett, and S. I. Miller. 1998. Lipid A acylation and bacterial resistance against vertebrate antimicrobial peptides. *Cell* **95:**189–198.

Hajjar, A. M., R. K. Ernst, J. H. Tsai, C. B. Wilson, and S. I. Miller. 2002. Human Toll-like receptor 4 recognizes host-specific LPS modifications. *Nat. Immunol.* **3:**354–359.

Hankins, J. V., J. A. Madsen, D. K. Giles, B. M. Childers, K. E. Klose, J. S. Brodbelt, and M. S. Trent. 2011. Elucidation of a novel *Vibrio cholerae* lipid A secondary hydroxy-acyltransferase and its role in innate immune recognition. *Mol. Microbiol.* **81:**1313–1329.

Hankins, J. V., and M. S. Trent. 2009. Secondary acylation of *Vibrio cholerae* lipopolysaccharide requires phosphorylation of Kdo. *J. Biol. Chem.* **284:**25804–25812.

Hegge, F. T., P. G. Hitchen, F. E. Aas, H. Kristiansen, C. Lovold, W. Egge-Jacobsen, M. Panico, W. Y. Leong, V. Bull, M. Virji, H. R. Morris, A. Dell, and M. Koomey. 2004. Unique modifications with phosphocholine and phosphoethanolamine define alternate antigenic forms of *Neisseria gonorrhoeae* type IV pili. *Proc. Natl. Acad. Sci. USA* **101:**10798–10803.

Heinrichs, D. E., M. A. Monteiro, M. B. Perry, and C. Whitfield. 1998a. The assembly system for the lipopolysaccharide R2 core-type of *Escherichia coli* is a hybrid of those found in *Escherichia coli* K-12 and *Salmonella enterica*. Structure and function of the R2 WaaK and WaaL homologs. *J. Biol. Chem.* **273:**8849–8859.

Heinrichs, D. E., J. A. Yethon, P. A. Amor, and C. Whitfield. 1998b. The assembly system for the outer core portion of R1- and R4-type lipopolysaccharides of *Escherichia coli*. The R1 core-specific beta-glucosyltransferase provides a novel attachment site for O-polysaccharides. *J. Biol. Chem.* **273:**29497–29505.

Heinrichs, D. E., J. A. Yethon, and C. Whitfield. 1998c. Molecular basis for structural diversity in the core regions of the lipopolysaccharides of *Escherichia coli* and *Salmonella enterica*. *Mol. Microbiol.* **30:**221–232.

Helander, I. M., I. Kilpelainen, and M. Vaara. 1994. Increased substitution of phosphate groups in lipopolysaccharides and lipid A of the polymyxin-resistant *pmrA* mutants of *Salmonella typhimurium*: a 31P-NMR study. *Mol. Microbiol.* **11:**481–487.

Herrera, C. M., J. V. Hankins, and M. S. Trent. 2010. Activation of PmrA inhibits LpxT-dependent phosphorylation of lipid A promoting resistance to antimicrobial peptides. *Mol. Microbiol.* **76:**1444–1460.

Hitchen, P. G., J. L. Prior, P. C. Oyston, M. Panico, B. W. Wren, R. W. Titball, H. R. Morris, and A. Dell. 2002. Structural characterization of lipo-oligosaccharide (LOS) from *Yersinia pestis*: regulation of LOS structure by the PhoPQ system. *Mol. Microbiol.* **44:**1637–1650.

Hobbs, M., and P. R. Reeves. 1994. The JUMPstart sequence: a 39 bp element common to several polysaccharide gene clusters. *Mol. Microbiol.* **12:**855–856.

Hong, W., B. Pang, S. West-Barnette, and W. E. Swords. 2007. Phosphorylcholine expression by nontypeable *Haemophilus influenzae* correlates with maturation of biofilm communities in vitro and in vivo. *J. Bacteriol.* **189:**8300–8307.

Hwang, P. M., W. Y. Choy, E. I. Lo, L. Chen, J. D. Forman-Kay, C. R. Raetz, G. G. Prive, R. E. Bishop, and L. E. Kay. 2002. Solution structure and dynamics of the outer membrane enzyme PagP by NMR. *Proc. Natl. Acad. Sci. USA* **99:**13560–13565.

Ingram, B. O., C. Sohlenkamp, O. Geiger, and C. R. Raetz. 2010. Altered lipid A structures and polymyxin hypersensitivity of *Rhizobium etli* mutants lacking the LpxE and LpxF phosphatases. *Biochim. Biophys. Acta* **1801:**593–604.

Isobe, T., K. A. White, A. G. Allen, M. Peacock, C. R. Raetz, and D. J. Maskell. 1999. *Bordetella pertussis waaA* encodes a monofunctional 2-keto-3-deoxy-D-manno-octulosonic acid transferase that can complement an *Escherichia coli waaA* mutation. *J. Bacteriol.* **181:**2648–2651.

Ivanov, I. E., E. N. Kintz, L. A. Porter, J. B. Goldberg, N. A. Burnham, and T. A. Camesano. 2011. Relating the physical properties of *Pseudomonas aeruginosa* lipopolysaccharides to virulence by atomic force microscopy. *J. Bacteriol.* **193:**1259–1266.

Jimenez, N., R. Canals, M. T. Salo, S. Vilches, S. Merino, and J. M. Tomas. 2008. The *Aeromonas hydrophila wb*O34* gene cluster: genetics and temperature regulation. *J. Bacteriol.* **190:**4198–4209.

Joiner, K. A., C. H. Hammer, E. J. Brown, R. J. Cole, and M. M. Frank. 1982. Studies on the mechanism of bacterial resistance to complement-mediated killing. I. Terminal complement components are deposited and released from *Salmonella minnesota* S218 without causing bacterial death. *J. Exp. Med.* **155:**797–808.

Kadam, S. K., A. Rehemtulla, and K. E. Sanderson. 1985. Cloning of *rfaG, B, I,* and *J* genes for glycosyltransferase enzymes for synthesis of the lipopolysaccharide core of *Salmonella typhimurium*. *J. Bacteriol.* **161:**277–284.

Kaluzny, K., P. D. Abeyrathne, and J. S. Lam. 2007. Coexistence of two distinct versions of O-antigen polymerase, Wzy-alpha and Wzy-beta, in *Pseudomonas aeruginosa* serogroup O2 and their contributions to cell surface diversity. *J. Bacteriol.* **189:**4141–4152.

Kalynych, S., X. Ruan, M. A. Valvano, and M. Cygler. 2011. Structure-guided investigation of lipopolysaccharide O-antigen chain length regulators reveals regions critical for modal length control. *J. Bacteriol.* **193:**3710–3721.

Kaniuk, N. A., E. Vinogradov, J. Li, M. A. Monteiro, and C. Whitfield. 2004. Chromosomal and plasmid-encoded enzymes are required for assembly of the R3-type core oligosaccharide in the lipopolysaccharide of *Escherichia coli* O157:H7. *J. Biol. Chem.* **279:**31237–31250.

Karbarz, M. J., S. R. Kalb, R. J. Cotter, and C. R. Raetz. 2003. Expression cloning and biochemical characterization of a *Rhizobium leguminosarum* lipid A 1-phosphatase. *J. Biol. Chem.* **278:**39269–39279.

Karow, M., O. Fayet, A. Cegielska, T. Ziegelhoffer, and C. Georgopoulos. 1991a. Isolation and characterization of the *Escherichia coli htrB* gene, whose product is essential for bacterial viability above 33°C in rich media. *J. Bacteriol.* **173:**741–750.

Karow, M., S. Raina, C. Georgopoulos, and O. Fayet. 1991b. Complex phenotypes of null mutations in the *htr* genes, whose products are essential for *Escherichia coli* growth at elevated temperatures. *Res. Microbiol.* **142:**289–294.

Karow, M., and C. Georgopoulos. 1991. Sequencing, mutational analysis, and transcriptional regulation of the *Escherichia coli htrB* gene. *Mol. Microbiol.* **5:**2285–2292.

Kato, A., and E. A. Groisman. 2004. Connecting two-component regulatory systems by a protein that protects a response regulator from dephosphorylation by its cognate sensor. *Genes Dev.* **18:**2302–2313.

Kawano, M., T. Manabe, and K. Kawasaki. 2010. *Salmonella enterica* serovar Typhimurium lipopolysaccharide deacylation enhances its intracellular growth within macrophages. *FEBS Lett.* **584:**207–212.

Kawasaki, K., K. China, and M. Nishijima. 2007. Release of the lipopolysaccharide deacylase PagL from latency compensates for a lack of lipopolysaccharide aminoarabinose modification-dependent resistance to the antimicrobial peptide polymyxin B in *Salmonella enterica*. *J. Bacteriol.* **189:**4911–4919.

Kawasaki, K., R. K. Ernst, and S. I. Miller. 2004a. 3-O-deacylation of lipid A by PagL, a PhoP/PhoQ-regulated deacylase of *Salmonella typhimurium*, modulates signaling through Toll-like receptor 4. *J. Biol. Chem.* **279:**20044–20048.

Kawasaki, K., R. K. Ernst, and S. I. Miller. 2004b. Deacylation and palmitoylation of lipid A by *Salmonellae* outer membrane enzymes modulate host signaling through Toll-like receptor 4. *J. Endotoxin Res.* **10:**439–444.

Kawasaki, K., R. K. Ernst, and S. I. Miller. 2005. Inhibition of *Salmonella enterica* serovar Typhimurium lipopolysaccharide deacylation by aminoarabinose membrane modification. *J. Bacteriol.* **187:**2448–2457.

Keenleyside, W. J., and C. Whitfield. 1996. A novel pathway for O-polysaccharide biosynthesis in *Salmonella enterica* serovar Borreze. *J. Biol. Chem.* **271:**28581–28592.

Keenleyside, W. J., and C. Whitfield. 1999. Genetics and biosynthesis of lipopolysaccharide O-antigens, p. 331–358. *In* C. M. H.

Brade, S. M. Opal, and S. N. Vogel (ed.), *Endotoxin in Health and Disease*. Marcel Dekker, Inc., New York, NY.

Kim, S. H., K. S. Kim, S. R. Lee, E. Kim, M. S. Kim, E. Y. Lee, Y. S. Gho, J. W. Kim, R. E. Bishop, and K. T. Chang. 2009. Structural modifications of outer membrane vesicles to refine them as vaccine delivery vehicles. *Biochim. Biophys. Acta* **1788:**2150–2159.

Kintz, E., and J. B. Goldberg. 2008. Regulation of lipopolysaccharide O antigen expression in *Pseudomonas aeruginosa. Future Microbiol.* **3:**191–203.

Kintz, E., J. M. Scarff, A. DiGiandomenico, and J. B. Goldberg. 2008. Lipopolysaccharide O-antigen chain length regulation in *Pseudomonas aeruginosa* serogroup O11 strain PA103. *J. Bacteriol.* **190:**2709–2716.

Knirel, Y. A., B. Lindner, E. Vinogradov, R. Z. Shaikhutdinova, S. N. Senchenkova, N. A. Kocharova, O. Holst, G. B. Pier, and A. P. Anisimov. 2005. Cold temperature-induced modifications to the composition and structure of the lipopolysaccharide of *Yersinia pestis. Carbohydr. Res.* **340:**1625–1630.

Koprivnjak, T., and A. Peschel. 2011. Bacterial resistance mechanisms against host defense peptides. *Cell. Mol. Life Sci.* **68:**2243–2254.

Koretke, K. K., A. N. Lupas, P. V. Warren, M. Rosenberg, and J. R. Brown. 2000. Evolution of two-component signal transduction. *Mol. Biol. Evol.* **17:**1956–1970.

Kox, L. F., M. M. Wosten, and E. A. Groisman. 2000. A small protein that mediates the activation of a two-component system by another two-component system. *EMBO J.* **19:**1861–1872.

Kulshin, V. A., U. Zahringer, B. Lindner, C. E. Frasch, C. M. Tsai, B. A. Dmitriev, and E. T. Rietschel. 1992. Structural characterization of the lipid A component of pathogenic *Neisseria meningitidis. J. Bacteriol.* **174:**1793–1800.

Kuzio, J., and A. M. Kropinski. 1983. O-antigen conversion in *Pseudomonas aeruginosa* PAO1 by bacteriophage D3. *J. Bacteriol.* **155:**203–212.

Lamarche, M. G., S. H. Kim, S. Crepin, M. Mourez, N. Bertrand, R. E. Bishop, J. D. Dubreuil, and J. Harel. 2008a. Modulation of hexaacyl pyrophosphate lipid A population under *Escherichia coli* phosphate (Pho) regulon activation. *J. Bacteriol.* **190:**5256–5264.

Lamarche, M. G., B. L. Wanner, S. Crepin, and J. Harel. 2008b. The phosphate regulon and bacterial virulence: a regulatory network connecting phosphate homeostasis and pathogenesis. *FEMS Microbiol. Rev.* **32:**461–473.

Larue, K., M. S. Kimber, R. Ford, and C. Whitfield. 2009. Biochemical and structural analysis of bacterial O-antigen chain length regulator proteins reveals a conserved quaternary structure. *J. Biol. Chem.* **284:**7395–7403.

Lee, H., F. F. Hsu, J. Turk, and E. A. Groisman. 2004. The PmrA-regulated *pmrC* gene mediates phosphoethanolamine modification of lipid A and polymyxin resistance in *Salmonella enterica. J. Bacteriol.* **186:**4124–4133.

Lee, J. H., K. L. Lee, W. S. Yeo, S. J. Park, and J. H. Roe. 2009a. SoxRS-mediated lipopolysaccharide modification enhances resistance against multiple drugs in *Escherichia coli. J. Bacteriol.* **191:**4441–4450.

Lee, S. R., S. H. Kim, K. J. Jeong, K. S. Kim, Y. H. Kim, S. J. Kim, E. Kim, J. W. Kim, and K. T. Chang. 2009b. Multi-immunogenic outer membrane vesicles derived from an MsbB-deficient *Salmonella enterica* serovar Typhimurium mutant. *J. Microbiol. Biotechnol.* **19:**1271–1279.

Lewis, L. A., B. Choudhury, J. T. Balthazar, L. E. Martin, S. Ram, P. A. Rice, D. S. Stephens, R. Carlson, and W. M. Shafer. 2009. Phosphoethanolamine substitution of lipid A and resistance of *Neisseria gonorrhoeae* to cationic antimicrobial peptides and complement-mediated killing by normal human serum. *Infect. Immun.* **77:**1112–1120.

Lippa, A. M., and M. Goulian. 2009. Feedback inhibition in the PhoQ/PhoP signaling system by a membrane peptide. *PLoS Genet.* **5:** e1000788.

Llama-Palacios, A., E. Lopez-Solanilla, and P. Rodriguez-Palenzuela. 2005. Role of the PhoP-PhoQ system in the virulence of *Erwinia chrysanthemi* strain 3937: involvement in sensitivity to plant antimicrobial peptides, survival at acid pH, and regulation of pectolytic enzymes. *J. Bacteriol.* **187:**2157–2162.

Loomis, W. F., A. Kuspa, and G. Shaulsky. 1998. Two-component signal transduction systems in eukaryotic microorganisms. *Curr. Opin. Microbiol.* **1:**643–648.

Lysenko, E. S., J. Gould, R. Bals, J. M. Wilson, and J. N. Weiser. 2000. Bacterial phosphorylcholine decreases susceptibility to the antimicrobial peptide LL-37/hCAP18 expressed in the upper respiratory tract. *Infect. Immun.* **68:**1664–1671.

Macarthur, I., J. W. Jones, D. R. Goodlett, R. K. Ernst, and A. Preston. 2011. The role of *pagL* and *lpxO* in *Bordetella bronchiseptica* lipid A biosynthesis. *J. Bacteriol.* **193:**4726–4735.

Macfarlane, E. L., A. Kwasnicka, and R. E. Hancock. 2000. Role of *Pseudomonas aeruginosa* PhoP-PhoQ in resistance to antimicrobial cationic peptides and aminoglycosides. *Microbiology* **146**(Pt. 10):2543–2554.

Macfarlane, E. L., A. Kwasnicka, M. M. Ochs, and R. E. Hancock. 1999. PhoP-PhoQ homologues in *Pseudomonas aeruginosa* regulate expression of the outer-membrane protein OprH and polymyxin B resistance. *Mol. Microbiol.* **34:**305–316.

Mackinnon, F. G., A. D. Cox, J. S. Plested, C. M. Tang, K. Makepeace, P. A. Coull, J. C. Wright, R. Chalmers, D. W. Hood, J. C. Richards, and E. R. Moxon. 2002. Identification of a gene (*lpt-3*) required for the addition of phosphoethanolamine to the lipopolysaccharide inner core of *Neisseria meningitidis* and its role in mediating susceptibility to bactericidal killing and opsonophagocytosis. *Mol. Microbiol.* **43:**931–943.

Marchal, K., S. De Keersmaecker, P. Monsieurs, B. van Boxel, K. Lemmens, G. Thijs, J. Vanderleyden, and B. De Moor. 2004. In silico identification and experimental validation of PmrAB targets in *Salmonella typhimurium* by regulatory motif detection. *Genome Biol.* **5:**R9.

Marolda, C. L., and M. A. Valvano. 1998. Promoter region of the *Escherichia coli* O7-specific lipopolysaccharide gene cluster: structural and functional characterization of an upstream untranslated mRNA sequence. *J. Bacteriol.* **180:**3070–3079.

Martorana, A. M., P. Sperandeo, A. Polissi, and G. Deho. 2011. Complex transcriptional organization regulates an *Escherichia coli* locus implicated in lipopolysaccharide biogenesis. *Res. Microbiol.* **162:**470–482.

McGee, D. J., A. E. George, E. A. Trainor, K. E. Horton, E. Hildebrandt, and T. L. Testerman. 2011. Cholesterol enhances *Helicobacter pylori* resistance to antibiotics and LL-37. *Antimicrob. Agents Chemother.* **55:**2897–2904.

Miller, S. I., A. M. Kukral, and J. J. Mekalanos. 1989. A two-component regulatory system (*phoP phoQ*) controls *Salmonella typhimurium* virulence. *Proc. Natl. Acad. Sci. USA* **86:**5054–5058.

Mohan, S., and C. R. Raetz. 1994. Endotoxin biosynthesis in *Pseudomonas aeruginosa:* enzymatic incorporation of laurate before 3-deoxy-D-*manno*-octulosonate. *J. Bacteriol.* **176:**6944–6951.

Moran, A. P., B. Lindner, and E. J. Walsh. 1997. Structural characterization of the lipid A component of *Helicobacter pylori* rough- and smooth-form lipopolysaccharides. *J. Bacteriol.* **179:**6453–6463.

Morona, R., L. Purins, A. Tocilj, A. Matte, and M. Cygler. 2009. Sequence-structure relationships in polysaccharide co-polymerase (PCP) proteins. *Trends Biochem. Sci.* **34:**78–84.

Morona, R., L. van den Bosch, and P. A. Manning. 1995. Molecular, genetic, and topological characterization of O-antigen chain length regulation in *Shigella flexneri*. *J. Bacteriol.* **177:**1059–1068.

Moss, J. E., P. E. Fisher, B. Vick, E. A. Groisman, and A. Zychlinsky. 2000. The regulatory protein PhoP controls susceptibility to the host inflammatory response in *Shigella flexneri*. *Cell. Microbiol.* **2:**443–452.

Mosser, J. L., and A. Tomasz. 1970. Choline-containing teichoic acid as a structural component of pneumococcal cell wall and its role in sensitivity to lysis by an autolytic enzyme. *J. Biol. Chem.* **245:**287–298.

Munford, R. S. 2008. Sensing gram-negative bacterial lipopolysaccharides: a human disease determinant? *Infect. Immun.* **76:**454–465.

Murata, T., W. Tseng, T. Guina, S. I. Miller, and H. Nikaido. 2007. PhoPQ-mediated regulation produces a more robust permeability barrier in the outer membrane of *Salmonella enterica* serovar Typhimurium. *J. Bacteriol.* **189:**7213–7222.

Murray, G. L., S. R. Attridge, and R. Morona. 2003. Regulation of *Salmonella typhimurium* lipopolysaccharide O antigen chain length is required for virulence; identification of FepE as a second Wzz. *Mol. Microbiol.* **47:**1395–1406.

Murray, G. L., S. R. Attridge, and R. Morona. 2005. Inducible serum resistance in *Salmonella typhimurium* is dependent on *wzz* (*fepE*)-regulated very long O antigen chains. *Microbes Infect.* **7:**1296–1304.

Murray, G. L., S. R. Attridge, and R. Morona. 2006. Altering the length of the lipopolysaccharide O antigen has an impact on the interaction of *Salmonella enterica* serovar Typhimurium with macrophages and complement. *J. Bacteriol.* **188:**2735–2739.

Nagy, G., T. Palkovics, A. Otto, H. Kusch, B. Kocsis, U. Dobrindt, S. Engelmann, M. Hecker, L. Emody, T. Pal, and J. Hacker. 2008. "Gently rough": the vaccine potential of a *Salmonella enterica* regulatory lipopolysaccharide mutant. *J. Infect. Dis.* **198:**1699–1706.

Newton, G. J., C. Daniels, L. L. Burrows, A. M. Kropinski, A. J. Clarke, and J. S. Lam. 2001. Three-component-mediated serotype conversion in *Pseudomonas aeruginosa* by bacteriophage D3. *Mol. Microbiol.* **39:**1237–1247.

Nikaido, H. 2003. Molecular basis of bacterial outer membrane permeability revisited. *Microbiol. Mol. Biol. Rev.* **67:**593–656.

Oertelt, C., B. Lindner, M. Skurnik, and O. Holst. 2001. Isolation and structural characterization of an R-form lipopolysaccharide from *Yersinia enterocolitica* serotype O:8. *Eur. J. Biochem.* **268:**554–564.

O'Neill, L. A. 2008a. The interleukin-1 receptor/Toll-like receptor superfamily: 10 years of progress. *Immunol. Rev.* **226:**10–18.

O'Neill, L. A. 2008b. When signaling pathways collide: positive and negative regulation of Toll-like receptor signal transduction. *Immunity* **29:**12–20.

Oyston, P. C., N. Dorrell, K. Williams, S. R. Li, M. Green, R. W. Titball, and B. W. Wren. 2000. The response regulator PhoP is important for survival under conditions of macrophage-induced stress and virulence in *Yersinia pestis*. *Infect. Immun.* **68:**3419–3425.

Papadopoulos, M., and R. Morona. 2010. Mutagenesis and chemical cross-linking suggest that Wzz dimer stability and oligomerization affect lipopolysaccharide O-antigen modal chain length control. *J. Bacteriol.* **192:**3385–3393.

Perez-Gutierrez, C., E. Llobet, C. M. Llompart, M. Reines, and J. A. Bengoechea. 2010. Role of lipid A acylation in *Yersinia enterocolitica* virulence. *Infect. Immun.* **78:** 2768–2781.

Pescaretti, M. M., F. E. Lopez, R. D. Morero, and M. A. Delgado. 2011. The PmrA/PmrB regulatory system controls the expression of *wzz*$_{fepE}$ gene involved in the O-antigen synthesis

of *Salmonella enterica* serovar Typhimurium. *Microbiology* **157:**2515–2521

Pilione, M. R., E. J. Pishko, A. Preston, D. J. Maskell, and E. T. Harvill. 2004. *pagP* is required for resistance to antibody-mediated complement lysis during *Bordetella bronchiseptica* respiratory infection. *Infect. Immun.* **72:**2837–2842.

Plested, J. S., K. Makepeace, M. P. Jennings, M. A. Gidney, S. Lacelle, J. Brisson, A. D. Cox, A. Martin, A. G. Bird, C. M. Tang, F. M. Mackinnon, J. C. Richards, and E. R. Moxon. 1999. Conservation and accessibility of an inner core lipopolysaccharide epitope of *Neisseria meningitidis*. *Infect. Immun.* **67:**5417–5426.

Preston, A., E. Maxim, E. Toland, E. J. Pishko, E. T. Harvill, M. Caroff, and D. J. Maskell. 2003. *Bordetella bronchiseptica* PagP is a Bvg-regulated lipid A palmitoyl transferase that is required for persistent colonization of the mouse respiratory tract. *Mol. Microbiol.* **48:**725–736.

Raetz, C. R., C. M. Reynolds, M. S. Trent, and R. E. Bishop. 2007. Lipid A modification systems in Gram-negative bacteria. *Annu. Rev. Biochem.* **76:**295–329.

Raetz, C. R., and C. Whitfield. 2002. Lipopolysaccharide endotoxins. *Annu. Rev. Biochem.* **71:**635–700.

Raina, S., and C. Georgopoulos. 1991. The *htrM* gene, whose product is essential for *Escherichia coli* viability only at elevated temperatures, is identical to the *rfaD* gene. *Nucleic Acids Res.* **19:**3811–3819.

Ranallo, R. T., R. W. Kaminski, T. George, A. A. Kordis, Q. Chen, K. Szabo, and M. M. Venkatesan. 2010. Virulence, inflammatory potential, and adaptive immunity induced by *Shigella flexneri msbB* mutants. *Infect. Immun.* **78:**400–412.

Ray, P. H., C. D. Benedict, and H. Grasmuk. 1981. Purification and characterization of cytidine 5′-triphosphate:cytidine 5′-monophosphate-3-deoxy-D-manno-octulosonate cytidylyltransferase. *J. Bacteriol.* **145:**1273–1280.

Raymond, C. K., E. H. Sims, A. Kas, D. H. Spencer, T. V. Kutyavin, R. G. Ivey, Y. Zhou, R. Kaul, J. B. Clendenning, and M. V. Olson. 2002. Genetic variation at the O-antigen biosynthetic locus in *Pseudomonas aeruginosa*. *J. Bacteriol.* **184:**3614–3622.

Rebeil, R., R. K. Ernst, B. B. Gowen, S. I. Miller, and B. J. Hinnebusch. 2004. Variation in lipid A structure in the pathogenic Yersiniae. *Mol. Microbiol.* **52:**1363–1373.

Reynolds, C. M., A. A. Ribeiro, S. C. McGrath, R. J. Cotter, C. R. Raetz, and M. S. Trent. 2006. An outer membrane enzyme encoded by *Salmonella typhimurium lpxR* that removes the 3′-acyloxyacyl moiety of lipid A. *J. Biol. Chem.* **281:**21974–21987.

Robey, M., W. O'Connell, and N. P. Cianciotto. 2001. Identification of *Legionella pneumophila rcp*, a *pagP*-like gene that confers resistance to cationic antimicrobial peptides and promotes intracellular infection. *Infect. Immun.* **69:**4276–4286.

Rojas, G., S. Saldias, M. Bittner, M. Zaldivar, and I. Contreras. 2001. The *rfaH* gene, which affects lipopolysaccharide synthesis in *Salmonella enterica* serovar Typhi, is differentially expressed during the bacterial growth phase. *FEMS Microbiol. Lett.* **204:**123–128.

Ruiz, N., D. Kahne, and T. J. Silhavy. 2009. Transport of lipopolysaccharide across the cell envelope: the long road of discovery. *Nat. Rev. Microbiol.* **7:**677–683.

Rutten, L., J. P. Mannie, C. M. Stead, C. R. Raetz, C. M. Reynolds, A. M. Bonvin, J. P. Tommassen, M. R. Egmond, M. S. Trent, and P. Gros. 2009. Active-site architecture and catalytic mechanism of the lipid A deacylase LpxR of *Salmonella typhimurium*. *Proc. Natl. Acad. Sci. USA* **106:**1960–1964.

Sarnacki, S. H., C. L. Marolda, M. Noto Llana, M. N. Giacomodonato, M. A. Valvano, and M. C. Cerquetti. 2009. Dam methylation controls O-antigen chain length in *Salmonella enterica* serovar Enteritidis by regulating the expression of Wzz protein. *J. Bacteriol.* **191:** 6694–6700.

Serino, L., and M. Virji. 2000. Phosphorylcholine decoration of lipopolysaccharide differentiates commensal *Neisseriae* from pathogenic strains: identification of *licA*-type genes in commensal *Neisseriae*. *Mol. Microbiol.* 35:1550–1559.

Serino, L., and M. Virji. 2002. Genetic and functional analysis of the phosphorylcholine moiety of commensal *Neisseria* lipopolysaccharide. *Mol. Microbiol.* 43:437–448.

Sevostyanova, A., V. Svetlov, D. G. Vassylyev, and I. Artsimovitch. 2008. The elongation factor RfaH and the initiation factor sigma bind to the same site on the transcription elongation complex. *Proc. Natl. Acad. Sci. USA* 105:865–870.

Silipo, A., and A. Molinaro. 2010. The diversity of the core oligosaccharide in lipopolysaccharides. *Subcell. Biochem.* 53:69–99.

Snyder, L. A., S. A. Butcher, and N. J. Saunders. 2001. Comparative whole-genome analyses reveal over 100 putative phase-variable genes in the pathogenic *Neisseria* spp. *Microbiology* **147**(Pt. 8):2321–2332.

Somerville, J. E., Jr., L. Cassiano, B. Bainbridge, M. D. Cunningham, and R. P. Darveau. 1996. A novel *Escherichia coli* lipid A mutant that produces an antiinflammatory lipopolysaccharide. *J. Clin. Investig.* 97:359–365.

Somerville, J. E., Jr., L. Cassiano, and R. P. Darveau. 1999. *Escherichia coli msbB* gene as a virulence factor and a therapeutic target. *Infect. Immun.* 67:6583–6590.

Soncini, F. C., E. Garcia Vescovi, F. Solomon, and E. A. Groisman. 1996. Molecular basis of the magnesium deprivation response in *Salmonella typhimurium*: identification of PhoP-regulated genes. *J. Bacteriol.* 178:5092–5099.

Sperandeo, P., R. Cescutti, R. Villa, C. Di Benedetto, D. Candia, G. Deho, and A. Polissi. 2007. Characterization of *lptA* and *lptB*, two essential genes implicated in lipopolysaccharide transport to the outer membrane of *Escherichia coli*. *J. Bacteriol.* 189:244–253.

Sperandeo, P., C. Pozzi, G. Deho, and A. Polissi. 2006. Nonessential KDO biosynthesis and new essential cell envelope biogenesis genes in the *Escherichia coli yrbG-yhbG* locus. *Res. Microbiol.* 157:547–558.

Sperandeo, P., R. Villa, A. M. Martorana, M. Samalikova, R. Grandori, G. Deho, and A. Polissi. 2011. New insights into the Lpt machinery for lipopolysaccharide transport to the cell surface: LptA-LptC interaction and LptA stability as sensors of a properly assembled transenvelope complex. *J. Bacteriol.* 193:1042–1053.

Spitznagel, J. K. 1990. Antibiotic proteins of human neutrophils. *J. Clin. Investig.* 86:1381–1386.

Stead, C. M., A. Beasley, R. J. Cotter, and M. S. Trent. 2008. Deciphering the unusual acylation pattern of *Helicobacter pylori* lipid A. *J. Bacteriol.* 190:7012–7021.

Stead, C. M., J. Zhao, C. R. Raetz, and M. S. Trent. 2010. Removal of the outer Kdo from *Helicobacter pylori* lipopolysaccharide and its impact on the bacterial surface. *Mol. Microbiol.* 78:837–852.

Steeghs, L., H. de Cock, E. Evers, B. Zomer, J. Tommassen, and P. van der Ley. 2001. Outer membrane composition of a lipopolysaccharide-deficient *Neisseria meningitidis* mutant. *EMBO J.* 20: 6937–6945.

Steeghs, L., M. P. Jennings, J. T. Poolman, and P. van der Ley. 1997. Isolation and characterization of the *Neisseria meningitidis lpxD-fabZ-lpxA* gene cluster involved in lipid A biosynthesis. *Gene* 190:263–270.

Stenutz, R., A. Weintraub, and G. Widmalm. 2006. The structures of *Escherichia coli* O-polysaccharide antigens. *FEMS Microbiol. Rev.* 30:382–403.

Stevenson, G., A. Kessler, and P. R. Reeves. 1995. A plasmid-borne O-antigen chain length determinant and its relationship to other chain length determinants. *FEMS Microbiol. Lett.* 125:23–30.

Strohmaier, H., P. Remler, W. Renner, and G. Hogenauer. 1995. Expression of genes *kdsA* and *kdsB* involved in 3-deoxy-D-manno-octulosonic acid metabolism and biosynthesis of enterobacterial lipopolysaccharide is growth phase regulated primarily at the transcriptional level in *Escherichia coli* K-12. *J. Bacteriol.* 177:4488–4500.

Svetlov, V., G. A. Belogurov, E. Shabrova, D. G. Vassylyev, and I. Artsimovitch. 2007. Allosteric control of the RNA polymerase by the elongation factor RfaH. *Nucleic Acids Res.* 35:5694–5705.

Swords, W. E., B. A. Buscher, K. Ver Steeg Ii, A. Preston, W. A. Nichols, J. N. Weiser, B. W. Gibson, and M. A. Apicella. 2000. Non-typeable *Haemophilus influenzae* adhere to and invade human bronchial epithelial cells via an interaction of lipooligosaccharide with the PAF receptor. *Mol. Microbiol.* 37:13–27.

Swords, W. E., M. R. Ketterer, J. Shao, C. A. Campbell, J. N. Weiser, and M. A. Apicella. 2001. Binding of the non-typeable *Haemophilus influenzae* lipooligosaccharide to the PAF receptor initiates host cell signalling. *Cell. Microbiol.* 3:525–536.

Tamayo, R., B. Choudhury, A. Septer, M. Merighi, R. Carlson, and J. S. Gunn. 2005. Identification of *cptA*, a PmrA-regulated locus required for phosphoethanolamine modification of the *Salmonella enterica* serovar Typhimurium lipopolysaccharide core. *J. Bacteriol.* 187:3391–3399.

Tamayo, R., S. S. Ryan, A. J. McCoy, and J. S. Gunn. 2002. Identification and genetic characterization of PmrA-regulated genes and genes involved in polymyxin B resistance in *Salmonella enterica* serovar Typhimurium. *Infect. Immun.* 70:6770–6778.

Tang, K. H., H. Guo, W. Yi, M. D. Tsai, and P. G. Wang. 2007. Investigation of the conformational states of Wzz and the Wzz.O-antigen complex under near-physiological conditions. *Biochemistry* 46:11744–11752.

Tocilj, A., C. Munger, A. Proteau, R. Morona, L. Purins, E. Ajamian, J. Wagner, M. Papadopoulos, L. Van Den Bosch, J. L. Rubinstein, J. Fethiere, A. Matte, and M. Cygler. 2008. Bacterial polysaccharide co-polymerases share a common framework for control of polymer length. *Nat. Struct. Mol. Biol.* 15:130–138.

Touze, T., A. X. Tran, J. V. Hankins, D. Mengin-Lecreulx, and M. S. Trent. 2008. Periplasmic phosphorylation of lipid A is linked to the synthesis of undecaprenyl phosphate. *Mol. Microbiol.* 67:264–277.

Tran, A. X., J. D. Whittimore, P. B. Wyrick, S. C. McGrath, R. J. Cotter, and M. S. Trent. 2006. The lipid A 1-phosphatase of *Helicobacter pylori* is required for resistance to the antimicrobial peptide polymyxin. *J. Bacteriol.* 188:4531–4541.

Trent, M. S., W. Pabich, C. R. Raetz, and S. I. Miller. 2001a. A PhoP/PhoQ-induced Lipase (PagL) that catalyzes 3-O-deacylation of lipid A precursors in membranes of *Salmonella typhimurium*. *J. Biol. Chem.* 276:9083–9092.

Trent, M. S., A. A. Ribeiro, S. Lin, R. J. Cotter, and C. R. Raetz. 2001b. An inner membrane enzyme in *Salmonella* and *Escherichia coli* that transfers 4-amino-4-deoxy-L-arabinose to lipid A: induction on polymyxin-resistant mutants and role of a novel lipid-linked donor. *J. Biol. Chem.* 276:43122–43131.

Trent, M. S., C. M. Stead, A. X. Tran, and J. V. Hankins. 2006. Diversity of endotoxin and its impact on pathogenesis. *J. Endotoxin Res.* 12:205–223.

Tzeng, Y. L., K. D. Ambrose, S. Zughaier, X. Zhou, Y. K. Miller, W. M. Shafer, and D. S. Stephens. 2005. Cationic antimicrobial peptide resistance in *Neisseria meningitidis*. *J. Bacteriol.* 187:5387–5396.

Tzeng, Y. L., A. Datta, V. K. Kolli, R. W. Carlson, and D. S. Stephens. 2002. Endotoxin of *Neisseria meningitidis* composed only of intact lipid A: inactivation of the meningococcal 3-deoxy-D-manno-octulosonic acid transferase. *J. Bacteriol.* 184:2379–2388.

Vaara, M., and M. Nurminen. 1999. Outer membrane permeability barrier in *Escherichia coli* mutants that are defective in the late acyltransferases of lipid A biosynthesis. *Antimicrob. Agents Chemother.* **43**:1459–1462.

Valvano, M. A. 2008. Undecaprenyl phosphate recycling comes out of age. *Mol. Microbiol.* **67**:232–235.

Viau, C., V. Le Sage, D. K. Ting, J. Gross, and H. Le Moual. 2011. Absence of PmrAB-mediated phosphoethanolamine modifications of *Citrobacter rodentium* lipopolysaccharide affects outer membrane integrity. *J. Bacteriol.* **193**:2168–2176.

Vines, E. D., C. L. Marolda, A. Balachandran, and M. A. Valvano. 2005. Defective O-antigen polymerization in *tolA* and *pal* mutants of *Escherichia coli* in response to extracytoplasmic stress. *J. Bacteriol.* **187**:3359–3368.

Vinogradov, E., M. B. Perry, and J. W. Conlan. 2002. Structural analysis of *Francisella tularensis* lipopolysaccharide. *Eur. J. Biochem.* **269**:6112–6118.

Vorachek-Warren, M. K., S. Ramirez, R. J. Cotter, and C. R. Raetz. 2002. A triple mutant of *Escherichia coli* lacking secondary acyl chains on lipid A. *J. Biol. Chem.* **277**:14194–14205.

Wang, L., S. Jensen, R. Hallman, and P. R. Reeves. 1998. Expression of the O antigen gene cluster is regulated by RfaH through the JUMPstart sequence. *FEMS Microbiol. Lett.* **165**:201–206.

Wang, X., M. J. Karbarz, S. C. McGrath, R. J. Cotter, and C. R. Raetz. 2004. MsbA transporter-dependent lipid A 1-dephosphorylation on the periplasmic surface of the inner membrane: topography of *Francisella novicida* LpxE expressed in *Escherichia coli*. *J. Biol. Chem.* **279**:49470–49478.

Warren, M. J., and M. P. Jennings. 2003. Identification and characterization of *pptA*: a gene involved in the phase-variable expression of phosphorylcholine on pili of *Neisseria meningitidis*. *Infect. Immun.* **71**:6892–6898.

Weiser, J. N., J. B. Goldberg, N. Pan, L. Wilson, and M. Virji. 1998. The phosphorylcholine epitope undergoes phase variation on a 43-kilodalton protein in *Pseudomonas aeruginosa* and on pili of *Neisseria meningitidis* and *Neisseria gonorrhoeae*. *Infect. Immun.* **66**:4263–4267.

Weiser, J. N., A. A. Lindberg, E. J. Manning, E. J. Hansen, and E. R. Moxon. 1989a. Identification of a chromosomal locus for expression of lipopolysaccharide epitopes in *Haemophilus influenzae*. *Infect. Immun.* **57**:3045–3052.

Weiser, J. N., J. M. Love, and E. R. Moxon. 1989b. The molecular mechanism of phase variation of *H. influenzae* lipopolysaccharide. *Cell* **59**:657–665.

Weiser, J. N., M. Shchepetov, and S. T. Chong. 1997. Decoration of lipopolysaccharide with phosphorylcholine: a phase-variable characteristic of *Haemophilus influenzae*. *Infect. Immun.* **65**:943–950.

West, A. H., and A. M. Stock. 2001. Histidine kinases and response regulator proteins in two-component signaling systems. *Trends Biochem. Sci.* **26**:369–376.

White, K. A., I. A. Kaltashov, R. J. Cotter, and C. R. Raetz. 1997. A mono-functional 3-deoxy-D-manno-octulosonic acid (Kdo) transferase and a Kdo kinase in extracts of *Haemophilus influenzae*. *J. Biol. Chem.* **272**:16555–16563.

Whitfield, C. 2006. Biosynthesis and assembly of capsular polysaccharides in *Escherichia coli*. *Annu. Rev. Biochem.* **75**:39–68.

Whitfield, C., P. A. Amor, and R. Koplin. 1997. Modulation of the surface architecture of gram-negative bacteria by the action of surface polymer:lipid A-core ligase and by determinants of polymer chain length. *Mol. Microbiol.* **23**:629–638.

Wilkinson, S. G., and L. Galbrath. 1975. Studies of lipopolysaccharides from *Pseudomonas aeruginosa*. *Eur. J. Biochem.* **52**:331–343.

Woisetschlager, M., and G. Hogenauer. 1986. Cloning and characterization of the gene encoding 3-deoxy-D-manno-octulosonate 8-phosphate synthetase from *Escherichia coli*. *J. Bacteriol.* **168**: 437–439.

Wollin, R., E. S. Creeger, L. I. Rothfield, B. A. Stocker, and A. A. Lindberg. 1983. *Salmonella typhimurium* mutants defective in UDP-D-galactose:lipopolysaccharide alpha 1,6-D-galactosyltransferase. Structural, immunochemical, and enzymologic studies of *rfaB* mutants. *J. Biol. Chem.* **258**:3769–3774.

Wong, S. M., and B. J. Akerley. 2005. Environmental and genetic regulation of the phosphorylcholine epitope of *Haemophilus influenzae* lipooligosaccharide. *Mol. Microbiol.* **55**:724–738.

Woodward, R., W. Yi, L. Li, G. Zhao, H. Eguchi, P. R. Sridhar, H. Guo, J. K. Song, E. Motari, L. Cai, P. Kelleher, X. Liu, W. Han, W. Zhang, Y. Ding, M. Li, and P. G. Wang. 2010. In vitro bacterial polysaccharide biosynthesis: defining the functions of Wzy and Wzz. *Nat. Chem. Biol.* **6**:418–423.

Wosten, M. M., L. F. Kox, S. Chamnongpol, F. C. Soncini, and E. A. Groisman. 2000. A signal transduction system that responds to extracellular iron. *Cell* **103**:113–125.

Wren, B. W. 2003. The *Yersiniae*—a model genus to study the rapid evolution of bacterial pathogens. *Nat. Rev. Microbiol.* **1**:55–64.

Yethon, J. A., D. E. Heinrichs, M. A. Monteiro, M. B. Perry, and C. Whitfield. 1998. Involvement of *waaY*, *waaQ*, and *waaP* in the modification of *Escherichia coli* lipopolysaccharide and their role in the formation of a stable outer membrane. *J. Biol. Chem.* **273**:26310–26316.

Zaslaver, A., A. Bren, M. Ronen, S. Itzkovitz, I. Kikoin, S. Shavit, W. Liebermeister, M. G. Surette, and U. Alon. 2006. A comprehensive library of fluorescent transcriptional reporters for *Escherichia coli*. *Nat. Methods* **3**:623–628.

Zhang, J. R., I. Idanpaan-Heikkila, W. Fischer, and E. I. Tuomanen. 1999. Pneumococcal *licD2* gene is involved in phosphorylcholine metabolism. *Mol. Microbiol.* **31**:1477–1488.

Zhao, J., and C. R. Raetz. 2010. A two-component Kdo hydrolase in the inner membrane of *Francisella novicida*. *Mol. Microbiol.* **78**:820–836.

Zhou, Z., S. Lin, R. J. Cotter, and C. R. Raetz. 1999. Lipid A modifications characteristic of *Salmonella typhimurium* are induced by NH4VO3 in *Escherichia coli* K12. Detection of 4-amino-4-deoxy-L-arabinose, phosphoethanolamine and palmitate. *J. Biol. Chem.* **274**:18503–18514.

Zhou, Z., A. A. Ribeiro, S. Lin, R. J. Cotter, S. I. Miller, and C. R. Raetz. 2001. Lipid A modifications in polymyxin-resistant *Salmonella typhimurium*: PMRA-dependent 4-amino-4-deoxy-L-arabinose, and phosphoethanolamine incorporation. *J. Biol. Chem.* **276**:43111–43121.

III. TOXINS AND ASSOCIATED VIRULENCE FACTOR PRODUCTION

Regulation of Bacterial Virulence
Edited by Michael L. Vasil and Andrew J. Darwin
© 2013 ASM Press, Washington, DC   doi:10.1128/9781555818524.ch12

Chapter 12

Toxin and Virulence Regulation in *Vibrio cholerae*

KAREN SKORUPSKI AND RONALD K. TAYLOR

OVERVIEW OF CHOLERA

Cholera is an acute intestinal infection caused by the bacterium *Vibrio cholerae*. It is estimated that cholera causes the deaths of more than 120,000 people worldwide every year (Sanchez and Holmgren, 2005). *V. cholerae* is a natural inhabitant of aquatic environments, where it is frequently found attached to the surface of marine organisms, and it is transmitted to humans by oral ingestion of contaminated water or food. It has a short incubation period (several hours to several days) and produces an enterotoxin (cholera toxin [CT]) that causes a copious diarrhea that can quickly lead to severe dehydration and death (reviewed in Kaper et al., 1995). CT is a bipartite toxin composed of a single active A subunit and five identical B subunits that bind the toxin to the intestinal epithelial cell surface receptor ganglioside GM_1 (reviewed in Sanchez and Holmgren, 2008). Once inside epithelial cells, the proteolytically cleaved A_1 subunit catalyzes the transfer of an ADP-ribose moiety to a regulatory G protein, resulting in the activation of adenylate cyclase and a subsequent increase in the intracellular levels of cyclic AMP (cAMP). This leads to massive secretion of chloride and water into the lumen of the intestine, which can result in loss of up to 30 liters of fluid per day. The most commonly used treatment for cholera is either oral or intravenous rehydration therapy.

Of the more than 200 different serogroups of *V. cholerae* that have been isolated, only two of these, O1 and O139, have been found to have epidemic and pandemic potential. The O1 serogroup is further divided into two biotypes, classical and El Tor, on the basis of their biochemical properties and phage susceptibilities (Kaper et al., 1995). The classical biotype was responsible for the first six pandemics, whereas the El Tor biotype is responsible for the seventh ongoing pandemic (Faruque et al., 1998). With respect to virulence, the disease caused by the classical biotype has a higher case fatality rate, whereas the El Tor biotype produces more asymptomatic carriers and is considered to be better adapted to survival in the environment. These properties are largely due to differences in the expression of important regulatory factors between the two biotypes (see below).

THE *V. CHOLERAE* VIRULENCE CASCADE

Strains of *V. cholerae* capable of causing the significant epidemics and pandemics of cholera that have occurred throughout history possess two genetic elements, the lysogenic bacteriophage CTXΦ (Waldor and Mekalanos, 1996) and the *Vibrio* pathogenicity island (VPI) (Karaolis et al., 1998), both of which were acquired by horizontal gene transfer (Fig. 1). CTXΦ carries the genes for CT (*ctxA* and *ctxB*) and the VPI contains the genes (*tcpA-E* and *tcpJ*) responsible for the synthesis and assembly of the essential colonization factor toxin-coregulated pilus (TCP) (Taylor et al., 1987; Herrington et al., 1988; Kirn et al., 2003). TCP is a type IV pilus that is assembled as a polymer of repeating subunits of TcpA pilin and forms long filaments that laterally associate into bundles capable of forming supertwisted structures (Craig et al., 2003; Jude and Taylor, 2011). The primary role of TCP in colonization is its ability to induce microcolony formation on the intestinal epithelium by promoting interactions between the bacteria (Kirn et al., 2000). TcpF is a secreted factor that is coexpressed with TcpA and which also has an essential role in colonization, but its mechanism of action remains unknown (Kirn et al., 2003; Megli et al., 2011). Other genes present on the VPI include the accessory colonization factors (*acfA* to -*D*) and *tagA*, *aldA*, and *tcpI*, which play poorly defined roles in intestinal colonization (Peterson and Mekalanos, 1988; Chaparro et al., 2010; Szabady et al., 2011) as well as several transcriptional regulators (see below).

The expression of the two primary virulence factors of *V. cholerae* O1, CT and TCP, is coordinately

Karen Skorupski and Ronald K. Taylor • Department of Microbiology and Immunology, Dartmouth Medical School, Hanover, NH 03755.

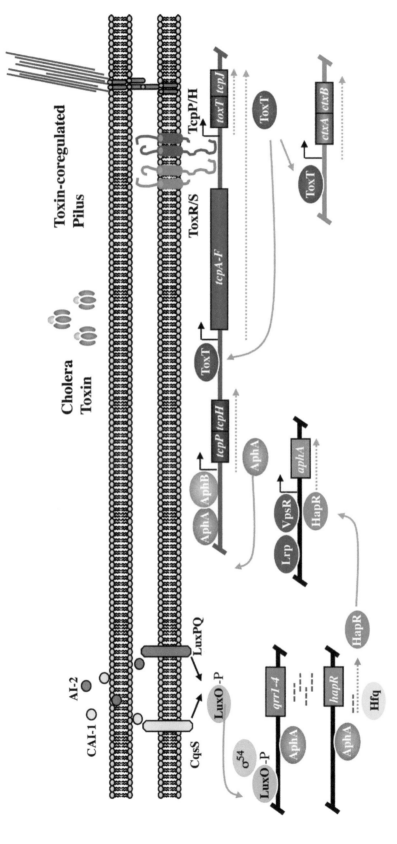

Figure 1. The *Vibrio cholerae* virulence cascade. The VPI is shown by a blue line, CTXΦ by a red line, and the ancestral genome by black lines. At low cell density, in the absence of CAI-1 and AI-2, CqsS and LuxPQ phosphorylate LuxO, which, together with σ⁵⁴, activates the expression of *qrr1* to *qrr4*. The resulting small RNAs, together with Hfq, destabilize the *hapR* message, leading to activation of *aphA* expression by Lrp and VpsR. AphA cooperates with AphB to activate the *tcpPH* promoter. TcpPH cooperates with ToxRS to activate the *toxT* promoter. ToxT then activates the *tcpA* and *ctx* promoters. At low cell density, AphA also represses the expression of *qrr2* to -4 and *hapR*. At high cell density, in the presence of CAI-1 and AI-2, LuxO is not phosphorylated and *qrr1* to *qrr4* are not expressed. This allows for the accumulation of HapR, which binds to a site in the *aphA* promoter overlapping the VpsR binding site, repressing its expression and turning off the virulence cascade.
doi:10.1128/9781555818524.ch12f1

regulated at the transcriptional level by a virulence cascade involving a number of activator proteins encoded both within the VPI and the ancestral genome (Fig. 1). The function of this regulation is to ensure that virulence factor expression occurs only under the appropriate conditions, such as when *V. cholerae* is in its proper niche within the host. ToxT, encoded on the VPI, is a member of the AraC-type regulator family that directly activates the expression of CT and TCP, as well as many of the other genes on the VPI (DiRita et al., 1991; Higgins et al., 1992; Champion et al., 1997; Hulbert and Taylor, 2002; Withey and DiRita, 2005a, 2005b). The activation of *toxT* expression requires cooperation between two homologous pairs of transmembrane proteins encoded by the *toxRS* operon in the ancestral genome (Miller et al., 1987; Miller et al., 1989; Higgins and DiRita, 1994; Krukonis et al., 2000) and the *tcpPH* operon contained on the VPI (Carroll et al., 1997; Häse and Mekalanos, 1998). ToxR and TcpP possess cytoplasmically localized DNA binding/transcription activation domains and periplasmic domains of unknown function. ToxS and TcpH are comprised predominately of a periplasmic domain and appear to interact with ToxR and TcpP, respectively (DiRita and Mekalanos, 1991; Häse and Mekalanos, 1998). The activation of the *tcpPH* operon by the regulatory proteins AphA and AphB, encoded in separate locations within the ancestral genome, initiates the expression of the virulence cascade (Skorupski and Taylor, 1999; Kovacikova and Skorupski, 1999). AphA is a winged-helix transcription factor (De Silva et al., 2005) that facilitates the binding of the LysR-type regulator

AphB to the promoter (Kovacikova and Skorupski, 2001, 2002a; Kovacikova et al., 2004).

The evolution of *V. cholerae* into a virulent organism appears to have occurred by independent horizontal gene transfer events that facilitated acquisition of the VPI and the CTX phage. Once the VPI was acquired, the elaboration of TCP, which serves as the high-affinity receptor for the CTX phage (Waldor and Mekalanos, 1996), allowed the CT genes to be acquired. The promoters within these elements then evolved to become regulated by the various transcriptional activators encoded within both the VPI and the ancestral genome to coordinate virulence gene expression and integrate it into the existing regulatory circuits of the cell.

INFLUENCE OF ENVIRONMENTAL STIMULI ON VIRULENCE GENE EXPRESSION

In order to optimize survival both inside and outside of the host, bacteria use multiple signal transduction pathways to respond to changes in their environment and regulate the expression of a wide variety of different genes, including those that play a role in virulence. In *V. cholerae*, the expression of the virulence cascade is influenced by a variety of environmental stimuli such as cell density, cyclic di-GMP (c-di-GMP) levels, pH, temperature, osmolarity, oxygen tension, nutrients, bile, fatty acids, bicarbonate, and phosphate (Miller et al., 2002; Tischler and Camilli, 2005; Miller and Mekalanos, 1988; Kovacikova et al., 2010; Chatterjee et al., 2007; Abuaita and Withey, 2009; Pratt et al., 2010) (Table 1). The mechanisms

Table 1. Environmental factors influencing virulence gene expression[a]

Signal	Factor/process	Reference(s)
Bicarbonate	ToxT activation of *tcpA*, *ctx*	Abuaita and Withey, 2009
Bile	ToxT activation of *tcpA*, *ctx*	Schuhmacher and Klose, 1999; Prouty et al., 2005
c-di-GMP	VieA activation of *toxT*	Tischler and Camilli, 2005
	CdgD activation of *frhA*	Syed et al., 2009
Cell density	HapR repression of *aphA*	Kovacikova and Skorupski, 2002b
	qrr activation of AphA	Rutherford et al., 2011
	HapR activation of T6SS genes	Zheng et al., 2010
	HapR activation of *hapA*	Jobling and Holmes, 1997
	HapA/PrtV proteolysis of GbpA	Jude et al., 2009
Fatty acids	ToxT activation of *tcpA*, *ctx*	Chatterjee et al., 2007; Lowden et al., 2010
Nutrients	cAMP-CRP repression of *tcpA*, *ctx*	Kovacikova and Skorupski, 2001; Nielsen et al., 2010
	Lrp activation of *aphA*	Lin et al., 2007
Oxygen	AphB activation of *tcpPH*	Kovacikova et al., 2010
pH	AphB activation of *tcpPH*	Kovacikova et al., 2010
	H-NS repression of *tcpA*, *ctx*	Nye et al., 2000
	YaeL proteolysis of TcpP	Matson and DiRita, 2005
Phosphate	PhoB repression of *tcpPH*	Pratt et al., 2010
Temperature	H-NS repression of *tcpA*, *ctx*	Nye et al., 2000
	YaeL proteolysis of TcpP	Matson and DiRita, 2005

[a]This table highlights the various signals and factors/processes discussed in this chapter. It is not inclusive of every signal or factor/process that has been reported to influence the expression of virulence genes in *V. cholerae*.

by which these various stimuli regulate virulence gene expression are of fundamental importance to pathogenesis, and they operate at different levels within the cascade, either directly influencing the activator proteins or functioning through other processes that control the expression or activities of the virulence activators, such as the involvement of global regulatory proteins, regulatory RNAs, and proteolysis (see below).

Quorum Sensing

Virulence gene expression in many pathogenic bacteria is influenced by cell density through quorum sensing systems. These are population-dependent regulatory systems that use extracellular signaling molecules referred to as autoinducers to control a wide range of processes (reviewed in Ng and Bassler, 2009). Such regulation is widespread among bacterial pathogens and allows the pathogenic response to occur only at the appropriate cell density. Although many quorum sensing systems function to activate virulence gene expression at high cell density, for *V. cholerae*, quorum sensing reduces virulence gene expression at high cell density (Miller et al., 2002; Zhu et al., 2002).

V. cholerae has at least two quorum sensing systems that feed into the response regulator LuxO (Fig.1). System 1 is comprised of the CqsA-dependent autoinducer CAI-1 and its cognate sensor CqsS, and system 2 is comprised of the LuxS-dependent autoinducer AI-2 and its cognate sensor LuxPQ (Miller et al., 2002). At low cell density, when the extracellular concentration of autoinducers is low, membrane-bound sensor kinases initiate a phosphorylation cascade, which results in activation of the response regulator LuxO. Phospho-LuxO, together with the alternative sigma factor σ^{54}, then activates the expression of four genes encoding small RNAs (*qrr1* to -4), which, in conjunction with the RNA binding protein Hfq, prevent expression of the LuxR homolog master regulator HapR by destabilizing its message (Lenz et al., 2004). At high cell density, when the extracellular concentration of autoinducers accumulates in the supernatant, the membrane-bound sensor kinases function as phosphatases, which render LuxO in an inactive, dephosphorylated state. This allows for an accumulation of HapR which, in turn, downregulates the virulence cascade by binding to a site centered at –71 in the *aphA* promoter and repressing its expression (Kovacikova and Skorupski, 2002b) (also see below). This reduces the intracellular levels of AphA and prevents activation of the *tcpPH* promoter. HapR also regulates other important processes in *V. cholerae* such as biofilm formation (Hammer

and Bassler, 2003) and production of the proteases HapA and PrtV (Jobling and Holmes, 1997; Zhu and Mekalanos, 2003) (also see below).

The HapR quorum sensing circuit is not functional in all toxigenic strains of *V. cholerae*, since a number of classical and El Tor biotype strains have a naturally occurring frameshift mutation in *hapR* that prevents HapR production regardless of the cell density. In these strains, AphA levels remain high and there is no decrease in virulence gene expression at high cell density (Kovacikova and Skorupski, 2002b; Zhu et al., 2002). Since these strains are still capable of causing severe disease, a functional quorum sensing system does not appear to be an absolute requirement for virulence in *V. cholerae*.

c-di-GMP Signaling and Biofilms

The intracellular second messenger molecule c-di-GMP is used by most bacteria to control the switch between motile, planktonic and sessile, biofilm-related lifestyles (reviewed in Hengge, 2009). c-di-GMP is produced from GTP by the activity of diguanylate cyclases (Ryjenkov et al., 2005), is degraded by phosphodiesterases (Schmidt et al., 2005; Ryan et al., 2006), and is sensed by c-di-GMP receptor proteins (Ryjenkov et al., 2006). The domains of proteins responsible for these activities are referred to as GGDEF, EAL or HD-GYP, and PilZ, respectively. *V. cholerae* encodes 62 proteins with domains predicted to be involved in influencing c-di-GMP levels (Galperin et al., 2001). Consistent with what is known in other bacteria, high levels of c-di-GMP in *V. cholerae* induce biofilm formation and repress motility and virulence factor expression, whereas low levels of c-di-GMP repress biofilm formation and induce motility and virulence factor expression (Tischler and Camilli, 2005; Tamayo et al., 2008; Liu et al., 2010). Thus, c-di-GMP is proposed to play a role in the transition of *V. cholerae* from the gastrointestinal tract of the host (low c-di-GMP) to the outside aquatic environment (high c-di-GMP), which is important for survival in both of these niches. The mechanisms by which it accomplishes this are not yet clearly understood (also see below).

Biofilm formation plays an important role in the survival of *V. cholerae* in the aquatic environment and for disease transmission (Zhu and Mekalanos, 2003). *V. cholerae* forms biofilms on the surfaces of chitinous organisms such as copepods and zooplankton, and this appears to contribute to the persistence of *V. cholerae* in environmental reservoirs between epidemics (Tamplin et al., 1990). This is due to the fact that bacteria within a biofilm are more resistant to chemical stresses (Yildiz and Schoolnik, 1999) and are efficient at forming conditionally viable environmental cells which are capable of long-term survival in aquatic environments (Kamruzzaman et al., 2010). Biofilm formation also helps *V. cholerae* resist the acidic environment of the stomach upon infection (Zhu and Mekalanos, 2003). Although strains defective for biofilm formation are still capable of causing severe disease, biofilm formation has been shown to enhance the infectivity of *V. cholerae* by forming multicellular clumps of bacteria shed in human stools that are physiologically proficient in inducing disease (Faruque et al., 2006; Tamayo et al., 2010).

Biofilm formation depends upon the expression of two linked operons, *vps*I and *vps*II, that encode proteins required for the production of exopolysaccharide (Yildiz and Schoolnik, 1999). The expression of the *vps* genes is controlled by two primary transcriptional activators: VpsT, a member of the LuxR family of transcriptional regulators, and VpsR, a member of the NtrC subclass of response regulators (Casper-Lindley and Yildiz, 2004; Yildiz et al., 2001). The induction of *vpsT* expression by high levels of c-di-GMP depends upon VpsR, and its repression at high cell density is due to the quorum sensing regulator HapR; both proteins recognize an overlapping binding site in the *vpsT* promoter (Hammer and Bassler, 2003; Waters et al., 2008; Srivastava et al., 2011). VpsT and VpsR directly bind c-di-GMP, resulting in induction of *vps* gene expression (Krasteva et al., 2010; Srivastava et al., 2011). The production of VPS appears to contribute to the in vivo fitness of *V. cholerae* (Fong et al., 2010).

VIRULENCE CASCADE ACTIVATOR PROTEINS

AphA

AphA is a member of a large family of at least 300 regulatory proteins with mostly unknown functions that show homology to PadR, a repressor that controls the detoxification of phenolic acids (Barthelmebs et al., 2000). AphA is a dimer with an N-terminal winged-helix DNA binding domain that is architecturally similar to that of the MarR family of transcriptional regulators (Alekshun et al., 2001) and a distinctive C-terminal dimerization domain comprising an extensive antiparallel coiled coil (De Silva et al., 2005). AphA activates the transcription of *tcpPH* by an unusual mechanism that appears to require a direct interaction with the LysR-type regulator AphB (Kovacikova et al., 2004). Upon binding of AphA to a site centered at –86 in the *tcpPH* promoter, the protein enhances the binding of AphB, the

primary transcriptional activator, to an adjacent and proximal site centered at –60 in the promoter (Fig. 1) (Kovacikova and Skorupski, 2001). Under the appropriate environmental conditions, this results in activation of the *tcpPH* promoter.

Although AphA activates the expression of the *tcpPH* promoter on the VPI, the *aphA* gene is located in the ancestral genome adjacent to genes with roles in cellular physiology. This suggested that AphA has other functions in *V. cholerae*, possibly regulating physiological processes. One such role for AphA is as a repressor of an operon encoding enzymes involved in the biosynthesis of acetoin (Kovacikova et al., 2005). The metabolic function of this pathway is to counteract lethal acidification when cells are grown in the presence of excess glucose by redirecting pyruvic acid into neutral rather than acidic end products. As part of the acetoin operon, AphA represses the expression of two PhoB-activated genes, *acgA* and *acgB*, that encode proteins which influence motility and biofilm formation by altering c-di-GMP levels in the cell (Kovacikova et al., 2005; Pratt et al., 2009). AphA also represses the expression of a gene encoding penicillin amidase (Kovacikova et al., 2003) and enhances biofilm formation by activating the expression of the biofilm regulator VpsT (Yang et al., 2010). Although specific signals influencing the activity of AphA have not yet been identified, a number of environmental signals influence the expression of *aphA* through other regulatory proteins. These include nutrient availability, cell density, and c-di-GMP through the regulators Lrp, HapR, and VpsR, respectively (see below). AphA also regulates its own expression (Lin et al., 2007).

The expression of *aphA* is activated by the leucine-responsive regulatory protein (Lrp) and modulated by the quorum sensing repressor HapR and the biofilm transcriptional activator VpsR, which compete for access to an overlapping binding site (Fig. 1) (Lin et al., 2007). Lrp is a small DNA binding protein that has a global role in regulating genes involved in amino acid metabolism and related processes in *Escherichia coli* (Brinkman et al., 2003). Metabolic cues, such as a decrease in nutrient availability, appear to stimulate transcriptional activation of genes positively regulated by Lrp (Calvo and Matthews, 1994). Lrp activates *aphA* expression by binding directly to a near-consensus site centered at –130 in the promoter (Lin et al., 2007). The molecular mechanisms involved in the transcriptional activation by Lrp proteins are not well understood. Since Lrp is known to significantly bend DNA and induce conformational changes in the promoter, Lrp may induce a bend in the *aphA* promoter that facilitates its interaction with RNA polymerase.

HapR is a member of the TetR family of transcriptional regulators (De Silva et al., 2007) and functions as both a transcriptional activator and repressor in *V. cholerae* (Jobling and Holmes, 1997; Hammer and Bassler, 2003; Lin et al., 2005). The location of the HapR binding site at –71 in the *aphA* promoter is somewhat unusual for a transcriptional repressor and suggested that HapR might function by a different mechanism than blocking access of RNA polymerase (Kovacikova and Skorupski, 2002b). It appears that the mechanism by which HapR represses the expression of the *aphA* promoter involves its ability to antagonize the functions of two different activators simultaneously (i) by interfering with Lrp-mediated activation initiated by binding of the protein to the site at –130 and (ii) by directly blocking access of VpsR to the promoter by competition for an overlapping binding site centered at –80 (Lin et al., 2007). In *V. cholerae*, the expression of *vpsR* is influenced by c-di-GMP levels (Tischler and Camilli, 2004), and VpsR also imparts this regulation on the expression of *aphA* (Srivastava et al., 2011). Although HapR and VpsR compete for access to the *aphA* promoter, HapR has a stronger affinity for it than does VpsR (Lin et al., 2007).

The above results suggest a model for the regulation of *aphA* expression by environmental signals (Lin et al., 2007). In response to a reduction in nutrients, such as the situation that might be encountered in the mucous lining of the small intestine, Lrp binding at –130 induces a conformational change in the promoter that allows the protein to interact with RNA polymerase and stimulate transcription. This activation is moderated by AphA binding to a site at –20, which likely interferes with RNA polymerase binding. In response to other environmental signals, such as cell density and c-di-GMP, VpsR and HapR compete for access to overlapping binding sites to induce opposing effects on the expression of *aphA*. At low cell density when levels of HapR are low, VpsR binding enhances Lrp-mediated activation of the promoter. As the cell density increases, HapR outcompetes VpsR for access to the promoter and also antagonizes Lrp-mediated activation. The regulation of the *aphA* promoter by Lrp, VpsR, HapR, and AphA appears to allow the integration of multiple physiological signals into the expression of *aphA* and its target genes.

AphA has recently been shown to be part of a reciprocal quorum sensing loop in *V. cholerae* due to its ability to repress the expression of the small RNA genes *qrr2*, *qrr3*, and *qrr4* as well as *hapR* at low cell density (Fig. 1) (Rutherford et al., 2011). The *qrr* small RNAs also activate the translation of AphA at low cell density (Rutherford et al., 2011). Once the cells have transitioned to high cell density, HapR represses the expression of *aphA*, which, in turn, derepresses the expression of *qrr2*, *qrr3*, and *qrr4* as well as *hapR*, further increasing HapR levels. This reciprocal quorum sensing loop thus ensures that at low cell density AphA levels are high and HapR levels are low, whereas at high cell density HapR levels are high and AphA levels are low.

AphB

AphB is a member of the LysR family of transcriptional regulators. LysR-type transcriptional regulators (LTTRs) comprise one of the largest families of transcriptional regulators in prokaryotes, and its members regulate an extremely diverse set of genes whose products are involved in a variety of biological processes such as metabolism, nitrogen fixation, oxidative stress responses, quorum sensing, and virulence (Maddocks and Oyston, 2008). Like other members of the LTTR family, AphB is a tetramer comprised of two distinct subunit conformations (Taylor et al., 2012). However, unlike most known LTTRs, AphB must undergo a significant conformational change in order to bind to DNA (also see below) (Taylor et al., 2012). At the *tcpPH* promoter, AphB plays the primary role in transcriptional activation, whereas AphA plays an accessory role by facilitating its interaction with the promoter from an upstream site (Fig. 1) (Kovacikova and Skorupski, 2001; Kovacikova et al., 2004). Although LTTRs typically require the binding of a metabolically important small molecule coinducer to a pocket region in the C-terminal regulatory domain of the protein in order to stimulate transcription, it is not yet known whether AphB requires a coinducer to activate *tcpPH* expression.

Similar to the situation with AphA, the *aphB* gene is not located on the VPI, nor is it located in the vicinity of *aphA*. Thus, in addition to activating the expression of *tcpPH*, it was anticipated that AphB has additional functions in *V. cholerae*. It has recently been shown that AphB directly activates the expression of a number of genes in *V. cholerae* that have roles in pH homeostasis (Kovacikova et al., 2010). One of these is *cadC*, which encodes a member of the ToxR-like family of transmembrane transcriptional regulators that directly activates the expression of the *cadBA* operon, encoding a lysine/cadaverine antiporter and lysine decarboxylase, respectively (Merrell and Camilli, 2000). Lysine decarboxylase plays an important role in an adaptive acid tolerance response in *V. cholerae* (Merrell and Camilli, 1999), and AphB plays a crucial role in this response by virtue of its ability to activate *cadC* expression (Kovacikova et al., 2010). Specific functions have not yet been defined for the other genes directly activated by AphB. These include a member of the poorly characterized Gpr1/Fun34/YaaH family, predicted to be transmembrane proteins; NhaB, which is a member of a group of Na^+/H^+ antiporters with roles in pH homeostasis and NaCl tolerance (Herz et al., 2003); carbonic anhydrase, which catalyzes the interconversion of carbon dioxide and bicarbonate and plays an important role in acid survival in *Helicobacter pylori* due to its ability to buffer the periplasm (Marcus et al., 2005); and ClC, which comprises a family of integral membrane proteins whose major action is to translocate chloride ions across cell membranes (Dutzler, 2006). This protein has previously been shown to confer mild resistance of *V. cholerae* to acid pH (Ding and Waldor, 2003).

The identification of additional AphB activated promoters in *V. cholerae* has begun to elucidate the specific signals that influence the virulence cascade through this regulator. The expression of all of the promoters activated by AphB was stimulated under aerobic conditions by lowering the pH and was stimulated at neutral pH by reducing the levels of oxygen (Kovacikova et al., 2010). These results indicate that all of the promoters directly activated by AphB are responsive to two different environmental signals that are encountered by *V. cholerae* during the early stages of the infection process: low pH and anaerobiosis. Precisely how this occurs is not yet known. Part of this response appears to involve an oxygen-dependent modification of a cysteine at position 227 that reduces the ability of AphB to activate transcription under aerobic conditions (Liu et al., 2011). Mutations have also been isolated in the putative ligand binding pocket region of the regulatory domain that allow AphB to constitutively activate *tcpPH* expression at high pH and in the presence of oxygen (Taylor et al., 2012). The crystal structure of one of these constitutive mutants reveals a pathway of conformational changes in the protein relative to the wild type, possibly mimicking a ligand-bound state, that reorients the DNA binding domain such that it becomes able to interact with the major groove of DNA (Taylor et al., 2012).

In vitro growth conditions differentially influence the expression of the virulence cascade in the two disease-causing biotypes of *V. cholerae*, classical and El Tor, and this is largely due to differences in the ability of AphB to activate the expression of *tcpPH* (Kovacikova and Skorupski, 2000). In the classical biotype, maximal virulence induction in vitro requires aerobic growth in LB medium, pH 6.5, at 30°C (Miller and Mekalanos, 1988), whereas the El Tor biotype requires a bicarbonate-containing medium (referred to as AKI) in which the cells are first incubated statically for 3.5 h before they are shifted to vigorous aeration (Iwanaga et al., 1986). The

molecular basis for this differential response of the virulence cascade to environmental signals is a single nucleotide change in the binding site for AphB (centered at –60) within the El Tor *tcpPH* promoter that disrupts the dyad symmetry of the site and reduces its affinity for AphB (Kovacikova and Skorupski, 2000, 2002a). This results in decreased expression of *tcpPH* and the rest of the virulence cascade in the El Tor biotype relative to the classical. The observation that low oxygen levels enhance the ability of AphB to activate the expression of its regulated genes (Kovacikova et al., 2010) may explain why the AKI method has been successful in inducing the expression of the virulence cascade in the El Tor biotype. The initial phase, in which the cells grow statically for 3.5 h, lowers the levels of oxygen, thereby increasing the expression of *tcpPH* sufficiently to induce the expression of the virulence cascade. Once induced, high-level expression of TCP and CT is obtained by vigorous aeration.

ToxR and TcpP

ToxR and TcpP are members of a group of unusual transcriptional activators that are localized to the inner membrane (Miller et al., 1987; Häse and Mekalanos, 1998). These proteins have N-terminal cytoplasmic DNA binding domains with a winged helix-turn-helix motif homologous to the OmpR/PhoB family of transcriptional activators (Martinez-Hackert and Stock, 1997), a single transmembrane domain, and C-terminal periplasmic domains. In contrast to typical two-component response regulators like OmpR and PhoB, in which signal transduction in response to environmental stimuli is accompanied by phosphorylation reactions, ToxR and TcpP do not contain phosphoacceptor domains and are not activated by phosphorylation. Whether these proteins directly sense environmental signals is not yet clear.

ToxR and TcpP cooperate to activate the expression of *toxT* (Higgins and DiRita, 1994; Häse and Mekalanos, 1998; Krukonis et al., 2000). TcpP appears to play the primary role in transcriptional activation and has a binding site centered around –40 in the *toxT* promoter, whereas ToxR plays an accessory role and has a more distal binding site centered around –80 (Fig. 1) (Krukonis et al., 2000). The working models for transcriptional activation of *toxT* involve ToxR binding to the promoter, directly interacting with the wing domain of TcpP, and recruiting TcpP to its binding site in the promoter to interact with RNA polymerase (Krukonis and DiRita, 2003; Goss et al., 2010). Since ToxR appears to be produced under a wide variety of environmental conditions (DiRita et al., 1996), whereas

TcpP is transcriptionally regulated in response to cell density, pH, and oxygen through AphA and AphB (Kovacikova and Skorupski, 2002b; Kovacikova et al., 2010), it is thought that ToxR constitutively binds the *toxT* promoter but activation does not occur until TcpP is produced (Krukonis et al., 2000).

In addition to its transcriptional control, the levels of TcpP in the cell are also regulated by proteolysis (Beck et al., 2004; Matson and DiRita, 2005). ToxR and TcpP both require the presence of membrane-bound effector proteins, ToxS, and TcpH, respectively, which are important for their functions (DiRita and Mekalanos 1991; Häse and Mekalanos, 1998). The exact role of ToxS is not clear, but it may enhance the dimerization of ToxR (DiRita and Mekalanos, 1991; Dziejman and Mekalanos, 1994). The role of TcpH appears to be to influence the stability of TcpP by protecting the periplasmic domain from degradation (Beck et al., 2004). Since loss of ToxS does not influence the stability of ToxR (Beck et al., 2004), the functions of the membrane-bound effector proteins are not equivalent. In the absence of TcpH, TcpP is degraded by two sequentially functioning proteases (Matson and DiRita, 2005). This process, referred to as regulated intramembrane proteolysis, occurs in both prokaryotes and eukaryotes (Makinoshima and Glickman, 2006). The initial proteolytic event is catalyzed by a currently unidentified protease, resulting in a TcpP species that is further degraded by the inner membrane-localized zinc metalloprotease YaeL (Matson and DiRita, 2005). The degradation of TcpP appears to be a control mechanism for rapidly shutting down virulence gene production under unfavorable conditions, such as when cultures are shifted to LB medium, pH 8.5, at 37°C (Matson and DiRita, 2005).

Independent of TcpP, ToxR also regulates the expression of other genes in the ancestral genome of *V. cholerae* (Bina et al., 2003). ToxR directly activates the expression of *ompU*, encoding an outer membrane porin (Crawford et al., 1998), and represses the expression of the alternative porin, OmpT (Li et al., 2000). At the *ompU* promoter, ToxR activates expression apparently without the need for coactivators other than ToxS and RNA polymerase (Crawford et al., 1998). At the *ompT* promoter, ToxR represses expression by binding to a region that interferes with CRP-mediated activation (Li et al., 2002). ToxR-mediated regulation of outer membrane porin expression is important for resistance to organic acids, bile and other related detergents, and antimicrobial peptides and for intestinal colonization in mice (Merrell et al., 2001; Provenzano and Klose, 2000; Mathur and Waldor, 2004). The role of OmpU in antimicrobial peptide resistance involves its ability to activate the σ^E-mediated periplasmic stress response (Mathur et al., 2007).

The localization of ToxR to the inner membrane is not an absolute requirement for it to function in gene regulation. Soluble forms of the cytoplasmic winged-helix domain of ToxR are capable of regulating the expression of *ompU* and *ompT*, whereas membrane localization of the winged helix is required for TcpP-dependent activation of *toxT* expression (Crawford et al., 2003). These results suggest that the major function of membrane localization of ToxR is to facilitate interaction with membrane-bound TcpP. Residues important for ToxR to activate *toxT* expression have been localized to the winged-helix domain, reflecting its role in promoter binding and possibly interaction with TcpP (Morgan et al., 2011). In contrast, residues important for ToxR to activate *ompU* have been localized to the α-loop between the two helices of the helix-turn-helix domain, which likely interacts with RNA polymerase (Morgan et al., 2011).

ToxT

ToxT is a member of the AraC/XylS family of transcriptional regulators, which are characterized by a conserved region of approximately 100 amino acids that comprises the DNA binding domain (Gallegos et al., 1997). Family members are involved in a variety of cellular processes such as carbon metabolism, stress responses, and virulence, and an increasing number of the virulence-associated members have been found to directly sense environmental effector molecules (Yang et al., 2011). The full-length crystal structure of ToxT has recently revealed that the protein is a monomer with N- and C-terminal domains resembling those of AraC (Lowden et al., 2010). The C-terminal domain contains two helix-turn-helix motifs with the DNA recognition helices oriented in a nonparallel manner, and the N-terminal domain contains eight antiparallel β-sheets, which form a pocket for the 16-carbon fatty acid effector *cis*-palmitoleate (see below).

ToxT is responsible for activating the transcription of most of the promoters within the VPI, as well as the *ctx* promoter on the lysogenic CTX phage, thus underscoring its importance to virulence in *V. cholerae* (Fig. 1) (DiRita et al., 1991; Higgins et al., 1992; Champion et al., 1997; Hulbert and Taylor, 2002; Withey and DiRita, 2005a, 2005b). ToxT regulates its own expression through activation of the *tcpA* promoter (Brown and Taylor, 1995; Yu and DiRita, 1999) and also activates the expression of two small regulatory RNAs, TarA and TarB (Richard et al., 2010; Bradley et al., 2011). TarA, located upstream of the ToxT-activated gene *tcpI*, influences glucose metabolism by negatively regulating a glucose transporter in the phosphotransferase system and plays a role in the in vivo fitness of *V. cholerae* (Richard et al., 2010). TarB, also located on the VPI, negatively regulates expression of the secreted colonization factor TcpF (Bradley et al., 2011). ToxT also contributes to fitness during infection by repressing the expression of the anticolonization pilus, mannose-sensitive hemagglutinin (Hsiao et al., 2006; Hsiao et al., 2009).

ToxT binds to DNA by recognizing degenerate 13-bp sequences in its regulated promoters referred to as toxboxes (Withey and DiRita, 2006). As toxboxes can be present either singly or as a pair of direct or inverted repeats, ToxT exhibits a large degree of flexibility in the way in which it activates transcription at its regulated promoters (Withey and DiRita, 2006; Bellair and Withey, 2008). At the *tcpA* and *ctx* promoters, the two toxboxes are positioned as direct repeats, whereas between the divergent *acfA* and *acfD* genes, the two toxboxes are organized as an inverted repeat (Withey and DiRita, 2005a). Although ToxT appears to bind independently as a monomer to each toxbox, full activation at these promoters apparently requires interactions between the monomers on adjacent toxboxes (Withey and DiRita, 2006). In contrast, at single toxboxes, such as the one upstream of *aldA*, a ToxT monomer activates alone (Withey and DiRita, 2005b). A synthetic inhibitor of ToxT, virstatin, prevents activation at many of the promoters where ToxT binds to adjacent toxboxes, suggesting that it interferes with interactions between ToxT monomers (Shakhnovich et al., 2007). The identification of small molecules that inhibit the activity of ToxT is a strategy for developing potential therapeutics for cholera.

Bile is a mixture of many molecules, including saturated fatty acids, unsaturated fatty acids, salts, and cholesterol, that is secreted into the intestine from the gallbladder (Chatterjee et al., 2007). Many enteric bacteria recognize bile as a signal that they have entered a human host, and it influences the expression of genes involved in resistance to its detergent properties as well as virulence (Gunn, 2000). In the case of *V. cholerae*, bile increases the expression of genes involved in processes that provide protection from it, such as motility, biofilm formation, efflux pumps, and the outer membrane protein OmpU, whereas bile decreases the expression of virulence factors and the outer membrane protein OmpT (Gupta and Chowdhury, 1997; Schuhmacher and Klose, 1999; Provenzano et al., 2000; Hung et al., 2006; Cerda-Maira et al., 2008). It is thought that bile present in the lumen of the intestine prevents premature expression of virulence genes by negatively regulating their expression until the bacterium penetrates the mucous

gel of the epithelium, where the bile concentration is sufficiently reduced (Schuhmacher and Klose, 1999).

The activity of ToxT is inhibited by bile and by unsaturated fatty acids, which are a component of bile (Schuhmacher and Klose, 1999; Chatterjee et al., 2007). The crystal structure of ToxT has revealed the presence of an almost completely buried and solvent inaccessible 16-carbon fatty acid, *cis*-palmitoleate, bound to the pocket at the interface between the N- and C-terminal domains (Lowden et al., 2010). In the presence of either *cis*-palmitoleate or oleate, the expression of the *tcp* and *ctx* operons is significantly reduced and ToxT is unable to bind to DNA (Lowden et al., 2010). Since oleate is present in a higher concentration in bile than *cis*-palmitoleate, oleate may be the natural ligand which occupies this pocket and inhibits ToxT in vivo. Residues that render ToxT insensitive to bile, fatty acids, and virstatin (see above) have been localized to the N-terminal domain of the protein and appear to be important for interactions between ToxT monomers (Prouty et al., 2005; Childers et al., 2011). A model for the regulation of ToxT function via fatty acid binding involves the head group of the fatty acid acting as a bridge between the N- and C-terminal domains, keeping the protein in a closed conformation that is not capable of binding to DNA (Lowden et al., 2010). Once the bacteria have penetrated the mucus of the intestine, where the concentrations of fatty acids are reduced, the fatty acid is removed from ToxT and it is able to form an open conformation that is competent for DNA binding and dimerization at adjacent toxboxes.

In contrast to bile, sodium bicarbonate appears to stimulate the activity of ToxT (Abuaita and Withey, 2009). Bicarbonate is secreted into the lumen of the small intestine to neutralize the acid that comes from the stomach. Thus, bicarbonate may be an important in vivo signal that increases the activity of ToxT during infection and induces virulence gene expression. Since bile and bicarbonate have competing effects on the activity of ToxT, this may serve to allow the proper spatiotemporal pattern of virulence gene expression within the host environment.

The activity of ToxT is also regulated by proteolysis. ToxT is rapidly degraded when *V. cholerae* is grown under environmental conditions unfavorable for virulence gene expression (pH 8.5 and 37°C) (Abuaita and Withey, 2011). This appears to be a control mechanism for rapidly shutting down virulence gene production prior to entry back into the aquatic environment after infection. Although the proteases responsible for this degradation are not known, an unstructured motif in ToxT between amino acids 100 and 109 is required (Abuaita and Withey, 2011).

MODULATORS OF VIRULENCE

H-NS

H-NS is an abundant, nucleid-associated global regulator that influences a variety of cellular processes, including virulence gene expression in many pathogens (reviewed in Dorman, 2007, and Fang and Rimsky, 2008). H-NS recognizes DNA with relatively low sequence specificity, preferentially binding to intrinsically curved, AT-rich DNA often associated with horizontally acquired elements. H-NS interacts with extensive regions on DNA and silences transcription by organizing promoter and regulatory regions into nucleoprotein complexes in response to environmental signals. Upon binding to high-affinity nucleation sites, H-NS polymerizes along DNA using sites of lower affinity to allow the formation of higher-order structures. Genes whose expression is influenced by H-NS are typically responsive to environmental parameters known to influence DNA topology, such as osmolarity, temperature, anaerobiosis, pH, and growth phase (Atlung and Ingmer, 1997).

H-NS directly represses expression from at least three promoters within the virulence cascade: $toxT$, ctx, and $tcpA$ (Nye et al., 2000; Stonehouse et al., 2008; Stonehouse et al., 2011). In the absence of H-NS, expression from each of these promoters was increased dramatically under virulence-repressing conditions (LB medium, pH 8.5 at 37°C) and, to a lesser degree, under virulence-inducing conditions (LB medium, pH 6.5 at 30°C), even in the absence of their cognate activator proteins (Nye et al., 2000). These results indicate that H-NS plays a significant role in repressing virulence gene expression under environmental conditions not normally permissive for their expression and that it represses expression even under inducing conditions. H-NS appears to exert its largest repressive effect on the $toxT$ promoter by influencing an extensive region from –172 to beyond –256 with respect to the start of transcription (Nye et al., 2000). H-NS also has a strong effect on the expression of the ctx promoter (Nye et al., 2000), where it binds to a region that overlaps the high-affinity ToxT binding site (–111 to –41; Yu and DiRita, 2002), the –35 element, and two regions downstream of the +1 site (Stonehouse et al., 2011). These results suggest that H-NS blocks transcription initiation by interfering with ToxT and RNA polymerase binding and activation. At the $tcpA$ promoter, the effect of H-NS is more moderate and the protein binds to a region that overlaps the ToxT binding site (–84 to –41; Yu and DiRita, 2002) as well as an upstream integration host factor (IHF) binding site (see below)

(Stonehouse et al., 2008). The overlapping binding sites for H-NS and ToxT at the ctx and $tcpA$ promoters suggest that ToxT functions as an antirepressor by displacing H-NS in order to bind as well as a direct activator that recruits RNA polymerase (Yu and DiRita, 2002; Stonehouse et al., 2008; Stonehouse et al., 2011).

The above results suggest a model in which the ctx, $tcpA$, and $toxT$ promoters are repressed by H-NS when $V.\ cholerae$ is in a nonintestinal environment. Once $V.\ cholerae$ has infected its host and virulence gene expression is induced by particular stimuli in the intestinal environment, ToxT, and possibly, ToxR/TcpP, functions at its respective promoter as antirepressor to alleviate the effects of H-NS in addition to functioning as a direct activator of transcription. Once $V.\ cholerae$ has disseminated out of the host and virulence gene expression is no longer required, TcpP and ToxT levels are decreased by proteolysis (Matson and DiRita, 2005; Abuaita and Withey, 2011) and H-NS reestablishes repression at the various promoters.

IHF

IHF is another nucleoid-associated global regulator that plays an important role in DNA bending and compaction and is important for the transcriptional regulation of many genes (Goosen and van de Putte, 1995). It is composed of two distinct subunits, IHFα and IHFβ, which form a heterodimer. Unlike H-NS, IHF recognizes a specific asymmetric site on DNA and typically functions as an architectural factor that leads to the correct promoter structure necessary for transcriptional activation. IHF also functions in antirepression of H-NS repressed promoters (van Ulsen et al., 1996).

In contrast to H-NS, IHF positively influences $V.\ cholerae$ virulence gene expression (Stonehouse et al., 2008). IHF was found to be required for expression of both the $tcpA$ and ctx promoters but not the $toxT$ promoter. IHF binds to the $tcpA$ promoter at a site centered at –162 that is upstream of the ToxT binding site and within the region protected from DNase I digestion by H-NS. In addition, the protein was found to bend the DNA 120°. Since IHF does not appear to bind to the ctx promoter, this suggests that the effect of IHF on ctx is indirect due to its influence on $toxT$ expression driven from the $tcpA$ promoter. The finding that activation by IHF appears to occur only when both H-NS and ToxT are present suggests that IHF functions to counteract H-NS repression, allowing for greater accessibility of ToxT to its binding site.

cAMP Receptor Protein

The cAMP receptor protein (cAMP-CRP) system functions as a global regulatory network that controls gene expression in response to carbon and energy sources in the environment (reviewed in Kolb et al., 1993). The intracellular levels of cAMP in enteric bacteria are regulated by the phosphoenolpyruvae:carbohydrate phosphotransferase system (PTS), a transport system that catalyzes the uptake and phosphorylation of numerous carbohydrates (reviewed in Deutscher et al., 2006). In the absence of PTS carbon sources, cAMP levels increase and the binding of the nucleotide to CRP results in a conformational change in the protein that induces sequence-specific binding to DNA. CRP functions as both a positive and negative regulator of gene expression and influences many different cellular processes. In *V. cholerae*, cAMP-CRP has been shown to influence the expression of more than 20% of its genome, with approximately equal numbers of genes positively and negatively regulated (Fong and Yildiz, 2008).

The cAMP-CRP system negatively regulates the expression of the virulence cascade in *V. cholerae*. A Δ*crp* mutation has been shown to cause significant derepression of TCP and CT expression under laboratory growth conditions not normally permissive for their expression (Skorupski and Taylor, 1997). This effect has largely been attributed to the presence of a near-consensus cAMP-CRP binding site in the *tcpPH* promoter that overlaps the binding sites for AphA and AphB (Kovacikova and Skorupski, 2001). The influence of a second binding site for cAMP-CRP that overlaps the −35 sequence of the *tcpA* promoter and which could compete for ToxT binding is less clear (Skorupski and Taylor, 1997; Kovacikova and Skorupski, 2001). The involvement of cAMP-CRP in the regulation of CT and TCP expression in *V. cholerae* suggests that carbon and energy sources in the environment influence virulence gene expression. The downregulation of virulence gene expression when cAMP levels are high may serve as a mechanism to limit the expression of virulence determinants in low-nutrient environments outside of the host and favor their expression in nutrient-rich environments such as in the lumen of the intestine.

VieSAB

The VieSAB locus encodes a three-component signal transduction system that is involved in regulating biofilm formation, motility, and virulence in the classical biotype of *V. cholerae* (Tischler et al., 2002; Tischler and Camilli, 2004, 2005; Martinez-Wilson et al., 2008). This system differs from conventional two-component signal transduction systems in that it encodes two response regulators, VieA and VieB. VieS is an inner membrane-localized hybrid sensor kinase that targets VieA for phosphorylation and controls its expression (Martinez-Wilson et al., 2008). VieA appears to be a typical response regulator with a helix-turn-helix domain for DNA binding, whereas VieB is unusual in that it contains a phosphoreceiver domain but lacks a recognizable DNA binding motif.

VieA has been shown to influence the levels of c-di-GMP in *V. cholerae* (Tischler and Camilli, 2004; Tamayo et al., 2005). By degrading c-di-GMP via its EAL-dependent phosphodiesterase activity, VieA promotes motility and virulence gene expression while repressing biofilm formation (Tischler and Camilli, 2004, 2005; Martinez-Wilson et al., 2008). Transcriptional profiling experiments have indicated that VieA is a major regulator of gene expression under classical biotype virulence gene-inducing conditions, influencing approximately 10% of the genes, whereas VieA does not significantly contribute to the regulation of gene expression in the El Tor biotype (Beyhan et al., 2006). The roles of VieB in *V. cholerae* have not yet been elucidated.

With respect to virulence, the c-di-GMP phosphodiesterase activity of VieA is required for maximal expression of *ctxAB* in classical biotype strains during growth under in vitro conditions, and this appears to be due to its ability to influence expression from the *toxT* promoter (Tischler and Camilli, 2005). Thus, VieA lowers the intracellular concentration of c-di-GMP, allowing for maximal expression of *toxT*, which, in turn, activates *ctxAB* transcription. Surprisingly, despite the reduction in *toxT* expression that is observed in the *vieA* mutant, the expression of *tcpA* does not appear to be affected (Tischler and Camilli, 2005). The mechanisms involved in this regulation are not yet understood. The phosphodiesterase activity of VieA is also necessary for classical biotype strains to colonize in the infant mouse model (Tischler and Camilli, 2005). By promoting motility and virulence gene expression and repressing biofilm formation, VieA may influence the transition from environmental reservoirs to the human host. Although VieA does not influence virulence in the El Tor biotype, artificially increasing the concentration of c-di-GMP reduces expression from the *toxT* promoter and decreases colonization by this biotype (Tamayo et al., 2008).

PhoB

PhoB is a response regulator that is activated by the histidine kinase PhoR under conditions of phosphate limitation and which regulates the transcription of a large set of genes involved in phosphate homeostasis known as the Pho regulon (reviewed in Hsieh and Wanner, 2010). The activation of PhoB is prevented by the high-affinity inorganic phosphate transporter Pst under conditions of high environmental phosphate by an unknown mechanism. The Pst system is composed of five components encoded within the *pstSCAB-phoU* operon. Null mutations in any of the *pst* genes disrupt regulation of PhoB activation and lead to constitutive expression of the Pho regulon independent of the environmental phosphate availability (Rao and Torriani, 1990).

Inactivation of the Pst system has been shown to decrease the expression of the *V. cholerae* virulence cascade and influence colonization in the infant mouse model (Pratt et al., 2010). Strains carrying a deletion of the *pst* operon exhibit decreased expression of CT and TCP, and this appears to be due to the ability of PhoB to repress expression of the *tcpPH* promoter. The *tcpPH* promoter contains a binding site for PhoB that is located downstream of the binding site for AphB in the region between −41 and +60 relative to the transcriptional start. However, since *V. cholerae* strains lacking Pst, PhoB, Pst and PhoB, or PhoR are all severely attenuated for colonization in the infant mouse model, these findings indicate that this regulatory system also plays other important roles during the infection process in addition to influencing the expression of *tcpPH* (Pratt et al., 2010). These roles appear to be dependent upon the expression of other genes in the Pho regulon and are important for the survival of *V. cholerae* when it is in the small intestine as well as when it is in a freshwater environment.

One operon under the control of PhoB, *acgAB*, has previously been shown to be negatively regulated by AphA from a promoter upstream of the acetoin operon and encodes proteins that influence motility and biofilm formation by altering c-di-GMP levels in the cell (Kovacikova et al., 2005; Pratt et al., 2009). PhoB positively regulates motility under low-phosphate conditions by indirectly activating the expression of *acgAB*, thus linking regulation by phosphate to c-di-GMP signaling (Pratt et al., 2009). A model for regulation by PhoB involves its activation during the late stages of infection when the concentration of exogenous phosphate becomes limiting where it functions to shut down virulence gene expression by repressing the expression of *tcpPH* and induces the expression of genes that play a role in survival in the aquatic environment upon dissemination (Pratt et al., 2009).

TsrA

TsrA is a global regulator that represses the expression of the primary virulence factors in *V.*

cholerae as well as a type VI secretion system (T6SS) (Zheng et al., 2010). TsrA is conserved in *Vibrio* species and shows a weak similarity to the N-terminal sequence of H-NS. In LB medium, a normally repressive condition for virulence factor expression in the El Tor biotype strain C6706, deletion of *tsrA* increased expression from both the *tcpA* and *ctx* promoters, as well as that of *toxT*, but did not influence either *tcpPH* or *toxRS* (Zheng et al., 2010). These results suggest that it may function along with H-NS at the *toxT* promoter, possibly interfering with the ability of the transmembrane transcriptional activators TcpP and ToxR to activate transcription. In the infant mouse, deletion of *tsrA* increased intestinal colonization almost 10-fold, indicating that TsrA is still capable of repressing virulence factor expression even in the host. Since TsrA does not appear to have a DNA binding domain, it may require another transcriptional activator to control its target genes.

In the El Tor biotype strain C6706, repression of the T6SS requires both TsrA and quorum sensing (Zheng et al., 2010). As mentioned earlier, LuxO is a response regulator that activates the expression of four small RNAs (*qrr1* to -4), which prevents expression of HapR at low cell density by destabilizing its mRNA (Lenz et al., 2004). Disruption of *luxO* stabilizes HapR message and HapR, in turn, induces expression of the T6SS genes *hcp*1 and *hcp*2 (hemolysin-coregulated proteins) (Zheng et al., 2010). However, full induction of these genes also requires disruption of *tsrA*. Once activated, the T6SS system is functional in this strain background and can translocate the effector protein VgrG-1 into host cells, resulting in diarrhea in the infant rabbit infection model. TsrA also plays a role in activating the expression of another quorum sensing-regulated factor, HapA (also known as HA/protease), which also requires HapR and appears to function as a detachase by releasing *V. cholerae* from the surface of intestinal cells (Finkelstein et al., 1992; Jobling and Holmes, 1997). These findings suggest that during the late stages of infection when *V. cholerae* has reached high cell densities, HapR simultaneously represses virulence gene expression and activates the expression of the T6SS and HapA protease, promoting detachment from the intestinal wall and contributing to the exit of *V. cholerae* from the host (also see below).

GbpA

GbpA is a secreted protein required for efficient intestinal colonization of *V. cholerae* (Kirn et al., 2005). GbpA mediates attachment of *V. cholerae* both to host epithelial cells and to the surface of zooplankton present in the environment by binding to a sugar residue, *N*-acetylglucosamine (GlcNAc), present on both surfaces (Kirn et al., 2005). GlcNAc is a common modification of glycoproteins and lipids present on the intestinal epithelium (Finne et al., 1989) and is the building block of the abundant aquatic carbon source chitin (Meibom et al., 2004). In the host environment, GbpA enhances colonization of *V. cholerae* in the gastrointestinal tract, whereas in the outside environment, GbpA facilitates colonization of the chitinous surfaces of copepods such as *Daphnia* spp., leading to improved survival (Kirn et al., 2005). GpbA thus plays an important role in *V. cholerae* in both the host and the aquatic environment.

The activity of GbpA is controlled by quorum sensing via a process involving the metalloproteases HapA and PrtV (Jude et al., 2009). HapA cleaves fibronectin, lactoferrin, and mucin and appears to function in the release of *V. cholerae* from the surface of intestinal cells (Finkelstein et al., 1983; Finkelstein et al., 1992). PrtV cleaves hemolysin A (also known as cytolysin) and has been shown to be essential for both *V. cholerae* killing of *Caenorhabditis elegans* and protecting *V. cholerae* from predator grazing by various flagellates (Ou et al., 2009; Vaitkevicius et al., 2006). The *hapA* and *prtV* genes are activated at high cell density by the quorum sensing regulator HapR (Jobling and Holmes, 1997; Zhu and Mekalanos, 2003). GbpA is present in cells at low cell density but not at high cell density in the presence of functional HapR, and this appears to be due to degradation by HapA and PrtV (Jude et al., 2009). The protease-mediated degradation of GbpA by HapA and PrtV may serve as a mechanism to facilitate release of *V. cholerae* from the surface of the intestinal epithelium once it has achieved high cell density and expedite its return to the aquatic environment. Although the presence of a functional quorum sensing system is not essential for the process of intestinal colonization, it appears to be important for enhancing the transition of *V. cholerae* from the host to the aquatic environment.

FrhA

FrhA (flagellum-regulated hemagglutinin A) is a novel adhesin that facilitates *V. cholerae* binding to human erythrocytes and epithelial cells and enhances intestinal colonization in the infant mouse (Syed et al., 2009). FrhA also plays a role in attachment to chitin and abiotic surfaces during biofilm development. Thus, like GbpA, FrhA plays an important role in *V. cholerae* in both the host and in the aquatic environment. FrhA is in the RTX (repeats-in-toxins) family of proteins (Satchell, 2011). It is a large protein containing four cadherin-like domains in its central

region. Cadherins comprise a family of eukaryotic transmembrane proteins that have distinctive repeat sequences in their extracellular regions that function in calcium-dependent cell-cell adhesion (Pokutta and Weis, 2007). The expression of FrhA is activated by the flagellar regulatory hierarchy (see below) through its effects on c-di-GMP via the diguanylate cyclase CdgD (Syed et al., 2009). It appears that FrhA is utilized to initiate colonization or attachment to abiotic surfaces when the bacteria are motile upon arrival at a surface structure.

Motility and Chemotaxis

V. cholerae has a single, polar flagellum and is highly motile. Flagellar motility facilitates attachment and subsequent colonization of *V. cholerae* on the intestinal epithelium, and in the infant mouse model it appears to play a more important role for strains of the El Tor biotype (Lee et al., 2001; Butler and Camilli, 2005; Liu et al., 2008; Martinez et al., 2009). Flagellar motility appears to be important for *V. cholerae* to penetrate through the mucosal layer of the epithelial cell surface, and nonmotile mutants colonize less efficiently than their motile counterparts because they do not make sufficient contact with the epithelium (Liu et al., 2008; Martinez et al., 2009). Nonmotile mutants have also been shown to express increased levels of the virulence factors TCP and CT, indicating that motility and expression of virulence genes are inversely regulated (Gardel and Mekalanos, 1996; Silva et al., 2006). A model for *V. cholerae* infection involves motile bacteria attaching to the intestinal cell surface, after which they upregulate virulence factor expression and downregulate motility.

The assembly of flagella is a highly ordered and regulated process. There are six chromosomal regions in the *V. cholerae* genome containing flagellum-specific genes, and their expression is regulated by a four-tiered transcriptional hierarchy (Prouty et al., 2001). The class I gene product FlrA is a master regulator that activates σ^{54}-dependent transcription of class II genes, which encode flagellar structural proteins as well as the two-component system FlrBC (Klose & Mekalanos, 1998; Correa et al., 2000). Phosphorylated FlrC activates σ^{54}-dependent transcription of class III genes, which encode structural proteins as well as the lipoprotein FlgP, which, in addition to its role in motility, is involved in epithelial cell attachment and enhances intestinal colonization (Morris et al., 2008; Martinez et al., 2009). Lastly, FliA is an alternate sigma factor (σ^{28}) that activates transcription of class IV genes, which encode additional structural components and the anti-sigma factor FlgM. By binding to FliA, FlgM prevents activation of σ^{28}-dependent genes until it is secreted through the flagellum (Correa et al., 2004).

Transcriptome analyses in the absence of the flagellar regulators σ^{54}, FlrA, FlrC, and FliA have demonstrated an upregulation of virulence gene expression, consistent with the inverse relationship between these two processes (Syed et al., 2009). The upregulation of *ctx*, *tcp*, and *toxT*, but not *tcpP* or *toxR*, suggests that the flagellar regulatory system influences expression from the *toxT* promoter. Since the genes encoding VieSAB are also upregulated in the flagellar regulatory mutants (Syed et al., 2009), this raises the possibility that the increased expression of *ctx*, *tcp*, and *toxT* is due to upregulation of *vieA* transcription. The flagellar regulatory system thus functions as an important signaling component of pathogenesis by positively regulating factors that promote arrival (motility) and attachment (FrhA; see above) to colonization sites while negatively regulating the expression of factors needed after the bacteria have attached (virulence).

The flagellar regulatory hierarchy also influences virulence gene expression in *V. cholerae* through the quorum sensing system. Passage of motile *V. cholerae* organisms through the mucous layer of the epithelium leads to breakage of the flagellum, which allows secretion of the anti-σ^{28} factor FlgM through the damaged flagellum (Correa et al., 2004; Liu et al., 2008). The decreased intracellular levels of FlgM increase the activity of FliA and induce the expression of σ^{28}-dependent genes. This, in turn, leads to decreased expression of the quorum sensing master regulator HapR (it is not yet known if this regulation is direct or indirect) and induces virulence gene expression by alleviating HapR-mediated repression of the cascade (Liu et al., 2008). These findings suggest that *V. cholerae* uses the flagellar machinery to enhance virulence gene expression by also influencing the quorum sensing system.

V. cholerae and other marine organisms utilize a sodium-driven motor to power their flagella instead of a proton-driven motor as in *Escherichia coli* and *Salmonella* (McCarter, 2001). Sodium-driven motors also require the Na$^+$ translocating NADH ubiquinone oxidoreductase (NQR), an enzyme involved in energy production, which drives the sodium motive force required to power the motor (Häse & Barquera, 2001). The enzyme is composed of six subunits encoded within an operon. Loss of NQR enzyme function by either specific inhibitors or mutations in the *nqr* gene cluster results in elevated expression from the *toxT* promoter (Häse and Mekalanos, 1999). Although the mechanisms responsible for this increase in virulence gene expression are not yet understood, they

appear to be related to the ability of *V. cholerae* to sense changes in the membrane energy levels and are independent of the flagellum (Häse, 2001).

Bacteria use chemotaxis to sense and respond to their surrounding environment by controlling the rotation of the flagella (reviewed in Porter et al., 2011). Chemoreceptors detect changes in favorable substances (attractants) and unfavorable ones (repellents) and signal through two-component systems to the flagellar motor to control the direction that they swim. This allows the bacteria to find better environments for growth. Chemotaxis has been shown to play a role in the virulence of *V. cholerae*. Nonchemotactic mutants of *V. cholerae* have a competitive advantage during infection of infant mice and outcompete the wild-type strain 70-fold in vivo (Lee et al., 2001). Part of the explanation for this is that nonchemotactic *V. cholerae* mutants show aberrant distribution within the infant mouse small intestine. Whereas wild-type strains colonize primarily in the distal part of the small intestine, nonchemotactic mutants are more evenly distributed throughout the small intestine, allowing a greater surface area to be colonized (Lee et al., 2001). *V. cholerae* thus appears to use chemotaxis to avoid colonizing the upper portion of the small intestine, possibly due to an unknown chemoattractant or chemorepellent gradient (Butler and Camilli, 2005).

The ability of nonchemotactic mutants to outcompete wild-type strains during infection is not actually due to a defect in chemotaxis per se but is due to an alteration in the bias of flagellar rotation, which influences colonization (Butler and Camilli, 2004). Strains with a counterclockwise-biased flagellar rotation exhibit enhanced colonization, whereas strains biased in the clockwise direction are actually attenuated 10-fold for infection (Butler and Camilli, 2004). Since clockwise-biased flagellar rotation mutants change the direction in which they swim extremely frequently, they are unable to make net progress in any single direction and as a result are attenuated for infection. In contrast, counterclockwise-biased mutants swim straight for relatively long periods, allowing them to colonize better by improving their contact with the mucous layer and villi.

Chemotaxis also appears to influence the infectivity of *V. cholerae* after passage through the host. *V. cholerae* organisms shed in rice-water stools from humans or recovered from the intestines of infant mice have been shown to be 10- to 100-fold more infectious than *V. cholerae* grown in vitro (Merrell et al., 2002; Alam et al., 2005). This hyperinfectivity appears to be at least partly due to a transient reduction in chemotaxis (Butler et al., 2006) and persisted for 5 h in pond water, but it was abolished by subsequent growth in vitro. Thus, passage of *V. cholerae* through the host imparts a transient competitive advantage on the organism by enhancing its transmissibility.

REGULATION OF VIRULENCE GENE EXPRESSION DURING INFECTION

Most of the studies on the regulation of virulence gene expression in *V. cholerae* have been carried out during growth under in vitro conditions. Within the host, *V. cholerae* encounters a variety of different microenvironments, each with its own set of signals that have the potential to influence virulence gene expression. In order to gain an understanding of how virulence genes are regulated during the infectious process, recombination-based in vivo expression technology (RIVET) was developed (Lee et al., 1999). Using RIVET, a number of differences were observed with regard to the roles of TcpP, ToxR, and TcpA in the expression of the *tcpA* and *ctx* promoters during infection of the infant mouse compared to in vitro conditions (Lee et al., 1999). In addition, the expression of *tcpA* was induced before *ctx*, suggesting that colonization of the intestinal epithelium by *V. cholerae* may produce a signal that induces the expression of *ctx*. A similar temporal pattern of expression has recently been shown for *tcpA* and *ctx* during in vitro growth in AKI medium (Kanjilal et al., 2010). The temporally distinct expression of *tcpA* and *ctx* in both of these studies may be a reflection of repression by the global regulator H-NS, since it has been shown that the *ctx* promoter is more strongly repressed by H-NS than *tcpA* (Yu and DiRita, 2002; Stonehouse et al., 2011).

The expression of *tcpA* and *ctx* has also been examined in the rabbit ileal loop model of cholera (Nielsen et al., 2010). Site-specific expression profiling revealed that the *tcpA* and *ctx* promoters are strongly upregulated early in the infectious cycle when *V. cholerae* is on the epithelial cell surface and in the overlying mucus compared to when it is in the luminal fluid. In addition, single-cell analyses using a destabilized variant of the green fluorescent protein that allows a real-time assessment of gene expression showed that the magnitude of *tcpA* expression is greater the closer *V. cholerae* is to the epithelial surface and paralleled its growth rate. These findings suggest the presence of some type of signal associated with the epithelial surface, but whether this signal is diffusible (such as a nutrient, fatty acid, or bicarbonate) or requires physical contact of the bacterium with either the mucus or cell surface is not yet clear.

The single cell gene expression studies of Nielsen et al. also revealed that during the transition into early stationary phase, the population of cells expressing *tcpA*, *ctx*, and *toxT* bifurcates into two nearly equal fractions (Nielsen et al., 2010). In one fraction, expression levels persist for at least 4 h, whereas in the other fraction, expression levels decline to preinduction levels. This bifurcation phenotype exhibits characteristics of a bistable switch and was shown to depend upon the ToxT autoregulatory circuit, since all of the cells lacking the ToxT-dependent promoter upstream of *tcpA* showed decreased *tcpA* expression during entry into stationary phase. In addition, the growth phase dependency of the bistable switch depends on the cAMP-CRP system, since none of the *crp* or *cya* mutant cells exhibit decreased *tcpA* expression during entry into stationary phase. It is thought that the bistable control of virulence gene expression could contribute to the transmission of cholera by generating a subpopulation of *V. cholerae* that continues to produce TCP and CT several hours after it has been shed into the environment.

Late in the infection process, the expression of *tcpA* and *ctx* is markedly reduced as bacteria detach from the epithelial cell surface and reenter the mucus as part of the mucosal escape response (Nielsen et al., 2006; Nielsen et al., 2010). This response is controlled by the stationary-phase sigma factor RpoS and is associated with an upregulation of motility and chemotaxis genes as bacteria migrate away from the epithelial surface into the lumen of the intestine and a downregulation of virulence gene expression as they prepare to exit from the host (Nielsen et al., 2006). The effect of RpoS on virulence gene expression may be due to its positive influence on the expression of *hapR* (Yildiz et al., 2004), which downregulates the virulence cascade and upregulates the production of HapA protease, which promotes the detachment of *V. cholerae* from epithelial surfaces (Finkelstein et al., 1992; Jude et al., 2009). The downregulation of virulence gene expression late in infection is consistent with a number of other studies that have shown that transcripts for *tcpA* and *ctx* are in relatively low abundance in human stool specimens (Merrell et al., 2002; Bina et al., 2003; LaRocque et al., 2005). Conversely, a program of gene expression is induced late in infection that facilitates survival in the aquatic environment (Schild et al., 2007).

Recent advances in the development of cDNA sequencing (RNA-seq) have facilitated the generation of comprehensive transcriptome profiles of *V. cholerae* during infection in both the rabbit and mouse models of cholera (Mandlik et al., 2011). Transcripts elevated in these models are derived from known factors involved in the virulence of *V. cholerae*, genes involved in adapting the metabolism of the organism to the host-specific conditions, and small RNAs not previously linked to infection. RNA-seq is a powerful approach that holds promise for further elucidating the complex patterns of bacterial gene expression that occur during infection.

REFERENCES

Abuaita, B. H., and J. H. Withey. 2009. Bicarbonate induces *Vibrio cholerae* virulence gene expression by enhancing ToxT activity. *Infect. Immun.* **77:**4111–4120.

Abuaita, B. H., and J. H. Withey. 2011. Termination of *Vibrio cholerae* virulence gene expression is mediated by proteolysis of the major virulence activator, ToxT. *Mol. Microbiol.* **81:**1640–1653.

Alam, A., R. C. LaRocque, J. B. Harris, C. Vanderspurt, E. T. Ryan, F. Qadri, and S. B. Calderwood. 2005. Hyperinfectivity of human-passaged *Vibrio cholerae* can be modeled by growth in the infant mouse. *Infect. Immun.* **73:**6674–6679.

Alekshun, M. N., S. B. Levy, T. R. Mealy, B. A. Seaton, and J. F. Head. 2001. The crystal structure of MarR, a regulator of multiple antibiotic resistance, at 2.3 Å resolution. *Nat. Struct. Biol.* **8:**710–714.

Atlung, T., and H. Ingmer. 1997. H-NS: a modulator of environmentally regulated gene expression. *Mol. Microbiol.* **24:**7–17.

Barthelmebs, L., B. Lecomte, C. Divies, and J.-F. Cavin. 2000. Inducible metabolism of phenolic acids in *Pediococcus pentosaceus* is encoded by an autoregulated operon which involves a new class of negative transcriptional regulator. *J. Bacteriol.* **182:**6724–6731.

Beck, N. A., E. S. Krukonis, and V. J. DiRita. 2004. TcpH influences virulence gene expression in *Vibrio cholerae* by inhibiting the degradation of the transcription activator TcpP. *J. Bacteriol.* **186:**8309–8316.

Bellair, M., and J. H. Withey. 2008. Flexibility of *Vibrio cholerae* ToxT in transcription activation of genes having altered promoter spacing. *J. Bacteriol.* **190:**7925-7931.

Beyhan, S., A. D. Tischler, A. Camilli, and F. H. Yildiz. 2006. Differences in gene expression between the classical and El Tor biotypes of *Vibrio cholerae* O1. *Infect. Immun.* **74:**3633–3642.

Bina, J., Z. Zhu, M. Dziejman, S. Faruque, S. Calderwood, and J. J. Mekalanos. 2003. ToxR regulon of *Vibrio cholerae* and its expression in vibrios shed by cholera patients. *Proc. Natl. Acad. Sci. USA* **100:**2801–2806.

Bradley, E. S., K. Bodi, A. M. Ismail, and A. Camilli. 2011. A genome-wide approach to discovery of small RNAs involved in regulation of virulence in *Vibrio cholerae*. *PLoS Pathog.* **7:**e1002126.

Brinkman, A. B., T. J. G. Ettema, W. M. de Vos, and J. van der Oost. 2003. The Lrp family of transcriptional regulators. *Mol. Microbiol.* **48:**287–294.

Brown, R. C., and R. K. Taylor. 1995. Organization of *tcp*, *acf*, and *toxT* genes within a ToxT-dependent operon. *Mol. Microbiol.* **16:**425–439.

Butler, S. M., and A. Camilli. 2004. Both chemotaxis and net motility greatly influence the infectivity of *Vibrio cholerae*. *Proc. Natl. Acad. Sci. USA* **101:**5018–5023.

Butler, S. M., and A. Camilli. 2005. Going against the grain: chemotaxis and infection in *Vibrio cholerae*. *Nat. Rev. Microbiol.* **3:**611–620.

Butler, S. M., E. J. Nelson, N. Chowdhury, S. M. Faruque, S. B. Calderwood, and A. Camilli. 2006. Cholera stool bacteria repress chemotaxis to increase infectivity. *Mol. Microbiol.* **60:**417–426.

Calvo, J. M., and R. G. Matthews. 1994. The leucine-responsive regulatory protein, a global regulator of metabolism in *Escherichia coli*. *Microbiol. Rev.* **58:**466–490.

Carroll, P. A., K. T. Tashima, M. B. Rogers, V. J. DiRita, and S. B. Calderwood. 1997. Phase variation in *tcpH* modulates expression of the ToxR regulon in *Vibrio cholerae*. *Mol. Microbiol.* **25:**1099–1111.

Casper-Lindley, C., and F. H. Yildiz. 2004. VpsT is a transcriptional regulator required for expression of *vps* biosynthesis genes and the development of rugose colonial morphology in *Vibrio cholerae* O1 El Tor. *J. Bacteriol.* **186:**1574–1578.

Cerda-Maira, F. A., C. S. Ringelberg, and R. K. Taylor. 2008. The bile response repressor BreR regulates expression of the *Vibrio cholerae breAB* efflux system operon. *J. Bacteriol.* **190:**7441–7452.

Champion, G. A., M. N. Neely, M. A. Brennan, and V. J. DiRita. 1997. A branch in the ToxR regulatory cascade of *Vibrio cholerae* revealed by characterization of *toxT* mutant strains. *Mol. Microbiol.* **23:**323–331.

Chaparro, A. P., S. K. Ali, and K. E. Klose. 2010. The ToxT-dependent methyl-accepting chemoreceptors AcfB and TcpI contribute to *Vibrio cholerae* intestinal colonization. *FEMS Microbiol. Lett.* **302:**99–105.

Chatterjee, A., P. K. Dutta, and R. Chowdhury. 2007. Effect of fatty acids and cholesterol present in bile on expression of virulence factors and motility of *Vibrio cholerae*. *Infect. Immun.* **75:**1946–1953.

Childers, B. M., X. Cao, G. G. Weber, B. Demeler, P. J. Hart, and K. E. Klose. 2011. N-terminal residues of the *Vibrio cholerae* virulence regulatory protein ToxT involved in dimerization and modulation by fatty acids. *J. Biol. Chem.* **286:**28644–28655.

Correa, N. E., J. R. Barker, and K. E. Klose. 2004. The *Vibrio cholerae* FlgM homologue is an anti-s[28] factor that is secreted through the sheathed polar flagellum. *J. Bacteriol.* **186:**4613–4619.

Correa, N. E., C. M. Lauriano, R. McGee, and K. E. Klose. 2000. Phosphorylation of the flagellar regulatory protein FlrC is necessary for *Vibrio cholerae* motility and enhanced colonization. *Mol. Microbiol.* **35:**743–755.

Craig, L., R. K. Taylor, M. E. Pique, B. D. Adair, A. S. Arvai, M. Singh, S. J. Lloyd, D. S. Shin, E. D. Getzoff, M. Yeager, K. T. Forest, and J. A. Tainer. 2003. Type IV pilin structure and assembly: X-ray and EM analyses of *Vibrio cholerae* toxin-coregulated pilus and *Pseudomonas aeruginosa* PAK pilin. *Mol. Cell* **11:**1139–1150.

Crawford, J. A., J. B. Kaper, and V. J. DiRita. 1998. Analysis of ToxR-dependent transcription activation of *ompU*, the gene encoding a major envelope protein in *Vibrio cholerae*. *Mol. Microbiol.* **29:**235–256.

Crawford, J. A., E. S. Krukonis, and V. J. DiRita. 2003. Membrane localization of the ToxR winged-helix domain is required for TcpP-mediated virulence gene activation in *Vibrio cholerae*. *Mol. Microbiol.* **47:**1459–1473.

De Silva, R. S., G. Kovacikova, W. Lin, R. K. Taylor, K. Skorupski, and F. J. Kull. 2005. Crystal structure of the virulence gene activator AphA from *Vibrio cholerae* reveals it is a novel member of the winged helix transcription factor superfamily. *J. Biol. Chem.* **280:**13779–13783.

De Silva, R. S., G. Kovacikova, W. Lin, R. K. Taylor, K. Skorupski, and F. J. Kull. 2007. Crystal structure of the *Vibrio cholerae* quorum-sensing regulatory protein HapR. *J. Bacteriol.* **189:**5683–5691.

Deutscher, J., C. Francke, and P. W. Postma. 2006. How phosphotransferase system-related protein phosphorylation regulates carbon metabolism in bacteria. *Microbiol. Mol. Biol. Rev.* **70:**939–1031.

Ding, Y., and M. K. Waldor. 2003. Deletion of a *Vibrio cholerae* ClC channel results in acid sensitivity and enhanced intestinal colonization. *Infect. Immun.* 71:4197–4200.

DiRita, V. J., and J. J. Mekalanos. 1991. Periplasmic interaction between two membrane regulatory proteins, ToxR and ToxS, results in signal transduction and transcriptional activation. *Cell* 64:29–37.

DiRita, V. J., C. Parsot, G. Jander, and J. J. Mekalanos. 1991. Regulatory cascade controls virulence in *Vibrio cholerae*. *Proc. Natl. Acad. Sci. USA* 88:5403–5407.

DiRita, V. J., M. Neely, R. K. Taylor, and P. M. Bruss. 1996. Differential expression of the ToxR regulon in classical and El Tor biotypes of *V. cholerae* is due to biotype-specific control over *toxT* expression. *Proc. Natl. Acad. Sci. USA* 93:7991–7995.

Dorman, C. J. 2007. H-NS, the genome sentinel. *Nat. Rev. Microbiol.* 5:157–161.

Dutzler, R. 2006. The ClC family of chloride channels and transporters. *Curr. Opin. Struct. Biol.* 16:439–446.

Dziejman, M., and J. J. Mekalanos. 1994. Analysis of membrane protein interaction: ToxR can dimerize the amino terminus of phage lambda repressor. *Mol. Microbiol.* 13:485–494.

Fang, F. C., and S. Rimsky. 2008. New insights into transcriptional regulation by H-NS. *Curr. Opin. Microbiol.* 11:113–120.

Faruque, S. M., M. J. Albert, and J. J. Mekalanos. 1998. Epidemiology, genetics, and ecology of toxigenic *Vibrio cholerae*. *Microbiol. Mol. Biol. Rev.* 62:1301–1314.

Faruque, S. M., K. Biswas, S. M. N. Udden, Q. S. Ahmad, D. A. Sack, G. B. Nair, and J. J. Mekalanos. 2006. Transmissibility of cholera: *in vivo*-formed biofilms and their relationship to infectivity and persistence in the environment. *Proc. Natl. Acad. Sci. USA* 103:6350–6355.

Finkelstein, R. A., M. Boesman-Finkelstein, Y. Chang, and C. C. Häse. 1992. *Vibrio cholerae* hemagglutinin/protease, colonial variation, virulence, and detachment. *Infect. Immun.* 60:472–478.

Finkelstein, R. A., M. Boesman-Finkelstein, and P. Holt. 1983. *Vibrio cholerae* hemagglutinin/lectin/protease hydrolyzes fibronectin and ovomucin: F. M. Burnet revisited. *Proc. Natl. Acad. Sci. USA* 80:1092–1095.

Finne, J., M. E. Breimer, G. C. Hansson, K.-A. Karlsson, H. Leffler, J. F. G. Vliegenthart, and H. van Halbeek. 1989. Novel polyfucosylated N-linked glycopeptides with blood group A, H, X, and Y determinants from human small intestinal epithelial cells. *J. Biol. Chem.* 264:5720–5735.

Fong, J. C. N., and F. H. Yildiz. 2008. Interplay between cyclic AMP-cyclic AMP receptor protein and cyclic di-GMP signaling in *Vibrio cholerae* biofilm formation. *J. Bacteriol.* 190:6646–6659.

Fong, J. C. N., K. A. Syed, K. E. Klose, and F. H. Yildiz. 2010. Role of *Vibrio* polysaccharide (*vps*) genes in VPS production, biofilm formation and *Vibrio cholerae* pathogenesis. *Microbiology* 156:2757–2769.

Gallegos, M.-T., R. Schleif, A. Bairoch, K. Hofmann, and J. L. Ramos. 1997. AraC/XylS family of transcriptional regulators. *Microbiol. Mol. Biol. Rev.* 61:393–410.

Galperin, M. Y., A. N. Nikolskaya, and E. V. Koonin. 2001. Novel domains of the prokaryotic two-component signal transduction systems. *FEMS Microbiol. Lett.* 203:11–21.

Gardel, C. L., and J. J. Mekalanos. 1996. Alterations in *Vibrio cholerae* motility phenotypes correlate with changes in virulence factor expression. *Infect. Immun.* 64:2246–2255.

Goosen, N., and P. van de Putte. 1995. The regulation of transcription initiation by integration host factor. *Mol. Microbiol.* 16:1–7.

Goss, T. J., C. P. Seaborn, M. D. Gray, and E. S. Krukonis. 2010. Identification of the TcpP binding site in the *toxT* promoter of *Vibrio cholerae* and the role of ToxR in TcpP-mediated activation. *Infect. Immun.* 78:4122–4133.

Gunn, J. S. 2000. Mechanisms of bacterial resistance and response to bile. *Microbes Infect.* 2:907–913.

Gupta, S., and R. Chowdhury. 1997. Bile affects production of virulence factors and motility of *Vibrio cholerae*. *Infect. Immun.* 65:1131–1134.

Hammer, B. K., and B. L. Bassler. 2003. Quorum sensing controls biofilm formation in *Vibrio cholerae*. *Mol. Microbiol.* 50:101–114.

Häse, C. C. 2001. Analysis of the role of flagellar activity in virulence gene expression in *Vibrio cholerae*. *Microbiology* 147:831–837.

Häse, C. C., and B. Barquera. 2001. Role of sodium bioenergetics in *Vibrio cholerae*. *Biochem. Biophys. Acta* 1505:169–178.

Häse, C. C., and J. J. Mekalanos. 1998. TcpP protein is a positive regulator of virulence gene expression in *Vibrio cholerae*. *Proc. Natl. Acad. Sci. USA* 95:730–734.

Häse, C. C., and J. J. Mekalanos. 1999. Effects of changes in membrane sodium flux on virulence gene expression in *Vibrio cholerae*. *Proc. Natl. Acad. Sci. USA* 96:3183–3187.

Hengge, R. 2009. Principles of c-di-GMP signalling in bacteria. *Nat. Rev. Microbiol.* 7:263–273.

Herrington, D. A., R. H. Hall, G. Losonsky, J. J. Mekalanos, R. K. Taylor, and M. M. Levine. 1988. Toxin, toxin-coregulated pili, and the *toxR* regulon are essential for *Vibrio cholerae* colonization in humans *J. Exp. Med.* 168:1487–1492.

Herz, K., S. Vimont, E. Padan, and P. Berche. 2003. Roles of NhaA, NhaB, and NhaD Na$^+$/H$^+$ antiporters in survival of *Vibrio cholerae* in a saline environment. *J. Bacteriol.* 185:1236–1244.

Higgins, D. E., and V. J. DiRita. 1994. Transcriptional control of *toxT*, a regulatory gene in the ToxR regulon of *Vibrio cholerae*. *Mol. Microbiol.* 14:17–29.

Higgins, D. E., E. Nazareno, and V. J. DiRita. 1992. The virulence gene activator ToxT from *Vibrio cholerae* is a member of the AraC family of transcriptional activators. *J. Bacteriol.* 174:6974–6980.

Hsiao, A., Z. Liu, A. Joelsson, and J. Zhu. 2006. *Vibrio cholerae* virulence regulator-coordinated evasion of host immunity. *Proc. Natl. Acad. Sci. USA* 103:14542–14547.

Hsiao, A., X. Xu, B. Kan, R. V. Kulkarni, and J. Zhu. 2009. Direct regulation by the *Vibrio cholerae* regulator ToxT to modulate colonization and anticolonization pilus expression. *Infect. Immun.* 77:1383–1388.

Hsieh, Y.-J., and B. L. Wanner. 2010. Global regulation by the seven-component P$_i$ signaling system. *Curr. Opin. Microbiol.* 13:198–203.

Hulbert, R. R., and R. K. Taylor. 2002. Mechanism of ToxT-dependent transcriptional activation at the *Vibrio cholerae* tcpA promoter. *J. Bacteriol.* 184:5533–5544.

Hung, D. T., J. Zhu, D. Sturtevant, and J. J. Mekalanos. 2006. Bile acids stimulate biofilm formation in *Vibrio cholerae*. *Mol. Microbiol.* 59:193–201.

Iwanaga, M., K. Yamamoto, N. Higa, Y. Ichinose, N. Nakasone, and M. Tanabe. 1986. Culture conditions for stimulating cholera toxin production by *Vibrio cholerae* O1 El Tor. *Microbiol. Immunol.* 30:1075–1083.

Jobling, M. G., and R. K. Holmes. 1997. Characterization of *hapR*, a positive regulator of the *Vibrio cholerae* HA/protease gene *hap*, and its identification as a functional homologue of the *Vibrio harveyi luxR* gene. *Mol. Microbiol.* 26:1023–1034.

Jude, B. A., R. M. Martinez, K. Skorupski, and R. K. Taylor. 2009. Levels of the secreted *Vibrio cholerae* attachment factor GbpA are modulated by quorum-sensing-induced proteolysis. *J. Bacteriol.* 191:6911–6917.

Jude, B. A., and R. K. Taylor. 2011. The physical basis of type 4 pilus-mediated microcolony formation by *Vibrio cholerae* O1. *J. Struct. Biol.* **175**:1–9.

Kamruzzaman, M., S. M. N. Udden, D. E. Cameron, S. B. Calderwood, G. B. Nair, J. J. Mekalanos, and S. M. Faruque. 2010. Quorum-regulated biofilms enhance the development of conditionally viable, environmental *Vibrio cholerae*. *Proc. Natl. Acad. Sci. USA* **107**:1588–1593.

Kanjilal, S., R. Citorik, R. C. LaRocque, M. F. Ramoni, and S. B. Calderwood. 2010. A systems biology approach to modeling *Vibrio cholerae* gene expression under virulence-inducing conditions. *J. Bacteriol.* **192**:4300–4310.

Kaper, J. B., J. G. Morris, Jr., and M. M. Levine. 1995. Cholera. *Clin. Microbiol. Rev.* **8**:48–86.

Karaolis, D. K. R., J. A. Johnson, C. C. Bailey, E. C. Boedeker, J. B. Kaper, and P. R. Reeves. 1998. A *Vibrio cholerae* pathogenicity island associated with epidemic and pandemic strains. *Proc. Natl. Acad. Sci. USA* **95**:3134–3139.

Kirn, T. J., N. Bose, and R. K. Taylor. 2003. Secretion of a soluble colonization factor by the TCP type 4 pilus biogenesis pathway for *V. cholerae*. *Mol. Microbiol.* **49**:81–92.

Kirn, T. J., B. A. Jude, and R. K. Taylor. 2005. A colonization factor links *Vibrio cholerae* environmental survival and human infection. *Nature* **438**:863–866.

Kirn, T. J., M. J. Lafferty, C. M. P. Sandoe, and R. K. Taylor. 2000. Delineation of pilin domains required for bacterial association into microcolonies and intestinal colonization by *Vibrio cholerae*. *Mol. Microbiol.* **35**:896–910.

Klose, K. E., and J. J. Mekalanos. 1998. Distinct roles of an alternative sigma factor during both free-swimming and colonizing phases of the *Vibrio cholerae* pathogenic cycle. *Mol. Microbiol.* **28**:501–520.

Kolb, A., S. Busby, H. Buc, S. Garges, and S. Adhya. 1993. Transcriptional regulation by cAMP and its receptor protein. *Annu. Rev. Biochem.* **62**:749–795.

Kovacikova, G., and K. Skorupski. 1999. A *Vibrio cholerae* LysR homolog, AphB, cooperates with AphA at the *tcpPH* promoter to activate expression of the ToxR virulence cascade. *J. Bacteriol.* **181**:4250–4256.

Kovacikova, G., and K. Skorupski. 2000. Differential activation of the *tcpPH* promoter by AphB determines biotype specificity of virulence gene expression in *Vibrio cholerae*. *J. Bacteriol.* **182**:3228–3238.

Kovacikova, G., and K. Skorupski. 2001. Overlapping binding sites for the virulence gene regulators AphA, AphB and cAMP-CRP at the *Vibrio cholerae tcpPH* promoter. *Mol. Microbiol.* **41**:393–407.

Kovacikova, G., and K. Skorupski. 2002a. Binding site requirements of the virulence gene regulator AphB: differential affinities for the *Vibrio cholerae* classical and El Tor *tcpPH* promoters. *Mol. Microbiol.* **44**:533–547.

Kovacikova, G., and K. Skorupski. 2002b. Regulation of virulence gene expression in *Vibrio cholerae* by quorum sensing: HapR functions at the *aphA* promoter. *Mol. Microbiol.* **46**:1135–1147.

Kovacikova, G., W. Lin, and K. Skorupski. 2003. The virulence activator AphA links quorum sensing to pathogenesis and physiology in *Vibrio cholerae* by repressing the expression of a penicillin amidase gene on the small chromosome. *J. Bacteriol.* **185**:4825–4836.

Kovacikova, G., W. Lin, and K. Skorupski. 2004. *Vibrio cholerae* AphA uses a novel mechanism for virulence gene activation that involves interaction with the LysR-type regulator AphB at the *tcpPH* promoter. *Mol. Microbiol.* **53**:129–142.

Kovacikova, G., W. Lin, and K. Skorupski. 2005. Dual regulation of genes involved in acetoin biosynthesis and motility/biofilm formation by the virulence activator AphA and the acetate-responsive LysR-type regulator AlsR in *Vibrio cholerae*. *Mol. Microbiol.* **57**:420–433.

Kovacikova, G., W. Lin, and K. Skorupski. 2010. The LysR-type virulence activator AphB regulates the expression of genes in *Vibrio cholerae* in response to low pH and anaerobiosis. *J. Bacteriol.* **192**:4181–4191.

Krasteva, P. V., J. C. N. Fong, N. J. Shikuma, S. Beyhan, M. V. A. S. Navarro, F. H. Yildiz, and H. Sondermann. 2010. *Vibrio cholerae* VpsT regulates matrix production and motility by directly sensing cyclic di-GMP. *Science* **327**:866–868.

Krukonis, E. S., R. R. Yu, and V. J. DiRita. 2000. The *Vibrio cholerae* ToxR/TcpP/ToxT virulence cascade: distinct roles for two membrane-localized transcriptional activators on a single promoter. *Mol. Microbiol.* **38**:67–84.

Krukonis, E. S., and V. J. DiRita. 2003. DNA binding and ToxR responsiveness by the wing domain of TcpP, an activator of virulence gene expression in *Vibrio cholerae*. *Mol. Cell* **12**:157–165.

LaRocque, R. C., J. B. Harris, M. Dziejman, X. Li, A. I. Khan, A. S. G. Faruque, S. M. Faruque, G. B. Nair, E. T. Ryan, F. Qadri, J. J. Mekalanos, and S. B. Calderwood. 2005. Transcriptional profiling of *Vibrio cholerae* recovered directly from patient specimens during early and late stages of human infection. *Infect. Immun.* **73**:4488–4493.

Lee, S. H., D. L. Hava, M. K. Waldor, and A. Camilli. 1999. Regulation and temporal expression patterns of *Vibrio cholerae* virulence genes during infection. *Cell* **99**:625–634.

Lee, S. H., S. M. Butler, and A. Camilli. 2001. Selection for in vivo regulators of bacterial virulence. *Proc. Natl. Acad. Sci. USA* **98**:6889–6894.

Lenz, D. H., K. C. Mok, B. N. Lilley, R. V. Kulkarni, N. S. Wingreen, and B. L. Bassler. 2004. The small RNA chaperone Hfq and multiple small RNAs control quorum sensing in *Vibrio harveyi* and *Vibrio cholerae*. *Cell* **118**:69–82.

Li, C. C., J. A. Crawford, V. J. DiRita, and J. B. Kaper. 2000. Molecular cloning and transcriptional regulation of *ompT*, a ToxR-repressed gene in *Vibrio cholerae*. *Mol. Microbiol.* **35**:189–203.

Li, C. C., D. S. Merrell, A. Camilli, and J. B. Kaper. 2002. ToxR interferes with CRP-dependent transcriptional activation of *ompT* in *Vibrio cholerae*. *Mol. Microbiol.* **43**:1577–1589.

Lin, W., G. Kovacikova, and K. Skorupski. 2005. Requirements for *Vibrio cholerae* HapR binding and transcriptional repression at the *hapR* promoter are distinct from those at the *aphA* promoter. *J. Bacteriol.* **187**:3013–3019.

Lin, W., G. Kovacikova, and K. Skorupski. 2007. The quorum sensing regulator HapR downregulates the expression of the virulence gene transcription factor AphA in *Vibrio cholerae* by antagonizing Lrp- and VpsR-mediated activation. *Mol. Microbiol.* **64**:953–967.

Liu, X., S. Beyhan, B. Lim, R. G. Linington, and F. H. Yildiz. 2010. Identification and characterization of a phosphodiesterase that inversely regulates motility and biofilm formation in *Vibrio cholerae*. *J. Bacteriol.* **192**:4541–4552.

Liu, Z., T. Miyashiro, A. Tsou, A. Hsiao, M. Goulian, and J. Zhu. 2008. Mucosal penetration primes *Vibrio cholerae* for host colonization by repressing quorum sensing. *Proc. Natl. Acad. Sci. USA* **105**:9769–9774.

Liu, Z., M. Yang, G. L. Peterfreund, A. M. Tsou, N. Selamoglu, F. Daldal, Z. Zhong, B. Kan, and J. Zhu. 2011. *Vibrio cholerae* anaerobic induction of virulence gene expression is controlled by thiol-based switches of virulence regulator AphB. *Proc. Natl. Acad. Sci. USA* **108**:810–815.

Lowden, M. J., K. Skorupski, M. Pellegrini, M. G. Chiorazzo, R. K. Taylor, and J. F. Kull. 2010. Structure of *Vibrio cholerae* ToxT reveals a mechanism for fatty acid regulation of virulence genes. *Proc. Natl. Acad. Sci. USA* **107:**2860–2865.

Maddocks, S. E., and P. C. F. Oyston. 2008. Structure and function of the LysR-type transcriptional regulator (LTTR) family proteins. *Microbiology* **154:**3609–3623.

Makinoshima, H., and M. S. Glickman. 2006. Site-2 proteases in prokaryotes: regulated intramembrane proteolysis expands to microbial pathogenesis. *Microbes Infect.* **8:**1882–1888.

Mandlik, A., J. Livny, W. P. Robins, J. M. Ritchie, J. J. Mekalanos, and M. K. Waldor. 2011. RNA-seq-based monitoring of infection-linked changes in *Vibrio cholerae* gene expression. *Cell Host Microbe* **10:**165–174.

Marcus, E. A., A. P. Moshfegh, G. Sachs, and D. R. Scott. 2005. The periplasmic α-carbonic anhydrase activity of *Helicobacter pylori* is essential for acid acclimation. *J. Bacteriol.* **187:**729–738.

Martinez, R. M., M. N. Dharmasena, T. J. Kirn, and R. K. Taylor. 2009. Characterization of two outer membrane proteins, FlgO and FlgP, that influence *Vibrio cholerae* motility. *J. Bacteriol.* **191:**5669–5679.

Martinez-Hackert, E., and A. M. Stock. 1997. The DNA-binding domain of OmpR: crystal structures of a winged helix transcription factor. *Structure* **5:**109–124.

Martinez-Wilson, H. F., R. Tamayo, A. D. Tischler, D. W. Lazinski, and A. Camilli. 2008. The *Vibrio cholerae* hybrid sensor kinase VieS contributes to motility and biofilm regulation by altering the cyclic diguanylate level. *J. Bacteriol.* **190:**6439–6447.

Mathur, J., and M. K. Waldor. 2004. The *Vibrio cholerae* ToxR-regulated porin OmpU confers resistance to antimicrobial peptides. *Infect. Immun.* **72:**3577–3583.

Mathur, J., B. M. Davis, and M. K. Waldor. 2007. Antimicrobial peptides activate the *Vibrio cholerae* σ^E regulon through an OmpU-dependent signalling pathway. *Mol. Microbiol.* **63:**848–858.

Matson, J. S., and V. J. DiRita. 2005. Degradation of the membrane-localized virulence activator TcpP by the YaeL protease in *Vibrio cholerae*. *Proc. Natl. Acad. Sci. USA* **102:**16403–16408.

McCarter, L. L. 2001. Polar flagellar motility of *Vibrionaceae*. *Microbiol. Mol. Biol. Rev.* **65:**445–462.

Megli, C. J., A. S. W. Yuen, S. Kolappan, M. R. Richardson, M. N. Dharmasena, S. J. Krebs, R. K. Taylor, and L. Craig. 2011. Crystal structure of the *Vibrio cholerae* colonization factor TcpF and identification of a functional immunogenic site. *J. Mol. Biol* **409:**146–158.

Meibom, K. L., X. B. Li, A. T. Nielsen, C.-Y. Wu, S. Roseman, and G. K. Schoolnik. 2004. The *Vibrio cholerae* chitin utilization program. *Proc. Natl. Acad. Sci. USA* **101:**2524–2529.

Merrell, D. S., S. M. Butler, F. Qadri, N. A. Dolganov, A. Alam, M. B. Cohen, S. B. Calderwood, G. K. Schoolnik, and A. Camilli. 2002. Host-induced epidemic spread of the cholera bacterium. *Nature* **417:**642–645.

Merrell, D. S., and A. Camilli. 1999. The *cadA* gene of *Vibrio cholerae* is induced during infection and plays a role in acid tolerance. *Mol. Microbiol.* **34:**836–849.

Merrell, D. S., and A. Camilli. 2000. Regulation of *Vibrio cholerae* genes required for acid tolerance by a member of the "ToxR-like" family of transcriptional regulators. *J. Bacteriol.* **182:**5342–5350.

Merrell, D. S., C. Bailey, J. B. Kaper, and A. Camilli. 2001. The ToxR-mediated organic acid tolerance response of *Vibrio cholerae* requires OmpU. *J. Bacteriol.* **183:**2746–2754.

Miller, M. B., K. Skorupski, D. H. Lenz, R. K. Taylor, and B. L. Bassler. 2002. Parallel quorum sensing systems converge to regulate virulence in *Vibrio cholerae*. *Cell* **110:**303–314.

Miller, V. L., R. K. Taylor, and J. J. Mekalanos. 1987. Cholera toxin transcriptional activator ToxR is a transmembrane DNA binding protein. *Cell* **48:**271–279.

Miller, V. L., and J. J. Mekalanos. 1988. A novel suicide vector and its use in construction of insertion mutations: osmoregulation of outer membrane proteins and virulence determinants in *Vibrio cholerae* requires *toxR*. *J. Bacteriol.* **170:**2575–2583.

Miller, V. L., V. J. DiRita, and J. J. Mekalanos. 1989. Identification of *toxS*, a regulatory gene whose product enhances ToxR-mediated activation of the cholera toxin promoter. *J. Bacteriol.* **171:**1288–1293.

Morgan, S. J., S. Felek, S. Gadwal, N. M. Koropatkin, J. W. Perry, A. B. Bryson, and E. S. Krukonis. 2011. The two faces of ToxR: activator of *ompU*, co-regulator of *toxT* in *Vibrio cholerae*. *Mol. Microbiol.* **81:**113–128.

Morris, D. C., F. Peng, J. R. Barker, and K. E. Klose. 2008. Lipidation of an FlrC-dependent protein is required for enhanced intestinal colonization by *Vibrio cholerae*. *J. Bacteriol.* **190:**231–239.

Ng, W.-L., and B. L. Bassler. 2009. Bacterial quorum-sensing network architectures. *Annu. Rev. Genet.* **43:**197–222.

Nielsen, A. T., N. A. Dolganov, G. Otto, M. C. Miller, C. Y. Wu, and G. K. Schoolnik. 2006. RpoS controls the *Vibrio cholerae* mucosal escape response. *PLoS Pathog.* **2:**e109.

Nielsen, A. T., N. A. Dolganov, T. Rasmussen, G. Otto, M. C. Miller, S. A. Felt, S. Torreilles, and G. K. Schoolnik. 2010. A bistable switch and anatomical site control *Vibrio cholerae* virulence gene expression in the intestine. *PLoS Pathog.* **6:**e1001102.

Nye, M. B., J. D. Pfau, K. Skorupski, and R. K. Taylor. 2000. *Vibrio cholerae* H-NS silences virulence gene expression at multiple steps in the ToxR regulatory cascade. *J. Bacteriol.* **182:**4295–4303.

Ou, G., P. K. Rompikuntal, A. Bitar, B. Lindmark, K. Vaitkevicius, S. N. Wai, and M.-L. Hammarstrom. 2009. *Vibrio cholerae* cytolysin causes an inflammatory response in human intestinal epithelial cells that is modulated by the PrtV protease. *PloS One* **4:**e7806.

Peterson, K. M., and J. J. Mekalanos. 1988. Characterization of the *Vibrio cholerae toxR* regulon: identification of novel genes involved in intestinal colonization. *Infect. Immun.* **56:**2822–2829.

Pokutta, S., and W. I. Weis. 2007. Structure and mechanism of cadherins and catenins in cell-cell contacts. *Annu. Rev. Cell Dev. Biol.* **23:**237–261.

Porter, S. L., G. H. Wadhams, and J. P. Armitage. 2011. Signal processing in complex chemotaxis pathways. *Nat. Rev. Microbiol.* **9:**153–165.

Pratt, J. T., A. M. Ismail, and A. Camilli. 2010. PhoB regulates both environmental and virulence gene expression in *Vibrio cholerae*. *Mol. Microbiol.* **77:**1595–1605.

Pratt, J. T., E. McDonough, and A. Camilli. 2009. PhoB regulates motility, biofilms and cyclic di-GMP in *Vibrio cholerae*. *J. Bacteriol.* **191:**6632–6642.

Prouty, M. G., N. E. Correa, and K. E. Klose. 2001. The novel σ^{54}- and σ^{28}-dependent flagellar gene transcription hierarchy of *Vibrio cholerae*. *Mol. Microbiol.* **39:**1595–1609.

Prouty, M. G., C. R. Osorio, and K. E. Klose. 2005. Characterization of functional domains of the *Vibrio cholerae* virulence regulator ToxT. *Mol. Microbiol.* **58:**1143–1156.

Provenzano, D., and K. E. Klose. 2000. Altered expression of the ToxR-regulated porins OmpU and OmpT diminishes *Vibrio cholerae* bile resistance, virulence factor expression, and intestinal colonization. *Proc. Natl. Acad. Sci. USA* **97:**10220–10224.

Provenzano, D., D. A. Schuhmacher, J. L. Barker, and K. E. Klose. 2000. The virulence regulatory protein ToxR mediates enhanced bile resistance in *Vibrio cholerae* and other pathogenic *Vibrio* species. *Infect. Immun.* **68:**1491–1497.

Rao, N. N., and A. Torriani. 1990. Molecular aspects of phosphate transport in *Escherichia coli*. *Mol. Microbiol.* **4:**1083–1090.

Richard, A. L., J. H. Withey, S. Beyhan, F. Yildiz, and V. J. DiRita. 2010. The *Vibrio cholerae* virulence regulatory cascade controls glucose uptake through activation of TarA, a small regulatory RNA. *Mol. Microbiol.* **78:**1171–1181.

Rutherford, S. T., J. C. van Kessel, Y. Shao, and B. L. Bassler. 2011. AphA and LuxR/HapR reciprocally control quorum sensing in vibrios. *Genes Dev.* **25:**397–408.

Ryan, R. P., Y. Fouhy, J. F. Lucey, L. C. Crossman, S. Spiro, Y.-W. He, L.-H. Zhang, S. Heeb, M. Camara, P. Williams, and J. M. Dow. 2006. Cell-cell signaling in *Xanthomonas campestris* involves an HD-GYP domain protein that functions in cyclic di-GMP turnover. *Proc. Natl. Acad. Sci. USA* **103:**6712–6717.

Ryjenkov, D. A., R. Simm, U. Romling, and M. Gomelsky. 2006. The PilZ domain is a receptor for the second messenger c-di-GMP: the PilZ domain protein YcgR controls motility in enterobacteria. *J. Biol. Chem.* **281:**30310–30314.

Ryjenkov, D. A., M. Tarutina, O. V. Moskvin, and M. Gomelsky. 2005. Cyclic diguanylate is a ubiquitous signalling molecule in bacteria: insights into biochemistry of the GGDEF protein domain. *J. Bacteriol.* **187:**1792–1798.

Sanchez, J., and J. Holmgren. 2005. Virulence factors, pathogenesis and vaccine protection in cholera and ETEC diarrhea. *Curr. Opin. Immunol.* **17:**388–398.

Sanchez, J., and J. Holmgren. 2008. Cholera toxin structure, gene regulation and pathophysiological and immunological aspects. *Cell. Mol. Life Sci.* **65:**1347–1360.

Satchell, K. J. F. 2011. Structure and function of MARTX toxins and other large repetitive RTX proteins. *Annu. Rev. Microbiol.* **65:**71–90.

Schild, S., R. Tamayo, E. J. Nelson, F. Qadri, S. B. Calderwood, and A. Camilli. 2007. Genes induced late in infection increase fitness of *Vibrio cholerae* after release into the environment. *Cell Host Microbe* **2:**264–277.

Schmidt, A. J., D. A. Ryjenkov, and M. Gomelsky. 2005. The ubiquitous protein domain EAL is a cyclic diguanylate-specific phosphodiesterase: enzymatically active and inactive EAL domains. *J. Bacteriol.* **187:**4774–4781.

Schuhmacher, D. A., and K. E. Klose. 1999. Environmental signals modulate ToxT-dependent virulence factor expression in *Vibrio cholerae. J. Bacteriol.* **181:**1508–1514.

Shakhnovich, E. A., D. T. Hung, E. Pierson, K. Lee, and J. J. Mekalanos. 2007. Virstatin inhibits dimerization of the transcriptional activator ToxT. *Proc. Natl. Acad. Sci. USA* **104:**2372–2377.

Silva, A. J., G. J. Leitch, A. Camilli, and J. A. Benitez. 2006. Contribution of hemagglutinin/protease and motility to the pathogenesis of El Tor biotype cholera. *Infect. Immun.* **74:**2072–2079.

Skorupski, K., and R. K. Taylor. 1997. Cyclic AMP and its receptor protein negatively regulate the coordinate expression of cholera toxin and toxin-coregulated pilus in *Vibrio cholerae. Proc. Natl. Acad. Sci. USA* **94:**265–270.

Skorupski, K., and R. K. Taylor. 1999. A new level in the *Vibrio cholerae* ToxR virulence cascade: AphA is required for transcriptional activation of the *tcpPH* operon. *Mol. Microbiol.* **31:**763–771.

Srivastava, D., R. C. Harris, and C. M. Waters. 2011. Integration of cyclic di-GMP and quorum sensing in the control of *vpsT* and *aphA* in *Vibrio cholerae. J. Bacteriol.* **193:**6331–6341.

Stonehouse, E., G. Kovacikova, R. K. Taylor, and K. Skorupski. 2008. Integration host factor positively regulates virulence gene expression in *Vibrio cholerae. J. Bacteriol.* **190:**4736–4748.

Stonehouse, E. A., R. R. Hulbert, M. B. Nye, K. Skorupski, and R. K. Taylor. 2011. H-NS binding and repression of the *ctx* promoter in *Vibrio cholerae. J. Bacteriol.* **193:**979–988.

Syed, K. A., S. Beyhan, N. Correa, J. Queen, J. Liu, F. Peng, K. J. F. Satchell, F. Yildiz, and K. E. Klose. 2009. The *Vibrio cholerae* flagellar regulatory hierarchy controls expression of virulence factors. *J. Bacteriol.* **191:**6555–6570.

Szabady, R. L., J. H. Yanta, D. K. Halladin, M. J. Schofield, and R. A. Welch. 2011. TagA is a secreted protease of *Vibrio cholerae* that specifically cleaves mucin glycoproteins. *Microbiology* **157:**516–525.

Tamayo, R., A. D. Tischler, and A. Camilli. 2005. The EAL domain protein VieA is a cyclic diguanylate phosphodiesterase. *J. Biol. Chem.* **280:**33324–33330.

Tamayo, R., B. Patimalla, and A. Camilli. 2010. Growth in a biofilm induces a hyperinfectious phenotype in *Vibrio cholerae. Infect. Immun.* **78:**3560–3569.

Tamayo, R., S. Schild, J. T. Pratt, and A. Camilli. 2008. Role of cyclic di-GMP during El Tor biotype *Vibrio cholerae* infection: characterization of the in vivo-induced cyclic di-GMP phosphodiesterase CdpA. *Infect. Immun.* **76:**1617–1627.

Tamplin, M. L., A. L. Gauzens, A. Huq, D. A. Sack, and R. R. Colwell. 1990. Attachment of *Vibrio cholerae* serogroup O1 to zooplankton and phytoplankton of Bangladesh waters. *Appl. Environ. Microbiol.* **56:**1977–1980.

Taylor, J. L., R. S. De Silva, G. Kovacikova, W. Lin, R. K. Taylor, K. Skorupski, and F. J. Kull. 2012. The crystal structure of AphB, a virulence gene activator from *Vibrio cholerae*, reveals residues that influence its response to oxygen and pH. *Mol. Microbiol.* **83:**457–470.

Taylor, R. K., V. L. Miller, D. B. Furlong, and J. J. Mekalanos. 1987. Use of *phoA* gene fusions to identify a pilus colonization factor coordinately regulated with cholera toxin. *Proc. Natl. Acad. Sci. USA* **84:**2833–2837.

Tischler, A. D., and A. Camilli. 2004. Cyclic diguanylate (c-di-GMP) regulates *Vibrio cholerae* biofilm formation. *Mol. Microbiol.* **53:**857–869.

Tischler, A. D., and A. Camilli. 2005. Cyclic diguanylate regulates *Vibrio cholerae* virulence gene expression. *Infect. Immun.* **73:**5873–5882.

Tischler, A. D., S. H. Lee, and A. Camilli. 2002. The *Vibrio cholerae vieSAB* locus encodes a pathway contributing to cholera toxin production. *J. Bacteriol.* **184:**4104–4113.

Vaitkevicius, K., B. Lindmark, G. Ou, T. Song, C. Toma, M. Iwanaga, J. Zhu, A. Andersson, M.-L. Hammarstrom, S. Tuck, and S. N. Wai. 2006. A *Vibrio cholerae* protease needed for killing of *Caenorhabditis elegans* has a role in protection from natural predator grazing. *Proc. Natl. Acad. Sci. USA* **103:**9280–9285.

van Ulsen, P., M. Hillebrand, L. Zulianello, P. van de Putte, and N. Goosen. 1996. Integration host factor alleviates the H-NS-mediated repression of the early promoter of bacteriophage Mu. *Mol. Microbiol.* **21:**567–578.

Waldor, M. K., and J. J. Mekalanos. 1996. Lysogenic conversion by a filamentous phage encoding cholera toxin. *Science* **272:**1910–1914.

Waters, C. M., W. Lu, J. D. Rabinowitz, and B. L. Bassler. 2008. Quorum sensing controls biofilm formation in *Vibrio cholerae* through modulation of cyclic di-GMP levels and repression of *vpsT. J. Bacteriol.* **190:**2527–2536.

Withey, J. H., and V. J. DiRita. 2005a. Activation of both *acfA* and *acfD* transcription by *Vibrio cholerae* ToxT requires binding to two centrally located DNA sites in an inverted repeat conformation. *Mol. Microbiol.* **56:**1062–1077.

Withey, J. H., and V. J. DiRita. 2005b. *Vibrio cholerae* ToxT independently activates the divergently transcribed *aldA* and *tagA* genes. *J. Bacteriol.* **187:**7890–7900.

Withey, J. H., and V. J. DiRita. 2006. The toxbox: specific DNA sequence requirements for activation of *Vibrio cholerae* virulence genes by ToxT. *Mol. Microbiol.* **59:**1779–1789.

Yang, J., M. Tauschek, and R. M. Robins-Browne. 2011. Control of bacterial virulence by AraC-like regulators that respond to chemical signals. *Trends Microbiol.* **19:**128–135.

Yang, M., E. M. Frey, Z. Liu, R. Bishar, and J. Zhu. 2010. The virulence transcriptional activator AphA enhances biofilm formation by *Vibrio cholerae* by activating the expression of the biofilm regulator VpsT. *Infect. Immun.* **78:**697–703.

Yildiz, F. H., and G. K. Schoolnik. 1999. *Vibrio cholerae* O1 El Tor: identification of a gene cluster required for the rugose colony type, exopolysaccharide production, chlorine resistance and biofilm formation. *Proc. Natl. Acad. Sci. USA* **96:**4028–4033.

Yildiz, F. H., N. A. Dolganov, and G. K. Schoolnik. 2001. VpsR, a member of the response regulators of the two-component regulatory systems, is required for expression of *vps* biosynthesis genes and EPS^ETr-associated phenotypes in *Vibrio cholerae* O1 El Tor. *J. Bacteriol.* **183:**1716–1726.

Yildiz, F. H., X. S. Liu, A. Heydorn, and G. K. Schoolnik. 2004. Molecular analysis of rugosity in a *Vibrio cholerae* O1 El Tor phase variant. *Mol. Microbiol.* **53:**497–515.

Yu, R. R., and V. J. DiRita. 1999. Analysis of an autoregulatory loop controlling ToxT, cholera toxin, and toxin-coregulated pilus production in *Vibrio cholerae*. *J. Bacteriol.* **181:**2584–2592.

Yu, R. R., and V. J. DiRita. 2002. Regulation of gene expression in *Vibrio cholerae* by ToxT involves both antirepression and RNA polymerase stimulation. *Mol. Microbiol.* **43:**119–134.

Zheng, J., O. S. Shin, D. E. Cameron, and J. J. Mekalanos. 2010. Quorum sensing and a global regulator TsrA control expression of type VI secretion and virulence in *Vibrio cholerae*. *Proc. Natl. Acad. Sci. USA* **107:**21128–21133.

Zhu, J., M. B. Miller, R. E. Vance, M. Dziejman, B. L. Bassler, and J. J. Mekalanos. 2002. Quorum sensing regulators control virulence gene expression in *Vibrio cholerae*. *Proc. Natl. Acad. Sci. USA* **99:**3129–3134.

Zhu, J., and J. J. Mekalanos. 2003. Quorum sensing-dependent biofilms enhance colonization in *Vibrio cholerae*. *Dev. Cell* **5:**647–656.

Regulation of Bacterial Virulence
Edited by Michael L. Vasil and Andrew J. Darwin
© 2013 ASM Press, Washington, DC   doi:10.1128/9781555818524.ch13

Chapter 13

Virulence Gene Regulation in *Bacillus anthracis* and Other *Bacillus cereus* Group Species

Jennifer L. Dale and Theresa M. Koehler

THE *BACILLUS CEREUS* GROUP SPECIES AS PATHOGENS

The genus *Bacillus* belongs to the phylum *Firmicutes* and includes a diverse group of gram-positive and gram-variable bacteria. *Bacillus* species are rod-shaped aerobic or facultative anaerobic bacteria that can form dormant endospores in response to nutrient deprivation and other environmental signals. The *Bacillus cereus* group species, also known as group 1 bacilli and *Bacillus cereus* sensu lato, are comprised of *B. cereus* sensu stricto, *B. anthracis*, *B. thuringiensis*, *B. weihenstephanensis*, *B. mycoides*, and *B. pseudomycoides* (Ash et al., 1991; Ash and Collins, 1992). These *Bacillus* species have similar physiology, cell structure, and genetic exchange systems. They can grow saprophytically under nutrient-rich conditions, including some soil environments, and are common inhabitants of the gut of invertebrates (Jensen et al., 2003).

The best-studied members of the *B. cereus* group, *B. anthracis*, *B. thuringiensis*, and *B. cereus* sensu stricto, are pathogens with common and unique features that facilitate their ability to cause disease. While *B. anthracis* is the causative agent of anthrax in mammals, *B. thuringiensis* is considered primarily to be a pathogen of insects, and some *B. cereus* sensu stricto strains produce toxins associated with mild food poisoning. *B. thuringiensis* and *B. cereus* sensu stricto can also cause local and systemic opportunistic infections in humans, some of which are hospital acquired (Craig et al., 1974; Colpin et al., 1981; Damgaard et al., 1997; Mahler et al., 1997; Hernandez et al., 1998; Al-Abri et al., 2011; Shimono et al., 2011).

B. anthracis and Anthrax Disease

As the etiological agent of anthrax, *B. anthracis* is the most renowned member of the *B. cereus* group. Inhalation or ingestion of *B. anthracis* spores can result in a lethal hemorrhagic septicemia. Entry of spores via a cutaneous lesion generally results in a localized infection that can become systemic if untreated. Inhalation anthrax is the best-studied form of anthrax disease. Immediately following entry into the host, spores are phagocytosed by resident macrophages and dendritic cells, which serve as vehicles for transit of the bacterium to regional lymph nodes (Barnes, 1947; Ross, 1957; Dixon et al., 1999; Mock and Fouet, 2001; Cleret et al., 2007). The nature of early intracellular *B. anthracis*-host cell interactions is unresolved, but the balance between spore clearance and escape from phagocytic cells is likely impacted by the multiplicity of infection, virulence of the strain, and other factors (Guidi-Rontani et al., 1999; Guidi-Rontani et al., 2001; Welkos et al., 2002; Cote et al., 2005; Hu et al., 2006; Cote et al., 2008). Nongerminating intracellular spores or vegetative cells that survive the initial host response ultimately escape the phagocytic cell and disseminate via a hematological route to various target tissues and organs throughout the host (Ribot et al., 2006). In late-stage anthrax disease, the vegetative cells can reach concentrations of 10^8 CFU/ml in blood and cerebrospinal fluid (Dixon et al., 1999).

B. anthracis evades host immune responses primarily by production of anthrax toxin and a poly-γ-D-glutamate capsule. Anthrax toxin represents an interesting variation on the classic A-B toxin model: one binding/translocating B component, protective antigen (PA), and two enzymatic A components, edema factor (EF) and lethal factor (LF). The combination of purified PA and EF is called edema toxin (ET), while purified PA plus LF is called lethal toxin (LT). LT and ET act on multiple cell types to disable the immune system (Baldari et al., 2006; Turk, 2007). LF is a metalloprotease that cleaves mitogen-activated protein

Jennifer L. Dale and Theresa M. Koehler • Department of Microbiology and Molecular Genetics, University of Texas—Houston Medical School, Houston, TX 77030.

kinase kinases (MAPKKs). This block in MAPK signaling significantly impairs cellular functions of the innate and adaptive immune systems. Immune response is also impaired by EF, a calmodulin-dependent adenylate cyclase that activates host cyclic AMP (cAMP)-dependent signaling pathways. Despite over 50 years of anthrax toxin investigation, much is still unknown regarding the multiple roles of the proteins in disease. Nevertheless, it is clear that low levels of toxin impact the early host responses to intracellular *B. anthracis*. In late-stage anthrax, the toxin reaches a critical threshold level, causing a broad spectrum of pathophysiologies even when bacterial proliferation is curtailed by antibiotics (Banks et al., 2006).

As is true for many other encapsulated pathogens, the *B. anthracis* capsule has an antiphagocytic role, and capsule synthesis is associated with dissemination during infection (Keppie et al., 1953; Makino et al., 1989a; Drysdale et al., 2005b; Ovodov, 2006; Glomski et al., 2008). Whereas most bacterial capsules are polysaccharide, the *B. anthracis* capsule composition is unique: it is comprised solely of D-glutamic acid residues that are gamma linked to form homopolymers of over 215 kDa (Record and Wallis, 1956). The *B. anthracis* capsule is relatively nonimmunogenic. Vegetative capsulated cells may avoid demise in the host because a strong humoral immune response is not generated against the outer surface of the bacterium.

B. thuringiensis, an Insect Pathogen

B. thuringiensis is widely known for its production of parasporal crystal toxins (δ-endotoxins) bearing insecticidal activity. This feature has led to use of natural and genetically modified strains as pesticides for control of the insect orders Lepidoptera, Diptera, and Coleoptera (Roh et al., 2007). Moreover, transfer of genes encoding the insecticidal proteins to plants such as corn, cotton, rice, and soybean has been successfully employed in agricultural biotechnology (Crickmore, 2006). In addition to the toxins, *B. thuringiensis* products have been investigated as agents for control of plant diseases. Zwittermicin A, a linear aminopolyol antibiotic, is active against oomycetes and gram-negative bacteria, and acyl homoserine lactone lactonases prevent bacterial quorum sensing (Silo-Suh et al., 1994; Dong et al., 2002; Zhou et al., 2008).

B. cereus Sensu Stricto and Diarrheal Disease

Although most *B. cereus* sensu stricto isolates are considered nonpathogens, some strains can cause food-borne illness. Diarrheal disease results from intestinal infection with enterotoxin-producing *B. cereus* sensu stricto strains. The pore-forming cytotoxins, hemolysin BL (Hbl), nonhemolytic enterotoxin (Nhe), and cytotoxin K (CytK), are synthesized by the bacterium during growth in the small intestine (Granum and Lund, 1997; Clavel et al., 2004; Stenfors Arnesen et al., 2008). An emetic syndrome results from ingestion of the preformed *B. cereus* sensu stricto toxin, cereulide (Agata et al., 1995; Ehling-Schulz et al., 2004).

B. thuringiensis and *B. cereus* Sensu Stricto as Opportunists

Opportunistic infections caused by *B. thuringiensis* and *B. cereus* sensu stricto are relatively uncommon, but they can have serious consequences whether local or systemic. The bacteria have been associated with bloodstream infections, endocarditis, respiratory tract infections, and endophthalmitis (Drobniewski, 1993; Stenfors Arnesen et al., 2008; Horii et al., 2011). There are few reports concerning association of specific virulence factors with these infections. Some reports implicate the membrane-damaging toxins of the species (Callegan et al., 2006; Ghelardi et al., 2007), and studies have suggested that phosphatidylinositol-specific phospholipase C (PLC) as well as sphingomyelinase may be related to *B. cereus* sensu stricto pathogenicity (Schoeni and Wong, 2005; Dohmae, et al., 2008; Stenfors Arnesen et al., 2008).

Nosocomial *B. cereus* sensu stricto infections have been described, but it can be difficult to confirm the species as a causative agent of infectious diseases because it is a common contaminant associated with clinical samples (Al-Abri et al., 2011; Shimono et al., 2011). Biofilm-producing strains of *B. thuringiensis* and *B. cereus* sensu stricto cause nosocomial bacteremia via catheter infection (Kuroki et al., 2009).

Virulence Plasmid Content of the *B. cereus* Group Species

The chromosomes of *B. anthracis*, *B. thuringiensis*, and *B. cereus* sensu stricto reveal striking sequence similarity and gene synteny (Rasko et al., 2005), but virulence-associated plasmid content can allow facile discrimination of the three species. *B. anthracis* is distinguished by its virulence plasmids, pXO1 (182 kb) and pXO2 (96 kb). The structural genes for the anthrax toxin proteins, *pagA* (PA), *lef* (LF), and *cya* (EF), are located on pXO1 (Koehler, 2002), while the biosynthetic operon for capsule synthesis, *capBCADE*, is located on pXO2 (Makino et al., 1988, 1989a; Okinaka et al., 1989a; Candela et al., 2005). The insecticidal toxins of *B. thuringiensis* are encoded by a number of large, sometimes self-transmissible plasmids (Gonzalez et al., 1982; Schnepf et al., 1998). In *B. cereus* sensu stricto, the genes for cereulide biosynthesis are plasmid borne, while the enterotoxin genes are located on the chromosome (Stenfors Arnesen et al., 2008).

MAJOR PLEIOTROPIC CONTROL SYSTEMS FOR VIRULENCE GENE EXPRESSION

As is true for many pathogens, certain virulence genes in *B. cereus* group species are transcribed coordinately in response to specific signals and regulators. Central players in coordinate control of gene expression are AtxA in *B. anthracis* and PlcR in *B. cereus* sensu stricto and *B. thuringiensis* (Fig. 1). Differential gene expression among the *B. cereus* group species can be attributed in part to PlcR and AtxA activities in the different species.

The *atxA* gene is located on *B. anthracis* plasmid pXO1 (Uchida et al., 1993; Koehler et al., 1994). The AtxA regulon includes structural genes for the anthrax toxin proteins, carried on pXO1, the biosynthetic operon for capsule synthesis, located on pXO2, and multiple other pXO1, pXO2, and chromosomal genes (Bourgogne et al., 2003). AtxA-regulated genes on the *B. anthracis* chromosome have homologues in the other *B. cereus* group species, but typical *B. cereus* sensu stricto and *B. thuringiensis* strains do not carry the plasmid harboring *atxA* and therefore

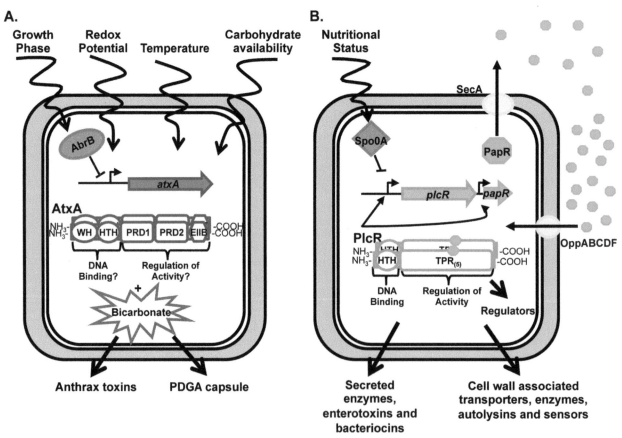

Figure 1. Models for AtxA and PlcR control of virulence gene expression. (A) *atxA* gene activation and AtxA function in *B. anthracis*. Multiple signals, including growth phase, redox potential, temperature, and carbohydrate availability, impact the transcription of *atxA*. The growth phase transition state regulator AbrB binds directly to the *atxA* promoter region to repress transcription. Predicted functional domains of AtxA are the winged helix (WH) and helix-turn-helix (HTH) for DNA binding, PTS domains (PRD1 and PRD2) for regulation of activity, and EIIB for multimerization. In the presence of elevated CO_2/bicarbonate, AtxA positively affects transcription of the anthrax toxin genes and the biosynthetic operon for synthesis of PDGA capsule. (B) *plcR* gene activation and PlcR-PapR function in *B. cereus* group members. Signals that impact *plcR/papR* transcription include nutritional status and cell density. The master response regulator Spo0A binds directly to the *plcR* promoter to repress transcription. PlcR contains a DNA-binding domain (HTH) and tetratricopeptide repeats (TPRs) that regulate activity. PapR is exported by the SecA machinery, proteolytically processed to a heptapeptide, and imported into the cell by the OppABCDF transport system. Mature processed PapR associates with PlcR, enabling dimerization and regulation of activity. The PlcR-PapR complex autogenously controls the *plcR/papR* bicistronic gene cluster in addition to multiple genes encoding secreted toxins and degradative enzymes, cell wall-associated proteins, and cytoplasmic regulatory proteins. doi:10.1128/9781555818524.ch13f1

exhibit differential expression of the AtxA-controlled chromosomal genes.

In *B. cereus* sensu stricto and *B. thuringiensis*, PlcR is a global transcriptional regulator, controlling expression of multiple genes, many associated with pathogenesis (Lereclus et al., 1996; Agaisse et al., 1999; Gohar et al., 2002). Unlike *atxA*, *plcR* is present on the chromosome in all three species. However, the *B. anthracis plcR* gene contains a nonsense mutation resulting in a nonfunctional protein product (Mignot et al., 2001). Thus, although homologues of many PlcR-regulated genes are present in the *B. anthracis* genome, they are not subject to PlcR control. Multiple phenotypic differences between *B. anthracis* and the other two species are attributed to low-level expression of the PlcR regulon in *B. anthracis* (Mignot et al., 2001).

Some studies have suggested that *plcR* and *atxA* are not compatible. Expression of a *B. thuringiensis plcR* gene in a *B. anthracis* strain containing *atxA* results in a significant decrease in sporulation, a phenotype that can be rescued by deletion of *atxA* (Mignot et al., 2001). It has been speculated that the nonsense mutation within *B. anthracis plcR* provided a selective advantage for evolution of the species.

AtxA: a Unique Regulator of *B. anthracis*

The *atxA* gene, originally named for its function in anthrax toxin gene activation, encodes a major positive regulator of the anthrax toxin genes, *pagA*, *lef*, and *cya*, located at distinct loci on pXO1. AtxA also positively affects transcription of the capsule biosynthetic operon, *capBCADE*, on pXO2 via control of the capsule gene regulator *acpA* (Drysdale et al., 2004; Uchida, et al., 1993; Koehler et al., 1994; Dai et al., 1995; Fouet and Mock, 1996; Guignot et al., 1997; Uchida et al., 1997; Sirard et al., 2000; Mignot et al., 2003). The AtxA regulon extends beyond the toxin and capsule genes. Transcriptional profiling and other experiments have revealed additional AtxA-regulated genes on the plasmids and chromosome (Hoffmaster and Koehler, 1997, 1999b; Bourgogne et al., 2003). AtxA function is critical for *B. anthracis* virulence; an *atxA* null mutant is attenuated in a murine model of anthrax disease (Dai et al., 1995).

The molecular basis for AtxA function has not been established, but features of potential domains have been described (Fig. 1A). AtxA is a 56-kDa soluble basic protein with amino acid sequence motifs suggestive of DNA-binding and regulatory domains. Amino-terminal winged-helix (WH) and helix-turn-helix (HTH) motifs are predicted to facilitate protein-DNA interaction, but specific DNA-binding activity has not

been reported (Uchida et al., 1997; Tsvetanova et al., 2007; Koehler, 2009). Centrally located amino acid sequences suggest two phosphoenolpyruvate:carbohydrate phosphotransferase system (PTS) regulation domains (PRDs). PRDs are common to transcription regulators that control genes associated with carbohydrate metabolism. Phosphorylation of specific histidine residues within PRDs affects protein oligomerization and function (Deutscher et al., 2006). AtxA can be phosphorylated at two histidines, one in each PRD. Phosphorylation at H199 appears to activate AtxA function, while phosphorylation at H379 is associated with decreased AtxA activity (Tsvetanova et al., 2007). Interestingly, transcriptional profiling studies have not revealed AtxA targets associated with catabolism (Bourgogne et al., 2003). Finally, sequences in the carboxy-terminal region of AtxA bear similarity to enzyme IIB (EIIB) proteins. EIIB proteins function within the PTS to phosphorylate carbohydrates as they pass through cognate EIIC permeases (Deutscher et al., 2006). The EIIB-like domain of AtxA has been shown to facilitate AtxA oligomerization (Hammerstrom et al., 2011).

Transcription of the toxin and capsule genes and many other AtxA-regulated genes is enhanced when *B. anthracis* cultures are incubated in 5% CO_2 in media containing 0.8% dissolved bicarbonate (Bartkus and Leppla, 1989; Cataldi et al., 1992; Koehler et al., 1994; Sirard et al., 1994; Dai et al., 1995; Fouet and Mock, 1996; Hoffmaster and Koehler, 1997, 1999b; Bourgogne et al., 2003; Mignot et al., 2003; Chitlaru et al., 2006). The CO_2/bicarbonate signal is considered to be physiologically significant for a mammalian pathogen. Growth in elevated CO_2/bicarbonate appears to enhance AtxA activity rather than AtxA protein levels (Dai and Koehler, 1997). There is a positive correlation between the CO_2/bicarbonate signal, the steady-state AtxA dimer/monomer ratio, and AtxA activity; however, a protein-ligand relationship between CO_2/bicarbonate and AtxA has not been established (Hammerstrom et al., 2011).

While AtxA activity is affected by CO_2/bicarbonate, AtxA transcript and protein levels vary under different growth conditions. In batch culture, *atxA* expression follows the growth curve; *atxA* transcript and AtxA protein levels are relatively low during exponential phase but increase as the culture approaches stationary phase (Saile and Koehler, 2002). Under some culture conditions, growth phase-dependent *atxA* transcription is affected by *trans*-acting factors. Mutation of genes involved in cytochrome *c* biogenesis result in increased *atxA* transcription during early exponential phase, suggesting a relationship between redox state and *atxA* expression. Nevertheless, the significance of this phenotype is uncertain because *atxA* transcript levels are not affected by these mutations when cells are cultured in elevated CO_2/bicarbonate (Wilson et al., 2009). Transcription of *atxA* is also increased in early exponential phase in a mutant lacking the transition state regulator AbrB (Saile and Koehler, 2002). AbrB is well characterized in the archetype *Bacillus* species, *B. subtilis*, as a pleiotropic transcriptional regulator that represses multiple genes during exponential growth phase. As a culture transitions to stationary phase, AbrB repression is relieved. The *B. anthracis* AbrB protein is the only *trans*-acting regulator that has been shown to interact directly and specifically with the *atxA* promoter (Strauch et al., 2005). Another *trans*-acting regulator, CodY, controls AtxA protein levels posttranslationally. CodY is a pleiotropic transcription regulator found in a number of gram-positive bacteria (Sonenshein, 2005). CodY senses intracellular GTP concentration as an indicator of nutritional conditions and regulates transcription of early-stationary-phase genes (Ratnayake-Lecamwasam et al., 2001). The influence of CodY on AtxA protein levels occurs by an unknown mechanism (van Schaik et al., 2009).

Other signals impacting *atxA* expression are temperature and the presence of glucose. In concordance with the significance of AtxA for *B. anthracis* pathogenesis, the optimal temperature for *atxA* transcription is 37°C (Dai and Koehler, 1997). The presence of glucose stimulates transcription of *atxA*, but the mechanism for this induction is not clear. The glucose effect requires the presence of the carbon catabolite protein CcpA; however, CcpA does not specifically bind the *atxA* promoter (Chiang et al., 2011). CcpA activity appears to be important for virulence. A *B. anthracis ccpA* null mutant is significantly attenuated in a murine model of infection (Chiang et al., 2011).

The PlcR Regulon of *B. cereus* Sensu Stricto and *B. thuringiensis*

The pleiotropic transcriptional regulator PlcR, initially discovered as a positive regulator of the phospholipase C gene in *B. thuringiensis* (Lereclus et al., 1996), controls multiple genes encoding secreted toxins and degradative enzymes, cell wall-associated proteins, and cytoplasmic regulatory proteins in *B. thuringiensis* and *B. cereus* sensu stricto (Lereclus et al., 1996; Agaisse et al., 1999; Gohar et al., 2008). PlcR regulation is not apparent in *B. anthracis*, because the *plcR* locus contains a nonsense mutation resulting in a truncated nonfunctional protein (Agaisse et al., 1999; Slamti et al., 2004). Proteomic studies, transcriptional profiling, and *in silico* analyses have been employed to determine PlcR regulons in several *B. thuringiensis* and

B. cereus sensu stricto strains (Agaisse et al., 1999; Okstad et al., 1999; Gohar et al., 2002; Gohar et al., 2008). Established PlcR-controlled virulence genes include those for enterotoxins, hemolysins, proteases, and phospholipases. These genes are spread throughout the genome and do not form pathogenic islands on the chromosome (Agaisse et al., 1999).

The 34-kDa PlcR protein contains an amino-terminal helix-turn-helix DNA-binding domain and a carboxy-terminal regulatory domain consisting of 11 helices that form five tetratricopeptide repeats (TPR) (Declerck et al., 2007). PlcR activity is dependent upon interaction with the quorum-sensing peptide PapR (peptide activating PlcR) (Slamti and Lereclus, 2002). In the current model for PlcR/PapR function (Fig. 1B), PapR is synthesized as a 48-amino-acid peptide and secreted by the SecA machinery. Once outside of the cell, PapR is proteolytically processed to a heptapeptide that is imported into the cell via the OppABCDF transport system (Gominet et al., 2001; Bouillaut et al., 2008). Inside the cell, the PapR heptapeptide associates with PlcR to activate target genes. The crystal structure of PlcR-PapR indicates that PapR binds to the concave side of PlcR TPR domain helices 5 and 7, triggering dimerization of two PlcR-PapR complexes via the TPR domains (Declerck et al., 2007).

The *plcR* and *papR* genes form a bicistronic cluster that is autogenously controlled (Lereclus et al., 1996; Agaisse et al., 1999), and *B. cereus* group members can be classified into four distinct groups based on the sequence and specificity of the PlcR-PapR pair. PlcR groups I, II, III, and IV are associated with PapR heptapeptides, including the carboxy-terminal sequences LPFE(F/Y), VP(F/Y)E(F/Y), MPFEF, and LPFEH, respectively (Slamti and Lereclus, 2005; Bouillaut et al., 2008). The first and last amino acids of these peptide sequences determine specificity of PlcR-PapR for its target genes (Slamti and Lereclus, 2002, 2005; Bouillaut et al., 2008). It has been suggested that each *B. cereus* sensu stricto PlcR-PapR group contains a group-dependent protease that is specific for PapR. The extracellular protease NprB has been implicated in PapR processing in group II strains (Pomerantsev et al., 2009). The PlcR-PapR complex binds to a consensus DNA sequence, the palindromic "PlcR box" (TATGNAN₄TNCATA), located up to 200 nucleotides upstream of the –10 box of a promoter region (Agaisse et al., 1999; Okstad, et al., 1999). *In silico* and genetic analyses have revealed variability in PlcR box sequences. The A+T content is higher in the vicinity of PlcR boxes that are active when grown under rich conditions (Luria-Bertani broth [LB], 30°C) (Gohar et al., 2008). PlcR target genes typically contain promoter regions that resemble the canonical –10 region of the housekeeping sigma factor, SigA, and a –35 recognition region that is slightly different than the typical SigA consensus sequence (Agaisse et al., 1999; Gohar et al., 2008).

plcR transcription is controlled by the developmental regulator Spo0A. Two Spo0A boxes flank the PlcR box upstream of *plcR*. Active phosphorylated Spo0A is thought to repress *plcR* transcription by competing with PlcR-PapR for binding to the *plcR* promoter region (Lereclus et al., 2000). During batch culture in rich media, transcription of *plcR* and PlcR-regulated genes increases at the transition from exponential to stationary phase of growth. When cells are cultured in sporulation media, phosphorylated Spo0A prevents *plcR* activation (Lereclus et al., 2000). Thus, elevated *plcR* transcription occurs when the nutrient status keeps Spo0A~P levels low, and when cells are at a high density due to quorum sensing (Lereclus et al., 2000; Gohar, et al., 2008).

Deletion of *plcR* in pathogenic *B. cereus* sensu stricto and *B. thuringiensis* strains decreases virulence in insect larvae, mice, and rabbit eye models (Salamitou et al., 2000; Slamti and Lereclus, 2002; Callegan et al., 2003). A majority of strains synthesize a functional PlcR protein, but a small proportion (1%) contain *plcR* or *papR* genes with mutations rendering nonfunctional proteins. *B. anthracis* harbors multiple orthologues of *plcR*-regulated genes but does not contain a functional PlcR due to a point mutation in *plcR* that results in a truncated protein (Agaisse et al., 1999; Slamti et al., 2004). Introduction of a functional *B. thuringiensis*-derived PlcR into *B. anthracis* facilitates expression of genes with PlcR boxes in their promoter regions, including genes encoding proteases, hemolysins, and phospholipases. However, increased expression of these PlcR-regulated virulence factors did not influence the virulence of *B. anthracis* in a murine model of anthrax infection (Mignot et al., 2001).

THE VIRULENCE ARSENAL

A large number of virulence factors have been established for the pathogenic *B. cereus* group species. Some of these factors are species specific and controlled by either AtxA or PlcR. Other virulence factors are shared among species and, in some cases, subject to additional levels of control.

Anthrax Toxin

Anthrax toxin is the best-studied and arguably the most important virulence factor produced by *B. anthracis*. The toxin is comprised of three proteins: PA, LF, and EF. Binary combination of PA and LF is termed LT, and combination of PA and EF is termed ET. Anthrax toxin entry is initiated when PA (85 kDa) binds to host cells via specific receptors (ANTXR1 and ANTXR2), is cleaved by a furin-like protease, and forms a multimeric prepore that is capable of binding LF (83 kDa) and/or EF (89 kDa). Upon protein-receptor complex endocytosis and endosomal acidification, the PA prepore undergoes a conformational change enabling insertion into the endosomal membrane and translocation of LT and ET into the host cell cytosol (Friedlander, 1986; Blaustein et al., 1989; Miller et al., 1999; Krantz et al., 2005; Young and Collier, 2007; Thoren and Krantz, 2011). LF is a zinc-dependent metalloprotease that inhibits the MAPK signal transduction pathway, ultimately resulting in host cell death (Duesbery et al., 1998; Koehler, 2000; Tournier et al., 2007). EF is a calmodulin-dependent adenylyl cyclase that elevates cellular levels of cAMP, causing host cell edema (Leppla, 1982; Tournier et al., 2007). Overall, production of anthrax toxin enables the bacterium to evade and escape the host immune response.

The structural genes for anthrax toxin include *cya*, *lef*, and *pagA*, which encode EF, LF, and PA, respectively. These genes are located noncontiguously on the virulence plasmid pXO1 within a 45-kb pathogenicity island that includes the positive regulator *atxA* (Mikesell et al., 1983; Okinaka et al., 1999a, 1999b). Transcription of the toxin genes is strongly affected by the *trans*-acting regulator AtxA. Thus, signals that impact *atxA* transcription and AtxA function, in particular CO_2/bicarbonate, impact steady-state levels of toxin gene transcripts and toxin proteins (Koehler et al., 1994; Sirard et al., 1994; Dai et al., 1995; Dai and Koehler, 1997). The monocistronic transcripts of the *cya* and *lef* genes map to single start sites, and transcription of both genes is AtxA dependent. The *pagA* gene is part of a bicistronic operon, *pagAR* (Hoffmaster and Koehler, 1999a). The *pagR* gene encodes PagR, a transcriptional repressor that autogenously controls expression of the operon by binding directly upstream of *pagA*. Interestingly, PagR has also been shown to repress transcription of the S-layer genes *sap* and *eag* by binding directly to their promoter regions. Sequence similarities in the promoter regions of PagR-regulated genes are not apparent, and a consensus sequence for PagR binding has not been identified (Mignot et al., 2003). Two transcriptional start sites in the *pagAR* promoter have been reported. The major start site (P1) is AtxA dependent. The minor start site (P2) is not dependent upon AtxA and is expressed constitutively at a relatively low level (Koehler et al., 1994; Dai et al., 1995). Common *cis*-acting regions of *atxA*-dependent promoters have not been identified. Nucleotide sequence similarities in promoter regions are not apparent. It has been suggested that DNA curvature plays a role in AtxA regulation of its target toxin genes (Hadjifrangiskou and Koehler, 2008).

Entomopathogenic Toxins

The entomopathogenesis of *B. thuringiensis* is dependent upon the production of characteristic insecticidal parasporal crystals called cryotoxins (Cry) and cytolysins (Cyt). The genes encoding Cry and Cyt are typically located on large, transmissible plasmids (Rasko et al., 2005). Distinctive crystalline aggregates of the large (~130- to 140-kDa) protoxin proteins are found in sporulating mother cells and are released at the completion of sporulation. These crystal inclusions may account for up to 25% of the dry weight of the sporulating cells (Lereclus et al., 2000; Rasko et al., 2005). Upon ingestion by insect larvae, the protoxin crystals are solubilized in the insect midgut and proteolytically cleaved by a host protease to form active toxins, termed δ-endotoxins, of about 60 kDa. Cry and Cyt are part of the pore-forming toxin family. These toxins undergo conformational changes in the insect facilitating insertion into midgut epithelial cells, with subsequent host cell lysis. Disruption of the midgut results in conditions favorable for spore germination and bacterial multiplication, ultimately causing septicemia and insect death (Hofte and Whiteley, 1989; Lereclus et al., 1996; Rasko et al., 2005; Raymond et al., 2010; Bravo et al., 2011). *B. thuringiensis* Cry and Cyt toxins exhibit highly specific insecticidal activity and a restricted host range. The toxins are categorized as specific for Lepidoptera, Lepidoptera and Diptera, Coleoptera, and Diptera. They are further grouped into subclasses according to both structural similarities and insecticidal specificity (Hofte and Whiteley, 1989).

The *cry* genes exhibit sporulation-dependent and sporulation-independent transcription during stationary-phase growth in batch culture. The best-studied sporulation-dependent *cry* gene, *cry1Aa* (Lepidoptera specific), contains two overlapping promoters, BtI and BtII, which are used sequentially for transcription (Wong et al., 1983). Transcription from BtI occurs between early sporulation and mid-sporulation, whereas transcription from BtII initiates during mid-sporulation and continues throughout the completion of sporulation (Brown and Whiteley, 1988, 1990; Lereclus et al., 2000). The BtI and BtII transcriptional start sites are recognized by the mother cell-specific sigma factor homologues of *B. subtilis* SigE (Brown and Whiteley, 1988) and SigK (Brown and Whiteley, 1990), respectively. SigE appears to be required for *cry1Aa* transcription, whereas SigK is needed for full

activation of the gene. Expression of a *cry1Aa-lacZ* transcriptional fusion is reduced in a *sigK* mutant and abolished in a *sigE* mutant (Adams et al., 1991; Bravo et al., 1996; Lereclus et al., 2000). Although only the *cry1Aa* promoter has been analyzed extensively, several *cry* genes contain similar SigE- and SigK-specific consensus sequences and are likely controlled in a similar manner.

The Coleoptera-specific *cry3A* gene represents a typical example of sporulation-independent expression. The *cry3A* promoter region resembles that recognized by the housekeeping sigma factor, SigA. Transcription of *cry3A* is weak during vegetative growth and is activated at the beginning of stationary phase. The mechanism by which *cry3A* is activated at the onset of stationary phase is not known, but transcription of *cry3A* is negatively controlled by *spo0A* and *sigE* (de Souza et al., 1993; Agaisse and Lereclus, 1994; Malvar and Baum, 1994; Baum and Malvar, 1995; Lereclus, et al.,1995; Salamitou et al., 1996, Lereclus et al., 2000).

In addition to control of transcription initiation, *cry* transcript stability is regulated. Increased *cry* mRNA stability has been attributed to 5′ and 3′ mRNA stabilizing factors. The 5′ stability determinant appears to be a Shine-Dalgarno consensus sequence that does not direct translation initiation but instead stabilizes the corresponding transcript (Agaisse and Lereclus, 1995). Transcription of a positive retroregulator (Schindler and Echols, 1981) sequence downstream of the respective *cry* gene is thought to protect the mRNA from exonucleolytic degradation (Wong and Chang, 1986; Agaisse and Lereclus, 1995).

In addition to Cry and Cyt proteins, *B. thuringiensis* expresses a vegetative insecticidal protein (VIP) that is expressed and secreted during vegetative growth. Transcription of the *vip* gene initiates during mid-log phase of growth and is actively expressed in sporulating cultures (Estruch et al., 1996), but little is known regarding *vip* gene control.

Enterotoxins

Diarrheal-type food poisoning caused by *B. cereus* sensu stricto and *B. thuringiensis* is attributed to the production of one or more heat-labile enterotoxins: hemolysin BL (Hbl), nonhemolytic enterotoxin (Nhe), and cytotoxin K (CytK) (Spira and Goepfert, 1972; Glatz et al., 1974; Beecher and Macmillan, 1991; Lund and Granum, 1996; Granum and Lund, 1997; Lund et al., 2000; Stenfors Arnesen et al., 2008). These pore-forming toxins are believed to disrupt the structural integrity of epithelial cell plasma membranes in the small intestine (Granum and Lund,

1997; Stenfors Arnesen et al., 2008). Hbl and Nhe are protein complexes composed of three protein subunits. They are part of the ClyA superfamily of toxins found in a number of gram-negative bacteria (Beecher and Macmillan, 1991; Beecher, et al., 1995; Lund and Granum, 1996; Lund et al., 2000; Lindback et al., 2004; Stenfors Arnesen et al., 2008). The Hbl protein subunits include a binding component, B (35 kDa), and two lytic components, L$_1$ and L$_2$ (35 and 45 kDa, respectively). Nhe is composed of NheA (41 kDa), NheB (40 kDa), and NheC (37 kDa), which have 18 to 44% amino acid identity to the Hbl proteins and are considered to have similar functions (Beecher and Macmillan, 1991; Beecher et al., 1995; Lund and Granum, 1996; Granum et al., 1999; Stenfors Arnesen et al., 2008). CytK (34 kDa) is a member of the β-barrel pore-forming toxin family and has little amino acid sequence similarity to Hbl and Nhe. Nevertheless, CytK, Hbl, and Nhe exhibit similar enterotoxin potencies (Lund et al., 2000).

Genes encoding the enterotoxins are located at multiple sites on the chromosome (Agaisse et al., 1999; Kolsto et al., 2009). The *hblC*, *hblD*, and *hblA* genes comprise an operon and encode Hbl proteins L$_2$, L$_1$, and B, respectively (Heinrichs et al., 1993; Lindback et al., 1999; Stenfors Arnesen et al., 2008). In some strains, an additional gene, *hblB*, is found downstream of *hblCDA*. *hblB* is likely a pseudogene that has arisen as a partial gene duplication of *hblA*. Detectable levels of *hblB* transcript are nonexistent, and the *hblCDA* mRNA transcript appears to terminate within *hblB* (Agaisse et al., 1999; Okstad et al., 1999; Stenfors Arnesen et al., 2008). The *nheA*, *nheB*, and *nheC* genes, encoding the Nhe toxin proteins, are also transcribed as an operon (Granum et al., 1999). It is notable that attempts to delete *nheA* have been unsuccessful, suggesting that it is an essential gene (Ramarao and Lereclus, 2006; Fagerlund et al., 2008; Stenfors Arnesen et al., 2008). Finally, CytK is encoded by *cytK* and bears no sequence similarity to either of the other enterotoxins Hbl and Nhe (Lund et al., 2000).

Several *B. cereus* group member strains have been examined for the presence of the enterotoxin genes. Fewer than 50% of the strains sequenced contain genes for Hbl and CytK, whereas Nhe genes are in 99% of the strains. Hbl and CytK genes are present at the highest frequencies in clinical and food-associated isolates (Ehling-Schulz et al., 2005a; Ehling-Schulz et al., 2006a, 2006b; Moravek et al., 2006).

The enterotoxin genes are part of the PlcR regulon, but they are also controlled by other transcriptional regulators in response to metabolic and environmental stimuli (Agaisse et al., 1999; Okstad et al., 1999; Lund et al., 2000; Gohar et al., 2008; Stenfors Arnesen et al., 2008). ResDE, the two-component signal transduction system that regulates metabolism in response to oxygen, positively controls enterotoxin production (Duport et al., 2006; Esbelin et al., 2009). Growth of *B. cereus* sensu stricto in low-oxidoreduction anaerobic environments such as those in the small intestine requires ResDE. These conditions also enhance production of Hbl and Nhe (Duport et al., 2006). Typical two-component systems contain a sensor histidine kinase (i.e., ResE) that autophosphorylates in response to a signal and further transfers its phosphate to a response regulator (i.e., ResD). DNA-binding activity of the response regulator ResD is modulated by phosphorylation, and its activity is highest under anaerobic growth conditions (Nakano and Zuber, 1998). ResD binds directly to the promoter regions of *plcR*, *hbl*, and *nhe*. Interestingly, ResD binding to the *plcR*, *hbl*, and *nhe* promoters is not dependent on its phosphorylation status (Esbelin et al., 2009). An essential function of ResDE is activation of the redox regulator Fnr upon oxygen limitation (Nakano and Zuber, 1998). Fnr is required for enterotoxin activation and was shown to bind directly to the *hbl* and *nhe* promoter regions (Zigha et al., 2007; Esbelin et al., 2008). Based on these data, it has been suggested that phosphorylated ResD interacts with Fnr to form a coactivator complex allowing active transcription of *hbl* and *nhe* operons; however, nonphosphorylated ResD is hypothesized to form an Fnr antiactivator complex, decreasing transcription of the operons (Esbelin et al., 2009). Since PlcR also binds to and controls expression of the *hbl* and *nhe* operons, the possibility of a quaternary complex of Fnr, ResD, and PlcR cannot be excluded.

CcpA, a transcriptional regulator of the LacI family associated with catabolite repression, may control *hbl* and *nhe* expression (Warner and Lolkema, 2003). In *B. subtilis*, CcpA binds to consensus sequences, called catabolite responsive elements (CRE sites), in the promoters of target genes (Fujita et al., 1995; Miwa et al., 2000; Stulke and Hillen, 2000). Putative CRE sites were identified in the *hbl* and *nhe* operons (van der Voort et al., 2008), but CcpA binding to these sites has not been reported. *hbl* transcription is repressed by increasing concentrations of glucose (Duport et al., 2004), and the expression of both *hbl* and *nhe* is increased in a *ccpA* null strain. *cytK* expression is not affected by CcpA (van der Voort et al., 2008).

There is a correlation between enterotoxin regulation and *B. cereus* sensu stricto motility. Initially, enterotoxin secretion was thought to be dependent upon the flagellar export apparatus, because flagellum mutants were deficient in Hbl secretion (Bouillaut et al., 2005; Ghelardi et al., 2007). However, Fagerlund et al. (Fagerlund et al., 2010) showed that Nhe, Hbl, and CytK harbor Sec-type signal peptide sequences enabling secretion by the Sec pathway. Nonflagellated mutants were shown to be unaffected for enterotoxin secretion but exhibited reduced enterotoxin transcription. The molecular basis for flagellar control of enterotoxin transcription is not understood, but several signs point to coregulation of motility and virulence. First, a *B. thuringiensis* nonflagellated mutant is avirulent and has reduced hemolytic activity (Zhang et al., 1993). Second, a *plcR* mutant exhibits reduced motility on agar plates and decreased flagellin expression (Gohar et al., 2002). Finally, *hbl* and flagellar genes are upregulated during swarming migration, whereas the majority of PlcR-regulated genes, including *nhe* and *cytK*, are downregulated (Ghelardi et al., 2007). There is speculation that MogR, a transcriptional repressor of flagellar motility found in *Listeria* and *B. cereus* group species, has a role in enterotoxin regulation. Four putative MogR binding sites have been identified within the *hbl* gene (Fagerlund et al., 2010).

Cholesterol-Dependent Cytolysins

B. anthracis, *B. thuringiensis*, and *B. cereus* senso stricto secrete pore-forming toxins of the cholesterol-dependent cytolysin (CDC) family. These toxins, anthrolysin O (Alo), cereolysin O (Clo), and thuringiolysin O (Tlo), bind cholesterol and form pores in cell membranes. This results in influx of calcium from outside the cell and release of calcium from intracellular stores such as the endoplasmic reticulum (Gekara et al., 2007; Gilbert, 2010). The toxins have hemolytic activity *in vitro* (Bernheimer and Grushoff, 1967; Shannon et al., 2003). Investigations regarding the role of the toxin in pathogenesis have centered on Alo. An *alo* null mutant is fully virulent in a mouse model for anthrax, but mutants lacking the *alo* gene in combination with the absence of certain phospholipase genes are attenuated (Heffernan et al., 2007). Antibodies against purified Alo can provide protection against *B. anthracis* intravenous challenges (Nakouzi et al., 2008). Finally, purified Alo kills human neutrophils, monocytes, and macrophages and disrupts barrier function in cultured gut epithelial cells (Mosser and Rest, 2006; Bourdeau et al., 2009). It has been proposed that Alo has a role in the establishment of gastrointestinal anthrax disease (Bishop et al., 2010).

The genes encoding the CDCs of the *B. cereus* group species contain PlcR recognition sites in their promoter regions, and in *B. cereus* sensu stricto and *B. thuringiensis*, the phospholipase genes are part of the PlcR regulon (Agaisse et al., 1999; Okstad et al., 1999; Gohar et al., 2002; Gohar et al., 2008). However, given that *B. anthracis* does not synthesize a functional PlcR, the *B. anthracis* gene encoding Alo is not subject to this control (Mignot et al., 2001; Ross and Koehler, 2006). In batch culture of *B. anthracis*, *alo* expression appears to be highest in rich media and ALO is produced in a growth phase-dependent manner, with highest expression during mid-exponential-phase growth (Shannon et al., 2003).

Emetic Toxin (Cereulide)

Two clonal clusters of *B. cereus* sensu stricto and a selected few *Bacillus weihenstephanensis* strains have been identified as emetic toxin producers that elicit emesis (vomiting) when ingested with contaminated food products (Ehling-Schulz et al., 2005a; Thorsen et al., 2006; Vassileva et al., 2007; Kolsto et al., 2009). Emetic toxin, also called cereulide, is a 1.2-kDa heat-stable acid-tolerant cyclic dodecadepsipeptide, [D-O-Leu-D-Ala-D-O-Val-D-Val]$_3$. The toxin is synthesized nonribosomally (Agata et al., 1995; Agata et al., 1996; Granum and Lund, 1997; Kotiranta et al., 2000), and the alternating peptide and ester bonds, presence of D-amino acids, and cyclic structure are characteristic of products produced by nonribosomal peptide synthetases (NRPSs) (Ehling-Schulz et al., 2005b). Cereulide resembles the potassium ionophore antibiotic valinomycin produced by *Streptomyces griseus* and, like valinomycin, is toxic for mitochondria (Agata et al., 1994; Shinagawa et al., 1996; Mikkola et al., 1999; Kroten et al., 2010).

Cereulide synthesis is dependent on a 24-kb biosynthetic gene cluster located on a pXO1-like plasmid of emetic strains (Ehling-Schulz et al., 2005a, 2006b; Stenfors Arnesen et al., 2008; Kolsto et al., 2009; Lucking et al., 2009). Synthesis by *B. cereus* sensu stricto strains occurs during growth in a wide range of temperatures, but maximal production occurs when cells are cultured between 12 and 22°C (Finlay et al., 2000). Optimal cereulide synthesis by *B. weihenstephanensis* has not been associated with a specific temperature range, but unlike *B. cereus* sensu stricto, *B. weihenstephanensis* produces cereulide during growth at temperatures as low as 8°C (Thorsen et al., 2006). The cereulide biosynthetic gene cluster includes seven genes, *cesH*, *cesP*, *cesT*, *cesA*, *cesB*, *cesC*, and *cesD*, encoding a putative hydrolase, 4'-phosphopantetheinyl transferase, a putative type II thioesterase, cereulide synthetase (structural genes), and a putative ABC transporter, respectively (Ehling-Schulz et al., 2005a, 2005b; Ehling-Schulz et al., 2006b). Together, these gene products are associated with activation and regeneration of the

NRPS, assembly of the peptide product, removal of misprimed monomers, and potentially either transport of cereulide out of the bacterial cell or self-resistance towards cereulide (Ehling-Schulz et al., 2006a).

The cereulide biosynthetic genes are not controlled by PlcR but are part of the Spo0A-AbrB regulon. *B. cereus* sensu stricto harbors two *abrB* genes: one located on the chromosome the other on a virulence plasmid. The chromosome-encoded AbrB binds specifically to the *cesP* promoter region of the biosynthetic gene cluster. Overexpression of the chromosomal AbrB exhibited a nontoxic phenotype in a HEp-2 cell assay, whereas overexpression of the plasmid-encoded AbrB remained cytotoxic, suggesting that the chromosome-encoded AbrB functions in repression of cereulide synthesis. A *spo0A* null mutant is cereulide deficient and exhibits a significant reduction in *cesA* (cereulide structural gene) transcripts associated with constant repression by AbrB (Lucking et al., 2009). It has been speculated that additional pathways control cereulide synthesis in *B. cereus* sensu stricto (Lucking et al., 2009).

Phospholipases

Phospholipases have been described or suggested as virulence factors for many gram-positive bacterial pathogens (Goldfine et al., 1998; Flores-Diaz and Alape-Giron, 2003). *B. cereus* sensu stricto, *B. thuringiensis*, and *B. anthracis* secrete three of these enzymes, PI-PLC (phosphatidylinositol-specific PLC, also known as PlcA), PC-PLC (phosphatidylcholine-preferring PLC, also known as PlcB), and sphingomyelinase (also known as SmcA). PLCs act by cleaving polar head groups from phospholipids in host cell membranes to lyse cells and release lipid second messengers (Titball, 1998; Schmiel and Miller, 1999). Sphingomyelinase hydrolyzes sphingomyelin to ceramide and phosphocholine. Amino acid sequence similarity and results of various biochemical studies indicate that the homologous enzymes from each of the species have common activities (Ikezawa et al., 1980; Oda et al., 2010). It has been suggested that PC-PLC may play a major role in *B. cereus* sensu stricto- and *B. thuringiensis*-mediated endophthalmitis (Titball, 1998). However, PC-PLC- and PI-PLC-deficient *B. thuringiensis* mutants are not attenuated in a rabbit eye model of infection (Callegan et al., 2002). In a murine model of anthrax, the three phospholipases appear to be functionally redundant. Virulence was attenuated only in a triple mutant, lacking *plcA*, *plcB*, and *smcA* (Heffernan et al., 2006; Heffernan et al., 2007).

There are few published studies concerning control of phospholipase gene expression. As is true for the cholesterol-dependent cytolysins, the phospholipase genes of *B. cereus* sensu stricto and *B. thuringiensis* are part of the PlcR regulon (Agaisse et al., 1999; Okstad et al., 1999; Gohar et al., 2002; Gohar et al., 2008), while the *B. anthracis* phospholipase genes are not PlcR controlled due to the absence of a functional PlcR in that species. In *B. anthracis*, the phospholipase genes are most highly expressed in cultures grown under anaerobic conditions with CO_2 (Klichko et al., 2003; Pomerantsev et al., 2003), but the mechanism for this control is not known.

Metalloproteases

B. cereus group species produce multiple secreted proteases, several of which are associated with virulence. Under some growth conditions, the metalloproteases InhA (for immune inhibitor A) and NprB (for neutral protease B, also known as NprA and Npr599) are among the most abundant proteins in culture supernatants (Donovan et al., 1997; Chitlaru, et al., 2006; Chung et al., 2006). InhA paralogues are present in some strains, and studies employing an insect model of *B. thuringiensis* infection show that multiple *inhA* genes impact pathogenicity (Fedhila et al., 2002; Guillemet et al., 2010). InhA1 from *B. thuringiensis* degrades insect antimicrobial peptides and enhances the ability of the bacterium to escape from macrophages (Ramarao and Lereclus, 2005). InhA1 from *B. anthracis* has been shown to target the host fibrinolytic system, increasing blood-brain barrier permeability and contributing to cerebral hemorrhages during systemic infection (Chung et al., 2006; Chung et al., 2008; Kastrup et al., 2008; Chung et al., 2009; Chung et al., 2011; Mukherjee et al., 2011).

InhA1 levels are controlled by SinR, a transition state regulator well characterized in *B. subtilis* (Mandic-Mulec et al., 1992; Chu et al., 2006; Chu et al., 2008). In *B. anthracis*, SinR, a DNA-binding repressor, and SinI, a SinR antagonist, have been shown to control transcription of *inhA1* (Pflughoeft et al., 2011). SinR directly regulates transcription of its target genes by binding to a consensus sequence (GTTCTYT, in which Y is C or T) in the promoter region (Chu et al., 2006; Pflughoeft et al., 2011). Although a similar sequence is present in the *inhA1* promoter, SinR binding has not been detected. The promoter of the structural gene for camelysin (*calY*), another metalloprotease (Grass et al., 2004), contains the SinR consensus sequence and is strongly repressed by SinR. Camelysin and InhA1 levels in *B. anthracis* culture supernatants from *sinR*, *inhA1*, and *calY* null mutants show that the concentration of InhA1 is inversely proportional to the concentration of camelysin. A model in which InhA1 protease levels are controlled at the transcriptional level by SinR and at the posttranslational level by camelysin has been proposed (Pflughoeft et al., 2011). Another report indicates that *inhA1* transcription in *B. thuringiensis* is repressed when SinR is present in high copy numbers (Grandvalet et al., 2001). Given the similarities in the *inhA1* loci of the *B. cereus* group species, it is likely that *inhA1* transcription is controlled in the same manner in the three species.

The metalloprotease NprB was first characterized in *B. thuringiensis* as an important degradative enzyme that functions in sporulation (Donovan et al., 1997). *B. anthracis* NprB shows activity against human tissue proteins, suggesting a role in virulence (Chung et al., 2006; Chung et al., 2011). A recent report reveals that *nprB* gene expression depends on a cell-cell communication system consisting of the regulatory protein NprR and the signaling peptide NprX (Perchat et al., 2011). Transcription of *nprB* is activated during sporulation and requires both NprR and NprX. In a mechanism resembling the PlcR-PapR regulatory system, NprX is secreted, processed to a heptapeptide, and reimported into the cell, where it binds to NprR, allowing transcription of *nprB*. In the *B. cereus* group species, seven strain-specific NprR-NprX pairs have been identified, and it has been suggested that the regulatory protein and its signaling protein coevolved (Perchat et al., 2011).

Capsules

Capsule synthesis by *B. anthracis* has long been considered a distinguishing feature for this *B. cereus* group species. With the exception of a few recent reports of unusual isolates, *B. anthracis* cells are capsulated, while *B. cereus* sensu stricto and *B. thuringiensis* are devoid of capsule material. Unlike most bacterial capsules, which are made of polysaccharide, the *B. anthracis* capsule is poly-γ-D-glutamate (PDGA) (Bruckner et al., 1953). The capsule is a major virulence factor of *B. anthracis*. Encapsulated cells are resistant to opsonophagocytic clearance during infection and critical for virulence in some animal models of anthrax (Drysdale et al., 2005b; Heninger et al., 2006).

The proteins encoded by the capsule biosynthetic operon *capBCADE*, located on pXO2, are necessary and sufficient for capsule synthesis, transport, attachment, and assembly (Green et al., 1985; Makino et al., 1988; Okinaka et al., 1999a; Candela and Fouet, 2005, 2006; Candela et al., 2005a). D-Glutamate is polymerized via γ-type amides formed between its α-amino and γ-carboxyl groups (Bruckner et al., 1953). In the current model for capsule synthesis, CapB is an ATP-dependent ligase that, together with CapC,

synthesizes PDGA. CapA and CapE transport PDGA across the membrane (Ashiuchi et al., 2001; Ashiuchi et al., 2004; Candela et al., 2005; Candela and Fouet, 2006). CapD cleaves PDGA and generates amide bonds with peptidoglycan cross bridges to anchor capsular material to the surface of the bacterium (Candela and Fouet, 2005; Richter et al., 2009).

Capsule synthesis, as quantified by the average thickness of the capsule on vegetative cells, is directly related to *capBCADE* transcript levels (Drysdale et al., 2004). Regulation of *cap* operon transcription presents an interesting example of cross talk between the two virulence plasmids. In fully virulent strains carrying pXO1 and pXO2, the pXO1-encoded regulator AtxA directs transcription of the *cap* operon via two pXO2-encoded regulators, AcpA (anthrax capsule activator A) and AcpB, a paralogue of AcpA (Vietri et al., 1995; Guignot et al., 1997; Uchida et al., 1997; Drysdale et al., 2004). AcpA and AcpB are positive regulators of the *cap* operon and can act independently to induce *capBCADE* transcription. AtxA induces transcription of the *acpA* gene, located upstream of *capBCADE*. The *acpA* transcript is monocistronic, while the *acpB* gene, located downstream from the weak transcriptional terminator of the *cap* operon, can be transcribed from its own promoter or via transcriptional read-through of *capBCADE*. Cotranscription of *capBCADE* and *acpB* creates a positive-feedback loop for *capBCADE* transcription (Drysdale et al., 2005a).

Detailed transcriptional analysis of the *cap* operon and the *acpA* and *acpB* genes reveals AtxA-dependent and AtxA-independent transcriptional start sites. The *acpA* gene has two apparent transcription start sites; one is activated approximately 20-fold by AtxA, and the other exhibits low-level constitutive expression. In addition to transcription via read-through of *capBCADE*, *acpB* transcription occurs via an AtxA-independent promoter immediately upstream of *acpB*. Thus, low-level expression of *acpA* and *acpB* from their constitutive promoters allows limited transcription of the *cap* operon and limited capsule synthesis in an *atxA* null mutant. A double mutant, with deletions of *acpA* and *acpB*, produces no detectable capsule. However, the presence of either *acpA* or *acpB* is sufficient for capsule synthesis that is comparable to that of the parent strain (Drysdale et al., 2004).

The functional similarity observed for AcpA and AcpB can be extended to AtxA. Transcription of *capBCADE* and capsule synthesis can occur in strains with deletions of *acpA* and *acpB* when the *atxA* gene is overexpressed, or when the CO_2/bicarbonate signal is increased to a level beyond that which is normally required to induce AtxA activity (Green et al.,

1985; Uchida, et al., 1997). The three regulators are of similar size (56 to 57 kDa) and exhibit amino acid sequence similarities including motifs for potential functional domains. As is true for AtxA, the amino-terminal amino acid sequences of AcpA and AcpB are suggestive of DNA-binding domains. Motifs for regulatory PRDs and EIIB-like domains are also apparent in the AcpA and AcpB amino acid sequences.

Despite their similarities, AtxA, AcpA, and AcpB are not complete functional homologues. AtxA has far-reaching effects on *B. anthracis* gene expression, while the AcpA regulon (and likely the AcpB regulon) is much more limited (Bourgogne et al., 2003). Also, in a murine model for inhalation anthrax, in which a pXO1$^+$ pXO2$^+$ parent strain has a 50% lethal dose of approximately 10^3 spores, a *cap* operon deletion mutant and an *acpAacpB* double mutant do not cause disease at doses up to 5 × 10^7 spores. Deletion of *acpA* alone does not result in attenuation, in agreement with the ability of this mutant to produce capsule in culture. However, an *acpB* null mutant, despite producing capsule in culture, is moderately attenuated in a murine model for inhalation anthrax. This suggests that the positive-feedback loop for *cap* operon expression may be critical for capsule synthesis during infection, and/or that *acpB* has an additional function(s) *in vivo* (Drysdale et al., 2005b).

The dogma that *B. anthracis* is the only encapsulated *B. cereus* group species has recently been challenged. A *B. thuringiensis* strain containing plasmid-borne genes with similarity to the capsule biosynthesis genes of *B. anthracis* produces a capsule that reacts with antibody raised against the PDGA capsule of *B. anthracis* (Cachat et al., 2008). Other reports have described six encapsulated *B. cereus* sensu stricto isolates that cause an inhalation anthrax-like disease in humans and great apes (Hoffmaster et al., 2004; Hoffmaster et al., 2006; Klee et al., 2006; Sue et al., 2006). The two encapsulated strains from the great apes harbor a pXO2-like plasmid (Klee et al., 2006). These strains and most others also carry pXO1-like plasmids and produce anthrax toxin. Yet four of the encapsulated strains do not carry the *B. anthracis* capsule biosynthetic genes. These isolates produce polysaccharide capsules.

Encapsulated *B. cereus* G9241, associated with an anthrax-like respiratory illness in humans (Hoffmaster et al., 2006), has been the most-studied *B. cereus* sensu stricto isolate with a polysaccharide capsule. This strain carries pBC218, encoding proteins responsible for synthesis of exopolysaccharide (Oh et al., 2011). The strain also carries pBCXO1, a plasmid similar to pXO1 of *B. anthracis* and harboring the anthrax toxin genes (Hoffmaster et al., 2004). The presence of both plasmids is required for virulence in a murine model (Wilson et al., 2011). Recently, strain G9241 was also shown to produce a hyaluronic acid capsule due to expression of the *hasACB* genes on pBCXO1. The hyaluronic capsule is not expressed by *B. anthracis* because the *hasACB* locus on pXO1 has a single nucleotide deletion that causes premature termination of *hasA* translation. Elaboration of both capsules by strain G9241 is essential for resistance of vegetative cells to phagocytic killing and for establishment of anthrax-like disease in an intraperitoneal murine model of infection (Oh et al., 2011). There are no reports describing regulation of capsule synthesis in this strain.

S-Layer Protein BslA

Adherence of bacteria to host cell surfaces is a common event for pathogenesis, but there are few published reports describing attachment of the *Bacillus* group species to host cells. Vegetative cells of *B. anthracis* can attach to and enter nonphagocytic cells (Russell et al., 2007; Kern and Schneewind, 2008; Russell et al., 2008), and candidate adhesins have been revealed in genomic and proteomic analyses. Only one of many surface-associated, or S-layer-type, proteins of *B. anthracis* has been shown to function as a true surface adhesin. The pXO1-encoded protein BslA mediates *B. anthracis* attachment and invasion of brain microvascular endothelial cells, facilitating penetration of the blood-brain barrier (Ebrahimi et al., 2009). A BslA-deficient mutant is highly attenuated in a guinea pig model (Kern and Schneewind, 2010). Detailed investigation of *bslA* gene expression has not been described, but transcription of the *bslA* gene is positively regulated by AtxA (Bourgogne et al., 2003).

SUMMARY AND PERSPECTIVES

The genome sequences and overall physiology of *B. anthracis*, *B. cereus* sensu stricto, and *B. thuringiensis* are remarkably similar. Nevertheless, a few crucial differences, mostly related to plasmid content and the presence of functional global regulators (AtxA and PlcR), result in the ability of these species to cause very different diseases. Recent discoveries of unusual isolates bearing genes and traits traditionally associated with more than one *B. cereus* group species blur the line of demarcation for these bacteria as causal agents of specific diseases. As the numbers of environmental surveys and genomic studies grow, it is likely that more "hybrid" strains will be found, making the evolution of the *B. cereus* group species increasingly intriguing.

The alternate lifestyles of the *B. cereus* group members—in the context of a host, or as saprophytic soil organisms—suggest a requirement for signaling pathways that allow discernment of signals specific for different environments. Certainly signals such as CO_2/bicarbonate, temperature, carbohydrate availability, and redox status have a profound influence on the overall metabolic activities of the species. Increased investigation of the mechanisms by which these signals impact virulence gene expression will enhance our understanding of relationships between virulence and overall fitness of the bacteria.

Finally, the developmental nature of the *B. cereus* group species is tied to establishment and transmission of disease. The morphogenic state of the bacterium plays a critical role in survival of the bacterium inside and outside of a host organism. Links between cell development and virulence gene expression have been demonstrated. AtxA and PlcR expression is tied to the developmental regulator Spo0A. Further, factors of the sporulation cascade of *B. anthracis* differ from those of the nonpathogenic *Bacillus* species *B. subtilis*. It has been proposed that acquisition of the virulence genes, via attainment of pXO1, is associated with loss of sporulation sensor histidine kinase activities (Brunsing et al., 2005).

Acknowledgments. We thank Didier Lereclus, Troy Hammerstrom, Michelle Swick, and Lori Horton for critical reading of the manuscript. Work in T.M.K.'s laboratory is supported by award number R01 AI033537 from the National Institute of Allergy and Infectious Diseases.

REFERENCES

Adams, L. F., K. L. Brown, and H. R. Whiteley. 1991. Molecular cloning and characterization of two genes encoding sigma factors that direct transcription from a *Bacillus thuringiensis* crystal protein gene promoter. *J. Bacteriol.* **173:**3846–3854.

Agaisse, H., M. Gominet, O. A. Okstad, A. B. Kolsto, and D. Lereclus. 1999. PlcR is a pleiotropic regulator of extracellular virulence factor gene expression in *Bacillus thuringiensis*. *Mol. Microbiol.* **32:**1043–1053.

Agaisse, H., and D. Lereclus. 1995. How does *Bacillus thuringiensis* produce so much insecticidal crystal protein? *J. Bacteriol.* **177:**6027–6032.

Agaisse, H., and D. Lereclus. 1994. Structural and functional analysis of the promoter region involved in full expression of the *cryIIIA* toxin gene of *Bacillus thuringiensis*. *Mol. Microbiol.* **13:**97–107.

Agata, N., M. Mori, M. Ohta, S. Suwan, I. Ohtani, and M. Isobe. 1994. A novel dodecadepsipeptide, cereulide, isolated from *Bacillus cereus* causes vacuole formation in HEp-2 cells. *FEMS Microbiol. Lett.* **121:**31–34.

Agata, N., M. Ohta, and M. Mori. 1996. Production of an emetic toxin, cereulide, is associated with a specific class of *Bacillus cereus*. *Curr. Microbiol.* **33:**67–69.

Agata, N., M. Ohta, M. Mori, and M. Isobe. 1995. A novel dodecadepsipeptide, cereulide, is an emetic toxin of *Bacillus cereus*. *FEMS Microbiol. Lett.* **129:**17–20.

Al-Abri, S. S., A. K. Al-Jardani, M. S. Al-Hosni, P. J. Kurup, S. Al-Busaidi, and N. J. Beeching. 2011. A hospital acquired outbreak of *Bacillus cereus* gastroenteritis, Oman. *J. Infect. Public Health* **4:**180–186.

Ash, C., and M. D. Collins. 1992. Comparative analysis of 23S ribosomal RNA gene sequences of *Bacillus anthracis* and emetic *Bacillus cereus* determined by PCR-direct sequencing. *FEMS Microbiol. Lett.* **73:**75–80.

Ash, C., J. A. Farrow, M. Dorsch, E. Stackebrandt, and M. D. Collins. 1991. Comparative analysis of *Bacillus anthracis*, *Bacillus cereus*, and related species on the basis of reverse transcriptase sequencing of 16S rRNA. *Int. J. Syst. Bacteriol.* **41:**343–346.

Ashiuchi, M., C. Nawa, T. Kamei, J. J. Song, S. P. Hong, M. H. Sung, K. Soda, and H. Misono. 2001. Physiological and biochemical characteristics of poly gamma-glutamate synthetase complex of *Bacillus subtilis*. *Eur. J. Biochem.* **268:**5321–5328.

Ashiuchi, M., K. Shimanouchi, H. Nakamura, T. Kamei, K. Soda, C. Park, M. H. Sung, and H. Misono. 2004. Enzymatic synthesis of high-molecular-mass poly-gamma-glutamate and regulation of its stereochemistry. *Appl. Environ. Microbiol.* **70:**4249–4255.

Baldari, C. T., F. Tonello, S. R. Paccani, and C. Montecucco. 2006. Anthrax toxins: a paradigm of bacterial immune suppression. *Trends Immunol.* **27:**434–440.

Banks, D. J., S. C. Ward, and K. A. Bradley. 2006. New insights into the functions of anthrax toxin. *Expert Rev. Mol. Med.* **8:**1–18.

Barnes, J. M. 1947. The development of anthrax following the administration of spores by inhalation. *Br. J. Exp. Pathol.* **28:**385–394.

Bartkus, J. M., and S. H. Leppla. 1989. Transcriptional regulation of the protective antigen gene of *Bacillus anthracis*. *Infect. Immun.* **57:**2295–2300.

Baum, J. A., and T. Malvar. 1995. Regulation of insecticidal crystal protein production in *Bacillus thuringiensis*. *Mol. Microbiol.* **18:**1–12.

Beecher, D. J., and J. D. Macmillan. 1991. Characterization of the components of hemolysin BL from *Bacillus cereus*. *Infect. Immun.* **59:**1778–1784.

Beecher, D. J., J. L. Schoeni, and A. C. Wong. 1995. Enterotoxic activity of hemolysin BL from *Bacillus cereus*. *Infect. Immun.* **63:**4423–4428.

Bernheimer, A. W., and P. Grushoff. 1967. Cereolysin: production, purification and partial characterization. *J. Gen. Microbiol.* **46:**143–150.

Bishop, B. L., J. P. Lodolce, L. E. Kolodziej, D. L. Boone, and W. J. Tang. 2010. The role of anthrolysin O in gut epithelial barrier disruption during *Bacillus anthracis* infection. *Biochem. Biophys. Res. Commun.* **394:**254–259.

Blaustein, R. O., T. M. Koehler, R. J. Collier, and A. Finkelstein. 1989. Anthrax toxin: channel-forming activity of protective antigen in planar phospholipid bilayers. *Proc. Natl. Acad. Sci. USA* **86:**2209–2213.

Bouillaut, L., S. Perchat, S. Arold, S. Zorrilla, L. Slamti, C. Henry, M. Gohar, N. Declerck, and D. Lereclus. 2008. Molecular basis for group-specific activation of the virulence regulator PlcR by PapR heptapeptides. *Nucleic Acids Res.* **36:**3791–3801.

Bouillaut, L., N. Ramarao, C. Buisson, N. Gilois, M. Gohar, D. Lereclus, and C. Nielsen-Leroux. 2005. FlhA influences *Bacillus thuringiensis* PlcR-regulated gene transcription, protein production, and virulence. *Appl. Environ. Microbiol.* **71:**8903–8910.

Bourdeau, R. W., E. Malito, A. Chenal, B. L. Bishop, M. W. Musch, M. L. Villereal, E. B. Chang, E. M. Mosser, R. F. Rest, and W. J. Tang. 2009. Cellular functions and X-ray structure of Anthrolysin O, a cholesterol-dependent cytolysin secreted by *Bacillus anthracis*. *J. Biol. Chem.* **284:**14645–14656.

Bourgogne, A., M. Drysdale, S. G. Hilsenbeck, S. N. Peterson, and T. M. Koehler. 2003. Global effects of virulence gene regulators in a *Bacillus anthracis* strain with both virulence plasmids. *Infect. Immun.* **71:**2736–2743.

Bravo, A., H. Agaisse, S. Salamitou, and D. Lereclus. 1996. Analysis of *cryIAa* expression in *sigE* and *sigK* mutants of *Bacillus thuringiensis*. *Mol. Gen. Genet.* **250:**734–741.

Bravo, A., S. Likitvivatanavong, S. S. Gill, and M. Soberon. 2011. *Bacillus thuringiensis*: a story of a successful bioinsecticide. *Insect Biochem. Mol. Biol.* **41:**423–431.

Brown, K. L., and H. R. Whiteley. 1988. Isolation of a *Bacillus thuringiensis* RNA polymerase capable of transcribing crystal protein genes. *Proc. Natl. Acad. Sci. USA* **85:**4166–4170.

Brown, K. L., and H. R. Whiteley. 1990. Isolation of the second *Bacillus thuringiensis* RNA polymerase that transcribes from a crystal protein gene promoter. *J. Bacteriol.* **172:**6682–6688.

Bruckner, V., J. Kovacs, and G. Denes. 1953. Structure of poly-D-glutamic acid isolated from capsulated strains of *B. anthracis*. *Nature* **172:**508.

Brunsing, R. L., C. La Clair, S. Tang, C. Chiang, L. E. Hancock, M. Perego, and J. A. Hoch. 2005. Characterization of sporulation histidine kinases of *Bacillus anthracis*. *J. Bacteriol.* **187:**6972–6981.

Cachat, E., M. Barker, T. D. Read, and F. G. Priest. 2008. A *Bacillus thuringiensis* strain producing a polyglutamate capsule resembling that of *Bacillus anthracis*. *FEMS Microbiol. Lett.* **285:**220–226.

Callegan, M. C., D. C. Cochran, S. T. Kane, M. S. Gilmore, M. Gominet, and D. Lereclus. 2002. Contribution of membrane-damaging toxins to *Bacillus* endophthalmitis pathogenesis. *Infect. Immun.* **70:**5381–5389.

Callegan, M. C., S. T. Kane, D. C. Cochran, M. S. Gilmore, M. Gominet, and D. Lereclus. 2003. Relationship of *plcR*-regulated factors to *Bacillus* endophthalmitis virulence. *Infect. Immun.* **71:**3116–3124.

Callegan, M. C., B. D. Novosad, R. Ramirez, E. Ghelardi, and S. Senesi. 2006. Role of swarming migration in the pathogenesis of *Bacillus* endophthalmitis. *Investig. Ophthalmol. Vis. Sci.* **47:**4461–4467.

Candela, T., and A. Fouet. 2005. *Bacillus anthracis* CapD, belonging to the gamma-glutamyltranspeptidase family, is required for the covalent anchoring of capsule to peptidoglycan. *Mol. Microbiol.* **57:**717–726.

Candela, T., and A. Fouet. 2006. Poly-gamma-glutamate in bacteria. *Mol. Microbiol.* **60:**1091–1098.

Candela, T., M. Mock, and A. Fouet. 2005. CapE, a 47-amino-acid peptide, is necessary for *Bacillus anthracis* polyglutamate capsule synthesis. *J. Bacteriol.* **187:**7765–7772.

Cataldi, A., A. Fouet, and M. Mock. 1992. Regulation of *pag* gene expression in *Bacillus anthracis*: use of a *pag-lacZ* transcriptional fusion. *FEMS Microbiol. Lett.* **77:**89–93.

Chiang, C., C. Bongiorni, and M. Perego. 2011. Glucose-dependent activation of *Bacillus anthracis* toxin gene expression and virulence requires the carbon catabolite protein CcpA. *J. Bacteriol.* **193:**52–62.

Chitlaru, T., O. Gat, Y. Gozlan, N. Ariel, and A. Shafferman. 2006. Differential proteomic analysis of the *Bacillus anthracis* secretome: distinct plasmid and chromosome CO_2-dependent cross talk mechanisms modulate extracellular proteolytic activities. *J. Bacteriol.* **188:**3551–3571.

Chu, F., D. B. Kearns, S. S. Branda, R. Kolter, and R. Losick. 2006. Targets of the master regulator of biofilm formation in *Bacillus subtilis*. *Mol. Microbiol.* **59:**1216–1228.

Chu, F., D. B. Kearns, A. McLoon, Y. Chai, R. Kolter, and R. Losick. 2008. A novel regulatory protein governing biofilm formation in *Bacillus subtilis*. *Mol. Microbiol.* **68:**1117–1127.

Chung, M.-C., T. G. Popova, B. A. Millis, D. V. Mukherjee, W. Zhou, L. A. Liotta, E. F. Petricoin, V. Chandhoke, C. Bailey, and S. G. Popov. 2006. Secreted neutral metalloproteases of *Bacillus anthracis* as candidate pathogenic factors. *J. Biol. Chem.* **281:**31408–31418.

Chung, M. C., S. C. Jorgensen, T. G. Popova, C. L. Bailey, and S. G. Popov. 2008. Neutrophil elastase and syndecan shedding contribute to antithrombin depletion in murine anthrax. *FEMS Immunol. Med. Microbiol.* **54:**309–318.

Chung, M. C., S. C. Jorgensen, T. G. Popova, J. H. Tonry, C. L. Bailey, and S. G. Popov. 2009. Activation of plasminogen activator inhibitor implicates protease InhA in the acute-phase response to *Bacillus anthracis* infection. *J. Med. Microbiol.* **58:**737–744.

Chung, M. C., S. C. Jorgensen, J. H. Tonry, F. Kashanchi, C. Bailey, and S. Popov. 2011. Secreted *Bacillus anthracis* proteases target the host fibrinolytic system. *FEMS Immunol. Med. Microbiol.* **62:**173–181.

Clavel, T., F. Carlin, D. Lairon, C. Nguyen-The, and P. Schmitt. 2004. Survival of *Bacillus cereus* spores and vegetative cells in acid media simulating human stomach. *J. Appl. Microbiol.* **97:**214–219.

Cleret, A., A. Quesnel-Hellmann, A. Vallon-Eberhard, B. Verrier, S. Jung, D. Vidal, J. Mathieu, and J. N. Tournier. 2007. Lung dendritic cells rapidly mediate anthrax spore entry through the pulmonary route. *J. Immunol.* **178:**7994–8001.

Colpin, G. G., H. F. Guiot, R. F. Simonis, and F. E. Zwaan. 1981. *Bacillus cereus* meningitis in a patient under gnotobiotic care. *Lancet* **ii:**694–695.

Cote, C. K., T. L. Dimezzo, D. J. Banks, B. France, K. A. Bradley, and S. L. Welkos. 2008. Early interactions between fully virulent *Bacillus anthracis* and macrophages that influence the balance between spore clearance and development of a lethal infection. *Microbes Infect.* **10:**613–619.

Cote, C. K., C. A. Rossi, A. S. Kang, P. R. Morrow, J. S. Lee, and S. L. Welkos. 2005. The detection of protective antigen (PA) associated with spores of *Bacillus anthracis* and the effects of anti-PA antibodies on spore germination and macrophage interactions. *Microb. Pathog.* **38:**209–225.

Craig, C. P., W. S. Lee, and M. Ho. 1974. Letter: *Bacillus cereus* endocarditis in an addict. *Ann. Intern. Med.* **80:**418–419.

Crickmore, N. 2006. Beyond the spore—past and future developments of *Bacillus thuringiensis* as a biopesticide. *J. Appl. Microbiol.* **101:**616–619.

Dai, Z., and T. M. Koehler. 1997. Regulation of anthrax toxin activator gene (*atxA*) expression in *Bacillus anthracis*: temperature, not CO_2/bicarbonate, affects AtxA synthesis. *Infect. Immun.* **65:**2576–2582.

Dai, Z., J. C. Sirard, M. Mock, and T. M. Koehler. 1995. The *atxA* gene product activates transcription of the anthrax toxin genes and is essential for virulence. *Mol. Microbiol.* **16:**1171–1181.

Damgaard, P. H., B. M. Hansen, J. C. Pedersen, and J. Eilenberg. 1997. Natural occurrence of *Bacillus thuringiensis* on cabbage foliage and in insects associated with cabbage crops. *J. Appl. Microbiol.* **82:**253–258.

Declerck, N., L. Bouillaut, D. Chaix, N. Rugani, L. Slamti, F. Hoh, D. Lereclus, and S. T. Arold. 2007. Structure of PlcR: insights into virulence regulation and evolution of quorum sensing in Gram-positive bacteria. *Proc. Natl. Acad. Sci. USA* **104:**18490–18495.

de Souza, M. T., M. M. Lecadet, and D. Lereclus. 1993. Full expression of the *cryIIIA* toxin gene of *Bacillus thuringiensis* requires a distant upstream DNA sequence affecting transcription. *J. Bacteriol.* **175:**2952–2960.

Deutscher, J., C. Francke, and P. W. Postma. 2006. How phosphotransferase system-related protein phosphorylation regulates carbohydrate metabolism in bacteria. *Microbiol. Mol. Biol. Rev.* **70:**939–1031.

Dixon, T. C., M. Meselson, J. Guillemin, and P. C. Hanna. 1999. Anthrax. *N. Engl. J. Med.* **341**:815–826.

Dohmae, S., T. Okubo, W. Higuchi, T. Takano, H. Isobe, T. Baranovich, S. Kobayashi, M. Uchiyama, Y. Tanabe, M. Itoh, and T. Yamamoto. 2008. *Bacillus cereus* nosocomial infection from reused towels in Japan. *J. Hosp. Infect.* **69**:361–367.

Dong, Y. H., A. R. Gusti, Q. Zhang, J. L. Xu, and L. H. Zhang. 2002. Identification of quorum-quenching N-acyl homoserine lactonases from *Bacillus* species. *Appl. Environ. Microbiol.* **68**:1754–1759.

Donovan, W. P., Y. Tan, and A. C. Slaney. 1997. Cloning of the *nprA* gene for neutral protease A of *Bacillus thuringiensis* and effect of in vivo deletion of *nprA* on insecticidal crystal protein. *Appl. Environ. Microbiol.* **63**:2311–2317.

Drobniewski, F. A. 1993. *Bacillus cereus* and related species. *Clin. Microbiol. Rev.* **6**:324–338.

Drysdale, M., A. Bourgogne, S. G. Hilsenbeck, and T. M. Koehler. 2004. *atxA* controls *Bacillus anthracis* capsule synthesis via *acpA* and a newly discovered regulator, *acpB*. *J. Bacteriol.* **186**:307–315.

Drysdale, M., A. Bourgogne, and T. M. Koehler. 2005a. Transcriptional analysis of the *Bacillus anthracis* capsule regulators. *J. Bacteriol.* **187**:5108–5114.

Drysdale, M., S. Heninger, J. Hutt, Y. Chen, C. R. Lyons, and T. M. Koehler. 2005b. Capsule synthesis by *Bacillus anthracis* is required for dissemination in murine inhalation anthrax. *EMBO J.* **24**:221–227.

Duesbery, N. S., C. P. Webb, S. H. Leppla, V. M. Gordon, K. R. Klimpel, T. D. Copeland, N. G. Ahn, M. K. Oskarsson, K. Fukasawa, K. D. Paull, and G. F. Vande Woude. 1998. Proteolytic inactivation of MAP-kinase-kinase by anthrax lethal factor. *Science* **280**:734–737.

Duport, C., S. Thomassin, G. Bourel, and P. Schmitt. 2004. Anaerobiosis and low specific growth rates enhance hemolysin BL production by *Bacillus cereus* F4430/73. *Arch. Microbiol.* **182**:90–95.

Duport, C., A. Zigha, E. Rosenfeld, and P. Schmitt. 2006. Control of enterotoxin gene expression in *Bacillus cereus* F4430/73 involves the redox-sensitive ResDE signal transduction system. *J. Bacteriol.* **188**:6640–6651.

Ebrahimi, C. M., J. W. Kern, T. R. Sheen, M. A. Ebrahimi-Fardooee, N. M. van Sorge, O. Schneewind, and K. S. Doran. 2009. Penetration of the blood-brain barrier by *Bacillus anthracis* requires the pXO1-encoded BslA protein. *J. Bacteriol.* **191**:7165–7173.

Ehling-Schulz, M., M. Fricker, H. Grallert, P. Rieck, M. Wagner, and S. Scherer. 2006a. Cereulide synthetase gene cluster from emetic *Bacillus cereus*: structure and location on a mega virulence plasmid related to *Bacillus anthracis* toxin plasmid pXO1. *BMC Microbiol.* **6**:20.

Ehling-Schulz, M., M. H. Guinebretiere, A. Monthan, O. Berge, M. Fricker, and B. Svensson. 2006b. Toxin gene profiling of enterotoxic and emetic *Bacillus cereus*. *FEMS Microbiol. Lett.* **260**:232–240.

Ehling-Schulz, M., M. Fricker, and S. Scherer. 2004. *Bacillus cereus*, the causative agent of an emetic type of food-borne illness. *Mol. Nutr. Food Res.* **48**:479–487.

Ehling-Schulz, M., B. Svensson, M. H. Guinebretiere, T. Lindback, M. Andersson, A. Schulz, M. Fricker, A. Christiansson, P. E. Granum, E. Martlbauer, C. Nguyen-The, M. Salkinoja-Salonen, and S. Scherer. 2005a. Emetic toxin formation of *Bacillus cereus* is restricted to a single evolutionary lineage of closely related strains. *Microbiology* **151**:183–197.

Ehling-Schulz, M., N. Vukov, A. Schulz, R. Shaheen, M. Andersson, E. Martlbauer, and S. Scherer. 2005b. Identification and partial characterization of the nonribosomal peptide synthetase gene responsible for cereulide production in emetic *Bacillus cereus*. *Appl. Environ. Microbiol.* **71**:105–113.

Esbelin, J., J. Armengaud, A. Zigha, and C. Duport. 2009. ResDE-dependent regulation of enterotoxin gene expression in *Bacillus cereus*: evidence for multiple modes of binding for ResD and interaction with Fnr. *J. Bacteriol.* **191**:4419–4426.

Esbelin, J., Y. Jouanneau, J. Armengaud, and C. Duport. 2008. ApoFnr binds as a monomer to promoters regulating the expression of enterotoxin genes of *Bacillus cereus*. *J. Bacteriol.* **190**:4242–4251.

Estruch, J. J., G. W. Warren, M. A. Mullins, G. J. Nye, J. A. Craig, and M. G. Koziel. 1996. Vip3A, a novel *Bacillus thuringiensis* vegetative insecticidal protein with a wide spectrum of activities against lepidopteran insects. *Proc. Natl. Acad. Sci. USA* **93**:5389–5394.

Fagerlund, A., T. Lindback, and P. E. Granum. 2010. *Bacillus cereus* cytotoxins Hbl, Nhe and CytK are secreted via the Sec translocation pathway. *BMC Microbiol.* **10**:304.

Fagerlund, A., T. Lindback, A. K. Storset, P. E. Granum, and S. P. Hardy. 2008. *Bacillus cereus* Nhe is a pore-forming toxin with structural and functional properties similar to the ClyA (HlyE, SheA) family of haemolysins, able to induce osmotic lysis in epithelia. *Microbiology* **154**:693–704.

Fedhila, S., T. Msadek, P. Nel, and D. Lereclus. 2002. Distinct *clpP* genes control specific adaptive responses in *Bacillus thuringiensis*. *J. Bacteriol.* **184**:5554–5562.

Finlay, W. J., N. A. Logan, and A. D. Sutherland. 2000. *Bacillus cereus* produces most emetic toxin at lower temperatures. *Lett. Appl. Microbiol.* **31**:385–389.

Flores-Diaz, M., and A. Alape-Giron. 2003. Role of *Clostridium perfringens* phospholipase C in the pathogenesis of gas gangrene. *Toxicon* **42**:979–986.

Fouet, A., and M. Mock. 1996. Differential influence of the two *Bacillus anthracis* plasmids on regulation of virulence gene expression. *Infect. Immun.* **64**:4928–4932.

Friedlander, A. M. 1986. Macrophages are sensitive to anthrax lethal toxin through an acid-dependent process. *J. Biol. Chem.* **261**:7123–7126.

Fujita, Y., Y. Miwa, A. Galinier, and J. Deutscher. 1995. Specific recognition of the *Bacillus subtilis gnt cis*-acting catabolite-responsive element by a protein complex formed between CcpA and seryl-phosphorylated HPr. *Mol. Microbiol.* **17**:953–960.

Gekara, N. O., K. Westphal, B. Ma, M. Rohde, L. Groebe, and S. Weiss. 2007. The multiple mechanisms of Ca^{2+} signalling by listeriolysin O, the cholesterol-dependent cytolysin of *Listeria monocytogenes*. *Cell. Microbiol.* **9**:2008–2021.

Ghelardi, E., F. Celandroni, S. Salvetti, M. Ceragioli, D. J. Beecher, S. Senesi, and A. C. Wong. 2007. Swarming behavior of and hemolysin BL secretion by *Bacillus cereus*. *Appl. Environ. Microbiol.* **73**:4089–4093.

Gilbert, R. J. 2010. Cholesterol-dependent cytolysins. *Adv. Exp. Med. Biol.* **677**:56–66.

Glatz, B. A., W. M. Spira, and J. M. Goepfert. 1974. Alteration of vascular permeability in rabbits by culture filtrates of *Bacillus cereus* and related species. *Infect. Immun.* **10**:299–303.

Glomski, I. J., F. Dumetz, G. Jouvion, M. R. Huerre, M. Mock, and P. L. Goossens. 2008. Inhaled non-capsulated *Bacillus anthracis* in A/J mice: nasopharynx and alveolar space as dual portals of entry, delayed dissemination, and specific organ targeting. *Microbes Infect.* **10**:1398–1404.

Gohar, M., K. Faegri, S. Perchat, S. Ravnum, O. A. Okstad, M. Gominet, A. B. Kolsto, and D. Lereclus. 2008. The PlcR virulence regulon of *Bacillus cereus*. *PLoS One* **3**:e2793.

Gohar, M., O. A. Okstad, N. Gilois, V. Sanchis, A. B. Kolsto, and D. Lereclus. 2002. Two-dimensional electrophoresis analysis of the extracellular proteome of *Bacillus cereus* reveals the importance of the PlcR regulon. *Proteomics* **2**:784–791.

Goldfine, H., T. Bannam, N. C. Johnston, and W. R. Zuckert. 1998. Bacterial phospholipases and intracellular growth: the two distinct phospholipases C of *Listeria monocytogenes*. *Symp. Ser. Soc. Appl. Microbiol.* 27:7S–14S.

Gominet, M., L. Slamti, N. Gilois, M. Rose, and D. Lereclus. 2001. Oligopeptide permease is required for expression of the *Bacillus thuringiensis* plcR regulon and for virulence. *Mol. Microbiol.* 40:963–975.

Gonzalez, J. M., Jr., B. J. Brown, and B. C. Carlton. 1982. Transfer of *Bacillus thuringiensis* plasmids coding for delta-endotoxin among strains of *B. thuringiensis* and *B. cereus*. *Proc. Natl. Acad. Sci. USA* 79:6951–6955.

Grandvalet, C., M. Gominet, and D. Lereclus. 2001. Identification of genes involved in the activation of the *Bacillus thuringiensis* inhA metalloprotease gene at the onset of sporulation. *Microbiology* 147:1805–1813.

Granum, P. E., and T. Lund. 1997. *Bacillus cereus* and its food poisoning toxins. *FEMS Microbiol. Lett.* 157:223–228.

Granum, P. E., K. O'Sullivan, and T. Lund. 1999. The sequence of the non-haemolytic enterotoxin operon from *Bacillus cereus*. *FEMS Microbiol. Lett.* 177:225–229.

Grass, G., A. Schierhorn, E. Sorkau, H. Muller, P. Rucknagel, D. H. Nies, and B. Fricke. 2004. Camelysin is a novel surface metalloproteinase from *Bacillus cereus*. *Infect. Immun.* 72:219–228.

Green, B. D., L. Battisti, T. M. Koehler, C. B. Thorne, and B. E. Ivins. 1985. Demonstration of a capsule plasmid in *Bacillus anthracis*. *Infect. Immun.* 49:291–297.

Guidi-Rontani, C., M. Levy, H. Ohayon, and M. Mock. 2001. Fate of germinated *Bacillus anthracis* spores in primary murine macrophages. *Mol. Microbiol.* 42:931–938.

Guidi-Rontani, C., M. Weber-Levy, E. Labruyere, and M. Mock. 1999. Germination of *Bacillus anthracis* spores within alveolar macrophages. *Mol. Microbiol.* 31:9–17.

Guignot, J., M. Mock, and A. Fouet. 1997. AtxA activates the transcription of genes harbored by both *Bacillus anthracis* virulence plasmids. *FEMS Microbiol. Lett.* 147:203–207.

Guillemet, E., C. Cadot, S. L. Tran, M. H. Guinebretiere, D. Lereclus, and N. Ramarao. 2010. The InhA metalloproteases of *Bacillus cereus* contribute concomitantly to virulence. *J. Bacteriol.* 192:286–294.

Hadjifrangiskou, M., and T. M. Koehler. 2008. Intrinsic curvature associated with the coordinately regulated anthrax toxin gene promoters. *Microbiology* 154:2501–2512.

Hammerstrom, T. G., J. H. Roh, E. P. Nikonowicz, and T. M. Koehler. 2011. *Bacillus anthracis* virulence regulator AtxA: oligomeric state, function and CO_2-signalling. *Mol. Microbiol.* 82:634–647.

Heffernan, B. J., B. Thomason, A. Herring-Palmer, and P. Hanna. 2007. *Bacillus anthracis* anthrolysin O and three phospholipases C are functionally redundant in a murine model of inhalation anthrax. *FEMS Microbiol. Lett.* 271:98–105.

Heffernan, B. J., B. Thomason, A. Herring-Palmer, L. Shaughnessy, R. McDonald, N. Fisher, G. B. Huffnagle, and P. Hanna. 2006. *Bacillus anthracis* phospholipases C facilitate macrophage-associated growth and contribute to virulence in a murine model of inhalation anthrax. *Infect. Immun.* 74:3756–3764.

Heinrichs, J. H., D. J. Beecher, J. D. MacMillan, and B. A. Zilinskas. 1993. Molecular cloning and characterization of the hblA gene encoding the B component of hemolysin BL from *Bacillus cereus*. *J. Bacteriol.* 175:6760–6766.

Heninger, S., M. Drysdale, J. Lovchik, J. Hutt, M. F. Lipscomb, T. M. Koehler, and C. R. Lyons. 2006. Toxin-deficient mutants of *Bacillus anthracis* are lethal in a murine model for pulmonary anthrax. *Infect. Immun.* 74:6067–6074.

Hernandez, E., F. Ramisse, J. P. Ducoureau, T. Cruel, and J. D. Cavallo. 1998. *Bacillus thuringiensis* subsp. *konkukian* (serotype H34) superinfection: case report and experimental evidence of pathogenicity in immunosuppressed mice. *J. Clin. Microbiol.* 36:2138–2139.

Hoffmaster, A. R., K. K. Hill, J. E. Gee, C. K. Marston, B. K. De, T. Popovic, D. Sue, P. P. Wilkins, S. B. Avashia, R. Drumgoole, C. H. Helma, L. O. Ticknor, R. T. Okinaka, and P. J. Jackson. 2006. Characterization of *Bacillus cereus* isolates associated with fatal pneumonias: strains are closely related to *Bacillus anthracis* and harbor *B. anthracis* virulence genes. *J. Clin. Microbiol.* 44:3352–3360.

Hoffmaster, A. R., and T. M. Koehler. 1997. The anthrax toxin activator gene atxA is associated with CO_2-enhanced nontoxin gene expression in *Bacillus anthracis*. *Infect. Immun.* 65:3091–3099.

Hoffmaster, A. R., and T. M. Koehler. 1999a. Autogenous regulation of the *Bacillus anthracis* pag operon. *J. Bacteriol.* 181:4485–4492.

Hoffmaster, A. R., and T. M. Koehler. 1999b. Control of virulence gene expression in *Bacillus anthracis*. *J. Appl. Microbiol.* 87:279–281.

Hoffmaster, A. R., J. Ravel, D. A. Rasko, G. D. Chapman, M. D. Chute, C. K. Marston, B. K. De, C. T. Sacchi, C. Fitzgerald, L. W. Mayer, M. C. Maiden, F. G. Priest, M. Barker, L. Jiang, R. Z. Cer, J. Rilstone, S. N. Peterson, R. S. Weyant, D. R. Galloway, T. D. Read, T. Popovic, and C. M. Fraser. 2004. Identification of anthrax toxin genes in a *Bacillus cereus* associated with an illness resembling inhalation anthrax. *Proc. Natl. Acad. Sci. USA* 101:8449–8454.

Hofte, H., and H. R. Whiteley. 1989. Insecticidal crystal proteins of *Bacillus thuringiensis*. *Microbiol. Rev.* 53:242–255.

Horii, T., S. Notake, K. Tamai, H. Yanagisawa, and P. Brennan. 2011. *Bacillus cereus* from blood cultures: virulence genes, antimicrobial susceptibility and risk factors for blood stream infection. *FEMS Immun. Med. Microbiol.* 63:202–209.

Hu, H., Q. Sa, T. M. Koehler, A. I. Aronson, and D. Zhou. 2006. Inactivation of *Bacillus anthracis* spores in murine primary macrophages. *Cell. Microbiol.* 8:1634–1642.

Ikezawa, H., M. Mori, and R. Taguchi. 1980. Studies on sphingomyelinase of *Bacillus cereus*: hydrolytic and hemolytic actions on erythrocyte membranes. *Arch. Biochem. Biophys.* 199:572–578.

Jensen, G. B., B. M. Hansen, J. Eilenberg, and J. Mahillon. 2003. The hidden lifestyles of *Bacillus cereus* and relatives. *Environ. Microbiol.* 5:631–640.

Kastrup, C. J., J. Q. Boedicker, A. P. Pomerantsev, M. Moayeri, Y. Bian, R. R. Pompano, T. R. Kline, P. Sylvestre, F. Shen, S. H. Leppla, W. J. Tang, and R. F. Ismagilov. 2008. Spatial localization of bacteria controls coagulation of human blood by 'quorum acting.' *Nat. Chem. Biol.* 4:742–750.

Keppie, J., H. Smith, and P. W. Harris-Smith. 1953. The chemical basis of the virulence of *Bacillus anthracis*. II. Some biological properties of bacterial products. *Br. J. Exp. Pathol.* 34:486–496.

Kern, J., and O. Schneewind. 2010. BslA, the S-layer adhesin of *B. anthracis*, is a virulence factor for anthrax pathogenesis. *Mol. Microbiol.* 75:324–332.

Kern, J. W., and O. Schneewind. 2008. BslA, a pXO1-encoded adhesin of *Bacillus anthracis*. *Mol. Microbiol.* 68:504–515.

Klee, S. R., M. Ozel, B. Appel, C. Boesch, H. Ellerbrok, D. Jacob, G. Holland, F. H. Leendertz, G. Pauli, R. Grunow, and M. Nattermann. 2006. Characterization of *Bacillus anthracis*-like bacteria isolated from wild great apes from Côte d'Ivoire and Cameroon. *J. Bacteriol.* 188:5333–5344.

Klichko, V. I., J. Miller, A. Wu, S. G. Popov, and K. Alibek. 2003. Anaerobic induction of *Bacillus anthracis* hemolytic activity. *Biochem. Biophys. Res. Commun.* 303:855–862.

Koehler, T. M. 2000. *Bacillus anthracis*, p. 519–528. *In* V. A. Fischetti (ed.), *Gram-Positive Pathogens*. American Society for Microbiology, Washington, DC.

Koehler, T. M. 2002. *Bacillus anthracis* genetics and virulence gene regulation. *Curr. Top. Microbiol. Immunol.* **271:**143–164.

Koehler, T. M. 2009. *Bacillus anthracis* physiology and genetics. *Mol. Aspects Med.* **30:**386–396.

Koehler, T. M., Z. Dai, and M. Kaufman-Yarbray. 1994. Regulation of the *Bacillus anthracis* protective antigen gene: CO_2 and a *trans*-acting element activate transcription from one of two promoters. *J. Bacteriol.* **176:**586–595.

Kolsto, A. B., N. J. Tourasse, and O. A. Okstad. 2009. What sets *Bacillus anthracis* apart from other *Bacillus* species? *Annu. Rev. Microbiol.* **63:**451–476.

Kotiranta, A., K. Lounatmaa, and M. Haapasalo. 2000. Epidemiology and pathogenesis of *Bacillus cereus* infections. *Microbes Infect.* **2:**189–198.

Krantz, B. A., R. A. Melnyk, S. Zhang, S. J. Juris, D. B. Lacy, Z. Wu, A. Finkelstein, and R. J. Collier. 2005. A phenylalanine clamp catalyzes protein translocation through the anthrax toxin pore. *Science* **309:**777–781.

Kroten, M. A., M. Bartoszewicz, and I. Swiecicka. 2010. Cereulide and valinomycin, two important natural dodecadepsipeptides with ionophoretic activities. *Pol. J. Microbiol.* **59:**3–10.

Kuroki, R., K. Kawakami, L. Qin, C. Kaji, K. Watanabe, Y. Kimura, C. Ishiguro, S. Tanimura, Y. Tsuchiya, I. Hamaguchi, M. Sakakura, S. Sakabe, K. Tsuji, M. Inoue, and H. Watanabe. 2009. Nosocomial bacteremia caused by biofilm-forming *Bacillus cereus* and *Bacillus thuringiensis*. *Intern. Med.* **48:**791–796.

Leppla, S. H. 1982. Anthrax toxin edema factor: a bacterial adenylate cyclase that increases cyclic AMP concentrations of eukaryotic cells. *Proc. Natl. Acad. Sci. USA* **79:**3162–3166.

Lereclus, D., H. Agaisse, M. Gominet, and J. Chaufaux. 1995. Overproduction of encapsulated insecticidal crystal proteins in a *Bacillus thuringiensis spo0A* mutant. *Nat. Biotechnol.* **13:**67-71.

Lereclus, D., H. Agaisse, M. Gominet, S. Salamitou, and V. Sanchis. 1996. Identification of a *Bacillus thuringiensis* gene that positively regulates transcription of the phosphatidylinositol-specific phospholipase C gene at the onset of the stationary phase. *J. Bacteriol.* **178:**2749–2756.

Lereclus, D., H. Agaisse, C. Grandvalet, S. Salamitou, and M. Gominet. 2000. Regulation of toxin and virulence gene transcription in *Bacillus thuringiensis*. *Int. J. Med. Microbiol.* **290:**295–299.

Lindback, T., A. Fagerlund, M. S. Rodland, and P. E. Granum. 2004. Characterization of the *Bacillus cereus* Nhe enterotoxin. *Microbiology* **150:**3959–3967.

Lindback, T., O. A. Okstad, A. L. Rishovd, and A. B. Kolsto. 1999. Insertional inactivation of *hblC* encoding the L2 component of *Bacillus cereus* ATCC 14579 haemolysin BL strongly reduces enterotoxigenic activity, but not the haemolytic activity against human erythrocytes. *Microbiology* **145:**3139–3146.

Lucking, G., M. K. Dommel, S. Scherer, A. Fouet, and M. Ehling-Schulz. 2009. Cereulide synthesis in emetic *Bacillus cereus* is controlled by the transition state regulator AbrB, but not by the virulence regulator PlcR. *Microbiology* **155:**922–931.

Lund, T., M. L. De Buyser, and P. E. Granum. 2000. A new cytotoxin from *Bacillus cereus* that may cause necrotic enteritis. *Mol. Microbiol.* **38:**254–261.

Lund, T., and P. E. Granum. 1996. Characterisation of a non-haemolytic enterotoxin complex from *Bacillus cereus* isolated after a foodborne outbreak. *FEMS Microbiol. Lett.* **141:**151–156.

Mahler, H., A. Pasi, J. M. Kramer, P. Schulte, A. C. Scoging, W. Bar, and S. Krahenbuhl. 1997. Fulminant liver failure in association with the emetic toxin of *Bacillus cereus*. *N. Engl. J. Med.* **336:**1142–1148.

Makino, S., C. Sasakawa, I. Uchida, N. Terakado, and M. Yoshikawa. 1988. Cloning and CO_2-dependent expression of the genetic region for encapsulation from *Bacillus anthracis*. *Mol. Microbiol.* **2:**371–376.

Makino, S., I. Uchida, N. Terakado, C. Sasakawa, and M. Yoshikawa. 1989a. Molecular characterization and protein analysis of the *cap* region, which is essential for encapsulation in *Bacillus anthracis*. *J. Bacteriol.* **171:**722–730.

Makino, Y., S. Negoro, I. Urabe, and H. Okada. 1989b. Stability-increasing mutants of glucose dehydrogenase from *Bacillus megaterium* IWG3. *J. Biol. Chem.* **264:**6381–6385.

Malvar, T., and J. A. Baum. 1994. Tn*5401* disruption of the *spo0F* gene, identified by direct chromosomal sequencing, results in CryIIIA overproduction in *Bacillus thuringiensis*. *J. Bacteriol.* **176:**4750–4753.

Mandic-Mulec, I., N. Gaur, U. Bai, and I. Smith. 1992. Sin, a stage-specific repressor of cellular differentiation. *J. Bacteriol.* **174:**3561–3569.

Mignot, T., M. Mock, and A. Fouet. 2003. A plasmid-encoded regulator couples the synthesis of toxins and surface structures in *Bacillus anthracis*. *Mol. Microbiol.* **47:**917–927.

Mignot, T., M. Mock, D. Robichon, A. Landier, D. Lereclus, and A. Fouet. 2001. The incompatibility between the PlcR- and AtxA-controlled regulons may have selected a nonsense mutation in *Bacillus anthracis*. *Mol. Microbiol.* **42:**1189–1198.

Mikesell, P., B. E. Ivins, J. D. Ristroph, and T. M. Dreier. 1983. Evidence for plasmid-mediated toxin production in *Bacillus anthracis*. *Infect. Immun.* **39:**371–376.

Mikkola, R., N. E. Saris, P. A. Grigoriev, M. A. Andersson, and M. S. Salkinoja-Salonen. 1999. Ionophoretic properties and mitochondrial effects of cereulide: the emetic toxin of *B. cereus*. *Eur. J. Biochem.* **263:**112–117.

Miller, C. J., J. L. Elliott, and R. J. Collier. 1999. Anthrax protective antigen: prepore-to-pore conversion. *Biochemistry* **38:**10432–10441.

Miwa, Y., A. Nakata, A. Ogiwara, M. Yamamoto, and Y. Fujita. 2000. Evaluation and characterization of catabolite-responsive elements (*cre*) of *Bacillus subtilis*. *Nucleic Acids Res.* **28:**1206–1210.

Mock, M., and A. Fouet. 2001. Anthrax. *Annu. Rev. Microbiol.* **55:**647–671.

Moravek, M., R. Dietrich, C. Buerk, V. Broussolle, M. H. Guinebretiere, P. E. Granum, C. Nguyen-The, and E. Martlbauer. 2006. Determination of the toxic potential of *Bacillus cereus* isolates by quantitative enterotoxin analyses. *FEMS Microbiol. Lett.* **257:**293–298.

Mosser, E. M., and R. F. Rest. 2006. The *Bacillus anthracis* cholesterol-dependent cytolysin, Anthrolysin O, kills human neutrophils, monocytes and macrophages. *BMC Microbiol.* **6:**56.

Mukherjee, D. V., J. H. Tonry, K. S. Kim, N. Ramarao, T. G. Popova, C. Bailey, S. Popov, and M. C. Chung. 2011. *Bacillus anthracis* protease InhA increases blood-brain barrier permeability and contributes to cerebral hemorrhages. *PLoS One* **6:**e17921.

Nakano, M. M., and P. Zuber. 1998. Anaerobic growth of a "strict aerobe" (*Bacillus subtilis*). *Annu. Rev. Microbiol.* **52:**165–190.

Nakouzi, A., J. Rivera, R. F. Rest, and A. Casadevall. 2008. Passive administration of monoclonal antibodies to anthrolysin O prolong survival in mice lethally infected with *Bacillus anthracis*. *BMC Microbiol.* **8:**159.

Oda, M., M. Takahashi, T. Matsuno, K. Uoo, M. Nagahama, and J. Sakurai. 2010. Hemolysis induced by *Bacillus cereus* sphingomyelinase. *Biochim. Biophys. Acta* **1798:**1073–1080.

Oh, S. Y., J. M. Budzik, G. Garufi, and O. Schneewind. 2011. Two capsular polysaccharides enable *Bacillus cereus* G9241 to cause anthrax-like disease. *Mol. Microbiol.* **80:**455–470.

Okinaka, R., K. Cloud, O. Hampton, A. Hoffmaster, K. Hill, P. Keim, T. Koehler, G. Lamke, S. Kumano, D. Manter, Y. Martinez, D. Ricke, R. Svensson, and P. Jackson. 1999a. Sequence, assembly and analysis of pX01 and pX02. *J. Appl. Microbiol.* **87:**261–262.

Okinaka, R. T., K. Cloud, O. Hampton, A. R. Hoffmaster, K. K. Hill, P. Keim, T. M. Koehler, G. Lamke, S. Kumano, J. Mahillon,

D. Manter, Y. Martinez, D. Ricke, R. Svensson, and P. J. Jackson. 1999b. Sequence and organization of pXO1, the large *Bacillus anthracis* plasmid harboring the anthrax toxin genes. *J. Bacteriol.* 181:6509–6515.

Okstad, O. A., M. Gominet, B. Purnelle, M. Rose, D. Lereclus, and A. B. Kolsto. 1999. Sequence analysis of three *Bacillus cereus* loci carrying PlcR-regulated genes encoding degradative enzymes and enterotoxin. *Microbiology* 145(Pt. 11):3129–3138.

Ovodov, Y. S. 2006. Capsular antigens of bacteria. Capsular antigens as the basis of vaccines against pathogenic bacteria. *Biochemistry* 71:955–961.

Perchat, S., T. Dubois, S. Zouhir, M. Gominet, S. Poncet, C. Lemy, M. Aumont-Nicaise, J. Deutscher, M. Gohar, S. Nessler, and D. Lereclus. 2011. A cell-cell communication system regulates protease production during sporulation in bacteria of the *Bacillus cereus* group. *Mol. Microbiol.* 82:619–633.

Pflughoeft, K. J., P. Sumby, and T. M. Koehler. 2011. *Bacillus anthracis sin* locus and regulation of secreted proteases. *J. Bacteriol.* 193:631–639.

Pomerantsev, A. P., K. V. Kalnin, M. Osorio, and S. H. Leppla. 2003. Phosphatidylcholine-specific phospholipase C and sphingomyelinase activities in bacteria of the *Bacillus cereus* group. *Infect. Immun.* 71:6591–6606.

Pomerantsev, A. P., O. M. Pomerantseva, A. S. Camp, R. Mukkamala, S. Goldman, and S. H. Leppla. 2009. PapR peptide maturation: role of the NprB protease in *Bacillus cereus* 569 PlcR/PapR global gene regulation. *FEMS Immunol. Med. Microbiol.* 55:361–377.

Ramarao, N., and D. Lereclus. 2006. Adhesion and cytotoxicity of *Bacillus cereus* and *Bacillus thuringiensis* to epithelial cells are FlhA and PlcR dependent, respectively. *Microbes Infect.* 8:1483–1491.

Ramarao, N., and D. Lereclus. 2005. The InhA1 metalloprotease allows spores of the *B. cereus* group to escape macrophages. *Cell. Microbiol.* 7:1357–1364.

Rasko, D. A., M. R. Altherr, C. S. Han, and J. Ravel. 2005. Genomics of the *Bacillus cereus* group of organisms. *FEMS Microbiol. Rev.* 29:303–329.

Ratnayake-Lecamwasam, M., P. Serror, K. W. Wong, and A. L. Sonenshein. 2001. *Bacillus subtilis* CodY represses early-stationary-phase genes by sensing GTP levels. *Genes Dev.* 15:1093–1103.

Raymond, B., P. R. Johnston, C. Nielsen-LeRoux, D. Lereclus, and N. Crickmore. 2010. *Bacillus thuringiensis*: an impotent pathogen? *Trends Microbiol.* 18:189–194.

Record, B. R., and R. G. Wallis. 1956. Physico-chemical examination of polyglutamic acid from *Bacillus anthracis* grown *in vivo*. *Biochem. J.* 63:443–447.

Ribot, W. J., R. G. Panchal, K. C. Brittingham, G. Ruthel, T. A. Kenny, D. Lane, B. Curry, T. A. Hoover, A. M. Friedlander, and S. Bavari. 2006. Anthrax lethal toxin impairs innate immune functions of alveolar macrophages and facilitates *Bacillus anthracis* survival. *Infect. Immun.* 74:5029–5034.

Richter, S., V. J. Anderson, G. Garufi, L. Lu, J. M. Budzik, A. Joachimiak, C. He, O. Schneewind, and D. Missiakas. 2009. Capsule anchoring in *Bacillus anthracis* occurs by a transpeptidation reaction that is inhibited by capsidin. *Mol. Microbiol.* 71:404–420.

Roh, J. Y., J. Y. Choi, M. S. Li, B. R. Jin, and Y. H. Je. 2007. *Bacillus thuringiensis* as a specific, safe, and effective tool for insect pest control. *J. Microbiol. Biotechnol.* 17:547–559.

Ross, C. L., and T. M. Koehler. 2006. *plcR papR*-independent expression of anthrolysin O by *Bacillus anthracis*. *J. Bacteriol.* 188:7823–7829.

Ross, J. M. 1957. The pathogenesis of anthrax following the administration of spores by the respiratory route. *J. Pathol. Bacteriol.* 73:485–494.

Russell, B. H., R. Vasan, D. R. Keene, T. M. Koehler, and Y. Xu. 2008. Potential dissemination of *Bacillus anthracis* utilizing human lung epithelial cells. *Cell. Microbiol.* 10:945–957.

Russell, B. H., R. Vasan, D. R. Keene, and Y. Xu. 2007. *Bacillus anthracis* internalization by human fibroblasts and epithelial cells. *Cell. Microbiol.* 9:1262–1274.

Saile, E., and T. M. Koehler. 2002. Control of anthrax toxin gene expression by the transition state regulator *abrB*. *J. Bacteriol.* 184:370–380.

Salamitou, S., H. Agaisse, A. Bravo, and D. Lereclus. 1996. Genetic analysis of *cryIIIA* gene expression in *Bacillus thuringiensis*. *Microbiology* 142:2049–2055.

Salamitou, S., F. Ramisse, M. Brehelin, D. Bourguet, N. Gilois, M. Gominet, E. Hernandez, and D. Lereclus. 2000. The *plcR* regulon is involved in the opportunistic properties of *Bacillus thuringiensis* and *Bacillus cereus* in mice and insects. *Microbiology* 146:2825–2832.

Schindler, D., and H. Echols. 1981. Retroregulation of the *int* gene of bacteriophage lambda: control of translation completion. *Proc. Natl. Acad. Sci. USA* 78:4475–4479.

Schmiel, D. H., and V. L. Miller. 1999. Bacterial phospholipases and pathogenesis. *Microbes Infect.* 1:1103–1112.

Schnepf, E., N. Crickmore, J. Van Rie, D. Lereclus, J. Baum, J. Feitelson, D. R. Zeigler, and D. H. Dean. 1998. *Bacillus thuringiensis* and its pesticidal crystal proteins. *Microbiol. Mol. Biol. Rev.* 62:775–806.

Schoeni, J. L., and A. C. Wong. 2005. *Bacillus cereus* food poisoning and its toxins. *J. Food Prot.* 68:636–648.

Shannon, J. G., C. L. Ross, T. M. Koehler, and R. F. Rest. 2003. Characterization of anthrolysin O, the *Bacillus anthracis* cholesterol-dependent cytolysin. *Infect. Immun.* 71:3183–3189.

Shimono, N., J. Hayashi, H. Matsumoto, N. Miyake, Y. Uchida, S. Shimoda, N. Furusyo, and K. Akashi. 26 October 2011. Vigorous cleaning and adequate ventilation are necessary to control an outbreak in a neonatal intensive care unit. *J. Infect. Chemother.* doi:10.1007/s10156-011-0326-y.

Shinagawa, K., Y. Ueno, D. Hu, S. Ueda, and S. Sugii. 1996. Mouse lethal activity of a HEp-2 vacuolation factor, cereulide, produced by *Bacillus cereus* isolated from vomiting-type food poisoning. *J. Vet. Med. Sci.* 58:1027–1029.

Silo-Suh, L. A., B. J. Lethbridge, S. J. Raffel, H. He, J. Clardy, and J. Handelsman. 1994. Biological activities of two fungistatic antibiotics produced by *Bacillus cereus* UW85. *Appl. Environ. Microbiol.* 60:2023–2030.

Sirard, J. C., C. Guidi-Rontani, A. Fouet, and M. Mock. 2000. Characterization of a plasmid region involved in *Bacillus anthracis* toxin production and pathogenesis. *Int. J. Med. Microbiol.* 290:313–316.

Sirard, J. C., M. Mock, and A. Fouet. 1994. The three *Bacillus anthracis* toxin genes are coordinately regulated by bicarbonate and temperature. *J. Bacteriol.* 176:5188–5192.

Slamti, L., and D. Lereclus. 2002. A cell-cell signaling peptide activates the PlcR virulence regulon in bacteria of the *Bacillus cereus* group. *EMBO J.* 21:4550–4559.

Slamti, L., and D. Lereclus. 2005. Specificity and polymorphism of the PlcR-PapR quorum-sensing system in the *Bacillus cereus* group. *J. Bacteriol.* 187:1182–1187.

Slamti, L., S. Perchat, M. Gominet, G. Vilas-Boas, A. Fouet, M. Mock, V. Sanchis, J. Chaufaux, M. Gohar, and D. Lereclus. 2004. Distinct mutations in PlcR explain why some strains of the *Bacillus cereus* group are nonhemolytic. *J. Bacteriol.* 186:3531–3538.

Sonenshein, A. L. 2005. CodY, a global regulator of stationary phase and virulence in Gram-positive bacteria. *Curr. Opin. Microbiol.* 8:203–207.

Spira, W. M., and J. M. Goepfert. 1972. *Bacillus cereus*-induced fluid accumulation in rabbit ileal loops. *Appl. Microbiol.* 24:341–348.

Stenfors Arnesen, L. P., A. Fagerlund, and P. E. Granum. 2008. From soil to gut: *Bacillus cereus* and its food poisoning toxins. *FEMS Microbiol. Rev.* 32:579–606.

Strauch, M. A., P. Ballar, A. J. Rowshan, and K. L. Zoller. 2005. The DNA-binding specificity of the *Bacillus anthracis* AbrB protein. *Microbiology* 151:1751–1759.

Stulke, J., and W. Hillen. 2000. Regulation of carbon catabolism in *Bacillus* species. *Annu. Rev. Microbiol.* 54:849–880.

Sue, D., A. R. Hoffmaster, T. Popovic, and P. P. Wilkins. 2006. Capsule production in *Bacillus cereus* strains associated with severe pneumonia. *J. Clin. Microbiol.* 44:3426–3428.

Thoren, K. L., and B. A. Krantz. 2011. The unfolding story of anthrax toxin translocation. *Mol. Microbiol.* 80:588–595.

Thorsen, L., B. M. Hansen, K. F. Nielsen, N. B. Hendriksen, R. K. Phipps, and B. B. Budde. 2006. Characterization of emetic *Bacillus weihenstephanensis*, a new cereulide-producing bacterium. *Appl. Environ. Microbiol.* 72:5118–5121.

Titball, R. W. 1998. Bacterial phospholipases. *Symp. Ser. Soc. Appl. Microbiol.* 27:127S–137S.

Tournier, J. N., A. Quesnel-Hellmann, A. Cleret, and D. R. Vidal. 2007. Contribution of toxins to the pathogenesis of inhalational anthrax. *Cell. Microbiol.* 9:555–565.

Tsvetanova, B., A. C. Wilson, C. Bongiorni, C. Chiang, J. A. Hoch, and M. Perego. 2007. Opposing effects of histidine phosphorylation regulate the AtxA virulence transcription factor in *Bacillus anthracis*. *Mol. Microbiol.* 63:644–655.

Turk, B. E. 2007. Manipulation of host signalling pathways by anthrax toxins. *Biochem. J.* 402:405–417.

Uchida, I., J. M. Hornung, C. B. Thorne, K. R. Klimpel, and S. H. Leppla. 1993. Cloning and characterization of a gene whose product is a *trans*-activator of anthrax toxin synthesis. *J. Bacteriol.* 175:5329–5338.

Uchida, I., S. Makino, T. Sekizaki, and N. Terakado. 1997. Cross-talk to the genes for *Bacillus anthracis* capsule synthesis by *atxA*, the gene encoding the *trans*-activator of anthrax toxin synthesis. *Mol. Microbiol.* 23:1229–1240.

van der Voort, M., O. P. Kuipers, G. Buist, W. M. de Vos, and T. Abee. 2008. Assessment of CcpA-mediated catabolite control of gene expression in *Bacillus cereus* ATCC 14579. *BMC Microbiol.* 8:62.

van Schaik, W., A. Chateau, M. A. Dillies, J. Y. Coppee, A. L. Sonenshein, and A. Fouet. 2009. The global regulator CodY regulates toxin gene expression in *Bacillus anthracis* and is required for full virulence. *Infect. Immun.* 77:4437–4445.

Vassileva, M., K. Torii, M. Oshimoto, A. Okamoto, N. Agata, K. Yamada, T. Hasegawa, and M. Ohta. 2007. A new phylogenetic cluster of cereulide-producing *Bacillus cereus* strains. *J. Clin. Microbiol.* 45:1274–1277.

Vietri, N. J., R. Marrero, T. A. Hoover, and S. L. Welkos. 1995. Identification and characterization of a *trans*-activator involved in the regulation of encapsulation by *Bacillus anthracis*. *Gene* 152:1–9.

Warner, J. B., and J. S. Lolkema. 2003. CcpA-dependent carbon catabolite repression in bacteria. *Microbiol. Mol. Biol. Rev.* 67:475–490.

Welkos, S., A. Friedlander, S. Weeks, S. Little, and I. Mendelson. 2002. *In-vitro* characterization of the phagocytosis and fate of anthrax spores in macrophage and the effects of anti-PA antibody. *J. Med. Microbiol.* 51:821–831.

Wilson, A. C., J. A. Hoch, and M. Perego. 2009. Two small c-type cytochromes affect virulence gene expression in *Bacillus anthracis*. *Mol. Microbiol.* 72:109–123.

Wilson, M. K., J. M. Vergis, F. Alem, J. R. Palmer, A. M. Keane-Myers, T. N. Brahmbhatt, C. L. Ventura, and A. D. O'Brien. 2011. *Bacillus cereus* G9241 makes anthrax toxin and capsule like highly virulent *B. anthracis* Ames but behaves like attenuated toxigenic nonencapsulated *B. anthracis* Sterne in rabbits and mice. *Infect. Immun.* 79:3012–3019.

Wong, H. C., and S. Chang. 1986. Identification of a positive retroregulator that stabilizes mRNAs in bacteria. *Proc. Natl. Acad. Sci. USA* 83:3233–3237.

Wong, H. C., H. E. Schnepf, and H. R. Whiteley. 1983. Transcriptional and translational start sites for the *Bacillus thuringiensis* crystal protein gene. *J. Biol. Chem.* 258:1960–1967.

Young, J. A., and R. J. Collier. 2007. Anthrax toxin: receptor binding, internalization, pore formation, and translocation. *Annu. Rev. Biochem.* 76:243–265.

Zhang, M. Y., A. Lovgren, M. G. Low, and R. Landen. 1993. Characterization of an avirulent pleiotropic mutant of the insect pathogen *Bacillus thuringiensis*: reduced expression of flagellin and phospholipases. *Infect. Immun.* 61:4947–4954.

Zhou, Y., Y. L. Choi, M. Sun, and Z. Yu. 2008. Novel roles of *Bacillus thuringiensis* to control plant diseases. *Appl. Microbiol. Biotechnol.* 80:563–572.

Zigha, A., E. Rosenfeld, P. Schmitt, and C. Duport. 2007. The redox regulator Fnr is required for fermentative growth and enterotoxin synthesis in *Bacillus cereus* F4430/73. *J. Bacteriol.* 189:2813–2824.

Regulation of Bacterial Virulence
Edited by Michael L. Vasil and Andrew J. Darwin
© 2013 ASM Press, Washington, DC   doi:10.1128/9781555818524.ch14

Chapter 14

Regulation of Extracellular Toxin Production in
Clostridium perfringens

JACKIE K. CHEUNG, LEE-YEAN LOW, THOMAS J. HISCOX, AND JULIAN I. ROOD

DISEASES CAUSED BY
CLOSTRIDIUM PERFRINGENS

Clostridium perfringens is a gram-positive, spore-forming anaerobic rod that is widespread in the environment and is commonly isolated from the gastrointestinal tract of humans and animals, as well as from soil and sewage (Rood and Cole, 1991; Rood, 1998). It is the causative agent of several diseases of both animals and humans. In humans, *C. perfringens* isolates causes gas gangrene (clostridial myonecrosis), food poisoning, and necrotizing enterocolitis (Rood and Cole, 1991; Songer, 2010). In animals, *C. perfringens* is responsible for diseases such as lamb dysentery and ovine enterotoxemia (also called pulpy kidney disease of sheep) (Rood and Cole, 1991; Songer, 1997) and avian necrotic enteritis (Van Immerseel et al., 2009; Timbermont et al., 2011). The pathogenesis of these diseases is mediated primarily by numerous potent extracellular toxins produced by *C. perfringens*. All *C. perfringens* strains produce α-toxin; however, different isolates can produce other toxins, such as θ-toxin (perfringolysin O), κ-toxin (collagenase), μ-toxin (hyaluronidase), λ-toxin (protease), ι-toxin (an actin-specific ADP-ribosyltransferase), sialidases, and the pore-forming toxins ε-toxin, β-toxin, β2-toxin, δ-toxin, NetB toxin, and *C. perfringens* enterotoxin (CPE) (Gibert et al., 1997; Rood, 1998; Keyburn et al., 2008; Popoff and Bouvet, 2009). No single strain is able to produce all the above-listed toxins; therefore, *C. perfringens* strains can be classified into five major toxinotypes (A to E) (Table 1) depending on the subset of toxins that are produced (Petit et al., 1999).

The most notable human disease caused by *C. perfringens* is gas gangrene or clostridial myonecrosis. The disease usually develops as a result of a traumatic injury that enables vegetative *C. perfringens* cells or viable spores to enter the body through contamination of a wound, either with soil or following perforation of the bowel (Titball, 2005). Damage to the body's vasculature associated with the traumatic injury provides the anaerobic conditions conducive to the growth of vegetative cells, which is accompanied by the production of extracellular toxins, in particular α-toxin and perfringolysin O (Awad et al., 1995; Awad, et al., 2001), and subsequently leads to massive tissue necrosis and gas production (Stevens et al., 1997). Treatment involves the use of a combination of antibiotic therapy and surgical intervention. If the infection is left untreated, the patient will almost invariably succumb to the disease as a result of generalized sepsis and toxemia.

C. perfringens is also one of the leading causes of human bacterial food poisoning (Brynestad and Granum, 2002; McClane, 2005). The disease develops as a result of ingestion of *C. perfringens*-contaminated food following improper reheating or storage of cooked meat (Brynestad and Granum, 2002). Heat-resistant *C. perfringens* spores are able to survive cooking, and the subsequent slow cooling of the meat allows for spore germination (McClane, 2005). Subsequent growth of the bacterium in the gastrointestinal tract leads to sporulation and CPE production (Skjelkvale and Uemura, 1977; Sarker et al., 1999; McClane, 2005). The disease results in diarrhea and acute abdominal pain (Johnson and Gerding, 1997; McClane, 2005) but tends to be self-limiting, and with the exception of debilitated patents and the elderly, very few patients require hospitalization (Brynestad and Granum, 2002). A more severe, but very rare, form of food poisoning is enteritis necroticans. This disease develops following the consumption of meat (commonly pork) that has been contaminated with *C. perfringens* type C isolates (Lawerence, 1997; Vidal et al., 2008). The primary virulence factor involved in the pathogenesis of the disease is β-toxin (Fisher et al., 2006; Sayeed et al., 2008; Vidal et al., 2008).

Jackie K. Cheung, Lee-Yean Low, Thomas J. Hiscox, and Julian I. Rood • Department of Microbiology, Monash University, Clayton, Victoria 3800, Australia.

Table 1. Toxins produced and diseases caused by *C. perfringens* toxinotypes

Toxinotype	Toxin				Primary disease(s)	
	α	β	ε	ι	Human	Animal
A	+	−	−	−	Gas gangrene, food poisoning	Necrotic enteritis in chickens
B	+	+	+	−		Lamb dysentery, enterotoxemia in sheep
C	+	+	−	−	Necrotizing enterocolitis (Darmbrand syndrome, or pig-bel)	Necrotic enteritis in pigs, lambs, calves, foals, and, rarely, in chickens
D	+	−	+	−		Enterotoxemia in sheep
E	+	−	−	+		Enterotoxemia in calves

C. perfringens is also a significant pathogen of domestic animals, with all of these diseases resulting from the production of potent extracellular toxins (Songer, 1996, 2010). For example, avian *C. perfringens* type A isolates produce the pore-forming toxin NetB, which is required for necrotic enteritis in chickens (Keyburn et al., 2008; Van Immerseel et al., 2009; Keyburn et al., 2010).

C. perfringens type B is the etiological agent of lamb dysentery (Songer, 1997). Infection occurs due to the rapid multiplication of *C. perfringens* type B and subsequent toxin production within the gastrointestinal tract of the newborn lamb (Fernandez-Miyakawa et al., 2007). Using the *C. perfringens* type B mouse intravenous injection model, both ε-toxin and β-toxin have been implicated in the pathogenesis of type B enterotoxemias (Fernandez-Miyakawa et al., 2007).

C. perfringens type C isolates are able to infect many different animals, causing similar diseases in cattle, lambs, and dogs (Songer, 2010). Infection of newborn piglets occurs from the mother (Songer and Uzal, 2005), with *C. perfringens* type C cells rapidly multiplying within the small intestine of the piglet and producing toxins (Songer, 1997). Using a newly developed mouse intraduodenal and intragastric challenge models of type C enterotoxemia, it was confirmed that β-toxin plays a key role in the pathogenesis of type C infections, although α-toxin and perfringolysin O may also be important (Uzal et al., 2009; Uzal and McClane, 2011).

Enterotoxemia of sheep (or pulpy kidney disease) is caused by *C. perfringens* type D (Songer, 1997; Sayeed et al., 2005; Uzal and Songer, 2008). Once more the organisms colonize the small intestine, resulting in the subsequent production of ε-toxin, the primary virulence factor (Fernandez Miyakawa and Uzal, 2003; Sayeed et al., 2005). Subsequent absorption of ε-toxin from the small intestine leads to fatal systemic intoxication and effects on the central nervous system (Fernandez Miyakawa and Uzal, 2003).

Finally, diseases caused by type E isolates are rare, but these isolates have been linked to enterotoxemia of lambs, calves, and rabbits (Songer and Miskimmins, 2004). It has been suggested that ι-toxin, an actin-specific ADP-ribosyltransferase (Tsuge et al., 2008), may play a role in the development of disease, but this has not been shown experimentally (Songer, 1996; Songer and Miskimmins, 2004).

Since protein toxins are crucial to the pathogenesis of *C. perfringens* infections, knowledge of how their production is regulated is essential if we are to understand host-pathogen interactions in the disease process. This chapter provides an overview of the regulatory systems and mechanisms involved in the control of toxin production in *C. perfringens*. To date, research has primarily focused on regulation by the VirSR two-component signal transduction system, but recent studies have identified other regulatory elements that also play a role in this important process.

THE GLOBAL VirSR TWO-COMPONENT SIGNAL TRANSDUCTION SYSTEM

Two-component signal transduction systems represent one of the most widespread mechanisms by which bacteria sense and respond to a diverse range of changes in both environmental stimuli and bacterial cell density (Casino et al., 2010). Many of these systems have been implicated in the regulation of virulence in pathogenic bacteria (Beier and Gross, 2006; Calva and Oropeza, 2006), and because they are not found in mammals they are attractive as potential drug targets (Gotoh et al., 2010). These regulatory networks generally consist of two components: a sensor histidine kinase and a cognate response regulator. These proteins comprise a phosphorelay cascade that plays a crucial role in translating an external stimulus into an adaptive response within the bacterial cell, often by differentially altering the expression of specific genes (Bourret and Silversmith, 2010).

Analysis of the genome sequences of three isolates of *C. perfringens* has shown that there are 24 sensor histidine kinases and 17 response regulators that are conserved in all three strains (Myers et al., 2006). The best-characterized and the most significant of these phosphorelay systems is the VirS/VirR

(VirSR) network from the gas gangrene-causing type A isolate, strain 13 (Lyristis et al., 1994; Shimizu et al., 1994). The VirSR system is a global regulatory network that regulates the expression of genes encoding extracellular toxins and enzymes (Lyristis et al., 1994; Shimizu et al., 1994; Ba-Thein et al., 1996; Ohtani et al., 2003; Cheung et al., 2010), as well as housekeeping genes (Ohtani et al., 2000; Kawsar et al., 2004; Andre et al., 2010; Yuan et al., 2011), genes encoding regulatory RNA molecules (Ohtani et al., 2002a; Shimizu et al., 2002c; Okumura et al., 2008), and genes involved in quorum sensing (Ohtani et al., 2002b; Ohtani et al., 2009) (Fig. 1). Furthermore, recent work has shown that regulation of toxin production by VirSR is not confined to human type A strains. This regulatory system is also involved in the

production of β-toxin by type C strains both in vitro, under aerobic (Vidal, et al., 2009b) and anaerobic conditions (Ma et al., 2011), and in vivo (Ma et al., 2011) and is required for optimal expression of the plasmid-encoded NetB toxin in avian necrotic enteritis isolates (Cheung et al., 2010).

In this regulatory network, the VirS sensor histidine kinase is predicted to contain seven transmembrane domains in its N-terminal region (Cheung et al., 2009), and it is this region that is postulated to detect the environmental or growth phase stimulus (Lyristis et al., 1994; Cheung et al., 2009). Mutagenesis studies have shown that several residues in these transmembrane domains are important to the function of this sensor (Cheung and Rood, 2000a; Cheung et al., 2009). Like most sensor histidine kinases, VirS has

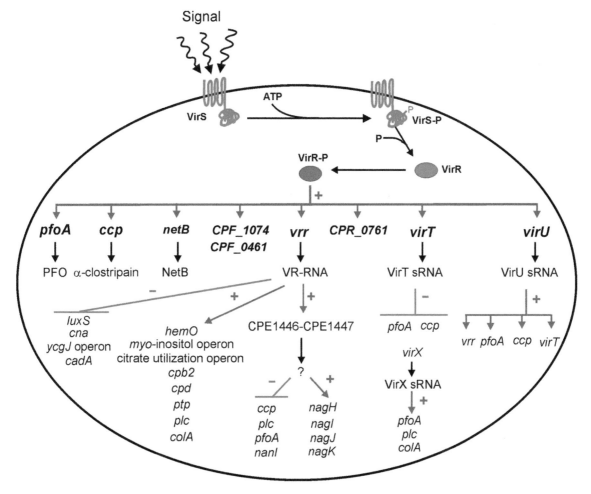

Figure 1. Regulation of gene expression by the VirSR two-component signal transduction system. The VirS sensor histidine kinase (shown in pink) autophosphorylates upon detection of an external signal. Phosphorylated VirS then donates the phosphoryl group to its cognate response regulator, VirR (shown in blue). Phosphorylated VirR directly regulates the expression of *pfoA*, *ccp*, *vrr*, *virT*, and *virU* in strain 13, *netB* in EHE-NE18, and presumably *CPF_1079* and *CPF_0461* in ATCC 13124 and *CPR_0761* in SM101. VirSR indirectly regulates the expression of the indicated genes through the VR-RNA, VirT, and VirU sRNAs. Positive regulation is denoted by the green arrows and plus symbols, while negative regulation is shown by the red lines and minus symbols. The VirX sRNA positively regulates the expression the *pfoA*, *plc*, and *colA* genes in a VirSR-independent manner.
doi:10.1128/9781555818524.ch14f1

a cytoplasmic C-terminal domain that contains the residues predicted to be involved in autophosphorylation. These residues include the proposed site of autophosphorylation, H255, and the GXGL (or G2) motif. When these residues were altered, a truncated form of the VirS protein was no longer able to autophosphorylate, indicating that these motifs were essential for this process (Cheung et al., 2009). Once phosphorylated, VirS acts as a phosphodonor for its cognate response regulator, VirR (Cheung et al., 2009).

VirR belongs to the AlgR/AgrA/LytR family of response regulators (Nikolskaya and Galperin, 2002). Like with other response regulators, the N-terminal region of VirR contains residues that are predicted to be involved in phosphorylation of the protein, including E8 and D9, which are postulated to form the phosphoacceptor pocket, and K105 and D57, with the latter residue being the proposed site of phosphorylation (Cheung et al., 2009). The importance of the D57 residue was demonstrated when substitution of this amino acid with asparagine eliminated phosphotransfer between a phosphorylated derivative of VirS and VirR (Cheung et al., 2009). Once activated, VirR is able to regulate transcription by binding to the promoter region of its target genes (Cheung and Rood, 2000b; Cheung et al., 2004). Two novel DNA-binding motifs, FxRxHrS and SKHR, which are located in the C-terminal region of VirR, have been shown to be essential for this function (McGowan et al., 2002; McGowan et al., 2003), and it is the FxRxHrS motif (or LytTR domain) that places VirR in the AlgR/AgrA/LytR family of response regulators. In silico analysis has shown that members of this family do not contain typical helix-turn-helix or winged-helix DNA binding domains (Nikolskaya and Galperin, 2002; Galperin, 2010). The crystal structure of the C-terminal DNA-binding region of *Staphylococcus aureus* AgrA complexed with its target DNA has revealed that the LytTR domain adopts a novel elongated β-β-β fold (Sidote et al., 2008). Loops within the domain insert into successive major grooves in the target DNA and cause DNA distortion. Interestingly, only two amino acid residues, H169 and R233, were shown to directly interact with the major groove and to be crucial for DNA binding. Mutation of these residues adversely affected binding efficiency (Sidote et al., 2008). Similar results were observed with VirR when the FxRxHrS and SKHR motifs were altered (McGowan et al., 2002; McGowan et al., 2003).

Genetic studies have shown that disruption of either *virR* (Shimizu et al., 1994) or *virS* (Lyristis et al., 1994) resulted in an altered toxin production profile. These mutants no longer produced perfringolysin O

or the extracellular protease α-clostripain and showed a significant reduction in α-toxin, sialidase, and collagenase activities (Lyristis et al., 1994; Shimizu et al., 1994; Okumura et al., 2008). Although VirSR was first identified as a positive regulator of extracellular toxin production, it is now considered a bifunctional system, as it has been demonstrated to positively and negatively regulate the expression of many genes at the transcriptional level (Banu et al., 2000; Shimizu et al., 2002b; Ohtani et al., 2010).

Comparison of gene expression profiles observed by microarray analysis in the wild-type strain and a *virR* mutant indicated that the genes in the VirSR regulon can be divided into two groups: genes that are directly regulated by VirSR and those that are regulated by the VirSR–VR-RNA cascade (Ohtani et al., 2010) (Fig. 1). To date, only six genes have been shown by in vivo studies to be directly activated by VirSR. Five of these genes, *pfoA* (encoding perfringolysin O), *ccp* (encoding α-clostripain), *vrr* (encoding the VR-RNA regulatory RNA molecule), *virT* (encoding the VirT regulatory RNA), and *virU* (encoding the VirU regulatory RNA), are from strain 13 (Okumura et al., 2008; Ohtani et al., 2010), while the *netB* gene (encoding NetB toxin) is from the avian necrotic enteritis type A strain, EHE-NE18 (Cheung et al., 2010). Direct regulation of the expression of these genes involves the binding of VirR to the promoter region. Specifically, VirR recognizes and binds to two imperfect direct repeats known as VirR boxes (Cheung and Rood, 2000b; Cheung et al., 2004; Cheung et al., 2010) (Fig. 2). These VirR boxes have also been identified upstream of several genes encoding hypothetical proteins in two sequenced type A *C. perfringens* isolates, ATCC 13124 and SM101 (Myers et al., 2006) (Fig. 2). Using a *pfoA* reporter system, it was demonstrated that these binding sites are essential for VirR-regulated transcription. The maintenance of the correct helical phasing, the correct spacing between the VirR boxes, and the correct distance between the VirR boxes and the –35 region were shown to be critical for optimal transcriptional activation (Cheung et al., 2004). These results, in addition to the finding that *C. perfringens* RNA polymerase binds to DNA more efficiently in the presence of VirR, suggest that VirR activates transcription by facilitating the binding of RNA polymerase to the promoter (Cheung et al., 2004).

The vast majority of genes that are regulated by the VirSR system fall into the second group; that is, genes that are regulated by the small regulatory RNA (sRNA) intermediate, VR-RNA (Ohtani et al., 2010) (Fig. 1). A total of 147 genes from strain 13 are in this group, comprising 30 single genes and 21 putative operons (Ohtani et al., 2010). Through the

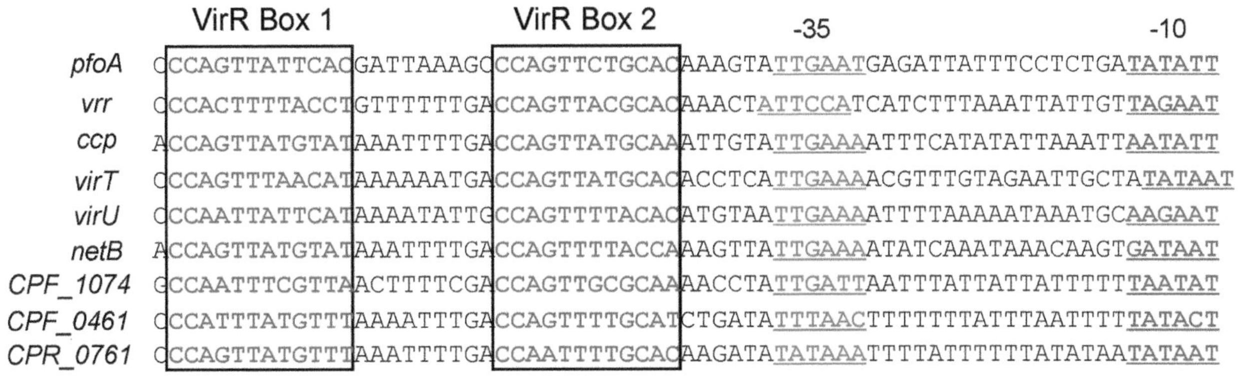

	VirR Box 1		VirR Box 2	-35	-10

```
         VirR Box 1              VirR Box 2            -35                      -10

 pfoA   CCCAGTTATTCAC GATTAAAGCCCAGTTCTGCAC AAAGTATTGAATGAGATTATTTCCTCTGATATATT
  vrr   CCCACTTTTACCTGTTTTTTGACCAGTTACGCAC AAACTATTCCATCATCTTTAAATTATTGTTAGAAT
  ccp   ACCAGTTATGTATAAATTTTGACCAGTTATGCAA ATTGTATTGAAAATTTCATATATTAAATTAATATT
 virT   CCCAGTTTAACATAAAAAATGACCAGTTATGCAC ACCTCATTGAAAACGTTTGTAGAATTGCTATATAAT
 virU   CCCAATTATTCATAAAATATTGCCAGTTTTACACATGTAATTGAAAATTTTAAAAATAAATGCAAGAAT
 netB   ACCAGTTATGTATAAATTTTGACCAGTTTTACCAAAGTATTGAAAATATCAAATAAACAAGTGATAAT
CPF_1074 GCCAATTTCGTTAACTTTTCGACCAGTTGCGCAAAACCTATTGATTAATTTATTATTATTTTTAATAT
CPF_0461 GCCATTTATGTTTAAAATTTGACCAGTTTTGCATCTGATATTTAACTTTTTTTTATTTAATTTTTATACT
CPR_0761 CCCAGTTATGTTTAAATTTTGACCAATTTTGCACAAGATATATAAATTTTTATTTTTTATATAATATAAT
```

Figure 2. VirR boxes in the *C. perfringens* genome. VirR boxes are written in blue and are boxed. The genes found downstream are indicated adjacent to the relevant VirR boxes. The putative –35 and –10 sequences are underlined and shown in pink and green, respectively.
doi:10.1128/9781555818524.ch14f2

VirSR–VR-RNA cascade, target genes can be either positively or negatively regulated. Of the virulence-related genes, *plc* (α-toxin), *colA* (collagenase) (Banu et al., 2000; Shimizu et al., 2002c), *cpb2* (β2-toxin) (Ohtani et al., 2003), and several other potential virulence factors (Ohtani et al., 2010) are positively regulated by this cascade, whereas *cna* (putative collagen adhesin) and *cadA* (cell wall-anchored DNase) (Okumura et al., 2005) are negatively regulated (Ohtani et al., 2003). Note that the *cpb2* and *cna* genes are located on pCP13 in strain 13, thereby demonstrating that the VirSR–VR-RNA cascade is able to regulate both chromosomal and plasmid-encoded genes (Ohtani et al., 2003). Many genes encoding proteins or enzymes that have housekeeping functions also fall into this group of genes (Ohtani et al., 2010). They include the *myo*-inositol operon, *ptp* (protein tyrosine phosphatase), and *cpd* (2′,3′-cyclic nucleotide 2′-phosphodiesterase), which are positively regulated (Banu et al., 2000; Kawsar et al., 2004). Recent studies have also implicated the VirSR–VR-RNA cascade in the partial regulation of the citrate utilization operon. In addition to negative regulation by VR-RNA, expression of this cluster of genes was shown to be regulated by two other two-component systems, CPE0518/CPE0519 and CPE0531/CPE0532, as well as two transcriptional regulators, CitI and CitG (Yuan et al., 2011). Finally, there are genes that encode proteins implicated in adaptive responses, such as *luxS* (autoinducer 2 [AI-2]), which is part of the *ycgJ-metB-cysK-ygaG* (or *ubiGmc-cBAluxS*) operon (Ohtani et al., 2000; Ohtani et al., 2002b; Andre et al., 2010), and *hemO* (heme oxygenase) (Hassan et al., 2010), which are negatively and positively controlled, respectively. VR-RNA also has recently been shown to positively regulate the expression of the *CPE1447-CPE1446* genes, which encode two novel toxin regulators (Obana and Nakamura,

2011). To increase the level of complexity, a subset of VirSR-regulated genes are also controlled by other sRNA molecules, namely, VirU, VirT (Okumura et al., 2008), and VirX (Ohtani et al., 2002a).

REGULATION BY sRNA MOLECULES

In recent years it has been shown that RNA molecules play an important role in gene regulation. Examples include mRNA leader sequences that affect the expression of genes in *cis* (riboswitches), sRNAs that bind to proteins or base-pair with target mRNAs, and CRISPR (clustered regularly interspaced short palindromic repeats) RNAs that inhibit the uptake of foreign DNA (Waters and Storz, 2009). sRNAs are the most extensively studied RNA regulators (Waters and Storz, 2009) and are defined as short, usually between 50 and 200 nucleotides in length (Gottesman, 2005), noncoding transcripts that control the translation or stability of target mRNAs (Liu and Camilli, 2010); they often regulate responses to changes in environmental conditions (Waters and Storz, 2009). Most sRNAs bind to ribosome binding sites in the 5′ untranslated region (UTR) of their target mRNAs, which causes translational repression and frequently destabilizes the mRNA (Waters and Storz, 2009). Conversely, some sRNAs activate the translation and stability of their target mRNAs by base-pairing far upstream of the ribosome binding site and promoting entry to the ribosome (Frohlich and Vogel, 2009).

Genomic analysis has predicted a number of sRNAs in the genomes of *C. perfringens* isolates. In the *C. perfringens* gas gangrene strains, 13 and ATCC 13124, 193 and 181 sRNAs have been predicted, respectively, whereas 131 sRNA were predicted in the food poisoning isolate, SM101 (Chen et al., 2011). To date, there have been very few functional studies carried out on any of these putative regulatory molecules. Most studies on *C. perfringens* sRNAs have involved VR-RNA (Banu et al., 2000), VirU (Okumura et al., 2008), VirT (Okumura et al., 2008) and VirX (Ohtani et al., 2002a). With the exception of VirX, three of these sRNA molecules have been associated with the VirSR regulatory system; the genes encoding these three sRNAs all contain consensus VirR-binding sites in their promoter regions and are directly regulated by the VirSR system (Cheung et al., 2004; Okumura et al., 2008).

VR-RNA was the first sRNA molecule identified in *C. perfringens* (Ohtani et al., 2010), and it is also the most extensively studied. It was initially found to be encoded by an open reading frame called *hyp7* (Banu et al., 2000), but the VR-RNA coding region was later renamed *vrr* (Okumura et al., 2008). Expression of *vrr* is mostly under the control of the VirSR system (Banu et al., 2000; Shimizu et al., 2002), but it is to a lesser extent regulated by the VirU sRNA molecule (Okumura et al., 2008).

Evidence that VR-RNA is a regulatory RNA molecule was obtained when a derivative carrying a nonsense mutation in the *hyp7* coding region was able to activate transcription of the *plc* and *colA* genes in the same way as the wild-type gene (Shimizu et al., 2002c). Similar mutations in the coding regions of *virU*, *virT* (Okumura et al., 2008), and *virX* (Ohtani et al., 2002a) indicated that these genes also encoded regulatory RNA molecules. Furthermore, deletion analysis showed that the region required for the regulatory function of VR-RNA was localized to a small portion at the 3' end of the molecule (Shimizu et al., 2002c). Recent work has shown that VR-RNA directly controls the stability of *colA* mRNA through an antisense mechanism (Obana et al., 2010). This base-pairing appears to be independent of the RNA chaperone, Hfq (Obana et al., 2010). The 3' region of VR-RNA base-pairs with the 5' UTR of *colA* mRNA. The binding of these two RNA species promotes cleavage of the *colA* mRNA at positions −79 and −78 (with respect to the ATG start codon) (Fig. 3A) and results in stabilization of the processed *colA* mRNA transcript and increased translation. This stability is suggested to be due to the binding of ribosomes to the ribosome binding site in the 5' UTR, which is normally sequestered in the unprocessed full-length transcript but is exposed upon cleavage of the *colA* mRNA (Fig. 3B) (Obana et al., 2010).

It seems that the same mechanism cannot be applied to VR-RNA-mediated regulation of the *plc* gene. Despite the fact that the 3' region of VR-RNA is important for *plc* expression (Shimizu et al., 2002c), the sequence of this region does not show any complementarity to the 5'

leader of *plc* mRNA (Obana et al., 2010). Therefore, the precise method by which VR-RNA regulates expression of such a diverse range of genes remains unclear and differs depending on the specific gene.

In addition to *vrr*, the transcription of two other sRNA-encoding genes, *virU* and *virT*, also is regulated directly by the VirSR system (Okumura et al., 2008). Initial database searches indicated that these open reading frames encoded hypothetical proteins, but experimental evidence showed that like with VR-RNA, the effects mediated by VirU and VirT were not altered by nonsense and frameshift mutations, suggesting that they encoded regulatory RNA molecules rather than proteins (Okumura et al., 2008). Analysis of a *virT* mutant showed that there was at least a 2.5-fold increase in the levels of both *pfoA* and *ccp* mRNA during logarithmic growth phase, whereas there was no significant change in *plc*, *colA*, *vrr*, or *virU* mRNA levels. These findings implied that the *virT* gene product was a negative regulator of *pfoA* and *ccp* expression (Okumura et al., 2008). For unknown reasons, it was not possible to construct a *virU* null mutant. The function of VirU was therefore determined by overexpression of the *virU* gene. In response to a higher *virU* copy number, *pfoA*, *ccp*, *vrr*, and *virT* mRNA levels were all increased, suggesting that the *virU* gene encodes a positive regulator of these genes (Okumura et al., 2008) or that it stabilizes the respective mRNA molecules. It is postulated that these sRNAs may function to fine-tune transcription of VirSR-regulated genes, thereby maintaining balanced gene expression (Okumura et al., 2008). The actual mechanisms by which VirU and VirT regulate their target genes are yet to be determined.

The final sRNA molecule to be studied is that encoded by *virX*. This open reading frame was identified from the screening of a chromosomal DNA library, with the insert carrying *virX* shown to modulate the expression of the *plc*, *colA*, and *pfoA* genes (Ohtani et al., 2002a). VirX sRNA was found to increase the levels of *plc* and *colA* mRNA independently of VirSR but was shown to activate *pfoA* expression in association with VirSR. Furthermore, this regulation was observed at different times during the growth of *C. perfringens*, with *pfoA*-specific mRNA increasing at mid-exponential phase and *plc* and *colA* mRNA increasing at late exponential phase (Ohtani et al., 2002a). Recent work has indicated that VirX also negatively regulates the expression of the *ubiG* operon, which is involved in the conversion of methionine to cysteine and in AI-2 production (Andre et al., 2010). However, like with the other sRNAs, the precise mechanism by which VirX regulates gene expression remains to be determined.

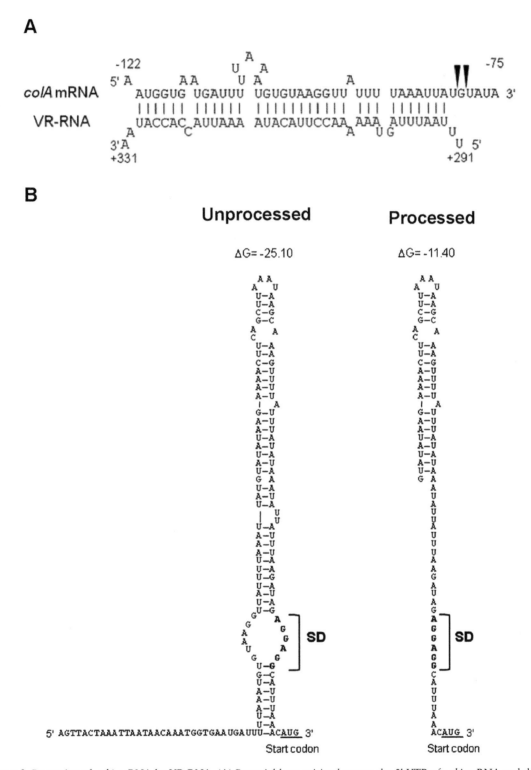

Figure 3. Processing of *colA* mRNA by VR-RNA. (A) Potential base-pairing between the 5′ UTR of *colA* mRNA and the 3′ region of VR-RNA. The numbers indicate the nucleotides relative to the *colA* ATG start, where A is +1. The triangles represent the processing sites on the *colA* mRNA. (B) Predicted structures of the unprocessed and processed 5′ UTR of *colA* mRNA from the transcription start site to the AUG start codon (underlined). The predicted ΔG values (in kilocalories per mole) are indicated above each structure. The ribosome binding sequences (SD) are shown in bold. Panels are reprinted from *Molecular Microbiology* (Obana et al., 2010) with the permission of the authors and the publisher.
doi:10.1128/9781555818524.ch14f3

THE ROLE OF CELL DENSITY AND QUORUM SENSING IN REGULATING TOXIN PRODUCTION

Early studies by Higashi et al. (Higashi et al., 1973), which were continued by Imagawa et al. (Imagawa et al., 1981; Imagawa and Higashi, 1992), demonstrated the production of a substance that could stimulate perfringolysin O production from a class of *C. perfringens* mutants (group b mutants) that could not produce this toxin. This complementation was initially achieved by cross-streaking this strain with another class of mutants (group a mutants) (Higashi, et al., 1973). The compound, later designated as substance A (Imagawa and Higashi, 1992), was diffusible, had a low molecular weight (Imagawa et al., 1981), and could be isolated from culture filtrates of group a mutants (Imagawa and Higashi, 1992). We now know that these characteristics are very similar to those of signal or quorum sensing peptides.

Recent studies suggest that VirSR-mediated gene regulation is part of an *agr*-like quorum sensing system (Ohtani et al., 2009). The *agr* (accessory gene regulator) system of *S. aureus* is the best-studied quorum sensing system in gram-positive bacteria and consists of a characteristic two-component signal transduction pathway, a regulatory RNA molecule and a small secreted polypeptide (Bronner et al., 2004). The system has been shown to regulate the expression of many virulence factors, including hemolysins (Novick et al., 1993) and surface proteins such as protein A and fibronectin binding protein (Vandenesch et al., 1991; Wolz et al., 2000; Otto, 2004).

The Agr system forms a quorum sensing network that is able to respond to changes in cell density. The system comprises the sensor histidine kinase AgrC and its cognate response regulator AgrA, which controls the expression of two transcripts (RNAII and RNAIII) from two different promoters (P1 and P2) (Novick et al., 1995). The RNAII transcript encompasses the *agrABCD* operon. AgrB is a membrane transporter that is involved in the export and proteolytic cleavage of the AgrD prepheromone to form a mature autoinducing peptide (AIP), which then acts as the ligand for the activation of AgrC (Ji et al., 1995; Novick, 2003). Phosphorylated AgrA is also able to bind to the P3 promoter, regulating the transcription of the regulatory RNA molecule, RNAIII (Janzon et al., 1989; Novick, 2003). RNAIII regulates the transcription and translation of several genes such as *hla* (α-hemolysin) and *hlb* (β-hemolysin) as well as *spa* (surface protein A) and the *agrABCD* operon (Bronner et al., 2004).

The identification of *agrBD* homologues on the *C. perfringens* chromosome has led to the discovery of a quorum sensing system that is similar to the *agr* system of *S. aureus* (Ohtani et al., 2009; Vidal et al., 2009a; Li et al., 2011). The products of the *C. perfringens agrBD* genes have 50 and 46% amino acid sequence similarity to their respective *S. aureus* counterparts (Ohtani et al., 2009). Consistent with the *S. aureus agr* system, the product of the *agrB* gene is predicted to function as a membrane transporter protein that is involved in the translocation and posttranslational modification of the *agrD* gene product to form a mature pheromone or AIP. Unlike with the *S. aureus* system, no potential AgrCA two-component signal transduction system was identified downstream of the *agrBD* genes (Ohtani et al., 2009). However, previous studies have reported similarity between VirR and AgrA both in terms of domain organization and in terms of the presence of the LytTR DNA-binding motif (Nikolskaya and Galperin, 2002; Sidote et al., 2008; Ohtani et al., 2009). An *agrBD* mutant does not produce perfringolysin O, and expression of the *plc* and *colA* genes is significantly reduced compared to that of the wild type (Ohtani et al., 2009). Expression of the *pfoA* gene in the *agrBD* mutant could be restored to levels similar to those of the wild type upon the addition of diluted stationary-phase culture supernatants harvested from the wild-type strain (Ohtani et al., 2009). Similarly, when an *agrB* mutant and a *pfoA* null mutant were grown together on the same culture but physically separated by a Transwell filter, perfringolysin O production was restored to wild-type levels (Vidal et al., 2009a). These observations suggest that the *agrBD* genes encode a signaling molecule that is required for *pfoA* expression (Ohtani et al., 2009). In an *agrBD-virR* double mutant, the expression of the *pfoA* gene could not be restored by the addition of stationary-phase culture supernatant, which suggests that the *agrBD* gene products regulate *pfoA* gene expression in a VirR-dependent manner (Ohtani et al., 2009). Note that the precise relationship between the Agr and VirSR systems in *C. perfringens* remains to be determined. In particular, it is not known if the AIP that is the processed product of the *agrD* gene actually binds to VirS and therefore is responsible for its autophosphorylation.

In addition to α-toxin and perfringolysin O, the *agr* system has been found to regulate the production of β2-toxin, CPE production, and sporulation in a *C. perfringens* type A food poisoning isolate (Li et al., 2011). The reduction in sporulation and CPE production was attributed, in part, to a considerable decrease in the levels of Spo0A and alternative sigma factors (Li et al., 2011), all of which have been shown to be essential for these processes (Huang et al., 2004; Harry et al., 2009; Li and McClane, 2010).

Although the *agr* system appears to play an important role in stimulating toxin production in response to quorum sensing in *C. perfringens*, it does not appear to be the only system involved in this process. An AI-2-based system also has been identified (Ohtani et al., 2002b). This type of system facilitates both intra- and interspecies communication among gram-negative and gram-positive bacteria (Galloway et al., 2011). Central to the synthesis of AI-2 is the *luxS* gene, with inactivation of *luxS* resulting in elimination of AI-2 production (Xavier and Bassler, 2003). In *C. perfringens* the *luxS* gene, initially known as *ygaG* (Ohtani et al., 2000), was identified as the last gene in the *ycgJ* operon (Ohtani et al., 2002b). The *luxS* gene appears to be involved in stimulating the production of α-toxin, perfringolysin O, and collagenase (κ-toxin). However, the mechanism of regulation seems to be different for the different toxins, with AI-2 stimulating the transcription of the *pfoA* gene but postulated to posttranscriptionally control the production of α-toxin and collagenase (Ohtani et al., 2002b). In addition, it was demonstrated that the culture supernatant derived from the wild-type *C. perfringens* strain could induce luminescence of a *Vibrio harveyi* indicator strain, an ability that is lost when *luxS* is mutated (Ohtani et al., 2002b). This result suggests that AI-2 from *C. perfringens* can be used for interspecies communication. In other species, AI-2 molecules bind to a sensor or receptor on the bacterial cell surface (Galloway et al., 2011). In *C. perfringens*, this sensor has not yet been identified, and although the VirSR–VR-RNA cascade negatively regulates the expression of *luxS* (Ohtani et al., 2000), this AI-2 molecule is not the VirSR effector (Ohtani et al., 2002b).

REGULATION OF TOXIN PRODUCTION IN RESPONSE TO HOST CELL CONTACT

Another means of activating toxin production in *C. perfringens* is by direct cell-to-cell contact. Recent work has shown that the production of α-toxin, β-toxin, β2-toxin, and perfringolysin O was rapidly upregulated when *C. perfringens* type C cells were used to infect intact enterocyte-like Caco-2 cells. This enhanced toxin production corresponded to an increase in transcription of the respective toxin genes, and expression occurred earlier in the growth cycle than with cells cultured in vitro (Vidal et al., 2009b). Supernatants of noninfected Caco-2 cell cultures could not induce β-toxin production, implying that stimulation was not due to a secreted host factor. In addition, induction of β-toxin production was eliminated when the two cell types were physically separated

by a Transwell membrane (Vidal et al., 2009b). Based on these data, it was concluded that contact between the *C. perfringens* and Caco-2 cells was necessary for this upregulation. The mechanism by which contact between these two cell types stimulates β-toxin and perfringolysin O production by *C. perfringens* is unclear, but the data suggest that it is dependent on the VirSR system but not the *luxS*-mediated quorum sensing system (Vidal et al., 2009b).

ALTERNATIVE REGULATORY MECHANISMS

Recent studies have found that some toxin genes are controlled in a VirSR-independent manner by several different regulatory systems (Hiscox et al., 2011; Obana and Nakamura, 2011). These systems are very different from each other, but a common feature is that they seem to regulate genes encoding some of the hyaluronidases and sialidases produced by *C. perfringens*. Previous work has shown that the two extracellular sialidases, NanI and NanJ, are not essential for virulence in the *C. perfringens* gas gangrene strain 13 (Chiarezza et al., 2009). Although the hyaluronidases are not as well studied as the other toxins in *C. perfringens*, they are postulated to be spreading factors that aid disease progression (Hynes and Walton, 2000). Strain 13 carries five genes that encode putative hyaluronidases (Shimizu et al., 2002a), while ATCC 13124 and SM101 contain four putative hyaluronidase genes (Myers et al., 2006). Of these genes, only *nagH* has been studied experimentally and shown to encode a functional hyaluronidase (Canard et al., 1994).

Expression of *nagH*, as well as another putative hyaluronidase gene, *nagL*, is regulated by an orphan response regulator called RevR (Hiscox et al., 2011). RevR has significant amino acid sequence identity to phosphate regulators in *Clostridium kluyveri* (PhoB) and *Bacillus subtilis* (PhoP), as well as VicR from *Streptococcus pneumoniae* and YycF (WalR) from *B. subtilis* (Hiscox et al., 2011). However, the *revR* gene is not located in the vicinity of a sensor histidine kinase gene. RevR also regulates the expression of two sialidase genes, *nanI* and *nanJ*, the *ccp* gene that encodes the cysteine protease α-clostripain, and several genes encoding proteins involved in sporulation (Hiscox et al., 2011). Enzymatic assays demonstrated that the changes in the production of hyaluronidase, sialidase, and protease correlated with the alterations in the transcription of these genes. This response regulator is important in regulating virulence, since a *revR* mutant was attenuated for virulence in the mouse myonecrosis model, an effect that was reversed upon complementation (Hiscox et al., 2011). Whether alterations in the production of these extracellular enzymes are responsible for the attenuation of virulence remains to be determined. The mechanism by which RevR controls transcription of its target genes is also yet to be elucidated, but it has a putative C-terminal winged helix-turn-helix motif and therefore appears to be a transcriptional regulator (Hiscox et al., 2011).

CPE1446 and CPE1447 comprise a novel regulatory system that also regulates extracellular toxin production (Obana and Nakamura, 2011). These open reading frames are cotranscribed and their expression is positively regulated by VR-RNA, but in a VirSR-independent manner (Obana and Nakamura, 2011). CPE1446 and CPE1447 appear to be transcriptional regulators that act by forming a protein complex that regulates the expression of the *ccp*, *plc*, and *pfoA* genes in a different manner from the VirSR–VR-RNA regulatory cascade. These proteins also positively control the expression of the *nagH*, *nagI*, *nagJ*, and *nagK* hyaluronidase genes but negatively regulate the *nanI* sialidase gene. The mechanism by which CPE1446 and CPE1447 regulate their target genes is unclear, and despite the presence of helix-turn-helix motifs in both proteins, they do not seem to bind upstream of the genes that they regulate (Obana and Nakamura, 2011).

Adding to the complexity of hyaluronidase and sialidase gene regulation is the finding that some of these genes are regulated at the posttranscriptional level by the protein encoded by the *CPE1268* open reading frame. Due to its amino acid sequence similarity to the Tex protein of *Bordetella pertussis*, CPE1268 was renamed as Tex (Abe et al., 2010). Sequence analysis of the *tex* region did not reveal VirR boxes, suggesting that this gene was not directly regulated by the VirSR system. Mutation of *tex* led to reduced expression of the *nagH*, *nagJ*, *nagL*, and *nanJ* genes, whereas other potential virulence genes were not affected (Abe et al., 2010). The Tex family of proteins all contain an RNA binding domain in the C-terminal region (Fuchs et al., 1996). RNA binding studies showed that purified *C. perfringens* Tex protein was able to bind specifically to the 5′ region of the *nagH*, *nagJ*, *nagL*, and *nanJ* mRNA molecules. It was postulated that Tex positively regulates these genes by binding to the target mRNA during transcription, thereby acting as an antiterminator (Abe et al., 2010).

CPE is the enterotoxin that is essential for *C. perfringens* type A food poisoning (McClane, 2005). Its regulation is different from that of other *C. perfringens* toxins. Production of CPE is linked to sporulation (Czeczulin et al., 1993), such that the *cpe* gene is expressed only in sporulating cells, not in vegetative cells (Czeczulin et al., 1993; Zhao and Melville, 1998). CPE is not actively secreted; it accumulates in the mother cell during sporulation and is released, along with the mature spore, when the mother cell lyses (Duncan, 1973; Duncan et al., 1973; Loffler and Labbe, 1986). Recent studies have shown that sporulation-specific sigma factors are involved in regulating the transcription of the *cpe* gene (Harry et al., 2009; Li and McClane, 2010). In *C. perfringens*, four sigma factors, SigE, SigF, SigG, and SigK, regulate the expression of genes involved in the formation of heat-resistant endospores (Harry et al., 2009; Li and McClane, 2010). When the genes encoding these sigma factors are disrupted, the resultant mutants either lose their ability to form spores (Li and McClane, 2010) or form spores at a severely reduced frequency (Harry et al., 2009). However, of these sigma factors, only SigE, SigF, and SigK are required for *cpe* expression and CPE production (Li and McClane, 2010). A model that illustrates sigma factor regulation of sporulation and CPE production depicts a hierarchy of control such that SigF regulates the production of SigE and SigK, which, in turn, regulate the transcription of *cpe* from SigE- and SigK-dependent *cpe* promoters (Li and McClane, 2010). In addition to these sigma factors, the catabolite repression transcriptional regulator CcpA has been shown to repress *cpe* expression during exponential phase but to activate expression during stationary phase (Varga et al., 2004). Taken together, it appears that sporulation and the concomitant CPE production in *C. perfringens* constitute a highly complex process where the correct timing of gene expression is paramount.

CONCLUDING REMARKS

C. perfringens is a prolific toxin and extracellular enzyme producer and as such employs many different regulatory systems and mechanisms to control the production of these proteins. It is evident that the VirSR two-component signal transduction system and VR-RNA are central to this regulatory process, but recent work has demonstrated that regulation extends beyond this global two-component system. In particular, an increasing number of regulatory sRNAs are being identified, and with the advent of whole-genome transcriptomic analysis, more sRNAs will undoubtedly be revealed in the not-too-distant future. Elucidating their targets and mechanism of action will prove to be a challenge. The presence of other regulatory systems, such as the RevR orphan response regulator, the CPE1446-CPE1447 protein complex, and the Tex RNA binding protein, illustrates the diverse mechanisms by which this organism controls the production of extracellular toxins and enzymes. There is no doubt that we still have a lot to learn about regulation of toxin production and virulence in this complex bacterial pathogen. The next few years promise to be very exciting as new regulatory networks are discovered and the mechanism of action of existing networks is elucidated.

Acknowledgments. Research in this laboratory was supported by grants from the Australian National Health and Medical Research Council. L.-Y.L. and T.J.H. were supported by a Monash Graduate Scholarship and an Australian Postgraduate Award, respectively.

REFERENCES

Abe, K., N. Obana, and K. Nakamura. 2010. Effects of depletion of RNA-binding protein Tex on the expression of toxin genes in *Clostridium perfringens*. *Biosci. Biotechnol. Biochem.* **74:**1564–1571.

Andre, G., E. Haudecoeur, M. Monot, K. Ohtani, T. Shimizu, B. Dupuy, and I. Martin-Verstraete. 2010. Global regulation of gene expression in response to cysteine availability in *Clostridium perfringens*. *BMC Microbiol.* **10:**234.

Awad, M. M., A. E. Bryant, D. L. Stevens, and J. I. Rood. 1995. Virulence studies on chromosomal α-toxin and θ-toxin mutants constructed by allelic exchange provide genetic evidence for the essential role of α-toxin in *Clostridium perfringens*-mediated gas gangrene. *Mol. Microbiol.* **15:**191–202.

Awad, M. M., D. M. Ellemor, R. L. Boyd, J. J. Emmins, and J. I. Rood. 2001. Synergistic effects of alpha-toxin and perfringolysin O in *Clostridium perfringens*-mediated gas gangrene. *Infect. Immun.* **69:**7904–7910.

Banu, S., K. Ohtani, H. Yaguchi, T. Swe, S. T. Cole, H. Hayashi, and T. Shimizu. 2000. Identification of novel VirR/VirS-regulated genes in *Clostridium perfringens*. *Mol. Microbiol.* **35:**854–864.

Ba-Thein, W., M. Lyristis, K. Ohtani, I. T. Nisbet, H. Hayashi, J. I. Rood, and T. Shimizu. 1996. The *virR/virS* locus regulates the transcription of genes encoding extracellular toxin production in *Clostridium perfringens*. *J. Bacteriol.* **178:**2514–2520.

Beier, D., and R. Gross. 2006. Regulation of bacterial virulence by two-component systems. *Curr. Opin. Microbiol.* **9:**143–152.

Bourret, R. B., and R. E. Silversmith. 2010. Two-component signal transduction. *Curr. Opin. Microbiol.* **13:**113–115.

Bronner, S., H. Monteil, and G. Prevost. 2004. Regulation of virulence determinants in *Staphylococcus aureus*: complexity and applications. *FEMS Microbiol. Rev.* **28:**183–200.

Brynestad, S., and P. E. Granum. 2002. *Clostridium perfringens* and foodborne infections. *Int. J. Food Microbiol.* **74:**195–202.

Calva, E., and R. Oropeza. 2006. Two-component signal transduction systems, environmental signals, and virulence. *Microb. Ecol.* **51:**166–176.

Canard, B., T. Garnier, B. Saint-Joanis, and S. T. Cole. 1994. Molecular genetic analysis of the *nagH* gene encoding a hyaluronidase of *Clostridium perfringens*. *Mol. Gen. Genet.* **243:**215–224.

Casino, P., V. Rubio, and A. Marina. 2010. The mechanism of signal transduction by two-component systems. *Curr. Opin. Struct. Biol.* **20:**763–771.

Chen, Y., D. C. Indurthi, S. W. Jones, and E. T. Papoutsakis. 2011. Small RNAs in the genus *Clostridium*. *mBio* **2:**e00340-00310.

Cheung, J. K., M. M. Awad, S. McGowan, and J. I. Rood. 2009. Functional analysis of the VirSR phosphorelay from *Clostridium perfringens*. *PLoS One* **4:**e5849.

Cheung, J. K., B. Dupuy, D. S. Deveson, and J. I. Rood. 2004. The spatial organization of the VirR boxes is critical for VirR-mediated expression of the perfringolysin O gene, *pfoA*, from *Clostridium perfringens*. *J. Bacteriol.* **186:**3321–3330.

Cheung, J. K., A. L. Keyburn, G. P. Carter, A. L. Lanckriet, F. Van Immerseel, R. J. Moore, and J. I. Rood. 2010. The VirSR two-component signal transduction system regulates NetB toxin production in *Clostridium perfringens*. *Infect. Immun.* **78:**3064–3072.

Cheung, J. K., and J. I. Rood. 2000a. Glutamate residues in the putative transmembrane region are required for the function of the VirS sensor histidine kinase from *Clostridium perfringens*. *Microbiology* **146**(Pt. 2):517–525.

Cheung, J. K., and J. I. Rood. 2000b. The VirR response regulator from *Clostridium perfringens* binds independently to two imperfect direct repeats located upstream of the *pfoA* promoter. *J. Bacteriol.* **182:**57–66.

Chiarezza, M., D. Lyras, S. J. Pidot, M. Flores-Diaz, M. M. Awad, C. L. Kennedy, L. M. Cordner, T. Phumoonna, R. Poon, M. L. Hughes, J. J. Emmins, A. Alape-Giron, and J. I. Rood. 2009. The NanI and NanJ sialidases of *Clostridium perfringens* are not essential for virulence. *Infect. Immun.* **77:**4421–4428.

Czeczulin, J. R., P. C. Hanna, and B. A. McClane. 1993. Cloning, nucleotide sequencing, and expression of the *Clostridium perfringens* enterotoxin gene in *Escherichia coli*. *Infect. Immun.* **61:**3429–3439.

Duncan, C. L. 1973. Time of enterotoxin formation and release during sporulation of *Clostridium perfringens* type A. *J. Bacteriol.* **113:**932–936.

Duncan, C. L., G. J. King, and W. R. Frieben. 1973. A paracrystalline inclusion formed during sporulation of enterotoxin-producing strains of *Clostridium perfringens* type A. *J. Bacteriol.* **114:**845–859.

Fernandez-Miyakawa, M. E., D. J. Fisher, R. Poon, S. Sayeed, V. Adams, J. I. Rood, B. A. McClane, and F. A. Uzal. 2007. Both epsilon-toxin and beta-toxin are important for the lethal properties of *Clostridium perfringens* type B isolates in the mouse intravenous injection model. *Infect. Immun.* **75:**1443–1452.

Fernandez Miyakawa, M. E., and F. A. Uzal. 2003. The early effects of *Clostridium perfringens* type D epsilon toxin in ligated intestinal loops of goats and sheep. *Vet. Res. Commun.* **27:**231–241.

Fisher, D. J., M. E. Fernandez-Miyakawa, S. Sayeed, R. Poon, V. Adams, J. I. Rood, F. A. Uzal, and B. A. McClane. 2006. Dissecting the contributions of *Clostridium perfringens* type C toxins to lethality in the mouse intravenous injection model. *Infect. Immun.* **74:**5200–5210.

Frohlich, K. S., and J. Vogel. 2009. Activation of gene expression by small RNA. *Curr. Opin. Microbiol.* **12:**674–682.

Fuchs, T. M., H. Deppisch, V. Scarlato, and R. Gross. 1996. A new gene locus of *Bordetella pertussis* defines a novel family of prokaryotic transcriptional accessory proteins. *J. Bacteriol.* **178:**4445–4452.

Galloway, W. R., J. T. Hodgkinson, S. D. Bowden, M. Welch, and D. R. Spring. 2011. Quorum sensing in Gram-negative bacteria: small-molecule modulation of AHL and AI-2 quorum sensing pathways. *Chem. Rev.* **111:**28–67.

Galperin, M. Y. 2010. Diversity of structure and function of response regulator output domains. *Curr. Opin. Microbiol.* **13:**150–159.

Gibert, M., C. Jolivert-Reynaud, and M. R. Popoff. 1997. Beta2 toxin, a novel toxin produced by *Clostridium perfringens*. *Gene* **203:**65–73.

Gotoh, Y., Y. Eguchi, T. Watanabe, S. Okamoto, A. Doi, and R. Utsumi. 2010. Two-component signal transduction as potential drug targets in pathogenic bacteria. *Curr. Opin. Microbiol.* **13:**232–239.

Gottesman, S. 2005. Micros for microbes: non-coding regulatory RNAs in bacteria. *Trends Genet.* **21**:399–404.

Harry, K. H., R. Zhou, L. Kroos, and S. B. Melville. 2009. Sporulation and enterotoxin (CPE) synthesis are controlled by the sporulation-specific sigma factors SigE and SigK in *Clostridium perfringens. J. Bacteriol.* **191**:2728–2742.

Hassan, S., K. Ohtani, R. Wang, Y. Yuan, Y. Wang, Y. Yamaguchi, and T. Shimizu. 2010. Transcriptional regulation of *hemO* encoding heme oxygenase in *Clostridium perfringens. J. Microbiol.* **48**:96–101.

Higashi, Y., M. Chazono, K. Inoue, Y. Yanagase, and T. Amano. 1973. Complementation of theta-toxinogenicity between mutants of two groups of *Clostridium perfringens. Biken J.* **16**:1–9.

Hiscox, T. J., A. Chakravorty, J. M. Choo, K. Ohtani, T. Shimizu, J. K. Cheung, and J. I. Rood. 2011. Regulation of virulence by the RevR response regulator in *Clostridium perfringens. Infect. Immun.* **79**:2145–2153.

Huang, I. H., M. Waters, R. R. Grau, and M. R. Sarker. 2004. Disruption of the gene (*spo0A*) encoding sporulation transcription factor blocks endospore formation and enterotoxin production in enterotoxigenic *Clostridium perfringens* type A. *FEMS Microbiol. Lett.* **233**:233–240.

Hynes, W. L., and S. L. Walton. 2000. Hyaluronidases of Grampositive bacteria. *FEMS Microbiol. Lett.* **183**:201–207.

Imagawa, T., and Y. Higashi. 1992. An activity which restores theta toxin activity in some theta toxin-deficient mutants of *Clostridium perfringens. Microbiol. Immunol.* **36**:523–527.

Imagawa, T., T. Tatsuki, Y. Higashi, and T. Amano. 1981. Complementation characteristics of newly isolated mutants from two groups of strains of *Clostridium perfringens. Biken J.* **24**:13–21.

Janzon, L., S. Lofdahl, and S. Arvidson. 1989. Identification and nucleotide sequence of the delta-lysin gene, *hld*, adjacent to the accessory gene regulator (*agr*) of *Staphylococcus aureus. Mol. Gen. Genet.* **219**:480–485.

Ji, G., R. C. Beavis, and R. P. Novick. 1995. Cell density control of staphylococcal virulence mediated by an octapeptide pheromone. *Proc. Natl. Acad. Sci. USA* **92**:12055–12059.

Johnson, S., and D. N. Gerding. 1997. Enterotoxemic infections, p. 117–140. *In* J. I. Rood, B. A. McClane, J. G. Songer, and R. W. Titball (ed.), *The Clostridia: Molecular Biology and Pathogenesis.* Academic Press, San Diego, CA.

Kawsar, H. I., K. Ohtani, K. Okumura, H. Hayashi, and T. Shimizu. 2004. Organization and transcriptional regulation of *myo*-inositol operon in *Clostridium perfringens. FEMS Microbiol. Lett.* **235**:289–295.

Keyburn, A. L., T. L. Bannam, J. Moore, and J. I. Rood. 2010. NetB, a pore-forming toxin from necrotic enteritis strains of *Clostridium perfringens. Toxins* **2**:1913–1927.

Keyburn, A. L., J. D. Boyce, P. Vaz, T. L. Bannam, M. E. Ford, D. Parker, A. Di Rubbo, J. I. Rood, and R. J. Moore. 2008. NetB, a new toxin that is associated with avian necrotic enteritis caused by *Clostridium perfringens. PLoS Pathog.* **4**:e26.

Lawerence, G. W. 1997. The pathogenesis of enteritis necroticans, p. 197–207. *In* J. I. Rood, B. A. McClane, J. G. Songer, and R. W. Titball (ed.), *The Clostridia: Molecular Biology and Pathogenesis.* Academic Press, San Diego, CA.

Li, J., J. Chen, J. E. Vidal, and B. A. McClane. 2011. The Agr-like quorum-sensing system regulates sporulation and production of enterotoxin and beta2 toxin by *Clostridium perfringens* type A non-food-borne human gastrointestinal disease strain F5603. *Infect. Immun.* **79**:2451–2459.

Li, J., and B. A. McClane. 2010. Evaluating the involvement of alternative sigma factors SigF and SigG in *Clostridium perfringens* sporulation and enterotoxin synthesis. *Infect. Immun.* **78**:4286–4293.

Liu, J. M., and A. Camilli. 2010. A broadening world of bacterial small RNAs. *Curr. Opin. Microbiol.* **13**:18–23.

Loffler, A., and R. Labbe. 1986. Characterization of a parasporal inclusion body from sporulating, enterotoxin-positive *Clostridium perfringens* type A. *J. Bacteriol.* **165**:542–548.

Lyristis, M., A. E. Bryant, J. Sloan, M. M. Awad, I. T. Nisbet, D. L. Stevens, and J. I. Rood. 1994. Identification and molecular analysis of a locus that regulates extracellular toxin production in *Clostridium perfringens. Mol. Microbiol.* **12**:761–777.

Ma, M., J. Vidal, J. Saputo, B. A. McClane, and F. Uzal. 2011. The VirS/VirR two-component system regulates the anaerobic cytotoxicity, intestinal pathogenicity, and enterotoxemic lethality of *Clostridium perfringens* type C isolate CN3685. *mBio* **2**:e00338-00310.

McClane, B. A. 2005. Clostridial enterotoxins, p. 385–406. *In* P. Dürre (ed.), *Handbook on Clostridia.* CRC Press, Boca Raton, FL.

McGowan, S., I. S. Lucet, J. K. Cheung, M. M. Awad, J. C. Whisstock, and J. I. Rood. 2002. The FxRxHrS motif: a conserved region essential for DNA binding of the VirR response regulator from *Clostridium perfringens. J. Mol. Biol.* **322**:997–1011.

McGowan, S., J. R. O'Connor, J. K. Cheung, and J. I. Rood. 2003. The SKHR motif is required for biological function of the VirR response regulator from *Clostridium perfringens. J. Bacteriol.* **185**:6205–6208.

Myers, G. S., D. A. Rasko, J. K. Cheung, J. Ravel, R. Seshadri, R. T. DeBoy, Q. Ren, J. Varga, M. M. Awad, L. M. Brinkac, S. C. Daugherty, D. H. Haft, R. J. Dodson, R. Madupu, W. C. Nelson, M. J. Rosovitz, S. A. Sullivan, H. Khouri, G. I. Dimitrov, K. L. Watkins, S. Mulligan, J. Benton, D. Radune, D. J. Fisher, H. S. Atkins, T. Hiscox, B. H. Jost, S. J. Billington, J. G. Songer, B. A. McClane, R. W. Titball, J. I. Rood, S. B. Melville, and I. T. Paulsen. 2006. Skewed genomic variability in strains of the toxigenic bacterial pathogen, *Clostridium perfringens. Genome Res.* **16**:1031–1040.

Nikolskaya, A. N., and M. Y. Galperin. 2002. A novel type of conserved DNA-binding domain in the transcriptional regulators of the AlgR/AgrA/LytR family. *Nucleic Acids Res.* **30**:2453–2459.

Novick, R. P. 2003. Autoinduction and signal transduction in the regulation of staphylococcal virulence. *Mol. Microbiol.* **48**:1429–1449.

Novick, R. P., S. J. Projan, J. Kornblum, H. F. Ross, G. Ji, B. Kreiswirth, F. Vandenesch, and S. Moghazeh. 1995. The *agr* P2 operon: an autocatalytic sensory transduction system in *Staphylococcus aureus. Mol. Gen. Genet.* **248**:446–458.

Novick, R. P., H. F. Ross, S. J. Projan, J. Kornblum, B. Kreiswirth, and S. Moghazeh. 1993. Synthesis of staphylococcal virulence factors is controlled by a regulatory RNA molecule. *EMBO J.* **12**:3967–3975.

Obana, N., and K. Nakamura. 2011. A novel toxin regulator, the CPE1446-CPE1447 protein heteromeric complex, controls toxin genes in *Clostridium perfringens. J. Bacteriol.* **193**:4417–4424.

Obana, N., Y. Shirahama, K. Abe, and K. Nakamura. 2010. Stabilization of *Clostridium perfringens* collagenase mRNA by VR-RNA-dependent cleavage in 5′ leader sequence. *Mol. Microbiol.* **77**:1416–1428.

Ohtani, K., S. K. Bhowmik, H. Hayashi, and T. Shimizu. 2002a. Identification of a novel locus that regulates expression of toxin genes in *Clostridium perfringens. FEMS Microbiol. Lett.* **209**:113–118.

Ohtani, K., H. Hayashi, and T. Shimizu. 2002b. The *luxS* gene is involved in cell-cell signalling for toxin production in *Clostridium perfringens. Mol. Microbiol.* **44**:171–179.

Ohtani, K., H. Hirakawa, K. Tashiro, S. Yoshizawa, S. Kuhara, and T. Shimizu. 2010. Identification of a two-component VirR/VirS regulon in *Clostridium perfringens. Anaerobe* **16**:258–264.

Ohtani, K., H. I. Kawsar, K. Okumura, H. Hayashi, and T. Shimizu. 2003. The VirR/VirS regulatory cascade affects transcription of plasmid-encoded putative virulence genes in *Clostridium perfringens* strain 13. *FEMS Microbiol. Lett.* **222:**137–141.

Ohtani, K., H. Takamura, H. Yaguchi, H. Hayashi, and T. Shimizu. 2000. Genetic analysis of the *ycgJ-metB-cysK-ygaG* operon negatively regulated by the VirR/VirS system in *Clostridium perfringens*. *Microbiol. Immunol.* **44:**525–528.

Ohtani, K., Y. Yuan, S. Hassan, R. Wang, Y. Wang, and T. Shimizu. 2009. Virulence gene regulation by the *agr* system in *Clostridium perfringens*. *J. Bacteriol.* **191:**3919–3927.

Okumura, K., H. I. Kawsar, T. Shimizu, T. Ohta, and H. Hayashi. 2005. Identification and characterization of a cell-wall anchored DNase gene in *Clostridium perfringens*. *FEMS Microbiol. Lett.* **242:**281–285.

Okumura, K., K. Ohtani, H. Hayashi, and T. Shimizu. 2008. Characterization of genes regulated directly by the VirR/VirS system in *Clostridium perfringens*. *J. Bacteriol.* **190:**7719–7727.

Otto, M. 2004. Quorum-sensing control in staphylococci—a target for antimicrobial drug therapy? *FEMS Microbiol. Lett.* **241:**135–141.

Petit, L., M. Gibert, and M. R. Popoff. 1999. *Clostridium perfringens*: toxinotype and genotype. *Trends Microbiol.* **7:**104–110.

Popoff, M. R., and P. Bouvet. 2009. Clostridial toxins. *Future Microbiol.* **4:**1021–1064.

Rood, J. I. 1998. Virulence genes of *Clostridium perfringens*. *Annu. Rev. Microbiol.* **52:**333–360.

Rood, J. I., and S. T. Cole. 1991. Molecular genetics and pathogenesis of *Clostridium perfringens*. *Microbiol. Rev.* **55:**621–648.

Sarker, M. R., R. J. Carman, and B. A. McClane. 1999. Inactivation of the gene (*cpe*) encoding *Clostridium perfringens* enterotoxin eliminates the ability of two *cpe*-positive *C. perfringens* type A human gastrointestinal disease isolates to affect rabbit ileal loops. *Mol. Microbiol.* **33:**946–958.

Sayeed, S., M. E. Fernandez-Miyakawa, D. J. Fisher, V. Adams, R. Poon, J. I. Rood, F. A. Uzal, and B. A. McClane. 2005. Epsilon-toxin is required for most *Clostridium perfringens* type D vegetative culture supernatants to cause lethality in the mouse intravenous injection model. *Infect. Immun.* **73:**7413–7421.

Sayeed, S., F. A. Uzal, D. J. Fisher, J. Saputo, J. E. Vidal, Y. Chen, P. Gupta, J. I. Rood, and B. A. McClane. 2008. Beta toxin is essential for the intestinal virulence of *Clostridium perfringens* type C disease isolate CN3685 in a rabbit ileal loop model. *Mol. Microbiol.* **67:**15–30.

Shimizu, T., W. Ba-Thein, M. Tamaki, and H. Hayashi. 1994. The *virR* gene, a member of a class of two-component response regulators, regulates the production of perfringolysin O, collagenase, and hemagglutinin in *Clostridium perfringens*. *J. Bacteriol.* **176:**1616–1623.

Shimizu, T., K. Ohtani, H. Hirakawa, K. Ohshima, A. Yamashita, T. Shiba, N. Ogasawara, M. Hattori, S. Kuhara, and H. Hayashi. 2002a. Complete genome sequence of *Clostridium perfringens*, an anaerobic flesh-eater. *Proc. Natl. Acad. Sci. USA* **99:**996–1001.

Shimizu, T., K. Shima, K. Yoshino, K. Yonezawa, and H. Hayashi. 2002b. Proteome and transcriptome analysis of the virulence genes regulated by the VirR/VirS system in *Clostridium perfringens*. *J. Bacteriol.* **184:**2587–2594.

Shimizu, T., H. Yaguchi, K. Ohtani, S. Banu, and H. Hayashi. 2002c. Clostridial VirR/VirS regulon involves a regulatory RNA molecule for expression of toxins. *Mol. Microbiol.* **43:**257–265.

Sidote, D. J., C. M. Barbieri, T. Wu, and A. M. Stock. 2008. Structure of the *Staphylococcus aureus* AgrA LytTR domain bound to DNA reveals a beta fold with an unusual mode of binding. *Structure* **16:**727–735.

Skjelkvale, R., and T. Uemura. 1977. Experimental diarrhoea in human volunteers following oral administration of *Clostridium perfringens* enterotoxin. *J. Appl. Bacteriol.* **43:**281–286.

Songer, J. G. 2010. Clostridia as agents of zoonotic disease. *Vet. Microbiol.* **140:**399–404.

Songer, J. G. 1997. Clostridial diseases of animals, p. 153–182. *In* J. I. Rood, B. A. McClane, J. G. Songer, and R. W. Titball (ed.), *The Clostridia: Molecular Biology and Pathogenesis*. Academic Press, San Diego, CA.

Songer, J. G. 1996. Clostridial enteric diseases of domestic animals. *Clin. Microbiol. Rev.* **9:**216–234.

Songer, J. G., and D. W. Miskimmins. 2004. *Clostridium perfringens* type E enteritis in calves: two cases and a brief review of the literature. *Anaerobe* **10:**239–242.

Songer, J. G., and F. A. Uzal. 2005. Clostridial enteric infections in pigs. *J. Vet. Diagn. Investig.* **17:**528–536.

Stevens, D. L., R. K. Tweten, M. M. Awad, J. I. Rood, and A. E. Bryant. 1997. Clostridial gas gangrene: evidence that alpha and theta toxins differentially modulate the immune response and induce acute tissue necrosis. *J. Infect. Dis.* **176:**189–195.

Timbermont, L., F. Haesebrouck, R. Ducatelle, and F. Van Immerseel. 2011. Necrotic enteritis in broilers: an updated review on the pathogenesis. *Avian Pathol.* **40:**341–347.

Titball, R. W. 2005. Gas gangrene: an open and closed case. *Microbiology* **151:**2821–2828.

Tsuge, H., M. Nagahama, M. Oda, S. Iwamoto, H. Utsunomiya, V. E. Marquez, N. Katunuma, M. Nishizawa, and J. Sakurai. 2008. Structural basis of actin recognition and arginine ADP-ribosylation by *Clostridium perfringens* iota-toxin. *Proc. Natl. Acad. Sci. USA* **105:**7399–7404.

Uzal, F., and B. A. McClane. 2011. Recent progress in understanding the pathogenesis of *Clostridium perfringens* type C infections. *Vet. Microbiol.* **153:**37–43.

Uzal, F. A., J. Saputo, S. Sayeed, J. E. Vidal, D. J. Fisher, R. Poon, V. Adams, M. E. Fernandez-Miyakawa, J. I. Rood, and B. A. McClane. 2009. Development and application of new mouse models to study the pathogenesis of *Clostridium perfringens* type C enterotoxemias. *Infect. Immun.* **77:**5291–5299.

Uzal, F. A., and J. G. Songer. 2008. Diagnosis of *Clostridium perfringens* intestinal infections in sheep and goats. *J. Vet. Diagn. Investig.* **20:**253–265.

Vandenesch, F., J. Kornblum, and R. P. Novick. 1991. A temporal signal, independent of *agr*, is required for *hla* but not *spa* transcription in *Staphylococcus aureus*. *J. Bacteriol.* **173:**6313–6320.

Van Immerseel, F., J. I. Rood, R. J. Moore, and R. W. Titball. 2009. Rethinking our understanding of the pathogenesis of necrotic enteritis in chickens. *Trends Microbiol.* **17:**32–36.

Varga, J., V. L. Stirewalt, and S. B. Melville. 2004. The CcpA protein is necessary for efficient sporulation and enterotoxin gene (*cpe*) regulation in *Clostridium perfringens*. *J. Bacteriol.* **186:**5221–5229.

Vidal, J. E., J. Chen, J. Li, and B. A. McClane. 2009a. Use of an EZ-Tn5-based random mutagenesis system to identify a novel toxin regulatory locus in *Clostridium perfringens* strain 13. *PLoS One* **4:**e6232.

Vidal, J. E., K. Ohtani, T. Shimizu, and B. A. McClane. 2009b. Contact with enterocyte-like Caco-2 cells induces rapid upregulation of toxin production by *Clostridium perfringens* type C isolates. *Cell. Microbiol.* **11:**1306–1328.

Vidal, J. E., B. A. McClane, J. Saputo, J. Parker, and F. A. Uzal. 2008. Effects of *Clostridium perfringens* beta-toxin on the rabbit small intestine and colon. *Infect. Immun.* **76:**4396–4404.

Waters, L. S., and G. Storz. 2009. Regulatory RNAs in bacteria. *Cell* **136:**615–628.

Wolz, C., P. Pohlmann-Dietze, A. Steinhuber, Y. T. Chien, A. Manna, W. van Wamel, and A. Cheung. 2000. Agr-independent regulation of fibronectin-binding protein(s) by the regulatory locus *sar* in *Staphylococcus aureus*. *Mol. Microbiol.* **36:**230–243.

Xavier, K. B., and B. L. Bassler. 2003. LuxS quorum sensing: more than just a numbers game. *Curr. Opin. Microbiol.* **6:**191–197.

Yuan, Y., K. Ohtani, S. Yoshizawa, and T. Shimizu. 21 September 2011. Complex transcriptional regulation of citrate metabolism in *Clostridium perfringens*. *Anaerobe* **18:**48–54. [Epub ahead of print.]

Zhao, Y., and S. B. Melville. 1998. Identification and characterization of sporulation-dependent promoters upstream of the enterotoxin gene (*cpe*) of *Clostridium perfringens*. *J. Bacteriol.* **180:**136–142.

Regulation of Bacterial Virulence
Edited by Michael L. Vasil and Andrew J. Darwin
© 2013 ASM Press, Washington, DC   doi:10.1128/9781555818524.ch15

Chapter 15

Regulation of Toxin Production in *Clostridium difficile*

GLEN P. CARTER, KATE E. MACKIN, JULIAN I. ROOD, AND DENA LYRAS

INTRODUCTION

Clostridium difficile is the causative agent of a range of intestinal diseases collectively referred to as *C. difficile* infections (CDI) or *C. difficile*-associated disease. Disease in humans normally presents as diarrhea, which can be mild to severe (Borriello, 1998). However, infection with *C. difficile* can result in more serious conditions, including pseudomembranous colitis, which is characterized by the presence of pseudomembranous plaques on the intestinal mucosa and severe inflammation of the colon, or, more rarely, toxic megacolon, which is a severe and acute form of colonic distension that is associated with high mortality rates (Dobson et al., 2003).

C. difficile is primarily a nosocomial pathogen, although community-acquired disease is being reported with increasing incidence (Baker et al., 2010). The major risk factors of CDI are advanced age (>65 years) and the prior use of antibiotics or other therapies that disrupt the normal intestinal microbiota (Borriello, 1998). In healthy individuals, the normal intestinal microbiota provides "colonization resistance" against *C. difficile*, which prevents the organism from becoming established in the gut (Reeves et al., 2011). Once this resistance is removed through the use of antibiotics, *C. difficile* is able to rapidly colonize the gut and cause disease.

C. difficile represents a significant health care problem and is the major cause of antibiotic-associated diarrhea in the developed world (Borriello, 1998). Estimates suggest that CDI costs the U.S. health care system approximately $3.2 billion per year (O'Brien et al., 2007). The problem has been exacerbated in the last decade with the emergence of so-called "hypervirulent" strains characterized as BI by restriction endonuclease analysis, NAP1 by North American pulsed-field gel electrophoresis, and 027 by ribotyping (McDonald et al., 2005; Warny et al., 2005). These strains reportedly produce more toxin (Warny et al., 2005) and have also developed resistance against moxifloxacin, gatifloxacin, and erythromycin (McDonald et al., 2005), which was not previously observed in *C. difficile*. In addition, some have been suggested to be proficient sporeformers (Merrigan et al., 2010). Importantly, infection with BI/NAP1/027 strains of *C. difficile* is associated with the onset of more severe disease and higher rates of morbidity and mortality (Loo et al., 2005; Stabler et al., 2009; Freeman et al., 2010).

CDI is a toxin-mediated disease, with most disease symptoms observed being directly linked to the production of two separate toxins known as toxin A (TcdA) and toxin B (TcdB) (Carter et al., 2010). Both toxins are monoglucosyltransferases that inactivate host cell Rho family GTPases, resulting in disruption of the actin cytoskeleton and cell death (Just et al., 1995). This process leads to a loss of tight junction integrity in the gut, followed by the movement of fluid into the intestinal lumen and the onset of diarrhea (Voth and Ballard, 2005). A number of strains, including BI/NAP1/027 isolates, also produce a third toxin known as CDT binary toxin (Perelle et al., 1997). This toxin is an ADP-ribosyltransferase that causes actin depolymerization, resulting in disruption of the host cell cytoskeleton and cell death (Perelle et al., 1997). However, the exact role of this toxin in disease is currently not known, as discussed later in this review.

The expression of toxins by *C. difficile* appears to be a highly regulated process, and it is well established that environmental stimuli such as nutrients (Mani et al., 2002), temperature (Karlsson et al., 2003), and growth phase (Hundsberger et al., 1997) play an important role in the control of toxin production. It is only recently, however, that a better understanding of how *C. difficile* regulates the production of toxins at the molecular level has emerged, primarily because the genetic tools required to make targeted genetic knockouts have been developed only in the last few years (O'Connor et al., 2006; Heap et al., 2007). This chapter discusses our current understanding of gene regulation in *C. difficile*, focusing on

Glen P. Carter, Kate E. Mackin, Julian I. Rood, and Dena Lyras • Department of Microbiology, Monash University, Clayton, Victoria 3800, Australia.

how toxin production is regulated, with a particular emphasis on the major toxins, toxin A and toxin B.

THE IMPACT OF ENVIRONMENT ON THE PRODUCTION OF TOXIN A AND TOXIN B

The onset of toxin synthesis is associated with entry into the stationary phase of growth (Hundsberger et al., 1997; Mani et al., 2002). The precise growth phase signals involved in the initiation of toxin production remain unknown, even though nutrient signals have clearly been shown to have a profound effect on toxin production by *C. difficile*. For example, the presence of glucose or another rapidly metabolizable sugar in the *C. difficile* growth medium strongly represses the production of the two major toxins, TcdA and TcdB (Mani et al., 2002), indicating that catabolite repression of toxin production occurs, as discussed below. Furthermore, the amino acids proline and, to a greater extent, cysteine have been shown to substantially reduce the amount of toxin produced by *C. difficile* (Karlsson et al., 2003). Conversely, increased levels of toxin production have been observed when the concentration of biotin is limited or when the bicarbonate concentration is increased, suggesting that the assimilation of CO_2 or HCO_3^- may be involved in toxin regulation (Karlsson et al., 1999).

Temperature has also been shown to influence toxin production, with maximal toxin expression detected at 37°C and lower levels occurring at 22 and 42°C (Karlsson et al., 2003). This observation is in keeping with the biological niche in which *C. difficile* produces toxins, namely, the mammalian host.

THE MOLECULAR MECHANISMS CONTROLLING TOXIN PRODUCTION

The Pathogenicity Locus

TcdA and TcdB are encoded by *tcdA* and *tcdB*, respectively, which are located in a region of the chromosome known as the pathogenicity locus (PaLoc) (Fig. 1) (Braun et al., 1996). In addition to the *tcdA* and *tcdB* genes, three other genes are located within the PaLoc: *tcdR*, *tcdC*, and *tcdE* (Braun et al., 1996). A transcriptional analysis of the PaLoc region showed that *tcdA*, *tcdB*, *tcdE*, and *tcdR* were all coordinately expressed during late logarithmic stages of growth (Hundsberger et al., 1997). In addition, they were transcribed as both mono- and polycistronic mRNAs, suggesting that the genes are arranged as a coordinately regulated toxin gene operon. Conversely, the *tcdC* gene was found to be transcribed primarily

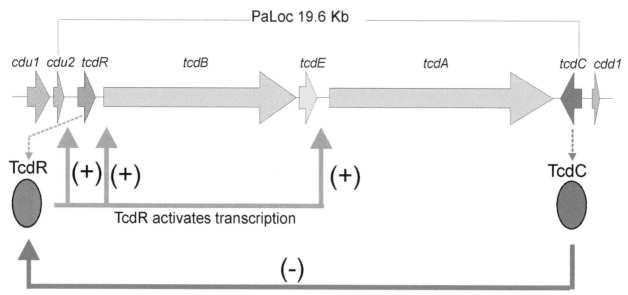

Figure 1. Structure of the PaLoc and flanking regions. Filled arrows indicate open reading frames, with arrows showing the direction of transcription. Toxin genes are shown in blue, regulatory genes are in orange and green, *tcdE* is in yellow, and genes located outside the PaLoc are in grey. The sigma factor TcdR interacts with the core RNA polymerase protein, facilitating recognition of the *tcdA*, *tcdB*, and *tcdR* promoters by the TcdR-RNA polymerase complex and promoting transcription from these promoters. The anti-sigma factor TcdC negatively regulates transcription by interacting with TcdR. doi:10.1128/9781555818524.ch15f1

during early phases of growth, with substantially reduced levels of transcription occurring during the stationary growth phase (Hundsberger et al., 1997), although a recent report suggests that transcription of *tcdC* may also occur during stationary-phase growth (Merrigan et al., 2010). The *tcdC* gene is transcribed as a monocistronic mRNA and appears to be expressed independently of the toxin gene operon. The genetic arrangement of the PaLoc and the temporal pattern of gene expression among the encoded genes led to the hypothesis that TcdR and TcdC may act as positive and negative regulators, respectively, of toxin production (Hundsberger et al., 1997). The role of TcdE remains to be determined; however, experiments conducted in *Escherichia coli* suggest that this protein may be involved in export and release of the toxins from *C. difficile* (Tan et al., 2001).

The Alternative Sigma Factor TcdR

The TcdR protein has a helix-turn-helix DNA binding motif (Moncrief et al., 1997) and has limited similarity to some clostridial transcriptional activator proteins, as well as to several families of eubacterial RNA polymerase sigma factors (Moncrief et al., 1997; Mani and Dupuy, 2001), which is consistent with the hypothesis that TcdR is a positive regulator of toxin production. The first experimental evidence to support this hypothesis was obtained by Moncrief et al. (Moncrief et al., 1997), who showed that the expression of a reporter gene transcriptionally fused to either the *tcdA* or *tcdB* promoter increased 500- or 800-fold, respectively, when TcdR was coexpressed in *trans* within *E. coli* cells. These experimental findings were subsequently corroborated in a second surrogate host, *Clostridium perfringens* (Mani and Dupuy, 2001). The latter study also showed that the TcdR protein functions as a sigma factor rather than as a transcriptional activator (Mani and Dupuy, 2001). Evidence for this conclusion came from the observations that TcdR could not directly bind to the *C. difficile* toxin promoters, that TcdR interacts with the core RNA polymerase protein, and, most importantly, that TcdR facilitates recognition of the *C. difficile* toxin promoters by RNA polymerase. This study also noted that the *tcdA* and *tcdB* promoter regions share a high degree of sequence similarity and do not resemble typical prokaryotic σ^{70} promoters, which is in keeping with the hypothesis that toxin gene expression is critically dependent on the presence of an alternative σ-factor, namely, TcdR (Mani and Dupuy, 2001). The development of genetic tools for *C. difficile* facilitated studies which showed that TcdR activates the expression of the toxin genes in the native *C. difficile* host (Mani et al., 2002), confirming the previous findings

on the role of TcdR in toxin gene expression carried out in heterologous hosts (Mani et al., 2002). This study also showed that TcdR regulates the expression of the *tcdR* gene; therefore, TcdR synthesis is positively autoregulated (Mani et al., 2002).

Homologues of TcdR have been identified in several other pathogenic clostridial species, including *C. botulinum*, *C. tetani*, and *C. perfringens* (Marvaud et al., 1998). These proteins, known as BotR, TetR, and UviA, have been shown to be critical for the expression of the botulinum (BoNT) and tetanus (TetX) toxins in *C. botulinum* and *C. tetani* (Raffestin et al., 2005) and for the synthesis of a UV-inducible bacteriocin (Bcn) in *C. perfringens* (Dupuy et al., 2005), respectively. TcdR and its homologues are functionally interchangeable in vitro and, to a limited extent, in vivo, with each protein able to facilitate binding of RNA polymerase and to promote transcription from the promoters for *tcdA*, *tcdB*, *ntnH*, the gene encoding the botulinum toxin, *tetX*, and *bcn*, albeit to various degrees (Dupuy et al., 2006). This finding suggests that the control of toxin and bacteriocin production in these species may occur via a conserved mechanism (Mani et al., 2002). The TcdR family of proteins is most similar to the extracytoplasmic function (ECF) sigma factors that are assigned to group 4 of the σ^{70} family; however, significant structural and functional differences exist between the TcdR-like proteins and the latter family. For this reason, the TcdR family of proteins comprises a new subgroup (group 5) of σ^{70} family RNA polymerase sigma factors (Dupuy et al., 2006).

As stated above, toxin production in *C. difficile* can be profoundly affected by environmental cues. Since toxin production relies on the presence of TcdR, it is not surprising that many of these stimuli influence toxin production by modulating the expression of TcdR. The presence of glucose in the growth medium, for example, substantially represses *tcdR* expression (Mani et al., 2002), as does growth of *C. difficile* at temperatures significantly above or below 37°C (Karlsson et al., 2003). Therefore, the negative impact of these cues on toxin production is likely to be mediated indirectly through the downregulation of *tcdR* expression, rather than via a direct effect on the toxin genes. Whether modulation of toxin production by other environmental stimuli also occurs in an indirect manner is currently not known and requires further investigation.

The Anti-Sigma Factor TcdC

Interest in TcdC and the putative role it plays in regulating toxin production increased when it was discovered that hypervirulent BI/NAP1/027 strains of

C. difficile have a naturally occurring mutation within the *tcdC* gene, resulting in the production of a truncated and nonfunctional TcdC protein (Warny et al., 2005). These strains were reported to make significantly more toxin than other clinical isolates, leading to the hypothesis that the mutation within *tcdC* was responsible for this phenotype, since TcdC was thought to act as a negative regulator of toxin production. Therefore, it was hypothesized that the *tcdC* mutation in these strains ultimately resulted in the increased severity and mortality associated with these strains. However, experimental evidence that TcdC acts as a negative regulator of toxin production, or that the *tcdC* mutation impacts toxin production or virulence of BI/NAP1/027 strains, was lacking until recently. The first evidence showing that TcdC has the capacity to act as a negative regulator was obtained through studies performed in the surrogate host *C. perfringens* (Matamouros et al., 2007). These studies showed that expression of TcdR significantly upregulates the expression of a *tcdA* promoter–β-glucuronidase reporter gene fusion and that coexpression of TcdC with TcdR reduces the level of β-glucuronidase expression from the toxin A promoter, providing convincing evidence that TcdC negatively regulates expression from the toxin gene promoter (Matamouros et al., 2007).

TcdC is a membrane-associated protein that is anchored to the cell membrane at the N terminus and is predicted to contain a transmembrane domain (Govind et al., 2006). The protein forms a dimer via a coiled-coil motif located within the middle of the protein (Matamouros et al., 2007). TcdC has little homology with any characterized proteins, and no DNA binding motif has been identified within the protein. Consequently, the mechanism by which TcdC represses toxin production in *C. difficile* was unclear, although the data suggest that TcdC functions as an anti-sigma factor (Dupuy et al., 2008). Anti-sigma factors generally interact with a cognate sigma factor, thereby preventing the formation of an active RNA polymerase holoenzyme and blocking transcription of the target genes. Importantly, anti-sigma factors are often membrane associated, particularly those that interact with the ECF family of sigma factors (Govind et al., 2006), to which TcdR is most closely related (Dupuy et al., 2006). Experimental evidence has confirmed this hypothesis by showing that TcdC interacts with TcdR and blocks TcdR-mediated transcription from the *C. difficile* toxin promoters. Pulldown experiments were used to confirm that TcdC associates with the TcdR protein, but confirmation of this result using surface plasmon resonance analysis was unable to be obtained (Matamouros et al., 2007). Importantly, in vitro transcription experiments were able to demonstrate that TcdC inhibited transcription from the

tcdA promoter by destabilizing the TcdR-containing holoenzyme-*tcdA* promoter complex. Subsequent experiments showed, however, that once a stable open complex was formed at the toxin gene promoter, TcdC no longer inhibited transcription, indicating that TcdC must act at an early stage of transcriptional initiation (Matamouros et al., 2007). Overall, these experiments clearly demonstrated that TcdC inhibits transcription from the *C. difficile tcdA* promoter.

The recent analysis of *tcdC* genes from a diverse group of clinical isolates has led to a reconsideration of the role of TcdC in toxin gene regulation and hypervirulence in BI/NAP1/027 strains of *C. difficile* (Curry et al., 2007; Murray et al., 2009). These studies found that mutations within the *tcdC* gene were widespread among clinical isolates and that the presence of these mutations was not an accurate indicator of high-level toxin production, such as that seen in many BI/NAP1/027 strains. The major limitation of these studies, however, was that these conclusions were drawn from the analysis of nonisogenic strains. It is very difficult to accurately assign gene function based on the analysis of nonisogenic strains, particularly in a genetically heterogeneous species like *C. difficile*, since the impact of secondary mutations and additional genetic variation cannot be assessed. Using a newly developed and more efficient method to genetically manipulate BI/NAP1/027 strains of *C. difficile*, the *tcdC* mutation in M7404, a Canadian BI/NAP1/027 isolate, was recently complemented with a wild-type copy of *tcdC* (Carter et al., 2011). In this study, isogenic strains were generated to definitively determine the role of TcdC in toxin production and virulence (Carter et al., 2011). Using these strains, the expression of TcdC within the native host was shown to significantly inhibit toxin production and to reduce the virulence of the *tcdC*-complemented strain in a hamster disease model. These experiments clearly demonstrate the importance of TcdC as a negative regulator of toxin production in *C. difficile* and provide the first experimental evidence that the *tcdC* mutation in BI/NAP1/027 strains contributes to the development of hypervirulence in *C. difficile* (Carter et al., 2011). Despite these advances in our understanding of the role played by this protein, the exact mechanism by which TcdC inhibits toxin production remains to be determined.

Other Regulators of Toxin A and Toxin B Production in *C. difficile*

The regulation of toxin production in *C. difficile* is clearly complex and relies on a number of different regulatory systems. Only now with the development of efficient genetic techniques which enable manipulation of a range of strain types can we begin to elucidate the mechanisms controlling these regulatory networks. *C. difficile* strain 630 is a nonhypervirulent, toxinotype 0 clinical isolate, which was the first *C. difficile* strain to have its complete genome sequence determined. Therefore, this strain is often used as a point of reference for *C. difficile* strain comparisons. Genomic analysis of strain 630 has led to the identification of 45 two-component systems as well as five orphan histidine kinases and six orphan response regulators (Sebaihia et al., 2006; Stabler et al., 2006). However, most of these putative regulators remain uncharacterized.

Three global regulators have been associated with toxin expression in *C. difficile*, described in detail below. Two of these proteins—Spo0A and CodY—are involved in the transition into stationary growth and sporulation. The involvement of SigH, a sigma factor associated with Spo0A, has also been described previously. The third regulator is CcpA, and it is responsible for glucose-mediated repression of toxin production.

The Global Regulator Spo0A

Spo0A is a global regulator that is the master regulator of sporulation, controlling the transition of the bacterial cell into the spore form. The sporulation pathway is best studied in the genus *Bacillus*, and homologues of a number of key genes in this pathway have also been identified in the clostridia (Paredes et al., 2005). In several species of *Bacillus* and *Clostridium*, a range of histidine kinases have been shown to detect environmental signals, ultimately leading to the phosphorylation and activation of Spo0A. This regulator is then able to exert an effect on a broad range of target genes.

In *Bacillus* species, activation of Spo0A occurs via a phosphorelay cascade that eventuates in the phosphorylation of Spo0A via Spo0F and Spo0B (Stephenson and Lewis, 2005). Once phosphorylated Spo0A reaches a critical level, it triggers transition from one cell behavior (such as motility and biofilm formation) to another (such as sporulation), as demonstrated in *Bacillus subtilis* (Lopez et al., 2009). In the clostridia, Spo0A activation appears to be controlled by an alternative mechanism, possibly relying on a two-component-like system, since no Spo0F or Spo0B homologues have been identified to date (Paredes et al., 2005).

Nevertheless, in both *Bacillus* and the clostridia, the phosphorylated form of Spo0A can act directly by binding upstream of a target gene, at consensus sequences known as 0A boxes, to produce a transcriptional response. Spo0A can also act indirectly by influencing the transcription of target genes that then themselves result in the up- or downregulation of other genes (Paredes et al., 2005).

In addition to its involvement in the sporulation cascade, Spo0A has also been shown to influence a range of other cell behaviors. In *B. subtilis*, for example, Spo0A is associated with the regulation of more than 500 genes, including those involved in motility and chemotaxis (Fawcett et al., 2000). Likewise, in the clostridia, Spo0A is known to regulate a number of non-sporulation-related genes, including those involved in virulence. For example, in a *C. perfringens spo0A* mutant, enterotoxin and TpeL toxin production was found to be markedly reduced (Huang et al., 2004; Paredes-Sabja et al., 2011). Interestingly, TpeL is a member of the large clostridial toxin family, with similarity to toxins A and B from *C. difficile*. Thus, it is possible that toxin production in *C. difficile* may also be regulated by Spo0A, with a possible association with sporulation. In keeping with this hypothesis, it has been suggested that sporulation and toxin production may represent different survival strategies for *C. difficile* and may therefore be subject to the same regulatory controls (Kamiya et al., 1992; Akerlund et al., 2006). Indeed, as discussed below, recent evidence suggests that Spo0A may play a regulatory role in toxin production (Underwood et al., 2009). However, since Spo0A is a pleiotropic regulator, it is not clear whether toxin production is associated with sporulation or another, as-yet-unidentified process that is also controlled by Spo0A in *C. difficile*.

Similar to the case with *Bacillus* species, mutagenesis of *spo0A* in *C. difficile* has confirmed the requirement of this regulator for sporulation (Heap et al., 2007). The role of Spo0A in toxin production has also subsequently been investigated, using an erythromycin-sensitive derivative of strain 630, defined as 630Δ*erm* (Underwood et al., 2009). As expected, a *spo0A* mutant in this strain was not able to produce detectable spores. This mutant also produced less toxin than the wild type, since Western blot analysis showed that toxin A levels were less than 10% of those produced by the wild-type strain during stationary phase. Furthermore, supernatant from the *spo0A* mutant showed a 1,000-fold decrease in activity compared to that of the wild-type strain in a Vero cell cytotoxicity assay. Thus, in addition to its role in controlling sporulation, Spo0A also seems to be involved in the regulation of toxin production in *C. difficile* (Underwood et al., 2009), although neither study complemented the *spo0A* mutations with a wild-type copy of the gene.

By interrogation of the strain 630 genome sequence, a putative sensor histidine kinase, CD2494, was also identified in this study and putatively found to be involved in regulation of toxin production in strain 630Δ*erm* (Underwood et al., 2009). Although not followed up by complementation analysis, mutagenesis of

CD2494 resulted in a 10-fold reduction in cytotoxicity compared to the wild type, with an estimated 50 to 60% reduction in toxin A levels. This mutant also showed a decreased sporulation capacity, suggestive of a role in spore formation. However, these properties were not completely abolished, suggesting that other kinases may also play a role in the signaling pathways leading to sporulation and toxin production (Underwood et al., 2009). Five orphan histidine kinases have been annotated in the genome sequence of strain 630 (Sebaihia et al., 2006). Three of these enzymes, including CD2494, were identified as putative sporulation-associated kinases (Underwood et al., 2009). Orthologues of these three putative kinases are found in all sequenced *C. difficile* strains examined thus far, but not in other clostridial species, suggesting that the signals required for activation of Spo0A in *C. difficile* are very specific (Underwood et al., 2009). Unfortunately, CD2494 could not be overexpressed and purified in order to test its ability to phosphorylate Spo0A. However, another putative kinase, CD1579, was shown to directly phosphorylate Spo0A in vitro (Underwood et al., 2009). This result lends weight to the suggestion that Spo0A is part of a two-component-like system and is directly phosphorylated by these kinases, in contrast to the phosphorelay systems of *Bacillus* species.

Together, these results permit an understanding of the initial events that occur in the activation of Spo0A in *C. difficile*. Once activated, Spo0A works either directly upon a target or indirectly. The control of toxin production by Spo0A is likely to be indirect, as upstream regions of *tcdA* and *tcdB* do not contain the 0A boxes required for binding of Spo0A to these regions (Underwood et al., 2009), nor does *tcdR* (Saujet et al., 2011). The signaling networks controlled by Spo0A are potentially vast and intricate. The major genes of the *Bacillus* sporulation cascade downstream of Spo0A have also been identified in the clostridia, including those encoding the key sigma factors (Paredes et al., 2005). The earliest sporulation-associated sigma factor, and the only such factor thus far examined in any detail in *C. difficile*, is SigH.

The Alternative Sigma Factor SigH

SigH is an alternative sigma factor that is involved in the transition to stationary phase and sporulation. In *Bacillus* species, SigH functions through a feedback control mechanism, via the transitional regulator AbrB. AbrB normally represses the expression of SigH in *B. subtilis* (Weir et al., 1991); AbrB is, in turn, repressed by phosphorylated Spo0A (Perego et al., 1988). Spo0A-mediated repression of AbrB results in the lifting of AbrB-mediated repression of SigH,

leading to further expression of Spo0A. In this way, phosphorylation of Spo0A results in a cycle which promotes further synthesis of Spo0A. Since *C. difficile* lacks the transition state regulator AbrB, it has been suggested that Spo0A directly controls the transcription of *sigH* (Saujet et al., 2011). In *C. difficile*, SigH also seems to positively regulate the transcription of *spo0A*, with a putative SigH recognition sequence located in the *spo0A* promoter region; consequently, regulation of Spo0A and that of SigH are intricately interrelated (Saujet et al., 2011).

Transcription of *sigH* in *C. difficile* is coordinated with transcription of *tcdA* and *tcdR* (Karlsson et al., 2008), suggesting a link between SigH and PaLoc expression. A *sigH* mutant of strain 630E, another erythromycin-sensitive derivative of strain 630, isolated independently of strain 630Δ*erm* (O'Connor et al., 2006), was recently constructed to further explore this potential association (Saujet et al., 2011). By comparing the expression profiles of the wild-type 630E and *sigH* mutant derivative, this study identified over 400 genes differentially regulated at the beginning of stationary phase, including many genes involved in sporulation, as has been observed in *B. subtilis*. The *sigH* mutant was unable to produce detectable spores, but sporulation was restored upon complementation (Saujet et al., 2011). Other genes altered in expression included those involved in motility, cellular division, metabolism, and virulence. Importantly, levels of expression of *tcdA*, *tcdB*, and *tcdR* were all increased in the *sigH* mutant compared to the wild-type strain, indicating that this protein plays a role in regulation of these genes. Protein dot blot analysis showed that the level of toxin A produced by the mutant was greater than that produced by the wild-type strain, confirming the results obtained from the transcriptional analysis (Saujet et al., 2011). The role of SigH in regulation of toxin production appears to be indirect, as no SigH recognition sequence could be identified upstream of the PaLoc genes (Saujet et al., 2011). The increased toxin production observed in the *sigH* mutant contrasts with the decreased toxin production seen in the previously described *spo0A* mutant; however, the way in which these regulators interact to control toxin production is still unknown, and more research is clearly needed to determine the complex networks involved.

The Global Transcriptional Regulator CodY

As discussed above, the expression of PaLoc genes is upregulated when cells enter stationary phase and nutrients become limited (Hundsberger et al., 1997; Dupuy and Sonenshein, 1998). CodY is a global transcriptional regulator involved in the bacterial response to nutrient limitation. Well conserved across the low-G+C gram-positive bacteria, including *C. difficile*, CodY acts as a dimer and consists of an N-terminal GAF family cofactor binding domain and a C-terminal winged helix-turn-helix DNA binding domain (Levdikov et al., 2006).

CodY represses transcription of many stationary-phase genes, preventing their expression during the exponential phase of growth. In *B. subtilis*, it has been shown that the affinity of CodY for its target sites is increased in the presence of GTP and branched-chain amino acids (BCAAs), either individually or in combination (Shivers and Sonenshein, 2004; Handke et al., 2008), since the presence of these cofactors is a signal of nutrient availability. When levels of GTP and BCAAs are reduced, CodY acts as a signal of nutrient limitation and is no longer able to bind to its target sites. Thus, when nutrients become limited CodY repression is lifted, allowing the expression of genes involved in bacterial adaptation to starvation. CodY is also a positive regulator of a small number of genes, such as those involved in the positive regulation of carbon overflow metabolism (Shivers et al., 2006). Overall, CodY is involved in the regulation of more than 100 genes in *B. subtilis* (Molle et al., 2003). These targets are associated with a range of adaptive behaviors, such as motility (Bergara et al., 2003), sugar and amino acid transport (Molle et al., 2003), and sporulation (Ratnayake-Lecamwasam et al., 2001). In some bacterial species, CodY is also involved in regulation of virulence factors (Pohl et al., 2009; van Schaik et al., 2009).

CodY is also associated with virulence and other phenotypes in *C. difficile* (Dineen et al., 2007; Dineen et al., 2010). Microarray analysis that compared strain 630E and an isogenic *codY* mutant showed that over 140 genes are under the control of CodY in *C. difficile* (Dineen et al., 2010). These include genes involved in metabolism, sporulation, and virulence. Regulation of binary toxin is not under the control of CodY (Dineen et al., 2010). However, the role of CodY in the regulation of toxin A and toxin B is becoming clearer, with CodY appearing to repress toxin gene expression in *C. difficile*. Phenotypic analysis of a *codY* mutant showed that it produced greater levels of toxin A in vitro than the wild-type strain, supporting transcriptional evidence showing that CodY represses toxin production (Dineen et al., 2007). Subsequent microarray analysis confirmed this result and showed that the expression of *tcdR*, *tcdA*, *tcdE*, and *tcdR* is altered in the *codY* mutant, in agreement with the earlier study (Dineen et al. 2007; Dineen et al., 2010).

The promoter regions of *tcdC*, *tcdA*, and *tcdB* have been shown to contain CodY binding sites (Dineen et al., 2007), although it is notable that CodY was able to bind only weakly to the two toxin promoters in vitro (Dineen et al., 2010). Regulation of toxin production seems to be mediated primarily through the interaction of CodY and the promoter region of *tcdR*, with CodY showing high-affinity binding to the promoter region of *tcdR* in vitro. In addition, GTP and BCAAs are cofactors for the repression of toxin production by CodY. Binding of *C. difficile* CodY to the *tcdR* promoter region was increased in the presence of GTP and BCAAs, in a similar way to *B. subtilis* CodY and its targets (Dineen et al., 2007). This supports previous evidence showing that a mixture of amino acids, including BCAAs, represses toxin production (Karlsson et al., 1999). When a *codY* mutant was grown in medium containing glucose, a known repressor of toxin production, toxin A levels were significantly lower than when the strain was grown in medium without glucose. This suggests that glucose repression of toxin production may be independent of CodY-mediated repression (Dineen et al., 2007) and that another regulatory mechanism may be involved.

The interplay of CodY with other regulatory networks is also becoming evident. Interestingly, several genes identified as regulated by CodY contain EAL and/or GGDEP domains (Dineen et al., 2010). These domains are associated with the secondary messenger cyclic di-GMP (c-di-GMP). *C. difficile* encodes a number of proteins potentially involved in the c-di-GMP signaling pathway, suggesting that it is important for the adaptation of this organism to environmental changes (Sudarsan et al., 2008; Bordeleau et al., 2011). In addition, the CodY and CcpA regulatory networks are intertwined. Carbon catabolite repression, mediated by the transcriptional regulator CcpA, is another mechanism involved in repression of toxin production (as discussed below). So, for example, under conditions of nutrient availability, CcpA stimulates production of BCAAs, promoting binding of CodY to its target sites (Sonenshein, 2007).

The Transcription Factor CcpA

The presence of glucose, or other rapidly metabolizable carbon sources, in vitro is known to inhibit the production of toxin A and toxin B, suggesting that production of these toxins is controlled by carbon catabolite repression (Dupuy and Sonenshein, 1998). In the low-G+C gram-positive bacteria, carbon catabolite repression is primarily mediated by CcpA. It is becoming increasingly clear that CcpA can also modulate virulence factor production in various species (Varga et al., 2004; Li et al., 2010; Chiang et al., 2011).

CcpA is a transcription factor belonging to the LacI-GalR family of proteins (Henkin et al., 1991).

Like the other global regulators discussed here, CcpA is best characterized in *Bacillus* species. CcpA binds to catabolite-responsive elements (*cre* sites) (Miwa et al., 2000), which are consensus DNA sequences found within or downstream of the promoter regions of target genes. CcpA is part of a complex regulatory pathway (Gorke and Stulke, 2008) and is activated by binding with the phosphorylated form of HPr, which is part of the phosphoenolpyruvate-dependent phosphotransferase system (PTS). In the presence of particular carbon sources, HPr is phosphorylated at a conserved serine residue (forming HPr-Ser-P) by the HPr kinase/phosphorylase (HprK/P). The activity of HprK/P is stimulated by fructose-1,6-biphosphate (FBP). The interaction between CcpA and HPr-Ser-P promotes CcpA binding to its target *cre* sites (Miwa et al., 2000). In *B. subtilis*, about 11% all of genes are regulated by CcpA; a similar result has been found for the *C. difficile* CcpA, with many of the affected genes being involved in carbon metabolism (Antunes et al., 2011).

A link between carbon catabolite repression and toxin production in *C. difficile* has recently been demonstrated (Antunes et al., 2011). In this study, mutant derivatives of strain 630E targeting various aspects of the carbon catabolite repression pathway were constructed, specifically, the PTS proteins enzyme I and HPr (encoded by the *ptsI* and *ptsH* genes), HprK, and CcpA. The results showed that PTS is required for glucose uptake and glucose-mediated toxin gene repression, since the *ptsI* and *ptsH* mutants produced the same amount of toxin whether or not the culture medium contained glucose. That is, derepression of toxin production was observed when these strains were cultured in the presence of glucose. By contrast, repression of toxin production in the presence of glucose was still observed in the *hprK* mutant. It appears that HprK-mediated phosphorylation of HPr is not required for repression of toxin production in *C. difficile* (Antunes et al., 2011). Depression of toxin production was also observed in the *ccpA* mutant grown in the presence of glucose, so CcpA is clearly required for glucose-mediated toxin repression. However, the *ccpA* mutant produced twofold-lower levels of toxin than the wild-type strain, suggesting that another regulator may be linked to CcpA and therefore involved in the regulation of toxin production.

As noted above, CcpA represses gene expression by binding to *cre* sites associated with its target genes (Miwa et al., 2000). Potential *cre* sites were identified in the promoter region of *tcdB* and the 5′ region of the *tcdA* coding sequence, and CcpA was found to interact with these regions in vitro (Antunes et al., 2011). Interestingly, FBP, glucose-6-phosphate, and NADP were all found to increase the in vitro affinity of CcpA for the regulatory regions of *tcdA* and *tcdB* in *C. difficile*. However, HPr and HPr-Ser-P did not promote CcpA binding to target DNA, in contrast to what has been documented in the *Bacillus* model. This result, and the fact that HprK-mediated phosphorylation of HPr is not required for repression of toxin production, suggests that *C. difficile* CcpA uses a different mechanism to respond to particular carbon sources (Antunes et al., 2011).

Bacteriophage-Mediated Regulation of Toxin Production

Recent studies suggest that prophages can influence the regulation of toxin A and B production in *C. difficile*. Initial studies suggested that toxin A and/or toxin B production was modified in a toxigenic *C. difficile* lysogen (Goh et al., 2005). More recent studies involving transcriptional analysis of PaLoc genes following lysogeny by ΦCD119, a *Myoviridae* family member, demonstrated a decrease in gene expression, which was probably mediated by the FCD119-encoded repressor gene *repR* (Govind et al., 2009). In this study, RepR was shown to downregulate *tcdA::gusA* and *tcdR::gusA* reporter fusions in *E. coli*, in the latter case by binding specifically to the region upstream of *tcdR* in the PaLoc (Govind et al., 2009). RepR was also shown to bind specifically to the *repR* upstream region from FCD119, and similarities between the RepR binding sites upstream of *tcdR* and *repR* were defined (Govind et al., 2009). The results from this study suggest that the presence of a temperate phage can influence toxin gene regulation in *C. difficile*, an effect that is mediated by a phage-encoded protein, RepR, most likely acting directly through *tcdR* to downregulate the expression of *tcdA* and *tcdB* (Govind et al., 2009). *C. difficile* phage infection studies involving a BI/NAP1/027 strain and the unrelated *Siphoviridae* family phage FCD38-2 have also shown that this phage modulates toxin production, but in this case immunoblots demonstrated 1.6- and 2.1-fold increases in toxin A and toxin B levels, respectively. This finding was supported by real-time quantitative reverse transcriptase PCR analysis that showed increased expression of all of the PaLoc genes; however, the mechanism responsible for this increased expression was not defined (Sekulovic et al., 2011). Undoubtedly, these recent studies provide convincing evidence that the presence of temperate phages can influence toxin production in *C. difficile*.

Interestingly, a c-di-GMP riboswitch has been identified within the lysis module of FCD119 (Sudarsan et al., 2008; Govind et al., 2009); riboswitches are mRNA domains involved in controlling gene expression in response to environmental or physiological changes. The presence of this riboswitch suggests that physiological changes in the host bacterium can be monitored by FCD119, which will then exert an effect on PaLoc genes in a cascade effect that results from multilayered and complex intertwined regulatory networks likely to involve both the phage and host bacterium (Sudarsan et al., 2008; Govind et al., 2009; Sekulovic et al., 2011).

Regulation of Binary Toxin Production

C. difficile transferase (CDT) is the third toxin produced by many strains of *C. difficile,* especially epidemic or hypervirulent strains. This binary toxin consists of two polypeptides: a binding component, CDTb, that is responsible for attachment of the toxin complex to the lipolysis-stimulated lipoprotein receptor on the host cell surface (Papatheodorou et al., 2011), and the active component, CDTa, which has actin-specific ADP-ribosyltransferase activity. The toxin is closely related to iota-toxin from *C. perfringens,* CST binary toxin from *Clostridium spiroforme,* and C2 toxin from *C. botulinum* (Barth et al., 2004). The role of CDT in the disease process remains to be elucidated. It clearly is not essential for disease, since many virulent strains do not produce CDT. However, it has been shown that CDT depolymerizes the actin cytoskeleton and also enhances the adhesion of *C. difficile* cells to the gastrointestinal epithelium by causing the formation of microtubule-based protrusions in the host cell membrane (Schwan et al., 2009; Schwan et al., 2011). Therefore, it has been postulated that CDT may augment the virulence of epidemic *C. difficile* strains (Schwan et al., 2009; Papatheodorou et al., 2011).

CDT is encoded by two genes, *cdtA* and *cdtB,* that are located in an operon at a defined locus on the *C. difficile* chromosome, the CdtLoc (Carter et al., 2007). This 6.2-kb locus includes a response regulator gene, *cdtR,* that is located upstream of the *cdt* operon. The CdtLoc seems to have been inserted just upstream of two open reading frames, *cd2602* and *cd2601,* that appear to encode a two-component signal transduction system (Fig 2). In most *C. difficile* strains that do not produce CDT, the CdtLoc is replaced by a 68-bp noncoding region. The sequenced *C. difficile* strain 630 carries this locus, but the *cdtA* and *cdtB* genes are pseudogenes in this strain (Carter et al., 2007). Initial analysis of 26 *C. difficile* isolates showed that all of the 18 strains that carried the *cdtAB* operon, as either intact genes or pseudogenes, also carried the *cdtR* gene as part of the CdtLoc. Analysis of genomic microarray data revealed that 82% of the strains that lacked the CdtLoc were in the A⁻B⁺ clade of *C. difficile* isolates (Carter et al., 2007). These findings were confirmed in subsequent studies, which identified five different *cdtR* alleles (Bouvet and Popoff, 2008). One of these alleles, which was present in four strains that also had a 39-bp deletion in *tcdC,* had an in-frame stop codon at codon 107, which would lead to the production of a truncated and nonfunctional CdtR protein. These strains all produced lower levels of ADP-ribosyltransferase activity than strains that did not have this internal *cdtR* stop codon. In a more recent study, a similar stop codon at position 107 of *cdtR* was observed in a ribotype 078 isolate (Metcalf and Weese, 2011).

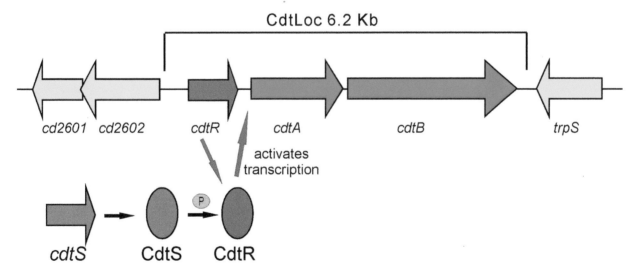

Figure 2. Activation of *cdtAB* by CdtR. An unknown signal activates the putative CdtS sensor histidine kinase, which then leads to the phosphorylation (P) of the response regulator CdtR. Phosphorylated CdtR activates the transcription of the *cdtAB* operon, which is part of the chromosomal CdtLoc, as shown.
doi:10.1128/9781555818524.ch15f2

CdtR is a LytTR-like positive transcriptional regulator of the Cdt operon and CDT production (Carter et al., 2007). When a plasmid that carries the *cdtR* gene is introduced into a strain that does not carry the CdtLoc but that has a recombinant plasmid that carries the *cdtAB* operon, significantly increased CDT production is observed (Carter et al., 2007). Increased toxin production is not observed when either a vector plasmid or a plasmid with a large in-frame deletion in the *cdtR* gene is introduced into the same strain. It is postulated that an unknown environmental signal leads to the autophosphorylation of an as-yet-unknown sensor histidine kinase, designated CdtS. Phosphorylated CdtS then binds to CdtR, leading to transfer of the phosphoryl group to CdtR and activation of the DNA binding domain of CdtR. CdtR then activates the transcription of the *cdtAB* operon, possibly directly, since we recently demonstrated binding of purified CdtR to the upstream *cdtA* promoter region using gel mobility shift assays (V. Rai, G. P. Carter, D. Lyras, and J. I. Rood, unpublished data). Clearly, there is a need for further research to identify CdtS and to determine the nature of the environmental or growth phase signal that leads to its activation as well as to identify the CdtR binding site.

CONCLUSIONS

In conclusion, the regulation of toxin production in *C. difficile* is complex, with multiple environmental and physiological factors involved in eliciting precise responses via specific regulatory proteins. The regulatory networks involved in triggering the resulting effects are intricate and interrelated in ways that are only now beginning to be understood. Recent breakthroughs in genetic manipulation technologies available for *C. difficile* (O'Connor et al., 2006; Heap et al., 2007), together with refined animal infection models, will facilitate future studies and will further define the complex regulatory networks involved in *C. difficile* toxin regulation.

Acknowledgments. Research in our laboratory was supported by grants from the Australian Research Council and the Australian National Health and Medical Research Council.

REFERENCES

Akerlund, T., B. Svenungsson, A. Lagergren, and L. G. Burman. 2006. Correlation of disease severity with fecal toxin levels in patients with *Clostridium difficile*-associated diarrhea and distribution of PCR ribotypes and toxin yields in vitro of corresponding isolates. *J. Clin. Microbiol.* **44:**353–358.

Antunes, A., I. Martin-Verstraete, and B. Dupuy. 2011. CcpA-mediated repression of *Clostridium difficile* toxin gene expression. *Mol. Microbiol.* **79:**882–899.

Baker, S. S., H. Faden, W. Sayej, R. Patel, and R. D. Baker. 2010. Increasing incidence of community-associated atypical *Clostridium difficile* disease in children. *Clin. Pediatr.* (Philadelphia) **49:**644–647.

Barth, H., K. Aktories, M. R. Popoff, and B. G. Stiles. 2004. Binary bacterial toxins: biochemistry, biology, and applications of common *Clostridium* and *Bacillus* proteins. *Microbiol. Mol. Biol. Rev.* **68:**373–402.

Bergara, F., C. Ibarra, J. Iwamasa, J. C. Patarroyo, R. Aguilera, and L. M. Marquez-Magana. 2003. CodY is a nutritional repressor of flagellar gene expression in *Bacillus subtilis*. *J. Bacteriol.* **185:**3118–3126.

Bordeleau, E., L. C. Fortier, F. Malouin, and V. Burrus. 2011. c-di-GMP turn-over in *Clostridium difficile* is controlled by a plethora of diguanylate cyclases and phosphodiesterases. *PLoS Genet.* **7:**e1002039.

Borriello, S. P. 1998. Pathogenesis of *Clostridium difficile* infection. *J. Antimicrob. Chemother.* **41:**13–19.

Bouvet, P. J., and M. R. Popoff. 2008. Genetic relatedness of *Clostridium difficile* isolates from various origins determined by triple-locus sequence analysis based on toxin regulatory genes *tcdC*, *tcdR*, and *cdtR*. *J. Clin. Microbiol.* **46:**3703–3713.

Braun, V., T. Hundsberger, P. Leukel, M. Sauerborn, and C. von Eichel-Streiber. 1996. Definition of the single integration site of the pathogenicity locus in *Clostridium difficile*. *Gene* **181:**29–38.

Carter, G. P., G. R. Douce, R. Govind, P. M. Howarth, K. E. Mackin, J. Spencer, A. M. Buckley, A. Antunes, D. Kotsanas, G. A. Jenkin, B. Dupuy, J. I. Rood, and D. Lyras. 2011. The anti-sigma factor TcdC modulates hypervirulence in an epidemic BI/NAP1/027 clinical isolate of *Clostridium difficile*. *PLoS Pathog.* **7:**e1002317.

Carter, G. P., D. Lyras, D. L. Allen, K. E. Mackin, P. M. Howarth, J. R. O'Connor, and J. I. Rood. 2007. Binary toxin production in *Clostridium difficile* is regulated by CdtR, a LytTR family response regulator. *J. Bacteriol.* **189:**7290–7301.

Carter, G. P., J. I. Rood, and D. Lyras. 2010. The role of toxin A and toxin B in *Clostridium difficile*-associated disease: past and present perspectives. *Gut Microbes* **1:**58–64.

Chiang, C., C. Bongiorni, and M. Perego. 2011. Glucose-dependent activation of *Bacillus anthracis* toxin gene expression and virulence requires the carbon catabolite protein CcpA. *J. Bacteriol.* **193:**52–62.

Curry, S. R., J. W. Marsh, C. A. Muto, M. M. O'Leary, A. W. Pasculle, and L. H. Harrison. 2007. *tcdC* genotypes associated with severe TcdC truncation in an epidemic clone and other strains of *Clostridium difficile*. *J. Clin. Microbiol.* **45:**215–221.

Dineen, S. S., S. M. McBride, and A. L. Sonenshein. 2010. Integration of metabolism and virulence by *Clostridium difficile* CodY. *J. Bacteriol.* **192:**5350–5362.

Dineen, S. S., A. C. Villapakkam, J. T. Nordman, and A. L. Sonenshein. 2007. Repression of *Clostridium difficile* toxin gene expression by CodY. *Mol. Microbiol.* **66:**206–219.

Dobson, G., C. Hickey, and J. Trinder. 2003. *Clostridium difficile* colitis causing toxic megacolon, severe sepsis and multiple organ dysfunction syndrome. *Intensive Care Med.* **29:**1030.

Dupuy, B., R. Govind, A. Antunes, and S. Matamouros. 2008. *Clostridium difficile* toxin synthesis is negatively regulated by TcdC. *J. Med. Microbiol.* **57:**685–689.

Dupuy, B., N. Mani, S. Katayama, and A. L. Sonenshein. 2005. Transcription activation of a UV-inducible *Clostridium perfringens* bacteriocin gene by a novel sigma factor. *Mol. Microbiol.* **55:**1196–1206.

Dupuy, B., S. Raffestin, S. Matamouros, N. Mani, M. R. Popoff, and A. L. Sonenshein. 2006. Regulation of toxin and bacteriocin gene expression in *Clostridium* by interchangeable RNA polymerase sigma factors. *Mol. Microbiol.* **60:**1044–1057.

Dupuy, B., and A. L. Sonenshein. 1998. Regulated transcription of *Clostridium difficile* toxin genes. *Mol. Microbiol.* **27:**107–120.

Fawcett, P., P. Eichenberger, R. Losick, and P. Youngman. 2000. The transcriptional profile of early to middle sporulation in *Bacillus subtilis. Proc. Natl. Acad. Sci. USA* **97:**8063–8068.

Freeman, J., M. P. Bauer, S. D. Baines, J. Corver, W. N. Fawley, B. Goorhuis, E. J. Kuijper, and M. H. Wilcox. 2010. The changing epidemiology of *Clostridium difficile* infections. *Clin. Microbiol. Rev.* **23:**529–549.

Goh, S., B. J. Chang, and T. V. Riley. 2005. Effect of phage infection on toxin production by *Clostridium difficile. J. Med. Microbiol.* **54:**129–135.

Gorke, B., and J. Stulke. 2008. Carbon catabolite repression in bacteria: many ways to make the most out of nutrients. *Nat. Rev. Microbiol.* **6:**613–624.

Govind, R., G. Vediyappan, R. D. Rolfe, B. Dupuy, and J. A. Fralick. 2009. Bacteriophage-mediated toxin gene regulation in *Clostridium difficile. J. Virol.* **83:**12037–12045.

Govind, R., G. Vediyappan, R. D. Rolfe, and J. A. Fralick. 2006. Evidence that *Clostridium difficile* TcdC is a membrane-associated protein. *J. Bacteriol.* **188:**3716–3720.

Handke, L. D., R. P. Shivers, and A. L. Sonenshein. 2008. Interaction of *Bacillus subtilis* CodY with GTP. *J. Bacteriol.* **190:**798–806.

Heap, J. T., O. J. Pennington, S. T. Cartman, G. P. Carter, and N. P. Minton. 2007. The ClosTron: a universal gene knock-out system for the genus *Clostridium. J. Microbiol. Methods* **70:**452–464.

Henkin, T. M., F. J. Grundy, W. L. Nicholson, and G. H. Chambliss. 1991. Catabolite repression of alpha-amylase gene expression in *Bacillus subtilis* involves a trans-acting gene product homologous to the *Escherichia coli lacI* and *galR* repressors. *Mol. Microbiol.* **5:**575–584.

Huang, I. H., M. Waters, R. R. Grau, and M. R. Sarker. 2004. Disruption of the gene (*spo0A*) encoding sporulation transcription factor blocks endospore formation and enterotoxin production in enterotoxigenic *Clostridium perfringens* type A. *FEMS Microbiol. Lett.* **233:**233–240.

Hundsberger, T., V. Braun, M. Weidmann, P. Leukel, M. Sauerborn, and C. von Eichel-Streiber. 1997. Transcription analysis of the genes *tcdA-E* of the pathogenicity locus of *Clostridium difficile. Eur. J. Biochem.* **244:**735–742.

Just, I., J. Selzer, M. Wilm, C. von Eichel-Streiber, M. Mann, and K. Aktories. 1995. Glucosylation of Rho proteins by *Clostridium difficile* toxin B. *Nature* **375:**500–503.

Kamiya, S., H. Ogura, X. Q. Meng, and S. Nakamura. 1992. Correlation between cytotoxin production and sporulation in *Clostridium difficile. J. Med. Microbiol.* **37:**206–210.

Karlsson, S., L. G. Burman, and T. Akerlund. 2008. Induction of toxins in *Clostridium difficile* is associated with dramatic changes of its metabolism. *Microbiology* **154:**3430–3436.

Karlsson, S., L. G. Burman, and T. Akerlund. 1999. Suppression of toxin production in *Clostridium difficile* VPI 10463 by amino acids. *Microbiology* **145:**1683–1693.

Karlsson, S., B. Dupuy, K. Mukherjee, E. Norin, L. G. Burman, and T. Akerlund. 2003. Expression of *Clostridium difficile* toxins A and B and their sigma factor TcdD is controlled by temperature. *Infect. Immun.* **71:**1784–1793.

Levdikov, V. M., E. Blagova, P. Joseph, A. L. Sonenshein, and A. J. Wilkinson. 2006. The structure of CodY, a GTP- and isoleucine-responsive regulator of stationary phase and virulence in gram-positive bacteria. *J. Biol. Chem.* **281:**11366–11373.

Li, C., F. Sun, H. Cho, V. Yelavarthi, C. Sohn, C. He, O. Schneewind, and T. Bae. 2010. CcpA mediates proline auxotrophy and is required for *Staphylococcus aureus* pathogenesis. *J. Bacteriol.* **192:**3883–3892.

Loo, V. G., L. Poirier, M. A. Miller, M. Oughton, M. D. Libman, S. Michaud, A. M. Bourgault, T. Nguyen, C. Frenette, M. Kelly, A.

Vibien, P. Brassard, S. Fenn, K. Dewar, T. J. Hudson, R. Horn, P. Rene, Y. Monczak, and A. Dascal. 2005. A predominantly clonal multi-institutional outbreak of *Clostridium difficile*-associated diarrhea with high morbidity and mortality. *N. Engl. J. Med.* **353:**2442–2449.

Lopez, D., H. Vlamakis, and R. Kolter. 2009. Generation of multiple cell types in *Bacillus subtilis. FEMS Microbiol. Rev.* **33:**152–163.

Mani, N., and B. Dupuy. 2001. Regulation of toxin synthesis in *Clostridium difficile* by an alternative RNA polymerase sigma factor. *Proc. Natl. Acad. Sci. USA* **98:**5844–5849.

Mani, N., D. Lyras, L. Barroso, P. Howarth, T. Wilkins, J. I. Rood, A. L. Sonenshein, and B. Dupuy. 2002. Environmental response and autoregulation of *Clostridium difficile* TxeR, a sigma factor for toxin gene expression. *J. Bacteriol.* **184:**5971–5978.

Marvaud, J., M. Gibert, K. Inoue, Y. Fujinaga, K. Oguma, and M. Popoff. 1998. *botR/A* is a positive regulator of botulinum neurotoxin and associated non-toxin protein genes in *Clostridium botulinum* A. *Mol. Microbiol.* **29:**1009–1018.

Matamouros, S., P. England, and B. Dupuy. 2007. *Clostridium difficile* toxin expression is inhibited by the novel regulator TcdC. *Mol. Microbiol.* **64:**1274–1288.

McDonald, L. C., G. E. Killgore, A. Thompson, R. C. Owens, S. V. Kazakova, S. P. Sambol, S. Johnson, and D. N. Gerding. 2005. An epidemic, toxin gene–variant strain of *Clostridium difficile. N. Engl. J. Med.* **353:**2433–2441.

Merrigan, M., A. Venugopal, M. Mallozzi, B. Roxas, V. K. Viswanathan, S. Johnson, D. N. Gerding, and G. Vedantam. 2010. Human hypervirulent *Clostridium difficile* strains exhibit increased sporulation as well as robust toxin production. *J. Bacteriol.* **192:**4904–4911.

Metcalf, D. S., and J. S. Weese. 2011. Binary toxin locus analysis in *Clostridium difficile. J. Med. Microbiol.* **60:**1137–1145.

Miwa, Y., A. Nakata, A. Ogiwara, M. Yamamoto, and Y. Fujita. 2000. Evaluation and characterization of catabolite-responsive elements (*cre*) of *Bacillus subtilis. Nucleic Acids Res.* **28:**1206–1210.

Molle, V., Y. Nakaura, R. P. Shivers, H. Yamaguchi, R. Losick, Y. Fujita, and A. L. Sonenshein. 2003. Additional targets of the *Bacillus subtilis* global regulator CodY identified by chromatin immunoprecipitation and genome-wide transcript analysis. *J. Bacteriol.* **185:**1911–1922.

Moncrief, J., L. Barroso, and T. Wilkins. 1997. Positive regulation of *Clostridium difficile* toxins. *Infect. Immun.* **65:**1105–1108.

Murray, R., D. Boyd, P. N. Levett, M. R. Mulvey, and M. J. Alfa. 2009. Truncation in the *tcdC* region of the *Clostridium difficile* PathLoc of clinical isolates does not predict increased biological activity of toxin B or toxin A. *BMC Infect. Dis.* **9:**103.

O'Brien, J. A., B. J. Lahue, J. J. Caro, and D. M. Davidson. 2007. The emerging infectious challenge of *Clostridium difficile*-associated disease in Massachusetts hospitals: clinical and economic consequences. *Infect. Control Hosp. Epidemiol.* **28:**1219–1227.

O'Connor, J. R., D. Lyras, K. A. Farrow, V. Adams, D. R. Powell, J. Hinds, J. K. Cheung, and J. I. Rood. 2006. Construction and analysis of chromosomal *Clostridium difficile* mutants. *Mol. Microbiol.* **61:**1335–1351.

Papatheodorou, P., J. E. Carette, G. W. Bell, C. Schwan, G. Guttenberg, T. R. Brummelkamp, and K. Aktories. 2011. Lipolysis-stimulated lipoprotein receptor (LSR) is the host receptor for the binary toxin *Clostridium difficile* transferase (CDT). *Proc. Natl. Acad. Sci. USA* **108:**16422–16427.

Paredes, C. J., K. V. Alsaker, and E. T. Papoutsakis. 2005. A comparative genomic view of clostridial sporulation and physiology. *Nat. Rev. Microbiol.* **3:**969–978.

Paredes-Sabja, D., N. Sarker, and M. R. Sarker. 2011. *Clostridium perfringens tpeL* is expressed during sporulation. *Microb. Pathog.* **51**:384–388.

Perego, M., G. B. Spiegelman, and J. A. Hoch. 1988. Structure of the gene for the transition state regulator, *abrB*: regulator synthesis is controlled by the *spo0A* sporulation gene in *Bacillus subtilis. Mol. Microbiol.* **2**:689–699.

Perelle, S., M. Gibert, P. Bourlioux, G. Corthier, and M. Popoff. 1997. Production of a complete binary toxin (actin-specific ADP-ribosyltransferase) by *Clostridium difficile* CD196. *Infect. Immun.* **65**:1402–1407.

Pohl, K., P. Francois, L. Stenz, F. Schlink, T. Geiger, S. Herbert, C. Goerke, J. Schrenzel, and C. Wolz. 2009. CodY in *Staphylococcus aureus*: a regulatory link between metabolism and virulence gene expression. *J. Bacteriol.* **191**:2953–2963.

Raffestin, S., B. Dupuy, J. C. Marvaud, and M. R. Popoff. 2005. BotR/A and TetR are alternative RNA polymerase sigma factors controlling the expression of the neurotoxin and associated protein genes in *Clostridium botulinum* type A and *Clostridium tetani. Mol. Microbiol.* **55**:235–249.

Ratnayake-Lecamwasam, M., P. Serror, K. W. Wong, and A. L. Sonenshein. 2001. *Bacillus subtilis* CodY represses early-stationary-phase genes by sensing GTP levels. *Genes Dev.* **15**:1093–1103.

Reeves, A. E., C. M. Theriot, I. L. Bergin, G. B. Huffnagle, P. D. Schloss, and V. B. Young. 2011. The interplay between microbiome dynamics and pathogen dynamics in a murine model of *Clostridium difficile* infection. *Gut Microbes* **2**:145–158.

Saujet, L., M. Monot, B. Dupuy, O. Soutourina, and I. Martin-Verstraete. 2011. The key sigma factor of transition phase, SigH, controls sporulation, metabolism, and virulence factor expression in *Clostridium difficile. J. Bacteriol.* **193**:3186–3196.

Schwan, C., T. Nolke, A. S. Kruppke, D. M. Schubert, A. E. Lang, and K. Aktories. 2011. Cholesterol- and sphingolipid-rich microdomains are essential for microtubule-based membrane protrusions induced by *Clostridium difficile* transferase (CDT). *J. Biol. Chem.* **286**:29356–29365.

Schwan, C., B. Stecher, T. Tzivelekidis, M. van Ham, M. Rohde, W. D. Hardt, J. Wehland, and K. Aktories. 2009. *Clostridium difficile* toxin CDT induces formation of microtubule-based protrusions and increases adherence of bacteria. *PLoS Pathog.* **5**:e1000626.

Sebaihia, M., B. W. Wren, P. Mullany, N. F. Fairweather, N. Minton, R. Stabler, N. R. Thomson, A. P. Roberts, A. M. Cerdeno-Tarraga, H. Wang, M. T. Holden, A. Wright, C. Churcher, M. A. Quail, S. Baker, N. Bason, K. Brooks, T. Chillingworth, A. Cronin, P. Davis, L. Dowd, A. Fraser, T. Feltwell, Z. Hance, S. Holroyd, K. Jagels, S. Moule, K. Mungall, C. Price, E. Rabbinowitsch, S. Sharp, M. Simmonds, K. Stevens, L. Unwin, S. Whithead, B. Dupuy, G. Dougan, B. Barrell, and J. Parkhill. 2006. The multidrug-resistant human pathogen *Clostridium difficile* has a highly mobile, mosaic genome. *Nat. Genet.* **38**:779–786.

Sekulovic, O., M. Meessen-Pinard, and L. C. Fortier. 2011. Prophage-stimulated toxin production in *Clostridium difficile* NAP1/027 lysogens. *J. Bacteriol.* **193**:2726–2734.

Shivers, R. P., S. S. Dineen, and A. L. Sonenshein. 2006. Positive regulation of *Bacillus subtilis ackA* by CodY and CcpA: establishing a potential hierarchy in carbon flow. *Mol. Microbiol.* **62**:811–822.

Shivers, R. P., and A. L. Sonenshein. 2004. Activation of the *Bacillus subtilis* global regulator CodY by direct interaction with branched-chain amino acids. *Mol. Microbiol.* **53**:599–611.

Sonenshein, A. L. 2007. Control of key metabolic intersections in *Bacillus subtilis. Nat. Rev. Microbiol.* **5**:917–927.

Stabler, R. A., D. N. Gerding, J. G. Songer, D. Drudy, J. S. Brazier, H. T. Trinh, A. A. Witney, J. Hinds, and B. W. Wren. 2006. Comparative phylogenomics of *Clostridium difficile* reveals clade specificity and microevolution of hypervirulent strains. *J. Bacteriol.* **188**:7297–7305.

Stabler, R. A., M. He, L. Dawson, M. Martin, E. Valiente, C. Corton, T. D. Lawley, M. Sebaihia, M. A. Quail, G. Rose, D. N. Gerding, M. Gibert, M. R. Popoff, J. Parkhill, G. Dougan, and B. W. Wren. 2009. Comparative genome and phenotypic analysis of *Clostridium difficile* 027 strains provides insight into the evolution of a hypervirulent bacterium. *Genome Biol.* **10**:R102.

Stephenson, K., and R. J. Lewis. 2005. Molecular insights into the initiation of sporulation in Gram-positive bacteria: new technologies for an old phenomenon. *FEMS Microbiol. Rev.* **29**:281–301.

Sudarsan, N., E. R. Lee, Z. Weinberg, R. H. Moy, J. N. Kim, K. H. Link, and R. R. Breaker. 2008. Riboswitches in eubacteria sense the second messenger cyclic di-GMP. *Science* **321**:411–413.

Tan, K. S., B. Y. Wee, and K. P. Song. 2001. Evidence for holin function of *tcdE* gene in the pathogenicity of *Clostridium difficile. J. Med. Microbiol.* **50**:613–619.

Underwood, S., S. Guan, V. Vijayasubhash, S. D. Baines, L. Graham, R. J. Lewis, M. H. Wilcox, and K. Stephenson. 2009. Characterization of the sporulation initiation pathway of *Clostridium difficile* and its role in toxin production. *J. Bacteriol.* **191**:7296–7305.

van Schaik, W., A. Chateau, M. A. Dillies, J. Y. Coppee, A. L. Sonenshein, and A. Fouet. 2009. The global regulator CodY regulates toxin gene expression in *Bacillus anthracis* and is required for full virulence. *Infect. Immun.* **77**:4437–4445.

Varga, J., V. L. Stirewalt, and S. B. Melville. 2004. The CcpA protein is necessary for efficient sporulation and enterotoxin gene (*cpe*) regulation in *Clostridium perfringens. J. Bacteriol.* **186**:5221–5229.

Voth, D. E., and J. D. Ballard. 2005. *Clostridium difficile* toxins: mechanism of action and role in disease. *Clin. Microbiol. Rev.* **18**:247–263.

Warny, M., J. Pepin, A. Fang, G. Killgore, A. Thompson, J. Brazier, E. Frost, and L. C. MacDonald. 2005. Toxin production by an emerging strain of *Clostridium difficile* associated with outbreaks of severe disease in North America and Europe. *Lancet* **366**:1079–1084.

Weir, J., M. Predich, E. Dubnau, G. Nair, and I. Smith. 1991. Regulation of *spo0H*, a gene coding for the *Bacillus subtilis* sigma H factor. *J. Bacteriol.* **173**:521–529.

Regulation of Bacterial Virulence
Edited by Michael L. Vasil and Andrew J. Darwin
© 2013 ASM Press, Washington, DC   doi:10.1128/9781555818524.ch16

Chapter 16

Anthrax and Iron

Paul E. Carlson, Jr., Shandee D. Dixon, and Philip C. Hanna

INTRODUCTION

In order to cause disease, pathogenic bacteria require specialized means to sense the various microenvironments presented to them in the context of a host and then to regulate the systems required for establishment, persistence, growth, and induction of the pathologies associated with infection. The example presented in this chapter focuses on one such system utilized by *Bacillus anthracis* during anthrax infections, iron acquisition. It is commonly believed that *B. anthracis* exists in the environment at-large, mostly to totally in its dormant spore morphotype. This implies that for its evolutionary success, there is a severe requirement for efficient, robust growth during the vegetative infectious cycle. After spores enter the body, *B. anthracis* will recognize its locale in the host and germinate (Dixon et al., 1999). The vegetative bacilli are then capable of replicating to extremely high titers within days, sometimes reaching >10^7 bacilli per ml of blood during severe systemic anthrax, and spreading to all organs of the body (Dixon et al., 1999). To accomplish this, *B. anthracis* needs to obtain, among other requirements, the nutrients essential to support that explosive growth. Recently, research focused on the acquisition of one such nutrient essential for successful anthrax infections, iron, has come to the fore. Additionally, the scarcity of available iron (a prevalent condition in the host) serves as an environmental cue to differentially regulate the expression of a defined subset of the *B. anthracis* genome (Carlson et al., 2009). Below, the various mechanisms used by *B. anthracis* for obtaining iron are summarized, the relative impact of each of these mechanisms on a successful anthrax infection is reviewed, and the transcriptome regulated by low concentrations of iron is presented.

IRON ACQUISITION DURING ANTHRAX

Iron is a required nutrient for most living organisms. Iron and iron-containing proteins are involved in a variety of metabolic and signaling functions, most notably those involving electron transport and transfer of electrons (Ratledge and Dover, 2000; Faraldo-Gomez and Sansom, 2003; Ratledge, 2004; Wandersman and Delepelaire, 2004). The low solubility and high toxicity of ferric iron (Fe^{3+}) have led to the majority of iron in eukaryotes being sequestered by iron-binding proteins, including hemoglobin, transferrin, and ferritin (Ratledge and Dover, 2000; Glanfield et al., 2007). The host's limitation of free iron ions also serves as a barrier to infection by pathogenic bacteria that require this metal for growth (Schaible and Kaufmann, 2004; Chu et al., 2010). For most bacterial pathogens, including *B. anthracis*, the ability to sequester and import iron is one of the major determinants of its survival and virulence, and these bacteria have evolved specialized mechanisms to acquire iron from various host sources (Ratledge and Dover, 2000; Faraldo-Gomez and Sansom, 2003; Glanfield et al., 2007). The primary mechanisms of iron acquisition in *B. anthracis* are believed to be the production of high-affinity, iron-binding siderophores and the breakdown and transport of host heme. The function and regulation of these systems are discussed in detail here.

Heme-Iron Acquisition

A high percentage (up to 80%) of the ferrous iron (Fe^{2+}) in a mammalian host is bound to heme, making this an obvious target molecule for bacterial iron acquisition mechanisms (Honsa and Maresso, 2011). The heme iron acquisition system is depicted in Fig. 1. Since heme is bound in hemoglobin, which

Paul E. Carlson, Jr., Shandee D. Dixon, and Philip C. Hanna • Department of Microbiology and Immunology, University of Michigan Medical School, Ann Arbor, MI 48108.

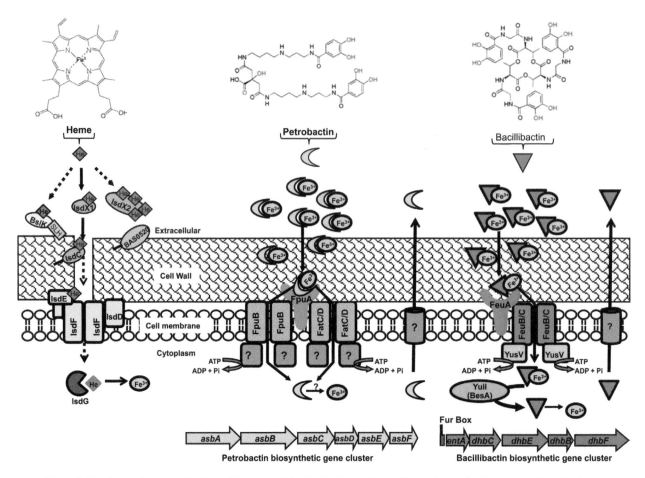

Figure 1. The three primary mechanisms of iron acquisition in *Bacillus anthracis*. The pathways for iron acquisition using heme (left), petrobactin (center), and bacillibactin (right) are shown. The structure of each molecule is shown above its respective transport system (red residues interact directly with iron). Iron-binding molecules are represented by red squares (heme), yellow crescents (petrobactin), and maroon triangles (bacillibactin). All known components of these three specific transport systems are represented. Specific components and functions of each iron acquisition system are discussed within the text. Representations of the genes in the biosynthetic operons for petrobactin and bacillibactin are shown beneath their respective transport systems. doi:10.1128/9781555818524.ch16f1

is packaged at high levels within red blood cells, a multistep process is required for *B. anthracis* to utilize this iron source. It is possible that during infection, *B. anthracis* may require release of heme from host cells for access of iron from this source to occur. When available, iron-bound heme is scavenged from hemoglobin by the secreted hemophores IsdX1 and IsdX2 (Maresso et al., 2008; Fabian et al., 2009). These proteins transport heme to the cell wall, where it is passed off the cell wall-anchored IsdC (Maresso et al., 2006). The proteins that bind and transfer heme contain near-iron transporter (NEAT) domains, which are conserved domains involved in binding of heme across many bacterial species (Gat et al., 2008; Tarlovsky et al., 2010; Honsa and Maresso, 2011). Once heme is localized to the bacterial cell membrane, it is imported into the bacterial cytoplasm through a heme-specific ATP binding cassette (ABC) transport system, IsdEFD (Maresso and Schneewind, 2006; Gat

et al., 2008; Honsa and Maresso, 2011). Finally, once inside the bacterial cytoplasm, iron is released from heme when the heme monooxygenase, IsdG, breaks down the molecule (Skaar et al., 2006).

Using this system, *B. anthracis* is able to grow in vitro with only heme as its iron source; however, the role of this system in the life and infectious cycles of this bacterium, and in its pathogenesis, remains to be elucidated. It is hypothesized that acquisition of iron from heme would occur during later stages of infection, when *B. anthracis* is growing extracellularly in the host bloodstream (Honsa and Maresso, 2011). When directly tested, some *B. anthracis* strains lacking the ability to utilize heme retain wild-type levels of virulence in mice, while mutant strains lacking other NEAT domain-containing proteins have exhibited only a mild attenuation (Gat et al., 2008; Carlson et al., 2009). While these data do not strongly support a major role for iron acquisition from heme during anthrax, it is possible that iron acquisition from heme is necessary for infection of mammalian hosts not used for laboratory experiments (i.e., common natural hosts such as large herbivores or rare hosts such as humans) or that the current animal models of infection do not accurately portray natural late-stage infection. Future experimentation with different experimental systems will be necessary to determine the true role of heme-iron acquisition in the pathogenesis of *B. anthracis*.

Bacillus anthracis siderophores

Siderophores are high-affinity iron-chelating molecules that are secreted into the extracellular environment, where they scavenge iron from a variety of host sources (Ratledge and Dover, 2000). After binding to host iron ions, the holo-iron-siderophore complexes are imported into the bacterial cytoplasm by specific ABC transport systems (Davidson et al., 2008). *B. anthracis* produces two siderophores, bacillibactin and petrobactin (May et al., 2001; Cendrowski et al., 2004). While both of these molecules exhibit high-affinity binding to iron (bacillibactin $K_f = 10^{48}$; petrobactin $K_f = 10^{23}$ where K_f = equilibrium constant for the formation of the ferri-siderophore complex over freely dissociated siderophore and ferric iron), they differ significantly in both structure and role in bacterial pathogenesis (Cendrowski et al., 2004; Honsa and Maresso, 2011).

Bacillibactin. Bacillibactin is a catecholate siderophore that contains 2,3-dihydroxybenzoyl moieties (May et al., 2001). This siderophore is synthesized by genes encoded in the *dhb* operon (Fig. 1, right), which is found in all strains of the *Bacillus cereus* sensu lato group (May et al., 2001). Most of the data regarding the function of this siderophore are inferred from studies using *Bacillus subtilis*. In *B. subtilis*, bacillibactin is secreted through the major facilitator superfamily protein YmfE (Miethke et al., 2008). Following secretion, iron-bound bacillibactin is specifically imported by an ABC transport system, which is encoded by *feuABC-yusV* (Fig. 1, right) (Dertz et al., 2006; Miethke et al., 2006; Ollinger et al., 2006). Iron-bound bacillibactin is then hydrolyzed in the cytoplasm by the YuiL esterase (encoded by *besA*), allowing release of imported iron ions in the cytoplasm for use by the bacterium (Fig. 1, right) (Miethke et al., 2006).

Although this siderophore is produced by many *Bacillus* species, it is not required for growth of *B. anthracis* under iron-limited conditions (Cendrowski et al., 2004). The *B. anthracis* genome encodes orthologs of all of the proteins for synthesis and uptake of bacillibactin found in other *Bacillus* spp.; however, strains lacking this siderophore retain wild-type

levels of virulence in laboratory models of inhalational anthrax (Cendrowski et al., 2004). One possible explanation for this could be altered patterns of bacillibactin secretion. In *B. anthracis*, compared to *B. subtilis*, secretion of bacillibactin in vitro appears to occur late in the life cycle, with measurable amounts being detected only after nearly 10 h of growth in iron-limited media (Wilson et al., 2009). This late secretion of this siderophore could be due to the fact that *B. anthracis* lacks a functional ortholog of YmfE, which is responsible for bacillibactin secretion in *B. subtilis* (Miethke et al., 2008). Similar to our current knowledge of heme-iron acquisition, there are currently no data associating bacillibactin with virulence, suggesting that it plays, at most, a minor role during infection. It is possible that bacillibactin plays some role in the pathogenesis of *B. anthracis*, perhaps in a host organism that is yet to be studied. This would not be unprecedented, as *Brucella abortus* produces two chatecholate siderophores, brucebactin and a monocatecholate 2,3-dihydroxybenzoic acid (Lopez-Goni et al., 1992; Gonzalez Carrero et al., 2002). Despite being produced under iron-starved conditions, the 2,3-dihydroxybenzoic acid siderophore is not required for virulence in mice (Bellaire et al., 1999; Bellaire et al., 2003a). However, it was later shown that this siderophore was required for *B. abortus* virulence in pregnant cattle, showing that these molecules can be species specific (Bellaire et al., 2003b). It may be that this difference is the result of growth using a different carbon source, erythritol, during bovine infection, which leads to an unusually high requirement for iron (Bellaire et al., 2003a).

Petrobactin. The second siderophore produced by *B. anthracis*, the one believed most relevant to pathogenesis, is petrobactin. This catacholate siderophore was originally identified and its chemical structure determined in the oil-degrading bacterium *Marinobacter hydrocarbonoclasticus* (Barbeau et al., 2002). Petrobactins from the two organisms are structurally identical. In *B. anthracis*, the enzymes required for biosynthesis of this siderophore are encoded in the six-gene *asb* operon (Fig. 1, center) (Cendrowski et al., 2004; Pfleger et al., 2007). Although petrobactin is a catecholate siderophore, like bacillibactin, it is structurally unique (see Fig. 1). Petrobactin is made up of one central citrate flanked by two spermidine molecules, each with terminal dihydroxybenzoic acid residues (Cendrowski et al., 2004; Pfleger et al., 2007). A key feature of the petrobactin structure is the presence of 3,4-dihydroxybenzoate moieties, the importance of which is discussed below (Abergel et al., 2006; Pfleger et al., 2007; Abergel et al., 2008; Koppisch et al., 2008; Pfleger et al., 2008).

Pathogenic *Bacillus* species, including *B. anthracis*, *B. cereus*, and some isolates of *B. thuringiensis*, can produce petrobactin (Koppisch et al., 2005). For *B. anthracis* under iron-starved conditions in vitro, the level of petrobactin secretion into culture supernatants is nearly five times greater than that of bacillibactin (Koppisch et al., 2005; Wilson et al., 2009). There is also a significant difference in the kinetics of siderophore secretion, with measureable levels of petrobactin present 5 h before detection of bacillibactin in vitro (Wilson et al., 2009).

Unlike bacillibactin, petrobactin is required for normal growth rates in low-iron media, growth in macrophages, and in vivo pathogenesis of *B. anthracis* in animal models (Cendrowski et al., 2004; Lee et al., 2007; Pfleger et al., 2007). Mutant strains lacking the ability to synthesize or import this siderophore are severely attenuated in murine models of anthrax infection (Cendrowski et al., 2004; Carlson et al., 2010). In fact, the level of attenuation observed for these mutants (approximately 1,000-fold) is similar to what has been observed for *B. anthracis* toxin mutants (Pezard et al., 1991). The importance of this siderophore in *B. anthracis* pathogenesis is thought to be, at least in part, due to the inability of the host immune system to recognize this molecule. The host immune system uses a protein called siderocalin to sequester certain bacterial siderophores, including bacillibactin (Goetz et al., 2002; Fischbach et al., 2006). The unique structure of petrobactin, specifically the presence of 3,4-dihydroxybenzoate moieties, prevents siderocalin binding, leaving the siderophore free to function in iron acquisition in vivo (see Fig. 1, center) (Abergel et al., 2006). This lack of immune recognition by a mammalian host led to petrobactin being defined as a "stealth" siderophore (Abergel et al., 2006; Pfleger et al., 2008; Zawadzka et al., 2009). This strategy of immune evasion is not without precedent, as glucosylation of the 2,3-dihydroxybenzoate moieties of salmochelin produced by *E. coli* also allows evasion of siderocalin (Muller et al., 2009).

Although the mechanism of petrobactin secretion is currently unknown, the import of petrobactin across the *B. anthracis* cell membrane appears to be mediated by a combination of ABC transporters and is depicted in Fig. 1. Two *B. cereus* proteins, FpuA and FatB, are able to bind tightly to petrobactin molecules (Zawadzka et al., 2009a). These proteins are annotated as the receptor components of iron-specific ABC transporters, and orthologs of each are encoded by the *B. anthracis* genome. Interestingly, only one of these proteins, FpuA, is required for *B. anthracis* petrobactin-based iron acquisition (Carlson et al., 2010). While the other receptor, FatB, does function in *B. subtilis*, genetic analyses in *B. anthracis* indicate

that it is completely dispensable for growth under low-iron conditions and murine virulence (Zawadzka et al., 2009b). *B. anthracis* strains lacking FatB retained the ability to grow to high levels in iron-limited media and also full virulence in a murine model of infection (Carlson et al., 2010). In sharp contrast, strains lacking only FpuA exhibited reduced growth rates and attenuated virulence nearly identical to what is observed for strains unable to synthesize petrobactin (Carlson et al., 2010). Although *B. anthracis* requires only one specific receptor protein for petrobactin uptake, it appears that multiple proteins can function as the permease and/or ATPase for this system. To date, two permeases (membrane-spanning component) have been identified as functioning in petrobactin import. FatCD and FpuB can each act, presumably in association with FpuA, to import this siderophore (unpublished data). Strains lacking either one of these proteins retain wild-type phenotypes, while double mutants ($\Delta fatCD$ $\Delta fpuB$) exhibit phenotypes identical to those of other petrobactin-deficient strains (unpublished data). The identities of the ATP binding proteins that provide energy for this transport to occur remain under investigation.

REGULATION OF IRON ACQUISITION MECHANISMS IN *B. ANTHRACIS*

B. anthracis transcriptional regulation during iron starvation has been well studied. Many genes are both induced and repressed during growth in iron-limited media, and these changes increase the longer the bacteria remain under these conditions (Carlson et al., 2009). Also, many genes that are induced under low-iron conditions are also coregulated in response to other host-specific cues, such as oxidative stress and growth in macrophages or host blood (Bergman et al., 2006; Passalacqua et al., 2007; Carlson et al., 2009). Many iron-specific genes are regulated by the transcription factor ferric uptake regulator (Fur). The presence of Fur recognition sites (Fur boxes) in the promoter region of genes may be indicative of altered regulation under iron-starved conditions (Baichoo and Helmann, 2002; Baichoo et al., 2002; Andrews et al., 2003; Ollinger et al., 2006). A known metabolic regulator, CodY, has also been associated with the *Bacillus* iron response. This transcription factor is required for heme-iron acquisition and virulence in *B. anthracis* and may play a role regulating expression of some of the iron-responsive genes that have been identified (Chateau et al., 2011). It is not surprising that the bacterium would have multiple mechanisms for regulating genes involved in iron acquisition. In the context of a low-iron environment, bacteria need to rapidly increase expression of the necessary mechanisms in order to survive. In contrast, in a high-iron environment, it would be necessary to stop expression of these genes, as high levels of intracellular iron are toxic to bacterial cells (Stauff and Skaar, 2009).

Transcription of the genes encoding the bacillibactin biosynthetic machinery (GBAA_2368 to GBAA_2376 [note that all gene designations are from NCBI genome NC_007530—*Bacillus anthracis* strain Ames Ancestor]) is highly induced during iron starvation (Carlson et al., 2009). These genes are also induced, to various degrees, in response to a range of host-specific signals, including growth in macrophages, growth in bovine blood, and oxidative stress (Bergman et al., 2006; Passalacqua et al., 2007). It is possible that growth in macrophages and growth in blood also represent iron-limiting conditions, further emphasizing the importance of this regulon. Although this siderophore has not been implicated in *B. anthracis* pathogenesis in mice, the fact that induction is observed across this range of host-specific signals implies that these proteins, and therefore this siderophore, may play some role in *B. anthracis* physiology, possibly during infection of other host species not used for laboratory experiments.

Genes that have been implicated in iron acquisition from heme can be induced by a variety of conditions. Proteins implicated in heme binding/scavenging (IsdC, IsdX1, and IsdX2) and transport of heme-iron (IsdEFD) are induced during growth under iron-limiting conditions, growth in macrophages, and growth in bovine blood and under oxidative stress (Bergman et al., 2006; Passalacqua et al., 2007; Carlson et al., 2009). Many genes containing NEAT domains are induced in response to iron starvation and growth in blood, including one novel family member of previously unknown function (encoded by GBAA_0520) (Carlson et al., 2009; Tarlovsky et al., 2010). Similarly, five other ABC transport systems that are annotated as iron transporters (encoded by GBAA_0615 to GBAA_0618, GBAA_3864 to GBAA_3867, GBAA_4595 to GBAA_4597, GBAA_5327 to GBAA_5330, and GBAA_5628 to GBAA_5630) also are induced under all of these conditions (Bergman et al., 2006; Passalacqua et al., 2007; Carlson et al., 2009). The regulation of these systems in response to low-iron conditions is not surprising, as each has a *fur* regulatory element in the promoter region of the first gene in the operon (Hotta et al., 2010).

Surprisingly, the iron acquisition genes most important for *B. anthracis* pathogenicity exhibit little to no differential regulation in response to iron starvation or any other conditions tested (Bergman et al., 2006; Passalacqua et al., 2007; Carlson et al., 2009). The genes encoding the petrobactin biosynthetic machinery (*asbABCDEF*) are not induced under these conditions (Bergman et al., 2006; Passalacqua et al., 2007; Carlson et al., 2009). It should be noted that unlike the genes discussed above, the promoter regions of these genes do not contain obvious Fur regulatory elements (Hotta et al., 2010). Despite this lack of transcriptional regulation, increased levels of petrobactin can be detected in *B. anthracis* culture supernatants grown under iron-limiting conditions, implying the presence of a posttranscriptional mechanism for the regulation of petrobactin production (Lee et al., 2011). Also, it is possible that expression of petrobactin biosynthetic genes is controlled by another host-specific signal, as oxygen level appears to alter expression of these genes (Lee et al., 2011). Similar to expression of petrobactin biosynthetic genes, the expression of the gene encoding the petrobactin receptor (*fpuA*) is not significantly induced in response to iron starvation (Carlson et al., 2009). While other components of the receptor system are induced in response to iron starvation, the level of induction (~2-fold) is much lower than what is observed for other iron-specific ABC transport systems in *B. anthracis* (Carlson et al., 2009). The minimal change in expression of these genes could indicate their importance to the survival of *B. anthracis*. These genes appear to be constitutively expressed, allowing for immediate activation of iron acquisition when the bacterium encounters an inhospitable environment. Given the recent findings on petrobactin biosynthesis, the presence of posttranscriptional mechanisms of regulation for these receptor proteins remains a possibility.

THE FUTURE OF IRON RESEARCH IN *B. ANTHRACIS*

Anthrax infections of mammalian hosts, which are often lethal and/or involve a massive septicemia, are believed to represent a major, if not the only, site for *B. anthracis* replication. After introduction into a host, *B. anthracis* establishment, vegetative growth, and serious pathologies are dependent upon a variety of abilities, including classical virulence traits. Recent advances clearly indicate that acquisition of iron from the host, an environment where iron sources are closely sequestered, also is a vital component involved in disease and, therefore, overall bacterial evolutionary success. Iron acquisition represents an important choke point in the *B. anthracis* infectious cycle. Consequently, molecules that inhibit or block iron acquisition, related membrane transport processes, or bacterial iron-based regulation, in general, may prove to be effective new medical countermeasures against anthrax and related infections. We predict that as more defined details of the genes, proteins, factors, and mechanisms that regulate iron metabolism become understood, reasonable countermeasure targets will be identified.

REFERENCES

Abergel, R. J., M. K. Wilson, J. E. Arceneaux, T. M. Hoette, R. K. Strong, B. R. Byers, and K. N. Raymond. 2006. Anthrax pathogen evades the mammalian immune system through stealth siderophore production. *Proc. Natl. Acad. Sci. USA* 103:18499–18503.

Abergel, R. J., A. M. Zawadzka, and K. N. Raymond. 2008. Petrobactin-mediated iron transport in pathogenic bacteria: coordination chemistry of an unusual 3,4-catecholate/citrate siderophore. *J. Am. Chem. Soc.* 130:2124–2125.

Andrews, S. C., A. K. Robinson, and F. Rodriguez-Quinones. 2003. Bacterial iron homeostasis. *FEMS Microbiol. Rev.* 27:215–237.

Baichoo, N., and J. D. Helmann. 2002. Recognition of DNA by Fur: a reinterpretation of the Fur box consensus sequence. *J. Bacteriol.* 184:5826–5832.

Baichoo, N., T. Wang, R. Ye, and J. D. Helmann. 2002. Global analysis of the *Bacillus subtilis* Fur regulon and the iron starvation stimulon. *Mol. Microbiol.* 45:1613–1629.

Barbeau, K., G. Zhang, D. H. Live, and A. Butler. 2002. Petrobactin, a photoreactive siderophore produced by the oil-degrading marine bacterium *Marinobacter hydrocarbonoclasticus. J. Am. Chem. Soc.* 124:378–379.

Bellaire, B. H., P. H. Elzer, C. L. Baldwin, and R. M. Roop II. 2003a. Production of the siderophore 2,3-dihydroxybenzoic acid is required for wild-type growth of *Brucella abortus* in the presence of erythritol under low-iron conditions in vitro. *Infect. Immun.* 71:2927–2932.

Bellaire, B. H., P. H. Elzer, S. Hagius, J. Walker, C. L. Baldwin, and R. M. Roop II. 2003b. Genetic organization and iron-responsive regulation of the *Brucella abortus* 2,3-dihydroxybenzoic acid biosynthesis operon, a cluster of genes required for wild-type virulence in pregnant cattle. *Infect. Immun.* 71:1794–1803.

Bellaire, B. H., P. H. Elzer, C. L. Baldwin, and R. M. Roop II. 1999. The siderophore 2,3-dihydroxybenzoic acid is not required for virulence of *Brucella abortus* in BALB/c mice. *Infect. Immun.* 67:2615–2618.

Bergman, N. H., E. C. Anderson, E. E. Swenson, M. M. Niemeyer, A. D. Miyoshi, and P. C. Hanna. 2006. Transcriptional profiling of the *Bacillus anthracis* life cycle in vitro and an implied model for regulation of spore formation. *J. Bacteriol.* 188:6092–6100.

Carlson, P. E., Jr., K. A. Carr, B. K. Janes, E. C. Anderson, and P. C. Hanna. 2009. Transcriptional profiling of *Bacillus anthracis* Sterne (34F2) during iron starvation. *PLoS One* 4:e6988.

Carlson, P. E., Jr., S. D. Dixon, B. K. Janes, K. A. Carr, T. D. Nusca, E. C. Anderson, S. E. Keene, D. H. Sherman, and P. C. Hanna. 2010. Genetic analysis of petrobactin transport in *Bacillus anthracis. Mol. Microbiol.* 75:900–909.

Cendrowski, S., W. MacArthur, and P. Hanna. 2004. *Bacillus anthracis* requires siderophore biosynthesis for growth in macrophages and mouse virulence. *Mol. Microbiol.* 51:407–417.

Chateau, A., W. van Schaik, A. Six, W. Aucher, and A. Fouet. 2011. CodY regulation is required for full virulence and heme iron acquisition in *Bacillus anthracis. FASEB J.* 25:4445–4456.

Chu, B. C., A. Garcia-Herrero, T. H. Johanson, K. D. Krewulak, C. K. Lau, R. S. Peacock, Z. Slavinskaya, and H. J. Vogel. 2010. Siderophore uptake in bacteria and the battle for iron with the host; a bird's eye view. *Biometals* 23:601–611.

Davidson, A. L., E. Dassa, C. Orelle, and J. Chen. 2008. Structure, function, and evolution of bacterial ATP-binding cassette systems. *Microbiol. Mol. Biol. Rev.* 72:317–364.

Dertz, E. A., J. Xu, A. Stintzi, and K. N. Raymond. 2006. Bacillibactin-mediated iron transport in *Bacillus subtilis. J. Am. Chem. Soc.* 128:22–23.

Dixon, T. C., M. Meselson, J. Guillemin, and P. C. Hanna. 1999. Anthrax. *N. Engl. J. Med.* 341:815–826.

Fabian, M., E. Solomaha, J. S. Olson, and A. W. Maresso. 2009. Heme transfer to the bacterial cell envelope occurs via a secreted hemophore in the Gram-positive pathogen *Bacillus anthracis. J. Biol. Chem.* 284:32138–32146.

Faraldo-Gomez, J. D., and M. S. Sansom. 2003. Acquisition of siderophores in gram-negative bacteria. *Nat. Rev. Mol. Cell Biol.* 4:105–116.

Fischbach, M. A., H. Lin, D. R. Liu, and C. T. Walsh. 2006. How pathogenic bacteria evade mammalian sabotage in the battle for iron. *Nat. Chem. Biol.* 2:132–138.

Gat, O., G. Zaide, I. Inbar, H. Grosfeld, T. Chitlaru, H. Levy, and A. Shafferman. 2008. Characterization of Bacillus anthracis iron-regulated surface determinant (Isd) proteins containing NEAT domains. *Mol. Microbiol.* 70:983–999.

Glanfield, A., D. P. McManus, G. J. Anderson, and M. K. Jones. 2007. Pumping iron: a potential target for novel therapeutics against schistosomes. *Trends Parasitol.* 23:583–588.

Goetz, D. H., M. A. Holmes, N. Borregaard, M. E. Bluhm, K. N. Raymond, and R. K. Strong. 2002. The neutrophil lipocalin NGAL is a bacteriostatic agent that interferes with siderophore-mediated iron acquisition. *Mol. Cell* 10:1033–1043.

Gonzalez Carrero, M. I., F. J. Sangari, J. Aguero, and J. M. Garcia Lobo. 2002. *Brucella abortus* strain 2308 produces brucebactin, a highly efficient catecholic siderophore. *Microbiology* 148:353–360.

Honsa, E. S., and A. W. Maresso. 2011. Mechanisms of iron import in anthrax. *Biometals* 24:533–545.

Hotta, K., C. Y. Kim, D. T. Fox, and A. T. Koppisch. 2010. Siderophore-mediated iron acquisition in *Bacillus anthracis* and related strains. *Microbiology* 156:1918–1925.

Koppisch, A. T., C. C. Browder, A. L. Moe, J. T. Shelley, B. A. Kinkel, L. E. Hersman, S. Iyer, and C. E. Ruggiero. 2005. Petrobactin is the primary siderophore synthesized by *Bacillus anthracis* str. Sterne under conditions of iron starvation. *Biometals* 18:577–585.

Koppisch, A. T., K. Hotta, D. T. Fox, C. E. Ruggiero, C. Y. Kim, T. Sanchez, S. Iyer, C. C. Browder, P. J. Unkefer, and C. J. Unkefer. 2008. Biosynthesis of the 3,4-dihydroxybenzoate moieties of petrobactin by *Bacillus anthracis. J. Org. Chem.* 73:5759–5765.

Lee, J. Y., B. K. Janes, K. D. Passalacqua, B. F. Pfleger, N. H. Bergman, H. Liu, K. Hakansson, R. V. Somu, C. C. Aldrich, S. Cendrowski, P. C. Hanna, and D. H. Sherman. 2007. Biosynthetic analysis of the petrobactin siderophore pathway from *Bacillus anthracis. J. Bacteriol.* 189:1698–1710.

Lee, J. Y., K. D. Passalacqua, P. C. Hanna, and D. H. Sherman. 2011. Regulation of petrobactin and bacillibactin biosynthesis in *Bacillus anthracis* under iron and oxygen variation. *PLoS One* 6:e20777.

Lopez-Goni, I., I. Moriyon, and J. B. Neilands. 1992. Identification of 2,3-dihydroxybenzoic acid as a *Brucella abortus* siderophore. *Infect. Immun.* 60:4496–4503.

Maresso, A. W., T. J. Chapa, and O. Schneewind. 2006. Surface protein IsdC and Sortase B are required for heme-iron scavenging of *Bacillus anthracis. J. Bacteriol.* 188:8145–8152.

Maresso, A. W., G. Garufi, and O. Schneewind. 2008. *Bacillus anthracis* secretes proteins that mediate heme acquisition from hemoglobin. *PLoS Pathog.* 4:e1000132.

Maresso, A. W., and O. Schneewind. 2006. Iron acquisition and transport in *Staphylococcus aureus. Biometals* 19:193–203.

May, J. J., T. M. Wendrich, and M. A. Marahiel. 2001. The *dhb* operon of *Bacillus subtilis* encodes the biosynthetic template for the catecholic siderophore 2,3-dihydroxybenzoate-glycine-threonine trimeric ester bacillibactin. *J. Biol. Chem.* 276:7209–7217.

Miethke, M., O. Klotz, U. Linne, J. J. May, C. L. Beckering, and M. A. Marahiel. 2006. Ferri-bacillibactin uptake and hydrolysis in *Bacillus subtilis. Mol. Microbiol.* **61:**1413–1427.

Miethke, M., S. Schmidt, and M. A. Marahiel. 2008. The major facilitator superfamily-type transporter YmfE and the multidrug-efflux activator Mta mediate bacillibactin secretion in *Bacillus subtilis. J. Bacteriol.* **190:**5143–5152.

Muller, S. I., M. Valdebenito, and K. Hantke. 2009. Salmochelin, the long-overlooked catecholate siderophore of *Salmonella. Biometals* **22:**691–695.

Ollinger, J., K. B. Song, H. Antelmann, M. Hecker, and J. D. Helmann. 2006. Role of the Fur regulon in iron transport in *Bacillus subtilis. J. Bacteriol.* **188:**3664–3673.

Passalacqua, K. D., N. H. Bergman, J. Y. Lee, D. H. Sherman, and P. C. Hanna. 2007. The global transcriptional responses of *Bacillus anthracis* Sterne (34F$_2$) and a ΔsodA1 mutant to paraquat reveal metal ion homeostasis imbalances during endogenous superoxide stress. *J. Bacteriol.* **189:**3996–4013.

Pezard, C., P. Berche, and M. Mock. 1991. Contribution of individual toxin components to virulence of *Bacillus anthracis. Infect. Immun.* **59:**3472–3477.

Pfleger, B. F., Y. Kim, T. D. Nusca, N. Maltseva, J. Y. Lee, C. M. Rath, J. B. Scaglione, B. K. Janes, E. C. Anderson, N. H. Bergman, P. C. Hanna, A. Joachimiak, and D. H. Sherman. 2008. Structural and functional analysis of AsbF: origin of the stealth 3,4-dihydroxybenzoic acid subunit for petrobactin biosynthesis. *Proc. Natl. Acad. Sci. USA* **105:**17133–17138.

Pfleger, B. F., J. Y. Lee, R. V. Somu, C. C. Aldrich, P. C. Hanna, and D. H. Sherman. 2007. Characterization and analysis of early enzymes for petrobactin biosynthesis in *Bacillus anthracis. Biochemistry* **46:**4147–4157.

Ratledge, C. 2004. Iron, mycobacteria and tuberculosis. *Tuberculosis* (Edinburgh) **84:**110–130.

Ratledge, C., and L. G. Dover. 2000. Iron metabolism in pathogenic bacteria. *Annu. Rev. Microbiol.* **54:**881–941.

Schaible, U. E., and S. H. Kaufmann. 2004. Iron and microbial infection. *Nat. Rev. Microbiol.* **2:**946–953.

Skaar, E. P., A. H. Gaspar, and O. Schneewind. 2006. *Bacillus anthracis* IsdG, a heme-degrading monooxygenase. *J. Bacteriol.* **188:**1071–1080.

Stauff, D. L., and E. P. Skaar. 2009. *Bacillus anthracis* HssRS signalling to HrtAB regulates haem resistance during infection. *Mol. Microbiol.* **72:**763–778.

Tarlovsky, Y., M. Fabian, E. Solomaha, E. Honsa, J. S. Olson, and A. W. Maresso. 2010. A *Bacillus anthracis* S-layer homology protein that binds heme and mediates heme delivery to IsdC. *J. Bacteriol.* **192:**3503–3511.

Wandersman, C., and P. Delepelaire. 2004. Bacterial iron sources: from siderophores to hemophores. *Annu. Rev. Microbiol.* **58:**611–647.

Wilson, M. K., R. J. Abergel, J. E. Arceneaux, K. N. Raymond, and B. R. Byers. 2009. Temporal production of the two *Bacillus anthracis* siderophores, petrobactin and bacillibactin. *Biometals* **23:**129–134.

Zawadzka, A. M., R. J. Abergel, R. Nichiporuk, U. N. Andersen, and K. N. Raymond. 2009a. Siderophore-mediated iron acquisition systems in *Bacillus cereus*: identification of receptors for anthrax virulence-associated petrobactin. *Biochemistry* **48:**3645–3657.

Zawadzka, A. M., Y. Kim, N. Maltseva, R. Nichiporuk, Y. Fan, A. Joachimiak, and K. N. Raymond. 2009b. Characterization of a *Bacillus subtilis* transporter for petrobactin, an anthrax stealth siderophore. *Proc. Natl. Acad. Sci. USA* **106:**21854–21859.

IV. PROTEIN EXPORT AND INTRACELLULAR LIFE WITHIN THE HOST

Regulation of Bacterial Virulence
Edited by Michael L. Vasil and Andrew J. Darwin
© 2013 ASM Press, Washington, DC   doi:10.1128/9781555818524.ch17

Chapter 17

Regulation of the Expression of Type III Secretion Systems: an Example from *Pseudomonas aeruginosa*

AUDREY LE GOUELLEC, BENOIT POLACK, DAKANG SHEN, AND BERTRAND TOUSSAINT

INTRODUCTION

Despite a long history of research that includes approximately 2,000 publications, the type III secretion systems (T3SSs) of gram-negative bacteria remain a major topic in the science of host-microbe interactions (for reviews, see Coburn et al., 2007; Galan and Wolf-Watz, 2006; Hauser, 2009; Troisfontaines and Cornelis, 2005; and Dean, 2011). Like the T4SS and T6SS, the T3SS is able to translocate effector proteins directly from the bacteria into the cytosol of the eukaryotic host cell. T3SSs are predominantly present in symbiotic and pathogenic bacteria associated with humans, animals, plants, and insects and can have diverse functions, such as bacterial internalization, apoptosis induction, and inflammation regulation (Coburn et al., 2007). The function depends on the nature of the translocated toxins, which are specific to each bacterial species (Pallen et al., 2005). T3SSs can also act on the host to reprogram host cellular processes to favor the infection. Lesser and Leong recently showed that type III effectors of pathogenic *Escherichia coli* and *Salmonella enterica* serovar Typhimurium can act as molecular scaffolds that not only recruit multiple eukaryotic proteins but also directly regulate their activity and localization (Lesser and Leong, 2011). This system of protein exportation indubitably plays a prominent role in human infection by gram-negative bacteria such as *Salmonella*, *Shigella*, *Yersinia*, *Escherichia*, and *Pseudomonas* species (Coburn et al., 2007), establishing it as a target for anti-infective drug development. However, T3SSs are very complex and are always composed of a large number of proteins; this system may also be a problem for the pathogen itself, either because it is metabolically costly or because it can become a target for host defenses, as exemplified by recent studies that revealed that the human immune system is able to detect the presence of T3SSs. For instance, Wangdi et al. have shown that pulmonary infection of mice with T3SS translocation-competent *Pseudomonas aeruginosa* triggers an inflammatory response through caspase 1 activation, while infection with a T3SS translocation-incompetent isogenic strain does not (Wangdi et al., 2010). This finding reveals that the highly complex interaction programs of the pathogens and hosts have coevolved to become more precise and that the bacterium in particular must have a fine-tuned regulatory system to ensure that the T3SS is activated at the correct time and location. In this review, after a general overview of T3SS regulation in the main pathogenic bacteria, we focus on the different aspects of the regulation of T3SS gene expression in *P. aeruginosa*.

COMMON ASPECTS OF T3SS REGULATION AMONG PATHOGENIC BACTERIA

Pathogenic bacteria occupy very different infection foci; for instance, the animal pathogens *Shigella* spp. and *Salmonella* spp. live intracellularly after successful invasion, whereas *Yersinia* spp., *P. aeruginosa*, and enteropathogenic *E. coli* (EPEC) predominantly remain extracellular (Francis et al., 2002; Nataro and Kaper, 1998). Therefore, different stimuli could be used to up- or downregulate the expression of T3SS genes: temperature, divalent cations, host cell contact, serum, or other factors. The regulation of secretion is controlled at several levels: transcriptional, posttranscriptional, and translational and via the conformational switching and secretion of T3SS components (for reviews, see Brutinel and Yahr, 2008, and Deane et al., 2010).

Audrey Le Gouellec, Benoit Polack, and Bertrand Toussaint • TheREx, TIMC-IMAG Laboratory, UMR 5525 CNRS, Université Joseph Fourier, Grenoble, France.   **Dakang Shen** • School of Cellular and Molecular Medicine, University of Bristol, University Walk, Bristol BS8 1TD, United Kingdom.

General Host Sensing

Most bacteria primarily reside outside their host. However, upon contact with an animal host, the growth temperature of the bacteria increases to 37°C. This temperature increase is known to induce T3SS gene expression (Table 1) in *Shigella* spp. (Maurelli et al., 1984), *Yersinia* spp. (Goguen et al., 1984; Michiels et al., 1990) and EPEC (Bustamante et al., 2001; Rosenshine et al., 1996; Umanski et al., 2002). In *Shigella*, when the temperature increases to 37°C, an AraC-like transcriptional activator, VirF, activates VirB, which subsequently activates T3SS gene expression (Beloin et al., 2002; Dorman et al., 2001). Because DNA supercoiling is temperature dependent, H-NS, a nucleoid-associated protein, is able to bind to and repress the promoter activity of VirF only at temperatures below 32°C (Falconi et al., 1998). Similarly, DNA topology has been implicated in the temperature-dependent expression of the T3SS genes of *Yersinia* spp., EPEC, and enterohemorrhagic *E. coli* (Cornelis et al., 1991; Rohde et al., 1994; Rohde et al., 1999). These four pathogens likely exhibit comparable physical bases for thermoregulation that probably involve the accessibility of the target DNA to H-NS (Francis et al., 2002). In addition to temperature, osmolarity, pH, and oxygen tension have been suggested to cause topological changes in local DNA structure. Not surprisingly, these environmental stimuli are all encountered by a pathogen during infection (Francis et al., 2002).

Although the physiological role of divalent cations is not clear, they can function as extracellular signals to regulate virulence gene expression (Garcia Vescovi et al., 1996). In *S.* *Typhimurium*, the two-component system PhoP/PhoQ senses the different Mg^{2+} concentrations outside and inside the host cell to inversely regulate two independent T3SSs: pathogenicity island I (SPI-1), which is important for ensuring the initial invasion, and SPI-2, which is important for intracellular survival and proliferation (Chamnongpol et al., 2003; Groisman, 2001). In *Yersinia* spp. and *P. aeruginosa*, T3SS gene expression

in vitro is induced by calcium depletion in the medium (Frank, 1997; Goguen et al., 1984; Michiels et al., 1990; Pollack et al., 1986; Yahr and Frank, 1994).

In *Yersinia* spp., growth in chemically defined media, for example, Dulbecco's minimal Eagle medium, does not lead to type III secretion, even under low-calcium conditions (Lee et al., 2001). Two additional signals, glutamate and host serum, must be provided in the chemically defined media to activate the type III pathway (Lee et al., 2001). Serum is able to activate the T3SSs of *Shigella* (Menard et al., 1994), *Salmonella* (Zierler and Galan, 1995), *Yersinia* (Lee et al., 2001), and *P. aeruginosa* (Hornef et al., 2000; Vallis et al., 1999). In *Shigella* spp., a group of chemical compounds, including Congo red, Evans blue, and direct orange, are able to induce the secretion of Ipa proteins by bacteria suspended in phosphate-buffered saline. In addition, activation of Ipa secretion by Congo red was observed in bacteria harvested throughout the exponential phase of growth but not in bacteria in the stationary phase (Bahrani et al., 1997). The interaction of bacteria with components of the extracellular matrix, such as fibronectin, laminin, or collagen type IV, can stimulate the release of Ipa effector proteins (Watarai et al., 1995). tRNA modification is also required for the full expression of the T3SS genes (Durand et al., 2000). In *Salmonella* spp., studies have shown the following: (i) neutral pH, low oxygen tension, and high-osmolarity conditions favor the maximal expression of T3SS genes (Bajaj et al., 1996); (ii) a rising concentration of acetate in the distal ileum (small intestine) provides a signal for T3SS gene expression, while propionate and butyrate, which are present in high concentrations in the cecum and colon (large intestine), inhibit the expression of T3SS genes (Gantois et al., 2006; Lawhon et al., 2002); (iii) cationic antimicrobial peptides, a conserved and highly effective component of innate immunity, repress the expression of T3SS genes while activating other virulence regulons (Bader et al., 2003); and (iv) the expression of *Salmonella* T3SS genes is repressed in the presence of bile (Prouty et al., 2004). In EPEC, maximal type III secretion occurs

Table 1. Summary of the main regulatory characteristics of the T3SSs of several animal pathogens

Organism(s)	Temp	Divalent cation	Cell contact	Serum	Stress and special factors[a]	Secreted factor	Interbacterial communication
P. aeruginosa		Ca^{2+}	+	+	+	ExsE	+
Yersinia spp.	+	Ca^{2+}	+	+	+	LcrQ (YscM)	
Shigella spp.	+		+	+	+	OspD1	+
Salmonella spp.[b]		Mg^{2+}		+	+		+
EPEC[c]	+	Ca^{2+}			+		+

[a]See text for details.
[b]SPI-1.

under conditions similar to those found in the gastro-intestinal tract, including the presence of sodium bicarbonate and millimolar concentrations of calcium and $Fe(NO_3)_3$ (Kenny et al., 1997). The negative regulation of T3SS gene expression was observed in the presence of ammonium or in LB (Bustamante et al., 2001). The changes in pH during the passage through the acidic stomach to the neutral small intestine determine the need for T3SS gene expression (Francis et al., 2002; Shin et al., 2001).

Host cell contact-dependent activation of T3SS gene expression has also been observed in *Shigella* spp. (Menard et al., 1994; Watarai et al., 1995), EPEC (Rosenshine et al., 1996), and *P. aeruginosa* (Hornef et al., 2000; Vallis et al., 1999).

Stress and Metabolic Signals

The number of stress and metabolic signals is too great for details to be given for each main gram-negative pathogen. In brief, T3SSs are often linked to the metabolic status of the bacterium. In *Shigella*, some metabolic intermediates involved in T3SS regulation have been identified, such as quinolinate, which is an intermediate in NAD biosynthesis that has recently been reported to inhibit T3SS-dependent invasion (Prunier et al., 2007), ornithine, and other amino acids (Durand and Bjork, 2009). Although the mechanism by which metabolic pathways influence virulence gene expression in bacteria is unclear, the example of *P. aeruginosa* detailed below could facilitate progress in that field.

Cell-to-Cell Communication in Bacteria

Quorum sensing (QS), cell-to-cell communication in bacteria, involves producing, releasing, detecting, and responding to small hormone-like molecules known as autoinducers, which allow the bacteria to monitor their environment and alter their behavior in a population density-dependent manner in response to changes in the number and/or species present in a community (Waters and Bassler, 2005). In *Y. enterocolitica*, the production of the Yop proteins in the wild-type strain and that in a QS mutant are indistinguishable (Throup et al., 1995). In *S. flexneri*, the maximal expression of the T3SS genes and the maximal activity of the type three secretion apparatus (TTSA) occur at high cell density. However, autoinducer-2 (AI-2), a QS signaling molecule active in *S. flexneri* in late log phase, does not influence T3SS gene expression (Day and Maurelli, 2001). In EPEC, QS controls T3SS gene transcription and protein secretion (Sperandio et al., 1999). Known regulators include AI-2 and AI-3 (Kendall et al., 2007; Sperandio et al., 2003); SdiA (Kanamaru et

al., 2000), an *E. coli* homologue of QS regulators; and QseA, QS *E. coli* regulator A (Sperandio et al., 2002). QseA shares homology with several hypothetical *P. aeruginosa* proteins and PtxR, which positively regulates exotoxin A production (Hamood et al., 1996). In *P. aeruginosa*, QS has an important impact on T3SS expression, as detailed below.

Cell Contact and Secretion/Activation Coupling

T3SSs are specifically regulated upon contact with the targeted host cell. One of the pathways involved in this particular activation allows the secretion function to be coupled to the activation of gene expression, and this "secretion/activation" coupling is encountered in all the main gram-negative bacteria that use T3SSs as a virulence factor. This pathway generally involves a transcriptional regulator of the AraC/XylS type, the activity of which is inhibited by a proteic ligand depending on the secretion function status of the bacteria.

In *Yersinia* spp., repression of T3SS gene expression is relieved when LcrQ in *Yersinia pseudotuberculosis* and two LcrQ homologues, YscM1 and YscM2 in *Yersinia enterocolitica*, are secreted by the TTSA (Cambronne et al., 2000; Cambronne et al., 2004; Pettersson et al., 1996; Stainier et al., 1997; Wulff-Strobel et al., 2002). In *Shigella* spp., the expression of late T3SS effectors, which is controlled by MxiE and IpgCa, is not initiated until OspD1, an anti-activator of MxiE, is secreted by the TTSA (Parsot et al., 2005). In *P. aeruginosa*, T3SS gene expression is activated when ExsE, a small inhibitor of the T3SS, is secreted through the opened type III secretion channel (Rietsch et al., 2005; Urbanowski et al., 2005).

Taken together, the expression of T3SS genes is controlled by multicomponent regulatory networks that integrate a diverse set of environmental cues, probably to restrict the energy-consuming synthesis of 30 or more proteins to the correct place and time (Hueck, 1998). Coupled secretion/activation, host environment sensing, bacterial metabolic status, and cell-to-cell communication in bacteria are examined in detail for *P. aeruginosa*.

P. AERUGINOSA IN DEPTH

In contrast to *Salmonella* or *Yersinia* spp., in which T3SSs are encoded by unstable genetic elements such as pathogenicity islands or plasmids, the entire set of genes necessary for the function and the regulation of the T3SS is stably located on the chromosome of *P. aeruginosa*.

This bacterium is particularly rich in regulatory protein-encoding genes, with 8% of genes involved in regulatory functions (Stover et al., 2000). Two-component regulatory systems are more highly represented than usual, with 64 sensor kinases and 72 response regulators (Rodrigue et al., 2000). These numerous regulatory circuits allow the bacterium to adapt through the coordinated regulation of virulence factors depending on the site or timing of infection, thus explaining the tight transcriptional and posttranscriptional regulation of the T3SS of *P. aeruginosa*. The T3SS is active in acute infection, allowing the bacteria to disseminate rapidly and resist immune cells (Dacheux et al., 1999; Engel and Balachandran, 2009; Hauser, 2009), but is less active during the chronic infection of cystic fibrosis (CF) patients. A multitude of fine-tuned systems for T3SS regulation exist, which consider the host environment, the metabolic status of the bacterium, and different signals from the host or other bacteria (Fig. 1).

The T3SS of *P. aeruginosa* is induced by at least three environmental signals (Frank, 1997; Hornef et al., 2000; Vallis et al., 1999; Yahr and Frank, 1994): (i) in vivo contact with eukaryotic host cells, (ii) in vitro removal of calcium from the medium, and (iii) the presence of serum (due to the existence of type III secretion factors, which have been identified as albumin and casein [Kim et al., 2005]). However, a subset of *P. aeruginosa* T3SS components, encoded by the operon *exsDPscB-L*, and the transcriptional activator ExsA are constitutively expressed, indicating the existence of an expression hierarchy similar to the ordered assembly of the closely related flagellar structure (Wolfgang et al., 2003). This type of hierarchy may facilitate the subsequent assembly and localization of additional T3SS components for the rapid secretion of presynthesized or cotranslational proteins under T3SS-inducing conditions. This type of T3SS regulation is referred to as secretion/activation coupling.

ExsA-Dependent Secretion/Activation Coupling

ExsA is the key regulator of the T3SS

ExsA, a member of the AraC/XylS family of transcriptional regulators, is the central regulator of T3SS gene expression. Because it is encoded by the last gene in the *exsCEBA* operon, it controls its own expression via the P_{exsC} promoter. By aligning the 10 identified ExsA-dependent promoters, Brutinel et al. (2008) identified a consensus sequence for ExsA binding: tAaAAAnwnMyGrCynnnmYTGayAk. This sequence is composed of three highly conserved regions: an

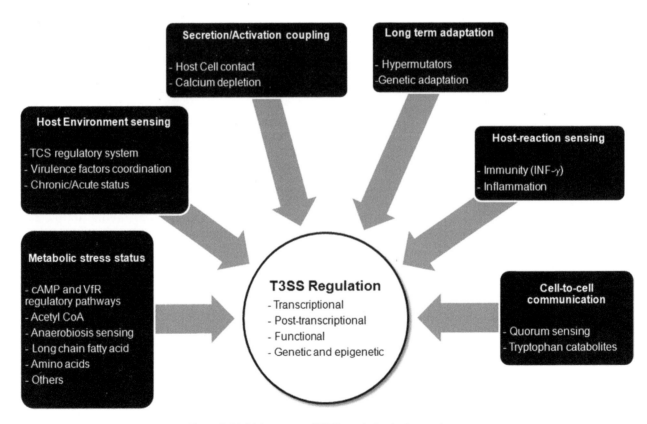

Figure 1. Multiple aspects of T3SS regulation in *P. aeruginosa*.
doi:10.1128/9781555818524.ch17f1

adenine-rich region centered at the −55 position from the transcriptional start site, a GxC sequence around the −44 position, and a TGxxA sequence close to the −35 position. These three regions are necessary for ExsA binding and activation of the P_{exoT} promoter (Brutinel et al., 2008). As a member of AraC/XylS family, ExsA is composed of an N-terminal domain involved in self-association and ligand binding and a C-terminal domain containing two helix-turn-helix motifs involved in DNA binding. In electrophoretic mobility shift assays, purified ExsA binds to the P_{exsC}, P_{exsD}, and P_{exoT} promoters with apparent equilibrium constants around 1, 4, and 5 nM, respectively (Brutinel et al., 2008). Two distinct shift products that correspond to the binding of one and two molecules of ExsA have been observed; binding at the first site of interaction is required for binding at the second site, but deletion of the second site has no effect on binding at the first site (Brutinel et al., 2008). Binding at the second site depends on monomer-monomer interactions through the association of the N-terminal domains (Brutinel et al., 2009). The Hill coefficient is twice as large for the P_{exsC} promoter as for the P_{exsD} and P_{exoT} promoters, suggesting that cooperative binding occurs only for the P_{exsC} promoter (Brutinel et al., 2008). This finding may reflect a difference in the action of ExsA on the promoter of this central regulator of the T3SS.

ExsA regulation

ExsA is fully connected to T3SS translocation/secretion and is triggered by cell contact or calcium lowering via function switching due to exchanges of protein partners, which involves all the proteins of the *exsCEDA* operon (Rietsch et al., 2005; Urbanowski et al., 2005).

ExsD was identified as the first negative regulator of the *P. aeruginosa* T3SS by McCaw et al. through a direct interaction with ExsA (McCaw et al., 2002). ExsD functions as an antiactivator. In vitro reconstitution experiments showed that ExsD disrupts the self-association and the DNA-binding ability of ExsA (Brutinel et al., 2010). ExsC, a T3SS chaperone for ExsE, acts as an antiantiactivator by binding and sequestering the antiactivator ExsD. Under T3SS-inducing conditions, ExsC forms a 2:2 complex with ExsD, thus freeing ExsA (Dasgupta et al., 2004; Lykken et al., 2006; Zheng et al., 2007). ExsC is also a type III chaperone, forming a 2:1 complex with ExsE (Zheng et al., 2007). Finally, ExsE is a secreted regulator that prevents the association between ExsC and ExsD (Rietsch et al., 2004; Urbanowski et al., 2005). Crystallographic analysis of the ExsC-ExsE complex revealed that ExsE is wrapped around one face of the ExsC homodimer (Vogelaar et al., 2010). ExsE contains two binding motifs for ExsC that are localized in the C-terminal and N-terminal extremities of the molecule. These domains are composed of an Arg-X-Val-X-Arg motif that partially overlaps the β motif previously described by Lilic et al. (Lilic et al., 2006) for the type III secretion chaperone in *Salmonella*, invasion protein A (SipA). In the ExsC-ExsE complex, each ExsE binding motif interacts with one ExsC molecule, explaining the 2:1 stoichiometry. In contrast to ExsE, ExsD has only one set of Arg-X-Val-X-Arg and β motifs located N terminally, which is sufficient to bind to ExsC and explains the 2:2 stoichiometry of the ExsC-ExsD complex (Vogelaar et al., 2010).

Taken together, under nonpermissive conditions, the ExsA-dependent transcription of the T3SS regulon is inactive because the binding equilibrium favors inhibitory complexes: ExsA-ExsD and ExsC-ExsE. Conversely, inducing conditions activate the secretion/translocation of ExsE, freeing ExsC to associate with ExsD and therefore liberating ExsA to activate the T3SS regulon. Under this scheme, termination of secretion would result from an increase in the cytoplasmic concentration of ExsE, which would sequestrate ExsD and/or dissociate the ExsC-ExsD complex, thus favoring the formation of the inhibitory complex ExsA-ExsD. However, this model does not seem possible, because the newly released ExsD molecules are unable to bind to ExsA (Brutinel et al., 2010). Although the proteins are purified as a complex when both proteins are expressed in *E. coli*, they are unable to form a complex when expressed and purified separately (Thibault et al., 2009). Because purified ExsD exists in a trimeric form (Zheng et al., 2007), its ExsA binding domain may be buried, prohibiting the formation of new ExsA-ExsD complexes. Therefore, the only way to switch off the T3SS would be to decrease the cytosolic ExsA concentration, thus generating an epigenetic regulation that is responsible for the bistability.

This exchange of partners would necessitate a hierarchy in the binding affinities of the different complexes: ExsC-ExsE > ExsC-ExsD > ExsD-ExsA. Measurement of the binding by isothermal titration calorimetry shows dissociation constants of 1 nM for ExsC-ExsE and 18 nM for ExsC-ExsD (Lykken et al., 2006; Zheng et al., 2007), but no binding was observed for the ExsD-ExsA complex.

Finally, the crystal structure of ExsD shows striking similarities with the transcriptional repressor KorB; although ExsD does not exhibit strong DNA binding affinity for the P_{exsD} promoter by electrophoretic mobility shift assay, ExsD does interact with DNA in differential scanning fluorimetry experiments, suggesting that the regulation may be even more complex (Bernhards et al., 2009).

PtrA, *Pseudomonas* type III repressor A, can also repress T3SS gene expression by inhibiting the function of ExsA through direct binding, as demonstrated in double hybrid and pulldown experiments (Ha et al., 2004). To shut down the energy-expensive T3SS, the expression of PtrA is highly and specifically induced by high copper, such as during the infection of a mouse burn wound, through the CopR-CopS two-component regulatory system. Interestingly, CopR is also a regulator of the OprD porin, and it links the Zn, Cu, and imipenem responses in *P. aeruginosa* by interacting with another two-component system, CzcRS (Caille et al., 2007). Although PtrA is a periplasmic protein that is specifically synthesized in the presence of copper and is involved in the tolerance of *P. aeruginosa* to copper, its role in the transcriptional inhibition of the T3SS is unclear (Elsen et al., 2011).

Epigenetic regulation: bistability of the T3SS

We have previously shown that despite a functionally wild-type genomic background, some *P. aeruginosa* strains are unable to develop type III-dependent cytotoxicity except when ExsA is overexpressed in *trans* (Dacheux et al., 2001). In addition, using green fluorescent protein reporter analysis of single cells, a cell-to-cell variation in the T3SS, which ranged from no expression to strong expression, was observed (Hornef et al., 2000; B. Polack and B. Toussaint, unpublished observations). These observations, together with the presence of a positive-feedback loop regulating the expression of *exsA*, led us to hypothesize the existence of an epigenetic switch, i.e., a bistability, between the inducible and noninducible T3SS in *P. aeruginosa* (Filopon et al., 2006). With the help of generalized logic and a formal computer approach, we designed a minimal model of *P. aeruginosa* T3SS regulation under activation and established the consistency of the hypothesis of an epigenetic acquisition of an inducible T3SS-activated phenotype. This model also showed that only an artificial (exogenous) transient increase in ExsA concentration would be sufficient to permanently switch a noninducible strain to an inducible one. We then demonstrated a stable acquisition of the secretory phenotype, in vitro and in vivo, after a transient increase in the ExsA signal, which was consistent with the model. Therefore, using in vitro and in vivo experiments according to the predictions of the formal model, we were the first to demonstrate that a stable acquisition (or loss) of a phenotypic trait involved in the pathogenicity of *P. aeruginosa*, i.e., a bistability, could arise from an epigenetic switch (Filopon et al., 2006).

Bistability, which now appears to be a fundamental trait in bacterial regulation (for examples, see Dubnau and Losick, 2006, or Turner et al., 2009), is epigenetic because it is not mediated through genetic changes. However, the major source of bistability may be transcriptional noise in key promoters, which generates transitions between alternative states of gene expression and greater diversity in phenotype expression, resulting in a higher fitness of the population (Mettetal and van Oudenaarden, 2007; Raser and O'Shea, 2005).

Host Environment Sensing

Two-component regulatory systems and small RNAs

P. aeruginosa uses different two-component regulatory pathways to regulate virulence or biofilm formation, which correspond to acute or chronic infections. The hybrid sensors RetS and LadS are involved in the transition between chronic and acute infections by antagonistically controlling the expression of genes involved in virulence, such as the T3SS, or genes that are required for biofilm formation, such as those involved in polysaccharide synthesis (Goodman et al., 2004; Laskowski et al., 2004; Ventre et al., 2006). These two sensors appear to overlap with another two-component system formed by the GacS/GacA pair, in which GacS is an unorthodox sensor and GacA is a response regulator. Signals transduced by the three sensor kinases, RetS, LadS, and GacS, are hypothesized to ultimately control the levels of the two small regulatory RNAs.

After the detection of an as-yet-unknown signal, the sensor kinase GacS autophosphorylates at a histidine residue (Heeb and Haas, 2001), and the phosphate is then transferred to GacA, which becomes activated. GacA is involved in the pathogenesis of numerous hosts (Gooderham and Hancock, 2009). Once activated, the GacA protein is able to bind specifically to the GacA box upstream of the *rsmZ* and *rsmY* genes (Brencic et al., 2009). The two small RNAs, *rsmZ* and *rsmY*, in turn, regulate the reciprocal expression of the genes involved in the T3SS and biofilm formation by titrating the amount of free or active RsmA protein in the cell (Burrowes et al., 2006; Goodman et al., 2004; Ventre et al., 2006).

RsmA, the "regulator of secondary metabolites," is a global regulator that negatively controls the expression of numerous genes at the posttranscriptional level and positively regulates T3SS gene expression (Burrowes et al., 2005; Heurlier et al., 2004; Mulcahy et al., 2006). More than 500 genes are under the control of RsmA (Brencic and Lory, 2009), which exerts its negative action by binding to the 5′ untranslated region of the RNA to inhibit translation initiation. Hence, RsmA is a small RNA-binding protein that could be considered to be the major posttranscriptional regulator of *P. aeruginosa*.

RsmA plays an important role in the posttranscriptional regulation of a number of virulence-related genes in *P. aeruginosa* (Burrowes et al., 2005; Heurlier et al., 2004; Pessi et al., 2001) and positively regulates T3SS gene expression by controlling other T3SS regulators (Mulcahy et al., 2006). Mutations in *rsmA* result in the decreased expression of five positive T3SS regulators, including RetS, ExsA, ExsC, CyaB, and Vfr, and one negative regulator, ExsD (Mulcahy et al., 2006), and the increased production of the QS signal C4-HSL (see below) (Pessi et al., 2001) and the multidrug resistance efflux pump MexEF-OprN (Burrowes et al., 2006). RsmA may act positively on T3SS by blocking ExsA RNA degradation by an endonuclease negatively controlled by RsmA (S. Lory, 2009, Pseudomonas Congress). All these effects, except that on ExsD, make RsmA a positive regulator of the expression of T3SS genes.

RsmA is highly homologous (85% identity) to *Salmonella* CsrA, which is a positive regulator of invasion genes and components of the *Salmonella* T3SS (Lawhon et al., 2003).

RetS, a regulator of exopolysaccharide and type III secretion (Goodman et al., 2004; Zolfaghar et al., 2005), and RtsM, a regulator of type III secretion (Laskowski et al., 2004), are the same protein, which was determined independently by two different groups. RetS, an unusual hybrid two-component signaling protein with a sensor kinase domain followed by two response regulator receiver domains in tandem, is required for the expression of T3SS genes (and also T2SS, lipase A, endotoxin A, and type IV pili) and for the repression of genes responsible for the exopolysaccharide components of the *P. aeruginosa* biofilm matrix and T6SS. An intact phosphoacceptor site within receiver domain 2, but not receiver domain 1, is absolutely required for RetS activity, and the periplasmic domain and the phosphorylation of receiver domain 1 may inhibit RetS activation. Furthermore, RetS may serve as a substrate for another sensor kinase (Laskowski and Kazmierczak, 2006). RetS acts in a fairly unique manner by forming heterodimers with GacS and preventing the activation of the GacS/GacA pathway (Goodman et al., 2009). LadS, another hybrid sensor kinase, and GacS/GacA, together with RetS, are all involved in the reciprocal regulation of T3SS gene expression and exopolysaccharide synthesis (Ventre et al., 2006).

LadS and RetS are hybrid sensors that may require a histidine phosphotransfer (Hpt) module (sometimes from another protein) to transfer phosphate to their cognate response regulator. In a recent study, an *hptB* mutant had a phenotype similar to that of a *retS* mutant, but despite these similarities, the HptB and RetS pathways are distinct. Both pathways terminate on the GacA response regulator, but HptB signaling controls the expression of *rsmY* only, whereas RetS signaling modulates both *rsmY* and *rsmZ* gene expression. HptB signaling is a complex regulatory cascade that involves a response regulator with an output domain belonging to the phosphatase 2C family and likely an anti-anti-σ factor (Bordi et al., 2010). This last work revealed that the initial input in the Gac system comes from several signaling pathways that have not yet been completely characterized. The final output is adjusted by the differential control of *rsmY* and *rsmZ*. This has some consequences for fine-tuning the regulation of different virulence factors, such as the RetS-dependent but HptB-independent control of the T6SS. Other outputs act directly on other small riboregulators, such as the global regulators MvaU and MvaT. MvaU and MvaT are H-NS-type transcriptional regulators that are able to form intra- or intermolecular DNA complex structures and inhibit the transcription of targeted genes by blocking access to the RNA polymerase (Dame et al., 2005; Dorman, 2007). In *P. aeruginosa*, MvaT and MvaU are involved in the regulation of the expression of more than 150 genes, the majority of which encode surface-associated proteins as well as the QS genes T6SS, *cupA* (biofilm), and *exsA* (Castang et al., 2008). The A+T-rich DNA segment upstream of the riboregulator *rsmZ* is bound and silenced by MvaT and MvaU.

This complex system, which includes different sensor molecules acting on different riboregulators, should allow *P. aeruginosa* to switch between patterns of gene expression that characterize acute infection versus chronic colonization (Brencic et al., 2009). However, the signals that are sensed by this system as well as the biochemical relationships between the various regulators remain mainly uncharacterized. RetS and LadS may respond to carbohydrates of host or bacterial origin because the periplasmic domains of both proteins belong to a class of bacterial periplasmic sensor modules, 7TMR-DISMED2, which has been identified in a variety of carbohydrate-binding proteins (Ventre et al., 2006).

However, virulence factor gene expression may not be exclusive to acute or chronic infections. An active secretion of T3SS toxins could occur during biofilm growth conditions (Mikkelsen et al., 2009). The system described above could contribute to the balance between acute and chronic phenotypes, but many other inputs are involved in the activation or repression of T3SSs, such as metabolic/stress status.

Bacterial Metabolic/Stress Status Sensing

Many metabolic or stress status signals that interfere with T3SSs have been discovered. Here we describe their principles.

cAMP and Vfr regulatory pathway

Cyclic AMP (cAMP) is a secondary messenger involved in numerous signal transduction pathways in *P. aeruginosa*. A membrane-associated adenylate cyclase (CyaB), responsible for cAMP synthesis, and a cAMP-binding cAMP receptor protein (CRP) homologue called Vfr are required for the expression of T3SS genes (Lory et al., 2004; Smith et al., 2004; Wolfgang et al., 2003). Modulation of intracellular cAMP seems to be an important mechanism for controlling T3SS expression. The catalytic domain of the CyaB enzyme appears to be regulated by inorganic carbons, such as bicarbonate ions (Linder and Schultz, 2003). By detecting HCO_3^-, CyaB allows the bacteria to respond to variations in inorganic carbon concentrations by modulating intracellular cAMP concentrations (Yahr and Wolfgang, 2006). Furthermore, low Ca^{2+} and high NaCl concentrations increase cAMP and T3SS activation (Rietsch and Mekalanos, 2006). cAMP levels act as an allosteric regulator of the protein Vfr, which is the homologue of the *E. coli* CRP protein in *P. aeruginosa* (Albus et al., 1997; Beatson et al., 2002; Wolfgang et al., 2003). Vfr is a transcriptional regulator that binds to a specific regulatory sequence in the upstream regions of target genes (*lasR*, *fleQ*, *regA*, and *toxA*), but it can also indirectly regulate genes, which should be the case for T3SS, but no binding of recombinant Vfr has been observed (Shen et al., 2006). However, in the absence of ExsD, Vfr is no longer required for T3SS gene expression, suggesting that Vfr acts to derepress ExsA through the antiactivator ExsD (Dasgupta et al., 2006). Vfr may be a central regulatory protein for T3SS on which many signals acting on the intracellular cAMP concentration converge. Some of these signals are not fully characterized, such as that involving FimL. FimL is homologous to the N-proximal domain of the complex chemosensory protein ChpA and regulates T3SS gene expression by intersecting with Vfr-modulated pathways (Whitchurch et al., 2005). FimL proteins have a phosphodiesterase activity that can modulate cAMP levels in *P. aeruginosa* (Inclan et al., 2011). VfR is probably more complicated because it has other allosteric regulators (such as cGMP) and is a global regulator for virulence factor expression through different mechanisms that may or may not involve cAMP (Fuchs et al., 2010). SadARS, a three-component regulatory system, is comprised of a putative sensor histidine kinase, SadS, and two response regulators, SadA and SadR. SadARS may repress T3SS gene expression by promoting biofilm formation (Kuchma et al., 2005). The presence of a Glu-Ala-Leu (EAL) domain in SadR suggests that this protein possesses a phosphodiesterase activity that can degrade the cyclic diguanylate monophosphate (di-GMPc). The di-GMPc could be another secondary messenger (with cAMP and cGMP) involved in T3SS regulation and may be involved in the regulation of other virulence factors (Lory et al., 2009).

Acetyl-CoA and other energy sensing

T3SS gene expression could be repressed by many metabolic stresses, but a partially common mechanism may use acetyl coenzyme A (acetyl-CoA) as part of the signal.

T3SS expression is abolished in mutants lacking pyruvate dehydrogenase (*aceA* or *aceB*), an enzyme involved in the production of acetyl-CoA from pyruvate when aerobic glycolysis occurs (Dacheux et al., 2002). The absence of the glucose transport regulator (*gltR*) also results in a failure to induce T3SS expression (Wolfgang et al., 2003). By contrast, after inactivation of citrate synthase (a tricarboxylic acid cycle enzyme involved in acetyl-CoA metabolism), an upregulation of T3SS occurs (Rietsch and Mekalanos, 2006). This finding strongly suggests a link between T3SS regulation and acetyl-CoA, as has also been proven in *Yersinia enterocolitica* (Schmid et al., 2009). Notably, acetyl-CoA is an essential metabolite for growth under aerobic conditions, and the anaerobic metabolism of *P. aeruginosa* is important for growth and biofilm formation during persistent infections (Platt et al., 2008).

Anaerobiosis sensing

In *P. aeruginosa*, the global anaerobic response regulator ANR senses low oxygen and subsequently triggers the expression of a range of target genes, including microaerobic induction of the T3SS through the RsmA protein (O'Callaghan et al., 2011). Low oxygen induces the T3SS in *P. aeruginosa* via modulation of the small RNAs *rsmZ* and *rsmY*. The role played by nitric oxide has recently been emphasized (Van Alst et al., 2009) by showing that T3SS expression is abolished in a mutant devoid of the nitric oxide-synthesizing enzyme nitrate reductase (NirS). Nitric oxide metabolism is also linked to anaerobiosis and its complex regulation through the ANR regulator.

The upregulation of the PsrA protein, a repressor of T3SS expression (see below), during anaerobiosis was recently identified by transcriptomic and proteomic analyses (Trunk et al., 2010).

Long-chain fatty acids

Another regulatory pathway linked to energy metabolism that involves long-chain fatty acid (LCFA) metabolism genes has been described previously (Kang et al., 2008). This pathway also involves the PsrA transcriptional regulator, which can respond to the LCFA concentration and hence activate or repress the transcription of the *exsCEBA* operon (Kang et al., 2009). PsrA also regulates RpoS expression (Kojic et al., 2002). Although PsrA has been shown to bind to the *exsCEBA* promoter by electrophoretic mobility shift assays (Shen et al., 2006), large concentrations of recombinant PsrA are required. Hence, it is possible that the LCFA and PsrA pathway indirectly modulate the RpoS protein, which is also known to negatively regulate the T3SS (Hogardt et al., 2004). The RpoS concentration increases greatly during stationary phase and is linked to the expression of QS genes known to inactivate the T3SS (Bleves et al., 2005; Hogardt et al., 2004). *phrS*, the first small RNA that provides a regulatory link between oxygen availability and the QS regulatory network known to interfere with T3SS expression, was recently identified (Sonnleitner and Haas, 2011).

Hence, the role of RpoS and QS and cell density regulation in general remains to be characterized.

Amino acids

When histidine is present in the medium, excessive uptake and degradation of histidine due to the overexpression of histidine utilization genes, such as *hutT*, abolish the ability to induce *exoS* expression and render the bacterium noncytotoxic (Rietsch et al., 2004). The cytotoxicity defect can be partially suppressed by an insertion in *cbrA*, a gene that encodes the sensor kinase of the CbrAB two-component regulatory system, which is involved in sensing and responding to a carbon-nitrogen imbalance. In minimal glucose media, the induction of *exoS* expression by calcium removal is ineffective, but the addition of Casamino Acids and glutamate is able to totally restore its induction (Rietsch and Mekalanos, 2006).

The role of tryptophan is discussed in "Bacterial Cell-to-Cell Communication" below because it has been implicated in the signaling for T3SS repression at high cell densities (Shen et al., 2008).

Other stress/metabolic signals

The *P. aeruginosa* magnesium transporter MgtE was recently shown to inhibit the transcription of the T3SS by acting directly or indirectly on the ExsECDA regulatory system, but while variations in calcium levels modulate T3SS gene expression, the addition of exogenous magnesium did not inhibit T3SS activity (Anderson et al., 2010).

The gene *truA*, which encodes a pseudouridinase enzyme, is required for the expression of the type III secretory genes. Pseudouridination of tRNAs is proposed to be critical for the translation of T3SS genes or their regulators (Ahn et al., 2004).

The gene *dsbA*, which encodes a periplasmic thiol/disulfide oxidoreductase, affects the expression of multiple virulence factors, including T3SS genes, probably due to a nonspecific effect resulting from abnormal protein folding caused by the lack of disulfide bonds (Ha et al., 2003).

Overexpression of the MexCD-OprJ and MexEF-OprN multidrug resistance efflux pumps is associated with a reduction of T3SS gene expression. Because these pumps can extrude a wide range of compounds belonging to different structural families, it is possible that the intracellular signal needed for the activation of T3SS gene expression is extruded (Linares et al., 2005).

By searching for molecular differences between T3SS-producing and -non-producing strains, Jin and coworkers observed that the *mexS/T* genes, which are regulators of the MexEF-OprN efflux systems, also regulate the T3SS, elastase, and pyocyanin. MexS/T from a T3SS-producing strain confers a high T3SS expression background on a PAO1 strain (Jin et al., 2011). DNA damage also has an influence on T3SS activation. Wu and Jin have shown that the PtrB protein is a repressor of T3SS that is regulated by the PtrR protein in response to mitomycin-induced DNA damage (Wu and Jin, 2005). We also confirm that UV-induced DNA damage reduced T3SS activation in an ExsA-dependent manner (unpublished data).

Molecules from host cells could act as inducers of the T3SS. This is the case for spermidine, an important molecule that also functions as an inducer for the T3SS, as demonstrated by spermidine uptake mutants (*spuE* gene) that showed decreased expression in T3SS genes in response to host cell extract and calcium depletion (Zhou et al., 2007).

All of the signals described above belong to the bacterium itself, but interbacterial communication also plays a prominent role in virulence factor regulation.

Bacterial Cell-to-Cell Communication

The synchronization of individual behavior is thought to aid bacterial infection by up- or down-regulating virulence determinants, like the T3SS, only when a sufficient bacterial cell density has been achieved to overwhelm host defenses. On the contrary, downregulation of the T3SS could benefit *P. aeruginosa* for the chronic infection of the host by conserving energy resources or by restricting antibody-mediated clearance within the host (Yahr and Wolfgang, 2006). In vitro, T3SS activation is regulated in a cell density-dependent manner (Shen et al., 2008). In correlation with growth phase-dependent internalization, the invasive strain PAK translocates much larger amounts of ExoS and ExoT into HeLa cells when it is in an exponential growth phase than when it is in a stationary growth phase (Ha and Jin, 2001). The regulation of type III secretion genes has previously been associated with the *las* and *rhl* QS systems, which are cell density dependent (Bleves et al., 2005; Hogardt et al., 2004; Juhas et al., 2005; Schaber et al., 2004; Yahr and Wolfgang, 2006). Many bacteria communicate using diffusible molecules known as autoinducers to coordinate population activities (de Kievit and Iglewski, 2000; Rumbaugh et al., 2000). *P. aeruginosa* possesses at least three main interrelated QS systems: the *las* and *rhl* systems, which employ acylated homoserine lactone signal molecules, and the 2-alkyl-4(1-H)-quinolone (AHQ) family. LasI is the synthase for the autoinducer N-(3-oxododecanoyl)-L-homoserine lactone ($3OC_{12}$-HSL), while RhlI synthesizes the autoinducer N-butanoyl-L-homoserine lactone (C_4-HSL). At high cell densities, $3OC_{12}$-HSL and C_4-HSL reach critical levels and activate their regulators, which, in turn, enhance the transcription of different virulence genes (de Kievit and Iglewski, 2000; Rumbaugh et al., 2000). Mutants lacking the *rhl* QS system show increased expression of T3SS genes and secretion of ExoS during exponential growth (Bleves et al., 2005; Hogardt et al., 2004; Shen et al., 2008). The stationary-phase sigma factor RpoS and the Rhl system are coordinately regulated (Latifi et al., 1996; Whiteley et al., 2000), and the expression of *exoS'-gfp* is upregulated in an *rpoS* mutant. Particularly in stationary phase, ExoS expression is downregulated by both the Rhl QS system and RpoS (Hogardt et al., 2004).

The AHQs include more than 50 quinolone compounds, 2 of which have been shown to act as QS signals, *Pseudomonas* quinolone signal (PQS; 2-heptyl-3-hydroxy-4-quinolone) and its immediate precursor, 4-hydroxy-2-heptylquinoline (HHQ) (Fletcher et al., 2007; Pesci et al., 1999). Singh et al. demonstrated a role for PQS in type III cytotoxin secretion changes by using clonal isolates from a patient who had developed ventilator-associated pneumonia. The clinical observations were mirrored by an in vitro temporal shift in the phenotype of the isolate from a nonsecreting to a type III cytotoxin-secreting (T3SS) phenotype (Singh et al., 2010).

We have shown that T3SS expression is dependent on cell density and that *exsCEBA* operon expression decreases rapidly in the second part of stationary phase (Shen et al., 2008). In fact, in stationary phase, *P. aeruginosa* secretes in the medium a signal that can inhibit T3SS expression independently of QS molecules such as $3OC_{12}$-HSL, C_4-HSL, and PQS and the stationary-phase sigma factor RpoS but is dependent on the tryptophan synthase TrpA (Shen et al., 2008). Tryptophan, which is not the signal molecule, is the precursor of known signaling molecules such as indole-3-acetic acid (IAA) (Cohen et al., 2003) in bacterium-plant symbiosis, kynurenine (Kurnasov et al., 2003) in immune regulation, and anthranilate (Chugani and Greenberg, 2010) and PQS (Farrow and Pesci, 2007). We have shown that IAA, naphthalenacetic acid (a compound similar to IAA), and 3-hydroxykynurenine inhibit *exsCEBA* operon expression at millimolar concentrations. Furthermore, inactivation of the *kynA* gene involved in kynurenine synthesis results in a decrease in the T3SS-inhibiting activity of stationary-phase culture supernatants. While a direct link has not been established between the tryptophan pathway and phenazine-1-carboxylic acid (PCA) synthesis, we found a role for PCA in the repression of T3SS gene expression during stationary-phase growth (unpublished data). Recently, Sonnleitner and Haas described the only known regulatory small RNA, PhrS, the transcription of which depends on the oxygen-responsive regulator ANR. When oxygen becomes limiting, as in stationary phase or in biofilms, PhrS is highly transcribed, leading to the increased translation of PqsR, a regulator of the PQS (Sonnleitner and Haas, 2011), and probably PCA, as has been demonstrated in *Pseudomonas* sp. strain M18 (Lu et al., 2009). This hypothesis has to be examined to better understand the molecular mechanism of the downregulation of the T3SS during stationary-phase growth. The cell density-dependent regulation of the T3SS is very complex. Because oxygen becomes limiting in stationary phase as in biofilm formation, we could assume that these two types of regulation work in parallel. Much progress has been made in understanding biofilm formation in vitro (Murray et al., 2007). It has been proposed that following the infection of CF patient airways, *P. aeruginosa* strains evolve to reduce T3SS expression (Lee et al., 2005), or populations of cells gradually change from a type III protein secretion-positive phenotype to a secretion-negative phenotype (Jain et al., 2004). However, these in vitro laboratory observations must be reconciled with the behavior of clinical isolates from a CF lung. The bacteria are able to integrate signals from themselves, from other bacteria, and also from the host. The dialogue between the bacteria and the host must also be considered.

Host Reaction Sensing (T3SS and Immunity)

In 2005, Wu et al. proposed that the opportunistic pathogen *P. aeruginosa* could adapt to the host by sensing alterations in the host immune function and respond by enhancing the virulence phenotype (Wu et al., 2005). Virulence in *P. aeruginosa* is highly regulated by the QS signaling system, a hierarchical system of virulence gene regulation that is dependent on bacterial cell density processes (Diggle et al., 2002; Smith and Iglewski, 2003; Wagner et al., 2003; Whiteley et al., 1999). The QS system is suggested to play a key role in the response of *P. aeruginosa* to the host cytokine gamma interferon (IFN-γ), and IFN-γ may shift the virulence of *P. aeruginosa* against epithelial cells. Interestingly, the effect is specific for IFN-γ, because other cytokines do not modify, for instance, the production of pyocyanin (Wu et al., 2005). *P. aeruginosa* binds specifically to IFN-γ through the outer membrane porin OprF (Wu et al., 2005). Hence, OprF is a host immune system sensor that modulates QS to enhance virulence. Production of *N*-acyl-L-homoserine lactone (AHL) signaling molecules, including 3OC$_{12}$-HSL and C$_4$-HSL, was reduced and delayed in an OprF mutant strain. The production of the non-AHL-signaling molecule PQS was decreased, while its precursor, HHQ, accumulated in the bacteria. The production of the phenazine pyocyanin, elastase, lectin PA-1L, and exotoxin A is clearly under the control of the QS network and is impaired in an *oprF* mutant (Fito-Boncompte et al., 2011). The production but likely not the secretion of ExoS and ExoT, the major toxin secreted by the T3SS, is impaired in the *oprF* mutant. However, this phenomenon can be overcome by overexpressing ExsA, the major activator of the T3SS system. Therefore, we propose that OprF may downregulate ExsA or maintain it in an inactive state. Together with previous findings that *P. aeruginosa* QS can alter the host immune response (e.g., by activating IFN-γ) and the work cited in the introduction (Wangdi et al., 2010), this illustrates an exciting new element of bacterium-host interactions in which *P. aeruginosa* both senses and modulates the host immune state but the immune system is also able to respond to the T3SS.

Long-Term Adaptation

It is now well known that the long-term adaptation of *P. aeruginosa* during CF infection is due to the selection of diverse genotypes and phenotypes from one infecting clone, a process termed adaptive radiation (Burns et al., 2001). During microevolution within CF lungs, *P. aeruginosa* populations with altered but not reduced virulence appear, and this adapted virulence plays a critical role in the pathogenesis of chronic infections (Bragonzi et al., 2009). This acquired diversity is due to loss-of-function mutations in selected *P. aeruginosa* genes. Loss-of-function mutations in *mucA* and *lasR* determine the phenotypic switch to mucoidy and a loss of QS (Hogardt and Heesemann, 2010; Winstanley and Fothergill, 2009). This genetic adaptation during CF infections is also catalyzed by hypermutation capacities due to DNA mismatch repair system-deficient hypermutable (or mutator) strains (Mena et al., 2008). The role of the T3SS during CF is not fully understood, but the percentage of T3SS-positive CF isolates diminishes drastically with the duration of PA infection, from 45% in first infections in children to 11% in adults (Jain et al., 2008). In an older study, 12 *P. aeruginosa* CF isolates that were unable to perform type III secretion were shown to possess T3SS genes and the *exsA* transcriptional activator but showed no transcriptional activation of the *exsCBA* regulatory operon (Dacheux et al., 2001). The expression of *exsA* in *trans* restored the in vitro secretion of the T3SS proteins, the first indication that CF adaptation could act on T3SS regulatory circuits.

A loss of function in *mucA*, which encodes an anti-sigma factor, allows the activation of the alternative sigma factor AlgU (also named AlgT) and hence the transcription of numerous genes, including alginate synthesis genes (Govan and Deretic, 1996). The activation of the AlgU regulon through mutation of *mucA* has been shown to provoke the inactivation of the T3SS by inhibiting the cAMP/Vfr signaling pathway (Jones et al., 2010). This work was the first demonstration that a single mutation could lead to the inverse regulation of virulence factors for acute or chronic infection. The selection for *mucA* mutations occurs under the pressure of the high energy cost of maintaining an active T3SS because during the biofilm mode of growth, there is no need for the T3SS. Intriguingly, this genetic adaptation corresponds to a stabilization of the classical transcriptional and post-transcriptional regulation described before that acts more rapidly but should be less stable. Otherwise, this kind of adaptation could also lead to different levels of active type III secretion between strains, a well-known phenomenon in T3SSs in *P. aeruginosa* strains.

CONCLUSION

Despite the large number of genes that influence the regulation of the expression of T3SS genes in *P. aeruginosa*, only a minority have been shown to have a precise role (Fig. 2). This is the case for

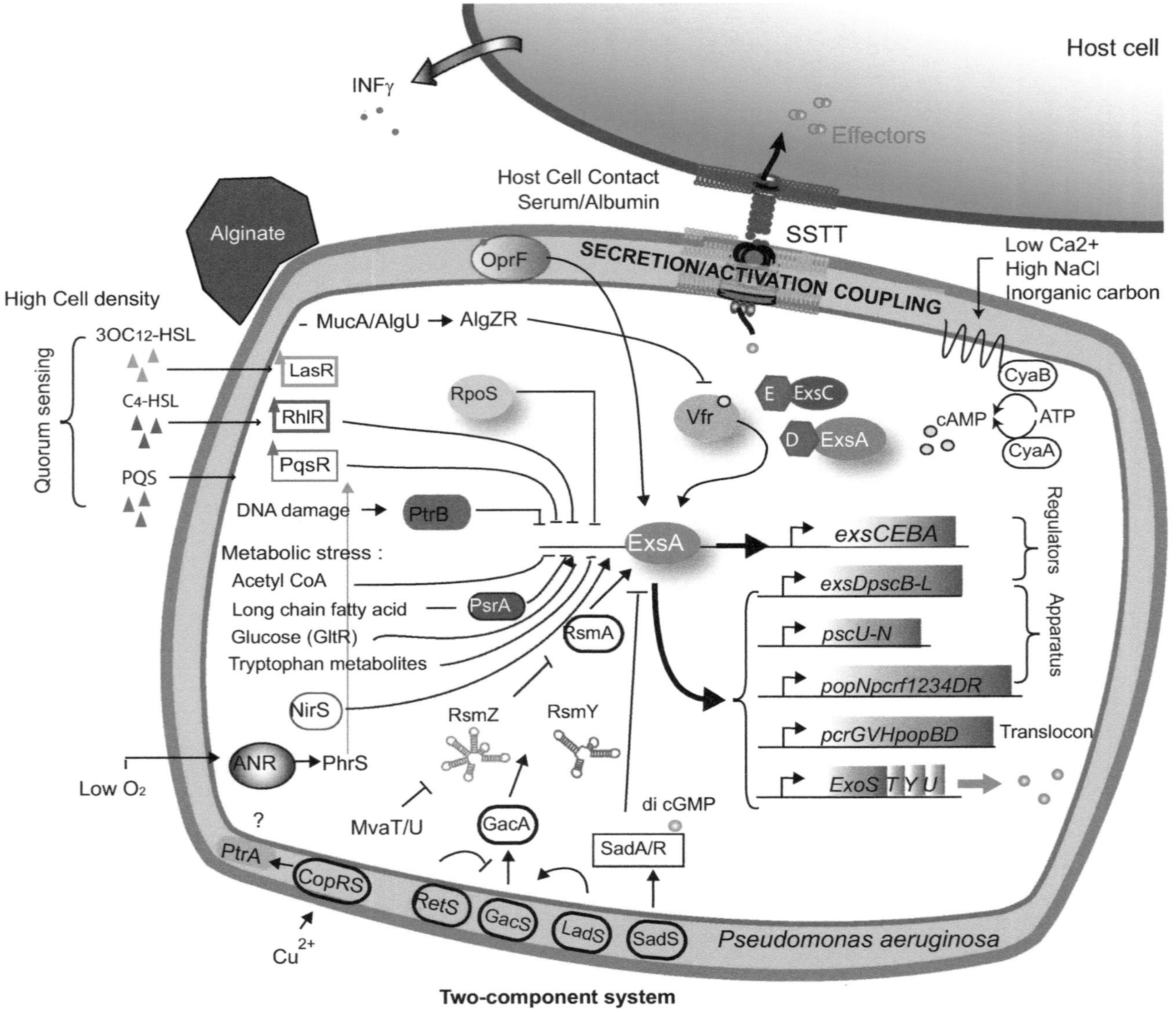

Figure 2. Schema of type III secretion system regulation in *P. aeruginosa*. doi:10.1128/9781555818524.ch17f2

the central core of T3SS regulation in *P. aeruginosa*, which is composed of the global regulator Vfr, the RNA-binding protein RsmA, the *exsCEBA* promoter-binding protein PsrA, and the ExsADCE secretion/regulation coupling cascade. Much work remains with respect to the other major components of the interfering pathways to have a fully integrated view of the physiological basis of this regulation.

Like in *P. aeruginosa*, T3SS regulation in other gram-negative pathogens probably acts at the following different levels: transcriptional, posttranscriptional, functional, genetic, and epigenetic (bistability). We attempted here to highlight the main factors governing T3SS regulation in *P. aeruginosa*.

(i) First, the bacteria have to induce T3SS expression by activating the secretion/activation ExsADCE system, which is primarily based on protein/protein inhibition. It is amazing that the signal(s) that govern this activation is unknown. This cascade is activated in vivo by host cell contact, which is only partially mimicked by calcium depletion in vitro. We are not sure that calcium depletion has the same physiological basis as natural host cell activation. Intriguingly, when the intracellular level of the ExsA protein is too low, calcium depletion is not able to activate the T3SS, which has been attributed to a bistability of the system by Filopon and coworkers (Filopon et al., 2006). This bistability could explain the occurrence of a T3SS-negative clone from a T3SS-positive strain after many generations without T3SS induction. (ii) A very complex network of metabolic/stress sensing genes is involved in the control of the ability of the given cell to activate this high-energy-consuming virulence factor. Energy control coupling has also been observed for other gram-negative pathogens, such as *Shigella*, which activates T3SS depending on the oxygenation level of the area of infection (Marteyn et al., 2010). Metabolic/stress sensing is probably devoted to the timing and topology of the infection. (iii) The bacteria also use a complete set of host or environment sensing systems, mainly to coordinate the expression of the different virulence factors depending on the type, time, or site of the infection. (iv) Special sensing/regulatory apparatuses may exist for detecting the host immune response, as has been suggested in two different studies by the intriguing although not fully elucidated role of the OprF protein (Wu et al., 2005). OprF could be a sensor of host immune activation (through IFN-γ) and seems to have a central role in the activation of different *P. aeruginosa* virulence factors. (v) *P. aeruginosa* also possesses a very efficient cell-to-cell communication system, which utilizes not only classic QS signals but probably also less well-characterized signals, some of which belong to tryptophan catabolism and could be a host cell-specific

signature in the case of immune cells (B. Toussaint, unpublished data). (vi) Finally, a more long-term adaptation could occur through the selection of loss-of-function mutations in key regulators.

Future research depends on the final objective. If the goal is to understand the physiology of *P. aeruginosa* infection, we must consider the following points. First, the data obtained must be progressively integrated into a complex model using modern computational biology tools, and the different models must be optimized by confrontation with high-quality in vivo observations. Therefore, tools for the precise measurement of gene activation and T3SS function in vivo are needed, which will become possible with in vivo imaging technology currently under development. We also need to integrate the different data obtained from different strains because important differences between strains sometimes yield contradictory results. We also need to understand why some strains (such as PAK, for example) have a very high level of T3SS expression, which may be addressed by studying T3SS regulation at the genetic level.

If the goal is to design therapeutic molecules that act on T3SSs, we will probably need to screen different target genes or proteins. Computational biology will reveal the main nodes to target to obtain a strong inhibition of T3SS without the possibility of being bypassed. One of the best targets is certainly the ExsA main transcriptional activator. The equivalent of ExsA in *Shigella* is the VirF transcription factor, which has been used with success as a target for high-throughput screening of antibacterial molecules (Hurt et al., 2010). We suppose that both fundamental and applied research will soon make the inhibition of T3SS expression one of the first new antibacterial therapeutics.

REFERENCES

Ahn, K. S., U. Ha, J. Jia, D. Wu, and S. Jin. 2004. The *truA* gene of *Pseudomonas aeruginosa* is required for the expression of type III secretory genes. *Microbiology* 150:539–547.

Albus, A. M., E. C. Pesci, L. J. Runyen-Janecky, S. E. West, and B. H. Iglewski. 1997. Vfr controls quorum sensing in *Pseudomonas aeruginosa*. *J. Bacteriol.* 179:3928–3935.

Anderson, G. G., T. L. Yahr, R. R. Lovewell, and G. A. O'Toole. 2010. The *Pseudomonas aeruginosa* magnesium transporter MgtE inhibits transcription of the type III secretion system. *Infect. Immun.* 78:1239–1249.

Bader, M. W., W. W. Navarre, W. Shiau, H. Nikaido, J. G. Frye, M. McClelland, F. C. Fang, and S. I. Miller. 2003. Regulation of *Salmonella typhimurium* virulence gene expression by cationic antimicrobial peptides. *Mol. Microbiol.* 50:219–230.

Bahrani, F. K., P. J. Sansonetti, and C. Parsot. 1997. Secretion of Ipa proteins by *Shigella flexneri*: inducer molecules and kinetics of activation. *Infect. Immun.* 65:4005–4010.

Bajaj, V., R. L. Lucas, C. Hwang, and C. A. Lee. 1996. Co-ordinate regulation of *Salmonella typhimurium* invasion genes by environmental and regulatory factors is mediated by control of *hilA* expression. *Mol. Microbiol.* **22**:703–714.

Beatson, S. A., C. B. Whitchurch, J. L. Sargent, R. C. Levesque, and J. S. Mattick. 2002. Differential regulation of twitching motility and elastase production by Vfr in *Pseudomonas aeruginosa*. *J. Bacteriol.* **184**:3605–3613.

Beloin, C., S. McKenna, and C. J. Dorman. 2002. Molecular dissection of VirB, a key regulator of the virulence cascade of *Shigella flexneri*. *J. Biol. Chem.* **277**:15333–15344.

Bernhards, R. C., X. Jing, N. J. Vogelaar, H. Robinson, and F. D. Schubot. 2009. Structural evidence suggests that antiactivator ExsD from *Pseudomonas aeruginosa* is a DNA binding protein. *Protein Sci.* **18**:503–513.

Bleves, S., C. Soscia, P. Nogueira-Orlandi, A. Lazdunski, and A. Filloux. 2005. Quorum sensing negatively controls type III secretion regulon expression in *Pseudomonas aeruginosa* PAO1. *J. Bacteriol.* **187**:3898–3902.

Bordi, C., M. C. Lamy, I. Ventre, E. Termine, A. Hachani, S. Fillet, B. Roche, S. Bleves, V. Mejean, A. Lazdunski, and A. Filloux. 2010. Regulatory RNAs and the HptB/RetS signalling pathways fine-tune *Pseudomonas aeruginosa* pathogenesis. *Mol. Microbiol.* **76**:1427–1443.

Bragonzi, A., M. Paroni, A. Nonis, N. Cramer, S. Montanari, J. Rejman, C. Di Serio, G. Doring, and B. Tummler. 2009. *Pseudomonas aeruginosa* microevolution during cystic fibrosis lung infection establishes clones with adapted virulence. *Am. J. Respir. Crit. Care Med.* **180**:138–145.

Brencic, A., and S. Lory. 2009. Determination of the regulon and identification of novel mRNA targets of *Pseudomonas aeruginosa* RsmA. *Mol. Microbiol.* **72**:612–632.

Brencic, A., K. A. McFarland, H. R. McManus, S. Castang, I. Mogno, S. L. Dove, and S. Lory. 2009. The GacS/GacA signal transduction system of *Pseudomonas aeruginosa* acts exclusively through its control over the transcription of the RsmY and RsmZ regulatory small RNAs. *Mol. Microbiol.* **73**:434–445.

Brutinel, E. D., C. A. Vakulskas, K. M. Brady, and T. L. Yahr. 2008. Characterization of ExsA and of ExsA-dependent promoters required for expression of the *Pseudomonas aeruginosa* type III secretion system. *Mol. Microbiol.* **68**:657–671.

Brutinel, E. D., C. A. Vakulskas, and T. L. Yahr. 2009. Functional domains of ExsA, the transcriptional activator of the *Pseudomonas aeruginosa* type III secretion system. *J. Bacteriol.* **191**:3811–3821.

Brutinel, E. D., C. A. Vakulskas, and T. L. Yahr. 2010. ExsD inhibits expression of the *Pseudomonas aeruginosa* type III secretion system by disrupting ExsA self-association and DNA binding activity. *J. Bacteriol.* **192**:1479–1486.

Brutinel, E. D., and T. L. Yahr. 2008. Control of gene expression by type III secretory activity. *Curr. Opin. Microbiol.* **11**:128–133.

Burns, J. L., R. L. Gibson, S. McNamara, D. Yim, J. Emerson, M. Rosenfeld, P. Hiatt, K. McCoy, R. Castile, A. L. Smith, and B. W. Ramsey. 2001. Longitudinal assessment of *Pseudomonas aeruginosa* in young children with cystic fibrosis. *J. Infect. Dis.* **183**:444–452.

Burrowes, E., A. Abbas, A. O'Neill, C. Adams, and F. O'Gara. 2005. Characterisation of the regulatory RNA RsmB from *Pseudomonas aeruginosa* PAO1. *Res. Microbiol.* **156**:7–16.

Burrowes, E., C. Baysse, C. Adams, and F. O'Gara. 2006. Influence of the regulatory protein RsmA on cellular functions in *Pseudomonas aeruginosa* PAO1, as revealed by transcriptome analysis. *Microbiology* **152**:405–418.

Bustamante, V. H., F. J. Santana, E. Calva, and J. L. Puente. 2001. Transcriptional regulation of type III secretion genes in enteropathogenic *Escherichia coli*: Ler antagonizes H-NS-dependent repression. *Mol. Microbiol.* **39**:664–678.

Caille, O., C. Rossier, and K. Perron. 2007. A copper-activated two-component system interacts with zinc and imipenem resistance in *Pseudomonas aeruginosa*. *J. Bacteriol.* **189**:4561–4568.

Cambronne, E. D., L. W. Cheng, and O. Schneewind. 2000. LcrQ/YscM1, regulators of the *Yersinia yop* virulon, are injected into host cells by a chaperone-dependent mechanism. *Mol. Microbiol.* **37**:263–273.

Cambronne, E. D., J. A. Sorg, and O. Schneewind. 2004. Binding of SycH chaperone to YscM1 and YscM2 activates effector *yop* expression in *Yersinia enterocolitica*. *J. Bacteriol.* **186**:829–841.

Castang, S., H. R. McManus, K. H. Turner, and S. L. Dove. 2008. H-NS family members function coordinately in an opportunistic pathogen. *Proc. Natl. Acad. Sci. USA* **105**:18947–18952.

Chamnongpol, S., M. Cromie, and E. A. Groisman. 2003. Mg^{2+} sensing by the Mg^{2+} sensor PhoQ of *Salmonella enterica*. *J. Mol. Biol.* **325**:795–807.

Chugani, S., and E. P. Greenberg. 2010. LuxR homolog-independent gene regulation by acyl-homoserine lactones in *Pseudomonas aeruginosa*. *Proc. Natl. Acad. Sci. USA* **107**:10673–10678.

Coburn, B., I. Sekirov, and B. B. Finlay. 2007. Type III secretion systems and disease. *Clin. Microbiol. Rev.* **20**:535–549.

Cohen, J. D., J. P. Slovin, and A. M. Hendrickson. 2003. Two genetically discrete pathways convert tryptophan to auxin: more redundancy in auxin biosynthesis. *Trends Plant Sci.* **8**:197–199.

Cornelis, G. R., C. Sluiters, I. Delor, D. Geib, K. Kaniga, C. Lambert de Rouvroit, M. P. Sory, J. C. Vanooteghem, and T. Michiels. 1991. *ymoA*, a *Yersinia enterocolitica* chromosomal gene modulating the expression of virulence functions. *Mol. Microbiol.* **5**:1023–1034.

Dacheux, D., I. Attree, C. Schneider, and B. Toussaint. 1999. Cell death of human polymorphonuclear neutrophils induced by a *Pseudomonas aeruginosa* cystic fibrosis isolate requires a functional type III secretion system. *Infect. Immun.* **67**:6164–6167.

Dacheux, D., I. Attree, and B. Toussaint. 2001. Expression of ExsA in *trans* confers type III secretion system-dependent cytotoxicity on noncytotoxic *Pseudomonas aeruginosa* cystic fibrosis isolates. *Infect. Immun.* **69**:538–542.

Dacheux, D., O. Epaulard, A. de Groot, B. Guery, R. Leberre, I. Attree, B. Polack, and B. Toussaint. 2002. Activation of the *Pseudomonas aeruginosa* type III secretion system requires an intact pyruvate dehydrogenase *aceAB* operon. *Infect. Immun.* **70**:3973–3977.

Dame, R. T., M. S. Luijsterburg, E. Krin, P. N. Bertin, R. Wagner, and G. J. Wuite. 2005. DNA bridging: a property shared among H-NS-like proteins. *J. Bacteriol.* **187**:1845–1848.

Dasgupta, N., A. Ashare, G. W. Hunninghake, and T. L. Yahr. 2006. Transcriptional induction of the *Pseudomonas aeruginosa* type III secretion system by low Ca^{2+} and host cell contact proceeds through two distinct signaling pathways. *Infect. Immun.* **74**:3334–3341.

Dasgupta, N., G. L. Lykken, M. C. Wolfgang, and T. L. Yahr. 2004. A novel anti-anti-activator mechanism regulates expression of the *Pseudomonas aeruginosa* type III secretion system. *Mol. Microbiol.* **53**:297–308.

Day, W. A., Jr., and A. T. Maurelli. 2001. *Shigella flexneri* LuxS quorum-sensing system modulates *virB* expression but is not essential for virulence. *Infect. Immun.* **69**:15–23.

Dean, P. 2011. Functional domains and motifs of bacterial type III effector proteins and their roles in infection. *FEMS Microbiol. Rev.* **35**:1100–1125.

Deane, J. E., P. Abrusci, S. Johnson, and S. M. Lea. 2010. Timing is everything: the regulation of type III secretion. *Cell. Mol. Life Sci.* **67**:1065–1075.

de Kievit, T. R., and B. H. Iglewski. 2000. Bacterial quorum sensing in pathogenic relationships. *Infect. Immun.* **68**:4839–4849.

Diggle, S. P., K. Winzer, A. Lazdunski, P. Williams, and M. Camara. 2002. Advancing the quorum in *Pseudomonas aeruginosa*: MvaT and the regulation of N-acylhomoserine lactone production and virulence gene expression. *J. Bacteriol.* **184**:2576–2586.

Dorman, C. J. 2007. H-NS, the genome sentinel. *Nat. Rev. Microbiol.* **5**:157–161.

Dorman, C. J., S. McKenna, and C. Beloin. 2001. Regulation of virulence gene expression in *Shigella flexneri*, a facultative intracellular pathogen. *Int. J. Med. Microbiol.* **291**:89–96.

Dubnau, D., and R. Losick. 2006. Bistability in bacteria. *Mol. Microbiol.* **61**:564–572.

Durand, J. M., and G. R. Bjork. 2009. Metabolic control through ornithine and uracil of epithelial cell invasion by *Shigella flexneri. Microbiology* **155**:2498–2508.

Durand, J. M., B. Dagberg, B. E. Uhlin, and G. R. Bjork. 2000. Transfer RNA modification, temperature and DNA superhelicity have a common target in the regulatory network of the virulence of *Shigella flexneri*: the expression of the *virF* gene. *Mol. Microbiol.* **35**:924–935.

Elsen, S., M. Ragno, and I. Attree. 2011. PtrA is a periplasmic protein involved in Cu tolerance in *Pseudomonas aeruginosa. J.Bacteriol.* **193**:3376–3378.

Engel, J., and P. Balachandran. 2009. Role of *Pseudomonas aeruginosa* type III effectors in disease. *Curr. Opin. Microbiol.* **12**:61–66.

Falconi, M., B. Colonna, G. Prosseda, G. Micheli, and C. O. Gualerzi. 1998. Thermoregulation of *Shigella* and *Escherichia coli* EIEC pathogenicity. A temperature-dependent structural transition of DNA modulates accessibility of *virF* promoter to transcriptional repressor H-NS. *EMBO J.* **17**:7033–7043.

Farrow, J. M., III, and E. C. Pesci. 2007. Two distinct pathways supply anthranilate as a precursor of the *Pseudomonas* quinolone signal. *J. Bacteriol.* **189**:3425–3433.

Filopon, D., A. Merieau, G. Bernot, J. P. Comet, R. Leberre, B. Guery, B. Polack, and J. Guespin-Michel. 2006. Epigenetic acquisition of inducibility of type III cytotoxicity in *P. aeruginosa. BMC Bioinformatics* **7**:272.

Fito-Boncompte, L., A. Chapalain, E. Bouffartigues, H. Chaker, O. Lesouhaitier, G. Gicquel, A. Bazire, A. Madi, N. Connil, W. Veron, L. Taupin, B. Toussaint, P. Cornelis, Q. Wei, K. Shioya, E. Deziel, M. G. Feuilloley, N. Orange, A. Dufour, and S. Chevalier. 2011. Full virulence of *Pseudomonas aeruginosa* requires OprF. *Infect. Immun.* **79**:1176–1186.

Fletcher, M. P., S. P. Diggle, M. Camara, and P. Williams. 2007. Biosensor-based assays for PQS, HHQ and related 2-alkyl-4-quinolone quorum sensing signal molecules. *Nat. Protoc.* **2**:1254–1262.

Francis, M. S., H. Wolf-Watz, and A. Forsberg. 2002. Regulation of type III secretion systems. *Curr. Opin. Microbiol.* **5**:166–172.

Frank, D. W. 1997. The exoenzyme S regulon of *Pseudomonas aeruginosa. Mol. Microbiol.* **26**:621–629.

Fuchs, E. L., E. D. Brutinel, A. K. Jones, N. B. Fulcher, M. L. Urbanowski, T. L. Yahr, and M. C. Wolfgang. 2010. The *Pseudomonas aeruginosa* Vfr regulator controls global virulence factor expression through cyclic AMP-dependent and -independent mechanisms. *J. Bacteriol.* **192**:3553–3564.

Galan, J. E., and H. Wolf-Watz. 2006. Protein delivery into eukaryotic cells by type III secretion machines. *Nature* **444**:567–573.

Gantois, I., R. Ducatelle, F. Pasmans, F. Haesebrouck, I. Hautefort, A. Thompson, J. C. Hinton, and F. Van Immerseel. 2006. Butyrate specifically down-regulates *Salmonella* pathogenicity island 1 gene expression. *Appl. Environ. Microbiol.* **72**:946–949.

Garcia Vescovi, E., F. C. Soncini, and E. A. Groisman. 1996. Mg²⁺ as an extracellular signal: environmental regulation of *Salmonella* virulence. *Cell* **84**:165–174.

Goguen, J. D., J. Yother, and S. C. Straley. 1984. Genetic analysis of the low calcium response in *Yersinia pestis* Mu d1(Ap *lac*) insertion mutants. *J. Bacteriol.* **160**:842–848.

Gooderham, W. J., and R. E. Hancock. 2009. Regulation of virulence and antibiotic resistance by two-component regulatory systems in *Pseudomonas aeruginosa. FEMS Microbiol. Rev.* **33**:279–294.

Goodman, A. L., B. Kulasekara, A. Rietsch, D. Boyd, R. S. Smith, and S. Lory. 2004. A signaling network reciprocally regulates genes associated with acute infection and chronic persistence in *Pseudomonas aeruginosa. Dev. Cell* **7**:745–754.

Goodman, A. L., M. Merighi, M. Hyodo, I. Ventre, A. Filloux, and S. Lory. 2009. Direct interaction between sensor kinase proteins mediates acute and chronic disease phenotypes in a bacterial pathogen. *Genes Dev.* **23**:249–259.

Govan, J. R., and V. Deretic. 1996. Microbial pathogenesis in cystic fibrosis: mucoid *Pseudomonas aeruginosa* and *Burkholderia cepacia. Microbiol. Rev.* **60**:539–574.

Groisman, E. A. 2001. The pleiotropic two-component regulatory system PhoP-PhoQ. *J. Bacteriol.* **183**:1835–1842.

Ha, U., and S. Jin. 2001. Growth phase-dependent invasion of *Pseudomonas aeruginosa* and its survival within HeLa cells. *Infect. Immun.* **69**:4398–4406.

Ha, U. H., J. Kim, H. Badrane, J. Jia, H. V. Baker, D. Wu, and S. Jin. 2004. An in vivo inducible gene of *Pseudomonas aeruginosa* encodes an anti-ExsA to suppress the type III secretion system. *Mol. Microbiol.* **54**:307–320.

Ha, U. H., Y. Wang, and S. Jin. 2003. DsbA of *Pseudomonas aeruginosa* is essential for multiple virulence factors. *Infect. Immun.* **71**:1590–1595.

Hamood, A. N., J. A. Colmer, U. A. Ochsner, and M. L. Vasil. 1996. Isolation and characterization of a *Pseudomonas aeruginosa* gene, *ptxR*, which positively regulates exotoxin A production. *Mol. Microbiol.* **21**:97–110.

Hauser, A. R. 2009. The type III secretion system of *Pseudomonas aeruginosa*: infection by injection. *Nat. Rev. Microbiol.* **7**:654–665.

Heeb, S., and D. Haas. 2001. Regulatory roles of the GacS/GacA two-component system in plant-associated and other gram-negative bacteria. *Mol. Plant-Microbe Interact.* **14**:1351–1363.

Heurlier, K., F. Williams, S. Heeb, C. Dormond, G. Pessi, D. Singer, M. Camara, P. Williams, and D. Haas. 2004. Positive control of swarming, rhamnolipid synthesis, and lipase production by the posttranscriptional RsmA/RsmZ system in *Pseudomonas aeruginosa* PAO1. *J. Bacteriol.* **186**:2936–2945.

Hogardt, M., and J. Heesemann. 2010. Adaptation of *Pseudomonas aeruginosa* during persistence in the cystic fibrosis lung. *Int. J. Med. Microbiol.* **300**:557–562.

Hogardt, M., M. Roeder, A. M. Schreff, L. Eberl, and J. Heesemann. 2004. Expression of *Pseudomonas aeruginosa exoS* is controlled by quorum sensing and RpoS. *Microbiology* **150**:843–851.

Hornef, M. W., A. Roggenkamp, A. M. Geiger, M. Hogardt, C. A. Jacobi, and J. Heesemann. 2000. Triggering the ExoS regulon of *Pseudomonas aeruginosa*: a GFP-reporter analysis of exoenzyme (Exo) S, ExoT and ExoU synthesis. *Microb. Pathog.* **29**:329–343.

Hueck, C. J. 1998. Type III protein secretion systems in bacterial pathogens of animals and plants. *Microbiol. Mol. Biol. Rev.* **62**:379–433.

Hurt, J. K., T. J. McQuade, A. Emanuele, M. J. Larsen, and G. A. Garcia. 2010. High-throughput screening of the virulence regulator VirF: a novel antibacterial target for shigellosis. *J. Biomol. Screen.* **15**:379–387.

Inclan, Y. F., M. J. Huseby, and J. N. Engel. 2011. FimL regulates cAMP synthesis in *Pseudomonas aeruginosa. PLoS One* **6**:e15867.

Jain, M., M. Bar-Meir, S. McColley, J. Cullina, E. Potter, C. Powers, M. Prickett, R. Seshadri, B. Jovanovic, A. Petrocheilou, J. D. King, and A. R. Hauser. 2008. Evolution of *Pseudomonas aeruginosa* type III secretion in cystic fibrosis: a paradigm of chronic infection. *Transl. Res.* **152**:257–264.

Jain, M., D. Ramirez, R. Seshadri, J. F. Cullina, C. A. Powers, G. S. Schulert, M. Bar-Meir, C. L. Sullivan, S. A. McColley, and A. R. Hauser. 2004. Type III secretion phenotypes of *Pseudomonas aeruginosa* strains change during infection of individuals with cystic fibrosis. *J. Clin. Microbiol.* **42**:5229–5237.

Jin, Y., H. Yang, M. Qiao, and S. Jin. 2011. MexT regulates the type III secretion system through MexS and PtrC in *Pseudomonas aeruginosa*. *J. Bacteriol.* **193**:399–410.

Jones, A. K., N. B. Fulcher, G. J. Balzer, M. L. Urbanowski, C. L. Pritchett, M. J. Schurr, T. L. Yahr, and M. C. Wolfgang. 2010. Activation of the *Pseudomonas aeruginosa* AlgU regulon through *mucA* mutation inhibits cyclic AMP/Vfr signaling. *J. Bacteriol.* **192**:5709–5717.

Juhas, M., L. Eberl, and B. Tummler. 2005. Quorum sensing: the power of cooperation in the world of *Pseudomonas*. *Environ. Microbiol.* **7**:459–471.

Kanamaru, K., I. Tatsuno, T. Tobe, and C. Sasakawa. 2000. SdiA, an *Escherichia coli* homologue of quorum-sensing regulators, controls the expression of virulence factors in enterohaemorrhagic *Escherichia coli* O157:H7. *Mol. Microbiol.* **38**:805–816.

Kang, Y., V. V. Lunin, T. Skarina, A. Savchenko, M. J. Schurr, and T. T. Hoang. 2009. The long-chain fatty acid sensor, PsrA, modulates the expression of *rpoS* and the type III secretion *exsCEBA* operon in *Pseudomonas aeruginosa*. *Mol. Microbiol.* **73**:120–136.

Kang, Y., D. T. Nguyen, M. S. Son, and T. T. Hoang. 2008. The *Pseudomonas aeruginosa* PsrA responds to long-chain fatty acid signals to regulate the *fadBA5* beta-oxidation operon. *Microbiology* **154**:1584–1598.

Kendall, M. M., D. A. Rasko, and V. Sperandio. 2007. Global effects of the cell-to-cell signaling molecules autoinducer-2, autoinducer-3, and epinephrine in a *luxS* mutant of enterohemorrhagic *Escherichia coli*. *Infect. Immun.* **75**:4875–4884.

Kenny, B., A. Abe, M. Stein, and B. B. Finlay. 1997. Enteropathogenic *Escherichia coli* protein secretion is induced in response to conditions similar to those in the gastrointestinal tract. *Infect. Immun.* **65**:2606–2612.

Kim, J., K. Ahn, S. Min, J. Jia, U. Ha, D. Wu, and S. Jin. 2005. Factors triggering type III secretion in *Pseudomonas aeruginosa*. *Microbiology* **151**:3575–3587.

Kojic, M., C. Aguilar, and V. Venturi. 2002. TetR family member PsrA directly binds the *Pseudomonas rpoS* and *psrA* promoters. *J. Bacteriol.* **184**:2324–2330.

Kuchma, S. L., J. P. Connolly, and G. A. O'Toole. 2005. A three-component regulatory system regulates biofilm maturation and type III secretion in *Pseudomonas aeruginosa*. *J. Bacteriol.* **187**:1441–1454.

Kurnasov, O., L. Jablonski, B. Polanuyer, P. Dorrestein, T. Begley, and A. Osterman. 2003. Aerobic tryptophan degradation pathway in bacteria: novel kynurenine formamidase. *FEMS Microbiol. Lett.* **227**:219–227.

Laskowski, M. A., and B. I. Kazmierczak. 2006. Mutational analysis of RetS, an unusual sensor kinase-response regulator hybrid required for *Pseudomonas aeruginosa* virulence. *Infect. Immun.* **74**:4462–4473.

Laskowski, M. A., E. Osborn, and B. I. Kazmierczak. 2004. A novel sensor kinase-response regulator hybrid regulates type III secretion and is required for virulence in *Pseudomonas aeruginosa*. *Mol. Microbiol.* **54**:1090–1103.

Latifi, A., M. Foglino, K. Tanaka, P. Williams, and A. Lazdunski. 1996. A hierarchical quorum-sensing cascade in *Pseudomonas aeruginosa* links the transcriptional activators LasR and RhlR (VsmR) to expression of the stationary-phase sigma factor RpoS. *Mol. Microbiol.* **21**:1137–1146.

Lawhon, S. D., J. G. Frye, M. Suyemoto, S. Porwollik, M. McClelland, and C. Altier. 2003. Global regulation by CsrA in *Salmonella typhimurium*. *Mol. Microbiol.* **48**:1633–1645.

Lawhon, S. D., R. Maurer, M. Suyemoto, and C. Altier. 2002. Intestinal short-chain fatty acids alter *Salmonella typhimurium* invasion gene expression and virulence through BarA/SirA. *Mol. Microbiol.* **46**:1451–1464.

Lee, V. T., S. K. Mazmanian, and O. Schneewind. 2001. A program of *Yersinia enterocolitica* type III secretion reactions is activated by specific signals. *J. Bacteriol.* **183**:4970–4978.

Lee, V. T., R. S. Smith, B. Tummler, and S. Lory. 2005. Activities of *Pseudomonas aeruginosa* effectors secreted by the type III secretion system in vitro and during infection. *Infect. Immun.* **73**:1695–1705.

Lesser, C. F., and J. M. Leong. 2011. Bacterial scaffolds assemble novel higher-order complexes to reengineer eukaryotic cell processes. *Sci. Signal.* **4**:pe32.

Lilic, M., M. Vujanac, and C. E. Stebbins. 2006. A common structural motif in the binding of virulence factors to bacterial secretion chaperones. *Mol. Cell* **21**:653–664.

Linares, J. F., J. A. Lopez, E. Camafeita, J. P. Albar, F. Rojo, and J. L. Martinez. 2005. Overexpression of the multidrug efflux pumps MexCD-OprJ and MexEF-OprN is associated with a reduction of type III secretion in *Pseudomonas aeruginosa*. *J. Bacteriol.* **187**:1384–1391.

Linder, J. U., and J. E. Schultz. 2003. The class III adenylyl cyclases: multi-purpose signalling modules. *Cell. Signal.* **15**:1081–1089.

Lory, S., M. Merighi, and M. Hyodo. 2009. Multiple activities of c-di-GMP in *Pseudomonas aeruginosa*. *Nucleic Acids Symp. Ser.* **53**:51–52.

Lory, S., M. Wolfgang, V. Lee, and R. Smith. 2004. The multitalented bacterial adenylate cyclases. *Int. J. Med. Microbiol.* **293**:479–482.

Lu, J., X. Huang, K. Li, S. Li, M. Zhang, Y. Wang, H. Jiang, and Y. Xu. 2009. LysR family transcriptional regulator PqsR as repressor of pyoluteorin biosynthesis and activator of phenazine-1-carboxylic acid biosynthesis in *Pseudomonas* sp. M18. *J. Biotechnol.* **143**:1–9.

Lykken, G. L., G. Chen, E. D. Brutinel, L. Chen, and T. L. Yahr. 2006. Characterization of ExsC and ExsD self-association and heterocomplex formation. *J. Bacteriol.* **188**:6832–6840.

Marteyn, B., N. P. West, D. F. Browning, J. A. Cole, J. G. Shaw, F. Palm, J. Mounier, M. C. Prevost, P. Sansonetti, and C. M. Tang. 2010. Modulation of *Shigella* virulence in response to available oxygen in vivo. *Nature* **465**:355–358.

Maurelli, A. T., B. Blackmon, and R. Curtiss, III. 1984. Temperature-dependent expression of virulence genes in *Shigella* species. *Infect. Immun.* **43**:195–201.

McCaw, M. L., G. L. Lykken, P. K. Singh, and T. L. Yahr. 2002. ExsD is a negative regulator of the *Pseudomonas aeruginosa* type III secretion regulon. *Mol. Microbiol.* **46**:1123–1133.

Mena, A., E. E. Smith, J. L. Burns, D. P. Speert, S. M. Moskowitz, J. L. Perez, and A. Oliver. 2008. Genetic adaptation of *Pseudomonas aeruginosa* to the airways of cystic fibrosis patients is catalyzed by hypermutation. *J. Bacteriol.* **190**:7910–7917.

Menard, R., P. Sansonetti, and C. Parsot. 1994. The secretion of the *Shigella flexneri* Ipa invasins is activated by epithelial cells and controlled by IpaB and IpaD. *EMBO J.* **13**:5293–5302.

Mettetal, J. T., and A. van Oudenaarden. 2007. Microbiology. Necessary noise. *Science* **317**:463–464.

Michiels, T., P. Wattiau, R. Brasseur, J. M. Ruysschaert, and G. Cornelis. 1990. Secretion of Yop proteins by yersiniae. *Infect. Immun.* **58**:2840–2849.

Mikkelsen, H., N. J. Bond, M. E. Skindersoe, M. Givskov, K. S. Lilley, and M. Welch. 2009. Biofilms and type III secretion are not mutually exclusive in *Pseudomonas aeruginosa*. *Microbiology* **155**:687–698.

Mulcahy, H., J. O'Callaghan, E. P. O'Grady, C. Adams, and F. O'Gara. 2006. The posttranscriptional regulator RsmA plays a role in the interaction between *Pseudomonas aeruginosa* and human airway epithelial cells by positively regulating the type III secretion system. *Infect. Immun.* **74:**3012–3015.

Murray, T. S., M. Egan, and B. I. Kazmierczak. 2007. *Pseudomonas aeruginosa* chronic colonization in cystic fibrosis patients. *Curr. Opin. Pediatr.* **19:**83–88.

Nataro, J. P., and J. B. Kaper. 1998. Diarrheagenic *Escherichia coli*. *Clin. Microbiol. Rev.* **11:**142–201.

O'Callaghan, J., F. J. Reen, C. Adams, and F. O'Gara. 26 August 2011. Low oxygen induces the T3SS in *Pseudomonas aeruginosa* via modulation of the small RNAs, *rsmZ* and *rsmY*. *Microbiology* doi:10.1099/mic.0.052050-0.

Pallen, M. J., S. A. Beatson, and C. M. Bailey. 2005. Bioinformatics, genomics and evolution of non-flagellar type-III secretion systems: a Darwinian perspective. *FEMS Microbiol. Rev.* **29:**201–229.

Parsot, C., E. Ageron, C. Penno, M. Mavris, K. Jamoussi, H. d'Hauteville, P. Sansonetti, and B. Demers. 2005. A secreted anti-activator, OspD1, and its chaperone, Spa15, are involved in the control of transcription by the type III secretion apparatus activity in *Shigella flexneri*. *Mol. Microbiol.* **56:**1627–1635.

Pesci, E. C., J. B. Milbank, J. P. Pearson, S. McKnight, A. S. Kende, E. P. Greenberg, and B. H. Iglewski. 1999. Quinolone signaling in the cell-to-cell communication system of *Pseudomonas aeruginosa*. *Proc. Natl. Acad. Sci. USA* **96:**11229–11234.

Pessi, G., F. Williams, Z. Hindle, K. Heurlier, M. T. Holden, M. Camara, D. Haas, and P. Williams. 2001. The global posttranscriptional regulator RsmA modulates production of virulence determinants and N-acylhomoserine lactones in *Pseudomonas aeruginosa*. *J. Bacteriol.* **183:**6676–6683.

Pettersson, J., R. Nordfelth, E. Dubinina, T. Bergman, M. Gustafsson, K. E. Magnusson, and H. Wolf-Watz. 1996. Modulation of virulence factor expression by pathogen target cell contact. *Science* **273:**1231–1233.

Platt, M. D., M. J. Schurr, K. Sauer, G. Vazquez, I. Kukavica-Ibrulj, E. Potvin, R. C. Levesque, A. Fedynak, F. S. Brinkman, J. Schurr, S. H. Hwang, G. W. Lau, P. A. Limbach, J. J. Rowe, M. A. Lieberman, N. Barraud, J. Webb, S. Kjelleberg, D. F. Hunt, and D. J. Hassett. 2008. Proteomic, microarray, and signature-tagged mutagenesis analyses of anaerobic *Pseudomonas aeruginosa* at pH 6.5, likely representing chronic, late-stage cystic fibrosis airway conditions. *J. Bacteriol.* **190:**2739–2758.

Pollack, C., S. C. Straley, and M. S. Klempner. 1986. Probing the phagolysosomal environment of human macrophages with a Ca²⁺-responsive operon fusion in *Yersinia pestis*. *Nature* **322:**834–836.

Prouty, A. M., I. E. Brodsky, J. Manos, R. Belas, S. Falkow, and J. S. Gunn. 2004. Transcriptional regulation of *Salmonella enterica* serovar Typhimurium genes by bile. *FEMS Immunol. Med. Microbiol.* **41:**177–185.

Prunier, A. L., R. Schuch, R. E. Fernandez, K. L. Mumy, H. Kohler, B. A. McCormick, and A. T. Maurelli. 2007. *nadA* and *nadB* of *Shigella flexneri* 5a are antivirulence loci responsible for the synthesis of quinolinate, a small molecule inhibitor of *Shigella* pathogenicity. *Microbiology* **153:**2363–2372.

Raser, J. M., and E. K. O'Shea. 2005. Noise in gene expression: origins, consequences, and control. *Science* **309:**2010–2013.

Rietsch, A., and J. J. Mekalanos. 2006. Metabolic regulation of type III secretion gene expression in *Pseudomonas aeruginosa*. *Mol. Microbiol.* **59:**807–820.

Rietsch, A., I. Vallet-Gely, S. L. Dove, and J. J. Mekalanos. 2005. ExsE, a secreted regulator of type III secretion genes in *Pseudomonas aeruginosa*. *Proc. Natl. Acad. Sci. USA* **102:**8006–8011.

Rietsch, A., M. C. Wolfgang, and J. J. Mekalanos. 2004. Effect of metabolic imbalance on expression of type III secretion genes in *Pseudomonas aeruginosa*. *Infect. Immun.* **72:**1383–1390.

Rodrigue, A., Y. Quentin, A. Lazdunski, V. Mejean, and M. Foglino. 2000. Two-component systems in *Pseudomonas aeruginosa*: why so many? *Trends Microbiol.* **8:**498–504.

Rohde, J. R., J. M. Fox, and S. A. Minnich. 1994. Thermoregulation in *Yersinia enterocolitica* is coincident with changes in DNA supercoiling. *Mol. Microbiol.* **12:**187–199.

Rohde, J. R., X. S. Luan, H. Rohde, J. M. Fox, and S. A. Minnich. 1999. The *Yersinia enterocolitica* pYV virulence plasmid contains multiple intrinsic DNA bends which melt at 37 degrees C. *J. Bacteriol.* **181:**4198–4204.

Rosenshine, I., S. Ruschkowski, and B. B. Finlay. 1996. Expression of attaching/effacing activity by enteropathogenic *Escherichia coli* depends on growth phase, temperature, and protein synthesis upon contact with epithelial cells. *Infect. Immun.* **64:**966–973.

Rumbaugh, K. P., J. A. Griswold, and A. N. Hamood. 2000. The role of quorum sensing in the in vivo virulence of *Pseudomonas aeruginosa*. *Microbes Infect.* **2:**1721–1731.

Schaber, J. A., N. L. Carty, N. A. McDonald, E. D. Graham, R. Cheluvappa, J. A. Griswold, and A. N. Hamood. 2004. Analysis of quorum sensing-deficient clinical isolates of *Pseudomonas aeruginosa*. *J. Med. Microbiol.* **53:**841–853.

Schmid, A., W. Neumayer, K. Trulzsch, L. Israel, A. Imhof, M. Roessle, G. Sauer, S. Richter, S. Lauw, E. Eylert, W. Eisenreich, J. Heesemann, and G. Wilharm. 2009. Cross-talk between type three secretion system and metabolism in *Yersinia*. *J. Biol. Chem.* **284:**12165–12177.

Shen, D. K., D. Filopon, H. Chaker, S. Boullanger, M. Derouazi, B. Polack, and B. Toussaint. 2008. High-cell-density regulation of the *Pseudomonas aeruginosa* type III secretion system: implications for tryptophan catabolites. *Microbiology* **154:**2195–2208.

Shen, D. K., D. Filopon, L. Kuhn, B. Polack, and B. Toussaint. 2006. PsrA is a positive transcriptional regulator of the type III secretion system in *Pseudomonas aeruginosa*. *Infect. Immun.* **74:**1121–1129.

Shin, S., M. P. Castanie-Cornet, J. W. Foster, J. A. Crawford, C. Brinkley, and J. B. Kaper. 2001. An activator of glutamate decarboxylase genes regulates the expression of enteropathogenic *Escherichia coli* virulence genes through control of the plasmid-encoded regulator, Per. *Mol. Microbiol.* **41:**1133–1150.

Singh, G., B. Wu, M. S. Baek, A. Camargo, A. Nguyen, N. A. Slusher, R. Srinivasan, J. P. Wiener-Kronish, and S. V. Lynch. 2010. Secretion of *Pseudomonas aeruginosa* type III cytotoxins is dependent on pseudomonas quinolone signal concentration. *Microb. Pathog.* **49:**196–203.

Smith, R. S., and B. H. Iglewski. 2003. *P. aeruginosa* quorum-sensing systems and virulence. *Curr. Opin. Microbiol.* **6:**56–60.

Smith, R. S., M. C. Wolfgang, and S. Lory. 2004. An adenylate cyclase-controlled signaling network regulates *Pseudomonas aeruginosa* virulence in a mouse model of acute pneumonia. *Infect. Immun.* **72:**1677–1684.

Sonnleitner, E., and D. Haas. 2011. Small RNAs as regulators of primary and secondary metabolism in *Pseudomonas* species. *Appl. Microbiol. Biotechnol.* **91:**63–79.

Sperandio, V., C. C. Li, and J. B. Kaper. 2002. Quorum-sensing *Escherichia coli* regulator A: a regulator of the LysR family involved in the regulation of the locus of enterocyte effacement pathogenicity island in enterohemorrhagic *E. coli*. *Infect. Immun.* **70:**3085–3093.

Sperandio, V., J. L. Mellies, W. Nguyen, S. Shin, and J. B. Kaper. 1999. Quorum sensing controls expression of the type III secretion gene transcription and protein secretion in enterohemorrhagic and enteropathogenic *Escherichia coli*. *Proc. Natl. Acad. Sci. USA* **96:**15196–15201.

Sperandio, V., A. G. Torres, B. Jarvis, J. P. Nataro, and J. B. Kaper. 2003. Bacteria-host communication: the language of hormones. *Proc. Natl. Acad. Sci. USA* **100:**8951–8956.

Stainier, I., M. Iriarte, and G. R. Cornelis. 1997. YscM1 and YscM2, two *Yersinia enterocolitica* proteins causing downregulation of *yop* transcription. *Mol. Microbiol.* **26:**833–843.

Stover, C. K., X. Q. Pham, A. L. Erwin, S. D. Mizoguchi, P. Warrener, M. J. Hickey, F. S. Brinkman, W. O. Hufnagle, D. J. Kowalik, M. Lagrou, R. L. Garber, L. Goltry, E. Tolentino, S. Westbrock-Wadman, Y. Yuan, L. L. Brody, S. N. Coulter, K. R. Folger, A. Kas, K. Larbig, R. Lim, K. Smith, D. Spencer, G. K. Wong, Z. Wu, I. T. Paulsen, J. Reizer, M. H. Saier, R. E. Hancock, S. Lory, and M. V. Olson. 2000. Complete genome sequence of *Pseudomonas aeruginosa* PAO1, an opportunistic pathogen. *Nature* **406:**959–964.

Thibault, J., E. Faudry, C. Ebel, I. Attree, and S. Elsen. 2009. Anti-activator ExsD forms a 1:1 complex with ExsA to inhibit transcription of type III secretion operons. *J. Biol. Chem.* **284:**15762–15770.

Throup, J. P., M. Camara, G. S. Briggs, M. K. Winson, S. R. Chhabra, B. W. Bycroft, P. Williams, and G. S. Stewart. 1995. Characterisation of the *yenI/yenR* locus from *Yersinia enterocolitica* mediating the synthesis of two N-acylhomoserine lactone signal molecules. *Mol. Microbiol.* **17:**345–356.

Troisfontaines, P., and G. R. Cornelis. 2005. Type III secretion: more systems than you think. *Physiology* (Bethesda) **20:**326–339.

Trunk, K., B. Benkert, N. Quack, R. Munch, M. Scheer, J. Garbe, L. Jansch, M. Trost, J. Wehland, J. Buer, M. Jahn, M. Schobert, and D. Jahn. 2010. Anaerobic adaptation in *Pseudomonas aeruginosa*: definition of the Anr and Dnr regulons. *Environ. Microbiol.* **12:**1719–1733.

Turner, K. H., I. Vallet-Gely, and S. L. Dove. 2009. Epigenetic control of virulence gene expression in *Pseudomonas aeruginosa* by a LysR-type transcription regulator. *PLoS Genet.* **5:**e1000779.

Umanski, T., I. Rosenshine, and D. Friedberg. 2002. Thermoregulated expression of virulence genes in enteropathogenic *Escherichia coli*. *Microbiology* **148:**2735–2744.

Urbanowski, M. L., G. L. Lykken, and T. L. Yahr. 2005. A secreted regulatory protein couples transcription to the secretory activity of the *Pseudomonas aeruginosa* type III secretion system. *Proc. Natl. Acad. Sci. USA* **102:**9930–9935.

Vallis, A. J., T. L. Yahr, J. T. Barbieri, and D. W. Frank. 1999. Regulation of ExoS production and secretion by *Pseudomonas aeruginosa* in response to tissue culture conditions. *Infect. Immun.* **67:**914–920.

Van Alst, N. E., M. Wellington, V. L. Clark, C. G. Haidaris, and B. H. Iglewski. 2009. Nitrite reductase NirS is required for type III secretion system expression and virulence in the human monocyte cell line THP-1 by *Pseudomonas aeruginosa*. *Infect. Immun.* **77:**4446–4454.

Ventre, I., A. L. Goodman, I. Vallet-Gely, P. Vasseur, C. Soscia, S. Molin, S. Bleves, A. Lazdunski, S. Lory, and A. Filloux. 2006. Multiple sensors control reciprocal expression of *Pseudomonas aeruginosa* regulatory RNA and virulence genes. *Proc. Natl. Acad. Sci. USA* **103:**171–176.

Vogelaar, N. J., X. Jing, H. H. Robinson, and F. D. Schubot. 2010. Analysis of the crystal structure of the ExsC·ExsE complex reveals distinctive binding interactions of the *Pseudomonas aeruginosa* type III secretion chaperone ExsC with ExsE and ExsD. *Biochemistry* **49:**5870–5879.

Wagner, V. E., D. Bushnell, L. Passador, A. I. Brooks, and B. H. Iglewski. 2003. Microarray analysis of *Pseudomonas aeruginosa* quorum-sensing regulons: effects of growth phase and environment. *J. Bacteriol.* **185:**2080–2095.

Wangdi, T., L. A. Mijares, and B. I. Kazmierczak. 2010. In vivo discrimination of type 3 secretion system-positive and -negative *Pseudomonas aeruginosa* via a caspase-1-dependent pathway. *Infect. Immun.* **78:**4744–4753.

Watarai, M., T. Tobe, M. Yoshikawa, and C. Sasakawa. 1995. Contact of *Shigella* with host cells triggers release of Ipa invasins and is an essential function of invasiveness. *EMBO J.* **14:**2461–2470.

Waters, C. M., and B. L. Bassler. 2005. Quorum sensing: cell-to-cell communication in bacteria. *Annu. Rev. Cell Dev. Biol.* **21:**319–346.

Whitchurch, C. B., S. A. Beatson, J. C. Comolli, T. Jakobsen, J. L. Sargent, J. J. Bertrand, J. West, M. Klausen, L. L. Waite, P. J. Kang, T. Tolker-Nielsen, J. S. Mattick, and J. N. Engel. 2005. *Pseudomonas aeruginosa fimL* regulates multiple virulence functions by intersecting with Vfr-modulated pathways. *Mol. Microbiol.* **55:**1357–1378.

Whiteley, M., K. M. Lee, and E. P. Greenberg. 1999. Identification of genes controlled by quorum sensing in *Pseudomonas aeruginosa*. *Proc. Natl. Acad. Sci. USA* **96:**13904–13909.

Whiteley, M., M. R. Parsek, and E. P. Greenberg. 2000. Regulation of quorum sensing by RpoS in *Pseudomonas aeruginosa*. *J. Bacteriol.* **182:**4356–4360.

Winstanley, C., and J. L. Fothergill. 2009. The role of quorum sensing in chronic cystic fibrosis *Pseudomonas aeruginosa* infections. *FEMS Microbiol. Lett.* **290:**1–9.

Wolfgang, M. C., V. T. Lee, M. E. Gilmore, and S. Lory. 2003. Coordinate regulation of bacterial virulence genes by a novel adenylate cyclase-dependent signaling pathway. *Dev. Cell* **4:**253–263.

Wu, L., O. Estrada, O. Zaborina, M. Bains, L. Shen, J. E. Kohler, N. Patel, M. W. Musch, E. B. Chang, Y. X. Fu, M. A. Jacobs, M. I. Nishimura, R. E. Hancock, J. R. Turner, and J. C. Alverdy. 2005. Recognition of host immune activation by *Pseudomonas aeruginosa*. *Science* **309:**774–777.

Wu, W., and S. Jin. 2005. PtrB of *Pseudomonas aeruginosa* suppresses the type III secretion system under the stress of DNA damage. *J. Bacteriol.* **187:**6058–6068.

Wulff-Strobel, C. R., A. W. Williams, and S. C. Straley. 2002. LcrQ and SycH function together at the Ysc type III secretion system in *Yersinia pestis* to impose a hierarchy of secretion. *Mol. Microbiol.* **43:**411–423.

Yahr, T. L., and D. W. Frank. 1994. Transcriptional organization of the *trans*-regulatory locus which controls exoenzyme S synthesis in *Pseudomonas aeruginosa*. *J. Bacteriol.* **176:**3832–3838.

Yahr, T. L., and M. C. Wolfgang. 2006. Transcriptional regulation of the *Pseudomonas aeruginosa* type III secretion system. *Mol. Microbiol.* **62:**631–640.

Zheng, Z., G. Chen, S. Joshi, E. D. Brutinel, T. L. Yahr, and L. Chen. 2007. Biochemical characterization of a regulatory cascade controlling transcription of the *Pseudomonas aeruginosa* type III secretion system. *J. Biol. Chem.* **282:**6136–6142.

Zhou, L., J. Wang, and L. H. Zhang. 2007. Modulation of bacterial type III secretion system by a spermidine transporter dependent signaling pathway. *PLoS One* **2:**e1291.

Zierler, M. K., and J. E. Galan. 1995. Contact with cultured epithelial cells stimulates secretion of *Salmonella typhimurium* invasion protein InvJ. *Infect. Immun.* **63:**4024–4028.

Zolfaghar, I., A. A. Angus, P. J. Kang, A. To, D. J. Evans, and S. M. Fleiszig. 2005. Mutation of *retS*, encoding a putative hybrid two-component regulatory protein in *Pseudomonas aeruginosa*, attenuates multiple virulence mechanisms. *Microbes Infect.* **7:**1305–1316.

Regulation of Bacterial Virulence
Edited by Michael L. Vasil and Andrew J. Darwin
© 2013 ASM Press, Washington, DC   doi:10.1128/9781555818524.ch18

Chapter 18

Regulation of Bacterial Type IV Secretion

JENNY A. LAVERDE-GOMEZ, MAYUKH SARKAR, AND PETER J. CHRISTIE

INTRODUCTION

Many bacterial pathogens rely on type IV secretion systems (T4SSs) for delivery of effector proteins to eukaryotic hosts during establishment of infection (Baron and Coombes, 2007; Alvarez-Martinez and Christie, 2009; Llosa et al., 2009; Backert and Clyne, 2011; Fischer, 2011; Nagai and Kubori, 2011). In their roles as protein delivery machines, T4SSs functionally resemble the type III secretion systems (T3SSs) discussed in the previous chapter (chapter 17) (see also Coburn et al., 2007, and Diaz et al., 2011). The T3SSs and T4SSs represent a fascinating example of surface machines that have evolved for similar purposes but from distinct ancestral organelles. The T3SSs evolved from the flagellum—a motility motor—and the T4SSs from the conjugation apparatus—a sex machine (Christie et al., 2005; Frank et al., 2005; Medini et al., 2006; Erhardt et al., 2010). Bacterial pathogens rely on both systems for delivery of effectors during infection, but the T3SSs and T4SSs also display important differences. Whereas the T3SSs translocate only monomeric proteins and are found almost exclusively in gram-negative species, the T4SSs translocate different types of substrates, including monomeric proteins, multimeric protein complexes, single-stranded DNA (ssDNA), and peptidoglycan. The T4SSs are also very broadly distributed among most, if not all, species of bacteria and some species of archaea. The T3SSs generally deliver substrates to specific eukaryotic target cells for a pathogenic outcome (Dean, 2011; Izore et al., 2011). While this is true for some T4SSs, others have the capacity to deliver DNA or protein substrates to a diverse range of bacterial, archaeal, and eukaryotic host cells (Alvarez-Martinez and Christie, 2009). A paradigmatic T4SS, the *Agrobacterium tumefaciens* VirB/VirD4 system, was adapted to deliver DNA and protein substrates to plant cells during infection, but in fact this system can

deliver substrates to a range of bacterial, fungal, plant, and human cell types (Christie, 2004).

In this chapter, we explore the question of how bacterial pathogens regulate the biogenesis and function of their T4SSs in pathogenic settings. First, we describe a regulatory cascade involving the perception of multiple signals exchanged between *A. tumefaciens* and its plant host. This signaling dialogue leads not only to infection of plant tissue but also to enhanced conjugative transfer of the virulence-associated tumor-inducing (pTi) plasmid. Second, we summarize the regulatory features of a large T4SS subfamily, the conjugation systems functioning in gram-negative and -positive species. This topic is timely and relevant to the chapter's focus on T4SSs and virulence because conjugation is intimately linked with pathogenesis as follows: (i) conjugative DNA transfer is the principal mechanism underlying the rapid dissemination of antibiotic resistance among and between pathogenic species in clinical settings, (ii) conjugation systems elaborate adhesive surface structures that promote attachment to and biofilm formation on abiotic (e.g., prosthetic devices) and biotic (e.g., human heart tissue) surfaces, and (iii) conjugation occurs in biofilms, resulting in propagation of fitness traits within and between coresident species. From a more basic science perspective, long-term studies of paradigmatic systems such as the *Escherichia coli* F and *Enterococcus faecalis* pCF10/pAD1 plasmid transfer systems have supplied many molecular details of regulatory mechanisms controlling conjugative DNA transfer. As detailed investigations of a more recently discovered T4SS subfamily, the effector translocators, proceed, it is increasingly evident that pathogens have evolved similar signaling/transduction pathways to regulate delivery of effector proteins to eukaryotic cells during infection.

In the third section of this chapter, we summarize regulatory features of the well-characterized

Jenny A. Laverde-Gomez, Mayukh Sarkar, and Peter J. Christie • Department of Microbiology and Molecular Genetics, University of Texas Medical School at Houston, 6431 Fannin St., Houston, TX 77030.

effector translocators of *Brucella* spp., *Legionella pneumophila*, and *Bartonella* spp., and also examine why and how these and other bacterial pathogens cross-regulate T4SSs and other surface motility or attachment devices such as flagella and type IV pili. Finally, we discuss posttranscriptional regulation of substrate-T4SS docking reactions and donor-target cell contacts. The overarching goal of this chapter is to identify mechanistic themes and variations that have evolved to regulate the myriad of T4SS activities exploited by bacterial pathogens during infection.

T4SS STRUCTURE AND FUNCTION

The T4SSs can be grouped into three subfamilies (Cascales and Christie, 2003). The conjugation systems are the largest subfamily, present in nearly all bacterial species and some archaeal species. These systems mediate the transfer of ssDNA from a bacterial donor to recipient cell through direct cell-to-contact and formation of mating pairs. These systems are responsible for rapid and widespread intra- and interspecies transmission of antibiotic resistance genes and

other virulence traits and, on a broader time scale, the plasticity and shaping of bacterial genomes. The "effector translocators" comprise a second subfamily, thus far identified only in gram-negative pathogens or symbionts. These systems translocate proteins and other effector molecules, including DNA and peptidoglycan, to eukaryotic target cells during the course of infection. Finally, the "DNA uptake/release" systems comprise a small subfamily of systems that exchange DNA with the extracellular milieu by mechanisms that do not require direct donor-target cell interactions.

In gram-negative bacteria, the T4SSs are composed of a dozen or more subunits synthesized from clustered and coexpressed gene sets (Alvarez-Martinez and Christie, 2009; Juhas et al., 2009). The conjugation systems elaborate a translocation channel and an extracellular filament termed the conjugative pilus (Fig. 1). The translocation channel spans both membranes and the intervening periplasm/cell wall, forming a gated channel through which DNA substrates pass. The conjugative pilus establishes initial contact with target cells. Following attachment, donor and target cells are brought into direct physical contact,

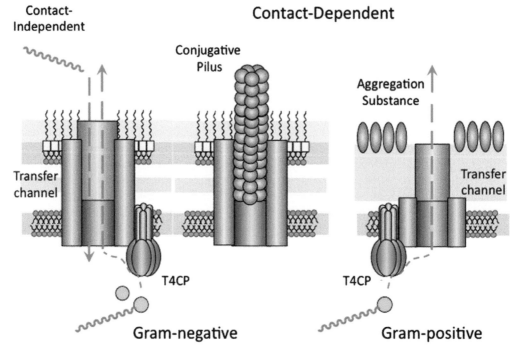

Figure 1. General architectures of T4SSs in gram-negative and -positive bacteria. The T4SSs of gram-negative bacteria are composed of a translocation channel and an extracellular filament, e.g., conjugative pilus, that may or may not be physically connected. These systems translocate substrates to prokaryotic or eukaryotic target cells by a contact-dependent mechanism. Conjugation systems translocate DNA substrates as ssDNA covalently bound at their 5′ ends to relaxase (yellow circle, green wavy line). Effector translocators deliver protein substrates (yellow circle) to target cells, often to aid in the infection process. Some T4SSs export DNA or protein substrates or take up DNA (green wavy line) from the extracellular milieu by a contact-independent mechanism. Gram-positive T4SSs elaborate surface adhesins, e.g., AS, rather than conjugative pili and conjugatively transfer DNA substrates to bacterial recipient cells through direct cell-to-cell contact.
doi:10.1128/9781555818524.ch18f1

through either pilus retraction or pilus-mediated aggregation (Lawley et al., 2003; Christie et al., 2005). In gram-positive bacteria, conjugation machines are composed of a lower number of subunits because the translocase spans only one membrane and these systems elaborate surface adhesins, as opposed to more structurally complex pili for mating pair formation (Grohmann et al., 2003; Alvarez-Martinez and Christie, 2009) (Fig. 1). The gene composition and genetic organization of effector translocator systems resemble those of conjugation machines, although novel adaptations also exist that account for structural and functional variations. As with the conjugation systems, most effector translocators deliver substrates to target cells by a cell contact-dependent mechanism (Christie et al., 2005). The *Bordetella pertussis* Ptl system is an exception in delivering its cargo, the multisubunit pertussis toxin, to the extracellular milieu independently of target cell contact (Locht et al., 2011). The DNA release/uptake systems also function independently of target cell contact. The *Neisseria gonorrhoeae* gonococcal genetic island (see below) system releases chromosomal DNA fragments to the extracellular milieu, whereas the *Helicobacter pylori* Com system imports DNA from the milieu (Karnholz et al., 2006; Ramsey et al., 2011).

The conjugation systems translocate mobile elements such as conjugative or mobilizable plasmids, or integrative and conjugative elements (ICEs), to bacterial recipient cells. Processing of DNA substrates for transfer is initiated by assembly of a protein termed the relaxase and one or more auxiliary proteins at the origin of transfer sequence (*oriT*), forming the relaxosome (Pansegrau et al., 1988). The relaxase nicks the DNA strand destined for transfer and remains covalently bound to the 5′ end of the translocated strand (ssDNA), so that the relaxase-ssDNA nucleoprotein complex corresponds to the transfer intermediate (Lanka and Wilkins, 1995; Byrd and Matson, 1997; Lee and Grossman, 2007). The relaxase carries internal or C-terminal translocation signals conferring substrate recognition by the T4SS, and the relaxase is also thought to pilot the ssDNA through the translocation channel and into the target cell (Vergunst et al., 2005; Parker and Meyer, 2007; Lang et al., 2010). Consequently, the conjugation machines can be viewed as protein translocation systems that evolved to recognize and translocate a DNA-nicking enzyme (e.g., a rolling-circle replicase) and, only coincidentally, the associated ssDNA (Fig. 1).

Most T4SSs employ common strategies for recruitment and translocation of substrates across the cell envelope (Fig. 1). Protein substrates of the effector translocators and the relaxase of conjugation systems generally carry C-terminal translocation signals

and can require auxiliary factors or chaperones for docking with the T4SS (Nagai et al., 2005; Schulein et al., 2005; Vergunst et al., 2005). Substrates first engage with a hexameric ATPase, termed the type IV coupling protein (T4CP) or substrate receptor, located at the entrance to the translocation channel (Alvarez-Martinez and Christie, 2009). The T4CP then delivers the substrate to a membrane translocase for delivery through an envelope-spanning proteinaceous channel composed of mating pair formation (Mpf) proteins. Although there is considerable variability in architecture and function of T4SS machines, this general model for substrate docking and translocation suffices for our discussion concerned with mechanisms of T4SS regulation.

THE *A. TUMEFACIENS* VirB/VirD4 T4SS: A CONJUGATION SYSTEM ADAPTED FOR USE AS AN EFFECTOR TRANSLOCATOR

Most conjugation and effector translocator systems convey either DNA substrates or effector proteins to target cells. A few systems function both as conjugation machines delivering plasmid DNA substrates to recipient bacteria and as effector translocators injecting protein or DNA substrates to eukaryotic cells. Such systems include (i) *A. tumefaciens* VirB/VirD4, which can inject effector proteins and DNA substrates to a wide range of plant, yeast, bacterial, and human cells (Christie, 2004); (ii) *L. pneumophila* Dot/Icm, which delivers upwards of 200 effector proteins to human cells during infection and an IncQ plasmid substrate to *L. pneumophila* recipients (Vogel et al., 1998; Nagai and Kubori, 2011); (iii) *Bartonella henselae* VirB/VirD4, which delivers effector proteins and DNA to mammalian cells and DNA to bacterial recipients (Dehio, 2008; Fernandez-Gonzalez et al., 2011; Schroder et al., 2011); and (iv) *E. coli* plasmid RP4, which delivers DNA to bacterial and yeast cells (Bates et al., 1998). How T4SSs are activated for transfer in response to perception of potential recipients is an interesting topic in general, but it is especially fascinating for systems displaying such versatility in substrate recognition and host range. The *A. tumefaciens* VirB/VirD4 T4SS has the broadest host range of all described T4SSs to date, and it is also unique in its use of a T4SS to deliver DNA to a eukaryotic target cell as a critical feature of infection. This system has been extensively studied, and below we describe molecular details of a regulatory network controlling T4SS biogenesis in response to a variety of plant host, environmental, and bacterial signals.

Upon encountering wounded plant tissue, *A. tumefaciens* integrates a number of extracellular signals into a regulatory cascade, with two dominant consequences: (i) infection of the susceptible tissue and (ii) propagation of the Ti plasmid among *A. tumefaciens* cells in the vicinity (White and Winans, 2007). Infection leads to a loss of plant cell growth control and the formation of tumors termed crown galls. *A. tumefaciens* incites disease by use of the VirB/VirD4 T4SS to deliver a segment of its genome, the transfer DNA (T-DNA), and several effector proteins to the nuclei of susceptible plant cells. When the T-DNA is integrated into the plant genome, products of the expressed genes disrupt plant auxin and cytokinin levels, resulting in loss of cell growth control and tumor production (Amor et al., 2005). The regulatory network controlling T-DNA transfer and subsequent events at the infection site is depicted in Fig. 2 (Christie, 2007).

Transcriptional Activation of the *vir* Regulon

The pTi plasmid carries genes whose products mediate T-DNA processing and transfer: (i) the VirA/VirG two-component regulatory system coordinately induces expression of the *vir* genes (the *vir* regulon) in response to perception of plant-derived signals, (ii) VirC and VirD proteins process T-DNA into a nucleoprotein particle for delivery to plant nuclei, and (iii) VirB proteins and VirD4 elaborate the T4SS for transfer of the T-DNA transfer intermediate and effector proteins across the bacterial envelope (Amor et al., 2005; White and Winans, 2007).

Infection is initiated when bacteria sense and respond to an array of signals, including specific classes of plant phenolic compounds, aldose monosaccharides, low PO$_4$, and an acidic pH that are present at a plant wound site (Fig. 2). The VirA/VirG signal transduction system—the first described two-component system in bacteria (Winans et al., 1986)—together with the ChvE sugar-binding protein senses plant phenolics and sugars (Winans et al., 1986; Winans et al., 1989; Winans, 1991). Upon sensory recognition, the VirA sensor kinase autophosphorylates and then phosphorylates the VirG response regulator. Phospho-VirG activates transcription of six essential *vir* operons within the *vir* regulon and other nonessential genes scattered on the pTi plasmid (Kalogeraki and

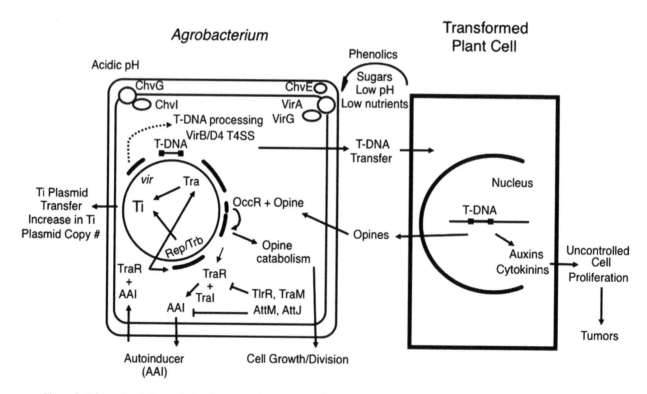

Figure 2. Schematic of chemical signaling events between *Agrobacterium* and the transformed plant cell. Signals released from wounded plant cells initiate the infection process through the VirA/VirG/ChvE and ChvG/ChvI sensory response systems, resulting in *vir* gene activation. The Vir proteins mediate T-DNA processing, assembly of the VirB/VirD4 T4SS, and T-DNA translocation to susceptible plant cells. VirA/VirG also induces expression of the Ti plasmid *rep* genes, resulting in elevated Ti plasmid copy number. Opines released from transformed plant cells activate opine catabolism functions for growth of infecting bacteria. Opines also activate synthesis of TraR for autoinducer (AAI) synthesis. TraR and AAI at a critical concentration activate the Ti plasmid conjugation functions. TlrR and TraM negatively regulate TraR activity, and AttM and AttJ negatively control AAI levels. doi:10.1128/9781555818524.ch18f2

Winans, 1998; Kalogeraki et al., 2000). VirA/VirG also enhances expression of the pTi *repABC* genes, resulting in an increase in plasmid copy number at the infection site (Cho and Winans, 2005). *A. tumefaciens* might have evolved this mode of copy number control as a means of enhancing virulence gene dosage and, hence, virulence potential upon perception of environmental conditions favorable for interkingdom DNA transfer.

The inducing phenolic compounds and monosaccharide sugars are intermediates of biosynthetic pathways involved in plant cell wall repair. Most plant wounds release these intermediates, which partly accounts for the capacity of *A. tumefaciens* to infect most plant species. VirA interacts with plant phenolics either directly or indirectly upon diffusion of the signal into the periplasm (Chang and Winans, 1992; Lee et al., 1992; Mukhopadhyay et al., 2004). ChvE binds plant sugars in the periplasm, and the ChvE-sugar complex then binds the periplasmic domain of VirA, inducing a conformational change that enhances VirA sensitivity to phenolic inducers (Cangelosi et al., 1990; Shimoda et al., 1993; Peng et al., 1998). The VirA periplasmic domain also is implicated in recognition of acidic pH, though the physical mechanism of pH perception is unknown. Upon activation, phospho-VirG binds *vir* boxes (TNCAATTGAAAPy) within the *vir* promoters to induce transcription (Pazour and Das, 1990; Han et al., 1992).

A second two-component system, ChvG/ChvI, senses acidic pH and, in turn, activates several of the *vir* genes and other chromosomal genes required for growth in acidic environments. ChvG/ChvI also induces *virG*, resulting in significantly elevated expression of the *vir* genes (Fig. 2) (Chang and Winans, 1992, 1996; Li et al., 2002; Yuan et al., 2008). The molecular basis of ChvG/ChvI-mediated acid sensing is unknown, but this system likely evolved for perception of the acidic environment of wounded plant tissue (Chang and Winans, 1996). The ChvG/ExoS two-component system of *Sinorhizobium meliloti* is highly related to the ChvG/ChvI system and required for establishment of plant symbiosis through regulation of succinoglycan and flagellum production (Cheng and Walker, 1998; Cheng and Yao, 2004; Yao et al., 2004; Wells et al., 2007). Other mammalian pathogens also rely on ChvG/ChvI-like systems to regulate expression of their T4SS genes (see below).

Plant-Derived Opines Induce Synthesis of LuxR Homologs

Following the transfer and integration of T-DNA into the plant nuclear genome, some of the transferred genes code for synthesis of novel amino acid condensation products termed opines (Fig. 2). Opines are released from the transformed cells, where they are then taken up by *A. tumefaciens* cells primarily for use as carbon and nitrogen sources (Valdivia et al., 1991). However, imported opines also bind and modulate the activities of cognate LysR transcription factors (Fuqua and Winans, 1994). For example, OccR positively regulates expression of the *occ* operon, whose genes are involved in octopine uptake and catabolism, by inducing a bend in the DNA at the OccR binding site. Octopine modulates OccR regulatory activity by altering both the affinity of OccR for its target site and the angle of the DNA bend (Wang and Winans, 1995). Although the majority of the *occ* operon codes for octopine transport and catabolism functions, the distal end of the *occR* operon contains a gene for a transcriptional activator termed TraR. TraR is related to LuxR, an activator shown nearly 30 years ago to regulate synthesis of an *N*-acylhomoserine lactone (AHL) termed autoinducer (Fuqua and Winans, 1994; White and Winans, 2007). As described in an earlier chapter (chapter 3) of this book, cells that synthesize autoinducer molecules secrete these molecules into the environment. At low cell densities, autoinducer is in low concentration, whereas at high cell densities, this substance accumulates in the surrounding environment and diffuses back into the bacterial cell to activate transcription of a defined set of genes. For *A. tumefaciens*, the autoinducer is an *N*-3-(oxo-octonoyl)-L-homoserine lactone termed *Agrobacterium* autoinducer (AAI or 3OC8-HSL (Zhu et al., 1998).

TraR and AAI Activate Transcription of the Ti Plasmid Trb/Tra T4SS

Upon diffusion of AAI across the *A. tumefaciens* envelope, AAI binds TraR, inducing a conformational change required for binding and activation of promoters for the pTi plasmid *tra/trb* genes, as well as *traI*, whose product mediates synthesis of AAI, creating a positive-feedback loop for AAI production (Fig. 2) (Fuqua and Winans, 1994; White and Winans, 2007). The *tra/trb* genes code for a second T4SS responsible for pTi conjugative transfer to bacterial recipient cells (Alt-Morbe et al., 1996). Sensing of the AAI quorum signal thus leads to enhanced transmission of the pTi plasmid and associated virulence traits among agrobacterial cells in the vicinity of the infection site.

The *tra/trb* T4SS genes are also negatively controlled (Fig. 2). For example, TraR activity is antagonized by two proteins, TraM and TrlR (Fuqua et al., 1995; Chai et al., 2001). TraM interacts with the carboxyl terminus of TraR, which inactivates TraR and disrupts TraR-DNA complexes. TrlR is a truncated form of TraR that suppresses TraR activity through formation of inactive heterodimers. In addition, an AAI signal turnover system is composed of the AttJ regulator and AttM (Chai et al., 2007). AttM is an AHL-lactonase that hydrolyzes the lactone ring of AAI (Zhang et al., 2002; Haudecoeur et al., 2009; Khan and Farrand, 2009; Haudecoeur and Faure, 2010). Interestingly, AAI stimulation of pTi plasmid transfer occurs only during exponential growth; upon entry into stationary phase, plasmid transfer is inhibited. This was shown to arise through induction of *attM* expression by the stationary-phase alarmone signal, (p)ppGpp, and RelA (p)ppGpp synthetase. The accumulated AttM lactonase degrades intracellular pools of the CO8-HSL quorum signal, resulting in shutdown of *tra/trb* gene expression (Zhang et al., 2004; Dalebroux et al., 2010). Overall, this complex network of negative regulatory controls might have evolved for fine-tuning of conjugative transfer in response to changes in cell growth/density and pTi plasmid status at the infection site (Khan and Farrand, 2009).

INDUCIBLE PLASMID TRANSFER: *ENTEROCOCCUS FAECALIS* pCF10 PHEROMONE-DEPENDENT TRANSFER

It is interesting that the activating AAI quorum signal for *A. tumefaciens* Ti plasmid transfer is released and sensed by Ti plasmid-carrying cells. In a homogeneous population, quorum sensing would thus appear to foster Ti plasmid transfer among donor cells. However, the infection site likely consists of a heterogeneous mix of plasmid-carrying and plasmid-free cells, the latter resulting from pTi plasmid curing at elevated temperatures (>28°C) or the fact that plant wound products serve as chemoattractants for other *A. tumefaciens* cells in the vicinity (Shaw et al., 1988). AHL-mediated plasmid transfer thus ensures maintenance of a Ti-plasmid-carrying population of cells, possibly to ensure a constant source of nutrients through continued delivery of opine-encoding T-DNA to susceptible plant tissue. Interestingly, the *E. faecalis* pCF10 conjugation system also has the potential to respond to a quorum signal released by donor cells, although several mechanisms have evolved to suppress donor-donor cell signaling while promoting the donor-recipient cell dialogue (Fig. 3) (Dunny et al., 1978; Dunny and Johnson, 2011).

Plasmid pCF10 is a member of a family of large, conjugative plasmids that code for tetracycline resistance, hemolysins, adhesins, and other virulence traits and whose transfer among enterococcal species is stimulated by small peptide pheromones. The pheromones are small (7- to 8-residue) peptides generated by intramembrane processing of signal sequences of chromosomally encoded lipoproteins of unknown function (Clewell et al., 2000). As depicted in Fig. 3, plasmid-free recipient cells release the pheromones, and the pheromones are taken up by donor cells, where they bind and activate a regulator to induce plasmid *tra* gene expression (Rocha-Estrada et al., 2010; Dunny and Johnson, 2011). Pheromone induction also results in synthesis of a surface glycoprotein, termed aggregation substance (AS), that binds to enterococcal binding substance present on the surface of recipient cells. This triggers bacterial cell clumping, establishment of donor-recipient cell mating junctions, and plasmid transfer (Dunny, 2007). AS is an important virulence factor by virtue of the fact it stimulates plasmid transfer and associated genes for antibiotic resistance, hemolysins, and other virulence determinants. AS also promotes adherence to and colonization of human heart valves and other tissues (Chuang et al., 2009; Chuang-Smith et al., 2010).

Two pheromone-responsive plasmids of *E. faecalis*, pAD1 and pCF10 (Fig. 3), have been extensively studied (Clewell, 2007; Wardal et al., 2010; Dunny and Johnson, 2011). Donor cells carrying pCF10 respond to the peptide pheromone cCF10 (LVTLVFV). cCF10 is encoded on the *E. faecalis* chromosome within the signal sequence of lipoprotein CcfA. Pre-cCF10 (residues 1 to 22) is processed by signal peptidase II cleavage at a conserved lipobox, removal of residues 1 to 12 by the metalloenzyme Eep, and removal of residues 20 to 22 by an unknown protease (Clewell et al., 2000; Antiporta and Dunny, 2002). Mature cCF10 (residues 13 to 19) is released from the cell and functions as a quorum signal (Antiporta and Dunny, 2002). It is imported by donor cells via the pCF10-encoded PrgZ receptor and a chromosomally encoded oligopeptide permease transporter (Leonard, 1996). Once in the cytosol, cCF10 binds and causes conformational changes in the regulator PrgX, resulting in activation of *tra* gene expression (Shi et al., 2005).

Although donor cells also produce and release cCF10, at least two mechanisms have evolved to prevent self-induction and donor-donor cell mating (Fig. 3). First, a plasmid-encoded membrane protein, PrgY, interacts with cCF10 upon release from the donor cell and sequesters or inactivates the pheromone (Chandler et al., 2005). Second, the pheromone-responsive plasmids code for an inhibitor peptide; for pCF10, the inhibitor is iCF10 (AITLIFI). This peptide also interacts with PrgX, but with an outcome opposite that of PrgX-cCF10 complex formation (Kozlowicz et al., 2006).

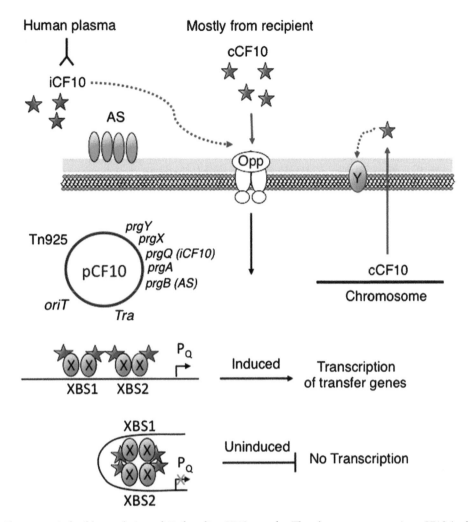

Figure 3. Pheromone-inducible regulation of *E. faecalis* pCF10 transfer. The pheromone-responsive pCF10 is shown, with locations for genes coding for regulation and transfer (*Tra*), the origin of transfer sequence (*oriT*), and Tn*925*. Chromosomally encoded peptide pheromone cCF10 released by donor cells is sequestered by *prgY*. cCF10 released by recipient cells is taken up through an oligopeptide permease transporter and when bound to PrgX stimulates transcription of the transfer genes. pCF10-encoded peptide inhibitor iCF10 negatively regulates transfer gene expression, limiting plasmid transfer potential in donor cell populations. Human plasma binds iCF10, indirectly promoting plasmid transfer in the human host. pCF10-encoded AS promotes bacterial aggregation and plasmid transfer and biofilm formation on human tissues. doi:10.1128/9781555818524.ch18f3

Transcriptional control of the *prgX*/*prgQ* region and adjoining *tra* (designated *prg* or *pcf*) genes has been extensively studied (Fig. 3; see also Rocha-Estrada et al., 2010, and Dunny and Johnson, 2011, for more details). Briefly, binding of cCF10 or iCF10 induces different conformational states of the regulator PrgX. Under noninducing conditions (lack of recipient cells and low cCF10/ iCF10 ratio), PrgX forms a tetramer that represses transcription initiation from the P_Q promoter through formation of a DNA loop; loop formation impedes access of RNA polymerase to the cognate promoter sequences. Under inducing conditions (presence of recipient cells and high cCF10/ iCF10 ratio), cCF10-bound PrgX fails to tetramerize and induce DNA looping; operator occupancy of

cCF10-PrgX dimers does not occlude binding of RNA polymerase, and transcription proceeds. Additional regulatory controls on iCF10 production and synthesis of the Tra proteins have been described, including modulation of transcription through binding of RNA polymerase to convergent promoters and antisense RNA anti-Q control (Johnson et al., 2010; Shokeen et al., 2010; Chatterjee et al., 2011). Ultimately, the induction state of an *E. faecalis* donor cell population appears to depend on the ratio of exogenous cCF10/ iCF10 pheromone (Chatterjee et al., 2011; Dunny and Johnson, 2011).

Of further interest, in the mammalian bloodstream a host factor selectively sequesters or degrades iCF10 (Fig. 3) (McCormick et al., 2000). In response to low iCF10 levels, the bacteria conjugatively transfer the pheromone-dependent plasmids and also produce elevated levels of AS, which promotes attachment and biofilm formation on host tissue. The iCF10-degrading component in human plasma has not been identified but is thought to be an albumin/lipid complex. iCF10 can be considered the functional analog to the *A. tumefaciens* AHL-lactonase AttM in that both inhibitors mitigate conjugative DNA transfer and transmission of virulence determinants. For *E. faecalis*, however, the inhibitor is sequestered by a host factor(s), thereby favoring bacterial colonization, plasmid transfer, and disease transmission.

ICEs: *B. SUBTILIS* ICE*Bs1*, *VIBRIO CHOLERAE* SXT, Tn*916*, AND CTnDOT

Integrative and conjugative elements (ICEs) are mobile elements residing on bacterial chromosomes (Alvarez-Martinez and Christie, 2009; Juhas et al., 2009; Wozniak and Waldor, 2010). These elements contribute to genome plasticity and are important determinants for bacterial pathogenesis by virtue of

their capacity to transmit antibiotic resistance and other virulence genes. In response to various environmental and physiological signals, ICEs excise from the chromosome and either reintegrate into the chromosome or undergo processing reactions analogous to the conjugative plasmids for transmission to bacterial target cells. The regulatory networks controlling excision and transmission are complex but can involve perception of signals found in the human host (Fig. 4) (Wozniak and Waldor, 2010). For example, DNA-damaging agents such as mitomycin C or quinolone antibiotics (e.g., ciprofloxacin) induce the SOS response, which can enhance transfer frequencies of *V. cholerae* SXT and *B. subtilis* ICE*Bs1* by orders of magnitude (Beaber et al., 2004; Auchtung et al., 2005; van der Veen and Abee, 2011). For SXT, the SOS response appears to involve RecA-assisted autocleavage of the SetR repressor, which is the main regulator of ICE transfer (Beaber et al., 2004). For ICE*Bs1*, the SOS response induces synthesis of the ImmA metallopeptidase, whose target, ImmR, is a repressor of ICE*Bs1* transfer gene expression (Fig. 4) (Auchtung et al., 2005; Bose et al., 2008; Bose and Grossman, 2010). Antibiotic-mediated induction

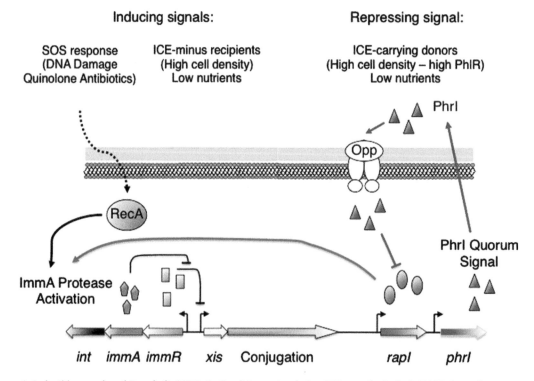

Figure 4. Inducible transfer of *B. subtilis* ICE*Bs1*. Conditions stimulating ICE transfer include DNA-damaging agents, e.g., quinolone antibiotics, and a high density of potential recipient cells lacking the ICE (whose presence is sensed by the ICE-encoded PhrL pheromone inhibitor [blue-shaded triangles]). Induction of the SOS response activates the ImmA protease, which, in turn, inactivates the ImmR repressor, leading to expression of the excision and transfer genes. At a high density of donor cells, PhlR accumulates in the extracellular milieu and upon internalization inactivates the RapI inducer (green-shaded ovals). Arrows denote activation of gene expression; bars denote inactivation/repression.
doi:10.1128/9781555818524.ch18f4

of ICE transfer also occurs among the Tn*916* and CTnDOT elements. In the case of Tn*916*, tetracycline relieves premature termination of a transcript so that the longer transcript formed in the presence of the antibiotic codes for regulatory proteins required for *tra* gene transcription (Roberts and Mullany, 2009; Wozniak and Waldor, 2010). For CTnDOT, tetracycline enables elongation of an mRNA through *rteA* and *rteB* through a translation attenuation mechanism. Then RteB-mediated activation of downstream regulators activates expression of the excision and *tra* genes (Jeters et al., 2009; Wozniak and Waldor, 2010). It is noteworthy that antibiotic treatment or exposure to other DNA-damaging reagents has the potential of inducing excision and transmission among pathogens of mobile elements carrying the genes for resistance to the antibiotic administered in the treatment regimen.

ICE*Bs1* excision and transmission are also regulated by a quorum signal produced by ICE-minus cells (Auchtung et al., 2005; Bose et al., 2008) (Fig. 4). Regulation by population density is mediated by the quorum-sensing regulator RapI and the pentapeptide PhrI (Auchtung et al., 2005). The Rap (aspartyl phosphate phosphatases) family of transcription factors control sporulation, competence development, and production of degradative enzymes and antibiotics (Auchtung et al., 2005; Rocha-Estrada et al., 2010). In *B. subtilis,* RapI activates expression of the ICE*Bs1* excision and transfer genes directly through promoter binding and indirectly by triggering ImmA-mediated cleavage of the ImmR repressor. The Phr pentapeptides are released by ICE*Bs1*-carrying cells and imported through an oligopeptide permease; in the bacterial host, they bind and inhibit RapI activity. Phr peptides thus functionally resemble the *E. faecalis* inhibitor peptide iCF10 by preventing donor-donor cell transmission. A homogeneous population of ICE*Bs1*-carrying cells expresses the *rapI* and *phrI* genes at high levels in response to a low-nutrient, high-cell-density environment; under these conditions, the accumulation of PhrI quorum signal inhibits ICE transmission. Conversely, when ICE*Bs1*-carrying cells are crowded by ICE*Bs1*-minus cells, the PhrI quorum signal does not accumulate, resulting in RapI-mediated activation of excision and ICE transmission (Auchtung et al., 2005).

REGULATION OF F PLASMID TRANSFER

The systems described above are regulated through quorum signals, and an important feature of their regulation is the coevolution of mechanisms for quorum sensing and discrimination of self (plasmid-carrying donors) from nonself (plasmid-free recipients) (Frost et al., 2005; Frost, 2009; Frost and Koraimann, 2010). Other mobile elements do not depend on quorum signals for activation but, instead, regulate transmission through sensing of environmental signals and physiological status. The best characterized of these systems is that encoded by the *E. coli* F plasmid (Lawley et al., 2003; Lawley et al., 2004; Frost and Koraimann, 2010). Long-term studies of the F transfer system have generated a fundamental understanding of how mobile genetic elements move and how through integration and excision from the chromosome they contribute over evolutionary time to the shaping of bacterial chromosomes. Furthermore, as discussed below, elements of the regulatory circuitry controlling F transfer are now known to control biogenesis and function of other T4SSs, including several effector translocator systems exploited by many pathogens. The general mechanistic themes include transcriptional regulation by sensing of extracytoplasmic stress and induction of the CpxAR regulon, stationary-phase-mediated silencing by H-NS and the involvement of other nucleoid-associated proteins in controlling transfer, and sensing of environmental stimuli and physiological status (Fig. 5).

The main regulators of T4SS *tra* gene expression are encoded by *traM, traJ*, and *traY* (Fig. 5) (Frost and Koraimann, 2010). Promoters are located upstream of each of these genes, wherein *traY* is the first gene in the *tra* operon (*traYALE…X*) encoding the T4SS system. Expression of the *traYALE…X* promoter is regulated by TraJ, TraY, and other proteins such as integration host factor (IHF) and the response regulator SfrA. TraM functions posttranslationally to regulate TraI/Y nicking at *oriT* and docking of the relaxosome at the TraD T4CP.

Two antisense RNAs, *utpR* and *finP*, along with their cognate RNA-binding proteins, Hfq and FinO, respectively, also regulate *tra* gene expression (Fig. 5) (Arthur et al., 2003; Will and Frost, 2006b). The global regulator Hfq appears to bind UtpR to promote degradation of *traJ* mRNA and transfer potential. The antisense RNA *finP* and FinO constitute a fertility inhibition system. *finP* represses *traJ* expression and plasmid transfer, but in the absence of FinO, the antisense RNA is rapidly degraded. F plasmids or other coresident plasmids encode FinO, which binds *finP* and prevents its degradation. Thus, in cells encoding the *finP*/FinO pair, F transfer is strongly repressed, such that only a few cells in a population are transfer competent (Frost and Koraimann, 2010). Interestingly, upon F transfer, the new transconjugants have low levels of *finP*/FinO for a brief period, and consequently, these cells transfer at high frequency to other plasmid-free cells in the vicinity. The net effect of high-frequency transfer is complete conversion of

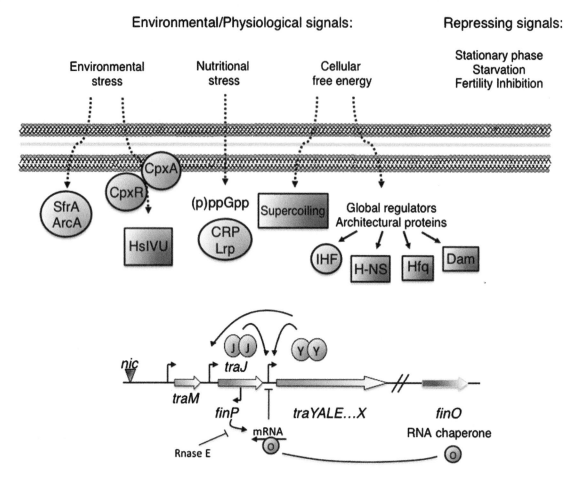

Figure 5. Transfer of the *E. coli* F factor is controlled by environmental and physiological signals acting through the regulatory factors shown. Factors that repress F *tra* gene expression are boxed and red shaded; those inducing *tra* gene expression are circled. The CpxRA two-component system is green shaded. The F-borne main regulators *traJ*, *traY*, and FinO/*finP* fertility inhibition system and their promoter targets are shown. RNase E degrades *finP* antisense RNA in the absence of bound RNA chaperone FinO. The regulators activated in response to environmental and physiological cues control F *tra* gene expression as described in the text (see also Frost and Koraimann, 2010). Arrows denote activation of gene expression; lines denote repression. Dotted arrows represent unspecified sensing or transduction mechanisms.
doi:10.1128/9781555818524.ch18f5

a recipient cell population to plasmid-bearing status. The fertility inhibition systems function generally to control plasmid transfer within donor populations, maximizing the chances for plasmid spread to new recipients and minimizing the metabolic burden accompanying T4SS assembly (Frost and Koraimann, 2010).

Gram-negative bacteria sense extracytoplasmic stresses through the CpxAR two-component system (Gubbins et al., 2002; Lau-Wong et al., 2008; MacRitchie et al., 2008). Cpx (conjugative plasmid expression) was so named from an early finding that certain *cpxA* mutations resulted in reduced levels of TraJ and reduced F plasmid transfer. More recent work established that CpxAR acts indirectly through effects on the HslVU protease-chaperone pair, which cleaves TraJ as part of the stress response (Fig. 5)

(Lau-Wong et al., 2008). Interestingly, whereas F transfer is inhibited upon sensing of extracytoplasmic stress, overexpression of F *tra* genes or expression at an inappropriate stage of cell growth also can activate the extracytoplasmic stress response. Transient upregulation of the CpxAR regulon in response to the loading of T4SS machine subunits into the cell envelope likely exerts a quantitative effect on F plasmid transfer and also controls the timing of plasmid transfer during cell growth cycle.

A number of global regulators affecting DNA architecture, e.g., IHF, H-NS, and FIS, also control F T4SS gene expression (Fig. 5) (Will et al., 2004; Will and Frost, 2006a; Frost and Koraimann, 2010). IHF binding contributes directly to transcription of *traY* and the T4SS genes. By contrast, H-NS silences *tra* gene expression mainly during stationary-phase growth. H-NS protein levels actually remain fairly constant during growth, and changes in gene expression are driven by nutritional and environmental signals that cause changes in chromosomal supercoiling, thereby affecting H-NS binding. H-NS-mediated gene silencing plays an important role in controlling *tra* gene expression during all stages of cell growth, but F has evolved several mechanisms to overcome H-NS silencing. Thus, it is thought that the complex F regulatory network ultimately serves to balance silencing and desilencing to achieve optimum levels of transfer potential. Other large conjugative plasmids also carry genes for H-NS as well as members of the related Hha/YmoA family that have been shown to regulate expression of genes required for horizontal DNA transfer in response to changes in environmental conditions, such as temperature. There is also increasing evidence that environmentally responsive global regulators have critical functions in controlling expression of virulence genes, including those encoding T4SS effector translocators or their substrates, during pathogenesis (see below).

REGULATION OF EFFECTOR TRANSLOCATOR SYSTEMS

Many mammalian pathogens have adapted a T4SS to deliver cargo substrates to the eukaryotic host to promote infection. Next, we summarize regulatory networks controlling type IV secretion for a few of the better-characterized systems, *Brucella* VirB, *L. pneumophila* Dot/Icm, and *Bartonella* VirB/VirD4. We also describe some interesting posttranscriptional modes of regulation exploited by *N. gonorrhoeae* for T4SS-mediated DNA release, *H. pylori* for CagA translocation, and *Pseudomonas aeruginosa* for transmission of the PAI-1 pathogenicity island.

Brucella VirB: Modulation of T4SS Assembly and Function at Different Stages during Intracellular Infection

Brucella spp. are intracellular pathogens that cause brucellosis in domestic and wild animals and humans (Pappas et al., 2005). Infection of humans is caused primarily by two species, *Brucella melitensis* and *Brucella abortus*, through direct contact with infected animals or carcasses or consumption of unpasteurized dairy products from goats and cattle. *Brucella* spp. have evolved to avoid the immune system of hosts, and infection is often characterized by long-term persistence of the bacteria. During infection, *Brucella* resides mainly intracellularly, sequestered from antibodies. Infected cells remain

viable because *Brucella* inhibits apoptosis with yet-unknown virulence factors. Cyclic β-1,2-glucan and the VirB T4SS are critical for intracellular survival, the former enabling avoidance of phagosome-lysosome fusion and the latter enabling survival and replication in host cells (Foulongne et al., 2000; Hong et al., 2000; Boschiroli et al., 2002a; Arellano-Reynoso et al., 2005; Roux et al., 2007; den Hartigh et al., 2008; Rambow-Larsen et al., 2009). Recent screens have identified more than six putative effectors whose translocation to human cells is dependent on a functional VirB T4SS (de Jong et al., 2008; de Barsy et al., 2011; Paschos et al., 2011).

Upon entry into the host cell, *Brucella* spp. are exposed to environmental conditions that stimulate expression of the T4SS genes (Fig. 6) (Rambow-Larsen et al., 2009). *Brucella* spp. require T4SS-mediated substrate delivery during a relatively confined period during intracellular infection, for several hours after uptake of the bacterium into a host cell until destruction in phagolysosomes has been avoided and a vacuole in the endoplasmic reticulum (ER) suitable for replication has been established. Upon entry into host cells, T4SS genes are induced upon acidification of the *Brucella*-containing vacuole, which occurs after the phagosome transiently fuses with early and late endosomes and lysosomes. Nutrient starvation also elicits expression of the *virB* genes. As also observed with the *A. tumefaciens* VirB/VirD4 T4SS, in culture the expression of *Brucella virB* genes is strongly induced upon switching cells from a rich to minimal medium at low pH (Boschiroli et al., 2002b; Rouot et al., 2003; Sieira et al., 2004).

In response to sensory perception of environmental, host, and, possibly, bacterial signals, several transcriptional regulators control expression of the *virB* genes in *Brucella* (Fig. 6). After phagocytosis, the first environment that the bacterium encounters is a nutrient-poor, acidic vacuole. Induction of the *virB* genes in nutrient starvation requires the stringent response regulator Rsh and IHF (Sieira et al., 2004). Interestingly, IHF is required for *virB* gene induction in medium at neutral pH and inside host cells. However, a second regulator, HutC, competes with IHF for binding to the same binding site in the *virB* promoter. HutC is required for *virB* activation under nutrient starvation and low-pH conditions (Dozot et al., 2006; Sieira et al., 2010). It is thought that IHF and HutC might act sequentially to activate *virB* gene expression early during intracellular infection when the *Brucella* phagosome is undergoing acidification, whereby IHF recruits or provides the correct promoter structure for binding of other transcription factors and HutC binding specifies correct induction of the T4SS *virB* operon under acidic conditions (Dozot

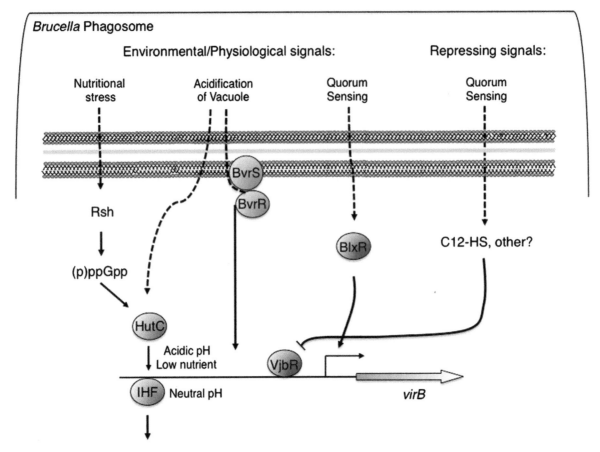

Figure 6. The *Brucella* VirB T4SS genes are expressed in the phagosome in response to environmental signals and quorum signals. At neutral pH, IHF activates *virB* expression, whereas nutritional stress resulting in elevated levels of the alarmone (p)ppGpp promotes displacement of HutC for IHF. The BvrS/BvrR two-component system (green shaded) regulates *virB* expression directly and indirectly by controlling VjbR activity. VjbR and BlxR (blue shaded) control *virB* expression in response to sensing of unknown quorum signals. Arrows denote activation of gene expression; lines denote repression. Dashed arrows represent unspecified sensing or transduction mechanisms.
doi:10.1128/9781555818524.ch18f6

et al., 2006; Rambow-Larsen et al., 2009; Sieira et al., 2010).

Three other transcriptional regulators modulate expression of the *virB* genes in response to extracellular signals (Fig. 6). VjbR is a LuxR homolog of demonstrated importance for activation of the *virB* promoter (Delrue et al., 2005; Uzureau et al., 2007). VjbR also regulates a large and diverse number of genes, many identified as virulence factors in other bacterial pathogens (Uzureau et al., 2010; Weeks et al., 2010). Correspondingly, mutations in *vjbR* highly attenuate intracellular survival of *B. melitensis*, suggestive of a role for quorum sensing in regulation of *Brucella* infections (Uzureau et al., 2007). Purified VjbR binds the putative quorum signal *N*-dodecanoyl homoserine lactone (C12-HSL) but interestingly, in *E. coli* cells VjbR binds the *virB* promoter region in the absence but not in the presence of the AHL (Uzureau et al., 2007). Thus, quorum sensing diminishes rather than enhances synthesis of the VirB T4SS. Of further

interest, *Brucella* does not contain any of the known genes for synthesis of C12-HSL, although small amounts of this AHL were isolated from *B. melitensis* culture supernatant (Taminiau et al., 2002). To account for the possible contribution of quorum sensing during *Brucella* infection, it was suggested that a low level of production of the AHL in the *Brucella* phagosome might permit sufficient VjbR-mediated expression of the *virB* genes. When *Brucella* then migrates to the ER, where the T4SS is no longer needed, C12-HSL concentrations might accumulate to levels sufficient to inhibit activity of VjbR and block *virB* expression. It is also possible that VjbR is sensing not an HSL of bacterial origin but, rather, a host-derived signal present only in the ER (Rambow-Larsen et al., 2009).

The BvrRS two-component system is also critical for virulence and expression of *virB* genes (Fig. 6) (Sola-Landa et al., 1998; Martinez-Nunez et al., 2010). BvrR/BvrS is highly related to the

A. tumefaciens ChvG/ChvI two-component system involved in mediating expression of the *virB* and other genes required for growth in acidic pH. Like the ChvG/ChvI system, BvrR/BvrS controls not only *virB* gene expression but also genes encoding proteins involved in outer membrane biogenesis/function, e.g., outer membrane proteins and proteins involved in lipopolysaccharide modifications (Viadas et al., 2010). During infection, BvrRS functions to prevent phagosome-lysosome fusion and promote replication in macrophages. BvrRS also positively controls expression of VjbR (Martinez-Nunez et al., 2010; Viadas et al., 2010). This two-component system therefore orchestrates critical control of T4SS *virB* expression through direct binding of activated phospho-BvrR to the *virB* promoter, possibly as a result of sensing the decrease in pH of the *Brucella* phagosome. BvrRS also acts indirectly on *virB* gene expression through induction of *vjbR* and the quorum sensing system.

Recently, a second quorum sensing regulator, BlxR (also called BabR), was shown to regulate *virB* genes (Fig. 6) (Rambow-Larsen et al., 2008). Unlike VjbR, however, BlxR does not appear to respond to C12-HSL, suggesting that another signaling molecule may regulate its activity (Weeks et al., 2010). VjbR and BlxR regulate overlapping sets of genes involved in virulence, stress responses, metabolism, and bacterial replication, and thus, the two quorum sensing regulators might exert their effects at different stages of the infection cycle (Rambow-Larsen et al., 2008). In conjunction with the BvrR/BvrS system, VjbR might activate *virB* gene expression early during formation of the *Brucella* phagosome when the T4SS is needed. Then, BlxR might function later to downregulate *virB* gene expression during establishment of the replication permissive niche in the ER when the T4SS is no longer important (Rambow-Larsen et al., 2009).

L. pneumophila Dot/Icm: Underscoring the Importance of Temporal Control of Machine Biogenesis, Effector Synthesis, and Substrate Transfer during Infection

L. pneumophila lives in a wide variety of aquatic habitats and also survives intracellularly in free-living amoebae. Upon inhalation of contaminated water sources, *L. pneumophila* also can infect mammalian phagocytes and cause a pneumonia termed Legionnaires' disease (Vogel and Isberg, 1999). *L. pneumophila* establishes an intracellular niche in host cells by forming *Legionella*-containing vacuoles (LCVs). The organism replicates in the LCVs and inhibits phagosome-lysosome fusion, instead recruiting early secretory vesicles and the ER. The Dot/Icm T4SS is essential for bacterial uptake and formation

of LCVs, and to date, this system has been shown to translocate as many as 200 effector proteins (Burstein et al., 2009; Nagai and Kubori, 2011; Zhu et al., 2011).

In LCVs, when conditions are favorable, *L. pneumophila* undergoes robust replication. When nutrients are exhausted, replication ceases and cells activate genes involved in promoting escape from the host cell, survival in the extracellular milieu, and infection of new phagocytic cells. This biphasic lifestyle, a replicative mode under nutrient-rich conditions and a transmissive mode allowing escape from the nutrient-depleted host cell, is controlled by differential expression of a large number of genes through a complex regulatory cascade (Molofsky and Swanson, 2004). Expression of genes encoding the Dot/Icm apparatus and its substrates is controlled through this regulatory cascade (Gal-Mor et al., 2002).

At least four interactive regulatory systems coordinate expression of genes coding for the Dot/Icm T4SS or protein effectors, although it is important to note that mutations of several of the regulatory components have only modest effects on *dot/icm* gene expression or virulence (Fig. 7). The envelope stress response two-component system CpxRA positively regulates the *icmR*, *icmWicmX*, and *icmV-dotA* transcriptional units as well as genes for several translocated effectors (Altman and Segal, 2008). CpxR also negatively regulates several effector genes, suggesting that distinct subsets of effectors are synthesized and translocated at different times during infection. An activating environmental signal for this two-component system has not been identified in *L. pneumophila*, but the inducing signals identified in other bacteria, e.g., periplasmic stress, accumulation of unfolded proteins in the periplasm, and pH sensing, could also be relevant signals in this pathogen (Altman and Segal, 2008).

A second two-component system, PmrAB, regulates a large number of effector genes (Fig. 7) (Zusman et al., 2007). Mutations in PmrAB exert a stronger effect on intracellular growth than mutations in CpxRA. The PmrA and CpxR regulators belong to the same family of DNA-binding proteins with winged helix-turn-helix motif, and both regulate expression of a number of virulence genes in *L. pneumophila* as well as *Shigella sonnei*, enteropathogenic *E. coli*, *Yersinia* spp., *Salmonella enterica*, *Pseudomonas aeruginosa*, and *Francisella* spp. By genome-wide microarray analyses, PmrA was found to regulate the expression of more than 270 genes in *L. pneumophila* (Al-Khodor et al., 2009). The nature of the environmental signal(s) sensed by the PmrAS system, the question of whether or how PmrAS temporally regulates effector synthesis/translocation, and the possibility of cross talk between

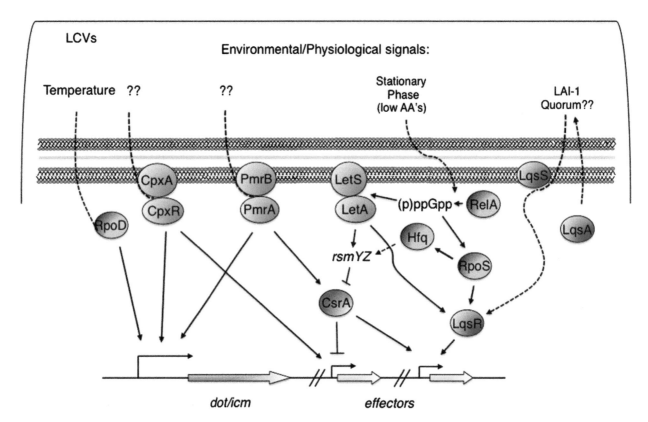

Figure 7. In the LCV, various signals regulate expression of genes coding for the Dot/Icm T4SS and the estimated 200 effectors. Signals are perceived by three two-component systems (green shaded), which act directly on gene expression or indirectly, e.g., through sRNAs *rsmY* and *rsmZ* and the RNA-binding protein CsrA. A complex regulatory network coordinates T4SS machine biogenesis with production of effectors for translocation at the appropriate stage of the infection cycle. A putative quorum sensing system (blue shaded) activates the LuxR homolog, LqsR. Arrows denote activation of gene expression; lines denote repression. Dashed arrows represent predicted activities.
doi:10.1128/9781555818524.ch18f7

PmrAS, CpxRA, and other regulatory mechanisms are intriguing areas for future study.

A third two-component system, LetAS, coordinates with two small RNAs (sRNAs), *rsmY* and *rsmZ*, and the RNA binding protein CsrA to regulate stationary-phase gene expression and activation of virulence traits, including some *icm/dot* effector genes (Fig. 7) (Molofsky and Swanson, 2003; Shi et al., 2006; Al-Khodor et al., 2008; Rasis and Segal, 2009; Edwards et al., 2010; Tiaden et al., 2010). In contrast, LetA activates expression of these genes through induction of the *rsmYZ* sRNAs, which inhibit CsrA activity. Two stationary-phase regulators, RpoS and Hfq, also positively regulate expression of these and other Dot/Icm effectors by enhancing the synthesis of the *rsmY* and *rsmZ* sRNAs (Bachman and Swanson, 2001, 2004; McNealy et al., 2005; Rasis and Segal, 2009; Sahr et al., 2009). The latter regulatory system mainly functions to control effector protein synthesis, but LetAS also activates transcription of *dotA*, whose product is both an essential Dot/Icm T4SS component and a substrate of this translocation system.

Finally, a putative quorum sensing system, LqsA/LqsS/LqsR, was recently identified (Tiaden et al., 2007; Tiaden et al., 2010). This system regulates expression of virulence genes, including flagellin genes and genes encoding Dot/Icm effectors SidB and SidD. The stationary-phase sigma factor RpoS and, possibly, the quorum signal 3-hydroxypentadecan-4-one promote LqsR activity. LqsR promotes phagocytosis, formation of the LCV, and intracellular replication. LqsR converges with the LetAS system, both through its activation by LetA and by antagonizing the activity of the RNA-binding CsrA global repressor. For T4SS function, in response to a quorum signal, LqsR might contribute to synthesis and translocation of specific Dot/Icm effectors at a specific stage of the infection cycle.

In sum, a complex regulatory circuitry controls expression of the *dot/icm* and effector genes at both the transcriptional and posttranscriptional levels. Although a number of possible signals and regulators have been identified, more rigorous studies are needed to discriminate direct from indirect effects on Dot/Icm machine assembly and function.

Bartonella: Coregulation of Two T4SSs Mediating Translocation and Adhesion

The alphaproteobacterium *Bartonella henselae* is a globally distributed zoonotic pathogen transmitted via a cat flea between cats and from cats to humans by scratches or bites (Dehio, 2005; Pulliainen and Dehio, 2009). *B. henselae* possesses two T4SSs, named VirB/VirD4 and Trw to reflect the phylogenetic relatedness of the systems with the *A. tumefaciens* and plasmid R388 T4SSs, respectively (Schmid et al., 2004; Vayssier-Taussat et al., 2010; Franz and Kempf, 2011). However, whereas the VirB/VirD4 T4SS translocates effector proteins and DNA (see below), the Trw system lacks a VirD4-like T4CP and is thought to function exclusively as an attachment organelle. Expression of both T4SS gene sets is under the regulatory control of the BatR/BatS two-component system (Quebatte et al., 2010). This system is closely related in sequence to the above-mentioned *A. tumefaciens* ChvG/ChvI and *Brucella* BvrR/BvrS systems, which activate T4SS gene expression through sensing of environmental signals, e.g., pH or envelope stress. BatR/BatS is activated in a neutral pH range (pH 7 to 7.8), with an optimum at the physiological pH of

blood (pH 7.4). Sensing of this physiological signal is thought to allow for discrimination of the mammalian host and arthropod vector environments.

The *Bartonella* VirB/VirD4 system recognizes and translocates effectors termed BEPs (*Bartonella* effector proteins) (Schulein et al., 2005). The BEPs carry a bipartite translocation signal composed of a C-terminal positively charged motif and internal sequences termed BID (BEP intracellular delivery) domains. Intriguingly, it was discovered that an *A. tumefaciens* relaxase, TraA, also carries a BID domain; when produced in *Bartonella*, TraA is recognized and translocated through the VirB/VirD4 system. This finding led to the notion that the *Bartonella* VirB/VirD4 might traffic DNA, and indeed this was recently shown. The *Bartonella* VirB/VirD4 system delivers a mobilizable plasmid into human cells by virtue of recognition of a BID-domain-containing relaxase (Schroder et al., 2011). Reminiscent of *A. tumefaciens*, therefore, *Bartonella* also has evolved as an effector translocator while retaining an ancestral function as a DNA translocation system.

The *Bartonella* VirB/VirD4 and BEP substrate genes are coregulated by the BatR/BatS system (Fig. 8), most probably for temporal control of machine

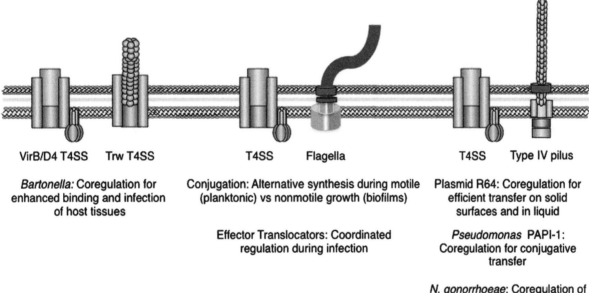

| VirB/D4 T4SS | Trw T4SS | | T4SS | Flagella | | T4SS | Type IV pilus |

Bartonella: Coregulation for enhanced binding and infection of host tissues

Conjugation: Alternative synthesis during motile (planktonic) vs nonmotile growth (biofilms)

Plasmid R64: Coregulation for efficient transfer on solid surfaces and in liquid

Effector Translocators: Coordinated regulation during infection

Pseudomonas PAPI-1: Coregulation for conjugative transfer

N. gonorrhoeae: Coregulation of DNA release and uptake systems

Figure 8. Coordinated regulation of T4SSs and other motility or attachment organelles. A pathogen may elaborate more than one T4SS for translocation of effectors to promote colonization at different stages of the infection cycle or for expansion of the infection niche. Conjugation systems function optimally in dense populations of nonmotile cells, e.g., biofilms; regulators induce *tra* genes and repress *fla* genes. Effector translocators coordinate the synthesis of T4SSs and flagella during the infection cycle through common environmentally responsive regulators. Most conjugation systems function efficiently among cells growing on solid surfaces; coregulation of type IV pili allows for sampling of the fluid environment for potential recipients for expanded transfer potential. In *N. gonorrhoeae*, the T4SS functions as a DNA release system; coregulation of a type IV pilus-mediated DNA uptake system promotes gene flux and genetic diversity.
doi:10.1128/9781555818524.ch18f8

assembly and substrate transfer (Quebatte et al., 2010). As noted above, BatR/BatS coregulation extends to the Trw system, which, interestingly, is dispensable for infection of endothelial cells but important for establishment of long-term chronic infections (Vayssier-Taussat et al. 2010). If, in fact, the Trw system functions exclusively to mediate binding to host cell receptors, BatR/BatS coregulation of both T4SSs might have evolved for expansion of the target cell repertoire in the human host.

COORDINATED REGULATION OF T4SS AND OTHER SURFACE ORGANELLES

The regulatory schemes discussed above evolved to optimize the timing and efficiency of substrate transfer in host environments suitable for growth and colonization by the infecting pathogen. Another interesting evolutionary adaptation for pathogenesis is the coordinated transcriptional control of genes encoding type IV secretion and other attachment or motility machines. As noted above, in *Bartonella* the coregulated VirB/VirD4 and Trw systems might function cooperatively during infection. Other such examples exist, although surface organelles also can be alternatively produced.

Regulation of T4SS and Flagella

Many bacterial pathogens rely on flagellar or type IV pilus-based motility to migrate to sites favorable for colonization within the host. Within the replication niche, free-swimming bacteria suppress their motility mechanisms and induce expression of other virulence genes, often forming biofilms at the infection site (Anderson et al., 2010; Mikkelsen et al., 2011; Ogasawara et al., 2011). It is now well established that bacterial conjugation machines also contribute to biofilm formation through elaboration of attachment factors, e.g., *E. coli* F plasmid-encoded pili or *E. faecalis* pCF10-encoded AS (Dudley et al., 2006; Ong et al., 2009; Mikkelsen et al., 2011; Ogasawara et al., 2011). Correspondingly, biofilms represent favorable microenvironments for horizontal gene transfer (Ghigo, 2001; Molin and Tolker-Nielsen, 2003; Cook et al., 2011). One regulatory mechanism controlling biofilm formation, motility, and type IV secretion involves synthesis in *E. coli* K-12 of the environmental response regulator Hha and another factor, YbaJ (Barrios et al., 2006). These factors control hemolysin production and biofilm formation but also stimulate expression of T4SS genes carried on a conjugative plasmid, resulting in enhanced transfer frequencies. Conversely, although plasmid-free cells are motile, cells carrying the conjugative plasmid are nonmotile and motility is restored by mutation of *hha* and *ybaJ*. Hha and YbaJ, therefore, coordinately control biofilm formation and conjugative DNA transfer and, in conjunction with a plasmid-encoded regulator, repress flagellar gene expression. Similarly, the growth conditions favorable for *A. tumefaciens* infection of plants—low pH and low nutrient content—induce synthesis of the VirB/VirD4 machine while repressing flagellar gene expression (Lai et al., 2000). The VirA/VirG two-component system regulates both activities, but it is not known whether the VirG response regulator acts directly or indirectly to control *fla* gene expression. In these instances, T4SS machine biogenesis is coupled with a sessile physiological state conducive for horizontal gene transfer.

Alternatively, some pathogens cross-regulate motility and type IV secretion. This is suggested by results of microarray and mutational studies of *Brucella* spp. and *L. pneumophila* (Fig. 8). In *L. pneumophila*, the biphasic life cycle consisting of a replicative growth phase and a transmissive phase requires activation of both motility and virulence genes upon entry into the latter (Hammer et al., 2002; Tiaden et al., 2007; Al-Khodor et al., 2009; Edwards et al., 2010). As mentioned above, this occurs during the transition into stationary phase in response to low nutrient content and other environmental signals that are integrated by a complex regulatory network involving two-component (PmrA/PmrB, CpxR/CpxA, LetA/LetS/*rsm*YZ/CsrA) and quorum sensing (LqsA/LqsS/LqsR) systems. Results of microarray analyses have shown that these sensory transduction systems coregulate expression of T4SS and flagellar genes. *Brucella*, on the other hand, is an apparently nonmotile bacterium, but it does carry *fla* genes and recently was shown to produce a flagellum. Flagellar mutants retain the capacity to infect macrophages but show reduced persistence upon infection of mice, indicating that the flagellum is important for pathogenesis but is not involved in entry into host cells. The BvrRS two-component system and both LuxR homologs, VlbR and BlxR, coordinately control flagellar and T4SS gene expression (Delrue et al., 2005; Uzureau et al., 2007; Rambow-Larsen et al., 2008; Viadas et al., 2010). The motility and effector translocation machines might both be required for formation of the *Brucella* phagosome but not subsequently for formation of the replication niche in the ER.

Coregulation of the T4SS and Type IV Pili

Several examples also are known in which conjugation systems are coregulated with type IV pili (Fig. 8). Type IV pili are phylogenetically unrelated to conjugative pili produced by T4SSs and instead share a common ancestry with T2SSs. Type IV pili contribute to virulence through a number of important functions, including twitching motility, cell adhesion, attachment to host cells, autoagglutination, microcolony formation, and DNA uptake (Craig and Li, 2008). *E. coli* plasmid R64 carries genes for a conjugation apparatus related to the *L. pneumophila* Dot/Icm system that elaborates a comparatively thick conjugative pilus (Komano et al., 2000). These types of pili are not thought to be dynamic, as has been reported for the F pilus (Clarke et al., 2008), and this T4SS mediates plasmid transfer to recipient cells only on solid surfaces (Komano et al., 2000). However, R64 also carries genes for a type IV pilus (Yoshida et al., 1999). This pilus functions as an attachment device and additionally as a dynamic organelle that can draw donor and recipients together by extension and retraction. The type IV pilus endows R64-carrying cells with the capacity to efficiently conjugate with recipient cells in broth matings. R64 thus appears to have co-opted a type IV pilus as a way of dynamically sampling the environment for potential recipients to enhance the probability of plasmid transmission.

Recently, it was shown that the *P. aeruginosa* PAI-1 pathogenicity island also carries R64-like T4SS and type IV pilus gene clusters that function coordinately to mediate its intercellular spread through conjugation (Fig. 8) (Carter et al., 2010). Both machines are also reduced in their structural complexity, with a reduced number of discernible T4SS homologs (e.g., VirD4 T4CP and VirB4 ATPase) and an incomplete type IV pilus cluster, resulting in dependence on other bacterial chromosomal factors for biogenesis of this organelle.

A third example of T4SS and type IV pilus coregulation occurs in *N. gonorrhoeae* (Fig. 8) (Ramsey et al., 2011). *N. gonorrhoeae* is an exclusively human pathogen and the causative agent of the sexually transmitted disease gonorrhea. About 80% of gonococcal strains carry the genes for a T4SS on a chromosomally borne genetic island (the gonococcal genetic island [GGI]). This T4SS functions differently than the conjugation systems by exporting chromosomal DNA to the extracellular milieu independently of target cell contact. Gonococci are also naturally competent, and DNA uptake occurs via a type IV pilus-mediated competence system (Ramsey et al., 2011). This combination of both a DNA release system and a DNA uptake system is thought to have evolved to facilitate acquisition of fitness traits and contribute to genetic variation, e.g., phase/antigenic variation, in the short term and genome plasticity over evolutionary time. The gonococcal T4SS is an F-like system; its GGI carries F-like T4SS genes that

for the most part are collinear with the F plasmid *tra* genes (Hamilton et al., 2005). Interestingly, however, the GGI does not encode homologs of the primary regulators for F plasmid transfer, including TraJ, TraY, and TraM, and therefore must be regulated differently than F. Recently, it was observed that piliated strains release appreciably more DNA into the milieu than nonpiliated strains (Salgado-Pabon et al., 2010). Follow-up studies established that expression of some of the T4SS genes is upregulated in strains producing type IV pili (Pil$^+$) compared with the isogenic pilus-minus (Pil$^-$) variants. The upregulated genes include *traD* (T4CP) and *traI* (relaxase), but not those encoding subunits of the transfer channel. These findings led the authors to suggest that gonococcal cells activate expression of T4SS genes specifically involved in substrate processing and recruitment as a response to perception of signals released by neighboring, pilus-producing cells. Perception of such signals would indicate the presence of cells in the vicinity that are competent for genetic transformation and activate the DNA release system. Another interesting finding that warrants follow-up study is that expression of T4SS genes is required for intracellular growth of some gonococcal strains by a mechanism linked to iron acquisition (Zola et al., 2010).

POSTTRANSCRIPTIONAL CONTROL OF T4SS BIOGENESIS OR FUNCTION

Regulation of Substrate-T4CP Docking Reactions

Type IV secretion is also regulated posttranscriptionally through mechanisms controlling T4SS biogenesis at specific sites in the cell, recruitment of secretion substrates to the translocation channel, or donor-target cell interactions (Fig. 9). Some T4SSs assemble at cell poles, including *A. tumefaciens* VirB/VirD4 (Kumar and Das, 2002; Atmakuri et al., 2003; Judd et al., 2005); the *Coxiella burnetii* Dot/Icm, which is similar to that of *L. pneumophila* (Morgan et al., 2010); and *B. subtilis* ICE*Bs1* (Berkmen et al., 2010). Other T4SSs, including the plasmid R27 and pCF10 conjugation systems, appear to form discrete foci at several locations around the cell (Gilmour et al., 2001; Chen et al., 2008). Although factors responsible for spatial positioning are not well defined, a few mechanisms have been shown to promote recruitment of substrates to spatially localized machines (Fig. 9). One such mechanism involves VirC1, a member of the ParA/MinD-like ATPase family required for efficient processing of the oncogenic T-DNA for translocation through the VirB/VirD4 T4SS (Atmakuri et al., 2007). VirC1 was shown to

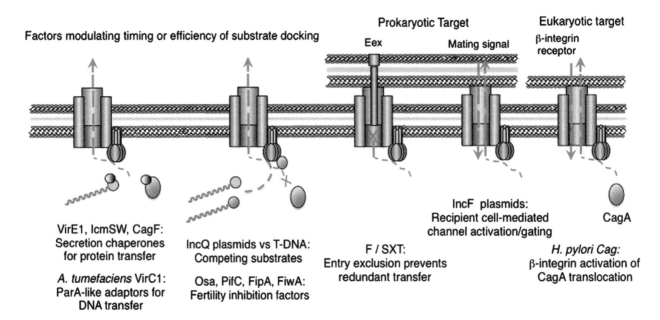

Figure 9. Posttranscriptional control of T4SS function. Intracellular factors can regulate the efficiency of timing of substrate translocation. Representative factors and their biological roles in modulating access of DNA or protein substrates with the T4CP receptor are listed. Target cell contact also regulates translocation. DNA transfer is inhibited by entry exclusion systems to prevent redundant transfer among equivalent donor cells. In F systems, a mating signal generated upon contact with a potential recipient cell activates the T4SS through an unknown mechanism. In *H. pylori*, binding of a T4SS adhesin serves to activate both β₁-integrin receptors on the mammalian cell and the Cag T4SS on the bacterial cell to stimulate CagA translocation. doi:10.1128/9781555818524.ch18f9

localize at the cell poles and interact with the polar VirD4 T4CP. Additionally, through interactions with the VirD2 relaxase and T-DNA, polar VirC1 was shown to mediate the polar positioning of the VirD2/ T-DNA transfer intermediate. Spatial positioning of the DNA substrate near the transfer machine is thought to promote the docking and translocation reactions. ParA-like proteins are commonly encoded by conjugative plasmids, and it is not known whether these proteins mediate one or both of the functions of plasmid partitioning and DNA substrate delivery to cognate T4SSs. However, ICEs also often carry genes for ParA-like functions despite the fact that these elements partition with the chromosome without the need of a dedicated partitioning function. These ICEs might rely on ParA functions for spatial coupling of the excised and circularized transfer intermediates with the T4SS apparatus. The *N. gonorrhoeae* GGI also encodes a ParA-like protein, and its production is necessary for delivery of chromosomal DNA through the GGI T4SS to the extracellular milieu, suggestive of a role in the coupling of DNA substrates with the T4SS (Hamilton et al., 2005).

Other adaptor or chaperone functions contribute to the docking of secretion substrates with their cognate transfer machines (Fig. 9). In *A. tumefaciens*, the VirE1 secretion chaperone is necessary for delivery of the VirE2 effector protein to the VirB/VirD4 T4SS.

VirE1 resembles secretion chaperones described for T3SSs in its physical properties (acidic pI and amphipathic helix) (Zhao et al., 2001). A VirE1-VirE2 crystal structure shows that VirE1 sequesters regions of VirE2 involved in self-association and interaction with DNA, and the C-terminal signal sequence is solvent exposed (Dym et al., 2008). Thus, VirE1 indirectly promotes VirE2 engagement with the T4SS by preventing premature protein-protein and -DNA interactions while enabling docking of the secretion signal with the VirD4 T4CP. In *L. pneumophila*, delivery of a subset of effectors to the Dot/Icm system is regulated by two chaperones, IcmS and IcmW, that also resemble the T3SS chaperones in physical properties and function (Fig. 9) (Bardill et al., 2005; Ninio et al., 2005; Vincent and Vogel, 2006; Cambronne and Roy, 2007). The IcmS and IcmW proteins are not essential components of the Dot/Icm secretion machinery, yet they interact with and mediate translocation of Sid and other effectors. Studies of SidG identified a central N-terminal domain where the IcmSW complex binds and a distinct C-terminal domain which carries the translocation signal. Interestingly, deletion of the IcmSW binding domain eliminated the requirement for the IcmSW complex to bind and trigger translocation. SidG therefore has an intrinsic regulatory function where an internal region inhibits recognition of the C-terminal translocation signal and

IcmSW binding relieves this inhibition. It is suggested that chaperone binding for translocation signal presentation represents a secondary level of regulation that contributes to appropriate timing of substrate delivery to the host. Similarly, in *Helicobacter pylori*, CagF is a chaperone required for CagA translocation through the Cag T4SS (Couturier et al., 2006; Pattis et al., 2007). A CagF-CagA complex associates predominantly with the membrane, and there is some evidence that CagF binds the membrane at or near the Cag T4SS, suggesting that it could coordinate docking of the CagA substrate with the translocation channel.

On the flip side, substrate docking can also be inhibited through binding of competing substrates or other factors (Fig. 9). In *A. tumefaciens*, IncQ plasmids are bona fide substrates of the VirB/VirD4 T4SS, but cells carrying these plasmids are suppressed in their transfer of the T-DNA and protein substrates to plants (Binns et al., 1995). IncQ plasmids interfere with docking of these substrates with the VirD4 T4CP, most probably through competitive inhibition (Cascales et al., 2005). Similarly, Osa, a protein encoded by IncW plasmids, strongly suppresses T-DNA and protein trafficking through the VirB/VirD4 T4SS, also by interfering with the docking of oncogenic substrates with VirD4 (Chen and Kado, 1994; Cascales et al., 2005). Osa is related to other fertility inhibitors—PifC (IncF), FipA (IncN), and FiwA (IncP)—that evolved to suppress transfer of coresident plasmids, most probably through interfering effects of substrate docking reactions (Cascales et al., 2005). Whether substrate-docking inhibitors modulate trafficking through effector translocators in pathogenic settings is not yet known.

Contact-Mediated T4SS Suppression or Activation

Conjugation systems have evolved several mechanisms to "sample" the environment for potential recipient cells and to prevent formation of donor-donor cell mating pairs (Fig. 9). As mentioned above, quorum signals and quorum signal-degrading factors or inhibitor analogs regulate *tra* gene expression and propagation of mobile elements. At the posttranscriptional level, a recent study showed that *E. coli* cells carrying IncF plasmids dynamically assemble and disassemble their F pili (Clarke et al., 2008). F pilus extension/retraction occurs stochastically and in the absence of a triggering event, suggesting that donor cells use pilus extension dynamics to randomly sample their environment for the presence of recipient cells.

In the event that cells in a donor population do form close contacts, some mobile elements prevent redundant transfer through a mechanism termed entry exclusion. IncF plasmid encodes an outer membrane surface lipoprotein, TraT, that might function to prevent pilus binding or channel activation (Lawley et al., 2003). Additionally, F plasmids code for two inner membrane components, TraS and TraG (Audette et al., 2007). TraG is related to *A. tumefaciens* VirB6, a polytopic inner membrane subunit thought to comprise part of the translocation channel. However, TraG carries a large C-terminal extension that is missing in VirB6. In donor-donor cell contacts, this C-terminal extension was shown to project across the donor cell envelope to establish a specific interaction with TraS in the partner cell. This TraG-TraS interaction blocks further mating pair formation and plasmid transfer. In donor-recipient mating pairs, by contrast, TraG is thought to project across the cell-cell interface and interact with unspecified inner membrane protein to establish a productive mating junction. The *V. cholerae* SXT ICE possesses a similar TraG-TraS Eex system, and homologs of these factors are present in other conjugation systems, suggesting that this exclusion mechanism is widely used to reduce redundant transfer (Marrero and Waldor, 2005, 2007).

Early studies of F also identified a "mating signal" propagated by recipient cells upon contact with plasmid-carrying donor cells (Lu and Frost, 2005). The nature of this mating signal is not known, but this signal supports the notion that T4SSs are gated and activated to translocate substrates only upon formation of productive mating junctions. The molecular details of donor-target cell contacts are also not well defined, but in the *H. pylori* Cag T4SS system, a putative adhesin, CagL, was shown to localize at the tip of a Cag-encoded pilus (Fig. 9). CagL binds to and activates integrin receptors on mammalian epithelial cells through an Arg-Gly-Asp (RGD) motif (Kwok et al., 2007; Backert et al., 2008; Tegtmeyer et al., 2010; Conradi et al., 2011). Then, integrin activation, in turn, stimulates the Cag T4SS for translocation of the CagA effector protein to the mammalian cell. Other T4SS subunits, including CagI and CagY, as well as CagA itself, bind the transmembrane receptor β_1 integrin, suggesting that a complex network of protein-protein interactions at the *H. pylori*-mammalian cell interface activates both the Cag T4SS and its integrin receptor on the mammalian host cell to effect CagA translocation (Tegtmeyer et al., 2011). It is interesting to speculate that T4SS-receptor binding activates the translocation channel by stimulating T4SS channel gating.

SUMMARY AND PERSPECTIVES

The long-term studies of conjugation systems and more recent investigations of the effector translocators

Table 1. Regulatory mechanisms controlling type IV secretion

Regulatory mechanism	Signal perceived	Response	Role in pathogenesis
Two-component systems			
A. tumefaciens VirA/VirG(ChvE)	Plant phenolics, sugars	Induction of vir genes	T-DNA/effector protein transfer, plant tumor production, opine release
	Acidic pH	Regulates VirB/VirD4 T4SS; stimulation of rep genes	
E. coli F plasmid CpxR/CpxA	Extracytoplasmic stress: pH, growth conditions	Upregulation of HslVU; TraJ cleavage, reduced F tra gene expression	Reduction in plasmid transfer and pilus-mediated attachment
Legionella CpxR/CpxA	Unknown stress signal	Induction of dot/icm genes, + or − control of effector genes	Effector translocation in postexponential (transmissive) phase
A. tumefaciens ChvG/ChvI	Acidic pH	Stimulation of some vir genes, induction of genes for growth in acid	Growth in plant wound environment, T-DNA transfer
Brucella BvrR/BvrS	Acidic pH of Brucella phagosome?	Induction of virB genes	Effector translocation in the Brucella phagosome
Bartonella BatR/BatS	Neutral pH of blood	Induction of VirB/VirD4 and BEP effector genes	Effector translocation in human epithelial cells
Sinorhizobium meliloti ExoS/ChvI	Acidic pH of plant host?	Induction of succinoglycan and flagellar genes; unknown effect on T4SS regulation	Sinorhizobium-plant symbiosis
Legionella PmrA/PmrB	Unknown	Induction of dot/icm genes, + or − control of effector genes	Effector translocation in postexponential (transmissive) phase
Legionella LetA/ LetS	Nutritional stress	Regulates effector genes	Effector translocation
Legionella LqsS/LqsR	Nutritional stress (low AAs)[a]	Regulates dot/icm effector genes, flagellar genes	Effector translocation, motility
Quorum-sensing systems			
Homoserine lactones			
A. tumefaciens LuxR	3OC8-HSL	pTi tra/trb upregulation, pTi rep gene upregulation	pTi dissemination at infection site, pTi copy number increase
Brucella VjbR	C12-HSL?, other	Blocks VjbR inducer activity, regulates flagellar genes	Might suppress virB expression in replication niche in ER
Brucella BlxR	Unknown	Regulates virB and flagellar genes	Unknown, but has overlapping gene targets with VjbR
Legionella LqsR	3-Hydroxypentadecan-4-one (LAI-1)?	Regulates some effector genes	Effector translocation
Peptide pheromones			
E. faecalis pCF10	cCF10; iCF10	pCF10 tra gene induction	Plasmid transfer (pheromone-dependent plasmids code for antibiotic resistance/hemolysins/other virulence traits)
		AS production	Bacterial aggregation, pCF10 transfer, biofilms on abiotic and biotic surfaces

System	Signal	Mechanism	Outcome
B. subtilis ICE*Bs1*	PhlR	Blocks expression of excision and transfer genes	Promotes ICE transfer to ICE-minus recipients
Environmentally activated systems			
SOS response			
B. subtilis ICE*Bs1*	DNA damage, e.g., oxidative compounds; quinolone antibiotics	Activates ImmA protease that degrades ImmR repressor	Stimulation of ICE transfer
V. cholerae SXT	DNA damage, e.g., oxidative compounds; quinolone antibiotics	RecA-stimulated autoproteolysis of SetR repressor	Stimulation of SXT transfer
Stringent response			
L. pneumophila	Nutrient stress	Coordination of LetA/S, RelA, and (p)ppGpp to relieve CsrA repression of effector gene expression	Modulates effector gene expression during transmissive phase
Brucella spp.	Nutrient stress	Coordination of Rsh and (p)ppGpp to activate HutC binding to *virB* promoter	HutC-mediated expression of *virB* genes in *Brucella* phagosome
A. tumefaciens	Nutrient stress	RelA and (p)ppGpp mediate activation of *attM* gene encoding lactonase	Degradation of 3OC8-HSL; reduction in Ti plasmid transfer in stationary phase
Responses to antibiotics			
Tn916, CTnDot	Tetracycline	Activates expression of excision and *tra* genes	Tetracycline-induced transmission of ICEs coding for Tet resistance
Regulation by small RNAs			
E. coli F plasmid			
Hfq/UtpR/*traJ* leader	Growth phase?	Repression of *traJ* expression	Reduced plasmid transfer
FinO/*finP*	Coresident plasmids	Repression of *traYALE-X* expression	Regulates F transfer, also can control transmission of coresident plasmids
E. faecalis pCF10	Pheromone, indirectly	Controls *tra* gene expression	Regulates pCF10 transfer by inhibiting *prgQ* transcription
Legionella Dot/Icm			
Hfq/*rsmYZ*/CsrA	Stationary-phase signals, e.g., low AAs	Modulation of CsrA RNA-binding activity, + or − regulation of effector genes	Dot/Icm-mediated translocation of effectors

*a*AAs, amino acids.

have identified two broad classes of regulatory mechanisms, those adapted specifically for activation of a given T4SS in response to bacterial or environmental cues and more general global systems adapted not only for controlling T4SS biogenesis but also for other cellular activities in response to sensing of the cell's physiological status. Table 1 lists common regulatory components of the complex regulatory circuitries reviewed in this chapter.

The inducible DNA transfer systems rely on two-component sensory/response systems or quorum sensing systems to respond to specific signals from neighboring bacteria or the environment. Among the best characterized of the inducible systems (*A. tumefaciens* VirB/VirD4 and Tra/Trb, *E. faecalis* pCF10, and *B. subtilis* ICE*Bs1*), the activating signals are known, as are mechanisms for signal uptake and modulation of T4SS gene transcription. Among the effector translocator systems, similar regulatory mechanisms are also employed, although our present understanding of transduction pathways is poor even for the better-characterized systems. ChvG/ChvI-like two-component systems have been adapted by *A. tumefaciens*, *Brucella*, *Bartonella*, and *S. meliloti*, most probably as a pH sensor. Other two-component systems have been adapted for sensory recognition of T4SS gene expression in response to a specific exogenous signal, e.g., *A. tumefaciens* VirA/VirG, or more general stresses/signals, e.g., CpxR/CpxA and LetA/LetS for mounting a more global response (Table 1).

In the well-characterized DNA transfer systems, at least three types of responses are coordinated upon perception of quorum signals released by neighboring cells. First, the T4SS and DNA processing machinery are synthesized. Second, various mechanisms are elaborated for discrimination of self, e.g., plasmid-carrying donors, from nonself, e.g., plasmid-free recipients. Finally, negative-feedback systems control the quorum response; for example, *A. tumefaciens* produces lactonase upon entry into stationary phase and plasmid-carrying *E. faecalis* produces competitive inhibitors. Among the pathogenic bacteria that utilize quorum signaling to control expression of genes for the effector translocator machines or their substrates, the inducing signals have yet to be identified or confirmed. Regardless, the evidence favors their involvement in regulating *Brucella* VirB and *L. pneumophila* Dot/Icm translocation.

Correspondingly, it can be predicted that these systems also have evolved mechanisms for discrimination of appropriate from inappropriate target cells. Discrimination of self from nonself might not be a problem for effector translocators that evolved specialized adhesins for binding to eukaryotic target cell receptors. However, the *L. pneumophila* and *Bartonella* T4SSs translocate effectors to human cells as well as DNA and protein substrates to bacterial recipients; for these pathogens, discrimination of self from nonself is critical for successful infection. Finally, an area of investigation currently receiving considerable attention is how effector protein trafficking is temporally and spatially controlled in the human host. Temporal control of both machine assembly and inactivation is probably critical, although at this time there is little knowledge of negative-feedback mechanisms operating during the infection cycle.

Another regulatory theme common among all T4SSs is their activation in response to environmental and physiological signals. Some environmental signals activate T4SSs through specific mechanisms, as illustrated by tetracycline-inducible transfer of Tn*916* and CTnDOT ICEs. The *A. tumefaciens* VirB/VirD4 system, various other mobile DNA systems, and probably all effector translocators are temperature dependent, although the mechanism of temperature sensing is generally poorly understood at this time. Two global stress responses, the SOS response and the stringent response, modulate T4SS biogenesis and function through well-characterized transduction networks that also control other cellular activities, including membrane biogenesis, and flagellar or type IV pilus-based motility. For the conjugation systems, such global regulatory controls appear to confine DNA transfer to metabolically active cells so that transmission is repressed upon entry into stationary-phase growth or exposure to nutrient-poor environments. For the effector translocators, sensing of the cell's physiological status restricts T4SS assembly and effector translocation to specific niches in the human host, e.g., the *Brucella* phagosome and *L. pneumophila* LCVs, wherein effector delivery promotes colonization and transmission. Results of microarray studies have also shown that global regulatory systems coordinate the assembly of T4SSs with other surface organelles, and here we highlighted the biological importance of cross-regulating T4SSs, flagella, and type IV pili during infection.

For all T4SSs, transduction of exogenous or physiological signals ultimately converges on the regulatory machinery controlling transcription of machine subunits, DNA processing enzymes, or protein effectors. Among the well-characterized conjugation machines, three main factors controlling transcription include (i) dedicated transcription factors, e.g., F plasmid TraJ; (ii) antisense RNAs and cognate RNA-binding proteins, e.g., F plasmid finP/FinO; and (iii) proteases, e.g., *B. subtilis* ICE*Bs1* ImmA. In general, these regulatory molecules are encoded within or near the genes coding for the T4SS machine and nic site. Studies of the effector translocator systems have not yet approached the level of resolution of the conjugation systems, preventing any conclusions about the role of these types of factors in T4SS biogenesis or function. However, recent work has shown the importance of global sRNAs and cognate CsrA, and possibly Hfq, RNA-binding proteins in regulating expression of *L. pneumophila dot/icm* effector genes. Identification of dedicated transcription factors, sRNAs, and proteases, and definition of their contributions to spatiotemporal control of T4SS assembly and function during infection, is a next critical phase of research on T4SS regulation in pathogenic bacteria.

Acknowledgments. We thank members of the Christie laboratory for helpful comments. J.A.L.-G. and M.S. contributed equally to this review.

Work in the Christie laboratory is supported by NIH grant GM48746 and BARD grant IS-4245-09.

REFERENCES

Al-Khodor, S., S. Kalachikov, I. Morozova, C. T. Price, and Y. Abu Kwaik. 2009. The PmrA/PmrB two-component system of *Legionella pneumophila* is a global regulator required for intracellular replication within macrophages and protozoa. *Infect. Immun.* 77:374–386.

Al-Khodor, S., C. T. Price, F. Habyarimana, A. Kalia, and Y. Abu Kwaik. 2008. A Dot/Icm-translocated ankyrin protein of *Legionella pneumophila* is required for intracellular proliferation within human macrophages and protozoa. *Mol. Microbiol.* 70:908–923.

Altman, E., and G. Segal. 2008. The response regulator CpxR directly regulates expression of several *Legionella pneumophila icm/dot* components as well as new translocated substrates. *J. Bacteriol.* 190:1985–1996.

Alt-Morbe, J., J. L. Stryker, C. Fuqua, P. L. Li, S. K. Farrand, and S. C. Winans. 1996. The conjugal transfer system of *Agrobacterium tumefaciens* octopine-type Ti plasmids is closely related to the transfer system of an IncP plasmid and distantly related to Ti plasmid *vir* genes. *J. Bacteriol.* 178:4248–4257.

Alvarez-Martinez, C. E., and P. J. Christie. 2009. Biological diversity of prokaryotic type IV secretion systems. *Microbiol. Mol. Biol. Rev.* 73:775–808.

Amor, J. C., J. Swails, X. Zhu, C. R. Roy, H. Nagai, A. Ingmundson, X. Cheng, and R. A. Kahn. 2005. The structure of RalF, an ADP-ribosylation factor guanine nucleotide exchange factor from *Legionella pneumophila*, reveals the presence of a cap over the active site. *J. Biol. Chem.* 280:1392–1400.

Anderson, J. K., T. G. Smith, and T. R. Hoover. 2010. Sense and sensibility: flagellum-mediated gene regulation. *Trends Microbiol.* 18:30–37.

Antiporta, M. H., and G. M. Dunny. 2002. *ccfA*, the genetic determinant for the cCF10 peptide pheromone in *Enterococcus faecalis* OG1RF. *J. Bacteriol.* 184:1155–1162.

Arellano-Reynoso, B., N. Lapaque, S. Salcedo, G. Briones, A. E. Ciocchini, R. Ugalde, E. Moreno, I. Moriyon, and J. P. Gorvel. 2005. Cyclic beta-1,2-glucan is a *Brucella* virulence factor required for intracellular survival. *Nat. Immunol.* 6:618–625.

Arthur, D. C., A. F. Ghetu, M. J. Gubbins, R. A. Edwards, L. S. Frost, and J. N. Glover. 2003. FinO is an RNA chaperone that facilitates sense-antisense RNA interactions. *EMBO J.* 22:6346–6355.

Atmakuri, K., E. Cascales, O. T. Burton, L. M. Banta, and P. J. Christie. 2007. *Agrobacterium* ParA/MinD-like VirC1 spatially coordinates early conjugative DNA transfer reactions. *EMBO J.* 26:2540–2551.

Atmakuri, K., Z. Ding, and P. J. Christie. 2003. VirE2, a type IV secretion substrate, interacts with the VirD4 transfer protein at cell poles of *Agrobacterium tumefaciens*. *Mol. Microbiol.* 49:1699–1713.

Auchtung, J. M., C. A. Lee, R. E. Monson, A. P. Lehman, and A. D. Grossman. 2005. Regulation of a *Bacillus subtilis* mobile genetic element by intercellular signaling and the global DNA damage response. *Proc. Natl. Acad. Sci. USA* 102:12554–12559.

Audette, G. F., J. Manchak, P. Beatty, W. A. Klimke, and L. S. Frost. 2007. Entry exclusion in F-like plasmids requires intact TraG in the donor that recognizes its cognate TraS in the recipient. *Microbiology* 153:442–451.

Bachman, M. A., and M. S. Swanson. 2004. Genetic evidence that *Legionella pneumophila* RpoS modulates expression of the transmission phenotype in both the exponential phase and the stationary phase. *Infect. Immun.* 72:2468–2476.

Bachman, M. A., and M. S. Swanson. 2001. RpoS co-operates with other factors to induce *Legionella pneumophila* virulence in the stationary phase. *Mol. Microbiol.* 40:1201–1214.

Backert, S., and M. Clyne. 2011. Pathogenesis of *Helicobacter pylori* infection. *Helicobacter* 16(Suppl. 1):19–25.

Backert, S., R. Fronzes, and G. Waksman. 2008. VirB2 and VirB5 proteins: specialized adhesins in bacterial type-IV secretion systems? *Trends Microbiol.* 16:409–413.

Bardill, J. P., J. L. Miller, and J. P. Vogel. 2005. IcmS-dependent translocation of SdeA into macrophages by the *Legionella pneumophila* type IV secretion system. *Mol. Microbiol.* 56:90–103.

Baron, C., and B. Coombes. 2007. Targeting bacterial secretion systems: benefits of disarmament in the microcosm. *Infect. Disord. Drug Targets* 7:19–27.

Barrios, A. F., R. Zuo, D. Ren, and T. K. Wood. 2006. Hha, YbaJ, and OmpA regulate *Escherichia coli* K12 biofilm formation and conjugation plasmids abolish motility. *Biotechnol. Bioeng.* 93:188–200.

Bates, S., A. M. Cashmore, and B. M. Wilkins. 1998. IncP plasmids are unusually effective in mediating conjugation of *Escherichia coli* and *Saccharomyces cerevisiae*. *J. Bacteriol.* 180:6538–6543.

Beaber, J. W., B. Hochhut, and M. K. Waldor. 2004. SOS response promotes horizontal dissemination of antibiotic resistance genes. *Nature* 427:72–74.

Berkmen, M. B., C. A. Lee, E. K. Loveday, and A. D. Grossman. 2010. Polar positioning of a conjugation protein from the integrative and conjugative element ICE*Bs1* of *Bacillus subtilis*. *J. Bacteriol.* 192:38–45.

Binns, A. N., C. E. Beaupre, and E. M. Dale. 1995. Inhibition of VirB-mediated transfer of diverse substrates from *Agrobacterium tumefaciens* by the IncQ plasmid RSF1010. *J. Bacteriol.* 177:4890–4899.

Boschiroli, M. L., S. Ouahrani-Bettache, V. Foulongne, S. Michaux-Charachon, G. Bourg, A. Allardet-Servent, C. Cazevieille, J. P. Lavigne, J. P. Liautard, M. Ramuz, and D. O'Callaghan. 2002a. Type IV secretion and *Brucella* virulence. *Vet. Microbiol.* 90:341–348.

Boschiroli, M. L., S. Ouahrani-Bettache, V. Foulongne, S. Michaux-Charachon, G. Bourg, A. Allardet-Servent, C. Cazevieille, J. P. Liautard, M. Ramuz, and D. O'Callaghan. 2002b. The *Brucella suis virB* operon is induced intracellularly in macrophages. *Proc. Natl. Acad. Sci. USA* 99:1544–1549.

Bose, B., J. M. Auchtung, C. A. Lee, and A. D. Grossman. 2008. A conserved anti-repressor controls horizontal gene transfer by proteolysis. *Mol. Microbiol.* 70:570–582.

Bose, B., and A. D. Grossman. 2011. Regulation of horizontal gene transfer in *Bacillus subtilis* by activation of a conserved site-specific protease. *J. Bacteriol.* **193**:22–29.

Burstein, D., T. Zusman, E. Degtyar, R. Viner, G. Segal, and T. Pupko. 2009. Genome-scale identification of *Legionella pneumophila* effectors using a machine learning approach. *PLoS Pathog.* **5**:e1000508.

Byrd, D. R., and S. W. Matson. 1997. Nicking by transesterification: the reaction catalysed by a relaxase. *Mol. Microbiol.* **25**:1011–1022.

Cambronne, E. D., and C. R. Roy. 2007. The *Legionella pneumophila* IcmSW complex interacts with multiple Dot/Icm effectors to facilitate type IV translocation. *PLoS Pathog.* **3**:e188.

Cangelosi, G. A., R. G. Ankenbauer, and E. W. Nester. 1990. Sugars induce the *Agrobacterium virulence* genes through a periplasmic binding protein and a transmembrane signal protein. *Proc. Natl. Acad. Sci. USA* **87**:6708–6712.

Carter, M. Q., J. Chen, and S. Lory. 2010. The *Pseudomonas aeruginosa* pathogenicity island PAPI-1 is transferred via a novel type IV pilus. *J. Bacteriol.* **192**:3249–3258.

Cascales, E., K. Atmakuri, Z. Liu, A. N. Binns, and P. J. Christie. 2005. *Agrobacterium tumefaciens* oncogenic suppressors inhibit T-DNA and VirE2 protein substrate binding to the VirD4 coupling protein. *Mol. Microbiol.* **58**:565–579.

Cascales, E., and P. J. Christie. 2003. The versatile bacterial type IV secretion systems. *Nat. Rev. Microbiol.* **1**:137–150.

Chai, Y., C. S. Tsai, H. Cho, and S. C. Winans. 2007. Reconstitution of the biochemical activities of the AttJ repressor and the AttK, AttL, and AttM catabolic enzymes of *Agrobacterium tumefaciens*. *J. Bacteriol.* **189**:3674–3679.

Chai, Y., J. Zhu, and S. C. Winans. 2001. TrlR, a defective TraR-like protein of *Agrobacterium tumefaciens*, blocks TraR function *in vitro* by forming inactive TrlR:TraR dimers. *Mol. Microbiol.* **40**:414–421.

Chandler, J. R., A. R. Flynn, E. M. Bryan, and G. M. Dunny. 2005. Specific control of endogenous cCF10 pheromone by a conserved domain of the pCF10-encoded regulatory protein PrgY in *Enterococcus faecalis*. *J. Bacteriol.* **187**:4830–4843.

Chang, C. H., and S. C. Winans. 1992. Functional roles assigned to the periplasmic, linker, and receiver domains. *J. Bacteriol.* **174**:7033–7039.

Chang, C. H., and S. C. Winans. 1996. Resection and mutagenesis of the acid pH-inducible P2 promoter of the *Agrobacterium tumefaciens virG* gene. *J. Bacteriol.* **178**:4717–4720.

Chatterjee, A., C. M. Johnson, C.-C. Shu, Y. N. Kaznessis, D. Ramkrishna, G. M. Dunny, and W.-S. Hu. 2011. Convergent transcription confers a bistable switch in *Enterococcus faecalis* conjugation. *Proc. Natl. Acad. Sci. USA* **108**:9721–9726.

Chen, C. Y., and C. I. Kado. 1994. Inhibition of *Agrobacterium tumefaciens* oncogenicity by the *osa* gene of pSa. *J. Bacteriol.* **176**:5697–5703.

Chen, Y., X. Zhang, D. Manias, H. J. Yeo, G. M. Dunny, and P. J. Christie. 2008. *Enterococcus faecalis* PcfC, a spatially localized substrate receptor for type IV secretion of the pCF10 transfer intermediate. *J. Bacteriol.* **190**:3632–3645.

Cheng, H. P., and G. C. Walker. 1998. Succinoglycan production by *Rhizobium meliloti* is regulated through the ExoS-ChvI two-component regulatory system. *J. Bacteriol.* **180**:20–26.

Cheng, H. P., and S. Y. Yao. 2004. The key *Sinorhizobium meliloti* succinoglycan biosynthesis gene *exoY* is expressed from two promoters. *FEMS Microbiol. Lett.* **231**:131–136.

Cho, H., and S. C. Winans. 2005. VirA and VirG activate the Ti plasmid *repABC* operon, elevating plasmid copy number in response to wound-released chemical signals. *Proc. Natl. Acad. Sci. USA* **102**:14843–14848.

Christie, P. J. 2009. *Agrobacterium* and plant cell transformation, p. 29–43. *In* M. Schaechter (ed.), *The Desk Encyclopedia of Microbiology*, 2nd ed. Academic Press, San Diego, CA.

Christie, P. J. 2004. Bacterial type IV secretion: the *Agrobacterium* VirB/D4 and related conjugation systems. *Biochim. Biophys. Acta* **1694**:219–234.

Christie, P. J., K. Atmakuri, V. Krishnamoorthy, S. Jakubowski, and E. Cascales. 2005. Biogenesis, architecture, and function of bacterial type IV secretion systems. *Annu. Rev. Microbiol.* **59**:451–485.

Chuang, O. N., P. M. Schlievert, C. L. Wells, D. A. Manias, T. J. Tripp, and G. M. Dunny. 2009. Multiple functional domains of *Enterococcus faecalis* aggregation substance Asc10 contribute to endocarditis virulence. *Infect. Immun.* **77**:539–548.

Chuang-Smith, O. N., C. L. Wells, M. J. Henry-Stanley, and G. M. Dunny. 2010. Acceleration of *Enterococcus faecalis* biofilm formation by aggregation substance expression in an *ex vivo* model of cardiac valve colonization. *PLoS One* **5**:e15798.

Clarke, M., L. Maddera, R. L. Harris, and P. M. Silverman. 2008. F-pili dynamics by live-cell imaging. *Proc. Natl. Acad. Sci. USA* **105**:17978–17981.

Clewell, D. B. 2007. Properties of *Enterococcus faecalis* plasmid pAD1, a member of a widely disseminated family of pheromone-responding, conjugative, virulence elements encoding cytolysin. *Plasmid* **58**:205–227.

Clewell, D. B., F. Y. An, S. E. Flannagan, M. Antiporta, and G. M. Dunny. 2000. Enterococcal sex pheromone precursors are part of signal sequences for surface lipoproteins. *Mol. Microbiol.* **35**:246–247.

Coburn, B., I. Sekirov, and B. B. Finlay. 2007. Type III secretion systems and disease. *Clin. Microbiol. Rev.* **20**:535–549.

Conradi, J., S. Huber, K. Gaus, F. Mertink, S. Royo Gracia, U. Strijowski, S. Backert, and N. Sewald. 2011. Cyclic RGD peptides interfere with binding of the *Helicobacter pylori* protein CagL to integrins alpha(V)beta (3) and alpha (5)beta (1). *Amino Acids* doi 10.1007/s00726-011-1066-0.

Cook, L., A. Chatterjee, A. Barnes, J. Yarwood, W.-S. Hu, and G. Dunny. 2011. Biofilm growth alters regulation of conjugation by a bacterial pheromone. *Mol. Microbiol.* doi:10.1111/j.1365-2958.2011.07786.x.

Couturier, M. R., E. Tasca, C. Montecucco, and M. Stein. 2006. Interaction with CagF is required for translocation of CagA into the host via the *Helicobacter pylori* type IV secretion system. *Infect. Immun.* **74**:273–281.

Craig, L., and J. Li. 2008. Type IV pili: paradoxes in form and function. *Curr. Opin. Struct. Biol.* **18**:267–277.

Dalebroux, Z. D., S. L. Svensson, E. C. Gaynor, and M. S. Swanson. 2010. ppGpp conjures bacterial virulence. *Microbiol. Mol. Biol. Rev.* **74**:171–199.

Dean, P. 2011. Functional domains and motifs of bacterial type III effector proteins and their roles in infection. *FEMS Microbiol. Rev.* doi:10.1111/j.1574-6976.2011.00271.x.

de Barsy, M., A. Jamet, D. Filopon, C. Nicolas, G. Laloux, J. F. Rual, A. Muller, J. C. Twizere, J. Nkengfac, J. Vandenhaute, D. E. Hill, S. P. Salcedo, J. P. Gorvel, J. J. Letesson, and X. De Bolle. 2011. Identification of a *Brucella* spp. secreted effector specifically interacting with human small GTPase Rab2. *Cell. Microbiol.* **13**:1044–1058.

Dehio, C. 2005. *Bartonella*-host-cell interactions and vascular tumour formation. *Nat. Rev. Microbiol.* **3**:621–631.

Dehio, C. 2008. Infection-associated type IV secretion systems of *Bartonella* and their diverse roles in host cell interaction. *Cell. Microbiol.* **10**:1591–1598.

de Jong, M. F., Y. H. Sun, A. B. den Hartigh, J. M. van Dijl, and R. M. Tsolis. 2008. Identification of VceA and VceC, two members of the VjbR regulon that are translocated into macrophages by the *Brucella* type IV secretion system. *Mol. Microbiol.* 70:1378–1396.

Delrue, R. M., C. Deschamps, S. Leonard, C. Nijskens, I. Danese, J. M. Schaus, S. Bonnot, J. Ferooz, A. Tibor, X. De Bolle, and J. J. Letesson. 2005. A quorum-sensing regulator controls expression of both the type IV secretion system and the flagellar apparatus of *Brucella melitensis. Cell. Microbiol.* 7:1151–1161.

den Hartigh, A. B., H. G. Rolan, M. F. de Jong, and R. M. Tsolis. 2008. VirB3 to VirB6 and VirB8 to VirB11, but not VirB7, are essential for mediating persistence of *Brucella* in the reticuloendothelial system. *J. Bacteriol.* 190:4427–4436.

Diaz, M. R., J. M. King, and T. L. Yahr. 2011. Intrinsic and extrinsic regulation of type III secretion gene expression in *Pseudomonas aeruginosa. Front. Microbiol.* 2:89.

Dozot, M., R. A. Boigegrain, R. M. Delrue, R. Hallez, S. Ouahrani-Bettache, I. Danese, J. J. Letesson, X. De Bolle, and S. Kohler. 2006. The stringent response mediator Rsh is required for *Brucella melitensis* and *Brucella suis* virulence, and for expression of the type IV secretion system *virB. Cell. Microbiol.* 8:1791–1802.

Dudley, E. G., C. Abe, J. M. Ghigo, P. Latour-Lambert, J. C. Hormazabal, and J. P. Nataro. 2006. An IncI1 plasmid contributes to the adherence of the atypical enteroaggregative *Escherichia coli* strain C1096 to cultured cells and abiotic surfaces. *Infect. Immun.* 74:2102–2114.

Dunny, G. M. 2007. The peptide pheromone-inducible conjugation system of *Enterococcus faecalis* plasmid pCF10: cell-cell signalling, gene transfer, complexity and evolution. *Philos. Trans. R. Soc. Lond. B* 362:1185–1193.

Dunny, G. M., B. L. Brown, and D. B. Clewell. 1978. Induced cell aggregation and mating in *Streptococcus faecalis*: evidence for a bacterial sex pheromone. *Proc. Natl. Acad. Sci. USA* 75:3479–3483.

Dunny, G. M., and C. M. Johnson. 2011. Regulatory circuits controlling enterococcal conjugation: lessons for functional genomics. *Curr. Opin. Microbiol.* 14:174–180.

Dym, O., S. Albeck, T. Unger, J. Jacobovitch, A. Branzburg, Y. Michael, D. Frenkiel-Krispin, S. G. Wolf, and M. Elbaum. 2008. Crystal structure of the *Agrobacterium virulence* complex VirE1-VirE2 reveals a flexible protein that can accommodate different partners. *Proc. Natl. Acad. Sci. USA* 105:11170–11175.

Edwards, R. L., M. Jules, T. Sahr, C. Buchrieser, and M. S. Swanson. 2010. The *Legionella pneumophila* LetA/LetS two-component system exhibits rheostat-like behavior. *Infect. Immun.* 78:2571–2583.

Erhardt, M., K. Namba, and K. T. Hughes. 2010. Bacterial nanomachines: the flagellum and type III injectisome. *Cold Spring Harbor Perspect. Biol.* 2:a000299.

Fernandez-Gonzalez, E., H. D. de Paz, A. Alperi, L. Agundez, M. Faustmann, F. J. Sangari, C. Dehio, and M. Llosa. 2011. Transfer of R388 derivatives by a pathogenesis-associated type IV secretion system into both bacteria and human cells. *J. Bacteriol.* 193:6257–6265.

Fischer, W. 2011. Assembly and molecular mode of action of the *Helicobacter pylori* Cag type IV secretion apparatus. *FEBS J.* 278:1203–1212.

Foulongne, V., G. Bourg, C. Cazevieille, S. Michaux-Charachon, and D. O'Callaghan. 2000. Identification of *Brucella suis* genes affecting intracellular survival in an in vitro human macrophage infection model by signature-tagged mutagenesis. *Infect. Immun.* 68:1297–1303.

Frank, A. C., C. M. Alsmark, M. Thollesson, and S. G. Andersson. 2005. Functional divergence and horizontal transfer of type IV secretion systems. *Mol. Biol. Evol.* 22:1325–1336.

Franz, B., and V. A. Kempf. 2011. Adhesion and host cell modulation: critical pathogenicity determinants of *Bartonella henselae. Parasites Vectors* 4:54.

Frost, L. S. 2009. Conjugation, bacterial, p. 294–308. *In* M. Schaechter (ed.), *The Desk Encyclopedia of Microbiology*, 2nd ed. Academic Press, San Diego, CA.

Frost, L. S., and G. Koraimann. 2010. Regulation of bacterial conjugation: balancing opportunity with adversity. *Future Microbiol.* 5:1057–1071.

Frost, L. S., R. Leplae, A. O. Summers, and A. Toussaint. 2005. Mobile genetic elements: the agents of open source genomics. *Nat. Rev. Microbiol.* 3:722–732.

Fuqua, C., M. Burbea, and S. C. Winans. 1995. Activity of the *Agrobacterium* Ti plasmid conjugal transfer regulator TraR is inhibited by the product of the *traM* gene. *J. Bacteriol.* 177:1367–1373.

Fuqua, W. C., and S. C. Winans. 1994. A LuxR-LuxI type regulatory system activates *Agrobacterium* Ti plasmid conjugal transfer in the presence of a plant tumor metabolite. *J. Bacteriol.* 176:2796–2806.

Gal-Mor, O., T. Zusman, and G. Segal. 2002. Analysis of DNA regulatory elements required for expression of the *Legionella pneumophila icm* and *dot* virulence genes. *J. Bacteriol.* 184:3823–3833.

Ghigo, J. M. 2001. Natural conjugative plasmids induce bacterial biofilm development. *Nature* 412:442–445.

Gilmour, M. W., T. D. Lawley, M. M. Rooker, P. J. Newnham, and D. E. Taylor. 2001. Cellular location and temperature-dependent assembly of IncHI1 plasmid R27-encoded TrhC-associated conjugative transfer protein complexes. *Mol. Microbiol.* 42:705–715.

Grohmann, E., G. Muth, and M. Espinosa. 2003. Conjugative plasmid transfer in Gram-positive bacteria. *Microbiol. Mol. Biol. Rev.* 67:277–301.

Gubbins, M. J., I. Lau, W. R. Will, J. M. Manchak, T. L. Raivio, and L. S. Frost. 2002. The positive regulator, TraJ, of the *Escherichia coli* F plasmid is unstable in a *cpxA** background. *J. Bacteriol.* 184:5781–5788.

Hamilton, H. L., N. M. Dominguez, K. J. Schwartz, K. T. Hackett, and J. P. Dillard. 2005. *Neisseria gonorrhoeae* secretes chromosomal DNA via a novel type IV secretion system. *Mol. Microbiol.* 55:1704–1721.

Hammer, B. K., E. S. Tateda, and M. S. Swanson. 2002. A two-component regulator induces the transmission phenotype of stationary-phase *Legionella pneumophila. Mol. Microbiol.* 44:107–118.

Han, D. C., C. Y. Chen, Y. F. Chen, and S. C. Winans. 1992. Altered-function mutations of the transcriptional regulatory gene *virG* of *Agrobacterium tumefaciens. J. Bacteriol.* 174:7040–7043.

Haudecoeur, E., and D. Faure. 2010. A fine control of quorum-sensing communication in *Agrobacterium tumefaciens. Commun. Integr. Biol.* 3:84–88.

Haudecoeur, E., M. Tannieres, A. Cirou, A. Raffoux, Y. Dessaux, and D. Faure. 2009. Different regulation and roles of lactonases AiiB and AttM in *Agrobacterium tumefaciens* C58. *Mol. Plant-Microbe Interact.* 22:529–537.

Hong, P. C., R. M. Tsolis, and T. A. Ficht. 2000. Identification of genes required for chronic persistence of *Brucella abortus* in mice. *Infect. Immun.* 68:4102–4107.

Izore, T., V. Job, and A. Dessen. 2011. Biogenesis, regulation, and targeting of the type III secretion system. *Structure* 19:603–612.

Jeters, R. T., G. R. Wang, K. Moon, N. B. Shoemaker, and A. A. Salyers. 2009. Tetracycline-associated transcriptional regulation of transfer genes of the *Bacteroides* conjugative transposon CTnDOT. *J. Bacteriol.* 191:6374–6382.

Johnson, C. M., D. A. Manias, H. A. Haemig, S. Shokeen, K. E. Weaver, T. M. Henkin, and G. M. Dunny. 2010. Direct evidence for control of the pheromone-inducible *prgQ* operon of *Enterococcus faecalis* plasmid pCF10 by a countertranscript-driven attenuation mechanism. *J. Bacteriol.* 192:1634–1642.

Judd, P. K., R. B. Kumar, and A. Das. 2005. Spatial location and requirements for the assembly of the *Agrobacterium tumefaciens* type IV secretion apparatus. *Proc. Natl. Acad. Sci. USA* 102:11498–11503.

Juhas, M., J. R. van der Meer, M. Gaillard, R. M. Harding, D. W. Hood, and D. W. Crook. 2009. Genomic islands: tools of bacterial horizontal gene transfer and evolution. *FEMS Microbiol. Rev.* **33**:376–393.

Kalogeraki, V. S., and S. C. Winans. 1998. Wound-released chemical signals may elicit multiple responses from an *Agrobacterium tumefaciens* strain containing an octopine-type Ti plasmid. *J. Bacteriol.* **180**:5660–5667.

Kalogeraki, V. S., J. Zhu, J. L. Stryker, and S. C. Winans. 2000. The right end of the *vir* region of an octopine-type Ti plasmid contains four new members of the *vir* regulon that are not essential for pathogenesis. *J. Bacteriol.* **182**:1774–1778.

Karnholz, A., C. Hoefler, S. Odenbreit, W. Fischer, D. Hofreuter, and R. Haas. 2006. Functional and topological characterization of novel components of the *comB* DNA transformation competence system in *Helicobacter pylori*. *J. Bacteriol.* **188**:882–893.

Khan, S. R., and S. K. Farrand. 2009. The BlcC (AttM) lactonase of *Agrobacterium tumefaciens* does not quench the quorum-sensing system that regulates Ti plasmid conjugative transfer. *J. Bacteriol.* **191**:1320–1329.

Komano, T., T. Yoshida, K. Narahara, and N. Furuya. 2000. The transfer region of IncI1 plasmid R64: similarities between R64 *tra* and *Legionella icm/dot* genes. *Mol. Microbiol.* **35**:1348–1359.

Kozlowicz, B. K., K. Shi, Z. Y. Gu, D. H. Ohlendorf, C. A. Earhart, and G. M. Dunny. 2006. Molecular basis for control of conjugation by bacterial pheromone and inhibitor peptides. *Mol. Microbiol.* **62**:958–969.

Kumar, R. B., and A. Das. 2002. Polar location and functional domains of the *Agrobacterium tumefaciens* DNA transfer protein VirD4. *Mol. Microbiol.* **43**:1523–1532.

Kwok, T., D. Zabler, S. Urman, M. Rohde, R. Hartig, S. Wessler, R. Misselwitz, J. Berger, N. Sewald, W. Konig, and S. Backert. 2007. *Helicobacter* exploits integrin for type IV secretion and kinase activation. *Nature* **449**:862–866.

Lai, E. M., O. Chesnokova, L. M. Banta, and C. I. Kado. 2000. Genetic and environmental factors affecting T-pilin export and T-pilus biogenesis in relation to flagellation of *Agrobacterium tumefaciens*. *J. Bacteriol.* **182**:3705–3716.

Lang, S., K. Gruber, S. Mihajlovic, R. Arnold, C. J. Gruber, S. Steinlechner, M. A. Jehl, T. Rattei, K. U. Frohlich, and E. L. Zechner. 2010. Molecular recognition determinants for type IV secretion of diverse families of conjugative relaxases. *Mol. Microbiol.* **78**:1539–1555.

Lanka, E., and B. M. Wilkins. 1995. DNA processing reactions in bacterial conjugation. *Annu. Rev. Biochem.* **64**:141–169.

Lau-Wong, I. C., T. Locke, M. J. Ellison, T. L. Raivio, and L. S. Frost. 2008. Activation of the Cpx regulon destabilizes the F plasmid transfer activator, TraJ, via the HslVU protease in *Escherichia coli*. *Mol. Microbiol.* **67**:516–527.

Lawley, T., B. M. Wilkins, and L. S. Frost. 2004. Bacterial conjugation in gram-negative bacteria, p. 203–226. *In* B. E. Funnell and G. J. Phillips (ed.), *Plasmid Biology*. ASM Press, Washington, DC.

Lawley, T. D., W. A. Klimke, M. J. Gubbins, and L. S. Frost. 2003. F factor conjugation is a true type IV secretion system. *FEMS Microbiol. Lett.* **224**:1–15.

Lee, C. A., and A. D. Grossman. 2007. Identification of the origin of transfer (*oriT*) and DNA relaxase required for conjugation of the integrative and conjugative element ICE*Bs1* of *Bacillus subtilis*. *J. Bacteriol.* **189**:7254–7261.

Lee, K., M. W. Dudley, K. M. Hess, D. G. Lynn, R. D. Joerger, and A. N. Binns. 1992. Mechanism of activation of *Agrobacterium* virulence genes: identification of phenol-binding proteins. *Proc. Natl. Acad. Sci. USA* **89**:8666–8670.

Leonard, B. A. B. 1996. *Enterococcus faecalis* pheromone binding protein, PrgZ, recruits a chromosomal oligopeptide permease system to import sex pheromone cCF10 for induction of conjugation. *Proc. Natl. Acad. Sci. USA* **93**:260–264.

Li, L., Y. Jia, Q. Hou, T. C. Charles, E. W. Nester, and S. Q. Pan. 2002. A global pH sensor: *Agrobacterium* sensor protein ChvG regulates acid- inducible genes on its two chromosomes and Ti plasmid. *Proc. Natl. Acad. Sci. USA* **99**:12369–12374.

Llosa, M., C. Roy, and C. Dehio. 2009. Bacterial type IV secretion systems in human disease. *Mol. Microbiol.* **73**:141–151.

Locht, C., L. Coutte, and N. Mielcarek. 2011. The ins and outs of pertussis toxin. *FEBS J.* doi:10.1111/j.1742-4658.2011.08237.x.

Lu, J., and L. S. Frost. 2005. Mutations in the C-terminal region of TraM provide evidence for in vivo TraM-TraD interactions during F-plasmid conjugation. *J. Bacteriol.* **187**:4767–4773.

MacRitchie, D. M., D. R. Buelow, N. L. Price, and T. L. Raivio. 2008. Two-component signaling and gram negative envelope stress response systems. *Adv. Exp. Med. Biol.* **631**:80–110.

Marrero, J., and M. K. Waldor. 2007. Determinants of entry exclusion within Eex and TraG are cytoplasmic. *J. Bacteriol.* **189**:6469–6473.

Marrero, J., and M. K. Waldor. 2005. Interactions between inner membrane proteins in donor and recipient cells limit conjugal DNA transfer. *Dev. Cell* **8**:963–970.

Martinez-Nunez, C., P. Altamirano-Silva, F. Alvarado-Guillen, E. Moreno, C. Guzman-Verri, and E. Chaves-Olarte. 2010. The two-component system BvrR/BvrS regulates the expression of the type IV secretion system VirB in *Brucella abortus*. *J. Bacteriol.* **192**:5603–5608.

McCormick, J. K., H. Hirt, G. M. Dunny, and P. M. Schlievert. 2000. Pathogenic mechanisms of enterococcal endocarditis. *Curr. Infect. Dis. Rep.* **2**:315–321.

McNealy, T. L., V. Forsbach-Birk, C. Shi, and R. Marre. 2005. The Hfq homolog in *Legionella pneumophila* demonstrates regulation by LetA and RpoS and interacts with the global regulator CsrA. *J. Bacteriol.* **187**:1527–1532.

Medini, D., A. Covacci, and C. Donati. 2006. Protein homology network families reveal step-wise diversification of type III and type IV secretion systems. *PLoS Comput. Biol.* **2**:e173.

Mikkelsen, H., M. Sivaneson, and A. Filloux. 2011. Key two-component regulatory systems that control biofilm formation in *Pseudomonas aeruginosa*. *Environ. Microbiol.* **13**:1666–1681.

Molin, S., and T. Tolker-Nielsen. 2003. Gene transfer occurs with enhanced efficiency in biofilms and induces enhanced stabilisation of the biofilm structure. *Curr. Opin. Biotechnol.* **14**:255–261.

Molofsky, A. B., and M. S. Swanson. 2003. *Legionella pneumophila* CsrA is a pivotal repressor of transmission traits and activator of replication. *Mol. Microbiol.* **50**:445–461.

Molofsky, A. B., and M. S. Swanson. 2004. Differentiate to thrive: lessons from the *Legionella pneumophila* life cycle. *Mol. Microbiol.* **53**:29–40.

Morgan, J. K., B. E. Luedtke, and E. I. Shaw. 2010. Polar localization of the *Coxiella burnetii* type IVB secretion system. *FEMS Microbiol. Lett.* **305**:177–183.

Mukhopadhyay, A., R. Gao, and D. G. Lynn. 2004. Integrating input from multiple signals: the VirA/VirG two-component system of *Agrobacterium tumefaciens*. *Chembiochem* **5**:1535–1542.

Nagai, H., E. D. Cambronne, J. C. Kagan, J. C. Amor, R. A. Kahn, and C. R. Roy. 2005. A C-terminal translocation signal required for Dot/Icm-dependent delivery of the *Legionella* RalF protein to host cells. *Proc. Natl. Acad. Sci. USA* **102**:826–831.

Nagai, H., and T. Kubori. 2011. Type IVB secretion systems of *Legionella* and other Gram-negative bacteria. *Front. Microbiol.* **2**:136.

Ninio, S., D. M. Zuckman-Cholon, E. D. Cambronne, and C. R. Roy. 2005. The *Legionella* IcmS-IcmW protein complex is important for Dot/Icm-mediated protein translocation. *Mol. Microbiol.* **55**:912–926.

Ogasawara, H., K. Yamamoto, and A. Ishihama. 2011. Role of the biofilm master regulator CsgD in cross-regulation between biofilm formation and flagellar synthesis. *J. Bacteriol.* **193**:2587–2597.

Ong, C. L., S. A. Beatson, A. G. McEwan, and M. A. Schembri. 2009. Conjugative plasmid transfer and adhesion dynamics in an *Escherichia coli* biofilm. *Appl. Environ. Microbiol.* **75:**6783–6791.

Pansegrau, W., G. Ziegelin, and E. Lanka. 1988. The origin of conjugative IncP plasmid transfer: interaction with plasmid-encoded products and the nucleotide sequence at the relaxation site. *Biochim. Biophys. Acta* **951:**365–374.

Pappas, G., N. Akritidis, M. Bosilkovski, and E. Tsianos. 2005. Brucellosis. *N. Engl. J. Med.* **352:**2325–2336.

Parker, C., and R. J. Meyer. 2007. The R1162 relaxase/primase contains two, type IV transport signals that require the small plasmid protein MobB. *Mol. Microbiol.* **66:**252–261.

Paschos, A., A. den Hartigh, M. A. Smith, V. L. Atluri, D. Sivanesan, R. M. Tsolis, and C. Baron. 2011. An *in vivo* high-throughput screening approach targeting the type IV secretion system component VirB8 identified inhibitors of *Brucella abortus* 2308 proliferation. *Infect. Immun.* **79:**1033–1043.

Pattis, I., E. Weiss, R. Laugks, R. Haas, and W. Fischer. 2007. The *Helicobacter pylori* CagF protein is a type IV secretion chaperone-like molecule that binds close to the C-terminal secretion signal of the CagA effector protein. *Microbiology* **153:**2896–2909.

Pazour, G. J., and A. Das. 1990. Characterization of the VirG binding site of *Agrobacterium tumefaciens*. *Nucleic Acids Res.* **18:**6909–6913. (Erratum, **19:**1358, 1991.)

Peng, W. T., Y. W. Lee, and E. W. Nester. 1998. The phenolic recognition profiles of the *Agrobacterium tumefaciens* VirA protein are broadened by a high level of the sugar binding protein ChvE. *J. Bacteriol.* **180:**5632–5638.

Pulliainen, A. T., and C. Dehio. 2009. *Bartonella henselae:* subversion of vascular endothelial cell functions by translocated bacterial effector proteins. *Int. J. Biochem. Cell Biol.* **41:**507–510.

Quebatte, M., M. Dehio, D. Tropel, A. Basler, I. Toller, G. Raddatz, P. Engel, S. Huser, H. Schein, H. L. Lindroos, S. G. Andersson, and C. Dehio. 2010. The BatR/BatS two-component regulatory system controls the adaptive response of *Bartonella henselae* during human endothelial cell infection. *J. Bacteriol.* **192:**3352–3367.

Rambow-Larsen, A. A., E. M. Petersen, C. R. Gourley, and G. A. Splitter. 2009. *Brucella* regulators: self-control in a hostile environment. *Trends Microbiol.* **17:**371–377.

Rambow-Larsen, A. A., G. Rajashekara, E. Petersen, and G. Splitter. 2008. Putative quorum-sensing regulator BlxR of *Brucella melitensis* regulates virulence factors including the type IV secretion system and flagella. *J. Bacteriol.* **190:**3274–3282.

Ramsey, M. E., K. L. Woodhams, and J. P. Dillard. 2011. The gonococcal genetic island and type IV secretion in the pathogenic *Neisseria*. *Front. Microbiol.* **2:**61.

Rasis, M., and G. Segal. 2009. The LetA-RsmYZ-CsrA regulatory cascade, together with RpoS and PmrA, post-transcriptionally regulates stationary phase activation of *Legionella pneumophila* Icm/Dot effectors. *Mol. Microbiol.* **72:**995–1010.

Roberts, A. P., and P. Mullany. 2009. A modular master on the move: the Tn*916* family of mobile genetic elements. *Trends Microbiol.* **17:**251–258.

Rocha-Estrada, J., A. E. Aceves-Diez, G. Guarneros, and M. de la Torre. 2010. The RNPP family of quorum-sensing proteins in Gram-positive bacteria. *Appl. Microbiol. Biotechnol.* **87:**913–923.

Rouot, B., M. T. Alvarez-Martinez, C. Marius, P. Menanteau, L. Guilloteau, R. A. Boigegrain, R. Zumbihl, D. O'Callaghan, N. Domke, and C. Baron. 2003. Production of the type IV secretion system differs among *Brucella species* as revealed with VirB5- and VirB8-specific antisera. *Infect. Immun.* **71:**1075–1082.

Roux, C. M., H. G. Rolan, R. L. Santos, P. D. Beremand, T. L. Thomas, L. G. Adams, and R. M. Tsolis. 2007. *Brucella* requires a functional type IV secretion system to elicit innate immune responses in mice. *Cell. Microbiol.* **9:**1851–1869.

Sahr, T., H. Bruggemann, M. Jules, M. Lomma, C. Albert-Weissenberger, C. Cazalet, and C. Buchrieser. 2009. Two small ncRNAs jointly govern virulence and transmission in *Legionella pneumophila*. *Mol. Microbiol.* **72:**741–762.

Salgado-Pabon, W., Y. Du, K. T. Hackett, K. M. Lyons, C. G. Arvidson, and J. P. Dillard. 2010. Increased expression of the type IV secretion system in piliated *Neisseria gonorrhoeae* variants. *J. Bacteriol.* **192:**1912–1920.

Schmid, M. C., R. Schulein, M. Dehio, G. Denecker, I. Carena, and C. Dehio. 2004. The VirB type IV secretion system of *Bartonella henselae* mediates invasion, proinflammatory activation and anti-apoptotic protection of endothelial cells. *Mol. Microbiol.* **52:**81–92.

Schroder, G., R. Schuelein, M. Quebatte, and C. Dehio. 2011. Conjugative DNA transfer into human cells by the VirB/VirD4 type IV secretion system of the bacterial pathogen *Bartonella henselae*. *Proc. Natl. Acad. Sci. USA* **108:**14643–14648.

Schulein, R., P. Guye, T. A. Rhomberg, M. C. Schmid, G. Schroder, A. C. Vergunst, I. Carena, and C. Dehio. 2005. A bipartite signal mediates the transfer of type IV secretion substrates of *Bartonella henselae* into human cells. *Proc. Natl. Acad. Sci. USA* **102:**856–861.

Shaw, C. H., A. M. Ashby, A. Brown, C. Royal, and G. J. Loake. 1988. *virA* and *virG* are the Ti-plasmid functions required for chemotaxis of *Agrobacterium tumefaciens* towards acetosyringone. *Mol. Microbiol.* **2:**413–417.

Shi, C., V. Forsbach-Birk, R. Marre, and T. L. McNealy. 2006. The *Legionella pneumophila* global regulatory protein LetA affects DotA and Mip. *Int. J. Med. Microbiol.* **296:**15–24.

Shi, K., C. K. Brown, Z. Y. Gu, B. K. Kozlowicz, G. M. Dunny, D. H. Ohlendorf, and C. A. Earhart. 2005. Structure of peptide sex pheromone receptor PrgX and PrgX/pheromone complexes and regulation of conjugation in *Enterococcus faecalis*. *Proc. Natl. Acad. Sci. USA* **102:**18596–18601.

Shimoda, N., A. Toyoda-Yamamoto, S. Aoki, and Y. Machida. 1993. Genetic evidence for an interaction between the VirA sensor protein and the ChvE sugar-binding protein of *Agrobacterium*. *J. Biol. Chem.* **268:**26552–26558.

Shokeen, S., C. M. Johnson, T. J. Greenfield, D. A. Manias, G. M. Dunny, and K. E. Weaver. 2010. Structural analysis of the Anti-Q-Qs interaction: RNA-mediated regulation of *E. faecalis* plasmid pCF10 conjugation. *Plasmid* **64:**26–35.

Sieira, R., G. M. Arocena, L. Bukata, D. J. Comerci, and R. A. Ugalde. 2010. Metabolic control of virulence genes in *Brucella abortus*: HutC coordinates *virB* expression and the histidine utilization pathway by direct binding to both promoters. *J. Bacteriol.* **192:**217–224.

Sieira, R., D. J. Comerci, L. I. Pietrasanta, and R. A. Ugalde. 2004. Integration host factor is involved in transcriptional regulation of the *Brucella abortus virB* operon. *Mol. Microbiol.* **54:**808–822.

Sola-Landa, A., J. Pizarro-Cerda, M. J. Grillo, E. Moreno, I. Moriyon, J. M. Blasco, J. P. Gorvel, and I. Lopez-Goni. 1998. A two-component regulatory system playing a critical role in plant pathogens and endosymbionts is present in *Brucella abortus* and controls cell invasion and virulence. *Mol. Microbiol.* **29:**125–138.

Taminiau, B., M. Daykin, S. Swift, M. L. Boschiroli, A. Tibor, P. Lestrate, X. De Bolle, D. O'Callaghan, P. Williams, and J. J. Letesson. 2002. Identification of a quorum-sensing signal molecule in the facultative intracellular pathogen *Brucella melitensis*. *Infect. Immun.* **70:**3004–3011.

Tegtmeyer, N., R. Hartig, R. M. Delahay, M. Rohde, S. Brandt, J. Conradi, S. Takahashi, A. J. Smolka, N. Sewald, and S. Backert. 2010. A small fibronectin-mimicking protein from bacteria induces cell spreading and focal adhesion formation. *J. Biol. Chem.* **285:**23515–23526.

Tegtmeyer, N., S. Wessler, and S. Backert. 2011. Role of the cag-pathogenicity island encoded type IV secretion system in *Helicobacter pylori* pathogenesis. *FEBS J.* **278:**1190–1202.

Tiaden, A., T. Spirig, T. Sahr, M. A. Walti, K. Boucke, C. Buchrieser, and H. Hilbi. 2010. The autoinducer synthase LqsA and putative sensor kinase LqsS regulate phagocyte interactions, extracellular filaments and a genomic island of *Legionella pneumophila*. *Environ. Microbiol.* **12**:1243–1259.

Tiaden, A., T. Spirig, S. S. Weber, H. Bruggemann, R. Bosshard, C. Buchrieser, and H. Hilbi. 2007. The *Legionella pneumophila* response regulator LqsR promotes host cell interactions as an element of the virulence regulatory network controlled by RpoS and LetA. *Cell. Microbiol.* **9**:2903–2920.

Uzureau, S., M. Godefroid, C. Deschamps, J. Lemaire, X. De Bolle, and J. J. Letesson. 2007. Mutations of the quorum sensing-dependent regulator VjbR lead to drastic surface modifications in *Brucella melitensis*. *J. Bacteriol.* **189**:6035–6047.

Uzureau, S., J. Lemaire, E. Delaive, M. Dieu, A. Gaigneaux, M. Raes, X. De Bolle, and J. J. Letesson. 2010. Global analysis of quorum sensing targets in the intracellular pathogen *Brucella melitensis* 16 M. *J. Proteome Res.* **9**:3200–3217.

Valdivia, R. H., L. Wang, and S. C. Winans. 1991. Characterization of a putative periplasmic transport system for octopine accumulation encoded by *Agrobacterium tumefaciens* Ti plasmid pTiA6. *J. Bacteriol.* **173**:6398–6405.

van der Veen, S., and T. Abee. 2011. Bacterial SOS response: a food safety perspective. *Curr. Opin. Biotechnol.* **22**:136–142.

Vayssier-Taussat, M., D. Le Rhun, H. K. Deng, F. Biville, S. Cescau, A. Danchin, G. Marignac, E. Lenaour, H. J. Boulouis, M. Mavris, L. Arnaud, H. Yang, J. Wang, M. Quebatte, P. Engel, H. Saenz, and C. Dehio. 2010. The Trw type IV secretion system of *Bartonella mediates* host-specific adhesion to erythrocytes. *PLoS Pathog.* **6**:e1000946.

Vergunst, A. C., M. C. van Lier, A. den Dulk-Ras, T. A. Grosse Stuve, A. Ouwehand, and P. J. Hooykaas. 2005. Positive charge is an important feature of the C-terminal transport signal of the VirB/D4-translocated proteins of *Agrobacterium*. *Proc. Natl. Acad. Sci. USA* **102**:832–837.

Viadas, C., M. C. Rodriguez, F. J. Sangari, J. P. Gorvel, J. M. Garcia-Lobo, and I. Lopez-Goni. 2010. Transcriptome analysis of the *Brucella abortus* BvrR/BvrS two-component regulatory system. *PloS One* **5**:e10216.

Vincent, C. D., and J. P. Vogel. 2006. The *Legionella pneumophila* IcmS-LvgA protein complex is important for Dot/Icm-dependent intracellular growth. *Mol. Microbiol.* **61**:596–613.

Vogel, J. P., H. L. Andrews, S. K. Wong, and R. R. Isberg. 1998. Conjugative transfer by the virulence system of *Legionella pneumophila*. *Science* **279**:873–876.

Vogel, J. P., and R. R. Isberg. 1999. Cell biology of *Legionella pneumophila*. *Curr. Opin. Microbiol.* **2**:30–34.

Wang, L., and S. C. Winans. 1995. High angle and ligand-induced low angle DNA bends incited by OccR lie in the same plane with OccR bound to the interior angle. *J. Mol. Biol.* **253**:32–38.

Wardal, E., E. Sadowy, and W. Hryniewicz. 2010. Complex nature of enterococcal pheromone-responsive plasmids. *Pol. J. Microbiol.* **9**:79–87.

Weeks, J. N., C. L. Galindo, K. L. Drake, G. L. Adams, H. R. Garner, and T. A. Ficht. 2010. *Brucella melitensis* VjbR and C12-HSL regulons: contributions of the N-dodecanoyl homoserine lactone signaling molecule and LuxR homologue VjbR to gene expression. *BMC Microbiol.* **10**:167.

Wells, D. H., E. J. Chen, R. F. Fisher, and S. R. Long. 2007. ExoR is genetically coupled to the ExoS-ChvI two-component system and located in the periplasm of *Sinorhizobium meliloti*. *Mol. Microbiol.* **64**:647–664.

White, C. E., and S. C. Winans. 2007. Cell-cell communication in the plant pathogen *Agrobacterium tumefaciens*. *Philos. Trans. R. Soc. Lond. B* **362**:1135–1148.

Will, W. R., and L. S. Frost. 2006a. Characterization of the opposing roles of H-NS and TraJ in transcriptional regulation of the F-plasmid *tra* operon. *J. Bacteriol.* **188**:507–514.

Will, W. R., and L. S. Frost. 2006b. Hfq is a regulator of F-plasmid TraJ and TraM synthesis in *Escherichia coli*. *J. Bacteriol.* **188**:124–131.

Will, W. R., J. Lu, and L. S. Frost. 2004. The role of H-NS in silencing F transfer gene expression during entry into stationary phase. *Mol. Microbiol.* **54**:769–782.

Winans, S. C. 1991. An *Agrobacterium* two-component regulatory system for the detection of chemicals released from plant wounds. *Mol. Microbiol.* **5**:2345–2350.

Winans, S. C., P. R. Ebert, S. E. Stachel, M. P. Gordon, and E. W. Nester. 1986. A gene essential for *Agrobacterium* virulence is homologous to a family of positive regulatory loci. *Proc. Natl. Acad. Sci. USA* **83**:8278–8282.

Winans, S. C., R. A. Kerstetter, J. E. Ward, and E. W. Nester. 1989. A protein required for transcriptional regulation of *Agrobacterium* virulence genes spans the cytoplasmic membrane. *J. Bacteriol.* **171**:1616–1622.

Wozniak, R. A., and M. K. Waldor. 2010. Integrative and conjugative elements: mosaic mobile genetic elements enabling dynamic lateral gene flow. *Nat. Rev. Microbiol.* **8**:552–563.

Yao, S. Y., L. Luo, K. J. Har, A. Becker, S. Ruberg, G. Q. Yu, J. B. Zhu, and H. P. Cheng. 2004. *Sinorhizobium meliloti* ExoR and ExoS proteins regulate both succinoglycan and flagellum production. *J. Bacteriol.* **186**:6042–6049.

Yoshida, T., S. R. Kim, and T. Komano. 1999. Twelve *pil* genes are required for biogenesis of the R64 thin pilus. *J. Bacteriol.* **181**:2038–2043.

Yuan, Z. C., P. Liu, P. Saenkham, K. Kerr, and E. W. Nester. 2008. Transcriptome profiling and functional analysis of *Agrobacterium tumefaciens* reveals a general conserved response to acidic conditions (pH 5.5) and a complex acid-mediated signaling involved in *Agrobacterium*-plant interactions. *J. Bacteriol.* **190**:494–507.

Zhang, H. B., C. Wang, and L. H. Zhang. 2004. The quormone degradation system of *Agrobacterium tumefaciens* is regulated by starvation signal and stress alarmone (p)ppGpp. *Mol. Microbiol.* **52**:1389–1401.

Zhang, H. B., L. H. Wang, and L. H. Zhang. 2002. Genetic control of quorum-sensing signal turnover in *Agrobacterium tumefaciens*. *Proc. Natl. Acad. Sci. USA* **99**:4638–4643.

Zhao, Z., E. Sagulenko, Z. Ding, and P. J. Christie. 2001. Activities of *virE1* and the VirE1 secretion chaperone in export of the multifunctional VirE2 effector via an *Agrobacterium* type IV secretion pathway. *J. Bacteriol.* **183**:3855–3865.

Zhu, J., J. W. Beaber, M. I. More, C. Fuqua, A. Eberhard, and S. C. Winans. 1998. Analogs of the autoinducer 3-oxooctanoyl-homoserine lactone strongly inhibit activity of the TraR protein of *Agrobacterium tumefaciens*. *J. Bacteriol.* **180**:5398–5405.

Zhu, W., S. Banga, Y. Tan, C. Zheng, R. Stephenson, J. Gately, and Z. Q. Luo. 2011. Comprehensive identification of protein substrates of the Dot/Icm type IV transporter of *Legionella pneumophila*. *PLoS One* **6**:e17638.

Zola, T. A., H. R. Strange, N. M. Dominguez, J. P. Dillard, and C. N. Cornelissen. 2010. Type IV secretion machinery promotes Ton-independent intracellular survival of *Neisseria gonorrhoeae* within cervical epithelial cells. *Infect. Immun.* **78**:2429–2437.

Zusman, T., G. Aloni, E. Halperin, H. Kotzer, E. Degtyar, M. Feldman, and G. Segal. 2007. The response regulator PmrA is a major regulator of the *icm/dot* type IV secretion system in *Legionella pneumophila* and *Coxiella burnetii*. *Mol. Microbiol.* **63**:1508–1523.

Regulation of Bacterial Virulence
Edited by Michael L. Vasil and Andrew J. Darwin
© 2013 ASM Press, Washington, DC   doi:10.1128/9781555818524.ch19

Chapter 19

PrfA and the *Listeria monocytogenes* Switch from Environmental Bacterium to Intracellular Pathogen

BOBBI XAYARATH AND NANCY E. FREITAG

Fundamental to the discovery and enhanced understanding of biological phenomena has been the use of model organisms in scientific research. Lessons learned from the detailed study of organisms that offer specific advantages, such as easy cultivation, rapid replication, and/or amenability to multiple genetic techniques, have provided insight into complex processes, including DNA replication and repair, cell division, cell differentiation, and population evolution (Baquero and Lemonnier, 2009; Goyal et al., 2011; Kaiser, 2008; Meisch and Prioleau, 2011). The facultative intracellular bacterium *Listeria monocytogenes* has been used for decades as a model infectious agent for the study of host innate and adaptive immunity. More recently, *L. monocytogenes* has served as a tool for the molecular investigation of eukaryotic cellular processes such as actin assembly, autophagy, and cytosolic immune surveillance pathways (Stavru et al., 2011). The ease of cultivation, the existence of numerous genetic tools, and the availability of excellent tissue culture and mouse infection models contribute to the establishment of *L. monocytogenes* as an excellent system for the exploration of host-pathogen interactions.

It is now becoming clear that in addition to the utility of *L. monocytogenes* as a probe for eukaryotic cell physiology and host immunity, the bacterium itself serves as a model for understanding how microorganisms that are adapted for life in the outside environment develop the capacity to cause disease in humans. Understanding the physiological mechanisms that enable *L. monocytogenes* to move from the outside environment to inside of a host is important for recognizing potential sources of emerging infections and for deciphering how relatively harmless microbes become human pathogens. This chapter focuses on describing what is currently known about how *L. monocytogenes* mediates the transition from saprophyte to mammalian pathogen and explores the regulatory circuit governed by a transcriptional activator known as PrfA that allows the bacterium to flourish in both soil and cytosol.

L. MONOCYTOGENES IN THE 'HOOD— NOT A DESIRABLE NEIGHBOR

L. monocytogenes is widely distributed in the outside environment and can be found in soil, silage, groundwater, sewage, and decaying vegetation, where it is believed to live as a saprophyte (Czuprynski, 2005; Fenlon, 1985; Freitag et al., 2009; Gandhi and Chikindas, 2007; Lecuit, 2007; Toledo-Arana et al., 2009). The bacterium does not form spores but can endure large fluctuations in temperature, salt concentration, and pH (Chaturongakul et al., 2008). This environmental resiliency provides *L. monocytogenes* with the means of contaminating and proliferating within food sources despite common methods used for food preservation that would quickly eradicate other harmful microorganisms (Farber and Peterkin, 1991; Gibbons et al., 2006; Gottlieb et al., 2006; Mead et al., 2006; Swaminathan et al., 2006). *L. monocytogenes* contamination of food products has resulted in thousands of reported cases of food-borne illness (referred to as listeriosis) as well as some of the most expensive food recalls in U.S. history (Centers for Disease Control and Prevention, 1998, 1999, 2002, 2004; Gottlieb et al., 2006; Lynch et al., 2006; Schwartz et al., 1988; Stone and Shoenberger, 2001).

Upon ingestion of contaminated foods by humans or other animals, *L. monocytogenes* transitions into a physiological state that promotes bacterial survival and replication within host cells (Cossart and Toledo-Arana, 2008; Freitag et al., 2009; Lecuit, 2007; Schlech et al., 1983). In healthy individuals, disease caused by *L. monocytogenes* is usually self-limiting and presents as a form of mild gastroenteritis (Drevets and Bronze, 2008; Ramaswamy et al., 2007;

Bobbi Xayarath and Nancy E. Freitag • Department of Microbiology and Immunology, University of Illinois at Chicago College of Medicine, Chicago, IL 60612.

Swaminathan and Gerner-Smidt, 2007). The infectious dose of *L. monocytogenes* is estimated to be 10^7 to 10^8 CFU for healthy persons but is approximately 10- to 100-fold less for highly susceptible individuals, which include the immunocompromised, the elderly, and pregnant women (Bortolussi, 2008; Farber et al. 1996). In these individuals, the bacterium is capable of causing systemic infections that lead to meningitis, encephalitis, bacteremia, and death. Pregnant women may experience a mild form of the disease, but the infection of the developing fetus can be severe and can lead to abortion, stillbirth, or neonatal infections with mortality rates as high as 50% (Bortolussi, 2008; Delgado, 2008; MacGowan et al., 1991; McClure and Goldenberg, 2009). In general, while the disease listeriosis is not generally as commonly reported as infections resulting from other food-borne pathogens, it has one of the highest case fatality rates, with approximately 30% of the known infected individuals succumbing to infection (Bortolussi, 2008; Gellin and Broome, 1989; Ramaswamy et al., 2007).

The primary route of *L. monocytogenes* infection is translocation across the intestinal epithelium following the consumption of contaminated food products (Gray and Killinger, 1966; Lecuit, 2007; Ramaswamy et al., 2007; Swaminathan and Gerner-Smidt, 2007). Once across the intestinal barrier, the bacteria enter into the bloodstream and are quickly taken up by resident macrophages that concentrate the bacteria within the liver and spleen, where they either are cleared by an effective host cell-mediated immune response or disseminate to other organs where they can further replicate (Portnoy and Jones, 1994). As mentioned, *L. monocytogenes* infections are often associated with central nervous system or brain stem infections (Drevets and Bronze, 2008), but bacteria are also known to target cardiac cells (Alonzo et al., 2011a; Antolin et al., 2008) and the bone marrow (Hardy et al., 2009), and *L. monocytogenes* has been reported to persist within the gallbladder of infected animals (Hardy et al., 2004). *L. monocytogenes* is thus capable of inhabiting a variety of environmental habitats, both inside and outside of mammalian hosts.

THE *L. MONOCYTOGENES* ARSENAL OF GENE PRODUCTS THAT PROMOTE BACTERIAL INFECTION

L. monocytogenes appears to maintain and express a variety of gene products that enable the bacterium to gain access to host cells, to replicate within the cytosol, and to spread to adjacent cells (Fig. 1) (Cossart and Lecuit, 1998; Kreft et al., 2002; Vazquez-Boland et al., 2001). To gain entry into nonprofessional phagocytic cells, such as epithelial cells, the bacterium expresses a number of surface proteins, including the internalins InlA and InlB, which trigger bacterial cell uptake (Bonazzi et al., 2009b; Ireton, 2007; Lecuit et al., 2001; Seveau et al., 2007). InlA binds E-cadherin, a calcium-dependent host cell adhesion molecule expressed predominantly in epithelial cells, whereas InlB binds to the hepatocyte growth factor receptor Met, a tyrosine kinase also found in epithelial cells and other cell types. Bacterial binding to these receptors induces the uptake of *L. monocytogenes* into the host cell through the exploitation of the host endocytic machinery (Ireton and Cossart, 1997; Pizarro-Cerda and Cossart, 2006). Once internalized, *L. monocytogenes* secretes a pore-forming cytolysin known as listeriolysin O (LLO) and two phospholipases, a phosphatidylinositol-specific phospholipase C (PI-PLC) encoded by *plcA* and a broad-range phospholipase C (PC-PLC) encoded by *plcB* (Gedde et al., 2000; Goldfine et al., 2000; Marquis et al., 1995; Portnoy et al., 1988; Schnupf and Portnoy, 2007; Vazquez-Boland et al., 1992; Wadsworth and Goldfine, 1999). These secreted products function together to disrupt the *Listeria*-containing phagosome or endosome to mediate bacterial escape from the vacuole. Bacterial replication occurs within the cell cytosol, with *L. monocytogenes* scavenging nutrients from the host to support replication (Chatterjee et al., 2006; Chico-Calero et al., 2002; Joseph and Goebel, 2007; Joseph et al., 2006; Keeney et al., 2009; Keeney et al., 2007; Marquis et al., 1993). Nutrients acquired from host cells include hexose phosphate sugars via the bacterial Hpt transporter, glycerol and other three-carbon sugars, lipoic acid, branched-chain amino acids, and peptides. Cytosolic bacteria move through the cytosol and into nearby host cells using a motile force generated by the polymerization of host actin (Tilney et al., 1990; Tilney and Portnoy, 1989). *L. monocytogenes* actin filament-based movement depends upon a bacterial membrane-anchored surface protein known as ActA (encoded by *actA*) that localizes to one end of the bacterial cells and serves as a scaffold for host proteins involved in actin filament nucleation and polymerization (Kocks et al., 1992; Kocks et al., 1993; Lambrechts et al., 2008). *L. monocytogenes* moves through the cell to spread into adjacent cells, a process facilitated through the relief of cell cortical tension via InlC (Rajabian et al., 2009), and the bacteria become encased in double-membrane secondary vacuoles formed as a result of cell-to-cell spread. Escape from the double-membrane vacuole is mediated by the activities of both LLO

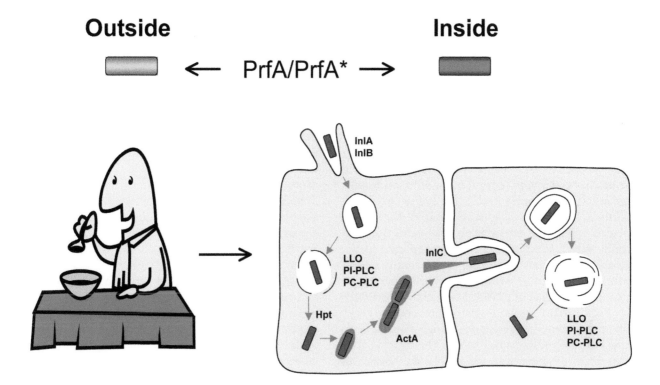

Outside

Inside

← PrfA/PrfA* →

Environment: vegetation, soil, water, food processing plants

Mammalian host cells

Figure 1. The *L. monocytogenes* transition from environmental bacterium to intracellular pathogen. *L. monocytogenes* survives under a number of diverse ecological conditions. It can be found growing within soil, decaying vegetation, water, silage, sewage, and food processing plants, but it also has the ability to adapt to life inside mammalian host cells. Central to the *L. monocytogenes* transition to life inside of host cells is the activity of a transcriptional regulator known as PrfA. Outside of host cells, PrfA exists in a low-activity state and directs low levels of virulence gene expression. Once inside a host, PrfA becomes highly activated (PrfA*) and dramatically increases the expression of a number of virulence gene products required for host cell invasion (internalins InlA and InlB), lysis of the phagosomal membrane (LLO, PI-PLC, and PC-PLC), intracellular growth (Hpt), cell-to-cell spread (actin assembly via ActA and relief of cell-cell cortical tension via InlC), and the dissolution of the double membrane resulting from cell-to-cell spread (LLO, PI-PLC, and PC-PLC). Adapted from Freitag et al. 2009. doi:10.1128/9781555818524.ch19f1

and the broad-specificity phospholipase PC-PLC (Marquis et al., 1995; Schnupf and Portnoy, 2007; Smith et al., 1995; Vazquez-Boland et al., 1992). Interestingly, PC-PLC is initially produced in an inactive proform that is sequestered at the bacterial membrane-cell wall interface until a drop in pH occurs as a result of vacuole acidification (Marquis and Hager, 2000). The correct compartmentalization of PC-PLC activity is important for avoiding damage to host cell membranes and is a critical aspect of *L. monocytogenes* virulence (Yeung et al., 2007). Mpl is a Zn-dependent metalloprotease encoded by *mpl* that processes the PC-PLC propeptide to its mature active form in response to a drop in vacuole pH (Forster et al., 2011; Poyart et al., 1993; Raveneau et al., 1992; Yeung et al., 2005). *L. monocytogenes* thus elaborates a number of bacterial gene products with fine-tuned activities that contribute to bacterial cell

invasion, phagosome escape, intracytosolic replication, and cell-to-cell spread.

PrfA, THE KEY TO *L. MONOCYTOGENES* PATHOGENESIS

As mentioned above, *L. monocytogenes* has developed the capacity to exploit a number of environmental niches, a talent that undoubtedly contributes to its widespread distribution. *L. monocytogenes* must thus be capable of discerning environment-specific cues so as to optimize the expression and activity of gene products that promote survival within disparate locations. Central to the ability of *L. monocytogenes* to transition from environments located outside to inside of a host is the master virulence regulatory protein, PrfA (<u>P</u>ositive <u>r</u>egulatory <u>f</u>actor <u>A</u>) (Freitag et al., 2009). PrfA is a 27-kDa transcriptional activator that is a member of the cyclic AMP (cAMP) receptor protein (CRP)/Fnr family of transcriptional regulators (Korner et al., 2003). PrfA activates transcription via recognition of a 14-bp palindromic DNA binding site, also known as the PrfA box, located in the –40 region of its target promoters (Freitag, 2006; Kreft and Vazquez-Boland, 2001). PrfA regulates the expression of a large number of gene products directly associated with bacterial virulence, including the products of *hly*, *plcA*, *hpt*, *actA*, *plcB*, *inlA*, *inlB*, and *inlC* (Chakraborty et al., 1992; Freitag, 2006; Scortti et al., 2007; Vazquez-Boland et al., 2001). Bacteria containing deletions or loss-of-function mutations within the *prfA* gene fail to replicate in the cytosol of host cells or spread to adjacent cells and are >100,000-fold less virulent in murine models of infection (Freitag et al., 1993; Mengaud et al., 1991).

PrfA further regulates the expression of genes that contribute to bile resistance, which may facilitate *L. monocytogenes* survival within the liver as well as bacterial persistence in the gallbladder (Begley et al., 2005; Dussurget et al., 2002; Hardy et al., 2004; Quillin et al., 2011; Sleator et al., 2005). PrfA induces the expression of a bile salt hydrolase (encoded by *bsh*) as well as a bile exclusion system (encoded by *bilE*), both of which have been shown to contribute to bacterial survival in the intestine in murine models of infection (Begley et al., 2005; Dussurget et al., 2002; Sleator et al., 2005). PrfA thus plays a pivotal role in regulating the expression of diverse factors associated with *L. monocytogenes* virulence and persistence. Numerous studies have indicated that the expression and activity of this central regulator itself are tightly regulated via mechanisms that encompass transcriptional, posttranscriptional, and posttranslational modes of control (Freitag, 2006; Gray et al., 2006). These multiple mechanisms of regulation act in concert to enable *L. monocytogenes* to fine-tune its pathogenic potential to its host without compromising bacterial survival in the outside environment.

THE FIRST MODE OF *prfA* REGULATION: TRANSCRIPTIONAL CONTROL OF *prfA* EXPRESSION

Transcriptional control of *prfA* is mediated by three separate promoter elements (Fig. 2A). Two of these promoters, *prfA*P1 and *prfA*P2, are located immediately upstream of the *prfA* translation initiation codon, and each directs the synthesis of monocistronic transcripts of *prfA* (Freitag and Portnoy, 1994).

These two promoters function to generate the initial basal levels of PrfA protein required for activation of *hly* and *plcA* expression and are needed for efficient escape of *L. monocytogenes* from host cell phagosomes. The *prfA*P1 promoter contains characteristics of an σ^A-dependent promoter, which is the primary sigma factor determining RNA polymerase specificity required for transcription in actively growing bacterial cells, such as during growth in rich broth culture media (Gray et al., 2006; Moran, 1993). The *prfA*P2 promoter region contains sequences that resemble a PrfA binding box as well as characteristics of both a σ^A-dependent and σ^B (the general stress response sigma factor)-dependent promoters (Freitag and Portnoy, 1994; Rauch et al., 2005). σ^B directs RNA polymerase to the promoter regions of a large number of genes whose products contribute to survival and adaptation to general environmental stresses, including conditions of low pH, high osmolarity, oxidative stress, and carbon starvation (Chaturongakul et al., 2008). Some of these stress response genes (*bilE*, *bsh*, and *inlA*) have promoters that are subject to both σ^B- and PrfA-dependent regulation and have been linked to both stress resistance and virulence (Dussurget et al., 2002; Kim et al., 2004; Kim et al., 2005; Sleator et al., 2005).

The third promoter lies immediately upstream of the *plcA* promoter and results in the production of a bicistronic *plcA-prfA* transcript as well as a monocistronic *plcA* transcript (Camilli et al., 1993; Freitag and Portnoy, 1994). Activation of this promoter results in an overall increase in the levels of *prfA* expression via the generation of the *plcA-prfA* transcript. This *plcA* promoter-directed increase in *prfA* expression is crucial for providing the high levels of PrfA required for stimulating *actA* expression, which is, in turn, needed for bacterial cell-to-cell spread (Camilli et al., 1993). While the *plcA* promoter is positively regulated by PrfA, the expression of both *prfA*P1 and *prfA*P2 is negatively influenced by PrfA, although it is unclear as to whether PrfA has a direct or indirect role in the downregulation of these two promoters (Freitag et al., 1993). Mutant strains containing deletions of either *prfA*P1 or *prfA*P2 are fully virulent in mouse infection models, indicating that these two promoters are functionally redundant in vivo (Freitag and Portnoy, 1994). Despite this functional redundancy, the presence of at least one of the two promoters is required to generate sufficient levels of PrfA for *hly*-dependent vacuole escape and *plcA* promoter-dependent increases in *prfA* expression. The *prfA*P1 and *prfA*P2 promoters thus function to "prime the pump" such that sufficient PrfA is generated for intracellular growth and cell-to-cell spread.

A

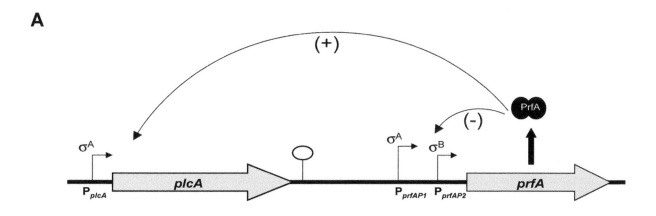

B

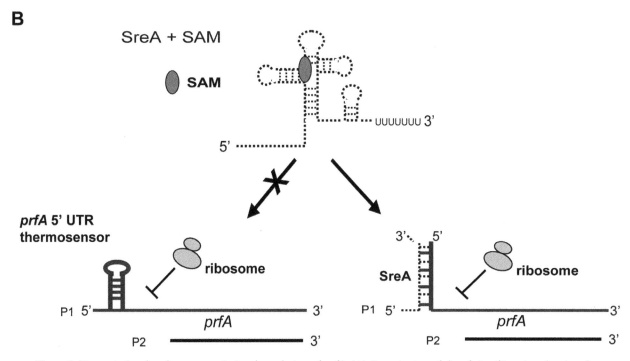

Figure 2. Transcriptional and posttranscriptional regulation of *prfA*. (A) Organization of the *plcA-prfA* region showing the location of all three promoter elements required for *prfA* expression. P$_{prfAP1}$ and P$_{prfAP2}$ are immediately upstream of *prfA*, while the third promoter, P$_{plcA}$, is located upstream of *plcA* and directs both a monocistronic *plcA* transcript and a bicistronic *plcA-prfA* transcript. Both P$_{plcA}$ and P$_{prfAP1}$ contain characteristics of an σ^A-dependent promoter, while P$_{prfAP2}$ contains characteristics of both σ^B- and σ^A-dependent promoters. The stem-loop structure indicates the *plcA* transcriptional terminator. The thermosensor is present within the 5′ UTR of the *prfA*P1 transcript and inhibits *prfA*P1 translation at lower temperatures. Both P$_{prfAP1}$ and P$_{prfAP2}$ are negatively [(−)] influenced by high levels of PrfA, whereas P$_{plcA}$ is positively [(+)] influenced, resulting in the production of the increased quantities of PrfA required for intracellular growth and spread. (B) Model of SreA *trans* regulation of *prfA* expression. SAM binding to the SAM-responsive riboswitch SreA results in a conformational change to the transcript and a terminator structure is formed, which then leads to the production of a short transcript lacking the downstream genes. Complementary nucleotide regions shared between this sRNA transcript and the 5′ UTR of the *prfA* mRNA result in a direct interaction between the two molecules that functions to reduce PrfA protein synthesis by blocking ribosome binding at 37°C. At lower temperatures (≤30°C), the thermosensor structure is formed at the 5′ UTR of the *prfA* mRNA and prevents both SreA binding and *prfA* mRNA translation.
doi:10.1128/9781555818524.ch19f2

THE SECOND MODE OF *prfA* REGULATION: POSTTRANSCRIPTIONAL CONTROL OF *prfA* EXPRESSION

RNA molecules have long been known to be important mediators of signal transduction and gene regulation, and in bacteria these molecules modulate gene expression in response to physiological cues (Winkler and Breaker, 2005). A number of RNA-based regulatory mechanisms have been described, including the presence of long mRNA leader sequences termed riboswitches that regulate the expression of genes immediately downstream, as well as small RNAs (sRNA) that bind proteins or base-pair with complementary target RNA molecules that can act either in *cis* or in *trans* (Waters and Storz, 2009). Most of these regulatory RNA molecules regulate gene expression by directly influencing mRNA transcription and/or transcript stability or by altering the efficiency of translational initiation. Long untranslated regions (UTRs) with extensive secondary structure are present in the mRNAs of multiple virulence genes, and studies demonstrating the importance of these regulatory RNAs in the modulation of *L. monocytogenes* gene expression have gained increasing attention (Shen and Higgins, 2005; Toledo-Arana et al., 2009; Wong et al., 2004).

Riboswitches present in mRNA of long UTRs have been viewed as one of the simplest forms of RNA regulatory elements (Waters and Storz, 2009). They are usually located within the 5′ UTRs of the mRNA molecules that they regulate and are normally *cis* encoded and *cis* acting. Acting in *cis* usually differentiates riboswitches from sRNAs, which often function in *trans* by base-pairing with specific target RNAs or through protein binding. The long leader sequences that make up riboswitches are capable of assuming structures that respond directly to temperature changes or physiological cues. Upon signal detection, a conformational shift occurs in the RNA secondary structure that serves to influence expression of the downstream genes.

Temperature has been shown to significantly influence the expression of virulence factors in *L. monocytogenes*, and Johansson et al. identified a thermosensor riboswitch present in the 5′ UTR of *prfA* that serves to regulate *prfA* expression as directed by the *prfA*P1 promoter (Johansson et al., 2002). The 5′ UTR of the *prfA*P1-directed mRNA forms a stem-loop structure at temperatures of 30°C or lower that effectively masks the ribosome binding region of *prfA* and inhibits *prfA* mRNA translation (Fig. 2B). This stem-loop becomes unstable at temperatures of 37°C or higher, thus enabling translation of *prfA*P1-directed mRNA and the production

of increased quantities of PrfA. In contrast, the *plcA* and *prfA*P2 promoters do not appear to be subject to this mode of thermoregulation, and therefore, transcripts from these promoters are likely to contribute to the expression of some PrfA-dependent virulence genes at temperatures at or below 30°C. While an increase in temperature has been suggested to signal the presence of *L. monocytogenes* within the environment of a mammalian host, increased temperature alone is not sufficient to induce PrfA-dependent gene expression. Bacteria grown at 37°C in broth culture still exhibit low to undetectable levels of PrfA activity in diverse media (Ripio et al., 1997b; Shetron-Rama et al., 2003; Wong and Freitag, 2004). In addition, PrfA activity is evident at 25°C in bacteria growing within infected insect cells, suggesting that a host cell-derived signal(s) can induce virulence gene expression in the absence of increased temperature (Cheng and Portnoy, 2003; Mansfield et al., 2003).

A novel role for a *cis*-acting riboswitch acting as a *trans* regulatory RNA for *prfA* expression was recently described by Loh *et al* (Loh et al., 2009). An intriguing connection was made between SreA, one of seven putative *S*-adenosylmethionine (SAM)-responsive riboswitches (SAM riboswitches) identified in the *L. monocytogenes* transcriptome, and *prfA*. The authors demonstrated a direct interaction between a defined region of *sreA* and the distal end of the *prfA*P1 5′ UTR. Experiments using an ectopic *Escherichia coli* expression system showed that *prf*-AP1 transcript-directed PrfA expression was reduced by the presence of *sreA*, and in vitro transcription/translation assays and electromobility shift assays demonstrated direct binding between SreA and the *prfA* 5′ UTR. Interestingly, SreA regulation of *prfA* was evident only at 37°C, suggesting that the *prfA* mRNA thermosensor represents the predominant regulation of *prfA*P1 transcripts at low temperatures, with SreA capable of functioning at higher temperatures that are relevant to bacterial infection of mammalian hosts (Fig. 2B).

THE THIRD MODE OF *prfA* REGULATION: POSTTRANSLATIONAL CONTROL OF PrfA

The third and least well-characterized yet perhaps most critical mechanism that exists for regulation of PrfA activity is posttranslational modification of PrfA. Based on structural and sequence homology, the PrfA protein belongs to the CRP-Fnr family of transcriptional regulators, of which there are approximately 400 members (Eiting et al., 2005; Korner et al., 2003). These regulators usually function as dimers and require the binding of small-molecule cofactors (for example, CRP with cAMP) or other forms of posttranslational modification (such as the binding of carbon monoxide by the heme moiety of CooA) to become fully activated. Previously published data had indicated that PrfA can be present within a cell but inactive (Renzoni et al., 1997). The identification of PrfA as a member of the CRP/Fnr family in combination with the high degree of structural similarity of PrfA to both CRP and CooA lent support to the premise that the binding of a small-molecule cofactor is likely to be important for PrfA activity (Eiting et al., 2005; Vega et al., 1998).

In addition to PrfA structural similarity to CRP and CooA, experimental evidence in support of PrfA posttranslational regulation includes the observation that PrfA-dependent gene expression is dramatically reduced in the presence of readily metabolized carbon sources, such as glucose and cellobiose, without a significant change in the amount of PrfA protein present in the bacterial cell (Milenbachs et al., 1997). Additional support for PrfA posttranslational regulation followed the identification of an *L. monocytogenes* strain that contained a single mutation within *prfA* coding sequences that resulted in constitutive expression of PrfA-dependent virulence genes in broth culture (Ripio et al., 1997b). The substitution of a serine for a glycine at position 145 within PrfA was analogous to an A144T mutation identified within CRP (CRP* mutants) that enabled CRP to become activated in the absence of its cAMP cofactor (Harman et al., 1986). CRP* mutations have been found to alter protein conformation in ways that mimic the allosteric changes that occur following the binding of cAMP (Garges and Adhya, 1985, 1988; Kim et al., 1992; Youn et al., 2006). These changes increase the DNA binding affinity of CRP for its target promoters. The PrfA* G145S mutant protein was found to have an approximately 18-fold increase in binding affinity for the *hly* promoter in comparison with wild-type PrfA, suggesting that the G145S amino acid substitution produced a conformational change in PrfA similar to those conferring the CRP* phenotype (Eiting et al., 2005; Vega et al., 1998). Although structural analysis of the PrfA* G145S mutant protein provided important insight into conformational changes that contribute to protein activation, studies of the mutant did not provide clues into the nature of the cofactor that may be required for full PrfA activity.

Based on homology with CRP, Eiting et al. identified a predicted cofactor binding site within the PrfA X-ray crystal structure consisting of a tunnel-like region located between the N-terminal β-barrel and C-terminal DNA-binding domains within the protein monomer (Eiting et al., 2005). While the identity of the PrfA cofactor is not known, the structure and chemical nature of the binding pocket may provide clues into the nature of the activating ligand. Electrostatic modeling of PrfA was recently used to probe the physical nature of the putative binding pocket (Xayarath et al., 2011b). This modeling revealed a high density of positive charge stemming from the presence of three lysine residues: K64, K122, and K130 (Fig. 3A). K64 and K122 are located on opposite sides of the opening of the pocket, while the K130 lysine is buried deep within the pocket interior. Charge neutralization of these lysine residues via glutamine substitution impaired PrfA DNA binding and activation of PrfA within the cytosol of infected host cells, whereas a K130 substitution completely abolished protein activity without affecting the protein levels (Xayarath et al., 2011b). Interestingly, the introduction of the *prfA** G145S mutation that constitutively activates PrfA in the absence of cofactor restored PrfA activity to the mutants containing the K64 or K122 substitution. These results suggested that the K64 and K122 mutations interfered with PrfA activation, presumably by reducing cofactor binding within the binding site pocket. PrfA K130Q activity could not be increased by the introduction of the PrfA* G145S mutation, suggesting that this substitution altered PrfA conformation such that the protein could no longer become activated (Xayarath et al., 2011b). While these studies still did not identify the activating cofactor, they implicated the positive charge of the PrfA binding pocket as contributing to the binding of a small anionic ligand, thereby leading to full activation of PrfA.

THE ISOLATION AND CHARACTERIZATION OF ADDITIONAL *prfA** MUTATIONS

Since the initial identification of the PrfA G145S mutant, a number of additional mutations that result in PrfA activation have been identified and characterized. *prfA** mutations identified to date include G145S, Y63C, S71C, E77K, A94T, L140F, Y154C, L148P, G155S, and PrfA P219S substitutions (Fig. 3B) (Miner et al., 2008a; Miner et al., 2008b; Monk et al., 2008; Mueller and Freitag, 2005; Shetron-Rama et al., 2003; Vega et al., 2004; Wong and Freitag, 2004; Xayarath et al., 2011a). The effects of these mutations on PrfA activity are not equivalent. Using PrfA-dependent *actA* gene expression as a readout, strains containing the different *prfA** alleles exhibited levels of *actA* expression in broth culture that were 4-fold to >200-fold greater than the levels of expression observed in wild-type bacteria (Miner et al., 2008a; Vega et al., 2004; Wong and Freitag, 2004; Xayarath et al., 2011a). *L. monocytogenes*

A

B

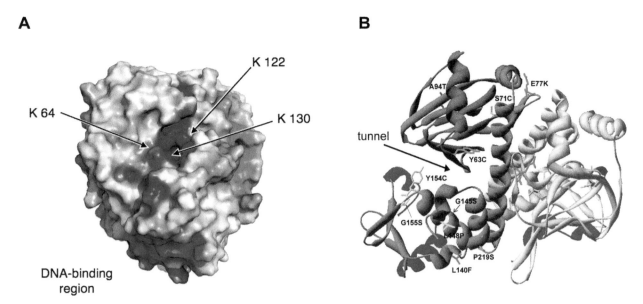

Figure 3. Model of the PrfA putative cofactor binding pocket and location of PrfA* amino acid substitutions. (A) Electrostatic modeling of PrfA protein showing the distribution of solvent-accessible surface charges on the protein dimer. Positive charge is shown in blue and negative charge in red, with electrostatic potentials ranging from −4kT/e (red) to +4kT/e (blue). Arrows point to the lysine residues that contribute to the positive charge of the putative cofactor binding pocket within PrfA. The positive charge of the DNA binding region is also highlighted at the bottom of the PrfA monomer. Reprinted from *PLoS One* (Xayarath et al., 2011b) with permission of the publisher. (B) Ribbon modeling of the PrfA dimer using DeepView-Swiss PdbViewer v4.0 (http://spdbv.vital-it.ch/) highlighting the tunnel region (black arrow) suggested by Eiting et al., 2005, as the cofactor binding site. The monomers that make up the dimer are colored either light or dark gray, and the DNA-binding helix-turn-helices are shown in blue. The locations of amino acid substitutions conferring the PrfA* phenotype are colored as follows: Y63C in green, S71C in tan, E77K in magenta, A94T in dark blue, L140F in red, G145S in black, L148P in yellow, Y154C in light pink, G155S in light blue, and P219S in orange. Residues G145 and G155 are difficult to see in this orientation and have been further highlighted by arrows pointing to their location.
doi:10.1128/9781555818524.ch19f3

strains containing *prfA** mutations also produced elevated levels of other PrfA-dependent gene products, including LLO, PI-PLC, PC-PLC, InlA, and InlB. While *prfA** mutations are thus defined by the resulting high-level expression of PrfA-dependent gene products, these mutations are located within very different regions of the protein (Fig. 3B).

Detailed structural information is thus far available only for the PrfA G145S mutant protein (Eiting et al., 2005). The G145 residue resides in αD, which is located at the C-terminal region of PrfA, and as previously described, the substitution of a serine at this position induces the repositioning of the PrfA helix-turn-helix DNA-binding domain that results in a substantial increase in PrfA DNA binding affinity. Substitutions at the nearby L140 residue (located in the same helix as G145) also dramatically increase PrfA DNA binding (Miner et al., 2008b; Wong and Freitag, 2004), presumably through a similar repositioning of the DNA-binding domain. In contrast, how the Y63C, S71C, E77K, A94T, L148P, Y154C, G155S, and P219S mutations lead to PrfA activation is less clear. Interestingly, while both the Y63C and P219S substitution mutants exhibited very high levels of PrfA-dependent virulence gene expression, neither

displayed a significant increase in their ability to bind DNA (Miner et al., 2008b; Xayarath et al., 2011a). The Y63 residue is located in a region of PrfA that forms part of the putative cofactor binding pocket; it is thus possible that the substitution of a tyrosine for a cysteine residue serves to enhance cofactor binding (Miner et al., 2008b). The P219 residue is located in αH, one of three alpha helices present within the extended C-terminal region which is unique to PrfA and absent from CRP (Fig. 3B) (Eiting et al., 2005; Xayarath et al., 2011a). These three C-terminal alpha helices have been proposed to stabilize the PrfA DNA-binding domain and to participate in homodimer interactions, such that residues within the three helices of one monomer form hydrogen bonds to β loops of the second monomer (Eiting et al., 2005; Herler et al., 2001). The *prfA* P219S mutation is thus not located near the putative cofactor binding region and would not be expected to enhance or influence cofactor binding unless through distal effects. Whatever the mechanism by which a P219S substitution activates PrfA, the change imposed by this specific mutation must be distinct from that conferred by G145S, as there is no evident increase in PrfA P219S DNA binding affinity (Xayarath et al., 2011a).

The *prfA* Y154C mutant has a phenotype that differs dramatically from that of the other *prfA** mutants (Miner et al., 2008b). The Y154 residue is located at the very end of αD, the same helix containing the G145 amino acid, but is oriented towards the PrfA tunnel that makes up the putative cofactor binding pocket (Fig. 3B). Interestingly, the substitution of cysteine for a tyrosine at position 154 resulted in modestly enhanced PrfA-dependent gene expression in broth culture as well as a modest enhancement in PrfA DNA binding affinity; however, the Y154C mutation inhibited full activation of PrfA within the cytosol, thereby significantly attenuating bacterial virulence (Miner et al., 2008b). Based on the apparent inhibition of intracellular PrfA activation, it was speculated that the Y154C mutation either interfered with cofactor binding or alternatively stabilized the low-activity form of PrfA, locking it in an "environmental" state. Overall, the variable phenotypes of the different *prfA** mutations indicate that PrfA activation can be achieved through a variety of structural modifications, all of which may have relevance to the structural changes that may be imposed by cofactor binding.

PrfA ACTIVATION AND THE *L. MONOCYTOGENES* BALANCE BETWEEN BACTERIAL LIFE INSIDE AND OUTSIDE OF A HOST

The activation of PrfA, triggered by *L. monocytogenes* entry into host cells, enhances the ability of the bacterium to adapt to life within host cells. *L. monocytogenes* strains containing *prfA** mutations are hyperinvasive, are quicker to mediate phagosome escape, and initiate bacterial actin-based motility more rapidly to promote cell-to-cell spread (Bruno and Freitag, 2010; Miner et al., 2008a; Mueller and Freitag, 2005; Wong and Freitag, 2004; Xayarath et al., 2011a). Activation of PrfA also appears to shift *L. monocytogenes* metabolism towards the use of three-carbon sugars, which have been demonstrated to be one of the principal carbon sources used by bacteria replicating within the cytosol (Eylert et al., 2008). *prfA** mutants are hypervirulent in mouse infection models and exhibit a competitive advantage over wild-type strains during infection (Bruno and Freitag, 2010; Shetron-Rama et al., 2003; Xayarath et al., 2011a). There thus appears to be no obvious disadvantages associated with constitutive activation of PrfA once bacteria have staked out a host. Why, then, have so many apparent regulatory mechanisms evolved to control PrfA activity?

The answer would appear to lie in the need for *L. monocytogenes* to balance life within a host with life in the outside environment. While physiological shifts resulting from the PrfA activation enhance the ability of *L. monocytogenes* to thrive in its intracellular replication niche, these physiological changes compromise bacterial fitness outside of host cells (Bruno and Freitag, 2010). Constitutively activated PrfA* mutants are impaired for flagellum-mediated swimming motility, and this defect would be anticipated to compromise bacterial fitness in environments where the bacteria must be able to detect and move towards available nutrient sources (O'Neil and Marquis, 2006; Port and Freitag, 2007; Shetron-Rama et al., 2003; Way et al., 2004; Wong and Freitag, 2004; Xayarath et al., 2011a). The swimming motility defect appears to be due to a defect not in flagellum assembly but, rather, in the ability of *prfA** mutants to detect and initiate movement towards nutrient sources (Xayarath et al., 2011a). It has, however, been reported that gene products that contribute to flagellum biosynthesis and chemotaxis, such as FlaA and MotA, are downregulated following PrfA activation (Port and Freitag, 2007; Toledo-Arana et al., 2009). When grown in rich broth culture media, monocultures of *prfA** mutants display no obvious growth defects; however, these mutants are rapidly outcompeted by wild-type bacteria when grown in mixed culture (Bruno and Freitag, 2010). Stress conditions of high osmolarity and low acidity exacerbate the competitive defects observed for *prfA** strains in a manner that appears independent of the stress-responsive sigma factor σ^B. *prfA** strains are also less efficient in their ability to utilize carbon sources such as glucose and cellobiose, preferring three-carbon sugars such as glycerol (Bruno and Freitag, 2010; Joseph and Goebel, 2007; Joseph et al., 2008; Marr et al., 2006; Stoll et al., 2008). It is thus becoming increasingly apparent that *L. monocytogenes* must carefully regulate the activity of PrfA in order to balance the needs of bacterial life in the outside environment with the needs required for life within host cells.

THE IMPACT OF PrfA ACTIVATION ON THE PROCESS AND PATTERN OF *L. MONOCYTOGENES* PROTEIN SECRETION

The core PrfA regulon encompasses 10 gene products directly regulated by PrfA that contribute to invasion, replication, and spread of *L. monocytogenes* to adjacent host cells (Scortti et al., 2007). Including *prfA* itself, six PrfA-dependent genes are located within a *Listeria* pathogenicity island referred to as the LIPI-1, which includes the genes *hly*, *plcA*, *prfA*, *mpl*, *actA*, and *plcB* (Table 1). Outside of the LIPI-1, *inlA*, *inlB*, *inlC*, *bsh*, and *hpt* are also directly

Table 1. Gene products directly or indirectly regulated by PrfA

Description	Gene no.	Gene name	Gene product	PrfA box (TTAACAnnTGTTAA)[c]	Reference(s)
Direct regulation					
LIPI-1	lmo0201	plcA[a,b]	Phosphatidylinositol-specific phospholipase C	TTAACAAATGTTAA	Mengaud et al., 1989
	lmo0200	prfA[a]	Listeriolysin-positive regulatory protein	TTAACAAATGTTAA	Freitag et al., 1993
		prfA[c]		CTAACATTTGTTGT	
	lmo0202	hly[b]	Listerolysin O	TTAACATTTGTTAA	Mengaud et al., 1989
	lmo0203	mpl	Zinc metalloproteinase precursor	TTAACAAATGTAAA	Mengaud et al., 1989
	lmo0204	actA[a]	Actin assembly-inducing protein	TTAACAAATGTTAG	Vazquez-Boland et al., 1992
	lmo0205	plcB[a]	Broad-range phospholipase C	TTAACAAATGTTAG	
Outside LIPI-1	lmo0433	inlA[a]	Internalin A	ATAACATAAGTTAA	Lingnau et al., 1995
	lmo0434	inlB[a]	Internalin B	ATAACATAAGTTAA	Lingnau et al., 1995
	lmo1786	inlC	Internalin C	TTAACGCTTGTTAA	Engelbrecht et al., 1996; Luo et al., 2004
	lmo0838	hpt	Hexose phosphate transport protein	ATAACAAGTGTTAA	Chico-Calero et al., 2002
	lmo2067	bsh	Bile salt hydrolase	TTAAAAATTATTAA	Dussurget et al., 2002
	lmo2219	prsA2	Posttranslocation molecular chaperone	TTTACAnnTATTAA	Port and Freitag, 2007
Indirect upregulation[d]					
ABC transporter	lmo0135	ctaP	Cysteine transport-associated protein	ND	Port and Freitag, 2007; Xayarath et al., 2009
	lmo2196	oppA	ABC transporter oligopeptide-binding protein	ND	Borezee et al., 2000; Port and Freitag, 2007
	lmo2417		Conserved lipoprotein (putative ABC transporter binding protein)	ND	Port and Freitag, 2007

Cell wall modifying	lmo1438		Similar to penicillin-binding protein	ND	Guinane et al., 2006; Port and Freitag, 2007
	lmo2691	*namA/ murA*	Similar to autolysin, N-acetylmuramidase	ND	Alonzo et al., 2011b; Carroll et al., 2003; Port and Freitag, 2007
	lmo2754	*pbp5*	Similar to D-alanyl-D-alanine carboxypeptidase	ND	Guinane et al., 2006; Port and Freitag, 2007
Antigenic lipoproteins	lmo1388	*tcsA*	CD4+ T-cell-stimulating antigen, lipoprotein	ND	Port and Freitag, 2007; Sanderson et al., 1995
	lmo2637		Conserved lipoprotein, putative pheromone	ND	Port and Freitag, 2007
	lmo0292	*htrA*	Similar to heat shock HtrA serine protease	ND	Port and Freitag, 2007; Stack et al., 2005; Wilson et al., 2006
	lmo1883	*chiA*	Chitinase A; hydrolyzes N-acetylglucosamine	ND	Chaudhuri et al., 2010; Larsen et al., 2010
Indirect downregulated[d]	lmo0013	*qoxA*	AA3-600 quinol oxidase subunit II	ND	Port and Freitag, 2007
	lmo00232	*clpC*	Endopeptidase Clp ATP binding chain C	ND	Nair et al., 2000
	lmo0690	*flaA*	Flagellin	ND	Dons et al., 2004; O'Neil and Marquis, 2006; Port and Freitag, 2007

[a]Bicistronic transcript, PrfA box shared with downstream gene(s).
[b]Divergently transcribed, PrfA box shared with divergently transcribed genes.
[c]PrfA box upstream of prfAP2, which negatively regulates PrfA expression.
[d]Proteins identified in a proteomics screen as upregulated secreted proteins in the presence of the constitutively active prfA* L140F mutant strain.
[e]n, any nucleotide. ND, not determined (nonrecognizable PrfA box).

regulated by PrfA (Table 1). Microarray analyses of *L. monocytogenes* wild-type and *prfA** mutants grown in brain heart infusion broth suggest the possibility of at least 145 or more additional genes associated with PrfA regulation (Marr et al., 2006; Milohanic et al., 2003; Scortti et al., 2007). Studies performed by Milohanic et al. featuring *L. monocytogenes* grown in the presence or absence of compounds known to either induce or repress PrfA activity identified 73 differentially regulated genes (Milohanic et al., 2003). The majority of these genes may be indirectly modulated by PrfA, as only a small number contained recognizable PrfA binding boxes within their promoter regions. Group I genes were shown to be positively regulated by PrfA, while group II genes were demonstrated to be repressed by PrfA under all conditions tested. Group III contained genes that were PrfA activated under conditions that repressed the expression of PrfA-dependent genes in group I or that were repressed when group I PrfA-dependent genes were activated. Interestingly, a subset of group III genes encoded proteins involved in many different types of stress response, and about half of this subset appeared to be subject to regulation by the stress-responsive sigma factor σ^B (Milohanic et al., 2003). However, an independent study by Ollinger et al. using reverse transcription-PCR showed that the transcript levels of some of genes identified by Milohanic et al. as group II or III were not significantly affected by the absence of *prfA* (Ollinger et al., 2008). The interplay between the regulatory networks that exist for PrfA in the presence of different sigma factors and thus different forms of RNA polymerase appears to be quite complex. These interactions may not be easily discerned under laboratory growth conditions, but in vitro studies may nonetheless provide some insight into the mechanisms used by *L. monocytogenes* to adapt and survive in its varied environments.

Not unexpectedly, PrfA activation also alters the patterns and processes of *L. monocytogenes* protein secretion. Bacterial virulence factors are often surface associated or secreted, providing a direct means for interactions with host cell components. Profiles of proteins isolated from the culture supernatants of the wild type and Δ*prfA* and *prfA** mutants identified 17 proteins that were differentially secreted following PrfA activation (Port and Freitag, 2007). The majority of the genes encoding these proteins did not contain recognizable PrfA binding sites in their upstream promoter regions, suggesting that the synthesis and/or secretion of these proteins was indirectly influenced by PrfA activation (Table 1). Fifteen of the secreted proteins were upregulated, and this group included some of the known virulence factors associated with uptake and proliferation within host cells,

while others identified were putative ABC transporter components, cell wall-modifying enzymes, antigenic lipoproteins, and chaperone proteins associated with protein secretion (Port and Freitag, 2007) (Table 1).

Several novel gene products whose secretion levels increased as a result of PrfA activation have been subsequently shown to contribute to *L. monocytogenes* virulence (Alonzo et al., 2011b; Chaudhuri et al., 2010; Port and Freitag, 2007; Xayarath et al., 2009). Examples include CtaP, a multifunctional cysteine-transport-associated Protein that contributes to many different aspects of *L. monocytogenes* pathogenesis, including adhesion to host cells, acid resistance, and bacterial membrane integrity during host infection (Xayarath et al., 2009). Insertion mutations within the gene encoding TcsA (Port and Freitag, 2007), a secreted protein first identified as a stimulating antigen for CD4+ T cells (Sanderson et al., 1995), also reduce bacterial virulence in mice, as does the loss of the *chiA*-encoded chitinase enzyme (Chaudhuri et al., 2010), recently shown to be PrfA regulated (Larsen et al., 2010). Mutational analysis of NamA, a murein hydrolase required for bacterial cell septation during logarithmic growth, indicates that while NamA is required for full virulence of *L. monocytogenes*, cytosolic bacteria lose the septation defect as the force of actin polymerization disrupts bacterial cell chains and enables cell-to-cell spread (Alonzo et al., 2011b; Carroll et al., 2003).

The identification of PrsA2 as a gene product whose expression and secretion are directly regulated by PrfA has resulted in a deeper appreciation of the costs of PrfA activation to *L. monocytogenes* membrane integrity and bacterial viability (Alonzo and Freitag, 2010; Alonzo et al., 2011c; Port and Freitag, 2007) (Fig. 4). PrsA2 is one of two posttranslocation secretion chaperones encoded by *L. monocytogenes* (Alonzo et al., 2009). It is upregulated in vivo and appears to have been specifically adapted to regulate the folding and activity of secreted virulence factors (Alonzo and Freitag, 2010; Alonzo et al., 2009; Alonzo et al., 2011c; Camejo et al., 2009; Chatterjee et al., 2006). *L. monocytogenes* mutants lacking PrsA2 exhibited normal growth characteristics in broth culture but had reduced viability when PrfA became activated (Alonzo and Freitag, 2010). The reduction in bacterial viability observed following PrfA activation was likely to reflect the accumulation of misfolded proteins at the membrane-cell wall interface, a situation that compromised bacterial survival in the cytosol of infected host cells. PrsA2 has also been found to contribute to the integrity of the bacterial cell wall: it appears to be necessary for the activity of proteins whose function is to modify the cell wall, presumably to promote protein secretion as well as bacterial survival in the cytosol (Alonzo et al., 2011c), where it has recently been shown that lysozyme can access *L. monocytogenes* (Rae et al., 2011). PrfA activation clearly has profound effects on multiple aspects of bacterial physiology that are integral to host infection.

In addition to the use of *prfA** mutations and in vitro methods of detecting genes and gene products influenced by PrfA, two independent in vivo transcription analyses have been carried out to identify changes in patterns of *L. monocytogenes* gene expression during intracellular bacterial replication within tissue culture cells, conditions under which PrfA is activated (Chatterjee et al., 2006; Joseph et al., 2006). By comparing the transcription profiles of broth-grown bacteria versus intracellular bacteria,

Outside Environment, Low PrfA Activity

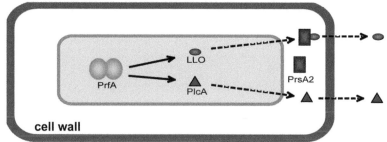

Host Cytosol, High PrfA Activity

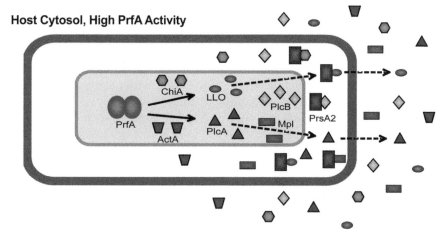

Host Cytosol, High PrfA Activity, No PrsA2

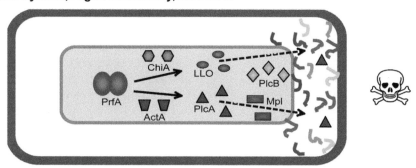

Figure 4. The posttranslocation secretion chaperone PrsA2 is critical for *L. monocytogenes* viability under conditions of PrfA activation. PrsA2 is a secreted chaperone that contributes to the folding and stability of *L. monocytogenes* secreted virulence factors and penicillin binding proteins at the membrane-cell wall interface. Outside of host cells and under conditions of low PrfA activity, PrsA2 is synthesized in small amounts and assists in the folding of virulence factors secreted at low levels, such as LLO (top). When PrfA becomes activated following the entry of *L. monocytogenes* into the cytosol, protein secretion increases dramatically, with a concurrent increase in PrsA2 levels to help mediate secreted protein folding and stability (middle). Under conditions of PrfA activation in the absence of PrsA2, misfolded proteins accumulate at the membrane-cell wall interface and compromise bacterial viability within the cytosol (bottom).
doi:10.1128/9781555818524.ch19f4

17 to 19% of the genes were found to be differentially expressed, including genes well associated with virulence (*prfA*, *hly*, *plcA*, *mpl*, *actA*, *plcB*, *inlA*, *inlB*, *inlC*, *hpt*, and *prsA2*). Upregulated genes also included those involved in general stress response, cell division, modification of the cell wall structure, and use of alternative carbon sources such as glycerol and phosphorylated sugars (Joseph et al., 2008; Joseph et al., 2006). The pentose phosphate pathway, not glycolysis, was therefore suggested to be the major pathway for sugar metabolism within the host cytosol (Joseph and Goebel, 2007; Joseph et al., 2008; Joseph et al., 2006). Another study examining the transcriptional profile observed between *L. monocytogenes* grown in brain heart infusion broth compared to in vivo growth in spleens isolated from mice revealed that approximately 20% of the genes were differentially expressed during infection, of which 80% were upregulated, while the remaining 20% were downregulated (Camejo et al., 2009). Similar to the observations made with bacteria grown in tissue culture cells, upregulated genes included those already known to be associated with virulence, stress response, cell wall metabolism, LPXTG surface proteins, lipoproteins, DNA metabolism, RNA/protein synthesis, and cell division. Interestingly, while this study also found that genes associated with the uptake of phosphorylated hexoses were highly induced, in vivo upregulation of genes encoding enzymes involved in glycolysis and downregulation of genes involved in the nonoxidative phase of the pentose phosphate pathway were also observed (Camejo et al., 2009). This is in contrast to the studies reported by Joseph et al. (Joseph and Goebel, 2007; Joseph et al., 2008; Joseph et al., 2006) demonstrating that the pentose phosphate pathway was the predominant method of carbon metabolism with the host cytosol. It is therefore possible that these discrepancies may reflect differences observed between bacterial growth in tissue culture cells versus growth in whole organs and animal tissues.

PUTTING IT ALL TOGETHER: UNLOCKING THE MECHANISMS LEADING TO PrfA ACTIVATION AND *L. MONOCYTOGENES* PATHOGENESIS

PrfA activation appears to function as the switch that enables *L. monocytogenes* to transition from saprophyte to pathogen (Freitag et al., 2009). To briefly summarize, when *L. monocytogenes* is living outside of a host, PrfA is predominantly found in its low-activity state, with modest amounts of protein provided via expression directed by the *prfA*P1 and *prfA*P2

promoters (Freitag and Portnoy, 1994; Gray et al., 2006) (Fig. 5). Transcripts directed by the *prfA*P1 promoter likely accumulate but are not translated due to the presence of the temperature-dependent thermosensor present in the *prfA*P1 5' UTR that prevents ribosome access (Johansson et al., 2002). Bacteria growing outside of host cells maintain flagellar motility to acquire nutrients and can also use swimming motility if ingested to more efficiently reach the intestinal epithelium (O'Neil and Marquis, 2006). Following host ingestion, elevated body temperature and the low acidity of the stomach increase the expression of stress-responsive genes, including *inlA*, *inlB*, and *prfA*, via the σ^B-dependent promoters (McGann et al., 2008; McGann et al., 2007). *L. monocytogenes* organisms that move from the stomach into the intestine are thus primed for host intestinal cell invasion and have accumulated increased amounts of nonactivated PrfA via σ^B-dependent *prfA*P2-directed transcription (Gray et al., 2006). PrfA protein levels may also have increased as a result of melting of the *prfA*P1 transcript thermosensor (Johansson et al., 2002); however, it is possible that the trans-acting riboswitch SreA continues to inhibit *prfA*P1 translation by preventing ribosome access (Loh et al., 2009). Once within the intestine, *L. monocytogenes* attaches to and invades epithelial cells through binding of InlA to host cell E-cadherin or via bacterial uptake through Peyer's patches (Bonazzi et al., 2009a; Ireton, 2007; Jensen et al., 1998). Following internalization by the host cell, a drop in pH within the phagosome triggers the activity of LLO, which has accumulated as a result of the low levels of PrfA generated via *prfA*P1 and *prfA*P2 expression and because of the presence of a high-affinity PrfA DNA binding site within the shared *hly*/*plcA* promoters (Freitag and Portnoy, 1994; Schnupf and Portnoy, 2007; Williams et al., 2000). LLO together with PlcA mediates bacterial escape into the cytosol (Goldfine et al., 1997). Once in the cytosol and through an as-yet-unknown signal, PrfA becomes fully activated and is synthesized at increased levels via the *plcA-prfA* read-through transcript, and the resulting accumulation of fully activated PrfA stimulates the expression of the multiple gene products required for *L. monocytogenes* acquisition of host cell nutrients, intracellular replication, and cell-to-cell spread (Freitag et al., 2009).

What remains elusive within this picture of the *L. monocytogenes* saprophyte-to-parasite transition is the nature of the signal(s) that triggers PrfA activation. Several lines of evidence suggest the importance of carbon sources and/or carbon metabolism to serve as a cue as to the presence of *L. monocytogenes* within a host habitat. It has been known for more than a decade that bacterial growth in the presence of readily metabolized carbohydrates such as glucose and cellobiose dramatically reduces the expression of PrfA-dependent gene products (Milenbachs et al., 1997; Milenbachs Lukowiak et al., 2004; Stoll et al., 2008), while the utilization of phosphorylated sugars such as glucose-1-phosphate or of glycerol does not lead to the repression of virulence gene expression (Chico-Calero et al., 2002; Joseph et al., 2008; Ripio et al., 1997a). *prfA** mutants are impaired for growth in the presence of glucose but more readily metabolize glycogen, glycerol, and other C_3 compounds that serve as intracellular carbon sources (Eisenreich et al., 2010; Joseph et al., 2008; Marr et al., 2006). Joseph et al. demonstrated that *L. monocytogenes*

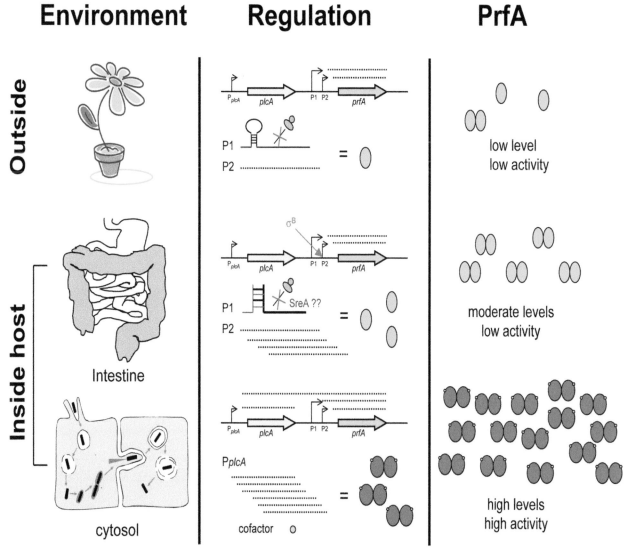

Figure 5. Model of *prfA* regulation during the *L. monocytogenes* transition from life in the outside environment to life inside the host. In the outside environment, low levels of PrfA protein are provided via transcripts directed from both the *prfA*P1 and *prfA*P2 promoters. The *prfA*P1 transcript contains a 5′ RNA thermosensor that serves to inhibit *prfA*P1 translation at temperatures below 30°C. Low levels of PrfA protein are provided from the *prfA*P2 transcripts (upper row). Following bacterial ingestion by a mammalian host, an increase in temperature and the acidic environment of the stomach induces the activity of the stress-responsive RNA polymerase alternative sigma factor σ^B, leading to increased expression of *prfA*, *inlA*, and *inlB* via their σ^B-regulated promoters. A low-activity form of PrfA accumulates as a result of the increase in *prfA*P2 promoter activity. While the increase in temperature relieves the inhibition of *prfA*P1 translation imposed by the *prfA* thermosensor, *prfA*P1 translation may still be inhibited via the direct binding of the SreA riboswitch to the 5′ UTR of *prfA*P1 transcripts (middle row). Once inside the host cell, PrfA-dependent production of LLO and PlcA mediates the efficient escape of *L. monocytogenes* from the phagosome into the cytosol, where unknown signals lead to the activation of PrfA and increased expression of *prfA* from the *plcA*P promoter. The *plcA*-*prfA* transcript results in the accumulation of increased levels of PrfA, which becomes activated via cofactor binding. High levels of activated PrfA increase the expression of a variety of gene products that contribute to *L. monocytogenes* intracellular replication and cell-to-cell spread (lower row).
doi:10.1128/9781555818524.ch19f5

switches from glycolysis to the pentose phosphate cycle when transferred from culture media to host cells, suggesting that the pentose phosphate cycle and not glycolysis is the predominant form of sugar metabolism within the host cytosol (Joseph et al., 2006). As previously mentioned, Camejo et al. reported that genes encoding enzymes involved in glycolysis were upregulated, while genes involved in the nonoxidative phase of the pentose phosphate pathway were repressed during in vivo growth in the spleens of mice (Camejo et al., 2009). Taken together, both studies indicate that

L. monocytogenes adjusts its metabolism in response to the host carbon sources it encounters and consumes during infection.

In contrast to carbon sources encountered within the host, PrfA-dependent gene expression is dramatically repressed in the presence of cellobiose and other readily metabolized sugars that are likely to be more prevalent in environments located outside of host cells (Milenbachs et al., 1997; Park and Kroll, 1993). These preferred carbon sources are transported into the bacterial cell via the phosphoenolpyruvate

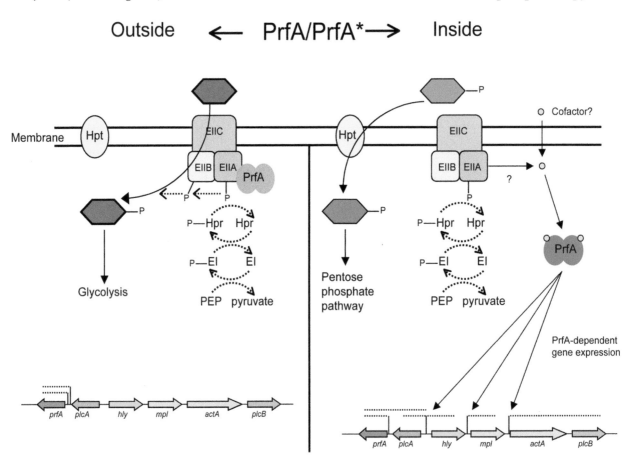

Figure 6. Model depicting the links between carbon transport and metabolism and PrfA-dependent virulence gene expression. The PEP-PTS is a multiprotein phosphorelay system that couples the transport of sugars across the bacterial membrane with their simultaneous phosphorylation. The PTS is comprised of three distinct proteins: EI, histidine protein (Hpr), and the permease EII. (Left) Outside of host cells, *L. monocytogenes* can be found in environments rich in the presence of PTS-dependent sugars, such as the plant sugar cellobiose. Transport of PTS-dependent sugars into the bacterial cell initiates with the autophosphorylation of EI using the phosphoryl group from PEP. EI subsequently transfers a phosphoryl group to Hpr, which then transfers it to the A domain of the EII permease group. As the sugar is transported into the bacterial cell, the phosphoryl group of EIIA is rapidly transferred to the EIIB domain and then to the incoming sugar as it passes through the EIIC domain of the permease. EIIA is therefore in a nonphosphorylated state during active PTS sugar transport, and it is this nonphosphorylated form of EIIA that has been proposed to sequester PrfA, thereby reducing or inhibiting its activity. (Right) Within the mammalian cell cytosol, alternative carbon sources such as hexose-phosphate sugars and C_3 sugars appear to be the predominant sources of carbohydrates used by *L. monocytogenes*. Transport of these non-PTS-dependent sugars into the bacterial cell occurs either via facilitated diffusion or through alternative transporters such as the Hpt system. Sugar transport via these alternative pathways leaves the EIIA component of the PTS in its phosphorylated state. It has been proposed that phosphorylated EIIA cannot sequester PrfA, which is then free to bind a cofactor that may be produced either by the bacterial cell or by the host cell. Activated PrfA binds target promoters and induces the expression of multiple bacterial gene products that contribute to host infection. Adapted from Freitag et al., 2009.

doi:10.1128/9781555818524.ch19f6

(PEP)-phosphotransferase system (PTS), a multiprotein phosphorelay system that couples the transport of sugars across the bacterial membrane with simultaneous phosphorylation of the incoming sugars (Barabote and Saier, 2005; Deutscher et al., 2006). Bacteria generally possess a number of PTS transporters which transport a variety of preferred carbon sources such as glucose, mannose, lactose, fructose, and galactose; *L. monocytogenes* appears to encode 29 complete PTSs (Glaser et al., 2001). There are seven families of PTSs based on the permeases that make up the individual PTSs, and each system is named accordingly, but functional redundancy does exist between the different systems (Barabote and Saier, 2005). The PTS is composed of three distinct proteins: enzyme I (EI), histidine protein (Hpr), and enzyme II (EII) permease. A functional PTS EII permease complex contains three or four components, including EIIA, EIIB, EIIC, and sometimes EIID. In this system, glycolytic intermediates signal the autophosphorylation of EI via the phosphoryl group from PEP. EI~P transfers its phosphate to a highly conserved His residue in Hpr, followed by the subsequent transfer to EIIA. EIIA rapidly transfers the phosphate to EIIB, which phosphorylates the incoming sugar as it passes through EIIC to enter the glycolytic pathway.

A model that describes the regulation of PrfA activity by carbon metabolism has evolved based on several independent reports that suggest that levels of PrfA-dependent gene expression correlate with the phosphorylation status of the components of the PTS permease complex (Joseph et al., 2008; Mertins et al., 2007; Stoll et al., 2008) (Fig. 6). In the presence of PTS-dependent sugars, rapid EII-dependent phosphorylation of the incoming sugar results in nonphosphorylated EII accumulating as the primary form of this protein in the cell. Nonphosphorylated EIIA correlates with a decrease in PrfA-dependent gene expression. In contrast, the phosphorylated form of EIIA accumulates in the presence of non-PTS-dependent carbon sources, and this form of EIIA is associated with high-level gene expression of PrfA-dependent gene products (Fig. 6). It has been proposed that one or more sugar-specific EIIA components of PTS in the nonphosphorylated form bind and sequester PrfA, keeping the regulator functionally inactive and preventing the induction of virulence gene expression. In the presence of non-PTS-dependent carbon sources such as hexose phosphates or glycerol, the lack of PTS-dependent sugar transport results in the accumulation of the phosphorylated form of EIIA and the release of PrfA, which is then fully active to induce target gene expression (Fig. 6) (Joseph et al., 2008; Joseph et al., 2006; Stoll et al., 2008).

While several features of this model seem plausible, it does raise a number of puzzling issues. A number of PTS-dependent sugars that require different EIIA molecules for transport have been associated with the repression of PrfA-dependent gene expression. This would suggest that each EIIA molecule must be capable of binding and sequestering PrfA and that the absence of a phosphate group on one EIIA is sufficient to sequester all of the available PrfA within the cell. This model would further suggest that PrfA may differ from other CRP/Fnr family members in not requiring the binding of a small-molecule signal or cofactor for full activity (Korner et al., 2003). The model proposed by Joseph et al. suggests that PrfA would be fully active following its release from EIIA without a requirement for the binding of a signal molecule or posttranslational modification (Joseph et al., 2008). However, both structural and functional analyses of wild-type and mutationally activated PrfA mutant proteins are consistent with increased PrfA activity as a result of conformational changes that promote higher-affinity DNA binding; it has been argued that these conformational changes occur as a result of cofactor binding within a structurally defined binding pocket (Eiting et al., 2005; Vega et al., 1998). Mutational analysis of the PrfA binding pocket supports a role for this region in PrfA activation (Xayarath et al., 2011b).

Alternatively, it is possible that *L. monocytogenes* phosphorylated PTS permeases function to stimulate the synthesis of a cofactor or second messenger that serves to activate PrfA (Fig. 6) (Freitag et al., 2009). EIIA-dependent stimulation of cofactor biosynthesis has been demonstrated in *E. coli* in that the glucose-specific PTS EIIA (EIIAGlc)-phosphate stimulates adenylate cyclase to produce the CRP cofactor cAMP (Gorke and Stulke, 2008). It is also possible that positive activation of PrfA by a cytosol-induced signal may be required as an additional step following the release of PrfA by EIIA or another PTS-related component to induce the fully activated PrfA conformational state (Fig. 6). The combined effects of PTS-dependent PrfA sequestration and activation of PrfA via cofactor binding combined with the multiple mechanisms of transcriptional and posttranscriptional regulation may thus fully integrate PrfA activity to the appropriate environmental location by virtue of the fact that activation requires multiple coordinated signals and events.

Finally, a very recent study that examined the *L. monocytogenes* transcriptome reported that the majority of the PrfA regulon was repressed during growth in the presence of bile (Quillin et al., 2011). It was speculated that the constituents of bile may negatively regulate PrfA activity, which, in turn, might represent a means of dialing down PrfA activity within host environments such as the gallbladder, where *L. monocytogenes* has been reported to persist in an extracellular form (Hardy et al., 2004). Interestingly, neither the expression of the PrfA-regulated bile salt hydrolase *bsh* nor the bile exclusion system *bilE* was repressed in the presence of bile; however, both of these genes contain promoters recognized by SigB (Milohanic et al., 2003). The regulation of PrfA activity thus appears to be quite complex, encompassing many regulatory networks and signals. This complexity no doubt stems from the need of *L. monocytogenes* to adapt to its exceedingly diverse and differentially challenging environmental niches.

CONCLUDING REMARKS

L. monocytogenes appears to be a master of diverse lifestyles, having developed the capacity to exist in habitats ranging from soil and groundwater to food processing plants and mammalian hosts. Many of the physiological changes underlying the transition of *L. monocytogenes* from soil bacterium to cytosolic pathogen appear to depend upon the activity of a single master regulatory switch: PrfA. While undoubtedly the premise of PrfA as the sole switch is simplistic, it remains striking that so many aspects of *L. monocytogenes* physiology and virulence can be traced back to the activation status of this single regulatory protein. The activity of this master switch must itself be carefully regulated in order to maximize *L. monocytogenes* survival and fitness in the diverse environments occupied by this bacterium.

The link between carbon source metabolism and the regulation of PrfA activity is intriguing given the structural similarities of PrfA with CRP, the central regulator of carbon metabolism in *Escherichia coli* (Eiting et al., 2005; Fic et al., 2009). It is possible that the original function of PrfA was tied to carbon metabolism and that the incorporation of virulence genes into the PrfA regulatory network evolved based on the ability of these gene products to promote bacterial survival in the presence of the carbohydrates found within mammalian cells. Indeed, available carbon sources may serve as a common theme for pathogens to identify a replication-permissive host environment. Links between carbon source utilization and bacterial virulence have been described for a variety of pathogens, including *Pseudomonas aeruginosa, Vibrio cholerae, Yersinia pestis, Yersinia pseudotuberculosis, Streptococcus pyogenes, Streptococcus pneumoniae, Streptococcus gordonii, Bacillus anthracis*, and *Staphylococcus aureus* (Chiang et al., 2011; Gorke and Stulke, 2008; Tsvetanova et al., 2007). Overall, the combined experimental evidence suggests that bacteria have coordinated carbon metabolism with virulence as a way of "tasting" their environment and then using the available flavor cues to determine which of the items in their arsenal of gene products are necessary for bacterial survival within that habitat. Environmental pathogens may thus be considered as having simply acquired a broader palate and better utensils for ultimately dining on a human host than those of their nonpathogenic microbial compatriots.

Acknowledgments. We thank Michael R. Campbell for his contributions to the writing of this chapter and also Nathan A. Styles for his helpful comments and suggestions.

REFERENCES

Alonzo, F., III, L. D. Bobo, D. J. Skiest, and N. E. Freitag. 2011a. Evidence for subpopulations of *Listeria monocytogenes* with enhanced invasion of cardiac cells. *J. Med. Microbiol.* 60:423–434.

Alonzo, F., III, P. D. McMullen, and N. E. Freitag. 2011b. Actin polymerization drives septation of *Listeria monocytogenes namA* hydrolase mutants, demonstrating host correction of a bacterial defect. *Infect. Immun.* 79:1458–1470.

Alonzo, F., III, B. Xayarath, J. C. Whisstock, and N. E. Freitag. 2011c. Functional analysis of the *Listeria monocytogenes* secretion chaperone PrsA2 and its multiple contributions to bacterial virulence. *Mol. Microbiol.* 80:1530–1548.

Alonzo, F., III, and N. E. Freitag. 2010. *Listeria monocytogenes* PrsA2 is required for virulence factor secretion and bacterial viability within the host cell cytosol. *Infect. Immun.* 78:4944–4957.

Alonzo, F., III, G. C. Port, M. Cao, and N. E. Freitag. 2009. The posttranslocation chaperone PrsA2 contributes to multiple facets of *Listeria monocytogenes* pathogenesis. *Infect. Immun.* 77:2612–2623.

Antolin, J., A. Gutierrez, R. Segoviano, R. Lopez, and R. Ciguenza. 2008. Endocarditis due to *Listeria*: description of two cases and review of the literature. *Eur. J. Intern. Med.* 19:295–296.

Baquero, F., and M. Lemonnier. 2009. Generational coexistence and ancestor's inhibition in bacterial populations. *FEMS Microbiol. Rev.* 33:958–967.

Barabote, R. D., and M. H. Saier, Jr. 2005. Comparative genomic analyses of the bacterial phosphotransferase system. *Microbiol. Mol. Biol. Rev.* 69:608–634.

Begley, M., R. D. Sleator, C. G. Gahan, and C. Hill. 2005. Contribution of three bile-associated loci, *bsh, pva*, and *btlB*, to gastrointestinal persistence and bile tolerance of *Listeria monocytogenes*. *Infect. Immun.* 73:894–904.

Bonazzi, M., M. Lecuit, and P. Cossart. 2009a. *Listeria monocytogenes* internalin and E-cadherin: from bench to bedside. *Cold Spring Harbor Perspect. Biol.* 1:a003087.

Bonazzi, M., M. Lecuit, and P. Cossart. 2009b. *Listeria monocytogenes* internalin and E-cadherin: from structure to pathogenesis. *Cell. Microbiol.* 11:693–702.

Borezee, E., E. Pellegrini, and P. Berche. 2000. OppA of *Listeria monocytogenes*, an oligopeptide-binding protein required for bacterial growth at low temperature and involved in intracellular survival. *Infect. Immun.* 68:7069–7077.

Bortolussi, R. 2008. Listeriosis: a primer. *Can. Med. Assoc. J.* 179:795–797.

Bruno, J. C., Jr., and N. E. Freitag. 2010. Constitutive activation of PrfA tilts the balance of *Listeria monocytogenes* fitness towards life within the host versus environmental survival. *PLoS One* 5:e15138.

Camejo, A., C. Buchrieser, E. Couve, F. Carvalho, O. Reis, P. Ferreira, S. Sousa, P. Cossart, and D. Cabanes. 2009. *In vivo* transcriptional profiling of *Listeria monocytogenes* and mutagenesis identify new virulence factors involved in infection. *PLoS Pathog.* 5:e1000449.

Camilli, A., L. G. Tilney, and D. A. Portnoy. 1993. Dual roles of *plcA* in *Listeria monocytogenes* pathogenesis. *Mol. Microbiol.* 8:143–157.

Carroll, S. A., T. Hain, U. Technow, A. Darji, P. Pashalidis, S. W. Joseph, and T. Chakraborty. 2003. Identification and characterization of a peptidoglycan hydrolase, MurA, of *Listeria monocytogenes*, a muramidase needed for cell separation. *J. Bacteriol.* 185:6801–6808.

Centers for Disease Control and Prevention. 1998. Multistate outbreak of listeriosis—United States, 1998. *MMWR Morb. Mortal. Wkly. Rep.* 47:1085–1086.

Centers for Disease Control and Prevention. 1999. Update: multistate outbreak of listeriosis—United States, 1998–1999. *MMWR Morb. Mortal. Wkly. Rep.* 47:1117–1118.

Centers for Disease Control and Prevention. 2002. Outbreak of listeriosis—northeastern United States, 2002. *MMWR Morb. Mortal. Wkly. Rep.* 51:950–951.

Centers for Disease Control and Prevention. 2004. Preliminary FoodNet data on the incidence of infection with pathogens transmitted commonly through food—selected sites, United States, 2003. *MMWR Morb. Mortal. Wkly. Rep.* 53:338–343.

Chakraborty, T., M. Leimeister-Wachter, E. Domann, M. Hartl, W. Goebel, T. Nichterlein, and S. Notermans. 1992. Coordinate regulation of virulence genes in *Listeria monocytogenes* requires the product of the *prfA* gene. *J. Bacteriol.* 174:568–574.

Chatterjee, S. S., H. Hossain, S. Otten, C. Kuenne, K. Kuchmina, S. Machata, E. Domann, T. Chakraborty, and T. Hain. 2006. Intracellular gene expression profile of *Listeria monocytogenes*. *Infect. Immun.* 74:1323–1338.

Chaturongakul, S., S. Raengpradub, M. Wiedmann, and K. J. Boor. 2008. Modulation of stress and virulence in *Listeria monocytogenes*. *Trends Microbiol.* 16:388–396.

Chaudhuri, S., J. C. Bruno, F. Alonzo III, B. Xayarath, N. P. Cianciotto, and N. E. Freitag. 2010. Contribution of chitinases to *Listeria monocytogenes* pathogenesis. *Appl. Environ. Microbiol.* 76:7302–7305.

Cheng, L. W., and D. A. Portnoy. 2003. *Drosophila* S2 cells: an alternative infection model for *Listeria monocytogenes*. *Cell. Microbiol.* 5:875–885.

Chiang, C., C. Bongiorni, and M. Perego. 2011. Glucose-dependent activation of *Bacillus anthracis* toxin gene expression and virulence requires the carbon catabolite protein CcpA. *J. Bacteriol.* 193:52–62.

Chico-Calero, I., M. Suarez, B. Gonzalez-Zorn, M. Scortti, J. Slaghuis, W. Goebel, and J. A. Vazquez-Boland. 2002. Hpt, a bacterial homolog of the microsomal glucose-6-phosphate translocase, mediates rapid intracellular proliferation in *Listeria*. *Proc. Natl. Acad. Sci. USA* 99:431–436.

Cossart, P., and M. Lecuit. 1998. Interactions of *Listeria monocytogenes* with mammalian cells during entry and actin-based movement: bacterial factors, cellular ligands and signaling. *EMBO J.* 17:3797–3806.

Cossart, P., and A. Toledo-Arana. 2008. *Listeria monocytogenes*, a unique model in infection biology: an overview. *Microbes Infect.* 10:1041–1050.

Czuprynski, C. J. 2005. *Listeria monocytogenes*: silage, sandwiches and science. *Anim. Health Res. Rev.* 6:211–217.

Delgado, A. R. 2008. Listeriosis in pregnancy. *J. Midwifery Womens Health* 53:255–259.

Deutscher, J., C. Francke, and P. W. Postma. 2006. How phosphotransferase system-related protein phosphorylation regulates carbohydrate metabolism in bacteria. *Microbiol. Mol. Biol. Rev.* 70:939–1031.

Dons, L., E. Eriksson, Y. Jin, M. E. Rottenberg, K. Kristensson, C. Larsen, J. Bresciani, and J. E. Olsen. 2004. Role of flagellin and the two-component CheA/CheY system of *Listeria monocytogenes* in host cell invasion and virulence. *Infect. Immun.* 72:3237–3244.

Drevets, D. A., and M. S. Bronze. 2008. *Listeria monocytogenes*: epidemiology, human disease, and mechanisms of brain invasion. *FEMS Immunol. Med. Microbiol.* 53:151–165.

Dussurget, O., D. Cabanes, P. Dehoux, M. Lecuit, C. Buchrieser, P. Glaser, and P. Cossart. 2002. *Listeria monocytogenes* bile salt hydrolase is a PrfA-regulated virulence factor involved in the intestinal and hepatic phases of listeriosis. *Mol. Microbiol.* 45:1095–1106.

Eisenreich, W., T. Dandekar, J. Heesemann, and W. Goebel. 2010. Carbon metabolism of intracellular bacterial pathogens and possible links to virulence. *Nat. Rev. Microbiol.* 8:401–412.

Eiting, M., G. Hageluken, W. D. Schubert, and D. W. Heinz. 2005. The mutation G145S in PrfA, a key virulence regulator of *Listeria monocytogenes*, increases DNA-binding affinity by stabilizing the HTH motif. *Mol. Microbiol.* 56:433–446.

Engelbrecht, F., S. K. Chun, C. Ochs, J. Hess, F. Lottspeich, W. Goebel, and Z. Sokolovic. 1996. A new PrfA-regulated gene of *Listeria monocytogenes* encoding a small, secreted protein which belongs to the family of internalins. *Mol. Microbiol.* 21:823–837.

Eylert, E., J. Schar, S. Mertins, R. Stoll, A. Bacher, W. Goebel, and W. Eisenreich. 2008. Carbon metabolism of *Listeria monocytogenes* growing inside macrophages. *Mol. Microbiol.* 69:1008–1017.

Farber, J. M., and P. I. Peterkin. 1991. *Listeria monocytogenes*, a food-borne pathogen. *Microbiol. Rev.* 55:476–511.

Farber, J. M., W. H. Ross, and J. Harwig. 1996. Health risk assessment of *Listeria monocytogenes* in Canada. *Int. J. Food Microbiol.* 30:145–156.

Fenlon, D. R. 1985. Wild birds and silage as reservoirs of *Listeria* in the agricultural environment. *J. Appl. Bacteriol.* 59:537–543.

Fic, E., P. Bonarek, A. Gorecki, S. Kedracka-Krok, J. Mikolajczak, A. Polit, M. Tworzydlo, M. Dziedzicka-Wasylewska, and Z. Wasylewski. 2009. cAMP receptor protein from *escherichia coli* as a model of signal transduction in proteins—a review. *J. Mol. Microbiol. Biotechnol.* 17:1–11.

Forster, B. M., A. P. Bitar, E. R. Slepkov, K. J. Kota, H. Sondermann, and H. Marquis. 2011. The metalloprotease of *Listeria monocytogenes* is regulated by pH. *J. Bacteriol.* 193:5090–5097.

Freitag, N. E. 2006. From hot dogs to host cells: how the bacterial pathogen *Listeria monocytogenes* regulates virulence gene expression. *Future Microbiol.* 1:89–101.

Freitag, N. E., G. C. Port, and M. D. Miner. 2009. *Listeria monocytogenes*—from saprophyte to intracellular pathogen. *Nat. Rev. Microbiol.* 7:623–628.

Freitag, N. E., and D. A. Portnoy. 1994. Dual promoters of the *Listeria monocytogenes prfA* transcriptional activator appear essential in vitro but are redundant in vivo. *Mol. Microbiol.* 12:845–853.

Freitag, N. E., L. Rong, and D. A. Portnoy. 1993. Regulation of the *prfA* transcriptional activator of *Listeria monocytogenes*: multiple promoter elements contribute to intracellular growth and cell-to-cell spread. *Infect. Immun.* 61:2537–2544.

Gandhi, M., and M. L. Chikindas. 2007. *Listeria*: a foodborne pathogen that knows how to survive. *Int. J. Food Microbiol.* 113:1–15.

Garges, S., and S. Adhya. 1985. Sites of allosteric shift in the structure of the cyclic AMP receptor protein. *Cell* 41:745–751.

Garges, S., and S. Adhya. 1988. Cyclic AMP-induced conformational change of cyclic AMP receptor protein (CRP): intragenic suppressors of cyclic AMP-independent CRP mutations. *J. Bacteriol.* **170**:1417–1422.

Gedde, M. M., D. E. Higgins, L. G. Tilney, and D. A. Portnoy. 2000. Role of listeriolysin O in cell-to-cell spread of *Listeria monocytogenes. Infect. Immun.* **68**:999–1003.

Gellin, B. G., and C. V. Broome. 1989. Listeriosis. *JAMA* **261**:1313–1320.

Gibbons, I. S., A. Adesiyun, N. Seepersadsingh, and S. Rahaman. 2006. Investigation for possible source(s) of contamination of ready-to-eat meat products with *Listeria* spp. and other pathogens in a meat processing plant in Trinidad. *Food Microbiol.* **23**:359–366.

Glaser, P., L. Frangeul, C. Buchrieser, C. Rusniok, A. Amend, F. Baquero, P. Berche, H. Bloecker, P. Brandt, T. Chakraborty, A. Charbit, F. Chetouani, E. Couve, A. de Daruvar, P. Dehoux, E. Domann, G. Dominguez-Bernal, A. Duchaud, L. Durant, O. Dussurget, K. D. Entian, H. Fsihi, F. Garcia-del Portillo, P. Garrido, L. Gautier, W. Goebel, N. Gomez-Lopez, T. Hain, J. Hauf, D. Jackson, L. M. Jones, U. Kaerst, J. Kreft, M. Kuhn, F. Kunst, G. Kurapkat, E. Madueno, A. Maitournam, J. M. Vicente, E. Ng, H. Nedjari, G. Nordsiek, S. Novella, B. de Pablos, J. C. Perez-Diaz, R. Purcell, B. Remmel, M. Rose, T. Schlueter, N. Simoes, A. Tierrez, J. A. Vazquez-Boland, H. Voss, J. Wehland, and P. Cossart. 2001. Comparative genomics of *Listeria* species. *Science* **294**:849–852.

Goldfine, H., C. Knob, D. Alford, and J. Bentz. 1997. Membrane permeabilization by *Listeria monocytogenes* phosphatidylinositol-specific phospholipase C is independent of phospholipid hydrolysis and cooperative with listeriolysin O. *Proc. Natl. Acad. Sci. USA* **94**:2772.

Goldfine, H., S. J. Wadsworth, and N. C. Johnston. 2000. Activation of host phospholipases C and D in macrophages after infection with *Listeria monocytogenes. Infect. Immun.* **68**:5735–5741.

Gorke, B., and J. Stulke. 2008. Carbon catabolite repression in bacteria: many ways to make the most out of nutrients. *Nat. Rev. Microbiol.* **6**:613–624.

Gottlieb, S. L., E. C. Newbern, P. M. Griffin, L. M. Graves, R. Hoekstra, N. L. Baker, S. B. Hunter, K. G. Holt, F. Ramsey, M. Head, P. Levine, G. Johnson, D. Schoonmaker-Bopp, V. Reddy, L. Kornstein, M. Gerwel, J. Nsubuga, L. Edwards, S. Stonecipher, S. Hurd, D. Austin, M. A. Jefferson, S. D. Young, K. Hise, E. D. Chernak, and J. Sobel. 2006. Multistate outbreak of listeriosis linked to turkey deli meat and subsequent changes in US regulatory policy. *Clin. Infect. Dis.* **42**:29–36.

Goyal, A., M. Takaine, V. Simanis, and K. Nakano. 2011. Dividing the spoils of growth and the cell cycle: the fission yeast as a model for the study of cytokinesis. *Cytoskeleton* (Hoboken) **68**:69–88.

Gray, M. J., N. E. Freitag, and K. J. Boor. 2006. How the bacterial pathogen *Listeria monocytogenes* mediates the switch from environmental Dr. Jekyll to pathogenic Mr. Hyde. *Infect. Immun.* **74**:2505–2512.

Gray, M. L., and A. H. Killinger. 1966. *Listeria monocytogenes* and listeric infections. *Bacteriol. Rev.* **30**:309–382.

Guinane, C. M., P. D. Cotter, R. P. Ross, and C. Hill. 2006. Contribution of penicillin-binding protein homologs to antibiotic resistance, cell morphology, and virulence of *Listeria monocytogenes* EGDe. *Antimicrob. Agents Chemother.* **50**:2824–2828.

Hardy, J., P. Chu, and C. H. Contag. 2009. Foci of *Listeria monocytogenes* persist in the bone marrow. *Dis. Model Mech.* **2**:39–46.

Hardy, J., K. P. Francis, M. DeBoer, P. Chu, K. Gibbs, and C. H. Contag. 2004. Extracellular replication of *Listeria monocytogenes* in the murine gall bladder. *Science* **303**:851–853.

Harman, J. G., M. McKenney, and A. Peterkofsky. 1986. Structure-function analysis of three cAMP-independent forms of the cAMP receptor protein. *J. Biol. Chem.* **261**:16332–16339.

Herler, M., A. Bubert, M. Goetz, Y. Vega, J. A. Vazquez-Boland, and W. Goebel. 2001. Positive selection of mutations leading to loss or reduction of transcriptional activity of PrfA, the central regulator of *Listeria monocytogenes* virulence. *J. Bacteriol.* **183**:5562–5570.

Ireton, K. 2007. Entry of the bacterial pathogen *Listeria monocytogenes* into mammalian cells. *Cell. Microbiol.* **9**:1365–1375.

Ireton, K., and P. Cossart. 1997. Host-pathogen interactions during entry and actin-based movement of *Listeria monocytogenes. Annu. Rev. Genet.* **31**:113–138.

Jensen, V. B., J. T. Harty, and B. D. Jones. 1998. Interactions of the invasive pathogens *Salmonella typhimurium*, *Listeria monocytogenes*, and *Shigella flexneri* with M cells and murine Peyer's patches. *Infect. Immun.* **66**:3758–3766.

Johansson, J., P. Mandin, A. Renzoni, C. Chiaruttini, M. Springer, and P. Cossart. 2002. An RNA thermosensor controls expression of virulence genes in *Listeria monocytogenes. Cell* **110**:551–561.

Joseph, B., and W. Goebel. 2007. Life of *Listeria monocytogenes* in the host cells' cytosol. *Microbes Infect.* **9**:1188–1195.

Joseph, B., S. Mertins, R. Stoll, J. Schar, K. R. Umesha, Q. Luo, S. Muller-Altrock, and W. Goebel. 2008. Glycerol metabolism and PrfA activity in *Listeria monocytogenes. J. Bacteriol.* **190**:5412–5430.

Joseph, B., K. Przybilla, C. Stuhler, K. Schauer, J. Slaghuis, T. M. Fuchs, and W. Goebel. 2006. Identification of *Listeria monocytogenes* genes contributing to intracellular replication by expression profiling and mutant screening. *J. Bacteriol.* **188**:556–568.

Kaiser, D. 2008. *Myxococcus*—from single-cell polarity to complex multicellular patterns. *Annu. Rev. Genet.* **42**:109–130.

Keeney, K., L. Colosi, W. Weber, and M. O'Riordan. 2009. Generation of branched-chain fatty acids through lipoate-dependent metabolism facilitates intracellular growth of *Listeria monocytogenes. J. Bacteriol.* **191**:2187–2196.

Keeney, K. M., J. A. Stuckey, and M. X. O'Riordan. 2007. LplA1-dependent utilization of host lipoyl peptides enables *Listeria* cytosolic growth and virulence. *Mol. Microbiol.* **66**:758–770.

Kim, H., K. J. Boor, and H. Marquis. 2004. *Listeria monocytogenes* σ^B contributes to invasion of human intestinal epithelial cells. *Infect. Immun.* **72**:7374–7378.

Kim, H., H. Marquis, and K. J. Boor. 2005. σ^B contributes to *Listeria monocytogenes* invasion by controlling expression of *inlA* and *inlB*. *Microbiology* **151**:3215–3222.

Kim, J., S. Adhya, and S. Garges. 1992. Allosteric changes in the cAMP receptor protein of *Escherichia coli*: hinge reorientation. *Proc. Natl. Acad. Sci. USA* **89**:9700–9704.

Kocks, C., E. Gouin, M. Tabouret, P. Berche, H. Ohayon, and P. Cossart. 1992. *L. monocytogenes*-induced actin assembly requires the *actA* gene product, a surface protein. *Cell* **68**:521–531.

Kocks, C., R. Hellio, P. Gounon, H. Ohayon, and P. Cossart. 1993. Polarized distribution of *Listeria monocytogenes* surface protein ActA at the site of directional actin assembly. *J. Cell Sci.* **05**(Pt. 3):699–710.

Korner, H., H. J. Sofia, and W. G. Zumft. 2003. Phylogeny of the bacterial superfamily of Crp-Fnr transcription regulators: exploiting the metabolic spectrum by controlling alternative gene programs. *FEMS Microbiol. Rev.* **27**:559–592.

Kreft, J., and J. A. Vazquez-Boland. 2001. Regulation of virulence genes in *Listeria. Int. J. Med. Microbiol.* **291**:145–157.

Kreft, J., J. A. Vazquez-Boland, S. Altrock, G. Dominguez-Bernal, and W. Goebel. 2002. Pathogenicity islands and other virulence elements in *Listeria. Curr. Top. Microbiol. Immunol.* **264**:109–125.

Lambrechts, A., K. Gevaert, P. Cossart, J. Vandekerckhove, and M. Van Troys. 2008. *Listeria* comet tails: the actin-based motility machinery at work. *Trends Cell Biol.* **18:**220–227.

Larsen, M. H., J. J. Leisner, and H. Ingmer. 2010. The chitinolytic activity of *Listeria monocytogenes* EGD is regulated by carbohydrates but also by the virulence regulator PrfA. *Appl. Environ. Microbiol.* **76:**6470–6476.

Lecuit, M. 2007. Human listeriosis and animal models. *Microbes Infect.* **9:**1216–1225.

Lecuit, M., S. Vandormael-Pournin, J. Lefort, M. Huerre, P. Gounon, C. Dupuy, C. Babinet, and P. Cossart. 2001. A transgenic model for listeriosis: role of internalin in crossing the intestinal barrier. *Science* **292:**1722–1725.

Lingnau, A., E. Domann, M. Hudel, M. Bock, T. Nichterlein, J. Wehland, and T. Chakraborty. 1995. Expression of the *Listeria monocytogenes* EGD *inlA* and *inlB* genes, whose products mediate bacterial entry into tissue culture cell lines, by PrfA-dependent and -independent mechanisms. *Infect. Immun.* **63:**3896–3903.

Loh, E., O. Dussurget, J. Gripenland, K. Vaitkevicius, T. Tiensuu, P. Mandin, F. Repoila, C. Buchrieser, P. Cossart, and J. Johansson. 2009. A *trans*-acting riboswitch controls expression of the virulence regulator PrfA in *Listeria monocytogenes*. *Cell* **139:**770–779.

Luo, Q., M. Rauch, A. K. Marr, S. Muller-Altrock, and W. Goebel. 2004. In vitro transcription of the *Listeria monocytogenes* virulence genes *inlC* and *mpl* reveals overlapping PrfA-dependent and -independent promoters that are differentially activated by GTP. *Mol. Microbiol.* **52:**39–52.

Lynch, M., J. Painter, R. Woodruff, and C. Braden. 2006. Surveillance for foodborne-disease outbreaks—United States, 1998–2002. *MMWR Surveill. Summ.* **55:**1–42.

MacGowan, A. P., P. H. Cartlidge, F. MacLeod, and J. McLaughlin. 1991. Maternal listeriosis in pregnancy without fetal or neonatal infection. *J. Infect.* **22:**53–57.

Mansfield, B. E., M. S. Dionne, D. S. Schneider, and N. E. Freitag. 2003. Exploration of host-pathogen interactions using *Listeria monocytogenes* and *Drosophila melanogaster*. *Cell. Microbiol.* **5:**901–911.

Marquis, H., H. G. Bouwer, D. J. Hinrichs, and D. A. Portnoy. 1993. Intracytoplasmic growth and virulence of *Listeria monocytogenes* auxotrophic mutants. *Infect. Immun.* **61:**3756–3760.

Marquis, H., V. Doshi, and D. A. Portnoy. 1995. The broad-range phospholipase C and a metalloprotease mediate listeriolysin O-independent escape of *Listeria monocytogenes* from a primary vacuole in human epithelial cells. *Infect. Immun.* **63:**4531–4534.

Marquis, H., and E. J. Hager. 2000. pH-regulated activation and release of a bacteria-associated phospholipase C during intracellular infection by *Listeria monocytogenes*. *Mol. Microbiol.* **35:**289–298.

Marr, A. K., B. Joseph, S. Mertins, R. Ecke, S. Muller-Altrock, and W. Goebel. 2006. Overexpression of PrfA leads to growth inhibition of *Listeria monocytogenes* in glucose-containing culture media by interfering with glucose uptake. *J. Bacteriol.* **188:**3887–3901.

McClure, E. M., and R. L. Goldenberg. 2009. Infection and stillbirth. *Semin. Fetal Neonatal Med.* **14:**182–189.

McGann, P., S. Raengpradub, R. Ivanek, M. Wiedmann, and K. J. Boor. 2008. Differential regulation of *Listeria monocytogenes* internalin and internalin-like genes by σ^B and PrfA as revealed by subgenomic microarray analyses. *Foodborne Pathog. Dis.* **5:**417–435.

McGann, P., M. Wiedmann, and K. J. Boor. 2007. The alternative sigma factor sigma B and the virulence gene regulator PrfA both regulate transcription of *Listeria monocytogenes* internalins. *Appl. Environ. Microbiol.* **73:**2919–2930.

Mead, P. S., E. F. Dunne, L. Graves, M. Wiedmann, M. Patrick, S. Hunter, E. Salehi, F. Mostashari, A. Craig, P. Mshar, T. Bannerman, B. D. Sauders, P. Hayes, W. Dewitt, P. Sparling, P. Griffin, D. Morse, L. Slutsker, and B. Swaminathan. 2006. Nationwide outbreak of listeriosis due to contaminated meat. *Epidemiol. Infect.* **134:**744–751.

Meisch, F., and M. N. Prioleau. 2011. Genomic approaches to the initiation of DNA replication and chromatin structure reveal a complex relationship. *Brief Funct. Genomics* **10:**30–36.

Mengaud, J., S. Dramsi, E. Gouin, J. A. Vazquez-Boland, G. Milon, and P. Cossart. 1991. Pleiotropic control of *Listeria monocytogenes* virulence factors by a gene that is autoregulated. *Mol. Microbiol.* **5:**2273–2283.

Mengaud, J., M. F. Vicente, and P. Cossart. 1989. Transcriptional mapping and nucleotide sequence of the *Listeria monocytogenes hlyA* region reveal structural features that may be involved in regulation. *Infect. Immun.* **57:**3695–3701.

Mertins, S., B. Joseph, M. Goetz, R. Ecke, G. Seidel, M. Sprehe, W. Hillen, W. Goebel, and S. Muller-Altrock. 2007. Interference of components of the phosphoenolpyruvate phosphotransferase system with the central virulence gene regulator PrfA of *Listeria monocytogenes*. *J. Bacteriol.* **189:**473–490.

Milenbachs, A. A., D. P. Brown, M. Moors, and P. Youngman. 1997. Carbon-source regulation of virulence gene expression in *Listeria monocytogenes*. *Mol. Microbiol.* **23:**1075–1085.

Milenbachs Lukowiak, A., K. J. Mueller, N. E. Freitag, and P. Youngman. 2004. Deregulation of *Listeria monocytogenes* virulence gene expression by two distinct and semi-independent pathways. *Microbiology* **150:**321–333.

Milohanic, E., P. Glaser, J. Y. Coppee, L. Frangeul, Y. Vega, J. A. Vazquez-Boland, F. Kunst, P. Cossart, and C. Buchrieser. 2003. Transcriptome analysis of *Listeria monocytogenes* identifies three groups of genes differently regulated by PrfA. *Mol. Microbiol.* **47:**1613–1625.

Miner, M. D., G. C. Port, H. G. Bouwer, J. C. Chang, and N. E. Freitag. 2008a. A novel *prfA* mutation that promotes *Listeria monocytogenes* cytosol entry but reduces bacterial spread and cytotoxicity. *Microb. Pathog.* **45:**273–281.

Miner, M. D., G. C. Port, and N. E. Freitag. 2008b. Functional impact of mutational activation on the *Listeria monocytogenes* central virulence regulator PrfA. *Microbiology* **154:**3579–3589.

Monk, I. R., C. G. Gahan, and C. Hill. 2008. Tools for functional postgenomic analysis of *Listeria monocytogenes*. *Appl. Environ. Microbiol.* **74:**3921–3934.

Moran, C. P. 1993. RNA polymerase and transcription factors, p. 653–657. *In* A. L. Sonenshein, J. A. Hoch, and R. Losick (ed.), Bacillus subtilis *and Other Gram-Positive Bacteria: Biochemistry, Physiology, and Molecular Genetics*. American Society for Microbiology, Washington, DC.

Mueller, K. J., and N. E. Freitag. 2005. Pleiotropic enhancement of bacterial pathogenesis resulting from the constitutive activation of the *Listeria monocytogenes* regulatory factor PrfA. *Infect. Immun.* **73:**1917–1926.

Nair, S., E. Milohanic, and P. Berche. 2000. ClpC ATPase is required for cell adhesion and invasion of *Listeria monocytogenes*. *Infect. Immun.* **68:**7061–7068.

Ollinger, J., M. Wiedmann, and K. J. Boor. 2008. σ^B- and PrfA-dependent transcription of genes previously classified as putative constituents of the *Listeria monocytogenes* PrfA regulon. *Foodborne Pathog. Dis.* **5:**281–293.

O'Neil, H. S., and H. Marquis. 2006. *Listeria monocytogenes* flagella are used for motility, not as adhesins, to increase host cell invasion. *Infect. Immun.* **74:**6675–6681.

Park, S. F., and R. G. Kroll. 1993. Expression of listeriolysin and phosphatidylinositol-specific phospholipase C is repressed by the plant-derived molecule cellobiose in *Listeria monocytogenes*. *Mol. Microbiol.* **8**:653–661.

Pizarro-Cerda, J., and P. Cossart. 2006. Subversion of cellular functions by *Listeria monocytogenes*. *J. Pathol.* **208**:215–223.

Port, G. C., and N. E. Freitag. 2007. Identification of novel *Listeria monocytogenes* secreted virulence factors following mutational activation of the central virulence regulator, PrfA. *Infect. Immun.* **75**:5886–5897.

Portnoy, D. A., P. S. Jacks, and D. J. Hinrichs. 1988. Role of hemolysin for the intracellular growth of *Listeria monocytogenes*. *J. Exp. Med.* **167**:1459–1471.

Portnoy, D. A., and S. Jones. 1994. The cell biology of *Listeria monocytogenes* infection (escape from a vacuole). *Ann. N. Y. Acad. Sci.* **730**:15–25.

Poyart, C., E. Abachin, I. Razafimanantsoa, and P. Berche. 1993. The zinc metalloprotease of *Listeria monocytogenes* is required for maturation of phosphatidylcholine phospholipase C: direct evidence obtained by gene complementation. *Infect. Immun.* **61**:1576–1580.

Quillin, S. J., K. T. Schwartz, and J. H. Leber. 2011. The novel *Listeria monocytogenes* bile sensor BrtA controls expression of the cholic acid efflux pump MdrT. *Mol. Microbiol.* **81**:129–142.

Rae, C. S., A. Geissler, P. C. Adamson, and D. A. Portnoy. 2011. Mutations of the *Listeria monocytogenes* peptidoglycan N-deacetylase and O-acetylase result in enhanced lysozyme sensitivity, bacteriolysis, and hyperinduction of innate immune pathways. *Infect. Immun.* **79**:3596–3606.

Rajabian, T., B. Gavicherla, M. Heisig, S. Muller-Altrock, W. Goebel, S. D. Gray-Owen, and K. Ireton. 2009. The bacterial virulence factor InlC perturbs apical cell junctions and promotes cell-to-cell spread of *Listeria*. *Nat. Cell Biol.* **11**:1212–1218.

Ramaswamy, V., V. M. Cresence, J. S. Rejitha, M. U. Lekshmi, K. S. Dharsana, S. P. Prasad, and H. M. Vijila. 2007. *Listeria*—review of epidemiology and pathogenesis. *J. Microbiol. Immunol. Infect.* **40**:4–13.

Rauch, M., Q. Luo, S. Muller-Altrock, and W. Goebel. 2005. SigB-dependent in vitro transcription of *prfA* and some newly identified genes of *Listeria monocytogenes* whose expression is affected by PrfA in vivo. *J. Bacteriol.* **187**:800–804.

Raveneau, J., C. Geoffroy, J. L. Beretti, J. L. Gaillard, J. E. Alouf, and P. Berche. 1992. Reduced virulence of a *Listeria monocytogenes* phospholipase-deficient mutant obtained by transposon insertion into the zinc metalloprotease gene. *Infect. Immun.* **60**:916–921.

Renzoni, A., A. Klarsfeld, S. Dramsi, and P. Cossart. 1997. Evidence that PrfA, the pleiotropic activator of virulence genes in *Listeria monocytogenes*, can be present but inactive. *Infect. Immun.* **65**:1515–1518.

Ripio, M. T., K. Brehm, M. Lara, M. Suarez, and J. A. Vazquez-Boland. 1997a. Glucose-1-phosphate utilization by *Listeria monocytogenes* is PrfA dependent and coordinately expressed with virulence factors. *J. Bacteriol.* **179**:7174–7180.

Ripio, M. T., G. Dominguez-Bernal, M. Lara, M. Suarez, and J. A. Vazquez-Boland. 1997b. A Gly145Ser substitution in the transcriptional activator PrfA causes constitutive overexpression of virulence factors in *Listeria monocytogenes*. *J. Bacteriol.* **179**:1533–1540.

Sanderson, S., D. J. Campbell, and N. Shastri. 1995. Identification of a CD4+ T cell-stimulating antigen of pathogenic bacteria by expression cloning. *J. Exp. Med.* **182**:1751–1757.

Schlech, W. F., III, P. M. Lavigne, R. A. Bortolussi, A. C. Allen, E. V. Haldane, A. J. Wort, A. W. Hightower, S. E. Johnson, S. H. King, E. S. Nicholls, and C. V. Broome. 1983. Epidemic listeriosis—evidence for transmission by food. *N. Engl. J. Med.* **308**:203–206.

Schnupf, P., and D. A. Portnoy. 2007. Listeriolysin O: a phagosome-specific lysin. *Microbes Infect.* **9**:1176–1187.

Schwartz, B., C. A. Ciesielski, C. V. Broome, S. Gaventa, G. R. Brown, B. G. Gellin, A. W. Hightower, and L. Mascola. 1988. Association of sporadic listeriosis with consumption of uncooked hot dogs and undercooked chicken. *Lancet* **2**:779–782.

Scortti, M., H. J. Monzo, L. Lacharme-Lora, D. A. Lewis, and J. A. Vazquez-Boland. 2007. The PrfA virulence regulon. *Microbes Infect.* **9**:1196–1207.

Seveau, S., J. Pizarro-Cerda, and P. Cossart. 2007. Molecular mechanisms exploited by *Listeria monocytogenes* during host cell invasion. *Microbes Infect.* **9**:1167–1175.

Shen, A., and D. E. Higgins. 2005. The 5′ untranslated region-mediated enhancement of intracellular listeriolysin O production is required for *Listeria monocytogenes* pathogenicity. *Mol. Microbiol.* **57**:1460–1473.

Shetron-Rama, L. M., K. Mueller, J. M. Bravo, H. G. Bouwer, S. S. Way, and N. E. Freitag. 2003. Isolation of *Listeria monocytogenes* mutants with high-level in vitro expression of host cytosol-induced gene products. *Mol. Microbiol.* **48**:1537–1551.

Sleator, R. D., H. H. Wemekamp-Kamphuis, C. G. Gahan, T. Abee, and C. Hill. 2005. A PrfA-regulated bile exclusion system (BilE) is a novel virulence factor in *Listeria monocytogenes*. *Mol. Microbiol.* **55**:1183–1195.

Smith, G. A., H. Marquis, S. Jones, N. C. Johnston, D. A. Portnoy, and H. Goldfine. 1995. The two distinct phospholipases C of *Listeria monocytogenes* have overlapping roles in escape from a vacuole and cell-to-cell spread. *Infect. Immun.* **63**:4231–4237.

Stack, H. M., R. D. Sleator, M. Bowers, C. Hill, and C. G. Gahan. 2005. Role for HtrA in stress induction and virulence potential in *Listeria monocytogenes*. *Appl. Environ. Microbiol.* **71**:4241–4247.

Stavru, F., C. Archambaud, and P. Cossart. 2011. Cell biology and immunology of *Listeria monocytogenes* infections: novel insights. *Immunol. Rev.* **240**:160–184.

Stoll, R., S. Mertins, B. Joseph, S. Muller-Altrock, and W. Goebel. 2008. Modulation of PrfA activity in *Listeria monocytogenes* upon growth in different culture media. *Microbiology* **154**:3856–3876.

Stone, S. C., and J. Shoenberger. 2001. Update: multistate outbreak of listeriosis—United States, 2000. *Ann. Emerg. Med.* **38**:339–341.

Swaminathan, B., and P. Gerner-Smidt. 2007. The epidemiology of human listeriosis. *Microbes Infect.* **9**:1236–1243.

Swaminathan, B., P. Gerner-Smidt, and J. M. Whichard. 2006. Foodborne disease trends and reports. *Foodborne Pathog. Dis.* **3**:316–318.

Tilney, L. G., P. S. Connelly, and D. A. Portnoy. 1990. Actin filament nucleation by the bacterial pathogen, *Listeria monocytogenes*. *J. Cell Biol.* **111**:2979–2988.

Tilney, L. G., and D. A. Portnoy. 1989. Actin filaments and the growth, movement, and spread of the intracellular bacterial parasite, *Listeria monocytogenes*. *J. Cell Biol.* **109**:1597–1608.

Toledo-Arana, A., O. Dussurget, G. Nikitas, N. Sesto, H. Guet-Revillet, D. Balestrino, E. Loh, J. Gripenland, T. Tiensuu, K. Vaitkevicius, M. Barthelemy, M. Vergassola, M. A. Nahori, G. Soubigou, B. Regnault, J. Y. Coppee, M. Lecuit, J. Johansson, and P. Cossart. 2009. The *Listeria* transcriptional landscape from saprophytism to virulence. *Nature* **459**:950–956.

Tsvetanova, B., A. C. Wilson, C. Bongiorni, C. Chiang, J. A. Hoch, and M. Perego. 2007. Opposing effects of histidine phosphorylation regulate the AtxA virulence transcription factor in *Bacillus anthracis*. *Mol. Microbiol.* **63**:644–655.

Vazquez-Boland, J. A., C. Kocks, S. Dramsi, H. Ohayon, C. Geoffroy, J. Mengaud, and P. Cossart. 1992. Nucleotide sequence of the lecithinase operon of *Listeria monocytogenes* and possible role of lecithinase in cell-to-cell spread. *Infect. Immun.* **60**:219–230.

Vazquez-Boland, J. A., M. Kuhn, P. Berche, T. Chakraborty, G. Dominguez-Bernal, W. Goebel, B. Gonzalez-Zorn, J. Wehland, and J. Kreft. 2001. *Listeria pathogenesis* and molecular virulence determinants. *Clin. Microbiol. Rev.* 14:584–640.

Vega, Y., C. Dickneite, M. T. Ripio, R. Bockmann, B. Gonzalez-Zorn, S. Novella, G. Dominguez-Bernal, W. Goebel, and J. A. Vazquez-Boland. 1998. Functional similarities between the *Listeria monocytogenes* virulence regulator PrfA and cyclic AMP receptor protein: the PrfA* (Gly145Ser) mutation increases binding affinity for target DNA. *J. Bacteriol.* 180:6655–6660.

Vega, Y., M. Rauch, M. J. Banfield, S. Ermolaeva, M. Scortti, W. Goebel, and J. A. Vazquez-Boland. 2004. New *Listeria monocytogenes prfA** mutants, transcriptional properties of PrfA* proteins and structure-function of the virulence regulator PrfA. *Mol. Microbiol.* 52:1553–1565.

Wadsworth, S. J., and H. Goldfine. 1999. *Listeria monocytogenes* phospholipase C-dependent calcium signaling modulates bacterial entry into J774 macrophage-like cells. *Infect. Immun.* 67:1770–1778.

Waters, L. S., and G. Storz. 2009. Regulatory RNAs in bacteria. *Cell* 136:615–628.

Way, S. S., L. J. Thompson, J. E. Lopes, A. M. Hajjar, T. R. Kollmann, N. E. Freitag, and C. B. Wilson. 2004. Characterization of flagellin expression and its role in *Listeria monocytogenes* infection and immunity. *Cell. Microbiol.* 6:235–242.

Williams, J. R., C. Thayyullathil, and N. E. Freitag. 2000. Sequence variations within PrfA DNA binding sites and effects on *Listeria monocytogenes* virulence gene expression. *J. Bacteriol.* 182:837–841.

Wilson, R. L., L. L. Brown, D. Kirkwood-Watts, T. K. Warren, S. A. Lund, D. S. King, K. F. Jones, and D. E. Hruby. 2006. *Listeria monocytogenes* 10403S HtrA is necessary for resistance to cellular stress and virulence. *Infect. Immun.* 74:765–768.

Winkler, W. C., and R. R. Breaker. 2005. Regulation of bacterial gene expression by riboswitches. *Annu. Rev. Microbiol.* 59:487–517.

Wong, K. K., H. G. Bouwer, and N. E. Freitag. 2004. Evidence implicating the 5′ untranslated region of *Listeria monocytogenes actA* in the regulation of bacterial actin-based motility. *Cell. Microbiol.* 6:155–166.

Wong, K. K., and N. E. Freitag. 2004. A novel mutation within the central *Listeria monocytogenes* regulator PrfA that results in constitutive expression of virulence gene products. *J. Bacteriol.* 186:6265–6276.

Xayarath, B., H. Marquis, G. C. Port, and N. E. Freitag. 2009. *Listeria monocytogenes* CtaP is a multifunctional cysteine transport-associated protein required for bacterial pathogenesis. *Mol. Microbiol.* 74:956–973.

Xayarath, B., J. I. Smart, K. J. Mueller, and N. E. Freitag. 2011a. A novel C-terminal mutation resulting in constitutive activation of the *Listeria monocytogenes* central virulence regulatory factor PrfA. *Microbiology* 157(Pt. 11):3138–3149.

Xayarath, B., K. W. Volz, J. I. Smart, and N. E. Freitag. 2011b. Probing the role of protein surface charge in the activation of PrfA, the central regulator of *Listeria monocytogenes* pathogenesis. *PLoS One* 6:e23502.

Yeung, P. S., Y. Na, A. J. Kreuder, and H. Marquis. 2007. Compartmentalization of the broad-range phospholipase C activity to the spreading vacuole is critical for *Listeria monocytogenes* virulence. *Infect. Immun.* 75:44–51.

Yeung, P. S., N. Zagorski, and H. Marquis. 2005. The metalloprotease of *Listeria monocytogenes* controls cell wall translocation of the broad-range phospholipase C. *J. Bacteriol.* 187:2601–2608.

Youn, H., R. L. Kerby, M. Conrad, and G. P. Roberts. 2006. Study of highly constitutively active mutants suggests how cAMP activates cAMP receptor protein. *J. Biol. Chem.* 281:1119–1127.

Regulation of Bacterial Virulence
Edited by Michael L. Vasil and Andrew J. Darwin
© 2013 ASM Press, Washington, DC   doi:10.1128/9781555818524.ch20

Chapter 20

The SsrAB Virulon of *Salmonella enterica*

SANDRA BILLIG, ALFONSO FELIPE-LÓPEZ, AND MICHAEL HENSEL

INTRODUCTION

The rapid adaptation to changing environments is an essential feature of free-living bacteria. Changing environments are specifically encountered by bacterial pathogens, and their presence in host tissues can impose a complex array of stress factors on the bacterial cell. In particular, the transition from an extracellular to an intracellular lifestyle requires a massive reprogramming of gene expression in order to cope with host defense mechanisms and altered nutritional conditions. This chapter focuses on the function of the regulatory system SsrAB of *Salmonella enterica*, which controls expression of virulence factors for the intracellular phase of the life of the pathogen.

PATHOGENESIS OF INFECTIONS BY *SALMONELLA ENTERICA*

Salmonella enterica is a common gastrointestinal pathogen but also serves as a well-established model system to study molecular mechanisms of bacterial virulence and regulation of virulence genes. *Salmonella* infections are one of the most common and widely distributed food-borne diseases and can be fatal during childhood and in elderly and immunocompromised patients. *Salmonella enterica* serovars Enteritidis and Typhimurium are the serotypes most frequently isolated from humans, causing gastroenteritis, which usually is a self-limiting disease but progresses to systemic disease in a small number of cases (reviewed in Gordon, 2008, and Raffatellu et al., 2008). Infections with *S. enterica* serovar Typhi can lead to a life-threatening condition known as typhoid fever. *S.* Typhi is still a major health problem in countries with low standards of hygiene and medical infrastructure. Typhoid fever is characterized by the systemic spread of the pathogen after oral uptake and proliferation in various host organs. *S.* Typhi is a highly human-adapted serovar, but many aspects of the infection are resembled in infection models with *S.* Typhimurium and *Salmonella*-susceptible inbred strains of mice.

Salmonella is an invasive, facultative intracellular pathogen. After invasion of host cells or phagocytic uptake, the bacteria can adopt an intracellular lifestyle within a membrane-bound compartment termed *Salmonella*-containing vacuole (SCV). In the course of the infection, the orally ingested salmonellae are exposed to changing environments in the various anatomical sites inside the host. These physiological stimuli include exposure to low pH in the stomach and the presence of digestive enzymes, antimicrobial peptides, and competition with an established microbial community in the intestine. After invasion of epithelial cells in the intestine, or phagocytosis by macrophages in the lamina propria of Peyer's patches, the entry into host cell provides signals for the intracellular lifestyle of *Salmonella*. Several sensor systems enable *Salmonella* to respond to the changing environment inside the host by activating specific sets of metabolic and virulence genes. Some of these virulence genes are organized in the *Salmonella* pathogenicity islands (SPIs), large blocks of genes horizontally acquired during the evolution of the bacteria. Depending on the type of *Salmonella* serovar and isolate, up to 20 distinct SPIs can be present in the genome (Gerlach and Hensel, 2007). So far, SPI1 and SPI2 have been extensively characterized; SPI1 controls invasion and inflammatory responses, and SPI2 controls intracellular replication (reviewed in Haraga et al., 2008). Both SPI1 and SPI2 encode type III secretion systems (T3SS) that are capable of directly translocating bacterial effector proteins into host cells, thereby altering physiological functions of mammalian cells. SPI1 effectors remodel the host cell cytoskeleton, thereby inducing the uptake of bacteria by nonphagocytic cells (Schlumberger and Hardt, 2006). Once the bacteria are intracellular, the activity

---

Sandra Billig, Alfonso Felipe-López, and Michael Hensel • Abteilung Mikrobiologie, Universität Osnabrück, Osnabrück, Germany.

of the SPI2-encoded T3SS and its effector proteins inhibits the maturation of the phagolysosome and enable intracellular bacterial survival and replication (Ibarra and Steele-Mortimer, 2009).

Here, we discuss the experimental data and current understanding of the function of the two-component system SsrAB. SsrAB is encoded by genes in SPI2 and is necessary to control expression of virulence genes during the intracellular stage of *Salmonella* infection.

SALMONELLA PATHOGENICITY ISLAND 2

Intracellular survival and replication is dependent on the function of the SPI2-encoded T3SS (SPI2-T3SS). SPI2 consists of a 40-kb region of the *Salmonella* chromosome and contains 32 genes with functions in intracellular pathogenesis. Another portion of SPI2 is not involved in intracellular replication but encodes

a metabolic pathway for anaerobic respiration using tetrathionate, a function recently reported to provide an advantage over the competing intestinal flora (Winter et al., 2010). The SPI2-T3SS translocates a cocktail of at least 20 effector proteins that jointly modify host cell vesicular transport and phagosomal maturation and subvert the presentation of antigens by the major histocompatibility complex II. Altogether, the control of all these individual processes enables the intracellular replication of *Salmonella* (reviewed in Ibarra and Steele-Mortimer, 2009, and Jantsch et al., 2011).

While all components of the type III secretion apparatus are encoded by genes within SPI2, the majority of the effector proteins are encoded by genes located on separate loci dispersed over the chromosome. Some of the effector genes are associated with bacteriophage genomes, and it has been considered that the effector genes are a part of a mobile pool of horizontally transferred virulence functions. The

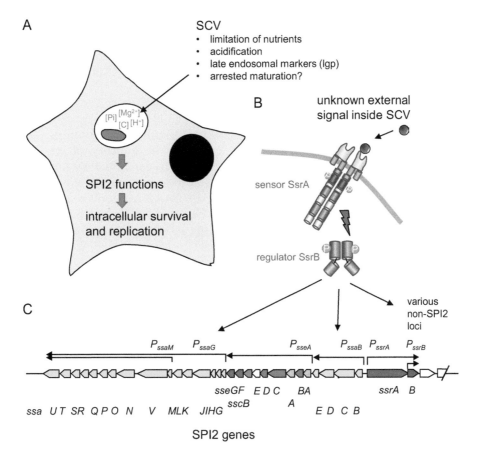

Figure 1. The SsrAB virulon in *Salmonella enterica*. (A) Intracellular salmonellae reside within the SCV, a membrane-bound compartment considered growth restricting due to nutritional limitations. By means of the SPI2-T3SS, intracellular salmonellae modify host cell functions, resulting in intracellular proliferation. (B) Unknown signals present in the SCV act on the SsrA sensor and, in turn, result in activation of the SsrAB virulon. (C) SsrB activates the transcription of several operons within SPI2 encoding the SPI2-T3SS, cognate effector proteins, and the SsrAB regulatory system, various loci outside of SPI2 encoding effector proteins of the SPI2-T3SS, and further loci with unknown contribution to the intracellular lifestyle. doi:10.1128/9781555818524.ch20f1

precise molecular function is known only for a subset of the individual SPI2-T3SS effector proteins, and the identification of host cell targets is still ongoing. The complex arrangement of a T3SS encoded by a horizontally acquired pathogenicity island and a large number of additional factors encoded by individual transcriptional units defines the requirement for a regulatory system that allows the precise temporal and spatial regulation of all components. This regulation is essential for successful adaptation to the intracellular lifestyle and for survival and replication within the host.

Of central importance for the coordinate regulation of SPI2-T3SS expression is the two-component system SsrAB (secretion system regulator) that is also encoded by SPI2 (Fig. 1). While SsrAB is of central importance for expression of all components of the T3SS and effectors, its activity is controlled by other regulatory systems encoded by the core genome of *Salmonella*. SsrA is the membrane-bound sensor of the system and SsrB the cytosolic response regulator. The structural and functional features of the SsrAB system are described in a later section.

ROLE OF SsrAB IN *SALMONELLA* VIRULENCE

The signature-tagged mutagenesis screen initially identified various transposon-insertion mutants with highly reduced virulence in the murine model of systemic infection (Hensel et al., 1995). The characterization of the mutant strains indicated insertions in various SPI2 genes, including *ssrA* and *ssrB*. Further analyses demonstrated that defects of *ssrA* or *ssrB* cause a high degree of attenuation similar to that observed for mutant strains with defects in structural components of the T3SS. Within infected host organisms, *ssrA/ssrB* mutant strains are unable to proliferate within host tissue and are eventually cleared by immune functions of the host (Shea et al., 1996). On the cellular level, the mutant strains fail to synthesize the SPI2-T3SS and to translocate effector proteins of the SPI2-T3SS into host cells (Cirillo et al., 1998). As observed for SPI2-T3SS-deficient strains, *ssrA/ssrB* mutant strains survive in epithelial cells and nonactivated phagocytes but are unable to proliferate within the host cells.

THE SsrAB VIRULON

SsrAB-Mediated Regulation

The SsrAB regulatory system controls a complex regulon with major functions in bacterial virulence.

Thus, we refer to the group of SsrAB-regulated genes as the virulon. The SsrAB virulon comprises the genes in SPI2 encoding the T3SS and a subset of effectors, the *ssrAB* genes, and various loci outside of SPI2 encoding the majority of the effector proteins (Fig. 1). In addition, a large number of SsrB-regulated genes without an obvious link to SPI2 function have been identified (Worley et al., 2000; Tomljenovic-Berube et al., 2010).

The function of the SPI2-encoded T3SS is specifically required for the intracellular lifestyle of *Salmonella*, and in line with this notion, the transcriptional activation of SsrAB-regulated genes occurs inside infected host cells and host tissue (Cirillo et al., 1998; Deiwick et al., 1999; Rollenhagen et al., 2004). Under experimental conditions, the expression can also be induced by growth in synthetic media that may mimic the intracellular environments of *Salmonella*. For example, minimal media with low concentrations of phosphate or magnesium or an acidic pH of 5.8 or below have been utilized to induce expression of the SsrAB virulon in vitro (Beuzon et al., 1999; Deiwick and Hensel, 1999; Coombes et al., 2004; Löber et al., 2006). Under in vitro conditions, expression of the SsrAB virulon is highest when *Salmonella* is reaching the late exponential growth phase.

Several studies reported the expression of the SsrAB virulon under extracellular conditions, especially in the intestinal environment (Brown et al., 2005; Osborne and Coombes, 2011). The stimuli leading to the induction of the SsrAB virulon under extracellular conditions are not known. However, we and others observed a low level of proteins encoded by the SsrAB virulon in bacterial cultures grown overnight in LB broth, indicating that acidification of the medium or nutritional limitation may occur under these conditions (Deiwick and Hensel, 1999; unpublished observations). Since the function of the SPI2-T3SS is restricted to the intracellular phase of pathogenesis, the expression under extracellular conditions might be a preparatory step for the subsequent intracellular phase (Brown et al., 2005).

SsrAB Target Genes within the SPI2 Locus

Genes controlled by SsrAB within SPI2 encode subunits of the T3SS (SsaBCDE and SsaGHIJ KLMVNOPQRSTU), components of the translocon (SseBCD), translocated effector proteins (SseFG), and dedicated chaperones of translocon and effector proteins (SscAB). Also, *ssaB/spiC* is regulated by SsrAB, but its role as a component of the T3SS or as a translocated effector is controversial (Uchiya et al., 1999; Freeman et al., 2002; Yu et al., 2002). The other *sse* genes have not been functionally analyzed. At least one further portion of the SPI2 locus has been acquired by an independent event of horizontal gene transfer. This portion encodes the tetrathionate reductase TtrABC required for anaerobic respiration, as well as a further two-component system, termed TtrRS (Hensel et al., 1999). The transcriptional control of *ttr* genes is independent of SsrAB but regulated by TtrSR. Other open reading frames in the SPI2 locus encode proteins without roles in intracellular pathogenesis or anaerobic respiration and await functional characterization.

The precise operon structure of the SPI2 genes encoding T3SS functions has not been demonstrated by molecular approaches. However, the combined information from bioinformatics prediction, reporter gene fusions, and chromatin immunoprecipitation (ChIP) allows rather detailed predictions about the promoters and transcripts. SsrB-controlled promoters in SPI2 are P_{ssaB}, P_{GssaG}, and P_{ssaM}, controlling expression of genes for T3SS subunits. The small (*ssaABCDE*) and large (*ssaGHIJKLMVNOPQRSTU*) apparatus operons encode proteins predicted or known to be components of the SPI2-T3SS apparatus. The full extent of the large apparatus operon has not been determined, but it is possible that a very large transcript consisting of *ssaG* to *ssaU* is synthesized. P_{sseA} controls genes encoding translocon and effector proteins with their cognate chaperones, and P_{ssrA} controls expression of *ssrAB*. "ChIP-on-chip" analyses also indicate the additional promoter P_{ssaR} (Tomljenovic-Berube et al., 2010), and previous characterization of SsrB binding sites revealed an additional promoter in the *ssrAB* operon, i.e., P_{ssrB} (Feng et al., 2003).

SsrAB Target Genes outside the SPI2 Locus

Various loci outside of SPI2 encode effector proteins of the SPI2-T3SS, including *gogB*, *pipB*, *pipB2*, *sifA*, *sifB*, *slrP*, *sopD2*, *sseI*, *sseJ*, *sseK1*, *sseK2*, *sseL*, *sspH1*, *sspH2*, and *steC* (Xu and Hensel, 2010). Only a few of these genes are not controlled by SsrB, namely, *sspH1*, *slrP*, and *gogB*. These loci either encode effectors that can be translocated by the SPI2-T3SS as well as by the SPI1-T3SS, with the corresponding genes being constitutively expressed (*sspH1* and *slrP*) or exhibiting autonomous control of expression, such as *gogB* (Coombes et al., 2005).

The function of most SsrAB-regulated effector proteins of the SPI2-T3SS is unknown, and here the focus is only on those that are well characterized. Detailed functional analyses of various other SsrAB-regulated effector proteins are beyond the scope of this chapter, but various recent reviews focus on this topic (Kuhle and Hensel, 2004; Haraga et al., 2008; Ibarra and Steele-Mortimer, 2009). SifA is probably the most relevant effector for intracellular proliferation and systemic pathogenesis in the murine model.

SifA function is required for induction of *Salmonella*-induced filament (SIF) (Stein et al., 1996) and to maintain the SCV membrane (Beuzon et al., 2000). SifA-deficient strains are highly attenuated (Beuzon et al., 2000). On the molecular level, SifA interacts with SifA kinesin interaction proteins (SKIP) in order to prevent recruitment of kinesin to the SCV (Boucrot et al., 2005). SseF and SseG contribute to SIF formation and are required to localize the SCV to the perinuclear, Golgi-associated position in host cells (Kuhle and Hensel, 2002; Salcedo and Holden, 2003). The lack of SseF or SseG results in mild attenuation of virulence (Hensel et al., 1998). SseJ has acyltransferase activity in vitro, and modification of the SCV membrane by SseJ has been proposed, but the lack of SseJ has resulted only in small virulence defects in the model systems investigated so far (Ohlson et al., 2005). PipB2 was shown to be a linker for host cell kinesin, and this function appears to be important to balance the function of SifA (Henry et al., 2006).

Other Loci Regulated by SsrB

In order to identify SsrAB-regulated genes, Worley et al. utilized arabinose-inducible overexpression of *ssrB* (Worley et al., 2000). This approach identified 10 loci outside of SPI2 termed *srf* (SsrB-regulated factor) which were mostly characterized by association with phage genes. Subsequent characterization of the loci identified *srfH* as encoding SPI2-T3SS effector protein SseI (Miao and Miller, 2000). The function of most of the other *srf* loci remains to be demonstrated.

A recent study used microarray analysis in combination with ChIP-on-chip analysis to identify SsrB-regulated genes. This approach identified 133 genes that are significantly upregulated by SsrB. In addition to the known members of the SsrAB regulon, further loci were identified that appear unlinked to the function of the SPI2-T3SS. These genes have predicted functions in transport, secretion, protein and membrane modification, and trafficking of cellular components (Tomljenovic-Berube et al., 2010) and await further characterization.

In general, those genes with strong levels of expression are predominantly localized to mobile genetic elements or recently acquired genomic islands; 56 of the 133 SsrB-regulated genes were present within genomic islands. Some weakly coexpressed genes represent ancestral *Salmonella* genes recruited into the SsrB regulon (Tomljenovic-Berube et al., 2010).

Expression Levels of SsrAB-Regulated Genes

A comparative analysis of expression levels under in vitro conditions indicated high diversity in expression levels of the various genes encoding structural components and effector proteins of the SPI2-T3SS. Expression levels were reported as follows: very high for *sseJ* and *sifA*, *ssaG*, *steC*, *sseL*, and *sopD2*; moderate for *ssaB*, *sseA*, *sseG*, *sifB*, *pipB2*, and *sspH1*; and low for *sspH2*, *sseI*, *slrP*, *sseK1*, *sseK2*, *pipB*, and *gogB* (Xu and Hensel, 2010). These data were obtained by analyses using in vitro conditions inducing the SsrAB regulon, but they were also in agreement with data obtained by analyses of intracellular *Salmonella* reporter strains in macrophages. Interestingly, the expression levels only partially corresponded to the role of the effector proteins as determined by experimental studies performed in the murine model or in cell culture models. There is no direct correlation between the amounts of effector proteins and their roles in the intracellular lifestyle of *Salmonella*. Effector proteins like SseJ or SseF are present in the host cell at high levels, while the translocated effectors SseK1, SseK2, and GogB could not be detected (Xu and Hensel, 2010). The time points of maximal expression levels are rather similar for SPI2 effector genes located within and outside of SPI2. The sequence of effector actions might depend not on a sequential expression of the specific effector genes but, instead, on differences in the half-lives of the translocated proteins or subtle differences in the kinetics of translocation (Xu and Hensel, 2010). As the quantity of secreted protein is not associated with its function, the affinity for its interaction partners should indicate how these proteins are able to control the intracellular traffic of the host cell.

STRUCTURE AND FUNCTION OF SsrA AND SsrB

Sensor Kinase SsrA

The *ssrA* gene encodes a large histidine sensor kinase composed of 920 amino acids (aa). SsrA is predicted to have two transmembrane domains in the inner membrane, which define a large periplasmic domain comprising presumably 250 aa. By analogy to other two-component system sensors, the periplasmic domain is considered the signal-sensing part of the molecule, although the molecular or physical nature of this stimulus is unknown (Fass and Groisman, 2009). In addition to the canonical ATP-binding domain and the receiver or phosphorylation domain, SsrA contains an unusually large number of protein modules such as response regulator and a histidyl phosphotransfer domain (Carroll et al., 2009). The predicted organization of modules in

SsrA and proposed membrane topology are shown in Fig. 2. Directly following the second transmembrane domain a HAMP domain (aa 317 to 369) was predicted. HAMP domains have been identified in histidine kinases, adenylate cyclases, methyl-accepting chemotaxis proteins, and phosphatases. The HAMP domain is followed by the HisKA (histidine kinase A, aa 395 to 460) and HATPase (histidine kinase-like ATPase, aa 507 to 614) domains. The canonical HisKA and HATPase domains are proposed to act as dimerization and phosphate-accepting and as ATP-binding and catalytic domains, respectively. As a further domain, a Rec domain spanning aa 689 to 804 is predicted. Rec domains are also found in other sensor proteins, as in SsrA, or in a separate protein. With lower probability, an HPt (histidine-containing phosphotransfer) domain was predicted for the C terminus of SsrA. The HPt domain may act as a further site for phosphorylation. The presence of the additional phosphor-accepting domains may classify SsrA as a hybrid kinase and indicates the presence of a complex phosphorelay within the protein. It has to be pointed out that these predictions are all sequence based and not confirmed by detailed experimental studies. Attempts to delete distinct domains in SsrA for subsequent functional characterization of the mutant strains have so far resulted in complete loss of SsrA function (L. Schwarzer and M. Hensel, unpublished observations).

SsrA is an unorthodox two-component sensor protein, since the C terminus contains a region similar to the receiver domain, which is usually found in the response regulators, and it comprises a histidine phosphotransfer domain, placing it in the BvgS family, with similarity to RcsC, LuxN, LuxQ, etc. (Garmendia et al., 2003). These domains suggest a phosphorelay, whereby the phosphoryl group is transferred via intramolecular reactions from the histidine to an aspartate to another histidine in SsrA and then onto the conserved aspartate of the SsrB response regulator (Carroll et al., 2009). The biological need for this putative four-step phosphorelay is still poorly understood. It might be that the additional phosphoreceiver domain plays an autoregulatory role for the sensor proteins as a second phosphorylation might exert a negative regulatory effect on the kinase activity. Such a mechanism was observed for the sensor BvgS of *Bordetella pertussis* (Beier et al., 1995).

Whether SsrA can directly sense different environmental signals is still an open question, as at least one of the signals could be sensed by another sensor-regulator which would then be linked through the SsrAB system. This is based on the observation of transcriptional activity of SsrB in the absence of SsrA (Carroll et al., 2009).

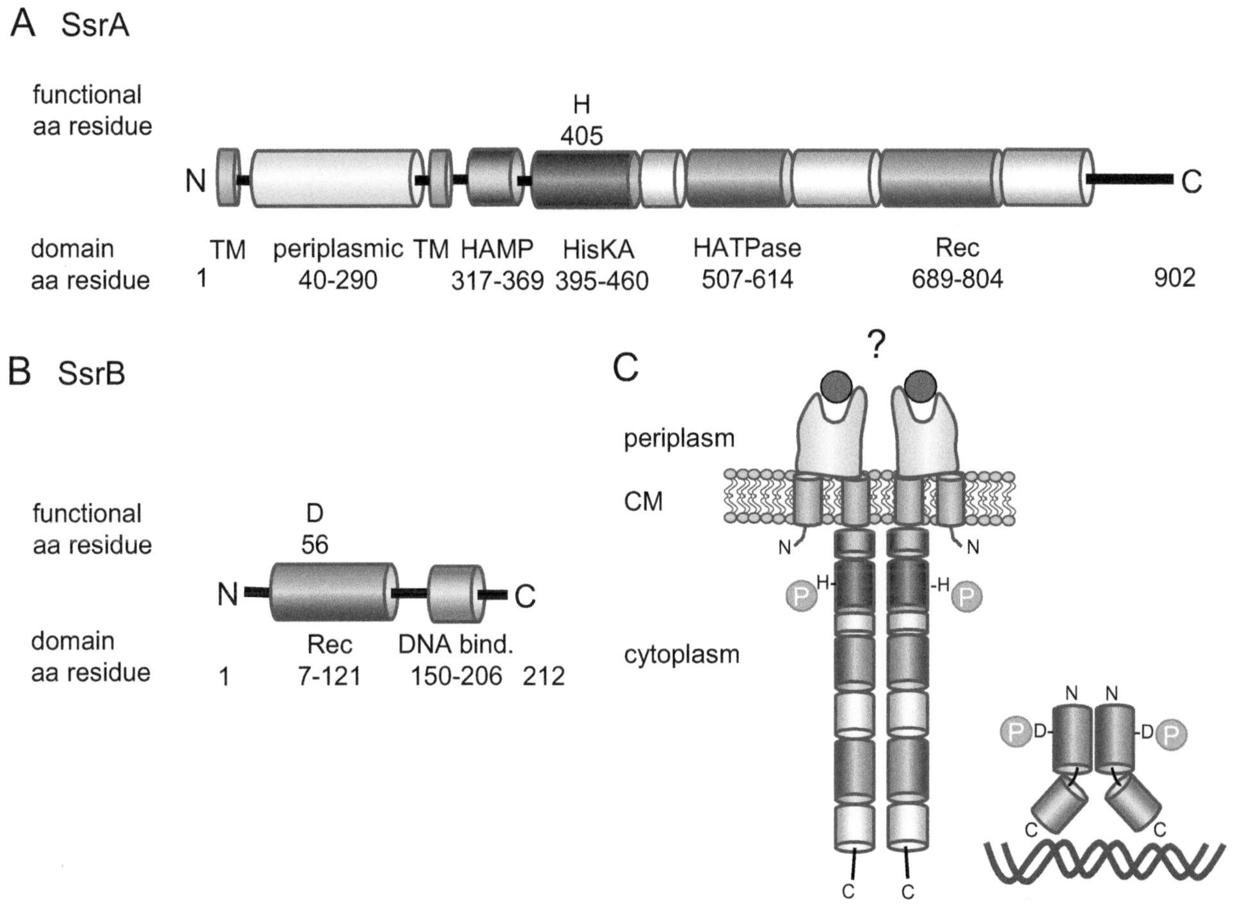

A SsrA

functional
aa residue

H
405

| N | | | | | | | | | C |

domain
aa residue

TM	periplasmic	TM	HAMP	HisKA	HATPase	Rec	
1	40-290	317-369	395-460	507-614		689-804	902

B SsrB

functional
aa residue

D
56

N ——[Rec]——[DNA bind.]—— C

domain
aa residue

	Rec	DNA bind.	
1	7-121	150-206	212

C

?

periplasm

CM

cytoplasm

Figure 2. Models for the domain organization of SsrA and SsrB. The predicted domain organization of SsrA (A) and SsrB (B) and positions of domains and functional residues are depicted. (C) Model for the topology of SsrA and SsrB. Red circle, unknown signal received by SsrA; green circle with P, phosphate group. CM, cytoplasmic membrane. doi:10.1128/9781555818524.ch20f2

Response Regulator SsrB

In contrast to the case with SsrA, several experimental approaches have been made towards understanding the function of SsrB, the response regulator of the SsrAB two-component system. SsrB is a typical two-domain response regulator of the NarL/FixJ subfamily (Feng et al., 2004), which binds to and controls the transcription of its own promoter and that corresponding to the upstream *ssrA* gene. SsrB consists of the N-terminal receiver domain with the highly conserved aspartic acid D56 residue and of the C-terminal DNA-binding and dimerization domains (Fig. 2B). The C terminus of SsrB folds into a four-helix bundle (Leu150–Ile161, α-H1; Gly169–Lys173, α-H2; Ser177–Met186, α-H3; and Val197–Arg205, α-H4). α-H2 and α-H3 contain the helix-turn-helix (HTH) DNA binding motif, while α-H4 forms the dimerization domain (Carroll et al., 2009). Dimerization and DNA binding of SsrB are both essential properties for transcriptional activity

of SsrB-dependent genes. The C terminus of SsrB is required for activation of gene transcription, while the N-terminal phosphorylation domain inhibits the C-terminal effector domain until it is phosphorylated. The inhibitory action by the nonphosphorylated receiver domain might be transferred by interactions of α-H5 in the N terminus with the C-terminal effector domain (Feng et al., 2004). Thereby, the N terminus physically blocks the C-terminal HTH motif, avoiding DNA binding activity in the absence of phosphorylation. Thus, SsrB is autoregulated similar to response regulators like Spo0A from *Bacillus subtilis* and PhoB from *Escherichia coli*, in which an isolated C terminus is required for transcriptional activity (Feng et al., 2004). Based on structural homology to NarL, SsrB supposedly binds as a dimer to DNA. Thereby, each monomer inserts its recognition helix into subsequent major grooves vertical to the axis of the DNA. Additional flanking interactions between SsrB and its cognate DNA occur along a total length of 20 nucleotides (Walthers et al., 2007). Like OmpR, a response regulator encoded outside of SPI2 on the *Salmonella* chromosome, SsrB makes only a few base contacts, binds to AT-rich DNA, and presumably forms numerous contacts with the phosphate backbone (Carroll et al., 2009). The critical residues in the HTH motif required for specific transcriptional activity by SsrB are conserved between members of the NarL family of response regulators. Lys179 and Met186 in the DNA recognition helix are important for SsrB transcriptional activation and DNA binding, while a Glu182 residue is predicted to make nonspecific contacts with the phosphate backbone (Carroll et al., 2009).

Dimerization of SsrB is required for the transcription of most SsrB-regulated genes. SsrB contains two adjacent cysteine residues (Cys45 and Cys46) within its N-terminal domain and a third cysteine (Cys203) within α-H4 of the C terminus, just downstream of the HTH motif (Walthers et al., 2007). The Cys45 residue might affect the dimerization, by remodeling the dimer interface from an inactive form involving the N terminus to an active form involving α-H4–α-H4 interactions, forming a disulfide bond through Cys203 in the C terminus (Carroll et al., 2009). In a Cys203 substitution the SsrB activity is reduced due to the modification of its dimer interface. This demonstrates that dimerization is essential for DNA binding and is required for the following activation of transcription. Val197 and Leu201 are required for DNA contact, although they are in the loop of the dimerization helix α-H4 and might be involved in dimerization of SsrB. Another residue, Val197, placed in dimerization helix α-H4, is predicted to contact DNA despite its position. Leu201 is also an essential hydrophobic residue localized in α-H4, which stabilizes the helix and maintains the ability to form a dimer (Carroll et al., 2009).

Promoter Specificity of SsrB

To bind to the DNA as a dimer, SsrB requires the presence of two tandem SsrB operator sites (Walthers et al., 2007). Sequence analysis showed that the SsrB binding sites are AT rich but not well conserved (Feng et al., 2004). From 189 unique sequences isolated by ChIP with SsrB, more than 80% contained only a degenerate consensus motif, ACCTyA (y is guanine or thymine) (Tomljenovic-Berube et al., 2010). It seems that SsrB does not recognize a precise sequence motif, but it enhances the DNA binding through the formation of more specific contacts with the phosphate backbone, which would not be apparent as a conserved binding site (Feng et al., 2004). The few highly conserved sites within the promoters were limited to regions of direct DNA-protein contact by SsrB,

whereas adjacent sequence showed substantial drift. The conserved sites are contained in a heptameric sequence in 7-4-7 tail-to-tail architecture that creates an 18-bp degenerate palindrome (see Fig. 3A). A single conserved heptamer is sufficient for transcriptional activation by SsrB, so degenerate palindromes with only one conserved half-site are also functional (Tomljenovic-Berube et al., 2010). As long as the orientation of one single heptamer is maintained with respect to the downstream gene, SsrB is tolerant of variations in the adjacent 4-bp spacer and heptamer sequences as well. This flexible palindrome sequence is the minimal architecture required for regulatory input by SsrB. Since dimerization is likely required for transcriptional activation by SsrB, the strong DNA contact by one monomer may stabilize the binding of a second monomer with a less-than-ideal sequence (Tomljenovic-Berube et al., 2010).

Besides the degenerate consensus sequences of the promoters, SsrB also shows a variable location of its binding sites with respect to the transcriptional start site, including sites upstream, overlapping, or downstream of the start of transcription, which indicates that the mechanism of SsrB activation is different from a classical transcriptional activator. It is likely that SsrB uses distinct mechanisms to activate transcription at different promoters (Walthers et al., 2007).

REGULATORY CONTROL OF *SsrAB* EXPRESSION

Autoregulation of the *ssrAB* Operon

The gene *ssrB* is located downstream of *ssrA*, separated by a 30-bp intergenic region. The transcriptional start site of *ssrA* is located 164 nucleotides upstream of the translational start, while the transcriptional start site of *ssrB* resides 150 nucleotides upstream from the *ssrB* initiating codon (Fig. 3). Depending on the environmental signal, *ssrB* could be expressed as a polycistronic transcript with *ssrA* or monocistronic from its own promoter (Feng et al., 2003). The expression of *ssrAB* is positively autoregulated, as the SsrB protein binds to *ssrA* and *ssrB* promoters and activates their transcription. Activation of SsrB itself is mediated by its sensor kinase SsrA, but the environmental signal sensed by SsrA is still unknown.

Core Genome-Encoded Regulators of *ssrAB* Expression

In addition to SsrAB, further global regulatory mechanisms also control the expression of the SsrAB

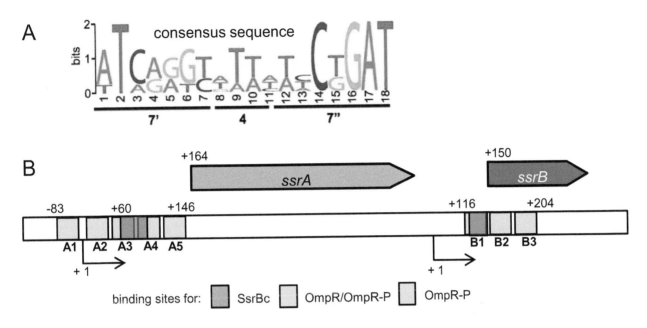

Figure 3. Binding specificity of SsrB and regulation of *ssrAB* expression. (A) The palindromic consensus motif of promoters of the SsrAB virulon as identified by one-hybrid screens and ChIP-on-chip experiments (modified from Tomljenovic-Berube et al., 2010). (B) The gene arrangement of *ssrAB* in SPI2 is depicted. The yellow box (A1) indicates the binding site location of OmpR (low affinity) and OmpR-P (high affinity) upstream of the transcriptional start site of *ssrA*. The light blue boxes (A2 to A5) indicate additional OmpR-P binding sites. Boxes B1 to B3 indicate the OmpR-P binding sites in the *ssrB* region. Overlapping binding sites for the C-terminal domain of SsrB are shown as red boxes (Feng et al., 2004). The transcriptional start sites of *ssrA* and *ssrB* are indicated by arrows and labeled +1, while the translational start sites are denoted by green arrows for *ssrA* and *ssrB*. The labels +164 and +150 indicate the presence of 164- and 150-nucleotide UTR in the *ssrA* and *ssrB* transcripts, respectively (Feng et al., 2003).
doi:10.1128/9781555818524.ch20f3

virulon. These include the two-component systems PhoPQ and OmpR/EnvZ, the transcriptional regulators SlyA and HilD, and the nucleoid-associated proteins Fis, histone-like nucleoid structuring protein (H-NS), and YdgT (Walthers et al., 2007; Xu and Hensel, 2010) (Fig. 4). Some of the SPI2 regulators depend on specific growth conditions the bacteria experience before entering a host cell; others may be required to optimize expression of SPI2 genes (Fass and Groisman, 2009).

OmpR/EnvZ

The ancestral OmpR/EnvZ two-component system regulates the porin genes *ompF* and *ompC* in response to changes in osmolarity (Kenney, 2002). OmpR/EnvZ is required for *Salmonella* replication and survival within macrophages and is essential for full virulence in the murine model (Garmendia et al., 2003).

The response regulator OmpR consists of two domains, the N-terminal phosphorylation domain and the C-terminal DNA-binding domain, which are separated by a linker. The C-terminal structure places OmpR in a family of winged HTH proteins. Its affinity

for DNA is increased when OmpR is phosphorylated and the C-terminal binding of a nonphosphorylated OmpR to DNA induces OmpR phosphorylation in the N terminus (Feng et al., 2003). OmpR is capable of a precise global regulation, because like SsrB, it makes only a few base contacts, binds to AT-rich DNA, and makes various phosphate backbone contacts (Carroll et al., 2009). OmpR is a typical class I activator that binds to the proximal sites of a promoter that terminate just upstream of the −35 element. These activators interact with the subunits of RNA polymerase, usually by a recruitment mechanism, and in this way directly activate transcription (Walthers et al., 2007).

Beside the diverse porin subtypes, OmpR regulates various additional genes, including the flagellar operon *flhDC*, *agfABCDEF* encoding curli fimbriae, and the cryptic porin-encoding *ompS1* and *ompS2* genes in *S.* Typhi and also acts at multiple steps in the regulatory cascade, activating SsrAB virulon genes both within and outside of SPI2. The regulation of *ssrAB* by OmpR is one of a few examples in which a two-component response regulator directly induces the expression of another two-component regulatory system (Feng et al., 2003). This shows that there is a

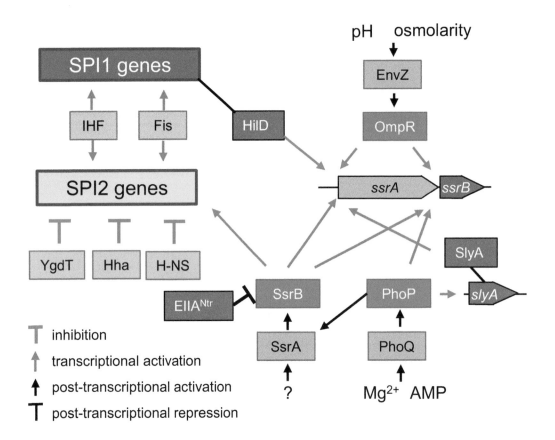

Figure 4. Model for regulatory circuits of the SsrAB virulon (modified from Fass and Groisman, 2009). The two-component systems (green boxes) SsrAB, EnvZ/OmpR, and PhoPQ are involved in SPI2 regulation. OmpR activates SPI2 genes by binding to the promoter and inducing the expression of *ssrA* and *ssrB*. PhoP directly binds the *ssrB* promoter and thereby indirectly induces *ssrA* transcription. SsrB binds to all SPI2 promoters, including *ssrA* and its own promoter, for which it is required for antagonizing the repression activity of H-NS. Transcription factors (red boxes) further affect the SsrAB regulon, with SlyA affecting *ssrAB* expression and EIIA^{Ntr} directly interacting with SsrB at the posttranscriptional level, thereby preventing SsrB-induced gene expression. The expression of *slyA* is controlled by PhoP. The SPI1-encoded HilD transcription factor is also involved (under certain conditions) in antagonizing H-NS activity. The NAPs (orange boxes) H-NS, Hha, and YdgT function as general negative SPI2 regulators, while Fis and IHF are NAPs that have a positive effect on SPI2 as well as SPI1 gene expression (Wilson et al., 2007; Fass and Groisman, 2009).
doi:10.1128/9781555818524.ch20f4

regulatory network between virulence genes and general regulators due to adaptation of *Salmonella* to a pathogenic lifestyle.

The phosphorylation state of OmpR is an essential factor for selective binding at the cognate sites in the *ssrA* promoter region and the 5′ untranslated regions (5′ UTR) of *ssrA* and *ssrB* (Feng et al., 2003; Feng et al., 2004) (Fig. 3). OmpR can bind to the *ssrA* sites in its phosphorylated and nonphosphorylated states, while it binds to the *ssrB* sites only in its phosphorylated form. The nonphosphorylated OmpR binds to the *ssrA* A1 site with an equilibrium dissociation constant (K_d) of about 260 nM and in the phosphorylated form OmpR-P with a K_d of 32 nM. The remaining OmpR-P low-affinity binding sites (A2 to A5) are downstream of the transcriptional start, suggesting a role for OmpR-P as a repressor (Feng et al.,

2003). Furthermore, OmpR binding to the *ssrA* A1 site causes a conformational change that stimulates phosphorylation. The phosphorylation of OmpR induces binding to *ssrA* sites A2 to A5. OmpR-P binds to three sites in the *ssrB*, named B1, B2, and B3. B1 is situated in the 5′ UTR of *ssrB*, while the B2 and B3 sites are located downstream from the translational start site of *ssrB*. It was observed that the phosphorylation of OmpR was absolutely required for the expression of *ssrB*, suggesting that binding of OmpR-P to the B1 site results in expression of the gene (Feng et al., 2004). *ssrA* and *ssrB* expression is also controlled by SsrB, and DNase I footprint analyses with the C-terminal portion of SsrB (SsrBc) demonstrated that the SsrBc binding sites of *ssrA* and *ssrB* overlap with the binding sites of OmpR-P (Fass and Groisman, 2009; Feng et al., 2004). Summarized, when OmpR-P levels are low, the A1 site would be occupied, leading to expression of *ssrA* and *ssrB*. An increase in OmpR-P levels would enable OmpR-P to bind to lower-affinity sites repressing *ssrA* and *ssrB* (Feng et al., 2003). This enables the bacteria to precisely regulate the expression of SsrA and SsrB by the amount of phosphorylated OmpR.

By this pathway, the signals that OmpR/EnvZ responds to are routed through SsrAB. EnvZ autokinase activity is assumed to be stimulated by an increasing potassium concentration, and the amount of phosphorylated OmpR in the following signal cascade increases over time in the presence of KCl. This suggests that the extracellular concentration of K^+ could act as the signal detected by OmpR/EnvZ when *Salmonella* senses changes in osmolarity (Garmendia et al., 2003). The observation that SsrB was present in an *ompR* null strain indicates that there are additional regulatory elements besides OmpR that affect SsrB levels (Feng et al., 2004).

PhoPQ Two-Component System

The PhoPQ two-component system is a conserved global regulatory system. In *Salmonella*, additional functions of PhoPQ were observed, suggesting that PhoPQ acts as a master regulator for the transition from the extracellular to the intracellular lifestyle of the pathogen (for examples, see Bader et al., 2005, and Zwir et al., 2005; for a review, see Groisman, 2001). The PhoPQ system is required for *Salmonella* virulence, as it controls the expression of more than 40 loci, including a number of pathogenicity determinants in SPI2, various effector gene loci, and SPI5, and it also represses genes of the HilA regulon, which incorporates genes involved in host cell invasion and adhesion to polarized cells (Bajaj et al., 1996; Gerlach et al., 2007; Thijs et al., 2007; Main-Hester et al., 2008). PhoP-activated genes are typically expressed by intracellular *Salmonella*, suggesting that this is the point in time when the PhoPQ system regulates expression of SPI2 genes (Fass and Groisman, 2009). The signals that stimulate the autophosphorylation of the sensor kinase PhoQ are known and include antimicrobial peptides and low concentrations of the divalent cations Mg^{2+} and Ca^{2+} (Zwir et al., 2005; Prost and Miller, 2008). PhoP directly binds to the *ssrB* promoter, where it activates *ssrB* transcription. *ssrA* gene expression is also controlled by PhoP at the posttranscriptional level in a process that might involve the 5′ leader of the *ssrA* mRNA and inhibition of translation (Fass and Groisman, 2009). A PhoPQ regulatory system defect does not reduce the maximum expression level of the SsrAB-controlled genes but causes highly delayed expression (Xu and Hensel,

2010). PhoP also regulates expression of the gene encoding SlyA, which is a regulator that activates the SPI2 expression (Fass and Groisman, 2009).

SlyA

The MarR-type regulator SlyA belongs to the terminal regulators in a cascade that integrates multiple signals to activate SPI2 gene expression. SlyA is suggested to control resistance to extracellular killing or to facilitate adhesion to macrophage receptors that induce phagocytic uptake. SlyA is also important for survival of *Salmonella* in the stationary phase. The SlyA protein is a homodimer that binds to five sites within the promoter region of the *slyA* gene, where it negatively regulates its own expression (Stapleton et al., 2002). Under specific conditions, SlyA and SsrAB can coordinate regulation of SPI2 and other virulence factors. The SlyA protein binds directly to the *ssrA* and *ssrB* promoters. SlyA shows a high affinity for inverted repeat sequences of two hexamers, TTAGCA-AGCTAA, as in the *ssrA* promoter region, but also to other binding sites that do not match the specific consensus sequence (Yoon et al., 2009). Deletion of the *slyA* gene had no effect on the expression of other SPI2-regulated genes but did affect *ssrA* and *ssrB* expression (Feng et al., 2004). Hence, the transcription of *ssrA* and *ssrB* is activated by SlyA, which itself depends on regulation by PhoP after upregulation in response to osmotic changes. SsrB and SlyA may deploy a similar mechanism to regulate SPI2, such as by counteracting the silencing activity of the nucleoid organizing proteins H-NS and YdgT (see "Role of NAPs" below for further details). Specifically, SsrB and SlyA compete against H-NS and YdgT for binding to the same target sequences, or they alter DNA structure to promote transcription of SPI2 by facilitating RNA polymerase access (Yoon et al., 2009).

EIIA^Ntr

The expression of *ssrB* is promoted by several transcription factors, including SsrB itself, OmpR, HilD, PhoPQ, and others. Therefore, the expression of SPI2 genes could increase to undesirable levels under activating conditions. To avoid the hyperactivation of SPI2 genes, *Salmonella* might use the *ptsN*-encoded EIIA^Ntr, a component of the nitrogen-metabolic phosphotransferase system (PTS). The PTS consists of the enzyme I^Ntr (EI^Ntr), the phosphoregulator Npr, and the enzyme IIA^Ntr (EIIA^Ntr). EIIA^Ntr interacts directly with the TrkA potassium transporter and the KdpD sensor kinase, which control cytoplasmic potassium levels. EIIA^Ntr also interacts directly with the SsrB at the posttranscriptional level, thereby preventing SsrB binding to its target promoters. The EIIA^Ntr-SsrB interaction is specific and might repress the phosphorylation of SsrB by SsrA or promote the dephosphorylation of phosphorylated SsrB. The SPI2 regulatory function of EIIA^Ntr by negative control of the stability or activity of the SsrB protein is independent of the phosphorylation state of EIIA^Ntr. A lack of EIIA^Ntr results in overexpression of the positive autoregulated *ssrB*. Thus, EIIA^Ntr negatively regulates SPI2 genes, resulting in their expression at appropriate levels under inducing conditions. Infection studies in the murine model of systemic *Salmonella* infection demonstrated the importance of EIIA^Ntr for virulence of *Salmonella* in vivo (Choi et al., 2010). The alternative sigma factor σ^E might enhance EIIA^Ntr-mediated SPI2 gene regulation inside macrophages, because it counteracts phagosomal defense molecules such as reactive oxygen species and antimicrobial peptides. The *ptsN* gene as part of the nitrogen-metabolic PTS is a member of the σ^E regulon (Rhodius et al., 2006). During *Salmonella* growth inside macrophages, σ^E, which is activated by the phagosomal stress, could enhance EIIA^Ntr expression, thereby controlling timing and/or levels of SPI2 expression (Choi et al., 2010). The control of the SsrAB virulon by EIIA^Ntr may imply that some nitrogen sources could inhibit the expression of SPI2 genes. It may be interesting to search for those SPI2-repressing nitrogen-containing components in the host cells.

Role of NAPs

The promoters of many genes have evolved to respond to specific perturbations in DNA supercoiling. Thus, supercoiling represents an important relay for environmental signals. The spatial organization of the transcription factors at target gene promoters is constricted by the degree of supercoiling in the underlying DNA scaffold. This implies that the topology of the bacterial chromosome imposes a global effect on gene expression. In *Salmonella*, the chromosome and plasmid molecules are generally negatively supercoiled. SPI1 gene products induce the invasion of intestinal epithelial cells when supercoiling is high, while SPI2 genes are expressed inside the cell when DNA topology transforms to a more relaxed conformation (Cameron et al., 2011). To modify their DNA topology, enteric bacteria deploy nucleoid-associated proteins (NAPs). These proteins do not covalently modify DNA but change the superhelical density of their binding regions (Cameron et al., 2011). Thereby, NAPs act as global regulators by controlling the expression of a large number of genes throughout their genomes (Fass and Groisman, 2009).

H-NS is a constitutive abundant NAP acting as a global transcriptional regulator and a genome-structuring protein that binds to AT-rich sequences commonly present in horizontally acquired DNA. It has been proposed that NAPs prevent the uncontrolled expression of SPI genes. The expression of *ssrAB* genes is downregulated by H-NS (Bustamante et al., 2008). Furthermore, in H-NS-defective *Salmonella*, a substantially smaller amount of the SsrB regulator is required to activate SPI2 genes, suggesting that SsrB counteracts the H-NS silencing. This relief of H-NS repression is necessary to activate the transcription of SPI2 genes (Walthers et al., 2007).

YdgT and Hha are two further NAPs which repress transcription of SPI2 genes. The *ydgT* gene is not expressed at early stages of infection, so SPI2 genes can be upregulated (Silphaduang et al., 2007). Hha is a homolog of YdgT. Expression of SPI2 genes during growth in LB medium, but not low-phosphate minimal medium, was increased in the Δ*hha* background. Thus, Hha silences expression when bacteria are exposed to an extracellular environment, and this mechanism enables *Salmonella* to express SPI2 genes under intracellular conditions (Silphaduang et al., 2007). YdgT and Hha can form heterodimers with H-NS by binding to its oligomerization domain (Stoebel et al., 2008). Thus, YdgT and Hha may form a repressor complex with H-NS and repress gene expression in concert with H-NS. It has also been suggested that YdgT and Hha associate with H-NS. These complexes stabilize the H-NS-promoted repression of SPI2 genes and limit the effect of SsrB in counteracting H-NS repression (Fass and Groisman, 2009). A defect in one of these components may result in the expression of SPI2 genes with unsuitable timing (i.e., under repressing conditions) and at inappropriate levels under inducing conditions (Choi et al., 2010).

There are also positive regulators of SPI2 expression that function to counteract the silencing effects of H-NS and other NAPs, thereby promoting the access of RNA polymerase to specific promoters (Lim et al., 2006; Silphaduang et al., 2007). Integration host factor (IHF) and the factor for inversion stimulation (Fis) are two NAPs which are essential for full expression of SPI2 genes. IHF is an important factor for DNA bending and compaction. IHF production is highest in the late exponential growth phase, coinciding with SPI2 expression (Fass and Groisman, 2009). SPI2 gene expression is reduced in the absence of IHF, but this effect may be indirect due to modulation of expression of *hilD* by IHF (Mangan et al., 2006).

Fis functions as a global regulator of supercoiling by stabilizing intermediate topologies. To maintain the homeostasis of DNA topology, Fis secludes the supercoiled DNA from DNA gyrases (GyrAB). As a further level of control, Fis affects the regulation of genes encoding topoisomerase (TopAB) and GyrAB (Keane and Dorman, 2003). Fis is a sequence-dependent DNA binding protein that activates gene expression by the recruitment of RNA polymerase, but Fis could also inhibit transcription by directly blocking a promoter. Thus, gene promoters may be fine-tuned by the Fis levels and topological conformation of their cognate DNA (Cameron et al., 2011). *ssrA* expression is optimal at a crucial concentration of Fis, and too-large amounts of Fis result in reduced *ssrA* expression (Lim et al., 2006). Fis also controls the expression of other regulatory genes such as *phoP* and SPI1 genes (Fass and Groisman, 2009).

Regulation by Other Horizontally Acquired Regulators

The SPI1-encoded HilD regulatory protein, which activates the expression of SPI1 genes, has been suggested to control the switch from SPI1 to SPI2 induction (Fass and Groisman, 2009). HilD functions by counteracting the H-NS-mediated repression of the *ssrAB* operon, subsequently enabling OmpR to activate transcription of SPI2 genes. HilD directly binds to the *ssrAB* regulatory region and thereby controls both the sensor kinase and the response regulator (Bustamante et al., 2008). The mechanism to differentially regulate the expression of SPI1 and SPI2 could be explained by a lower DNA binding constant at the *ssrA* site than at SPI1 binding sites. So a larger amount of HilD is required for *ssrA* expression than for SPI1 genes. Thus, the regulatory shift from SPI1 to SPI2 gene expression could be explained by the requirement of a higher level of HilD for binding to P_{ssrA} (Bustamante et al., 2008). Apart from the higher affinity for the *hilA* promoter region versus the *ssrAB* promoter region, the mechanism by which HilD controls *ssrAB* expression appears to depend on a number of different factors. These include the parallel growth phase-dependent expression of other key regulators of SPI2, such as OmpR, or the competitive action of other regulators. If SPI1 expression is turned off, HilD is redirected toward *ssrAB* activation (Bustamante et al., 2008). HilD is not required for SPI2 regulon expression under growth conditions that are thought to resemble the intracellular environment, including a slightly acidic pH and low-nutrient concentrations. The HilD-dependent activation of SPI2 expression could be observed only in high-osmolarity and nutrient-rich media, a milieu that is believed to give some approximation of the appropriate in vivo environment

of the intestinal lumen (Bustamante et al., 2008). Based on these findings, it was proposed that two independent pathways exist to activate genes of the SsrAB virulon in different host cell environments. The HilD-dependent mechanism is active in the intestinal lumen and another regulatory circuit specifically induced in the SCV during the intracellular lifestyle; these respond to different environmental conditions (Bustamante et al., 2008). While the biological role of the activation of the SsrAB virulon during intracellular life of *Salmonella* is obvious, the functional importance of observed expression under extracellular conditions needs further experimental investigation.

HOST FACTORS ACTING ON SsrAB

Upon stimulation by gamma interferon (IFN-γ), macrophages generate nitric oxide (NO), a highly reactive radical with antimicrobial activity. Work of the Vazques-Torres group showed that NO also abrogates expression of the SsrAB virulon (McCollister et al., 2005). Production of NO and its reaction products represses SPI2-T3SS effector genes both inside and outside of SPI2, suggesting a reduced level of SsrA or SsrB which controls this virulon. The inhibition of SPI2 gene expression by NO facilitates the interaction of SCVs with degradative compartments required for antimicrobial activity. A functional SPI2-T3SS would prevent trafficking of vesicles containing the NADPH oxidase (Vazquez-Torres et al., 2000) or inducible nitric oxide synthase (iNOS) (Chakravortty et al., 2002) enzymatic complexes and thereby reduce the contact of SCVs with these vesicles to approximately 20%. The iNOS enzymatic complex oxidates the guanidino group of L-arginine by generating L-citrulline and the radical diatomic NO. Even with its unpaired electron, NO can diffuse freely through membranes and react with cellular macromolecules. This makes dehydratases, membranes, and enzymes involved in DNA synthesis and replication potential targets of NO and its derivatives (reviewed in Bogdan et al., 2000). *Salmonella* resists this NO-mediated antimicrobial effect by an antinitrosative response, which repairs the radical-mediated damages and detoxifies a variety of NO derivatives. The ability to avoid contact with iNOS-containing vesicles represents another important protection mechanism against NO exposure during the innate host response (reviewed in Chakravortty and Hensel, 2003). IFN-γ is secreted in response to lipopolysaccharide of *Salmonella* and enhances the transcription of iNOS, thereby increasing the antimicrobial activity of the macrophages. IFN-γ also augments NADPH oxidase activity by increasing the levels of transcription of cognate genes. The high levels of iNOS and NADPH oxidase produce a high NO and peroxynitrite output in IFN-γ-treated macrophages, which inhibits SPI2 transcription (McCollister et al., 2005). NO represses *ssrA* transcription in a concentration-dependent manner. In nonstimulated macrophages, only small amounts of NO are generated. *Salmonella* can oppose these low rates of NO by the NO dioxygenase Hmp and other antinitrosative defenses such as glutathione, homocysteine, and glutathione reductase (Chakravortty and Hensel, 2003). The resistance mechanisms allow SPI2 transcription at small NO amounts, while at higher rates of NO synthesis, generated by IFN-γ-activated macrophages, SPI2 transcription is inhibited. It is still unknown if the inhibition of *ssrA* expression by NO proceeds in a direct or indirect manner. The cysteine residues in SsrB and OmpR, which directly control *ssrA* transcription, may represent potential targets for NO-mediated inhibition, resulting in a general downregulation of SPI2 expression (McCollister et al., 2005). Whether other OmpR-regulated functions, such as expression of porin genes, are also affected by NO has not been reported.

POTENTIAL SIGNALS FOR SsrAB ACTIVATION

So far, the specific signal or signals which are received by the sensor kinase SsrA leading to the phosphorylation of the response regulator SsrB and expression of SPI2 genes are unknown. Indirect evidence indicates that growth in media with low osmolarity, nutritional limitations, and slightly acidic conditions activates SPI2 gene expression (Deiwick et al., 1999; Lee et al., 2000; Coombes et al., 2004). The conditions for SsrAB virulon induction in vitro may well correspond to the intracellular environment experienced by *Salmonella* during life within macrophages (Xu and Hensel, 2010). The induction of SPI2 gene expression in phosphate-limiting media suggests that the availability of inorganic phosphate (P_i) within the SCV could be a signal affecting SPI2 expression. These P_i limiting conditions are encountered by intracellular *Salmonella* and induce an overproduction of sensor protein SsrB in vitro (Löber et al., 2006). More evidence that P_i is strictly limited and absolutely required in SCV is that the gene *pstS*, which encodes a part of the high-affinity P_i uptake system Pst, is highly induced by *Salmonella* within the SCV (Löber et al., 2006).

It has also been shown that the SCV undergoes acidification to a pH between 4.5 and 5.0. A rapid increase in SsrB levels was induced when bacterial cultures were shifted to media with pHs between 4.6 and 6.2, and increased levels of SPI2 effector proteins were detected shortly after shift to acidic pH (Garmendia et al., 2003). The slightly acidic pH of about 6.0 in early endosomes seems to be sufficient to activate SPI2 gene expression, and such slight acidification is achieved rapidly after phagosome formation. Besides SPI2 gene expression, the acidic pH within the SCV of infected host cells has been shown to affect the assembly of the SPI2-T3SS and to activate the secretion of the translocon and the effector proteins (Rappl et al., 2003; Löber et al., 2006; Yu et al., 2010). The induction of SPI2 expression by acidic pH inside the SCV is mediated by both SsrAB and OmpR/EnvZ (Garmendia et al., 2003). *ssrB* expression was found to be OmpR dependent at pH 7.4 but not at pH 5.7. In addition, at pH 5.7, SsrA can still be detected in the absence of OmpR/EnvZ or SsrB via immunoblotting (Feng et al., 2004). This demonstrates that the pH dependence of *ssrA* and *ssrB* expression is distinct and that expression of both components is uncoupled (Feng et al. 2004). The pH-dependent effect on SPI2 gene expression is distinct from that mediated by the amount of available inorganic phosphate and other nutrients. It is not known to date if pH represents the relevant signal to activate the SsrAB system, while P_i limitation leads only to an indirect induction of SPI2 gene expression, or if the rapid gene expression induced by low pH and the delayed expression induced by P_i starvation are equally required signals which are combined by SsrAB (Löber et al., 2006).

Deiwick et al., 1999, reported that in the presence of low Mg^{2+} concentrations SPI2 expression is induced, while high Mg^{2+} concentrations inhibit the expression of SPI2 effectors, and this effect could be reverted by lowering the phosphate concentration. In contrast to these results, Garmendia et al. showed by flow cytometry and by immunoblotting of green fluorescent protein-labeled SPI2 effectors that low and high $MgCl_2$ concentrations lead to no significant differences in levels of proteins encoded by these SPI2 genes (Garmendia et al., 2003). The calcium concentration is about five times lower (400 μM) in macrophage lysosomes than the concentration outside the cell (2 mM) (Vieira et al., 2002). Garmendia et al. suggested that low Ca^{2+} concentration affects gene expression mediated mainly through SsrAB and also considered that the osmosensitive OmpR/EnvZ is essential for full SPI2 gene expression in environments with low calcium concentrations (Garmendia et al., 2003). By contrast, Miao and Miller presented data showing that SsrAB-dependent activation of SPI2 genes could be detected in the presence of low and moderate concentrations of magnesium or calcium, indicating that SsrAB induction is independent of Mg^{2+} or Ca^{2+} limitation (Miao and Miller, 2002). Hence, the response in SPI2 gene expression generated by low osmolarity in the absence of Ca^{2+} or at low concentrations of Mg^{2+} is controversial.

As a result of this controversy, the influence of other osmotic substrates as a potential signal to the SsrAB virulon was examined. It has been shown that high osmolarity induces SPI1 gene expression, while it represses expression from the SPI2 promoters. This inhibitory effect could be partially prevented by the addition of the osmoprotectant glycine betaine (Garmendia et al., 2003). The response to the environmental stimulus osmolarity could explain why SPI1 is activated in the high-osmolarity intestinal lumen, while SPI2 gene expression is activated in the low-osmolarity intracellular environment. An inverse regulation of SPI1 and SPI2 genes in response to osmolarity provides Salmonella with the opportunity to express these functionally distinct systems independently and only in the environment where they are required during infection. The effect of osmolarity on SPI2 gene expression is mostly mediated through SsrAB, whereas OmpR/EnvZ, which is also activated by low osmolarity, seems to play only a minor role in sensing osmotic pressure (Garmendia et al., 2003). S. Typhimurium compensates for a sudden osmotic upshift by taking up large amounts of K^+ from the environment. Yet, the K^+ concentration has no effect on gene expression comparable to that caused by changes in osmolarity. This indicates that the intracellular potassium level is not the signal sensed by SsrA. Despite these facts, the potassium concentration might play a role in mediating osmolarity-dependent activation of OmpR/EnvZ. Therefore, the extracellular increase of potassium might not result in its uptake into the periplasm but might be sensed as osmotic pressure by OmpR/EnvZ (Garmendia et al., 2003).

When Salmonella cultures have reached the late stationary phase and SPI1 expression has declined, the SPI2 genes are expressed in a growth phase-dependent manner in vitro. The signals for this growth phase-dependent gene expression are unknown, but regulatory proteins acting primarily in response to population density could be excluded (Bustamante et al., 2008). The expression of genes for the SPI1-encoded T3SS and cognate effectors is controlled by a complex regulatory cascade and a complex array of environmental stimuli (reviewed in Ellermeier and Slauch, 2007). A similar complexity may be found for the SsrAB virulon. Still, a lot of unanswered questions remain related to the signals sensed by SsrA. It will be a challenge to identify those signals to provide a better understanding of this regulatory process during Salmonella infection.

CONCLUSIONS AND OUTLOOK

The gene products of the SsrAB virulon have crucial roles in the intracellular lifestyle of Salmonella and the progression of systemic infections. This complex virulon comprises genes acquired by horizontal gene transfer of the SPI2 locus and various additional loci from successive independent events of gene transfer. There are indications that the SsrAB virulon extends beyond genes directly related to the SPI2-T3SS. Significant progress has been made in understanding the various components of the SsrAB virulon and the molecular basis of interaction of regulator SsrB with the promoters of the virulon. In contrast, the identities of the physicochemical signals recognized by sensor SsrA are unknown. Understanding these stimuli would provide a major step towards a comprehensive view of intracellular pathogenesis. The SsrAB virulon is a paradigm for the complex wiring of a regulatory circuit for expression of virulence genes in bacteria. The levels of regulation range from the SPI2-encoded regulatory system SsrAB and core genome-encoded regulators to global control elements such as DNA topology regulators. An integrated understanding of this complex, multilevel virulence determinant remains a challenge for future investigations.

Acknowledgments. Work in our group was supported by grants from the Deutsche Forschungsgemeinschaft and a fellowship of Deutscher Akademischer Austauschdienst to A. F.-L.

REFERENCES

Bader, M. W., S. Sanowar, M. E. Daley, A. R. Schneider, U. Cho, W. Xu, R. E. Klevit, H. Le Moual, and S. I. Miller. 2005. Recognition of antimicrobial peptides by a bacterial sensor kinase. *Cell* **122**:461–472.

Bajaj, V., R. L. Lucas, C. Hwang, and C. A. Lee. 1996. Co-ordinate regulation of *Salmonella typhimurium* invasion genes by environmental and regulatory factors is mediated by control of *hilA* expression. *Mol. Microbiol.* **22**:703–714.

Beier, D., B. Schwarz, T. M. Fuchs, and R. Gross. 1995. In vivo characterization of the unorthodox BvgS two-component sensor protein of *Bordetella pertussis*. *J. Mol. Biol.* **248**:596–610.

Beuzon, C. R., G. Banks, J. Deiwick, M. Hensel, and D. W. Holden. 1999. pH-dependent secretion of SseB, a product of the SPI-2 type III secretion system of *Salmonella typhimurium*. *Mol. Microbiol.* **33**:806–816.

Beuzon, C. R., S. Meresse, K. E. Unsworth, J. Ruiz-Albert, S. Garvis, S. R. Waterman, T. A. Ryder, E. Boucrot, and D. W. Holden. 2000. *Salmonella* maintains the integrity of its intracellular vacuole through the action of SifA. *EMBO J.* **19**:3235–3249.

Bogdan, C., M. Rollinghoff, and A. Diefenbach. 2000. The role of nitric oxide in innate immunity. *Immunol. Rev.* **173**:17–26.

Boucrot, E., T. Henry, J. P. Borg, J. P. Gorvel, and S. Meresse. 2005. The intracellular fate of *Salmonella* depends on the recruitment of kinesin. *Science* **308**:1174–1178.

Brown, N. F., B. A. Vallance, B. K. Coombes, Y. Valdez, B. A. Coburn, and B. B. Finlay. 2005. *Salmonella* pathogenicity island 2 is expressed prior to penetrating the intestine. *PLoS Pathog.* 1:e32.

Bustamante, V. H., L. C. Martinez, F. J. Santana, L. A. Knodler, O. Steele-Mortimer, and J. L. Puente. 2008. HilD-mediated transcriptional cross-talk between SPI-1 and SPI-2. *Proc. Natl. Acad. Sci. USA* 105:14591–14596.

Cameron, A. D., D. M. Stoebel, and C. J. Dorman. 2011. DNA supercoiling is differentially regulated by environmental factors and FIS in *Escherichia coli* and *Salmonella enterica*. *Mol. Microbiol.* 80:85–101.

Carroll, R. K., X. Liao, L. K. Morgan, E. M. Cicirelli, Y. Li, W. Sheng, X. Feng, and L. J. Kenney. 2009. Structural and functional analysis of the C-terminal DNA binding domain of the *Salmonella typhimurium* SPI-2 response regulator SsrB. *J. Biol. Chem.* 284:12008–12019.

Chakravortty, D., I. Hansen-Wester, and M. Hensel. 2002. *Salmonella* pathogenicity island 2 mediates protection of intracellular *Salmonella* from reactive nitrogen intermediates. *J. Exp. Med.* 195:1155–1166.

Chakravortty, D., and M. Hensel. 2003. Inducible nitric oxide synthase and control of intracellular bacterial pathogens. *Microbes Infect.* 5:621–627.

Choi, J., D. Shin, H. Yoon, J. Kim, C. R. Lee, M. Kim, Y. J. Seok, and S. Ryu. 2010. *Salmonella* pathogenicity island 2 expression negatively controlled by EIIANtr-SsrB interaction is required for *Salmonella* virulence. *Proc. Natl. Acad. Sci. USA* 107:20506–20511.

Cirillo, D. M., R. H. Valdivia, D. M. Monack, and S. Falkow. 1998. Macrophage-dependent induction of the *Salmonella* pathogenicity island 2 type III secretion system and its role in intracellular survival. *Mol. Microbiol.* 30:175–188.

Coombes, B. K., N. F. Brown, Y. Valdez, J. H. Brumell, and B. B. Finlay. 2004. Expression and secretion of *Salmonella* pathogenicity island-2 virulence genes in response to acidification exhibit differential requirements of a functional type III secretion apparatus and SsaL. *J. Biol. Chem.* 26:49804–49815.

Coombes, B. K., M. E. Wickham, N. F. Brown, S. Lemire, L. Bossi, W. W. Hsiao, F. S. Brinkman, and B. B. Finlay. 2005. Genetic and molecular analysis of GogB, a phage-encoded type III-secreted substrate in *Salmonella enterica* serovar Typhimurium with autonomous expression from its associated phage. *J. Mol. Biol.* 348:817–830.

Deiwick, J., and M. Hensel. 1999. Regulation of virulence genes by environmental signals in *Salmonella typhimurium*. *Electrophoresis* 20:813–817.

Deiwick, J., T. Nikolaus, S. Erdogan, and M. Hensel. 1999. Environmental regulation of *Salmonella* pathogenicity island 2 gene expression. *Mol. Microbiol.* 31:1759–1773.

Ellermeier, J. R., and J. M. Slauch. 2007. Adaptation to the host environment: regulation of the SPI1 type III secretion system in *Salmonella enterica* serovar Typhimurium. *Curr. Opin. Microbiol.* 10:24–29.

Fass, E., and E. A. Groisman. 2009. Control of *Salmonella* pathogenicity island-2 gene expression. *Curr. Opin. Microbiol.* 12:199–204.

Feng, X., R. Oropeza, and L. J. Kenney. 2003. Dual regulation by phospho-OmpR of *ssrA/B* gene expression in *Salmonella* pathogenicity island 2. *Mol. Microbiol.* 48:1131–1143.

Feng, X., D. Walthers, R. Oropeza, and L. J. Kenney. 2004. The response regulator SsrB activates transcription and binds to a region overlapping OmpR binding sites at *Salmonella* pathogenicity island 2. *Mol. Microbiol.* 54:823–835.

Freeman, J. A., C. Rappl, V. Kuhle, M. Hensel, and S. I. Miller. 2002. SpiC is required for translocation of *Salmonella* pathogenicity island 2 effectors and secretion of translocon proteins SseB and SseC. *J. Bacteriol.* 184:4971–4980.

Garmendia, J., C. R. Beuzon, J. Ruiz-Albert, and D. W. Holden. 2003. The roles of SsrA-SsrB and OmpR-EnvZ in the regulation of genes encoding the *Salmonella typhimurium* SPI-2 type III secretion system. *Microbiology* 149:2385–2396.

Gerlach, R. G., and M. Hensel. 2007. *Salmonella* pathogenicity islands in host specificity, host pathogen-interactions and antibiotics resistance of *Salmonella enterica*. *Berl. Munch. Tierarztl. Wochenschr.* 120:317–327.

Gerlach, R. G., D. Jäckel, N. Geymeier, and M. Hensel. 2007. *Salmonella* pathogenicity island 4-mediated adhesion is coregulated with invasion genes in *Salmonella enterica*. *Infect. Immun.* 75:4697–4709.

Gordon, M. A. 2008. *Salmonella* infections in immunocompromised adults. *J. Infect.* 56:413–422.

Groisman, E. A. 2001. The pleiotropic two-component regulatory system PhoP-PhoQ. *J. Bacteriol.* 183:1835–1842.

Haraga, A., M. B. Ohlson, and S. I. Miller. 2008. Salmonellae interplay with host cells. *Nat. Rev. Microbiol.* 6:53–66.

Henry, T., C. Couillault, P. Rockenfeller, E. Boucrot, A. Dumont, N. Schroeder, A. Hermant, L. A. Knodler, P. Lecine, O. Steele-Mortimer, J. P. Borg, J. P. Gorvel, and S. Meresse. 2006. The *Salmonella* effector protein PipB2 is a linker for kinesin-1. *Proc. Natl. Acad. Sci. USA* 103:13497–13502.

Hensel, M., A. P. Hinsley, T. Nikolaus, G. Sawers, and B. C. Berks. 1999. The genetic basis of tetrathionate respiration in *Salmonella typhimurium*. *Mol. Microbiol.* 32:275–288.

Hensel, M., J. E. Shea, C. Gleeson, M. D. Jones, E. Dalton, and D. W. Holden. 1995. Simultaneous identification of bacterial virulence genes by negative selection. *Science* 269:400–403.

Hensel, M., J. E. Shea, S. R. Waterman, R. Mundy, T. Nikolaus, G. Banks, A. Vazquez-Torres, C. Gleeson, F. Fang, and D. W. Holden. 1998. Genes encoding putative effector proteins of the type III secretion system of *Salmonella* pathogenicity island 2 are required for bacterial virulence and proliferation in macrophages. *Mol. Microbiol.* 30:163–174.

Ibarra, J. A., and O. Steele-Mortimer. 2009. *Salmonella*—the ultimate insider. *Salmonella* virulence factors that modulate intracellular survival. *Cell. Microbiol.* 11:1579–1586.

Jantsch, J., D. Chikkaballi, and M. Hensel. 2011. Cellular aspects of immunity to intracellular *Salmonella enterica*. *Immunol. Rev.* 240:185–195.

Keane, O. M., and C. J. Dorman. 2003. The *gyr* genes of *Salmonella enterica* serovar Typhimurium are repressed by the factor for inversion stimulation, Fis. *Mol. Genet. Genomics* 270:56–65.

Kenney, L. J. 2002. Structure/function relationships in OmpR and other winged-helix transcription factors. *Curr. Opin. Microbiol.* 5:135–141.

Kuhle, V., and M. Hensel. 2004. Cellular microbiology of intracellular *Salmonella enterica*: functions of the type III secretion system encoded by *Salmonella* pathogenicity island 2. *Cell. Mol. Life Sci.* 61:2812–2826.

Kuhle, V., and M. Hensel. 2002. SseF and SseG are translocated effectors of the type III secretion system of *Salmonella* pathogenicity island 2 that modulate aggregation of endosomal compartments. *Cell. Microbiol.* 4:813–824.

Lee, A. K., C. S. Detweiler, and S. Falkow. 2000. OmpR regulates the two-component system SsrA-SsrB in *Salmonella* pathogenicity island 2. *J. Bacteriol.* 182:771–781.

Lim, S., B. Kim, H. S. Choi, Y. Lee, and S. Ryu. 2006. Fis is required for proper regulation of *ssaG* expression in *Salmonella enterica* serovar Typhimurium. *Microb. Pathog.* 41:33–42.

Löber, S., D. Jäckel, N. Kaiser, and M. Hensel. 2006. Regulation of *Salmonella* pathogenicity island 2 genes by independent environmental signals. *Int. J. Med. Microbiol.* 296:435–447.

Main-Hester, K. L., K. M. Colpitts, G. A. Thomas, F. C. Fang, and S. J. Libby. 2008. Coordinate regulation of *Salmonella* pathogenicity island 1 (SPI1) and SPI4 in *Salmonella enterica* serovar Typhimurium. *Infect. Immun.* **76**:1024–1035.

Mangan, M. W., S. Lucchini, V. Danino, T. O. Croinin, J. C. Hinton, and C. J. Dorman. 2006. The integration host factor (IHF) integrates stationary-phase and virulence gene expression in *Salmonella enterica* serovar Typhimurium. *Mol. Microbiol.* **59**:1831–1847.

McCollister, B. D., T. J. Bourret, R. Gill, J. Jones-Carson, and A. Vazquez-Torres. 2005. Repression of SPI2 transcription by nitric oxide-producing, IFNγ-activated macrophages promotes maturation of *Salmonella* phagosomes. *J. Exp. Med.* **202**:625–635.

Miao, E. A., and S. I. Miller. 2000. A conserved amino acid sequence directing intracellular type III secretion by *Salmonella typhimurium*. *Proc. Natl. Acad. Sci. USA* **97**:7539–7544.

Ohlson, M. B., K. Fluhr, C. L. Birmingham, J. H. Brumell, and S. I. Miller. 2005. SseJ deacylase activity by *Salmonella enterica* serovar Typhimurium promotes virulence in mice. *Infect. Immun.* **73**:6249–6259.

Osborne, S. E., and B. K. Coombes. 2011. Transcriptional priming of *Salmonella* pathogenicity island-2 precedes cellular invasion. *PLoS One* **6**:e21648.

Prost, L. R., and S. I. Miller. 2008. The *Salmonellae* PhoQ sensor: mechanisms of detection of phagosome signals. *Cell. Microbiol.* **10**:576–582.

Raffatellu, M., R. P. Wilson, S. E. Winter, and A. J. Bäumler. 2008. Clinical pathogenesis of typhoid fever. *J. Infect. Dev. Ctries.* **2**:260–266.

Rappl, C., J. Deiwick, and M. Hensel. 2003. Acidic pH is required for the functional assembly of the type III secretion system encoded by *Salmonella* pathogenicity island 2. *FEMS Microbiol. Lett.* **226**:363–372.

Rhodius, V. A., W. C. Suh, G. Nonaka, J. West, and C. A. Gross. 2006. Conserved and variable functions of the σ^E stress response in related genomes. *PLoS Biol.* **4**:e2.

Rollenhagen, C., M. Sorensen, K. Rizos, R. Hurvitz, and D. Bumann. 2004. Antigen selection based on expression levels during infection facilitates vaccine development for an intracellular pathogen. *Proc. Natl. Acad. Sci. USA* **101**:8739–8744.

Salcedo, S. P., and D. W. Holden. 2003. SseG, a virulence protein that targets *Salmonella* to the Golgi network. *EMBO J.* **22**:5003–5014.

Schlumberger, M. C., and W. D. Hardt. 2006. *Salmonella* type III secretion effectors: pulling the host cell's strings. *Curr. Opin. Microbiol.* **9**:46–54.

Shea, J. E., M. Hensel, C. Gleeson, and D. W. Holden. 1996. Identification of a virulence locus encoding a second type III secretion system in *Salmonella typhimurium*. *Proc. Natl. Acad. Sci. USA* **93**:2593–2597.

Silphaduang, U., M. Mascarenhas, M. Karmali, and B. K. Coombes. 2007. Repression of intracellular virulence factors in *Salmonella* by the Hha and YdgT nucleoid-associated proteins. *J. Bacteriol.* **189**:3669–3673.

Stapleton, M. R., V. A. Norte, R. C. Read, and J. Green. 2002. Interaction of the *Salmonella typhimurium* transcription and virulence factor SlyA with target DNA and identification of members of the SlyA regulon. *J. Biol. Chem.* **277**:17630–17637.

Stein, M. A., K. Y. Leung, M. Zwick, F. Garcia-del Portillo, and B. B. Finlay. 1996. Identification of a *Salmonella* virulence gene

required for formation of filamentous structures containing lysosomal membrane glycoproteins within epithelial cells. *Mol. Microbiol.* **20**:151–164.

Stoebel, D. M., A. Free, and C. J. Dorman. 2008. Anti-silencing: overcoming H-NS-mediated repression of transcription in Gram-negative enteric bacteria. *Microbiology* **154**:2533–2545.

Thijs, I. M., S. C. De Keersmaecker, A. Fadda, K. Engelen, H. Zhao, M. McClelland, K. Marchal, and J. Vanderleyden. 2007. Delineation of the *Salmonella enterica* serovar Typhimurium HilA regulon through genome-wide location and transcript analysis. *J. Bacteriol.* **189**:4587–4596.

Tomljenovic-Berube, A. M., D. T. Mulder, M. D. Whiteside, F. S. Brinkman, and B. K. Coombes. 2010. Identification of the regulatory logic controlling *Salmonella* pathoadaptation by the SsrA-SsrB two-component system. *PLoS Genet.* **6**:e1000875.

Uchiya, K., M. A. Barbieri, K. Funato, A. H. Shah, P. D. Stahl, and E. A. Groisman. 1999. A *Salmonella* virulence protein that inhibits cellular trafficking. *EMBO J.* **18**:3924–3933.

Vazquez-Torres, A., Y. Xu, J. Jones-Carson, D. W. Holden, S. M. Lucia, M. C. Dinauer, P. Mastroeni, and F. C. Fang. 2000. *Salmonella* pathogenicity island 2-dependent evasion of the phagocyte NADPH oxidase. *Science* **287**:1655–1658.

Vieira, O. V., R. J. Botelho, and S. Grinstein. 2002. Phagosome maturation: aging gracefully. *Biochem. J.* **366**:689–704.

Walthers, D., R. K. Carroll, W. W. Navarre, S. J. Libby, F. C. Fang, and L. J. Kenney. 2007. The response regulator SsrB activates expression of diverse *Salmonella* pathogenicity island 2 promoters and counters silencing by the nucleoid-associated protein H-NS. *Mol. Microbiol.* **65**:477–493.

Wilson, J. W., C. Coleman, and C. A. Nickerson. 2007. Cloning and transfer of the *Salmonella* pathogenicity island 2 type III secretion system for studies of a range of gram-negative genera. *Appl. Environ. Microbiol.* **73**:5911–5918.

Winter, S. E., P. Thiennimitr, M. G. Winter, B. P. Butler, D. L. Huseby, R. W. Crawford, J. M. Russell, C. L. Bevins, L. G. Adams, R. M. Tsolis, J. R. Roth, and A. J. Baumler. 2010. Gut inflammation provides a respiratory electron acceptor for *Salmonella*. *Nature* **467**:426–429.

Worley, M. J., K. H. Ching, and F. Heffron. 2000. *Salmonella* SsrB activates a global regulon of horizontally acquired genes. *Mol. Microbiol.* **36**:749–761.

Xu, X., and M. Hensel. 2010. Systematic analysis of the SsrAB virulon of *Salmonella enterica*. *Infect. Immun.* **78**:49–58.

Yoon, H., J. E. McDermott, S. Porwollik, M. McClelland, and F. Heffron. 2009. Coordinated regulation of virulence during systemic infection of *Salmonella enterica* serovar Typhimurium. *PLoS Pathog.* **5**:e1000306.

Yu, X. J., K. McGourty, M. Liu, K. E. Unsworth, and D. W. Holden. 2010. pH sensing by intracellular *Salmonella* induces effector translocation. *Science* **328**:1040–1043.

Yu, X. J., J. Ruiz-Albert, K. E. Unsworth, S. Garvis, M. Liu, and D. W. Holden. 2002. SpiC is required for secretion of *Salmonella* pathogenicity island 2 type III secretion system proteins. *Cell. Microbiol.* **4**:531–540.

Zwir, I., D. Shin, A. Kato, K. Nishino, T. Latifi, F. Solomon, J. M. Hare, H. Huang, and E. A. Groisman. 2005. Dissecting the PhoP regulatory network of *Escherichia coli* and *Salmonella enterica*. *Proc. Natl. Acad. Sci. USA* **102**:2862–2867.

Regulation of Bacterial Virulence
Edited by Michael L. Vasil and Andrew J. Darwin
© 2013 ASM Press, Washington, DC   doi:10.1128/9781555818524.ch21

Chapter 21

Francisella tularensis: Regulation of Gene Expression, Intracellular Trafficking, and Subversion of Host Defenses

Nrusingh P. Mohapatra, Shipan Dai, and John S. Gunn

FRANCISELLA TULARENSIS AND TULAREMIA

Francisella tularensis is a gram-negative coccobacillus that is found in diverse environments, including animal, protozoan, and insect hosts. *F. tularensis* is the causative agent of the zoonotic disease tularemia, which can also be transmitted to humans through arthropod bites, consumption of contaminated food or water, or handling of infected animal carcasses (Oyston et al., 2004; Thomas and Schaffner, 2010). Tularemia can have different clinical presentations depending on the route of transmission and the infecting strain, including ulceroglandular, glandular, oropharyngeal, pneumonic, oculoglandular, and typhoidal, with ulceroglandular tularemia the most common type, representing 75% of all cases. Pneumonic tularemia is the most fatal form of the disease, with a mortality rate of 30 to 60% if untreated, and is the most likely form in the event of an act of bioterror (Ellis et al., 2002; Evans et al., 1985).

F. tularensis can be divided into the subspecies *F. tularensis* subsp. *tularensis*, *F. tularensis* subsp. *horlarctica*, and *F. tularensis* subsp. *mediasiatica* (Kugeler et al., 2009; Staples et al., 2006). *F. novicida* is normally considered the fourth subspecies of *F. tularensis*; however, recent genome-wide polymorphism analysis places it as an independent species (Johansson et al., 2010). Every subspecies has its own unique genomic distribution and virulence capabilities. *F. tularensis* subsp. *tularensis*, also called type A, is the most virulent subspecies, with as few as 10 bacteria able to cause disease in humans through the respiratory route (Dienst, 1963). It is exclusively found in North America. *F. tularensis* subsp. *horlarctica* is also called type B. It is more widely distributed in the Northern Hemisphere, less virulent in humans than type A, but highly virulent in mice. A *Francisella* live vaccine strain (LVS) was developed from the type B subspecies by repeated passaging,

and it contains an unknown mutation(s). It is being considered but is currently not licensed for use in the United States. *F. tularensis* subsp. *mediasiatica* is not virulent in humans but is recovered primarily from ticks and animals in prescribed regions of Central Asia. *F. novicida* does not cause significant disease in humans but is virulent in mice. *F. novicida* has been widely used in mouse models of tularemia but is distinct, as described later, from *F. tularensis* in several ways. It is found in North America and has recently been found also in Australia (Whipp et al., 2003). *F. tularensis* has attracted significant attention over the past years because of its high virulence and potential to be used as a bioweapon. In fact, it was used to develop bioweapons by the former Soviet Union and Japan during World War II (Dennis et al., 2001). Because of its high virulence, ease of dissemination, and ability to cause public panic and social disruption, the Centers for Disease Control and Prevention (CDC) has listed *F. tularensis* type A and type B as category A select agents. The virulent strain Schu S4 is the prototypical type A strain used in laboratory research. It demonstrates almost 99% sequence identity with the LVS (type B) strain, yet there are numerous rearrangements and pseudogene content disparities that likely distinguish the virulence phenotypes and geographic and host distributions of type A and B strains.

ENVIRONMENT AND HOST ADAPTATION

I know of no other infection of animals communicable to man that can be acquired from sources so numerous and so diverse. In short, one can but feel that the status of Tularemia, both as a disease in nature and of man, is one of potentiality.

R. R. Parker (Parker, 1934)

Nrusingh P. Mohapatra, Shipan Dai, and John S. Gunn • Department of Microbial Infection and Immunity, Center for Microbial Interface Biology, The Ohio State University, Columbus, OH 43210.

One distinct feature of *F. tularensis* as a zoonotic and human pathogen is its wide distribution in the environment, as *F. tularensis* is found in more than 250 animal species, including mammals, insects, arthropods, and freshwater protozoans (Ellis et al., 2002; Keim et al., 2007). Small mammals, including lagomorphs, voles, squirrels, muskrats, and various rodents, serve as hosts for *F. tularensis*. Arthropods, including ticks, biting flies, fleas, lice, and mosquitoes, play important roles in maintaining *F. tularensis* as a reservoir by transfer between small mammals, but they can also transmit the disease directly to humans (Keim et al., 2007). *Francisella* spp. can survive in water, wet soil, and animal carcasses for extended periods.

To be able to persist in numerous hosts and in such various environments, *F. tularensis* must be able to adapt, and like most microbes, it does so via gene regulation. In fact, the environment plays a very important role in regulating the virulence of *F. tularensis*, especially those environmental cues it encounters during infection. Global transcriptional profiling of *F. tularensis* within infected macrophages has shown a wide range of changes in gene expression (Wehrly et al., 2009). Protein profiles of *F. tularensis* grown in broth versus inside macrophages also revealed differential expression of several virulence proteins (Golovliov et al., 1997). Although the molecular mechanisms of gene regulation are not fully understood, environmental and host adaptation likely play important roles in regulating *F. tularensis* virulence.

ANIMAL MODELS FOR *F. TULARENSIS* INFECTION

Several animal models have been used for the study of *F. tularensis* infection, primarily rodents (including mice, rats, and guinea pigs) as well as rabbits (Cowley and Elkins, 2011). For all the obvious reasons, the majority of studies have been carried out in mouse models. The mouse model has been a reasonable model for in vitro and in vivo studies of *F. tularensis* intracellular life cycles, molecular mechanisms for pathogenesis, and immune responses. However, there are huge discrepancies between mice and humans in terms of the virulence and lethality of *F. tularensis* infection. Humans are most susceptible to type A infection. Type B is less virulent and rarely lethal in humans. *F. novicida* and LVS do not cause disease in healthy humans. In contrast, infection with type A, type B, or *F. novicida* will kill mice essentially with any dose and through any route (Cowley and Elkins, 2011). Nevertheless, *F. novicida* and LVS infections of mice still serve as the most utilized animal model and

have produced valuable data (Conlan et al., 2011). Recent studies using Fischer 344 rats suggested that infection with different *F. tularensis* subspecies in this rat strain yielded sensitivity phenotypes similar to those with infection in humans. Thus, this might serve as a better model for infection, immunological studies, and vaccine development (Ray et al., 2010; Raymond and Conlan, 2009; Wu et al., 2009). Little or no investigation of the molecular mechanism of pathogenesis has been carried out in vertebrate models other than mice.

Nonhuman primates, such as rhesus macaques, grivet monkeys, and African green monkeys (Twenhafel et al., 2009), have also been used in the study of *F. tularensis* pathogenesis and vaccine development (Rick Lyons and Wu, 2007). Most of these studies dated back to the 1960s and 1970s. Aerosol, intranasal, intracutaneous, and oral infections with type A, type B, and LVS were used in these early studies. Respiratory infections led to acute bronchiolitis to bronchopneumonia, with lesions in spleens, livers, and regional lymph nodes (Hall et al., 1973; Schricker et al., 1972). The dose of infection, disease progression, and mortality rates were similar to those with the disease in humans. A more recent study of pneumonic tularemia in African green monkeys infected with Schu S4 also showed many features in common with human tularemia (Twenhafel et al., 2009). Additionally, numerous and widespread necrotizing pyogranulomatous lesions were seen in lung and lymphoid tissues, and bacteria were also frequently found in alveolar and other tissue macrophages, indicating the important role of macrophages in the pathogenesis of tularemia. Based on the studies described thus far, the nonhuman primate model is currently the best in recapitulating the disease in humans; however, this model is not cost-effective, which limits its use in research (Rick Lyons and Wu, 2007).

Because of the importance of arthropods in the persistence of *F. tularensis* and the spread of disease, insect models of *F. tularensis* infection have also been recently investigated. Researchers have used *Drosophila melanogaster*, which is genetically tractable and has been widely used as a model system for almost every aspect of eukaryotic biology, including infectious diseases (Akimana and Kwaik, 2011). All studies of *F. tularensis* infection of adult flies indicate that *Francisella* grows to high numbers and causes lethal infection by extracellular replication and entrance into phagocytic hemocytes (Ahlund et al., 2010; Moule et al., 2010; Vonkavaara et al., 2008). Studies have suggested that the virulence factors required for infection of mammalian hosts are often also required for *F. tularensis* to survive in the fly. The intracellular trafficking and intracellular replication of *F. novicida* and LVS in S2 and SualB *Drosophila* cells are similar to those observed in mouse and human macrophages, including a transient phagosomal maturation process without direct fusion with lysosomes, phagosomal escape within 30 to 60 min postinfection, and replication within host cell cytosol (Santic et al., 2009). *Francisella* pathogenicity island (FPI)-encoded proteins as well as regulatory factors that impact the expression of FPI-associated genes, including IglA, IglC, IglD, PdpA, PdpB, and MglA, are also required for *F. tularensis* replication inside mammalian and arthropod cell lines. This suggests some common mechanisms utilized by this pathogen to infect both mammals and arthropods. A recent genome-wide screening for bacterial factors important for intracellular proliferation and in vivo growth and survival within the fly showed that approximately 400 genes (22% of the bacterial genome) are important for successful infection in flies (Asare et al., 2010). In another screen, among the nearly 250 *F. novicida* mutants that are attenuated in mice, 20% of them are also attenuated in flies (Ahlund et al., 2010), suggesting that while *Francisella* uses some common virulence mechanisms for infection of both mammals and *Drosophila*, there are also notable differences demonstrated by the large proportion of virulence factors active in mammals that are dispensable in insect models.

A recent study also extended *Francisella* infection to a zebra fish model (Vojtech et al., 2009). This is another genetically tractable system aimed at studying *Francisellla* subspecies that infect fish. The innate immune response in zebra fish during acute *Francisella* infection shares many features with the infection in mammals, including upregulation of tumor necrosis factor alpha (TNF-α), interleukin 1β (IL-1β), and gamma interferon (IFN-γ). Pathogenesis of *Francisella* infection in this model is largely unknown and awaits further investigation.

Similarities and differences of the molecular mechanisms of infection and pathogenesis among various hosts imply an important role for the host environment and host adaptation in determining the pathogenicity of *F. tularensis*. In addition, it has come to light that lab-to-lab variability of what should be identical strains, as well as how these strains are grown, can have dramatic implications regarding pathogenesis.

ENVIRONMENTAL STIMULI

The host response can be unique regarding *F. tularensis* infection. This can be largely due to the different in vivo environments this bacterial pathogen encounters during infection (e.g., temperature changes when *F. tularensis* is transmitted from arthropods to mammals). In addition, iron sources, stress, and host intramacrophage (IMQ) components can all lead to altered expression profiles.

Temperature

One distinct environmental change during *F. tularensis* (and many other pathogens) infection of mammals is the in vitro to in vivo temperature shift. A transcriptomic study was carried out recently to compare the genome-wide expression profile of LVS grown in defined media at an environmental temperature of 26°C with that of LVS grown at 37°C (mammalian body temperature) (Horzempa et al., 2008). In this study, around 11% of the genes were differentially regulated, and 40% of the protein-coding genes induced at 37°C have been previously shown to be involved in virulence or intracellular growth of *F. tularensis*. Another recent study monitoring alterations in proteomes of LVS and Schu S4 associated with a temperature shift from 25 to 37°C also found a significant increase in the level of FPI-encoded proteins, including IglC, IglD, and PdpC (Lenco et al., 2009). In addition, *F. novicida* changes its outer surface by modifying lipid A of lipopolysaccharide (LPS) in response to temperature changes from 25 to 37°C (Shaffer et al., 2007).

It is not clear how *F. tularensis* senses and responds to temperature changes, but alternative sigma factors and the induction of heat shock proteins are likely, as they are utilized by many other bacteria. However, in addition to the main sigma factor, σ^{70} (RpoD), *F. tularensis* has only one alternative sigma factor, σ^{32} (RpoH), in contrast to most other bacteria, which have multiple alternative sigma factors (Grall et al., 2009). Transcription profiling under heat stress conditions showed that a number of heat shock proteins and proteins involved in virulence are upregulated (Grall et al., 2009). Mutations in an RpoH-regulated locus (genes *FTL0200* to *FTL0209* in LVS) revealed that it is involved in the response to various stress conditions, including heat shock, oxidative, and pH stresses, and is involved in intracellular replication and survival (Dieppedale et al., 2011). Another possible regulator is the RNA chaperone Hfq, which has recently been shown to be involved in the stress tolerance of *F. novicida*, including heat, oxidative, and pH stresses (Chambers and Bender, 2011).

Iron

Iron is essential for many metabolic processes for both host and pathogen. Sequestration of iron is an important mechanism utilized by mammalian hosts to restrain pathogenic organisms. On the other hand, the ability to compete for iron is critical for microbes because of the limited freely available iron inside or outside mammalian host cells. Thus, iron availability has a great impact on bacteria pathogen virulence and also host immune responses (Skaar, 2010; Wang and Cherayil, 2009). The successful pathogen is able to respond to iron-limiting conditions by upregulating genes involved in both iron acquisition and virulence (Skaar, 2010).

Unlike many other human pathogens, such as *Bacillus anthacis* and *Yersinia pestis*, *F. tularensis* does not seem to have a mechanism for the acquisition of iron through heme uptake systems (Lindgren et al., 2009). *F. tularensis* secretes a polycarboxylate siderophore to acquire ferric iron (Sullivan et al., 2006). Essential genes for siderophore production and iron acquisition are located in an operon designated *fsl* (*F. tularensis* siderophore locus) or *fig* (*F. tularensis* iron genes) that is conserved among different *F. tularensis* strains (Kiss et al., 2008; Sullivan et al., 2006). One of the gene products, FslE, is responsible for siderophore uptake in both *F. novicida* and Schu S4 and was suggested to function as a siderophore receptor (Ramakrishnan et al., 2008; Sen et al., 2010). Recent studies also identified FupA, a homologue of FslE in both Schu S4 and LVS, as a protein involved in iron uptake. Notably, a *fupA* mutant showed a defect in intracellular replication and decreased virulence in mice (Lindgren et al., 2009; Sen et al., 2010), suggesting the importance of iron acquisition in *F. tularensis* pathogenesis.

In bacteria, iron sensing is generally controlled by ferric uptake regulator, Fur, which forms a complex with ferrous iron and a specific DNA sequence upstream of iron-regulated genes, called the Fur box. Fur then represses downstream gene expression (Escolar et al., 1999). In *F. tularensis*, the *fsl*/*fig* operon is located downstream of *fur* and is induced under iron restriction conditions in a Fur-dependent manner (Deng et al., 2006; Sullivan et al., 2006). In addition, a Fur box is also found upstream of *iglC* and *pdpB*, two genes located within the FPI (Deng et al., 2006). A microarray assay found that nearly 80 genes, including FPI genes, are differentially expressed in iron-starved LVS (Deng et al., 2006). Another proteomic study found that the levels of a significant number of proteins, including FPI proteins encoded in the *igl* operon, were changed, with 52 increased and 89 decreased, under iron-restricted conditions (Lenco et al., 2007). In addition, PdpA, another FPI-encoded protein, was also shown to be upregulated under iron-limiting conditions (Schmerk et al., 2009b). Moreover, a recent study of the iron

content in different *F. tularensis* subspecies showed a direct link between iron metabolism and the ability of the bacteria to resist oxidative stress and cause disease (Lindgren et al., 2011). All of the above indicate an important role for iron in regulating *F. tularensis* virulence factors.

OXIDATIVE, pH, AND NUTRIENT STRESSES

Production of reactive oxygen species and reactive nitrogen species is an essential innate immune defense mechanism against invading microorganisms (Fang, 2004). Many pathogens have evolved ways to combat this host defense mechanism. A screen for bacterial proteins responding to heat and hydrogen peroxide has identified several chaperones, including DnaK, GroEL, and GroES, that are involved in oxidative responses of *F. tularensis* (Ericsson et al., 1994). In addition, around 20 proteins, including the previously mentioned chaperones and some virulence factors (encoded by *katG*, *ahpC*, *ahpF*, *gorA*, *grxA*, *sodA*, and *fumC*) were also upregulated upon exposure to oxidative stress (Lenco et al., 2005). Among the very few regulators present in *Francisella* spp., MglA, the transcriptional regulator of *F. tularensis* virulence, has been demonstrated to also play a role in regulating the response to oxidative stress (Guina et al., 2007). Another posttranscriptional regulator, RNA-binding protein Hfq, is involved in the oxidative and pH stress responses, in addition to its role in regulating virulence by negatively regulating the *pdp* operon in the FPI (Chambers and Bender, 2011). Notably, besides sensing oxidative stress and responding by regulating gene expression, *F. tularensis* is able to inhibit the host oxidative burst by directly inactivating the host NADPH oxidase (McCaffrey and Allen, 2006; McCaffrey et al., 2010; Mohapatra et al., 2010; Schulert et al., 2009). This is likely through dephosphorylation of host kinases and NADPH oxidase subunits by bacterial factors, such as the family of acid phosphatases (Mohapatra et al., 2010).

In addition to oxidative stress, during its intracellular life, especially when residing within phagosomes, *F. tularensis* faces changes in the environmental pH. After phagocytosis and before escaping into the host cytosol, *F. tularensis* resides in the phagosome, which goes through a transient maturation and acidification process. FPI genes are induced during this transient phagosomal stage, suggesting that the phagosomal environment is involved in regulating gene expression (Chong et al., 2008). However, studies of the roles of pH in virulence gene induction and bacterial phagosomal escape have been controversial (Chong et al., 2008; Clemens et al., 2009; Santic et al., 2008).

Nevertheless, at least one gene encoding the cytoplasmic membrane protein, RipA, which is involved in *F. tularensis* intracellular replication and host immune evasion, is regulated by pH, with increased expression in neutral rather than acidic environments (Fuller et al., 2009).

Also involved in virulence induction during the intracellular life of *F. tularensis* might be the limited source of nutrients inside host cells and within phagosomes. Some of the *F. tularensis* virulence regulators have also been implicated in the stringent response (bacterial reaction to starvation). For example, guanosine tetraphosphate (ppGpp), a global regulator that inhibits RNA synthesis during the bacterial stringent response, can promote the interaction of FevR with the MglA/SspA/RNA polymerase (RNAP) complex and regulate virulence gene expression (Charity et al., 2009). Its synthase, RelA, is also involved in *F. tularensis* virulence and intracellular survival (Dean et al., 2009). SspA, which interacts with MglA and regulates virulence gene expression, is also a stringent starvation transcriptional regulator (Charity et al., 2007). The involvement of these regulators suggests a critical role for the stringent response in *F. tularensis* virulence induction.

Intracellular Components

In addition to all the stresses an intracellular pathogen encounters when infecting host cells, certain intracellular components might also trigger the changes in bacterial virulence gene expression, such as polyamines. These are small polycationic molecules involved in a wide range of biological processes, including transcription and translation regulation. A recent study has identified spermine, a polyamine produced only in eukaryotic cells, as one such intracellular component sensed by *F. tularensis* (Carlson et al., 2009; Russo et al., 2011). Infection with *F. tularensis* cultured in the presence or absence of spermine induced altered patterns of cytokine production in macrophages. Consistent with this, as shown by a microarray assay, a significant portion of the *F. tularensis* transcriptome was affected in the presence or absence of spermine. This is in part mediated by the induction of *Francisella* insertion sequence (IS) elements ISFtu1 and ISFtu2, which are sufficient to induce the expression of downstream genes in response to spermine (Carlson et al., 2009). A more recent study identified, through screening of a transposon library, a gene in LVS and also its homologue in the prototypical virulent type A strain Schu S4 that were responsible for the spermine responses. Mutations in these two genes resulted in an increased level of cytokine production, and the corresponding mutants were attenuated in vivo (Russo et al., 2011). This observation suggested a novel model of gene regulation in which IMQ compounds elicit substantial changes in *Francisella* gene expression and result in low cytokine production and immune evasion.

FRANCISELLA INTRACELLULAR TRAFFICKING AND REPLICATION

In vivo, *F. tularensis* is able to infect a variety of cells, including alveolar macrophages, monocytes, neutrophils, hepatocytes, and lung epithelial cells. However, it primarily replicates inside macrophages (Hall et al., 2008). Several receptors have been implicated in the phagocytosis of different subspecies of *F. tularensis* by macrophages, including complement receptors, Fcγ receptors, scavenger receptors, and the mannose receptor (Balagopal et al., 2006; Clemens et al., 2005; Pierini, 2006; Schulert and Allen, 2006). *F. tularensis* is engulfed by macrophages by inducing unique asymmetric spacious pseudopod loops (Clemens et al., 2005).

Once inside macrophages, *F. tularensis* is able to avoid normal bactericidal processes. It can inhibit host oxidative burst, in part by dephosphorylation of host kinases and NADPH oxidase subunits (McCaffrey and Allen, 2006; McCaffrey et al., 2010; Mohapatra et al., 2010; Schulert et al., 2009). After entering macrophages, *F. tularensis* resides in the phagosome (or FCP, for *Francisella*-containing phagosomes). The FCP will go through a transient maturation process by acquiring early endosomal and late endosomal markers, such as EEA1, LAMP1, LAMP2, and CD63. However, it avoids direct fusion with lysosomes and does not acquire hydrolases, such as cathepsin D (Checroun et al., 2006; Chong et al., 2008; Clemens et al., 2004; Santic et al., 2010; Santic et al., 2005a). The FCP also acidifies during this transient maturation process. As mentioned, it remains controversial regarding the extent to which the FCP acidifies and whether it plays any role in *F. tularensis* phagosomal escape. In addition to altered phagosomal trafficking, as early as 30 to 60 min postinfection, *F. tularensis* is able to escape the phagosome and replicate within the host cytosol (Checroun et al., 2006; Chong et al., 2008; Clemens et al., 2004; Santic et al., 2010; Santic et al., 2005a). The time of escape varies in the literature due to differences in bacterial strains, host cells, and infection conditions. Phagosomal escape requires viable bacteria and occurs via an unknown mechanism. Mutants that are defective in phagosomal escape are also impaired in intracellular replication, suggesting that this is a critical process for *F. tularensis* pathogenesis (Meibom and Charbit, 2010).

After a short delay following phagosomal escape, *F. tularensis* starts replicating inside the host cell cytosol to high numbers. This intracellular replication can be detected by host cells through inflammasomes, which activate caspase-1, induce IL-1β and IL-18 processing, and also lead to host cell apoptosis or pyroptosis (Fernandes-Alnemri et al., 2010; Gavrilin and Wewers, 2011; Jones et al., 2010; Monack, 2008). *F. tularensis* can then be released to initiate another cycle of infection. At this late stage of infection, prior to host cell death, cytosolic *F. tularensis* was also found to reenter the endocytic system by inducing autophagy in mouse macrophages (Checroun et al., 2006). The exact molecular mechanism and the significance of this process for either *Francisella* infection or host immune responses are not clear; however, autophagy-inducing compounds have been shown to kill *Francisella* within host macrophages (Chiu et al., 2009).

FRANCISELLA MODULATION OF HOST IMMUNE RESPONSES

Host proinflammatory responses, including IFN-γ, TNF-α, and IL-1β production, are critical for controlling *F. tularensis* infection (Mariathasan et al., 2005; Metzger et al., 2007). However, *F. tularensis* has developed ways to avoid host killing and evade or subvert host immune responses. Although they are not fully understood yet, *F. tularensis* achieves this through several mechanisms.

F. tularensis resists complement-mediated killing and can survive extracellularly in whole blood in vitro and in vivo in the mouse model (Forestal et al., 2007; Yu et al., 2008). Complement deposition was extensively studied, and complement resistance was found to depend on LPS O antigen (Clay et al., 2008). Complement factor C3, which is critical for membrane attack complex formation and complement-mediated lysis, is deposited on the surface of all *F. tularensis* strains. However, it is quickly converted to C3b and then quickly inactivated to C3bi, in an O-antigen-dependent way (Ben Nasr and Klimpel, 2008; Clay et al., 2008). The deposition of C3bi on the bacterial surface also serves as the ligand for CR3, which is critical for the phagocytosis of *F. tularensis* by human macrophages (Clemens et al., 2005).

In addition to its role in complement resistance, *F. tularensis* LPS has another unique feature: compared to LPSs of other gram-negative bacteria, it is underacylated (tetra) and only monophosphorylated. Because of this, it does not interact, or only very poorly interacts, with Toll-like receptor 4 (TLR4) and is unable to induce any immune response in all cell types

tested (Gunn and Ernst, 2007; Hajjar et al., 2006). Instead, studies so far have suggested that most host immune responses are driven through TLR2. MyD88 and TIRAP are both required for full TLR2-mediated signaling (Cole et al., 2010). TLR2 is required for murine dendritic cells to activate NF-κB, to produce TNF-α (Katz et al., 2006), and for the production of proinflammatory cytokines in LVS-infected mouse macrophages, including TNF-α, IL-6, IL-1β, IFN-γ, IFN-β, and IL-12 (Abplanalp et al., 2009; Cole et al., 2007). It can also be activated by *F. tularensis* from within phagosomes. Consistently, *F. tularensis* mutants that cannot escape to host cytosol also produce a higher level of proinflammatory cytokines (Cole et al., 2008). As with other intracellular pathogens, this initial proinflammatory cytokine production is critical for controlling *F. tularensis* infection.

In vivo studies in the mouse model suggested that upon *F. tularensis* infection, there was a lack of immune responses during the first two days of infection, followed with overwhelming upregulation in proinflammatory cytokine productions by day 3, which was detrimental to the host (Andersson et al., 2006; Bosio et al., 2007; Conlan et al., 2008). Consistent with the in vivo observations, in vitro studies also suggested an inhibition of proinflammatory responses upon *F. tularensis* infection (Telepnev et al., 2003; Telepnev et al., 2005) with the highly virulent type A but not with the nonpathogenic *F. novicida* (Butchar et al., 2008; Cremer et al., 2009). In addition, infected host cells are no longer responsive to secondary stimuli (Telepnev et al., 2003). Several TLRs are also downregulated following *F. tularensis* infection in monocytes (Butchar et al., 2008; Telepnev et al., 2003). The ability of *F. tularensis* to down-modulate the host immune response is critical for the pathogenesis of tularemia. The underlying molecular mechanism is not clear. A recent study suggested that RipA, an *F. tularensis* cytosolic protein, is involved in inhibiting mitogen-activated protein kinase (MAPK) and inflammasome activation (Huang et al., 2010). However, this cannot explain the difference between the abilities of *F. novicida* and Schu S4 to suppress host responses. RipA is conserved in all subspecies (Fuller et al., 2008), including the avirulent *F. novicida*. Another study demonstrated that the difference in the immune suppression following *F. novicida* and Schu S4 infection might be due to their ability to differentially target the PI3K/Akt eukaryotic pathway through regulation of the inositol phosphatase SHIP-1, an important negative regulator, by differentially regulating miR-155 (Cremer et al., 2009). Several studies indicated that phosphatidylinositol 3-kinase (PI3K)/Akt pathway plays a central role in host responses following *F. tularensis* infection (Cremer et al., 2011). The activation of the PI3K/Akt pathway during *F. novicida* infection is critical for proinflammatory cytokine production, whereas in LVS infection, it leads to immune suppression through an increased level of MKP-1 (MAPK phosphatase-1), which inhibits MAPK activation (Medina et al., 2010). In Schu S4 infection, its activation is avoided by activating SHIP-1 (Cremer et al., 2009). It is not clear how the different subspecies can induce these different responses. However, bacterial surface structures, which will determine the initial interaction with macrophages, and putative secreted

bacterial virulence proteins, such as those encoded in FPI, and their regulators might play important roles by interacting with or hijacking host signaling. See Fig. 1 for a depiction of the intracellular *Francisella*-mediated events.

FRANCISELLA PATHOGENICITY ISLAND

Pathogenicity correlates with the expression of virulence factors. The genomes of bacterial pathogens are composed of conserved essential/critical genes

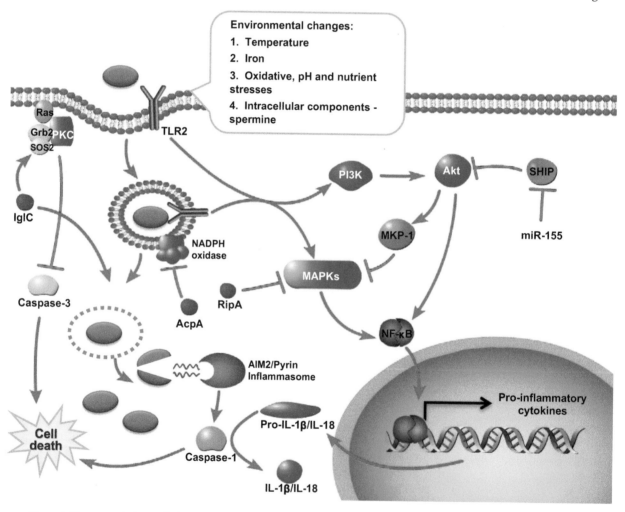

Figure 1. Phagocytosis, intracellular trafficking, and host immune responses during *F. tularensis* infection of host macrophages. *F. tularensis* enters host cells by utilizing different receptors. It resides in FCPs transiently after phagocytosis and escapes into the host cell cytosol shortly after infection. In the host cytosol, *F. tularensis* replicates to high numbers before inducing pyroptosis and apoptosis. FPI-encoded IglC, an important player in *F. tularensis* phagosomal escape and intracellular replication, is able to activate Ras and inhibit caspase-3 activation. The host cell detects cytosolic *F. tularensis* by AIM2 or pyrin inflammasomes and leads to IL-1β and IL-18 maturation and secretion. *F. tularensis* is recognized by the TLR2 signaling pathway at the cell surface and also while within phagosomes, followed by activation of MAPK and NF-κB pathways and proinflammatory cytokine production. *F. tularensis* is unique in its ability to suppress host immune responses by inhibiting MAPK activation through either a bacterial factor, RipA, host MAPK phosphatase (MKP-1), or other unknown mechanisms. Specific to low-virulence *F. novicida*, but not the highly virulent Schu S4, is the increased level of miR-155, which downregulates SHIP-1, a negative regulator of the PI3K/Akt pathway, and results in high levels of proinflammatory responses. In addition, AcpA can also dephosphorylate NADPH oxidase components and contribute to respiratory burst suppression following *F. tularensis* infection. doi:10.1128/9781555818524.ch21f1

and flexible additional pools of genes that encode supplementary traits that are beneficial for bacteria under certain environmental circumstances. These genes confer resistance to antibiotics and toxic compounds and produce virulence factors. The flexible gene pool represents variable chromosomal regions and includes mobile and accessory genetic elements, such as plasmids, pathogenicity islands (PIs), IS elements, transposons, and integrons. PIs are generally characterized by distinct G+C content compared to the overall content of the genome in which they are located (suggesting horizontal transmission) and are frequently flanked by tRNA genes or direct repeat sequences. Virulence genes that are located on PIs can be divided into several groups according to their associated phenotypes, such as expression of adherence factors, siderophores, capsules, endotoxin of gram-negative organisms, exotoxins, secretion systems, and endotoxin (LPS).

The 33-kbp FPI was first identified in 2004 by Nano (Nano et al., 2004). This FPI fulfills the required characteristics of PIs and encodes genes necessary for IMQ growth and survival, as well as a putative type VI secretion system (Barker et al., 2009; Ludu et al., 2008). Two identical copies of the FPI are found in the virulent *F. tularensis* subsp. *tularensis* and *F. tularensis* subsp. *holarctica*, while *F. tularensis* subsp. *novicida* has a single FPI that shares 97% nucleotide identity with that found in *F. tularensis* subsp. *tularensis* (Nano and Schmerk, 2007). The FPI of *F. tularensis* subsp. *holarctica* lacks the *pdpD* gene present in *F. tularensis* subsp. *tularensis* and *F. tularensis* subsp. *novicida* (Barker et al., 2009; Nano and Schmerk, 2007). The *pdpD* gene product is an outer membrane protein suggested to be required for full virulence (Ludu et al., 2008). It is not clear if these differences in FPI copy number and gene content explain some of the differences in virulence between the different subspecies. The FPI consists primarily of two large convergently transcribed operons, the *pdpD* and *iglABCD* operon and the *pdpA* operon, which contains 11 additional genes (Barker and Klose, 2007; Nano and Schmerk, 2007).

The pathogenesis of tularemia is dependent upon modulation of the host immune system by *F. tularensis* (Broms et al., 2010). Several of the immune alteration mechanisms are directly or indirectly linked to FPI proteins. Every gene within the FPI has been inactivated in *F. novicida*, and all are required for IMQ survival and virulence in mice except *pdpD* (Broms et al., 2010). The inactivation of *pdpD* has resulted in different findings by various groups. These findings might be due to polarity of the *pdpD* mutation or the origin of the strain (Ludu et al., 2008). *iglC* was the first FPI gene identified to be upregulated

during IMQ infection (Golovliov et al., 2003; Lai et al., 2004; Lauriano et al., 2003; Lauriano et al., 2004). Its protein product is essential for IMQ survival and virulence in mice, amoebae (*Acanthamoeba* and *Hartmannella*), fish (*Tilapia*), and mosquitoes in *F. tularensis* subsp. *novicida*, *F. tularensis* subsp. *holarctica*, and *F. tularensis* subsp. *tularensis* (Carlson et al., 2007; Lai et al., 2004; Lauriano et al., 2004; Read et al., 2008; Santic et al., 2010; Santic et al., 2008; Santic et al., 2007; Santic et al., 2011; Schulert et al., 2009; Soto et al., 2009; Soto et al., 2011). IglC is important for *Francisella* phagosome escape and avoidance of phagosome-lysosome fusion in macrophages (Lai et al., 2004; Lauriano et al., 2004; Santic et al., 2007; Santic et al., 2009; Soto et al., 2011). The cytotoxic effect of *Francisella* infection of macrophages is dependent on IglC, and it modulates the host immune response by downregulating TLR signaling (Lindgren et al., 2004; Santic et al., 2005b; Soto et al., 2011). Macrophages infected with an *iglC* mutant showed increased expression of IL-1β and decreased expression of IFN-γ, inducible nitric oxide synthase, and IP10, and these cytokines are required for the macrophage proinflammatory response (Carlson et al., 2007; Lindgren et al., 2004; Santic et al., 2005b). Infection of monocytes with *F. tularensis* subsp. *novicida* triggers *iglC*-dependent temporal activation of Ras through recruitment of protein kinase Cα (PKCα) and PKCβI to the SOS2/GrB2 complex. However, the silencing of SOS2, GrB2, and PKCα and PKCβI by RNA interference in monocytes has no effect on evasion of lysomosal fusion and bacterial escape into the cytosol, but it prevented replication of bacteria in the cytosol (Al-Khodor and Abu Kwaik, 2010).

A partial deletion of *iglA* in *F. novicida* resulted in avirulence in a chicken embryo model and an inability to grow in macrophages (Nix et al., 2006). As previously mentioned, *F. tularensis* LVS grown under iron limitation conditions revealed that more than 140 proteins were differentially expressed. For example, IglA, IglB, and IglC protein production increased significantly under iron-limited growth compared to when *F. tularensis* LVS was grown under iron-replete conditions (Lenco et al., 2007). IglA is localized to the cytoplasm of *Francisella*, and its expression and stability are dependent on the presence of IglB (de Bruin et al., 2007). Similarly, deletion of *iglB* and *iglD* in *F. novicida* and LVS induces protective immunity against homotypic and heterotypic challenge in the mouse model, and both of these mutants are unable to grow in murine macrophages (Cong et al., 2009; Santic et al., 2008; Santic et al., 2007). However, deletion of IglB and IglD in the Schu S4 strain showed marked attenuation but provided marginal protection against homologous challenge (Kadzhaev et al.,

2009). IglB was one of the three proteins that induced cellular activity during a *Francisella*-specific T-cell epitope screening (Valentino et al., 2011).

The nomenclature of the *pdpA* operon is currently being adjusted according to increasing knowledge of the function of the individual operon proteins (Broms et al., 2010). This operon encodes proteins with homology to rhoptry proteins in protists (e.g., *Plasmodium* and *Toxoplasma gondii*), which are involved in secretion and invasion of host cells (Kats et al., 2006). As mentioned earlier, *pdpD* is not essential for IMQ growth, while *pdpB* shares homology with *icmF* paralogs potentially involved in secretion (Barker et al., 2009; Deng et al., 2006; Read et al., 2008). However, *pdpA* and *pdpC* have been shown to be important for IMQ growth and virulence in mice (Read et al., 2008; Schmerk, 2009a; Schmerk, 2009b). Four genes in this operon are core components of type VI secretory systems: *pigB*, *pigC*, *pigF*, and *pigI* (recently changed to *vgrG*, *clpV*, *dotU*, and *hcp*, respectively) (Barker et al., 2009; Broms et al., 2010). The remaining *pig* genes, i.e., *pigA*, *pigD*, *pigE*, *pigG*, and *pigH*, have been renamed *iglE*, *iglG*, *iglH*, *iglI*, and *iglJ*, respectively, suggesting that these genes are genetically similar to *iglABCD*, which are required for intracellular growth (Barker et al., 2009; Barker and Klose, 2007; Bröms et al., 2011; Broms et al., 2010; Ludu et al., 2008; Nano and Schmerk, 2007).

TRANSCRIPTIONAL REGULATORY FACTORS

Transcription is an essential step in gene expression, and its understanding has been one of the major interests in molecular and cellular biology. Transcriptional regulation typically emerges from the interaction between *trans* factors that bind to *cis*-regulatory elements in the context of a particular chromatin/chromosome structure (e.g., promoter) (Browning and Busby, 2004). In prokaryotes, these *cis* regions occupy up to ~400 bp (Collado-Vides et al., 1991). Transcription initiation in bacteria requires sigma factors that are essential for proper promoter recognition by RNAP (Helmann, 2002; Kazmierczak et al., 2005; Maeda et al., 2000a; Maeda et al., 2000b; Paget and Helmann, 2003). In bacteria, sigma factors are divided into two main families: σ^{70} and σ^{54}. The σ^{70} family includes the housekeeping genes. *Escherichia coli* and *Bacillus subtilis* have six to eight alternative sigma factors that are used under different environmental circumstances; however, *Francisella*, as mentioned above, has only two sigma factors (σ^{70} and σ^{32} [stress/temperature response]). *Francisella* RNAP is unique, as it contains two distinct alpha

subunits that are not identical (Charity et al., 2007; Charity et al., 2009). Thus, as discussed below, multiple distinct RNAPs could be made (combinations of the alpha subunits), and because alpha subunits are frequently interaction points for *trans*-acting factors, it is unclear if these may have distinct roles in transcription or regulation.

Transcription factors are classified into several families based on at least two domains that allow them to function as regulatory switches (Ruger, 1972). One domain functions as a signal sensor by ligand binding or protein-protein interaction (Martinez-Antonio et al., 2006; Ptashne, 1984). The other domain is the responsive element that directly interacts with a target DNA sequence. In bacteria, the helix-turn-helix domain is the most common (Madan Babu et al., 2006; Seshasayee et al., 2006). Also, in bacteria, most of these domains are present in one single protein, except for two-component systems (TCS) (Ulrich et al., 2005). Classically, in TCS, the sensor protein senses an exogenous condition; it then autophosphorylates and transfers this phosphate to a cytoplasmic partner, which has a transcriptional regulatory activity (Mascher, 2006). These TCS work as a unit: in *Escherichia coli*, 26 of the 29 pairs are encoded in the same operon (Janga et al., 2007). Negative regulators bind to the promoter, interfering directly with RNAP, whereas positive regulators bind to the upstream region of the promoter, helping to recruit the polymerase and initiate transcription (Collado-Vides et al., 1991; Madan Babu and Teichmann, 2003). Transcription factors usually work as homodimers, tetramers, hexamers, and even, in a few cases, heterodimers (Goulian, 2004).

MglA

There are four known major regulators of virulence in *F. tularensis*: MglA, SspA, PmrA, and FevR, which positively control the expression of FPI genes (Baron and Nano, 1998; Brotcke and Monack, 2008; Charity et al., 2007; Mohapatra et al., 2007). Each of these proteins also regulates the expression of genes outside the FPI. Virulence gene regulation in *F. tularensis* was first identified during a screen for acid phosphatase mutants; a strain was isolated that carried a spontaneous point mutation that resulted in reduced cleavage of the chromogenic phosphatase substrate 5-bromo-4-chloro-3-indolylphosphate (XP) (Baron and Nano, 1998). This mutant was defective in growth and survival in several types of macrophages compared to the parental strain. Subsequent complementation studies and homology analysis led to the identification of the macrophage growth loci A and B (*mglAB*) as the site of mutation in strain GB2 (Baron and Nano, 1998). The amino acid sequence of MglA was 20% identical to that of stringent starvation protein, SspA, of *Escherichia coli*, and that of MglB was 30% identical to that of *E. coli* SspB. Transposon mutagenesis of these genes appeared to result in strains defective for growth in macrophages, having growth kinetics similar to those of the initial spontaneous mutant (Baron and Nano, 1998).

The cross talk between MglA and FPI was discovered by Francis Nano and Karl Klose in 2004 (Lauriano et al., 2004). Real-time reverse transcriptase PCR experiments demonstrated that loss of MglA affects the expression of several FPI genes, and microarray analysis implicated MglA in regulation of more than 100 genes throughout the *Francisella* genome (Lauriano et al., 2004). Not only does mutation of *mglA* have pleiotropic effects, but it also causes a reduced in vitro growth rate, inability to replicate within host cells, and a general sensitivity to cellular stresses (Brotcke et al., 2006; Santic et al., 2006; Santic et al., 2005b). Similarly, two-dimensional gel analysis by Lauriano et al. demonstrated that MglA positively regulates the expression of seven different proteins of the FPI in *F. tularensis* subsp. *novicida*, including IglC (Lauriano et al., 2004). IglC was one of the four most prominently induced proteins after *F. tularensis* entry into macrophages (Lauriano et al., 2004). Microarray-based transcriptional analysis of the genes surrounding *iglC* suggested that *pdpD* and *iglABCD* might be arranged in a single MglA-dependent transcriptional unit (Brotcke et al., 2006). Furthermore, *mglA* and some of the genes in the FPI are required for virulence in flies and for replication in mosquitoes and in amoebae (Lauriano et al., 2004; Santic et al., 2009). An *F. novicida* Δ*mglA* mutant is attenuated when delivered via the aerosol route but does not provide protection against homologous or heterologous challenge (West et al., 2008). The *mglA* mutant does replicate within the lungs, and control of an infection with this mutant involves MyD88 and IFN-γ (West et al., 2008). In addition to the FPI, other virulence factors such as the metalloprotease PepO are regulated by MglA (Guina et al., 2007).

PepO

PepO is a homologue for a proendothelin, which is a eukaryotic precursor for a potent vasoconstrictor, endothelin. The increased secretion of PepO results in increased vasoconstriction at the site of infection that may limit the spread of the bacteria (Guina et al., 2007; Hager et al., 2006). Mutations in some of the proteins comprising the type IV secretion

machinery likely responsible for secretion of PepO in the more virulent *F. tularensis* subspecies result in lower levels of PepO secretion and increased dissemination compared to those of *F. novicida* that does not have such mutations (Hager et al., 2006; Oyston, 2008).

SspA

As mentioned earlier, MglA is 20% identical to SspA from other gram-negative bacteria, including *E. coli*. MglA and SspA regulate similar sets of genes, including positive regulation of the FPI-associated genes (Charity et al., 2007). MglA and SspA interact with one another, and both can bind *Francisella* RNAP. Research suggests that MglA and SspA cooperate to regulate genes important for *Francisella* virulence. SspA is maximally expressed during stationary phase, while MglA is maximally expressed during the lag and exponential phases of growth (Brotcke et al., 2006).

RNA POLYMERASE

Strikingly, the study of Charity et al. also revealed the existence of two alpha (designated α1 and α2) subunits in the *F. tularensis* RNAP, encoded by two *rpoA* alleles and sharing only 32% amino acid identity (Charity et al., 2007). Since the alpha subunit participates in the recognition of the promoter region, the nature of the functional RNAP protein complex in *F. tularensis* appears to be particularly complex. Indeed, the combination of two SspA variants and two alpha subunit variants creates up to 16 possible permutations, and the combination (MglA-SspA) + (α1, α2, β, β', and ω) + σ[70] might be only one of several that function in *F. tularensis* gene regulation (Charity et al., 2007).

FevR/PigR

Brotcke and Monack devised a genetic screen in *F. tularensis* subsp. *novicida* to identify additional genes involved in the expression of the MglA-SspA regulon (Brotcke et al., 2006). The authors used the MglA-regulated promoter of *pepO* fused to the *lacZ* gene as a reporter system to screen for transposon mutants unable to induce *lacZ* transcription. This procedure led to the identification of five genes, including *FTN_0480* (designated *Francisella* effector of virulence regulation [FevR]). This gene is essential for IMQ replication and virulence in the mouse model

(Brotcke and Monack, 2008). The fact that FevR itself does not positively regulate the expression of MglA-SspA, while MglA-SspA positively regulates the expression of FevR, suggests that FevR might act downstream of MglA-SspA in the regulatory cascade and could act independently of MglA-SspA. To test this hypothesis, the authors devised an elegant procedure where they expressed *fevR* from a constitutive MglA-SspA-independent promoter in different wild-type and mutant (*mglA* or *fevR*) backgrounds (Brotcke and Monack, 2008). The *mglA* mutant strain expressing FevR constitutively failed to induce the entire MglA regulon, demonstrating that FevR is not sufficient to induce the MglA regulon. Finally, coimmunoprecipitation experiments determined that FevR physically binds to MglA-SspA and acts in parallel to regulate gene expression. Hence, it is likely that upon activation of FevR expression by MglA-SspA, the three proteins work in concert to initiate transcription (Brotcke and Monack, 2008; Charity et al., 2007). As pointed out by Brotcke and Monack, this type of regulation, where one activator induces the expression of a second activator and they act together as coactivators of downstream genes, corresponds to a "feed-forward loop" model, a theoretical regulatory network initially proposed by Mangan and Alon (Mangan and Alon, 2003). Feed-forward loops are expected to accelerate the time of response to persistent stimuli. In the case of MglA-SspA and FevR, the exact nature of the stimuli remains to be discovered. The fact that FevR is weakly homologous to transcription factors suggests that it is a DNA-binding protein; however, FevR binding to a FevR box remains to be experimentally demonstrated.

TWO-COMPONENT REGULATORY SYSTEMS

TCS are widely employed by gram-negative bacteria to monitor and respond to environmental signals. Virulent strains of *F. tularensis* are devoid of classical, tandemly arranged TCS, but orphaned members exist in the genome, such as the response regulator PmrA and sensor histidine kinases KdpD and QseC. These two kinases were identified in an in vivo negative screening and hence suspected to be involved in virulence (Weiss et al., 2007). The deletion of *pmrA* in *F. novicida* and LVS results in a strain deficient for IMQ growth and avirulent in the mouse model (Mohapatra et al., 2007). PmrA regulates the expression of the genes within the FPI and *fevR*. The phosphorylation of PmrA aids binding its own promoter and the FPI-borne *pdpD*, suggesting that phosphorylation was required for activity despite the lack of an associated sensor kinase. KdpD

was subsequently found to be the histidine kinase primarily responsible for phosphorylation of PmrA, targeting the aspartic acid at position 51 (D51) (Bell et al., 2010). In a strain expressing PmrA D51A, this mutant regulator still retains some DNA binding to its target genes, but there is reduced expression of the genes in the PmrA regulon. Moreover, the PmrA D51A mutant strain is deficient for IMQ replication and is attenuated in the mouse model. PmrA coprecipitates with the FPI transcription factors MglA and SspA, which, as discussed above, bind RNAP (Bell et al., 2010). Together, these data suggest a model of *Francisella* gene regulation that includes a TCS consisting of KdpD and PmrA. Once phosphorylated, PmrA binds to regulated gene promoters, thereby recruiting free or RNA P-bound MglA and SspA to initiate FPI gene transcription (Bell et al., 2010).

In another study, the screening of a library of organic molecules led to the identification of a specific inhibitor (LED209) of QseC, which, upon binding to QseC, prevented its autophosphorylation and its capacity to transduce a signal (Rasko et al., 2008). In particular, treatment with LED209 reduced the efficiency of intracellular survival and virulence of the Schu S4 strain of *F. tularensis* subsp. *tularensis*. Quantitative PCR assays revealed that *qseC* expression was upregulated during infection of mice with *F. tularensis* Schu S4 and that the inhibitor LED209 decreased the expression of *qseC* itself, as well as *iglC* and *pdpA* (Rasko et al., 2008). This suggests that QseC may be directly or indirectly involved in FPI gene regulation. *Francisella* QseC also contributes to the modification of LPS structure (Mokrievich et al., 2010). Deletion of quorum sensing gene *qseC* in *Francisella tularensis* 15 by site-directed mutagenesis resulted in a low growth rate on solid nutrient medium and abolished resistance to rabbit serum and survival in macrophages. In addition, the strain lost virulence and a significant phenotypic alteration was observed in the LPS. That is, the mutant strain was unable to synthesize significant amounts of the LPS with the high-molecular-mass O polysaccharide, presumably due to the loss of the ability to incorporate part of the repeating O polysaccharide (4-amino-4,6-dideoxy-D-glucose) (Mokrievich et al., 2010).

MigR/CiaC

A number of genetic screens have been set up to identify loci critical to the intracellular survival and multiplication of *F. tularensis*. Buchan et al. (Buchan et al., 2009) constructed an *F. tularensis* LVS derivative containing a chromosomal copy of the *lacZ* reporter gene under the control of the *iglB* promoter.

Screening of a library of transposon-insertion mutants identified one transposon-insertion mutant that demonstrated reduced expression of genes in the *igl* operon. Loss of the gene identified, *FTL_1542* (designated *migR/ciaC*), also led to a drastic reduction in *fevR* expression. Altogether, these data suggest that (i) the regulatory effect of MigR on the *igl* operon is likely to be indirect, (ii) it is mediated by the downregulation of FevR, and (iii) MigR is a novel major regulator of FevR expression (Buchan et al., 2009). MigR may be involved in the production of the alarmone ppGpp, thereby stabilizing the MglA-SspA-FevR-RNAP complex, which activates *fevR* transcription.

Hfq

Regulation of gene expression by noncoding RNAs or small RNAs (sRNAs) is prevalent in both prokaryotes and eukaryotes. Most bacterial sRNAs frequently require the Hfq chaperones for function, and they most commonly act through base-pairing with their mRNA targets, which are usually encoded in *trans* (Chao and Vogel, 2010). Hfq is a bacterial RNA-binding protein that was initially recognized as a microbial factor for the replication of the QB RNA phage in *E. coli* (Chao and Vogel, 2010). Hfq is a posttranscriptional regulator that binds sRNAs and mRNA and facilitates RNA-RNA interaction. The sRNA-mRNA-Hfq interactions frequently result in mRNA degradation (e.g., via RNase E) and/or inhibition of translation, but in some cases, they can also result in increased translation. Deletion of *hfq* has been correlated with considerable changes in transcript levels or protein levels, and *hfq* mutants generally exhibit pleiotropic phenotypes. Recently, the role of Hfq in pathogenesis has been examined in several bacterial species (Bearson et al., 2008; Bohn et al., 2007; Chambers and Bender, 2011; Christiansen et al., 2004; Christiansen et al., 2006; Crabbe et al., 2011; Havlasova et al., 2005; Jones et al., 2008; Kadzhaev et al., 2009; Koo et al., 2011; Liu et al., 2011; Meibom et al., 2009; Pannekoek et al., 2009; Robertson and Roop, 1999; Schiano et al., 2010; Sittka et al., 2007; Viegas et al., 2007). *hfq* mutants are often severely attenuated for intracellular replication in vitro and for virulence in mice.

The mutation of *hfq* in *F. tularensis* LVS, *F. novicida*, and the human pathogenic clinical isolate FSC200 contributes to stress resistance, IMQ survival, and the ability to cause disease in mice (Chambers and Bender, 2011; Kadzhaev et al., 2009; Meibom et al., 2009). Hfq directly or indirectly affects the expression of more than 100 genes in LVS

and *F. novicida*. Interestingly, as mentioned above, Hfq negatively regulates a subset of FPI genes (one of the two operons), including *pdpA* and *pdpB*.

SMALL RNA

A recent study by Postic et al. identified 24 sRNAs in the LVS genome by in silico prediction (Postic et al., 2010). However, they identified only two sRNAs by experimental assays (FtrA and FtrB). Deletion of these sRNAs did not have any virulence defect in macrophages or in the mouse model. Given the dearth of transcriptional factors and TCS in *Francisella* spp., future research on sRNAs in combination with Hfq may find an abundance of non-protein-based virulence gene regulation. See Fig. 2 for a depiction of FPI regulation mediated by the factors discussed above.

OTHER VIRULENCE FACTORS

Acid phosphatases

Acid phosphatases are ubiquitous in nature, hydrolyze phosphomonoesters at acidic pH, and have been associated with pathogen survival inside the phagosome through the inhibition of respiratory burst (Baca et al., 1993; Carlin et al., 2009; Sato et al., 1988). In our previous studies, we observed that the combined deletion of AcpA with AcpB, AcpC, and Hap in *F. novicida* resulted in an attenuated strain that was 100% protective against homologous challenge in the mouse model (Mohapatra et al., 2008). Additionally, this mutant did not escape from the macrophage phagosome, had negligible phosphatase activity, and failed to suppress the oxidative burst in human phagocytes. Transcriptional analysis demonstrated that expression of *acpA* and *hap* increased at the initial stage of macrophage infection. However, a study performed by Child et al. (Child et al., 2010) illustrated that deletion of AcpA, AcpB, and AcpC (ΔABC) in *F. tularensis* SchuS4 did not affect virulence in mouse and human macrophages. Similar findings were obtained with the ΔABC *F. novicida* mutant as reported by Mohapatra et al. (Mohapatra et al., 2008), but the additional deletion of Hap in *F. novicida* (ΔABCH) resulted in the attenuated phenotypes described above. The *hap* gene was not deleted in the ΔABC *F. tularensis* SchuS4 mutant because it already contains a C-terminal disruption by an insertion element in this strain. However, it is unclear if the truncated Hap protein is active or inactive.

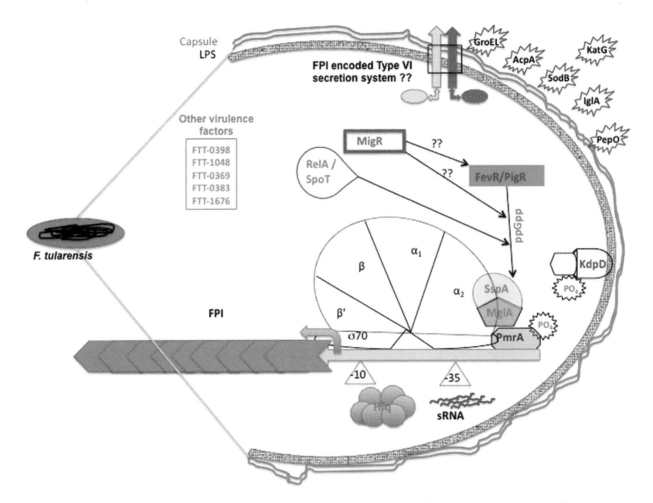

Figure 2. *Francisella* virulence factors and regulation of virulence genes. The *Francisella* TCS sensor kinase KdpD is phosphorylated due to unknown environmental signals and phosphorylates the response regulator PmrA, allowing PmrA to bind to the promoter regions of different target genes. PmrA binding recruits the MglA-SspA complex or MglA-SspA-RNAP, helping to initiate transcription. In the presence of the alarmone ppGpp, FevR binds to MglA-SspA to form a complex and FevR expression requires PmrA, MglA, SspA, and MigR. However, PmrA, MglA, SspA, and FevR regulate many additional genes outside of the FPI. Hfq regulates protein expression under various external conditions (e.g., peroxide, temperature, salt, and detergent) and inhibits regulation of the *pdpA-pdpE* operon of the FPI. More studies are required in the future to better understand the role of Hfq in aiding *Francisella* sRNAs to control virulence gene regulation. *Francisella* spp. secrete several proteins (e.g., AcpA, GroEL, KatG, PepO, SodB, and IglA, plus 26 more hypothetical proteins) in culture supernatants. The *Francisella* capsule and LPS have proved to be only mildly antigenic and to induce proinflammatory cytokines.
doi:10.1128/9781555818524.ch21f2

Unknown Virulence Factors

Whole-genome genetic screens have been used to identify *Francisella* virulence factors. An in vivo negative selection technique identified some new and reaffirmed many of the known *F. tularensis* virulence factors (Weiss et al., 2007). In this study, a *F. novicida* transposon library was constructed, pooled, and used to infect mice. Using microarray analysis, Weiss et al. identified 27 known virulence-associated genes and 44 previously unidentified genes of unknown function. Of these, null mutants of FTT0398 and FTT01048 were shown to have a significant virulence

defect yet retained their ability to replicate within macrophages in vitro (Weiss et al., 2007).

A global transcriptional profiling study was performed by Wehrly et al. (Wehrly et al., 2009) during the *F. tularensis* intracellular life cycle within primary murine macrophages. The goal of this study was to characterize the intracellular biology of the bacterium and to identify pathogenic determinants based on their intracellular expression profiles. Differential gene expression profiles were correlated with the intracellular niche of the bacteria. Expression of the FPI genes increased during the intracellular cycle. Twenty-seven putative hypothetical, secreted, outer membrane proteins or transcriptional regulators were identified. Among these, deletion of FTT0383 (FevR), FTT0369c (Sel I family of repeat-containing proteins), or FTT1676 (an outer membrane protein) abolished the ability of Schu S4 to survive or proliferate intracellularly and cause lethality in mice.

Francisella LPS

LPS is the major component of the outer membrane of gram-negative bacteria, and the interaction of LPS/MD-2 complexes with TLR4 on macrophages and endothelial cells activates a signaling cascade that results in the release of proinflammatory cytokine (Raetz et al., 2009). The structure of LPS includes a lipid portion (lipid A) that anchors it into the membrane, a polysaccharide core, and an oligo- or polysaccharide extending from the core beyond the bacterial surface (Raetz et al., 2009). Alterations in LPS structure, including changes in phosphorylation or fatty acid chain length, position, and number, significantly affect endotoxin bioactivity. The LPS of *Francisella* can be modified by various carbohydrates, including glucose, mannose, and galactosamine, which affect various aspects of virulence. However, *F. tularensis* LPS is 1,000-fold less potent than the LPS of enteric bacteria and does not activate cells via TLR4 (Zarrella et al., 2011). Recent study showed that the major lipid A species of *F. tularensis* LVS is composed of a glucosamine disaccharide backbone substituted with four fatty acyl groups and a phosphate (1-position) with a molecular mass of 1,505 Da. The major lipid A component contained 18:0 [3-OH (16:0)] in the distal subunit and two 18:0 (3-OH) fatty acyl chains at the 2- or 3-position of the reducing subunit. Additional variations in the fatty acyl groups of the lipid A species include a phosphate or a phosphoryl galactosamine at the 1-position and a hexose at the 4' or 6' position (Beasley et al., 2011; Soni et al., 2011). The lipid A moiety of *Francisella* LPS is linked to the core domain by a single 2-keto-3-deoxy-D-manno-octulosonic acid (Kdo) residue. *F. novicida* KdtA is bifunctional, but *F. novicida* contains a membrane-bound Kdo hydrolase that removes the outer Kdo unit. The hydrolase consists of two proteins (KdoH1 and KdoH2), which are expressed from adjacent, cotranscribed genes. KdoH1 has a single predicted N-terminal transmembrane segment. KdoH2 contains seven putative transmembrane sequences (Zhao and Raetz, 2010). There are multiple loci and genes identified in the *Francisella* genome that encode enzymes required for the synthesis of lipid A. These enzymes include two phosphatases (LpxE and LpxF), two dolichyl phosphate mannose synthase-like enzymes

(PmrF1 and PmrF2), one glycosyltransferase (PmrK), and a second hydroxyacyl acetyltransferase (LpxD) (Gunn and Ernst, 2007). The deletion of LpxF in *F. novicida* rendered it unable to grow in macrophages and highly attenuated in mice. The attenuation of this strain was due to enhanced cytokine production and infiltration of neutrophils (Wang et al., 2007).

Francisella Capsules

Capsular polysaccharides are important factors in bacterial pathogenesis and have been the target of a number of successful vaccines. *Francisella tularensis* has been considered to express a capsular antigen. The purified capsule contains a tetrasaccharide repeat, 2-acetamido-2,6-dideoxy-O-D-glucose (O-QuiNAc), 4,6-dideoxy-4-formamido-D-glucose (O-Qui4NFm), and 2-acetamido-2-deoxy-O-D-galacturonamide (O-GalNAcAN), which is similar, but not identical, to the LPS O antigen (called O-antigen capsule). Passive immunization with an O-antigen capsule-specific monoclonal antibody and active immunization of BALB/c mice with capsule each was protective against a 150-fold lethal challenge of *F. tularensis* LVS by the peritoneal route (Apicella et al., 2010). The disruption of glycosyltransferase genes involved in LPS O-antigen biosynthesis (*wbtI*, *wbtA1*, *wbtA2*, *wbtM*, *wbtI*, and *wbtC*) resulted in the loss of capsule expression, demonstrating that the LPS O-antigen sugars are used to build the O-antigen capsule. Mutations in *F. tularensis* LVS *capB* and *capC*, which are similar to *Bacillus anthracis* capsule genes, had no effect on capsule expression (Apicella et al., 2010). However, Jia et al. reported that the deletion of *capB* produced a strain that was significantly attenuated and induced potent protective immunity against homologous and heterologous challenge (Jia et al., 2010).

Another study identified a capsule-like complex in the *F. tularensis* LVS strain grown in Chamberlain's medium (Bandara et al., 2011). The putative 12.5-kb locus contains 12 genes (FTL_1432 to FTL_1421) involved in capsule synthesis. Deletion of FTL_1422 and FTL_1423 in the LVS resulted in mutants that lacked capsule-like complex, still maintained wild-type serum resistance, were negatively affected in terms of IMQ survival, and were attenuated in mice. Furthermore, inactivation of *wbtA* (LPS O-antigen biosynthesis) led to the complete loss of the O antigen, conferred serum sensitivity, impaired intracellular replication, and severely attenuated virulence in the mouse model. Notably, this mutant afforded protection against a challenge with virulent LVS (Bandara et al., 2011).

Acknowledgments. This work was supported by funding from The Region V "Great Lakes" Regional Center of Excellence in Biodefense and Emerging Infectious Diseases Consortium (NIH award 1-U54-AI-057153).

We thank Larry Schlesinger, Susheela Tridandapani, Mark Wewers, Tom Zahrt, and Shilpa Soni for their support on this work. N.P.M. and S.D. contributed equally to this work.

REFERENCES

Abplanalp, A. L., I. R. Morris, B. K. Parida, J. M. Teale, and M. T. Berton. 2009. TLR-dependent control of *Francisella tularensis* infection and host inflammatory responses. *PLoS One* **4:** e7920.

Ahlund, M. K., P. Ryden, A. Sjostedt, and S. Stoven. 2010. Directed screen of *Francisella novicida* virulence determinants using *Drosophila melanogaster*. *Infect. Immun.* **78:**3118–3128.

Akimana, C., and Y. A. Kwaik. 2011. *Francisella*-arthropod vector interaction and its role in patho-adaptation to infect mammals. *Front. Microbiol.* **2:**34.

Al-Khodor, S., and Y. Abu Kwaik. 2010. Triggering Ras signalling by intracellular *Francisella tularensis* through recruitment of PKCα and βI to the SOS2/GrB2 complex is essential for bacterial proliferation in the cytosol. *Cell. Microbiol.* **12:**1604–1621.

Andersson, H., B. Hartmanova, R. Kuolee, P. Ryden, W. Conlan, W. Chen, and A. Sjostedt. 2006. Transcriptional profiling of host responses in mouse lungs following aerosol infection with type A *Francisella tularensis*. *J. Med. Microbiol.* **55:**263–271.

Apicella, M. A., D. M. Post, A. C. Fowler, B. D. Jones, J. A. Rasmussen, J. R. Hunt, S. Imagawa, B. Choudhury, T. J. Inzana, T. M. Maier, D. W. Frank, T. C. Zahrt, K. Chaloner, M. P. Jennings, M. K. McLendon, and B. Gibson. 2010. Identification, characterization and immunogenicity of an O-antigen capsular polysaccharide of *Francisella tularensis*. *PLoS One* **5:**e11060.

Asare, R., C. Akimana, S. Jones, and Y. Abu Kwaik. 2010. Molecular bases of proliferation of *Francisella tularensis* in arthropod vectors. *Environ. Microbiol.* **12:**2587–2612.

Baca, O. G., M. J. Roman, R. H. Glew, R. F. Christner, J. E. Buhler, and A. S. Aragon. 1993. Acid phosphatase activity in *Coxiella burnetii*: a possible virulence factor. *Infect. Immun.* **61:**4232–4239.

Balagopal, A., A. S. MacFarlane, N. Mohapatra, S. Soni, J. S. Gunn, and L. S. Schlesinger. 2006. Characterization of the receptor-ligand pathways important for entry and survival of *Francisella tularensis* in human macrophages. *Infect. Immun.* **74:**5114–5125.

Bandara, A. B., A. E. Champion, X. Wang, G. Berg, M. A. Apicella, M. McLendon, P. Azadi, D. S. Snyder, and T. J. Inzana. 2011. Isolation and mutagenesis of a capsule-like complex (CLC) from *Francisella tularensis*, and contribution of the CLC to *F. tularensis* virulence in mice. *PLoS One* **6:**e19003.

Barker, J. R., A. Chong, T. D. Wehrly, J. J. Yu, S. A. Rodriguez, J. Liu, J. Celli, B. P. Arulanandam, and K. E. Klose. 2009. The *Francisella tularensis* pathogenicity island encodes a secretion system that is required for phagosome escape and virulence. *Mol. Microbiol.* **74:**1459–1470.

Barker, J. R., and K. E. Klose. 2007. Molecular and genetic basis of pathogenesis in *Francisella tularensis*. *Ann. N. Y. Acad. Sci.* **1105:**138–159.

Baron, G. S., and F. E. Nano. 1998. MglA and MglB are required for the intramacrophage growth of *Francisella novicida*. *Mol. Microbiol.* **29:**247–259.

Bearson, B. L., S. M. Bearson, J. J. Uthe, S. E. Dowd, J. O. Houghton, I. Lee, M. J. Toscano, and D. C. Lay, Jr. 2008. Iron regulated genes of *Salmonella enterica* serovar Typhimurium in response to norepinephrine and the requirement of *fepDGC* for norepinephrine-enhanced growth. *Microbes Infect.* **10:**807–816.

Beasley, A. S., R. J. Cotter, S. N. Vogel, T. J. Inzana, A. A. Qureshi, and N. Qureshi. 2012. A variety of novel lipid A structures obtained from *Francisella tularensis* live vaccine strain. *Innate Immun.* **18:**268–278.

Bell, B. L., N. P. Mohapatra, and J. S. Gunn. 2010. Regulation of virulence gene transcripts by the *Francisella novicida* orphan response regulator PmrA: role of phosphorylation and evidence of MglA/SspA interaction. *Infect. Immun.* **78:**2189–2198.

Ben Nasr, A., and G. R. Klimpel. 2008. Subversion of complement activation at the bacterial surface promotes serum resistance and opsonophagocytosis of *Francisella tularensis*. *J. Leukoc. Biol.* **84:**77–85.

Bohn, C., C. Rigoulay, and P. Bouloc. 2007. No detectable effect of RNA-binding protein Hfq absence in *Staphylococcus aureus*. *BMC Microbiol.* **7:**10.

Bosio, C. M., H. Bielefeldt-Ohmann, and J. T. Belisle. 2007. Active suppression of the pulmonary immune response by *Francisella tularensis* Schu4. *J. Immunol.* **178:**4538–4547.

Bröms, J. E., M. Lavander, L. Meyer, and A. Sjöstedt. 2011. IglG and IglI of the *Francisella* pathogenicity island are important virulence determinants of *Francisella tularensis* LVS. *Infect. Immun.* **79:**3683–3696.

Broms, J. E., A. Sjostedt, and M. Lavander. 2010. The role of the *Francisella tularensis* pathogenicity island in type VI secretion, intracellular survival, and modulation of host cell signaling. *Front. Microbiol.* **1:**136.

Brotcke, A., and D. M. Monack. 2008. Identification of *fevR*, a novel regulator of virulence gene expression in *Francisella novicida*. *Infect. Immun.* **76:**3473–3480.

Brotcke, A., D. S. Weiss, C. C. Kim, P. Chain, S. Malfatti, E. Garcia, and D. M. Monack. 2006. Identification of MglA-regulated genes reveals novel virulence factors in *Francisella tularensis*. *Infect. Immun.* **74:**6642–6655.

Browning, D. F., and S. J. Busby. 2004. The regulation of bacterial transcription initiation. *Nat. Rev. Microbiol.* **2:**57–65.

Buchan, B. W., R. L. McCaffrey, S. R. Lindemann, L. A. Allen, and B. D. Jones. 2009. Identification of *migR*, a regulatory element of the *Francisella tularensis* live vaccine strain *iglABCD* virulence operon required for normal replication and trafficking in macrophages. *Infect. Immun.* **77:**2517–2529.

Butchar, J. P., T. J. Cremer, C. D. Clay, M. A. Gavrilin, M. D. Wewers, C. B. Marsh, L. S. Schlesinger, and S. Tridandapani. 2008. Microarray analysis of human monocytes infected with *Francisella tularensis* identifies new targets of host response subversion. *PLoS One* **3:**e2924.

Carlin, A. F., Y. C. Chang, T. Areschoug, G. Lindahl, N. Hurtado-Ziola, C. C. King, A. Varki, and V. Nizet. 2009. Group B *Streptococcus* suppression of phagocyte functions by protein-mediated engagement of human Siglec-5. *J. Exp. Med.* **206:**1691–1699.

Carlson, P. E., Jr., J. A. Carroll, D. M. O'Dee, and G. J. Nau. 2007. Modulation of virulence factors in *Francisella tularensis* determines human macrophage responses. *Microb. Pathog.* **42:**204–214.

Carlson, P. E., Jr., J. Horzempa, D. M. O'Dee, C. M. Robinson, P. Neophytou, A. Labrinidis, and G. J. Nau. 2009. Global transcriptional response to spermine, a component of the intramacrophage environment, reveals regulation of *Francisella* gene expression through insertion sequence elements. *J. Bacteriol.* **191:**6855–6864.

Chambers, J. R., and K. S. Bender. 2011. The RNA chaperone Hfq is important for growth and stress tolerance in *Francisella novicida*. *PLoS One* **6:**e19797.

Chao, Y., and J. Vogel. 2010. The role of Hfq in bacterial pathogens. *Curr. Opin. Microbiol.* **13**:24–33.

Charity, J. C., L. T. Blalock, M. M. Costante-Hamm, D. L. Kasper, and S. L. Dove. 2009. Small molecule control of virulence gene expression in *Francisella tularensis*. *PLoS Pathog.* **5**:e1000641.

Charity, J. C., M. M. Costante-Hamm, E. L. Balon, D. H. Boyd, E. J. Rubin, and S. L. Dove. 2007. Twin RNA polymerase-associated proteins control virulence gene expression in *Francisella tularensis*. *PLoS Pathog.* **3**:e84.

Checroun, C., T. D. Wehrly, E. R. Fischer, S. F. Hayes, and J. Celli. 2006. Autophagy-mediated reentry of *Francisella tularensis* into the endocytic compartment after cytoplasmic replication. *Proc. Natl. Acad. Sci. USA* **103**:14578–14583.

Child, R., T. D. Wehrly, D. Rockx-Brouwer, D. W. Dorward, and J. Celli. 2010. Acid phosphatases do not contribute to the pathogenesis of type A *Francisella tularensis*. *Infect. Immun.* **78**:59–67.

Chiu, H. C., S. Soni, S. K. Kulp, H. Curry, D. Wang, J. S. Gunn, L. S. Schlesinger, and C. S. Chen. 2009. Eradication of intracellular *Francisella tularensis* in THP-1 human macrophages with a novel autophagy inducing agent. *J. Biomed. Sci.* **16**:110.

Chong, A., T. D. Wehrly, V. Nair, E. R. Fischer, J. R. Barker, K. E. Klose, and J. Celli. 2008. The early phagosomal stage of *Francisella tularensis* determines optimal phagosomal escape and *Francisella* pathogenicity island protein expression. *Infect. Immun.* **76**:5488–5499.

Christiansen, J. K., M. H. Larsen, H. Ingmer, L. Sogaard-Andersen, and B. H. Kallipolitis. 2004. The RNA-binding protein Hfq of *Listeria monocytogenes*: role in stress tolerance and virulence. *J. Bacteriol.* **186**:3355–3362.

Christiansen, J. K., J. S. Nielsen, T. Ebersbach, P. Valentin-Hansen, L. Sogaard-Andersen, and B. H. Kallipolitis. 2006. Identification of small Hfq-binding RNAs in *Listeria monocytogenes*. *RNA* **12**:1383–1396.

Clay, C. D., S. Soni, J. S. Gunn, and L. S. Schlesinger. 2008. Evasion of complement-mediated lysis and complement C3 deposition are regulated by *Francisella tularensis* lipopolysaccharide O antigen. *J. Immunol.* **181**:5568–5578.

Clemens, D. L., B. Y. Lee, and M. A. Horwitz. 2004. Virulent and avirulent strains of *Francisella tularensis* prevent acidification and maturation of their phagosomes and escape into the cytoplasm in human macrophages. *Infect. Immun.* **72**:3204–3217.

Clemens, D. L., B. Y. Lee, and M. A. Horwitz. 2005. *Francisella tularensis* enters macrophages via a novel process involving pseudopod loops. *Infect. Immun.* **73**:5892–5902.

Clemens, D. L., B. Y. Lee, and M. A. Horwitz. 2009. *Francisella tularensis* phagosomal escape does not require acidification of the phagosome. *Infect. Immun.* **77**:1757–1773.

Cole, L. E., M. H. Laird, A. Seekatz, A. Santiago, Z. Jiang, E. Barry, K. A. Shirey, K. A. Fitzgerald, and S. N. Vogel. 2010. Phagosomal retention of *Francisella tularensis* results in TIRAP/Mal-independent TLR2 signaling. *J. Leukoc. Biol.* **87**:275–281.

Cole, L. E., A. Santiago, E. Barry, T. J. Kang, K. A. Shirey, Z. J. Roberts, K. L. Elkins, A. S. Cross, and S. N. Vogel. 2008. Macrophage proinflammatory response to *Francisella tularensis* live vaccine strain requires coordination of multiple signaling pathways. *J. Immunol.* **180**:6885–6891.

Cole, L. E., K. A. Shirey, E. Barry, A. Santiago, P. Rallabhandi, K. L. Elkins, A. C. Puche, S. M. Michalek, and S. N. Vogel. 2007. Toll-like receptor 2-mediated signaling requirements for *Francisella tularensis* live vaccine strain infection of murine macrophages. *Infect. Immun.* **75**:4127–4137.

Collado-Vides, J., B. Magasanik, and J. D. Gralla. 1991. Control site location and transcriptional regulation in *Escherichia coli*. *Microbiol. Rev.* **55**:371–394.

Cong, Y., J.-J. Yu, M. N. Guentzel, M. T. Berton, J. Seshu, K. E. Klose, and B. P. Arulanandam. 2009. Vaccination with a defined *Francisella tularensis* subsp. *novicida* pathogenicity island mutant (Δ*iglB*) induces protective immunity against homotypic and heterotypic challenge. *Vaccine* **27**:5554–5561.

Conlan, J. W., W. Chen, C. M. Bosio, S. C. Cowley, and K. L. Elkins. 2011. Infection of mice with *Francisella* as an immunological model. *Curr. Protoc. Immunol.* **93**:19.14.1–19.14.16.

Conlan, J. W., X. Zhao, G. Harris, H. Shen, M. Bolanowski, C. Rietz, A. Sjostedt, and W. Chen. 2008. Molecular immunology of experimental primary tularemia in mice infected by respiratory or intradermal routes with type A *Francisella tularensis*. *Mol. Immunol.* **45**:2962–2969.

Cowley, S. C., and K. L. Elkins. 2011. Immunity to *Francisella*. *Front. Microbiol.* **2**:26.

Crabbe, A., M. J. Schurr, P. Monsieurs, L. Morici, J. Schurr, J. W. Wilson, C. M. Ott, G. Tsaprailis, D. L. Pierson, H. Stefanyshyn-Piper, and C. A. Nickerson. 2011. Transcriptional and proteomic responses of *Pseudomonas aeruginosa* PAO1 to spaceflight conditions involve Hfq regulation and reveal a role for oxygen. *Appl. Environ. Microbiol.* **77**:1221–1230.

Cremer, T. J., J. P. Butchar, and S. Tridandapani. 2011. *Francisella* subverts innate immune signaling: focus on PI3K/Akt. *Front. Microbiol.* **5**:13.

Cremer, T. J., D. H. Ravneberg, C. D. Clay, M. G. Piper-Hunter, C. B. Marsh, T. S. Elton, J. S. Gunn, A. Amer, T. D. Kanneganti, L. S. Schlesinger, J. P. Butchar, and S. Tridandapani. 2009. MiR-155 induction by *F. novicida* but not the virulent *F. tularensis* results in SHIP down-regulation and enhanced pro-inflammatory cytokine response. *PLoS One* **4**:e8508.

Dean, R. E., P. M. Ireland, J. E. Jordan, R. W. Titball, and P. C. Oyston. 2009. RelA regulates virulence and intracellular survival of *Francisella novicida*. *Microbiology* **155**:4104–4113.

de Bruin, O. M., J. S. Ludu, and F. E. Nano. 2007. The *Francisella* pathogenicity island protein IglA localizes to the bacterial cytoplasm and is needed for intracellular growth. *BMC Microbiol.* **7**:1.

Deng, K., R. J. Blick, W. Liu, and E. J. Hansen. 2006. Identification of *Francisella tularensis* genes affected by iron limitation. *Infect. Immun.* **74**:4224–4236.

Dennis, D. T., T. V. Inglesby, D. A. Henderson, J. G. Bartlett, M. S. Ascher, E. Eitzen, A. D. Fine, A. M. Friedlander, J. Hauer, M. Layton, S. R. Lillibridge, J. E. McDade, M. T. Osterholm, T. O'Toole, G. Parker, T. M. Perl, P. K. Russell, and K. Tonat. 2001. Tularemia as a biological weapon: medical and public health management. *JAMA* **285**:2763–2773.

Dienst, F. T., Jr. 1963. Tularemia: a perusal of three hundred thirty-nine cases. *J. La. State Med. Soc.* **115**:114–127.

Dieppedale, J., D. Sobral, M. Dupuis, I. Dubail, J. Klimentova, J. Stulik, G. Postic, E. Frapy, K. L. Meibom, M. Barel, and A. Charbit. 2011. Identification of a putative chaperone involved in stress resistance and virulence in *Francisella tularensis*. *Infect. Immun.* **79**:1428–1439.

Ellis, J., P. C. Oyston, M. Green, and R. W. Titball. 2002. Tularemia. *Clin. Microbiol. Rev.* **15**:631–646.

Ericsson, M., A. Tarnvik, K. Kuoppa, G. Sandstrom, and A. Sjostedt. 1994. Increased synthesis of DnaK, GroEL, and GroES homologs by *Francisella tularensis* LVS in response to heat and hydrogen peroxide. *Infect. Immun.* **62**:178–183.

Escolar, L., J. Perez-Martin, and V. de Lorenzo. 1999. Opening the iron box: transcriptional metalloregulation by the Fur protein. *J. Bacteriol.* **181**:6223–6229.

Evans, M. E., D. W. Gregory, W. Schaffner, and Z. A. McGee. 1985. Tularemia: a 30-year experience with 88 cases. *Medicine* (Baltimore) **64**:251–269.

Fang, F. C. 2004. Antimicrobial reactive oxygen and nitrogen species: concepts and controversies. *Nat. Rev. Microbiol.* **2**:820–832.

Fernandes-Alnemri, T., J. W. Yu, C. Juliana, L. Solorzano, S. Kang, J. Wu, P. Datta, M. McCormick, L. Huang, E. McDermott, L. Eisenlohr, C. P. Landel, and E. S. Alnemri. 2010. The AIM2 inflammasome is critical for innate immunity to *Francisella tularensis. Nat. Immunol.* 11:385–393.

Forestal, C. A., M. Malik, S. V. Catlett, A. G. Savitt, J. L. Benach, T. J. Sellati, and M. B. Furie. 2007. *Francisella tularensis* has a significant extracellular phase in infected mice. *J. Infect. Dis.* 196:134–137.

Fuller, J. R., R. R. Craven, J. D. Hall, T. M. Kijek, S. Taft-Benz, and T. H. Kawula. 2008. RipA, a cytoplasmic membrane protein conserved among *Francisella* species, is required for intracellular survival. *Infect. Immun.* 76:4934–4943.

Fuller, J. R., T. M. Kijek, S. Taft-Benz, and T. H. Kawula. 2009. Environmental and intracellular regulation of *Francisella tularensis ripA. BMC Microbiol.* 9:216.

Gavrilin, M. A., and M. D. Wewers. 2011. *Francisella* recognition by inflammasomes: differences between mice and men. *Front. Microbiol.* 2:11.

Golovliov, I., V. Baranov, Z. Krocova, H. Kovarova, and A. Sjostedt. 2003. An attenuated strain of the facultative intracellular bacterium *Francisella tularensis* can escape the phagosome of monocytic cells. *Infect. Immun.* 71:5940–5950.

Golovliov, I., M. Ericsson, G. Sandstrom, A. Tarnvik, and A. Sjostedt. 1997. Identification of proteins of *Francisella tularensis* induced during growth in macrophages and cloning of the gene encoding a prominently induced 23-kilodalton protein. *Infect. Immun.* 65:2183–2189.

Goulian, M. 2004. Robust control in bacterial regulatory circuits. *Curr. Opin. Microbiol.* 7:198–202.

Grall, N., J. Livny, M. Waldor, M. Barel, A. Charbit, and K. Meibom. 2009. Pivotal role of the *Francisella tularensis* heat-shock sigma factor RpoH. *Microbiology* 155:2560–2572.

Guina, T., D. Radulovic, A. J. Bahrami, D. L. Bolton, L. Rohmer, K. A. Jones-Isaac, J. Chen, L. A. Gallagher, B. Gallis, S. Ryu, G. K. Taylor, M. J. Brittnacher, C. Manoil, and D. R. Goodlett. 2007. MglA regulates *Francisella tularensis* subsp. *novicida* (*Francisella novicida*) response to starvation and oxidative stress. *J. Bacteriol.* 189:6580–6586.

Gunn, J. S., and R. K. Ernst. 2007. The structure and function of *Francisella* lipopolysaccharide. *Ann. N. Y. Acad. Sci.* 1105:202–218.

Hager, A. J., D. L. Bolton, M. R. Pelletier, M. J. Brittnacher, L. A. Gallagher, R. Kaul, S. J. Skerrett, S. I. Miller, and T. Guina. 2006. Type IV pili-mediated secretion modulates *Francisella* virulence. *Mol. Microbiol.* 62:227–237.

Hajjar, A. M., M. D. Harvey, S. A. Shaffer, D. R. Goodlett, A. Sjostedt, H. Edebro, M. Forsman, M. Bystrom, M. Pelletier, C. B. Wilson, S. I. Miller, S. J. Skerrett, and R. K. Ernst. 2006. Lack of in vitro and in vivo recognition of *Francisella tularensis* subspecies lipopolysaccharide by Toll-like receptors. *Infect. Immun.* 74:6730–6738.

Hall, J. D., M. D. Woolard, B. M. Gunn, R. R. Craven, S. Taft-Benz, J. A. Frelinger, and T. H. Kawula. 2008. Infected-host-cell repertoire and cellular response in the lung following inhalation of *Francisella tularensis* Schu S4, LVS, or U112. *Infect. Immun.* 76:5843–5852.

Hall, W. C., R. M. Kovatch, and R. L. Schricker. 1973. Tularaemic pneumonia: pathogenesis of the aerosol-induced disease in monkeys. *J. Pathol.* 110:193–201.

Havlasova, J., L. Hernychova, M. Brychta, M. Hubalek, J. Lenco, P. Larsson, M. Lundqvist, M. Forsman, Z. Krocova, J. Stulik, and A. Macela. 2005. Proteomic analysis of anti-*Francisella tularensis* LVS antibody response in murine model of tularemia. *Proteomics* 5:2090–2103.

Helmann, J. D. 2002. The extracytoplasmic function (ECF) sigma factors. *Adv. Microb. Physiol.* 46:47–110.

Horzempa, J., P. E. Carlson, Jr., D. M. O'Dee, R. M. Shanks, and G. J. Nau. 2008. Global transcriptional response to mammalian temperature provides new insight into *Francisella tularensis* pathogenesis. *BMC Microbiol.* 8:172.

Huang, M. T., B. L. Mortensen, D. J. Taxman, R. R. Craven, S. Taft-Benz, T. M. Kijek, J. R. Fuller, B. K. Davis, I. C. Allen, W. J. Brickey, D. Gris, H. Wen, T. H. Kawula, and J. P. Ting. 2010. Deletion of *ripA* alleviates suppression of the inflammasome and MAPK by *Francisella tularensis. J. Immunol.* 185:5476–5485.

Janga, S. C., H. Salgado, J. Collado-Vides, and A. Martinez-Antonio. 2007. Internal versus external effector and transcription factor gene pairs differ in their relative chromosomal position in *Escherichia coli. J. Mol. Biol.* 368:263–272.

Jia, Q., B. Y. Lee, R. Bowen, B. J. Dillon, S. M. Som, and M. A. Horwitz. 2010. A *Francisella tularensis* live vaccine strain (LVS) mutant with a deletion in *capB*, encoding a putative capsular biosynthesis protein, is significantly more attenuated than LVS yet induces potent protective immunity in mice against *F. tularensis* challenge. *Infect. Immun.* 78:4341–4355.

Johansson, A., J. Celli, W. Conlan, K. L. Elkins, M. Forsman, P. S. Keim, P. Larsson, C. Manoil, F. E. Nano, J. M. Petersen, and A. Sjostedt. 2010. Objections to the transfer of *Francisella novicida* to the subspecies rank of *Francisella tularensis. Int. J. Syst. Evol. Microbiol.* 60:1717–1718.

Jones, J. W., N. Kayagaki, P. Broz, T. Henry, K. Newton, K. O'Rourke, S. Chan, J. Dong, Y. Qu, M. Roose-Girma, V. M. Dixit, and D. M. Monack. 2010. Absent in melanoma 2 is required for innate immune recognition of *Francisella tularensis. Proc. Natl. Acad. Sci. USA* 107:9771–9776.

Jones, M. K., E. Warner, and J. D. Oliver. 2008. Survival of and in situ gene expression by *Vibrio vulnificus* at varying salinities in estuarine environments. *Appl. Environ. Microbiol.* 74:182–187.

Kadzhaev, K., C. Zingmark, I. Golovliov, M. Bolanowski, H. Shen, W. Conlan, and A. Sjostedt. 2009. Identification of genes contributing to the virulence of *Francisella tularensis* SCHU S4 in a mouse intradermal infection model. *PLoS One* 4:e5463.

Kats, L. M., C. G. Black, N. I. Proellocks, and R. L. Coppel. 2006. *Plasmodium rhoptries*: how things went pear-shaped. *Trends Parasitol.* 22:269–276.

Katz, J., P. Zhang, M. Martin, S. N. Vogel, and S. M. Michalek. 2006. Toll-like receptor 2 is required for inflammatory responses to *Francisella tularensis* LVS. *Infect. Immun.* 74:2809–2816.

Kazmierczak, M. J., M. Wiedmann, and K. J. Boor. 2005. Alternative sigma factors and their roles in bacterial virulence. *Microbiol. Mol. Biol. Rev.* 69:527–543.

Keim, P., A. Johansson, and D. M. Wagner. 2007. Molecular epidemiology, evolution, and ecology of *Francisella. Ann. N. Y. Acad. Sci.* 1105:30–66.

Kiss, K., W. Liu, J. F. Huntley, M. V. Norgard, and E. J. Hansen. 2008. Characterization of *fig* operon mutants of *Francisella novicida* U112. *FEMS Microbiol. Lett.* 285:270–277.

Koo, J. T., T. M. Alleyne, C. A. Schiano, N. Jafari, and W. Lathem. 29 August 2011. Global discovery of small RNAs in *Yersinia pseudotuberculosis* identifies *Yersinia*-specific small, noncoding RNAs required for virulence. *Proc. Natl. Acad. Sci. USA* doi:10.1073/pnas.1101655108.

Kugeler, K. J., P. S. Mead, A. M. Janusz, J. E. Staples, K. A. Kubota, L. G. Chalcraft, and J. M. Petersen. 2009. Molecular epidemiology of *Francisella tularensis* in the United States. *Clin. Infect. Dis.* 48:863–870.

Lai, X. H., I. Golovliov, and A. Sjostedt. 2004. Expression of IglC is necessary for intracellular growth and induction of apoptosis in murine macrophages by *Francisella tularensis. Microb. Pathog.* 37:225–230.

Lauriano, C. M., J. R. Barker, F. E. Nano, B. P. Arulanandam, and K. E. Klose. 2003. Allelic exchange in *Francisella tularensis* using PCR products. *FEMS Microbiol. Lett.* 229:195–202.

Lauriano, C. M., J. R. Barker, S. S. Yoon, F. E. Nano, B. P. Arulanandam, D. J. Hassett, and K. E. Klose. 2004. MglA regulates transcription of virulence factors necessary for *Francisella tularensis* intraamoebae and intramacrophage survival. *Proc. Natl. Acad. Sci. USA* 101:4246–4249.

Lenco, J., Hubalek, M., P. Larsson, A. Fucikova, M. Brychta, A. Macela, and J. Stulik. 2007. Proteomics analysis of the *Francisella tularensis* LVS response to iron restriction: induction of the *F. tularensis* pathogenicity island proteins IglABC. *FEMS Microbiol. Lett.* 269:11–21.

Lenco, J., M. Link, V. Tambor, J. Zakova, L. Cerveny, and A. J. Stulik. 2009. iTRAQ quantitative analysis of *Francisella tularensis* ssp. *holarctica* live vaccine strain and *Francisella tularensis* ssp. *tularensis* SCHU S4 response to different temperatures and stationary phases of growth. *Proteomics* 9:2875–2882.

Lenco, J., I. Pavkova, M. Hubalek, and J. Stulik. 2005. Insights into the oxidative stress response in *Francisella tularensis* LVS and its mutant DeltaiglC1+2 by proteomics analysis. *FEMS Microbiol. Lett.* 246:47–54.

Lindgren, H., I. Golovliov, V. Baranov, R. K. Ernst, M. Telepnev, and A. Sjostedt. 2004. Factors affecting the escape of *Francisella tularensis* from the phagolysosome. *J. Med. Microbiol.* 53:953–958.

Lindgren, H., M. Honn, I. Golovlev, K. Kadzhaev, W. Conlan, and A. Sjostedt. 2009. The 58-kilodalton major virulence factor of *Francisella tularensis* is required for efficient utilization of iron. *Infect. Immun.* 77:4429–4436.

Lindgren, H., M. Honn, E. Salomonsson, K. Kuoppa, A. Forsberg, and A. Sjostedt. 2011. Iron content differs between *Francisella tularensis* subspecies *tularensis* and subspecies *holarctica* strains and correlates to their susceptibility to H_2O_2-induced killing. *Infect. Immun.* 79:1218–1224.

Liu, H., Q. Wang, Q. Liu, X. Cao, C. Shi, and Y. Zhang. 2011. Roles of Hfq in the stress adaptation and virulence in fish pathogen *Vibrio alginolyticus* and its potential application as a target for live attenuated vaccine. *Appl. Microbiol. Biotechnol.* 91:353–364.

Ludu, J. S., O. M. de Bruin, B. N. Duplantis, C. L. Schmerk, A. Y. Chou, K. L. Elkins, and F. E. Nano. 2008. The *Francisella* pathogenicity island protein PdpD is required for full virulence and associates with homologues of the type VI secretion system. *J. Bacteriol.* 190:4584–4595.

Madan Babu, M., and S. A. Teichmann. 2003. Evolution of transcription factors and the gene regulatory network in *Escherichia coli*. *Nucleic Acids Res.* 31:1234–1244.

Madan Babu, M., S. A. Teichmann, and L. Aravind. 2006. Evolutionary dynamics of prokaryotic transcriptional regulatory networks. *J. Mol. Biol.* 358:614–633.

Maeda, H., N. Fujita, and A. Ishihama. 2000a. Competition among seven *Escherichia coli* sigma subunits: relative binding affinities to the core RNA polymerase. *Nucleic Acids Res.* 28:3497–3503.

Maeda, H., M. Jishage, T. Nomura, N. Fujita, and A. Ishihama. 2000b. Two extracytoplasmic function sigma subunits, σ^E and σ^{FecI}, of *Escherichia coli*: promoter selectivity and intracellular levels. *J. Bacteriol.* 182:1181–1184.

Mangan, S., and U. Alon. 2003. Structure and function of the feed-forward loop network motif. *Proc. Natl. Acad. Sci. USA* 100:11980–11985.

Mariathasan, S., D. S. Weiss, V. M. Dixit, and D. M. Monack. 2005. Innate immunity against *Francisella tularensis* is dependent on the ASC/caspase-1 axis. *J. Exp. Med.* 202:1043–1049.

Martinez-Antonio, A., S. C. Janga, H. Salgado, and J. Collado-Vides. 2006. Internal-sensing machinery directs the activity of the regulatory network in *Escherichia coli*. *Trends Microbiol.* 14:22–27.

Mascher, T. 2006. Intramembrane-sensing histidine kinases: a new family of cell envelope stress sensors in *Firmicutes* bacteria. *FEMS Microbiol. Lett.* 264:133–144.

McCaffrey, R. L., and L. A. Allen. 2006. *Francisella tularensis* LVS evades killing by human neutrophils via inhibition of the respiratory burst and phagosome escape. *J. Leukoc. Biol.* 80:1224–1230.

McCaffrey, R. L., J. T. Schwartz, S. R. Lindemann, J. G. Moreland, B. W. Buchan, B. D. Jones, and L. A. Allen. 2010. Multiple mechanisms of NADPH oxidase inhibition by type A and type B *Francisella tularensis*. *J. Leukoc. Biol.* 88:791–805.

Medina, E. A., I. R. Morris, and M. T. Berton. 2010. Phosphatidylinositol 3-kinase activation attenuates the TLR2-mediated macrophage proinflammatory cytokine response to *Francisella tularensis* live vaccine strain. *J. Immunol.* 185:7562–7572.

Meibom, K. L., and A. Charbit. 2010. The unraveling panoply of *Francisella tularensis* virulence attributes. *Curr. Opin. Microbiol.* 13:11–17.

Meibom, K. L., A. L. Forslund, K. Kuoppa, K. Alkhuder, I. Dubail, M. Dupuis, A. Forsberg, and A. Charbit. 2009. Hfq, a novel pleiotropic regulator of virulence-associated genes in *Francisella tularensis*. *Infect. Immun.* 77:1866–1880.

Metzger, D. W., C. S. Bakshi, and G. Kirimanjeswara. 2007. Mucosal immunopathogenesis of *Francisella tularensis*. *Ann. N. Y. Acad. Sci.* 1105:266–283.

Mohapatra, N. P., S. Soni, B. L. Bell, R. Warren, R. K. Ernst, A. Muszynski, R. W. Carlson, and J. S. Gunn. 2007. Identification of an orphan response regulator required for the virulence of *Francisella* spp. and transcription of pathogenicity island genes. *Infect. Immun.* 75:3305–3314.

Mohapatra, N. P., S. Soni, M. V. Rajaram, P. M. Dang, T. J. Reilly, J. El-Benna, C. D. Clay, L. S. Schlesinger, and J. S. Gunn. 2010. *Francisella* acid phosphatases inactivate the NADPH oxidase in human phagocytes. *J. Immunol.* 184:5141–5150.

Mohapatra, N. P., S. Soni, T. J. Reilly, J. Liu, K. E. Klose, and J. S. Gunn. 2008. Combined deletion of four *Francisella novicida* acid phosphatases attenuates virulence and macrophage vacuolar escape. *Infect. Immun.* 76:3690–3699.

Mokrievich, A. N., A. N. Kondakova, E. Valade, M. E. Platonov, G. M. Vakhrameeva, R. Z. Shaikhutdinova, R. I. Mironova, D. Blaha, I. V. Bakhteeva, G. M. Titareva, T. B. Kravchenko, T. I. Kombarova, D. Vidal, V. M. Pavlov, B. Lindner, I. A. Dyatlov, and Y. A. Knirel. 2010. Biological properties and structure of the lipopolysaccharide of a vaccine strain of *Francisella tularensis* generated by inactivation of a quorum sensing system gene *qseC*. *Biochemistry* (Moscow) 75:443–451.

Monack, D. M. 2008. The inflammasome: a key player in the inflammation triggered in response to bacterial pathogens. *J. Pediatr. Gastroenterol. Nutr.* 46(Suppl. 1): E14.

Moule, M. G., D. M. Monack, and D. S. Schneider. 2010. Reciprocal analysis of *Francisella novicida* infections of a *Drosophila melanogaster* model reveal host-pathogen conflicts mediated by reactive oxygen and imd-regulated innate immune response. *PLoS Pathog.* 6:e1001065.

Nano, F. E., and C. Schmerk. 2007. The *Francisella* pathogenicity island. *Ann. N. Y. Acad. Sci.* 1105:122–137.

Nano, F. E., N. Zhang, S. C. Cowley, K. E. Klose, K. K. Cheung, M. J. Roberts, J. S. Ludu, G. W. Letendre, A. I. Meierovics, G. Stephens, and K. L. Elkins. 2004. A *Francisella tularensis* pathogenicity island required for intramacrophage growth. *J. Bacteriol.* 186:6430–6436.

Nix, E. B., K. K. Cheung, D. Wang, N. Zhang, R. D. Burke, and F. E. Nano. 2006. Virulence of *Francisella* spp. in chicken embryos. *Infect. Immun.* 74:4809–4816.

Oyston, P. C. 2008. *Francisella tularensis*: unravelling the secrets of an intracellular pathogen. *J. Med. Microbiol.* **57:**921–930.

Oyston, P. C., A. Sjostedt, and R. W. Titball. 2004. Tularaemia: bioterrorism defence renews interest in *Francisella tularensis*. *Nat. Rev. Microbiol.* **2:**967–978.

Paget, M. S., and J. D. Helmann. 2003. The σ^{70} family of sigma factors. *Genome Biol.* **4:**203.

Pannekoek, Y., R. Huis in't Veld, C. T. Hopman, A. A. Langerak, D. Speijer, and A. van der Ende. 2009. Molecular characterization and identification of proteins regulated by Hfq in *Neisseria meningitidis*. *FEMS Microbiol. Lett.* **294:**216–224.

Parker, R. R. 1934. Recent studies of tick-borne diseases made at the United States Public Health Service Laboratory at Hamilton, Montana. *Proc. Fifth Pacific Congr.*, p. 3367–3374.

Pierini, L. M. 2006. Uptake of serum-opsonized *Francisella tularensis* by macrophages can be mediated by class A scavenger receptors. *Cell. Microbiol.* **8:**1361–1370.

Postic, G., E. Frapy, M. Dupuis, I. Dubail, J. Livny, A. Charbit, and K. L. Meibom. 2010. Identification of small RNAs in *Francisella tularensis*. *BMC Genomics* **11:**625.

Ptashne, M. 1984. DNA-binding proteins. *Nature* **308:**753–754.

Raetz, C. R., Z. Guan, B. O. Ingram, D. A. Six, F. Song, X. Wang, and J. Zhao. 2009. Discovery of new biosynthetic pathways: the lipid A story. *J. Lipid Res.* **50**(Suppl.)**:**S103–S108.

Ramakrishnan, G., A. Meeker, and B. Dragulev. 2008. *fslE* is necessary for siderophore-mediated iron acquisition in *Francisella tularensis* Schu S4. *J. Bacteriol.* **190:**5353–5361.

Rasko, D. A., C. G. Moreira, C. G., D. R. Li, N. C. Reading, J. M. Ritchie, M. K. Waldor, N. Williams, R. Taussig, S. Wei, M. Roth, D. T. Hughes, J. F. Huntley, M. W. Fina, J. R. Falck, and V. Sperandio. 2008. Targeting QseC signaling and virulence for antibiotic development. *Science* **321:**1078–1080.

Ray, H. J., P. Chu, T. H. Wu, C. R. Lyons, A. K. Murthy, M. N. Guentzel, K. E. Klose, and B. P. Arulanandam. 2010. The Fischer 344 rat reflects human susceptibility to *Francisella* pulmonary challenge and provides a new platform for virulence and protection studies. *PLoS One* **5:**e9952.

Raymond, C. R., and J. W. Conlan. 2009. Differential susceptibility of Sprague-Dawley and Fischer 344 rats to infection by *Francisella tularensis*. *Microb. Pathog.* **46:**231–244.

Read, A., S. J. Vogl, K. Hueffer, L. A. Gallagher, and G. M. Happ. 2008. *Francisella* genes required for replication in mosquito cells. *J. Med. Entomol.* **45:**1108–1116.

Rick Lyons, C., and T. H. Wu. 2007. Animal models of *Francisella tularensis* infection. *Ann. N. Y. Acad. Sci.* **1105:**238–265.

Robertson, G. T., and R. M. Roop, Jr. 1999. The *Brucella abortus* host factor I (HF-I) protein contributes to stress resistance during stationary phase and is a major determinant of virulence in mice. *Mol. Microbiol.* **34:**690–700.

Ruger, W. 1972. Transcription of genetic information and its regulation by protein factors. *Angew Chem. Int. Ed. Engl.* **11:**883–893.

Russo, B. C., J. Horzempa, D. M. O'Dee, D. M. Schmitt, M. J. Brown, P. E. Carlson, Jr., R. J. Xavier, and G. J. Nau. 2011. A *Francisella tularensis* locus required for spermine responsiveness is necessary for virulence. *Infect. Immun.* **79:**3665–3676.

Santic, M., C. Akimana, R. Asare, J. C. Kouokam, S. Atay, and Y. Abu Kwaik. 2009. Intracellular fate of *Francisella tularensis* within arthropod-derived cells. *Environ. Microbiol.* **11:**1473–1481.

Santic, M., S. Al-Khodor, and Y. Abu Kwaik. 2010. Cell biology and molecular ecology of *Francisella tularensis*. *Cell. Microbiol.* **12:**129–139.

Santic, M., R. Asare, I. Skrobonja, S. Jones, and Y. Abu Kwaik. 2008. Acquisition of the vacuolar ATPase proton pump and phagosome acidification are essential for escape of *Francisella*

tularensis into the macrophage cytosol. *Infect. Immun.* **76:**2671–2677.

Santic, M., M. Molmeret, J. R. Barker, K. E. Klose, A. Dekanic, M. Doric, and Y. Abu Kwaik. 2007. A *Francisella tularensis* pathogenicity island protein essential for bacterial proliferation within the host cell cytosol. *Cell. Microbiol.* **9:**2391–2403.

Santic, M., M. Molmeret, K. E. Klose, and Y. Abu Kwaik. 2006. *Francisella tularensis* travels a novel, twisted road within macrophages. *Trends Microbiol.* **14:**37–44.

Santic, M., M. Molmeret, and Y. Abu Kwaik. 2005a. Modulation of biogenesis of the *Francisella tularensis* subsp. *novicida*-containing phagosome in quiescent human macrophages and its maturation into a phagolysosome upon activation by IFN-gamma. *Cell. Microbiol.* **7:**957–967.

Santic, M., M. Molmeret, K. E. Klose, S. Jones, and Y. A. Kwaik. 2005b. The *Francisella tularensis* pathogenicity island protein IglC and its regulator MglA are essential for modulating phagosome biogenesis and subsequent bacterial escape into the cytoplasm. *Cell. Microbiol.* **7:**969–979.

Santic, M., M. Ozanic, V. Semic, G. Pavokovic, V. Mrvcic, and Y. A. Kwaik. 2011. Intra-vacuolar proliferation of *F. novicida* within *H. vermiformis*. *Front. Microbiol.* **2:**78.

Sato, K., H. Saito, and H. Tomioka. 1988. Enhancement of host resistance against *Listeria* infection by *Lactobacillus casei*: activation of liver macrophages and peritoneal macrophages by *Lactobacillus casei*. *Microbiol. Immunol.* **32:**689–698.

Schiano, C. A., L. E. Bellows, and W. W. Lathem. 2010. The small RNA chaperone Hfq is required for the virulence of *Yersinia pseudotuberculosis*. *Infect. Immun.* **78:**2034–2044.

Schmerk, C. L., B. N. Duplantis, P. L. Howard, and F. E. Nano. 2009a. A *Francisella novicida pdpA* mutant exhibits limited intracellular replication and remains associated with the lysosomal marker LAMP-1. *Microbiology* **155:**1498–1504.

Schmerk, C. L., B. N. Duplantis, D. Wang, R. D. Burke, A. Y. Chou, K. L. Elkins, J. S. Ludu, and F. E. Nano. 2009b. Characterization of the pathogenicity island protein PdpA and its role in the virulence of *Francisella novicida*. *Microbiology* **155:**1489–1497.

Schricker, R. L., H. T. Eigelsbach, J. Q. Mitten, and W. C. Hall. 1972. Pathogenesis of tularemia in monkeys aerogenically exposed to *Francisella tularensis* 425. *Infect. Immun.* **5:**734–744.

Schulert, G. S., and L. A. Allen. 2006. Differential infection of mononuclear phagocytes by *Francisella tularensis*: role of the macrophage mannose receptor. *J. Leukoc. Biol.* **80:**563–571.

Schulert, G. S., R. L. McCaffrey, B. W. Buchan, S. R. Lindemann, C. Hollenback, B. D. Jones, and L. A. Allen. 2009. *Francisella tularensis* genes required for inhibition of the neutrophil respiratory burst and intramacrophage growth identified by random transposon mutagenesis of strain LVS. *Infect. Immun.* **77:**1324–1336.

Sen, B., A. Meeker, and G. Ramakrishnan. 2010. The *fslE* homolog, FTL_0439 (*fupA/B*), mediates siderophore-dependent iron uptake in *Francisella tularensis* LVS. *Infect. Immun.* **78:**4276–4285.

Seshasayee, A. S., P. Bertone, G. M. Fraser, and N. M. Luscombe. 2006. Transcriptional regulatory networks in bacteria: from input signals to output responses. *Curr. Opin. Microbiol.* **9:**511–519.

Shaffer, S. A., M. D. Harvey, D. R. Goodlett, and R. K. Ernst. 2007. Structural heterogeneity and environmentally regulated remodeling of *Francisella tularensis* subspecies *novicida* lipid A characterized by tandem mass spectrometry. *J. Am. Soc. Mass Spectrom.* **18:**1080–1092.

Sittka, A., V. Pfeiffer, K. Tedin, and J. Vogel. 2007. The RNA chaperone Hfq is essential for the virulence of *Salmonella typhimurium*. *Mol. Microbiol.* **63:**193–217.

Skaar, E. P. 2010. The battle for iron between bacterial pathogens and their vertebrate hosts. *PLoS Pathog.* **6:**e1000949.

Soni, S., R. K. Ernst, A. Muszynski, N. P. Mohapatra, M. B. Perry, E. Vinogradov, R. W. Carlson, and J. S. Gunn. 2011. *Francisella tularensis* blue-gray phase variation involves structural modifications of lipopolysaccharide O-antigen, core and lipid A and affects intramacrophage survival and vaccine efficacy. *Front. Microbiol.* 1:129.

Soto, E., D. Fernandez, and J. P. Hawke. 2009. Attenuation of the fish pathogen *Francisella* sp. by mutation of the *iglC** gene. *J. Aquat. Anim. Health* 21:140–149.

Soto, E., D. Fernandez, R. Thune, and J. P. Hawke. 2011. Interaction of *Francisella asiatica* with tilapia (*Oreochromis niloticus*) innate immunity. *Infect. Immun.* 78:2070–2078.

Staples, J. E., K. A. Kubota, L. G. Chalcraft, P. S. Mead, and J. M. Petersen. 2006. Epidemiologic and molecular analysis of human tularemia, United States, 1964–2004. *Emerg. Infect. Dis.* 12:1113–1118.

Sullivan, J. T., E. F. Jeffery, J. D. Shannon, and G. Ramakrishnan. 2006. Characterization of the siderophore of *Francisella tularensis* and role of *fslA* in siderophore production. *J. Bacteriol.* 188:3785–3795.

Telepnev, M., I. Golovliov, T. Grundstrom, A. Tarnvik, and A. Sjostedt. 2003. *Francisella tularensis* inhibits Toll-like receptor-mediated activation of intracellular signalling and secretion of TNF-alpha and IL-1 from murine macrophages. *Cell. Microbiol.* 5:41–51.

Telepnev, M., I. Golovliov, and A. Sjostedt. 2005. *Francisella tularensis* LVS initially activates but subsequently down-regulates intracellular signaling and cytokine secretion in mouse monocytic and human peripheral blood mononuclear cells. *Microb. Pathog.* 38:239–247.

Thomas, L. D., and W. Schaffner. 2010. Tularemia pneumonia. *Infect. Dis. Clin. N. Am.* 24:43–55.

Twenhafel, N. A., D. A. Alves, and B. K. Purcell. 2009. Pathology of inhalational *Francisella tularensis* spp. *tularensis* SCHU S4 infection in African green monkeys (*Chlorocebus aethiops*). *Vet. Pathol.* 46:698–706.

Ulrich, L. E., E. V. Koonin, and I. B. Zhulin. 2005. One-component systems dominate signal transduction in prokaryotes. *Trends Microbiol.* 13:52–56.

Valentino, M. D., C. S. Abdul-Alim, Z. J. Maben, D. Skrombolas, L. L. Hensley, T. H. Kawula, M. Dziejman, E. M. Lord, J. A. Frelinger, and J. G. Frelinger. 2011. A broadly applicable approach to T cell epitope identification: application to improving tumor associated epitopes and identifying epitopes in complex pathogens. *J. Immunol. Methods* 373:111–126.

Viegas, S. C., V. Pfeiffer, A. Sittka, I. J. Silva, J. Vogel, and C. M. Arraiano. 2007. Characterization of the role of ribonucleases in *Salmonella* small RNA decay. *Nucleic Acids Res.* 35:7651–7664.

Vojtech, L. N., G. E. Sanders, C. Conway, V. Ostland, and J. D. Hansen. 2009. Host immune response and acute disease in a zebrafish model of *Francisella* pathogenesis. *Infect. Immun.* 77:914–925.

Vonkavaara, M., M. V. Telepnev, P. Ryden, A. Sjostedt, and S. Stoven. 2008. *Drosophila melanogaster* as a model for elucidating the pathogenicity of *Francisella tularensis*. *Cell. Microbiol.* 10:1327–1338.

Wang, L., and B. J. Cherayil. 2009. Ironing out the wrinkles in host defense: interactions between iron homeostasis and innate immunity. *J. Innate Immun.* 1:455–464.

Wang, X., A. A. Ribeiro, Z. Guan, S. N. Abraham, and C. R. Raetz. 2007. Attenuated virulence of a *Francisella* mutant lacking the lipid A 4'-phosphatase. *Proc. Natl. Acad. Sci. USA* 104:4136–4141.

Wehrly, T. D., A. Chong, K. Virtaneva, D. E. Sturdevant, R. Child, J. A. Edwards, D. Brouwer, V. Nair, E. R. Fischer, L. Wicke, A. J. Curda, J. J. Kupko III, C. Martens, D. D. Crane, C. M. Bosio, S. F. Porcella, and J. Celli. 2009. Intracellular biology and virulence determinants of *Francisella tularensis* revealed by transcriptional profiling inside macrophages. *Cell. Microbiol.* 11:1128–1150.

Weiss, D. S., A. Brotcke, T. Henry, J. J. Margolis, K. Chan, and D. M. Monack. 2007. In vivo negative selection screen identifies genes required for *Francisella* virulence. *Proc. Natl. Acad. Sci. USA* 104:6037–6042.

West, T. E., M. R. Pelletier, M. C. Majure, A. Lembo, A. M. Hajjar, and S. J. Skerrett. 2008. Inhalation of *Francisella novicida* Δ*mglA* causes replicative infection that elicits innate and adaptive responses but is not protective against invasive pneumonic tularemia. *Microbes Infect.* 10:773–780.

Whipp, M. J., J. M. Davis, G. Lum, J. de Boer, Y. Zhou, S. W. Bearden, J. M. Petersen, M. C. Chu, and G. Hogg. 2003. Characterization of a *novicida*-like subspecies of *Francisella tularensis* isolated in Australia. *J. Med. Microbiol.* 52:839–842.

Wu, T. H., J. L. Zsemlye, G. L. Statom, J. A. Hutt, R. M. Schrader, A. A. Scrymgeour, and C. R. Lyons. 2009. Vaccination of Fischer 344 rats against pulmonary infections by *Francisella tularensis* type A strains. *Vaccine* 27:4684–4693.

Yu, J. J., E. K. Raulie, A. K. Murthy, M. N. Guentzel, K. E. Klose, and B. P. Arulanandam. 2008. The presence of infectious extracellular *Francisella tularensis* subsp. *novicida* in murine plasma after pulmonary challenge. *Eur. J. Clin. Microbiol. Infect. Dis.* 27:323–325.

Zarrella, T. M., A. Singh, C. Bitsaktsis, T. Rahman, B. Sahay, P. J. Feustel, E. J. Gosselin, T. J. Sellati, and K. R. Hazlett. 2011. Host-adaptation of *Francisella tularensis* alters the bacterium's surface-carbohydrates to hinder effectors of innate and adaptive immunity. *PLoS One* 6:e22335.

Zhao, J., and C. R. Raetz. 2010. A two-component Kdo hydrolase in the inner membrane of *Francisella novicida*. *Mol. Microbiol.* 78:820–836.

V. STRESS RESPONSE DURING INFECTION

Regulation of Bacterial Virulence
Edited by Michael L. Vasil and Andrew J. Darwin
© 2013 ASM Press, Washington, DC   doi:10.1128/9781555818524.ch22

Chapter 22

Regulation of *Salmonella* Resistance to Oxidative and Nitrosative Stress

CALVIN A. HENARD AND ANDRÉS VÁZQUEZ-TORRES

INTRODUCTION

Many of the over 2,500 serovars of *Salmonella enterica* cause significant morbidity and mortality worldwide. Nontyphoidal salmonellae residing in the gastrointestinal tract of wild and domestic animals are transmitted to humans after the consumption of contaminated food or water. Nontyphoidal salmonellae primarily cause gastroenteritis; however, these serovars can disseminate to systemic sites in immunocompromised individuals with low CD4+ T-cell counts or defects in gamma interferon (IFN-γ) signaling (de Jong et al., 1998; Rongkavilit et al., 2000). The species *S. enterica* also includes the human-adapted serovars Typhi and Paratyphi, which cause life-threatening typhoid fever. Upon reaching the intestine, salmonellae invade enterocytes and specialized M cells in Peyer's patches or are taken up from the lumen by phagocytic cells (Jones et al., 1994; Vazquez-Torres et al., 1999). Salmonellae have affinity for macrophages populating gut-associated tissue and systemic sites. The ability of salmonellae to survive and replicate within macrophages is a hallmark of the pathogenesis of these intracellular bacteria (Fields et al., 1986). Intracellular salmonellae are exposed to a plethora of toxic species derived from superoxide ($O_2^{\cdot-}$) and NO, which are synthesized by NADPH phagocyte oxidase (Phox or NOX2) and inducible NO synthase (iNOS) enzymatic complexes, respectively. The role that the NADPH oxidase plays in resistance to salmonellosis is underscored by the fact that chronic granulomatous disease patients with a nonfunctional NADPH phagocyte oxidase are hypersusceptible to *Salmonella* infection (Mouy et al., 1989). Immunocompromised mice lacking NADPH oxidase or iNOS have experimentally established the importance of these hemoproteins, and the myriad of reactive species they produce, in host defense against *Salmonella*. Although many bacteria are killed in the harsh intracellular environment of macrophages, some salmonellae survive the onslaught. How do salmonellae survive inside cells of the innate host defense system under such hostile conditions? Mounting evidence indicates that the concerted action of redox sensors of reactive oxygen and nitrogen species (ROS and RNS, respectively) enhances the intracellular fitness of *Salmonella* in macrophages. This chapter presents the molecular mechanisms used by *Salmonella* to sense and respond to reactive species encountered at various phases during the infectious cycle.

EXPOSURE OF *SALMONELLA* TO ROS

Salmonellae are exposed to ROS produced endogenously through the univalent or divalent reduction of O_2 by enzymes of the electron transport chain or cytoplasmic flavoproteins. Intracellular salmonellae are also exposed to high fluxes of ROS generated exogenously by the enzymatic activity of the NADPH phagocyte oxidase during the innate response of phagocytic cells.

Endogenous Sources of Oxidative Stress

O_2 is the most energetically favorable terminal electron acceptor of the electron transport chain. O_2 is completely reduced to H_2O by Cu^{2+} and heme-based terminal quinol cytochrome oxidases. Although the terminal step in respiration is remarkably efficient, electrons can leak from upstream steps in the electron transport chain. For example, electrons at the Q site of NADH dehydrogenases univalently reduce O_2 to $O_2^{\cdot-}$ (Boveris and Chance, 1973). While the respiratory chain is an important source of $O_2^{\cdot-}$, other cellular activities also generate $O_2^{\cdot-}$ and hydrogen peroxide (H_2O_2). Flavoenzymes in the cytoplasm can

Calvin A. Henard and Andrés Vázquez-Torres • Department of Microbiology, University of Colorado Denver School of Medicine, Aurora, CO 80045.

adventitiously transfer one or two electrons to O_2, producing steady fluxes of $O_2^{\cdot-}$ and H_2O_2 (Messner and Imlay, 1999). The flavin-dependent dehydrogenase NadB in the nicotinamide biosynthetic pathway is a primary source of H_2O_2 in the cytoplasm of *Escherichia coli* (Korshunov and Imlay, 2010). All in all, actively metabolizing *E. coli* generates 5 µM/s $O_2^{\cdot-}$ and 15 µM/s H_2O_2 (Seaver and Imlay, 2004; Korshunov and Imlay, 2006). However, only a third of the intracellular H_2O_2 produced during metabolic processes has been accounted for (Korshunov and Imlay, 2010). Therefore, identifying possible sources of endogenous oxidative stress remains an interesting and active line of research.

ROS Produced by NADPH Phagocyte Oxidase

As an intracellular pathogen, *Salmonella* must deal with a burst of oxyradicals produced by the NADPH phagocyte oxidase in the innate response of macrophages (Vazquez-Torres et al., 2000a). The NADPH phagocyte oxidase is an enzymatic complex consisting of the membrane-associated subunits gp-91*phox* and p22*phox*, comprising cytochrome b_{558}, and the cytosolic components p40*phox*, p47*phox*, p67*phox*, and the GTPase Rac. The phagocytic process stimulates assembly of cytosolic components and the cytochrome b_{558} in the membrane to form a highly active NADPH oxidase enzymatic complex. Recognition of pathogen-associated molecular patterns, such as outer membrane proteins and lipopolysaccharide in *Salmonella*'s cell envelope, induces the assembly of an active NADPH phagocyte oxidase (Chateau and Caravano, 1997; Berger et al., 2010). NADPH phagocyte oxidase can also be activated by proinflammatory cytokines such as tumor necrosis factor alpha and interleukin 1. Transfer of electrons from NADPH to molecular O_2 results in the production of high concentrations of $O_2^{\cdot-}$ in a process termed the "respiratory burst" (reviewed in Bylund et al., 2010). The NADPH oxidase delivers >100 µM $O_2^{\cdot-}$ in the phagosome, unleashing the generation of H_2O_2 and other reactive species (Winterbourn et al., 2006).

The respiratory burst of phagocytes plays a critical role in controlling *Salmonella* infection. Recessive X-linked gp91*phox* or autosomal p22*phox*, p47*phox*, or p67*phox* mutations dramatically increase the susceptibility of chronic granulomatous disease patients to *Salmonella* (Mouy et al., 1989). Mice lacking either gp91*phox* or p47*phox* are excellent experimental animal models of chronic granulomatous disease in humans, and particularly gp91*phox*-deficient mice accurately recapitulate the important role that the NADPH phagocyte oxidase plays in resistance to

Salmonella (Mastroeni et al., 2000). Mice deficient in gp91*phox* succumb to *Salmonella* infection within 5 days after intraperitoneal inoculation. Systemic viscera of gp91*phox* knockout mice are loaded with a 10^3- to 10^4-fold-higher *Salmonella* burden than wild-type congenic mice a few days after challenge. The hypersusceptibility of gp91*phox*-deficient mice correlates with the poor anti-*Salmonella* activity of their macrophages (Mastroeni et al., 2000; Vazquez-Torres et al., 2000a). Interestingly, *Salmonella* can exploit the NADPH oxidase-dependent production of ROS in the gastrointestinal tract (Winter et al., 2010). Oxyradicals generated by the NADPH phagocyte oxidase react with sulfur compounds in the gut lumen, generating the alternative electron acceptor tetrathionate. The availability of tetrathionate allows *Salmonella* to outcompete the normal gut microbiota, thereby enhancing the virulence of this enteropathogen in the intestinal phase of the infection.

Antioxidant Defenses of *Salmonella*

Given the substantial generation of ROS, aerobic organisms express several superoxide dismutases (SODs)/reductases, catalases/peroxidases, thioredoxins, and glutaredoxins to detoxify $O_2^{\cdot-}$ and H_2O_2 or to repair lesions incurred by these toxic species. The cytoplasm of *Salmonella* is also rich in low-molecular-weight thiols and NADPH, which provide reducing power to thioredoxin reductase, glutaredoxin, and glutathione oxidoreductase detoxifying systems. These antioxidant defenses are remarkably efficient at protecting bacteria from steady-state production of ROS during endogenous metabolism. However, if concentrations of reactive species exceed the activities of the aforementioned enzymes, other detoxifying and repair mechanisms are often evoked.

MOLECULAR TARGETS OF ROS

The molecular targets of ROS are depicted in Fig. 1. As a negatively charged molecule, $O_2^{\cdot-}$ does not freely cross membranes, but it can enter the periplasmic space of gram-negative organisms through porins. On the other hand, H_2O_2 freely diffuses through bacterial cell membranes. The molecular targets of $O_2^{\cdot-}$ and H_2O_2 started to be elucidated in seminal work by Touati's group (Carlioz and Touati, 1986). Aerobically grown *E. coli* lacking SODs or catalases/peroxidases displays growth defects associated with auxotrophies for branched-chain, aromatic, and sulfur-containing amino acids (Boehm et al., 1976; Brown et al., 1979; Carlioz and Touati, 1986; Jang

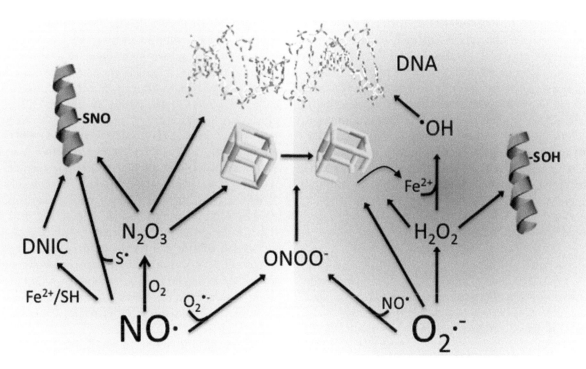

Figure 1. Molecular targets of RNS- and ROS-mediated anti-*Salmonella* activity. N₂O₃ formed in the reaction of NO with molecular oxygen O₂ is one of the indirect means by which NO causes cytotoxicity (blue box). NO forms ONOO⁻ through its interactions with superoxide anion (O₂⁻), and dinitrosyl-iron complexes (DNIC) by reacting with iron and low-molecular-weight thiols (-SH). The strong oxidant ONOO⁻ targets [4Fe-4S] clusters of dehydratases. The NO radical can also react directly with the sulfenyl radical (-S˙) to form S-nitrosylated protein derivatives. Moreover, N₂O₃ and DNIC are common sources of transnitrosation reactions and nitrosative stress. NO₂˙, N₂O₃, ONOO⁻, O₂⁻, and H₂O₂ are common sources of oxidative stress (purple box). These species damage [Fe-S] clusters, liberating catalytically active Fe²⁺. In turn, Fe²⁺ reduces H₂O₂ to the highly reactive hydroxyl radical (˙OH), which causes extensive DNA damage. H₂O₂ also oxidizes reactive cysteine residues in proteins to form sulfenic acid derivatives (-SOH). O₂ and NO also target copper and heme cofactors in terminal cytochromes of the electron transport chain; however, hemoprotein targets are not depicted because heme-based sensors of O₂ or NO have not yet been identified in *Salmonella*.
doi:10.1128/9781555818524.ch22f1

and Imlay, 2007). Dihydroxyacid dehydratase, an enzyme of the branched-chain amino acid anabolic pathway, was the first identified target of O_2^{-} (Boehm et al., 1976). The inhibition of dihydroxyacid dehydratase by O_2^{-} and H_2O_2 occurs through the inactivation of the solvent-exposed Fe_α in the catalytic $[4Fe-4S]^{2+}$ cluster (Kuo et al., 1987). O_2^{-} and H_2O_2 oxidize $[4Fe-4S]^{2+}$ clusters of dehydratases to an inactive $[3Fe-4S]^{+}$ state (Jang and Imlay, 2007). The biosynthetic defects in aromatic and sulfur-containing amino acids exhibited by SOD- and catalase/peroxidase-deficient *E. coli* are less clear. Aromatic amino acid biosynthesis is dependent on the production of erythrose-4-phosphate by transketolase as part of the nonoxidative branch of the pentose phosphate pathway. O_2^{-} may interfere with the catalytic function of transketolase, thereby limiting carbon flow for aromatic amino acid biosynthesis, but strong evidence for this is still lacking (Benov and Fridovich, 1999). Sulfur metabolism is

clearly interrupted in *E. coli* lacking SOD enzymatic activity (Carlioz and Touati, 1986). SOD mutants excrete sulfite, suggesting a blockage in sulfate assimilation (Benov and Fridovich, 1997). Based on sequence homology to spinach nitrate reductase, sulfite reductases from *E. coli* and *Salmonella* are suspected to contain a $[4Fe-4S]^{2+}$ cluster in their active site (Ostrowski et al., 1989). Diminished sulfite reductase activity provides a rationale for the observed defects in sulfur metabolism in strains lacking proper antioxidant defenses. The inhibition of synthesis of amino acids and nonfermentable metabolites can cause growth arrest. The effect of ROS on *Salmonella* central metabolism may be especially pertinent in phagosomes of macrophages, where nutrients might be a limited resource.

H_2O_2 oxidizes labile ferrous iron (Fe^{2+}) generating hydroxyl radical (OH˙) (equations 1 and 2), a species that reacts at diffusion-limiting rates with most biomolecules encountered near the site of its formation (Imlay et al., 1988). Because some free iron is associated with DNA, the chromosome is a preferred target of OH· (Levin et al., 1982; Farr et al., 1986). In this model, $O_2^{·-}$ supports DNA damage by releasing iron from [Fe-S] clusters of dehydratases (Kuo et al., 1987). The importance of DNA as a target of H_2O_2 is shown by the fact that mutations in *recA* of the SOS response render *Salmonella* avirulent, while a catalase-deficient strain remains virulent (Buchmeier et al., 1993; Buchmeier et al., 1995). Therefore, repair of DNA damage appears to be an integral component of antioxidant defense.

$$Fe^{2+} + H_2O_2 \rightarrow FeO_2^+ + H_2O \qquad (1)$$
$$FeO_2^+ + H^+ \rightarrow Fe^{3+} + OH· \qquad (2)$$

H_2O_2 also has a high affinity for redox-active sulfhydryl groups. Interestingly, the posttranslational oxidation of a thiol group to a sulfenic acid derivative can affect protein function (Hoffmann et al., 2004). Sulfenic acid formation is reversible, making this modification a signaling mechanism comparable to phosphorylation of aspartate in prokaryotes and serine, threonine, and tyrosine in eukaryotes. The importance of thiol-mediated sensing of ROS and RNS has been established in both prokaryotes and eukaryotes. A role of thiol oxidation in *Salmonella* signaling is discussed below.

REGULATORY NETWORKS THAT COORDINATE ANTIOXIDANT DEFENSES

The adaptation of *Salmonella* to $O_2^{·-}$ and H_2O_2 involves different regulatory systems, emphasizing differences in reactivity and molecular targets between these two ROS. Given the aforementioned affinity of $O_2^{·-}$ and H_2O_2 for sulfur in the side chain of cysteine and [Fe-S] clusters, it is not surprising that OxyR (oxidative stress response), SoxR (superoxide response), and FNR (fumarate/nitrate reduction) have incorporated these moieties as sensors of oxidative stress (Fig. 2). Many of the mechanisms mediating $O_2^{·-}$ and H_2O_2 sensing have been elucidated in *E. coli*, but these basic principles appear to be conserved in many prokaryotes, including the closely related *Salmonella*. OxyR, SoxR, and FNR are for most part dedicated sensors of H_2O_2, $O_2^{·-}$, and O_2, respectively. However, these regulators can also sense RNS, an aspect that is discussed separately.

Sensing of $O_2^{·-}$ by SoxRS

$O_2^{·-}$ activates the transcription of ~100 genes, many of which are under the control of the SoxRS

system in *E. coli* and *Salmonella* (Pomposiello and Demple, 2000; Pomposiello et al., 2001). SoxR acts as a homodimer containing two redox-active [2Fe-2S]$^+$ clusters (Hidalgo et al., 1995). As a negatively charged diatomic radical, $O_2^{·-}$ is electrostatically attracted to the positively charged [2Fe-2S]$^+$ cluster of SoxR. The low redox potential of SoxR (−290 mV) ensures its quick oxidation by $O_2^{·-}$ (Ding et al., 1996). SoxR is oxidized in less than 2 min of exposure to as little as 100 μM of the -generating, redox-cycling paraquat (Ding and Demple, 1997). However, given its low redox potential, several oxidants should theoretically react with SoxR, and thus it has been proposed that SoxR may not recognize $O_2^{·-}$ per se but instead respond to the redox balance of the cell (Krapp et al., 2011). Accordingly, perturbation of the NAD(P)H/NAD(P)$^+$ ratio by redox cycling drugs (e.g., menadione and paraquat) has been noted to activate SoxR (Dietrich and Kiley, 2011; Gu and Imlay, 2011).

Reduced and oxidized SoxR bind to the promoter of the *soxS* gene; however, only the oxidized SoxR recruits the RNA polymerase to the promoter of the *soxS* gene, its only known target (Gaudu et al., 1997). The oxidized [2Fe-2S]$^{2+}$ cluster induces conformational changes in the dimer and thereby triggers the formation of an open complex at the *soxS* promoter. The SoxS protein enhances the transcription of several genes in the SoxRS regulon, some of which directly protect the cell against $O_2^{·-}$ stress. For instance, the *sodA*-encoded cytoplasmic Mn-containing SOD detoxifies $O_2^{·-}$ to H_2O_2. Of note, SodA is required for the intracellular survival of *Salmonella* within J774 macrophage-like cells (Tsolis et al., 1995; Pomposiello and Demple, 2000). Other components regulated by the system include the *nfo*-encoded endonuclease involved in DNA repair, *zwf*-encoded glucose-6-phosphate dehydrogenase that generates NADPH reducing equivalents, and oxidant-resistant fumarase C and aconitase A of central metabolic pathways (Greenberg et al., 1990; Tsaneva and Weiss, 1990; Liochev and Fridovich, 1992; Gruer and Guest, 1994). The SoxRS-dependent regulation of the *zwf*-encoded glucose-6-phosphate dehydrogenase helps recover redox homeostasis. Interestingly, the reductive power provided through glucose-6-phosphate dehydrogenase enzymatic activity mediates resistance of *Salmonella* against the respiratory burst of macrophages and optimizes virulence of this enteropathogen in mice (Lundberg et al., 1999). The SoxRS-dependent induction of *fur*, which encodes a global repressor of ferric iron uptake, may lessen the formation of OH· from H_2O_2 by limiting iron availability (Zheng et al., 1999). The roles of other members of the SoxRS regulon are less clear. Expression

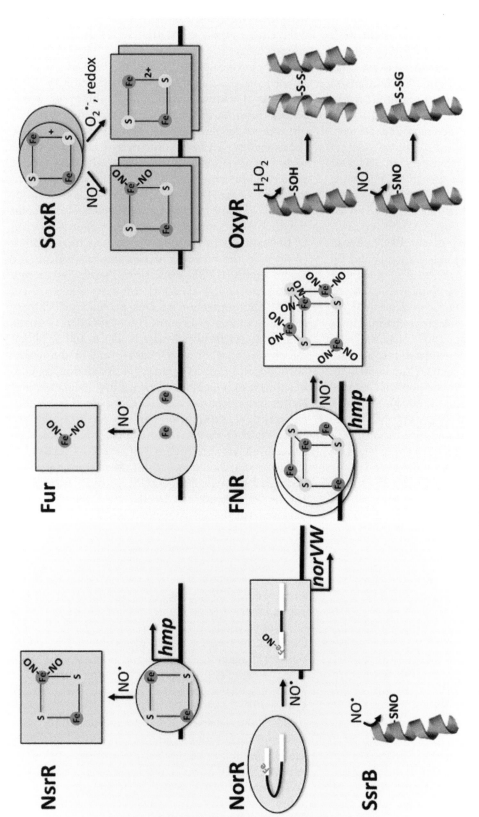

Figure 2. Sensors of oxidative and nitrosative stress. The dedicated NO sensors NsrR and NorR react with NO (yellow box). Dinitrosyl-iron complex formation in the NsrR [2Fe-2S] cluster derepresses transcription of target genes such as *hmp*, encoding a flavohemoglobin that detoxifies NO to NO_3^-. The NorR metalloprotein containing a nonheme iron center is also activated by NO. The N-terminal regulatory domain of NorR represses *norVW*, encoding flavorubredoxin and associated oxidoreductase. The formation of a mononitrosyl-iron species in NorR activates transcription. The redox-active thiol of Cys[203] of the SsrB response regulator that controls SPI2 gene transcription is the first thiol-based sensor of RNS to be identified in *Salmonella*. Some sensors such as Fur, FNR, SoxR, and OxyR can respond to both oxidative and nitrosative stress (green box). The transcriptional repressors Fur and FNR bind to DNA as homodimers. Dinitrosyl-iron complexes disrupt the DNA binding activity of Fur and FNR, derepressing transcription. Fur can be indirectly activated by oxidative stress-mediated disruption of iron homeostasis (not shown). O_2 and $O_2^{\cdot-}$ oxidize the [4Fe-4S] cluster of FNR (not shown). The [2Fe-2S] cluster of SoxR is primarily dedicated to sensing and redox changes in the cell. Conformational changes associated with the oxidation or nitrosylation of SoxR [2Fe-2S][+] activate *soxS* transcription. OxyR Cys[199] is a primary sensor of H_2O_2. H_2O_2 oxidizes the Cys[199] thiolate to sulfenic acid, which condenses with Cys[208] to form an intramolecular disulfide. OxyR Cys[199] can also be S nitrosylated and form a mixed disulfide with glutathione (-SG). Both oxidized and RNS-modified OxyR are transcriptionally active. doi:10.1128/9781555818524.ch22f2

of *tolC* and the *acrAB*-encoded efflux pump possibly enhances the exclusion of biomolecules damaged by O_2^--derived reactive species (Chou et al., 1993; Ma et al., 1996; Aono et al., 1998). Consistent with the notion that [Fe-S] clusters are preferred targets of O_2^-, the SoxRS system also induces expression of flavodoxin A, flavodoxin B, and YggX involved in iron trafficking and [Fe-S] cluster repair (Liochev and Fridovich, 1992; Gaudu and Weiss, 2000; Gralnick and Downs, 2001; Gralnick and Downs, 2003).

Collectively, the SoxRS system responds to, and protects enteric bacteria against, O_2^- stress. However, this regulator is dispensable for *Salmonella* pathogenesis and does not contribute to the survival of these intracellular bacteria within NADPH oxidase-expressing macrophages (Fang et al., 1997). These seemingly paradoxical findings may be explained by the fact that several members of the SoxRS regulon respond to other regulatory systems as well, many of which are important for *Salmonella* virulence (e.g., σ^s and FNR) (Fang et al., 1992; Hassan and Sun, 1992; Koh and Roe, 1996; Fink et al., 2007). Alternatively, the dispensability of the SoxRS-regulated cytoplasmic SodA Mn-SOD might reflect that O_2^- generated by the host NADPH phagocyte oxidase poisons extracytoplasmic targets (Craig and Slauch, 2009). Thus, the bacteriophage-encoded Cu-Zn periplasmic SOD contributes to *Salmonella* virulence in mice by detoxifying O_2^- in the periplasm (De Groote et al., 1997). The protective actions of Cu-Zn SOD would render cytoplasmic SodA and SodB dispensable for resistance to NADPH oxidase-derived O_2^- (De Groote et al., 1997; Uzzau et al., 2002; Krishnakumar et al., 2004).

Sensing of H_2O_2 by OxyR

OxyR belongs to the LysR family of transcriptional regulators. OxyR contains a redox-active cysteine residue at position 199 that actively responds to H_2O_2 (Zheng et al., 1998). What makes a particular cysteine reactive while the majority of them are not? The accessibility and the chemical environment of the cysteine, particularly the surrounding amino acid residues, can increase the reactivity of its sulfur with H_2O_2. Charged amino acids in the neighborhood of OxyR Cys^{199} lower the pK_a of the thiol side chain, keeping it in the thiolate ionic form at physiological pH (Kullik et al., 1995). Furthermore, hydrophobicity and allosteric interactions that stabilize the thiolate are important determinants for the reactivity of cysteines. The Cys^{199} thiolate group in the OxyR protein reacts with H_2O_2 with second-order rate kinetics of ~1.1×10^5 $M^{-1}s^{-1}$ to form a sulfenic acid (SOH) derivative (Lee et al., 2004). Biochemical and structural

evidence suggest that the sulfenylated Cys^{199} condenses with Cys^{208} to form an intramolecular disulfide bond (Zheng et al., 1998; Choi et al., 2001). The disulfide bond changes the conformation of OxyR, which binds cooperatively as a tetramer to four adjacent major grooves of target genes (Choi et al., 2001).

Oxidized OxyR activates transcription of about 30 genes (Christman et al., 1985), many of which encode enzymes that have antioxidant function. For instance, hydroperoxidase I and alkyl hydroperoxide reductase detoxify H_2O_2, organic hydroperoxides, and peroxynitrite ($ONOO^-$) (Bryk et al., 2000; Zheng et al., 2001; McLean et al., 2010a). The OxyR-dependent induction of the [Fe-S] cluster repair/assembly *suf* operon helps repair oxidized metalloproteins (Jang and Imlay, 2010). Similar to SoxS-dependent activation of *fur* expression, the OxyR-mediated transcription of *fur* and the consequent decrease in iron uptake likely diminish OH^- formation, thereby protecting DNA from Fenton-mediated toxicity (Zheng et al., 1999). The expression of Dps protein further protects DNA from H_2O_2- and Fe^{2+}-dependent damage by (i) nonspecifically binding to DNA and shielding the chromosome from OH^- generated in the neighborhood and (ii) abrogating the formation of OH^- in situ through its ability to bind Fe^{2+} (Martinez and Kolter, 1997; Zhao et al., 2002; Pacello et al., 2008). Dps-deficient salmonellae are hypersusceptible to authentic H_2O_2 in vitro and display attenuated virulence in an acute model of salmonellosis (Halsey et al., 2004; Pacello et al., 2008). It remains to be investigated whether Dps actually protects *Salmonella* against ROS generated by the NADPH phagocyte oxidase. The OxyR-dependent induction of glutathione reductase, glutaredoxin 1, and thioredoxin 2 ensures that the cell maintains cellular thiol/disulfide exchange balance. Interestingly, the Cys^{199}-Cys^{208} disulfide of OxyR is a substrate of glutaredoxin-1 (Aslund et al., 1999). This finding has two important implications. (i) Cys^{199} and Cys^{208} constitute a bona fide thiol switch adapted to sensing H_2O_2 and (ii) OxyR is the target of feedback regulation by members of the regulon. Although the expression of many genes of the OxyR regulon, such as *dps* and *ahpC*, is induced in macrophages, OxyR is not required for *Salmonella* virulence (Francis et al., 1997; Taylor et al., 1998; Aussel et al., 2011). In addition, both *katG* and *ahpC* null mutants are as virulent as wild-type *Salmonella* in murine models of infection (Buchmeier et al., 1995; Taylor et al., 1998). The dispensability of OxyR may be attributed to the compensatory regulation of catalases/peroxidases (Robbe-Saule et al., 2001). Salmonellae express at least five catalases/peroxidases with overlapping contributions to virulence in mice and macrophages (Hebrard et al., 2009). The attenuation of a *Salmonella* strain deficient in all catalase/peroxidase enzymatic activity is nonetheless difficult to interpret, because the growth of this strain is inhibited by atmospheric O_2.

Collectively, it is clear that the OxyR- and SoxRS-mediated responses to H_2O_2 and O_2^- help maintain redox and metabolic homeostasis in the presence of endogenous ROS. However, these regulators appear to play secondary roles in the antioxidant defenses of *Salmonella* against reactive species generated in the inflammatory response. Other mechanisms must assume a protective role when salmonellae are exposed to the oxidative stress associated with the respiratory burst of macrophages. For example, the FNR metalloprotein appears to play an essential role coordinating the antioxidant defense of *Salmonella* in an acute model of infection (Fink et al., 2007). Moreover, in coordination with classical antioxidant defenses, the type III secretion system encoded within *Salmonella* pathogenicity island 2 (SPI2) provides *Salmonella* an additional strategy of defense against the cytotoxicity associated with the enzymatic activity of the NADPH phagocyte oxidase.

Sensing of O_2 and O_2^- by FNR

FNR is an O_2-sensitive global transcriptional regulator. The FNR protein contains a $[4Fe-4S]^{2+}$ cluster that is converted to a $[2Fe-2S]^{2+}$ upon exposure to O_2 or O_2^- (Khoroshilova et al., 1997; Crack et al., 2004). Reduced $[4Fe-4S]^{2+}$ FNR binds as a dimer to consensus target DNA sites, acting as a repressor or an activator. Under anaerobic conditions, FNR controls the expression of >300 genes, many of which are involved in metabolism (Fink et al., 2007). Oxidation of the $[4Fe-4S]^{2+}$ cluster disrupts the dimer and diminishes the affinity of FNR for cognate promoters. FNR-dependent regulation is important for *Salmonella* virulence in the gastrointestinal phase of infection (Fink et al., 2007). Surprisingly, FNR is needed for *Salmonella* intracellular survival and for virulence in a systemic model of acute infection dominated by the antimicrobial actions of the NADPH phagocyte oxidase (Fink et al., 2007). This is unexpected since studies suggest that phagosomes harbor adequate levels of O_2 for *Salmonella* to perform aerobic metabolism (Eriksson et al., 2003). Salmonellae lacking *fnr* become virulent in gp91*phox*-deficient mice, suggesting that FNR-regulated gene products antagonize the cytotoxicity of ROS generated by the NADPH phagocyte oxidase (Fink et al., 2007). The susceptibility of *fnr* mutant *Salmonella* to ROS might be associated with the general effects of FNR on metabolic pathways (Hassan and Sun, 1992; Fang et al., 1997).

SALMONELLA EVASION OF THE NADPH PHAGOCYTE OXIDASE BY SPI2

The ability of *Salmonella* to survive the harsh intracellular environment of professional phagocytes is greatly influenced by the expression of the horizontally acquired, 25-kb SPI2 gene cluster in the *Salmonella* chromosome (Cirillo et al., 1998). The SPI2 locus encodes a type III secretion system that translocates effector proteins into the host cell cytoplasm. SPI2 effectors interfere with a variety of host cell functions. Of interest to this chapter, SPI2 lessens the oxidative and nitrosative stress that *Salmonella* must endure within macrophages (Vazquez-Torres et al., 2000b; Gallois et al., 2001; Vazquez-Torres et al., 2001; Chakravortty et al., 2002; Suvarnapunya and Stein, 2005; Berger et al., 2010). SPI2 prevents the contact of phagosomes with NADPH phagocyte oxidase enzymatic complexes by excluding both cytochrome b_{558} from the phagosomal membrane and vesicles harboring preformed NADPH oxidase (Vazquez-Torres et al., 2000b; Gallois et al., 2001). Recent evidence indicates that SPI2 protects intraphagosomal *Salmonella* from the respiratory burst triggered by the signalling lymphocyte-activation molecule family (SLAM)-dependent activation of intracellular class III phosphatidylinositol kinase Vps34 and its regulatory protein kinase Vps15 (Berger et al., 2010). SPI2 mutant salmonellae are exposed to higher levels of ROS than isogenic wild-type controls, as indicated by the extent of DNA damage (Suvarnapunya and Stein, 2005). Therefore, the SPI2 type III secretion system minimizes exposure of *Salmonella* to O_2^- while fostering its replication in a protected intracellular niche.

EXPOSURE OF *SALMONELLA* TO RNS

Salmonellae have several regulators that control responses to host-derived nitrogen oxides. Salmonellae are exposed to NO in the gastrointestinal tract and within the intracellular niche of macrophages. In this section, we briefly discuss the sources of NO and the chemistry of RNS relevant to *Salmonella* pathogenesis. The reader is directed to Henard and Vazquez-Torres, 2011, to find more about this fascinating aspect of *Salmonella* pathogenesis.

Nonenzymatic Production of RNS in the Stomach

The contents of the gastric lumen and, in particular, nitrogen oxides generated in the low pH of the stomach represent the first serious barrier encountered by enteropathogenic *Salmonella* in vertebrate hosts. Nitrate (NO_3^-) concentrated in the salivary glands through enterosalivary circulation is converted to nitrite (NO_2^-) by the resident microbiota of the tongue (Duncan et al., 1995). In the low pH of the stomach, NO_2^- is protonated to nitrous acid (HNO_2), a precursor for several RNS, including NO, nitrogen dioxide (NO_2), and dinitrogen trioxide (N_2O_3) (Lundberg et al., 2004). The battery of nitrogen oxides produced in the stomach increases the antimicrobial defenses of the gastric lumen against enteric pathogens such as *Salmonella* (Bourret et al., 2008).

Enzymatic Production of RNS

NO is produced in large amounts through the enzymatic activity of iNOS as part of the innate immune response. *Salmonella* infection of the gastrointestinal mucosa elicits an inflammatory response rich in NO synthesis. The NO produced in the gastrointestinal tract adds to host defense against *Salmonella*, as suggested by the increased localization of salmonellae to Peyer's patches and increased invasion of gut tissue in iNOS-deficient mice (Ackermann et al., 2008; Alam et al., 2008). Tissue macrophages and epithelial cells are the sources of RNS in the gastrointestinal mucosa of mice and humans infected with *Salmonella* (Witthoft et al., 1998; Giacomodonato et al., 2003; Enocksson et al., 2004).

Macrophages infected with *Salmonella* generate an arsenal of RNS. The iNOS hemoprotein is expressed in response to *Salmonella*'s lipopolysaccharide, fimbriae, and porins (Vazquez-Torres et al., 2004; Vitiello et al., 2008; Buzzo et al., 2010). Moreover, flagellin, SipB, and SopE secreted by the SPI1 type III secretion system also induce expression of iNOS (Cherayil et al., 2000; Muller et al., 2009). *Salmonella* can handle most of the NO produced in the innate response of macrophages; however, high NO fluxes generated in response to IFN-γ enhance the anti-*Salmonella* activity of macrophages (Vazquez-Torres et al., 2000a). NO exerts bacteriostasis in IFN-γ-activated macrophages. In addition, high NO fluxes generated by IFN-γ-activated macrophages promote the maturation of the *Salmonella* phagosome along the degradative pathway. The fusion of *Salmonella* phagosomes with lysosomes in IFN-γ-activated macrophages is a manifestation of the NO-mediated inhibition of SPI2 transcription (McCollister et al., 2005). The importance of NO in resistance to experimental murine salmonellosis is demonstrated by increased mortality of iNOS knockout mice challenged with *Salmonella* (Shiloh et al., 1999; Mastroeni et al., 2000).

MOLECULAR TARGETS OF RNS

The biological chemistry of NO is derived from both direct and indirect effects. NO can react quite specifically with a limited number of molecular targets, but the presence of O_2 enriches the oxidative and nitrosative chemistry of this diatomic radical (Fig. 1). O_2^- produced endogenously during respiration, or exogenously in the respiratory burst, reacts with NO to generate $ONOO^-$. Collectively, these RNS modify a variety of biomolecules. NO readily reacts with metal prosthetic groups such as Cu^{2+} and heme in the active site of terminal quinol cytochrome oxidases of the electron transport chain. Nitrosylation of metal centers in quinol cytochrome oxidases is fast and specific. For example, NO interacts with *bd* quinol cytochrome oxidase with a second-order rate constant of 3×10^8 $M^{-1}s^{-1}$ (Mason et al., 2009). Consequently, even at low fluxes of NO, quinol cytochrome oxidases are a prime target of NO (Husain et al., 2008). NO also reacts with the $Fe\alpha$ of [Fe-S] clusters in dehydratases in several central metabolic pathways (Keyer and Imlay, 1997; Brandes et al., 2007). Similar to ROS-mediated disruption of the [Fe-S] cluster of dihydroxyacid dehydratase, RNS inhibit branched-chain amino acid synthesis and aconitases of the tricarboxylic acid cycle (Hausladen and Fridovich, 1994). The generation of $ONOO^-$ enhances the NO-dependent oxidation of [Fe-S] clusters (Keyer and Imlay, 1997).

RNS react with redox active sulfhydryls in cysteines to produce *S*-nitrosothiols, disulfide bonds, and *S*-thiolated and sulfenylated derivatives (A. Vazquez-Torres, unpublished data). *S*-nitrosylation occurs not from the direct reaction of NO with thiols but, rather, through nucleophilic attack of a thiolate anion ($-S^-$) to a nitrosonium cation (NO^+). However, NO^+ does not occur free in solution. N_2O_3 produced in the autoxidation of NO with O_2 and transnitrosation reactions from *S*-nitrosothiols and dinitrosyl iron complexes are sources of NO^+-like electrophiles. Alternatively, S-nitrosylation can occur through the direct reaction of a sulfenyl radical ($-S^-$) with the diatomic NO radical. In addition to *S*-nitrosylation, RNS can engender a variety of S-thiolated products or oxidative products such as sulfenic acid and disulfide bonds, all of which contribute to important aspects of the biological chemistry of NO congeners. As it will be seen below, direct and indirect reactions of NO with metalloproteins and redox-active cysteines are common themes for how *Salmonella*, *E. coli*, and other bacteria sense RNS.

SALMONELLA REGULATORY NETWORKS THAT COORDINATE DEFENSES AGAINST RNS

NO is an intermediate of denitrification and chemical reactions in the gut lumen, and it is one of the toxic species produced by macrophages and epithelial cells.

Salmonella possesses three devoted NO sensors: the global regulator NsrR, the ATPase NorR, and the two-component regulatory protein SsrB (Fig. 2). Furthermore, *Salmonella* uses SoxR, OxyR, FNR, and Fur transcriptional regulators to sense both ROS and RNS.

Dedicated NO Sensors

Dedicated NO sensors are activated through the direct nitrosylation of iron cofactors by NO itself. Given the high affinity and reactivity of NO for metals, the dedicated sensors NsrR and NorR have incorporated an [Fe-S] cluster and a nonheme iron center, respectively, as sensing devices. Alternatively, a redox-active cysteine in the SPI2 regulator SsrB senses oxidative products of NO.

Direct sensing of NO by NsrR

NsrR belongs to the Rrf2 family of transcription repressors widespread in prokaryotes. NsrR contains an N-terminal helix-turn-helix domain responsible for DNA binding, and three cysteines implicated in ligating a [2Fe-2S] cluster in its C terminus (Tucker et al., 2008). NsrR binds specifically to DNA as a dimer, repressing the transcription of ~20 genes (Bodenmiller and Spiro, 2006; Filenko et al., 2007; Tucker et al., 2008). The NsrR operator region overlaps with the transcriptional start site or the –10 region of the promoter, suggesting that NsrR represses transcription by blocking the binding of RNA polymerase (Bodenmiller and Spiro, 2006). The [2Fe-2S] cluster of NsrR directly senses NO. The formation of a di-nitrosyl-iron complex in the [2Fe-2S] cluster inhibits binding of NsrR to DNA, thereby derepressing the transcription of target genes (Tucker et al., 2008).

The nitrosylation of NsrR upregulates transcription of *hmpA*. The *hmpA*-encoded flavohemoprotein serves as the primary means of NO detoxification in *Salmonella* (Bang et al., 2006). The flavohemoprotein Hmp contributes to the inducible antinitrosative response of *Salmonella* and many other bacteria by di-nitrosylating NO to NO_3^-, utilizing O_2 and NADH in the process (Crawford and Goldberg, 1998; Gardner et al., 1998). Hmp protects the respiratory activity of terminal quinol cytochrome oxidases from NO toxic effects and reduces the accumulation of low-molecular-weight *S*-nitrosothiols in *Salmonella*-infected macrophages (Stevanin et al., 2000; Laver et al., 2010). Hmp protects *Salmonella* against the bacteriostatic effects of NO and defends this intracellular pathogen against RNS-mediated cytotoxicity in murine and human macrophages (Stevanin et al., 2002; Bang et al., 2006; Gilberthorpe et al., 2007). Additionally, Hmp contributes to the antinitrosative

defenses of *Salmonella* in an Nramp1[+] murine model of salmonellosis (Stevanin et al., 2002; Bang et al., 2006). Although Hmp can consume some NO in the absence of O_2, the enzyme does not detoxify significant amounts of NO in anoxic environments (Mills et al., 2008). The expression of *hmpA* must be tightly regulated, because Hmp can catalyze the production of cytotoxic O_2^- (McLean et al., 2010b). Possibly as a consequence of the O_2^- produced by Hmp and the NO-mediated damage of [Fe-S] clusters, nitrosylated NsrR also derepresses the expression of the *ytfE* gene product implicated in [Fe-S] cluster repair (Gilberthorpe et al., 2007). The NO-mediated derepression of NsrR also induces the *nrfA*-encoded nitrite reductase in *E. coli*. However, recent evidence suggests that the *Salmonella nrfA* promoter region does not contain an NsrR binding site, and its induction might instead depend on FNR (see below) (Browning et al., 2010).

Sensing of NO by NorR

NorR is a member of the bacterial enhancer-binding protein family of transcriptional regulators that serves as a dedicated NO sensor in several bacterial species, including *Salmonella* (Gilberthorpe and Poole, 2008). NorR contains three structural domains dedicated to NO sensing and activation of transcription. The N-terminal regulatory domain harbors a nonheme iron center that represses a central ATPase domain. Moreover, NorR has a helix-turn-helix domain in the C terminus that mediates binding of hexamers to three enhancer sites (Justino et al., 2005; Tucker et al., 2006). The formation of a mononitrosyl-iron species in NO-treated NorR activates transcription by influencing two independent events (D'Autreaux et al., 2005). First, nitrosylation of the iron cofactor in NorR exposes a conserved GAFTGA domain that stimulates recruitment of the alternative sigma factor σ^{54} (i.e., RpoN) (Bush et al., 2010). Second, ATP hydrolysis is stimulated in the nitrosylated protein. The activated ATPase provides the energy needed to drive conformational changes in the RNA polymerase holoenzyme and the consequent activation of an open complex at the promoter. Interaction between NorR at enhancer sites upstream of the promoter and the holoenzyme at the transcription start site is made possible by the integration host factor-mediated DNA bending (Hoover et al., 1990).

Nitrosylated NorR activates the transcription of the *norVW* gene cluster, encoding a flavorubredoxin and its associated oxidoreductase (Hutchings et al., 2002). NorV is an O_2-sensitive NO reductase that catalyzes the reduction of NO to nitrous oxide (N_2O) under anaerobic conditions (Gardner et al., 2002;

Gomes et al., 2002). The transcription of *norV* is activated in *Salmonella* residing within macrophages (Eriksson et al., 2003); however, *norV* mutant salmonellae are virulent in an acute model of salmonellosis (Bang et al., 2006). Two factors may influence the apparent dispensability of NorV in this model of infection. First, a functional Hmp may have complemented the lack of NorV. Second, iNOS does not play much of a role in this acute model of *Salmonella* infection. Consequently, as described for NrfA, the flavorubredoxin may be important in other facets of the *Salmonella* infection such as the gastrointestinal tract, where NO is produced in copious amounts.

Sensing of RNS by the SPI2 SsrB response regulator

As described above, redox-active thiols are preferred targets of RNS. The thiol of Cys^{203} in the dimerization domain of the SPI2 master regulator SsrB is an example of an S-nitrosylated *Salmonella* molecular target. SsrB was recently recognized to contain a redox-active cysteine at position 203 that senses nitrosative stress (Husain et al., 2010). The relevance of the NO-sensing activity of SsrB is manifested by the fact that a strain of *Salmonella* expressing a redox-resistant SsrB C203S variant is attenuated in the context of nitrogen oxides generated in an Nramp1$^+$ model of oral salmonellosis. Nramp1$^+$ is a metal transporter expressed in phagosomal membranes that has been associated with resistance to *Salmonella* and increased production of RNS by the iNOS hemoprotein (Govoni and Gros, 1998; Fritsche et al., 2003; Fritsche et al., 2008; Nairz et al., 2009). The fact that SsrB Cys^{203} is conserved in both typhoidal and nontyphoidal strains of *Salmonella* further indicates the importance of this residue in the function of SsrB. It is possible that by decreasing *Salmonella*-induced apoptosis and reducing recognition by T and B lymphocytes, the tight control of SPI2 expression exerted by a redox-active SsrB may increase intracellular survival, limit the specific immune response, and thus enhance *Salmonella* fitness. The idea that the downregulation of SPI2 by the NO-sensing activity of SsrB is key to some aspect of *Salmonella* pathogenesis is in keeping with the fact that both positive and negative regulation of SPI2 are critical for *Salmonella* virulence (Coombes et al., 2005; Choi et al., 2010). Given the reactivity of thiols with ROS as well as RNS, it remains to be investigated whether the regulatory activity of SsrB Cys^{203} can also be modulated by ROS.

Indirect NO Sensors

Because of the overlap in some of the molecular targets of ROS and RNS, it is not surprising that ROS and RNS can modulate some common bacterial sensors. For instance, SoxR, OxyR, FNR, and Fur are responsive to both ROS and RNS.

Sensing of NO by SoxR

Elegant experiments by Nunoshiba et al. demonstrated that the $O_2^{·-}$ sensor SoxR can be activated by NO (Nunoshiba et al., 1995). Dinitrosyl iron adducts are formed through the nitrosylation of the $[2Fe-2S]^+$ cluster of SoxR (Ding and Demple, 2000). In contrast to NsrR, the nitrosylated SoxR protein is transcriptionally active. The SoxRS-mediated regulation of the Zwf glucose-6-phosphate dehydrogenase that generates NADPH-reducing power likely explains why SoxRS senses both $O_2^{·-}$ and NO. Interestingly, a *Salmonella* glucose-6-phosphate dehydrogenase mutant is sensitive to the NO donor S-nitrosoglutathione and is attenuated in mice capable of producing NO (Lundberg et al., 1999).

Sensing of RNS by OxyR

Work by the Stamler group demonstrated that in addition to sensing H_2O_2, OxyR senses RNS (Hausladen et al., 1996). The redox-active residue Cys^{199} can be S nitrosylated, S glutathionylated, and sulfenylated in response to NO (Hausladen et al., 1996; McLean et al., 2010). Similar to the oxidized protein, OxyR bearing these covalent modifications is transcriptionally active. The hypersusceptibility of *E. coli oxyR* mutants to the NO donor S-nitrosocysteine suggests that OxyR coordinates antinitrosative defenses (Hausladen et al., 1996). The OxyR-dependent induction of glutathione reductase, glutaredoxin 1, and thioredoxin 2 may be important to reestablish thiol/disulfide homeostasis after NO-mediated modification of cysteines. In addition, the expression of the OxyR-regulated *ahpC*-encoded alkyl hydroperoxide reductase may be important for the detoxification of $ONOO^-$ to NO_2^- (Chen et al., 1998).

Sensing of NO by FNR

In addition to responding to O_2 and $O_2^{·-}$, the $[4Fe-4S]^{2+}$ cluster of FNR is also modified by NO (Hassan and Sun, 1992). The formation of dinitrosyl-iron complexes in FNR derepresses genes involved in the antinitrosative response of *Salmonella*. FNR represses *hmpA* gene transcription (Cruz-Ramos et al., 2002). Moreover, *nrfA* expression is positively regulated by FNR, which is in contrast to *E. coli*, in which *nrfA* gene expression is independently controlled by NsrR (Browning et al., 2010). The *nrfA* gene encodes a bifunctional respiratory nitrite reductase that reduces

NO_2^- to NH_3 and catalyzes the anaerobic, five-electron reduction of NO to NO_3^- (Poock et al., 2002). NrfA activity is important for the resistance of *Salmonella* to NO under anaerobic conditions in vitro (Mills et al., 2008). However, this enzyme appears to play a minimal role in protecting *Salmonella* against NO produced in the course of the infection, as a *Salmonella nrfA* mutant is virulent when inoculated intraperitoneally (Bang et al., 2006). As discussed above for NorV, the lack of attenuation of *nrfA*-deficient *Salmonella* in this model of infection does not rule out the possibility that NrfA increases *Salmonella* fitness in the gastrointestinal tract, where O_2 is limited.

Sensing of NO by Fur

Fur (ferric uptake regulator) has been extensively studied for its role in iron homeostasis. Fur functions as an iron-dependent repressor, controlling the expression of genes involved in iron acquisition and utilization. Each Fur monomer is capable of binding an iron molecule. When iron is available, ferrous iron-containing Fur binds to operator regions of target genes. Under iron-limiting conditions, apo-Fur cannot bind to DNA, and thus repression is relieved. NO can react with the iron center of Fur, forming a dinitrosyl-iron complex (D'Autreaux et al., 2002). The ferrous-dinitrosyl form of Fur is unable to bind to DNA in vitro, and treatment of *E. coli* and *Salmonella* with NO causes derepression of Fur-regulated promoters (D'Autreaux et al., 2002; Mukhopadhyay et al., 2004; Bourret et al., 2008). Fur is required for *Salmonella* to cause systemic disease in an Nramp1⁺ mouse model of infection, but not in Nramp1⁻ mice (Troxell et al., 2011). This finding suggests that Fur may play a role in the antinitrosative defenses of *Salmonella* in response to the high NO synthesis associated with a functional NRAMP1 allele (Nairz et al., 2009). The iron overload and consequent increase in oxidative stress may contribute to the attenuation of *fur* mutant *Salmonella*. The induction of *hmpA* in response to NO is slightly increased in *fur*-deficient *Salmonella*, suggesting that Fur may repress *hmpA* transcription (Hernandez-Urzua et al., 2007). That the *Salmonella hmpA* promoter does not contain a predicted Fur box suggests that the regulation of *hmpA* by Fur is probably indirect. Further investigations delineating a possible role for Fur in the antinitrosative defenses of *Salmonella* are warranted.

CONCLUSIONS

Intracellular *Salmonella* must adapt to challenges generated as part of endogenous metabolism or the host response. Salmonellae encounter ROS generated as by-products of respiration and diverse metabolic pathways and are confronted with a burst of oxyradicals synthesized by NADPH phagocyte oxidase in the innate response of phagocytes. Macrophages also count with nitrogen oxides in their anti-*Salmonella* arsenal. Given the extraordinary metabolic constraints imposed by these reactive species, salmonellae use a network of regulators to sense ROS and RNS, thereby coordinating a variety of antioxidant and antinitrosative defenses.

ROS and RNS have distinct biological chemistries, but they also share some common molecular targets. For example, O_2^-, H_2O_2, and several RNS disrupt enzymes with [Fe-S] clusters or reactive cysteine residues. Many *Salmonella* transcriptional regulators have adopted the redox capacity of [Fe-S] cluster cofactors, nonheme iron centers, and the thiol group in the side chain of cysteine in order to sense and respond to ROS and RNS (Fig. 2). [Fe-S] clusters in SoxR, NsrR and FNR, nonheme iron cofactors in NorR and Fur, and the redox-active thiols of OxyR and SsrB react with O_2^-, NO, O_2, and/or H_2O_2. The fast kinetics of the interaction between the reactive species and regulatory proteins allows for the quick expression of antioxidant and antinitrosative defenses before reactive species rise to dangerous levels in the cell. For instance, the reaction of H_2O_2 and OxyR Cys¹⁹⁹ occurs at a second-order rate constant of 10^5 $M^{-1}s^{-1}$, compared to 15 $M^{-1}s^{-1}$ for the reaction of H_2O_2 with free cysteine (Luo et al., 2005). The high reactivity of OxyR Cys¹⁹⁹ with H_2O_2 is a consequence of the low pK_a of the thiol group and its steric interactions with acidic and hydrophobic amino acids. SoxR, OxyR, and FNR are also readily modified by NO or reactive species generated from the reaction of this diatomic radical with O_2, iron, and/or cysteine. Delineating the mechanisms that contribute to the reactivity of redox-active moieties in regulatory proteins is a fascinating line of research. It is likely that additional redox sensors responsive to RNS and/or ROS will be identified in *Salmonella* in the years to come.

Salmonella's ability to survive inside the hostile environment of macrophages differentiates this gram-negative bacterium from most strains of its ancestral cousin *E. coli*. Nonetheless, *Salmonella* and *E. coli* express many of the same SODs and catalases/peroxidases and use OxyR, SoxR, FNR, and Fur to sense ROS and RNS. Of these, sensing of reactive species by FNR and Fur is critical for the full expression of *Salmonella* virulence. In addition to commonly expressed antioxidant and antinitrosative defenses, *Salmonella enterica* uniquely expresses the SPI2 type III secretion system. This specialized delivery system decreases exposure of *Salmonella* to reactive species generated by NADPH phagocyte oxidase and iNOS hemoproteins. The realization that the SPI2 master regulator SsrB can be a sensor of RNS illustrates the complex strategies used by intracellular *Salmonella* to sense reactive species engendered in the course of the infection.

Acknowledgments. This review was supported by grants from the Burroughs Wellcome Fund, NIH project AI54959, and Institutional Training grant T32 AI052066.

REFERENCES

Ackermann, M., B. Stecher, N. E. Freed, P. Songhet, W. D. Hardt, and M. Doebeli. 2008. Self-destructive cooperation mediated by phenotypic noise. *Nature* **454:**987–990.

Alam, M. S., M. H. Zaki, T. Sawa, S. Islam, K. A. Ahmed, S. Fujii, T. Okamoto, and T. Akaike. 2008. Nitric oxide produced in Peyer's patches exhibits antiapoptotic activity contributing to an antimicrobial effect in murine salmonellosis. *Microbiol. Immunol.* **52:**197–208.

Aono, R., N. Tsukagoshi, and M. Yamamoto. 1998. Involvement of outer membrane protein TolC, a possible member of the *mar-sox* regulon, in maintenance and improvement of organic solvent tolerance of *Escherichia coli* K-12. *J. Bacteriol.* **180:**938–944.

Aslund, F., M. Zheng, J. Beckwith, and G. Storz. 1999. Regulation of the OxyR transcription factor by hydrogen peroxide and the cellular thiol-disulfide status. *Proc. Natl. Acad. Sci. USA* **96:**6161–6165.

Aussel, L., W. Zhao, M. Hebrard, A. A. Guilhon, J. P. Viala, S. Henri, L. Chasson, J. P. Gorvel, F. Barras, and S. Meresse. 2011. *Salmonella* detoxifying enzymes are sufficient to cope with the host oxidative burst. *Mol. Microbiol.* **80:**628–640.

Bang, I. S., L. Liu, A. Vazquez-Torres, M. L. Crouch, J. S. Stamler, and F. C. Fang. 2006. Maintenance of nitric oxide and redox homeostasis by the *Salmonella* flavohemoglobin Hmp. *J. Biol. Chem.* **281:**28039–28047.

Benov, L., and I. Fridovich. 1997. Superoxide imposes leakage of sulfite from *Escherichia coli*. *Arch. Biochem. Biophys.* **347:**271–274.

Benov, L., and I. Fridovich. 1999. Why superoxide imposes an aromatic amino acid auxotrophy on *Escherichia coli*. The trans ketolase connection. *J. Biol. Chem.* **274:**4202–4206.

Berger, S. B., X. Romero, C. Ma, G. Wang, W. A. Faubion, G. Liao, E. Compeer, M. Keszei, L. Rameh, N. Wang, M. Boes, J. R. Regueiro, H. C. Reinecker, and C. Terhorst. 2010. SLAM is a microbial sensor that regulates bacterial phagosome functions in macrophages. *Nat. Immunol.* **11:**920–927.

Bodenmiller, D. M., and S. Spiro. 2006. The *yjeB* (*nsrR*) gene of *Escherichia coli* encodes a nitric oxide-sensitive transcriptional regulator. *J. Bacteriol.* **188:**874–881.

Boehm, D. E., K. Vincent, and O. R. Brown. 1976. Oxygen and toxicity inhibition of amino acid biosynthesis. *Nature* **262:**418–420.

Bourret, T. J., S. Porwollik, M. McClelland, R. Zhao, T. Greco, H. Ischiropoulos, and A. Vazquez-Torres. 2008. Nitric oxide antagonizes the acid tolerance response that protects *Salmonella* against innate gastric defenses. *PLoS One* **3:**e1833.

Boveris, A., and B. Chance. 1973. The mitochondrial generation of hydrogen peroxide. General properties and effect of hyperbaric oxygen. *Biochem. J.* **134:**707–716.

Brandes, N., A. Rinck, L. I. Leichert, and U. Jakob. 2007. Nitrosative stress treatment of *E. coli* targets distinct set of thiol-containing proteins. *Mol. Microbiol.* **66:**901–914.

Brown, O., F. Yein, D. Boehme, L. Foudin, and C. S. Song. 1979. Oxygen poisoning of NAD biosynthesis: a proposed site of cellular oxygen toxicity. *Biochem. Biophys. Res. Commun.* **91:**982–990.

Browning, D. F., D. J. Lee, S. Spiro, and S. J. Busby. 2010. Down-regulation of the *Escherichia coli* K-12 *nrf* promoter by binding of the NsrR nitric oxide-sensing transcription repressor to an upstream site. *J. Bacteriol.* **192:**3824–3828.

Bryk, R., P. Griffin, and C. Nathan. 2000. Peroxynitrite reductase activity of bacterial peroxiredoxins. *Nature* **407:**211–215.

Buchmeier, N. A., S. J. Libby, Y. Xu, P. C. Loewen, J. Switala, D. G. Guiney, and F. C. Fang. 1995. DNA repair is more important than catalase for *Salmonella* virulence in mice. *J. Clin. Investig.* **95:**1047–1053.

Buchmeier, N. A., C. J. Lipps, M. Y. So, and F. Heffron. 1993. Recombination-deficient mutants of *Salmonella typhimurium* are avirulent and sensitive to the oxidative burst of macrophages. *Mol. Microbiol.* **7:**933–936.

Bush, M., T. Ghosh, N. Tucker, X. Zhang, and R. Dixon. 2010. Nitric oxide-responsive interdomain regulation targets the σ^{54}-interaction surface in the enhancer binding protein NorR. *Mol. Microbiol.* **77:**1278–1288.

Buzzo, C. L., J. C. Campopiano, L. M. Massis, S. L. Lage, A. A. Cassado, R. Leme-Souza, L. D. Cunha, M. Russo, D. S. Zamboni, G. P. Amarante-Mendes, and K. R. Bortoluci. 2010. A novel pathway for inducible nitric-oxide synthase activation through inflammasomes. *J. Biol. Chem.* **285:**32087–32095.

Bylund, J., K. L. Brown, C. Movitz, C. Dahlgren, and A. Karlsson. 2010. Intracellular generation of superoxide by the phagocyte NADPH oxidase: how, where, and what for? *Free Radic. Biol. Med.* **49:**1834–1845.

Carlioz, A., and D. Touati. 1986. Isolation of superoxide dismutase mutants in *Escherichia coli*: is superoxide dismutase necessary for aerobic life? *EMBO J.* **5:**623–630.

Chakravortty, D., I. Hansen-Wester, and M. Hensel. 2002. *Salmonella* pathogenicity island 2 mediates protection of intracellular *Salmonella* from reactive nitrogen intermediates. *J. Exp. Med.* **195:**1155–1166.

Chateau, M. T., and R. Caravano. 1997. The oxidative burst triggered by *Salmonella typhimurium* in differentiated U937 cells requires complement and a complete bacterial lipopolysaccharide. *FEMS Immunol. Med. Microbiol.* **17:**57–66.

Chen, L., Q. W. Xie, and C. Nathan. 1998. Alkyl hydroperoxide reductase subunit C (AhpC) protects bacterial and human cells against reactive nitrogen intermediates. *Mol. Cell* **1:**795–805.

Cherayil, B. J., B. A. McCormick, and J. Bosley. 2000. *Salmonella enterica* serovar Typhimurium-dependent regulation of inducible nitric oxide synthase expression in macrophages by invasins SipB, SipC, and SipD and effector SopE2. *Infect. Immun.* **68:**5567–5574.

Choi, H., S. Kim, P. Mukhopadhyay, S. Cho, J. Woo, G. Storz, and S. E. Ryu. 2001. Structural basis of the redox switch in the OxyR transcription factor. *Cell* **105:**103–113.

Choi, J., D. Shin, H. Yoon, J. Kim, C. R. Lee, M. Kim, Y. J. Seok, and S. Ryu. 2010. *Salmonella* pathogenicity island 2 expression negatively controlled by EIIANtr-SsrB interaction is required for *Salmonella* virulence. *Proc. Natl. Acad. Sci. USA* **107:**20506–20511.

Chou, J. H., J. T. Greenberg, and B. Demple. 1993. Posttranscriptional repression of *Escherichia coli* OmpF protein in response to redox stress: positive control of the *micF* antisense RNA by the *soxRS* locus. *J. Bacteriol.* **175:**1026–1031.

Christman, M. F., R. W. Morgan, F. S. Jacobson, and B. N. Ames. 1985. Positive control of a regulon for defenses against oxidative stress and some heat-shock proteins in *Salmonella typhimurium*. *Cell* 41:753–762.

Cirillo, D. M., R. H. Valdivia, D. M. Monack, and S. Falkow. 1998. Macrophage-dependent induction of the *Salmonella* pathogenicity island 2 type III secretion system and its role in intracellular survival. *Mol. Microbiol.* 30:175–188.

Coombes, B. K., M. E. Wickham, M. J. Lowden, N. F. Brown, and B. B. Finlay. 2005. Negative regulation of *Salmonella* pathogenicity island 2 is required for contextual control of virulence during typhoid. *Proc. Natl. Acad. Sci. USA* 102:17460–17465.

Crack, J., J. Green, and A. J. Thomson. 2004. Mechanism of oxygen sensing by the bacterial transcription factor fumarate-nitrate reduction (FNR). *J. Biol. Chem.* 279:9278–9286.

Craig, M., and J. M. Slauch. 2009. Phagocytic superoxide specifically damages an extracytoplasmic target to inhibit or kill *Salmonella*. *PLoS One* 4:e4975.

Crawford, M. J., and D. E. Goldberg. 1998. Role for the *Salmonella* flavohemoglobin in protection from nitric oxide. *J. Biol. Chem.* 273:12543–12547.

Cruz-Ramos, H., J. Crack, G. Wu, M. N. Hughes, C. Scott, A. J. Thomson, J. Green, and R. K. Poole. 2002. NO sensing by FNR: regulation of the *Escherichia coli* NO-detoxifying flavohaemoglobin, Hmp. *EMBO J.* 21:3235–3244.

D'Autreaux, B., D. Touati, B. Bersch, J. M. Latour, and I. Michaud-Soret. 2002. Direct inhibition by nitric oxide of the transcriptional ferric uptake regulation protein via nitrosylation of the iron. *Proc. Natl. Acad. Sci. USA* 99:16619–16624.

D'Autreaux, B., N. P. Tucker, R. Dixon, and S. Spiro. 2005. A non-haem iron centre in the transcription factor NorR senses nitric oxide. *Nature* 437:769–772.

De Groote, M. A., U. A. Ochsner, M. U. Shiloh, C. Nathan, J. M. McCord, M. C. Dinauer, S. J. Libby, A. Vazquez-Torres, Y. Xu, and F. C. Fang. 1997. Periplasmic superoxide dismutase protects *Salmonella* from products of phagocyte NADPH-oxidase and nitric oxide synthase. *Proc. Natl. Acad. Sci. USA* 94:13997–14001.

de Jong, R., F. Altare, I. A. Haagen, D. G. Elferink, T. Boer, P. J. van Breda Vriesman, P. J. Kabel, J. M. Draaisma, J. T. van Dissel, F. P. Kroon, J. L. Casanova, and T. H. Ottenhoff. 1998. Severe mycobacterial and *Salmonella* infections in interleukin-12 receptor-deficient patients. *Science* 280:1435 1438.

Dietrich, L. E., and P. J. Kiley. 2011. A shared mechanism of SoxR activation by redox-cycling compounds. *Mol. Microbiol.* 79:1119–1122.

Ding, H., and B. Demple. 2000. Direct nitric oxide signal transduction via nitrosylation of iron-sulfur centers in the SoxR transcription activator. *Proc. Natl. Acad. Sci. USA* 97:5146–5150.

Ding, H., and B. Demple. 1997. In vivo kinetics of a redox-regulated transcriptional switch. *Proc. Natl. Acad. Sci. USA* 94:8445–8449.

Ding, H., E. Hidalgo, and B. Demple. 1996. The redox state of the [2Fe-2S] clusters in SoxR protein regulates its activity as a transcription factor. *J. Biol. Chem.* 271:33173–33175.

Duncan, C., H. Dougall, P. Johnston, S. Green, R. Brogan, C. Leifert, L. Smith, M. Golden, and N. Benjamin. 1995. Chemical generation of nitric oxide in the mouth from the enterosalivary circulation of dietary nitrate. *Nat. Med.* 1:546–551.

Enocksson, A., J. Lundberg, E. Weitzberg, A. Norrby-Teglund, and B. Svenungsson. 2004. Rectal nitric oxide gas and stool cytokine levels during the course of infectious gastroenteritis. *Clin. Diagn. Lab. Immunol.* 11:250–254.

Eriksson, S., S. Lucchini, A. Thompson, M. Rhen, and J. C. Hinton. 2003. Unravelling the biology of macrophage infection by gene expression profiling of intracellular *Salmonella enterica*. *Mol. Microbiol.* 47:103–118.

Fang, F. C., S. J. Libby, N. A. Buchmeier, P. C. Loewen, J. Switala, J. Harwood, and D. G. Guiney. 1992. The alternative sigma factor KatF (RpoS) regulates *Salmonella* virulence. *Proc. Natl. Acad. Sci. USA* 89:11978–11982.

Fang, F. C., A. Vazquez-Torres, and Y. Xu. 1997. The transcriptional regulator SoxS is required for resistance of *Salmonella typhimurium* to paraquat but not for virulence in mice. *Infect. Immun.* 65:5371–5375.

Farr, S. B., R. D'Ari, and D. Touati. 1986. Oxygen-dependent mutagenesis in *Escherichia coli* lacking superoxide dismutase. *Proc. Natl. Acad. Sci. USA* 83:8268–8272.

Fields, P. I., R. V. Swanson, C. G. Haidaris, and F. Heffron. 1986. Mutants of *Salmonella typhimurium* that cannot survive within the macrophage are avirulent. *Proc. Natl. Acad. Sci. USA* 83:5189–5193.

Filenko, N., S. Spiro, D. F. Browning, D. Squire, T. W. Overton, J. Cole, and C. Constantinidou. 2007. The NsrR regulon of *Escherichia coli* K-12 includes genes encoding the hybrid cluster protein and the periplasmic, respiratory nitrite reductase. *J. Bacteriol.* 189:4410–4417.

Fink, R. C., M. R. Evans, S. Porwollik, A. Vazquez-Torres, J. Jones-Carson, B. Troxell, S. J. Libby, M. McClelland, and H. M. Hassan. 2007. FNR is a global regulator of virulence and anaerobic metabolism in *Salmonella enterica* serovar Typhimurium (ATCC 14028s). *J. Bacteriol.* 189:2262–2273.

Francis, K. P., P. D. Taylor, C. J. Inchley, and M. P. Gallagher. 1997. Identification of the *ahp* operon of *Salmonella typhimurium* as a macrophage-induced locus. *J. Bacteriol.* 179:4046–4048.

Fritsche, G., M. Dlaska, H. Barton, I. Theurl, K. Garimorth, and G. Weiss. 2003. Nramp1 functionality increases inducible nitric oxide synthase transcription via stimulation of IFN regulatory factor 1 expression. *J. Immunol.* 171:1994–1998.

Fritsche, G., M. Nairz, E. R. Werner, H. C. Barton, and G. Weiss. 2008. Nramp1-functionality increases iNOS expression via repression of IL-10 formation. *Eur. J. Immunol.* 38:3060–3067.

Gallois, A., J. R. Klein, L. A. Allen, B. D. Jones, and W. M. Nauseef. 2001. *Salmonella* pathogenicity island 2-encoded type III secretion system mediates exclusion of NADPH oxidase assembly from the phagosomal membrane. *J. Immunol.* 166:5741–5748.

Gardner, A. M., R. A. Helmick, and P. R. Gardner. 2002. Flavorubredoxin, an inducible catalyst for nitric oxide reduction and detoxification in *Escherichia coli*. *J. Biol. Chem.* 277:8172–8177.

Gardner, P. R., A. M. Gardner, L. A. Martin, and A. L. Salzman. 1998. Nitric oxide dioxygenase: an enzymic function for flavohemoglobin. *Proc. Natl. Acad. Sci. USA* 95:10378–10383.

Gaudu, P., N. Moon, and B. Weiss. 1997. Regulation of the SoxRS oxidative stress regulon. Reversible oxidation of the Fe-S centers of SoxR in vivo. *J. Biol. Chem.* 272:5082–5086.

Gaudu, P., and B. Weiss. 2000. Flavodoxin mutants of *Escherichia coli* K-12. *J. Bacteriol.* 182:1788–1793.

Giacomodonato, M. N., N. B. Goren, D. O. Sordelli, M. I. Vaccaro, D. H. Grasso, A. J. Ropolo, and M. C. Cerquetti. 2003. Involvement of intestinal inducible nitric oxide synthase (iNOS) in the early stages of murine salmonellosis. *FEMS Microbiol. Lett.* 223:231–238.

Gilberthorpe, N. J., M. E. Lee, T. M. Stevanin, R. C. Read, and R. K. Poole. 2007. NsrR: a key regulator circumventing *Salmonella enterica* serovar Typhimurium oxidative and nitrosative stress in vitro and in IFN-γ-stimulated J774.2 macrophages. *Microbiology* 153:1756–1771.

Gilberthorpe, N. J., and R. K. Poole. 2008. Nitric oxide homeostasis in *Salmonella typhimurium*: roles of respiratory nitrate reductase and flavohemoglobin. *J. Biol. Chem.* **283**:11146–11154.

Gomes, C. M., A. Giuffre, E. Forte, J. B. Vicente, L. M. Saraiva, M. Brunori, and M. Teixeira. 2002. A novel type of nitric-oxide reductase. *Escherichia coli* flavorubredoxin. *J. Biol. Chem.* **277**:25273–25276.

Govoni, G., and P. Gros. 1998. Macrophage NRAMP1 and its role in resistance to microbial infections. *Inflamm. Res.* **47**:277–284.

Gralnick, J., and D. Downs. 2001. Protection from super oxide damage associated with an increased level of the YggX protein in *Salmonella enterica*. *Proc. Natl. Acad. Sci. USA* **98**:8030–8035.

Gralnick, J. A., and D. M. Downs. 2003. The YggX protein of *Salmonella enterica* is involved in Fe(II) trafficking and minimizes the DNA damage caused by hydroxyl radicals: residue CYS-7 is essential for YggX function. *J. Biol. Chem.* **278**:20708–20715.

Greenberg, J. T., P. Monach, J. H. Chou, P. D. Josephy, and B. Demple. 1990. Positive control of a global antioxidant defense regulon activated by superoxide-generating agents in *Escherichia coli*. *Proc. Natl. Acad. Sci. USA* **87**:6181–6185.

Gruer, M. J., and J. R. Guest. 1994. Two genetically-distinct and differentially-regulated aconitases (AcnA and AcnB) in *Escherichia coli*. *Microbiology* **140**:2531–2541.

Gu, M., and J. A. Imlay. 2011. The SoxRS response of *Escherichia coli* is directly activated by redox-cycling drugs rather than by superoxide. *Mol. Microbiol.* **79**:1136–1150.

Halsey, T. A., A. Vazquez-Torres, D. J. Gravdahl, F. C. Fang, and S. J. Libby. 2004. The ferritin-like Dps protein is required for *Salmonella enterica* serovar Typhimurium oxidative stress resistance and virulence. *Infect. Immun.* **72**:1155–1158.

Hassan, H. M., and H. C. Sun. 1992. Regulatory roles of Fnr, Fur, and Arc in expression of manganese-containing superoxide dismutase in *Escherichia coli*. *Proc. Natl. Acad. Sci. USA* **89**:3217–3221.

Hausladen, A., and I. Fridovich. 1994. Superoxide and peroxynitrite inactivate aconitases, but nitric oxide does not. *J. Biol. Chem.* **269**:29405–29408.

Hausladen, A., C. T. Privalle, T. Keng, J. DeAngelo, and J. S. Stamler. 1996. Nitrosative stress: activation of the transcription factor OxyR. *Cell* **86**:719–729.

Hebrard, M., J. P. Viala, S. Meresse, F. Barras, and L. Aussel. 2009. Redundant hydrogen peroxide scavengers contribute to *Salmonella* virulence and oxidative stress resistance. *J. Bacteriol.* **191**:4605–4614.

Henard, C. A., and A. Vazquez-Torres. 2011. Nitric oxide and *Salmonella* pathogenesis. *Front. Microbiol.* **2**:84.

Hernandez-Urzua, E., D. S. Zamorano-Sanchez, J. Ponce-Coria, E. Morett, S. Grogan, R. K. Poole, and J. Membrillo-Hernandez. 2007. Multiple regulators of the flavohaemoglobin (*hmp*) gene of *Salmonella enterica* serovar Typhimurium include RamA, a transcriptional regulator conferring the multidrug resistance phenotype. *Arch. Microbiol.* **187**:67–77.

Hidalgo, E., J. M. Bollinger, Jr., T. M. Bradley, C. T. Walsh, and B. Demple. 1995. Binuclear [2Fe-2S] clusters in the *Escherichia coli* SoxR protein and role of the metal centers in transcription. *J. Biol. Chem.* **270**:20908–20914.

Hoffmann, J. H., K. Linke, P. C. Graf, H. Lilie, and U. Jakob. 2004. Identification of a redox-regulated chaperone network. *EMBO J.* **23**:160–168.

Hoover, T. R., E. Santero, S. Porter, and S. Kustu. 1990. The integration host factor stimulates interaction of RNA polymerase with NIFA, the transcriptional activator for nitrogen fixation operons. *Cell* **63**:11–22.

Husain, M., T. J. Bourret, B. D. McCollister, J. Jones-Carson, J. Laughlin, and A. Vazquez-Torres. 2008. Nitric oxide evokes an adaptive response to oxidative stress by arresting respiration. *J. Biol. Chem.* **283**:7682–7689.

Husain, M., J. Jones-Carson, M. Song, B. D. McCollister, T. J. Bourret, and A. Vazquez-Torres. 2010. Redox sensor SsrB Cys[203] enhances *Salmonella* fitness against nitric oxide generated in the host immune response to oral infection. *Proc. Natl. Acad. Sci. USA* **107**:14396–14401.

Hutchings, M. I., N. Mandhana, and S. Spiro. 2002. The NorR protein of *Escherichia coli* activates expression of the flavorubredoxin gene *norV* in response to reactive nitrogen species. *J. Bacteriol.* **184**:4640–4643.

Imlay, J. A., S. M. Chin, and S. Linn. 1988. Toxic DNA damage by hydrogen peroxide through the Fenton reaction *in vivo* and *in vitro*. *Science* **240**:640–642.

Jang, S., and J. A. Imlay. 2010. Hydrogen peroxide inactivates the *Escherichia coli* Isc iron-sulphur assembly system, and OxyR induces the Suf system to compensate. *Mol. Microbiol.* **78**:1448–1467.

Jang, S., and J. A. Imlay. 2007. Micromolar intracellular hydrogen peroxide disrupts metabolism by damaging iron-sulfur enzymes. *J. Biol. Chem.* **282**:929–937.

Jones, B. D., N. Ghori, and S. Falkow. 1994. *Salmonella typhimurium* initiates murine infection by penetrating and destroying the specialized epithelial M cells of the Peyer's patches. *J. Exp. Med.* **180**:15–23.

Justino, M. C., V. M. Goncalves, and L. M. Saraiva. 2005. Binding of NorR to three DNA sites is essential for promoter activation of the flavorubredoxin gene, the nitric oxide reductase of *Escherichia coli*. *Biochem. Biophys. Res. Commun.* **328**:540–544.

Keyer, K., and J. A. Imlay. 1997. Inactivation of dehydratase [4Fe-4S] clusters and disruption of iron homeostasis upon cell exposure to peroxynitrite. *J. Biol. Chem.* **272**:27652–27659.

Khoroshilova, N., C. Popescu, E. Munck, H. Beinert, and P. J. Kiley. 1997. Iron-sulfur cluster disassembly in the FNR protein of *Escherichia coli* by O$_2$: [4Fe-4S] to [2Fe-2S] conversion with loss of biological activity. *Proc. Natl. Acad. Sci. USA* **94**:6087–6092.

Koh, Y. S., and J. H. Roe. 1996. Dual regulation of the paraquat-inducible gene *pqi-5* by SoxS and RpoS in *Escherichia coli*. *Mol. Microbiol.* **22**:53–61.

Korshunov, S., and J. A. Imlay. 2006. Detection and quantification of superoxide formed within the periplasm of *Escherichia coli*. *J. Bacteriol.* **188**:6326–6334.

Korshunov, S., and J. A. Imlay. 2010. Two sources of endogenous hydrogen peroxide in *Escherichia coli*. *Mol. Microbiol.* **75**:1389–1401.

Krapp, A. D., M. V. Humbert, and N. Carrillo. 2011. The *soxRS* response of *Escherichia coli* can be induced in the absence of oxidative stress and oxygen by modulation of NADPH contents. *Microbiology* **157**:957–965.

Krishnakumar, R., M. Craig, J. A. Imlay, and J. M. Slauch. 2004. Differences in enzymatic properties allow SodCI but not SodCII to contribute to virulence in *Salmonella enterica* serovar Typhimurium strain 14028. *J. Bacteriol.* **186**:5230–5238.

Kullik, I., M. B. Toledano, L. A. Tartaglia, and G. Storz. 1995. Mutational analysis of the redox-sensitive transcriptional regulator OxyR: regions important for oxidation and transcriptional activation. *J. Bacteriol.* **177**:1275–1284.

Kuo, C. F., T. Mashino, and I. Fridovich. 1987. α,β-Dihydroxyisovalerate dehydratase. A superoxide-sensitive enzyme. *J. Biol. Chem.* **262**:4724–4727.

Laver, J. R., T. M. Stevanin, S. L. Messenger, A. D. Lunn, M. E. Lee, J. W. Moir, R. K. Poole, and R. C. Read. 2010. Bacterial nitric oxide detoxification prevents host cell S-nitrosothiol formation: a novel mechanism of bacterial pathogenesis. *FASEB J.* 24:286–295.

Lee, C., S. M. Lee, P. Mukhopadhyay, S. J. Kim, S. C. Lee, W. S. Ahn, M. H. Yu, G. Storz, and S. E. Ryu. 2004. Redox regulation of OxyR requires specific disulfide bond formation involving a rapid kinetic reaction path. *Nat. Struct. Mol. Biol.* 11:1179–1185.

Levin, D. E., M. Hollstein, M. F. Christman, E. A. Schwiers, and B. N. Ames. 1982. A new *Salmonella* tester strain (TA102) with A X T base pairs at the site of mutation detects oxidative mutagens. *Proc. Natl. Acad. Sci. USA* 79:7445–7449.

Liochev, S. I., and I. Fridovich. 1992. Fumarase C, the stable fumarase of *Escherichia coli*, is controlled by the SoxRS regulon. *Proc. Natl. Acad. Sci. USA* 89:5892–5896.

Lundberg, B. E., R. E. Wolf, Jr., M. C. Dinauer, Y. Xu, and F. C. Fang. 1999. Glucose 6-phosphate dehydrogenase is required for *Salmonella typhimurium* virulence and resistance to reactive oxygen and nitrogen intermediates. *Infect. Immun.* 67:436–438.

Lundberg, J. O., E. Weitzberg, J. A. Cole, and N. Benjamin. 2004. Nitrate, bacteria and human health. *Nat. Rev. Microbiol.* 2:593–602.

Luo, D., S. W. Smith, and B. D. Anderson. 2005. Kinetics and mechanism of the reaction of cysteine and hydrogen peroxide in aqueous solution. *J. Pharm. Sci.* 94:304–316.

Ma, D., M. Alberti, C. Lynch, H. Nikaido, and J. E. Hearst. 1996. The local repressor AcrR plays a modulating role in the regulation of *acrAB* genes of *Escherichia coli* by global stress signals. *Mol. Microbiol.* 19:101–112.

Martinez, A., and R. Kolter. 1997. Protection of DNA during oxidative stress by the nonspecific DNA-binding protein Dps. *J. Bacteriol.* 179:5188–5194.

Mason, M. G., M. Shepherd, P. Nicholls, P. S. Dobbin, K. S. Dodsworth, R. K. Poole, and C. E. Cooper. 2009. Cytochrome *bd* confers nitric oxide resistance to *Escherichia coli. Nat. Chem. Biol.* 5:94–96.

Mastroeni, P., A. Vazquez-Torres, F. C. Fang, Y. Xu, S. Khan, C. E. Hormaeche, and G. Dougan. 2000. Antimicrobial actions of the NADPH phagocyte oxidase and inducible nitric oxide synthase in experimental salmonellosis. II. Effects on microbial proliferation and host survival *in vivo. J. Exp. Med.* 192: 237–248.

McCollister, B. D., T. J. Bourret, R. Gill, J. Jones-Carson, and A. Vazquez-Torres. 2005. Repression of SPI2 transcription by nitric oxide-producing, IFNγ-activated macrophages promotes maturation of *Salmonella* phagosomes. *J. Exp. Med.* 202:625–635.

McLean, S., L. A. Bowman, and R. K. Poole. 2010a. KatG from *Salmonella typhimurium* is a peroxynitritase. *FEBS Lett.* 584:1628–1632.

McLean, S., L. A. Bowman, and R. K. Poole. 2010b. Peroxynitrite stress is exacerbated by flavohaemoglobin-derived oxidative stress in *Salmonella* Typhimurium and relieved by NO. *Microbiology* 156:3556–3565.

McLean, S., L. A. Bowman, G. Sanguinetti, R. C. Read, and R. K. Poole. 2010c. Peroxynitrite toxicity in *Escherichia coli* K12 elicits expression of oxidative stress responses and protein nitration and nitrosylation. *J. Biol. Chem.* 285:20724–20731.

Messner, K. R., and J. A. Imlay. 1999. The identification of primary sites of superoxide and hydrogen peroxide formation in the aerobic respiratory chain and sulfite reductase complex of *Escherichia coli. J. Biol. Chem.* 274:10119–10128.

Mills, P. C., G. Rowley, S. Spiro, J. C. Hinton, and D. J. Richardson. 2008. A combination of cytochrome c nitrite reductase (NrfA) and flavorubredoxin (NorV) protects *Salmonella enterica* serovar Typhimurium against killing by NO in anoxic environments. *Microbiology* 154:1218–1228.

Mouy, R., A. Fischer, E. Vilmer, R. Seger, and C. Griscelli. 1989. Incidence, severity, and prevention of infections in chronic granulomatous disease. *J. Pediatr.* 114:555–560.

Mukhopadhyay, P., M. Zheng, L. A. Bedzyk, R. A. LaRossa, and G. Storz. 2004. Prominent roles of the NorR and Fur regulators in the *Escherichia coli* transcriptional response to reactive nitrogen species. *Proc. Natl. Acad. Sci. USA* 101:745–750.

Muller, A. J., C. Hoffmann, M. Galle, A. Van Den Broeke, M. Heikenwalder, B. Falter, B. Misselwitz, M. Kremer, R. Beyaert, and W. D. Hardt. 2009. The *S.* Typhimurium effector SopE induces caspase-1 activation in stromal cells to initiate gut inflammation. *Cell Host Microbe* 6:125–136.

Nairz, M., G. Fritsche, M. L. Crouch, H. C. Barton, F. C. Fang, and G. Weiss. 2009. Slc11a1 limits intracellular growth of *Salmonella enterica* sv. Typhimurium by promoting macrophage immune effector functions and impairing bacterial iron acquisition. *Cell. Microbiol.* 11:1365–1381.

Nunoshiba, T., T. DeRojas-Walker, S. R. Tannenbaum, and B. Demple. 1995. Roles of nitric oxide in inducible resistance of *Escherichia coli* to activated murine macrophages. *Infect. Immun.* 63:794–798.

Ostrowski, J., J. Y. Wu, D. C. Rueger, B. E. Miller, L. M. Siegel, and N. M. Kredich. 1989. Characterization of the *cysJIH* regions of *Salmonella typhimurium* and *Escherichia coli* B. DNA sequences of *cysI* and *cysH* and a model for the siroheme-Fe4S4 active center of sulfite reductase hemoprotein based on amino acid homology with spinach nitrite reductase. *J. Biol. Chem.* 264:15726–15737.

Pacello, F., P. Ceci, S. Ammendola, P. Pasquali, E. Chiancone, and A. Battistoni. 2008. Periplasmic Cu,Zn superoxide dismutase and cytoplasmic Dps concur in protecting *Salmonella enterica* serovar Typhimurium from extracellular reactive oxygen species. *Biochim. Biophys. Acta* 1780:226–232.

Pomposiello, P. J., M. H. Bennik, and B. Demple. 2001. Genome-wide transcriptional profiling of the *Escherichia coli* responses to superoxide stress and sodium salicylate. *J. Bacteriol.* 183:3890–3902.

Pomposiello, P. J., and B. Demple. 2000. Identification of SoxS-regulated genes in *Salmonella enterica* serovar Typhimurium. *J. Bacteriol.* 182:23–29.

Poock, S. R., E. R. Leach, J. W. Moir, J. A. Cole, and D. J. Richardson. 2002. Respiratory detoxification of nitric oxide by the cytochrome c nitrite reductase of *Escherichia coli. J. Biol. Chem.* 277:23664–23669.

Robbe-Saule, V., C. Coynault, M. Ibanez-Ruiz, D. Hermant, and F. Norel. 2001. Identification of a non-haem catalase in *Salmonella* and its regulation by RpoS (σ^s). *Mol. Microbiol.* 39: 1533–1545.

Rongkavilit, C., Z. M. Rodriguez, O. Gomez-Marin, G. B. Scott, C. Hutto, D. M. Rivera-Hernandez, and C. D. Mitchell. 2000. Gram-negative bacillary bacteremia in human immunodeficiency virus type 1-infected children. *Pediatr. Infect. Dis. J.* 19:122–128.

Seaver, L. C., and J. A. Imlay. 2004. Are respiratory enzymes the primary sources of intracellular hydrogen peroxide? *J. Biol. Chem.* 279:48742–48750.

Shiloh, M. U., J. D. MacMicking, S. Nicholson, J. E. Brause, S. Potter, M. Marino, F. Fang, M. Dinauer, and C. Nathan. 1999. Phenotype of mice and macrophages deficient in both phagocyte oxidase and inducible nitric oxide synthase. *Immunity* 10:29–38.

Stevanin, T. M., N. Ioannidis, C. E. Mills, S. O. Kim, M. N. Hughes, and R. K. Poole. 2000. Flavohemoglobin Hmp affords inducible protection for *Escherichia coli* respiration, catalyzed by cytochromes *bo'* or *bd*, from nitric oxide. *J. Biol. Chem.* 275:35868–35875.

Stevanin, T. M., R. K. Poole, E. A. Demoncheaux, and R. C. Read. 2002. Flavohemoglobin Hmp protects *Salmonella enterica* serovar Typhimurium from nitric oxide-related killing by human macrophages. *Infect. Immun.* **70**:4399–4405.

Suvarnapunya, A. E., and M. A. Stein. 2005. DNA base excision repair potentiates the protective effect of *Salmonella* pathogenicity island 2 within macrophages. *Microbiology* **151**: 557–567.

Taylor, P. D., C. J. Inchley, and M. P. Gallagher. 1998. The *Salmonella typhimurium* AhpC polypeptide is not essential for virulence in BALB/c mice but is recognized as an antigen during infection. *Infect. Immun.* **66**:3208–3217.

Troxell, B., M. L. Sikes, R. C. Fink, A. Vazquez-Torres, J. Jones-Carson, and H. M. Hassan. 2011. Fur negatively regulates *hns* and is required for the expression of HilA and virulence in *Salmonella enterica* serovar Typhimurium. *J. Bacteriol.* **193**:497–505.

Tsaneva, I. R., and B. Weiss. 1990. *soxR*, a locus governing a superoxide response regulon in *Escherichia coli* K-12. *J. Bacteriol.* **172**:4197–4205.

Tsolis, R. M., A. J. Baumler, and F. Heffron. 1995. Role of *Salmonella typhimurium* Mn-superoxide dismutase (SodA) in protection against early killing by J774 macrophages. *Infect. Immun.* **63**:1739–1744.

Tucker, N. P., B. D'Autreaux, S. Spiro, and R. Dixon. 2006. Mechanism of transcriptional regulation by the *Escherichia coli* nitric oxide sensor NorR. *Biochem. Soc. Trans.* **34**:191–194.

Tucker, N. P., M. G. Hicks, T. A. Clarke, J. C. Crack, G. Chandra, N. E. Le Brun, R. Dixon, and M. I. Hutchings. 2008. The transcriptional repressor protein NsrR senses nitric oxide directly via a [2Fe-2S] cluster. *PLoS One* **3**:e3623.

Uzzau, S., L. Bossi, and N. Figueroa-Bossi. 2002. Differential accumulation of *Salmonella* [Cu, Zn] superoxide dismutases SodCI and SodCII in intracellular bacteria: correlation with their relative contribution to pathogenicity. *Mol. Microbiol.* **46**:147–156.

Vazquez-Torres, A., G. Fantuzzi, C. K. Edwards III, C. A. Dinarello, and F. C. Fang. 2001. Defective localization of the NADPH phagocyte oxidase to *Salmonella*-containing phagosomes in tumor necrosis factor p55 receptor-deficient macrophages. *Proc. Natl. Acad. Sci. USA* **98**:2561–2565.

Vazquez-Torres, A., J. Jones-Carson, A. J. Baumler, S. Falkow, R. Valdivia, W. Brown, M. Le, R. Berggren, W. T. Parks, and F. C. Fang. 1999. Extraintestinal dissemination of *Salmonella* by CD18-expressing phagocytes. *Nature* **401**:804–808.

Vazquez-Torres, A., J. Jones-Carson, P. Mastroeni, H. Ischiropoulos, and F. C. Fang. 2000a. Antimicrobial actions of the NADPH phagocyte oxidase and inducible nitric oxide synthase in experimental salmonellosis. I. Effects on microbial killing by activated peritoneal macrophages *in vitro*. *J. Exp. Med.* **192**:227–236.

Vazquez-Torres, A., Y. Xu, J. Jones-Carson, D. W. Holden, S. M. Lucia, M. C. Dinauer, P. Mastroeni, and F. C. Fang. 2000b. *Salmonella* pathogenicity island 2-dependent evasion of the phagocyte NADPH oxidase. *Science* **287**:1655–1658.

Vazquez-Torres, A., B. A. Vallance, M. A. Bergman, B. B. Finlay, B. T. Cookson, J. Jones-Carson, and F. C. Fang. 2004. Toll-like receptor 4 dependence of innate and adaptive immunity to *Salmonella*: importance of the Kupffer cell network. *J. Immunol.* **172**:6202–6208.

Vitiello, M., M. D'Isanto, E. Finamore, R. Ciarcia, A. Kampanaraki, and M. Galdiero. 2008. Role of mitogen-activated protein kinases in the iNOS production and cytokine secretion by *Salmonella enterica* serovar Typhimurium porins. *Cytokine* **41**:279–285.

Winter, S. E., P. Thiennimitr, M. G. Winter, B. P. Butler, D. L. Huseby, R. W. Crawford, J. M. Russell, C. L. Bevins, L. G. Adams, R. M. Tsolis, J. R. Roth, and A. J. Baumler. 2010. Gut inflammation provides a respiratory electron acceptor for *Salmonella*. *Nature* **467**:426–429.

Winterbourn, C. C., M. B. Hampton, J. H. Livesey, and A. J. Kettle. 2006. Modeling the reactions of superoxide and myeloperoxidase in the neutrophil phagosome: implications for microbial killing. *J. Biol. Chem.* **281**:39860–39869.

Witthoft, T., L. Eckmann, J. M. Kim, and M. F. Kagnoff. 1998. Enteroinvasive bacteria directly activate expression of iNOS and NO production in human colon epithelial cells. *Am. J. Physiol.* **275**:G564–G571.

Zhao, G., P. Ceci, A. Ilari, L. Giangiacomo, T. M. Laue, E. Chiancone, and N. D. Chasteen. 2002. Iron and hydrogen peroxide detoxification properties of DNA-binding protein from starved cells. A ferritin-like DNA-binding protein of *Escherichia coli*. *J. Biol. Chem.* **277**:27689–27696.

Zheng, M., F. Aslund, and G. Storz. 1998. Activation of the OxyR transcription factor by reversible disulfide bond formation. *Science* **279**:1718–1721.

Zheng, M., B. Doan, T. D. Schneider, and G. Storz. 1999. OxyR and SoxRS regulation of *fur*. *J. Bacteriol.* **181**:4639–4643.

Zheng, M., X. Wang, L. J. Templeton, D. R. Smulski, R. A. LaRossa, and G. Storz. 2001. DNA microarray-mediated transcriptional profiling of the *Escherichia coli* response to hydrogen peroxide. *J. Bacteriol.* **183**:4562–4570.

Regulation of Bacterial Virulence
Edited by Michael L. Vasil and Andrew J. Darwin
© 2013 ASM Press, Washington, DC   doi:10.1128/9781555818524.ch23

Chapter 23

Regulation of Vesicle Formation

Aimee K. Wessel, Gregory C. Palmer, and Marvin Whiteley

INTRODUCTION

In 1966, electron microscopy gave direct micrographic evidence of cell envelope blebbing, showing the outer membrane (OM) of *Escherichia coli* bulging outwards upon starvation of lysine (Knox et al., 1966; Work et al., 1966) (Fig. 1A). A year later, it was documented that during exponential phase growth of *Vibrio cholerae*, the OM bulged outward and presumably pinched off from the cell (Chatterjee and Das, 1967) (Fig. 1B). Additional experiments elucidated that these blebs, called outer membrane vesicles (OMVs), were not products of cell lysis but, rather, a membrane shedding event conserved among growing bacteria (Mayrand and Grenier, 1989; Pettit and Judd, 1992; Kadurugamuwa and Beveridge, 1995; Li et al., 1998; Zhou et al., 1998). Through multiple studies of gram-negative bacteria, OMVs were determined to range in size from 20 to 500 nm and include OM proteins, with very few inner membrane (IM) and cytoplasmic components (Hoekstra et al., 1976; Loeb and Kilner, 1979; Katsui et al., 1982; Kato et al., 2002). From 1966 to 1990, reports of bacterial vesiculation have focused on the blebbing of the OM of Gram-negative bacteria, though eventually, increased interest in Gram-positive vesiculation occurred in the 1990s and the late 2000s. This chapter addresses the regulation of OMV production in Gram-negative bacteria; however, a brief section is dedicated to summarizing current knowledge of Gram-positive membrane vesicles (MVs).

WHO MAKES MVs?

Secretion of vesicles is a highly conserved process occurring in all domains of life, in archaea, mammalian cells, and bacteria (Knox et al., 1966; Work et al., 1966; Dorward and Garon, 1990; Thery et al., 2002) (Fig. 1). The production of MVs serves many universal purposes: vesicles can package, protect, and transport numerous components that may travel long distances before reaching a target cell. Bacteria produce OMVs capable of trafficking a variety of components to both eukaryotes and prokaryotes (Dorward et al., 1989; Kadurugamuwa and Beveridge, 1996, 1997, 1999). It has been reported that all Gram-negative bacteria and some Gram-positive bacteria are capable of producing MVs, and numerous OMV-producing species are introduced throughout this chapter.

As would be expected, not all bacterial species produce identical vesicles, and individual vesicles produced by the same species are not always uniform in size and content. The heterogeneous distribution of envelope components, differences in lipopolysaccharide (LPS) structure, lipids, peptidoglycan (PG), protein production, and stages of growth have all been shown to affect bacterial vesiculation and are reviewed in this chapter. Additionally, changes encountered in the environment can dramatically alter cellular membranes and OMV production, due to physical and nutritional fluctuations and subsequent genetic regulation. OMVs can package soluble and insoluble components inside membrane-bound nanocompartments, enabling the delivery of proteins, DNA, RNA, lipids, and small molecules. Upon delivering these components to target cells, OMVs can affect cellular processes across long distances. OMV production has been observed in a variety of growth environments, including planktonic and biofilm laboratory cultures as well as natural environments such as sewage and rivers (Beveridge, 1999; Schooling and Beveridge, 2006).

WHY MAKE OMVs?

OMVs have been shown to interact with eukaryotic cells as well as Gram-negative and Gram-positive

---

Aimee K. Wessel • Section of Molecular Genetics and Microbiology, The University of Texas at Austin, Austin, TX 78712.
Gregory C. Palmer • Institute for Cellular and Molecular Biology, The University of Texas at Austin, Austin, TX 78712.
Marvin Whiteley • Institute for Cellular and Molecular Biology and Section of Molecular Genetics and Microbiology, The University of Texas at Austin, Austin, TX 78712.

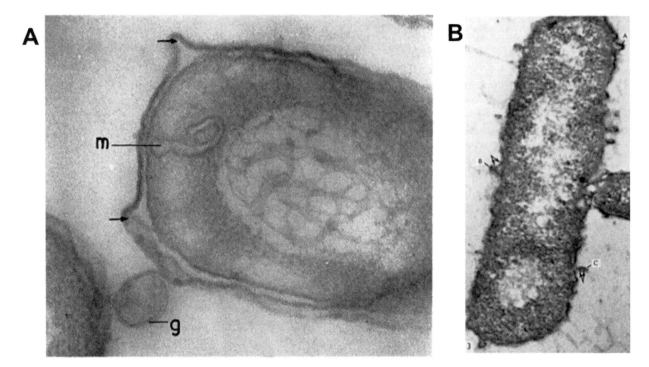

Figure 1. Two papers published in 1966 and 1967 give visible evidence of bacterial membrane blebbing. (A) An electron micrograph of lysine-limited *E. coli* cells, illustrating the OM blebbing away from the IM (at arrows). g, extracellular globule, presumably an OMV; m, intracytoplasmic membranous organelle. Reprinted from the *Journal of Bacteriology* (Knox et al., 1966) with permission of the publisher. (B) Exponential-phase *Vibrio cholerae* grown in peptone water, exhibiting multiple areas of membrane blebbing. Reprinted from the *Journal of General Microbiology* (Chatterjee and Das, 1967) with permission of the publisher.
doi:10.1128/9781555818524.ch23f1

bacteria (Kadurugamuwa and Beveridge, 1996, 1999; Kesty et al., 2004; Bomberger et al., 2009). While the ability to deliver contents via OMVs to other cells can be useful, what is additionally important to note is that OMVs can protect their cargo en route. Under conditions present in many common growth environments, proteins can be degraded by proteases, DNA degraded by DNases, and small molecules degraded or taken up by surrounding cells (Dorward and Garon, 1990; Lin et al., 2003; Wang and Leadbetter, 2005). By surrounding contents with a structured membrane, OMVs protect their contents from environmental degradation. Additionally, OMVs can package and traffic hydrophobic molecules (Mashburn and Whiteley, 2005), which normally cannot easily disperse during growth in aqueous environments.

WHAT IS THE MOLECULAR MECHANISM OF OMV FORMATION?

To understand the molecular mechanisms of OMV formation, it is important to first review the structural differences between the cell envelopes of Gram-negative and Gram-positive bacteria. While both classes of bacteria contain a semipermeable membrane and PG layer surrounding the cytosol, only Gram-negative bacteria contain an additional protein-studded OM, consisting of an asymmetric bilayer with an outer leaflet of LPS and an inner leaflet of phospholipids. LPS generally consists of three components: the hydrophobic lipid A domain, core oligosaccharide, and O antigen. Some OM lipoproteins interact with the underlying PG, linking the OM to the PG. Between the OM and IM is the periplasmic space, which contains proteins in addition to the relatively thin PG layer. The Gram-positive cell envelope contains teichoic acids and lipoteichoic acids, polymers of glycerol, phosphates, and ribitol covalently bound to a thick layer of PG.

Despite intense interest and research in the field since the discovery of OMVs, the molecular mechanism of OMV formation has not been completely elucidated. Three main models for the mechanism of OMV formation have been proposed, which are not mutually exclusive. Within these models, five factors are proposed to contribute to OMV formation: (i) expansion of the OM in areas that lack protein linkages to the PG layer (Hoekstra et al., 1976; Wensink and Witholt, 1981), (ii) cell division and cell shape (Burdett and Murray, 1974a, 1974b), (iii) OM stress caused by accumulation of misfolded proteins within the periplasmic space (Zhou et al., 1998; Hayashi et al., 2002; McBroom et al., 2006; McBroom and Kuehn, 2007; Tashiro et al., 2009), (iv) charge-charge repulsion of LPS (Kadurugamuwa and Beveridge, 1995, 1996; Li et al., 1996), and (v) change in the rigidity and curvature of the membrane due to insertion or association of small molecules in the membrane (Kadurugamuwa and Beveridge, 1995; Mashburn-Warren et al., 2008). In the following subsections, the contributions of these five factors to the molecular mechanism of OMV formation and release are discussed in detail (Fig. 2).

Model 1: OM Anchored to the PG

Model 1 outline

As the OM grows, it can over time bulge, particularly in areas with few covalent protein linkages to the PG, and pinch off from the cell. There are multiple ways in which the OM can become detached from the PG layer. The association between the OM and PG decreases (i) in areas where few OM-PG- and OM-PG-IM-associated proteins exist, (ii) when the OM grows, and (iii) during cell division and in spaces where the shape of the cell is highly curved.

OM-PG linkage. The OM is anchored to the PG layer by covalent and noncovalent protein interactions. Interestingly, significantly fewer lipoproteins and their associated proteins are present in OMVs than in the OM (Hoekstra et al., 1976; Wensink and Witholt, 1981). A study of lipoprotein association to PG differentiated free lipoprotein from PG-bound lipoprotein (Wensink and Witholt, 1981), indicating that PG-bound lipoprotein was mostly excluded from *E. coli* OMVs (Wensink and Witholt, 1981). Many groups have hypothesized that OM detachment occurs in areas of localized decreased lipoprotein connections to the underlying PG layer, leading to blebbing of the OM and OMV release (Wensink and Witholt, 1981; Zhou et al., 1998; Mashburn and Whiteley, 2005; Mashburn-Warren and Whiteley, 2006; Deatherage et al., 2009; Tashiro et al., 2009). Mutations in genes encoding PG-associated lipoproteins OmpA, Pal, and Lpp have been shown to significantly affect OMV formation in *E. coli*, *V. cholerae*, and *Salmonella*. Deleting or truncating these anchoring proteins decreases OM association to the PG, increasing vesiculation events (Sonntag et al., 1978; Yem and Wu, 1978; Bernadac et al., 1998; Cascales et al., 2002; Song et al., 2008; Deatherage et al., 2009). Additionally, Bernadac and

colleagues demonstrated that *E. coli tol-pal* mutants produce high levels of OMVs (Bernadac et al., 1998). The Tol/Pal complex spans the entire envelope, including Pal in the OM, TolB in the periplasm, and several other Tol IM proteins. This complex connects large portions of the cell envelope together and is known to be crucial for maintaining OM integrity (Lazzaroni et al., 1999; Llamas et al., 2000). However, in a mutant screen of *E. coli* vesiculation, several hypervesiculators maintained OM stability, giving evidence that OM integrity is not the only variable involved in OMV production (McBroom et al., 2006). Additionally, some have questioned the validity of conclusions drawn from Bernadac's experiments (Deatherage et al., 2009), due to strains and methods used. Regardless, it has been demonstrated that physical contacts mediated by protein interactions occurring between the OM, PG, and IM maintain OM stability and affect vesiculation.

OM growth. Wensick and Witholt proposed that OMVs form when the OM enlarges faster than the PG layer (Wensink and Witholt, 1981). It is hypothesized that vesicles form in zones of OM growth, as *E. coli* vesicles are enriched in newly synthesized proteins (Mug-Opstelten and Witholt, 1978). Mug-Opstelten and Witholt performed experiments using both ^{3}H- and ^{14}C-radiolabeled leucine and demonstrated that *E. coli* preferentially releases newly radiolabeled protein into what was referred to as "outer membrane fragments." Overnight cultures that had been labeled with [^{3}H]leucine were subcultured and subsequently given [^{14}C]leucine. Newly synthesized protein could be distinguished from old protein, and newly labeled protein was released into membrane fragments, indicating that OMVs release where new proteins have inserted into the OM (Mug-Opstelten and Witholt, 1978).

Cell division and cell shape. As bacterial populations grow, cells increase in mass, replicate DNA, and synthesize new cell envelope components. In binary fission, after chromosome replication, the two chromosomes are partitioned by the formation of a septum. As the IM and PG invaginate, the OM remains excluded from the septum, and PG synthesis occurs along the wall of the newly formed septum. Eventually, the OM grows and the new daughter cells separate (Burdett and Murray, 1974) (Fig. 3). At the division septum, the OM is highly detached from the PG, a physical occurrence which leads to the formation of OMVs (Fig. 3) (Burdett and Murray, 1974).

Deatherage and colleagues proposed that the three-dimensional shape of cells affects OMV formation, as membrane surfaces of various degrees of curvature have different physical properties. The

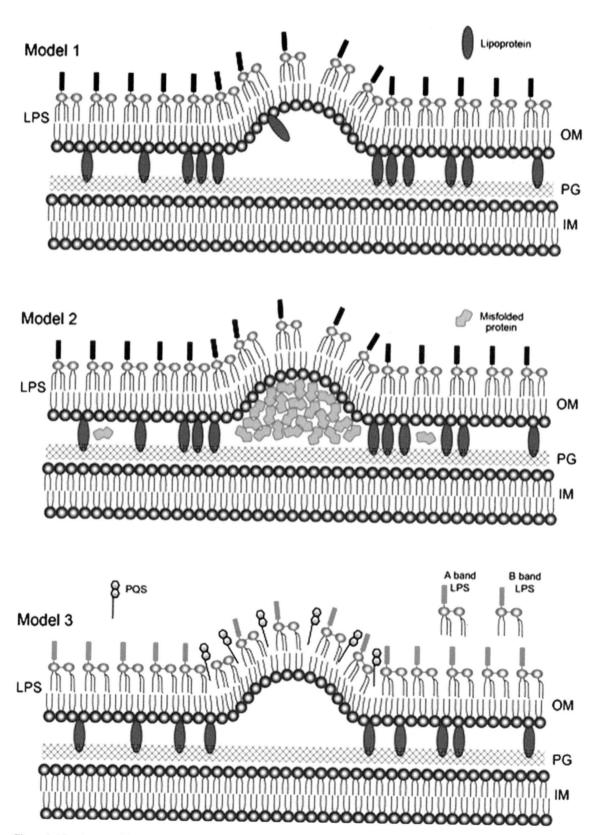

Figure 2. The three models of OMV formation. The cell envelope contains an OM with an outer leaflet of LPS. In models 1 and 2, LPS is hexa-acylated to illustrate a common *E. coli* structure, while in model 3, LPS is penta-acylated to illustrate a common *P. aeruginosa* structure. In model 1, the membrane can bleb in areas where the OM is not well anchored to the PG. In model 2, pressure on the OM caused by accumulation of proteins in the periplasmic space is relieved by membrane blebbing (figure adapted from Kulp and Kuehn, 2010). In model 3, charge-charge repulsion of LPS and PQS insertion is shown. In *P. aeruginosa*, in areas containing large amounts of B band LPS, charge-charge interactions induce curvature of the membrane. Membrane curvature is additionally enhanced by PQS preferentially inserting into the outer leaflet of the OM.
doi:10.1128/9781555818524.ch23f2

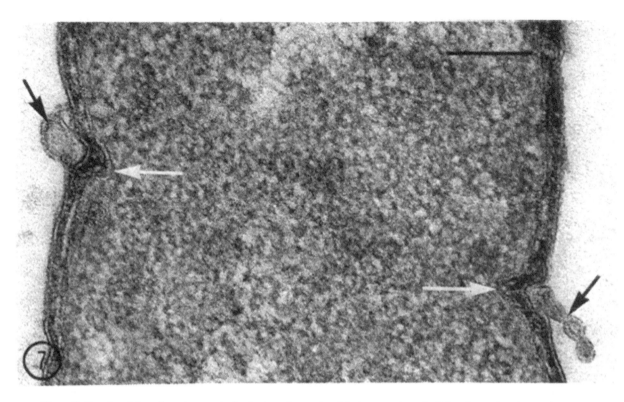

Figure 3. *E. coli* vesicles released upon the beginning of septation (black arrows), as the PG and cytoplasmic membrane grow inward toward the center of the cell (white arrows). Scale bar, 100 nm. Reprinted from the *Journal of Bacteriology* (Burdett and Murray, 1974) with permission of the publisher.
doi:10.1128/9781555818524.ch23f3

differences in biophysical properties of membranes of cocci and bacilli have yet to be studied. However, since it is known that membrane curvature during septation induces OMV formation, it is possible that preexisting degrees of membrane curvature would also similarly affect degrees of membrane blebbing. Though hypotheses for this potential OMV inducing factor have been proposed, to date no experimental evidence has been reported.

Model 2: Misfolded Proteins in the Periplasm

Model 2 outline

It is assumed that the shape of the prokaryotic cell is determined by tension as a result of hydrostatic pressure, expanding and contracting with changes in osmolarity (Koch et al., 1982). The presence of excess proteins in the cell envelope can also contribute to pressure on the OM, and under conditions where protein levels are increased, proteins often become misfolded. The cell responds with a variety of stress responses that degrade or remove excess proteins (reviewed in Ades, 2004). Recently, McBroom and Kuehn proposed a model (here named model 2) where an additional stress response pathway responds to

cell envelope stress by removing excess and misfolded cell envelope proteins via OMV secretion (McBroom and Kuehn, 2007).

Accumulation of misfolded proteins

The presence of misfolded proteins in the periplasm induces the production of proteases and chaperones through several well-characterized stress response pathways (Raivio, 2005; Rhodius et al., 2006). McBroom and Kuehn found that vesicle production is an alternate stress response pathway that is not directly regulated by previously identified stress response systems; however, it was shown that impairing the sigma E stress response pathway increased OMV production (McBroom and Kuehn, 2007). By depleting DegS, a periplasmic sensor protease responsible for sigma E activation, misfolded proteins accumulated in the periplasm, and vesicle production increased (Mogensen and Otzen, 2005; McBroom and Kuehn, 2007). To further elucidate the role of misfolded proteins in OMV formation, a construct was designed to mimic a misfolded envelope protein, and the mimic was preferentially sorted in OMVs. Additionally, when periplasmic proteins were overexpressed, OMV production increased (McBroom and Kuehn, 2007). With these findings, McBroom and Kuehn provide evidence for the second model, demonstrating that accumulation of periplasmic protein causes increased blebbing of the OM. A more detailed review of the regulation of OMV formation due to misfolded proteins can be found in the section "Cell Envelope Stress."

Model 3: Anionic Charge Repulsion of LPS and Outer Leaflet Expansion

Model 3 outline

LPS normally provides a structural barrier to the environment, with divalent cations forming cross bridges between adjacent LPS molecules. Not all LPS barriers are identical: LPS structure not only varies from species to species but also varies within species due to changes in physical, chemical, and nutrient conditions (Kropinski et al., 1987; McGroarty and Rivera, 1990; Makin and Beveridge, 1996). For example, *Pseudomonas aeruginosa* produces both A-band and B-band LPS, which differ in the structure of the O polysaccharide: A-band LPS contains mostly uncharged, short sugar chains, while B-band LPS is larger and highly anionic (Lam et al., 1989; Rivera and McGroarty, 1989). Mg^{2+} and Ca^{2+} salt bridges neutralize the charge-charge repulsions that occur between B-band LPS molecules in the OM.

LPS structure

Kadurugamuwa, Beveridge, and colleagues hypothesized that *P. aeruginosa* OMVs form due to anionic charge repulsion in the OM, specifically due to the presence of B-band LPS (Kadurugamuwa et al., 1993b; Kadurugamuwa and Beveridge, 1995; Kadurugamuwa and Beveridge, 1996, Li et al., 1996; Beveridge, 1999). Compared to A-band LPS, B-band LPS has longer, more negatively charged side chains, which causes B-band LPS side chains to repel each other due to neighboring terminal phosphate groups (Beveridge, 1999). It has been proposed that areas containing high levels of B-band LPS produce a charge-charge repulsion causing outward blebbing of the OM. As the B-band-rich OM blebs, periplasmic components can become packaged within blebs as they pinch off and leave the cell (Kadurugamuwa and Beveridge, 1995). Evidence for this model includes the fact that OMVs isolated from *P. aeruginosa* almost exclusively contain B-band LPS (Kadurugamuwa and Beveridge, 1995; Nguyen et al., 2003; Sabra et al., 2003) as well as additional experimental evidence discussed in the following paragraph. It should

also be noted that LPS molecules with long polysaccharide chains were found to be more abundant in *Porphyromonas gingivalis* OMVs than in the OM; however, the O antigen has been shown not to play a key role in the mechanism of *P. gingivalis* OMV formation, as strains that lack O antigen are able to produce OMVs (Haurat et al., 2011).

Using transmission electron microscopy (TEM), Sabra and colleagues showed that *P. aeruginosa* produced more OMVs under conditions of oxidative stress (pO_2, ~350% of air saturation). Under oxidative stress, cells became enriched in B-band LPS, which enhanced OMV formation (Sabra et al., 2003). However, in oxygen-limited environments (pO_2, ~0%), B-band LPS was weakly detected in the OM, and OMVs were rarely observed. Direct quantification of OMVs produced under these conditions was not shown; however, a threefold increase of mannuronic acid, a main component of B-band LPS, was detected in the culture supernatants of cells grown under oxidative stress (Sabra et al., 2003). Additional support for the LPS charge-charge repulsion model is discussed below; however, it should be noted that oxygen-limited cells experience global changes in genetic regulation, which could also affect OMV formation (see "Environmental Effects on OMV Formation").

OMV-inducing molecules

The order and fluidity of membranes are determined by the molecular structure and chemical interactions of membrane lipids, proteins, and small molecules. Mashburn-Warren et al. hypothesized that the bacterial OM can develop curvature at loci with decreased membrane fluidity, which can eventually lead to blebbing and OMV formation (Mashburn-Warren et al., 2008). Different lipids and molecules can decrease the fluidity of a membrane, and when these molecules bind or interact with the OM, they can presumably affect OMV formation.

Several studies have demonstrated that aminoglycoside antibiotics weaken the cell surface through ionic binding to the membrane (Hancock, 1984; Martin and Beveridge, 1986; Walker and Beveridge, 1988). For example, the bactericidal activity of gentamicin is not limited to inhibition of the 30S ribosome but is also due to disruption of the cell surface (Kadurugamuwa et al., 1993a, 1993b). Using strains that varied in their A- and B-band LPS contents, Beveridge's group investigated the nature of the ionic binding of gentamicin to the OM and found that the degree of gentamicin binding varied by LPS content: strains with more B-band LPS had a higher affinity for gentamicin than strains with A-band LPS alone (Kadurugamuwa et al., 1993a, 1993b). This high affinity also made B-band strains more susceptible to gentamicin. Gentamicin, a polycationic antibiotic, can replace structurally important cations like Mg^{2+} and Ca^{2+}, which normally function to cross bridge LPS molecules together (Hancock, 1984; Nikaido and Vaara, 1985; Peterson et al., 1985). Gentamicin binds and alters the LPS packing order, causing the membrane to bleb, and at higher concentrations forms transient holes in the membrane (Martin and Beveridge, 1986). Additional studies found that gentamicin stimulated OMV formation three- to fivefold, supporting the idea that gentamicin destabilizes the OM by altering the packing of LPS (Kadurugamuwa and Beveridge, 1995).

Similar to LPS interactions with gentamicin, a hydrophobic cell-cell signaling molecule, *Pseudomonas* quinolone signal (2-heptyl-3-hydroxy-4-quinolone [PQS]; reviewed in Pesci et al., 1999, and discussed in the section "QS and OMV Formation"), was shown to interact with and alter LPS packing. It was proposed that PQS destabilizes Mg^{2+} and Ca^{2+} salt bridges that normally neutralize the charge-charge repulsions of B-band LPS molecules, as it is known that quinolones can interact with cations (Marshall and Piddock, 1994). Further experimentation demonstrated not only that PQS is packaged into OMVs but also that PQS production is required and sufficient for OMV formation in *P. aeruginosa* (Mashburn and Whiteley, 2005). More specifically, a successive study elucidated the interaction between PQS and the OM, showing that PQS strongly interacts with the 4'-phosphate group, as well as the acyl chains of LPS (Mashburn-Warren et al., 2008).

An exciting new addition to this model was very recently proposed (Schertzer and Whiteley, 2012). It has been hypothesized that PQS inserts preferentially into the outer leaflet of the OM, due to its higher affinity for LPS over inner-leaflet phospholipids (Mashburn-Warren et al., 2008). Insertion of PQS primarily into the LPS would result in an asymmetric expansion of the outer leaflet, which has previously been proposed as the mechanism behind membrane blebbing in red blood cells (RBCs) exposed to certain membrane-active small molecules (Sheetz and Singer, 1974; Lim et al., 2002). Sheetz and Singer developed the "bilayer-couple hypothesis," which states that lateral expansion of one leaflet relative to the other causes membrane curvature, analogous to how a bimetallic thermocouple responds to heating. Though PQS was shown to have a higher affinity for LPS over phospholipids, HHQ, the precursor to PQS that lacks only the 3-hydroxyl group, does not have a higher affinity for LPS over phospholipids and does not induce OMV formation (Mashburn-Warren et al., 2008). Because HHQ lacks direct interaction with

LPS, it can likely easily flip-flop between leaflets and therefore does not contribute to asymmetric lateral growth of the outer leaflet. It was shown that PQS induces membrane curvature in RBCs and that the curvature was dependent upon the same characteristics of the molecule as are required for OMV formation in *P. aeruginosa* (Schertzer and Whiteley, 2012). Accordingly, neither HHQ nor other analogs of PQS tested had this effect. This model is consistent with work showing that PQS physically promotes OMV formation (Mashburn and Whiteley, 2005) and raises the possibility that other organisms might regulate OMV formation through the secretion of membrane-active small molecules.

Summary

Together, all of the parameters mentioned within the models above may have a collective effect on the production of OMVs. According to the first model, membrane blebbing likely occurs in areas where the OM does not contain strong association with the PG, specifically in areas where the PG is being cleaved or reorganized and in areas where the OM is growing. In the second model, blebbing occurs in areas where pressure builds due to accumulation of excess and misfolded proteins. The third model proposes that the membrane blebs due to LPS charge-charge repulsion and in areas where insertion of molecules in the OM changes the fluidity and length ratio, and therefore curvature, of the inner and outer leaflets. Because these changes in molecular interactions of the OM have all been shown to affect OMV production, they may all together alter the OM structure and impact membrane vesiculation.

WHAT REGULATES OMV FORMATION?

A significant, albeit elusive, goal in the OMV field has been to identify genes whose products are involved in regulation of OMV production. Given the importance of OMVs in sharing of genetic material, pathogenicity, and host-pathogen relationships, understanding what controls vesiculation opens the door for new antimicrobial therapeutic targets and has implications for environmental science and agriculture. Unfortunately, no one mode of genetic regulation of OMV formation has been discovered; however, many inroads have been made into understanding the cellular components that regulate OMV formation, as well as identifying proteins and RNA that may function as regulators. As mentioned in a previous section (see "Model 1: OM Anchored to the PG"), loss of OM integrity can enhance vesiculation and has been associated with altered protein-lipid interactions that tether the OM to the PG layer. More recent results have called into question whether membrane integrity is the only factor affecting OMV formation. The presence of misfolded proteins in the OM, the cell envelope stress response, quorum sensing (QS), and other mechanisms are all thought to control OMV formation. In this section we summarize the current work on the regulation of OMV production.

Cell Envelope Stress

As the barrier to the outside world, the cell envelope is frequently exposed to varying and harsh environments and the cell must maintain membrane integrity to survive. The structure of the cell envelope and particularly the OM are also critical to OMV formation, which suggests that the mechanisms cells utilize for coping with cell envelope stress could also directly or indirectly affect OMV formation. Evidence in support of this comes from a recent genetic screen by McBroom and colleagues for *E. coli* mutants that over- and underproduced OMVs. Among the genes identified in the screen were several whose products are involved in the cell envelope stress response and also increased OMV production (McBroom et al., 2006). The cell envelope stress response, which has been best characterized for *E. coli*, is regulated by an alternative sigma factor encoded by the *rpoE* gene, sigma E. Regulation of sigma E is mediated through its sequestration at the membrane by a transmembrane anti-sigma factor, RseA. When OM proteins become misfolded, a protease, DegS, is activated, which begins a sequential process resulting in the cleavage of RseA and release of sigma E. Free sigma E induces transcription of genes involved in the cell envelope stress response, including a downstream effector protease, DegP, which degrades misfolded proteins (Chaba et al. 2007; reviewed in Alba and Gross, 2004, and Ades, 2008). It was noted in a previous section (see "Model 2: Misfolded Proteins in the Periplasm") that the presence of misfolded proteins enhances OMV production. Considering that sigma E and OMV formation are both activated in the presence of membrane stress, it is logical to hypothesize that the sigma E pathway may induce OMV formation. However, this is not strictly the case, as mutations that both diminish (*degS*) and enhance (*rseA*) sigma E activity increased OMV formation more than 100-fold compared to that of the wild type (McBroom et al., 2006). The McBroom screen also identified *degP* as a hypervesiculating mutant, indicating that abolish-

ing downstream products of sigma E activation also enhances OMV formation.

How can OMV production be positively affected by both activation and deactivation of sigma E? McBroom and colleagues proposed in subsequent work that OMV formation is an alternative cell envelope stress response pathway meant to release misfolded proteins in the cell envelope regardless of sigma E activity. The strongest evidence presented for this hypothesis was the presence of overexpressed proteins in OMV preparations and the preferential sorting of a peptide that mimics misfolded envelope proteins into OMVs (McBroom and Kuehn, 2007). Thus, the presence of misfolded proteins in the cell envelope induces not only canonical stress-coping mechanisms like the sigma E pathway but also the production of OMVs.

The fact that misfolded proteins induce OMV formation is also consistent with an early observation that heat induces OMV formation (Katsui et al., 1982) (see "Environmental Effects on OMV Formation"). Prior to studies of misfolded proteins enhancing OMV production, it was believed that heat stress induced OMV formation by increasing the likelihood of protein misfolding and perturbing interactions between lipids, proteins, and PG, which all, in turn, disrupt membrane integrity (Hoekstra et al., 1976; Katsui et al., 1982). Consistent with this hypothesis, the same screen for mutants that over- and underproduced OMVs identified several loci that when mutated both decrease membrane integrity, as determined by detergent sensitivity, and increase OMV formation (McBroom et al., 2006). For the most part, these mutants contained defects in proteins critical for maintaining cell envelope structure, including constituents of the Tol-Pal system; OmpC, a porin; OmpR, a regulator of porin production; and PonB, a protein involved in growth and cross-linking of the PG layer (McBroom et al., 2006). The identification of *tolA*, *tolB*, and *pal* confirmed the utility of the screen, as the Tol-Pal system had been previously identified as affecting OMV formation (Bernadac et al., 1998). It has also been reported that expression of recombinant soluble protein fragments that interact with the Tol-Pal system can induce OMV production when present in the periplasm (Bernadac et al., 1998; Henry et al., 2004). The role these proteins play in membrane structure and OMV formation is discussed more thoroughly in another section (see "Model 1: OM Anchored to the PG"). However, the fact that Tol-Pal proteins are known to affect OMV formation is consistent with the hypothesis that perturbation of the Tol-Pal system during envelope stress enhances OMV production. It is worth noting that cell envelope stress generated by misfolded proteins and loss of important structural components within the cell envelope encompass proposed mechanisms of OMV formation regulation that are not mutually exclusive.

A final example of OMV formation as a stress-coping mechanism comes from *Pseudomonas putida*, which has been shown to increase OMV formation in the presence of the organic solvent toluene (Kobayashi et al., 2000). A study by Kobayashi and colleagues demonstrated that less toluene remained cell associated in a toluene-resistant strain than in a toluene-sensitive strain and that this coincided with an increase in toluene associated with OMVs produced by the resistant strain (Kobayashi et al., 2000). The authors presented this as evidence of a novel toluene resistance mechanism that involves shedding of cell-associated toluene via OMVs, and it is reminiscent of the proposed mechanism for release of misfolded proteins through OMV production (Kobayashi et al., 2000).

The Role of Proteins in Regulation of Vesiculation

In addition to the broad regulatory effects of membrane stress on OMV formation, several examples of proteins involved in regulation of OMV formation have also been found. The McBroom screen for mutants that over- and underproduced OMVs isolated a mutant in *nlpA* that underproduced vesicles, suggesting that it may play a role in positively regulating OMV formation. NlpA, a IM-bound periplasmic protein, had no known function until its vesiculation phenotype (McBroom et al., 2006). More recent work has demonstrated that NlpA transcription is negatively controlled by the enterotoxigenic

E. coli (ETEC) virulence regulator Rns and related *E. coli* proteins CfaD and AggR (Bodero et al., 2007). These proteins also control production of the ETEC heat-labile toxin, which is transmitted on the surface of OMVs. Rns, CfaD, and AggR may broadly regulate virulence by controlling both production of toxins and their release in OMVs (Bodero et al., 2007).

Interactions between protein and OMVs could be an important regulatory scheme in newly discovered OMV-secreting organelles called nanopods (Shetty et al., 2011) (Fig. 4). First identified in micrographs of *Delftia* strain Cs1-4, nanopods are filamentous structures that can be up to 6 µm long and appear under TEM as a layer of crystalline protein enclosing contents that resemble OMVs (Shetty et al., 2011). Analysis of the macromolecular contents of the external and internal structures revealed the former to be primarily composed of a previously uncharacterized protein, NpdA, while the latter contained LPS, periplasmic proteins, and OM proteins typically associated with OMVs (Shetty et al., 2011). Little is known about how the formation of these structures is regulated or what role they may play in regulation of OMV formation. However, the identification of NpdA as the primary constituent of the bacteria's surface layer protein (SLP) indicates a possible regulatory mechanism, as extracted SLP has been previously shown to assemble complex structures in the presence of native membrane (Engelhardt et al., 1990; Shetty et al., 2011). This raises the possibility that mechanisms of OMV formation may be involved in formation of the nanopod secretion apparatus, though further analysis of the interplay between NpdA and OMVs is required. The existence of an OMV-secreting organelle is an intriguing, unique

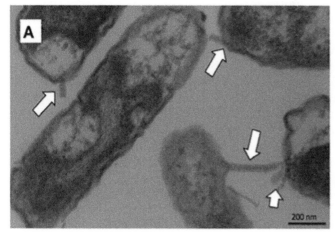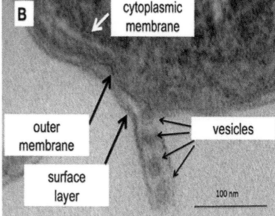

Figure 4. Thin-section TEM of *Delftia* producing nanopods, an organelle which secretes OMVs. (A) Cell-attached nanopods (white arrows) can be up to 6 µm in length. Scale bar, 200 nm. (B) The cell-nanopod junction. Scale bar, 100 nm. Reprinted from *PloS One* (Shetty et al. 2011) with permission of the publisher.
doi:10.1128/9781555818524.ch23f4

method for regulating vesicle production and trafficking. While this particular structure seems to be unique to the *Comamonadaceae* family, the existence of a similar tube structure for transfer of cellular components in *Bacillus subtilis* raises the possibility that this phenomenon may be prevalent in nature, though the latter structure was not associated with OMVs (Dubey and Ben-Yehuda, 2011).

As mentioned in a recent review, the existence of remarkably different OMV formation phenotypes between two strains of *Lysobacter* may indicate new regulatory mechanisms for OMV formation (Vasilyeva et al., 2009; Kulp and Kuehn, 2010). A single parent *Lysobacter* strain differentiates into two distinct strains, XL1 and XL2, upon long-term culturing (Vasilyeva et al., 2009). In a nutrient-poor medium that induces secretion of lytic enzymes, quantification of OMV production by assaying total protein in the vesicle preparation demonstrated approximately 60-fold more protein in OMVs from XL1 than from XL2 (Vasilyeva et al., 2009). It was proposed that this was due to the presence of larger vesicles in XL1 than in XL2, and this difference in vesicle size may help explain any OMV formation regulatory difference between the strains (Vasilyeva et al., 2009). The fact that these strains arose from the same progenitor and are presumably nearly genetically identical suggests that the differences between these two strains should be studied for insight into regulatory mechanisms of OMV production (Vasilyeva et al., 2009; Kulp and Kuehn, 2010).

sRNAs and OMV Formation

Small RNAs (sRNAs) are regulatory RNA molecules that can affect protein expression levels in many cases by binding the 5′ untranslated region of a messenger RNA and either altering interactions with the ribosome or targeting the RNA for degradation. Consequently, these interactions can either positively or negatively affect expression of the target protein, and they are often facilitated by a protein chaperone, Hfq. Reports have emerged of sigma E-regulated sRNAs in *E. coli* (RseX and MicA) and *Salmonella* (RybB and MicA) that decrease the abundance of OM proteins by degrading their mRNA transcripts (Udekwu et al., 2005; Douchin et al., 2006; Figueroa-Bossi et al., 2006; Johansen et al., 2006; Papenfort et al., 2006; Thompson et al., 2007; Udekwu and Wagner, 2007; Song et al., 2008). While these reports did not specifically address the effect of sRNAs on vesiculation, it follows that their ability to alter expression of proteins known to affect vesiculation suggests that the sRNAs are capable of regulating OMV production. This hypothesis was recently confirmed

by studies of *Vibrio cholerae*, which showed that an sRNA directly affects OMV formation (Song et al., 2008). Similar to the sRNAs described previously, the *V. cholerae* sRNA VrrA binds to the 5′ region of the OmpA transcript in a manner that blocks access to the ribosome binding site and inhibits translation of the porin protein. As with OmpC discussed above (see "Cell Envelope Stress"), OmpA is believed to be an important OM protein involved in maintenance of OM stability and tethering the OM to the PG layer. Deletion of VrrA decreased OMV production, corresponding to increased levels of OmpA protein in the membrane; overexpression of VrrA had the opposite effect (Song et al., 2008). The presence of sigma E-regulated sRNAs that affect vesiculation provides further evidence in support of the hypothesis that OMV formation is inducible as part of the normal cellular response to OM stress. With the help of techniques like RNA-Seq, novel sRNAs are rapidly being discovered, and as the population of known sRNAs increases, it will be interesting to uncover the role they play in OMV formation.

QS and OMV Formation

QS is a means of bacterial cell-to-cell communication that results in coordinated gene expression and group behavior. The canonical QS scheme requires production and sensing of small-molecule (gram-negative bacteria) or peptide (gram-positive bacteria) signals called autoinducers. When concentrations of constitutively produced autoinducer signals reach a threshold level, the signal molecules interact with a transcriptional regulator(s) and alter gene expression in a cell density-dependent manner. Processes regulated by QS include production of secondary metabolites and virulence factors, light production, biofilm formation, and, importantly for this chapter, OMV formation. In *P. aeruginosa*, a model organism for QS studies, production of OMVs is substantially diminished in the absence of an autoinducer called PQS (Mashburn and Whiteley, 2005). It has been shown that OMVs package and disseminate PQS, a highly hydrophobic signal whose diffusion would be significantly curtailed in the aqueous extracellular environment. Subsequent work is consistent with a model that PQS inserts into the outer leaflet of the OM, interacts with specific chemical moieties in LPS, and potentially stabilizes membrane blebs, leading to OMV formation (see "Curvature-Inducing Molecules") (Mashburn-Warren et al., 2009; Schertzer and Whiteley, 2012). Consistent with other autoinducer molecules, once the concentration of PQS reaches a threshold concentration, it interacts with the transcriptional regulator MvfR (PqsR) and induces expression of a range of genes, including its own biosynthetic operon (*pqsABCD*). This represents a positive-feedback loop, which ultimately results in greater production of OMVs. This example of direct gene regulation of OMV formation has implications for multispecies interactions mediated by vesicle trafficking, as well as host-pathogen interactions in this clinically relevant organism.

Summary

While no broad genetic regulation scheme for OMV production has been determined for all OMV-producing bacteria, there have been many discoveries that point to possible cell components and regulatory pathways that are likely involved. The observation that OMV production is increased in the presence of heat, misfolded proteins, and toxic compounds like toluene in a manner independent of known cell envelope stress pathways suggests that OMV formation may be an alternative stress-coping mechanism, though the mechanism of genetic regulation of this phenomenon remains elusive. Further, recent reports of sRNA and QS regulation of OMV formation raise intriguing new prospects for genetic regulation of OMV formation. The ubiquity of OMV production among gram-negative bacteria implies that it is a critical process for growth and survival. This, combined with the ability of OMVs to traffic important compounds for virulence and interspecies interactions, makes OMV production an attractive target for curtailing the growth and/or pathogenicity of bacteria. Thus, it becomes clear that understanding the mechanisms for regulation of OMV formation could lead to novel antimicrobial targets. Finally, understanding genetic regulation of OMV formation could also allow investigators to overproduce OMVs, which could be useful for industrial applications and vaccine development (Roy et al., 2010; Zollinger et al., 2010; McConnell et al., 2011).

WHAT IS PACKAGED IN OMVs?

OMVs package proteins, signaling molecules, and genetic material within a bilayered membrane derived from the OM, consisting of an outer leaflet of LPS and an inner leaflet of phospholipids. OMV contents can be trafficked to neighboring cells over long distances, influencing the environment in a variety of ways. Specific proteins and lipids are packaged into OMVs, while others are excluded, suggesting the existence of a sorting mechanism, as opposed to indiscriminate blebbling of the OM. The contents of OMVs are not identical among bacteria and differ greatly due to the diversity of bacterial species.

Initial experiments were performed to investigate whether OMVs originated from a pinching off of the OM or were merely artifacts of cell lysis. In studies of naturally produced OMVs, components of OMVs were compared to components found in the OM, periplasm, IM, and cell cytoplasm. Experiments concluded that proteins and lipids found in MVs were indeed most similar to the protein content of the OM, containing some periplasmic components and no cytoplasmic components (Beveridge, 1999; Horstman and Kuehn, 2000; Bauman and Kuehn, 2006). This confirmed the theory that OMVs originated from OM blebs of living cells. However, there is still much debate over whether OMVs package cytoplasmic components, as some proteomic analysis suggests that OMVs originate from the OM and periplasm and contain cytoplasmic components (Lee et al., 2007; Lee et al., 2008; Choi et al., 2011).

An important distinction of OMV studies should be made: many studies of OMV contents have examined detergent-extracted OMVs, which are synthetically produced with detergent treatment (Nally et al., 2005; Ferrari et al., 2006; Uli et al., 2006; Vipond et al., 2006), while other researchers have studied "native" OMVs, naturally produced by cells (Ferrari et al., 2006; Lee et al., 2007; Berlanda Scorza et al., 2008; Kwon et al., 2009). Additionally, within native-OMV studies, some researchers have examined OMVs purified from mutants; mutant OMV content may contain minor, though possibly significant, differences from wild-type OMVs.

Proteins

The protein content of vesicles has been studied using sodium dodecyl sulfate-polyacrylamide gel electrophoresis (SDS-PAGE) with Coomassie or silver staining, Western blotting, biochemical analysis, and mass spectroscopy (Horstman and Kuehn, 2000; Ferrari et al., 2006; Lee et al., 2007; Kwon et al., 2009). Proteomic studies using SDS-PAGE have examined total protein recovered from whole cells, purified OM, periplasm, and OMVs. OMVs had a banding pattern distinct from the rest of the cell components, suggesting the presence of a protein sorting method during OMV biogenesis (Renelli et al., 2004; Bauman and Kuehn, 2006; Tashiro et al., 2010). While several abundant OM proteins were demonstrated to also be present in OMVs (Kesty and Kuehn, 2004), there is evidence that some OM proteins are largely excluded from OMVs (Horstman and Kuehn, 2000) (Fig. 5). For example, the lipoprotein Lpp, which connects the OM to the PG, was not detected in *E. coli* OMVs (Hoekstra et al., 1976; Wensink and Witholt, 1981b). Additionally, biochemical studies have provided evidence that OMVs

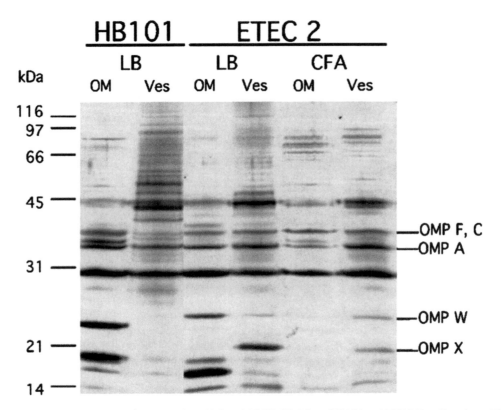

Figure 5. Protein banding patterns of OM proteins (OM) and OMVs (Ves) from HB101 and ETEC *E. coli* strains, with varying growth conditions. OM and Ves (0.5 µg) were applied to 12.5% SDS-PAGE gels and silver stained. Banding patterns for Ves look similar to the OM protein banding; however, some proteins appear to be preferentially sorted into OMVs. Molecular mass standards are indicated on the left, in kilodaltons. Reprinted from *Journal of Biological Chemistry* (Horstman and Kuehn, 2000) with permission of the publisher.
doi:10.1128/9781555818524.ch23f5

contain both outer membrane and periplasmic components (Horstman and Kuehn, 2000).

OMVs produced by several gram-negative bacteria have been shown to contain virulence factors, enabling bacteria to kill both prokaryotic and eukaryotic cells, even from a great distance. Several bacteria produce OMVs with antimicrobial activity, including *Citrobacter*, *Enterobacter*, *Escherichia*, *Klebsiella*, *Morganella*, *Proteus*, *Pseudomonas*, *Salmonella*, and *Shigella* (Kadurugamuwa and Beveridge, 1996; Li et al., 1998). Murein hydrolases (Li et al., 1998), phospholipase C, alkaline phosphatase, proelastase, and hemolysin (Kadurugamuwa and Beveridge, 1996) have all been shown to be packaged within OMVs. Additionally, β-lactamase can be packaged into OMVs, allowing populations of cells to lower the antibiotic concentration present in an environment (Ciofu et al., 2000), as well as potentially transferring active β-lactamase to β-lactam-sensitive bacteria.

Several toxins have also been detected within OMVs. The oral pathogen *Aggregatibacter actinomycetemcomitans* produces a leukotoxin, LtxA, which, interestingly, does not contain a type II signal peptide sequence but is still translocated across the IM

into the periplasm by LtxB and LtxD and then integrates into the OM (Lally et al., 1989). Once in the OM, LtxA is packaged into OMVs more frequently than other *A. actinomycetemcomitans* OM proteins, lending further evidence for the presence of an OMV sorting mechanism (Kato et al., 2002). More recently, Haurat and colleagues reported additional evidence for the existence of a protein sorting mechanism in a study of *P. gingivalis* OMVs (Haurat et al., 2011). *P. gingivalis* produces gingipains, a group of proteases that serve as a major virulence factor during infection. Gingipains are selectively sorted into OMVs, while some abundant OM proteins are excluded (Haurat et al., 2011). Understanding OMV protein sorting mechanisms may provide insight into the molecular mechanism of OMV formation and may better elucidate how to engineer bacteria to package and deliver specific cargo.

Previous work has shown that changes in the growth medium affect protein composition of the OM (Alphen and Lugtenberg, 1977; Horstman and Kuehn, 2000), which suggests that OMV protein composition also changes due to environmental perturbations. As seen in Fig. 5, Horstman and Kuehn report that changing the growth medium moderately affected the OMV protein profile; the banding patterns between the two medium conditions do not appear identical but are somewhat similar. Keenan and Allardyce give stronger evidence for OMV protein composition change due to alterations in the growth medium; the protein composition of *Helicobacter pylori* OMVs was altered under iron-limiting conditions, as less of the vacuolating cytotoxin VacA was observed, while two new proteases were detected. These phenotypes were reversed upon addition of iron (Keenan and Allardyce, 2000). Similarly, the specific activity of proteins in OMVs has been shown to be altered by environmental changes, as greater proteolytic activity was observed in OMVs obtained from hemin-limited *P. gingivalis* cells, while the total protein content remained the same (Smalley et al., 1991).

While the protein profiles of OMVs have been shown to vary due to environmental fluctuations, it is clear that many OMV proteins are conserved among species. Several groups have used a proteomics approach to examine total protein content of native OMVs isolated from both pathogenic and nonpathogenic bacteria, including *Neisseria meningitidis*, *P. aeruginosa*, *Pseudoalteromonas antarctica* NF$_3$, and *E. coli* (Post et al., 2005; Bauman and Kuehn, 2006; Ferrari et al., 2006; Nevot et al., 2006; Lee et al. 2007; Berlanda Scorza et al., 2008; Kwon et al., 2009; Choi et al., 2011). With these studies, researchers found a diverse array of proteins that differed between species; however, the presence of several protein families was found to be conserved: porins and OM proteins, murein hydrolases, multidrug efflux pumps, ABC transporters, protease/chaperone proteins, and motility proteins were are all common constituents of gram-negative OMVs. These proteomic studies were recently reviewed by Lee and colleagues (Lee et al., 2008).

Some researchers have challenged the convention that OMVs exclusively contain OM and periplasmic components, arguing that OMVs also package IM and cytoplasmic proteins (Lee et al., 2007; Kwon et al., 2009). Cytoplasmic and IM proteins may be packaged in OMVs; however, it is also possible that the detection of IM and cytoplasmic proteins could be due to contamination during the preparation of OMVs, as proteins originating from trace amounts of cells or lysed cell material may be detected by mass spectrometry. Most studies of native OMV protein content determined that IM and cytoplasmic proteins are not included in OMVs, an idea which is also supported by TEM studies that do not show an additional IM layer within OMVs (Knox et al., 1966; Bauman and Kuehn, 2006).

Nucleic Acids

In addition to protein, OMVs also enable the transfer of genetic material in a mechanism substantially different from the canonical means of horizontal gene transfer (transfection, transformation, and conjugation). Within OMVs, genetic material can be moved over long distances, transferring genetic information without requiring a physical attachment of cells. DNA associates with OMV surfaces but may also be enclosed within OMVs, which provide protection from DNases and other means of degradation in the extracellular environment. Double-stranded DNA has been identified in OMVs produced by *Neisseria gonorrhoeae*, *E. coli* O157:H7, and *P. aeruginosa* (Dorward et al., 1989; Kolling and Matthews, 1999; Yaron et al., 2000; Renelli et al., 2004). Successful transfer of plasmid DNA was demonstrated in *N. gonorrhoeae*, where a donor strain was able to transfer a plasmid containing a β-lactamase gene into recipient cells. In addition, a plasmid containing replication, mobilization, and partitioning genes was isolated from *E. coli* O157:H7 OMVs. OMVs facilitated the transfer of genetic material to both *Salmonella enterica* serovar Enteritidis and *E. coli* JM109 recipient cells. Like the previously mentioned species, *P. aeruginosa* packages chromosomal and plasmid DNA in OMVs; however, the transfer of DNA to a recipient has been unsuccessful under experimental conditions tested thus far (Renelli et al., 2004), suggesting that the mechanisms of DNA transfer via OMVs vary among species. Renelli and colleagues proposed two mechanisms of DNA packaging and transfer via OMVs in *P. aeruginosa* (Renelli et al., 2004): (i) DNA moves from the cytoplasm to the periplasm, where it incorporates into OMVs, or (ii) extracellular DNA moves through the OM and into the periplasm, where it then becomes packaged in OMVs. The mechanism for DNA movement into the periplasm in each model is unknown. Based on the experiments of Renelli et al., it seems likely that both models are relevant to DNA transfer via OMVs. In addition to DNA, RNA has been identified in OMVs in smaller amounts, but this phenomenon has not yet been widely investigated (Dorward et al., 1989).

LPS and Phospholipid Content of OMVs

The OM of Gram-negative bacteria is an asymmetric bilayer, consisting of an outer layer of LPS and an inner layer of phospholipids. As stated above, generally, LPS is comprised of three components: the hydrophobic lipid A domain, core oligosaccharide, and O antigen. Similarly, OMVs contain LPS, glycerophospholipids, OM proteins, and periplasmic components. Several studies have shown that *P. aeruginosa* OMVs primarily contain B-band LPS, despite B-band LPS being less prevalent in the OM than A-band LPS. It had been proposed that repulsive forces between adjacent B-band LPS molecules cause membrane curvature, leading to blebbing and eventual OMV release; this model is reviewed in the previous section (see "Model 3: Anionic Charge Repulsion of LPS").

In studies of *E. coli*, Hoekstra et al. found that phospholipid and fatty acid contents of the OM and OMVs appeared to be mostly similar (Hoekstra et al., 1976). Adding to those studies, Horstman and Kuehn studied ETEC OMVs using thin-layer chromatography and found that the lipid content of vesicles was very similar to that of the OM, containing LPS, phosphatidylethanolamine, phosphatidylglycerol, and cardiolipin, though they did not discuss the enrichment of specific lipids (Horstman and Kuehn, 2000).

In more recent studies of *P. aeruginosa*, Tashiro et al. examined differences in the phospholipid and fatty acid content in OMVs, compared to the content of the OM. The major phospholipid found in OMVs was phosphatidylglycerol, while the major OM phospholipid was phosphatidylethanolamine (Tashiro et al., 2011). In addition, OMVs contained more saturated fatty acids and longer-chain fatty acids than the OM, demonstrating that *P. aeruginosa* OMVs are more rigid than the OM. The differences Tashiro et al. point out may be species specific; therefore, additional studies are needed to conclude whether OMVs are universally enriched for specific lipids.

Small Molecules

Lipids and proteins of the OM interact with the external environment, and often small molecules associate with the OM due to their hydrophobic or cationic character. For example, in a study by Goedhart and colleagues, it was shown that Nod factor, a small molecule produced by nitrogen-fixing bacteria that induces nodulation in their leguminous symbiotes, associates with artificial membranes, suggesting that Nod factor may also be packaged into bacterial OMVs (Goedhart et al., 1999). Interestingly, artificially produced membranes containing Nod factor could transfer Nod factor to a host plant (Goedhart et al., 1999). In addition, the signaling molecule PQS (see "Model 3: Anionic Charge Repulsion of LPS") associates with OMVs due to its hydrophobic nature and ability to specifically interact with LPS. It is likely that many other hydrophobic molecules, including antimicrobial and signaling molecules, are packaged and trafficked in OMVs (Kadurugamuwa and Beveridge, 1996; Li et al., 1996, 1998); in

P. aeruginosa, several quinolones have been identified in OMVs and are thought to contribute to their antimicrobial activity (Mashburn and Whiteley, 2005).

Summary

OMVs have been shown to package a range of substrates. Proteomic analysis of MV contents has confirmed their OM-derived nature, as the protein profiles of OMVs and the OM are often similar. Additionally, non-OM proteins are found in OMVs, including virulence factors and toxins. OMVs also package nucleic acids, including antibiotic resistance genes, which suggests that the contribution of OMVs to pathogenicity goes beyond delivery of toxins. The lipid content of OMVs is also similar to that of the OM; however, differences in lipids present in OMVs may represent a regulatory mechanism for OMV formation. Finally, the fact that intercellular signaling molecules have also been associated with OMVs indicates a role for OMVs in intra- and interspecies communication. OMVs provide cells a means of concentrating and protecting secreted cellular components; continued characterization of these components and the mechanisms by which they are secreted will yield greater knowledge of the evolutionary purpose of OMVs and possibly lead to the exploitation of OMVs in industry and medicine.

ENVIRONMENTAL EFFECTS ON OMV FORMATION

The greatest obstacle that bacteria must overcome is variability between and within the environments in which they live, and their ability to cope with and adapt to the environment is critical to successful reproduction and propagation. As OMV formation has been associated with both normal growth and the stress response, it is not surprising that the environment can often profoundly affect OMV formation. Beginning with the aforementioned observation that heat induces vesicle formation (see "Cell Envelope Stress"), there has been much progress toward understanding environmental effects on OMV formation, and this section summarizes the effects of environmental features like temperature, salt/osmolarity, and available nutrients on OMV formation. Because of the well-documented role OMVs play in the pathogenicity of some organisms, much of the OMV field has focused on how the host responds to bacterially derived OMVs (Bauman and Kuehn, 2006; Bomberger et al., 2009; Park et al., 2010; Vidakovics et al., 2010). Of note, the use of vesicles as novel vaccines has become one area of widespread interest (Roy et al., 2010; Zollinger et al., 2010; McConnell et al., 2011). While the effect of OMVs on the host is outside the scope of a chapter on regulation of OMV formation, the reciprocal effect of both antibiotic treatment and the host on OMV formation is discussed.

Temperature

It has been known since the early 1980s that heat induces vesicle formation (Katsui et al., 1982). Work by Katsui and colleagues demonstrated that OM blebs were present when *E. coli* cells were incubated at 55°C for 30 min (Katsui et al., 1982), and these blebs were found to have protein and lipid contents similar to those of the OM. Thus, they argued these were not an artifact of cell lysis. Katsui and colleagues were also able to image membrane blebs that appear similar to vesicles and occurred primarily at division septa (see "Model 1: OM Anchored to the PG" and Fig. 3) (Katsui et al., 1982). Importantly, the vesicles contained periplasmic alkaline phosphatase activity; however, no activity was demonstrated for the cytoplasmic enzyme glucose-6-phosphate dehydrogenase, confirming that the blebs were indeed genuine vesicles rather than the result of dead or lysing cells (Katsui et al., 1982).

Available Nutrients

The nutritional environment in which a bacterium finds itself can often profoundly affect the physiology and behavior of the organism. This is also true for OMV formation, though there is no known strict rule for how available nutrients affect OMV formation. Some bacteria display enhanced OMV formation when nutrients are low. For example, in early studies, OMVs were observed in a lysine-limited medium (Knox et al., 1966) and under starvation conditions (Marden et al., 1985). Additionally, inoculation of *Haemophilus influenzae* into a competence-inducing medium that is missing several nutrients critical for growth was shown to also induce OMV formation (Deich and Hoyer, 1982), and the previously discussed strains of *Lysobacter* (see "The Role of Proteins in Regulation of Vesiculation") also made more OMVs in a nutrient-poor medium (Vasilyeva et al., 2009). On the other hand, *Pseudomonas fragi* makes OMVs when grown in rich media but not in a nutrient-poor, minimal medium (Thompson et al., 1985). Consistent with the former idea, it has been suggested that some organisms use vesicles to help acquire or degrade sources of carbon and energy, and the lysis of other organisms is one means to accomplish this. Enzymes capable of such nutrient acquisition by degradation have been associated with *P. aeruginosa* OMVs (Li et al., 1996). Further, the concentration

of proteolytic enzymes in *H. pylori* vesicles that may aid in nutrient acquisition was increased in an iron-limited medium, though the number of vesicles was unaffected (Keenan and Allardyce, 2000). Finally, production of recently discovered OMV-secreting structures called nanopods (see "The Role of Proteins in Regulation of Vesiculation" and Fig. 4) was induced in the presence of the polycyclic aromatic hydrocarbon phenanthrene (Shetty et al., 2011). The *Delftia* strain in which these structures were discovered was isolated from an oil-contaminated area, and it was proposed that nanopods and the vesicles they secrete may have a role in phenanthrene metabolism (Shetty et al., 2011). The presence of degradative enzymes in OMVs combined with their upregulation in nutrient-poor environments in many species suggests that nutrient acquisition may be one of the more significant roles OMVs play in bacterial growth and development.

The ability of the QS molecule PQS (see "QS and OMV Formation") to regulate OMV formation in *P. aeruginosa* (Mashburn and Whiteley, 2005; Mashburn-Warren et al., 2009) also raises an interesting possibility for nutritional effects on OMV formation. *P. aeruginosa* is well known to cause chronic infections in the lungs of individuals with the genetic disease cystic fibrosis (CF). When *P. aeruginosa* was grown in lung secretions from individuals with CF, PQS production was found to be increased approximately fivefold, raising the possibility that a nutrient(s) within CF lung fluids could induce PQS production (Palmer et al., 2005). Subsequent work confirmed that aromatic amino acids, specifically phenylalanine and tyrosine, are responsible for enhanced PQS production in CF lung secretions (Palmer et al., 2007). In addition to aromatic amino acids, oxygen has also been shown to be critical for PQS production and consequently OMV formation in *P. aeruginosa* (Schertzer et al., 2010). The ability of available nutrients to affect OMV formation and virulence phenotypes associated with OMVs underscores the need to characterize nutrients at infection sites, as this could lead to the ability to manipulate the behavior of bacteria during disease.

Salt

As discussed in a previous section (see "Model 3: Anionic Charge Repulsion of LPS"), concentration of ions, such as Mg^{2+} and Ca^{2+}, can affect interactions between LPS molecules in the OM of gram-negative bacteria. Additionally, osmotic stress derived from the presence of high salts in an aqueous environment has also been shown to affect OMV formation. When *Pseudomonas fluorescens* was exposed to osmotic stress by addition of 0.5 M NaCl to its growth medium, an increase in OMV formation was observed (Guyard-Nicodeme et al., 2008). It is not clear whether this response was due specifically to the osmotic stress or downregulation of the Pal homolog, OprL, which affects OMV formation (Bernadac et al., 1998; Guyard-Nicodeme et al., 2008).

Antibiotics

One of the classic targets for antimicrobial therapeutics has been the unique cell surface structures found in bacteria. Not surprisingly, antibiotics that disrupt the cell envelope, such as antimicrobial peptides, have also been shown to affect vesiculation (van der Kraan et al., 2005). Other antibiotics that do not affect the cell envelope have also been shown to affect vesiculation, as production of Shiga toxin and its release in OMVs by *Shigella dysenteriae* is increased in the presence of the antitumor drug mitomycin C; whether toxin release or OMV formation is the target of this regulation remains unknown (Dutta et al., 2004). Finally, exposing *P. aeruginosa* cells to the aminoglycoside gentamicin was demonstrated to increase OMV formation approximately threefold (Kadurugamuwa and Beveridge, 1995). This was hypothesized to be due to the ability of aminoglycosides to replace charge-charge interactions within the LPS of the OM (see "Curvature-Inducing Molecules"), and the unique macromolecular composition of the vesicles compared to that of non-gentamicin-induced vesicles suggests that they may be formed in a mechanistically distinct manner (Kadurugamuwa and Beveridge, 1995).

Host and Other Organisms

The utility of vesicles for trafficking molecules to other bacteria and even organisms in other domains of life suggests that target organisms can be affected by OMVs. Indeed, vesicles from a range of pathogenic organisms have been observed at infection sites, and they can be isolated from infected tissues and fluids (Dorward et al., 1991; Solcia et al., 1999; Namork and Brandtzaeg, 2002; Irazoqui et al., 2010). Many studies outside the scope of this chapter have examined the effects of OMVs on other organisms, and one particularly interesting thread of research investigates the immunogenic properties of vesicles and their utility as vaccines (Roy et al., 2010; Zollinger et al., 2010; McConnell et al., 2011). If the presence of vesicles can affect host behavior, it follows that OMV formation could be modulated in the presence of other organisms including the host. It has been argued that pathogens have co-opted the natural process of vesicle shedding for delivery of toxins and other virulence factors, and it is true that pathogenic bacteria are generally found to produce more OMVs than nonpathogenic bacteria (Kuehn and Kesty, 2005). Examples of this include the fact that both an ETEC strain and a leukotoxin-producing strain of *A. actinomycetemcomitans* produced more vesicles than strains that did not make the toxins (Lai et al., 1981; Horstman and Kuehn, 2000). Additionally, *H. influenzae* grown in either rats or human cerebrospinal fluid produced OMV-like structures within 2 h of growth, in contrast to cells grown in a standard culture medium that needed to reach stationary phase to produce OMVs (Dargis et al., 1992).

Obstacles for bacterial growth within the host often affect OMV formation. Altering LPS and other surface molecules within the host has been a survival strategy for several pathogens (Moxon et al., 2003; Cigana et al., 2009). It has also been reported that modifying such molecules can result in OMV formation phenotypes; thus, phenomena like serotype switching can affect OMV formation (Nguyen et al., 2003). Another obstacle to growth within the host is low levels of free iron, which is generally sequestered by host iron-binding molecules. Iron limitation decreased the number of vesiculating cells in a commercially significant rainbow trout pathogen, *Flavobacterium psychrophilum* (Moller et al., 2005). Host-derived substances also affect OMV formation, as the presence of bile salts induced production of vesicle-like structures in an anaerobe found in the intestine, *Bacteroides fragilis* (Pumbwe et al., 2007). It is possible that the presence of bile salts reflects a host environment, which alters the behavior of the organism by inducing OMV formation. Karavolos and colleagues recently reported an additional example of an OMV-inducing host molecule, as exposure of *Salmonella enterica* serovar Typhi to host neuroendocrine hormones increased OMV production (Karavolos et al., 2011). The authors observed that this was likely due to increased expression of an sRNA, *micA*, which represses expression of *ompA* (Karavolos et al., 2011) (see "sRNAs and OMV Formation"), representing another example of genetic control of OMV formation through sRNAs.

In a recent microscopic analysis of *P. aeruginosa* infecting the nematode *Caenorhabditis elegans*, the pathogen was found to produce an extracellular matrix that included putative OMVs (Fig. 6A), while a noninvasive *E. coli* strain (OP50) did not produce OMVs in the nematode (Irazoqui et al., 2010). Most pathogens exist within microcolony or biofilm structures inside the host, which, among other advantages, often confer resistance to host-derived antimicrobial

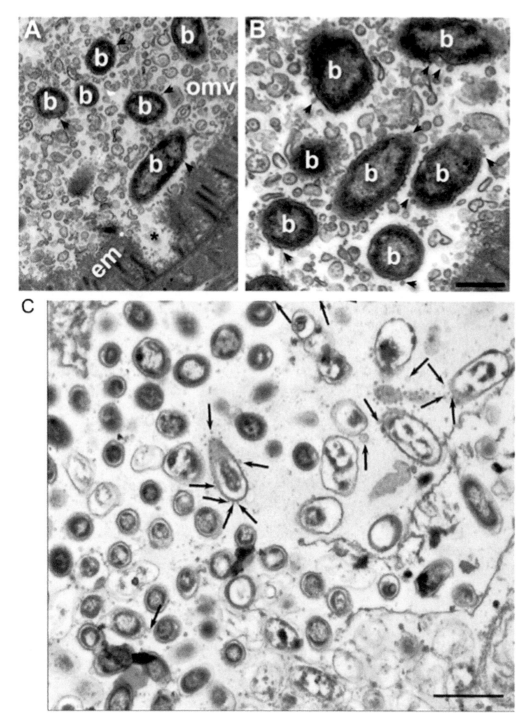

Figure 6. TEM of OMVs produced in biofilms. (A and B) *P. aeruginosa* (b) infects *C. elegans* in the lumen of the intestine, producing what appears to be OMVs (omv), in the presence of extracellular material (em). Scale bar, 0.5 μm. Reprinted from *PLoS Pathogens* (Irazoqui et al., 2010) with permission of the publisher. (C) TEM of a biofilm isolated from a domestic bathroom drain, indicating the presence of OMVs between cells (arrows), as well as blebbing off the cell surface (arrows). Scale bar, 1 μm. Reprinted from *Journal of Bacteriology* (Schooling and Beveridge, 2006) with permission of the publisher. doi:10.1128/9781555818524.ch23f6

agents as well as exogenous antibiotics. The contribution of OMVs to biofilm structures is discussed more in the section below; however, specifically within the host, the propensity to form microcolonies and the role OMVs play during infection may reveal novel biofilm-related regulatory mechanisms of OMV formation within the host.

OMVs Produced in Biofilms

In nature, bacteria commonly grow as biofilms, populations of cells attached to surfaces. Biofilm communities produce an extracellular matrix (ECM), which serves to protect the biofilm from harsh environments and can also help to create a three-dimensional colony structure. The ECM commonly consists of exopolysaccharides, proteins, lipids, nucleic acids, and, of note for this chapter, OMVs (Costerton et al., 1978; Costerton et al., 1981; Schooling and Beveridge, 2006; Schooling et al., 2009; Shetty et al., 2011). Interestingly, *P. aeruginosa* biofilm-derived OMVs have been shown to be distinct from planktonically derived OMVs, as they displayed a unique protein profile and greater proteolytic activity (Schooling and Beveridge, 2006). Laboratory-grown biofilms of *Shewanella oneidensis*, *E. coli*, *Azotobacter*, and *H. pylori* have all been shown to contain OMVs as a part of the ECM (Schooling and Beveridge, 2006; Yonezawa et al., 2011). This, combined with many studies that have also demonstrated that OMVs are produced in naturally occurring biofilms (Beveridge et al., 1997; Schooling and Beveridge, 2006) (Fig. 6), supports the hypothesis that OMVs are functionally and structurally important components of bacterial communities.

Summary

Bacteria exist in widely varying environments, and many alter their physiology and behavior as part of a coping mechanism to mitigate this stress and successfully adapt to new or changing conditions. Production of OMVs is one of the behaviors altered as the environment changes and may provide bacteria with an advantage as growth conditions change. As discussed, heat stress induces OMV production, and the ability to scavenge for nutrients or degrade nutrients present is often mediated by OMVs and their contents. Bacteria also respond to the challenges of antibiotic treatment and the host environment by altering OMV production. Finally, OMVs are a component of biofilms, which are likely the most common lifestyle for bacteria in the environment and one that is often associated with protection from harsh conditions. The differences observed in OMV formation in different environments underscores the importance of studying bacteria in situ, as the specific environment in which bacteria are growing dictates these phenotypes.

GRAM-POSITIVE MVs

Several decades after Gram-negative OMVs were reported, Dorward and Garon noted in 1990 that two Gram-positive species produced MVs: *B. subtilis* and *Bacillus cereus* (Dorward and Garon, 1990). Reports that Gram-positive bacteria indeed produce MVs provide evidence that MV formation is an evolutionarily conserved process. In a study of *Staphylococcus aureus*, Lee and colleagues examined culture supernatants and visualized purified vesicles using TEM (Fig. 7A) (Lee et al., 2009). Their analysis identified 90 vesicular proteins and provided their theoretical cellular localization, based on SDS-PAGE, tryptic digestion, nano-LC-ESI-MS/MS, and *in silico* analysis. According to their proteomic studies, 56.7% were classified as cytoplasmic, 16.7% membrane, and 23.3% extracellular (Lee et al., 2009). It is logical that gram-positive MVs contain a high number of cytoplasmic proteins, as cytoplasmic membrane blebbing is required for MV formation in Gram-positive organisms. Lee et al. also reported that a large amount of ribosomal and metabolic proteins are packaged in MVs, as well as the inclusion of transporter proteins, antibiotic resistance proteins, and numerous virulence factors. One year after Lee and colleagues published evidence for MV production in *B. subtilis*, Rivera and colleagues reported that *Bacillus anthracis*, the etiological agent of anthrax disease, produces MVs both in the environment and within macrophages (Fig. 7B) (Rivera et al., 2010). Using immunoelectron microscopy and enzyme-linked immunosorbent assay analyses, Rivera showed that *B. anthracis* vesicles contain multiple toxin components and sometimes have double membranes. Though some remain skeptical over whether gram-positive bacteria indeed produce MVs, several control experiments have been performed. For example, vesicles could not be isolated from heat-killed *B. anthracis* incubated under regular MV culture conditions (Rivera et al., 2010). Most recently, it was also reported that the Gram-positive *Streptomyces coelicolor* produces MVs (Schrempf et al., 2011).

In 2007, Marsollier and colleagues showed that the acid-fast, Gram-positive causative agent of Buruli ulcers, *Mycobacterium ulcerans*, produces MVs within its ECM (Marsollier et al., 2007). *M. ulcerans* MVs package a large number of membrane proteins (51 of 57 total proteins identified). Additionally, isolated *M. ulcerans* MVs have been shown to package two polyketide synthases required for production of

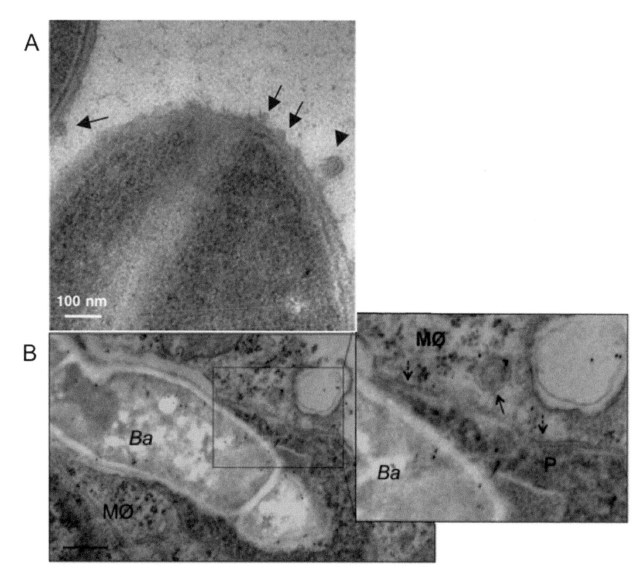

Figure 7. Production of MVs is not limited to Gram-negative bacteria. (A) A thin-section TEM of the gram-positive bacterium *S. aureus*. MV formation occurs at the cell surface (arrows), and a secreted MV is shown nearby (arrowhead). Scale bar, 100 nm. Reprinted from *Proteomics* (Lee et al., 2009) with permission of the publisher. (B) The Gram-positive bacterium *B. anthracis* (Ba) produces MVs within macrophages (MØ). A disrupted phagosome (P) double membrane is visible (two dashed arrows), and a *B. anthracis* vesicle is present in the macrophage cytoplasm (solid arrow). Scale bar, 500 nm. Reprinted from *Proceedings of the National Academy of Sciences of the United States of America* (Rivera et al., 2010) with permission of the publisher.
doi:10.1128/9781555818524.ch23f7

the toxin mycolactone, the only known virulence factor responsible for Buruli ulcers (George et al., 1999; Marsollier et al., 2007). Mycolactone was detected within vesicles, and purified vesicles contained cytotoxic activity. Mycolactone packaged within vesicles was more cytotoxic to host cells than equal levels of purified toxin alone (Marsollier et al., 2007). Though only a few Gram-positive species have been reported to produce MVs, the number of reported species will likely increase as research in the field progresses. The molecular mechanism of Gram-positive MV formation is also unknown; however, curiosity remains in

the field as to how MVs can form and detach from the cell while in the presence of a thick outer layer of PG.

CONCLUSION

Through the combined efforts of many investigators over the course of decades of research, much light has been shed on the highly conserved process of bacterial MV formation, though several questions remain unanswered. Every Gram-negative species tested has demonstrated the ability to produce OMVs; however, it is still unclear whether MV formation is as common among Gram-positive bacteria. The field has also delineated several models describing the molecular mechanisms of OMV formation, but more effort is needed to determine which models apply to different species and growth conditions. Regulatory schemes for OMV formation are actively being determined, and some of the future progress here could be derived from collaboration with other research areas like QS and regulatory RNAs. As for the contents of OMVs, the packaging of cytoplasmic components into Gram-negative OMVs is still a highly debated topic and yet another active area of investigation. Finally, the presence of OMVs within biofilms and the evidence that biofilm-derived OMVs are distinct from planktonically derived OMVs suggest that the OMV field will continue to investigate these clinically and industrially relevant structures as well. The importance of OMVs for normal growth and development, virulence, and a range of other processes will ensure that these and many other questions will soon be answered.

Acknowledgments. A.K.W. and G.C.P. contributed equally to this work.

REFERENCES

Ades, S. E. 2004. Control of the alternative sigma factor σ^E in *Escherichia coli. Curr. Opin. Microbiol.* **7:**157–162.

Ades, S. E. 2008. Regulation by destruction: design of the σ^E envelope stress response. *Curr. Opin. Microbiol.* **11:**535–540.

Alba, B. M., and C. A. Gross. 2004. Regulation of the *Escherichia coli* sigma-dependent envelope stress response. *Mol. Microbiol.* **52:**613–619.

Alphen, W. V., and B. Lugtenberg. 1977. Influence of osmolarity of the growth medium on the outer membrane protein pattern of *Escherichia coli. J. Bacteriol.* **131:**623–630.

Bauman, S. J., and M. J. Kuehn. 2006. Purification of outer membrane vesicles from *Pseudomonas aeruginosa* and their activation of an IL-8 response. *Microbes Infect.* **8:**2400–2408.

Berlanda Scorza, F., F. Doro, M. J. Rodriguez-Ortega, M. Stella, S. Liberatori, A. R. Taddei, L. Serino, D. Gomes Moriel, B. Nesta, M. R. Fontana, A. Spagnuolo, M. Pizza, N. Norais, and G. Grandi. 2008. Proteomics characterization of outer membrane vesicles from the extraintestinal pathogenic *Escherichia coli* ∆tolR IHE$_{3034}$ mutant. *Mol. Cell. Proteomics* **7:**473–485.

Bernadac, A., M. Gavioli, J. C. Lazzaroni, S. Raina, and R. Lloubes. 1998. *Escherichia coli* tol-pal mutants form outer membrane vesicles. *J. Bacteriol.* **180:**4872–4878.

Beveridge, T. J. 1999. Structures of gram-negative cell walls and their derived membrane vesicles. *J. Bacteriol.* **181:**4725–4733.

Beveridge, T. J., S. A. Makin, J. L. Kadurugamuwa, and Z. Li. 1997. Interactions between biofilms and the environment. *FEMS Microbiol. Rev.* **20:**291–303.

Bodero, M. D., M. C. Pilonieta, and G. P. Munson. 2007. Repression of the inner membrane lipoprotein NlpA by Rns in enterotoxigenic *Escherichia coli. J. Bacteriol.* **189:**1627–1632.

Bomberger, J. M., D. P. Maceachran, B. A. Coutermarsh, S. Ye, G. A. O'Toole, and B. A. Stanton. 2009. Long-distance delivery of bacterial virulence factors by *Pseudomonas aeruginosa* outer membrane vesicles. *PLoS Pathog.* **5:**e1000382.

Burdett, I. D. J., and R. G. E. Murray. 1974a. Electron microscope study of septum formation in *Escherichia coli* strains B and B/r during synchronous growth. *J. Bacteriol.* **119:**1039–1056.

Burdett, I. D. J., and R. G. E. Murray. 1974b. Septum formation in *Escherichia coli*: characterization of septal structure and the effects of antibiotics on cell division. *J. Bacteriol.* **119:**303–324.

Cascales, E., A. Bernadac, M. Gavioli, J. C. Lazzaroni, and R. Lloubes. 2002. Pal lipoprotein of *Escherichia coli* plays a major role in outer membrane integrity. *J. Bacteriol.* **184:**754–759.

Chaba, R., I. L. Grigorova, J. M. Flynn, T. A. Baker, and C. A. Gross. 2007. Design principles of the proteolytic cascade governing the σ^E-mediated envelope stress response in *Escherichia coli*: keys to graded, buffered, and rapid signal transduction. *Genes Dev.* **21:**124–136.

Chatterjee, S. N., and J. Das. 1967. Electron microscopic observations on the excretion of cell-wall material by *Vibrio cholerae. J. Gen. Microbiol.* **49:**1–11.

Choi, D. S., D. K. Kim, S. J. Choi, J. Lee, J. P. Choi, S. Rho, S. H. Park, Y. K. Kim, D. Hwang, and Y. S. Gho. 2011. Proteomic analysis of outer membrane vesicles derived from *Pseudomonas aeruginosa. Proteomics* **11:**3424–3429.

Cigana, C., L. Curcuru, M. R. Leone, T. Ierano, N. I. Lore, I. Bianconi, A. Silipo, F. Cozzolino, R. Lanzetta, A. Molinaro, M. L. Bernardini, and A. Bragonzi. 2009. *Pseudomonas aeruginosa* exploits lipid A and muropeptides modification as a strategy to lower innate immunity during cystic fibrosis lung infection. *PLoS One* **4:**e8439.

Ciofu, O., T. J. Beveridge, J. Kadurugamuwa, J. Walther-Rasmussen, and N. Hoiby. 2000. Chromosomal beta-lactamase is packaged into membrane vesicles and secreted from *Pseudomonas aeruginosa. J. Antimicrob. Chemother.* **45:**9–13.

Costerton, J. W., G. G. Geesey, and K. J. Cheng. 1978. How bacteria stick. *Sci. Am.* **238:**86–95.

Costerton, J. W., R. T. Irvin, and K. J. Cheng. 1981. The bacterial glycocalyx in nature and disease. *Annu. Rev. Microbiol.* **35:**299–324.

Dargis, M., P. Gourde, D. Beauchamp, B. Foiry, M. Jacques, and F. Malouin. 1992. Modification in penicillin-binding proteins during in vivo development of genetic competence of *Haemophilus influenzae* is associated with a rapid change in the physiological state of cells. *Infect. Immun.* **60:**4024–4031.

Deatherage, B. L., J. C. Lara, T. Bergsbaken, S. L. Rassoulian Barrett, S. Lara, and B. T. Cookson. 2009. Biogenesis of bacterial membrane vesicles. *Mol. Microbiol.* **72:**1395–1407.

Deich, R. A., and L. C. Hoyer. 1982. Generation and release of DNA-binding vesicles by *Haemophilus influenzae* during induction and loss of competence. *J. Bacteriol.* **152:**855–864.

Dorward, D. W., and C. F. Garon. 1990. DNA is packaged within membrane-derived vesicles of gram-negative but not gram-positive bacteria. *Appl. Environ. Microbiol.* **56:**1960–1962.

Dorward, D. W., C. F. Garon, and R. C. Judd. 1989. Export and intercellular transfer of DNA via membrane blebs of *Neisseria gonorrhoeae. J. Bacteriol.* **171:**2499–2505.

Dorward, D. W., T. G. Schwan, and C. F. Garon. 1991. Immune capture and detection of *Borrelia burgdorferi* antigens in urine, blood, or tissues from infected ticks, mice, dogs, and humans. *J. Clin. Microbiol.* **29:**1162–1170.

Douchin, V., C. Bohn, and P. Bouloc. 2006. Down-regulation of porins by a small RNA bypasses the essentiality of the regulated intramembrane proteolysis protease RseP in *Escherichia coli. J. Biol. Chem.* **281:**12253–12259.

Dubey, G. P., and S. Ben-Yehuda. 2011. Intercellular nanotubes mediate bacterial communication. *Cell* **144:**590–600.

Dutta, S., K. Iida, A. Takade, Y. Meno, G. B. Nair, and S. Yoshida. 2004. Release of Shiga toxin by membrane vesicles in *Shigella dysenteriae* serotype 1 strains and in vitro effects of antimicrobials on toxin production and release. *Microbiol. Immunol.* 48:965–969.

Engelhardt, H., S. Gerblrieger, D. Krezmar, S. Schneidervoss, A. Engel, and W. Baumeister. 1990. Structural properties of the outer membrane and the regular surface protein of *Comamonas acidovorans. J. Struct. Biol.* 105:92–102.

Ferrari, G., I. Garaguso, J. Adu-Bobie, F. Doro, A. R. Taddei, A. Biolchi, B. Brunelli, M. M. Giuliani, M. Pizza, N. Norais, and G. Grandi. 2006. Outer membrane vesicles from group B *Neisseria meningitidis* Δgna33 mutant: proteomic and immunological comparison with detergent-derived outer membrane vesicles. *Proteomics* 6:1856–1866.

Figueroa-Bossi, N., S. Lemire, D. Maloriol, R. Balbontin, J. Casadesus, and L. Bossi. 2006. Loss of Hfq activates the σ^E-dependent envelope stress response in *Salmonella enterica. Mol. Microbiol.* 62:838–852.

George, K. M., D. Chatterjee, G. Gunawardana, D. Welty, J. Hayman, R. Lee, and P. L. Small. 1999. Mycolactone: a polyketide toxin from *Mycobacterium ulcerans* required for virulence. *Science* 283:854–857.

Goedhart, J., H. Rohrig, M. A. Hink, A. van Hoek, A. J. Visser, T. Bisseling, and T. W. Gadella, Jr. 1999. Nod factors integrate spontaneously in biomembranes and transfer rapidly between membranes and to root hairs, but transbilayer flip-flop does not occur. Biochemistry 38:10898–10907.

Guyard-Nicodeme, M., A. Bazire, G. Hemery, T. Meylheuc, D. Molle, N. Orange, L. Fito-Boncompte, M. Feuilloley, D. Haras, A. Dufour, and S. Chevalier. 2008. Outer membrane modifications of *Pseudomonas fluorescens* MF37 in response to hyperosmolarity. *J. Proteome Res.* 7:1218–1225.

Hancock, R. E. 1984. Alterations in outer membrane permeability. *Annu. Rev. Microbiol.* 38:237–264.

Haurat, M. F., J. Aduse-Opoku, M. Rangarajan, L. Dorobantu, M. R. Gray, M. A. Curtis, and M. F. Feldman. 2011. Selective sorting of cargo proteins into bacterial membrane vesicles. *J. Biol. Chem.* 286:1269–1276.

Hayashi, J., N. Hamada, and H. K. Kuramitsu. 2002. The autolysin of *Porphyromonas gingivalis* is involved in outer membrane vesicle release. *FEMS Microbiol. Lett.* 216:217–222.

Henry, T., S. Pommier, L. Journet, A. Bernadac, J. P. Gorvel, and R. Lloubes. 2004. Improved methods for producing outer membrane vesicles in Gram-negative bacteria. *Res. Microbiol.* 155:437–446.

Hoekstra, D., J. W. van der Laan, L. de Leij, and B. Witholt. 1976. Release of outer membrane fragments from normally growing *Escherichia coli*. Biochim. Biophys. Acta 455:889–899.

Horstman, A. L., and M. J. Kuehn. 2000. Enterotoxigenic *Escherichia coli* secretes active heat-labile enterotoxin via outer membrane vesicles. *J. Biol. Chem.* 275:12489–12496.

Irazoqui, J. E., E. R. Troemel, R. L. Feinbaum, L. G. Luhachack, B. O. Cezairliyan, and F. M. Ausubel. 2010. Distinct pathogenesis and host responses during infection of C. *elegans* by P. *aeruginosa* and S. *aureus*. PLoS Pathog. 6:e1000982.

Johansen, J., A. A. Rasmussen, M. Overgaard, and P. Valentin-Hansen. 2006. Conserved small non-coding RNAs that belong to the σ^E regulon: role in down-regulation of outer membrane proteins. *J. Mol. Biol.* 364:1–8.

Kadurugamuwa, J. L., and T. J. Beveridge. 1996. Bacteriolytic effect of membrane vesicles from *Pseudomonas aeruginosa* on other bacteria including pathogens: conceptually new antibiotics. *J. Bacteriol.* 178:2767–2774.

Kadurugamuwa, J. L., and T. J. Beveridge. 1999. Membrane vesicles derived from *Pseudomonas aeruginosa* and *Shigella flexneri* can be integrated into the surfaces of other gram-negative bacteria. *Microbiology* 145(Pt. 8):2051–2060.

Kadurugamuwa, J. L., and T. J. Beveridge. 1997. Natural release of virulence factors in membrane vesicles by *Pseudomonas aeruginosa* and the effect of aminoglycoside antibiotics on their release. *J. Antimicrob. Chemother.* 40:615–621.

Kadurugamuwa, J. L., and T. J. Beveridge. 1995. Virulence factors are released from *Pseudomonas aeruginosa* in association with membrane vesicles during normal growth and exposure to gentamicin: a novel mechanism of enzyme secretion. *J. Bacteriol.* 177:3998–4008.

Kadurugamuwa, J. L., A. J. Clarke, and T. J. Beveridge. 1993a. Surface action of gentamicin on *Pseudomonas aeruginosa*. *J. Bacteriol.* 175:5798–5805.

Kadurugamuwa, J. L., J. S. Lam, and T. J. Beveridge. 1993b. Interaction of gentamicin with the A band and B band lipopolysaccharides of *Pseudomonas aeruginosa* and its possible lethal effect. *Antimicrob. Agents Chemother.* 37:715–721.

Karavolos, M. H., D. M. Bulmer, H. Spencer, G. Rampioni, I. Schmalen, S. Baker, D. Pickard, J. Gray, M. Fookes, K. Winzer, A. Ivens, G. Dougan, P. Williams, and C. M. Khan. 2011. *Salmonella* Typhi sense host neuroendocrine stress hormones and release the toxin haemolysin E. *EMBO Rep.* 12:252–258.

Kato, S., Y. Kowashi, and D. R. Demuth. 2002. Outer membrane-like vesicles secreted by *Actinobacillus actinomycetemcomitans* are enriched in leukotoxin. *Microb. Pathog.* 32:1–13.

Katsui, N., T. Tsuchido, R. Hiramatsu, S. Fujikawa, M. Takano, and I. Shibasaki. 1982. Heat-induced blebbing and vesiculation of the outer membrane of *Escherichia coli. J. Bacteriol.*151:1523–1531.

Keenan, J. I., and R. A. Allardyce. 2000. Iron influences the expression of *Helicobacter pylori* outer membrane vesicle-associated virulence factors. *Eur. J. Gastroenterol. Hepatol.* 12:1267–1273.

Kesty, N. C., and M. J. Kuehn. 2004. Incorporation of heterologous outer membrane and periplasmic proteins into *Escherichia coli* outer membrane vesicles. *J. Biol. Chem.* 279:2069–2076.

Kesty, N. C., K. M. Mason, M. Reedy, S. E. Miller, and M. J. Kuehn. 2004. Enterotoxigenic *Escherichia coli* vesicles target toxin delivery into mammalian cells. *EMBO J.* 23:4538–4549.

Knox, K. W., M. Vesk, and E. Work. 1966. Relation between excreted lipopolysaccharide complexes and surface structures of a lysine-limited culture of *Escherichia coli. J. Bacteriol.* 92:1206–1217.

Kobayashi, H., K. Uematsu, H. Hirayama, and K. Horikoshi. 2000. Novel toluene elimination system in a toluene-tolerant microorganism. *J. Bacteriol.* 182:6451–6455.

Koch, A. L., M. L. Higgins, and R. J. Doyle. 1982. The role of surface stress in the morphology of microbes. *J. Gen. Microbiol.* 128:927–945.

Kolling, G. L., and K. R. Matthews. 1999. Export of virulence genes and Shiga toxin by membrane vesicles of *Escherichia coli* O157:H7. *Appl. Environ. Microbiol.* 65:1843–1848.

Kropinski, A. M., V. Lewis, and D. Berry. 1987. Effect of growth temperature on the lipids, outer membrane proteins, and lipopolysaccharides of *Pseudomonas aeruginosa* PAO. *J. Bacteriol.* 169:1960–1966.

Kuehn, M. J., and N. C. Kesty. 2005. Bacterial outer membrane vesicles and the host-pathogen interaction. *Genes Dev.* 19:2645–2655.

Kulp, A., and M. J. Kuehn. 2010. Biological functions and biogenesis of secreted bacterial outer membrane vesicles. *Annu. Rev. Microbiol.* 64:163–184.

Kwon, S. O., Y. S. Gho, J. C. Lee, and S. I. Kim. 2009. Proteome analysis of outer membrane vesicles from a clinical *Acinetobacter baumannii* isolate. *FEMS Microbiol. Lett.* 297:150–156.

Lai, C. H., M. A. Listgarten, and B. F. Hammond. 1981. Comparative ultrastructure of leukotoxic and non-leukotoxic strains of *Actinobacillus actinomycetemcomitans. J. Periodontal Res.* 16:379–389.

Lally, E. T., E. E. Golub, I. R. Kieba, N. S. Taichman, J. Rosenbloom, J. C. Rosenbloom, C. W. Gibson, and D. R. Demuth. 1989. Analysis of the *Actinobacillus actinomycetemcomitans* leukotoxin gene. Delineation of unique features and comparison to homologous toxins. *J. Biol. Chem.* **264:**15451–15456.

Lam, M. Y., E. J. McGroarty, A. M. Kropinski, L. A. MacDonald, S. S. Pedersen, N. Hoiby, and J. S. Lam. 1989. Occurrence of a common lipopolysaccharide antigen in standard and clinical strains of *Pseudomonas aeruginosa. J. Clin. Microbiol.* **27:**962–967.

Lazzaroni, J. C., P. Germon, M. C. Ray, and A. Vianney. 1999. The Tol proteins of *Escherichia coli* and their involvement in the uptake of biomolecules and outer membrane stability. *FEMS Microbiol. Lett.* **177:**191–197.

Lee, E. Y., J. Y. Bang, G. W. Park, D. S. Choi, J. S. Kang, H. J. Kim, K. S. Park, J. O. Lee, Y. K. Kim, K. H. Kwon, K. P. Kim, and Y. S. Gho. 2007. Global proteomic profiling of native outer membrane vesicles derived from *Escherichia coli. Proteomics* **7:**3143–3153.

Lee, E. Y., D. S. Choi, K. P. Kim, and Y. S. Gho. 2008. Proteomics in gram-negative bacterial outer membrane vesicles. *Mass Spectrom. Rev.* **27:**535–555.

Lee, E. Y., D. Y. Choi, D. K. Kim, J. W. Kim, J. O. Park, S. Kim, S. H. Kim, D. M. Desiderio, Y. K. Kim, K. P. Kim, and Y. S. Gho. 2009. Gram-positive bacteria produce membrane vesicles: proteomics-based characterization of *Staphylococcus aureus*-derived membrane vesicles. *Proteomics* **9:**5425–5436.

Li, Z., A. J. Clarke, and T. J. Beveridge. 1998. Gram-negative bacteria produce membrane vesicles which are capable of killing other bacteria. *J. Bacteriol.* **180:**5478–5483.

Li, Z., A. J. Clarke, and T. J. Beveridge. 1996. A major autolysin of *Pseudomonas aeruginosa*: subcellular distribution, potential role in cell growth and division and secretion in surface membrane vesicles. *J. Bacteriol.* **178:**2479–2488.

Lim, H. W. G., M. Wortis, and R. Mukhopadhyay. 2002. Stomatocyte-discocyte-echinocyte sequence of the human red blood cell: evidence for the bilayer-couple hypothesis from membrane mechanics. *Proc. Natl. Acad. Sci. USA* **99:**16766–16769.

Lin, Y. H., J. L. Xu, J. Hu, L. H. Wang, S. L. Ong, J. R. Leadbetter, and L. H. Zhang. 2003. Acyl-homoserine lactone acylase from *Ralstonia* strain XJ12B represents a novel and potent class of quorum-quenching enzymes. *Mol. Microbiol.* **47:**849–860.

Llamas, M. A., J. L. Ramos, and J. J. Rodriguez-Herva. 2000. Mutations in each of the *tol* genes of *Pseudomonas putida* reveal that they are critical for maintenance of outer membrane stability. *J. Bacteriol.* **182:**4764–4772.

Loeb, M. R., and J. Kilner. 1979. Effect of growth medium on the relative polypeptide composition of cellular outer membrane and released outer membrane material in *Escherichia coli. J. Bacteriol.* **137:**1031–1034.

Makin, S. A., and T. J. Beveridge. 1996. *Pseudomonas aeruginosa* PAO1 ceases to express serotype-specific lipopolysaccharide at 45 degrees C. *J. Bacteriol.* **178:**3350–3352.

Marden, P., A. Tunlid, K. Malmcronafriberg, G. Odham, and S. Kjelleberg. 1985. Physiological and morphological changes during short-term starvation of marine bacterial isolates. *Arch. Microbiol.* **142:**326–332.

Marshall, A. J., and L. J. Piddock. 1994. Interaction of divalent cations, quinolones and bacteria. *J. Antimicrob. Chemother.* **34:**465–483.

Marsollier, L., P. Brodin, M. Jackson, J. Kordulakova, P. Tafelmeyer, E. Carbonnelle, J. Aubry, G. Milon, P. Legras, J. P. Andre, C. Leroy, J. Cottin, M. L. Guillou, G. Reysset, and S. T. Cole. 2007. Impact of *Mycobacterium ulcerans* biofilm on transmissibility to ecological niches and Buruli ulcer pathogenesis. *PLoS Pathog.* **3:**e62.

Martin, N. L., and T. J. Beveridge. 1986. Gentamicin interaction with *Pseudomonas aeruginosa* cell envelope. *Antimicrob. Agents Chemother.* **29:**1079–1087.

Mashburn, L. M., and M. Whiteley. 2005. Membrane vesicles traffic signals and facilitate group activities in a prokaryote. *Nature* **437:**422–425.

Mashburn-Warren, L., J. Howe, K. Brandenburg, and M. Whiteley. 2009. Structural requirements of the *Pseudomonas* quinolone signal for membrane vesicle stimulation. *J. Bacteriol.* **191:**3411–3414.

Mashburn-Warren, L., J. Howe, P. Garidel, W. Richter, F. Steiniger, M. Roessle, K. Brandenburg, and M. Whiteley. 2008. Interaction of quorum signals with outer membrane lipids: insights into prokaryotic membrane vesicle formation. *Mol. Microbiol.* **69:**491–502.

Mashburn-Warren, L. M., and M. Whiteley. 2006. Special delivery: vesicle trafficking in prokaryotes. *Mol. Microbiol.* **61:**839–846.

Mayrand, D., and D. Grenier. 1989. Biological activities of outer membrane vesicles. *Can. J. Microbiol.* **35:**607–613.

McBroom, A. J., A. P. Johnson, S. Vemulapalli, and M. J. Kuehn. 2006. Outer membrane vesicle production by *Escherichia coli* is independent of membrane instability. *J. Bacteriol.* **188:**5385–5392.

McBroom, A. J., and M. J. Kuehn. 2007. Release of outer membrane vesicles by Gram-negative bacteria is a novel envelope stress response. *Mol. Microbiol.* **63:**545–558.

McConnell, M. J., C. Rumbo, G. Bou, and J. Pachon. 2011. Outer membrane vesicles as an acellular vaccine against *Acinetobacter baumannii. Vaccine* **29:**5705–5710.

McGroarty, E. J., and M. Rivera. 1990. Growth-dependent alterations in production of serotype-specific and common antigen lipopolysaccharides in *Pseudomonas aeruginosa* PAO1. *Infect. Immun.* **58:**1030–1037.

Mogensen, J. E., and D. E. Otzen. 2005. Interactions between folding factors and bacterial outer membrane proteins. *Mol. Microbiol.* **57:**326–346.

Moller, J. D., A. C. Barnes, I. Dalsgaard, and A. E. Ellis. 2005. Characterisation of surface blebbing and membrane vesicles produced by *Flavobacterium psychrophilum. Dis. Aquat. Organ.* **64:**201–209.

Moxon, E. R., V. Bouchet, D. W. Hood, J. J. Li, J. R. Brisson, G. A. Randle, A. Martin, Z. Li, R. Goldstein, E. K. H. Schweda, S. I. Pelton, and J. C. Richards. 2003. Host-derived sialic acid is incorporated into *Haemophilus influenzae* lipopolysaccharide and is a major virulence factor in experimental otitis media. *Proc. Natl. Acad. Sci. USA* **100:**8898–8903.

Mug-Opstelten, D., and B. Witholt. 1978. Preferential release of new outer membrane fragments by exponentially growing *Escherichia coli. Biochim Biophys Acta* **508:**287–295.

Nally, J. E., J. P. Whitelegge, R. Aguilera, M. M. Pereira, D. R. Blanco, and M. A. Lovett. 2005. Purification and proteomic analysis of outer membrane vesicles from a clinical isolate of *Leptospira interrogans* serovar *Copenhageni. Proteomics* **5:**144–152.

Namork, E., and P. Brandtzaeg. 2002. Fatal meningococcal septicaemia with "blebbing" meningococcus. *Lancet* **360:**1741.

Nevot, M., V. Deroncele, P. Messner, J. Guinea, and E. Mercade. 2006. Characterization of outer membrane vesicles released by the psychrotolerant bacterium *Pseudoalteromonas antarctica* NF3. *Environ. Microbiol.* **8:**1523–1533.

Nguyen, T. T., A. Saxena, and T. J. Beveridge. 2003. Effect of surface lipopolysaccharide on the nature of membrane vesicles liberated from the Gram-negative bacterium *Pseudomonas aeruginosa. J. Electron Microsc.* (Tokyo) **52:**465–469.

Nikaido, H., and M. Vaara. 1985. Molecular basis of bacterial outer membrane permeability. *Microbiol. Rev.* **49:**1–32.

Palmer, K. L., L. M. Aye, and M. Whiteley. 2007. Nutritional cues control *Pseudomonas aeruginosa* multicellular behavior in cystic fibrosis sputum. *J. Bacteriol.* **189:**8079–8087.

Palmer, K. L., L. M. Mashburn, P. K. Singh, and M. Whiteley. 2005. Cystic fibrosis sputum supports growth and cues key aspects of *Pseudomonas aeruginosa* physiology. *J. Bacteriol.* **187**:5267–5277.

Papenfort, K., V. Pfeiffer, F. Mika, S. Lucchini, J. C. Hinton, and J. Vogel. 2006. σ^E-Dependent small RNAs of *Salmonella* respond to membrane stress by accelerating global *omp* mRNA decay. *Mol. Microbiol.* **62**:1674–1688.

Park, K. S., K. H. Choi, Y. S. Kim, B. S. Hong, O. Y. Kim, J. H. Kim, C. M. Yoon, G. Y. Koh, Y. K. Kim, and Y. S. Gho. 2010. Outer membrane vesicles derived from *Escherichia coli* induce systemic inflammatory response syndrome. *PLoS One* **5**:e11334.

Pesci, E. C., J. B. Milbank, J. P. Pearson, S. McKnight, A. S. Kende, E. P. Greenberg, and B. H. Iglewski. 1999. Quinolone signaling in the cell-to-cell communication system of *Pseudomonas aeruginosa. Proc. Natl. Acad. Sci. USA* **96**:11229–11234.

Peterson, A. A., R. E. Hancock, and E. J. McGroarty. 1985. Binding of polycationic antibiotics and polyamines to lipopolysaccharides of *Pseudomonas aeruginosa. J. Bacteriol.* **164**:1256–1261.

Pettit, R. K., and R. C. Judd. 1992. The interaction of naturally elaborated blebs from serum-susceptible and serum-resistant strains of *Neisseria gonorrhoeae* with normal human serum. *Mol. Microbiol.* **6**:729–734.

Post, D. M., D. Zhang, J. S. Eastvold, A. Teghanemt, B. W. Gibson, and J. P. Weiss. 2005. Biochemical and functional characterization of membrane blebs purified from *Neisseria meningitidis* serogroup B. *J. Biol. Chem.* **280**:38383–38394.

Pumbwe, L., C. A. Skilbeck, V. Nakano, M. J. Avila-Campos, R. M. Piazza, and H. M. Wexler. 2007. Bile salts enhance bacterial co-aggregation, bacterial-intestinal epithelial cell adhesion, biofilm formation and antimicrobial resistance of *Bacteroides fragilis. Microb. Pathog.* **43**:78–87.

Raivio, T. L. 2005. Envelope stress responses and Gram-negative bacterial pathogenesis. *Mol. Microbiol.* **56**:1119–1128.

Renelli, M., V. Matias, R. Y. Lo, and T. J. Beveridge. 2004. DNA-containing membrane vesicles of *Pseudomonas aeruginosa* PAO1 and their genetic transformation potential. *Microbiology* **150**:2161–2169.

Rhodius, V. A., W. C. Suh, G. Nonaka, J. West, and C. A. Gross. 2006. Conserved and variable functions of the σ^E stress response in related genomes. *PLoS Biol.* **4**:e2.

Rivera, J., R. J. Cordero, A. S. Nakouzi, S. Frases, A. Nicola, and A. Casadevall. 2010. *Bacillus anthracis* produces membrane-derived vesicles containing biologically active toxins. *Proc. Natl. Acad. Sci. USA* **107**:19002–19007.

Rivera, M., and E. J. McGroarty. 1989. Analysis of a common-antigen lipopolysaccharide from *Pseudomonas aeruginosa. J. Bacteriol.* **171**:2244–2248.

Roy, N., S. Barman, A. Ghosh, A. Pal, K. Chakraborty, S. S. Das, D. R. Saha, S. Yamasaki, and H. Koley. 2010. Immunogenicity and protective efficacy of *Vibrio cholerae* outer membrane vesicles in rabbit model. *FEMS Immunol. Med. Microbiol.* **60**:18–27.

Sabra, W., H. Lunsdorf, and A. P. Zeng. 2003. Alterations in the formation of lipopolysaccharide and membrane vesicles on the surface of *Pseudomonas aeruginosa* PAO1 under oxygen stress conditions. *Microbiology* **149**:2789–2795.

Schertzer, J. W., S. A. Brown, and M. Whiteley. 2010. Oxygen levels rapidly modulate *Pseudomonas aeruginosa* social behaviours via substrate limitation of PqsH. *Mol. Microbiol.* **77**:1527–1538.

Schertzer, J. W., and M. Whiteley. 2012. A bilayer-couple model of bacterial outer membrane vesicle biogenesis. *MBio* Mar 13;3(2). pii: e00297-11. doi: 10.1128/mBio.00297-11.

Schooling, S. R., and T. J. Beveridge. 2006. Membrane vesicles: an overlooked component of the matrices of biofilms. *J. Bacteriol.* **188**:5945–5957.

Schooling, S. R., A. Hubley, and T. J. Beveridge. 2009. Interactions of DNA with biofilm-derived membrane vesicles. *J. Bacteriol.* **191**:4097–4102.

Schrempf, H., I. Koebsch, S. Walter, H. Engelhardt, and H. Meschke. 2011. Extracellular *Streptomyces* vesicles: amphorae for survival and defence. *Microb. Biotechnol.* **4**:286–299.

Sheetz, M. P., and S. J. Singer. 1974. Biological membranes as bilayer couples. A molecular mechanism of drug-erythrocyte interactions. *Proc. Natl. Acad. Sci. USA* **71**:4457–4461.

Shetty, A., S. Chen, E. I. Tocheva, G. J. Jensen, and W. J. Hickey. 2011. Nanopods: a new bacterial structure and mechanism for deployment of outer membrane vesicles. *PLoS One* **6**:e20725.

Smalley, J. W., A. J. Birss, A. S. McKee, and P. D. Marsh. 1991. Haemin-restriction influences haemin-binding, haemagglutination and protease activity of cells and extracellular membrane vesicles of *Porphyromonas gingivalis* W50. *FEMS Microbiol. Lett.* **69**:63–67.

Solcia, E., R. Fiocca, V. Necchi, P. Sommi, V. Ricci, J. Telford, and T. L. Cover. 1999. Release of *Helicobacter pylori* vacuolating cytotoxin by both a specific secretion pathway and budding of outer membrane vesicles. Uptake of released toxin and vesicles by gastric epithelium. *J. Pathol.* **188**:220–226.

Song, T., F. Mika, B. Lindmark, Z. Liu, S. Schild, A. Bishop, J. Zhu, A. Camilli, J. Johansson, J. Vogel, and S. N. Wai. 2008. A new *Vibrio cholerae* sRNA modulates colonization and affects release of outer membrane vesicles. *Mol. Microbiol.* **70**:100–111.

Sonntag, I., H. Schwarz, Y. Hirota, and U. Henning. 1978. Cell envelope and shape of *Escherichia coli*: multiple mutants missing the outer membrane lipoprotein and other major outer membrane proteins. *J. Bacteriol.* **136**:280–285.

Tashiro, Y., S. Ichikawa, M. Shimizu, M. Toyofuku, N. Takaya, T. Nakajima-Kambe, H. Uchiyama, and N. Nomura. 2010. Variation of physiochemical properties and cell association activity of membrane vesicles with growth phase in *Pseudomonas aeruginosa. Appl. Environ. Microbiol.* **76**:3732–3739.

Tashiro, Y., A. Inagaki, M. Shimizu, S. Ichikawa, N. Takaya, T. Nakajima-Kambe, H. Uchiyama, and N. Nomura. 2011. Characterization of phospholipids in membrane vesicles derived from *Pseudomonas aeruginosa. Biosci. Biotechnol. Biochem.* **75**:605–607.

Tashiro, Y., R. Sakai, M. Toyofuku, I. Sawada, T. Nakajima-Kambe, H. Uchiyama, and N. Nomura. 2009. Outer membrane machinery and alginate synthesis regulators control membrane vesicle production in *Pseudomonas aeruginosa. J. Bacteriol.* **191**:7509–7519.

Thery, C., L. Zitvogel, and S. Amigorena. 2002. Exosomes: composition, biogenesis and function. *Nat. Rev. Immunol.* **2**:569–579.

Thompson, K. M., V. A. Rhodius, and S. Gottesman. 2007. σ^E regulates and is regulated by a small RNA in *Escherichia coli. J. Bacteriol.* **189**:4243–4256.

Thompson, S. S., Y. M. Naidu, and J. J. Pestka. 1985. Ultrastructural localization of an extracellular protease in *Pseudomonas fragi* by using the peroxidase-antiperoxidase reaction. *Appl. Environ. Microbiol.* **50**:1038–1042.

Udekwu, K. I., F. Darfeuille, J. Vogel, J. Reimegard, E. Holmqvist, and E. G. Wagner. 2005. Hfq-dependent regulation of OmpA synthesis is mediated by an antisense RNA. *Genes Dev.* **19**:2355–2366.

Udekwu, K. I., and E. G. Wagner. 2007. Sigma E controls biogenesis of the antisense RNA MicA. *Nucleic Acids Res.* **35**:1279–1288.

Uli, L., L. Castellanos-Serra, L. Betancourt, F. Dominguez, R. Barbera, F. Sotolongo, G. Guillen, and R. Pajon Feyt. 2006. Outer membrane vesicles of the VA-MENGOC-BC vaccine against serogroup B of *Neisseria meningitidis*: analysis of protein components by two-dimensional gel electrophoresis and mass spectrometry. *Proteomics* **6**:3389–3399.

van der Kraan, M. I., J. van Marle, K. Nazmi, J. Groenink, W. van't Hof, E. C. Veerman, J. G. Bolscher, and A. V. Nieuw Amerongen.

2005. Ultrastructural effects of antimicrobial peptides from bovine lactoferrin on the membranes of *Candida albicans* and *Escherichia coli*. *Peptides* **26:**1537–1542.

Vasilyeva, N. V., I. M. Tsfasman, N. E. Suzina, O. A. Stepnaya, and I. S. Kulaev. 2009. Outer membrane vesicles of *Lysobacter* sp. *Dokl. Biochem. Biophys.* **426:**139–142.

Vidakovics, M. L., J. Jendholm, M. Morgelin, A. Mansson, C. Larsson, L. O. Cardell, and K. Riesbeck. 2010. B cell activation by outer membrane vesicles—a novel virulence mechanism. *PLoS Pathog.* **6:**e1000724.

Vipond, C., J. Suker, C. Jones, C. Tang, I. M. Feavers, and J. X. Wheeler. 2006. Proteomic analysis of a meningococcal outer membrane vesicle vaccine prepared from the group B strain NZ98/254. *Proteomics* **6:**3400–3413.

Walker, S. G., and T. J. Beveridge. 1988. Amikacin disrupts the cell envelope of *Pseudomonas aeruginosa* ATCC 9027. *Can. J. Microbiol.* **34:**12–18.

Wang, Y. J., and J. R. Leadbetter. 2005. Rapid acyl-homoserine lactone quorum signal biodegradation in diverse soils. *Appl. Environ. Microbiol.* **71:**1291–1299.

Wensink, J., and B. Witholt. 1981a. Identification of different forms of the murein-bound lipoprotein found in isolated outer membranes of *Escherichia coli*. *Eur. J. Biochem.* **113:**349–357.

Wensink, J., and B. Witholt. 1981b. Outer-membrane vesicles released by normally growing *Escherichia coli* contain very little lipoprotein. *Eur. J. Biochem.* **116:**331–335.

Work, E., K. W. Knox, and M. Vesk. 1966. The chemistry and electron microscopy of an extracellular lipopolysaccharide from *Escherichia coli*. *Ann. N. Y. Acad. Sci.* **133:**438–449.

Yaron, S., G. L. Kolling, L. Simon, and K. R. Matthews. 2000. Vesicle-mediated transfer of virulence genes from *Escherichia coli* O157:H7 to other enteric bacteria. *Appl. Environ. Microbiol.* **66:**4414–4420.

Yem, D. W., and H. C. Wu. 1978. Physiological characterization of an *Escherichia coli* mutant altered in the structure of murein lipoprotein. *J. Bacteriol.* **133:**1419–1426.

Yonezawa, H., T. Osaki, T. Woo, S. Kurata, C. Zaman, F. Hojo, T. Hanawa, S. Kato, and S. Kamiya. 2011. Analysis of outer membrane vesicle protein involved in biofilm formation of *Helicobacter pylori*. *Anaerobe* **17:**388–390.

Zhou, L., R. Srisatjaluk, D. E. Justus, and R. J. Doyle. 1998. On the origin of membrane vesicles in gram-negative bacteria. *FEMS Microbiol. Lett.* **163:**223–228.

Zollinger, W. D., M. A. Donets, D. H. Schmiel, V. B. Pinto, J. Labrie, E. E. Moran, B. L. Brandt, B. Ionin, R. Marques, M. Wu, P. Chen, M. B. Stoddard, and P. B. Keiser. 2010. Design and evaluation in mice of a broadly protective meningococcal group B native outer membrane vesicle vaccine. *Vaccine* **28:**5057–5067.

Regulation of Bacterial Virulence
Edited by Michael L. Vasil and Andrew J. Darwin
© 2013 ASM Press, Washington, DC   doi:10.1128/9781555818524.ch24

Chapter 24

Regulation of Envelope Stress Responses
by *Mycobacterium tuberculosis*

DANIEL J. BRETL AND THOMAS C. ZAHRT

INTRODUCTION

The ability of bacteria to recognize and readily adapt to potentially adverse stimuli is an important characteristic that contributes to the overall survival and/or maintenance of these organisms within their specific environment. Bacteria that are pathogens of humans must resist various insults that are generated as part of the host's innate and adaptive immune response. *Mycobacterium tuberculosis* is a strict human pathogen that infects a large number of individuals and is a significant cause of morbidity and mortality throughout the world. One factor that has contributed to its success as a pathogen is its ability to persist for decades in the human host without killing or subsequent elimination by the host's immune system. Following its transmission to a susceptible individual, *M. tuberculosis* encounters numerous stressors that are capable of inducing potentially lethal insult at its cellular envelope. To counteract the effects of these stressors, *M. tuberculosis* expresses a variety of regulatory determinants, including extracytoplasmic function sigma factors, two-component signal transduction systems (TCSSs), and serine-threonine protein kinases (STPKs) that mediate adaptation programs through alterations in gene expression and/or protein modification. Here we discuss some of the stressors likely to target the cell envelope of *M. tuberculosis* during infection, and the corresponding regulatory elements expressed by the bacterium to counteract this stress.

MYCOBACTERIUM TUBERCULOSIS CHARACTERISTICS

Mycobacterium tuberculosis is an aerobic, rod-shaped, nonmotile, slow-growing bacterium and the etiological agent of the respiratory disease tuberculosis (TB). The bacillus measures ~0.2 to 0.6 µm by 1 to 10 µm and often grows in clumps or cords due to the presence of hydrophobic mycolic acids and other long-chain fatty acids in its cell wall. While the organism is Gram positive, its cell wall stains poorly with the Gram stain, requiring microscopic identification by alternative means such as acid-fast staining. *M. tuberculosis* is a member of a genetically closely related group of *Mycobacterium* species called the *M. tuberculosis* complex. In addition to *M. tuberculosis*, this group includes *M. africanum*, *M. bovis*, "*M. canetti*," *M. caprae*, *M. microti*, *M. pinnipedii*, and "*M. mungi*." Collectively, these organisms cause a TB-like disease in humans and other animals (Brosch et al., 2002; Cousins et al., 2003; Reed et al., 2009; Alexander et al., 2010).

Phylogenetically, *M. tuberculosis* is a member of the phylum Actinobacteria, which also includes several notable human pathogens, including species of the genera *Streptomyces*, *Corynebacterium*, *Nocardia*, and *Rhodococcus*. Molecular evidence indicates that *M. tuberculosis* is an ancient organism, and it has been detected in the preserved remains of mummified Egyptians dating between ~3000 and 500 BCE (Zink et al., 2001; Zink et al., 2003). However, *M. tuberculosis* was not described as the causative agent of TB until 1884, when the German physician-scientist Robert Koch made this demonstration by fulfilling several criteria that we now know as "Koch's postulates" (Koch, 1884).

TUBERCULOSIS

Prior to the confirmation of *M. tuberculosis* as the causative agent of TB, the disease was known as consumption, wasting disease, or white plague due to the chronic cough, productive sputum, fever, night sweats, and weight loss that are typical symptoms associated with active disease. TB is primarily

Daniel J. Bretl and Thomas C. Zahrt   •   Department of Microbiology and Molecular Genetics, Center for Infectious Disease Research, Medical College of Wisconsin, Milwaukee, WI 53226.

an infection of the lungs; however, other disease forms are also common and are generally associated with dissemination of the bacterium to secondary infection sites. *M. tuberculosis* is primarily spread through the air from an infected individual with active disease by coughing and/or sneezing, generating respiratory droplet nuclei that contain between ~1 and 3 bacilli. Approximately 10% of infected individuals develop a primary, acute disease that is characterized by uncontrolled bacterial replication within the lungs and failure of the host's adaptive immune response to control and/or contain the infection. The robust proliferation of *M. tuberculosis* during this infection state induces extensive tissue damage, resulting in some of the characteristic radiological hallmarks associated with this disease. If active TB is left untreated, the death rate associated with it can be greater than 50% (Tiemersma et al., 2011). In contrast, infection of an otherwise immunocompetent individual most often (~90%) leads to initial containment of the bacteria and establishment of an asymptomatic, latent infection state. During latency, bacterial replication is controlled, primarily through the active recruitment of immune cells to sites of infection and subsequent formation of granulomatous structures which act to wall off and compartmentalize the bacteria within defined regions. It is thought that within granulomas *M. tuberculosis* enters a state of nonreplicating persistence (NRP) (Wayne 1994) during which the bacterium is able to persist for extended periods without sterilization or elimination. Latently infected individuals harbor an ~10% risk of developing reactivated disease at some point during their lifetime. This risk increases to 10% annually following immunosuppression, such as that observed following chemotherapy or coinfection with the human immunodeficiency virus (Luetkemeyer et al., 2010). Currently, it is estimated that roughly one in every three people worldwide is infected by *M. tuberculosis* and suffer from latent TB. Significantly, there were 9 million new cases of active TB (many resulting from reactivation) and more than 1.7 million deaths due to *M. tuberculosis* in 2009 alone (World Health Organization, 2010). Thus, active and latent TB continues to be a source of significant morbidity and mortality worldwide.

THE LIFE CYCLE OF *M. TUBERCULOSIS*

M. tuberculosis is a facultative intracellular pathogen, and its host range is restricted to humans. The bacterium is not normally found free within the environment, so its continued survival within the human population requires that it be transmitted directly from an infected individual with active disease to one that is susceptible to infection. Following inhalation by a new host, *M. tuberculosis* becomes deposited in the alveolus, where it is subsequently phagocytosed by resident macrophages and dendritic cells of the lung. Within macrophages, the bacterium is exposed to a variety of potentially bactericidal conditions; however, the bacterium resists killing and establishes a productive intracellular replication niche by delaying and/or blocking key steps in the phagosome maturation process, including the inhibition of phagosome-lyosome fusion (Armstrong and Hart, 1971; Russell et al., 2002). Prior to development of acquired immunity, *M. tuberculosis* replicates within these cell types uncontrolled. However, generation of a protective adaptive immune response leads to macrophage activation and a corresponding reduction in bacterial replication. This response ultimately culminates in the recruitment of various activated immune cells to sites of infection and the encasement of *M. tuberculosis*-infected macrophages within granulomas. These organized structures function to wall off the bacteria and prevent subsequent dissemination to distant infection foci. However, recent evidence obtained using a zebrafish model system of TB and *Mycobacterium marinum*, a close relative of *M. tuberculosis*, indicates that these structures are highly dynamic and may also mediate the active transport of *M. tuberculosis* to distant infection sites (Davis and Ramakrishnan, 2009). Furthermore, though secretion of cytokines and other proinflammatory mediators by host cells comprising or surrounding the granuloma limits proliferation of *M. tuberculosis*, continued secretion of these factors over time damages host tissues and leads to necrotic, caseous lesions. In vitro, exposure of *M. tuberculosis* to specific environmental stimuli likely to be present within the granuloma, including low oxygen tension, elevated nitric oxide levels, and nutrient depletion, promotes the transition of the bacterium from a state of active replication to NRP or "dormancy" (Wayne and Hayes, 1996; Betts et al., 2002; Voskuil et al., 2003). Significantly, *M. tuberculosis* is capable of maintaining this persistent physiological state for the entire lifetime of its host and then reactivating at a later date to initiate a secondary acute infection. While little is known about the process of reactivation, this secondary disease state is often associated with generation of cavitary lesions in the lung(s) and subsequent transmission of infectious, viable bacilli that are present within the sputum.

THE MYCOBACTERIAL CELL ENVELOPE

Mycobacterium species, including *M. tuberculosis*, possess an unusual and highly complex cell envelope (Fig. 1). It is well recognized that this determinant plays an important role in resistance to both innate and adaptive immune responses, as well as other potentially damaging host-derived compounds. In *M. tuberculosis*, the cell envelope consists of a plasma membrane, a cell wall, an outer membrane (OM), and a capsular-like structure (Brennan and Nikaido, 1995; Kaur et al., 2009). The plasma membrane appears to be a typical bacterial membrane (Daffe and Etienne, 1999). Similarly, the cell wall is comprised of a core peptidoglycan (PG) layer that is typical of Gram-positive organisms; however, it exhibits a relatively high degree of cross-linking (70 to 80%) between peptide side chains, making the bacterium structurally rigid (Matsuhashi, 1966; Kaur et al., 2009). Covalently attached to the PG is the major polysaccharide of the cell wall, arabinogalactan (AG) (Kaur et al., 2009). The main purpose of AG is to serve as a bridge between the PG core and the long-chain lipids (C_{70} to C_{90}), including mycolic acids, which are found exterior to the AG. Interestingly, these lipids represent a major component of the cell envelope, estimated to comprise ~30 to 40% of the mass of the envelope (Barry, 2001; Rastogi et al., 2001). External to the AG-mycolic acid layer is an OM which exhibits characteristics similar to that of Gram-negative bacteria, including the ability to act as a permeability barrier to the outside environment. The OM of *Mycobacterium* is unusual in that it contains an abundance of mannosylated constituents, including phosphatidyl-*myo*-inositol mannosides (PIMs), lipomannan, and lipoarabinomannan (LAM), which are noncovalently anchored to the OM through their phosphatidyl-*myo*-inositol moiety. In addition, some *Mycobacterium* species, including *M. tuberculosis*, contain LAM that is "capped" with mannose (i.e., ManLAM) (Kaur et al., 2009). The OM of *M. tuberculosis* also contains numerous acyltrehaloses which are esterified to different fatty acyl groups in the OM. These include sulfatides (SLs), 2,3-di-O-acyltrehaloses (DATs), triacyltrehaloses, polyacyltrehaloses (PATs), trehalose monomycolate, and trehalose dimycolate (Kaur et al., 2009). Lastly, the loosely associated capsular-like surface of *M. tuberculosis* is composed primarily of proteins and polysaccharides, with only a minor amount of lipids (Kaur et al., 2009). There is mounting evidence that the polysaccharides within the capsular-like material contribute to pathogenesis of the bacterium, having both immunomodulatory and antiphagocytic activities (Cywes et al., 1997; Stokes et al., 2004; Gagliardi et al., 2007).

THE *M. TUBERCULOSIS* GENOME

Prototypical laboratory strain *M. tuberculosis* H37Rv was the first *Mycobacterium* isolate to be completely sequenced (Cole et al., 1998). Since then, the complete sequences of more than 20 *Mycobacterium* genomes, including 5 *M. tuberculosis* isolates, have been delineated. The genomes of most *M. tuberculosis* isolates are between 4 and 5 million base pairs and have the capacity to encode ~4,000 proteins (Cole et al., 1998). The *M. tuberculosis* genome is highly G/C rich (~65%) and contains numerous repetitive DNA sequence elements, including insertion sequences and multigene families (Cole et al., 1998). When compared against available protein databases, it can be seen that ~85% of the coding sequences are homologous to known or predicted proteins from other organisms. The remaining coding sequences are unique and may encode proteins with novel functions. *M. tuberculosis* encodes proteins known to generate all of the essential amino acids, vitamins, and enzyme cofactors required for its own survival (Cole et al., 1998). The bacterium also encodes determinants allowing growth on a variety of sugars, hydrocarbons, alcohols, ketones, and carboxylic acids (Cole et al., 1998). Genes devoted to lipid metabolism, glycolysis, the pentose phosphate pathway, the tricarboxylic acid cycle, and the glyoxylate shunt are also represented. Importantly, the tubercle bacillus is predicted to encode a large number of regulatory proteins (>150), many of which are known or predicted to mediate adaptation processes in response to stress encountered during infection. Regulatory proteins from *M. tuberculosis* that have been best characterized include those encoding sigma factors, TCSSs, STPKs, and other evolutionarily conserved regulatory protein families.

EXPOSURE OF *M. TUBERCULOSIS* TO ENVELOPE STRESS

Given the complex life cycle of *M. tuberculosis,* the unusual composition of its cell envelope, and the relatively large size of its genome, it is not surprising that the bacterium is capable of resisting and/or surviving exposure to the variety of adverse environmental stimuli both exterior to and within the host (Fig. 2). Following expulsion from an infected individual, single water droplet nuclei containing bacilli

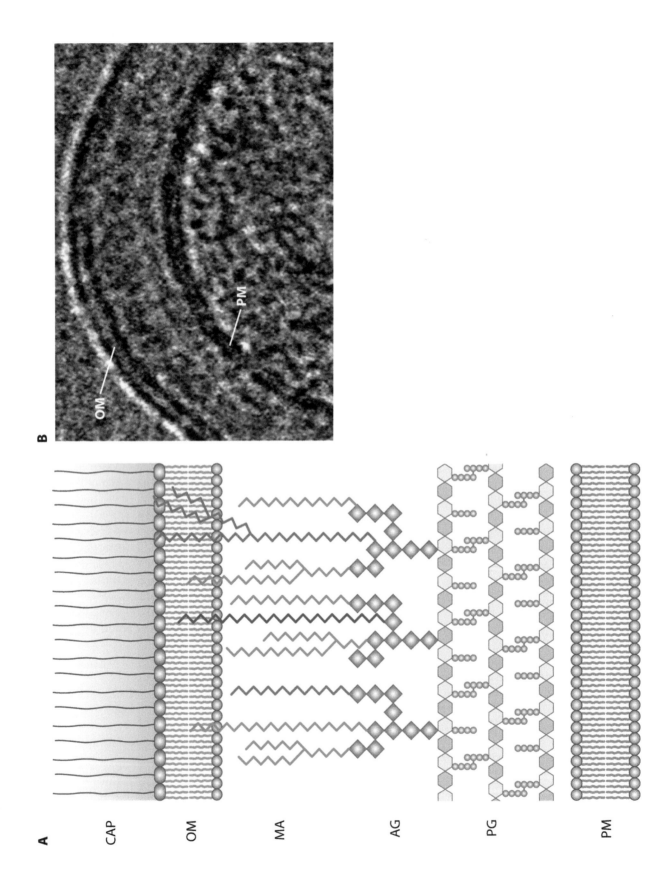

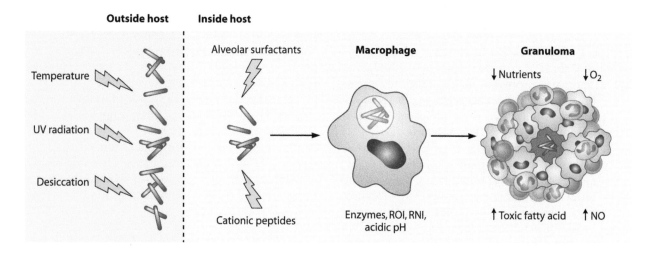

Figure 2. *M. tuberculosis* is exposed to a variety of environmental stressors found both outside and inside the host. M. tuberculosis is exposed to alterations in temperature, UV radiation, and desiccation during transmission from an infected individual to a susceptible host. Following inhalation and deposition in lung tissues, the bacterium encounters host alveolar surfactant proteins and cationic peptides which may perturb the bacterial cell envelope. Macrophages that ingest the tubercle bacillus generate bactericidal products, including ROI and RNI. Finally, M. tuberculosis establishes persistent infection within granulomatous structures generated following an adaptive host immune response. Within this environment, the bacterium must adapt to a variety of potentially adverse conditions, including altered nutrients, elevated levels of NO, toxic fatty acids, and low oxygen concentrations. doi:10.1128/9781555818524.ch24f2

are transmitted into the external atmosphere. Within this environment, *M. tuberculosis* must deal with desiccation, ionizing radiation, and fluctuations in temperature. Following its deposition into the terminal bronchioles and alveoli of the next host, *M. tuberculosis* first encounters mucosal cells of the airway epithelium, including alveolar epithelial type I (ATI) and type II (ATII) pneumocytes. ATII cells, along with Clara cells of the bronchi, secrete pulmonary surfactant (PS) molecules, which have been shown to exhibit antibacterial activity. Normally, PS coats the surface of the alveolus, where it helps maintain alveolus stability by minimizing surface tension during respiration. PS is a mixture of lipids (90%) and proteins (10%). Dipalmitoylphosphatidylcholine is the most abundant lipid that comprises the PS, and it associates with one of four surfactant proteins (SP-A, SP-B, SP-C, and SP-D) to produce functional PS. In addition to their function in reducing surface tension of alveoli, a subset of surfactant proteins are also

antibacterial. For example, SP-A and SP-D bind surface-exposed carbohydrate residues present in the *M. tuberculosis* envelope. Once bound, these molecules promote bacterial agglutination, mediate the association of *M. tuberculosis* with components of the innate and adaptive immune system, and alter aspects of the phagocytic response to infection (Downing et al., 1995; Pasula et al., 1997; Ferguson et al., 2002; Hall-Stoodley et al., 2006). Interestingly, recent evidence indicates that certain lipids present in the *M. tuberuclosis* cell envelope may limit the antimycobacterial activity of these peptides (Wang et al., 2008).

Additional products of the innate immune system are also likely to be encountered by *M. tuberculosis* either prior to or following ingestion by resident phagocytes of the lung. These determinants include antimicrobial peptides such as α-defensins, β-defensins, cathelicidins, and ubiquinated host peptides (Liu et al., 2006; Alonso et al., 2007; Wang et al., 2008; Sonawane et al., 2011). The cationic nature of

Figure 1. (A) The mycobacterial cell envelope is composed of four main compartments. The first is a typical bacterial plasma membrane (PM). Exterior to the PM is the cell wall, composed of several biomolecular moieties, including PG, AG, and mycolic acids (MA). Noncovalently attached to MA and other lipids of the cell wall layer is the OM. Finally, the entire cell is surrounded by a loosely associated capsule-like structure (CAP) composed primarily of proteins and carbohydrates. Lipids making up the MA layer are depicted as different lengths and colors to illustrate the variety of long-chain fatty acids found in the mycobacterial cell envelope. The image is not to scale. (B) Cryo-electron micrograph of the *M. smegmatis* cell envelope. The PM and OM are visible and enclose an area of different electron density that is presumed to contain PG, AG, and MA. This image is modified from that taken by Hoffmann et al. (Hoffmann et al., 2008) and is used with permission from the National Academy of Sciences. doi:10.1128/9781555818524.ch24f1

defensins allows them to associate with the anionic phospholipids comprising the plasma membrane and/or OM, promoting the permeabilization and subsequent leakage of intracellular contents from the bacterium (Lehrer et al., 1993). Evidence also indicates that defensins are present in the bronchoalveolar lavage fluid of individuals with active TB (Ashitani et al., 2002). Importantly, these molecules are biologically active against *M. tuberculosis* (Lehrer et al., 1993). Some antibacterial peptides also possess chemotactic activity on immune cells (Territo et al., 1989; Yang et al., 1999), compounding the potential antibacterial nature of these molecules. More recently, host-derived ubiquitinated peptides initially present in the cytosol of macrophages have also been shown to be translocated into vacuolar compartments that are accessed by *M. tuberculosis* during infection via the autophagy pathway (Alonso et al., 2007). These peptides are further processed by cathepsins into cationic peptides that are bactericidal against the bacterium during growth both in vitro and in vivo (Alonso et al., 2007).

Ingestion of *M. tuberculosis* by resident alveolar macrophages, dendritic cells, and neutrophils exposes *M. tuberculosis* to numerous environmental conditions or compounds that may induce stress at the cellular envelope. Internalization of *M. tuberculosis* into a phagosome results in the upregulated production of reactive oxygen intermediates (ROI) within the vacuolar compartment, including superoxide ($\cdot O_2^-$), and its various associated by-products such as hydrogen peroxide (H_2O_2). Similarly, infection of macrophages by *M. tuberculosis*, or activation of macrophages via other mechanisms, leads to the upregulation of *NOS2*. The corresponding enzyme is responsible for production of nitric oxide (NO) and other reactive nitrogen intermediates (RNI) such as peroxynitrate ($ONOO^-$), which is formed following the interaction of NO with $\cdot O_2^-$ (MacMicking et al., 1997). While ROI and RNI are not thought to directly target the cell envelope per se, exposure to these compounds can lead to perturbations and/or modifications in proteins and lipids comprising the cellular envelope, subsequently altering their characteristics. To combat ROI and RNI stress, *M. tuberculosis* encodes genetic determinants that are directly or indirectly involved in their detoxification, for example, superoxide dismutase and catalase (Diaz and Wayne, 1974; Kusunose et al., 1976; Zhang et al., 1992; Cole et al., 1998), or proteins that scavenge these toxic molecules (Ouellet et al., 2002; Pathania et al., 2002).

Maturation of *M. tuberculosis*-containing vacuoles through the endocytic pathway exposes the bacterium to an increasingly acidic environment brought forth by the acquisition of vacuolar ATPases on the phagosomal membrane. *M. tuberculosis* normally delays and/or blocks phagosome maturation in resting macrophages, limiting phagosome acidification to a pH of 6.4 (Sturgill-Koszycki et al., 1994). However, *M. tuberculosis*-containing phagosomes become more acidic (pH = 5.2) in macrophages that have been activated (Desjardins et al., 1994; Vergne et al., 2004; Russell et al., 2010; Welin et al., 2011). Additionally, fusion of the *M. tuberculosis* phagosome with the lysosome exposes the bacterium to an environment that is highly acidic and rich in acid hydrolases, lipases, proteases, and other potentially degradative factors.

RESPONSE TO CELL ENVELOPE STRESS IN *M. TUBERCULOSIS*

Due to the multitude of conditions found within the human host that could induce damage at the cell envelope, it is not surprising that *M. tuberculosis* employs several regulatory systems to respond to or mediate resistance against potentially damaging cell envelope stress. Though the regulation of cell envelope constituents, biosynthetic enzymes, and stress-responsive genes is only beginning to be understood, it is clear that signaling in response to cell envelope damage is complex, multifactorial, and highly regulated. In particular, the response of *M. tuberculosis* to cell envelope stress includes induction of coordinated gene expression programs that result from activation of sigma factors, TCSSs, and STPKs.

SIGMA FACTORS SigE AND SigB MEDIATE TRANSCRIPTIONAL REGULATION IN RESPONSE TO CELL ENVELOPE DAMAGE

Prokaryotic transcription is carried out by an RNA polymerase containing five necessary subunits: β, β', two α subunits, and ω. Transcription initiation is dependent on an additional, dissociable subunit, the σ factor. Sigma factors bind to specific DNA sequences upstream of transcriptional start sites of genes, and they allow discrimination between promoter elements whose activation leads to specific gene expression. Sigma factors can be classified into two distinct families based on structural and functional homology to the σ^{70} and σ^{54} sigma factors of *Escherichia coli* (Gruber and Gross, 2003). σ^{54} sigma factors are rare and not found in mycobacteria. In contrast, σ^{70} sigma factors are found in all bacteria, including *M. tuberculosis*, and are subdivided into additional categories: principal sigma factors responsible for transcription of the majority of housekeeping genes, principal-like sigma factors that are closely related to principal factors but nonessential for bacterial growth, sporulation-associated sigma factors, and extracytoplasmic function (ECF) sigma factors. The presence of multiple ECF sigma factors, in particular, allows bacteria to regulate transcription of specific gene programs in response to diverse environmental signals. *M. tuberculosis* contains a total of 13 sigma factors, including the principal sigma factor, SigA; the principal-like sigma factor, SigB; the sporulation-associated sigma factor, SigF; and 10 ECF sigma factors (SigC to SigE and SigG to SigM) (reviewed in Rodrigue et al., 2006). Individual ECF sigma factors are dispensable for *M. tuberculosis* growth in laboratory medium but become essential for survival under various stresses in vitro, for growth in macrophages, and/or for virulence in vivo (Rodrigue et al., 2006).

M. tuberculosis SigE

The ECF sigma factor SigE is responsive to conditions that directly or indirectly induce insult at the mycobacterial cell envelope (Table 1). Although this sigma factor is nonessential for *M. tuberculosis* growth under physiological conditions (Manganelli

et al., 2001), it is conserved across different species of *Mycobacterium*, including *Mycobacterium leprae*. *M. leprae* is a species that has undergone extensive genomic decay, suggesting that the function of SigE is essential to the stress response of the bacterium. The conserved function of SigE in *Mycobacterium* is supported by the fact that expression of *M. leprae sigE* can genetically complement and restore phenotypes to an *M. tuberculosis* strain harboring a *sigE* mutation (Williams et al., 2007). Disruption of *sigE* in both *Mycobacterium smegmatis* and *M. tuberculosis* results in increased sensitivity to oxidative stress, heat, and the ionic detergent sodium dodecyl sulfate (SDS) (Wu et al., 1997; Manganelli et al., 2001). Furthermore, deletion of *sigE* attenuates *M. tuberculosis* growth in macrophages (Manganelli et al., 2001), alters the host immune response to *M. tuberculosis* infection (Giacomini et al., 2006; Fontan et al., 2008; Hernandez Pando et al., 2010), and attenuates virulence of *M. tuberculosis* in mice (Ando et al., 2003; Manganelli et al., 2004; Hernandez Pando et al., 2010). Accordingly, SigE signaling is upregulated in response to heat shock (Manganelli et al., 1999), oxidative stress (Jensen-Cain and

Table 1. Regulators of cell envelope and environmental stresses known to activate their signaling

Transcriptional regulator	No. of genes in regulon	Reference(s)	Environmental stress	Reference(s)
MprA	320	He et al., 2006; Pang et al., 2007	SDS	Manganelli et al., 2001; He et al., 2006
			Triton X-100	He et al., 2006
			Alkaline pH	He et al., 2006
			Nutrient limitation	Betts et al., 2002
			Hollow-fiber granuloma	Karakousis et al., 2004
			In vivo (mice)	Talaat et al., 2004; Talaat et al., 2007
SigE	76	Manganelli et al., 2001; Fontan et al., 2008	SDS	Manganelli et al., 2001
			Oxidative stress	Jensen-Cain and Quinn, 2001
			Heat	Manganelli et al., 1999
			Macrophage	Graham and Clark-Curtiss, 1999; Jensen-Cain and Quinn, 2001
SigB	170	Lee et al., 2008; Fontan et al., 2009	SDS	Manganelli et al., 2001
			Heat shock	Manganelli et al., 2001
			Oxidative stress	Mehra and Kaushal, 2009
			Nutrient limitation	Betts et al., 2002
			Stationary phase	Hu and Coates, 1999
ClgR	14	Mehra and Kaushal, 2009; Estorninho et al., 2010; Manganelli and Provvedi, 2010	SDS	Manganelli et al., 2001; He et al., 2006
			Oxidative stress	Mehra and Kaushal 2009
			Heat shock	Stewart et al., 2002
			Macrophage	Schnappinger et al., 2003
PknB	—	—	PG fragments	Mir et al., 2011
			SDS, vancomycin	Barik et al., 2010

Quinn, 2001), and SDS (Manganelli et al., 1999) and during growth within macrophages (Graham and Clark-Curtiss, 1999; Jensen-Cain and Quinn, 2001). Interestingly, *sigE* is upregulated by another ECF sigma factor, SigH, in response to heat shock and oxidative stress (Raman et al., 2001; Manganelli et al., 2002; Dona et al., 2008), suggesting that the detection of heat shock and oxidative stress is distinct from the detection of stress leading to cell envelope damage (Manganelli et al., 2001; Song et al., 2008).

Approximately 25 genes are regulated in *M. tuberculosis* by SigE in response to cell envelope stress induced by SDS exposure (Manganelli et al., 2001). Interestingly, unlike most ECF sigma factors, SigE is not autoregulatory (Manganelli et al., 2001). Rather, the increased expression of *sigE* in response to SDS, Triton X-100, and alkaline pH is dependent on positive regulation by the MprAB TCSS. Details of this interaction are discussed in greater detail below (He et al., 2006; Dona et al., 2008). In a positive-feedback manner, SigE directly regulates expression of the entire *mprAB-pepD-moaB2* operon from a promoter positioned upstream of *mprA* (Manganelli et al., 2001) (Fig. 3). SigE also directly regulates expression of an operon that includes the transcriptional regulator *clgR*; a heat shock chaperone, *acr2*; and four genes involved in mycolic acid biosynthesis: *fbpC2*, *fbpB*, *fabD*, and *acpM* (Manganelli et al., 2001) (Fig. 3). In addition to in vitro stress, SigE-mediated regulation of these genes also occurs following infection of *M. tuberculosis* within macrophages (Fontan et al., 2008). Under these conditions, SigE also regulates genes involved in mycolic acid biosynthesis, genes encoding determinants linking the cell wall and mycolic acids, and genes mediating the translocation of folded proteins across the cell membrane (Fontan et al., 2008). Thus, SigE participates in a variety of fundamental processes linked in some way to the cell envelope.

Like some ECF sigma factors, SigE is regulated posttranscriptionally by binding to a cognate anti-sigma factor, RseA (Dona et al., 2008; Barik et al., 2010). Binding of RseA to SigE limits the ability of SigE to interact with the RNA polymerase and initiate transcription. Following surface stress caused by SDS or vancomycin, RseA becomes phosphorylated by the STPK PknB (Barik et al., 2010) (Fig. 3). Phosphorylation of RseA marks it for degradation by a specific protease complex, ClpC1P2. Significantly, phosphorylation and degradation of RseA result in dissociation of SigE from its anti-sigma factor without degradation of SigE itself. SigE then contributes to a second positive-feedback loop, upregulating expression of the aforementioned transcriptional regulator ClgR, which, in turn, upregulates the Clp proteases responsible for degradation of RseA (Mehra and Kaushal, 2009; Barik et al., 2010).

M. tuberculosis SigB

In addition to SigE, the principal-like sigma factor SigB contributes to the cell envelope damage response. SigB shares significant homology with the principal sigma factor SigA (Doukhan et al., 1995; Beggs et al., 1996; Hu and Coates, 1999), but unlike SigA, SigB is dispensable for growth (Gomez et al., 1998; Fontan et al., 2008). *sigB* is autoregulatory and is also positively regulated by multiple transcriptional regulators, including the sigma factors SigE, SigF, SigH, and SigL, as well as the response regulator MprA (Manganelli et al., 2001; Dainese et al., 2006; He et al., 2006; Pang et al., 2007; Lee, et al., 2008). Interestingly, overexpression of *sigB* does not result in increased expression of any other sigma factors (Lee et al., 2008); yet, the convergence of multiple transcriptional factors upon *sigB* suggests that this sigma factor is a central determinant that propagates signaling under multiple stress conditions. Indeed, expression of *sigB* is upregulated following exposure of *M. tuberculosis* to a multitude of stressors, including SDS, heat shock, nutrient starvation, oxidative stress, and stationary-phase growth (Hu and Coates, 1999; Manganelli et al., 1999; Betts et al., 2002). However, SigB regulates a unique set of genes compared to SigE or SigH, suggesting that the response mediated by SigB may be related to but distinct from that of ECF sigma factors (Mukherjee and Chatterji, 2005; Lee et al., 2008; Fontan et al., 2009). Several of the genes regulated by SigB are important in cell wall maintenance (*wag31*, etc.) and lipid biosynthesis (*fadD31*, *kasA*, *pks2*, *ppsD*, etc.) (Lee et al., 2008; Fontan et al., 2009). Additionally, overexpression of *M. tuberculosis sigB* in *M. smegmatis* results in altered colony morphology and accumulation of hyperglycosylated glycopeptidolipids (GPLs), complex lipids found at the outermost layer of the cell envelope (Mukherjee et al., 2005). Interestingly, GPLs are not present in *M. tuberculosis*. Rather, they are replaced with phenolic glycolipids (PGLs) (Ortalo-Magne et al., 1996). Therefore, the contribution of SigB in regulation of these complex cell envelope lipids remains undefined. Finally, an *M. tuberculosis sigB* mutant has increased sensitivity to SDS and is compromised for growth under hypoxic conditions (Fontan et al., 2009); however,

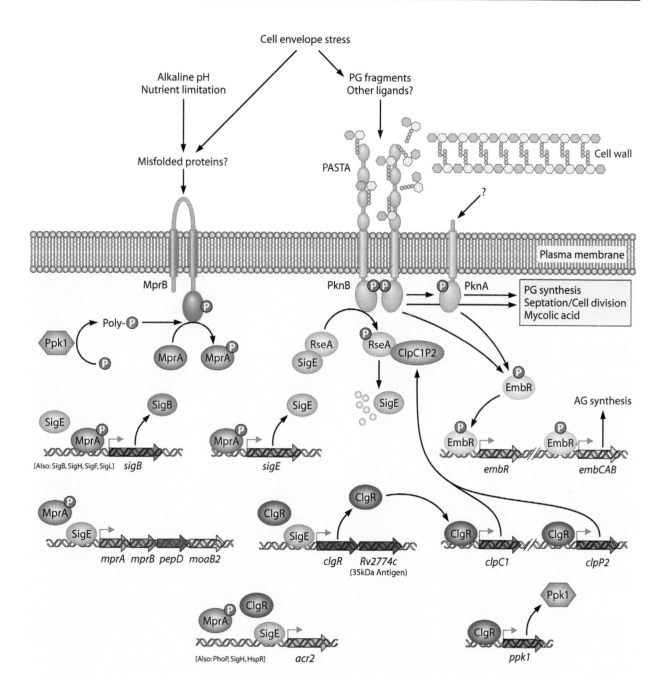

Figure 3.   Cell envelope stress response network of *M. tuberculosis.* Cell envelope stress that generates PG fragments is sensed by PknB via its PASTA domains. Activation of PknB results in phosphorylation of other STPKs, including PknA and several determinants involved in PG synthesis, cell division, and mycolic acid production. Additionally, PknB and other STPKs phosphorylate EmbR, leading to increased expression of the embCAB operon and increased AG synthesis. PknB phosphorylation of the SigE anti-sigma factor, RseA, results in RseA degradation by the ClpC1P2 protease complex. Dissociated SigE directs upregulation of several genes, including *mprAB*, *sigB*, *acr2*, and *clgR*. Cell envelope damage is also sensed by the extracytoplasmic domain of MprB. Activation of MprB leads to phosphorelay to MprA, and phosphorylated MprA directs transcription of *sigB*, *sigE acr2*, and several other genes. Finally, ClgR directs expression of the genes encoding the ClpC1P2 protease, as well as Ppk1, which is responsible for generation of polyphosphate molecules that can be utilized by MprB to alternatively phosphorylate MprA. doi:10.1128/9781555818524.ch24f3

it is not attenuated for growth and/or survival in macrophages, mice, or guinea pigs (Fontan et al., 2009). Therefore, the role of this sigma factor in the context of an infection remains undefined.

Other *M. tuberculosis* sigma factors

While information regarding the importance of other *M. tuberculosis* sigma factors in the cell envelope damage response is lacking, at least three additional sigma factors may play a role. SigL is regulated by a membrane-localized anti-sigma factor, RslA, which contains an extracytoplasmic domain, suggesting that it may be directly regulated by extracytoplasmic signals (Hahn et al., 2005; Dainese et al., 2006). While SigL is not activated by oxidative stress or exposure to SDS (Hahn et al., 2005), it is autoregulated and controls expression of genes involved in cell envelope processes, including determinants required for production of mycolic acid and other lipids (including *pks10* and *pks7*) (Hahn et al., 2005; Dainese et al., 2006). *sigL* mutants of *M. tuberculosis* are not attenuated in organ burden within a mouse model system of infection; however, these mutants do exhibit a delayed time to death. These results suggest a role for SigL-dependent gene regulation in aspects of virulence (Dainese et al., 2006). The sporulation-like sigma factor SigF may also play a role in the cell envelope damage response. While *M. tuberculosis* does not form spores (Traag et al., 2010), SigF regulates numerous genes in *M. tuberculosis* during stationary phase, including some important for cell envelope synthesis and structure (Geiman et al., 2004). Additionally, disruption of *sigF* in *M. tuberculosis* results in a mutant derivative that is unable to stain with neutral red (Chen et al., 2000). This strain is also more permeable to hydrophobic solutes, suggesting that SigF may regulate genes important for cell wall structure (Chen et al., 2000). Finally, SigM negatively regulates the mycolic acid biosynthetic enzymes KasA and KasB, as well as enzymes for phthiocerol dimycocerosate (PDIM) biosynthesis and sulfolipids (Raman et al., 2006). However, the signal(s) recognized by SigM, or the mechanism(s) by which SigM mediates this regulation, remains unclear.

M. tuberculosis encodes a complex network of sigma factors that are involved in various aspects of the cell envelope stress response. It is clear that the ECF sigma factor SigE, as well as SigB, plays a central role in this response. Examination of the genes regulated by these ECF sigma factors indicates that there is extensive cross talk with other signaling pathways, including those regulated by MprAB and ClgR (Table 2). Regardless, additional research will be required to define the role of these, and other, sigma factor determinants in the response of *M. tuberculosis* to conditions inducing cell envelope stress.

The TCSS MprAB Responds to Cell Envelope Stress

TCSSs are a dominant mechanism utilized by bacteria to sense their environment. These systems are ubiquitous in prokaryotes but are absent in higher eukaryotes, including humans. Prototypical TCSSs contain a membrane-localized histidine sensor kinase (SK) and a cytoplasmically localized response regulator (RR). SKs are often comprised of an extracytoplasmic sensor domain, one or more transmembrane domains, and a cytoplasmic region containing a dimerization motif and a kinase domain. Activation of SKs leads to autophosphorylation of SK dimers in *trans* on conserved histidine residues. The phosphate is then transferred via a unique phosphorelay reaction to a conserved aspartic acid residue on a cognate RR. The transfer of phosphate to the RR induces conformational changes that promote DNA binding, transcriptional regulation, and/or other enzymatic activities. *M. tuberculosis* encodes eleven genetically linked TCSSs, 2 orphaned SKs, and 5 orphaned RRs (Cole et al., 1998). It is beginning to be appreciated that the TCSSs of *M. tuberculosis* respond to a variety of environmental stresses. For example, the SenX3-RegX3 system responds to inorganic phosphate (Glover et al., 2007) and the DosRS-DosT system responds to limited oxygen environments, nitric oxide, and other conditions (Park et al., 2003; Voskuil et al., 2003; Kumar et al., 2008). A subset of *M. tuberculosis* TCSSs are activated within macrophages and/or are upregulated during the course of infection within both mice and humans (Via et al., 1996; Graham and Clark-Curtiss, 1999; Perez et al., 2001; Parish et al., 2003; Haydel and Clark-Curtiss, 2004; Leyten et al., 2006). Consistent with this observation, the deletion of specific *M. tuberculosis* TCSSs results in various in vivo phenotypes that include both attenuation and hypervirulence in mouse models of TB (Parish et al., 2003; Rickman et al., 2004; Walters et al., 2006; Converse et al., 2009; Luetkemeyer et al., 2010). Collectively, research on *M. tuberculosis* TCSSs indicates that these signaling systems are critical for aspects of survival and virulence within the host.

The MprAB system of *M. tuberculosis*

The MprAB TCSS of *M. tuberculosis* has been shown to respond and mediate resistance to cell envelope stress. RR MprA was first identified as being necessary for *M. tuberculosis* persistence within mice (Zahrt and Deretic, 2001), as reflected in its designation as a mycobacterium persistence regulator (Zahrt and Deretic, 2001). Expression of *mprAB* occurs in vivo during both acute and latent infection stages within mice (Talaat et al., 2004; Talaat et al., 2007).

Table 2. Genes coregulated by transcriptional regulators associated with cell envelope stress

Transcriptional regulators[a]	Gene number[b]	Gene product[b]	Function[b]
SigE/SigB	Rv0685	Tuf	Elongation factor
	Rv0863	Rv0863	Unknown
	Rv1398c	VapB10	Possible antitoxin
	Rv2204c	Rv2204c	Unknown
MprA/SigB	Rv0464c	Rv0464c	Unknown
	Rv1361c	PPE19	PPE family protein
	Rv1792	EsxM	ESAT-6-like protein
	Rv2625c	Rv2625c	Unknown
	Rv2287	YjcE	Possible Na+/H+ transporter
	Rv2712c	Rv2712c	Unknown
	Rv2989	Rv2989	Transcriptional regulator
	Rv3136	PPE5	PPE family protein
	Rv3180c	Rv3180c	Unknown
	Rv3767c	Rv3767c	Methyl transferase
	Rv3825c	Pks2	Polyketide synthase
MprA/SigE	Rv0129c	FbpC	Antigen 85 complex
	Rv0467	Icl1 (AceA)	Isocitrate lyase
	Rv0563	HtpX	Probable heat shock protein
	Rv0981	MprA	Two-component response regulator
	Rv0982	MprB	Two-component SK
	Rv0983	PepD	Serine protease
	Rv0984	MoaB2	Molybdopterin biosynthesis
	Rv1057	Rv1057	Unknown
	Rv1197	EsxK	ESAT-6-like protein
	Rv1884c	RpfC	Resuscitation-promoting factor
	Rv2052c	Rv2052c	Unknown
	Rv2053c	Rv2053c	Unknown
	Rv2244	AcpM	Acyl carrier protein
	Rv2710	SigB	RNA polymerase sigma factor
	Rv2743c	Rv2743c	Unknown
	Rv3139	FadE24	Acyl-CoA dehydrogenase
	Rv3140	FadE23	Acyl-CoA dehydrogenase
	Rv3614c	EspD	ESX-1 secretion-associated protein
	Rv3615c	EspC	ESX-1 secretion-associated protein
	Rv3616c	EspA	ESX-1 secretion-associated protein
MprA/SigB/SigE	Rv0465c	Rv0465c	Transcriptional regulator
	Rv1129c	Rv1129c	Transcriptional regulator
	Rv1130	PrpD	Methyl citrate dehydratase
	Rv1131	PrpC	Methyl citrate synthase
MprA/ClgR	Rv2460c	ClpP2	Protease
	Rv2461c	ClpP1	Protease
MprA/ClgR/SigB	Rv0384c	ClpB	Chaperone/protease
MprA/SigE/SigB/ClgR	Rv0249c	Rv0249c	Unknown
	Rv0250c	Rv0250c	Unknown
	Rv0251c	Acr2 (Hsp)	Heat shock protein
	Rv2744c	Rv2744c	35-kDa antigen
	Rv2745c	ClgR	Transcriptional regulator

[a]Genes coregulated by indicated transcriptional regulators were determined by comparison of available microarray data: MprA (He et al., 2006; Pang et al., 2007), SigE (Manganelli et al., 2001; Fontan et al., 2008), SigB (Lee et al., 2008; Fontan et al., 2009), and ClgR (Estorninho et al., 2010; Sherrid et al., 2010).
[b]Gene number, gene product, and gene function as described by the TubercuList (http://tuberculist.epfl.ch).

mprAB expression is also upregulated during growth of *M. tuberculosis* within implanted artificial hollow fibers, an infection model that simulates growth within granulomatous lesions (Karakousis et al., 2004). Interestingly, disruption of *mprA* does not limit the ability of *M. tuberculosis* to replicate during acute stages of infection within mice, suggesting that the environmental signal(s) detected by this system in vivo is likely to occur only following the onset of adaptive immunity (Zahrt and Deretic, 2001). It is well established that SK MprB and RR MprA form a functional TCSS (Zahrt et al., 2003). Activation of this TCSS occurs in vitro in response to agents that perturb the cell envelope, including SDS, the nonionic detergent TritonX-100, and alkaline pH (Table 1) (Manganelli et al., 2001; He et al., 2006). Interestingly, additional environmental stresses, including nutrient limitation, have also been shown to activate MprAB signaling (Betts et al., 2002). Therefore, MprAB may be activated by multiple signaling inputs. More likely, exposure to several different environmental stressors may manifest in a common signal that activates MprAB. Currently, this is hypothesized to be misfolded or aggregated proteins that accumulate in the extracytoplasmic space following stress exposure (Pang et al., 2007; White et al., 2010). Regardless of the specific signal detected following cell envelope stress, signaling through MprB requires the extracytoplasmic sensing domain, lending further support to the idea that MprAB processes signals that initiate exterior to the plasma membrane (He et al., 2006). As one would predict, the transcriptional cascade initiated by MprAB following cell envelope stress is necessary for survival under these conditions. For example, *mprAB* deletion mutants of *M. tuberculosis* and *M. smegmatis* exhibit increased sensitivity to SDS (White et al., 2010). *mprAB* deletion mutants in *M. smegmatis* also exhibit increased sensitivity to cycloserine, cefuroxime, and vancomycin, antibiotics that perturb the cell envelope by inhibiting steps in PG synthesis (White et al., 2010; White et al., 2011).

Microarray analyses have revealed that MprA both positively and negatively regulates a diverse set of greater than 300 genes (Table 1) (He et al., 2006; Pang et al., 2007). A subset of these genes is directly regulated through binding by MprA to DNA sequences in the promoter region of these genes (He and Zahrt, 2005; Pang and Howard, 2007). Other genes are likely regulated via indirect mechanisms, such as through SigE (Manganelli et al., 2001) or DosRST (Pang et al., 2007). MprA autoregulates expression of its own operon, a region that includes *mprA*, *mprB*, and the downstream genes *pepD* and *moaB2*. PepD is an HtrA-like serine protease, while MoaB2 is predicted to be a molybdopterin-biosynthetic

protein (He and Zahrt, 2005). HtrA-like proteases are a well-conserved family of proteins that carry out multiple functions in both Gram-positive and Gram-negative bacteria, including the refolding or degradation of misfolded proteins in the extracellular milieu (Alba et al., 2001; Farn and Roberts, 2004; Buelow and Raivio, 2005; Kim and Kim, 2005; Stack et al., 2005; Flannagan et al., 2007). In *M. smegmatis*, deletion of *pepD* results in increased sensitivity to SDS. Interestingly, deletion of *pepD* in *M. tuberculosis* does not increase sensitivity to SDS, due most likely to the compensatory upregulation of *sigE* (White et al., 2010). Regardless, deletion of *pepD* delays time to death in severe combined immunodeficient (SCID) mice infected with *M. tuberculosis* (Mohamedmohaideen et al., 2008). These mice also exhibit reduced tissue pathology despite similar organ burden (Mohamedmohaideen et al., 2008). PepD exhibits both protease and chaperone activity in vitro, dual functions that are consistent with those of other HtrA family members (Mohamedmohaideen et al., 2008; White et al., 2010). Proteomic studies have indicated that PepD interacts with and cleaves an abundant *M. tuberculosis* protein (Rv2744c) known as the 35-kDa antigen (White et al., 2011). This protein localizes to the cell wall fraction (White et al., 2011) and has homology to the PspA family of phage shock response proteins. These determinants are thought to participate in multiple processes, including those involving cell envelope stress responses (reviewed in Joly et al., 2010). Interestingly, over expression of the 35-kDa antigen in an *M. smegmatis* Δ*pepD* mutant rescues sensitivity of this strain to vancomycin, cycloserine, and cefuroxime (White et al., 2011). Thus, the interaction between PepD and the 35-kDa antigen may be important for maintaining cell envelope homeostasis under conditions of surface stress (White et al., 2011). Importantly, the gene encoding the 35-kDa antigen lies immediately downstream of and in an operon with the transcriptional regulator ClgR, previously discussed as being regulated by SigE. ClgR itself upregulates several proteases and chaperones that may be important under stress conditions, including another HtrA-like protease, Rv1043c (Estorninho et al., 2010).

Another important stress-induced gene directly regulated by MprA is *acr2* (Pang and Howard, 2007; Pang et al., 2007). Expression of *acr2* increases dramatically in an MprA-dependent manner following exposure of *M. tuberculosis* to SDS (Pang and Howard, 2007). In *M. tuberculosis*, *acr2* is also upregulated following various other environmental stresses; significantly, it is the gene most highly upregulated in response to heat stress (Stewart et al., 2002). Interestingly, *acr2* is also regulated by the ECF sigma factors SigH (Manganelli et al., 2002) and SigE (Manganelli et al., 2001), the RR PhoP (Walters et al., 2006), ClgR (Estorninho et al., 2010), and a transcriptional inhibitor, HspR (Stewart et al., 2002). HspR regulates additional heat shock proteins, including the essential *M. tuberculosis* Hsp70 homolog, DnaK (Stewart et al., 2002). All of these transcriptional regulators compete for binding within the *acr2* promoter, therefore tightly regulating expression of this important stress protein in *M. tuberculosis*. As might be expected given the extensive transcriptional regulation at this locus, *acr2* is also expressed within macrophages (Schnappinger et al., 2003; Wilkinson et al., 2005) and mice (Stewart et al., 2005). Deletion of *acr2* does not reduce the organ burden of *M. tuberculosis* in infected mice, but it does reduce the ability of *M. tuberculosis* to induce tissue pathology (Stewart et al., 2005). Functionally, Acr2 is a heat shock protein with homology to the alpha-crystallin family of proteins found in both prokaryotes and eukaryotes, including humans (Stewart et al., 2005). Acr2 has chaperone activity in vitro (Tabira et al., 2000) and is found in the ribosomal fraction of heat-stressed mycobacteria (Ohara et al., 1997). Thus, Acr2 is predicted to interact with and stabilize the 30S ribosome to help initiate translation under conditions of stress (Ohara et al., 1997). Direct regulation of *acr2* by MprA indicates that this function of Acr2, or some other, unknown function of this protein, is also important under conditions of envelope stress.

Other *M. tuberculosis* TCSSs regulating expression of cell envelope products

While it is clear that MprAB regulates a number of genes in response to cell envelope perturbation, other TCSSs may also indirectly play a role in this response. For example, deletion of the RR *phoP* in *M. tuberculosis* results in altered colony morphology. While the environmental signal(s) leading to PhoP-PhoR activation is not known, this mutant strain forms smaller colonies and exhibits an inability to form serpentine cords on solid medium, fix neutral red, and stain acid-fast (Perez et al., 2001; Gonzalo Asensio et al., 2006; Walters et al., 2006). These phenotypes can be explained in part by a lack of specific lipids within the cell envelope of the *phoP* mutant, including 2,3-di-*O*-acyltrehaloses (DATs), polyacyltrehaloses (PATs), and sulfolipids (SLs) (Gonzalo Asensio et al., 2006; Walters et al., 2006). PhoP directly regulates expression of the *msl3* operon, necessary for proper synthesis of DATs and PATS, and *pks2*, encoding the enzyme for SL synthesis (Gonzalo Asensio et al., 2006; Walters et al., 2006). These observations suggest that PhoP-PhoR may be important in regulating

the lipid repertoire of the cell envelope. Indeed, PhoP also regulates expression of several additional lipid biosynthetic genes or lipid transporters, including *fadD26*, *ppsA*, and *ppsB* (PDIM-biosynthetic genes); *pks5*, *pks6*, and *pks11* (polyketide biosynthesis); and *mmpL5*, *mmpS5*, and *mmpL12* (putative lipid transporters) (Gonzalo Asensio et al., 2006; Walters et al., 2006). Finally, in addition to other environmental stressors, *M. tuberculosis phoP* mutants are more sensitive to several cell wall-perturbing antibiotics, including vancomycin and cloxacillin (Walters et al., 2006), and these strains are attenuated in vivo (Perez et al., 2001). Undoubtedly, as the research on *M. tuberculosis* continues, it is likely that additional TCSSs will prove to be involved in responding to cell envelope stress or regulating genes that are important in cell envelope biosynthesis and maintenance.

Serine-threonine protein kinases

Posttranslational phosphorylation of proteins was traditionally thought to be limited to eukaryotic cells. However, the discovery of TCSSs in bacteria shifted this paradigm to include phosphorylation of prokaryotic SKs on a conserved histidine residue and phosphorelay to a conserved aspartic acid residue on the cognate RRs. Recently, the paradigm has again shifted to include "eukaryotic-like" STPKs, tyrosine kinases, and even serine-threonine phosphatases and tyrosine phosphatases. *M. tuberculosis* has served as a model organism for the study of STPKs; its genome contains 11 STPKs (named PknA, PknB, and PknD to -L) and a single serine-threonine phosphatase, PstP (Cole et al., 1998). In contrast, STPKs have not been described for *E. coli*. *M. tuberculosis* also encodes a single tyrosine kinase and two tyrosine phosphatases, though very little is currently known about these proteins (Cole et al., 1998; Bach et al., 2009).

All *M. tuberculosis* STPKs have the ability to phosphorylate themselves on several threonine and serine residues (Av-Gay et al., 1999; Chaba et al., 2002; Molle et al., 2003; Young et al., 2003; Cowley et al., 2004; Duran et al., 2005; Kang et al., 2005; Molle et al., 2003a; Molle et al., 2006b; Singh et al., 2006; Canova et al., 2008; Mieczkowski et al., 2008; Thakur et al., 2008; Kumar et al., 2009; Jang et al., 2010). In general, autophosphorylation increases the kinase activity of STPKs, as the prominent phosphorylation sites within these proteins are positioned within the kinase activation loop (Boitel et al., 2003; Ortiz-Lombardia et al., 2003; Young et al., 2003; Duran et al., 2005; Kang et al., 2005; Mieczkowski et al., 2008). In addition, many *M. tuberculosis* STPKs are also able to cross-phosphorylate each other, suggesting significant overlap in signaling among these kinases (Kang et al., 2005). Nearly all of the identified substrates of *M. tuberculosis* STPKs are dephosphorylated by PstP, a serine-threonine-specific phosphatase (Boitel et al., 2003; Chopra et al., 2003; Pullen et al., 2004). Intriguingly, PstP is itself a substrate of at least PknA and PknB, and phosphorylation by these STPKs increases the phosphatase activity of PstP (Boitel et al., 2003; Duran et al., 2005; Sajid et al., 2011). In turn, PstP can dephosphorylate itself (Sajid et al., 2011), as well as several STPKs, including PknA and PknB (Boitel et al., 2003; Chopra et al., 2003; Duran et al., 2005; Sharma et al., 2006). Therefore, STPK and PstP activities likely temper the activity of each other, allowing regulation of protein function on the basis of phosphorylation. Interestingly, proteomic studies have indicated that there are approximately 300 phosphorylated proteins present in lysates of *M. tuberculosis* (Prisic et al., 2010). This represents approximately 7% of the total protein content of the cell (Prisic et al., 2010). These proteins are known or predicted to possess a broad range of functional activities (Prisic et al., 2010). Importantly, unlike sigma factors and TCSSs, STPKs regulate protein activity at a posttranslational level. However, not surprisingly, transcriptional regulators are also substrates for STPKs (Molle et al., 2003; Kumar et al., 2009; Chao et al., 2010).

Most STPKs of *M. tuberculosis* localize to the cell membrane via a transmembrane domain flanked by an extracytoplasmic domain and a cytoplasmic kinase domain (Narayan et al., 2007). This general topology predicts that many of the STPKs of *M. tuberculosis* may respond to environmental signals. In contrast, PknG and PknK are soluble proteins (Koul et al., 2001; Kumar et al., 2009). PstP contains a transmembrane domain and is also localized to the cell membrane (Chopra et al., 2003). PknA and PknB have served as models for the study of STPKs in *M. tuberculosis*. PknA shares predicted homology with STPKs from other bacterial species important for sporulation and cell division, especially Pkn1 from *Myxococcus xanthus* (Chaba et al., 2002). PknB also shares overall prototypical STPK structure but, unlike other STPKs of *M. tuberculosis*, contains four PASTA (penicillin-binding protein and serine-threonine kinase-associated) domains

in its extracytoplasmic region (Boitel et al., 2003; Ortiz-Lombardia et al., 2003; Young et al., 2003; Kang et al., 2005). Proteins containing PASTA domains bind to PG subunits and have been implicated in such processes as germination of *Bacillus subtilis* endospores (Shah et al., 2008). Recently, it has been confirmed that the PASTA domains of PknB bind PG fragments and are essential for localization of PknB to mid-cell and polar locations (Mir et al., 2011). Importantly, the generation of PG fragments by Rpf proteins has been shown to be involved in promoting *M. tuberculosis* resuscitation and growth following dormancy (Mukamolova et al., 2002; Keep et al., 2006). Therefore, PknB may be directly involved in this response. The PASTA domains of PknB are arranged in a unique linear conformation and are less conserved than PASTA domains of penicillin-binding proteins, suggesting that PknB may have additional unidentified ligands (Jones and Dyson, 2006; Barthe et al., 2010). Regardless, the presence of the PASTA domains within the extracytoplasmic region of PknB suggests that this STPK may also directly respond to cell surface stress that perturbs cell wall integrity. Supporting this hypothesis, structural analysis has shown that PknB dimerization is required for its activation (Lombana et al., 2010). However, dimerization of PknB is weak and may require a ligand (i.e., PG fragments) to promote the association, localization, and activation of PknB (Mieczkowski et al., 2008; Barthe et al., 2010; Lombana et al., 2010; Mir et al., 2011). Interestingly, once activated, PknB dimerization is no longer required for this kinase to phosphorylate additional substrates (Lombana et al., 2010).

PknA and PknB are essential STPKs in *M. tuberculosis*

PknA and PknB are the only STPKs essential for *M. tuberculosis* survival (Sassetti et al., 2001; Fernandez et al., 2006), and it has been established that they function as critical regulatory elements important for cell envelope homeostasis and cell division (Fig. 4). *pknA*, *pknB*, and *pstP* are in an operon (Kang et al., 2005) that is positioned near the mycobacterial origin of replication, *oriC* (Cole, et al. 1998; Chaba et

Figure 4. PknA and PknB phosphorylate several proteins involved in biosynthesis or maintenance of the *M. tuberculosis* cell envelope. PknA phosphorylates MurD and FtsZ, while PknB phosphorylates PbpA, Rv0020c, and PapA5. Both kinases have been shown to phosphorylate Wag31, GlmU, and FipA. PknA and PknB also phosphorylate all of the enzymes of the core mycolic acid synthesis machinery. Phosphorylation of these enzymes is generally inhibitory, with the exception of KasB (stimulatory) and Pks13 (unknown). It is important to note that the substrates shown in this figure are not an exhaustive list, and other substrates having functions not directly related to cell envelope maintenance exist. Furthermore, other STPKs not shown are able to phosphorylate several of these substrates. Finally, the phosphorylation of the kinase substrates and the kinases themselves is reversible via dephosphorylation by PstP, the *M. tuberculosis* serine-threonine phosphatase (not shown). doi:10.1128/9781555818524.ch24f4

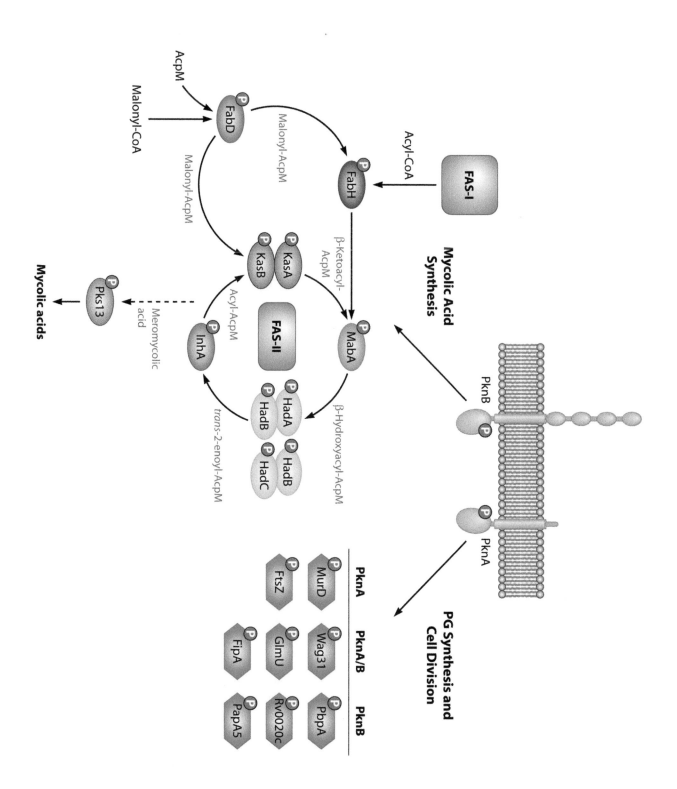

al., 2002). Also present within this operon are genes of the SEDS (shape, elongation, division, and sporulation) family (Henriques et al., 1998). *pknB*, and by extension the entire operon, is highly expressed during exponential growth but is downregulated approximately 10-fold upon reaching stationary phase (Kang et al., 2005), suggesting the importance of this operon during active growth. Additionally, expression of the operon increases in human monocyte-like THP-1 cells, mouse macrophage-like J774A.1 cells, and primary human alveolar macrophages, indicating the importance of these kinases in the context of cellular infection (Av-Gay et al., 1999; Singh et al., 2006).

Overexpression of *M. tuberculosis pknA* or *pknB* in *E. coli*, *M. smegmatis*, or *Mycobacterium bovis* BCG drastically changes the cellular morphology of these strains (Chaba et al., 2002; Kang et al., 2005). *pknA* overexpression results in elongated cells, at times 30- to 60-fold longer than wild-type cells (Chaba et al., 2002; Kang et al., 2005). In contrast, *pknB* overexpression results in short, "fat" cells (Chaba et al., 2002; Kang et al., 2005). Furthermore, overexpression of these kinases decreases growth rate and viability of *M. smegmatis* and *M. bovis* BCG (Kang et al., 2005). Importantly, these phenotypes are dependent on kinase activity, as overexpression of kinases defective in catalytic activity results in no observable cell morphological alterations (Kang et al., 2005). Depletion of PknA or PknB by antisense RNA also decreases the growth rate of *M. smegmatis* and *M. bovis* BCG and changes cellular morphology (Kang et al., 2005). Collectively, these data suggest that both the relative abundance and the activity of these kinases are tightly regulated and critical to their function.

Subsequent genetic and biochemical analyses have confirmed the role of PknA and PknB in cell envelope synthesis, septation, and cellular division. PknA and PknB phosphorylate Wag31 (Rv2145c), a homolog of *Streptomyces coelicolor* DivIVA (Kang et al., 2005). This protein regulates cell division and cell shape (Flärdh, 2003; Kang et al., 2005). *wag31* is an essential gene (Sassetti et al., 2001; Kang et al., 2008) that is part of the conserved division of cell wall (*dcw*) cluster of genes (Pucci et al., 1997). This gene family also includes *ftsZ*, important for septum formation, and *mur* genes, encoding PG-biosynthetic enzymes (Cole et al., 1998; Kang et al., 2008). Overexpression of *wag31* results in short, fat cells, reminiscent of PknB overexpression (Nguyen et al., 2007; Kang et al., 2008), and a reduction of Wag31 causes *M. tuberculosis* cells to disproportionally bulge at one end of the cell and apparently fail to divide, forming diplococci (Kang et al., 2008). Wag31 localizes to the poles of elongating cells, where the large percentage of nascent PG synthesis occurs during active mycobacterial growth (Nguyen et al., 2007; Kang et al., 2008). This protein then relocalizes to mid-cell just before division and likely plays a significant role in appropriate PG biosynthesis at this location as well (Nguyen et al., 2007; Kang et al., 2008). Importantly, phosphorylation of Wag31 increases its oligomerization, enhances its polar localization, and results in increased PG synthesis (Nguyen et al., 2007; Jani et al., 2010). Furthermore, Wag31 interacts with the penicillin-binding protein PbpB (Pbp3, FtsI) (Mukherjee et al., 2009). Penicillin-binding proteins cross-link PG subunits during the final stages of PG biosynthesis (Pratt, 2008). Interestingly, PbpB is cleaved by a membrane-localized protease during conditions of oxidative stress and elevated temperatures (Jani et al., 2010). The interaction of Wag31 prevents this cleavage event, presumably to maintain PbpB activity under stress conditions (Mukherjee et al., 2009; Jani et al., 2010). It is not yet known if the phosphorylation of Wag31 affects its interaction with PbpB.

Interestingly, another penicillin-binding protein of *M. tuberculosis*, PbpA, is a substrate of PknB (Dasgupta et al., 2006). *pbpA* (*Rv0016c*) is found in the same operon as *pknA*, *pknB*, and *pstP*, and deletion of *pbpA* in *M. smegmatis* results in elongated cells, defective septum formation, and defective PG synthesis at the mid-cell location (Dasgupta et al., 2006). The phenotype can be complemented by expression of *M. tuberculosis pbpA* but is not complemented by expression of a mutant *pbpA* that cannot be phosphorylated by PknB at a key residue, T437 (Dasgupta et al., 2006). This observation emphasizes the importance of PbpA phosphorylation as essential to its function. Another PG synthesis protein, MurD, interacts with and is phosphorylated by PknA (Thakur and Chakraborti, 2008). However, the impact of this phosphorylation event is currently unknown. Finally, both PknA and PknB phosphorylate GlmU (Parikh et al., 2009). *glmU* is an essential gene involved in PG synthesis, producing the key PG precursor UDP-GlcNac (Zhang et al., 2008). GlmU contains two well-defined domains, an acetyltransferase domain that acetylates glucosamine-1-phosphate and a uridyltransferase domain that converts this intermediate to UDP-N-acetylglucosamine. Interestingly, phosphorylation of this dual-function enzyme inhibits the acetyltransferase domain without affecting the uridyltransferase activity (Parikh et al., 2009). Additionally, phosphorylation of the acetyltransferase domain occurs in a 30-amino-acid region that is found only in *M. tuberculosis*, suggesting that *M. tuberculosis* has evolved key regulatory elements within this protein (Parikh et al., 2009).

In addition to directly regulating the activity of various PG-biosynthetic enzymes, PknA and PknB also play a direct role in septation and cellular division. As previously mentioned, overexpression of PknA results in elongated cells likely due to a septation defect (Thakur and Chakraborti, 2006). In bacterial cells, the formation of a septum is dependent on initial formation of a Z-ring at the mid-cell. FtsZ is a bacterial homolog of eukaryotic tubulin and is necessary for Z-ring formation (Adams and Errington, 2009). Like tubulin, FtsZ displays GTPase activity that allows for polymerization and dynamic polymers (Adams and Errington, 2009). In *M. tuberculosis* and *M. smegmatis*, *ftsZ* is essential (Dziadek et al., 2002; Dziadek et al., 2003). Phosphorylation of FtsZ by PknA inhibits its GTPase activity, thus preventing polymerization and proper Z-ring formation (Thakur and Chakraborti, 2006), though others have found that GTPase activity of FtsZ is not affected by phosphorylation (Sureka et al., 2010). However, under conditions of oxidative stress, phosphorylation of FtsZ is necessary for interaction with FipA (Rv0019c, FhaA). Therefore, Z-ring formation does not occur without FipA under oxidative stress (Sureka et al., 2010). FipA is phosphorylated by both PknA and PknB (Gupta et al., 2009; Sureka et al., 2010). Another septum-forming associated protein, FtsQ, is also complexed with FtsZ and FipA, but only under conditions of stress (Sureka et al., 2010). These results suggest that interactions between FtsZ and FipA, which are dependent on phosphorylation, promote localization of additional proteins to the mid-cell location. These interactions are important for promoting appropriate cellular division during periods of oxidative stress (Sureka et al., 2010). Intriguingly, FtsZ also complexes with FtsW and PbpB, drawing further connections between STPK regulation of septum formation and cellular division (Datta et al., 2006). Furthermore, PknB also phosphorylates a protein of unknown function encoded by *Rv0020c*, the gene directly downstream of *fipA* (Grundner et al., 2005). Finally, FipA interacts with PapA5 (Gupta et al., 2009). PapA5 generates lipids important for *M. tuberculosis* virulence by catalyzing the final step of PDIM synthesis (Trivedi et al., 2005). PapA5 itself is also phosphorylated by PknB (Gupta et al., 2009), though the implications of this modification for in vivo growth are unknown at this time.

M. tuberculosis STPKs regulate synthesis of AG

STPKs also regulate the production of determinants that are exterior to the PG layer within the mycobacteria. In particular, these determinants regulate the synthesis of AG, an abundant moiety that is connected to the PG layer via lipids. Synthesis of AG is indirectly regulated by PknA and PknB via phosphorylation of the transcriptional regulator EmbR (Sharma et al., 2006). EmbR was originally described as a substrate of PknH, an STPK encoded in the same operon as *embR* (Molle et al., 2003). Additionally, EmbR is also phosphorylated by PknJ, suggesting that this determinant is subject to multiple levels of regulation (Jang et al., 2010). EmbR regulates expression of several genes, notably the *embCAB* operon. Proteins encoded by *embCAB* are involved in the synthesis of AG components and were identified, along with EmbR, in genetic screens for determinants mediating ethambutol resistance (Belanger et al., 1996). Phosphorylation of EmbR results in the upregulated expression of *embCAB* (Sharma et al., 2006). Therefore, signals transmitted through PknA, PknB, PknH, and PknJ all converge at EmbR, subsequently influencing AG production and resistance to ethambutol in *M. tuberculosis* (Sharma et al., 2006). Not surprisingly, PstP serves to fine-tune this response by dephosphorylating the STPKs or, alternatively, directly dephosphorylating EmbR.

Mycolic acid synthesis in *M. tuberculosis* is also regulated by STPKs

In addition to regulating aspects of cell wall biosynthesis and cellular division, PknA, PknB, and other *M. tuberculosis* STPKs have also been shown to regulate mycolic acid synthesis (Fig. 4). Production of these determinants involves two systems, the eukaryotic-like fatty acid synthase type I (FAS-I), responsible for de novo lipid synthesis, and the FAS-II system (Dover et al., 2007; Kaur et al., 2009). The initial condensation reaction leading to mycolic acid synthesis is carried out by FabH, an important enzyme that bridges both FAS-I and the FAS-II system. FabH utilizes acyl coenzyme A (acyl-CoA) primers generated by FAS-I and malonyl-AcpM generated by another enzyme, FabD, to generate β-ketoacyl-AcpM products. These products are then fed into FAS-II and elongated through a series of sequential and cyclic enzymatic reactions. Core enzymes involved in mycolic acid synthesis include KasA/KasB, MabA, HadAB/HadBC, and InhA (Dover et al., 2007; Sacco et al., 2007). The contribution of these enzymes to mycolic acid production is described below. Additional modifications to maturing mycolic acids are also mediated by Pks13 (Portevin et al., 2004). Interestingly, all of the mycolic acid-biosynthetic enzymes are phosphorylated by *M. tuberculosis* STPKs. FabD and FabH are phosphorylated by several STPKs, including PknA and PknB (Molle et al., 2006; Veyron-Churlet et al., 2009). Phosphorylation of FabH is inhibitory and may fine-tune the activity of this enzyme (Veyron-Churlet et al., 2009).

A key step in mycolic acid synthesis is catalyzed by MabA, which converts β-ketoacyl-AcpM molecules to β-hydroxyacyl-AcpM products. MabA is phosphorylated by PknA and PknB as well as PknD, PknE, and PknL (Veyron-Churlet et al., 2010). One of the phosphorylation sites on MabA lies close to the substrate/cofactor binding site (Veyron-Churlet et al., 2010). Phosphorylation of this site inhibits MabA activity (Veyron-Churlet et al., 2010). The enzymatic reaction following MabA converts β-hydroxyacyl-AcpM to trans-2-enoyl-AcpM. This step is carried out by the hydratase complexes, HadAB and HadBC. These enzymes are also phosphorylated by all STPKs tested, and phosphorylation decreases the enzymatic activity of these enzymes (Slama et al., 2011). The next core FAS-II enzyme, InhA, converts trans-2-enoyl-AcpM to acyl-CoA products. InhA is phosphorylated by PknA and PknB, as well as PknE, PknH, and PknL (Khan et al., 2010; Molle et al., 2010). Phosphorylation of InhA lowers the affinity of this enzyme for NADH, a cofactor necessary for its enzymatic activity. Therefore, phosphorylation inhibits the enzyme and decreases mycobacterial growth (Khan et al., 2010; Molle et al., 2010). Interestingly, overexpression of an InhA allele containing a phosphomimetic amino acid substitution at the site of phosphorylation results in lethality when expressed in M. smegmatis due to a nearly complete abrogation of mycolic acid synthesis (Molle et al., 2010). Subsequent elongation of these fatty acids is facilitated by condensation reactions mediated by KasA and KasB, which convert acyl-CoA chains and malonyl-AcpM (generated by FabD) into further β-ketoacyl-AcpM products. Both KasA and KasB are phosphorylated by PknA and PknB, as well as PknD, PknE, PknF, PknH, and PknL (Molle et al., 2006). Interestingly, although KasA and KasB carry out similar yet distinct enzymatic reactions, phosphorylation of KasA is inhibitory, while phosphorylation of KasB increases its catalytic activity (Molle et al., 2006). In addition, KasB-generated products are longer than those generated by KasA. Therefore, increasing activity of KasB has been hypothesized to increase the longer, more mature mycolic acids (Molle et al., 2006). Finally, full-length mycolic acid precursors generated by the FAS-II elongation cycle, referred to as meromycolic acids, are condensed further with additional products of FAS-I to produce mature mycolic acids. This final reaction is catalyzed by Pks13, and like all of the other enzymes discussed, Pks13 is phosphorylated by STPKs (Veyron-Churlet et al., 2009).

However, the effect of this modification on enzymatic activity has not yet been established.

Posttranslation modification of anti-sigma factors by STPKs

There is mounting evidence that PknB, and perhaps other STPKs, may also play an important regulatory role in the response of M. tuberculosis to environmental stress by directly regulating the activity of anti-sigma factors. As discussed previously, ECF sigma factors are often regulated posttranslationally by binding to another protein, the anti-sigma factor. Interaction of a sigma factor with its anti-sigma counterpart inhibits the ability to mediate transcriptional regulation. It was recently demonstrated that PknB, as well as PknE, phosphorylates Rv3221A (RshA), the anti-sigma factor of SigH (Song et al., 2003; Greenstein et al., 2007; Park et al., 2008). Phosphorylation of RshA by PnkB decreases the affinity of this anti-sigma factor for SigH. The ECF sigma factor SigH is itself also phosphorylated by PknB; however, the impact of this event is unknown (Park et al., 2008). Similarly, PknB-dependent phosphorylation of RseA occurs following exposure of M. tuberculosis to cell envelope damage induced by vancomycin or SDS (Barik et al., 2010). Phosphorylation of RseA by PknB marks the anti-sigma factor for degradation by ClpC1P2, allowing for SigE-dependent gene regulation (Barik et al., 2010). How PknB distinguishes between different environmental signals and between specific anti-sigma factors is unknown.

Integration of the Cell Envelope Damage Response in M. tuberculosis

A large body of work has helped to shape the current model of how M. tuberculosis senses cell envelope damage, including the regulatory mechanisms by which the bacterium responds to these stimuli (Fig. 3). Conditions that directly (e.g., antibiotics and SDS) or indirectly (oxidative stress and cellular division) disrupt components of the cell envelope may lead to the release of PG fragments, allowing for dimerization and activation of PknB. This STPK phosphorylates numerous protein substrates within the tubercle bacillus, including PknA and other STPKs and proteins implicated in aspects of cell wall synthesis and cellular division, as well as the anti-sigma factors of SigE and SigH. The ensuing release of SigE and SigH following anti-sigma factor phosphorylation allows for the association of these determinants with RNA polymerase, where it is able to direct specific gene transcription. Meanwhile, cell envelope stress also activates the TCSS MprAB via an as-yet-undefined mechanism, possibly the accumulation of unfolded or misfolded proteins in the extracytoplasmic space. Detection of the signal by the MprB sensor kinase is likely mediated via the extracytoplasmic domain of this protein, leading to its autophosphorylation and subsequent relay of this phosphate to response regulator MprA. Phosphorylated MprA upregulates expression of several genes, including *sigE*, increasing the pool of free SigE, which further perpetuates the response by upregulating *mprA* in a positive-feedback manner. MprA and SigE also upregulate stress-induced genes, including *pepD* and *acr2*, which seem to play various roles in the stress response. Additionally, SigE promotes enhanced expression of the transcriptional regulator *clgR*. This determinant, in turn, upregulates several proteases and chaperones, including the Clp proteases responsible for degradation of RseA. ClgR also upregulates the enzyme Ppk1, which catalyzes the formation of polyphosphate molecules that can be used by MprB as an alternative phosphate source for phosphorelay to MprA (Sureka et al., 2007) and even greater potential MprA signaling.

CONCLUDING REMARKS

The cell envelope damage response of *M. tuberculosis* is complex, and undoubtedly significant gaps in our knowledge of this network remain. The genome of this highly successful pathogen contains at least 150 regulatory proteins, and the contribution of these proteins in the cell envelope damage response is only now beginning to be understood. Based on studies from a variety of laboratories, it is becoming increasingly clear that the genetic response to cell envelope stress involves a large and functionally diverse group of determinants whose expression is regulated by a small number of regulatory factors. Interestingly, it appears that overlap exists among determinants comprising these networks, and that a core set of stress responsive elements whose expression is required to counteract cell envelope stress in *M. tuberculosis* may exist (Table 2). Additional research into the regulatory determinants and downstream targets will be critical to better understanding the unusual life cycle of *M. tuberculosis* and, importantly, aid in development of novel antimycobacterial drugs and vaccines.

Acknowledgments. We thank current and previous members of the laboratory for useful discussions and work pertaining to the MprAB system from *M. tuberculosis*. We apologize to those laboratories that have made contributions to our understanding of stress response pathways in *M. tuberculosis* which we have not been able to acknowledge here. Work in the laboratory is supported by RO1 AI51669 from the National Institute of Allergy and Infectious Diseases (NIH) and an Investigator in the Pathogenesis of Infectious Disease Award from the Burroughs Wellcome Fund.

REFERENCES

Adams, D. W., and J. Errington. 2009. Bacterial cell division: assembly, maintenance and disassembly of the Z ring. *Nat. Rev. Microbiol.* **7:**642–653.

Alba, B. M., H. J. Zhong, J. C. Pelayo, and C. A. Gross. 2001. *degS* (*hhoB*) is an essential *Escherichia coli* gene whose indispensable function is to provide sigma (E) activity. *Mol. Microbiol.* **40:**1323–1333.

Alexander, K. A., P. N. Laver, A. L. Michel, M. Williams, P. D. van Helden, R. M. Warren, and N. C. Gey van Pittius. 2010. Novel *Mycobacterium tuberculosis* complex pathogen, *M. mungi. Emerg. Infect. Dis.* **16:**1296–1299.

Alonso, S., K. Pethe, D. G. Russell, and G. E. Purdy. 2007. Lysosomal killing of *Mycobacterium* mediated by ubiquitin-derived peptides is enhanced by autophagy. *Proc. Natl. Acad. Sci. USA* **104:**6031–6036.

Ando, M., T. Yoshimatsu, C. Ko, P. J. Converse, and W. R. Bishai. 2003. Deletion of *Mycobacterium tuberculosis* sigma factor E results in delayed time to death with bacterial persistence in the lungs of aerosol-infected mice. *Infect. Immun.* **71:**7170–7172.

Armstrong, J. A., and P. D. Hart. 1971. Response of cultured macrophages to *Mycobacterium tuberculosis*, with observations on fusion of lysosomes with phagosomes. *J. Exp. Med.* **134:**713–740.

Ashitani, J., H. Mukae, T. Hiratsuka, M. Nakazato, K. Kumamoto, and S. Matsukura. 2002. Elevated levels of alpha-defensins in plasma and BAL fluid of patients with active pulmonary tuberculosis. *Chest* **121:**519–526.

Av-Gay, Y., S. Jamil, and S. J. Drews. 1999. Expression and characterization of the *Mycobacterium tuberculosis* serine/threonine protein kinase PknB. *Infect. Immun.* **67:**5676–5682.

Bach, H., D. Wong, and Y. Av-Gay. 2009. *Mycobacterium tuberculosis* PtkA is a novel protein tyrosine kinase whose substrate is PtpA. *Biochem. J.* **420:**155–160.

Barik, S., K. Sureka, P. Mukherjee, J. Basu, and M. Kundu. 2010. RseA, the SigE specific anti-sigma factor of *Mycobacterium tuberculosis*, is inactivated by phosphorylation-dependent ClpC1P2 proteolysis. *Mol. Microbiol.* **75:**592–606.

Barry, C. E., III. 2001. Interpreting cell wall 'virulence factors' of *Mycobacterium tuberculosis. Trends Microbiol.* **9:**237–241.

Barthe, P., G. V. Mukamolova, C. Roumestand, and M. Cohen-Gonsaud. 2010. The structure of PknB extracellular PASTA domain from *Mycobacterium tuberculosis* suggests a ligand-dependent kinase activation. *Structure* **18:**606–615.

Beggs, M. L., M. D. Cave, and K. D. Eisenach. 1996. Isolation and sequence of a *Mycobacterium tuberculosis* sigma factor. *Gene* **174:**285–287.

Belanger, A. E., G. S. Besra, M. E. Ford, K. Mikusova, J. T. Belisle, P. J. Brennan, and J. M. Inamine. 1996. The *embAB* genes of *Mycobacterium avium* encode an arabinosyl transferase involved in cell wall arabinan biosynthesis that is the target for the antimycobacterial drug ethambutol. *Proc. Natl. Acad. Sci. USA* **93:**11919–11924.

Betts, J. C., P. T. Lukey, L. C. Robb, R. A. McAdam, and K. Duncan. 2002. Evaluation of a nutrient starvation model of *Mycobacterium tuberculosis* persistence by gene and protein expression profiling. *Mol. Microbiol.* **43:**717–731.

Boitel, B., M. Ortiz-Lombardia, R. Duran, F. Pompeo, S. T. Cole, C. Cervenansky, and P. M. Alzari. 2003. PknB kinase activity is regulated by phosphorylation in two Thr residues and dephosphorylation by PstP, the cognate phospho-Ser/Thr phosphatase, in *Mycobacterium tuberculosis. Mol. Microbiol.* **49:**1493–1508.

Brennan, P. J., and H. Nikaido. 1995. The envelope of mycobacteria. *Annu. Rev. Biochem.* **64:**29–63.

Brosch, R., S. V. Gordon, M. Marmiesse, P. Brodin, C. Buchrieser, K. Eiglmeier, T. Garnier, C. Gutierrez, G. Hewinson, K. Kremer, L. M. Parsons, A. S. Pym, S. Samper, D. van Soolingen, and S. T. Cole. 2002. A new evolutionary scenario for the *Mycobacterium tuberculosis* complex. *Proc. Natl. Acad. Sci. USA* **99:**3684–3689.

Buelow, D. R., and T. L. Raivio. 2005. Cpx signal transduction is influenced by a conserved N-terminal domain in the novel inhibitor CpxP and the periplasmic protease DegP. *J. Bacteriol.* **187:**6622–6630.

Canova, M. J., R. Veyron-Churlet, I. Zanella-Cleon, M. Cohen-Gonsaud, A. J. Cozzone, M. Becchi, L. Kremer, and V. Molle. 2008. The *Mycobacterium tuberculosis* serine/threonine kinase PknL phosphorylates Rv2175c: mass spectrometric profiling of the activation loop phosphorylation sites and their role in the recruitment of Rv2175c. *Proteomics* **8:**521–533.

Chaba, R., M. Raje, and P. K. Chakraborti. 2002. Evidence that a eukaryotic-type serine/threonine protein kinase from *Mycobacterium tuberculosis* regulates morphological changes associated with cell division. *Eur. J. Biochem.* **269:**1078–1085.

Chao, J. D., K. G. Papavinasasundaram, X. Zheng, A. Chavez-Steenbock, X. Wang, G. Q. Lee, and Y. Av-Gay. 2010. Convergence of Ser/Thr and two-component signaling to coordinate expression of the dormancy regulon in *Mycobacterium tuberculosis. J. Biol. Chem.* **285:**29239–29246.

Chen, P., R. E. Ruiz, Q. Li, R. F. Silver, and W. R. Bishai. 2000. Construction and characterization of a *Mycobacterium tuberculosis* mutant lacking the alternate sigma factor gene, *sigF. Infect. Immun.* **68:**5575–5580.

Chopra, P., B. Singh, R. Singh, R. Vohra, A. Koul, L. S. Meena, H. Koduri, M. Ghildiyal, P. Deol, T. K. Das, A. K. Tyagi, and Y. Singh. 2003. Phosphoprotein phosphatase of *Mycobacterium tuberculosis* dephosphorylates serine-threonine kinases PknA and PknB. *Biochem. Biophys. Res. Commun.* **311:**112–120.

Cole, S. T., R. Brosch, J. Parkhill, T. Garnier, C. Churcher, D. Harris, S. V. Gordon, K. Eiglmeier, S. Gas, C. E. Barry III, F. Tekaia, K. Badcock, D. Basham, D. Brown, T. Chillingworth, R. Connor, R. Davies, K. Devlin, T. Feltwell, S. Gentles, N. Hamlin, S. Holroyd, T. Hornsby, K. Jagels, A. Krogh, J. McLean, S. Moule, L. Murphy, K. Oliver, J. Osborne, M. A. Quail, M. A. Rajandream, J. Rogers, S. Rutter, K. Seeger, J. Skelton, R. Squares, S. Squares, J. E. Sulston, K. Taylor, S. Whitehead, and B. G. Barrell. 1998. Deciphering the biology of *Mycobacterium tuberculosis* from the complete genome sequence. *Nature* **393:**537–544.

Converse, P. J., P. C. Karakousis, L. G. Klinkenberg, A. K. Kesavan, L. H. Ly, S. S. Allen, J. H. Grosset, S. K. Jain, G. Lamichhane, Y. C. Manabe, D. N. McMurray, E. L. Nuermberger, and W. R. Bishai. 2009. Role of the *dosR-dosS* two-component regulatory system in *Mycobacterium tuberculosis* virulence in three animal models. *Infect. Immun.* **77:**1230–1237.

Cousins, D. V., R. Bastida, A. Cataldi, V. Quse, S. Redrobe, S. Dow, P. Duignan, A. Murray, C. Dupont, N. Ahmed, D. M. Collins, W. R. Butler, D. Dawson, D. Rodriguez, J. Loureiro, M. I. Romano, A. Alito, M. Zumarraga, and A. Bernardelli. 2003. Tuberculosis in seals caused by a novel member of the *Mycobacterium tuberculosis* complex: *Mycobacterium pinnipedii* sp. nov. *Int. J. Syst. Evol. Microbiol.* **53:**1305–1314.

Cowley, S., M. Ko, N. Pick, R. Chow, K. J. Downing, B. G. Gordhan, J. C. Betts, V. Mizrahi, D. A. Smith, R. W. Stokes, and Y. Av-Gay. 2004. The *Mycobacterium tuberculosis* protein serine/threonine kinase PknG is linked to cellular glutamate/glutamine levels and is important for growth in vivo. *Mol. Microbiol.* **52:**1691–1702.

Cywes, C., H. C. Hoppe, M. Daffe, and M. R. Ehlers. 1997. Nonopsonic binding of *Mycobacterium tuberculosis* to complement receptor type 3 is mediated by capsular polysaccharides and is strain dependent. *Infect. Immun.* **65:**4258–4266.

Daffe, M., and G. Etienne. 1999. The capsule of *Mycobacterium tuberculosis* and its implications for pathogenicity. *Tuber. Lung Dis.* **79:**153–169.

Dainese, E., S. Rodrigue, G. Delogu, R. Provvedi, L. Laflamme, R. Brzezinski, G. Fadda, I. Smith, L. Gaudreau, G. Palu, and R. Manganelli. 2006. Posttranslational regulation of *Mycobacterium tuberculosis* extracytoplasmic-function sigma factor sigma L and roles in virulence and in global regulation of gene expression. *Infect. Immun.* **74:**2457–2461.

Dasgupta, A., P. Datta, M. Kundu, and J. Basu. 2006. The serine/threonine kinase PknB of *Mycobacterium tuberculosis* phosphorylates PBPA, a penicillin-binding protein required for cell division. *Microbiology* **152:**493–504.

Datta, P., A. Dasgupta, A. K. Singh, P. Mukherjee, M. Kundu, and J. Basu. 2006. Interaction between FtsW and penicillin-binding protein 3 (PBP3) directs PBP3 to mid-cell, controls cell septation and mediates the formation of a trimeric complex involving FtsZ, FtsW and PBP3 in mycobacteria. *Mol. Microbiol.* **62:**1655–1673.

Davis, J. M., and L. Ramakrishnan. 2009. The role of the granuloma in expansion and dissemination of early tuberculous infection. *Cell* **136:**37–49.

Desjardins, M., J. E. Celis, G. van Meer, H. Dieplinger, A. Jahraus, G. Griffiths, and L. A. Huber. 1994. Molecular characterization of phagosomes. *J. Biol. Chem.* **269:**32194–32200.

Diaz, G. A., and L. G. Wayne. 1974. Isolation and characterization of catalase produced by *Mycobacterium tuberculosis. Am. Rev. Respir. Dis.* **110:**312–319.

Dona, V., S. Rodrigue, E. Dainese, G. Palu, L. Gaudreau, R. Manganelli, and R. Provvedi. 2008. Evidence of complex transcriptional, translational, and posttranslational regulation of the extracytoplasmic function sigma factor σ^E in *Mycobacterium tuberculosis. J. Bacteriol.* **190:**5963–5971.

Doukhan, L., M. Predich, G. Nair, O. Dussurget, I. Mandic-Mulec, S. T. Cole, D. R. Smith, and I. Smith. 1995. Genomic organization of the mycobacterial sigma gene cluster. *Gene* **165:**67–70.

Dover, L. G., L. J. Alderwick, A. K. Brown, K. Futterer, and G. S. Besra. 2007. Regulation of cell wall synthesis and growth. *Curr. Mol. Med.* **7:**247–276.

Downing, J. F., R. Pasula, J. R. Wright, H. L. Twigg III, and W. J. Martin II. 1995. Surfactant protein A promotes attachment of *Mycobacterium tuberculosis* to alveolar macrophages during infection with human immunodeficiency virus. *Proc. Natl. Acad. Sci. USA* **92:**4848–4852.

Duran, R., A. Villarino, M. Bellinzoni, A. Wehenkel, P. Fernandez, B. Boitel, S. T. Cole, P. M. Alzari, and C. Cervenansky. 2005. Conserved autophosphorylation pattern in activation loops and juxtamembrane regions of *Mycobacterium tuberculosis* Ser/Thr protein kinases. *Biochem. Biophys. Res. Commun.* **333:**858–867.

Dziadek, J., M. V. Madiraju, S. A. Rutherford, M. A. Atkinson, and M. Rajagopalan. 2002. Physiological consequences associated with overproduction of *Mycobacterium tuberculosis* FtsZ in mycobacterial hosts. *Microbiology* **148:**961–971.

Dziadek, J., S. A. Rutherford, M. V. Madiraju, M. A. Atkinson, and M. Rajagopalan. 2003. Conditional expression of *Mycobacterium smegmatis ftsZ*, an essential cell division gene. *Microbiology* **149:**1593–1603.

Estorninho, M., H. Smith, J. Thole, J. Harders-Westerveen, A. Kierzek, R. E. Butler, O. Neyrolles, and G. R. Stewart. 2010. ClgR regulation of chaperone and protease systems is essential for *Mycobacterium tuberculosis* parasitism of the macrophage. *Microbiology* **156:**3445–3455.

Farn, J., and M. Roberts. 2004. Effect of inactivation of the HtrA-like serine protease DegQ on the virulence of *Salmonella enterica* serovar Typhimurium in mice. *Infect. Immun.* **72:**7357–7359.

Ferguson, J. S., D. R. Voelker, J. A. Ufnar, A. J. Dawson, and L. S. Schlesinger. 2002. Surfactant protein D inhibition of human macrophage uptake of *Mycobacterium tuberculosis* is independent of bacterial agglutination. *J. Immunol.* **168:**1309–1314.

Fernandez, P., B. Saint-Joanis, N. Barilone, M. Jackson, B. Gicquel, S. T. Cole, and P. M. Alzari. 2006. The Ser/Thr protein kinase PknB is essential for sustaining mycobacterial growth. *J. Bacteriol.* **188:**7778–7784.

Flannagan, R. S., D. Aubert, C. Kooi, P. A. Sokol, and M. A. Valvano. 2007. *Burkholderia cenocepacia* requires a periplasmic HtrA protease for growth under thermal and osmotic stress and for survival in vivo. *Infect. Immun.* **75:**1679–1689.

Flärdh, K. 2003. Essential role of DivIVA in polar growth and morphogenesis in *Streptomyces coelicolor* A3(2). *Mol. Microbiol.* **49:**1523–1536.

Fontan, P. A., V. Aris, M. E. Alvarez, S. Ghanny, J. Cheng, P. Soteropoulos, A. Trevani, R. Pine, and I. Smith. 2008. *Mycobacterium tuberculosis* sigma factor E regulon modulates the host inflammatory response. *J. Infect. Dis.* **198:**877–885.

Fontan, P. A., M. I. Voskuil, M. Gomez, D. Tan, M. Pardini, R. Manganelli, L. Fattorini, G. K. Schoolnik, and I. Smith. 2009. The *Mycobacterium tuberculosis* sigma factor σB is required for full response to cell envelope stress and hypoxia in vitro, but it is dispensable for in vivo growth. *J. Bacteriol.* **191:**5628–5633.

Gagliardi, M. C., A. Lemassu, R. Teloni, S. Mariotti, V. Sargentini, M. Pardini, M. Daffe, and R. Nisini. 2007. Cell wall-associated alpha-glucan is instrumental for *Mycobacterium tuberculosis* to block CD1 molecule expression and disable the function of dendritic cell derived from infected monocyte. *Cell. Microbiol.* **9:**2081–2092.

Geiman, D. E., D. Kaushal, C. Ko, S. Tyagi, Y. C. Manabe, B. G. Schroeder, R. D. Fleischmann, N. E. Morrison, P. J. Converse, P. Chen, and W. R. Bishai. 2004. Attenuation of late-stage disease in mice infected by the *Mycobacterium tuberculosis* mutant lacking the SigF alternate sigma factor and identification of SigF-dependent genes by microarray analysis. *Infect. Immun.* **72:**1733–1745.

Giacomini, E., A. Sotolongo, E. Iona, M. Severa, M. E. Remoli, V. Gafa, R. Lande, L. Fattorini, I. Smith, R. Manganelli, and E. M. Coccia. 2006. Infection of human dendritic cells with a *Mycobacterium tuberculosis* sigE mutant stimulates production of high levels of interleukin-10 but low levels of CXCL10: impact on the T-cell response. *Infect. Immun.* **74:**3296–3304.

Glover, R. T., J. Kriakov, S. J. Garforth, A. D. Baughn, and W. R. Jacobs, Jr. 2007. The two-component regulatory system senX3-regX3 regulates phosphate-dependent gene expression in *Mycobacterium smegmatis*. *J. Bacteriol.* **189:**5495–5503.

Gomez, M., L. Doukhan, G. Nair, and I. Smith. 1998. *sigA* is an essential gene in *Mycobacterium smegmatis*. *Mol. Microbiol.* **29:**617–628.

Gonzalo Asensio, J., C. Maia, N. L. Ferrer, N. Barilone, F. Laval, C. Y. Soto, N. Winter, M. Daffe, B. Gicquel, C. Martin, and M. Jackson. 2006. The virulence-associated two-component PhoP-PhoR system controls the biosynthesis of polyketide-derived lipids in *Mycobacterium tuberculosis*. *J. Biol. Chem.* **281:**1313–1316.

Graham, J. E., and J. E. Clark-Curtiss. 1999. Identification of *Mycobacterium tuberculosis* RNAs synthesized in response to phagocytosis by human macrophages by selective capture of transcribed sequences (SCOTS). *Proc. Natl. Acad. Sci. USA* **96:**11554–11559.

Greenstein, A. E., J. A. MacGurn, C. E. Baer, A. M. Falick, J. S. Cox, and T. Alber. 2007. *M. tuberculosis* Ser/Thr protein kinase D phosphorylates an anti-anti-sigma factor homolog. *PLoS Pathog.* **3:**e49.

Gruber, T. M., and C. A. Gross. 2003. Multiple sigma subunits and the partitioning of bacterial transcription space. *Annu. Rev. Microbiol.* **57:**441–466.

Grundner, C., L. M. Gay, and T. Alber. 2005. *Mycobacterium tuberculosis* serine/threonine kinases PknB, PknD, PknE, and PknF phosphorylate multiple FHA domains. *Protein Sci.* **14:**1918–1921.

Gupta, M., A. Sajid, G. Arora, V. Tandon, and Y. Singh. 2009. Forkhead-associated domain-containing protein Rv0019c and polyketide-associated protein PapA5, from substrates of serine/threonine protein kinase PknB to interacting proteins of *Mycobacterium tuberculosis*. *J. Biol. Chem.* **284:**34723–34734.

Hahn, M. Y., S. Raman, M. Anaya, and R. N. Husson. 2005. The *Mycobacterium tuberculosis* extracytoplasmic-function sigma factor SigL regulates polyketide synthases and secreted or membrane proteins, and is required for virulence. *J. Bacteriol.* **187:**7062–7071.

Hall-Stoodley, L., G. Watts, J. E. Crowther, A. Balagopal, J. B. Torrelles, J. Robison-Cox, R. F. Bargatze, A. G. Harmsen, E. C. Crouch, and L. S. Schlesinger. 2006. *Mycobacterium tuberculosis* binding to human surfactant proteins A and D, fibronectin, and small airway epithelial cells under shear conditions. *Infect. Immun.* **74:**3587–3596.

Haydel, S. E., and J. E. Clark-Curtiss. 2004. Global expression analysis of two-component system regulator genes during *Mycobacterium tuberculosis* growth in human macrophages. *FEMS Microbiol. Lett.* **236:**341–347.

He, H., R. Hovey, J. Kane, V. Singh, and T. C. Zahrt. 2006. MprAB is a stress-responsive two-component system that directly regulates expression of sigma factors SigB and SigE in *Mycobacterium tuberculosis*. *J. Bacteriol.* **188:**2134–2143.

He, H., and T. C. Zahrt. 2005. Identification and characterization of a regulatory sequence recognized by *Mycobacterium tuberculosis* persistence regulator MprA. *J. Bacteriol.* **187:**202–212.

Henriques, A. O., P. Glaser, P. J. Piggot, and C. P. Moran, Jr. 1998. Control of cell shape and elongation by the *rodA* gene in *Bacillus subtilis*. *Mol. Microbiol.* **28:**235–247.

Hernandez Pando, R., L. D. Aguilar, I. Smith, and R. Manganelli. 2010. Immunogenicity and protection induced by a *Mycobacterium tuberculosis* sigE mutant in a BALB/c mouse model of progressive pulmonary tuberculosis. *Infect. Immun.* **78:**3168–3176.

Hoffmann, C., A. Leis, M. Niederweis, J. M. Plitzko, and H. Engelhardt. 2008. Disclosure of the mycobacterial outer membrane: cryo-electron tomography and vitreous sections reveal the lipid bilayer structure. *Proc. Natl. Acad. Sci. USA* **105:**3963–3967.

Hu, Y., and A. R. Coates. 1999. Transcription of two sigma 70 homologue genes, *sigA* and *sigB*, in stationary-phase *Mycobacterium tuberculosis*. *J. Bacteriol.* **181:**469–476.

Jang, J., A. Stella, F. Boudou, F. Levillain, E. Darthuy, J. Vaubourgeix, C. Wang, F. Bardou, G. Puzo, M. Gilleron, O. Burlet-Schiltz, B. Monsarrat, P. Brodin, B. Gicquel, and O. Neyrolles. 2010. Functional characterization of the *Mycobacterium tuberculosis* serine/threonine kinase PknJ. *Microbiology* **156:**1619–1631.

Jani, C., H. Eoh, J. J. Lee, K. Hamasha, M. B. Sahana, J. S. Han, S. Nyayapathy, J. Y. Lee, J. W. Suh, S. H. Lee, S. J. Rehse, D. C. Crick, and C. M. Kang. 2010. Regulation of polar peptidoglycan biosynthesis by Wag31 phosphorylation in mycobacteria. *BMC Microbiol.* **10:**327.

Jensen-Cain, D. M., and F. D. Quinn. 2001. Differential expression of *sigE* by *Mycobacterium tuberculosis* during intracellular growth. *Microb. Pathog.* **30:**271–278.

Joly, N., C. Engl, G. Jovanovic, M. Huvet, T. Toni, X. Sheng, M. P. Stumpf, and M. Buck. 2010. Managing membrane stress: the phage shock protein (Psp) response, from molecular mechanisms to physiology. *FEMS Microbiol. Rev.* 34:797–827.

Jones, G., and P. Dyson. 2006. Evolution of transmembrane protein kinases implicated in coordinating remodeling of gram-positive peptidoglycan: inside versus outside. *J. Bacteriol.*188:7470–7476.

Kang, C. M., D. W. Abbott, S. T. Park, C. C. Dascher, L. C. Cantley, and R. N. Husson. 2005. The *Mycobacterium tuberculosis* serine/threonine kinases PknA and PknB: substrate identification and regulation of cell shape. *Genes Dev.* 19:1692–1704.

Kang, C. M., S. Nyayapathy, J. Y. Lee, J. W. Suh, and R. N. Husson. 2008. Wag31, a homologue of the cell division protein DivIVA, regulates growth, morphology and polar cell wall synthesis in mycobacteria. *Microbiology* 154:725–735.

Karakousis, P. C., T. Yoshimatsu, G. Lamichhane, S. C. Woolwine, E. L. Nuermberger, J. Grosset, and W. R. Bishai. 2004. Dormancy phenotype displayed by extracellular *Mycobacterium tuberculosis* within artificial granulomas in mice. *J. Exp. Med.* 200:647–657.

Kaur, D., M. E. Guerin, H. Skovierova, P. J. Brennan, and M. Jackson. 2009. Chapter 2: biogenesis of the cell wall and other glycoconjugates of *Mycobacterium tuberculosis. Adv. Appl. Microbiol.* 69:23–78.

Keep, N. H., J. M. Ward, M. Cohen-Gonsaud, and B. Henderson. 2006. Wake up! Peptidoglycan lysis and bacterial non-growth states. *Trends Microbiol.* 14:271–276.

Khan, S., S. N. Nagarajan, A. Parikh, S. Samantaray, A. Singh, D. Kumar, R. P. Roy, A. Bhatt, and V. K. Nandicoori. 2010. Phosphorylation of enoyl-acyl carrier protein reductase InhA impacts mycobacterial growth and survival.*J. Biol. Chem.* 285:37860–37871.

Kim, D. Y., and K. K. Kim. 2005. Structure and function of HtrA family proteins, the key players in protein quality control. *J. Biochem. Mol. Biol.* 38:266–274.

Koch, R. 1884. 2 Die Aetiologie der Tuberkulose. *Mitt Kaiser Gesundh.* 1884:1–88.

Koul, A., A. Choidas, A. K. Tyagi, K. Drlica, Y. Singh, and A. Ullrich. 2001. Serine/threonine protein kinases PknF and PknG of *Mycobacterium tuberculosis*: characterization and localization. *Microbiology* 147:2307–2314.

Kumar, A., J. S. Deshane, D. K. Crossman, S. Bolisetty, B. S. Yan, I. Kramnik, A. Agarwal, and A. J. Steyn. 2008. Heme oxygenase-1-derived carbon monoxide induces the *Mycobacterium tuberculosis* dormancy regulon. *J. Biol. Chem.* 283:18032–18039.

Kumar, P., D. Kumar, A. Parikh, D. Rananaware, M. Gupta, Y. Singh, and V. K. Nandicoori. 2009. The *Mycobacterium tuberculosis* protein kinase K modulates activation of transcription from the promoter of mycobacterial monooxygenase operon through phosphorylation of the transcriptional regulator VirS. *J. Biol. Chem.* 284:11090–11099.

Kusunose, E., K. Ichihara, Y. Noda, and M. Kusunose. 1976. Superoxide dismutase from *Mycobacterium tuberculosis. J. Biochem.* 80:1343–1352.

Lee, J. H., P. C. Karakousis, and W. R. Bishai. 2008. Roles of SigB and SigF in the *Mycobacterium tuberculosis* sigma factor network. *J. Bacteriol.* 190:699–707.

Lehrer, R. I., A. K. Lichtenstein, and T. Ganz. 1993. Defensins: antimicrobial and cytotoxic peptides of mammalian cells. *Annu. Rev. Immunol.* 11:105–128.

Leyten, E. M., M. Y. Lin, K. L. Franken, A. H. Friggen, C. Prins, K. E. van Meijgaarden, M. I. Voskuil, K. Weldingh, P. Andersen, G. K. Schoolnik, S. M. Arend, T. H. Ottenhoff, and M. R. Klein. 2006. Human T-cell responses to 25 novel antigens encoded by genes of the dormancy regulon of *Mycobacterium tuberculosis. Microbes Infect.* 8:2052–2060.

Liu, P. T., S. Stenger, H. Li, L. Wenzel, B. H. Tan, S. R. Krutzik, M. T. Ochoa, J. Schauber, K. Wu, C. Meinken, D. L. Kamen, M. Wagner, R. Bals, A. Steinmeyer, U. Zugel, R. L. Gallo, D.

Eisenberg, M. Hewison, B. W. Hollis, J. S. Adams, B. R. Bloom, and R. L. Modlin. 2006. Toll-like receptor triggering of a vitamin D-mediated human antimicrobial response. *Science* 311:1770–1773.

Lombana, T. N., N. Echols, M. C. Good, N. D. Thomsen, H. L. Ng, A. E. Greenstein, A. M. Falick, D. S. King, and T. Alber. 2010. Allosteric activation mechanism of the *Mycobacterium tuberculosis* receptor Ser/Thr protein kinase, PknB. *Structure* 18:1667–1677.

Luetkemeyer, A. F., D. V. Havlir, and J. S. Currier. 2010. Complications of HIV disease and antiretroviral treatment. *Top. HIV Med.* 18:57–65.

MacMicking, J., Q. W. Xie, and C. Nathan. 1997. Nitric oxide and macrophage function. *Annu. Rev. Immunol.* 15:323–350.

Manganelli, R., E. Dubnau, S. Tyagi, F. R. Kramer, and I. Smith. 1999. Differential expression of 10 sigma factor genes in *Mycobacterium tuberculosis. Mol. Microbiol.* 31:715–724.

Manganelli, R., L. Fattorini, D. Tan, E. Iona, G. Orefici, G. Altavilla, P. Cusatelli, and I. Smith. 2004. The extra cytoplasmic function sigma factor σ^E is essential for *Mycobacterium tuberculosis* virulence in mice. *Infect. Immun.* 72:3038–3041.

Manganelli, R., and R. Provvedi. 2010. An integrated regulatory network including two positive feedback loops to modulate the activity of σ^E in mycobacteria. *Mol. Microbiol.* 75:538–542.

Manganelli, R., M. I. Voskuil, G. K. Schoolnik, E. Dubnau, M. Gomez, and I. Smith. 2002. Role of the extracytoplasmic-function sigma factor σ^H in *Mycobacterium tuberculosis* global gene expression. *Mol. Microbiol.* 45:365–374.

Manganelli, R., M. I. Voskuil, G. K. Schoolnik, and I. Smith. 2001. The *Mycobacterium tuberculosis* ECF sigma factor σ^F: role in global gene expression and survival in macrophages. *Mol. Microbiol.* 41:423–437.

Matsuhashi, M. 1966. [Biosynthesis in the bacterial cell wall]. *Tanpakushitsu Kakusan Koso* 11:875–886.

Mehra, S., and D. Kaushal. 2009. Functional genomics reveals extended roles of the *Mycobacterium tuberculosis* stress response factor σ^H. *J. Bacteriol.* 191:3965–3980.

Mieczkowski, C., A. T. Iavarone, and T. Alber. 2008. Auto-activation mechanism of the *Mycobacterium tuberculosis* PknB receptor Ser/Thr kinase. *EMBO J.* 27:3186–3197.

Mir, M., J. Asong, X. Li, J. Cardot, G. J. Boons, and R. N. Husson. 2011. The extracytoplasmic domain of the *Mycobacterium tuberculosis* Ser/Thr kinase PknB binds specific muropeptides and is required for PknB localization. *PLoS Pathog.* 7:e1002182.

Mohamedmohaideen, N. N., S. K. Palaninathan, P. M. Morin, B. J. Williams, M. Braunstein, S. E. Tichy, J. Locker, D. H. Russell, W. R. Jacobs, Jr., and J. C. Sacchettini. 2008. Structure and function of the virulence-associated high-temperature requirement A of *Mycobacterium tuberculosis. Biochemistry* 47:6092–6102.

Molle, V., A. K. Brown, G. S. Besra, A. J. Cozzone, and L. Kremer. 2006a. The condensing activities of the *Mycobacterium tuberculosis* type II fatty acid synthase are differentially regulated by phosphorylation. *J. Biol. Chem.* 281:30094–30103.

Molle, V., I. Zanella-Cleon, J. P. Robin, S. Mallejac, A. J. Cozzone, and M. Becchi. 2006b. Characterization of the phosphorylation sites of *Mycobacterium tuberculosis* serine/threonine protein kinases, PknA, PknD, PknE, and PknH by mass spectrometry. *Proteomics* 6:3754–3766.

Molle, V., C. Girard-Blanc, L. Kremer, P. Doublet, A. J. Cozzone, and J. F. Prost. 2003a. Protein PknE, a novel transmembrane eukaryotic-like serine/threonine kinase from *Mycobacterium tuberculosis. Biochem. Biophys. Res. Commun.* 308:820–825.

Molle, V., L. Kremer, C. Girard-Blanc, G. S. Besra, A. J. Cozzone, and J. F. Prost. 2003b. An FHA phosphoprotein recognition domain mediates protein EmbR phosphorylation by PknH, a Ser/Thr protein kinase from *Mycobacterium tuberculosis. Biochemistry* 42:15300–15309.

Molle, V., G. Gulten, C. Vilcheze, R. Veyron-Churlet, I. Zanella-Cleon, J. C. Sacchettini, W. R. Jacobs, Jr., and L. Kremer. 2010. Phosphorylation of InhA inhibits mycolic acid biosynthesis and growth of *Mycobacterium tuberculosis. Mol. Microbiol.* 78:1591–1605.

Mukamolova, G. V., O. A. Turapov, D. I. Young, A. S. Kaprelyants, D. B. Kell, and M. Young. 2002. A family of autocrine growth factors in *Mycobacterium tuberculosis. Mol. Microbiol.* 46:623–635.

Mukherjee, P., K. Sureka, P. Datta, T. Hossain, S. Barik, K. P. Das, M. Kundu, and J. Basu. 2009. Novel role of Wag31 in protection of mycobacteria under oxidative stress. *Mol. Microbiol.* 73:103–119.

Mukherjee, R., and D. Chatterji. 2005. Evaluation of the role of sigma B in *Mycobacterium smegmatis. Biochem. Biophys. Res. Commun.* 338:964–972.

Mukherjee, R., M. Gomez, N. Jayaraman, I. Smith, and D. Chatterji. 2005. Hyperglycosylation of glycopeptidolipid of *Mycobacterium smegmatis* under nutrient starvation: structural studies. *Microbiology* 151:2385–2392.

Narayan, A., P. Sachdeva, K. Sharma, A. K. Saini, A. K. Tyagi, and Y. Singh. 2007. Serine threonine protein kinases of mycobacterial genus: phylogeny to function. *Physiol. Genomics* 29:66–75.

Nguyen, L., N. Scherr, J. Gatfield, A. Walburger, J. Pieters, and C. J. Thompson. 2007. Antigen 84, an effector of pleiomorphism in *Mycobacterium smegmatis. J. Bacteriol.* 189:7896–7910.

Ohara, N., M. Naito, C. Miyazaki, S. Matsumoto, Y. Tabira, and T. Yamada. 1997. HrpA, a new ribosome-associated protein which appears in heat-stressed *Mycobacterium bovis* bacillus Calmette-Guerin. *J. Bacteriol.* 179:6495–6498.

Ortalo-Magne, A., A. Lemassu, M. A. Laneelle, F. Bardou, G. Silve, P. Gounon, G. Marchal, and M. Daffe. 1996. Identification of the surface-exposed lipids on the cell envelopes of *Mycobacterium tuberculosis* and other mycobacterial species. *J. Bacteriol.* 178:456–461.

Ortiz-Lombardia, M., F. Pompeo, B. Boitel, and P. M. Alzari. 2003. Crystal structure of the catalytic domain of the PknB serine/threonine kinase from *Mycobacterium tuberculosis. J. Biol. Chem.* 278:13094–13100.

Ouellet, H., Y. Ouellet, C. Richard, M. Labarre, B. Wittenberg, J. Wittenberg, and M. Guertin. 2002. Truncated hemoglobin HbN protects *Mycobacterium bovis* from nitric oxide. *Proc. Natl. Acad. Sci. USA* 99:5902–5907.

Pang, X., and S. T. Howard. 2007. Regulation of the alpha-crystallin gene *acr2* by the MprAB two-component system of *Mycobacterium tuberculosis. J. Bacteriol.* 189:6213–6221.

Pang, X., P. Vu, T. F. Byrd, S. Ghanny, P. Soteropoulos, G. V. Mukamolova, S. Wu, B. Samten, and S. T. Howard. 2007. Evidence for complex interactions of stress-associated regulons in an *mprAB* deletion mutant of *Mycobacterium tuberculosis. Microbiology* 153:1229–1242.

Parikh, A., S. K. Verma, S. Khan, B. Prakash, and V. K. Nandicoori. 2009. PknB-mediated phosphorylation of a novel substrate, N-acetylglucosamine-1-phosphate uridyltransferase, modulates its acetyltransferase activity. *J. Mol. Biol.* 386:451–464.

Parish, T., D. A. Smith, S. Kendall, N. Casali, G. J. Bancroft, and N. G. Stoker. 2003. Deletion of two-component regulatory systems increases the virulence of *Mycobacterium tuberculosis. Infect. Immun.* 71:1134–1140.

Park, H. D., K. M. Guinn, M. I. Harrell, R. Liao, M. I. Voskuil, M. Tompa, G. K. Schoolnik, and D. R. Sherman. 2003. Rv3133c/dosR is a transcription factor that mediates the hypoxic response of *Mycobacterium tuberculosis. Mol. Microbiol.* 48:833–843.

Park, S. T., C. M. Kang, and R. N. Husson. 2008. Regulation of the SigH stress response regulon by an essential protein

kinase in *Mycobacterium tuberculosis. Proc. Natl. Acad. Sci. USA* 105:13105–13110.

Pasula, R., J. F. Downing, J. R. Wright, D. L. Kachel, T. E. Davis, Jr., and W. J. Martin II. 1997. Surfactant protein A (SP-A) mediates attachment of *Mycobacterium tuberculosis* to murine alveolar macrophages. *Am. J. Respir. Cell Mol. Biol.* 17:209–217.

Pathania, R., N. K. Navani, A. M. Gardner, P. R. Gardner, and K. L. Dikshit. 2002. Nitric oxide scavenging and detoxification by the *Mycobacterium tuberculosis* haemoglobin, HbN in *Escherichia coli. Mol. Microbiol.* 45:1303–1314.

Perez, E., S. Samper, Y. Bordas, C. Guilhot, B. Gicquel, and C. Martin. 2001. An essential role for *phoP* in *Mycobacterium tuberculosis* virulence. *Mol. Microbiol.* 41:179–187.

Portevin, D., C. De Sousa-D'Auria, C. Houssin, C. Grimaldi, M. Chami, M. Daffe, and C. Guilhot. 2004. A polyketide synthase catalyzes the last condensation step of mycolic acid biosynthesis in mycobacteria and related organisms. *Proc. Natl. Acad. Sci. USA* 101:314–319.

Pratt, R. F. 2008. Substrate specificity of bacterial DD-peptidases (penicillin-binding proteins). *Cell. Mol. Life Sci.* 65:2138–2155.

Prisic, S., S. Dankwa, D. Schwartz, M. F. Chou, J. W. Locasale, C. M. Kang, G. Bemis, G. M. Church, H. Steen, and R. N. Husson. 2010. Extensive phosphorylation with overlapping specificity by *Mycobacterium tuberculosis* serine/threonine protein kinases. *Proc. Natl. Acad. Sci. USA* 107:7521–7526.

Pucci, M. J., J. A. Thanassi, L. F. Discotto, R. E. Kessler, and T. J. Dougherty. 1997. Identification and characterization of cell wall-cell division gene clusters in pathogenic gram-positive cocci. *J. Bacteriol.* 179:5632–5635.

Pullen, K. E., H. L. Ng, P. Y. Sung, M. C. Good, S. M. Smith, and T. Alber. 2004. An alternate conformation and a third metal in PstP/Ppp, the *M. tuberculosis* PP2C-family Ser/Thr protein phosphatase. *Structure* 12:1947–1954.

Raman, S., X. Puyang, T. Y. Cheng, D. C. Young, D. B. Moody, and R. N. Husson. 2006. *Mycobacterium tuberculosis* SigM positively regulates Esx secreted protein and nonribosomal peptide synthetase genes and down regulates virulence-associated surface lipid synthesis. *J. Bacteriol.* 188:8460–8468.

Raman, S., T. Song, X. Puyang, S. Bardarov, W. R. Jacobs, Jr., and R. N. Husson. 2001. The alternative sigma factor SigH regulates major components of oxidative and heat stress responses in *Mycobacterium tuberculosis. J. Bacteriol.* 183:6119–6125.

Rastogi, N., E. Legrand, and C. Sola. 2001. The mycobacteria: an introduction to nomenclature and pathogenesis. *Rev. Sci. Tech.* 20:21–54.

Reed, M. B., V. K. Pichler, F. McIntosh, A. Mattia, A. Fallow, S. Masala, P. Domenech, A. Zwerling, L. Thibert, D. Menzies, K. Schwartzman, and M. A. Behr. 2009. Major *Mycobacterium tuberculosis* lineages associate with patient country of origin. *J. Clin. Microbiol.* 47:1119–1128.

Rickman, L., J. W. Saldanha, D. M. Hunt, D. N. Hoar, M. J. Colston, J. B. Millar, and R. S. Buxton. 2004. A two-component signal transduction system with a PAS domain-containing sensor is required for virulence of *Mycobacterium tuberculosis* in mice. *Biochem. Biophys. Res. Commun.* 314:259–267.

Rodrigue, S., R. Provvedi, P. E. Jacques, L. Gaudreau, and R. Manganelli. 2006. The sigma factors of *Mycobacterium tuberculosis. FEMS Microbiol. Rev.* 30:926–941.

Russell, D. G., H. C. Mwandumba, and E. E. Rhoades. 2002. *Mycobacterium* and the coat of many lipids. *J. Cell Biol.* 158:421–426.

Russell, D. G., B. C. VanderVen, W. Lee, R. B. Abramovitch, M. J. Kim, S. Homolka, S. Niemann, and K. H. Rohde. 2010. *Mycobacterium tuberculosis* wears what it eats. *Cell Host Microbe* 8:68–76.

Sacco, E., A. S. Covarrubias, H. M. O'Hare, P. Carroll, N. Eynard, T. A. Jones, T. Parish, M. Daffe, K. Backbro, and A. Quemard. 2007. The missing piece of the type II fatty acid synthase system from *Mycobacterium tuberculosis*. *Proc. Natl. Acad. Sci. USA* 104:14628–14633.

Sajid, A., G. Arora, M. Gupta, S. Upadhyay, V. K. Nandicoori, and Y. Singh. 2011. Phosphorylation of *Mycobacterium tuberculosis* Ser/Thr phosphatase by PknA and PknB. *PLoS One* 6:e17871.

Sassetti, C. M., D. H. Boyd, and E. J. Rubin. 2001. Comprehensive identification of conditionally essential genes in mycobacteria. *Proc. Natl. Acad. Sci. USA* 98:12712–12717.

Schnappinger, D., S. Ehrt, M. I. Voskuil, Y. Liu, J. A. Mangan, I. M. Monahan, G. Dolganov, B. Efron, P. D. Butcher, C. Nathan, and G. K. Schoolnik. 2003. Transcriptional adaptation of *Mycobacterium tuberculosis* within macrophages: insights into the phagosomal environment. *J. Exp. Med.* 198:693–704.

Shah, I. M., M. H. Laaberki, D. L. Popham, and J. Dworkin. 2008. A eukaryotic-like Ser/Thr kinase signals bacteria to exit dormancy in response to peptidoglycan fragments. *Cell* 135:486–496.

Sharma, K., M. Gupta, A. Krupa, N. Srinivasan, and Y. Singh. 2006. EmbR, a regulatory protein with ATPase activity, is a substrate of multiple serine/threonine kinases and phosphatase in *Mycobacterium tuberculosis*. *FEBS J.* 273:2711–2721.

Sherrid, A. M., T. R. Rustad, G. A. Cangelosi, and D. R. Sherman. 2010. Characterization of a Clp protease gene regulator and the reaeration response in *Mycobacterium tuberculosis*. *PLoS One* 5:e11622.

Singh, A., Y. Singh, R. Pine, L. Shi, R. Chandra, and K. Drlica. 2006. Protein kinase I of *Mycobacterium tuberculosis*: cellular localization and expression during infection of macrophage-like cells. *Tuberculosis* (Edinburgh) 86:28–33.

Slama, N., J. Leiba, N. Eynard, M. Daffe, L. Kremer, A. Quemard, and V. Molle. 2011. Negative regulation by Ser/Thr phosphorylation of HadAB and HadBC dehydratases from *Mycobacterium tuberculosis* type II fatty acid synthase system. *Biochem. Biophys. Res. Commun.* 412:401–406.

Sonawane, A., J. C. Santos, B. B. Mishra, P. Jena, C. Progida, O. E. Sorensen, R. Gallo, R. Appelberg, and G. Griffiths. 2011. Cathelicidin is involved in the intracellular killing of mycobacteria in macrophages. *Cell. Microbiol.* 13:1601–1617.

Song, T., S. L. Dove, K. H. Lee, and R. N. Husson. 2003. RshA, an anti-sigma factor that regulates the activity of the mycobacterial stress response sigma factor SigH. *Mol. Microbiol.* 50:949–959.

Song, T., S. E. Song, S. Raman, M. Anaya, and R. N. Husson. 2008. Critical role of a single position in the −35 element for promoter recognition by *Mycobacterium tuberculosis* SigE and SigH. *J. Bacteriol.* 190:2227–2230.

Stack, H. M., R. D. Sleator, M. Bowers, C. Hill, and C. G. Gahan. 2005. Role for HtrA in stress induction and virulence potential in *Listeria monocytogenes*. *Appl. Environ. Microbiol.* 71:4241–4247.

Stewart, G. R., S. M. Newton, K. A. Wilkinson, I. R. Humphreys, H. N. Murphy, B. D. Robertson, R. J. Wilkinson, and D. B. Young. 2005. The stress-responsive chaperone alpha-crystallin 2 is required for pathogenesis of *Mycobacterium tuberculosis*. *Mol. Microbiol.* 55:1127–1137.

Stewart, G. R., L. Wernisch, R. Stabler, J. A. Mangan, J. Hinds, K. G. Laing, D. B. Young, and P. D. Butcher. 2002. Dissection of the heat-shock response in *Mycobacterium tuberculosis* using mutants and microarrays. *Microbiology* 148:3129–3138.

Stokes, R. W., R. Norris-Jones, D. E. Brooks, T. J. Beveridge, D. Doxsee, and L. M. Thorson. 2004. The glycan-rich outer layer of the cell wall of *Mycobacterium tuberculosis* acts as an antiphagocytic capsule limiting the association of the bacterium with macrophages. *Infect. Immun.* 72:5676–5686.

Sturgill-Koszycki, S., P. H. Schlesinger, P. Chakraborty, P. L. Haddix, H. L. Collins, A. K. Fok, R. D. Allen, S. L. Gluck, J. Heuser, and D. G. Russell. 1994. Lack of acidification in *Mycobacterium* phagosomes produced by exclusion of the vesicular proton-ATPase. *Science* 263:678–681.

Sureka, K., S. Dey, P. Datta, A. K. Singh, A. Dasgupta, S. Rodrigue, J. Basu, and M. Kundu. 2007. Polyphosphate kinase is involved in stress-induced *mprAB-sigE-rel* signalling in mycobacteria. *Mol. Microbiol.* 65:261–276.

Sureka, K., T. Hossain, P. Mukherjee, P. Chatterjee, P. Datta, M. Kundu, and J. Basu. 2010. Novel role of phosphorylation-dependent interaction between FtsZ and FipA in mycobacterial cell division. *PLoS One* 5:e8590.

Tabira, Y., N. Ohara, and T. Yamada. 2000. Identification and characterization of the ribosome-associated protein, HrpA, of Bacillus Calmette-Guerin. *Microb. Pathog.* 29:213–222.

Talaat, A. M., R. Lyons, S. T. Howard, and S. A. Johnston. 2004. The temporal expression profile of *Mycobacterium tuberculosis* infection in mice. *Proc. Natl. Acad. Sci. USA* 101:4602–4607.

Talaat, A. M., S. K. Ward, C. W. Wu, E. Rondon, C. Tavano, J. P. Bannantine, R. Lyons, and S. A. Johnston. 2007. Mycobacterial bacilli are metabolically active during chronic tuberculosis in murine lungs: insights from genome-wide transcriptional profiling. *J. Bacteriol.* 189:4265–4274.

Territo, M. C., T. Ganz, M. E. Selsted, and R. Lehrer. 1989. Monocyte-chemotactic activity of defensins from human neutrophils. *J. Clin. Investig.* 84:2017–2020.

Thakur, M., R. Chaba, A. K. Mondal, and P. K. Chakraborti. 2008. Interdomain interaction reconstitutes the functionality of PknA, a eukaryotic type Ser/Thr kinase from *Mycobacterium tuberculosis*. *J. Biol. Chem.* 283:8023–8033.

Thakur, M., and P. K. Chakraborti. 2008. Ability of PknA, a mycobacterial eukaryotic-type serine/threonine kinase, to transphosphorylate MurD, a ligase involved in the process of peptidoglycan biosynthesis. *Biochem. J.* 415:27–33.

Thakur, M., and P. K. Chakraborti. 2006. GTPase activity of mycobacterial FtsZ is impaired due to its transphosphorylation by the eukaryotic-type Ser/Thr kinase, PknA. *J. Biol. Chem.* 281:40107–40113.

Tiemersma, E. W., M. J. van der Werf, M. W. Borgdorff, B. G. Williams, and N. J. Nagelkerke. 2011. Natural history of tuberculosis: duration and fatality of untreated pulmonary tuberculosis in HIV negative patients: a systematic review. *PLoS One* 6:e17601.

Traag, B. A., A. Driks, P. Stragier, W. Bitter, G. Broussard, G. Hatfull, F. Chu, K. N. Adams, L. Ramakrishnan, and R. Losick. 2010. Do mycobacteria produce endospores? *Proc. Natl. Acad. Sci. USA* 107:878–881.

Trivedi, O. A., P. Arora, A. Vats, M. Z. Ansari, R. Tickoo, V. Sridharan, D. Mohanty, and R. S. Gokhale. 2005. Dissecting the mechanism and assembly of a complex virulence mycobacterial lipid. *Mol. Cell* 17:631–643.

Vergne, I., J. Chua, S. B. Singh, and V. Deretic. 2004. Cell biology of *Mycobacterium tuberculosis* phagosome. *Annu. Rev. Cell Dev. Biol.* 20:367–394.

Veyron-Churlet, R., V. Molle, R. C. Taylor, A. K. Brown, G. S. Besra, I. Zanella-Cleon, K. Futterer, and L. Kremer. 2009. The *Mycobacterium tuberculosis* beta-ketoacyl-acyl carrier protein synthase III activity is inhibited by phosphorylation on a single threonine residue. *J. Biol. Chem.* 284:6414–6424.

Veyron-Churlet, R., I. Zanella-Cleon, M. Cohen-Gonsaud, V. Molle, and L. Kremer. 2010. Phosphorylation of the *Mycobacterium tuberculosis* beta-ketoacyl-acyl carrier protein reductase MabA regulates mycolic acid biosynthesis. *J. Biol. Chem.* 285:12714–12725.

Via, L. E., R. Curcic, M. H. Mudd, S. Dhandayuthapani, R. J. Ulmer, and V. Deretic. 1996. Elements of signal transduction in *Mycobacterium tuberculosis*: in vitro phosphorylation and in vivo expression of the response regulator MtrA. *J. Bacteriol.* 178:3314–3321.

Voskuil, M. I., D. Schnappinger, K. C. Visconti, M. I. Harrell, G. M. Dolganov, D. R. Sherman, and G. K. Schoolnik. 2003. Inhibition of respiration by nitric oxide induces a *Mycobacterium tuberculosis* dormancy program. *J. Exp. Med.* **198:**705–713.

Walters, S. B., E. Dubnau, I. Kolesnikova, F. Laval, M. Daffe, and I. Smith. 2006. The *Mycobacterium tuberculosis* PhoPR two-component system regulates genes essential for virulence and complex lipid biosynthesis. *Mol. Microbiol.* **60:**312–330.

Wang, Z., U. Schwab, E. Rhoades, P. R. Chess, D. G. Russell, and R. H. Notter. 2008. Peripheral cell wall lipids of *Mycobacterium tuberculosis* are inhibitory to surfactant function. *Tuberculosis* (Edinburgh) **88:**178–186.

Wayne, L. G. 1994. Dormancy of *Mycobacterium tuberculosis* and latency of disease. *Eur. J. Clin. Microbiol. Infect. Dis.* **13:**908–914.

Wayne, L. G., and L. G. Hayes. 1996. An in vitro model for sequential study of shiftdown of *Mycobacterium tuberculosis* through two stages of nonreplicating persistence. *Infect. Immun.* **64:**2062–2069.

Welin, A., J. Raffetseder, D. Eklund, O. Stendahl, and M. Lerm. 2011. Importance of phagosomal functionality for growth restriction of *Mycobacterium tuberculosis* in primary human macrophages. *J. Innate Immun.* **3:**508–518.

White, M. J., H. He, R. M. Penoske, S. S. Twining, and T. C. Zahrt. 2010. PepD participates in the mycobacterial stress response mediated through MprAB and SigE. *J. Bacteriol.* **192:**1498–1510.

White, M. J., J. P. Savaryn, D. J. Bretl, H. He, R. M. Penoske, S. S. Terhune, and T. C. Zahrt. 2011. The HtrA-like serine protease PepD interacts with and modulates the *Mycobacterium tuberculosis* 35-kDa antigen outer envelope protein. *PLoS One* **6:** e18175.

Wilkinson, K. A., G. R. Stewart, S. M. Newton, H. M. Vordermeier, J. R. Wain, H. N. Murphy, K. Horner, D. B. Young, and R. J. Wilkinson. 2005. Infection biology of a novel alpha-crystallin of *Mycobacterium tuberculosis*: Acr2. *J. Immunol.* **174:** 4237–4243.

Williams, D. L., T. L. Pittman, M. Deshotel, S. Oby-Robinson, I. Smith, and R. Husson. 2007. Molecular basis of the defective heat stress response in *Mycobacterium leprae*. *J. Bacteriol.* **189:**8818–8827.

World Health Organization. 2010. *Global Tuberculosis Control.* World Health Organization, Geneva, Switzerland.

Wu, Q. L., D. Kong, K. Lam, and R. N. Husson. 1997. A mycobacterial extracytoplasmic function sigma factor involved in survival following stress. *J. Bacteriol.* **179:**2922–2929.

Yang, D., O. Chertov, S. N. Bykovskaia, Q. Chen, M. J. Buffo, J. Shogan, M. Anderson, J. M. Schroder, J. M. Wang, O. M. Howard, and J. J. Oppenheim. 1999. Beta-defensins: linking innate and adaptive immunity through dendritic and T cell CCR6. *Science* **286:**525–528.

Young, T. A., B. Delagoutte, J. A. Endrizzi, A. M. Falick, and T. Alber. 2003. Structure of *Mycobacterium tuberculosis* PknB supports a universal activation mechanism for Ser/Thr protein kinases. *Nat. Struct. Biol.* **10:**168–174.

Zahrt, T. C., and V. Deretic. 2001. *Mycobacterium tuberculosis* signal transduction system required for persistent infections. *Proc. Natl. Acad. Sci. USA* **98:**12706–12711.

Zahrt, T. C., C. Wozniak, D. Jones, and A. Trevett. 2003. Functional analysis of the *Mycobacterium tuberculosis* MprAB two-component signal transduction system. *Infect. Immun.* **71:**6962–6970.

Zhang, W., V. C. Jones, M. S. Scherman, S. Mahapatra, D. Crick, S. Bhamidi, Y. Xin, M. R. McNeil, and Y. Ma. 2008. Expression, essentiality, and a microtiter plate assay for mycobacterial GlmU, the bifunctional glucosamine-1-phosphate acetyltransferase and N-acetylglucosamine-1-phosphate uridyltransferase. *Int. J. Biochem. Cell Biol.* **40:**2560–2571.

Zhang, Y., B. Heym, B. Allen, D. Young, and S. Cole. 1992. The catalase-peroxidase gene and isoniazid resistance of *Mycobacterium tuberculosis*. *Nature* **358:**591–593.

Zink, A., C. J. Haas, U. Reischl, U. Szeimies, and A. G. Nerlich. 2001. Molecular analysis of skeletal tuberculosis in an ancient Egyptian population. *J. Med. Microbiol.* **50:**355–366.

Zink, A. R., C. Sola, U. Reischl, W. Grabner, N. Rastogi, H. Wolf, and A. G. Nerlich. 2003. Characterization of *Mycobacterium tuberculosis* complex DNAs from Egyptian mummies by spoligotyping. *J. Clin. Microbiol.* **41:**359–367.

VI. EMERGING REGULATORY MECHANISMS OF SPECIAL SIGNIFICANCE

Regulation of Bacterial Virulence
Edited by Michael L. Vasil and Andrew J. Darwin
© 2013 ASM Press, Washington, DC   doi:10.1128/9781555818524.ch25

Chapter 25

Regulatory Mechanisms of Special Significance: Role of Small RNAs in Virulence Regulation

Kai Papenfort, Colin P. Corcoran, Sanjay K. Gupta, Masatoshi Miyakoshi,
Nadja Heidrich, Yanjie Chao, Kathrin S. Fröhlich, Cynthia M. Sharma,
Wilma Ziebuhr, Alex Böhm, and Jörg Vogel

INTRODUCTION

Pathogenic bacteria express a myriad of systems aimed at subversion and exploitation of a host to promote proliferation and survival. Naturally, host cells have evolved equally complex defensive mechanisms which the bacterial pathogen must overcome in order to successfully establish an infection. Successful pathogenesis therefore depends on the correct temporal expression of virulence traits in response to host-related environmental cues, such as the increase in temperature associated with ingestion of a waterborne pathogen. These responses have traditionally been associated with the activity of transcription factors, whereas the extent and importance of posttranscriptional regulation in virulence control have only recently been recognized. This paradigm shift was undoubtedly aided by advances of biocomputational predictions and the advent of high-throughput RNA sequencing (RNA-seq) technologies which precipitated an explosion of small RNA (sRNA) discovery (Sharma and Vogel, 2009). The association of many of these sRNAs with the ubiquitous RNA-binding proteins Hfq and CsrA, whose deletion results in reduced virulence in many bacteria (Lucchetti-Miganeh et al., 2008; Chao and Vogel, 2010), was an early indication that sRNAs form integral components of bacterial pathogenesis. Extensive functional characterization of these new sRNAs has confirmed both their central role in fine-tuning gene expression and more dramatic roles during stress response or virulence gene expression.

The ever-expanding discovery and characterization of sRNAs in both gram-positive and gram-negative pathogens have provided a wealth of information on posttranscriptional virulence control, which is summarized in this chapter using the best-characterized examples.

Regulatory RNAs can operate at all layers of gene regulation, ranging from transcriptional initiation to protein activity (Waters and Storz, 2009; Papenfort and Vogel, 2010). Small regulatory RNAs can be divided into those elements encoded intrinsically within the regulated mRNA or those that are encoded distinctly from their targets. Intrinsically encoded RNA regulatory elements which directly respond to environmental signals are collectively termed riboswitches. Regulatory RNAs that are encoded distinctly from their targets are collectively termed small noncoding regulatory RNAs (sRNAs) and can be further classified based on their mode of action into regulatory RNAs which act by base-pairing with target mRNAs to modulate translation and/or stability of the mRNA or those which act posttranslationally to regulate protein activity.

Riboswitches are *cis*-acting regulatory RNAs encoded within the 5′ untranslated region (5′ UTR) of an mRNA that are directly responsive to a wide range of important environmental signals, including levels of specific metabolites, pH, and temperature (Grundy and Henkin, 2006). For many pathogenic bacteria, host body temperature is an important signal that triggers the expression of virulence genes. Temperature sensing occurs at every level of prokaryotic gene regulation, including genome-wide alterations in transcription induced by global changes in DNA supercoiling (Dorman and Corcoran, 2009), and more local effects such as altered translation

Kai Papenfort, Colin P. Corcoran, Sanjay K. Gupta, Masatoshi Miyakoshi, Nadja Heidrich, Yanjie Chao, Kathrin S. Fröhlich, Wilma Ziebuhr, Alex Böhm, and Jörg Vogel • Institute for Molecular Infection Biology, University of Würzburg, D-97080 Würzburg, Germany. Cynthia M. Sharma • Research Centre of Infectious Diseases, University of Würzburg, D-97080 Würzburg, Germany.

of a thermosensitive mRNA or altered activity of a thermosensitive protein (Hurme and Rhen, 1998). Thermosensitive riboswitches, or RNA thermometers, typically consist of a "closed" structure overlapping the ribosome binding site (RBS) which "melts" upon an increase in temperature (Klinkert and Narberhaus, 2009). For metabolite-sensing riboswitches, the presence of the metabolite specifically bound by the riboswitch typically results in formation of either a transcription terminator or an inhibitory structure overlapping the RBS. Upon intracellular depletion of the metabolite, an antiterminator is formed, allowing read-through of downstream genes, or if the RBS is liberated, resulting in increased translation of the mRNA (Henkin, 2008).

Other *cis*-acting RNAs include antisense-encoded RNAs, which are transcribed from the opposite strand within or near the target coding sequence and therefore contain perfect sequence complementarity to the target mRNA (Thomason and Storz, 2010). While these *cis*-encoded RNAs are generally considered *cis* acting, early studies involving the transposable element Tn*10* (which is controlled by an antisense-encoded RNA) revealed that *cis*-encoded sRNAs can also regulate *trans*-encoded targets as a mechanism for copy number control (Simons and Kleckner, 1983). Unlike the *cis*-acting elements described above, the majority of sRNAs described to date are encoded distally from their targets and are thus considered *trans* acting.

The best-characterized and largest group of *trans*-acting sRNAs modulate mRNA translation by base-pairing with limited sequence homology to their target mRNA. There are very few unifying features among these sRNAs, which vary in size (from 50 to 500 nucleotides [nt]), sequence, and secondary structure (Waters and Storz, 2009). The defining factor that links many of these sRNAs, particularly in gram-negative bacteria, is their association with the RNA chaperone Hfq. Hfq-dependent sRNAs constitute a considerable portion of all characterized sRNAs (Corcoran et al., 2011). Hfq was originally discovered in the 1960s as a host factor of an RNA phage, Qβ (hence the name), in nonpathogenic *Escherichia coli* (Franze de Fernandez et al., 1968). Its role as a pleiotropic regulator of gene expression became apparent much later through the broad phenotypes of an *E. coli* K-12 *hfq* mutant (Tsui et al., 1994). Hfq is a bacterial member of the large Sm/LSm family of RNA-binding proteins that assemble into a ring-like multimeric quaternary structure, such as a doughnut-shaped homohexamer in the case of Hfq. Hfq binds tightly to many sRNAs and is required for both intracellular stability (primarily through occlusion of binding sites for the major endoribonuclease

RNase E) and target mRNA pairing of sRNAs in many bacteria (Vogel and Luisi, 2011). The extent of posttranscriptional control by Hfq in a pathogenic bacterium is best illustrated in *Salmonella enterica* serovar Typhimurium, in which Hfq coimmunoprecipitation followed by RNA sequencing implicated over 60 sRNAs in the regulation of several hundred mRNAs, arguing that posttranscriptional regulation by Hfq and sRNAs affects almost 20% of all genes in this gram-negative model pathogen (Sittka et al., 2008; Ansong et al., 2009; Sittka et al., 2009). One important feature of many Hfq-dependent sRNAs is the ability to directly control multiple genes through a single short (<16-nt) targeting region, which is often located at the 5′ end of the sRNA (Papenfort et al., 2010). These globally acting posttranscriptional regulators often act in concert with transcriptional regulators to improve the fidelity of the response to a single environmental cue, such as envelope stress (Gogol et al., 2011), altered nutrient conditions (Beisel and Storz, 2011), or quorum sensing (QS) control (Sonnleitner et al., 2009).

The vast majority of Hfq-dependent sRNAs characterized to date act by base-pairing within the 5′ UTR to inhibit translation initiation. This translational block often results in the rapid degradation of the target mRNA due to reduced protection from ribonucleases in the absence of translating ribosomes. However, sRNAs can also inhibit translation without resulting in mRNA destruction (Moller et al., 2002). This can serve to discoordinate gene expression within an operon or to reversibly silence an mRNA, effectively priming the gene for rapid induction upon relief of sRNA-based repression. sRNAs are also capable of direct stimulation of gene expression, which is best illustrated by the regulation of *rpoS* translation by three sRNAs (DsrA, RprA, and ArcZ). The 5′ UTR of *rpoS* contains an inhibitory stem-loop, which sequesters the SD (Shine-Dalgarno sequence) and thus prevents efficient translation. DsrA, RprA, and ArcZ each target different regions in the 5′ UTR to disrupt the stem-loop structure, liberating the SD and allowing efficient translation (Sledjeski et al., 1996; Lease et al., 1998; Majdalani et al., 1998; Majdalani et al., 2001; Mandin and Gottesman, 2010; Soper et al., 2010). This "anti-antisense" mechanism is the most common mode of gene expression stimulation by an sRNA and, at least in the case of *rpoS*, is fully dependent on Hfq (Frohlich and Vogel, 2009).

A more recently discovered class of *trans*-acting sRNAs is involved in an adaptive response that confers immunity to bacteriophage and conjugative plasmids. These so-called CRISPR RNAs (clustered regularly interspaced short palindromic repeats) act through specific base-pairing with invading nucleic acids and target the duplex for destruction by the conserved CRISPR-associated proteins (Deveau et al., 2010). CRISPR RNAs are ubiquitous in bacteria and contribute to bacterial survival of predation by phage. Since CRISPR RNAs are not specifically associated with pathogenesis but represent a basic survival function, CRISPR loci in pathogenic bacteria are not discussed in this review unless of special interest.

RNA regulators of protein activity represent a distinct group of sRNAs that act not by base-pairing with target mRNAs but posttranslationally by mimicking the target of a protein and thus sequestering it in an unproductive complex (Babitzke and Romeo, 2007). The best-characterized member of this family is 6S RNA, which mimics an open promoter and sequesters RNA polymerase bound to the housekeeping sigma factor, σ^{70} (Wassarman and Storz, 2000). The widely distributed RNA-binding protein CsrA (also called RsmA in *Pseudomonas* species) is also subject to posttranslational control by the CsrB family of sRNAs. CsrA proteins target many mRNAs, including the *hfq* mRNA (Baker et al., 2007), at GGA-rich elements to inhibit ribosome binding, which often results in mRNA decay (Babitzke et al., 2009). CsrB-like sRNAs, which can be induced in response to nutrient limitation, consist of multiple high-affinity sites containing the GGA motif and thus titrate CsrA away from regulatory targets. Similar to *hfq* mutation, deletion of *csrA/rsmA* often results in pleiotropic changes in gene expression (Timmermans and Van Melderen, 2010), including reduced virulence in a diverse range of pathogens such as *Pseudomonas aeruginosa*, *Vibrio cholerae*, *Helicobacter pylori*, *Legionella pneumophila*, and *Salmonella enterica* serovar Typhimurium (Lucchetti-Miganeh et al., 2008).

In this chapter, we selectively detail the best-characterized examples of sRNAs and their molecular function in bacterial pathogens of special interest.

GRAM-NEGATIVE PATHOGENS

Salmonella

S. enterica serovar Typhimurium is a gram-negative pathogen causing gastroenteritis in humans. Upon ingestion, the bacterium encounters various stressful environments before it finally invades the epithelial lining. The ability of *Salmonella* to invade and survive inside the host mainly relies on a range of horizontally acquired virulence regions called *Salmonella* pathogenicity islands (SPIs). Of these, SPI-1 and SPI-2 encode type III secretion systems (T3SS), which allow the injection of effector proteins to the host cell. SPI-1 is required for invasion of nonphagocytic cells,

whereas SPI-2 is involved in survival under harsh intracellular conditions (Haraga et al., 2008).

When it comes to numbers and functions of studied RNA regulators, *Salmonella* constitutes one of the best-understood bacterial pathogens. There are two main reasons for this advanced knowledge. First, *Salmonella* is a long-standing genetically tractable model organism providing many genetic tools. Second (and maybe more relevant), its close phylogenetic relation to *E. coli* K-12, the "pioneer organism" for bacterial sRNA screens, allowed the study of many conserved sRNAs (Hershberg et al., 2003), some of which evolved *Salmonella*-specific functions (see below). As in *E. coli*, most sRNAs in *Salmonella* bind to Hfq, and the relevance of this interplay is reflected by the pleiotropic nature of phenotypes discovered for *Salmonella hfq* mutants. The most prominent virulence-related phenotypes of this gene defect include decreased motility, attenuation of epithelial cell invasion, and decreased survival within cultured macrophages (Sittka et al., 2007; Ansong et al., 2009). Following these pronounced phenotypes, the quest for noncoding RNAs in *Salmonella* has been strongly influenced by Hfq coimmuniprecipitation studies combined with microarray or high-throughput sequencing techniques to globally detect Hfq-binding partners (Sittka et al., 2008; Sittka et al., 2009). These studies validated many previously predicted sRNA candidates of two bioinformatics screens focusing on either the entire *Salmonella* chromosome (Pfeiffer et al., 2007) or horizontally acquired sRNAs only (Padalon-Brauch et al., 2008). The current count of validated *Salmonella* sRNAs runs at ~140, derived from samples of 20 different conditions (Kröger et al., 2012).

As outlined above, many characterized *Salmonella* sRNAs were originally discovered in *E. coli*, and sequence conservation analysis or, in some cases, wet-lab studies suggest that their biological function has been maintained. The study of these conserved sRNAs (also called core sRNAs) predicts the discovery of functions that reach beyond a single species (Papenfort et al., 2009). A prominent example of such conserved sRNA activity is illustrated by two sRNAs, i.e., RybB and MicA (Table 1), both of which are transcriptionally controlled by the alternative sigma factor σ^{E} (Johansen et al., 2006; Papenfort et al., 2006; Thompson et al., 2007). Induction of σ^{E} triggers RybB and MicA expression, which, in turn, downregulates a large suite of target mRNAs, many of which encode outer membrane proteins (OMPs) (Papenfort et al., 2006; Bossi and Figueroa-Bossi, 2007; Papenfort et al., 2010; Gogol et al., 2011). Control of OMP mRNAs is a recurring theme in sRNA-mediated gene regulation

Table 1. sRNAs in bacterial pathogens

Organism	Name	Alternative name(s)	Size (nt)	Adjacent genes	Role(s) in pathogenesis	Reference(s)
Salmonella	GcvB	IS 145	209	*gcvA/ygdI*	Induced under nutrient-rich conditions to repress ABC transporter systems	Sharma et al., 2007; Sharma et al., 2011
	RybB	P25	78	STM0869/STM0870	Expressed in response to envelope stress and represses many major and minor OMPs, including *ompA/C/D/F*	Papenfort et al., 2006; Papenfort et al., 2010; Gogol et al., 2011
	SgrS	ryaA	240	*yabN/leuD*	Represses *ptsG, manXYZ,* and *sopD* mRNAs	Vanderpool and Gottesman, 2004; Rice and Vanderpool, 2011
	CyaR	RyeE	87	*yegQ*/STM2137	Activated by Crp; repressor of *ompX* mRNA	Papenfort et al., 2008
	IsrE	*RybB2/RfrB*	97	STM1273/STM1274	Induced during macrophage infection; repressor of *sodB* mRNA	Padalon-Brauch et al., 2008
	MicC	IS063/*Tke8*	110	*nifJ/ynaF*	Represses *ompC* and *ompD*	Pfeiffer et al., 2009
	SdsR	*RyeB,* *Tpke79*	104	STM1871/STM1873	Controlled by σ^s; repressor of *ompD*; hot spot for integration of foreign DNA	Balbontin et al., 2008; Frohlich et al., 2012
	InvR	STnc270	80	*invH*/STM2901	Under invasive conditions, induced by HilD and represses *ompD*	Pfeiffer et al., 2007
	ChiX	MicM/*RybC*/SroB	80	*ybaK/ybaP*	Represses *chiP* mRNA	Figueroa-Bossi et al., 2009
	MicA	SraD	70	*luxS/gshA*	Induced by σ^E, under envelope stress conditions; represses *ompA* and *lamB*	Papenfort, et al., 2006; Bossi and Figueroa-Bossi, 2007
	CsrB		360	*yqcC/syd*	Antagonizes CsrA activity	Altier et al., 2000; Fortune et al., 2006
	CsrC	*SraK/RyiB/Tpk2*	240	*yihA/yihI*	Antagonizes CsrA activity	Altier et al., 2000; Fortune et al., 2006
	IsrJ		74	STM2614/STM2616	Induced under low-oxygen and -magnesium conditions; involved in translocation of virulence-associated proteins	Padalon-Brauch et al., 2008
	ArcZ	*sraH/ryhA*	120	*yhbL/arcB*	Activates *rpoS* mRNA; represses *sdaC, tpx,* and STM3216 mRNAs	Papenfort et al., 2009; Soper et al., 2010
	IsrM		328	STM2762/STM2763	Induced upon cold shock; repressor of *hilE* and *sopA* mRNAs	Padalon-Brauch et al., 2008; Gong et al., 2011
	AmgR		1,200	Antisense to *mgtC*	Mutants of *amgR* display enhanced survival in a murine infection model	Lee and Groisman, 2010
Vibrio	VrrA		140	VC1741/VC1743	Represses outer membrane porins OmpA and OmpT	Song and Wai, 2009; Song et al., 2010

Continued on following page

Table 1. *Continued*

Organism	Name	Alternative name(s)	Size (nt)	Adjacent genes	Role(s) in pathogenesis	Reference(s)
	Qrr1 to -4		96–108	VC1020/VC1021 VCA0040/ VCA0041 VCA0826/ VCA0827 VCA0958/ VCA0959	Activated by phosphorylated LuxO; repressor of *hapR* and *luxO*; activator of *aphA*, VCA0939	Lenz et al., 2004; Hammer and Bassler, 2007; Rutherford et al., 2011
	CsrB/C/D		~300–400	VC0190/ VC0191, VC0882/ VC0883, VC0839/ VC0840	Antagonists of CsrA function	Lenz et al., 2005
	TarA		72	VC0825/VC0826	Negatively regulates glucose transporter, *ptsG*	Richard et al., 2010
	TarB		77	VC0845/VC0846	Repressor of the TcpF colonization factor	Bradley et al., 2011
	RyhB		225	VC0106/VC0107	Regulated by Fur; represses *sodB* mRNA	Davis et al., 2005
Pseudomonas	PhrS		212	PA3305/PA3306	Induced by oxygen-responsive regulator ANR; induces expression of PqsR	Sonnleitner et al., 2008; Sonnleitner et al., 2011
	PrrF1/ PrrF2		116/114	PA4704/*phuW*	Induced under iron-limiting conditions; degrades *sodB*, *sdh*, and PA4880 sRNA	Wilderman et al., 2004
	RgsA		120	PA2958/PA2959	Induced by RpoS; required for H_2O_2 resistance	Gonzalez et al., 2008
	PrrH		325	PA4704/*phuW*	Repressed by iron or heme, leading to increased expression of PrrF-regulated genes	Oglesby-Sherrouse and Vasil, 2010
	RsmY/Z		123, 115	*dnr*/PA0528, *fdxA*/*rpoS*	Regulated by the GacS/ GacA TCS; modulates QS	Brencic et al., 2009
	CrcZ		407	*cbrB*/*pcnB*	Controls selection of most favorable carbon source	Sonnleitner et al., 2009
Yersinia	GcvB	Ysr45	130, 206	*gcvA*/y3155	Negatively regulates expression of *dppA* mRNA	McArthur et al., 2006
	CsrB/C		282, 255	*syd*/YPTB3010 *engB*/YPTB0020	Antagonists of CsrA function	Heroven et al., 2008
	RybB	Ysr48	~80	YPTB1356/ YPTB01326	Hfq-dependent sRNA at 37°C	Koo et al., 2011
	Ysr23		~200	*ybcI*/YPTB2678	Mild attenuation in a murine infection model	Koo et al., 2011
	Ysr29		NA[a]	*cyaA*/YPTB0186	Mild attenuation in a murine infection model	Koo et al., 2011
	Ysr35		339	*nrdA*/*ubiG*	Mild attenuation in a murine infection model	Koo et al., 2011
Francisella	FtrA		111	FTL_1319/ FTL_1320	Increased expression under iron-depleted conditions	Postic et al., 2010
	FtrB		115	FTL_0035/ FTL_0036	No effect on virulence in mouse model	Postic et al., 2010

Continued on following page

Table 1. sRNAs in bacterial pathogens (*Continued*)

Organism	Name	Alternative name(s)	Size (nt)	Adjacent genes	Role(s) in pathogenesis	Reference(s)
Shigella	RyhB		90	*yhhY/yhhX*	Represses master regulator of virulence, *virB*	Murphy and Payne, 2007
	RnaG		450	*virA/icsA*	*cis*-encoded repressor of the *icsA* invasion factor	Giangrossi et al., 2010
Bordetella	BprJ2		220	BP3395/BP3396	Controlled by the TCS BvgA/BvgS	Hot et al., 2011
	BprC		190	BP1878/BP1879	Likely to act on adjacent mRNA	Scarlato et al., 1991
Borrelia	DsrA$_{Bb}$		213–352	bb0577/bb0578	Regulates *rpoS* in a temperature-dependent manner	Lybecker and Samuels, 2007
Neisseria	NrrF		195	NGO2002/ NGO2004	Repressor of the *sdhCDAB* operon under iron starvation	Mellin et al., 2007
	AniS		147	NMB1206/ NMB1204	Anaerobically induced sRNA; downregulates NMB1468/NMB0214	Fantappie et al., 2011
	FnrS		108	*bfrB*/NGO0797	Induced under anaerobic conditions	Isabella and Clark, 2011
Chlamydia	IhtA		120	CT674/CT675	Inhibits HC1 translation	Grieshaber et al., 2006
	ctrR0332		80, 240	CTLon_0331/ CTLon_0332	Differentially expressed in EBs and RBs	Albrecht et al., 2010
Helicobacter	IG-443		60	*fur*/HP1033	Biological function unclear	Xiao et al., 2009a
	IG-524		60	HP1322/HP1326	Biological function unclear	Xiao et al., 2009a
	HPnc5490		87	HP1043/HP1044	Represses chemotaxis receptor TlpB	Sharma et al., 2010
	5' ureB- sRNA		292	Antisense to *ureAB*	Induced by ArsR response regulator under neutral conditions; represses urease	Wen et al., 2010
Staphylococcus	RNAIII		514	*agrB*/SA1841	QS-controlled sRNA; encodes the *hld* ORF; direct regulator of *rot*, *spa*, *hla*, *coa*, SA1000, and SA2353	Novick et al., 1993; Morfeldt et al., 1995; Boisset et al., 2007; Chevalier et al., 2010
	RsaE		100	SA0859/SA0860	Regulates several metabolic pathways involved in amino acid and peptide transport	Geissmann et al., 2009; Bohn et al., 2010
	SprD		140	*scn/chp*	Represses translation initiation of *sbi*	Chabelskaya et al., 2010
Listeria	LhrA		268	lmo2256/ lmo2258	Controls the expression of lmo0850, lmo0302, and *chiA*	Christiansen et al., 2006; Nielsen et al., 2011
	RliB		360	lmo0509/ lmo0510	Induced during infection; affects colonization of mice	Mandin et al., 2007; Toledo-Arana et al., 2009
	Rli31		144	lmo0558/ lmo0559	Important for colonization inside macrophages	Mraheil et al., 2011b
	Rli33-1,2		186, 274	lmo0671/ lmo0672	Important for colonization inside macrophages	Mraheil et al., 2011b
	Rli50		350	lmo2709/ lmo2710	Important for colonization inside macrophages	Mraheil et al., 2011b
	SbrA		70	lmo1374/ lmo1375	Activated by SigB	Nielsen et al., 2008
	SbrE	Rli47	514	lmo2141/ lmo2142	Activated by SigB	Oliver et al., 2009

Continued on following page

Table 1. *Continued*

Organism	Name	Alternative name(s)	Size (nt)	Adjacent genes	Role(s) in pathogenesis	Reference(s)
	SreA		229	lmo2419/ lmo2417	Activated by PrfA; represses *prfA* expression	Loh et al., 2009
	SreB		180	lmo0595/ lmo0594	Activated by PrfA; represses *prfA* expression	Loh et al., 2009
Streptococcus	FasX		250	*fasA/rnpA*	Major effector molecule of the *fasBCA* operon	Kreikemeyer et al., 2001
	Pel		459	*sagA/sagB*	Crucial for regulation of several virulence factors	Mangold et al., 2004
	RivX		180/220	*rivR/pgi*	Activates the genes of the *mga* regulon	Roberts and Scott, 2007
	csRNA4/5		87–151	Spr0239/*uppS* Spr0218/*gpmB*	Stimulates stationary-phase autolysis	Halfmann et al., 2007
Bacillus	RatA		222	*txpA/yqbM*	Antisense RNA of a type I TA system	Silvaggi et al., 2005
	SurA		~300	*yndK/yndL*	Differentially regulated during sporulation	Silvaggi et al., 2006
	SurC		199	*dnaJ/dnaK*	Induced by the mother cell-specific alternative sigma factor (σ^K)	Silvaggi et al., 2006
	FsrA		84	*ykuI/ykuJ*	Fur-regulated sRNA; facilitates adaptation to low-iron conditions	Gaballa et al., 2008
	BsrF		115	*yobO/csaA*	Decreased expression during sporulation; positively regulated by CodY	Preis et al., 2009; Saito et al., 2009
	SR1		205	*slp/speA*	Inhibits translation of the *ahrC* mRNA; encodes a small ORF	Heidrich et al., 2006
Mycobacterium	Mcr16		~100	Antisense to *fabD*	*cis*-encoded antisense RNA	DiChiara et al., 2010
	ASdes		75, 110	Antisense to *desA1*	*cis*-encoded antisense RNA	Arnvig and Young, 2009
	B11	MTS2822	93	Rv3660c/Rv3661	Overexpression impairs growth	Arnvig and Young, 2009
	G2	MTS1310	65	Rv1689/Rv1690	Overexpression impairs growth	Arnvig and Young, 2009
	F6	Mcr14, Mpr13, MTS0194	55, 100	Rv0243/Rv0244c	Induced by oxidative and acid stress; overexpression impairs growth	Arnvig and Young, 2009; DiChiara et al., 2010
	MTS0997	Mcr11	~110	Rv1264/Rv1265	Induced during infection in mice	DiChiara et al., 2010; Arnvig et al., 2011
	MTS1338		~100	Rv1733c/ Rv1735c	Induced in a murine infection model; regulated by DosR	Arnvig et al., 2011
	MTS2823		250, 300	Rv3661/Rv3662c	Induced in a murine infection model; regulates *prpC* and *prpD* mRNAs	Arnvig et al., 2011
Clostridia	VR-RNA		~400	*mutM/clg*	Promotes production of α-toxin, collagenase, and β2 toxin	Okumura et al., 2008

[a]NA, not applicable.

of *Salmonella* and *E. coli* (Vogel and Papenfort, 2006; Papenfort et al., 2010). For example, the CyaR sRNA is conserved among many enterobacterial species (including *Salmonella* and *E. coli*), as are its transcriptional control by the global transcriptional regulator Crp and repression of the OmpX protein (Johansen et al., 2008; Papenfort et al., 2008). Similarly, regulation of the YbfM chitin porin by the ChiX (also called MicM) sRNA was discovered concurrently in *Salmonella* and *E. coli* (Figueroa-Bossi et al., 2009; Overgaard et al., 2009).

In contrast to these conserved cases, OMP regulation by sRNAs also evolved diversifications. Posttranscriptional regulation of OmpD, a major porin of *Salmonella* which is silent in *E. coli* K-12 (Lee and Schnaitman, 1980; Hindahl et al., 1984), was observed for four different sRNAs, three of which are conserved, i.e., RybB, MicC, and SdsR (Papenfort et al., 2006; Pfeiffer et al., 2009; Frohlich et al., 2012), as well as the *Salmonella*-specific, SPI-1-encoded InvR (Pfeiffer et al., 2007). InvR was the first Hfq-binding sRNA discovered that originates from a horizontally acquired locus, and its transcriptional regulation is controlled by the major SPI-1 transcription factor HilD (Pfeiffer et al., 2007). Numerous other

sRNAs from SPIs have been described subsequently (Padalon-Brauch et al., 2008; Sittka et al., 2008; Sittka et al., 2009), of which IsrJ and IsrM have been shown to contribute to the *Salmonella* virulence program. Mutation of *isrJ* reduces effector protein translocation into host cells by an unknown mechanism (Padalon-Brauch et al., 2008), whereas IsrM affects *Salmonella* pathogenicity transcriptionally by targeting the *hilE* mRNA, encoding a global repressor of SPI-1 functions (Gong et al., 2011). Disruption of the *isrM* locus results in a strong defect in colonization of liver and spleen in a murine infection model, and this effect seems to result partially from lack of *hilE* repression. In addition to *hilE* mRNA, IsrM also downregulates expression of the *sopA* mRNA, encoding a secreted effector protein of the SPI-1 regulon (Gong et al., 2011). Transcriptional regulators driving IsrM expression are yet to be determined, and it is not clear if there are auxiliary protein factors, e.g., Hfq, which may be required for regulation of *hilE* and *sopA* mRNAs (Fig. 1). Another aspect of sRNA-mediated regulation in the context of the *Salmonella* virulence regions is those sRNAs that have been conserved in many enterobacteria but whose functions expanded to control horizontally acquired genes. For example,

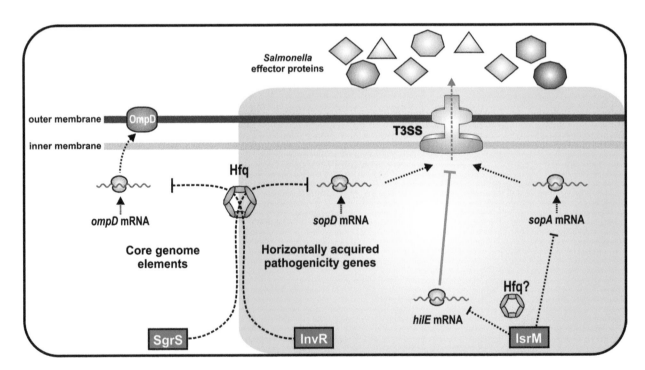

Figure 1. Cross talk of core and horizontally acquired genomic elements at the posttranscriptional level. The core genome-encoded SgrS sRNA downregulates expression of the *sopD* mRNA, encoding a virulence factor secreted by the T3SS, at the posttranscriptional level. InvR, encoded on the horizontally acquired SPI-1, represses the core genome-encoded *ompD* mRNA. Both sRNAs require the RNA chaperone Hfq for target regulation. IsrM, also acquired by horizontal gene transfer, reduces the expression of the *sopA* and *hilE* mRNAs. Thereby, IsrM indirectly induces T3SS secretion, as *hilE* encodes a global transcriptional regulator inhibiting virulence factor expression. It is currently unclear if IsrM requires Hfq for target regulation. doi:10.1128/9781555818524.ch25f1

the conserved ArcZ sRNA posttranscriptionally reduces the expression of the *sdaCB* and *tpx* mRNAs, both of which are present in *Salmonella* and *E. coli*. In addition, ArcZ also downregulates STM3216, a gene specific to *Salmonella* and predicted to act in chemotaxis control (Papenfort et al., 2009). Similarly, the conserved SgrS sRNA, known to control carbohydrate uptake (Vanderpool and Gottesman, 2007), employs a conserved domain at its 3′ end that serves to reduce expression of the *sopD* effector protein (Fig. 1) (Papenfort et al., 2012).

GcvB and RyhB constitute two additional Hfq-binding sRNAs which are conserved among *E. coli* and *Salmonella*. Both sRNAs control large regulons: GcvB reduces the expression of amino acid and oligopeptide uptake genes when nutrients are plentiful (Urbanowski et al., 2000; Sharma et al., 2007) and other genes involved in nitrogen homeostasis, such as the transcriptional regulator *lrp* (Sharma et al., 2011). The function of RyhB is equally specific but is dedicated to the control of intracellular iron levels (Masse et al., 2007). Interestingly, *Salmonella* harbors two *ryhB* loci: one homologous to the *ryhB* gene of *E. coli* and another copy, i.e., IsrE (also called RyhB-2 or RfrB), which is located on a horizontally acquired island of *Salmonella* (Ellermeier and Slauch, 2008; Padalon-Brauch et al., 2008; Sittka et al., 2008). While these two sRNAs evolved different patterns of transcriptional control, each of them contributes to growth inhibition under iron-limiting conditions, suggesting that their cellular activities are not fully redundant (Padalon-Brauch et al., 2008).

As outlined above, sRNA-mediated control of CsrA-like proteins is a highly conserved mechanism (Babitzke et al., 2009). *Salmonella csrA* mutants as well as overproducers of CsrA are defective for SPI-1 gene expression, which, in turn, causes attenuation of epithelial host cell invasion (Altier et al., 2000). Not surprisingly, combinatorial mutation of the two sRNA antagonists of CsrA, i.e., CsrB and CsrC, also affects SPI-1-mediated gene expression and results in decreased invasion rates (Fortune et al., 2006).

Finally, *Salmonella* also encodes a number of *cis*-regulatory RNA elements; however, evidence for direct roles in virulence is scarce. One example is the "fourU" type RNA thermometers upstream of the *Salmonella agsA* gene (Waldminghaus et al., 2007). Similarly, several riboswitches have been predicted in *Salmonella*. Among these, magnesium sensing by the *mgtA* leader probably constitutes one of the best-studied systems. The *mgtA* locus encodes a magnesium transporter, and binding of Mg^{2+} to its 5′ UTR results in transcription attenuation, a process which reduces overflowing Mg^{2+} import when concentrations are high (Cromie et al., 2006). Recently, the

same 5′ UTR was also reported to encode a short (18-amino-acid-long) proline-rich peptide which controls MgtA production via translation attenuation (Park et al., 2010). The biological role of combined Mg^{2+} and proline sensing has yet to be fully determined; however, multiple signal integration via the *mgtA* leader might indicate a physiologically relevant connection. In addition, Mg^{2+} homeostasis is also controlled by the *cis*-encoded AmgR sRNA. AmgR promotes specific degradation of *mgtC* in the polycistronic *mgtCBR* mRNA. Both *mgtCBR* and *amgR* are activated by the PhoP transcription factor (Fig. 2A), indicating that AmgR functions as a temporal regulator to alter MgtC and MgtB levels after the onset of PhoP-inducing conditions (Lee and Groisman, 2010).

Vibrio

Vibrio is a gram-negative marine bacterial genus which can grow planktonically in bodies of water or in association with marine animals and plants. This genus harbors commensals as well as pathogenic species, some of which are specific to marine animals (e.g., *Vibrio harveyi* and *Vibrio anguillarum*), whereas others cause disease in humans. Among the latter, *Vibrio cholerae*, the causative agent of the food- and waterborne disease cholera, constitutes one of the best-studied organisms of this genus and has a major impact on global health. Upon ingestion, *V. cholerae* colonizes the small intestine and can cause fatal diarrhea (Sack et al., 2004).

Vibrio species have seen a number of screens for sRNA regulators using either bioinformatics (Livny et al., 2005; Livny et al., 2008) or wet-lab-based approaches (Liu et al., 2009; Mandlik et al., 2011). However, knowledge on *Vibrio* sRNA regulators mainly stems from two, partially overlapping, areas of research: first, sRNAs that control or contribute to virulence and, second, sRNAs that act in QS. QS is a major regulator of *V. cholerae* pathogenesis, and therefore QS deficiency often results in attenuation of virulence (Higgins et al., 2007). Likewise, mutation of *hfq* abrogates QS signaling (Lenz et al., 2004) as well as colonization of the murine small intestine (Ding et al., 2004). While the sRNA determinants of the latter process are largely unknown, QS control via 4 or 5 paralogous sRNAs (Qrr1 to -5) is a well-studied process (Ng and Bassler, 2009). *Vibrio* QS involves up to three independent diffusible signaling molecules (autoinducers), which convey information on population density and species composition. At the molecular level, autoinducer detection triggers a complex signaling cascade at the center of which is the LuxO transcription factor. At low cell density

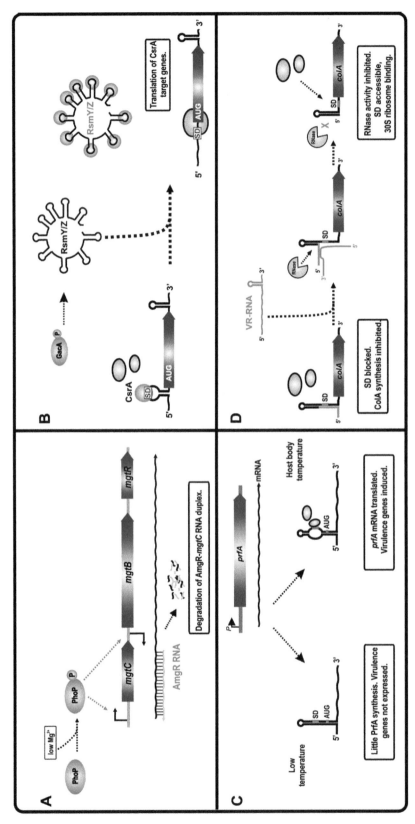

Figure 2. Repertoire of regulatory mechanisms employed by bacterial RNA regulators. (A) AmgR is a *cis*-encoded RNA that is expressed convergent to the *mgtC* ORF in *Salmonella*. The PhoPQ TCS activates expression of AmgR and *mgtC*, whereas interaction of both RNAs results in degradation of the RNA duplex. (B) CsrA is an RNA-binding protein that modulates expression by antagonizing translational initiation. CsrB-like RNAs (RsmY/Z) carry multiple CsrA binding sites and counteract CsrA activity via a titration mechanism. (C) The mRNA of the PrfA virulence factor of *L. monocytogenes* carries an RNA thermometer in its 5′ UTR. This regulatory structure supports translation initiation at high temperatures of (e.g., the mammalian host) but inhibits ribosome binding at lower temperatures. (D) The VR-RNA activates *colA* expression in *C. perfringens*. In the absence of VR-RNA, *colA* translation is inhibited by an internal stem-loop structure, blocking ribosome binding. VR-RNA binding to the 5′ UTR of *colA* induces an endonucleolytic cleavage which opens the RBS and generates a novel 5′ end which is resistant to RNase activity.
doi:10.1128/9781555818524.ch25f2

(LCD), LuxO is continuously phosphorylated and in conjunction with RNA polymerase-bound σ^{54} drives the transcription of the Qrr sRNAs which base-pair to and downregulate the *hapR* mRNA (Lenz et al., 2004; Bardill et al., 2011). Active HapR inhibits the expression of virulence genes and type III secretion, while it induces the production of proteases and several other genes. In contrast, at high cell density (HCD), LuxO phosphorylation is inhibited and HapR is expressed. Interestingly, Qrr4 also represses the expression of the *luxO* mRNA, giving rise to a posttranscriptional feedback loop which facilitates precise gene regulation during transition from LCD to HCD (Tu et al., 2010). Complexity in sRNA-mediated QS control is further increased by the activity of Qrr2 to –5, all of which are capable of activating *aphA* expression (Rutherford et al., 2011). AphA is a well-known transcriptional repressor of the *hapR* gene and the dominant QS regulator at LCD. The regulatory architecture linking the Qrr sRNAs, HapR, and AphA allows maximal AphA levels at LCD, while HapR production is strongest at HCD, with the Qrr sRNAs acting as the switch determining which of these transcription factors is expressed.

Finally, sRNA-mediated control of QS signaling may also function independently of *hapR* regulation. Some *V. cholerae* pathovars, including *V. cholerae* El Tor strain N16961 (the strain responsible for the current cholera pandemic), carry a frameshift mutation in the *hapR* gene, suggesting deficiency in QS regulation (Joelsson et al., 2006). The Qrr sRNAs were also found to directly activate the *vca0939* mRNA encoding a GGDEF domain-containing protein responsible for the production of the second messenger cyclic di-GMP (Hammer and Bassler, 2007). Therefore, sRNA-mediated regulation of QS also functions outside the AphA/HapR transcriptional network.

Vibrio also possesses regulatory RNAs antagonizing CsrA activity. The VarS/A two-component system (TCS) controls the expression of the CsrB, CsrC, and CsrD sRNAs and thereby indirectly controls the expression of target genes of CsrA. The CsrA protein also indirectly increases *luxO* expression and thus facilitates Qrr sRNA-mediated *hapR* repression. At LCD, the VarS/A TCS is inactive, and high CsrA activity will increase Qrr expression via LuxO production. By contrast, higher cell concentrations induce the VarS/A TCS and trigger CsrB/C/D expression, which will repress CsrA activity. This finally channels into QS by decreasing *luxO* levels and reducing Qrr expression, resulting in elevated *hapR* levels (Lenz et al., 2005).

Another *Vibrio*-specific RNA regulator is the σ^E-controlled VrrA sRNA. Similarly to MicA and RybB of *E. coli* and *Salmonella*, VrrA downregulates porin

production (OmpA and OmpT) upon damage to the outer membrane (Song et al., 2008; Song et al., 2010). In addition, VrrA also affects the production of outer membrane vesicles, and *vrrA* mutants display increased levels of intestinal colonization in an infant mice model. The latter effect might be partially based on increased production of the toxin-coregulated pilus protein, TcpA, the main colonization factor of *V. cholerae* (Song et al., 2008).

The two most recent additions to the list of *Vibrio* sRNA regulators affecting pathogenicity are TarA and TarB, both of which are controlled by the master virulence regulator ToxT (Richard et al., 2010; Bradley et al., 2011). In analogy to SgrS of *E. coli* (Vanderpool and Gottesman, 2004), TarA negatively regulates the major *Vibrio* glucose transporter *ptsG* (Richard et al., 2010), while TarB downregulates the secreted colonization factor TcpF (Bradley et al., 2011). Mutations of *tarA* and *tarB* impair virulence, though to different extents in the classical biotype strain of *V. cholerae* and the current pandemic El Tor biotype (Richard et al., 2010; Bradley et al., 2011). Finally, *Vibrio* also harbors a homolog of the widespread type of the Fur-regulated RyhB sRNA; however, iron does not seem limiting in a murine infection model, and *ryhB* mutation does not impair colonization of mammalian hosts (Davis et al., 2005).

Information on *cis*-acting RNA regulators affecting *Vibrio* pathogenesis is scarce. However, the identification of a riboswitch that modulates gene expression upon binding of cyclic di-GMP suggests that RNA-mediated regulators might play a role in biofilm formation, as well as in the expression of virulence determinants (Sudarsan et al., 2008). In particular, the identification of two cyclic di-GMP riboswitches upstream of the VC1722 and *gpbA* genes of *V. cholerae* highlights potential roles of these elements in cholera pathogenesis. The sugar-binding capacity of GbpA facilitates the attachment to human epithelial cells as well as to zooplankton, making it an important bifunctional molecule in the *Vibrio* infection process (Kirn et al., 2005).

Pseudomonas

Pseudomonas aeruginosa is a gram-negative opportunistic pathogen that causes serious infections in immunocompromised individuals and cystic fibrosis patients. This pathogen contains an arsenal of virulence factors, including cell association factors such as flagellum, lipopolysaccharide, alginate, and type IV pili, as well as extracellular factors such as exotoxin A, elastase, pyoverdine, and pyocyanin.

Hfq plays an important role in the expression of virulence factors in *P. aeruginosa*, and deletion of *hfq* in this pathogen causes pleiotropic effects (Sonnleitner et al., 2003), including decreased fitness on agar plates and rich medium, inhibited adaptation to stress-inducing conditions, and reduced motility and biofilm formation. In addition, deletion of *hfq* reduces virulence of *P. aeruginosa* PAO1 in both mice and *Galleria mellonella* larvae (Sonnleitner et al., 2003). Deletion of *hfq* affects ~5% of all transcripts in an RpoS-independent manner, and many of these are also regulated by QS systems (Sonnleitner et al., 2006).

Three sRNA screens have been conducted analyzing the intergenic regions of the *P. aeruginosa* genome (Livny et al., 2006; Gonzalez et al., 2008; Sonnleitner et al., 2008). These searches yielded a total number of ~40 sRNAs, including the four previously characterized sRNAs, PrrF1/2, RsmY, and RsmZ (see below).

Virulence of *P. aeruginosa* is modulated by QS systems that control the production of several virulence factors in a cell density-dependent manner. QS is governed by three interconnected regulatory systems: the LasI/LasR and RhlI/RhlR systems synthesize and recognize acylated homoserine lactones, whereas the third QS system functions through the action of the *Pseudomonas* quinolone signal (PQS). PQS acts as a coinducer of the LysR-type transcriptional regulator PqsR, which activates the transcription of several virulence factors along with the PQS-synthetic genes (Venturi, 2006). While the transcription of *pqsR* is positively regulated by LasR and negatively regulated by RhlR, PqsR is also subject to posttranscriptional regulation by the Hfq binding sRNA, PhrS (Sonnleitner et al., 2011). Under anaerobic conditions, PhrS is induced by the global transcriptional regulator ANR and activates the translation of PqsR by positively affecting the translation of the small open reading frame (ORF) upstream of *pqsR*. In addition to acting as a base-pairing sRNA, PhrS encodes a 37-amino-acid peptide whose function is still unknown.

The paralogous sRNAs, PrrF1 (116 nt) and PrrF2 (114 nt), are also implicated in the PQS-mediated QS of *P. aeruginosa*. Their expression is regulated by the ferric uptake regulator Fur and induced under iron limitation (Wilderman et al., 2004). As in the case of the Fur-regulated RyhB sRNA of *E. coli*, the targets of PrrF sRNAs include the superoxide dismutase gene *sodB*, the succinate dehydrogenase gene *sdhC*, and the bacterioferritin gene *bfrB*. Unexpectedly, the most highly regulated genes are *antABC* and *antR*, which are divergently transcribed and encode anthranilate dioxygenase and an AraC family transcriptional regulator controlling *antABC*, respectively. Since anthranilate is a precursor

of PQS, the level of PQS is significantly altered by PrrF1 and PrrF2, and therefore, these sRNAs provide a regulatory link between iron homeostasis and QS (Oglesby et al., 2008). Specifically in *P. aeruginosa*, a third, longer (325-nt) sRNA is encoded in the *prrF* locus, designated PrrH (Oglesby-Sherrouse and Vasil, 2010). Transcription of PrrH initiates at the 5′ end of *prrF1*, proceeds through *prrF1* terminator and *prrF1-prrF2* intergenic region, and terminates at the 3′ end of *prrF2*, suggesting that PrrH is produced by an antitermination mechanism. The expression of PrrH is repressed by free iron as well as heme, whereas only the latter is available in the host during infection. Using the unique 95-nt region that is absent from PrrF1/2, PrrH regulates the *nirL* gene, which encodes nitrite reductase.

Posttranscriptional control by the RNA-binding protein RsmA (CsrA) regulates many virulence genes of *P. aeruginosa* (Pessi et al., 2001). The activity of RsmA is antagonized by two sRNAs, RsmY and RsmZ (Fig. 2B), whose expression is exclusively regulated by the GacS/GacA TCS, the master regulator of virulence in *P. aeruginosa*. The GacS/GacA system is required for pathogenicity in plants, nematodes, and insects, as well as for both acute and chronic infections in murine infection models. The GacA response regulator transduces external regulatory signals and binds exclusively to two chromosomal loci to activate the expression of RsmY and RsmZ (Brencic et al., 2009). Both RsmY and RsmZ contain multiple GGA motifs to sequester RsmA (Fig. 2B) (Heurlier, et al., 2004; Kay et al., 2006; Sonnleitner et al., 2006). RsmA itself binds to the conserved GGA motifs located in the 5′ UTR of target mRNAs, resulting in translational repression. A structural analysis using a RsmA homolog from plant-pathogenic *P. fluorescens* revealed that RsmA binds specifically to the 5′-(A/U)CANGGANG(U/A)-3′ consensus sequence, which resembles the ideal SD (Schubert et al., 2007).

Two studies have suggested that the RsmA regulon in *P. aeruginosa* comprises >500 genes (Burrowes et al., 2006; Brencic and Lory, 2009), although there is little overlap between the genes proposed to be RsmA regulated in these two studies. This notwithstanding, it has been proposed that RsmA switches the expression of two mutually exclusive sets of virulence factors. The acute disease-associated genes, the T3SS, and the type IV pili, are positively regulated, while the chronic infection-associated genes, the type VI secretion system (T6SS), and exopolysaccharide components for biofilm production, are negatively regulated. All of the direct targets of RsmA, including the T6SS genes, are negatively regulated and contain a GGA motif adjacent to the RBS, indicating that RsmA translationally represses these genes and that positive regulation by RsmA is indirect (Brencic and Lory, 2009).

Although GacA activates only the transcription of RsmY and RsmZ, other factors also control the levels of these two sRNAs. The HptB phosphorelay participates in the control of biofilm and T3SS through negative regulation of *rsmY* (Bordi et al., 2010). The BfiSR two-component regulatory system is also implicated in biofilm development (Petrova and Sauer, 2010): BfiSR activates the transcription of *cafA*, encoding RNase G, which modulates the level of RsmZ to induce biofilm development.

Recently, expression profiling of sRNAs in a *gacA* mutant has identified a 120-nt sRNA, RgsA (Gonzalez et al., 2008). The expression of RgsA is under the direct control of the alternative sigma factor σ^S. Unlike RsmY and RsmZ, RgsA contains only a single GGA motif and is unable to sequester RsmA. Although RgsA has been shown to be involved in modulating the stress response by contributing to H_2O_2 resistance in *Pseudomonas fluorescens*, its role in the opportunistic pathogen *P. aeruginosa* is yet to be addressed.

One of the most exciting recent discoveries in *Pseudomonas* is CrcZ, a 407-nt-long sRNA that is regulated by RpoN and the CbrA/CbrB signal transduction pathway. CrcZ controls the selection of the energetically most favorable carbon source by antagonizing the activity of the Crc RNA-binding protein (Sonnleitner et al., 2009). Crc employs a mechanism similar to CsrA/RsmA but binds to CrcZ containing a pentarepeat of the sequence 5'-AANAANAA-3' with high affinity. In the presence of the preferred carbon source, i.e., succinate, the level of CrcZ is low, allowing Crc to bind the *amiE* mRNA (encoding an aliphatic amidase) and inhibit its translation. When nonpreferred carbon sources are present, the TCS CbrA/CbrB induces the CrcZ sRNA, resulting in sequestration of Crc and *amiE* translation. A *crc* mutant shows defects in biofilm formation, stress response, antibiotic resistance, and virulence (Linares et al., 2010). Proteomics analyses have identified 65 proteins belonging to the Crc regulon, 15 of which contain a predicted Crc-binding site in their RBS. Similar to the redundancy of CsrA/RsmA-binding sRNAs, several *Pseudomonas* species, e.g., *P. putida*, *P. fluorescens*, and *P. syringae*, but not *P. aeruginosa*, contain an additional CrcZ-like sRNA CrcY, the level of which is not significantly affected by the CbrA/CbrB TCS in *P. putida* (Moreno et al., 2011).

Yersinia

Among the 11 species included in the genus *Yersinia*, *Y. pestis*, *Y. pseudotuberculosis*, and *Y.* *enterocolitica* cause disease in humans. These three species differ extremely in their pathogenesis. The food- and waterborne enteropathogens *Y. enterocolitica* and *Y. pseudotuberculosis* cause a mild, self-limiting gastrointestinal disease referred to as yersiniosis (Heroven et al., 2008; Koo et al., 2011). In contrast, *Y. pestis* is the causative agent of both the bubonic and pneumonic forms of plague, which are fatal if not treated early. This divergence is particularly striking considering that *Y. pestis* has only evolved within the last 20,000 years from *Y. pseudotuberculosis* and their chromosomal genomes share more than 97% identity in 75% of all chromosomal genes (Achtman et al., 1999; Chain et al., 2004).

Deletion of *hfq* in *Yersinia* is accompanied by multiple defects, including increased sensitivity towards heat, osmotic, envelope, and oxidative stresses (Geng et al., 2009; Chao and Vogel, 2010). Interestingly, when comparing the impact of Hfq on several sRNAs at different temperatures, the RNA chaperone seems to be more relevant at elevated temperature in *Y. pestis* than in *Y. pseudotuberculosis* (Koo et al., 2011).

Common to all studied *Yersinia* species is a critical role for Hfq in pathogenesis: both *Y. pestis* and *Y. pseudotuberculosis hfq* mutants are strongly attenuated in animal infection models as well as in survival inside macrophage-like cells (Schiano et al., 2010). In *Y. enterocolitica*, Hfq promotes expression of the secreted toxin YstA (Nakao et al., 1995). Similar to that of other pathogens, like *Salmonella* or *Pseudomonas*, virulence of *Yersinia* depends on the translocation of bacterial effector proteins into host cells via a T3SS. Consequently, the defects in pathogenicity observed for *Y. pseudotuberculosis hfq* mutants can in part be ascribed to the reduced expression of several effector protein-coding genes (e.g., *yopE*, *yopH*, *yopJ*, and *yopT*) (Schiano et al., 2010).

A recent deep-sequencing approach identified in total ~50 orthologs of enterobacterial sRNAs in *Yersinia*, and there is evidence that their biological functions are at least partially conserved as well (Koo et al., 2011). These include GcvB, which had been shown previously to negatively regulate production of the dipeptide transporter subunit DppA in *Y. pestis* (McArthur et al., 2006), a known GcvB target in *E. coli* and *Salmonella* (Urbanowski et al., 2000; Sharma et al., 2007). Similarly, *Yersinia* possesses a Csr-like system comprising the regulatory protein CsrA as well as two sRNAs antagonizing its function, CsrB and CsrC (Heroven et al., 2008). Again similar to *Salmonella* and *E. coli*, *Y. pseudotuberculosis csrA* mutants display decreased motility and impaired adhesion and invasion in an *in vitro* cell culture model (Wei et al., 2001; Lawhon et al., 2003;

Heroven et al., 2008). A trait specific to *Yersinia* is the control of the global transcription factor RovA by CsrA. RovA activity contributes to host cell adhesion and colonization, and its expression is increased upon induction of CsrB and CsrC sRNAs, which function as molecular sponges for CsrA (Heroven et al., 2008).

Of the total of ~150 sRNAs identified by deep sequencing, only a minor fraction, termed Ysrs, was solely present in *Y. pseudotuberculosis,* while most candidates were found to be specific for the genus *Yersinia* (Koo et al., 2011). The role of sRNAs in *Y. pseudotuberculosis* virulence, in particular the contribution of the two Hfq-dependent sRNAs RybB (Ysr48) and Ysr29 as well as the Hfq-independent Ysr23 and Ysr35, was assessed in a murine infection model. None of the sRNA mutants displayed defects during *in vitro* growth, and deletion of RybB or Ysr23 also did not significantly alter virulence of the respective mutant strains. In contrast, mice infected with strains lacking Ysr29 or Ysr35 displayed increased survival compared to that with the wild type, and in a murine model of pneumonic plague, the *ysr35* deletion mutant of *Y. pestis* showed significant attenuation. However, little is known about the cellular functions of *Yersinia*-specific sRNAs, in particular about their targets. A proteomics approach identified eight proteins to be deregulated in the absence of *Y. pseudotuberculosis*-specific Ysr29, but whether this effect can be ascribed to a direct activity of the sRNA remains to be shown (Koo et al., 2011).

Francisella

Francisella tularensis is a gram-negative, highly infective, airborne pathogen of humans and animals and the etiological agent of the fatal disease tularemia. The relatively small (~2-Mb) genome encodes very few classical transcription regulators, such as TCSs or alternative sigma factors, raising the possibility that this organism might rely heavily on riboregulation. Using a combination of *in silico* predictions, cDNA cloning, and sequencing techniques, several sRNAs that are conserved across species as well as two highly expressed nonconserved sRNAs, FtrA and FtrB, were detected in *Francisella* (Postic et al., 2010). Individual deletion of the latter two sRNAs impacted the cellular levels of several *Francisella* mRNAs. However, the *ftrA/ftrB* deletion mutants did not show any phenotype in various *in vivo* infection models (Postic et al., 2010). This is in marked contrast to *hfq* mutants, which display attenuation in a mouse model for acute infection as well as for persistence (Meibom et al., 2009). Part of the attenuation phenotype can be attributed to aberrant expression of genes in the *Francisella* pathogenicity

island, but the exact regulatory mechanism remains to be determined.

Bordetella

Bordetella pertussis is a gram-negative human pathogen and the causative agent of whooping cough. Its virulence factors include adhesins (filamentous hemagglutinin, fimbriae, and pertactin) and toxins (pertussis toxin, adenylate cyclase-hemolysin bifunctional protein, dermonecrotic toxin, and tracheal cytotoxin). Expression of nearly all known virulence factors is regulated by the BvgAS TCS in response to environmental stimuli such as temperature, $MgSO_4$, or nicotinic acid (Cotter and Jones, 2003). The level of the BvgA protein is increased ~50-fold under inducing conditions, while the level of *bvgA* mRNA shows only 3-fold upregulation (Scarlato et al., 1991). The *bvgAS* genes are transcribed from at least three promoters, whereas another promoter is located ~20 nt from the start codon of *bvgA* in the opposite direction. This divergent promoter drives transcription of an ~190-nt antisense RNA designated BprC (Hot et al., 2011). Since BprC is complementary to the 5′ UTR of *bvgA*, this antisense RNA may account for the posttranscriptional regulation of *bvgA* mRNA, possibly by altering a secondary structure upstream of the RBS (Scarlato et al., 1991).

The only genome-wide search for *Bordetella* sRNAs was a bioinformatics approach to predict sRNA genes in intergenic regions of *B. pertussis* that are conserved among the other *Bordetella* species, *B. bronchiseptica, B. parapertussis,* and *B. avium* (Hot et al., 2011). Of the 20 candidate sRNAs, 13 were confirmed by Northern analysis and designated BprA to BprN. The *bprA, bprD, bprE, bprJ, bprL,* and *brpN* loci each encode two overlapping transcripts of different lengths. Among them, BprJ2 sRNA is transcribed only under BvgAS-inducing conditions in *B. pertussis,* implying that BprJ2 belongs to the BvgAS regulon (Hot et al., 2011). While targets of the Bpr sRNAs have remained elusive, there might be more sRNAs that mediate species-specific control of the BvgAS regulon (Cummings et al., 2006). It is also noteworthy that *B. pertussis* has a predicted Hfq homolog (Chao and Vogel, 2010), which has yet to be investigated experimentally.

Helicobacter and *Campylobacter*

Helicobacter pylori colonizes the stomachs of more than 50% of the world's population and is known as the causative agent of gastritis, peptic ulcer, and gastric cancer (Cover and Blaser, 2009). Many studies have focused on the virulence and acid resistance mechanisms of *H. pylori*, but almost nothing is known about posttranscriptional regulation and sRNAs in this microaerophilic, gram-negative Epsilonproteobacterium. Because of its small genome size (1.67 Mb, *H. pylori* 26695), low number of transcriptional regulators, including only three sigma factors (RpoD, FliA, and RpoN) and four TCSs (Tomb et al., 1997), and the lack of the RNA chaperone Hfq (Chao and Vogel, 2010), it was suggested that *H. pylori* has only basic regulatory circuits (Mitarai et al., 2007). Moreover, as none of the enterobacterial sRNAs, except for the housekeeping RNAs (tmRNA, SRP RNA, and M1 RNA), are conserved in *H. pylori*, it was considered to completely lack riboregulation (Mitarai et al., 2007). However, already initial studies based on bioinformatics predictions and a small-scale cDNA library cloning approach predicted the existence of several potential sRNAs in *H. pylori* (Livny et al., 2006; Xiao et al., 2009a; Xiao et al., 2009b). Moreover, a novel differential RNA sequencing approach (dRNA-seq) selective for the 5′ end of primary transcripts recently revealed an unexpected complex and compact transcriptional output from the small *Helicobacter* genome (Sharma et al., 2010). This included the discovery of more than 60 small RNAs, including potential regulators of *cis-* and *trans*-encoded mRNA targets.

Among these novel sRNAs, the *H. pylori* dRNA-seq study revealed a homolog of 6S RNA which was previously undiscovered in the Epsilon-subdivision (Barrick et al., 2005). This 180-nt-long RNA from *H. pylori* can fold into the characteristic long hairpin structure with a central bulge by which *E. coli* 6S RNA titrates RNA polymerase in stationary phase (Wassarman and Storz, 2000). In addition, the presence of 6S RNA-associated pRNAs (product RNAs) (Wassarman and Saecker, 2006) in the dRNA-seq data indicated that the 6S homolog of *Helicobacter* is functional. It still needs to be investigated whether *H. pylori* 6S RNA also has an impact on virulence like in *Legionella* (Faucher et al., 2010). Among the other housekeeping RNAs, tmRNA was shown to be essential and to be required for stress response in *H. pylori* (Thibonnier et al., 2008).

Several of the abundant sRNAs identified by dRNA-seq in intergenic regions of *Helicobacter* are potential candidates for *trans*-encoded antisense RNAs. A prominent example is the 87-nt-long HPnc5490 RNA, which was predicted to interact with the 5′ UTR of *tlpB* mRNA, encoding one of the chemotaxis receptors in *Helicobacter* (Sharma et al., 2010). Analyses of whole-cell protein fractions as well as quantitative real-time PCR of wild-type and HPnc5490 mutant strains confirmed downregulation of the TlpB protein and the *tlpB* mRNA by HPnc5490 sRNA.

Besides *trans*-encoded RNAs, a massive number of *cis*-encoded antisense RNAs (>900) have been identified in *H. pylori* (Sharma et al., 2010). For almost half of all ORFs, including several housekeeping genes, at least one associated antisense transcriptional start site was detected, suggesting genome-wide antisense regulation. Some of these *cis*-encoded RNAs are located on the opposite strand of small hydrophobic proteins, which resemble small bacterial toxins or antimicrobial peptides. For example, a family of six structurally related ~80-nt sRNAs, IsoA1 to -6 (RNA inhibitor of small ORF family A), is expressed antisense to small ORFs, AapA1 to -6 (antisense RNA-associated peptide family A), of homologous 22- to 30-amino-acid peptides (Sharma et al., 2010). *In vitro* translation assays indicated that translation of the small ORFs is strongly and specifically inhibited by the cognate antisense RNAs, and thus, the *aapA-isoA* loci might constitute the first examples of class I toxin-antitoxin systems in *H. pylori*.

An increasing number of antisense RNAs have also been detected in several other bacteria (Thomason and Storz, 2010). Since *H. pylori* lacks homologs of the endonucleolytic RNases E/G enzymes (Tomb et al., 1997; Parkhill et al., 2000), antisense-mediated processing by the double-strand-specific RNase III could be a major mechanism of gene regulation in this pathogen. Expression of an artificial antisense RNA has been successfully used to repress the essential *ahpC* gene, encoding alkyl hydroperoxide reductase (Croxen et al., 2007). Furthermore, a *cis*-encoded 292-nt-long antisense sRNA encoded opposite to the *ureAB* operon in *Helicobacter pylori* has recently been demonstrated to negatively regulate urease activity by binding to the 5′ end of *ureB* and mediating truncation of the *ureAB* mRNA (Wen et al., 2010). *In vitro* and *in vivo* approaches demonstrated that the 5′ *ureB* sRNA is induced under neutral pH conditions by the unphosphorylated ArsR response regulator of the ArsRS TCS, which activates transcription of the urease operon in its phosphorylated form in response to acid (Pflock et al., 2005).

Since there is no recognizable Hfq homolog in the Epsilonproteobacteria, the question arises whether riboregulation is independent of an RNA chaperone or whether a different protein replaces the function of Hfq in *H. pylori*. The finding that the RpoN chaperone HP0958 posttranscriptionally modulates the amount of *flaA* mRNA in *H. pylori* indicates that there might be Hfq-unrelated proteins involved in riboregulation (Douillard et al., 2008). In addition, *H. pylori* carries a homolog of the RNA-binding protein CsrA which was shown to be required for full motility, biofilm formation, survival under oxidative stress, and infection of epithelial cells and mice by *H. pylori* and the related pathogen

Campylobacter jejuni (Barnard et al., 2004; Fields and Thompson, 2008). However, homologs of the CsrB/C RNAs which have been shown to regulate the activity of CsrA are still unknown for both pathogens.

Similar to the case with *H. pylori*, almost nothing is known about posttranscriptional regulation in the related Epsilonproteobacterium *Campylobacter jejuni,* a leading cause of bacterial foodborne gastroenteritis (Young et al., 2007). Except for the housekeeping RNAs and 6S RNA, none of the *H. pylori* sRNAs are conserved in *Campylobacter*. Recently, conventional RNA-seq analysis of *Campylobacter jejuni* NCTC11168 revealed five intergenic regions which could harbor sRNA genes (Chaudhuri et al., 2011), and a comparative dRNA-seq analysis of multiple strains indicates that the *Campylobacter* genome encodes a variety of conserved and strain-specific *cis*- and *trans*-encoded sRNAs (G. Dugar and C. Sharma, unpublished data). Considering their small genome size compared to that of *E. coli, Helicobacter* and *Campylobacter* express an equal number of sRNAs, and the first examples of functional sRNAs are now being reported for these clinically relevant human pathogens.

Legionella

Legionella pneumophila is a gram-negative opportunistic human pathogen that causes Legionnaires' disease, a severe progressive pneumonia with significant mortality (Shands and Fraser, 1980). *L. pneumophila* can be found in almost all natural and engineered water systems and is transmitted via contaminated water aerosols. As for many intracellular pathogens, the bacterium's survival and spread depends on its ability to replicate inside eukaryotic phagocytic cells. In humans *L. pneumophila* replicates inside macrophages within a modified vacuole, whereas outside this host many different phagocytic protozoa such as *Acanthamoeba castellanii* and *Hartmannella vermiformis* support *L. pneumophila* replication (Faucher and Shuman, 2011).

Bioinformatic, microarray, and deep-sequencing approaches identified 79 sRNAs in *L. pneumophila* (Faucher et al., 2010). Up to now, for three sRNAs, namely, RsmY, RsmZ, and 6S RNA, essential roles in controlling virulence-related properties have been reported.

6S RNA in *L. pneumophila* is present in two copies (6S-1 RNA and 6S-2 RNA) and seems to have a different function in gene regulation (Weissenmayer et al., 2011). In *Legionella*, many genes are positively regulated by 6S RNA, and the majority of these genes are σ^S independent. This is in contrast to the case with *E. coli*, where 6S RNA mainly acts as a negative regulator of gene expression during stationary phase (Trotochaud and Wassarman, 2004). Deletion of 6S-1 RNA significantly inhibited the bacterial growth inside host cells, indicating that this sRNA plays an essential role for *L. pneumophila* to survive as an intracellular pathogen (Faucher et al., 2010).

In *L. pneumophila*, several TCSs have been shown to play major roles in regulatory processes that ensure a precise timing of virulence pathways. One is the LetA/LetS (*Legionella* transmission activator and sensor, respectively) system, which is necessary for efficient host transmission and survival in the environment. This TCS has been shown to regulate the CsrB homologs RsmY/RsmZ, which sequester CsrA and thereby abolish its activity as a translational repressor of genes required for intracellular multiplication (Rasis and Segal, 2009). Thus, as for many species covered in this chapter, these sRNAs have a key function in regulating the switch from a replicative/nonvirulent to a transmissive/virulent pathogen (Cazalet and Buchrieser, 2010).

Borrelia

Borrelia burgdorferi belongs to a distinct phylum of gram-negative bacteria, the spirochetes, which are characterized by their long helically coiled shape. Occasionally transmitted to humans, *B. burgdorferi* can cause Lyme disease (Anguita et al., 2003).

The relatively small genome of *B. burgdorferi* consists of a linear chromosome of ~1 Mb and multiple plasmids combined of 0.6 Mb (Fraser et al., 1997). To identify noncoding RNAs other than the housekeeping 4.5S RNA, RNaseP RNA, and tmRNA (as well as predicted homologs thereof), the *B. burgdorferi* genome was screened using different bioinformatics approaches (Ostberg et al., 2004). Two sRNA candidates could be verified by Northern blot analysis: BsrA, a small RNA previously misannotated as a tmRNA homolog, and a transcript from an intergenic region (*bb0577-bb0578*) displaying complementarity to the 5′ UTR of *rpoS*, which is now referred to as DsrA$_{Bb}$ (Lybecker and Samuels, 2007).

While *E. coli* RpoS is considered a global regulator of the general stress response (Hengge-Aronis, 2002), its *B. burgdorferi* homolog mainly controls virulence gene expression (Caimano et al., 2004). *B. burgdorferi rpoS* is expressed from two distinct promoter elements: a proximal, RpoN-dependent site which accounts for mRNA production under HCD conditions (Studholme and Buck, 2000; Samuels, 2011) and a more distal site giving rise to a longer transcript predicted to fold into a self-inhibitory stem-loop structure (Lybecker and Samuels, 2007).

DsrA$_{Bb}$ was demonstrated to stimulate RpoS production upon temperature increase under LCD conditions, presumably by a base-pairing mechanism as described for *E. coli* (Sledjeski et al., 1996). For its functional homology to the *E. coli* regulator, the RNA was named DsrA$_{Bb}$ (Lybecker and Samuels, 2007), although it is noteworthy that the two sRNAs do not share sequence or structural similarity (Repoila and Gottesman, 2001; Lybecker and Samuels, 2007).

For a long time, the surprisingly low number of verified sRNAs in *B. burgdorferi* was assigned to the obvious lack of both the RNA chaperone Hfq and the endonuclease RNase E in this organism (Ostberg et al., 2004). A seminal study queried the genome for small ORFs carrying the characteristic Sm1 and Sm2 RNA-binding motifs and identified the previously uncharacterized BB0268 as an atypical homolog of Hfq in *B. burgdorferi*, referred to as Hfq$_{Bb}$ (Lybecker et al., 2010). Although sharing only 33% similarity and 12% identity with *E. coli* Hfq, the predicted tertiary structure of Hfq$_{Bb}$ can be superimposed onto the experimentally solved structures of *S. aureus* and *E. coli* Hfq homologs (Schumacher et al., 2002; Link et al., 2009). Furthermore, Hfq$_{Bb}$ was shown to partially cross-complement an *E. coli* mutant with regard to sRNA-mediated translational activation of an *rpoS::lacZ* reporter fusion. In addition, ectopic expression of *E. coli* Hfq could rescue both a growth defect and abnormal cell shape phenotype of *hfq$_{Bb}$* mutants. Interestingly, virulence attenuation in a mouse model of *Borrelia* infection was restored only by complementation with the *B. burgdorferi* and not the *E. coli hfq* gene. Apart from the already-mentioned sRNA DsrA$_{Bb}$, which controls RpoS synthesis in early growth phases upon temperature increase, no other sRNAs or mRNAs have so far been identified to interact with Hfq$_{Bb}$. However, the various phenotypes of the *hfq$_{Bb}$* mutant provide substantial evidence that further sRNAs exist which might play a role in the virulence mechanisms of *B. burgdorferi*.

Neisseria

Neisseria spp. are gram-negative cocci which belong to the most common colonists of the human upper respiratory tract. The genus *Neisseria* encompasses two pathogenic species, namely, *N. gonorrhoeae*, the causative agent of gonorrhea, and *N. meningitidis*, which causes severe systemic infections leading to meningitis and deadly sepsis (Johnson, 1983). *N. meningitidis* is a major cause of morbidity and mortality during childhood in industrialized countries and accounts for epidemics in Africa and in Asia (Genco and Wetzler, 2010). *Neisseriae*

spp. include ~50 species of bacteria that are naturally competent for transformation during all phases of growth, with the capacity to incorporate naked DNA from the extracellular environment (Chen and Dubnau, 2004).

The first reported sRNA in the genus *Neisseria* is NrrF (for neisserial regulatory RNA responsive to iron [Fe]), which is involved in iron homeostasis. This Fur-regulated sRNA downregulates the expression of *sdhCDAB* operon, coding for succinate dehydrogenase, under iron starvation and has a well-conserved ortholog in *N. gonorrhoeae* (Mellin et al., 2007; Metruccio et al., 2009). Interestingly, neither the stability nor the regulation of the *sdhCDAB* operon was affected in an *hfq* deletion strain, suggesting that NrrF-mediated regulation is Hfq independent (Mellin et al., 2010). Further studies pointing to an involvement of other sRNAs in the regulation of *N. meningitidis* virulence were obtained by the analysis of *hfq* deletion strains. Those strains displayed significant attenuation and showed altered expression levels of more than 25 different proteins, including major virulence factors (Fantappie et al., 2009; Pannekoek et al., 2009).

The second sRNA identified in meningococci is AniS, which has predicted homologs in all currently available *Neisseria* genomes. AniS is an anaerobically induced sRNA whose expression is strictly dependent on the FNR transcriptional regulator. Hfq is required for the downregulation of AniS target mRNAs, although in contrast to its well-characterized role in promoting stability of many sRNAs, Hfq promotes the decay of AniS (Fantappie et al., 2011).

Application of RNA-seq to *N. gonorrhoeae* cultivated under anaerobic conditions identified another anaerobically induced sRNA termed FnrS (Isabella and Clark, 2011). The discovery of anaerobically induced sRNAs in *N. meningitidis* and *N. gonorrhoeae* suggests a major role for sRNAs to quickly sense changing oxygen conditions.

Chlamydia

The genus *Chlamydia* comprises gram-negative, obligate intracellular bacteria, among which *Chlamydia trachomatis* and *Chlamydia pneumoniae* constitute important human pathogens. *C. trachomatis* gives rise to several distinct clinical syndromes, including trachoma and urogenital infections, whereas *C. pneumoniae* causes atypical pneumonia. Upon intracellular uptake, chlamydiae undergo a drastic structural and physiological transformation from elementary bodies (EBs), representing the extracellular and metabolically inert stage of their life cycle, to larger and metabolically active reticulate bodies

(RBs) (Moulder, 1991). The infection process begins with the engulfment of EBs by the eukaryotic cell, where they are internalized in vacuole-bound membranes called inclusion bodies. Here, the core of condensed chromatin disperses, leading EBs to differentiate into active RBs. The RBs multiply by binary fission for about 18 to 48 h postinfection, after which they begin to reorganize and condense into inert EBs for release from infected cells (Shaw et al., 2000). In addition to significant changes in size and membrane permeability from EBs to RBs, chromatin organization varies significantly between the two phases.

Transition from RB to EB requires the expression of two histone-like proteins, Hc1 and Hc2, which bind to and densely compact the chromatin, a process that excludes transcriptional regulators from DNA and thereby renders the EB transcriptionally inactive (Perara et al., 1992). Overexpression of Hc1 (encoded by the *hctA* gene) is toxic to *E. coli*, a fact that allowed a heterologous genetic screen searching for chlamydial genes that can act as suppressors of Hc1 lethality (Grieshaber et al., 2004). Indeed, two genetic loci have been selected that rescued *E. coli* growth, one of which was subsequently identified to encode an sRNA, termed IhtA (inhibitor of hctA translation). IhtA seems to act at the level of translation inhibition, since overexpression did not alter transcription or transcript stability of *hctA*. Expression studies further indicate that IhtA is present in RBs but repressed in EBs, where Hc1 becomes upregulated, suggesting that IhtA restricts Hc1 expression in RBs but not in EBs (Grieshaber et al., 2006).

A deep-sequencing approach to discover noncoding RNAs identified 42 genomic and one plasmid-encoded novel sRNAs in *C. trachomatis* (Albrecht et al., 2010). Sixteen putative sRNAs were found to be located in intergenic regions, suggesting that they are *trans*-acting sRNAs, and 9 of these were confirmed by Northern blotting (Albrecht et al., 2010). One of these nine novel sRNAs, ctrR0332, was expressed abundantly and differentially in EBs and RBs, indicating a possible role in the developmental cycle of *C. trachomatis*. Extension of the RNA-seq approach to *Chlamydia pneumoniae* revealed 75 novel putative noncoding RNAs, including 20 *trans*-encoded RNAs, 47 *cis*-encoded RNAs, and 8 candidates encoded in sense direction with respect to annotated ORFs. Of the 13 sRNAs which were confirmed by Northern blotting, only 3 have been conserved in *C. trachomatis*, whereas most of the remaining novel sRNAs are specific to *C. pneumoniae* (Albrecht et al., 2011). Considering the wealth of sRNAs discovered in *Chlamydia* (Albrecht et al., 2010; Abdelrahman et al., 2011; Albrecht et al., 2011), it is interesting that no Hfq homolog has yet been found.

Shigella and Pathogenic *E. coli*

Shigella and pathogenic *Escherichia coli* are members of a closely related group of gram-negative bacteria that cause a wide range of human diseases, from different forms of intestinal infections associated with diarrhea to urinary tract infections or neonatal meningitis. A recent systematic bioinformatics analysis of several *Shigella* genomes complemented by microarray and Northern blot experiments identified at least nine novel sRNAs encoded on the 230-kbp *Shigella* virulence plasmid (Peng, et al., 2011). However, at present the roles of these sRNAs and any potential link to *Shigella* virulence are unclear.

A prominent example for a *trans*-acting sRNA which has been linked to *Shigella* virulence is RyhB. Apart from controlling iron-responsive genes in *E. coli* and *Salmonella* as described in the section on *Salmonella* above and by Massé and Gottesman (Massé and Gottesman, 2002). Overexpression of RyhB in *Shigella dysenteriae* has been found to repress the gene encoding the master regulator of *Shigella* virulence, VirB, which, in turn, is essential for expression of many genes encoding virulence or stress tolerance functions (Murphy and Payne, 2007). This intricate connection might serve to coordinate gene expression when *Shigella* enters host cells.

A well-characterized example from *Shigella flexneri* is the *cis*-encoded antisense RNA RnaG, which represses expression of the invasion factor-encoding gene *icsA* (*virG*) by a transcriptional attenuation mechanism, rather than by interfering with translation initiation or mRNA stability (Giangrossi et al., 2010). This 450-nt-long regulatory RNA makes extensive base-pairing with the nascent *icsA* transcript and triggers the formation of an intrinsic terminator structure, which, in turn, causes premature transcription termination. An 80-nt stretch at the 5′ end of RnaG is essential and sufficient for this effect, and a model for the interaction sites between RnaG and its *icsA* target has been suggested (Tran et al., 2011). Apart from these two examples, little is known about the roles of other sRNAs for virulence.

In sharp contrast to the vast knowledge about regulatory RNAs in nonpathogenic *E. coli* K-12, very little is known about riboregulation in the diverse pathogenic *E. coli* strains. Nevertheless, an important role of *hfq* has been observed in adherent invasive (Simonsen et al., 2011) or uropathogenic *E. coli* (Kulesus et al., 2008), as well as enterohemorrhagic *E. coli* O157:H7 (EHEC). For the latter group of pathogenic *E. coli*, the influence of Hfq on virulence gene expression appears to be strongly dependent on the exact strain background. Although it has been shown in two publications that Hfq represses expression of virulence genes encoded in the locus of enterocyte effacement of the widely studied EHEC strain EDL93 (Hansen and Kaper, 2009; Shakhnovich et al., 2009), a more recent article reported that Hfq in EHEC strain 86-24 is a positive regulator of locus of enterocyte effacement genes (Kendall et al., 2011). The molecular basis of opposite roles of Hfq for virulence regulation in these related EHEC strains is currently unclear.

GRAM-POSITIVE PATHOGENS

Bacillus

Bacilli are rod-shaped, gram-positive bacteria that are ubiquitous in nature. One defining characteristic of bacilli is their ability to form spores in response to nutrient starvation which are highly resistant to antimicrobial agents, including antibiotics and UV light (Kunst et al., 1997). The nonpathogenic soil bacterium *Bacillus subtilis* is historically the model organism for gram-positive bacteria, but the group also contains important pathogens such as *B. anthracis* and *B. cereus*. *B. anthracis* is one of the most potent human pathogens and became notorious after its exploitation as a bioterrorism agent in 2001 (Cummings and Relman, 2002). *B. cereus* causes self-limiting foodborne gastrointestinal infections and, rarely, chronic skin infections. Pathogenesis of both *B. anthracis* and *B. cereus* infections is associated with the production of potent toxins and other virulence factors (e.g., capsules; see chapter 13).

Almost 70 sRNAs have been predicted in *Bacillus subtilis*, primarily through high-throughput screens using tiling microarrays (Rasmussen et al., 2009) and transcriptional start site mapping by dRNA-seq (Irnov et al., 2010). Regulation by sRNAs in *Bacillus* is poorly characterized, and the role for Hfq is as yet unclear. Deletion of the Hfq homolog (YmaH), the crystal structure of which has recently been identified (Someya et al., 2012), had no effect on growth or sporulation (Silvaggi et al., 2005), and a number of studies have shown sRNA activity to be independent of Hfq. For example, Hfq binds to both the sRNA SR1 and its mRNA target, *ahrC*, and association of SR1 sRNA with *ahrC* mRNA causes the inhibition of *ahrC* mRNA translation. However, contrary to the well-defined chaperone activity of Hfq in gram-negative bacteria, Hfq neither stabilizes SR1 nor promotes duplex formation between SR1 and *ahrC*; instead, it was required for activation of *ahrC* translation (Heidrich et al., 2006; Waldminghaus et al., 2007). Interestingly, comparison of known Hfq binding sites and the use of systematic evolution of ligands by exponential enrichment (SELEX) to identify preferred

substrates of *B. subtilis* Hfq indicated a preference for AG repeats over the AU-rich sequences typical *in E. coli* Hfq-binding sites (Someya et al., 2012). This altered motif has also been identified in additional gram-positive bacteria (Nielsen et al., 2010), suggesting a general distinction between Hfq targets in gram-positive and gram-negative bacteria.

Iron acquisition is an essential component of bacterial pathogenesis. In many bacteria, the iron-sensing ferric uptake repressor (Fur) acts as a global regulator to maintain iron homeostasis. Gram-negative bacteria often employ a Fur-controlled Hfq-dependent sRNA (e.g., RyhB in *E. coli*) to rapidly downregulate nonessential iron utilizing proteins while simultaneously upregulating the machinery required for iron uptake (Masse et al., 2007). In *B. subtilis*, a similar iron-saving response is mediated by the Fur-regulated sRNA FsrA (Gaballa et al., 2008). FsrA is strongly induced in response to iron limitation and is essential for growth under iron-limiting conditions. Deletion of *fsrA* dramatically limits the ability of the cell to prevent expression of nonessential iron-utilizing proteins, thus depleting the iron available for essential iron-dependent proteins. Many of the phenotypes associated with *fur* deletion in *B. subtilis*, such as inability to grow on succinate or fumarate as a sole carbon source or ammonium as a sole nitrogen source, are linked to expression of FsrA, indicating an important role for this sRNA in posttranscriptional control in *Bacillus* (Gaballa et al., 2008). The action of FsrA is independent of the *Bacillus* Hfq homolog (YmaH) but involves three Fur-regulated small basic proteins (FbpA to -C), which are predicted to act as RNA chaperones and modulators of FsrA function (Gaballa et al., 2008).

One focus of sRNA research in *Bacillus* has been spore formation. Shortly after the start of sporulation, the developing cell forms compartments called the forespore and the mother cell, which contain a copy of the chromosome but undergo dramatically different regulatory cascades. Considering that over 500 genes are differentially regulated during sporulation, it is not surprising that a number of sRNAs have been identified which alter their expression pattern during this morphogenesis. A seminal study (Silvaggi et al., 2006) identified a number of sRNAs (termed Sur for small untranslated RNA) which were differentially regulated in response to sporulation. The expression of one sRNA, SurC, which is conserved in *B. anthracis*, is controlled by the mother cell-specific sigma factor σ^K, which triggers the expression of SurC in the mother cell (but not the forespore) during sporulation (Silvaggi et al., 2006). Unfortunately, no functional studies were performed to characterize the role of SurC, and thus, its contribution to sporulation remains unclear. Expression of the sRNA BsrF decreases ~8-fold during sporulation, tentatively indicating a role in repression of proteins required for spore formation (Preis et al., 2009). Interestingly, BsrF expression is activated by the pleiotropic global transcriptional regulator CodY, which coordinates expression of diverse regulons, including nutrient acquisition, cell differentiation, and virulence pathways (Sonenshein, 2005). Considering that the regulatory role for CodY is negative for the vast majority of target genes, the positive regulation of BsrF by CodY is an interesting aspect of this sRNA and potentially allows more rapid downregulation of CodY-repressed genes by the coupled posttranscriptional targeting by BsrF (Preis et al., 2009; Saito et al., 2009). BsrF is therefore an interesting candidate for further investigation.

Toxin-antitoxin (TA) systems were originally thought to function simply as selfish elements and are known to ensure the maintenance of plasmids or other mobile genetic elements. More recently, a diverse range of functions have been identified for TA systems, such as adaptation to stress and pathogenicity (Blower et al., 2011). RNA antitoxin A (RatA), which was identified in a screen for sRNAs encoded in intergenic regions of *Bacillus subtilis* (Silvaggi et al., 2005), is a 222-nt transcript that forms part of a type I TA system. In this system, expression of *tpxA*, which encodes a 59-amino-acid toxic peptide that causes cell lysis, is controlled by an antisense-encoded RNA (RatA). Interestingly, the *txpA-ratA* module is located within the ~42-kb *skin* (σ^K intervening) element that, during sporulation, is excised by recombination to form circular DNA. The concomitant juxtaposition of the flanking chromosomal DNA segments restores the coding sequence for the mother cell-specific alternative sigma factor (σ^K) (Kunkel et al., 1990). This genomic rearrangement is specific to the mother cell and results in terminal differentiation and ultimately cell lysis. Considering the regulatory significance of the *skin* region (Kunkel et al., 1990), it is possible that the repression of TpxA synthesis by RatA is eventually overcome on the excised circular DNA in the mother cell, promoting cell lysis by the toxic molecule and release of the spore.

Listeria

Listeria monocytogenes is a gram-positive non-spore-forming facultative intracellular pathogen. It is one of the most virulent foodborne pathogens causing fatal infection in humans and animals, with a mortality rate approaching 30% (Izar et al., 2011). *L. monocytogenes* is also one of the best-studied pathogens and serves as a classic model pathogen for investigation

of bacterial pathogenesis and host-pathogen interactions (Cossart, 2007). *Listeria* is able to thrive under a variety of conditions, including the low temperature in refrigerators, and can invade various types of host cells. Upon encountering stressful environments or the hosts, *Listeria* primarily uses an alternative sigma factor, σ^B, and a master virulence regulator, PrfA, to coordinate gene expression (Nielsen et al., 2008). PrfA is a transcriptional regulator encoded on a 10-kb genomic island which controls the expression of additional essential virulence factors encoded by the same gene cluster or elsewhere, including the well-known pore-forming toxin, listeriolysin (Cossart, 2007; Cossart and Toledo-Arana, 2008). In addition, VirR, a two-component response regulator, has also been found to play a critical role in *Listeria* pathogenesis (Mandin et al., 2005).

The housekeeping 4.5S RNA of the signal recognition particle was the first sRNA identified in *L. monocytogenes* (Barry et al., 1999). The first *cis*-regulatory RNA was discovered in *Listeria* in 2002, a thermometer located in the 5′ UTR of *prfA* which senses elevated temperature (37°C) in order to initiate the virulence gene cascade (Fig. 1C) (Johansson et al., 2002). The first *trans*-encoded sRNAs were identified by coimmunoprecipitation of Hfq (Christiansen et al., 2006) and by *in silico* searches (Mandin et al., 2007). Subsequently, more than 180 *trans*-acting sRNAs, as well as *cis*-encoded antisense RNAs and riboswitches, were discovered in genome-wide analyses of transcriptomes from *in vitro*- and *in vivo*-grown *Listeria* by tiling array and deep-sequencing approaches (Oliver et al., 2009; Toledo-Arana et al., 2009; Izar et al., 2011; Mraheil et al., 2011a; Mraheil et al., 2011b).

A function for *trans*-acting sRNAs in *Listeria* virulence was implicated by studying the role of Hfq in the virulence of *Listeria*. Disruption of *hfq* reduced bacterial counts in mouse organs upon infection (Christiansen et al., 2004). Interestingly, while in many bacteria the Hfq levels are constant throughout growth, in *Listeria* Hfq expression is regulated by σ^B. The same group also identified three Hfq-associated sRNAs (LhrA, -B, and -C) (Christiansen et al., 2006) and showed that Hfq was required for the stability of LhrA and the regulation of one of its targets (Nielsen et al., 2010; Nielsen et al., 2011), making *Listeria* a paradigm for Hfq-mediated riboregulation in gram-positive species. We note, however, that many other sRNAs of *Listeria* do not seem to be dependent on Hfq, or at least their cellular abundance is not influenced by Hfq (Mandin et al., 2007; Toledo-Arana et al., 2009).

The Hfq-dependent LhrA sRNA has been characterized in detail (Nielsen et al., 2010; Nielsen et al.,

2011). Almost 300 genes were affected by deletion of *lhrA* in *Listeria*. Interestingly, half of them also belong to the σ^B regulon, indicating a functional overlap or putative link between *lhrA* and σ^B. At least three of these genes, i.e., lmo0850, lmo0302, and *chiA* (encoding a chitinase), were verified to be directly targeted by LhrA via base-pairing. Interestingly, *chiA* is regulated by both PrfA and σ^B, and it has been shown to be critical for *Listeria* infection in mice (Chaudhuri et al., 2010; Larsen et al., 2010). However, the physiological role of LhrA sRNA in *Listeria* virulence has not been investigated.

RliB is encoded in the intergenic region between lmo0509 and lmo0510 in *L. monocytogenes*, with five copies of 29-nt-long repeat sequences (Mandin et al., 2007). Only two copies of the repeat sequences are present in the animal-pathogenic strain, *Listeria ivanovii*, while the sRNA is fully absent in nonpathogenic strains (Mandin et al., 2007). RliB is strongly induced when *Listeria* is exposed to human blood and intestinal lumen in mice. Therefore, RliB might be a very promising candidate sRNA involved in pathogenesis. The *rliB* mutant was tested in a mouse infection model and colonized the liver faster than the wild-type strain (Toledo-Arana et al., 2009). However, mRNA targets of RliB are still unknown.

Rli38, which is absent in nonpathogenic *Listeria innocua*, is induced ~25-fold when cells are cultivated in blood. The *rli38* deletion mutant was highly attenuated, with 10 times fewer bacteria recovered from multiple organs in a murine infection model (Toledo-Arana et al., 2009). The targets of Rli38 are not yet experimentally tested, although Rli38 was predicted to target the global iron uptake regulator Fur, deletion of which resulted in a similar attenuation (Rea et al., 2004; Toledo-Arana et al., 2009).

Three sRNAs, Rli31, Rli33-1, and Rli50, were identified in a study in which sRNAs induced upon growth of *Listeria* in macrophages were screened by a deep-sequencing-based approach (Mraheil et al., 2011b). All three sRNAs are required for full *Listeria* virulence, since disrupting any of these three sRNAs caused strong attenuation in all tested infection models (cell culture, insect, and mouse). Other interesting sRNAs of *Listeria* include SbrA and SbrE, which are directly regulated by σ^B and might play a role in virulence (Nielsen et al., 2008; Oliver et al., 2009).

Work from the *Listeria* field also revealed an interesting twist with respect to the origin of sRNAs. The *cis*-acting SAM riboswitches SreA and SreB (sensing the metabolite *S*-adenosylmethionine) were found to additionally function in *trans* and act as sRNAs, which base-pair with the mRNA encoding the master virulence regulator PrfA (Loh et al., 2009). Thus, the number of existing

sRNAs could be higher than previously expected assuming that many *cis*-regulatory elements might also function in *trans*. At this point, it is not clear how SreA/B affect *Listeria* virulence. However, it was shown that SreA/B sRNA expression depends on PrfA and is highly induced in host cells, thus forming a negative-feedback loop where high levels of PrfA activate SreA/B and which, in turn, repress *prfA* expression (Loh et al., 2009).

Streptococcus

Streptococci are typical commensals of human skin and the mucosae of the respiratory, gastrointestinal, and genitourinary tracts. Some species, such as *Streptococcus pyogenes* (group A streptococcus [GAS]) and *Streptococcus pneumoniae* (pneumococcus), are also important human pathogens. In the pre-antibiotic era, GAS was a very common cause of death. The bacterium typically causes purulent infections at the site of entry such as wound infections, abscesses, pharyngitis, and necrotizing fasciitis, whereas other GAS infections (e.g., scarlet fever and streptococcal toxic shock syndrome) are due to the production of potent streptococcal toxins and superantigens (Cunningham, 2000). Also, GAS gives rise to immunopathological disorders (e.g., acute rheumatic fever, endocarditis, and glomerulonephritis) by eliciting autoreactive antibodies against various streptococcal structures. *S. pneumoniae*, another streptococcal species with strong pathogenic potential, causes mainly infections of the respiratory tract and pneumonia, as well as severe and generalized infections like meningitis and septicemia (Weiser, 2010).

Understanding the coordinated expression and regulation of the numerous virulence-associated factors is an important focus of *Streptococcus* research. In GAS, TCSs and a range of so-called stand-alone transcription factors build the scaffold of gene regulation (Kreikemeyer et al., 2003). The first sRNA identified in GAS was FasX, which is 205 nt in size and was found to be located in the vicinity of the *fasBCA* locus (Kreikemeyer et al., 2001; Ramirez-Pena et al., 2010). Although not representing a QS system, the *fasBCA* locus exhibits certain organizational and functional similarities to the staphylococcal *agr* locus by coordinating the expression of matrix protein-binding adhesins and secreted virulence factors (e.g., streptokinase and streptolysin S) and by employing FasX to act as the sRNA effector molecule of the system (Kreikemeyer et al., 2001). A detailed study of FasX-mediated regulation of the streptokinase-encoding *ska* mRNA recently established a new mechanism of posttranscriptional activation. FasX binds to the 5′ portion of this transcript by an antisense mechanism. Due to its secondary structure, the FasX/ *ska* complex is protected from degradation and the *ska* mRNA is stabilized, resulting in a 10-fold increase in streptokinase expression (Ramirez-Pena et al., 2010).

Pel (pleiotropic effect locus) is another sRNA involved in GAS virulence control (Mangold et al., 2004). The 459-bp transcript, which also contains the coding sequence for streptolysin S, acts both at the transcriptional and posttranscriptional levels and activates expression of a number of virulence factors, such as the surface-associated M protein, the immune evasion molecule Sic, and the SpeB protease (Mangold et al., 2004). During study of the regulatory networks of the Mga stand-alone transcription factor and the TCS CovRS, the RivX sRNA was discovered downstream of the gene *rivR*, which encodes a RofA-like transcription factor (Roberts and Scott, 2007). RivX was shown to act as an independent regulatory RNA and to participate in the positive control of Mga-controlled virulence factors (Roberts and Scott, 2007).

A comprehensive approach combining bioinformatics and tiling microarrays increased the number of sRNAs in GAS to 75 (Perez et al., 2009). Many of these candidate sRNAs were found to be differentially expressed during growth, and for a number of them transcription start sites and terminator signals as well as stabilities were determined. Differential RNA sequencing recently identified approximately 140 novel sRNA candidates in GAS (Deltcheva et al., 2011). One of these sRNAs is tracrRNA, a *trans*-encoded small RNA which is a functional component of the CRISPR locus of *S. pyogenes* (Deltcheva et al., 2011). These CRISPR arrays are transcribed as a single long RNA molecule that needs to be processed to generate mature and functionally active CRISPR RNAs from the spacer regions. In GAS, tracrRNA was found to exhibit complementarities to the repeat regions, and tracrRNA binding to the long CRISPR RNA triggered processing and maturation of the CRISPR RNAs by RNase III (Deltcheva et al., 2011).

With respect to pneumococci, natural competence is a characteristic feature of *S. pneumoniae*, and uptake and incorporation of DNA are regarded as a major driving force for generating genetic diversity and the dissemination of virulence and antibiotic resistance traits in this pathogen (Johnsborg and Havarstein, 2009). Similar to that in GAS, pneumococcal gene regulation is mainly accomplished through TCSs, which are known to be involved in the control of pneumococcal autolysis, competence, bacteriocin release, and the regulation of basic metabolic functions (Johnsborg and Havarstein, 2009). Among these two-component regulators, the CiaRH

system has recently been shown not only to govern a range of cellular functions but also to drive expression of at least five sRNAs, named csRNAs1 to -5 (for *cia*-dependent small RNAs) (Halfmann et al., 2007). Interestingly, csRNA4 and csRNA5 were demonstrated to regulate stationary-phase autolysis, a process which mediates DNA release and transformation (Halfmann et al., 2007).

Thanks to a recent screen using high-resolution tiling arrays, the numbers of sRNAs in *S. pneumoniae* has recently grown to >50 (Kumar et al., 2010). Based on previous bioinformatics predictions (Livny et al., 2006), expression of a range of candidate sRNAs was validated and three novel sRNAs (i.e., Spd-sr17, Spd-sr37, and CcnA) were investigated in more detail in this study (Tsui et al., 2010). Spd-sr37 was found to adopt a base-paired structure, while Spd-sr17 and CcnA exhibit single-stranded stretches between hairpins. Deletion and overexpression of either of these sRNAs, however, did not result in fundamental effects on global transcription patterns under various growth conditions or virulence in a mouse model.

Staphylococcus

Staphylococcus aureus is a highly versatile microorganism living on the edge between commensalism and pathogenicity. On the one hand, the species occurs in 20% of healthy individuals as a harmless skin and mucosal commensal. On the other hand, it is one of the most significant human pathogens causing both health care- and community-acquired infections. *S. aureus* infections comprise a broad range of diseases which either are directly associated with the presence of the bacteria at the site of infection (e.g., skin and wound infections, pneumonia, and septicemia) or are mediated by the production of distinct staphylococcal toxins (e.g., food poisoning and toxic shock syndrome) (Tristan et al., 2007). *S. aureus* has additional natural reservoirs in mammals and birds and plays an important role in veterinary medicine. It is also notorious for the rapid evolution of antibiotic-resistant strains (Chambers and DeLeo, 2009).

Due to its multifaceted lifestyle, *S. aureus* must be able to cope with a wide range of environmental conditions. Well-studied players in staphylococcal gene regulation are classical regulatory proteins such as TCS proteins, DNA-binding proteins, and the alternative sigma factor σ^B (Cheung et al., 2004). More recent additions to this complex network are regulatory RNAs. The first sRNA to be identified in *S. aureus* was RNAIII, which represents the effector of the staphylococcal Agr system (Novick et al., 1993). Agr elegantly combines a classical TCS with a QS system and a multifunctional regulatory RNA (Novick and Geisinger, 2008). The *agr* locus consists of two adjacent transcriptional units from which the *agrBDCA* operon mediates both production and sensing of small circularized peptides which are continuously secreted during growth. Once a sufficient density (quorum) of these signaling molecules is reached, the system becomes activated, leading first to further autoinduction of *agrBDCA* and then to induction of the second transcriptional unit encoding RNAIII (Fig. 3). The Agr system mainly coordinates the expression of surface-associated and extracellular proteins during transition from exponential to stationary growth (Novick, 2003) but has also been demonstrated to influence a range of additional functions, such as biofilm formation (Kong et al., 2006), amino acid biosynthesis and transport (Batzilla et al., 2006), and the expression of other global regulators (Dunman et al., 2001). Many of these Agr-controlled factors have a significant impact on virulence (e.g., adhesins and toxins) (Ziebandt et al., 2004). Interestingly, RNAIII fulfills a bifunctional role as an mRNA encoding the delta-hemolysin peptide, which is involved in virulence, and as a regulatory RNA molecule (Verdon et al., 2009).

RNAIII adopts a highly structured conformation whereby different domains influence different mRNA targets (Benito et al., 2000). RNAIII represses translation of its target mRNAs upon binding, mostly by sequestering the ribosomal binding site and triggering mRNA degradation through RNase III (Huntzinger et al., 2005; Boisset et al., 2007; Chevalier et al., 2010). One notable exception is the translation activation of alpha toxin-encoding *hla* mRNA (Fig. 3) (Morfeldt et al., 1995).

Agr and RNAIII have taken an eminent position in staphylococcal virulence control and gene regulation, but the search for other staphylococcal sRNAs recently moved into focus. A first systematic screening of the *S. aureus* genome by bioinformatics and experimental verification identified several sRNAs on pathogenicity islands (Pichon and Felden, 2005). More recent studies utilizing combinations of bioinformatics (Pichon and Felden, 2005; Geissmann et al., 2009; Marchais et al., 2009), microarray hybridization (Anderson et al., 2006; Roberts et al., 2006), next-generation sequencing approaches (Beaume et al., 2010; Bohn et al., 2010), and Northern blot analyses (Abu-Qatouseh et al., 2010) detected hundreds of novel sRNAs in *S. aureus*. A wide diversity

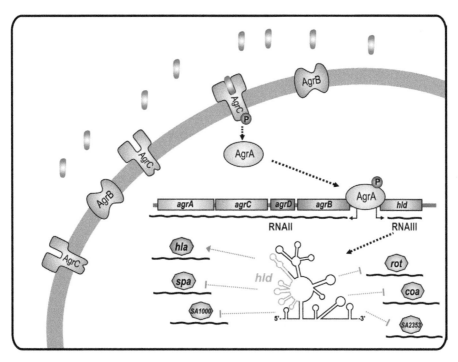

Figure 3. QS- and RNAIII-controlled gene expression. *S. aureus* produces an autoinducing peptide which is sensed by a histidine kinase (AgrC). Sensing of the autoinducing peptide by AgrC leads to phosphorylation of the response regulator AgrA, which, in turn, is a transcriptional activator of the bifunctional RNAIII. RNAIII encodes the *hld* gene (coding for δ-hemolysin) but also acts as a posttranscriptional regulator of several target mRNAs, most of which have profound impact on virulence. Whereas *spa*, *coa*, *rot*, SA1000, and SA2353 mRNAs are inhibited, the *hla* mRNA is induced by RNAIII. doi:10.1128/9781555818524.ch25f3

of sRNAs have been uncovered both in the core genome and on mobile genetic elements, comprising *trans*-encoded sRNAs, *cis*-encoded antisense sRNAs, and *cis*-acting RNA leaders in the 5′ UTRs of mRNAs (Felden et al., 2011).

Naturally, the sRNAs encoded on horizontally acquired virulence elements were expected to regulate mainly factors encoded within these genetic elements (Pichon and Felden, 2005). Interestingly, such sRNAs were also shown to exert effects on virulence-associated factors of the core genome. A good example is SprD, a *trans*-encoded sRNA which directly inhibits translation initiation of the immune evasion molecule Sbi (Chabelskaya et al., 2010). Other recently discovered sRNAs comprise antisense RNAs, often as components of type I TA modules (e.g., SprG and SprF) (Pichon and Felden, 2005; Fozo et al., 2010; Felden et al., 2011). Another sRNA, which encodes a small phenol-soluble modulin (PSM), was identified on a staphylococcal SCC *mec* genomic island conferring methicillin resistance (Queck et al., 2009). As this PSM plays a role in immune evasion, a link between methicillin resistance and virulence in *S. aureus* was postulated.

Dependence of toxin and other virulence factor expression on environmental stress and growth conditions is a common theme in bacterial pathogens, including *S. aureus*. Recent research has revealed fascinating and so-far-neglected links between stress, virulence, and metabolism in this organism (Somerville and Proctor, 2009). Interestingly, numerous sRNAs were found to be embedded in these complex regulatory circuits as well. Thus, a recent study detected 11 novel sRNAs (RsaA to -K) from which some (i.e., RsaA, -D, and -F) were found to be under the control of the stress-induced alternative σ^B factor, while the expression of others (i.e., RsaE) depends on Agr (Geissmann et al., 2009). All of these sRNAs make use of a conserved sequence motif to interact with their target mRNAs, suggesting that RsaA to -K represent a novel class of sRNAs that repress translation initiation (Geissmann et al., 2009). For RsaE, it was shown that this sRNA interacts with mRNAs of central metabolic pathways such as the tricarboxylic acid cycle, one-carbon metabolism, and ABC transporters (Geissmann et al., 2009; Bohn et al., 2010). Since staphylococci central carbon metabolism is linked to virulence via, for example, biofilm formation (Zhu et al., 2007; Zhu et al., 2009), RsaE might represent a good candidate to interfere with staphylococcal pathogenesis.

RsaA to -K were all shown to be independent of the RNA chaperone Hfq, at least as judged by intracellular stability (Geissmann et al., 2009). Hfq is well conserved in all strains and species of the genus *Staphylococcus* (Chao and Vogel, 2010). There has been a major debate as to the importance of Hfq for virulence and stress control in *S. aureus*. A pioneering study utilizing phenotypic microarrays failed to detect any effect of an *hfq*-null mutant on sRNA-mRNA interactions, stress and antibiotic resistance, or metabolic pathway control (Bohn et al., 2007). In contrast, another study identified Hfq as an important factor in *S. aureus* stress resistance and virulence control (Liu et al., 2010). The authors detected the Hfq protein only in some *S. aureus* strains, suggesting a strain-specific production of the chaperone.

Mycobacterium

The slow-growing facultative intracellular pathogen *Mycobacterium tuberculosis* is the causative agent of tuberculosis. It is one of the world's most successful pathogens, estimated to have infected one-third of the world's population (Mathema et al., 2006). *M. tuberculosis* employs a thick lipid-rich cell wall to cope with hostile environmental and intracellular conditions, to escape from host immune surveillance and establish a long-term, usually lifetime infection (Ehrt and Schnappinger, 2007). The regulation of virulence genes and lipid metabolism in *Mycobacterium* is achieved by a large and complex regulatory network consisting of 13 sigma factors, 11 TCSs, dozens of TA pairs, and over 100 annotated transcriptional regulators encoded in its 4.4-Mb genome. Even though the *M. tuberculosis* genome was among the first to be sequenced (Cole et al., 1998), riboregulation by sRNAs was a largely unexplored area until very recently.

An RNomics screen, which involved the cloning of size-fractionated short transcripts (20 to ~75 nt), produced a list of nine sRNAs in *M. tuberculosis* (Arnvig and Young, 2009). Subsequently, more comprehensive RNomics combined with biocomputational predictions identified an additional 34 sRNAs in the *M. tuberculosis* complex, which consists of virulent *M. tuberculosis*, *Mycobacterium bovis*, and the vaccine strain *M. bovis* BCG (DiChiara et al., 2010). The latest screen involved RNA-seq of *in vitro*-cultured *M. tuberculosis* and identified an additional 6 sRNAs, as well as a large number of abundant antisense RNAs (Arnvig et al., 2011), which brings the total number of sRNAs in *M. tuberculosis* to 88, close to the numbers known for *E. coli* and *Salmonella* (Waters and Storz, 2009; Papenfort and Vogel, 2010).

Intriguingly, more than half (48 out of 88) of the above-mentioned sRNAs are *cis*-encoded antisense RNAs. Many of them are located adjacent to or overlapping with the 3′ ends of genes and are complementary to the 3′ UTRs of genes (Arnvig and Young, 2009). Some *cis*-encoded RNAs are located antisense to important lipid metabolic and/or virulence-related genes, such as Mcr16 and ASdes. Mcr16 is antisense to the *fabD* gene, which is involved in fatty acid synthesis (DiChiara et al., 2010). ASdes RNA is located antisense to the *desA1* gene (Rv0824c), which encodes an essential lipid fatty acid desaturase induced during infection (Arnvig and Young, 2009). Furthermore, due to the high sequence identity of a second desaturase gene (*desA2/Rv1094*) with *desA1*, it will be exciting to see whether the *cis*-encoded ASdes RNA also regulates the *trans*-encoded *desA2* gene, in order to achieve a tight control of desaturase levels in *M. tuberculosis* during infection.

trans-encoded sRNAs also appear to regulate important metabolic pathways during *M. tuberculosis* infection, although their direct target genes have not yet been fully elucidated. Overexpression of three sRNAs, B11 (Mpr19), G2, and F6 (Mcr14, Mpr13), is lethal or causes growth deficits, and it was suggested that B11 overexpression affects cell wall synthesis and/or cell division (Arnvig and Young, 2009). Furthermore, three sRNAs, MTS0997 (Mcr11), MTS1338, and MTS2823, were shown to be induced in mycobacteria recovered from lungs of chronically infected mice, indicating their potential regulatory functions during mouse infection. Despite having only a slight growth defect, constitutive overexpression of MTS2823 causes downregulation of ~300 genes. Two adjacent genes, *prpC* and *prpD*, encoding methyl citrate synthase and dehydratase, respectively, are the most downregulated genes (Arnvig and Young, 2009).

In addition, the expression of MTS0997 and MTS1338 is regulated by the global transcriptional regulator CRP and dormancy survival regulator DosR, respectively (Arnvig et al., 2011). σ^B recognition sites have already been found upstream of the F6 sRNA (Arnvig and Young, 2009). Therefore, it can be expected that these new sRNAs will soon be integrated into the global regulatory network of mycobacteria.

Although its function was previously unknown, the CRISPR system has been used for mycobacterial genotyping (spoligotyping) for ~15 years (Kamerbeek et al., 1997). Interestingly, an abundant antisense RNA was found to be complementary to Cas2, the key gene for CRISPR function, and to be present in all CRISPR-Cas systems (Makarova et al., 2011). A better understanding of mycobacterial CRISPR may help to improve the current mycobacteriophage-based clinical tests (Kalantri et al., 2005).

Clostridia

Clostridia are gram-positive bacteria which belong to the *Firmicutes*. Their obligate anaerobe lifestyle discriminates them from the *Bacilli*, and their abilities to form spores and to produce toxins make them potent human pathogens (Rood, 1998; Bruggemann, 2005). Among these, *Clostridium botulinum*, *Clostridium tetani*, and *Clostridum perfringens* constitute the most-studied organisms; however, analysis of RNA-mediated gene expression has so far been limited to *C. perfringens*, the causative agent of gas gangrene in humans.

The VirR/S TCS system is an important mediator of *C. perfringens* toxin production because it activates expression of the genes encoding perfringolysin O and the cysteine protease α-clostripain as well as at least three regulatory RNAs, i.e., VR-RNA, VirT, and VirU (Banu et al., 2000; Shimizu et al., 2002a; Okumura et al., 2008). Whereas the function of the latter two riboregulators has not been studied in detail, VR-RNA has been observed to induce the production of α-toxin, collagenase (κ-toxin), and β2-toxin. Similar to RNAIII of *S. aureus*, VR-RNA is part of an extended RNA molecule that encodes a small ORF (*hyp7*) of yet-unknown function with its regulatory portion located in the 3′ end of the transcript (Shimizu et al., 2002b). Binding of VR-RNA to *colA* (encoding collagenase) relieves a *cis*-inhibitory structure in the 5′ UTR of the mRNA and induces an endonucleolytic cleavage downstream of the binding site. This dual mechanism increases *colA* mRNA stability efficiently as it increases *colA* translation and blocks further endonucleolytic decay (Fig. 1D) (Obana et al., 2010).

OUTLOOK

The rapid expansion of sRNA discovery that began in 2001 has continued unabated through the application of innovative new techniques and the expansion beyond typical model organisms such as *E. coli* and *Salmonella*. Intricate sRNA-mediated posttranscriptional regulons are continuously being discovered in organisms previously presumed to be devoid of sRNAs, highlighting the ubiquity of posttranscriptional control by small regulatory RNAs (Storz et al., 2011). Initial characterization of sRNAs uncovered their integral roles in a multitude of essential cellular pathways such as membrane homeostasis, metabolism, and iron acquisition. More recently, the focus has shifted towards identifying the roles of sRNAs in virulence gene control. These studies are often hampered by the lack of drastic phenotypes after deletion of the sRNA. This might reflect the general role of sRNAs in fine-tuning regulatory responses rather than the essential roles common for virulence-associated transcription factors. There are a few notable exceptions, including RNAIII in *Staphylococcus*, IsrM in *Salmonella*, and several sRNAs in *Listeria*, deletion of which leads to clear virulence phenotypes.

The roles of many sRNAs in virulence gene control are most often uncovered through broad screens. The affordability of new high-throughput sequencing technologies will certainly aid the discovery of new virulence-associated sRNAs, e.g., by facilitating the discovery of those sRNAs that are differentially expressed under virulence-inducing conditions. While these conditions are often mimicked through modification in culture media, more precise data will be derived from the deep sequencing of bacteria isolated at various stages of interaction with the host. Monitoring changes of expression levels of virulence-related genes in response to deletion or overexpression of a suite of sRNAs can also be effectively applied to identify posttranscriptional control in targets of interest (Papenfort et al., 2008; Mandin and Gottesman, 2009, 2010).

While many functionally related virulence factors are often clustered in horizontally acquired pathogenicity islands, *trans*-acting sRNAs located in the ancestral genome can be co-opted into regulation of horizontally acquired genes, thus linking expression of virulence factors with regulation of the core genome. Equally, horizontally acquired pathogenicity islands are capable of manipulating the core genome. One example of this occurs in *Salmonella*, where the horizontally acquired InvR downregulates expression of the ancestral *ompD* gene during expression of the type III secretion apparatus required for invasion of host cells (Pfeiffer et al., 2007). How newly acquired sRNAs are successfully incorporated into an existing posttranscriptional network is poorly understood. It has recently become clear that Hfq levels are limiting in the cell, and thus, there is strong competition between sRNAs for association with Hfq (Hussein and Lim, 2011; Moon and Gottesman, 2011). Induction of one sRNA (ArcZ) to ~200 copies per cell, which is comparable to the levels of many natively expressed sRNAs, was sufficient to titrate Hfq, depleting the pool of available Hfq and causing pleiotropic effects reminiscent of an *hfq* deletion (Papenfort et al., 2009). This raises the question of whether mobile genetic elements such as large self-transmissible virulence plasmids encode RNA chaperones specifically to reduce sequestration of core genome-encoded Hfq and thus lower the fitness cost of plasmid acquisition. Exactly such a mechanism is used in an analogous system by the large virulence plasmid of *S. flexneri*, which encodes a homolog to the core-encoded global repressor protein, H-NS. This prevents sequestration of H-NS by the newly acquired virulence plasmid and thus reduces pleiotropic changes in transcription that occur upon plasmid acquisition in the absence of the H-NS homolog (Doyle et al., 2007). The evolution of this regulatory cross talk within bacterial chromosomes, which are often a mosaic of core and horizontally acquired genes, is an intriguing aspect of posttranscriptional networks that awaits further characterization. The identification of RNA-binding proteins that facilitate horizontal gene transfer might provide a key insight into the evolution of posttranscriptional networks.

Possibly the most intriguing question remaining to date is whether a bacterial sRNA can be transferred to the eukaryotic host cell to directly facilitate pathogenesis, similarly to the many characterized translocated effector proteins.

REFERENCES

Abdelrahman, Y. M., L. A. Rose, and R. J. Belland. 2011. Developmental expression of non-coding RNAs in *Chlamydia trachomatis* during normal and persistent growth. *Nucleic Acids Res.* **39**:1843–1854.

Abu-Qatouseh, L. F., S. V. Chinni, J. Seggewiss, R. A. Proctor, J. Brosius, T. S. Rozhdestvensky, G. Peters, C. von Eiff, and K. Becker. 2010. Identification of differentially expressed small non-protein-coding RNAs in *Staphylococcus aureus* displaying both the normal and the small-colony variant phenotype. *J. Mol. Med.* (Berlin) **88**:565–575.

Achtman, M., K. Zurth, G. Morelli, G. Torrea, A. Guiyoule, and E. Carniel. 1999. *Yersinia pestis*, the cause of plague, is a recently emerged clone of *Yersinia pseudotuberculosis*. *Proc. Natl. Acad. Sci. USA* **96**:14043–14048.

Albrecht, M., C. M. Sharma, M. T. Dittrich, T. Muller, R. Reinhardt, J. Vogel, and T. Rudel. 2011. The transcriptional landscape of *Chlamydia pneumoniae*. *Genome Biol.* **12**:R98.

Albrecht, M., C. M. Sharma, R. Reinhardt, J. Vogel, and T. Rudel. 2010. Deep sequencing-based discovery of the *Chlamydia trachomatis* transcriptome. *Nucleic Acids Res.* **38**:868–877.

Altier, C., M. Suyemoto, and S. D. Lawhon. 2000. Regulation of *Salmonella enterica* serovar Typhimurium invasion genes by csrA. *Infect. Immun.* **68**:6790–6797.

Anderson, K. L., C. Roberts, T. Disz, V. Vonstein, K. Hwang, R. Overbeek, P. D. Olson, S. J. Projan, and P. M. Dunman. 2006. Characterization of the *Staphylococcus aureus* heat shock, cold shock, stringent, and SOS responses and their effects on log-phase mRNA turnover. *J. Bacteriol.* **188**:6739–6756.

Anguita, J., M. N. Hedrick, and E. Fikrig. 2003. Adaptation of *Borrelia burgdorferi* in the tick and the mammalian host. *FEMS Microbiol. Rev.* **27**:493–504.

Ansong, C., H. Yoon, S. Porwollik, H. Mottaz-Brewer, B. O. Petritis, N. Jaitly, J. N. Adkins, M. McClelland, F. Heffron, and R. D. Smith. 2009. Global systems-level analysis of Hfq and SmpB deletion mutants in *Salmonella*: implications for virulence and global protein translation. *PLoS One* **4**:e4809.

Arnvig, K. B., I. Comas, N. R. Thomson, J. Houghton, H. I. Boshoff, N. J. Croucher, G. Rose, T. T. Perkins, J. Parkhill, G. Dougan, and D. B. Young. 2011. Sequence-based analysis uncovers an abundance of non-coding RNA in the total transcriptome of *Mycobacterium tuberculosis*. *PLoS Pathog.* 7:e1002342.

Arnvig, K. B., and D. B. Young. 2009. Identification of small RNAs in *Mycobacterium tuberculosis*. *Mol. Microbiol.* 73:397–408.

Babitzke, P., C. S. Baker, and T. Romeo. 2009. Regulation of translation initiation by RNA binding proteins. *Annu. Rev. Microbiol.* 63:27–44.

Babitzke, P., and T. Romeo. 2007. CsrB sRNA family: sequestration of RNA-binding regulatory proteins. *Curr. Opin. Microbiol.* 10:156–163.

Baker, C. S., L. A. Eory, H. Yakhnin, J. Mercante, T. Romeo, and P. Babitzke. 2007. CsrA inhibits translation initiation of *Escherichia coli hfq* by binding to a single site overlapping the Shine-Dalgarno sequence. *J. Bacteriol.* 189:5472–5481.

Balbontin, R., N. Figueroa-Bossi, J. Casadesus, and L. Bossi. 2008. Insertion hot spot for horizontally acquired DNA within a bidirectional small-RNA locus in *Salmonella enterica*. *J. Bacteriol.* 190:4075–4078.

Banu, S., K. Ohtani, H. Yaguchi, T. Swe, S. T. Cole, H. Hayashi, and T. Shimizu. 2000. Identification of novel VirR/VirS-regulated genes in *Clostridium perfringens*. *Mol. Microbiol.* 35:854–864.

Bardill, J. P., X. Zhao, and B. K. Hammer. 2011. The *Vibrio cholerae* quorum sensing response is mediated by Hfq-dependent sRNA/mRNA base pairing interactions. *Mol. Microbiol.* 80:1381–1394.

Barnard, F. M., M. F. Loughlin, H. P. Fainberg, M. P. Messenger, D. W. Ussery, P. Williams, and P. J. Jenks. 2004. Global regulation of virulence and the stress response by CsrA in the highly adapted human gastric pathogen *Helicobacter pylori*. *Mol. Microbiol.* 51:15–32.

Barrick, J. E., N. Sudarsan, Z. Weinberg, W. L. Ruzzo, and R. R. Breaker. 2005. 6S RNA is a widespread regulator of eubacterial RNA polymerase that resembles an open promoter. *RNA* 11:774–784.

Barry, T., M. Kelly, B. Glynn, and J. Peden. 1999. Molecular cloning and phylogenetic analysis of the small cytoplasmic RNA from *Listeria monocytogenes*. *FEMS Microbiol. Lett.* 173:47–53.

Batzilla, C. F., S. Rachid, S. Engelmann, M. Hecker, J. Hacker, and W. Ziebuhr. 2006. Impact of the accessory gene regulatory system (Agr) on extracellular proteins, *codY* expression and amino acid metabolism in *Staphylococcus epidermidis*. *Proteomics* 6:3602–3613.

Beaume, M., D. Hernandez, L. Farinelli, C. Deluen, P. Linder, C. Gaspin, P. Romby, J. Schrenzel, and P. Francois. 2010. Cartography of methicillin-resistant *S. aureus* transcripts: detection, orientation and temporal expression during growth phase and stress conditions. *PLoS One* 5:e10725.

Beisel, C. L., and G. Storz. 2011. The base-pairing RNA spot 42 participates in a multioutput feedforward loop to help enact catabolite repression in *Escherichia coli*. *Mol. Cell* 41:286–297.

Benito, Y., F. A. Kolb, P. Romby, G. Lina, J. Etienne, and F. Vandenesch. 2000. Probing the structure of RNAIII, the *Staphylococcus aureus agr* regulatory RNA, and identification of the RNA domain involved in repression of protein A expression. *RNA* 6:668–679.

Blower, T. R., G. P. C. Salmond, and B. Luisi. 2011. Balancing at survival's edge: the structure and adaptive benefits of prokaryotic toxin-antitoxin partners. *Curr. Opin. Struct. Biol.* 21:109–118.

Bohn, C., C. Rigoulay, and P. Bouloc. 2007. No detectable effect of RNA-binding protein Hfq absence in *Staphylococcus aureus*. *BMC Microbiol.* 7:10.

Bohn, C., C. Rigoulay, S. Chabelskaya, C. M. Sharma, A. Marchais, P. Skorski, E. Borezee-Durant, R. Barbet, E. Jacquet, A. Jacq, D. Gautheret, B. Felden, J. Vogel, and P. Bouloc. 2010. Experimental discovery of small RNAs in *Staphylococcus aureus*

reveals a riboregulator of central metabolism. *Nucleic Acids Res.* 38:6620–6636.

Boisset, S., T. Geissmann, E. Huntzinger, P. Fechter, N. Bendridi, M. Possedko, C. Chevalier, A. C. Helfer, Y. Benito, A. Jacquier, C. Gaspin, F. Vandenesch, and P. Romby. 2007. *Staphylococcus aureus* RNAIII coordinately represses the synthesis of virulence factors and the transcription regulator Rot by an antisense mechanism. *Genes Dev.* 21:1353–1366.

Bordi, C., M. C. Lamy, I. Ventre, E. Termine, A. Hachani, S. Fillet, B. Roche, S. Bleves, V. Mejean, A. Lazdunski, and A. Filloux. 2010. Regulatory RNAs and the HptB/RetS signalling pathways fine-tune *Pseudomonas aeruginosa* pathogenesis. *Mol. Microbiol.* 76:1427–1443.

Bossi, L., and N. Figueroa-Bossi. 2007. A small RNA downregulates LamB maltoporin in *Salmonella*. *Mol. Microbiol.* 65:799–810.

Bradley, E. S., K. Bodi, A. M. Ismail, and A. Camilli. 2011. A genome-wide approach to discovery of small RNAs involved in regulation of virulence in *Vibrio cholerae*. *PLoS Pathog.* 7:e1002126.

Brencic, A., and S. Lory. 2009. Determination of the regulon and identification of novel mRNA targets of *Pseudomonas aeruginosa* RsmA. *Mol. Microbiol.* 72:612–632.

Brencic, A., K. A. McFarland, H. R. McManus, S. Castang, I. Mogno, S. L. Dove, and S. Lory. 2009. The GacS/GacA signal transduction system of *Pseudomonas aeruginosa* acts exclusively through its control over the transcription of the RsmY and RsmZ regulatory small RNAs. *Mol. Microbiol.* 73:434–445.

Bruggemann, H. 2005. Genomics of clostridial pathogens: implication of extrachromosomal elements in pathogenicity. *Curr. Opin. Microbiol.* 8:601–605.

Burrowes, E., C. Baysse, C. Adams, and F. O'Gara. 2006. Influence of the regulatory protein RsmA on cellular functions in *Pseudomonas aeruginosa* PAO1, as revealed by transcriptome analysis. *Microbiology* 152:405–418.

Caimano, M. J., C. H. Eggers, K. R. Hazlett, and J. D. Radolf. 2004. RpoS is not central to the general stress response in *Borrelia burgdorferi* but does control expression of one or more essential virulence determinants. *Infect. Immun.* 72:6433–6445.

Cazalet, C., L. Gomez-Valero, C. Rusniok, M. Lomma, D. Dervins-Ravault, H. J. Newton, F. M. Sansom, S. Jarraud, N. Zidane, L. Ma, C. Bouchier, J. Etienne, E. L. Hartland, and C. Buchrieser. 2010. Analysis of the *Legionella longbeachae* genome and transcriptome uncovers unique strategies to cause Legionnaires' disease. *PLoS Genet.* 6:e1000851.

Chabelskaya, S., O. Gaillot, and B. Felden. 2010. A *Staphylococcus aureus* small RNA is required for bacterial virulence and regulates the expression of an immune-evasion molecule. *PLoS Pathog.* 6:e1000927.

Chain, P. S., E. Carniel, F. W. Larimer, J. Lamerdin, P. O. Stoutland, W. M. Regala, A. M. Georgescu, L. M. Vergez, M. L. Land, V. L. Motin, R. R. Brubaker, J. Fowler, J. Hinnebusch, M. Marceau, C. Medigue, M. Simonet, V. Chenal-Francisque, B. Souza, D. Dacheux, J. M. Elliott, A. Derbise, L. J. Hauser, and E. Garcia. 2004. Insights into the evolution of *Yersinia pestis* through whole-genome comparison with *Yersinia pseudotuberculosis*. *Proc. Natl. Acad. Sci. USA* 101:13826–13831.

Chambers, H. F., and F. R. Deleo. 2009. Waves of resistance: *Staphylococcus aureus* in the antibiotic era. *Nat. Rev. Microbiol.* 7:629–641.

Chao, Y., and J. Vogel. 2010. The role of Hfq in bacterial pathogens. *Curr. Opin. Microbiol.* 13:24–33.

Chaudhuri, R. R., L. Yu, A. Kanji, T. T. Perkins, P. P. Gardner, J. Choudhary, D. J. Maskell, and A. J. Grant. 2011. Quantitative RNA-seq analysis of the transcriptome of *Campylobacter jejuni*. *Microbiology* 157:2922–2932.

Chaudhuri, S., J. C. Bruno, F. Alonzo III, B. Xayarath, N. P. Cianciotto, and N. E. Freitag. 2010. Contribution of chitinases to *Listeria monocytogenes* pathogenesis. *Appl. Environ. Microbiol.* **76:**7302–7305.

Chen, I., and D. Dubnau. 2004. DNA uptake during bacterial transformation. *Nat. Rev. Microbiol.* **2:**241–249.

Cheung, A. L., A. S. Bayer, G. Zhang, H. Gresham, and Y. Q. Xiong. 2004. Regulation of virulence determinants in vitro and in vivo in *Staphylococcus aureus*. *FEMS Immunol. Med. Microbiol.* **40:**1–9.

Chevalier, C., S. Boisset, C. Romilly, B. Masquida, P. Fechter, T. Geissmann, F. Vandenesch, and P. Romby. 2010. *Staphylococcus aureus* RNAIII binds to two distant regions of *coa* mRNA to arrest translation and promote mRNA degradation. *PLoS Pathog.* **6:**e1000809.

Christiansen, J. K., M. H. Larsen, H. Ingmer, L. Sogaard-Andersen, and B. H. Kallipolitis. 2004. The RNA-binding protein Hfq of *Listeria monocytogenes*: role in stress tolerance and virulence. *J. Bacteriol.* **186:**3355–3362.

Christiansen, J. K., J. S. Nielsen, T. Ebersbach, P. Valentin-Hansen, L. Sogaard-Andersen, and B. H. Kallipolitis. 2006. Identification of small Hfq-binding RNAs in *Listeria monocytogenes*. *RNA* **12:**1383–1396.

Cole, S. T., R. Brosch, J. Parkhill, T. Garnier, C. Churcher, D. Harris, S. V. Gordon, K. Eiglmeier, S. Gas, C. E. Barry III, F. Tekaia, K. Badcock, D. Basham, D. Brown, T. Chillingworth, R. Connor, R. Davies, K. Devlin, T. Feltwell, S. Gentles, N. Hamlin, S. Holroyd, T. Hornsby, K. Jagels, A. Krogh, J. McLean, S. Moule, L. Murphy, K. Oliver, J. Osborne, M. A. Quail, M. A. Rajandream, J. Rogers, S. Rutter, K. Seeger, J. Skelton, R. Squares, S. Squares, J. E. Sulston, K. Taylor, S. Whitehead, and B. G. Barrell. 1998. Deciphering the biology of *Mycobacterium tuberculosis* from the complete genome sequence. *Nature* **393:**537–544.

Corcoran, C. P., K. Papenfort, and J. Vogel. 2011. Hfq-associated regulatory small RNAs, p. 15–50. *In* W. R. Hess and A. Marchfelder (ed.), *Regulatory RNAs in Prokaryotes*. Springer, New York, NY.

Cossart, P. 2007. Listeriology (1926–2007): the rise of a model pathogen. *Microbes Infect.* **9:**1143–1146.

Cossart, P., and A. Toledo-Arana. 2008. *Listeria monocytogenes*, a unique model in infection biology: an overview. *Microbes Infect.* **10:**1041–1050.

Cotter, P. A., and A. M. Jones. 2003. Phosphorelay control of virulence gene expression in *Bordetella*. *Trends Microbiol.* **11:**367–373.

Cover, T. L., and M. J. Blaser. 2009. *Helicobacter pylori* in health and disease. *Gastroenterology* **136:**1863–1873.

Cromie, M. J., Y. Shi, T. Latifi, and E. A. Groisman. 2006. An RNA sensor for intracellular Mg^{2+}. *Cell* **125:**71–84.

Croxen, M. A., P. B. Ernst, and P. S. Hoffman. 2007. Antisense RNA modulation of alkyl hydroperoxide reductase levels in *Helicobacter pylori* correlates with organic peroxide toxicity but not infectivity. *J. Bacteriol.* **189:**3359–3368.

Cummings, C. A., H. J. Bootsma, D. A. Relman, and J. F. Miller. 2006. Species- and strain-specific control of a complex, flexible regulon by *Bordetella* BvgAS. *J. Bacteriol.* **188:**1775–1785.

Cummings, C. A., and D. A. Relman. 2002. Microbial forensics—"cross-examining pathogens." *Science* **296:**1976–1979.

Cunningham, M. W. 2000. Pathogenesis of group A streptococcal infections. *Clin. Microbiol. Rev.* **13:**470–511.

Davis, B. M., M. Quinones, J. Pratt, Y. Ding, and M. K. Waldor. 2005. Characterization of the small untranslated RNA RyhB and its regulon in *Vibrio cholerae*. *J. Bacteriol.* **187:**4005–4014.

Deltcheva, E., K. Chylinski, C. M. Sharma, K. Gonzales, Y. Chao, Z. A. Pirzada, M. R. Eckert, J. Vogel, and E. Charpentier. 2011.

CRISPR RNA maturation by *trans*-encoded small RNA and host factor RNase III. *Nature* **471:**602–607.

Deveau, H., J. E. Garneau, and S. Moineau. 2010. CRISPR/Cas system and its role in phage-bacteria interactions. *Annu. Rev. Microbiol.* **64:**475–493.

DiChiara, J. M., L. M. Contreras-Martinez, J. Livny, D. Smith, K. A. McDonough, and M. Belfort. 2010. Multiple small RNAs identified in *Mycobacterium bovis* BCG are also expressed in *Mycobacterium tuberculosis* and *Mycobacterium smegmatis*. *Nucleic Acids Res.* **38:**4067–4078.

Ding, Y., B. M. Davis, and M. K. Waldor. 2004. Hfq is essential for *Vibrio cholerae* virulence and downregulates sigma expression. *Mol. Microbiol.* **53:**345–354.

Dorman, C. J., and C. P. Corcoran. 2009. Bacterial DNA topology and infectious disease. *Nucleic Acids Res.* **37:**672–678.

Douillard, F. P., K. A. Ryan, D. L. Caly, J. Hinds, A. A. Witney, S. E. Husain, and P. W. O'Toole. 2008. Posttranscriptional regulation of flagellin synthesis in *Helicobacter pylori* by the RpoN chaperone HP0958. *J. Bacteriol.* **190:**7975–7984.

Doyle, M., M. Fookes, A. Ivens, M. W. Mangan, J. Wain, and C. J. Dorman. 2007. An H-NS-like stealth protein aids horizontal DNA transmission in bacteria. *Science* **315:**251–252.

Dunman, P. M., E. Murphy, S. Haney, D. Palacios, G. Tucker-Kellogg, S. Wu, E. L. Brown, R. J. Zagursky, D. Shlaes, and S. J. Projan. 2001. Transcription profiling-based identification of *Staphylococcus aureus* genes regulated by the *agr* and/or *sarA* loci. *J. Bacteriol.* **183:**7341–7353.

Ehrt, S., and D. Schnappinger. 2007. *Mycobacterium tuberculosis* virulence: lipids inside and out. *Nat. Med.* **13:**284–285.

Ellermeier, J. R., and J. M. Slauch. 2008. Fur regulates expression of the *Salmonella* pathogenicity island 1 type III secretion system through HilD. *J. Bacteriol.* **190:**476–486.

Fantappie, L., M. M. Metruccio, K. L. Seib, F. Oriente, E. Cartocci, F. Ferlicca, M. M. Giuliani, V. Scarlato, and I. Delany. 2009. The RNA chaperone Hfq is involved in stress response and virulence in *Neisseria meningitidis* and is a pleiotropic regulator of protein expression. *Infect. Immun.* **77:**1842–1853.

Fantappie, L., F. Oriente, A. Muzzi, D. Serruto, V. Scarlato, and I. Delany. 2011. A novel Hfq-dependent sRNA that is under FNR control and is synthesized in oxygen limitation in *Neisseria meningitidis*. *Mol. Microbiol.* **80:**507–523.

Faucher, S. P., G. Friedlander, J. Livny, H. Margalit, and H. A. Shuman. 2010. *Legionella pneumophila* 6S RNA optimizes intracellular multiplication. *Proc. Natl. Acad. Sci. USA* **107:**7533–7538.

Faucher, S. P., and H. A. Shuman. 2011. Small regulatory RNA and *Legionella pneumophila*. *Front. Microbiol.* **2:**98.

Felden, B., F. Vandenesch, P. Bouloc, and P. Romby. 2011. The *Staphylococcus aureus* RNome and its commitment to virulence. *PLoS Pathog.* **7:**e1002006.

Fields, J. A., and S. A. Thompson. 2008. *Campylobacter jejuni* CsrA mediates oxidative stress responses, biofilm formation, and host cell invasion. *J. Bacteriol.* **190:**3411–3416.

Figueroa-Bossi, N., M. Valentini, L. Malleret, F. Fiorini, and L. Bossi. 2009. Caught at its own game: regulatory small RNA inactivated by an inducible transcript mimicking its target. *Genes Dev.* **23:**2004–2015.

Fortune, D. R., M. Suyemoto, and C. Altier. 2006. Identification of CsrC and characterization of its role in epithelial cell invasion in *Salmonella enterica* serovar Typhimurium. *Infect. Immun.* **74:**331–339.

Fozo, E. M., K. S. Makarova, S. A. Shabalina, N. Yutin, E. V. Koonin, and G. Storz. 2010. Abundance of type I toxin-antitoxin systems in bacteria: searches for new candidates and discovery of novel families. *Nucleic Acids Res.* **38:**3743–3759.

Franze de Fernandez, M. T., L. Eoyang, and J. T. August. 1968. Factor fraction required for the synthesis of bacteriophage Qbeta-RNA. *Nature* 219:588–590.

Fraser, C. M., S. Casjens, W. M. Huang, G. G. Sutton, R. Clayton, R. Lathigra, O. White, K. A. Ketchum, R. Dodson, E. K. Hickey, M. Gwinn, B. Dougherty, J. F. Tomb, R. D. Fleischmann, D. Richardson, J. Peterson, A. R. Kerlavage, J. Quackenbush, S. Salzberg, M. Hanson, R. van Vugt, N. Palmer, M. D. Adams, J. Gocayne, J. Weidman, T. Utterback, L. Watthey, L. McDonald, P. Artiach, C. Bowman, S. Garland, C. Fuji, M. D. Cotton, K. Horst, K. Roberts, B. Hatch, H. O. Smith, and J. C. Venter. 1997. Genomic sequence of a Lyme disease spirochaete, *Borrelia burgdorferi. Nature* 390:580–586.

Fröhlich, K. S., K. Papenfort, A. A. Berger, and J. Vogel. 2012. A conserved RpoS-dependent small RNA controls the synthesis of major porin OmpD. *Nucleic Acids Res.* 40:3623–3640.

Fröhlich, K. S., and J. Vogel. 2009. Activation of gene expression by small RNA. *Curr. Opin. Microbiol.* 12:674–682.

Gaballa, A., H. Antelmann, C. Aguilar, S. K. Khakh, K. B. Song, G. T. Smaldone, and J. D. Helmann. 2008. The *Bacillus subtilis* iron-sparing response is mediated by a Fur-regulated small RNA and three small, basic proteins. *Proc. Natl. Acad. Sci. USA* 105:11927–11932.

Geissmann, T., C. Chevalier, M. J. Cros, S. Boisset, P. Fechter, C. Noirot, J. Schrenzel, P. Francois, F. Vandenesch, C. Gaspin, and P. Romby. 2009. A search for small noncoding RNAs in *Staphylococcus aureus* reveals a conserved sequence motif for regulation. *Nucleic Acids Res.* 37:7239–7257.

Genco, C., and L. Wetzler. 2010. *Neisseria: Molecular Mechanisms of Pathogenesis*. Caister Academic Press, Wymondham, United Kingdom.

Geng, J., Y. Song, L. Yang, Y. Feng, Y. Qiu, G. Li, J. Guo, Y. Bi, Y. Qu, W. Wang, X. Wang, Z. Guo, R. Yang, and Y. Han. 2009. Involvement of the post-transcriptional regulator Hfq in *Yersinia pestis* virulence. *PLoS One* 4:e6213.

Giangrossi, M., G. Prosseda, C. N. Tran, A. Brandi, B. Colonna, and M. Falconi. 2010. A novel antisense RNA regulates at transcriptional level the virulence gene *icsA* of *Shigella flexneri. Nucleic Acids Res.* 38:3362–3375.

Gogol, E. B., V. A. Rhodius, K. Papenfort, J. Vogel, and C. A. Gross. 2011. Small RNAs endow a transcriptional activator with essential repressor functions for single-tier control of a global stress regulon. *Proc. Natl. Acad. Sci. USA* 108:12875–12880.

Gong, H., G. P. Vu, Y. Bai, E. Chan, R. Wu, E. Yang, F. Liu, and S. Lu. 2011. A *Salmonella* small non-coding RNA facilitates bacterial invasion and intracellular replication by modulating the expression of virulence factors. *PLoS Pathog.* 7:e1002120.

Gonzalez, N., S. Heeb, C. Valverde, E. Kay, C. Reimmann, T. Junier, and D. Haas. 2008. Genome-wide search reveals a novel GacA-regulated small RNA in *Pseudomonas* species. *BMC Genomics* 9:167.

Grieshaber, N. A., E. R. Fischer, D. J. Mead, C. A. Dooley, and T. Hackstadt. 2004. Chlamydial histone-DNA interactions are disrupted by a metabolite in the methylerythritol phosphate pathway of isoprenoid biosynthesis. *Proc. Natl. Acad. Sci. USA* 101:7451–7456.

Grieshaber, N. A., S. S. Grieshaber, E. R. Fischer, and T. Hackstadt. 2006. A small RNA inhibits translation of the histone-like protein Hc1 in *Chlamydia trachomatis. Mol. Microbiol.* 59:541–550.

Grundy, F. J., and T. M. Henkin. 2006. From ribosome to riboswitch: control of gene expression in bacteria by RNA structural rearrangements. *Crit. Rev. Biochem. Mol. Biol.* 41:329–338.

Halfmann, A., M. Kovacs, R. Hakenbeck, and R. Bruckner. 2007. Identification of the genes directly controlled by the response regulator CiaR in *Streptococcus pneumoniae*: five out of 15 promoters drive expression of small non-coding RNAs. *Mol. Microbiol.* 66:110–126.

Hammer, B. K., and B. L. Bassler. 2007. Regulatory small RNAs circumvent the conventional quorum sensing pathway in pandemic *Vibrio cholerae. Proc. Natl. Acad. Sci. USA* 104:11145–11149.

Hansen, A. M., and J. B. Kaper. 2009. Hfq affects the expression of the LEE pathogenicity island in enterohaemorrhagic *Escherichia coli. Mol. Microbiol.* 73:446–465.

Haraga, A., M. B. Ohlson, and S. I. Miller. 2008. Salmonellae interplay with host cells. *Nat. Rev. Microbiol.* 6:53–66.

Heidrich, N., A. Chinali, U. Gerth, and S. Brantl. 2006. The small untranslated RNA SR1 from the *Bacillus subtilis* genome is involved in the regulation of arginine catabolism. *Mol. Microbiol.* 62:520–536.

Hengge-Aronis, R. 2002. Recent insights into the general stress response regulatory network in *Escherichia coli. J. Mol. Microbiol. Biotechnol.* 4:341–346.

Henkin, T. M. 2008. Riboswitch RNAs: using RNA to sense cellular metabolism. *Genes Dev.* 22:3383–3390.

Heroven, A. K., K. Bohme, M. Rohde, and P. Dersch. 2008. A Csr-type regulatory system, including small non-coding RNAs, regulates the global virulence regulator RovA of *Yersinia pseudotuberculosis* through RovM. *Mol. Microbiol.* 68:1179–1195.

Hershberg, R., S. Altuvia, and H. Margalit. 2003. A survey of small RNA-encoding genes in *Escherichia coli. Nucleic Acids Res.* 31:1813–1820.

Heurlier, K., F. Williams, S. Heeb, C. Dormond, G. Pessi, D. Singer, M. Camara, P. Williams, and D. Haas. 2004. Positive control of swarming, rhamnolipid synthesis, and lipase production by the posttranscriptional RsmA/RsmZ system in *Pseudomonas aeruginosa* PAO1. *J. Bacteriol.* 186:2936–2945.

Higgins, D. A., M. E. Pomianek, C. M. Kraml, R. K. Taylor, M. F. Semmelhack, and B. L. Bassler. 2007. The major *Vibrio cholerae* autoinducer and its role in virulence factor production. *Nature* 450:883–886.

Hindahl, M. S., G. W. Crockford, and R. E. Hancock. 1984. Outer membrane protein NmpC of *Escherichia coli*: pore-forming properties in black lipid bilayers. *J. Bacteriol.* 159:1053–1055.

Hot, D., S. Slupek, B. Wulbrecht, A. D'Hondt, C. Hubans, R. Antoine, C. Locht, and Y. Lemoine. 2011. Detection of small RNAs in *Bordetella pertussis* and identification of a novel repeated genetic element. *BMC Genomics* 12:207.

Huntzinger, E., S. Boisset, C. Saveanu, Y. Benito, T. Geissmann, A. Namane, G. Lina, J. Etienne, B. Ehresmann, C. Ehresmann, A. Jacquier, F. Vandenesch, and P. Romby. 2005. *Staphylococcus aureus* RNAIII and the endoribonuclease III coordinately regulate *spa* gene expression. *EMBO J.* 24:824–835.

Hurme, R., and M. Rhen. 1998. Temperature sensing in bacterial gene regulation—what it all boils down to. *Mol. Microbiol.* 30:1–6.

Hussein, R., and H. N. Lim. 2011. Disruption of small RNA signaling caused by competition for Hfq. *Proc. Natl. Acad. Sci. USA* 108:1110–1115.

Irnov, I., C. M. Sharma, J. Vogel, and W. C. Winkler. 2010. Identification of regulatory RNAs in *Bacillus subtilis. Nucleic Acids Res.* 38:6637–6651.

Isabella, V. M., and V. L. Clark. 2011. Deep sequencing-based analysis of the anaerobic stimulon in *Neisseria gonorrhoeae. BMC Genomics* 12:51.

Izar, B., M. A. Mraheil, and T. Hain. 2011. Identification and role of regulatory non-coding RNAs in *Listeria monocytogenes. Int. J. Mol. Sci.* 12:5070–5079.

Joelsson, A., Z. Liu, and J. Zhu. 2006. Genetic and phenotypic diversity of quorum-sensing systems in clinical and environmental isolates of *Vibrio cholerae. Infect. Immun.* 74:1141–1147.

Johansen, J., M. Eriksen, B. Kallipolitis, and P. Valentin-Hansen. 2008. Down-regulation of outer membrane proteins by noncoding RNAs: unraveling the cAMP-CRP- and sigmaE-dependent CyaR-*ompX* regulatory case. *J. Mol. Biol.* 383:1–9.

Johansen, J., A. A. Rasmussen, M. Overgaard, and P. Valentin-Hansen. 2006. Conserved small non-coding RNAs that belong to the sigmaE regulon: role in down-regulation of outer membrane proteins. *J. Mol. Biol.* 364:1–8.

Johansson, J., P. Mandin, A. Renzoni, C. Chiaruttini, M. Springer, and P. Cossart. 2002. An RNA thermosensor controls expression of virulence genes in *Listeria monocytogenes*. *Cell* 110:551–561.

Johnsborg, O., and L. S. Havarstein. 2009. Regulation of natural genetic transformation and acquisition of transforming DNA in *Streptococcus pneumoniae*. *FEMS Microbiol. Rev.* 33:627–642.

Johnson, A. P. 1983. The pathogenic potential of commensal species of *Neisseria*. *J. Clin. Pathol.* 36:213–223.

Kalantri, S., M. Pai, L. Pascopella, L. Riley, and A. Reingold. 2005. Bacteriophage- based tests for the detection of *Mycobacterium tuberculosis* in clinical specimens: a systematic review and meta-analysis. *BMC Infect. Dis.* 5:59.

Kamerbeek, J., L. Schouls, A. Kolk, M. van Agterveld, D. van Soolingen, S. Kuijper, A. Bunschoten, H. Molhuizen, R. Shaw, M. Goyal, and J. van Embden. 1997. Simultaneous detection and strain differentiation of *Mycobacterium tuberculosis* for diagnosis and epidemiology. *J. Clin. Microbiol.* 35:907–914.

Kay, E., B. Humair, V. Denervaud, K. Riedel, S. Spahr, L. Eberl, C. Valverde, and D. Haas. 2006. Two GacA-dependent small RNAs modulate the quorum-sensing response in *Pseudomonas aeruginosa*. *J. Bacteriol.* 188:6026–6033.

Kendall, M. M., C. C. Gruber, D. A. Rasko, D. T. Hughes, and V. Sperandio. 2011. Hfq virulence regulation in enterohemorrhagic *Escherichia coli* O157:H7 strain 86-24. *J. Bacteriol.* 193:6843–6851.

Kirn, T. J., B. A. Jude, and R. K. Taylor. 2005. A colonization factor links *Vibrio cholerae* environmental survival and human infection. *Nature* 438:863–866.

Klinkert, B., and F. Narberhaus. 2009. Microbial thermosensors. *Cell. Mol. Life Sci.* 66:2661–2676.

Kong, K. F., C. Vuong, and M. Otto. 2006. *Staphylococcus* quorum sensing in biofilm formation and infection. *Int. J. Med. Microbiol.* 296:133–139.

Koo, J. T., T. M. Alleyne, C. A. Schiano, N. Jafari, and W. W. Lathem. 2011. Global discovery of small RNAs in *Yersinia pseudotuberculosis* identifies *Yersinia*-specific small, noncoding RNAs required for virulence. *Proc. Natl. Acad. Sci. USA* 108:E709–E717.

Kreikemeyer, B., M. D. Boyle, B. A. Buttaro, M. Heinemann, and A. Podbielski. 2001. Group A streptococcal growth phase-associated virulence factor regulation by a novel operon (Fas) with homologies to two-component-type regulators requires a small RNA molecule. *Mol. Microbiol.* 39:392–406.

Kreikemeyer, B., K. S. McIver, and A. Podbielski. 2003. Virulence factor regulation and regulatory networks in *Streptococcus pyogenes* and their impact on pathogen-host interactions. *Trends Microbiol.* 11:224–232.

Kröger, C., S. C. Dillon, A. D. Cameron, K. Papenfort, S. K. Sivasankaran, K. Hokamp, Y. Chao, A. Sittka, M. Hébrard, K. Händler, A. Colgan, P. Leekitcharoenphon, G. C. Langridge, A. J. Lohan, B. Loftus, S. Lucchini, D. W. Ussery, C. J. Dorman, N. R. Thomson, J. Vogel, and J. C. Hinton. 15 May 2012. The transcriptional landscape and small RNAs of *Salmonella enterica* serovar Typhimurium. *Proc. Natl. Acad. Sci. USA* 109(20):E1277–E1286.

Kulesus, R. R., K. Diaz-Perez, E. S. Slechta, D. S. Eto, and M. A. Mulvey. 2008. Impact of the RNA chaperone Hfq on the fitness and virulence potential of uropathogenic *Escherichia coli. Infect. Immun.* 76:3019–3026.

Kumar, R., P. Shah, E. Swiatlo, S. C. Burgess, M. L. Lawrence, and B. Nanduri. 2010. Identification of novel non-coding small RNAs from *Streptococcus pneumoniae* TIGR4 using high-resolution genome tiling arrays. *BMC Genomics* 11:350.

Kunkel, B., R. Losick, and P. Stragier. 1990. The *Bacillus subtilis* gene for the development transcription factor sigma K is generated by excision of a dispensable DNA element containing a sporulation recombinase gene. *Genes Dev.* 4:525–535.

Kunst, F., N. Ogasawara, I. Moszer, et al. 1997. The complete genome sequence of the Gram-positive bacterium *Bacillus subtilis. Nature* 390:249–256.

Larsen, M. H., J. J. Leisner, and H. Ingmer. 2010. The chitinolytic activity of *Listeria monocytogenes* EGD is regulated by carbohydrates but also by the virulence regulator PrfA. *Appl. Environ. Microbiol.* 76:6470–6476.

Lawhon, S. D., J. G. Frye, M. Suyemoto, S. Porwollik, M. McClelland, and C. Altier. 2003. Global regulation by CsrA in *Salmonella typhimurium. Mol. Microbiol.* 48:1633–1645.

Lease, R. A., M. E. Cusick, and M. Belfort. 1998. Riboregulation in *Escherichia coli*: DsrA RNA acts by RNA:RNA interactions at multiple loci. *Proc. Natl. Acad. Sci. USA* 95:12456–12461.

Lee, D. R., and C. A. Schnaitman. 1980. Comparison of outer membrane porin proteins produced by *Escherichia coli* and *Salmonella typhimurium. J. Bacteriol.* 142:1019–1022.

Lee, E. J., and E. A. Groisman. 2010. An antisense RNA that governs the expression kinetics of a multifunctional virulence gene. *Mol. Microbiol.* 76:1020–1033.

Lenz, D. H., M. B. Miller, J. Zhu, R. V. Kulkarni, and B. L. Bassler. 2005. CsrA and three redundant small RNAs regulate quorum sensing in *Vibrio cholerae. Mol. Microbiol.* 58:1186–1202.

Lenz, D. H., K. C. Mok, B. N. Lilley, R. V. Kulkarni, N. S. Wingreen, and B. L. Bassler. 2004. The small RNA chaperone Hfq and multiple small RNAs control quorum sensing in *Vibrio harveyi* and *Vibrio cholerae. Cell* 118:69–82.

Linares, J. F., R. Moreno, A. Fajardo, L. Martinez-Solano, R. Escalante, F. Rojo, and J. L. Martinez. 2010. The global regulator Crc modulates metabolism, susceptibility to antibiotics and virulence in *Pseudomonas aeruginosa. Environ. Microbiol.* 12:3196–3212.

Link, T. M., P. Valentin-Hansen, and R. G. Brennan. 2009. Structure of *Escherichia coli* Hfq bound to polyriboadenylate RNA. *Proc. Natl. Acad. Sci. USA* 106:19292–19297.

Liu, J. M., J. Livny, M. S. Lawrence, M. D. Kimball, M. K. Waldor, and A. Camilli. 2009. Experimental discovery of sRNAs in *Vibrio cholerae* by direct cloning, 5S/tRNA depletion and parallel sequencing. *Nucleic Acids Res.* 37:e46.

Liu, Y., N. Wu, J. Dong, Y. Gao, X. Zhang, C. Mu, N. Shao, and G. Yang. 2010. Hfq is a global regulator that controls the pathogenicity of *Staphylococcus aureus. PLoS One* 5:e13069.

Livny, J., A. Brencic, S. Lory, and M. K. Waldor. 2006. Identification of 17 *Pseudomonas aeruginosa* sRNAs and prediction of sRNA-encoding genes in 10 diverse pathogens using the bioinformatic tool sRNAPredict2. *Nucleic Acids Res.* 34:3484–3493.

Livny, J., M. A. Fogel, B. M. Davis, and M. K. Waldor. 2005. sRNAPredict: an integrative computational approach to identify sRNAs in bacterial genomes. *Nucleic Acids Res.* 33:4096–4105.

Livny, J., H. Teonadi, M. Livny, and M. K. Waldor. 2008. High-throughput, kingdom-wide prediction and annotation of bacterial non-coding RNAs. *PLoS One* 3:e3197.

Loh, E., O. Dussurget, J. Gripenland, K. Vaitkevicius, T. Tiensuu, P. Mandin, F. Repoila, C. Buchrieser, P. Cossart, and J. Johansson. 2009. A *trans*-acting riboswitch controls expression of the virulence regulator PrfA in *Listeria monocytogenes. Cell* 139:770–779.

Lucchetti-Miganeh, C., E. Burrowes, C. Baysse, and G. Ermel. 2008. The post-transcriptional regulator CsrA plays a central role in the adaptation of bacterial pathogens to different stages of infection in animal hosts. *Microbiology* 154:16–29.

Lybecker, M. C., C. A. Abel, A. L. Feig, and D. S. Samuels. 2010. Identification and function of the RNA chaperone Hfq in the Lyme disease spirochete *Borrelia burgdorferi*. *Mol. Microbiol.* 78:622–635.

Lybecker, M. C., and D. S. Samuels. 2007. Temperature-induced regulation of RpoS by a small RNA in *Borrelia burgdorferi*. *Mol. Microbiol.* 64:1075–1089.

Majdalani, N., S. Chen, J. Murrow, K. St John, and S. Gottesman. 2001. Regulation of RpoS by a novel small RNA: the characterization of RprA. *Mol. Microbiol.* 39:1382–1394.

Majdalani, N., C. Cunning, D. Sledjeski, T. Elliott, and S. Gottesman. 1998. DsrA RNA regulates translation of RpoS message by an anti-antisense mechanism, independent of its action as an antisilencer of transcription. *Proc. Natl. Acad. Sci. USA* 95:12462–12467.

Makarova, K. S., D. H. Haft, R. Barrangou, S. J. Brouns, E. Charpentier, P. Horvath, S. Moineau, F. J. Mojica, Y. I. Wolf, A. F. Yakunin, J. van der Oost, and E. V. Koonin. 2011. Evolution and classification of the CRISPR-Cas systems. *Nat. Rev. Microbiol.* 9:467–477.

Mandin, P., H. Fsihi, O. Dussurget, M. Vergassola, E. Milohanic, A. Toledo-Arana, I. Lasa, J. Johansson, and P. Cossart. 2005. VirR, a response regulator critical for *Listeria monocytogenes* virulence. *Mol. Microbiol.* 57:1367–1380.

Mandin, P., and S. Gottesman. 2009. A genetic approach for finding small RNAs regulators of genes of interest identifies RybC as regulating the DpiA/DpiB two-component system. *Mol. Microbiol.* 72:551–565.

Mandin, P., and S. Gottesman. 2010. Integrating anaerobic/aerobic sensing and the general stress response through the ArcZ small RNA. *EMBO J.* 29:3094–3107.

Mandin, P., F. Repoila, M. Vergassola, T. Geissmann, and P. Cossart. 2007. Identification of new noncoding RNAs in *Listeria monocytogenes* and prediction of mRNA targets. *Nucleic Acids Res.* 35:962–974.

Mandlik, A., J. Livny, W. P. Robins, J. M. Ritchie, J. J. Mekalanos, and M. K. Waldor. 2011. RNA-Seq-based monitoring of infection-linked changes in *Vibrio cholerae* gene expression. *Cell Host Microbe* 10:165–174.

Mangold, M., M. Siller, B. Roppenser, B. J. Vlaminckx, T. A. Penfound, R. Klein, R. Novak, R. P. Novick, and E. Charpentier. 2004. Synthesis of group A streptococcal virulence factors is controlled by a regulatory RNA molecule. *Mol. Microbiol.* 53:1515–1527.

Marchais, A., M. Naville, C. Bohn, P. Bouloc, and D. Gautheret. 2009. Single-pass classification of all noncoding sequences in a bacterial genome using phylogenetic profiles. *Genome Res.* 19:1084–1092.

Masse, E., and S. Gottesman. 2002. A small RNA regulates the expression of genes involved in iron metabolism in *Escherichia coli*. *Proc. Natl. Acad. Sci. USA* 99:4620–4625.

Masse, E., H. Salvail, G. Desnoyers, and M. Arguin. 2007. Small RNAs controlling iron metabolism. *Curr. Opin. Microbiol.* 10:140–145.

Mathema, B., N. E. Kurepina, P. J. Bifani, and B. N. Kreiswirth. 2006. Molecular epidemiology of tuberculosis: current insights. *Clin. Microbiol. Rev.* 19:658–685.

McArthur, S. D., S. C. Pulvermacher, and G. V. Stauffer. 2006. The *Yersinia pestis gcvB* gene encodes two small regulatory RNA molecules. *BMC Microbiol.* 6:52.

Meibom, K. L., A. L. Forslund, K. Kuoppa, K. Alkhuder, I. Dubail, M. Dupuis, A. Forsberg, and A. Charbit. 2009. Hfq, a novel pleiotropic regulator of virulence-associated genes in *Francisella tularensis*. *Infect. Immun.* 77:1866–1880.

Mellin, J. R., S. Goswami, S. Grogan, B. Tjaden, and C. A. Genco. 2007. A novel Fur- and iron-regulated small RNA, NrrF, is required for indirect Fur-mediated regulation of the *sdhA* and *sdhC* genes in *Neisseria meningitidis*. *J. Bacteriol.* 189:3686–3694.

Mellin, J. R., R. McClure, D. Lopez, O. Green, B. Reinhard, and C. Genco. 2010. Role of Hfq in iron-dependent and -independent gene regulation in *Neisseria meningitidis*. *Microbiology* 156:2316–2326.

Metruccio, M. M., L. Fantappie, D. Serruto, A. Muzzi, D. Roncarati, C. Donati, V. Scarlato, and I. Delany. 2009. The Hfq-dependent small noncoding RNA NrrF directly mediates Fur-dependent positive regulation of succinate dehydrogenase in *Neisseria meningitidis*. *J. Bacteriol.* 191:1330–1342.

Mitarai, N., A. M. Andersson, S. Krishna, S. Semsey, and K. Sneppen. 2007. Efficient degradation and expression prioritization with small RNAs. *Phys. Biol.* 4:164–171.

Moller, T., T. Franch, C. Udesen, K. Gerdes, and P. Valentin-Hansen. 2002. Spot 42 RNA mediates discoordinate expression of the *E. coli* galactose operon. *Genes Dev.* 16:1696–1706.

Moon, K., and S. Gottesman. 2011. Competition among Hfq-binding small RNAs in *Escherichia coli*. *Mol. Microbiol.* 82:1545–1562.

Moreno, R., P. Fonseca, and F. Rojo. 2012. Two small RNAs, CrcY and CrcZ, act in concert to sequester the Crc global regulator in *Pseudomonas putida*, modulating catabolite repression. *Mol. Microbiol.* 83:24–40.

Morfeldt, E., D. Taylor, A. von Gabain, and S. Arvidson. 1995. Activation of alpha-toxin translation in *Staphylococcus aureus* by the *trans*-encoded antisense RNA, RNAIII. *EMBO J.* 14:4569–4577.

Moulder, J. W. 1991. Interaction of chlamydiae and host cells in vitro. *Microbiol. Rev.* 55:143–190.

Mraheil, M. A., A. Billion, C. Kuenne, J. Pischimarov, B. Kreikemeyer, S. Engelmann, A. Hartke, J. C. Giard, M. Rupnik, S. Vorwerk, M. Beier, J. Retey, T. Hartsch, A. Jacob, F. Cemic, J. Hemberger, T. Chakraborty, and T. Hain. 2011a. Comparative genome-wide analysis of small RNAs of major Gram-positive pathogens: from identification to application. *Microb. Biotechnol.* 3:658–676.

Mraheil, M. A., A. Billion, W. Mohamed, K. Mukherjee, C. Kuenne, J. Pischimarov, C. Krawitz, J. Retey, T. Hartsch, T. Chakraborty, and T. Hain. 2011b. The intracellular sRNA transcriptome of *Listeria monocytogenes* during growth in macrophages. *Nucleic Acids Res.* 39:4235–4248.

Murphy, E. R., and S. M. Payne. 2007. RyhB, an iron-responsive small RNA molecule, regulates *Shigella dysenteriae* virulence. *Infect. Immun.* 75:3470–3477.

Nakao, H., H. Watanabe, S. Nakayama, and T. Takeda. 1995. *yst* gene expression in *Yersinia enterocolitica* is positively regulated by a chromosomal region that is highly homologous to *Escherichia coli* host factor 1 gene (*hfq*). *Mol. Microbiol.* 18:859–865.

Ng, W. L., and B. L. Bassler. 2009. Bacterial quorum-sensing network architectures. *Annu. Rev. Genet.* 43:197–222.

Nielsen, J. S., M. H. Larsen, E. M. Lillebaek, T. M. Bergholz, M. H. Christiansen, K. J. Boor, M. Wiedmann, and B. H. Kallipolitis. 2011. A small RNA controls expression of the chitinase ChiA in *Listeria monocytogenes*. *PLoS One* 6:e19019.

Nielsen, J. S., L. K. Lei, T. Ebersbach, A. S. Olsen, J. K. Klitgaard, P. Valentin-Hansen, and B. H. Kallipolitis. 2010. Defining a role for Hfq in Gram-positive bacteria: evidence for Hfq-dependent antisense regulation in *Listeria monocytogenes*. *Nucleic Acids Res.* 38:907–919.

Nielsen, J. S., A. S. Olsen, M. Bonde, P. Valentin-Hansen, and B. H. Kallipolitis. 2008. Identification of a sigma B-dependent small noncoding RNA in *Listeria monocytogenes*. *J. Bacteriol.* 190:6264–6270.

Novick, R. P. 2003. Autoinduction and signal transduction in the regulation of staphylococcal virulence. *Mol. Microbiol.* 48:1429–1449.

Novick, R. P., and E. Geisinger. 2008. Quorum sensing in staphylococci. *Annu. Rev. Genet.* 42:541–564.

Novick, R. P., H. F. Ross, S. J. Projan, J. Kornblum, B. Kreiswirth, and S. Moghazeh. 1993. Synthesis of staphylococcal virulence factors is controlled by a regulatory RNA molecule. *EMBO J.* 12:3967–3975.

Obana, N., Y. Shirahama, K. Abe, and K. Nakamura. 2010. Stabilization of *Clostridium perfringens* collagenase mRNA by VR-RNA-dependent cleavage in 5' leader sequence. *Mol. Microbiol.* 77:1416–1428.

Oglesby, A. G., J. M. Farrow III, J. H. Lee, A. P. Tomaras, E. P. Greenberg, E. C. Pesci, and M. L. Vasil. 2008. The influence of iron on *Pseudomonas aeruginosa* physiology: a regulatory link between iron and quorum sensing. *J. Biol. Chem.* 283:15558–15567.

Oglesby-Sherrouse, A. G., and M. L. Vasil. 2010. Characterization of a heme-regulated non-coding RNA encoded by the *prrF* locus of *Pseudomonas aeruginosa*. *PLoS One* 5:e9930.

Okumura, K., K. Ohtani, H. Hayashi, and T. Shimizu. 2008. Characterization of genes regulated directly by the VirR/VirS system in *Clostridium perfringens*. *J. Bacteriol.* 190:7719–7727.

Oliver, H. F., R. H. Orsi, L. Ponnala, U. Keich, W. Wang, Q. Sun, S. W. Cartinhour, M. J. Filiatrault, M. Wiedmann, and K. J. Boor. 2009. Deep RNA sequencing of *L. monocytogenes* reveals overlapping and extensive stationary phase and sigma B-dependent transcriptomes, including multiple highly transcribed noncoding RNAs. *BMC Genomics* 10:641.

Ostberg, Y., I. Bunikis, S. Bergstrom, and J. Johansson. 2004. The etiological agent of Lyme disease, *Borrelia burgdorferi*, appears to contain only a few small RNA molecules. *J. Bacteriol.* 186:8472–8477.

Overgaard, M., J. Johansen, J. Moller-Jensen, and P. Valentin-Hansen. 2009. Switching off small RNA regulation with trap-mRNA. *Mol. Microbiol.* 73:790–800.

Padalon-Brauch, G., R. Hershberg, M. Elgrably-Weiss, K. Baruch, I. Rosenshine, H. Margalit, and S. Altuvia. 2008. Small RNAs encoded within genetic islands of *Salmonella typhimurium* show host-induced expression and role in virulence. *Nucleic Acids Res.* 36:1913–1927.

Pannekoek, Y., R. Huis in't Veld, C. T. Hopman, A. A. Langerak, D. Speijer, and A. van der Ende. 2009. Molecular characterization and identification of proteins regulated by Hfq in *Neisseria meningitidis*. *FEMS Microbiol. Lett.* 294:216–224.

Papenfort, K., M. Bouvier, F. Mika, C. M. Sharma, and J. Vogel. 2010. Evidence for an autonomous 5' target recognition domain in an Hfq-associated small RNA. *Proc. Natl. Acad. Sci. USA* 107:20435–20440.

Papenfort, K., V. Pfeiffer, S. Lucchini, A. Sonawane, J. C. Hinton, and J. Vogel. 2008. Systematic deletion of *Salmonella* small RNA genes identifies CyaR, a conserved CRP-dependent riboregulator of OmpX synthesis. *Mol. Microbiol.* 68:890–906.

Papenfort, K., V. Pfeiffer, F. Mika, S. Lucchini, J. C. Hinton, and J. Vogel. 2006. SigmaE-dependent small RNAs of *Salmonella* respond to membrane stress by accelerating global *omp* mRNA decay. *Mol. Microbiol.* 62:1674–1688.

Papenfort, K., D. Podkaminski, J. C. Hinton, and J. Vogel. 2012. The ancestral SgrS RNA discriminates horizontally acquired *Salmonella* mRNAs through a single G-U wobble pair. *Proc. Natl. Acad. Sci. USA* 109(13):E757–E764.

Papenfort, K., N. Said, T. Welsink, S. Lucchini, J. C. Hinton, and J. Vogel. 2009. Specific and pleiotropic patterns of mRNA regulation by ArcZ, a conserved, Hfq-dependent small RNA. *Mol. Microbiol.* 74:139–158.

Papenfort, K., and J. Vogel. 2010. Regulatory RNA in bacterial pathogens. *Cell Host Microbe* 8:116–127.

Park, S. Y., M. J. Cromie, E. J. Lee, and E. A. Groisman. 2010. A bacterial mRNA leader that employs different mechanisms to sense disparate intracellular signals. *Cell* 142:737–748.

Parkhill, J., B. W. Wren, K. Mungall, J. M. Ketley, C. Churcher, D. Basham, T. Chillingworth, R. M. Davies, T. Feltwell, S. Holroyd, K. Jagels, A. V. Karlyshev, S. Moule, M. J. Pallen, C. W. Penn, M. A. Quail, M. A. Rajandream, K. M. Rutherford, A. H. van Vliet, S. Whitehead, and B. G. Barrell. 2000. The genome sequence of the food-borne pathogen *Campylobacter jejuni* reveals hypervariable sequences. *Nature* 403:665–668.

Peng, J., J. Yang, and Q. Jin. 2011. An integrated approach for finding overlooked genes in *Shigella*. *PLoS One* 6:e18509.

Perara, E., D. Ganem, and J. N. Engel. 1992. A developmentally regulated chlamydial gene with apparent homology to eukaryotic histone H1. *Proc. Natl. Acad. Sci. USA* 89:2125–2129.

Perez, N., J. Trevino, Z. Liu, S. C. Ho, P. Babitzke, and P. Sumby. 2009. A genome-wide analysis of small regulatory RNAs in the human pathogen group A *Streptococcus*. *PLoS One* 4:e7668.

Pessi, G., F. Williams, Z. Hindle, K. Heurlier, M. T. Holden, M. Camara, D. Haas, and P. Williams. 2001. The global posttranscriptional regulator RsmA modulates production of virulence determinants and N-acylhomoserine lactones in *Pseudomonas aeruginosa*. *J. Bacteriol.* 183:6676–6683.

Petrova, O. E., and K. Sauer. 2010. The novel two-component regulatory system BfiSR regulates biofilm development by controlling the small RNA *rsmZ* through CafA. *J. Bacteriol.* 192:5275–5288.

Pfeiffer, V., K. Papenfort, S. Lucchini, J. C. Hinton, and J. Vogel. 2009. Coding sequence targeting by MicC RNA reveals bacterial mRNA silencing downstream of translational initiation. *Nat. Struct. Mol. Biol.* 16:840–846.

Pfeiffer, V., A. Sittka, R. Tomer, K. Tedin, V. Brinkmann, and J. Vogel. 2007. A small non-coding RNA of the invasion gene island (SPI-1) represses outer membrane protein synthesis from the *Salmonella* core genome. *Mol. Microbiol.* 66:1174–1191.

Pflock, M., S. Kennard, I. Delany, V. Scarlato, and D. Beier. 2005. Acid-induced activation of the urease promoters is mediated directly by the ArsRS two-component system of *Helicobacter pylori*. *Infect. Immun.* 73:6437–6445.

Pichon, C., and B. Felden. 2005. Small RNA genes expressed from *Staphylococcus aureus* genomic and pathogenicity islands with specific expression among pathogenic strains. *Proc. Natl. Acad. Sci. USA* 102:14249–14254.

Postic, G., E. Frapy, M. Dupuis, I. Dubail, J. Livny, A. Charbit, and K. L. Meibom. 2010. Identification of small RNAs in *Francisella tularensis*. *BMC Genomics* 11:625.

Preis, H., R. A. Eckart, R. K. Gudipati, N. Heidrich, and S. Brantl. 2009. CodY activates transcription of a small RNA in *Bacillus subtilis*. *J. Bacteriol.* 191:5446–5457.

Queck, S. Y., B. A. Khan, R. Wang, T. H. Bach, D. Kretschmer, L. Chen, B. N. Kreiswirth, A. Peschel, F. R. Deleo, and M. Otto. 2009. Mobile genetic element-encoded cytolysin connects virulence to methicillin resistance in MRSA. *PLoS Pathog.* 5:e1000533.

Ramirez-Pena, E., J. Trevino, Z. Liu, N. Perez, and P. Sumby. 2010. The group A *Streptococcus* small regulatory RNA FasX enhances streptokinase activity by increasing the stability of the *ska* mRNA transcript. *Mol. Microbiol.* 78:1332–1347.

Rasis, M., and G. Segal. 2009. The LetA-RsmYZ-CsrA regulatory cascade, together with RpoS and PmrA, post-transcriptionally regulates stationary phase activation of *Legionella pneumophila* Icm/Dot effectors. *Mol. Microbiol.* 72:995–1010.

Rasmussen, S., H. B. Nielsen, and H. Jarmer. 2009. The transcriptionally active regions in the genome of *Bacillus subtilis*. *Mol. Microbiol.* 73:1043–1057.

Rea, R. B., C. G. Gahan, and C. Hill. 2004. Disruption of putative regulatory loci in *Listeria monocytogenes* demonstrates a significant role for Fur and PerR in virulence. *Infect. Immun.* 72:717–727.

Repoila, F., and S. Gottesman. 2001. Signal transduction cascade for regulation of RpoS: temperature regulation of DsrA. *J. Bacteriol.* 183:4012–4023.

Rice, J. B., and C. K. Vanderpool. 2011. The small RNA SgrS controls sugar-phosphate accumulation by regulating multiple PTS genes. *Nucleic Acids Res.* 39:3806–3819.

Richard, A. L., J. H. Withey, S. Beyhan, F. Yildiz, and V. J. DiRita. 2010. The *Vibrio cholerae* virulence regulatory cascade controls glucose uptake through activation of TarA, a small regulatory RNA. *Mol. Microbiol.* 78:1171–1181.

Roberts, C., K. L. Anderson, E. Murphy, S. J. Projan, W. Mounts, B. Hurlburt, M. Smeltzer, R. Overbeek, T. Disz, and P. M. Dunman. 2006. Characterizing the effect of the *Staphylococcus aureus* virulence factor regulator, SarA, on log-phase mRNA half-lives. *J. Bacteriol.* 188:2593–2603.

Roberts, S. A., and J. R. Scott. 2007. RivR and the small RNA RivX: the missing links between the CovR regulatory cascade and the Mga regulon. *Mol. Microbiol.* 66:1506–1522.

Rood, J. I. 1998. Virulence genes of *Clostridium perfringens*. *Annu. Rev. Microbiol.* 52:333–360.

Rutherford, S. T., J. C. van Kessel, Y. Shao, and B. L. Bassler. 2011. AphA and LuxR/HapR reciprocally control quorum sensing in vibrios. *Genes Dev.* 25:397–408.

Sack, D. A., R. B. Sack, G. B. Nair, and A. K. Siddique. 2004. Cholera. *Lancet* 363:223–233.

Saito, S., H. Kakeshita, and K. Nakamura. 2009. Novel small RNA-encoding genes in the intergenic regions of *Bacillus subtilis*. *Gene* 428:2–8.

Samuels, D. S. 2011. Gene regulation in *Borrelia burgdorferi*. *Annu. Rev. Microbiol.* 65:479–499.

Scarlato, V., B. Arico, A. Prugnola, and R. Rappuoli. 1991. Sequential activation and environmental regulation of virulence genes in *Bordetella pertussis*. *EMBO J.* 10:3971–3975.

Schiano, C. A., L. E. Bellows, and W. W. Lathem. 2010. The small RNA chaperone Hfq is required for the virulence of *Yersinia pseudotuberculosis*. *Infect. Immun.* 78:2034–2044.

Schubert, M., K. Lapouge, O. Duss, F. C. Oberstrass, I. Jelesarov, D. Haas, and F. H. Allain. 2007. Molecular basis of messenger RNA recognition by the specific bacterial repressing clamp RsmA/CsrA. *Nat. Struct. Mol. Biol.* 14:807–813.

Schumacher, M. A., R. F. Pearson, T. Moller, P. Valentin-Hansen, and R. G. Brennan. 2002. Structures of the pleiotropic translational regulator Hfq and an Hfq-RNA complex: a bacterial Sm-like protein. *EMBO J.* 21:3546–3556.

Shakhnovich, E. A., B. M. Davis, and M. K. Waldor. 2009. Hfq negatively regulates type III secretion in EHEC and several other pathogens. *Mol. Microbiol.* 74:347–363.

Shands, K. N., and D. W. Fraser. 1980. Legionnaires' disease. *Dis. Mon.* 27:1–59.

Sharma, C. M., F. Darfeuille, T. H. Plantinga, and J. Vogel. 2007. A small RNA regulates multiple ABC transporter mRNAs by targeting C/A-rich elements inside and upstream of ribosome-binding sites. *Genes Dev.* 21:2804–2817.

Sharma, C. M., S. Hoffmann, F. Darfeuille, J. Reignier, S. Findeiss, A. Sittka, S. Chabas, K. Reiche, J. Hackermuller, R. Reinhardt, P. F. Stadler, and J. Vogel. 2010. The primary transcriptome of the major human pathogen *Helicobacter pylori*. *Nature* 464:250–255.

Sharma, C. M., K. Papenfort, S. R. Pernitzsch, H. J. Mollenkopf, J. C. Hinton, and J. Vogel. 2011. Pervasive post-transcriptional control of genes involved in amino acid metabolism by the Hfq-dependent GcvB small RNA. *Mol. Microbiol.* 81:1144–1165.

Sharma, C. M., and J. Vogel. 2009. Experimental approaches for the discovery and characterization of regulatory small RNA. *Curr. Opin. Microbiol.* 12:536–546.

Shaw, E. I., C. A. Dooley, E. R. Fischer, M. A. Scidmore, K. A. Fields, and T. Hackstadt. 2000. Three temporal classes of gene expression during the *Chlamydia trachomatis* developmental cycle. *Mol. Microbiol.* 37:913–925.

Shimizu, T., K. Shima, K. Yoshino, K. Yonezawa, and H. Hayashi. 2002a. Proteome and transcriptome analysis of the virulence genes regulated by the VirR/VirS system in *Clostridium perfringens*. *J. Bacteriol.* 184:2587–2594.

Shimizu, T., H. Yaguchi, K. Ohtani, S. Banu, and H. Hayashi. 2002b. Clostridial VirR/VirS regulon involves a regulatory RNA molecule for expression of toxins. *Mol. Microbiol.* 43:257–265.

Silvaggi, J. M., J. B. Perkins, and R. Losick. 2006. Genes for small, noncoding RNAs under sporulation control in *Bacillus subtilis*. *J. Bacteriol.* 188:532–541.

Silvaggi, J. M., J. B. Perkins, and R. Losick. 2005. Small untranslated RNA antitoxin in *Bacillus subtilis*. *J. Bacteriol.* 187:6641–6650.

Simons, R. W., and N. Kleckner. 1983. Translational control of Is10 transposition. *Cell* 34:683–691.

Simonsen, K. T., G. Nielsen, J. V. Bjerrum, T. Kruse, B. H. Kallipolitis, and J. Moller-Jensen. 2011. A role for the RNA chaperone Hfq in controlling adherent-invasive *Escherichia coli* colonization and virulence. *PLoS One* 6:e16387.

Sittka, A., S. Lucchini, K. Papenfort, C. M. Sharma, K. Rolle, T. T. Binnewies, J. C. Hinton, and J. Vogel. 2008. Deep sequencing analysis of small noncoding RNA and mRNA targets of the global post-transcriptional regulator, Hfq. *PLoS Genet.* 4:e1000163.

Sittka, A., V. Pfeiffer, K. Tedin, and J. Vogel. 2007. The RNA chaperone Hfq is essential for the virulence of *Salmonella typhimurium*. *Mol. Microbiol.* 63:193–217.

Sittka, A., C. M. Sharma, K. Rolle, and J. Vogel. 2009. Deep sequencing of *Salmonella* RNA associated with heterologous Hfq proteins in vivo reveals small RNAs as a major target class and identifies RNA processing phenotypes. *RNA Biol.* 6:266–275.

Sledjeski, D. D., A. Gupta, and S. Gottesman. 1996. The small RNA, DsrA, is essential for the low temperature expression of RpoS during exponential growth in *Escherichia coli*. *EMBO J.* 15:3993–4000.

Somerville, G. A., and R. A. Proctor. 2009. At the crossroads of bacterial metabolism and virulence factor synthesis in staphylococci. *Microbiol. Mol. Biol. Rev.* 73:233–248.

Someya, T., S. Baba, M. Fujimoto, G. Kawai, T. Kumasaka, and K. Nakamura. 2012. Crystal structure of Hfq from *Bacillus subtilis* in complex with SELEX-derived RNA aptamer: insight into RNA-binding properties of bacterial Hfq. *Nucleic Acids Res.* 40:1856–1867.

Sonenshein, A. L. 2005. CodY, a global regulator of stationary phase and virulence in Gram-positive bacteria. *Curr. Opin. Microbiol.* 8:203–207.

Song, T., F. Mika, B. Lindmark, Z. Liu, S. Schild, A. Bishop, J. Zhu, A. Camilli, J. Johansson, J. Vogel, and S. N. Wai. 2008. A new *Vibrio cholerae* sRNA modulates colonization and affects release of outer membrane vesicles. *Mol. Microbiol.* 70:100–111.

Song, T., D. Sabharwal, and S. N. Wai. 2010. VrrA mediates Hfq-dependent regulation of OmpT synthesis in *Vibrio cholerae*. *J. Mol. Biol.* 400:682–688.

Song, T., and S. N. Wai. 2009. A novel sRNA that modulates virulence and environmental fitness of *Vibrio cholerae*. *RNA Biol.* 6:254–258.

Sonnleitner, E., L. Abdou, and D. Haas. 2009. Small RNA as global regulator of carbon catabolite repression in *Pseudomonas aeruginosa*. *Proc. Natl. Acad. Sci. USA* 106:21866–21871.

Sonnleitner, E., N. Gonzalez, T. Sorger-Domenigg, S. Heeb, A. S. Richter, R. Backofen, P. Williams, A. Huttenhofer, D. Haas, and U. Blasi. 2011. The small RNA PhrS stimulates synthesis of the *Pseudomonas aeruginosa* quinolone signal. *Mol. Microbiol.* 80:868–885.

Sonnleitner, E., S. Hagens, F. Rosenau, S. Wilhelm, A. Habel, K. E. Jager, and U. Blasi. 2003. Reduced virulence of a *hfq* mutant of *Pseudomonas aeruginosa* O1. *Microb. Pathog.* 35:217–228.

Sonnleitner, E., M. Schuster, T. Sorger-Domenigg, E. P. Greenberg, and U. Blasi. 2006. Hfq-dependent alterations of the transcriptome profile and effects on quorum sensing in *Pseudomonas aeruginosa*. *Mol. Microbiol.* 59:1542–1558.

Sonnleitner, E., T. Sorger-Domenigg, M. J. Madej, S. Findeiss, J. Hackermuller, A. Huttenhofer, P. F. Stadler, U. Blasi, and I. Moll. 2008. Detection of small RNAs in *Pseudomonas aeruginosa* by RNomics and structure-based bioinformatic tools. *Microbiology* 154:3175–3187.

Soper, T., P. Mandin, N. Majdalani, S. Gottesman, and S. A. Woodson. 2010. Positive regulation by small RNAs and the role of Hfq. *Proc. Natl. Acad. Sci. USA* 107:9602–9607.

Storz, G., J. Vogel, and K. M. Wassarman. 2011. Regulation by small RNAs in bacteria: expanding frontiers. *Mol. Cell* 43:880–891.

Studholme, D. J., and M. Buck. 2000. Novel roles of σ^N in small genomes. *Microbiology* 146(Pt. 1):4–5.

Sudarsan, N., E. R. Lee, Z. Weinberg, R. H. Moy, J. N. Kim, K. H. Link, and R. R. Breaker. 2008. Riboswitches in eubacteria sense the second messenger cyclic di-GMP. *Science* 321:411–413.

Thibonnier, M., J. M. Thiberge, and H. De Reuse. 2008. *trans*-Translation in *Helicobacter pylori*: essentiality of ribosome rescue and requirement of protein tagging for stress resistance and competence. *PLoS One* 3:e3810.

Thomason, M. K., and G. Storz. 2010. Bacterial antisense RNAs: how many are there, and what are they doing? *Annu. Rev. Genet.* 44:167–188.

Thompson, K. M., V. A. Rhodius, and S. Gottesman. 2007. σ^E regulates and is regulated by a small RNA in *Escherichia coli*. *J. Bacteriol.* 189:4243–4256.

Timmermans, J., and L. Van Melderen. 2010. Post-transcriptional global regulation by CsrA in bacteria. *Cell. Mol. Life Sci.* 67:2897–2908.

Toledo-Arana, A., O. Dussurget, G. Nikitas, N. Sesto, H. Guet-Revillet, D. Balestrino, E. Loh, J. Gripenland, T. Tiensuu, K. Vaitkevicius, M. Barthelemy, M. Vergassola, M. A. Nahori, G. Soubigou, B. Regnault, J. Y. Coppee, M. Lecuit, J. Johansson, and P. Cossart. 2009. The *Listeria* transcriptional landscape from saprophytism to virulence. *Nature* 459:950–956.

Tomb, J. F., O. White, A. R. Kerlavage, R. A. Clayton, G. G. Sutton, R. D. Fleischmann, K. A. Ketchum, H. P. Klenk, S. Gill, B. A. Dougherty, K. Nelson, J. Quackenbush, L. Zhou, E. F. Kirkness, S. Peterson, B. Loftus, D. Richardson, R. Dodson, H. G. Khalak, A. Glodek, K. McKenney, L. M. Fitzegerald, N. Lee, M. D. Adams, E. K. Hickey, D. E. Berg, J. D. Gocayne, T. R. Utterback, J. D. Peterson, J. M. Kelley, M. D. Cotton, J. M. Weidman, C. Fujii, C. Bowman, L. Watthey, E. Wallin, W. S. Hayes, M. Borodovsky, P. D. Karp, H. O. Smith, C. M. Fraser, and J. C. Venter. 1997. The complete genome sequence of the gastric pathogen *Helicobacter pylori*. *Nature* 388:539–547.

Tran, C. N., M. Giangrossi, G. Prosseda, A. Brandi, M. L. Di Martino, B. Colonna, and M. Falconi. 2011. A multifactor regulatory circuit involving H-NS, VirF and an antisense RNA modulates transcription of the virulence gene *icsA* of *Shigella flexneri*. *Nucleic Acids Res.* 39:8122–8134.

Tristan, A., T. Ferry, G. Durand, O. Dauwalder, M. Bes, G. Lina, F. Vandenesch, and J. Etienne. 2007. Virulence determinants in community and hospital meticillin-resistant *Staphylococcus aureus*. *J. Hosp. Infect.* 65(Suppl. 2):105–109.

Trotochaud, A. E., and K. M. Wassarman. 2004. 6S RNA function enhances long-term cell survival. *J. Bacteriol.* 186:4978–4985.

Tsui, H. C., D. Mukherjee, V. A. Ray, L. T. Sham, A. L. Feig, and M. E. Winkler. 2010. Identification and characterization of noncoding small RNAs in *Streptococcus pneumoniae* serotype 2 strain D39. *J. Bacteriol.* 192:264–279.

Tsui, H. C. T., H. C. E. Leung, and M. E. Winkler. 1994. Characterization of broadly pleiotropic phenotypes caused by an Hfq insertion mutation in *Escherichia coli* K-12. *Mol. Microbiol.* 13:35–49.

Tu, K. C., T. Long, S. L. Svenningsen, N. S. Wingreen, and B. L. Bassler. 2010. Negative feedback loops involving small regulatory RNAs precisely control the *Vibrio harveyi* quorum-sensing response. *Mol. Cell* 37:567–579.

Urbanowski, M. L., L. T. Stauffer, and G. V. Stauffer. 2000. The *gcvB* gene encodes a small untranslated RNA involved in expression of the dipeptide and oligopeptide transport systems in *Escherichia coli*. *Mol. Microbiol.* 37:856–868.

Vanderpool, C. K., and S. Gottesman. 2004. Involvement of a novel transcriptional activator and small RNA in post-transcriptional regulation of the glucose phosphoenolpyruvate phosphotransferase system. *Mol. Microbiol.* 54:1076–1089.

Vanderpool, C. K., and S. Gottesman. 2007. The novel transcription factor SgrR coordinates the response to glucose-phosphate stress. *J. Bacteriol.* 189:2238–2248.

Venturi, V. 2006. Regulation of quorum sensing in *Pseudomonas*. *FEMS Microbiol. Rev.* 30:274–291.

Verdon, J., N. Girardin, C. Lacombe, J. M. Berjeaud, and Y. Hechard. 2009. δ-Hemolysin, an update on a membrane-interacting peptide. *Peptides* 30:817–823.

Vogel, J., and B. F. Luisi. 2011. Hfq and its constellation of RNA. *Nat. Rev. Microbiol.* 9:578–589.

Vogel, J., and K. Papenfort. 2006. Small non-coding RNAs and the bacterial outer membrane. *Curr. Opin. Microbiol.* 9:605–611.

Waldminghaus, T., N. Heidrich, S. Brantl, and F. Narberhaus. 2007. FourU: a novel type of RNA thermometer in *Salmonella*. *Mol. Microbiol.* 65:413–424.

Wassarman, K. M., and R. M. Saecker. 2006. Synthesis-mediated release of a small RNA inhibitor of RNA polymerase. *Science* 314:1601–1603.

Wassarman, K. M., and G. Storz. 2000. 6S RNA regulates *E. coli* RNA polymerase activity. *Cell* 101:613–623.

Waters, L. S., and G. Storz. 2009. Regulatory RNAs in bacteria. *Cell* 136:615–628.

Wei, B. L., A. M. Brun-Zinkernagel, J. W. Simecka, B. M. Pruss, P. Babitzke, and T. Romeo. 2001. Positive regulation of motility and *flhDC* expression by the RNA-binding protein CsrA of *Escherichia coli*. *Mol. Microbiol.* 40:245–256.

Weiser, J. N. 2010. The pneumococcus: why a commensal misbehaves. *J. Mol. Med.* (Berlin) 88:97–102.

Weissenmayer, B. A., J. G. Prendergast, A. J. Lohan, and B. J. Loftus. 2011. Sequencing illustrates the transcriptional response of *Legionella pneumophila* during infection and identifies seventy novel small non-coding RNAs. *PLoS One* 6:e17570.

Wen, Y., J. Feng, D. R. Scott, E. A. Marcus, and G. Sachs. 2010. A *cis*-encoded antisense small RNA regulated by the HP0165-HP0166 two-component system controls expression of *ureB* in *Helicobacter pylori*. *J. Bacteriol.* **193:**40–51.

Wilderman, P. J., N. A. Sowa, D. J. FitzGerald, P. C. FitzGerald, S. Gottesman, U. A. Ochsner, and M. L. Vasil. 2004. Identification of tandem duplicate regulatory small RNAs in *Pseudomonas aeruginosa* involved in iron homeostasis. *Proc. Natl. Acad. Sci. USA* **101:**9792–9797.

Xiao, B., W. Li, G. Guo, B. Li, Z. Liu, K. Jia, Y. Guo, X. Mao, and Q. Zou. 2009a. Identification of small noncoding RNAs in *Helicobacter pylori* by a bioinformatics-based approach. *Curr. Microbiol.* **58:**258–263.

Xiao, B., W. Li, G. Guo, B. S. Li, Z. Liu, B. Tang, X. H. Mao, and Q. M. Zou. 2009b. Screening and identification of natural antisense transcripts in *Helicobacter pylori* by a novel approach based on RNase I protection assay. *Mol. Biol. Rep.* **36:**1853–1858.

Young, K. T., L. M. Davis, and V. J. Dirita. 2007. *Campylobacter jejuni*: molecular biology and pathogenesis. *Nat. Rev. Microbiol.* **5:**665–679.

Zhu, Y., E. C. Weiss, M. Otto, P. D. Fey, M. S. Smeltzer, and G. A. Somerville. 2007. *Staphylococcus aureus* biofilm metabolism and the influence of arginine on polysaccharide intercellular adhesin synthesis, biofilm formation, and pathogenesis. *Infect. Immun.* **75:**4219–4226.

Zhu, Y., Y. Q. Xiong, M. R. Sadykov, P. D. Fey, M. G. Lei, C. Y. Lee, A. S. Bayer, and G. A. Somerville. 2009. Tricarboxylic acid cycle-dependent attenuation of *Staphylococcus aureus* in vivo virulence by selective inhibition of amino acid transport. *Infect. Immun.* **77:**4256–4264.

Ziebandt, A. K., D. Becher, K. Ohlsen, J. Hacker, M. Hecker, and S. Engelmann. 2004. The influence of *agr* and σB in growth phase dependent regulation of virulence factors in *Staphylococcus aureus*. *Proteomics* **4:**3034–3047.

Regulation of Bacterial Virulence
Edited by Michael L. Vasil and Andrew J. Darwin
© 2013 ASM Press, Washington, DC   doi:10.1128/9781555818524.ch26

Chapter 26

Negative Regulation during Bacterial Infection

Andrew M. Stern, Ansel Hsiao, and Jun Zhu

INTRODUCTION

One simple, alluring goal of molecular bacterial pathogenesis research has been to reduce the bacterial cell's activity during infection to a series of regulatory events, whereby the microbe senses the presence of the host and responds accordingly. One difficulty in studying this is that bacterial regulation in the host depends on the local microenvironment of microbes and can vary dramatically depending on the stage of colonization or infection. Movement between and survival in these very different situations may require completely different transcriptional programs.

Some microbial adaptations to survival in the face of host immunity or nutrient deprivation are due to random mutation or phase variation (Virgin, 2007). But pathogens are also able to employ a series of transcriptional regulators that can tightly control the expression of virulence determinants and survival factors at different stages of infection. As bacteria move from one compartment of the host to the next, a broad range of genes may be required to colonize, proliferate, communicate, and survive in each new niche. Regulation by transcription factors can also be positive or negative. Genes that are required for one phase of infection may be irrelevant or even deleterious to survival in another. Just as virulence determinants may need to be upregulated for colonization, invasion, or disease, "antivirulence factors" that may reduce pathogen fitness inside the host must be suppressed. This regulatory flexibility is especially important for microorganisms whose life cycles involve different hosts and different environmental reservoirs. Given this, the set of upregulated virulence factors that are necessary during infection but not under fixed in vitro growth conditions may only be the proverbial tip of the iceberg compared to what occurs during the course of infection. Historically, most of the research on bacterial pathogenesis has focused on which genes are upregulated during infection, but it has become increasingly clear that gene repression is just as important for successful pathogenesis. In this chapter, we attempt to highlight some of the many interesting cases of negative regulation at work during different stages of infection for several pathogens, focusing on the diverse molecular mechanisms involved in gene repression and the selective pressures that led to their evolution.

NEGATIVE REGULATION TO PROMOTE VECTOR-HOST ZOONOTIC TRANSMISSION

Some pathogenic bacteria cycle through multiple hosts; for example, some pathogens use arthropod vectors to invade human populations, an extreme transition that demands a great deal of regulatory flexibility. One example is the spirochete *Borrelia burgdorferi*, the etiologic agent of Lyme disease, which follows a zoonotic cycle of transmission between *Ixodes* ticks and small mammals in the wild and can be transmitted to humans by infected ticks while feeding (Burgdorfer et al., 1982; Lane et al., 1991). To be a successful pathogen, *B. burgdorferi* must carefully manage the expression of outer surface proteins (Osps) on its coat, which requires timely transcriptional repression during its life cycle.

Interestingly, the discovery of this genetic regulation began with the search for antigens for a Lyme disease vaccine. OspA was first characterized by raising murine antibodies against spirochetes isolated from ticks (Barbour et al., 1983). OspC was identified as a protein expressed in bacteria isolated from infected patients and ticks; it was encoded by a plasmid maintained across strains from many *B. burgdorferi* isolates (Fuchs et al., 1992; Tilly et al., 1997). OspA was chosen as the target of the first Lyme disease vaccine (Steere et al., 1998). Significantly, the

Andrew M. Stern and Jun Zhu • Department of Microbiology, Perelman School of Medicine, University of Pennsylvania, Philadelphia, PA 19104.   Ansel Hsiao • Center for Genome Sciences & Systems Biology, Washington University School of Medicine, Saint Louis, MO 63110.

antibody responses of infected mice to OspA differed depending on how the initial inoculum was prepared: inoculation with broth-cultured *B. burgdorferi* led to the production of anti-OspA antibodies, but animals inoculated directly by ticks exhibited an attenuated immune response (Roehrig et al., 1992). In addition, the levels of OspA present in infected *Ixodes scapularis* ticks varied during the course of feeding, with quantities of OspA dropping during the last 12 hours of tick feeding and then recovering after feeding (Burkot et al., 1994). These observations suggested that *B. burgdorferi* actively regulated the composition of its outer surface proteins during transmission from ticks to humans.

Work by the Rosa laboratory subsequently confirmed the differential expression of OspC and OspA, this being regulated in response to the feeding status of the infected tick (Schwan et al., 1995). *B. burgdorferi* in the midgut of infected ticks produced OspA but no detectable OspC, yet mouse hosts produced antibodies to OspC far more readily than they produced antibodies to OspA, suggesting that the host is exposed to the opposite combination of proteins when the spirochete enters their bodies, which in turn would mean that the spirochete upregulates OspC and downregulates OspA during transmission. Indeed, OspC-negative bacteria taken from ticks and passaged at higher temperatures or lower pH, conditions mimicking those found during transmission, readily expressed OspC. Likewise, bacteria taken from ticks that had been allowed to feed on mice began to produce OspC (Schwan et al., 1995). In contrast, OspA levels in bacteria in the tick dropped dramatically during feeding but returned within hours of its completion, and spirochetes recently acquired by ticks feeding on an infected animal rapidly began to express OspA (Schwan and Piesman, 2000). These results have been supported by microarray analysis of the transcriptome of *B. burgdorferi* in fed and unfed ticks, as compared to cells grown in vivo in dialysis membrane compartments implanted in rats (Revel et al., 2002). These results imply accurate and timely downregulation of OspA and OspC at different times: OspA during transmission to the host, and OspC during infection of the tick midgut. It is now known that OspA acts as an adhesin in the *Ixodes* midgut, which suggests that this downregulation of OspA expression during transmission from tick to mammal host is actually a mechanism for escape from the tick reservoir (Pal et al., 2004a, 2004b; Yang et al., 2004).

OspC, however, plays a critical role in the movement of *B. burgdorferi* from the tick midgut reservoir to the tick salivary gland, a necessary step for zoonotic transmission (Schwan, 2003; Pal et al., 2004b). *B. burgdorferi* organisms that have been exposed to tick hemolymph upregulate OspC production (Johns et al., 2000), and OspC is responsible for spirochete binding to the tick salivary gland (Pal et al., 2004b). OspC has also been shown to be important for resisting the initial innate immune response by the host after tick bite or needle inoculation (Antonara et al., 2010), and is proposed to be a factor promoting dissemination from the initial site of infection to distant host tissues (Roehrig et al., 1992; Seemanapalli et al., 2011). OspC may, however, be detrimental for *B. burgdorferi* if it is expressed in mammalian hosts later in infection, by presenting a target for the host humoral immune response. OspC expression is often lost in infected immunocompetent mice but not immunodeficient hosts such as SCID mice (Liang et al., 2002, 2004). Furthermore, passive immunization of SCID mice with anti-OspC antibody is sufficient to abrogate OspC production, suggesting that OspC is actively downregulated during infection in response to host immune pressure (Liang et al., 2002).

Although the complete mechanism of regulation of *ospA* and *ospC* is not understood, it has been shown that *ospC* expression and *ospA* expression are reciprocally regulated via the same regulatory cascade. In this cascade, the sigma factor RpoN leads to production of a second sigma factor, RpoS, which in turn activates *ospC* and represses *ospA* (Hübner et al., 2001; Yang et al., 2005; Caimano et al., 2007). Activation of the RpoN-RpoS cascade itself is mediated by acetyl phosphate-dependent phosphorylation of the Rrp2 response regulator (Yang et al., 2003; Xu et al., 2010) and by BosR, a homologue of the iron-sensor Fur (Hyde et al., 2009, 2010; Ouyang et al., 2009, 2011). Although temperature, pH, and oxygen concentration are thought to play a role, the host-derived signals that lead to OspA downregulation are still not precisely understood (Burtnick et al., 2007), nor is the mechanism by which they lead to activation of the Rrp2/BosR-RpoN-RpoS system known. It is also not yet understood what genetic programs the spirochete uses to downregulate OspC late in infection, but it is clear that the repression of these two factors, OspA and OspC, at different time points during the zoonotic life cycle is critical for successful pathogenesis.

REPRESSION OF FLAGELLA AND PILI TO EVADE IMMUNE RECOGNITION

During infection, pathogenic bacteria must contend with an in vivo environment that is under the surveillance of immune mechanisms capable of rapidly identifying and eliminating foreign microorganisms. Successful pathogens often have mechanisms to survive the recognition of conserved pathogen-associated molecular patterns (PAMPs) by the innate immune system of the host. This is often accomplished by the selective repression of antivirulence factors that are highly antigenic during the infection. As can be seen from the following examples, this regulation can be tightly controlled transcriptionally or posttranslationally and driven by diverse signals.

In the related respiratory pathogens *Bordetella bronchiseptica* and *B. pertussis*, the latter being the etiologic agent of whooping cough in humans, virulence and colonization genes are controlled by the two-component BvgAS regulatory system (Fig. 1) (Arico et al., 1989, 1991; Scarlato et al., 1991; Roy et al., 1989, 1990; Roy and Falkow, 1991). BvgS is an inner membrane protein that autophosphorylates in response to a nebulously defined set of environmental stimuli. BvgS in turn transfers the phosphate to the DNA-binding cytosolic protein BvgA, changing the affinity of BvgA for promoters (Locht et al., 2001). This phosphorelay positively and negatively regulates transcription in bordetellae in several distinct phases (Yuk et al., 1996). In the Bvg⁺ phase, both *B. bronchiseptica* and *B. pertussis* activate the production of numerous virulence factors (Weiss and Hewlett, 1986); these include adhesins such as fimbriae and the filamentous hemagglutinin and toxins such as pertussis toxin and adenylate cyclase toxin. Also, under these conditions, motility genes like those required for the expression of flagellar biosynthesis components are repressed (Akerley et al., 1992, 1995). The BvgAS system thus represents a regulator that controls a spectrum of transcriptional activities for a large number of genes, which are activated or repressed in stages as bacteria move from Bvg⁻ phase conditions to Bvg⁺ (Cotter and Jones, 2003). The in

vivo environmental conditions during infection that influence BvgAS activity are not known; however, in vitro, growth at 26°C instead of 37°C, or the addition of millimolar quantities of sulfate ions or nicotinic acid to the growth medium (Miller et al., 1989a; Cummings et al., 2006), can mimic Bvg⁻ regulation.

The importance of gene repression during infection was demonstrated by Akerley et al. while studying the regulation of flagellar biosynthesis genes, which are normally repressed by BvgAS in the Bvg⁺ phase (Akerley et al., 1995). The authors placed the flagellar biosynthesis regulator locus *frlAB* under the control of the Bvg⁺ phase BvgAS-inducible promoter of *fhaB*, the gene encoding the structural subunit of filamentous hemagglutin (Jacob-Dubuisson et al., 2000). This strain thus expressed flagella during the Bvg⁺ phase, whereas in wild-type cells, flagella would be expressed only in the Bvg⁻ phase. When flagellar genes were expressed out of sequence in this manner, colonization of rats was inhibited relative to that by wild-type *B. bronchiseptica*. Intriguingly, nonmotile suppressor mutants of the initial Bvg⁺ flagellated strain were isolated from infected animals. These were shown to be nonmotile even in Bvg⁻ growth, and thus were not revertants to wild type. These data suggested that strong selective pressure exists against flagellar synthesis during infection and that repression of these genes is critical for success of the pathogen.

It is thought that *B. bronchiseptica* represses flagellum synthesis during infection to evade host immune recognition. Initial host responses to pathogens are mediated by molecules that recognize PAMPs such as flagellin, peptidoglycan, and lipopolysaccharide. Flagellin is one of the primary PAMPs sensed by Toll-like receptor 5 (TLR5), which is one of the family of innate immune sensors conserved from *Drosophila melanogaster* to humans that serves to detect the presence of microbial products (Honko and Mizel,

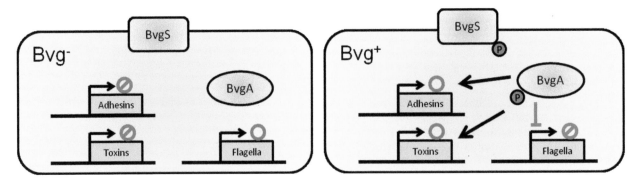

Figure 1. The BvgAS two-component system of *Bordetella* controls multiple genes. Under Bvg⁻ conditions, the membrane protein BvgS and cytosolic DNA-binding protein BvgA are unphosphorylated. This allows transcription of flagellar genes but not virulence factors such as adhesins and toxins. In response to certain environmental stimuli, the cell switches to a Bvg⁺ state, which is characterized by a phosphorylated BvgS and BvgA. This activated form of BvgA represses flagellar gene expression and activates the expression of adhesins and toxins. The timely repression of flagellar genes in the Bvg⁺ state is critical for infection. doi:10.1128/9781555818524.ch26f1

2005; Leulier and Lemaitre, 2008). *Bordetella* flagellin, the major component of flagellar structure, has been shown to activate signaling by both human and mouse TLR5 and thereby to elicit the production of proinflammatory cytokines and chemokines (Lopez-Boado et al., 2005). From these results, it seems that the function of negative regulation of flagellar components in *B. bronchiseptica* is to suppress the production of flagellin, which would otherwise act as an antivirulence factor by eliciting host immune responses during infection. In fact, flagellar biosynthesis is thought to be inactive in the genomes of the related pathogens *B. pertussis* and *B. parapertussis*, due to the presence of numerous insertion elements and pseudogenes within the biosynthesis operon (Parkhill et al., 2003). That this has occurred in these more host-adapted organisms suggests that host pressures such as immune recognition may have selected for a more permanent suppression of flagellar biosynthesis than regulation of transcription by BvgAS.

In addition to *Bordetella*, which must repress flagellar synthesis to avoid the immune system in the respiratory tract, the downregulation of outer surface structures is important in pathogens of the intestinal tract. An example exists in *Vibrio cholerae*, the causative agent of the epidemic diarrheal disease cholera (Hsiao et al. 2006). The structure examined was the type IV mannose-sensitive hemagglutinin (MSHA), which is produced in vitro but not during early colonization of the mouse intestinal tract. A wide variety of *V. cholerae* strains are able to produce MSHA pili (Svennerholm et al., 1991; Faruque et al., 1997), which are thought to aid *V. cholerae* persistence in the aquatic environment in between human hosts. MSHA pili mediate the attachment of *V. cholerae* cells to chitin, the basis of the exoskeletons of many small marine organisms in the *V. cholerae* reservoir (Chiavelli et al., 2001; Meibom et al., 2004). In addition, MSHA is required for the formation of biofilms, three-dimensional matrices of bacterial cells and secreted exopolysaccharide in which *V. cholerae* is often found (Watnick et al., 1999; Zhu and Mekalanos, 2003; Moorthy and Watnick, 2004, 2005).

MSHA pili are in fact not required for colonization in humans or the infant mouse model of colonization (Attridge et al., 1996; Thelin and Taylor, 1996), in accordance with the finding that *msh* genes are repressed during early infection. The reason behind this is that host-secreted immunoglobulin A (S-IgA) nonspecifically binds to *V. cholerae* cells in an MSHA-dependent manner. S-IgA is a dimeric antibody (Phalipon and Corthesy, 2003) that is heavily modified by N- and O-linked glycosylation. This glycosylation expands the binding range of IgA beyond the specific antigen recognized by the Fab regions of the IgA monomers,

as the glycan moieties can interact with bacterial carbohydrate-binding structures (Royle et al., 2003). Bacteria bound by S-IgA in the gut can become entrapped in the mucus layer of the intestine, excluding them from the epithelium and leading to clearance by bulk flow (Dickinson et al., 1998; Macpherson et al., 2001; Royle et al., 2003). Antibody-bound antigen is also preferentially taken up by M-cells in the intestinal Peyer's patches, which can promote the generation of a specific host immune response in adults (Apodaca et al., 1994; Royle et al., 2003; Blanco and DiRita, 2006). Indeed, when the genes responsible for MSHA biogenesis were artificially expressed constitutively in *V. cholerae*, these bacteria failed to colonize wild-type but not $IgA^{-/-}$ infant mice (Hsiao et al., 2006).

V. cholerae therefore must coordinate MSHA pilus biogenesis such that it is expressed in the aquatic environment, where it is important for survival, but repressed in the human host, where it is detrimental. To accomplish this, *V. cholerae* uses a similar mechanism to *Bordetella*: repression of MSHA is coupled to activation of virulence by the same transcriptional regulator. The transcription factor ToxT, which directly binds to and activates the promoter of the *ctx* and *tcp* operons required for intestinal colonization, also binds to and represses the promoter of the MSHA operon (DiRita et al., 1991; Hsiao et al., 2009). Thus, this pathogen very efficiently combines its responses to host stimuli in order to simultaneously upregulate factors necessary for virulence (toxin genes), while using the same factors to transcriptionally repress the biogenesis of detrimental factors such as MSHA.

The repression of MSHA goes beyond transcription. ToxT not only represses expression of the *msh* genes but also activates enzymes that degrade MSHA pili proteins posttranslationally. Within the *tcp* operon is the *tcpJ* gene, which encodes a peptidase that is normally required for proper assembly of the virulence pilus TCP (Kaufman et al., 1991; DiRita et al., 1991; LaPointe and Taylor, 2000; Hsiao et al., 2006). However, *tcpJ* was shown to cleave an additional target, the main MSHA subunit MshA, and prepare it for degradation (Hsiao et al., 2008). Thus, the repression of MSHA to avoid host immunity is so critical for *V. cholerae* that two mechanisms, one transcriptional and the other posttranslational, evolved to ensure repression of MSHA biogenesis early in infection.

Listeria monocytogenes, a gram-positive bacterium that causes the systemic infection listeriosis as well as some cases of meningitis, provides another example of the downregulation of flagella to avoid immune recognition. *L. monocytogenes* downregulates the production of its flagella at the higher temperatures found in its mammalian host (Peel et al., 1988; Way et al., 2004). Furthermore, *Listeria* flagellin is known to activate TLR5 and potentiate an immune response to the bacteria (Dons et al., 2004; Way et al., 2004). Noting this, Higgins et al. sought to determine how *L. monocytogenes* suppresses flagellar synthesis at higher temperatures (Fig. 2). They discovered that MogR is a transcriptional regulator that not only represses the transcription of the central flagellar subunit *flaA* but also binds to the promoters of several components of the flagellar synthesis system and represses them, suggesting that MogR is the central repressor of flagellar synthesis (Gründling et al., 2004; Shen and Higgins, 2006). Intriguingly, however, neither the levels of MogR protein nor the ability of MogR to bind the promoters of flagellar genes appeared temperature dependent (Shen and Higgins, 2006). The mystery was resolved by the discovery of another regulator, GmaR, which binds to and titrates MogR from flagellar promoters at lower temperatures, allowing transcription (Shen et al., 2006). Surprisingly, the temperature dependence of the system relies on a conformational change in GmaR which leads to its dissociation from MogR and its degradation (Kamp and Higgins, 2009, 2011). Thus, *L. monocytogenes* uses a unique and elegant system to repress the synthesis of its flagella to avoid recognition by the immune system and promote disease.

Immune recognition presents a significant challenge to pathogenic bacteria. A central mechanism to evade host defenses is to stop producing the structures, such as flagella and pili, that are recognized by host antibodies and TLRs. A broad variety of bacteria require the selective downregulation of these surface apparatuses to successfully invade a host. *Bordetella*, *Vibrio*, and *Listeria* are just some of the bacteria that are able to use a remarkably diverse set of molecular mechanisms to accomplish this stealthy maneuver.

GENE REPRESSION TO MEDIATE TRANSITIONS FROM ENVIRONMENT TO HOST AND BACK

Some bacterial pathogens use their human hosts as substrates for replication but enter an environmental reservoir between hosts. This type of life cycle requires dramatic changes in gene expression between the host and the environment. These switches have been well studied in *V. cholerae* and have again revealed that gene repression plays a critical role.

After *V. cholerae* organisms have established colonization of the small intestines, replicated, and caused disease, they must find their way out of the host in order to infect other individuals. Although the voluminous diarrhea characteristic of cholera is obviously a mechanism for exit, it has become clear through recent work that exit from the host intestinal tract is not only accomplished by a passive shedding of bacterial cells but is also mediated by careful transcriptional events known as the "mucosal escape response," discussed in detail in Chapter 28. In brief, virulence factors such as the cholera toxin (CT) and toxin-coregulated pilus (TCP) are downregulated in most *V. cholerae* cells late in infection, whereas stress-responsive genes are upregulated (Nielsen et al., 2006). This shift in gene expression, along with detachment from the gut epithelium and an increase in motility, requires the alternative sigma factor RpoS. It is thought that RpoS might exert its effects through the quorum-sensing regulator HapR. HapR is expressed at high cell density to repress virulence gene expression (Kovacikova and Skorupski, 2002; Miller et al., 2002; Zhu et al., 2002), which is thought to occur in a stationary phase-like cellular state shortly before exit from the host.

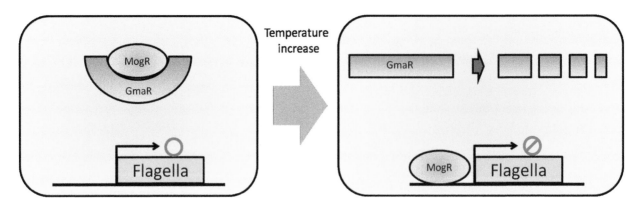

Figure 2. Temperature-dependent repression of flagellar synthesis in *Listeria monocytogenes*. Flagellar genes are repressed by the DNA-binding protein MogR, but under low temperatures, the anti-repressor GmaR binds to and titrates MogR away from the flagellar promoter, allowing expression of flagellar genes. As the temperature increases, a conformational change in GmaR takes place, targeting it for degradation. This allows MogR to bind flagellar promoters and repress transcription. doi:10.1128/9781555818524.ch26f2

However, *V. cholerae* also exists at high cell density as it enters the host before it establishes its niche at the epithelial surface, due to the high infectious dose and likely presence in dense biofilms (Leclerc et al., 2002). Thus, once the bacteria have entered a new host, *hapR* must be repressed in order to begin another cycle of *V. cholerae* pathogenesis. Interestingly, the repression of *hapR* early in infection is tied to the presence of flagella on the surface of the *V. cholerae* cell (Fig. 3). As *V. cholerae* moves toward the epithelium of the human small intestine, it must pass through a glycocalyx of mucins. It has been thought that this process requires flagellum-based motility, as mutants with mutations in flagellar assembly have decreased fitness in colonization (Freter et al., 1981b; Butler and Camilli, 2005) and are also limited in their ability to penetrate artificial mucin barriers (Freter et al., 1981a). Initial studies of the relationship between motility and virulence noted an inverse correlation between flagellar motility and virulence gene expression (Gardel and Mekalanos, 1996; Häse and Mekalanos, 1999; Häse, 2001). However, it was not until 2008 that a causative relationship between

flagellar loss and virulence gene expression was discovered by Liu et al. (2008) to be mediated through HapR. The authors performed a genetic screen to find *V. cholerae* mutants that could overcome HapR-mediated repression of virulence genes. They used a transposon to generate a library of mutants with a *luxO* mutant background, which normally constitutively overexpresses *hapR* and cannot activate virulence even under virulence-inducing conditions. Interestingly, several mutants were isolated that contained transposon insertions in flagellar synthesis genes, suggesting that flagella and virulence are inversely regulated. Further analysis revealed that *hapR* expression is controlled by the same regulators that control the final steps in flagellar synthesis. In the normal flagellar assembly pathway, the outer structural proteins of the flagella are not produced until the basal-body-hook structure, which is able to secrete proteins outside the cell, is assembled (Correa et al., 2004). One of the proteins secreted by this apparatus is FlgM, which normally binds to and sequesters the sigma factor FliA. Once FliA is free to bind promoters, it activates expression of the late flagellar

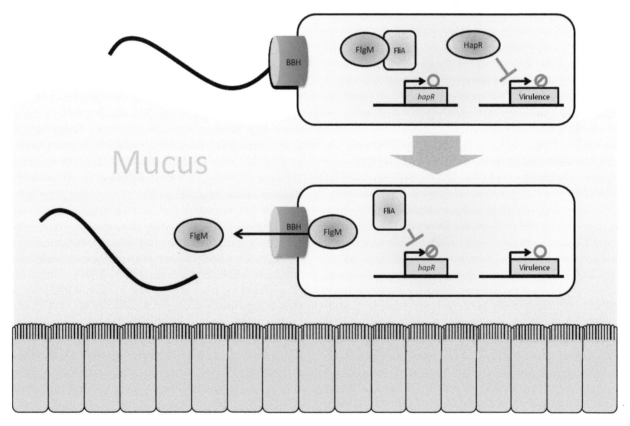

Figure 3. Flagellar breakage signals repression of *hapR* to allow virulence gene expression in *V. cholerae*. At the high cell densities present in a *V. cholerae* inoculum, virulence genes are repressed by the quorum-sensing regulator HapR. As the vibrios penetrate the mucus layer of the host small intestines, the flagella break off, allowing the basal-body-hook complex (BBH) to export the anti-sigma factor FlgM. This frees the sigma factor FliA to repress *hapR*, which in turn allows virulence gene expression.
doi:10.1128/9781555818524.ch26f3

genes so that the structural proteins they encode can be assembled into a flagellum. The authors found that FliA represses *hapR* expression, implying that if FlgM is exported from the cell, FliA would then be free to repress *hapR* (Liu et al., 2008). Knowing that *V. cholerae* must penetrate a layer of mucus to reach its desired niche in the intestinal crypts, the authors hypothesized that the viscous mucus might cause the flagellum to be lost from the surface of the bacteria while the secretion system underlying the flagellum remains intact. Indeed, electron microscopy demonstrated that the bacteria lost their flagella after swimming through mucus, that FlgM is then exported from the cells, and that this mechanism leads to repression of *hapR* by FliA.

In summary, *V. cholerae* possesses mechanisms capable of promoting virulence and escape responses at appropriate times. As discussed in Chapter 28, during infection the virulence program represses quorum sensing and biofilms to allow replication and disease progression. Late in infection, quorum sensing represses virulence and promotes cellular detachment and biofilm formation to allow escape. During infection of a new host, flagellar breakage allows the derepression of virulence again. At the heart of these mechanisms is a complex gene regulation apparatus, primarily controlled by a multilevel network of gene repression events. Thus, *V. cholerae* is a particularly good example of how bacteria use negative regulation to control gene expression during host-environment transitions.

GENE REPRESSION TO ESTABLISH AN INTRACELLULAR NICHE

Negative regulation in transcriptional programs in vivo has also been found in *Salmonella enterica*, a gram-negative pathogen that is the cause of disease in humans and other mammalian hosts (Haraga et al., 2008). Full virulence in *S. enterica* requires the products of genes on at least two major pathogenicity islands, *Salmonella* pathogenicity islands 1 and 2 (SPI-1 and SPI-2) (Galan and Curtiss, 1989; Shea et al., 1996). Each island contains genes encoding a type III secretion system (T3SS), which is capable of translocating a number of effectors into host cells to mediate numerous roles, including rearrangement of the host actin cytoskeleton (Zhou et al., 1999) and disruption of intracellular trafficking (Uchiya et al., 1999), electrolyte loss (Wood et al., 1998), and invasion of host cells (Hardt et al., 1998).

Salmonellae are able to invade host epithelial and phagocytic cells in the gut (Jones et al., 1994; Vazquez-Torres et al., 1999), and can survive and replicate within macrophages (Richter-Dahlfors et al., 1997). Indeed, the ability to survive intracellularly is critical for systemic virulence (Fields et al., 1986). The T3SSs and their effectors are critical to this process but are not constitutively active: salmonellae are able to control the expression of these genes in response to a variety of environmental cues (Lucas and Lee, 2000). The genes of SPI-1, for example, are thought to be important largely during *Salmonella* invasion of the intestinal epithelium, as SPI-1 mutants are attenuated for virulence after oral infection but not intraperitoneal infection (Galan and Curtiss, 1989; Hansen-Wester and Hensel, 2001). In contrast, SPI-2 mutants are deficient for survival within host macrophages (Hensel et al., 1998) and for virulence when administered intraperitoneally (Hensel et al., 1995; Shea et al., 1996). SPI-2 expression can occur in the lumen of the intestine before invasion, perhaps acting to prime *Salmonella* for intracellular survival within host macrophages (Brown et al., 2005). However, the ordered expression of these secretion systems is critical for successful pathogenesis, and gene repression events play an important role in ensuring the timeliness of this process.

There are several intertwined regulatory systems governing SPI expression in vivo. The PhoPQ two-component regulatory system consists of the membrane-bound PhoQ and the cytoplasmic DNA-binding protein PhoP, similar to the BvgAS system of *Bordetella* described above (Miller et al., 1989b; Prost and Miller, 2008). The PhoPQ system is active—and PhoP is phosphorylated—in the acidified phagosomes of macrophages and inactive outside. In this way, the PhoPQ system is thought to act as a sensor of the location of salmonellae vis-à-vis the interior of the host cell and as the key upstream regulator of SPI virulence genes in this pathogen (Cotter and DiRita, 2000). Outside of host cells, the PhoPQ system is inactive, allowing for the expression of SPI-1 genes under the control of the SPI-1-encoded regulator HilA (Bajaj et al., 1995, 1996). Within the phagosome of host phagocytic cells, the PhoPQ system is activated (Alpuche Aranda et al., 1992) and leads to the activation of SPI-2 genes through the SsrAB two-component system encoded within SPI-2 (Deiwick et al., 1999). Furthermore, under these conditions, the invasion genes encoded within SPI-1 are no longer necessary; PhoPQ then acts to repress SPI-1 gene transcription (Pegues et al., 1995). Thus, this regulatory system optimizes the virulence gene transcription of *Salmonella*, depending on the stage of infection in which the microbe finds itself, by repressing one set of genes and activating another. The repression of SPI-1 before activation in the gut lumen is mediated transcriptionally by H-NS (described below) and posttranscriptionally by the small RNA CsrA (Martínez et al., 2011).

Transcription of the genes of the SPI-2 system is also negatively regulated during some stages of infection. Coombes et al. (2005) were able to identify the small protein YdgT, which bears some homology to the nucleoid-associated protein H-NS, as being able to negatively regulate the transcription of genes within SPI-2. *Salmonella ydgT* mutants displayed enhanced survival within murine macrophages, as the SPI-2 genes were derepressed. However, contrary to what might be expected after deletion of a negative regulator of virulence, *ydgT* mutants were attenuated for virulence in systemic infections of mice. Interestingly, competition in a mouse model showed higher colonization for *ydgT* mutants compared to wild type within 24 hours of infection but lower colonization after a further 31 hours of infection. *S. enterica* constitutively expressing *ydgT* was attenuated for colonization at both time points. These data suggested that although virulence genes must be activated to survive as a pathogen, negative regulation of the same genes might be required until bacteria enter an appropriate stage of infection.

REPRESSION OF HORIZONTALLY ACQUIRED VIRULENCE GENES AS FOREIGN DNA

A common feature of many pathogens is that they acquired their virulence genes through lateral gene transfer (Juhas et al., 2009). Virulence genes are frequently encoded in clusters on the genomes of pathogens, and these clusters are termed "pathogenicity islands." The pathogenicity islands often differ from the host genome in their G+C content and are frequently flanked by repeat sequences, indicating that they were inherited from outside sources, such as bacteriophages or conjugative plasmids from other bacteria. Even more obvious examples of lateral transfer are plasmids which encode virulence genes. This horizontal gene transfer has resulted in rapid increases in antibiotic resistance and the emergence of new pathogens. However, not all horizontally transferred genes benefit the recipient: there are parasitic genes, lytic phages, and otherwise toxic DNA sequences against which the bacterium must protect itself. To do this, there are multiple described defense systems, such as the recently identified CRISPR system for phage immunity (Deveau et al., 2011). However, one of the first global repressor proteins identified, H-NS, is thought to serve as a first line of defense against unwanted gene expression from horizontally acquired DNA (Fig. 4). Work by the Fang group and others have identified a model whereby H-NS represses newly inherited genes in a process

termed "xenogeneic silencing" to ensure the health of the host bacterium in the presence of its new genomic content (Navarre et al., 2006). Furthermore, H-NS has been implicated in the repression of virulence genes in multiple pathogens, making it a particularly interesting transcriptional repressor.

The role of H-NS in *V. cholerae* was first identified by complementing the *hns* mutation in *E. coli* with genomic DNA from *V. cholerae* and finding that tampering with levels of H-NS protein in *V. cholerae* disrupted phenotypes such as colony morphology and motility (Tendeng et al., 2000). The link with virulence was established in the same year, when Nye et al. (2000) isolated an *hns* mutant strain of *V. cholerae* and found that H-NS repressed more than one step in the virulence transcription cascade. This study produced two interesting results regarding *V. cholerae* pathogenesis. First, H-NS represses two different pathogenicity islands on the genome: the *ctx* locus, encoding the cholera toxin that was inherited from the CTXΦ phage, and the *tcp* locus, which was inherited through an unknown mechanism. Second, *hns* mutants exhibited virulence gene expression under non-virulence-inducing conditions that was just as high as what was normally seen when virulence was induced in wild-type bacteria. It is now thought that positive transcriptional regulators must displace H-NS from the promoters of virulence genes to fully activate transcription (Stonehouse et al., 2008, 2011). This was surprising, in that the entire inducibility of the virulence program, and likely the bacterium's ability to cause disease, was dependent on relieving multilocus repression by H-NS. The importance of this regulation was underscored by studies that demonstrated that *hns* mutants displayed a colonization defect in mice despite higher levels of TcpA production (Ghosh et al., 2006).

V. cholerae is not the only bacterium in which H-NS has been shown to be a critical repressor of virulence. Extensive studies have focused on the role of H-NS in gene regulation in *Salmonella enterica*, the cause of enteritis and typhoid fever. As discussed above, *Salmonella* possesses horizontally acquired virulence genes on several *Salmonella* pathogenicity islands (SPIs). The first genome-wide screens of H-NS-binding sites revealed a dramatic enrichment at low-G+C loci, in particular the SPIs (Navarre et al., 2006; Lucchini et al., 2006). The importance of H-NS-mediated repression is highlighted by the reduced viability of the *hns* mutant, as well as the complementation of this low-viability phenotype by deletion of the critical *Salmonella* virulence regulator SsrA (Lucchini et al., 2006), suggesting that the repression of virulence genes by H-NS is important for bacterial survival under nonvirulent conditions. The

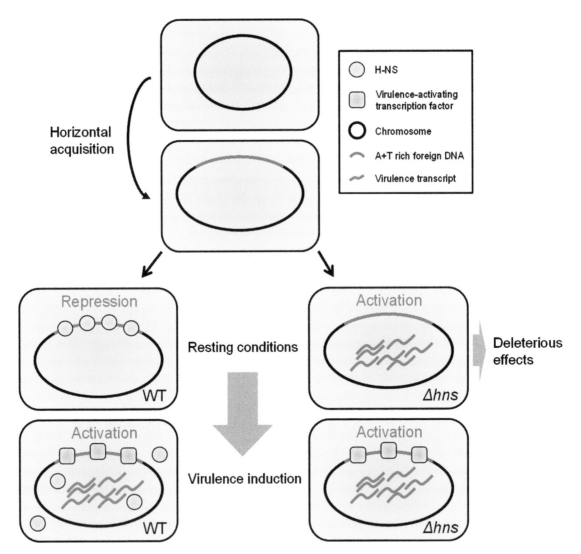

Figure 4. H-NS-mediated repression of transcription of foreign DNA. The nucleoid-associated protein H-NS represses transcription of A+T-rich foreign DNA, a process known as "xenogeneic silencing." H-NS-mediated gene repression has evolved to become a timing mechanism for the expression of virulence genes, which are commonly inherited horizontally. After acquisition of new A+T-rich DNA, such as might encode virulence genes, the H-NS protein represses transcription of this DNA. Under virulence-inducing conditions, a virulence-activating transcription factor may displace H-NS from the DNA to allow transcription. In an *hns* mutant, however, virulence genes may be active under nonvirulent conditions, which can be deleterious to the bacteria. The presence of H-NS or other repressive nucleoid-associated proteins thus ensures timely activation of virulence programs for a wide variety of bacteria.
doi:10.1128/9781555818524.ch26f4

mechanism by which H-NS mediates virulence gene repression in *Salmonella* is similar to that in *V. cholerae*. Studies have shown that deletion of *hns* results in abrogation of the inducibility of SPI-2 (Walthers et al., 2007, 2011) and SPI-1 (Olekhnovich and Kadner, 2006, 2007), leaving the virulence genes highly expressed under noninducing conditions. Furthermore, the inducibility of SPI-2 gene expression is dependent on physical displacement of H-NS from the virulence genes by the transcription factor SsrB (Walthers et al., 2011). However, in *Salmonella*, this is not the only mechanism by which H-NS-mediated repression is reversed; the *Salmonella* regulator SlyA binds to the

pagC and *ugtL* promoters concurrently with H-NS and overcomes their repression without displacing H-NS from the DNA molecule (Perez et al., 2008). H-NS has also been shown to repress horizontally acquired virulence genes in pathogenic *E. coli*, *Yersinia*, and *Shigella* (Porter and Dorman, 1994; Bustamante et al., 2001; Goldberg et al., 2001; Ellison and Miller, 2006; Torres et al., 2007; Turner and Dorman, 2007; Castellanos et al., 2009), highlighting how the effects of global repression are widespread among the domain of pathogenic bacteria.

In addition to H-NS, other nucleoid-associated proteins have been demonstrated to have a global effect on A+T-rich DNA, including virulence genes. YdgT is a nucleoid-associated protein from *Salmonella* that negatively regulates SPI-2 (Coombes et al., 2005). In *Mycobacterium tuberculosis*, which lacks an H-NS homologue, the protein Lsr2 possesses several histone-like properties similar to H-NS, such as DNase I protection, oligomerization on DNA, DNA bridging, and a preference for A+T-rich binding (Colangeli et al., 2007; Gordon et al., 2008, 2010). This protein represses virulence genes and antibiotic resistance genes in a manner similar to H-NS (Colangeli et al., 2007, 2009; Gordon et al., 2008, 2010). Lsr2 also complements phenotypes in an *E. coli hns* mutant (Gordon et al., 2008). All this suggests that although pathogenic mycobacteria do not possess an exact H-NS homologue, the requirement for a global transcriptional repressor to mediate appropriate timing of virulence may be much more widespread than H-NS.

From these studies, one could conclude that repression by nucleoid-associated proteins is, in fact, the "default" state for the bacteria and that the regulatory effect of these proteins has evolved from a self-defense mechanism preventing deleterious overexpression of foreign genes to a global switching mechanism ensuring temporally appropriate virulence gene expression.

GENE REPRESSION TO MAINTAIN COMMENSALITY

For some pathogens, causing disease in a host is not always the most beneficial outcome for the bacteria. Unlike *V. cholerae* or *S. enterica,* which tend to cause only one or two disease syndromes, certain bacteria such as *Streptococcus pyogenes*, also known as group A *Streptococcus* (GAS), cause a wide range of diseases. These range from benign skin infections to rheumatic fever and necrotizing fasciitis. Considerable research has been done to address the question of what determines whether GAS will switch from a harmless colonizer to an invasive pathogen. As it turns out, negative regulation of virulence factors plays a major role in restricting GAS to its noninvasive state (Fig. 5).

Interestingly, the central pathway for reducing virulence potential is by posttranslational cleavage mediated by the cysteine protease SpeB. SpeB is capable of cleaving both host and bacterial proteins during the course of infection (Kapur et al., 1994; Berge and Björck, 1995). Among the bacterial proteins cleaved by this enzyme are several that are critical for virulence, including the anti-phagocytic fibrinogen- and IgG-binding protein M1, the plasminogen activator streptokinase that associates with M1, the

bacteriophage-encoded DNase I Sda1, and various GAS superantigens (Whitnack and Beachey, 1982; Raeder et al., 1998; Kansal et al., 2003; Rezcallah et al., 2004; Buchanan et al., 2006; Walker et al., 2007). Modification of the M1 surface protein by SpeB involves the removal of a 24-amino-acid portion of the N terminus. The resulting truncated form of M1 is unable to bind to as wide a range of human IgG types as is the full-length form (Raeder et al., 1998). Though type IIo GAS, which express full-length M1, and type IIb GAS, which express SpeB and thus a truncated M1, are both resistant to phagocytosis in vitro, type IIo GAS are much more virulent in mouse models of infection, suggesting that a full binding range of M1 is important for the severity of GAS infections (Raeder and Boyle, 1996). In addition, the ability of M1 to bind in a complex with streptokinase and host plasminogen has been shown to be critical to the GAS invasion. SpeB thus acts at each of these steps to inhibit maximal virulence in invasive GAS disease (Cunningham, 2000; McKay et al., 2004). While to some extent these effects on GAS virulence are reversed by the ability of cleaved IgG-binding proteins to dissociate from the cell and induce complement activation away from the streptococci themselves, as well as some activity of this protease on host IgG itself, the presence of this protease appears to be inversely related to the severity of invasive disease caused by these pathogens (von Pawel-Rammingen and Bjorck, 2003; Hollands et al., 2010). These effects together can have a significant impact on the propensity of GAS to cause invasive disease (Cole et al., 2006). Thus, although not a transcriptional repressor, SpeB is an inhibitor of the virulent state of the organism and restricts the capacity of GAS to cause invasive, potentially fatal, disease to its host.

However, many infections do progress to invasive disease. In these cases, some selective pressure should exist during infection that causes GAS to downregulate SpeB expression. To investigate the genetic basis for the switch from noninvasive to invasive GAS, Sumby et al. (2006) took noninvasive and invasive GAS clinical isolates and used the noninvasive GAS to infect mice subcutaneously. From the resulting lesions, invasive streptococci were recovered. These, as well as the initial non-invasive GAS used for the starting inoculum, were then sequenced and compared. From this analysis, the only difference detected was a frameshift mutation in the *covRS* ("control of virulence") two-component system. Analogous to the BvgAS and PhoPQ systems described above, CovS is a membrane protein that can sense external cues and transfer a phosphate group to the cytoplasmic DNA-binding protein CovR, altering its affinity for

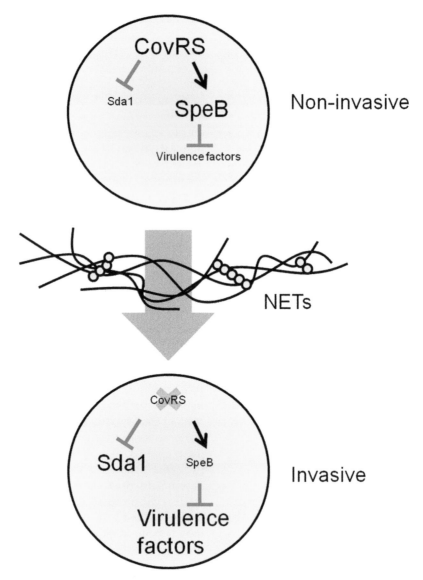

Figure 5. Selection of invasive mutants in group A *Streptococcus*. In the colonizing but noninvasive state, the protease SpeB is expressed, causing degradation or truncation of numerous virulence factors. The expression of SpeB is controlled by the two-component system CovRS. In the presence of neutrophil extracellular traps, there is a selective pressure for expression of the DNase Sda1, which is normally repressed by CovRS. This pressure causes the outgrowth of *covRS* mutants, which has the effect of coselecting for strains with decreased SpeB expression, leading to increased virulence factor production and an invasive phenotype.
doi:10.1128/9781555818524.ch26f5

promoters. This system is involved in the regulation of several genes associated with pathogenesis in GAS, including *speB* and the genes encoding streptokinase, capsule production, and the toxin streptolysin (Federle et al., 1999; Heath et al., 1999) . A 7-bp insertion into *covS* generated a truncated form of this histidine kinase, and this was determined to be responsible for the increased severity of phenotype in invasive strains isolated after infection in mice (Kansal et al., 2000; Sumby et al., 2006). Interestingly, this form of *covS* containing the 7-bp insertion also caused abrogation of *speB* expression.

Walker et al. (2007) were able to demonstrate that the acquisition of a phage containing the genes expressing Sda1, a GAS DNase I enzyme that is also a target of SpeB, selected for *covRS* mutants that did not express SpeB. Sda1 allows GAS to escape neutrophil extracellular traps (NETs). These are networks of fibers secreted by activated neutrophils consisting of chromatin and granular components that can effect the killing of extracellular bacteria independently of phagocytosis (Brinkmann et al., 2004). Sda1 allows GAS to escape extracellular killing by activated neutrophils by degrading the DNA components of these NETs (Buchanan et al., 2006); given the critical role played by neutrophils in the control of many bacterial infections, it is not surprising that a strong selective force would be exerted on a GAS population to abrogate SpeB production, allowing Sda1 to act on NETs and assist in GAS escape from neutrophils. This also enables the production of the full-length form of M1 that allows for optimal virulence in this pathogen. Thus, selection by innate neutrophil-mediated immune responses drives the mutation of *covS* in GAS populations during infection, a process that simultaneously derepresses the expression of several virulence genes and overcomes the posttranslational effects of SpeB to allow the maximal production of virulence factors, and consequently a more severe outcome for the host.

The *covRS* two-component system has also been studied in the group B *Streptococcus agalactiae*, which is a major cause of sepsis and meningitis in newborns. Similar to CovRS in GAS, GBS CovRS negatively regulates a broad array of genes which includes virulence genes such as those encoding hemolysin and C5a peptidase (Lamy et al., 2004; Jiang et al., 2005). Although inactivation of *covR* produces different effects on animal killing depending on the model used (Lamy et al., 2004; Jiang et al., 2005; Lembo et al., 2010), it has been shown that *covR* mutants cause increased permeability of the blood-brain barrier and activation of chemokine production by the brain epithelium (Lembo et al., 2010). This suggests a mechanism for how the normally commensal GBS can switch from colonizing the vaginal epithelium to causing invasive, life-threatening disease in the newborn. Direct evidence of CovR derepression during GBS pathogenesis has not been shown, but there is some limited evidence that GBS may possess such a switching mechanism. Like the response regulators in most two-component systems, the DNA-binding activity of CovR is determined by its phosphorylation state. Interestingly, CovR appears to be the target not only of the histidine kinase CovS but also of the serine/threonine kinase Stk1 (Lin et al., 2009), suggesting that multiple pathways exist to collectively cause the derepression of virulence genes.

Both GAS and GBS exist in states that are innocuous to their hosts, and yet both possess the ability to cause severe disease. At the fulcrum of this switch appears to be the two-component system CovRS and the protease SpeB, deactivating the virulence genes and their products until the organism is ready to cause invasive disease. The deactivation of CovR has been shown to be mediated through selection of CovS mutants in the presence of innate immune responses, but it also appears to be regulated by as-yet-unknown signals sensed by CovS. Nonetheless, GAS and GBS are examples of how bacteria may use negative regulation to remain in a non-disease-causing state.

CONCLUSION

In the rapidly advancing field of bacterial pathogenesis, the primary focus of research has been on the identification of virulence factors, from secretion systems to toxins and adherence factors, and the mechanisms by which these bacteria activate their expression. But an increasing number of studies of negative regulation have revealed that these infectious organisms possess exquisitely tuned genetic programs that activate and repress their virulence factors at precisely timed junctions during pathogenesis. Pathogenic bacteria usually travel through numerous potentially hostile environments during the course of infecting a host as well as in their nonhost habitats, and each of these environments requires its own genetic response from the bacteria to ensure survival. This entails both positive regulation of beneficial genes and negative regulation of genes for which expression would be deleterious. In this chapter, we have highlighted several such pressures: arthropod vectors, parasitic DNA, intracellular versus extracellular conditions, and the mammalian immune system, and examined the various transcriptional and posttranslational mechanisms that bacteria use to coordinate negative regulation. Further research is likely to solidify our understanding of negative regulation in bacterial diseases and add to the complex and astounding array of genetic systems already documented.

REFERENCES

Akerley, B. J., P. A. Cotter, and J. F. Miller. 1995. Ectopic expression of the flagellar regulon alters development of the Bordetella-host interaction. *Cell* 80:611–620.

Akerley, B. J., D. M. Monack, S. Falkow, and J. F. Miller. 1992. The *bvgAS* locus negatively controls motility and synthesis of flagella in *Bordetella bronchiseptica*. *J. Bacteriol.* 174:980–990.

Alpuche Aranda, C. M., J. A. Swanson, W. P. Loomis, and S. I. Miller. 1992. Salmonella typhimurium activates virulence gene transcription within acidified macrophage phagosomes. *Proc. Natl. Acad. Sci. USA* 89:10079–10083.

Antonara, S., L. Ristow, J. McCarthy, and J. Coburn. 2010. Effect of *Borrelia burgdorferi* OspC at the site of inoculation in mouse skin. *Infect. Immun.* 78:4723–4733.

Apodaca, G., L. A. Katz, and K. E. Mostov. 1994. Receptor-mediated transcytosis of IgA in MDCK cells is via apical recycling endosomes. *J. Cell Biol.* 125:67–86.

Arico, B., J. F. Miller, C. Roy, S. Stibitz, D. Monack, S. Falkow, R. Gross, and R. Rappuoli. 1989. Sequences required for expression of Bordetella pertussis virulence factors share homology with prokaryotic signal transduction proteins. *Proc. Natl. Acad. Sci. USA* 86:6671–6675.

Arico, B., V. Scarlato, D. M. Monack, S. Falkow, and R. Rappuoli. 1991. Structural and genetic analysis of the *bvg* locus in *Bordetella* species. *Mol. Microbiol.* **5**:2481–2491.

Attridge, S. R., P. A. Manning, J. Holmgren, and G. Jonson. 1996. Relative significance of mannose-sensitive hemagglutinin and toxin-coregulated pili in colonization of infant mice by *Vibrio cholerae* El Tor. *Infect. Immun.* **64**:3369–3373.

Bajaj, V., C. Hwang, and C. A. Lee. 1995. HilA is a novel OmpR/ToxR family member that activates the expression of Salmonella typhimurium invasion genes. *Mol. Microbiol.* **18**:715–727.

Bajaj, V., R. L. Lucas, C. Hwang, and C. A. Lee. 1996. Co-ordinate regulation of *Salmonella* Typhimurium invasion genes by environmental and regulatory factors is mediated by control of *hilA* expression. *Mol. Microbiol.* **22**:703–714.

Barbour, A. G., S. L. Tessier, and W. J. Todd. 1983. Lyme disease spirochetes and ixodid tick spirochetes share a common surface antigenic determinant defined by a monoclonal antibody. *Infect. Immun.* **41**:795–804.

Berge, A., and L. Björck. 1995. Streptococcal cysteine proteinase releases biologically active fragments of streptococcal surface proteins. *J. Biol. Chem.* **270**:9862–9867.

Blanco, L. P., and V. J. DiRita. 2006. Antibodies enhance interaction of *Vibrio cholerae* with intestinal M-like cells. *Infect. Immun.* **74**:6957–6964.

Brinkmann, V., U. Reichard, C. Goosmann, B. Fauler, Y. Uhlemann, D. S. Weiss, Y. Weinrauch, and A. Zychlinsky. 2004. Neutrophil extracellular traps kill bacteria. *Science* **303**:1532–1535.

Brown, N. F., B. A. Vallance, B. K. Coombes, Y. Valdez, B. A. Coburn, and B. B. Finlay. 2005. *Salmonella* pathogenicity island 2 is expressed prior to penetrating the intestine. *PLoS Pathog.* **1**:e32.

Buchanan, J. T., A. J. Simpson, R. K. Aziz, G. Y. Liu, S. A. Kristian, M. Kotb, J. Feramisco, and V. Nizet. 2006. DNase expression allows the pathogen Group A *Streptococcus* to escape killing in neutrophil extracellular traps. *Curr. Biol.* **16**:396–400.

Burgdorfer, W., A. G. Barbour, S. F. Hayes, J. L. Benach, E. Grunwaldt, and J. P. Davis. 1982. Lyme disease—a tick-borne spirochetosis? *Science* **216**:1317–1319.

Burkot, T. R., J. Piesman, and R. A. Wirtz. 1994. Quantitation of the *Borrelia burgdorferi* outer surface protein A in *Ixodes scapularis*: fluctuations during the tick life cycle, doubling times, and loss while feeding. *J. Infect. Dis.* **170**:883–889.

Burtnick, M. N., J. S. Downey, P. J. Brett, J. A. Boylan, J. G. Frye, T. R. Hoover, and F. C. Gherardini. 2007. Insights into the complex regulation of *rpoS* in *Borrelia burgdorferi*. *Mol Microbiol.* **65**:277–293.

Bustamante, V. H., F. J. Santana, E. Calva, and J. L. Puente. 2001. Transcriptional regulation of type III secretion genes in enteropathogenic *Escherichia coli*: Ler antagonizes H-NS-dependent repression. *Mol. Microbiol.* **39**:664–678.

Butler, S. M., and A. Camilli. 2005. Going against the grain: chemotaxis and infection in *Vibrio cholerae*. *Nat. Rev. Microbiol.* **3**:611–620.

Caimano, M. J., R. Iyer, C. H. Eggers, C. Gonzalez, E. A. Morton, M. A. Gilbert, I. Schwartz, and J. D. Radolf. 2007. Analysis of the RpoS regulon in *Borrelia burgdorferi* in response to mammalian host signals provides insight into RpoS function during the enzootic cycle. *Mol. Microbiol.* **65**:1193–1217.

Castellanos, M. I., D. J. Harrison, J. M. Smith, S. K. Labahn, K. M. Levy, and H. J. Wing. 2009. VirB Alleviates H-NS repression of the *icsP* promoter in *Shigella flexneri* from sites more than one kilobase upstream of the transcription start site. *J. Bacteriol.* **191**:4047–4050.

Chiavelli, D. A., J. W. Marsh, and R. K. Taylor. 2001. The mannose-sensitive hemagglutinin of *Vibrio cholerae* promotes adherence to zooplankton. *Appl. Environ. Microbiol.* **67**:3220–3225.

Colangeli, R., A. Haq, V. L. Arcus, E. Summers, R. S. Magliozzo, A. McBride, A. K. Mitra, M. Radjainia, A. Khajo, W. R. Jacobs, P. Salgame, and D. Alland. 2009. The multifunctional histone-like protein Lsr2 protects mycobacteria against reactive oxygen intermediates. *Proc. Natl. Acad. Sci. USA* **106**:4414–4418.

Colangeli, R., D. Helb, C. Vilchèze, M. H. Hazbón, C.-G. Lee, H. Safi, B. Sayers, I. Sardone, M. B. Jones, R. D. Fleischmann, S. N. Peterson, W. R. Jacobs, and D. Alland. 2007. Transcriptional regulation of multi-drug tolerance and antibiotic-induced responses by the histone-like protein Lsr2 in *M. tuberculosis*. *PLoS Pathog.* **3**:e87.

Cole, J. N., J. D. McArthur, F. C. McKay, M. L. Sanderson-Smith, A. J. Cork, M. Ranson, M. Rohde, A. Itzek, H. Sun, D. Ginsburg, M. Kotb, V. Nizet, G. S. Chhatwal, and M. J. Walker. 2006. Trigger for group A streptococcal M1T1 invasive disease. *FASEB J.* **20**:1745–1747.

Coombes, B. K., M. E. Wickham, M. J. Lowden, N. F. Brown, and B. B. Finlay. 2005. Negative regulation of *Salmonella* pathogenicity island 2 is required for contextual control of virulence during typhoid. *Proc. Natl. Acad. Sci. USA* **102**:17460–17465.

Correa, N. E., J. R. Barker, and K. E. Klose. 2004. The *Vibrio cholerae* FlgM homologue is an anti-σ^{28} factor that is secreted through the sheathed polar flagellum. *J. Bacteriol.* **186**:4613–4619.

Cotter, P. A., and V. J. DiRita. 2000. Bacterial virulence gene regulation: an evolutionary perspective. *Annu. Rev. Microbiol.* **54**:519–565.

Cotter, P. A., and A. M. Jones. 2003. Phosphorelay control of virulence gene expression in *Bordetella*. *Trends Microbiol.* **11**:367–373.

Cummings, C. A., H. J. Bootsma, D. A. Relman, and J. F. Miller. 2006. Species- and strain-specific control of a complex, flexible regulon by *Bordetella* BvgAS. *J. Bacteriol.* **188**:1775–1785.

Cunningham, M. W. 2000. Pathogenesis of group A streptococcal infections. *Clin. Microbiol. Rev.* **13**:470–511.

Deiwick, J., T. Nikolaus, S. Erdogan, and M. Hensel. 1999. Environmental regulation of *Salmonella* pathogenicity island 2 gene expression. *Mol. Microbiol.* **31**:1759–1773.

Deveau, H., J. E. Garneau, and S. Moineau. 2011. CRISPR/Cas system and its role in phage-bacteria interactions. *Annu. Rev. Microbiol.* **64**:475–493.

Dickinson, E. C., J. C. Gorga, M. Garrett, R. Tuncer, P. Boyle, S. C. Watkins, S. M. Alber, M. Parizhskaya, M. Trucco, M. I. Rowe, and H. R. Ford. 1998. Immunoglobulin A supplementation abrogates bacterial translocation and preserves the architecture of the intestinal epithelium. *Surgery* **124**:284–290.

DiRita, V. J., C. Parsot, G. Jander, and J. J. Mekalanos. 1991. Regulatory cascade controls virulence in *Vibrio cholerae*. *Proc. Natl. Acad. Sci. USA* **88**:5403–5407.

Dons, L., E. Eriksson, Y. Jin, M. E. Rottenberg, K. Kristensson, C. N. Larsen, J. Bresciani, and J. E. Olsen. 2004. Role of flagellin and the two-component CheA/CheY system of *Listeria monocytogenes* in host cell invasion and virulence. *Infect. Immun.* **72**:3237–3244.

Ellison, D. W., and V. L. Miller. 2006. H-NS represses *inv* transcription in *Yersinia enterocolitica* through competition with RovA and interaction with YmoA. *J. Bacteriol.* **188**:5101–5112.

Faruque, S. M., K. M. Ahmed, A. K. Siddique, K. Zaman, A. R. Alim, and M. J. Albert. 1997. Molecular analysis of toxigenic *Vibrio cholerae* O139 Bengal strains isolated in Bangladesh between 1993 and 1996: evidence for emergence of a new clone of the Bengal vibrios. *J. Clin. Microbiol.* **35**:2299–2306.

Federle, M. J., K. S. McIver, and J. R. Scott. 1999. A response regulator that represses transcription of several virulence operons in the group A *Streptococcus*. *J. Bacteriol.* **181**:3649–3657.

Fields, P. I., R. V. Swanson, C. G. Haidaris, and F. Heffron. 1986. Mutants of Salmonella typhimurium that cannot survive within the macrophage are avirulent. *Proc. Natl. Acad. Sci. USA* 83:5189–5193.

Freter, R., B. Allweiss, P. C. O'Brien, S. A. Halstead, and M. S. Macsai. 1981a. Role of chemotaxis in the association of motile bacteria with intestinal mucosa: in vitro studies. *Infect. Immun.* 34:241–249.

Freter, R., P. C. O'Brien, and M. S. Macsai. 1981b. Role of chemotaxis in the association of motile bacteria with intestinal mucosa: in vivo studies. *Infect. Immun.* 34:234–240.

Fuchs, R., S. Jauris, F. Lottspeich, V. Preac-Mursic, B. Wilske, and E. Soutschek. 1992. Molecular analysis and expression of a Borrelia burgdorferi gene encoding a 22 kDa protein (pC) in Escherichia coli. *Mol. Microbiol.* 6:503–509.

Galan, J. E., and R. Curtiss. 1989. Cloning and molecular characterization of genes whose products allow *Salmonella* Typhimurium to penetrate tissue culture cells. *Proc. Natl. Acad. Sci. USA* 86:6383–6387.

Gardel, C., and J. Mekalanos. 1996. Alterations in *Vibrio cholerae* motility phenotypes correlate with changes in virulence factor expression. *Infect. Immun.* 64:2246–2255.

Ghosh, A., K. Paul, and R. Chowdhury. 2006. Role of the histone-like nucleoid structuring protein in colonization, motility, and bile-dependent repression of virulence gene expression in *Vibrio cholerae*. *Infect. Immun.* 74:3060–3064.

Goldberg, M. D., M. Johnson, J. C. D. Hinton, and P. H. Williams. 2001. Role of the nucleoid-associated protein Fis in the regulation of virulence properties of enteropathogenic *Escherichia coli*. *Mol. Microbiol.* 41:549–559.

Gordon, B. R. G., R. Imperial, L. Wang, W. W. Navarre, and J. Liu. 2008. Lsr2 of *Mycobacterium* represents a novel class of H-NS-like proteins. *J. Bacteriol.* 190:7052–7059.

Gordon, B. R. G., Y. Li, L. Wang, A. Sintsova, H. van Bakel, S. Tian, W. W. Navarre, B. Xia, and J. Liu. 2010. Lsr2 is a nucleoid-associated protein that targets AT-rich sequences and virulence genes in *Mycobacterium tuberculosis*. *Proc. Natl. Acad. Sci. USA* 107:5154–5159.

Gründling, A., L. S. Burrack, H. G. A. Bouwer, and D. E. Higgins. 2004. *Listeria monocytogenes* regulates flagellar motility gene expression through MogR, a transcriptional repressor required for virulence. *Proc. Natl. Acad. Sci. USA* 101:12318–12323.

Hansen-Wester, I., and M. Hensel. 2001. *Salmonella* pathogenicity islands encoding type III secretion systems. *Microbes Infect.* 3:549–559.

Haraga, A., M. B. Ohlson, and S. I. Miller. 2008. Salmonellae interplay with host cells. *Nat. Rev. Microbiol.* 6:53–66.

Hardt, W. D., L. M. Chen, K. E. Schuebel, X. R. Bustelo, and J. E. Galan. 1998. *Salmonella* Typhimurium encodes an activator of Rho GTPases that induces membrane ruffling and nuclear responses in host cells. *Cell* 93:815–826.

Häse, C. C. 2001. Analysis of the role of flagellar activity in virulence gene expression in *Vibrio cholerae*. *Microbiology* 147:831–837.

Häse, C. C., and J. J. Mekalanos. 1999. Effects of changes in membrane sodium flux on virulence gene expression in *Vibrio cholerae*. *Proc. Natl. Acad. Sci. USA* 96:3183–3187.

Heath, A., V. J. DiRita, N. L. Barg, and N. C. Engleberg. 1999. A two-component regulatory system, CsrR-CsrS, represses expression of three *Streptococcus pyogenes* virulence factors, hyaluronic acid capsule, streptolysin S, and pyrogenic exotoxin B. *Infect. Immun.* 67:5298–5305.

Hensel, M., J. E. Shea, C. Gleeson, M. D. Jones, E. Dalton, and D. W. Holden. 1995. Simultaneous identification of bacterial virulence genes by negative selection. *Science* 269:400–403.

Hensel, M., J. E. Shea, S. R. Waterman, R. Mundy, T. Nikolaus, G. Banks, A. Vazquez-Torres, C. Gleeson, F. C. Fang, and D. W. Holden. 1998. Genes encoding putative effector proteins of the

type III secretion system of *Salmonella* pathogenicity island 2 are required for bacterial virulence and proliferation in macrophages. *Mol. Microbiol.* 30:163–174.

Hollands, A., M. A. Pence, A. M. Timmer, S. R. Osvath, L. Turnbull, C. B. Whitchurch, M. J. Walker, and V. Nizet. 2010. Genetic switch to hypervirulence reduces colonization phenotypes of the globally disseminated Group A *Streptococcus* M1T1 clone. *J. Infect. Dis.* 202:11–19.

Honko, A. N., and S. B. Mizel. 2005. Effects of flagellin on innate and adaptive immunity. *Immunol. Res.* 33:83–101.

Hsiao, A., Z. Liu, A. Joelsson, and J. Zhu. 2006. *Vibrio cholerae* virulence regulator-coordinated evasion of host immunity. *Proc. Natl. Acad. Sci. USA* 103:14542–14547.

Hsiao, A., K. Toscano, and J. Zhu. 2008. Post-transcriptional cross-talk between pro- and anti-colonization pili biosynthesis systems in *Vibrio cholerae*. *Mol. Microbiol.* 67:849–860.

Hsiao, A., X. Xu, B. Kan, R. V. Kulkarni, and J. Zhu. 2009. Direct regulation by the *Vibrio cholerae* regulator ToxT to modulate colonization and anticolonization pilus expression. *Infect. Immun.* 77:1383–1388.

Hübner, A., X. Yang, D. M. Nolen, T. G. Popova, F. C. Cabello, and M. V. Norgard. 2001. Expression of *Borrelia burgdorferi* OspC and DbpA is controlled by a RpoN-RpoS regulatory pathway. *Proc. Natl. Acad. Sci. USA* 98:12724–12729.

Hyde, J. A., D. K. Shaw, R. Smith III, J. P. Trzeciakowski, and J. T. Skare. 2009. The BosR regulatory protein of *Borrelia burgdorferi* interfaces with the RpoS regulatory pathway and modulates both the oxidative stress response and pathogenic properties of the Lyme disease spirochete. *Mol. Microbiol.* 74:1344–1355.

Hyde, J. A., D. K. Shaw, R. Smith, J. P. Trzeciakowski, and J. T. Skare. 2010. Characterization of a conditional *bosR* mutant in *Borrelia burgdorferi*. *Infect. Immun.* 78:265–274.

Jacob-Dubuisson, F., B. Kehoe, E. Willery, N. Reveneau, C. Locht, and D. A. Relman. 2000. Molecular characterization of *Bordetella bronchiseptica* filamentous haemagglutinin and its secretion machinery. *Microbiology* 146:1211–1221.

Jiang, S.-M., M. J. Cieslewicz, D. L. Kasper, and M. R. Wessels. 2005. Regulation of virulence by a two-component system in group B *Streptococcus*. *J. Bacteriol.* 187:1105–1113.

Johns, R. H., D. E. Sonenshine, and W. L. Hynes. 2000. Enhancement of OspC expression by *Borrelia burgdorferi* in the presence of tick hemolymph. *FEMS Microbiol. Lett.* 193:137–141.

Jones, B. D., N. Ghori, and S. Falkow. 1994. *Salmonella* Typhimurium initiates murine infection by penetrating and destroying the specialized epithelial M cells of the Peyer's patches. *J. Exp. Med.* 180:15–23.

Juhas, M., J. R. Van Der Meer, M. Gaillard, R. M. Harding, D. W. Hood, and D. W. Crook. 2009. Genomic islands: tools of bacterial horizontal gene transfer and evolution. *FEMS Microbiol. Rev.* 33:376–393.

Kamp, H. D., and D. E. Higgins. 2011. A protein thermometer controls temperature-dependent transcription of flagellar motility genes in *Listeria monocytogenes*. *PLoS Pathog.* 7:e1002153.

Kamp, H. D., and D. E. Higgins. 2009. Transcriptional and post-transcriptional regulation of the GmaR antirepressor governs temperature-dependent control of flagellar motility in *Listeria monocytogenes*. *Mol. Microbiol.* 74:421–435.

Kansal, R. G., V. Nizet, A. Jeng, W. J. Chuang, and M. Kotb. 2003. Selective modulation of superantigen-induced responses by streptococcal cysteine protease. *J. Infect. Dis.* 187:398–407.

Kansal, R. G., A. McGeer, D. E. Low, A. Norrby-Teglund, and M. Kotb. 2000. Inverse relation between disease severity and expression of the streptococcal cysteine protease, SpeB, among clonal M1T1 isolates recovered from invasive group A streptococcal infection cases. *Infect. Immun.* 68:6362–6369.

Kapur, V., J. T. Maffei, R. S. Greer, L.-L. Li, G. J. Adams, and J. M. Musser. 1994. Vaccination with streptococcal extracellular cysteine protease (interleukin-1β convertase) protects mice against challenge with heterologous group A streptococci. *Microb. Pathog.* **16**:443–450.

Kaufman, M. R., J. M. Seyer, and R. K. Taylor. 1991. Processing of TCP pilin by TcpJ typifies a common step intrinsic to a newly recognized pathway of extracellular protein secretion by gramnegative bacteria. *Genes. Dev.* **5**:1834–1846.

Kovacikova, G., and K. Skorupski. 2002. Regulation of virulence gene expression in *Vibrio cholerae* by quorum sensing: HapR functions at the *aphA* promoter. *Mol. Microbiol.* **46**:1135–1147.

Lamy, M.-C., M. Zouine, J. Fert, M. Vergassola, E. Couve, E. Pellegrini, P. Glaser, F. Kunst, T. Msadek, P. Trieu-Cuot, and C. Poyart. 2004. CovS/CovR of group B streptococcus: a twocomponent global regulatory system involved in virulence. *Mol Microbiol.* **54**:1250–1268.

Lane, R. S., J. Piesman, and W. Burgdorfer. 1991. Lyme borreliosis: relation of its causative agent to its vectors and hosts in North America and Europe. *Annu. Rev. Entomol.* **36**:587–609.

LaPointe, C. F., and R. K. Taylor. 2000. The type 4 prepilin peptidases comprise a novel family of aspartic acid proteases. *J. Biol. Chem.* **275**:1502–1510.

Leclerc, H., L. Schwartzbrod, and E. Dei-Cas. 2002. Microbial agents associated with waterborne diseases. *Crit. Rev. Microbiol.* **28**:371–409.

Lembo, A., M. A. Gurney, K. Burnside, A. Banerjee, M. De Los Reyes, J. E. Connelly, W.-J. Lin, K. A. Jewell, A. Vo, C. W. Renken, K. S. Doran, and L. Rajagopal. 2010. Regulation of CovR expression in Group B *Streptococcus* impacts blood–brain barrier penetration. *Mol. Microbiol.* **77**:431–443.

Leulier, F., and B. Lemaitre. 2008. Toll-like receptors—taking an evolutionary approach. *Nat. Rev. Genet.* **9**:165–178.

Liang, F. T., J. Yan, M. L. Mbow, S. L. Sviat, R. D. Gilmore, M. Mamula, and E. Fikrig. 2004. *Borrelia burgdorferi* changes its surface antigenic expression in response to host immune responses. *Infect. Immun.* **72**:5759–5767.

Liang, F. T., M. B. Jacobs, L. C. Bowers, and M. T. Philipp. 2002. An immune evasion mechanism for spirochetal persistence in Lyme borreliosis. *J. Exp. Med.* **195**:415–422.

Lin, W.-J., D. Walthers, J. E. Connelly, K. Burnside, K. A. Jewell, L. J. Kenney, and L. Rajagopal. 2009. Threonine phosphorylation prevents promoter DNA binding of the Group B *Streptococcus* response regulator CovR. *Mol. Microbiol.* **71**:1477–1495.

Liu, Z., T. Miyashiro, A. Tsou, A. Hsiao, M. Goulian, and J. Zhu. 2008. Mucosal penetration primes *Vibrio cholerae* for host colonization by repressing quorum sensing. *Proc. Natl. Acad. Sci. USA* **105**:9769–9774.

Locht, C., R. Antoine, and F. Jacob-Dubuisson. 2001. *Bordetella pertussis*, molecular pathogenesis under multiple aspects. *Curr. Opin. Microbiol.* **4**:82–89.

Lopez-Boado, Y. S., L. M. Cobb, and R. Deora. 2005. *Bordetella bronchiseptica* flagellin is a proinflammatory determinant for airway epithelial cells. *Infect. Immun.* **73**:7525–7534.

Lucas, R. L., and C. A. Lee. 2000. Unravelling the mysteries of virulence gene regulation in *Salmonella* Typhimurium. *Mol. Microbiol.* **36**:1024–1033.

Lucchini, S., G. Rowley, M. D. Goldberg, D. Hurd, M. Harrison, and J. C. D. Hinton. 2006. H-NS mediates the silencing of laterally acquired genes in bacteria. *PLoS Pathog.* **2**:e81.

Macpherson, A. J., L. Hunziker, K. McCoy, and A. Lamarre. 2001. IgA responses in the intestinal mucosa against pathogenic and non-pathogenic microorganisms. *Microbes Infect.* **3**:1021–1035.

Martínez, L. C., H. Yakhnin, M. I. Camacho, D. Georgellis, P. Babitzke, J. L. Puente, and V. H. Bustamante. 2011. Integration of a complex regulatory cascade involving the SirA/BarA and Csr global regulatory systems that controls expression of the *Salmonella* SPI-1 and SPI-2 virulence regulons through HilD. *Mol. Microbiol.* **80**:1637–1656.

McKay, F. C., J. D. McArthur, M. L. Sanderson-Smith, S. Gardam, B. J. Currie, K. S. Sriprakash, P. K. Fagan, R. J. Towers, M. R. Batzloff, G. S. Chhatwal, M. Ranson, and M. J. Walker. 2004. Plasminogen binding by group A streptococcal isolates from a region of hyperendemicity for streptococcal skin infection and a high incidence of invasive infection. *Infect. Immun.* **72**:364–370.

Meibom, K. L., X. B. Li, A. T. Nielsen, C. Y. Wu, S. Roseman, and G. K. Schoolnik. 2004. The *Vibrio cholerae* chitin utilization program. *Proc. Natl. Acad. Sci. USA* **101**:2524–2529.

Miller, J. F., C. R. Roy, and S. Falkow. 1989a. Analysis of *Bordetella pertussis* virulence gene regulation by use of transcriptional fusions in *Escherichia coli*. *J. Bacteriol.* **171**:6345–6348.

Miller, M. B., K. Skorupski, D. H. Lenz, R. K. Taylor, and B. L. Bassler. 2002. Parallel quorum sensing systems converge to regulate virulence in *Vibrio cholerae*. *Cell* **110**:303–314.

Miller, S. I., A. M. Kukral, and J. J. Mekalanos. 1989b. A two-component regulatory system (*phoP phoQ*) controls *Salmonella* Typhimurium virulence. *Proc. Natl. Acad. Sci. USA* **86**:5054–5058.

Moorthy, S., and P. I. Watnick. 2004. Genetic evidence that the *Vibrio cholerae* monolayer is a distinct stage in biofilm development. *Mol. Microbiol.* **52**:573–587.

Moorthy, S., and P. I. Watnick. 2005. Identification of novel stagespecific genetic requirements through whole genome transcription profiling of *Vibrio cholerae* biofilm development. *Mol. Microbiol.* **57**:1623–1635.

Navarre, W. W., S. Porwollik, Y. Wang, M. McClelland, H. Rosen, S. J. Libby, and F. C. Fang. 2006. Selective silencing of foreign DNA with low GC content by the H-NS protein in *Salmonella*. *Science* **313**:236–238.

Nielsen, A. T., N. A. Dolganov, G. Otto, M. C. Miller, C. Y. Wu, and G. K. Schoolnik. 2006. RpoS controls the *Vibrio cholerae* mucosal escape response. *PLoS Pathog.* **2**:e109.

Nye, M. B., J. D. Pfau, K. Skorupski, and R. K. Taylor. 2000. *Vibrio cholerae* H-NS silences virulence gene expression at multiple steps in the ToxR regulatory cascade. *J. Bacteriol.* **182**:4295–4303.

Olekhnovich, I. N., and R. J. Kadner. 2006. Crucial roles of both flanking sequences in silencing of the *hilA* promoter in *Salmonella enterica*. *J. Mol. Biol.* **357**:373–386.

Olekhnovich, I. N., and R. J. Kadner. 2007. Role of nucleoidassociated proteins Hha and H-NS in expression of *Salmonella enterica* activators HilD, HilC, and RtsA required for cell invasion. *J. Bacteriol.* **189**:6882–6890.

Ouyang, Z., R. K. Deka, and M. V. Norgard. 2011. BosR (BB0647) controls the RpoN-RpoS regulatory pathway and virulence expression in *Borrelia burgdorferi* by a vovel DNA-binding mechanism. *PLoS Pathog.* **7**:e1001272.

Ouyang, Z., M. Kumar, T. Kariu, S. Haq, M. Goldberg, U. Pal, and M. V. Norgard. 2009. BosR (BB0647) governs virulence expression in *Borrelia burgdorferi*. *Mol. Microbiol.* **74**:1331–1343.

Pal, U., X. Li, T. Wang, R. R. Montgomery, N. Ramamoorthi, A. M. Desilva, F. Bao, X. Yang, M. Pypaert, D. Pradhan, F. S. Kantor, S. Telford, J. F. Anderson, and E. Fikrig. 2004a. TROSPA, an *Ixodes scapularis* receptor for *Borrelia burgdorferi*. *Cell* **119**:457–468.

Pal, U., X. Yang, M. Chen, L. K. Bockenstedt, J. F. Anderson, R. A. Flavell, M. V. Norgard, and E. Fikrig. 2004b. OspC facilitates *Borrelia burgdorferi* invasion of *Ixodes scapularis* salivary glands. *J. Clin. Investig.* **113**:220–230.

Parkhill, J., M. Sebaihia, A. Preston, L. D. Murphy, N. Thomson, D. E. Harris, M. T. Holden, C. M. Churcher, S. D. Bentley,

K. L. Mungall, A. M. Cerdeno-Tarraga, L. Temple, K. James, B. Harris, M. A. Quail, M. Achtman, R. Atkin, S. Baker, D. Basham, N. Bason, I. Cherevach, T. Chillingworth, M. Collins, A. Cronin, P. Davis, J. Doggett, T. Feltwell, A. Goble, N. Hamlin, H. Hauser, S. Holroyd, K. Jagels, S. Leather, S. Moule, H. Norberczak, S. O'Neil, D. Ormond, C. Price, E. Rabbinowitsch, S. Rutter, M. Sanders, D. Saunders, K. Seeger, S. Sharp, M. Simmonds, J. Skelton, R. Squares, S. Squares, K. Stevens, L. Unwin, S. Whitehead, B. G. Barrell, and D. J. Maskell. 2003. Comparative analysis of the genome sequences of *Bordetella pertussis*, *Bordetella parapertussis*, and *Bordetella bronchiseptica*. *Nat. Genet.* **35**:32–40.

Peel, M., W. Donachie, and A. Shaw. 1988. Temperature-dependent expression of flagella of *Listeria monocytogenes* studied by electron microscopy, SDS-PAGE and Western blotting. *J. Gen. Microbiol.* **134**:2171–2178.

Pegues, D. A., M. J. Hantman, I. Behlau, and S. I. Miller. 1995. PhoP/PhoQ transcriptional repression of *Salmonella* Typhimurium invasion genes: evidence for a role in protein secretion. *Mol. Microbiol.* **17**:169–181.

Perez, J. C., T. Latifi, and E. A. Groisman. 2008. Overcoming H-NS-mediated transcriptional silencing of horizontally acquired genes by the PhoP and SlyA Proteins in *Salmonella enterica*. *J. Biol. Chem.* **283**:10773–10783.

Phalipon, A., and B. Corthesy. 2003. Novel functions of the polymeric Ig receptor: well beyond transport of immunoglobulins. *Trends Immunol.* **24**:55–58.

Porter, M. E., and C. J. Dorman. 1994. A role for H-NS in the thermo-osmotic regulation of virulence gene expression in *Shigella flexneri*. *J. Bacteriol.* **176**:4187–4191.

Prost, L. R., and S. I. Miller. 2008. The *Salmonella* PhoQ sensor: mechanisms of detection of phagosome signals. *Cell. Microbiol.* **10**:576–582.

Raeder, R., and M. D. Boyle. 1996. Properties of IgG-binding proteins expressed by *Streptococcus pyogenes* isolates are predictive of invasive potential. *J. Infect. Dis.* **173**:888–895.

Raeder, R., M. Woischnik, A. Podbielski, and M. D. Boyle. 1998. A secreted streptococcal cysteine protease can cleave a surface-expressed M1 protein and alter the immunoglobulin binding properties. *Res. Microbiol.* **149**:539–548.

Revel, A. T., A. M. Talaat, and M. V. Norgard. 2002. DNA microarray analysis of differential gene expression in *Borrelia burgdorferi*, the Lyme disease spirochete. *Proc. Natl. Acad. Sci. USA* **99**:1562–1567.

Rezcallah, M. S., M. D. P. Boyle, and D. D. Sledjeski. 2004. Mouse skin passage of *Streptococcus pyogenes* results in increased streptokinase expression and activity. *Microbiology* **150**:365–371.

Richter-Dahlfors, A., A. M. Buchan, and B. B. Finlay. 1997. Murine salmonellosis studied by confocal microscopy: *Salmonella* Typhimurium resides intracellularly inside macrophages and exerts a cytotoxic effect on phagocytes in vivo. *J. Exp. Med.* **186**:569–580.

Roehrig, J. T., J. Piesman, A. R. Hunt, M. G. Keen, C. M. Happ, and B. J. Johnson. 1992. The hamster immune response to tick-transmitted *Borrelia burgdorferi* differs from the response to needle-inoculated, cultured organisms. *J. Immunol.* **149**:3648–3653.

Roy, C. R., and S. Falkow. 1991. Identification of *Bordetella pertussis* regulatory sequences required for transcriptional activation of the *fhaB* gene and autoregulation of the *bvgAS* operon. *J. Bacteriol.* **173**:2385–2392.

Roy, C. R., J. F. Miller, and S. Falkow. 1990. Autogenous regulation of the *Bordetella pertussis bvgABC* operon. *Proc. Natl. Acad. Sci. USA* **87**:3763–3767.

Roy, C. R., J. F. Miller, and S. Falkow. 1989. The *bvgA* gene of *Bordetella pertussis* encodes a transcriptional activator required for coordinate regulation of several virulence genes. *J. Bacteriol.* **171**:6338–6344.

Royle, L., A. Roos, D. J. Harvey, M. R. Wormald, D. van Gijlswijk-Janssen, el-R. M. Redwan, I. A. Wilson, M. R. Daha, R. A. Dwek, and P. M. Rudd. 2003. Secretory IgA N- and O-glycans provide a link between the innate and adaptive immune systems. *J. Biol. Chem.* **278**:20140–20153.

Scarlato, V., A. Prugnola, B. Arico, and R. Rappuoli. 1991. The bvg-dependent promoters show similar behaviour in different *Bordetella* species and share sequence homologies. *Mol. Microbiol.* **5**:2493–2498.

Schwan, T. G. 2003. Temporal regulation of outer surface proteins of the Lyme-disease spirochaete *Borrelia burgdorferi*. *Biochem. Soc. Trans.* **31**:108–112.

Schwan, T. G., J. Piesman, W. T. Golde, M. C. Dolan, and P. A. Rosa. 1995. Induction of an outer surface protein on *Borrelia burgdorferi* during tick feeding. *Proc. Natl. Acad. Sci. USA* **92**:2909–2913.

Schwan, T. G., and J. Piesman. 2000. Temporal changes in outer surface proteins A and C of the Lyme disease-associated spirochete, *Borrelia burgdorferi*, during the chain of infection in ticks and mice. *J. Clin. Microbiol.* **38**:382–388.

Seemanapalli, S. V., Q. Xu, K. McShan, and F. T. Liang. 2011. Outer surface protein C is a dissemination-facilitating factor of *Borrelia burgdorferi* during mammalian infection. *PLoS ONE* **5**:e15830.

Shea, J. E., M. Hensel, C. Gleeson, and D. W. Holden. 1996. Identification of a virulence locus encoding a second type III secretion system in *Salmonella* Typhimurium. *Proc. Natl. Acad. Sci. USA* **93**:2593–2597.

Shen, A., and D. E. Higgins. 2006. The MogR transcriptional repressor regulates nonhierarchal expression of flagellar motility genes and virulence in *Listeria monocytogenes*. *PLoS Pathog.* **2**:e30.

Shen, A., H. D. Kamp, A. Gründling, and D. E. Higgins. 2006. A bifunctional O-GlcNAc transferase governs flagellar motility through anti-repression. *Genes Dev.* **20**:3283–3295.

Steere, A. C., V. K. Sikand, F. Meurice, D. L. Parenti, E. Fikrig, R. T. Schoen, J. Nowakowski, C. H. Schmid, S. Laukamp, C. Buscarino, and D. S. Krause. 1998. Vaccination against Lyme disease with recombinant *Borrelia burgdorferi* outer-surface lipoprotein A with adjuvant. *N. Engl. J. Med.* **339**:209–215.

Stonehouse, E. A., R. R. Hulbert, M. B. Nye, K. Skorupski, and R. K. Taylor. 2011. H-NS binding and repression of the *ctx* promoter in *Vibrio cholerae*. *J. Bacteriol.* **193**:979–988.

Stonehouse, E., G. Kovacikova, R. K. Taylor, and K. Skorupski. 2008. Integration host factor positively regulates virulence gene expression in *Vibrio cholerae*. *J. Bacteriol.* **190**:4736–4748.

Sumby, P., A. R. Whitney, E. A. Graviss, F. R. DeLeo, and J. M. Musser. 2006. Genome-wide analysis of Group A streptococci reveals a mutation that modulates global phenotype and disease specificity. *PLoS Pathog.* **2**:e5.

Svennerholm, A. M., G. Johnson, and C. Yan. 1991. A method for studies of an El Tor-associated antigen of *Vibrio cholerae* O1. *FEMS Microbiol. Lett.* **63**:179–185.

Tendeng, C., C. Badaut, E. Krin, P. Gounon, S. Ngo, A. Danchin, S. Rimsky, and P. Bertin. 2000. Isolation and characterization of *vicH*, encoding a new pleiotropic regulator in *Vibrio cholerae*. *J. Bacteriol.* **182**:2026–2032.

Thelin, K. H., and R. K. Taylor. 1996. Toxin-coregulated pilus, but not mannose-sensitive hemagglutinin, is required for colonization by *Vibrio cholerae* O1 El Tor biotype and O139 strains. *Infect. Immun.* **64**:2853–2856.

Tilly, K., S. Casjens, B. Stevenson, J. L. Bono, D. S. Samuels, D. Hogan, and P. Rosa. 1997. The *Borrelia burgdorferi* circular plasmid cp26: conservation of plasmid structure and targeted inactivation of the ospC gene. *Mol. Microbiol.* **25**:361–373.

Torres, A. G., G. N. Lopez-Sanchez, L. Milflores-Flores, S. D. Patel, M. Rojas-Lopez, C. F. Martinez de la Pena, M. M. P. Arenas-Hernandez, and Y. Martinez-Laguna. 2007. Ler and H-NS, regulators controlling expression of the long polar fimbriae of *Escherichia coli* O157:H7. *J. Bacteriol.* **189:**5916–5928.

Turner, E. C., and C. J. Dorman. 2007. H-NS antagonism in *Shigella flexneri* by VirB, a virulence gene transcription regulator that is closely related to plasmid partition factors. *J. Bacteriol.* **189:**3403–3413.

Uchiya, K., M. A. Barbieri, K. Funato, A. H. Shah, P. D. Stahl, and E. A. Groisman. 1999. A *Salmonella* virulence protein that inhibits cellular trafficking. *EMBO J.* **18:**3924–3933.

Vazquez-Torres, A., J. Jones-Carson, A. J. Baumler, S. Falkow, R. Valdivia, W. Brown, M. Le, R. Berggren, W. T. Parks, and F. C. Fang. 1999. Extraintestinal dissemination of *Salmonella* by CD18-expressing phagocytes. *Nature* **401:**804–808.

Virgin, H. W. 2007. In vivo veritas: pathogenesis of infection as it actually happens. *Nat. Immunol.* **8:**1143–1147.

von Pawel-Rammingen, U., and L. Bjorck. 2003. IdeS and SpeB: immunoglobulin-degrading cysteine proteinases of *Streptococcus pyogenes*. *Curr. Opin. Microbiol.* **6:**50–55.

Walker, M. J., A. Hollands, M. L. Sanderson-Smith, J. N. Cole, J. K. Kirk, A. Henningham, J. D. McArthur, K. Dinkla, R. K. Aziz, R. G. Kansal, A. J. Simpson, J. T. Buchanan, G. S. Chhatwal, M. Kotb, and V. Nizet. 2007. DNase Sda1 provides selection pressure for a switch to invasive Group A streptococcal infection. *Nat. Med.* **13:**981–985.

Walthers, D., R. K. Carroll, W. W. Navarre, S. J. Libby, F. C. Fang, and L. J. Kenney. 2007. The response regulator SsrB activates expression of diverse *Salmonella* pathogenicity island 2 promoters and counters silencing by the nucleoid-associated protein H-NS. *Mol. Microbiol.* **65:**477–493.

Walthers, D., Y. Li, Y. Liu, G. Anand, J. Yan, and L. J. Kenney. 2011. *Salmonella enterica* response regulator SsrB relieves H-NS silencing by displacing H-NS bound in polymerization mode and directly activates transcription. *J. Biol. Chem.* **286:**1895–1902.

Watnick, P. I., K. J. Fullner, and R. Kolter. 1999. A role for the mannose-sensitive hemagglutinin in biofilm formation by *Vibrio cholerae* El Tor. *J. Bacteriol.* **181:**3606–3609.

Way, S. S., L. J. Thompson, J. E. Lopes, A. M. Hajjar, T. R. Kollmann, N. E. Freitag, and C. B. Wilson. 2004. Characterization of flagellin expression and its role in *Listeria monocytogenes* infection and immunity. *Cell. Microbiol.* **6:**235–242.

Weiss, A. A., and E. L. Hewlett. 1986. Virulence factors of *Bordetella pertussis*. *Annu. Rev. Microbiol.* **40:**661–686.

Whitnack, E., and E. H. Beachey. 1982. Antiopsonic activity of fibrinogen bound to M protein on the surface of Group A streptococci. *J. Clin. Investig.* **69:**1042–1045.

Wood, M. W., M. A. Jones, P. R. Watson, S. Hedges, T. S. Wallis, and E. E. Galyov. 1998. Identification of a pathogenicity island required for *Salmonella* enteropathogenicity. *Mol. Microbiol.* **29:**883–891.

Xu, H., M. J. Caimano, T. Lin, M. He, J. D. Radolf, S. J. Norris, F. Gheradini, A. J. Wolfe, and X. F. Yang. 2010. Role of acetyl-phosphate in activation of the Rrp2-RpoN-RpoS pathway in *Borrelia burgdorferi*. *PLoS Pathog.* **6:**e1001104.

Yang, X. F., U. Pal, S. M. Alani, E. Fikrig, and M. V. Norgard. 2004. Essential role for OspA/B in the life cycle of the Lyme disease spirochete. *J. Exp. Med.* **199:**641–648.

Yang, X. F., S. M. Alani, and M. V. Norgard. 2003. The response regulator Rrp2 is essential for the expression of major membrane lipoproteins in *Borrelia burgdorferi*. *Proc. Natl. Acad. Sci. USA* **100:**11001–11006.

Yang, X. F., M. C. Lybecker, U. Pal, S. M. Alani, J. Blevins, A. T. Revel, D. S. Samuels, and M. V. Norgard. 2005. Analysis of the *ospC* regulatory element controlled by the RpoN-RpoS regulatory pathway in *Borrelia burgdorferi*. *J. Bacteriol.* **187:**4822–4829.

Yuk, M. H., P. A. Cotter, and J. F. Miller. 1996. Genetic regulation of airway colonization by *Bordetella* species. *Am. J. Respir. Crit. Care Med.* **154:**S150–154.

Zhou, D., M. S. Mooseker, and J. E. Galan. 1999. Role of the *Salmonella* Typhimurium actin-binding protein SipA in bacterial internalization. *Science* **283:**2092–2095.

Zhu, J., and J. J. Mekalanos. 2003. Quorum sensing-dependent biofilms enhance colonization in *Vibrio cholerae*. *Dev. Cell* **5:**647–656.

Zhu, J., M. B. Miller, R. E. Vance, M. Dziejman, B. L. Bassler, and J. J. Mekalanos. 2002. Quorum-sensing regulators control virulence gene expression in *Vibrio cholerae*. *Proc. Natl. Acad. Sci. USA* **99:**3129–3134.

Regulation of Bacterial Virulence
Edited by Michael L. Vasil and Andrew J. Darwin
© 2013 ASM Press, Washington, DC   doi:10.1128/9781555818524.ch27

Chapter 27

Regulation in Response to Host-Derived Signaling Molecules

CHARLEY GRUBER AND VANESSA SPERANDIO

INTRODUCTION

To survive in their environment, bacteria must sense the conditions in which they live and adapt accordingly. From the perspective of a bacterium, the host is little more than another environment in which to grow and thrive, although potentially one actively trying to kill it. So it is in the best interest of bacteria to know where they are and the state of their host. The best way to acquire this information is to sense various molecules produced by the host. This interaction, along with the host sensing bacterium-derived molecules, has been termed interkingdom signaling (Sperandio, 2004).

While the term may be fairly recent, the evidence for this interaction has been mounting for some time. Stress has long been known to increase the susceptibility of animals to infectious diseases (Peterson et al., 1991). This was long thought to be due to the effect of stress on the functioning of the immune system (Khansari et al., 1990). However, it was shown that social stress can actually boost the acute immune response in mice (Lyte et al., 1990). If stress was not weakening the immune system under these circumstances, then it might have been increasing the virulence of the bacteria. This changes the paradigm. Instead of opportunistic pathogens passively taking advantage of a person's weakened immune system, they may actively sense their host's stress and exploit its weakness by increasing their virulence.

While the potential from this observation was revolutionary, the exact mechanism used by bacteria to sense the state of their host was at first elusive. Over the next few years, a large number of examples were identified in many different organisms responding to many varying signals, but it was only in the last decade or so that the mechanisms of these responses were elucidated.

SENSING OF HOST-DERIVED HORMONES

Hormones, the principal signaling molecules in multicellular organisms, are present in every environment where bacteria interact with the host. Their ubiquity and the potential information they provide on the status of the host make them a useful signal for bacteria to detect.

Aside from sensing host hormones, many species of bacteria themselves have been shown to produce their own hormone and hormone-like molecules, primarily those derived from amino acids. The catecholamines dopamine and norepinephrine (NE) have been detected in several species of bacteria including *Bacillus subtilis* and *Serratia marcescens* (Tsavkelova et al., 2000). Many species have also been observed to produce the tryptophan-derived neurotransmitter serotonin (Hsu et al., 1986). Bacterial production of the immune-modulating molecule histamine is well documented (Fernandez et al., 2006), and bacterial production of large amounts of histamine in food is a major cause of food intoxication (Bodmer et al., 1999).

The microbial production of molecules that are classically considered to be produced by the host potentially complicates interkingdom signaling, since in some cases the origin of the molecule may be ambiguous. However, the shared production of the same signaling molecules across kingdoms is intriguing and suggests that the use of these "host" hormones as signaling molecules may be far older than anticipated.

Effect of Catecholamines on Growth

Catecholamines are a family of three neuroendocrine hormones that are produced from the amino acid tyrosine. Tyrosine is hydroxylated to produce L-DOPA, which is converted into dopamine, which can be further modified into NE and then epinephrine

Charley Gruber and Vanessa Sperandio • Department of Microbiology, UT Southwestern Medical Center, Dallas, TX 75390.

(Epi). Dopamine and NE are present throughout the body in blood and other tissues, particularly the intestines (Aneman et al., 1996). Stress and tissue damage greatly increase the amount of these two hormones in circulation and tissues, as well as causing the systemic release of Epi. Severe systemic infections such as sepsis greatly increase the concentrations of all three catecholamines (Hahn et al., 1995).

The connection to stress made this class of hormones one of the prime suspects for the increased virulence of bacteria under stress conditions. It was reported that these catecholamines increased the growth rate of several species of bacteria including *Escherichia coli* and *Yersinia enterocolitica* in serum-SAPI medium, a minimal salts medium supplemented with 30% bovine serum (Lyte and Ernst, 1992). This provided one of the first pieces of evidence that bacteria are capable of sensing host hormones (Lyte, 1992).

Serum normally contains the iron-binding protein transferrin (Tf) while other secretions contain lactoferrin (Lf), both of which sequester iron to prevent the growth of bacteria. Different species of bacteria have evolved numerous mechanisms for acquiring iron, such as the production of iron-scavenging siderophores (Ratledge and Dover, 2000). Many of these siderophores, such as enterobactin, contain catechol groups that bind iron ions with high affinity, allowing them to acquire iron from Tf and Lf. Bacteria can then take up these molecules to harvest iron. As their name suggests, catecholamines themselves have a catechol group which is able to bind iron on its own (Siraki et al., 2000; Paris et al., 2005). It has been proposed that the host NE is involved in iron uptake. NE is proposed to assist with the removal and transfer of iron from Tf and Lf to siderophores. For *E. coli*, this process requires both the siderophore enterobactin and its TonB-dependent transport system (Freestone, 2000; Freestone, 2003).

Which catecholamines are able to induce growth seems to vary depending on the species of bacteria. All three hormones are able to assist in growth enhancement in *E. coli* and *Salmonella enterica*, but Epi has little effect on the growth of *Y. enterocolitica*. Epi is still able to increase iron uptake by *Y. enterocolitica*, but it does not increase its growth rate. However, it can antagonize the action of NE and dopamine. Addition of α- but not β-adrenergic antagonists blocked NE- and Epi-dependent growth induction in *E. coli* and *S. enterica* and NE-dependent growth induction in *Y. enterocolitica* as well as the uptake of NE in all three species. The addition of dopamine receptor antagonists blocks the dopamine-dependent growth induction of all three species. None of the antagonists decreased the catecholamine-induced increase in iron uptake (Freestone et al., 2007). The effect of antagonists and

the observation that Epi acts as an antagonist to NE and dopamine in *Y. enterocolitica* suggest that there is more to growth induction in these species in serum-SAPI medium than iron acquisition.

It has also been reported that *E. coli* grown in serum-based media with NE leads to the production of an autoinducer (NE-AI). Addition of preconditioned media containing this autoinducer is able to increase the growth rate at a level similar to that caused by catecholamines (Lyte et al., 1996). This AI is produced by many different species of bacteria and is able to induce growth across species and in the absence of Tf, Lf, or catecholamines (Freestone et al., 1999). It has been proposed that this molecule is the siderophore enterobactin or its breakdown product 2,3-dihydroxybenzoylserine. The proposed AI was not produced in an enterobactin mutant in M9 minimal medium (Burton, 2002). However, it has also been reported that the AI was not able to acquire iron from Tf in an NE-dependent manner like enterobactin (Freestone, 2003). Further work must be done to characterize this molecule and determine its structure.

Bordetella pertussis and *B. bronchiseptica* are obligate pathogens of the respiratory tract in mammals, with *B. pertussis* being the causative agent of whooping cough (Bordet and Gengou, 1906; Goodnow, 1980). Like many other species, their growth in serum-containing media is enhanced by the presence of catecholamines, and this is due to the increased availability of iron. Among their three major iron acquisition pathways, *Bordetella* spp. make use of siderophores produced by other species. In the presence of enterobactin or other structurally similar siderophores, the AraC family regulator BfeR activates the transcription of the enterobactin receptor BfeA (Beall and Sanden, 1995; Anderson and Armstrong, 2004, 2006). However, catecholamines alone are able to induce *bfeA* in the absence of any xenosiderophore, and neither *bfeA* nor its regulator is required for the catecholamine-induced growth enhancement. This activation likely stems from the structural similarity of catecholamines and catechol-containing siderophores. Similar to previous examples, NE is able to harvest iron from Tf, but instead of transferring the iron to a siderophore, *Bordetella* directly takes up the iron-bound NE through an unknown TonB-dependent receptor (Anderson and Armstrong, 2008).

Sensing of Epinephrine and Norepinephrine in Enterohemorrhagic *E. coli*

Enterohemorrhagic *E. coli* (EHEC) is a major food-borne pathogen that causes outbreaks of disease throughout the world, with serotype O157:H7 being the most prevalent agent. It colonizes the large intestine, and causes bloody diarrhea and hemolytic-uremic syndrome (HUS). The production of the phage-encoded Shiga toxin (Stx) is responsible for HUS, which leads to the majority of the mortality associated with EHEC (Kaper et al., 2004).

EHEC attaches to the intestinal lining and forms an attaching and effacing (AE) lesion. These lesions are characterized by the destruction of microvilli and the formation of a pedestal-like structure beneath each individual bacterium (Staley et al., 1969). The virulence factors responsible for this phenotype include a type III secretion system (Jarvis et al., 1995) several of its effectors encoded in a 41-gene pathogenicity island called the locus of enterocyte effacement (LEE) (McDaniel, 1995), which contains five major operons (Mellies et al., 1999), and another effector encoded in a cryptic prophage, EspFu/TccP (Campellone et al., 2004; Garmendia et al., 2004). The major transcriptional regulator of the LEE is Ler, the LEE-encoded regulator, which is encoded within the *LEE1* operon and is required for the expression of all genes within the LEE (Mellies et al., 1999; Bustamante et al., 2001).

AE lesion formation, motility, and Stx are all regulated by quorum sensing (QS) in EHEC through the bacterial signaling molecule autoinducer 3 (AI-3) (Sperandio et al., 1999, 2001, 2002). Additionally, the catecholamine hormones Epi and NE have the same effect as AI-3, linking QS and interkingdom signaling (Sperandio et al., 2003). All three signals are sensed by the QseC histidine sensor kinase (Clarke et al., 2006). Upon sensing its signals, the histidine kinase QseC autophosphorylates and transfers the phosphate to its cognate response regulator QseB. Phosphorylated QseB is able to activate the transcription of *fhlDC*, the master regulators of the flagella, which leads to the production of flagella and an increase in motility, while unphosphorylated QseB binds to a different site to repress the expression of *fhlDC*. QseC is also able to activate expression of the LEE genes by phosphorylating the response regulator KdpE, which activates Ler and thus the rest of the LEE (Fig. 1) (Hughes et al., 2009).

Downstream of the QseBC system there is another two-component system, QseEF, whose regulation is activated by QseBC (Reading et al., 2007). The histidine kinase QseE senses Epi as well as sulfate and phosphate (Fig. 1) (Reading et al., 2009). QseF can be phosphorylated by multiple histidine kinases—BaeS, EnvZ, RstB, UhpB—as well as by both bacterial adrenergic sensors, QseC and QseE (Yamamoto et al., 2005; Hughes et al., 2009). QseF is required for the expression of the non-LEE-encoded effector EspFu that is required for pedestal formation (Reading et al., 2007), and it activates the expression of the sRNA

GlmY through which the sRNA GlmZ regulates glucosamine synthetase (GlmS) (Gorke and Vogel, 2008; Reichenbach et al., 2009).

Homologues of the QseBC and QseEF systems are present in other strains of *E. coli* (Kostakioti et al., 2009) and many other species of bacteria (Rasko et al., 2008). QseBC has been reported to be vital to the virulence of uropathogenic *E. coli*. The QseBC system regulates flagella as well as its major virulence factors, curli and type 1 fimbriae, though the posphorylation state of QseB in a manner similar to that used in EHEC (Kostakioti et al., 2009). QseBC is known to be involved in virulence in *Salmonella enterica* serovar Typhimurium (Merighi et al., 2006, 2009; Bearson and Bearson, 2008; Bearson et al., 2010; Moreira et al., 2010) and *Francisella tularensis* (Weiss et al., 2007). The conservation of these proteins and their role in virulence have made them an attractive target for drug discovery. Since they are not required for *in vitro* growth, such drugs are thought to create less selection for resistant organisms, which is an extremely attractive prospect in the era of antibiotics resistance. A high-throughput screen has uncovered a small molecule, LED209, which binds to QseC, preventing it from recognizing AI-3/Epi and NE to activate virulence gene expression. LED209 proved effective in blocking the pathogenesis of EHEC, *S. enterica* serovar Typhimurium and *F. tularensis* in vitro and in animal models (Rasko et al., 2008). This shows that the study of interkingdom signaling may provide new drug targets.

Sensing of Norepinephrine by *Pseudomonas*

Pseudomonas aeruginosa, one of the most clinically important opportunistic pathogens, is able to infect many different sites of the host and cause a variety of diseases. Additionally, *Pseudomonas* is present at low levels in the intestinal tract of healthy individuals as well as being prevalent in the environment (Alverdy et al., 2000).

P. aeruginosa encodes a slew of virulence factors. It encodes a type III secretion system that damages immune cells and assists in breaching the mucosal barrier (Engel and Balachandran, 2009). It also encodes a large variety of exotoxins and degradative enzymes such as elastases that damage cells and their extracellular matrix (Wretlind et al., 1987). Pyocyanin, a secondary metabolite, also disrupts host cell function, as well as acting as an antibiotic against other bacteria, potentially killing off its competitors (Baron and Rowe, 1981; Lau et al., 2004). *Pseudomonas* is also capable of forming resistant biofilms that are greatly refractory to antibiotic treatment (Harmsen et al., 2010). The biosurfactant rhamnolipid is a major

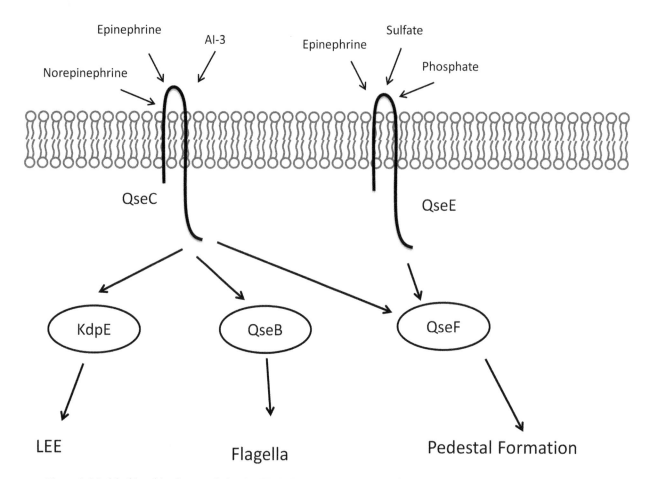

Figure 1. Model of interkingdom regulation in EHEC. QseC senses AI-3/epinephrine and NE to increase its autophosphorylation. It then transfers this phosphate to three response regulators: QseB (controls flagellar expression), KdpE (controls LEE gene expression), and QseF (controls pedestal formation). QseBC activate the expression of *qseEF*, where QseE senses Epi, sulfate, and phosphate and transfers its phosphate to QseF.
doi:10.1128/9781555818524.ch27f1

component of these biofilms and has been shown to act as a barrier against the immune system by directly killing cells such as neutrophils and macrophages (Van Gennip et al., 2009).

QS plays a role in regulating all of these virulence factors (Smith and Iglewski, 2003; de Kievit, 2009). *P. aeruginosa* encodes two separate acyl homoserine lactone (AHL) systems. AHL QS systems consist of a LuxI homolog that produces the signal and a LuxR homolog that senses it (Shadel et al., 1990). LasI is responsible for the production of 3-oxo-C12-AHL that is sensed by LasR, while RhlI produces C4-AHL that is sensed by RhlR (Gambello and Iglewski, 1991; Pearson et al., 1995). RhlR and LasR are specific to the AHL produced by their respective synthase, and they also create a positive feedback loop by activating their AHL-respective synthase (Seed et al., 1995; Latifi et al., 1996). The two systems exist in a hierarchy, as LasR is able to activate RhlR (Fig. 2) (Latifi et al., 1996).

In addition to these two systems, *Pseudomonas* produces a quinolone signaling molecule, 2-heptyl-3-hydroxy-4-quinolone, or *Pseudomonas* quinolone signal (PQS). The production of this signal requires LasR-AHL (Pesci et al., 1999). PQS also activates RhlI (McKnight et al., 2000), and RhlR-AHL in turn then represses PQS synthesis (Wade et al., 2005). In addition, there are numerous other regulators in this system (Fig. 2) (Schuster and Greenberg, 2006).

Besides these bacterium-derived signals, *Pseudomonas* also senses the host hormone NE. Exposure to NE increases the expression of several virulence factors such as elastase and pyocyanin as well as the signal PQS. It also increases the adherence and invasion of epithelial cells by *Pseudomonas* (Hegde et al., 2009). NE has no effect on *rhlR* expression but greatly increases that of *lasR*. This, coupled with the activation of PQS, suggests that NE acts through the *lasI* QS pathway. The expression of GacA, an activator of the *lasIR* system, is

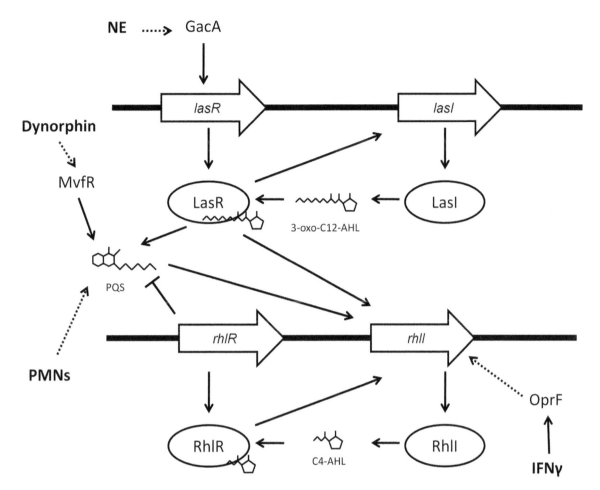

Figure 2. Model of interkingdom signaling in *Pseudomonas aeruginosa*. NE influences the expression of GacA, which in turn influences the expression of *lasR*, encoding the receptor for the 3-oxo-C12-AHL quorum-sensing signal. LasR controls the expression of *lasI* that encodes the synthase of the 3-oxo-C12-AHL, forming a positive auto-regulatory loop. LasR also activates expression of the RlhRI quorum-sensing system, which represses expression of the PQS signal. Expression of PQS is positively activated by the presence of PMNs, and through MvfR by dynorphin. IFN-γ through OprF activates the expression of *rlhI*. doi:10.1128/9781555818524.ch27f2

also increased, which suggests that the GacAS two-component system may play a role in the NE signaling pathway (Fig. 2) (Hegde et al., 2009).

Pseudomonas also represses some virulence factors when grown in serum-containing medium with NE. The increased availability of iron due to NE treatment in this medium decreases the expression of the iron starvation sigma factor PvdS, which is an activator of the gene encoding the exotoxin ToxA and siderophore synthesis genes (Li et al., 2009). This suggests that NE plays a dual role in iron availability and in indirect regulation, as well as the more direct regulation of virulence genes.

Sensing of Catecholamines by Other Organisms

While the QseBC and QseEF systems are the best characterized, and a significant amount of information about the effect of NE on *Pseudomonas* is

known, these catecholamines have also been reported to affect the virulence of many other organisms.

Borrelia burgdorferi, the causative agent of Lyme disease, cycles through mammalian and tick hosts. Outer surface protein A (OspA) is required for the colonization of its tick host (Pal et al., 2000; Yang et al., 2004), and its expression is upregulated by both Epi and NE (Scheckelhoff et al., 2007). The treatment of *Borrelia*-infected mice with the β-adrenergic antagonist propranolol greatly reduced the ability of the bacteria to be taken up by feeding ticks. The inflammation caused by the tick bite may result in the release of catecholamines, which prepares *Borrelia* for survival in its arthropod host. Alternatively, catecholamines may be present in the tick's saliva. Either way, *Borrelia* uses host hormones to regulate the expression of an important tick colonization factor.

The intestinal pathogen *Campylobacter jejuni* is one of the leading causes of food-borne illness in the world and causes enteritis and diarrhea (Butzler, 2004). The primary virulence factor of *Campylobacter* is its flagella, which are required to penetrate the mucosal layer (Lee et al., 1986) and invade the cells of the intestinal lining (Grant et al., 1993). The exposure of *Campylobacter* to NE increases its growth rate by supplying iron, as with other organisms, as well as increasing its motility. The ability of the bacteria to invade monolayers of intestinal epithelial cells was also greatly increased by NE treatment (Cogan et al., 2007). The invasion of the intestinal lining by *Campylobacter* triggers a cytokine response (Abram et al., 2000) that recruits CD8[+] T cells which would destroy the infected intestinal cells (Vuckovic et al., 2006). Since this T-cell-mediated response is thought to be responsible for the destruction of the intestinal lining during infection, increased NE levels in the intestinal lumen potentially lead to a more severe infection. The mechanism by which *Campylobacter* responds to NE is not yet known, and treatment with α- and β-adrenergic antagonists does not have an effect (Cogan et al., 2007).

Given the prevalence and importance of catecholamines in animals, many other pathogens are likely able to respond to these host-derived hormones.

Sensing of Opioids

Opioids are neurotransmitters that play multiple roles in host responses such as stress, tissue damage, and regulation of the immune system (Peterson et al., 1998; Neudeck and Loeb, 2002; Neudeck et al., 2003; Vallejo et al., 2004). This diverse group of hormones is classifed by whether they act on the δ-, κ-, or μ-opioid receptors. Aside from the nervous system, they are particularly abundant in the gastrointestinal tract, which is highly innervated (Sternini et al., 2004), and immune cells such as macrophages have been reported to secrete opioids in response to inflammation. This suggests that intestinal bacteria are exposed to opioids, which led to the investigation into their role in virulence regulation.

The opportunistic pathogen *P. aeruginosa* has been shown to respond to host stress and tissue damage by increasing its virulence (Wu et al., 2005). Intestinal damage causes the κ-opioid dynorphin to be released into the intestinal lumen. Dynorphin is able to enter the *Pseudomonas* cytoplasm, where it activates the PQS QS signal. This greatly upregulates various virulence factors including the compound 2-heptyl-4-hydroxy quinoline-N-oxide (HQNQ) that *Pseudomonas* uses to kill colonizing bacteria such as lactobacilli that are normally considered protective. The activation of PQS by dynorphin requires the multiple virulence factor regulator (*mvfR*), which is

a known regulator of the PQS system (Fig. 2) (Deziel et al., 2005). Like NE, *Pseudomonas* uses dynorphin to sense host stress and respond by increasing its virulence.

Sensing of Peptide Hormones

Another class of hormones used by eukaryotes is peptide hormones. These signaling molecules are translated from an mRNA to produce a prohormone, which is processed and exported out of the cell. This is an extremely diverse group of hormones that have many roles in regulating physiology.

Helicobacter pylori colonizes the stomach and is suspected of infecting 50% of the human population. Without treatment, this bacterium potentially colonizes its host for life and may cause severe diseases including duodenal ulcers and gastric cancer (Brown, 2000). It is notable for surviving the harsh environment of the stomach, where it is exposed to extremely low pH as well as several host hormones. One of these is the peptide hormone gastrin, which is secreted by cells of the stomach and is responsible for stimulating the release of gastric acid (Gregory and Tracy, 1964). Gastrin has been shown to increase the growth rate of *H. pylori* and is taken up by the bacteria (Chowers et al., 1999). This effect can be competed using the synthetic peptide analog pentagastrin and another peptide hormone, cholecystokinin, which is involved in stimulating bile release and shares a similar C-terminus with gastrin. This suggests that gastrin acts through a receptor that can be blocked by similar molecules. Additionally, this growth response is specific to human gastrin and may help explain the specificity of *H. pylori* to humans (Chowers et al., 2002). Another peptide hormone present in the stomach is somatostatin, which functions to inhibit the secretion of gastric acid (Johansson et al., 1978). This hormone has been reported to suppress the growth of *H. pylori in vitro*. The peptide binds to the bacteria with high affinity, which can be blocked through an α-somatostatin antibody, although the receptor has not been identified. Somatostatin greatly increases the *H. pylori* intracellular concentrations of both cGMP and cAMP, and increased concentrations of cGMP are sufficient for growth inhibition (Yamashita et al., 1998). *H. pylori* infection also changes the levels of these two hormones, increasing the amount of gastrin and decreasing the amount of somatostatin to increase its own growth (Joseph and Kirschner, 2004; Maciorkowska et al., 2006). Given the ability of *Helicobacter* to thrive in extremely low-pH environments, it appears to sense these peptide hormones to maximize its growth rate under these conditions.

Burkholderia pseudomallei is an environmental organism most commonly found in Southeast Asia; it is the causative agent of the severe disease melioidosis (Vuddhakul et al., 1999). Since the most common risk factor of melioidosis is diabetes mellitus, the effect of insulin on *B. pseudomallei* was investigated. Insulin is the peptide hormone secreted by the pancreas that is responsible for the regulation of blood glucose levels by causing liver and muscle cells to take up glucose. It was reported that *B. pseudomallei* and the related species *B. cepacia* are both capable of binding insulin but only *B. pseudomallei* had reduced activity of phospholipase C and acid phosphatase (Kanai et al., 1996). *B. pseudomallei* grew better in sera collected from diabetic rats, and this could be complemented with the addition of human insulin. Additionally, the bacterium was found to be more virulent in a diabetic mouse model. The addition of purified human insulin to growth media inhibited its growth (Woods et al., 1993). However, this was later shown to be the result of the preservative *m*-cresol that was present in the commercially available human serum (Simpson and Wuthiekanun, 2000). The epidemiology on which the research was based has also been challenged. While diabetes is a major risk factor for melioidosis, most cases involved type II diabetics who were mostly not insulin dependent (Currie, 1995). While *B. pseudomallei* does appear to bind and respond to insulin, more work must be done to further characterize this interaction and identify the insulin receptor.

Natriuretic peptides are a class of peptide hormones involved in the osmoregulation of blood. This family contains three structurally related peptides, atrial natriuretic peptide (ANP), brain natriuretic peptide (BNP), and C-type natriuretic peptide (CNP), which signal through membrane-bound guanylyl cyclase receptors (Potter et al., 2009). Due to their structural similarity to cyclic antimicrobial peptides, the antimicrobial properties of human natriuretic peptides were investigated and they were reported to have this activity (Kourie, 1999; Krause et al., 2001), although not at concentrations found in sera (Kalra et al., 2003). Treatment of a *P. aeruginosa* strain with BNP and CNP has been reported to increase its cytotoxicity *in vitro* and activate modifications of its lipopolysaccharide (LPS). As in animals, treatment with these peptides increased the levels of cGMP as well as cAMP (Veron et al., 2007). Cyclic nucleotides are known to be involved in the regulation of virulence genes in *Pseudomonas* through two cytoplasmic adenylate cyclases (Smith et al., 2004), and the cAMP-binding protein Vfr is required for the natriuretic peptide response (Veron et al., 2007). The sensor for these peptides has not yet been identified, but the shared use of cyclic nucleotides as second messengers

in this response in both prokaryotes and eukaryotes is an interesting observation.

Since peptide hormones are one of the most diverse classes of eukaryotic signaling molecules, there are certainly many systems that have not yet been discovered.

SENSING OF THE IMMUNE SYSTEM

Since the immune system directly interacts with pathogens and actively tries to eliminate them, it would be in the best interest of any pathogen to gauge the immune system. The major signaling molecules used by the immune system are cytokines, which makes them a prime target for being eavesdropped on by bacteria.

The first indication that bacteria were capable of responding to cytokines comes from the observation that several different cytokines increased the growth rate of pathogenic *E. coli* (Porat et al., 1991). Interleukin 1 (IL-1) is a pro-inflammatory cytokine released by activated macrophages that bacteria would likely encounter during an infection. Several virulent clinical isolates grew faster in the presence of this cytokine, while avirulent strains did not. The effect of IL-1β is greater than that of IL-1α and can be decreased by the addition of IL-1 receptor antagonist (IL-1ra). This suggests that whatever is sensing IL-1 was also able to bind to but not recognize IL-1ra. Two cytokine growth factors, IL-2 and granulocyte macrophage colony-stimulating factor, have also been shown to stimulate the growth of pathogenic *E. coli* (Denis et al., 1991). The mechanism of action of these three cytokines in bacteria has not yet been elucidated.

After the discovery that IL-1 binds to pathogenic *E. coli*, it was reported that IL-1 also binds to *Yersinia pestis*, the causative agent of plague (Zav'yalov et al., 1995). One of the *Y. pestis* virulence factors is its capsule (Brubaker, 1991). The genes encoding the F1 antigen of this capsule were cloned in *E. coli*, which then gained the ability to bind IL-1. The protein responsible for this was determined to be capsule antigen F1 assembly (Caf1A), an outer membrane protein that is responsible for the assembly of the F1 antigen (Karlyshev et al., 1992). The binding of IL-1 and the Caf1 protein that makes up the F1 antigen was competitive, suggesting that they share a similar binding site. Sequence analysis of the Caf1A protein indicated that it has significant homology to the human IL-1β receptor, which raises the possibility that this gene was acquired through horizontal gene transfer from humans (Zav'yalov et al., 1995). The effects of IL-1 binding to *Y. pestis* are unknown. It may act to sop up IL-1 produced by immune cells in a manner similar to soluble cytokine receptors produced by viruses, or it may be used to detect macrophages that still have membrane-anchored IL-1 (Fuhlbrigge et al., 1988).

Tumor necrosis factor α (TNF-α) is another pro-inflammatory cytokine that is released during infection in response to various microbe-associated molecular patterns. Several Gram-negative bacteria including *E. coli*, *Shigella*, and *Salmonella* are known to bind TNF-α (Luo et al., 1993). *Shigella flexneri* is an intracellular pathogen that is capable of invading epithelial cells and escaping into the cytoplasm (Sansonetti and Egile, 1998). *Shigella* treated with TNF-α shows increased ability to invade both epithelial cells and macrophages. However, such treatment also decreased its ability to grow in macrophages, possibly due to residual TNF activating the macrophages (Luo et al., 1993). Again, the exact mechanisms through which TNF-α acts are still unknown.

Another key cytokine is gamma interferon (IFN-γ). IFN-γ activates the immune system and is a major factor in clearance of bacteria. The opportunistic pathogen *Pseudomonas aeruginosa* possesses the ability to sense this cytokine. IFN-γ binds to the outer membrane porin F (OprF), which leads to the activation of the QS gene *rhlI*; however, the exact mechanism and whether other QS systems are involved are still not known (Fig. 2). RhlI then produces the autoinducer C4-HSL, resulting in the expression of virulence genes including lectin PA-I, which is vital to the disruption of the intestinal barrier in *Pseudomonas* infections by allowing it to adhere the intestinal cells (Laughlin et al., 2000; Wu et al., 2003) and pyocyanin (Wu et al., 2005). Since IFN-γ is normally used to help clear bacterial infections, this implies that the ability to sense this cytokine may allow *Pseudomonas* to survive the immune system by preparing its weaponry as a countermeasure. This would explain the observation that IFN-γ treatment actually worsens *P. aeruginosa* infections in several mouse models and clinical studies (Babalola et al., 2004).

Additionally, *Pseudomonas* biofilms alter gene expression in response to the presence of neutrophils (PMNs). Normally, *in vitro P. aeruginosa* biofilms do not produce large amounts of rhamnolipids (Morici et al., 2007). However, upon exposure to PMNs, *Pseudomonas* greatly increases the expression of rhamnolipids that coat the biofilm and provide protection from the neutrophils (Alhede et al., 2009). This response is QS dependent and leads to the activation of PQS and then RhlR, which regulate rhamnolipid production (Fig. 2). The molecule that *P. aeruginosa* is sensing is unknown, but it is not dynorphin. Once again, the response of *Pseudomonas*

to an activated immune system is to take measures to oppose it.

Sensing of Antimicrobial Peptides by Gram-Negative Bacteria

The immune system produces many other molecules aside from cytokines. Among these are cationic antimicrobial peptides (CAMPs). These peptides have a wide array of diversity but share a positive charge and amphipathic regions that allow them to interact with and disrupt the negatively charged bacterial membrane (Dathe and Wieprecht, 1999). Bacteria have evolved various mechanisms to survive the onslaught of these peptides. One mechanism is best characterized by the intracellular pathogen *S. enterica* serovar Typhimurium. Upon entry into cells, *Salmonella* modifies the lipid A portion of its LPS to prevent binding of the CAMPs to its outer membrane (Peschel, 2002). These modifications of LPS are controlled by the PhoPQ two-component system (Guo et al., 1997). Additionally, these two genes have been found to be vital for the virulence of *Salmonella* as well as other pathogens (Fields et al., 1989), and it has been shown that the histidine kinase PhoQ is activated within the phagosome of a macrophage (Alpuche Aranda et al., 1992).

PhoQ is known to be repressed by divalent metal ions such as Ca^{2+} and Mg^{2+}, and their depletion in media is often used to induce virulence gene expression *in vitro*. These ions act as a bridge between acidic residues in the periplasmic domain of the protein and the negatively charged phospholipids on the membrane. These metal bridges lock the protein in a repressed state (Cho et al., 2006). However, the absence of these metal ions is not responsible for the activation of PhoQ, as the concentration of Mg^{2+} is high enough in the phagosome to repress its activity (Martin-Orozco et al., 2006). Instead, it is the presence of CAMPs in the activated phagosome that activates PhoQ. These antimicrobial peptides bind directly to the same metal-binding sites on PhoQ even in the presence of physiologically relevant concentrations of divalent cations (Fig. 3) (Bader et al., 2005). The positive charge and amphipathic nature of CAMPs allow them to bind to both the membrane and the protein in a similar manner to the metals. However, the larger size of the peptide alters the conformation of PhoQ and activates its kinase activity. This leads to the transfer of phosphate to PhoP and the activation of downstream genes.

In addition to CAMPs, PhoQ is able to respond to the low pH present in the phagosome. This response is also independent of the presence of metal ions; however, unlike CAMPs, low pH does not affect

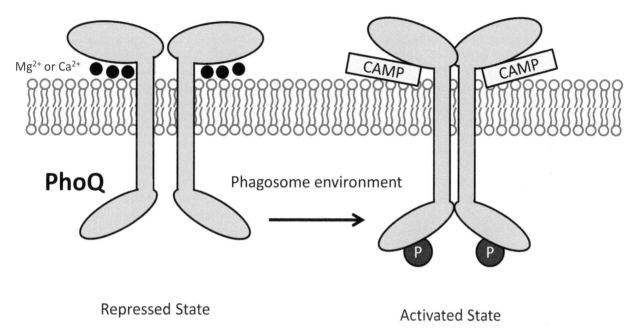

Figure 3. Model of activation of the *S. enterica* serovar Typhimurium PhoQ sensor by CAMPs. doi:10.1128/9781555818524.ch27f3

the binding of the metal ions (Prost et al., 2007). Additionally, low pH and CAMPs have an additive effect on the activation of the kinase. The *Salmonella* PhoQ protein acts as a sensor for the environment of the phagosome and is vital to the survival of serovar Typhimurium *in vivo*. The PhoQ from *P. aeruginosa* still senses divalent metal cations although via a different mechanism, but the pH-sensing region is not conserved and it lacks the ability to sense the two host-derived signals. While it can functionally complement for the *Salmonella* PhoQ *in vitro*, it cannot restore full virulence in an animal model (Prost et al., 2008). The PhoQ histidine kinase of *Salmonella* has specifically evolved to sense the environment of the phagosome by using the host's own bactericidal tools and to respond in a manner that thwarts the host's attempts to kill it.

Additionally, a protein that acts as a negative regulator of this system has been identified. MgrB is an inner membrane protein that is positively regulated by the PhoPQ system and is capable of binding PhoQ. Deletion of the *mgrB* gene results in significantly high expression of PhoP target genes, while overexpression of this protein results in the downregulation of PhoP targets even under PhoQ-inducing conditions. MgrB is thought to bind and inactivate the PhoQ kinase, creating a negative feedback loop on the PhoPQ regulatory system (Lippa and Goulian, 2009).

While the *Salmonella* PhoQ is the best studied, the PhoPQ system is known to regulate virulence and antimicrobial resistance in a wide array of organisms

including *Shigella flexneri*, *Yersinia pestis*, and the plant pathogens *Erwinia amylovora* and *Erwinia chrysanthemi* (Moss et al., 2000; Oyston et al., 2000; Llama-Palacios et al., 2003; Nakka et al., 2010). However, there is little direct evidence that these homologs sense the same signals.

Sensing of Antimicrobial Peptides by Gram-Positive Bacteria

Gram-positive bacteria also must face the wide array of antimicrobial peptides produced by the host. These bacteria prevent the action of these peptides by the enzyme Dlt adding D-alanine to their teichoic acids (Abachin et al., 2002; Kovacs et al., 2006), and by the enzyme MprF modifying their cell membrane phospholipids through the addition of lysine (Peschel et al., 2001; Thedieck et al., 2006).

It was reported that *Staphylococcus epidermidis* exposed to the cationic antimicrobial peptide human beta-defensin-3 (hBD-3) increased expression of the *dlt* and *mprF* as well as the ABC transport system VraFG and the adjacent two-component system *graRS/apsRS* that also contains another protein ApsX (Li et al., 2007). VraFG is required for the intermediate resistance of *Staphylococcus aureus* to the glycopeptide antibiotic vancomycin, and mutations in the response regulator GraR have been identified in vancomycin-intermediate *S. aureus* (VISA) strains (Kuroda et al., 2000; Meehl et al., 2007; Neoh et al., 2008). Mutants with mutations of both genes of the two-component system as well as *apsX* showed greatly decreased expression of *dlt* and *mprF*, which was blind to hBD-3; however, the regulation pattern of *vraFG* suggests that other factors are involved. These mutants were also more susceptible to killing by hBD-3 (Li et al., 2007).

The ApsS histidine sensor kinase contains a small, strongly negatively charged periplasmic loop thought to be the peptide-binding site. An antibody directed against this region blocks the activation of *dlt* expression by hBD-3. The sensor is also capable of responding to various other cationic peptides but not anionic peptides (Li et al., 2007). It is not known if this protein is also able to respond to vancomycin or if the VraFG function in removing CAMPs, although it appears that the same system in *Staphylococcus* is involved in resistance to eukaryotic and prokaryotic antimicrobials.

SENSING OF BILE SALTS

Any pathogen that infects the host through the fecal-oral route will encounter various hazards such as the low pH of the stomach, proteolytic enzymes, and bile salts. Bile is a heterogeneous mixture of various bile salts, cholesterol, and bilirubin that serves to solubilize lipids to aid in digestion. Bile salts also function to eliminate many bacteria, given that their surfactant ability disrupts bacterial membranes (Hofmann, 1998).

Vibrio cholerae, the causative agent of the severe diarrheal disease cholera, normally inhabits aquatic environments and upon ingestion is able to cause disease (Mekalanos et al., 1997). While in the intestine, *V. cholerae* inevitably encounters bile and responds to it in several ways. It alters the expression of porins to increase its survival (Provenzano et al., 2000). Bile also modulates the expression of cholera toxin and the toxin-coregulated pilus, two virulence factors vital for survival in the host (Gupta and Chowdhury, 1997; Schuhmacher and Klose, 1999; Hung and Mekalanos, 2005).

V. cholerae is also able to form a biofilm, which has been shown to be important in the spread of the disease (Watnick and Kolter, 1999). This process is regulated by the orphan response regulators VpsT and VpsR (Yildiz et al., 2001; Casper-Lindley and Yildiz, 2004; Rashid et al., 2004), which activate the *vps* genes that produce the exopolysaccharides that form the biofilm. Crude bile and the purified bile salt sodium cholate induce the formation of biofilms in *V. cholerae* through VpsR, although the upstream regulators are still unknown. Biofilm formation has been reported to protect *V. cholerae* from bile salts

in vitro. Bacteria that are part of the biofilm are able to survive, while planktonic bacteria are eventually killed (Hung et al., 2006). *V. cholerae* senses bile and responds to maximize its survival as well as modulating its virulence gene expression.

SENSING OF HOST-DERIVED METABOLITES

Sensing of Host Signals by Gram-Negative Bacteria

In addition to hormones and various uniquely host-derived molecules, bacteria are capable of mounting a specific response to relatively common molecules that are released by their host during infection.

Enteropathogenic *E. coli* (EPEC) is a major cause of watery diarrhea in the developing world. Like its relative EHEC, EPEC is an AE pathogen that contains a type III secretion system encoded by the LEE (McDaniel et al., 1995). EPEC adheres to the epithelial cells in the intestine and secretes effectors that damage and destroy the intestinal lining. This has been shown to release various cellular contents including ATP, which rapidly degrades into adenosine when exposed to the environment of the intestinal lumen. This adenosine has been shown to contribute to the watery diarrhea associated with EPEC infection by stimulating the secretion of chloride by the intestinal cells (Crane et al., 2002).

In addition to having an effect on host cells, the exposure of EPEC to adenosine has a significant effect on the expression of virulence genes. EPEC normally adheres to cells by forming localized clusters of cells called microcolonies (Cravioto et al., 1991). This pattern is dependent on the function of the bundle-forming pilus (Bfp) (Vuopio-Varkila and Schoolnik, 1991), whose expression is regulated by the plasmid-encoded regulator (Per) (Kenny and Finlay, 1995). The exposure of EPEC to adenosine greatly alters its attachment pattern from microcolonies to a much more diffuse pattern by downregulating Per, which results in lower levels of Bfp (Crane and Shulgina, 2009).

Adenosine has a different effect on the expression of the various proteins (Esps) that are secreted through EPEC's type III secretion system. Adenosine concentrations of 20 to 40 µM have been reported to increase the expression of Esps, while higher concentrations (60 to 80 µM) result in a repression of these genes. The exact mechanism of this regulation is still unknown, but it is not through either Per or Ler, the major regulators of the LEE genes (Crane and Shulgina, 2009). It is hypothesized that the massive disruption of the intestinal lining during

EPEC infection releases large amounts of adenosine, which acts as a signal that the bacteria should start dispersing in order to infect more distal regions of the intestine or to simply leave and infect the next host.

EPEC is not the only bacterium that senses adenosine; the opportunistic pathogen *P. aeruginosa* shares this ability. Adenosine functions as a signaling molecule that the host releases under hypoxic conditions or in response to tissue damage (Volmer et al., 2006). It then binds to the A2b receptor to enhance epithelial barrier function and promote wound healing (Zhou et al., 2009). It was discovered that cells expressing the hypoxia-inducible factor 1 (HIF-1α), which produce large quantities of adenosine, greatly increased the expression of the *Pseudomonas* lectin PA-I. While adenosine increased the expression of this virulence factor, its downstream product inosine was a far more potent activator (Patel et al., 2007). Interestingly, the cultured epithelial cells were not producing any inosine; instead, *Pseudomonas* was metabolizing the adenosine to make inosine, which is essentially functioning as a QS molecule, although the mechanism of inosine-induced activation of PA-I has not yet been elucidated. Additionally, the presence of the epithelial cells increases the rate of conversion, suggesting that there is another signaling molecule sensed by *Pseudomonas* (Patel et al., 2007). This is yet another example of an opportunistic pathogen sensing the host's signaling molecules and taking advantage of a potential weakness.

Sensing of Host Signals by Staphylococci

The ability to respond to host-derived molecules is not limited to Gram-negative bacteria. Gram-positive bacteria including *Staphylococcus aureus* are also known to accomplish this feat. *S. aureus* is one of the most significant human pathogens. It can cause a wide array of disease from minor skin infections to septicemia (Brook, 2002), and a large percentage of the population are long-term carriers (Wertheim et al., 2004). *S. aureus* encodes numerous virulence factors such as exotoxins and adhesins whose regulation is likely influenced by host factors (Yarwood et al., 2002; Cheung et al., 2004). One of the better studied regulatory systems is the *agr* system, which produces the QS signal autoinducer peptide (AIP) (George and Muir, 2007).

S. aureus uses heme from its host as its preferred iron source (Skaar et al., 2004). The presence of heme activates the two-component system HssRS, which activates the heme transporters HrtAB that remove excess heme to prevent any toxic effects (Friedman et al., 2006; Torres et al., 2007). The absence of either the two-component system or the heme transporter

leads to increased secretion of virulence factors and an increase in liver damage in a mouse model (Torres et al., 2007). Without these systems, *S. aureus* would be unable to remove excess heme, which would activate the stress response and thus many virulence factors. This system may exist to limit the amount of cellular damage and heme released during an infection. Since too much damage would elicit a stronger immune response or potentially kill the host, this would increase the survival of *S. aureus*.

Other components of blood are also known to play a role in the virulence regulation of *S. aureus*. The α and β chains of hemoglobin downregulate exotoxin production through the *ssrAB* two-component system which regulates the Agr QS system (Yarwood et al., 2001; Schlievert et al., 2007). Additionally, a component of low-density lipoprotein (LDL) and very low-density lipoprotein (VLDL), apolipoprotein B, disrupts *S. aureus* signaling by binding AIP and preventing activation of the *agr* system (Peterson et al., 2008). This may be a mechanism used by the host to thwart *S. aureus* signaling, or this binding could be exploited by the bacteria to downregulate its QS system under conditions where it is not advantageous.

SENSING OF HOST-DERIVED SIGNALS IN PLANTS

The sensing of host-derived molecules is not a phenomenon found only in animal pathogens. Plant pathogens have also evolved mechanisms by which they can sense the status of their host. One of the oldest and best-studied examples is *Agrobacterium tumefaciens*. This bacterium can induce crown gall tumors at injury sites on a wide variety of different plant species. It accomplishes this by transferring the Ti plasmid through a type IV secretion system into its host. The region of the plasmid that is transferred to the host contains genes that induce the host plant to produce opines that the bacterium uses as nutrients. The untransferred region contains the genes that encode this transfer system as well as their regulators. These genes are induced by phenolic compounds which are released by the plant at the site of injury as well as several different monosaccharides (Ankenbauer and Nester, 1990).

These signals are detected and transduced by the VirAG two-component system. This system consists of the VirG, the response regulator, and VirA, its cognate histidine sensor kinase (Stachel and Zambryski, 1986). This protein can be divided into different domains that function to sense different signals. The periplasmic domain is required for VirA to sense monosaccharides; however, it does not directly interact with them. Instead, the periplasmic sugar-binding protein ChvE (Huang et al., 1990) binds to various simple sugars that are either plant metabolites or components of the plant cell wall (Ankenbauer and Nester, 1990). The ChvE-monosaccharide complex then interacts with this domain of VirA to activate its kinase activity. A wound would be the site of highest concentrations of these monosaccharides due to either leaking of cytoplasmic contents or restructuring of a damaged cell wall. Additionally, the activity of VirA is increased by low pH.

The cytoplasmic linker domain of VirA is directly responsible for the detection of the plant-derived phenolic compounds (Lee et al., 1996). These compounds are released by plants at the site of injury. However, the presence of phenolic compounds is not sufficient to activate VirA. Either low pH or the presence of monosaccharide bound ChvE is needed. The ability and requirement for VirA to sense multiple signals keeps its virulence genes tightly regulated until it is at a wound site where it can start an infection (Fig. 4).

The virulence mechanism in *A. tumefaciens* is also under the control of a QS system. It contains a LuxRI system, TraRI, that produces the AHL *N*-(3-oxooctanoyl) homoserine lactone (OC8-HSL), which is known to control the conjugation of the Ti plasmid (Piper et al., 1993; Zhang et al., 1993), as well as plasmid copy number and severity of the disease (Pappas and Winans, 2003). This system is positively regulated by the opines produced by the infected plant (Piper et al., 1999). *Agrobacterium* is also able to downregulate its QS system through the production of two lactonases, AttK and AiiA, which are able to inactivate AHLs by cleaving the lactone ring (Zhang et al., 2002; Carlier et al., 2003).

Gamma-aminobutyric acid (GABA) is a major inhibitory neurotransmitter in animals (Owens and Kriegstein, 2002), but it is also a signaling molecule in plants and is known to accumulate at sites of stress (Shelp et al., 1999; Roberts, 2007). *Agrobacterium* contains an ABC transporter for GABA (Morera et al., 2008), and once in the cytoplasm, GABA activates

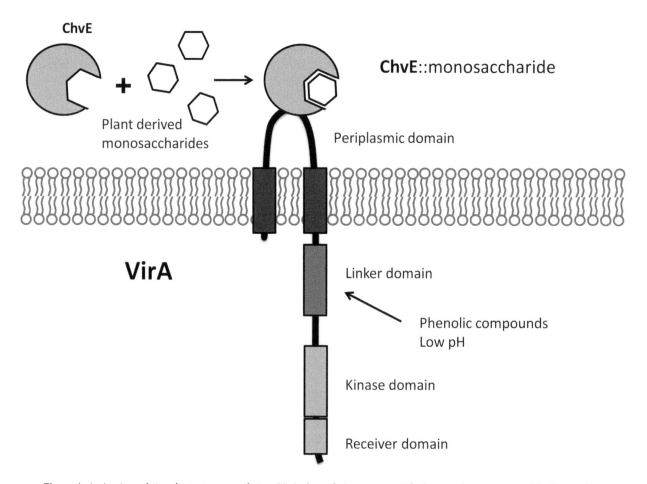

Figure 4. Activation of *Agrobacterium tumefaciens* VirA through interaction with the periplasmic protein ChvE bound to plant-derived monosaccharides.
doi:10.1128/9781555818524.ch27f4

the expression of AttK. Bacteria treated with GABA do not accumulate OC8-HSL, and plants producing large amounts of GABA were protected from *Agrobacterium* infection (Chevrot et al., 2006). *Agrobacterium* appears to be using this plant signal to downregulate its QS system in order to regulate its conjugative system and virulence. Since GABA is also an important signaling molecule in animals and is produced and secreted by many bacteria (Minuk, 1986; Mountfort and Pybus, 1992; Higuchi et al., 1997), it is possible that it may serve as an important signaling molecule in both inter- and intrakingdom signaling.

This ability to sense plant-derived molecules is not unique to *Agrobacterium*. Other plant pathogens such as *Pseudomonas syringae* are also able to sense plant-derived phenolic compounds and various monosaccharides through different systems. Additionally, symbiotic relationships such as those of *Rhizobium*, a nodule-forming, nitrogen-fixing bacterium, also depend on interkingdom signaling (Gottfert, 1993).

SENSING OF BACTERIUM-DERIVED MOLECULES BY THE HOST

Just as bacteria are capable of sensing host-derived molecules, the host is also capable of sensing bacterium-derived signaling molecules. The bacterial QS molecules AHLs are known to elicit a response in mammalian cells. The earliest evidence of this was the observation that respiratory epithelial cells produced IL-8 when exposed to the *Pseudomonas* AHL produced by LasI, 3-oxo-C12-AHL (OdDHL) (DiMango, 1995). Further studies showed that this AHL, but not C4-AHL produced by RhlI, has immunomodulatory effects. OdDHL is capable of inhibiting cytokine production by activated macrophages and lymphocyte proliferation. Low concentrations stimulate antibody production, while higher concentrations suppress it (Telford et al., 1998). The structural requirements of this inhibition of cytokine production were determined. Acylated homoserine lactones with an 11- to 13-C side chain with either a 3-oxo or a 3-hydroxy group have the highest immune-suppressive activity (Chhabra et al., 2003). While in some studies IL-4 is less inhibited than IFN-γ, OdDHL is capable of inhibiting the proliferation of both Th1 and Th2 cells (Ritchie et al., 2003, 2005). *In vivo* data from *Pseudomonas* sepsis patients with detectable AHL in their sera mirror these patterns, suggesting that these effects are clinically relevant (Boontham et al., 2008).

While AHLs suppress some parts of the immune system, OdDHL also elicits a pro-inflammatory response. OdDHL induces IL-8 production through the MAPK pathway via the transcription factors nuclear factor NF-κB and AP-2 (Smith et al., 2001). While IL-8 is a potent chemokine for neutrophils, OdDHL itself is also able to act as a chemotactic signal for PMNs (Zimmermann et al., 2006). The activation of NF-κB also activates cyclooxygenase (Cox)-2, a pro-inflammatory enzyme responsible for the production of the eicosanoid prostaglandin (Smith et al., 2002). Injection of OdDHL under the skin of mice triggers a strong IL-1α, IL-6, and chemokine response (Smith et al., 2002). This strong induction of inflammation would assist in the clearance of a *Pseudomonas* infection in a healthy individual. However, in a cystic fibrosis patient this hyperinflammatory response contributes to the disease (Mayer et al., 2011).

In addition to their immune modulator role, AHLs are capable of inducing apoptosis in various mammalian cell types. OdDHL, but again not C4-AHL, was reported to induce apoptosis in macrophages and neutrophils but not in some epithelial cell lines (Tateda et al., 2003). Breast cell carcinoma cell lines were also shown to be susceptible to OdDHL-induced apoptosis. This was reported to be due to OdDHL-induced suppression of signaling through the signal transducer and activator of transcription 3 (STAT3) pathway. Inhibiting this pathway had the same effect, and a constitutively active STAT3 protected the cells from OdDHL-induced apoptosis (Li et al., 2004). It was later reported that OdDHL causes a release of calcium from the endoplasmic reticulum in murine fibroblasts. This release and subsequent apoptosis could be blocked by inhibitors of calcium signaling. However, this did not alter the modulation of the immune response (Shiner et al., 2006), which implies that AHLs act through at least two signaling pathways in mammalian cells.

Given these effects induced by OdDHL in mammalian cells, it should not be surprising that animals have evolved a means of protecting themselves (Yang et al., 2005). A family of enzymes called paraoxonases (PONs) act as lactonases (Khersonsky and Tawfik, 2005). Since these enzymes are not secreted in the extracellular environment, they are unable to subvert QS, but they are able to detoxify any AHL that enters the cell (Chun et al., 2004). Knockout studies with mice have determined that PON2 is the most likely candidate for an active AHL-degrading enzyme (Stoltz et al., 2007). Interestingly, OdDHL also has a mechanism to shut down PON2. The increase in cytoplasmic calcium it induces causes the PON2 mRNA to degrade and decrease the hydrolytic activity of the enzyme (Horke et al., 2010). Since bacteria evolve far faster than mammals, it also should not be surprising that they have evolved around the protection provided by PON2.

Table 1. Summary of interkingdom signaling systems

Compound	Organism	Sensor(s)	Effect	Reference(s)
Catecholamines				
NE, Epi, Dopamine	Various	N/A	Increased growth due to availability of iron	Lyte, 1992; Freestone et al., 2000; Anderson and Armstrong, 2004
NE, Epi	EHEC	QseC, QseE	Increased expression of virulence genes	Clarke et al., 2006; Reading et al., 2007
NE	Pseudomonas aeruginosa	?	Increases LasIR QS pathway and downstream virulence genes	Hegde et al., 2009
NE	Borrelia burgdorferi	?	Increases expression of tick colonization factor OspA	Scheckelhoff et al., 2007
NE	Campylobacter jejuni	?	Increased growth rate and motility	Cogan et al., 2007
Opioids				
Dynorphin	Pseudomonas aeruginosa	?	Increased virulence gene expression through PQS QS system	Wu et al., 2005
Peptide hormones				
Gastrin	Helicobacter pylori	?	Increased growth	Chowers et al., 1999
Somatostatin	Helicobacter pylori	?	Decreased growth	Yamashita et al., 1998
Natriuretic peptides	Pseudomonas aeruginosa	?	Increased cytotoxicity and modification of LPS through cyclic nucleotides	Veron et al., 2007
Insulin	Burkholderia pseudomallei	?	Reduced activity of phospholipase C and acid phosphatase	Woods et al., 1993; Kanai et al., 1996; Simpson and Wuthiekanun, 2000
Cytokines				
IL-1	Escherichia coli	?	Increased growth	Porat et al., 1991
IL-1	Yersinia pestis	Caf1A	?	Zav'yalov et al., 1995
IL-2	Escherichia coli	?	Increased growth	Denis et al., 1991
GM-CSF	Escherichia coli	?	Increased growth	Denis et al., 1991
TNF-α	Shigella flexneri	?	Increased invasion of epithelial cells and macrophages	Luo et al., 1993
IFN-γ	Pseudomonas aeruginosa	OprF	Increased virulence gene expression through RhlIR QS system	Wu et al., 2003
Antimicrobials				
CAMPs	Salmonella	PhoQ	Modification of Lipid A and virulence genes	Bader et al., 2005
CAMPs	Staphylococcus epidermidis	GraS/ApsS	Modification of teichoic acids and phospholipids	Li et al., 2007
Bile Salts	Vibrio cholerae	?	Regulates porins, virulence genes, and biofilm formation	Gupta and Chowdhury, 1997; Provenzano et al., 2000; Hung et al., 2006
Metabolites				
Adenosine	EPEC	?	Alters expression of virulence genes and attachment pattern	Crane and Shulgina, 2009
Adenosine	Pseudomonas aeruginosa	?	Converted to inosine, which activates expression of PA-1 lectin	Patel et al., 2007
Heme	Staphylococcus aureus	HssS	Heme export, stress, virulence factors	Torres et al., 2007
Hemoglobin	Staphylococcus aureus	AgrC	Repression of virulence genes through AIP QS system	Schlievert et al., 2007
Apolipoprotein B	Staphylococcus aureus	AIP	Repression of virulence genes by sequestering AIP signal	Peterson et al., 2008
Plants				
Phenolics	Agrobacterium tumefaciens	VirA	Regulation of Ti plasmid and virulence genes	Lee et al., 1996
Monosaccharides	Agrobacterium tumefaciens	VirA	Regulation of Ti plasmid and virulence genes	Huang et al., 1990
GABA	Agrobacterium tumefaciens	?	Downregulation of QS and conjugation	Chevrot et al., 2006

AHLs play many, often conflicting roles in mammalian signaling. Due to the abundance of bacterial signaling molecules, it is likely that animals are able to sense other classes of molecules.

CONCLUSION

While these systems are extremely diverse, there is a recurrent trend. Interkingdom signaling intersects with QS, either directly as in the case of EHEC (Clarke et al., 2006) or indirectly as with *Pseudomonas* (Zaborina et al., 2007; Hegde et al., 2009). This may simply reflect the central role that QS plays in regulating virulence in many bacteria. Given that bacteria can produce many of the same small molecules and in other cases similar molecules to eukaryotes, these behaviors may be an example of convergent evolution or an indication that they share a common origin. After all, hormones and other signaling molecules used by multicellular eukaryotes serve the same basic purpose as QS molecules: intercellular communication that mediates cooperative behaviors.

In any case, the ability to sense and respond to host signals has been identified in most major pathogens (Table 1). The prevalence of this ability suggests that interkingdom signaling may be extremely common and that many more systems will be uncovered in the future. Given the central role they play in virulence regulation, these interkingdom signaling pathways make for potential targets for new drugs to treat bacterial diseases.

REFERENCES

Abachin, E., C. Poyart, E. Pellegrini, E. Milohanic, F. Fiedler, P. Berche, and P. Trieu-Cuot. 2002. Formation of D-alanyl-lipoteichoic acid is required for adhesion and virulence of Listeria monocytogenes. *Mol. Microbiol.* **43:**1–14.

Abram, M., D. Vuckovic, B. Wraber, and M. Doric. 2000. Plasma cytokine response in mice with bacterial infection. *Mediators Inflamm.* **9:**229–234.

Alhede, M., T. Bjarnsholt, P. O. Jensen, R. K. Phipps, C. Moser, L. Christophersen, L. D. Christensen, M. van Gennip, M. Parsek, N. Hoiby, T. B. Rasmussen, and M. Givskov. 2009. Pseudomonas aeruginosa recognizes and responds aggressively to the presence of polymorphonuclear leukocytes. *Microbiology* **155:**3500–3508.

Alpuche Aranda, C. M., J. A. Swanson, W. P. Loomis, and S. I. Miller. 1992. *Salmonella typhimurium* activates virulence gene transcription within acidified macrophage phagosomes. *Proc. Natl. Acad. Sci. USA* **89:**10079–10083.

Alverdy, J., C. Holbrook, F. Rocha, L. Seiden, R. L. Wu, M. Musch, E. Chang, D. Ohman, and S. Suh. 2000. Gut-derived sepsis occurs when the right pathogen with the right virulence genes meets the right host: evidence for in vivo virulence expression in Pseudomonas aeruginosa. *Ann. Surg.* **232:**480–489.

Anderson, M. T., and S. K. Armstrong. 2004. The BfeR regulator mediates enterobactin-inducible expression of *Bordetella* enterobactin utilization genes. *J. Bacteriol.* **186:**7302–7311.

Anderson, M. T., and S. K. Armstrong. 2006. The *Bordetella* Bfe system: growth and transcriptional response to siderophores, catechols, and neuroendocrine catecholamines. *J. Bacteriol.* **188:**5731–5740.

Anderson, M. T., and S. K. Armstrong. 2008. Norepinephrine mediates acquisition of transferrin-iron in *Bordetella bronchiseptica. J. Bacteriol.* **190:**3940–3947.

Aneman, A., G. Eisenhofer, L. Olbe, J. Dalenback, P. Nitescu, L. Fandriks, and P. Friberg. 1996. Sympathetic discharge to mesenteric organs and the liver. Evidence for substantial mesenteric organ norepinephrine spillover. *J. Clin. Investig.* **97:**1640–1646.

Ankenbauer, R. G., and E. W. Nester. 1990. Sugar-mediated induction of *Agrobacterium tumefaciens* virulence genes: structural specificity and activities of monosaccharides. *J. Bacteriol.* **172:**6442–6446.

Babalola, C. P., C. H. Nightingale, and D. P. Nicolau. 2004. Effect of adjunctive treatment with gamma interferon against Pseudomonas aeruginosa pneumonia in neutropenic and nonneutropenic hosts. *Int. J. Antimicrob. Agents* **24:**219–225.

Bader, M. W., S. Sanowar, M. E. Daley, A. R. Schneider, U. Cho, W. Xu, R. E. Klevit, H. Le Moual, and S. I. Miller. 2005. Recognition of antimicrobial peptides by a bacterial sensor kinase. *Cell* **122:**461–472.

Baron, S. S., and J. J. Rowe. 1981. Antibiotic action of pyocyanin. *Antimicrob. Agents Chemother.* **20:**814–820.

Beall, B., and G. N. Sanden. 1995. A Bordetella pertussis fepA homologue required for utilization of exogenous ferric enterobactin. *Microbiology* **141:**3193–3205.

Bearson, B. L., and S. M. Bearson. 2008. The role of the QseC quorum-sensing sensor kinase in colonization and norepinephrine-enhanced motility of Salmonella enterica serovar Typhimurium. *Microb. Pathog.* **44:**271–278.

Bearson, B. L., S. M. Bearson, I. S. Lee. and B. W. Brunelle. 2010. The Salmonella enterica serovar Typhimurium QseB response regulator negatively regulates bacterial motility and swine colonization in the absence of the QseC sensor kinase. *Microb. Pathog.* **48:**214-219.

Bodmer, S., C. Imark, and M. Kneubuhl. 1999. Biogenic amines in foods: histamine and food processing. *Inflamm. Res.* **48:**296–300.

Boontham, P., A. Robins, P. Chandran, D. Pritchard, M. Camara, P. Williams, S. Chuthapisith, A. McKechnie, B. J. Rowlands, and O. Eremin. 2008. Significant immunomodulatory effects of Pseudomonas aeruginosa quorum-sensing signal molecules: possible link in human sepsis. *Clin. Sci.* (London) **115:**343–351.

Bordet, J., and O. Gengou. 1906. Le microbe de la coqueluche. *Ann. Inst. Pasteur* (Paris) **20:**731–741.

Brook, I. 2002. Secondary bacterial infections complicating skin lesions. *J. Med. Microbiol.* **51:**808–812.

Brown, L. M. 2000. Helicobacter pylori: epidemiology and routes of transmission. *Epidemiol. Rev.* **22:**283–297.

Brubaker, R. R. 1991. Factors promoting acute and chronic diseases caused by yersiniae. *Clin. Microbiol. Rev.* **4:**309–324.

Burton, C. L., S. R. Chhabra, S. Swift, T. J. Baldwin, H. Withers, S. J. Hill, and P. Williams. 2002. The growth response of *Escherichia coli* to neurotransmitters and related catecholamine drugs requires a functional enterobactin biosynthesis and uptake system. *Infect. Immun.* **70:**5913–5923.

Bustamante, V. H., F. J. Santana, E. Calva, and J. L. Puente. 2001. Transcriptional regulation of type III secretion genes in enteropathogenic Escherichia coli: Ler antagonizes H-NS-dependent repression. *Mol. Microbiol.* **39:**664–678.

Butzler, J. P. 2004. Campylobacter, from obscurity to celebrity. *Clin. Microbiol. Infect.* **10:**868–876.

Campellone, K. G., D. Robbins, and J. M. Leong. 2004. EspFU is a translocated EHEC effector that interacts with Tir and N-WASP and promotes Nck-independent actin assembly. *Dev. Cell* **7:**217–228.

Carlier, A., S. Uroz, B. Smadja, R. Fray, X. Latour, Y. Dessaux, and D. Faure. 2003. The Ti plasmid of *Agrobacterium tumefaciens* harbors an *attM*-paralogous gene, *aiiB*, also encoding N-acyl homoserine lactonase activity. *Appl. Environ. Microbiol.* **69**:4989–4993.

Casper-Lindley, C., and F. H. Yildiz. 2004. VpsT is a transcriptional regulator required for expression of *vps* biosynthesis genes and the development of rugose colonial morphology in *Vibrio cholerae* O1 El Tor. *J. Bacteriol.* **186**:1574–1578.

Cheung, A. L., A. S. Bayer, G. Zhang, H. Gresham, and Y. Q. Xiong. 2004. Regulation of virulence determinants in vitro and in vivo in Staphylococcus aureus. *FEMS Immunol. Med. Microbiol.* **40**:1–9.

Chevrot, R., R. Rosen, E. Haudecoeur, A. Cirou, B. J. Shelp, E. Ron, and D. Faure. 2006. GABA controls the level of quorum-sensing signal in *Agrobacterium tumefaciens*. *Proc. Natl. Acad. Sci. USA* **103**:7460–7464.

Chhabra, S. R., C. Harty, D. S. Hooi, M. Daykin, P. Williams, G. Telford, D. I. Pritchard, and B. W. Bycroft. 2003. Synthetic analogues of the bacterial signal (quorum sensing) molecule N-(3-oxododecanoyl)-L-homoserine lactone as immune modulators. *J. Med. Chem.* **46**:97–104.

Cho, U. S., M. W. Bader, M. F. Amaya, M. E. Daley, R. E. Klevit, S. I. Miller, and W. Xu. 2006. Metal bridges between the PhoQ sensor domain and the membrane regulate transmembrane signaling. *J. Mol. Biol.* **356**:1193–1206.

Chowers, M. Y., N. Keller, S. Bar-Meir, and Y. Chowers. 2002. A defined human gastrin sequence stimulates the growth of Helicobacter pylori. *FEMS Microbiol. Lett.* **217**:231–236.

Chowers, M. Y., N. Keller, R. Tal, I. Barshack, R. Lang, S. Bar-Meir, and Y. Chowers. 1999. Human gastrin: a *Helicobacter pylori*-specific growth factor. *Gastroenterology* **117**:1113–1118.

Chun, C. K., E. A. Ozer, M. J. Welsh, J. Zabner, and E. P. Greenberg. 2004. Inactivation of a *Pseudomonas aeruginosa* quorum-sensing signal by human airway epithelia. *Proc. Natl. Acad. Sci. USA* **101**:3587–3590.

Clarke, M. B., D. T. Hughes, C. Zhu, E. C. Boedeker, and V. Sperandio. 2006. The QseC sensor kinase: a bacterial adrenergic receptor. *Proc. Natl. Acad. Sci. USA* **103**:10420–10425.

Cogan, T. A., A. O. Thomas, L. E. Rees, A. H. Taylor, M. A. Jepson, P. H. Williams, J. Ketley, and T. J. Humphrey. 2007. Norepinephrine increases the pathogenic potential of Campylobacter jejuni. *Gut* **56**:1060–1065.

Crane, J. K., R. A. Olson, H. M. Jones, and M. E. Duffey. 2002. Release of ATP during host cell killing by enteropathogenic *E. coli* and its role as a secretory mediator. *Am. J. Physiol. Gastrointest. Liver Physiol.* **283**:G74–G86.

Crane, J. K., and I. Shulgina. 2009. Feedback effects of host-derived adenosine on enteropathogenic *Escherichia coli*. *FEMS Immunol. Med. Microbiol.* **57**:214–228.

Cravioto, A., A. Tello, A. Navarro, J. Ruiz, H. Villafan, F. Uribe, and C. Eslava. 1991. Association of *Escherichia coli* HEp-2 adherence patterns with type and duration of diarrhoea. *Lancet* **337**:262–264.

Currie, B. 1995. *Pseudomonas pseudomallei*-insulin interaction. *Infect. Immun.* **63**:3745.

Dathe, M., and T. Wieprecht. 1999. Structural features of helical antimicrobial peptides: their potential to modulate activity on model membranes and biological cells. *Biochim. Biophys. Acta* **1462**:71–87.

de Kievit, T. R. 2009. Quorum sensing in Pseudomonas aeruginosa biofilms. *Environ. Microbiol.* **11**:279–288.

Denis, M., D. Campbell, and E. O. Gregg. 1991. Interleukin-2 and granulocyte-macrophage colony-stimulating factor stimulate growth of a virulent strain of *Escherichia coli*. *Infect. Immun.* **59**:1853–1856.

Deziel, E., S. Gopalan, A. P. Tampakaki, F. Lepine, K. E. Padfield, M. Saucier, G. Xiao, and L. G. Rahme. 2005. The contribution of MvfR to Pseudomonas aeruginosa pathogenesis and quorum sensing circuitry regulation: multiple quorum sensing-regulated genes are modulated without affecting lasRI, rhlRI or the production of N-acyl-L-homoserine lactones. *Mol. Microbiol.* **55**:998–1014.

DiMango, E., H. J. Zar, R. Bryan, and A. Prince. 1995. Diverse Pseudomonas aeruginosa gene products stimulate respiratory epithelial cells to produce interleukin-8. *J. Clin. Investig.* **96**:2204–2210.

Engel, J., and P. Balachandran. 2009. Role of *Pseudomonas aeruginosa* type III effectors in disease. *Curr. Opin. Microbiol.* **12**:61–66.

Fernandez, M., B. del Rio, D. M. Linares, M. C. Martin, and M. A. Alvarez. 2006. Real-time polymerase chain reaction for quantitative detection of histamine-producing bacteria: use in cheese production. *J. Dairy Sci.* **89**:3763–3769.

Fields, P. I., E. A. Groisman, and F. Heffron. 1989. A Salmonella locus that controls resistance to microbicidal proteins from phagocytic cells. *Science* **243**:1059–1062.

Freestone, P. P., R. D. Haigh, and M. Lyte. 2007. Blockade of catecholamine-induced growth by adrenergic and dopaminergic receptor antagonists in *Escherichia coli* O157:H7, *Salmonella enterica* and *Yersinia enterocolitica*. *BMC Microbiol.* **7**:8.

Freestone, P. P., R. D. Haigh, P. H. Williams, and M. Lyte. 1999. Stimulation of bacterial growth by heat-stable, norepinephrine-induced autoinducers. *FEMS Microbiol. Lett.* **172**:53–60.

Freestone, P. P., R. D. Haigh, P. H. Williams, and M. Lyte. 2003. Involvement of enterobactin in norepinephrine-mediated iron supply from transferrin to enterohaemorrhagic Escherichia coli. *FEMS Microbiol. Lett.* **222**:39–43.

Freestone, P. P., M. Lyte, C. P. Neal, A. F. Maggs, R. D. Haigh, and P. H. Williams. 2000. The mammalian neuroendocrine hormone norepinephrine supplies iron for bacterial growth in the presence of transferrin or lactoferrin. *J. Bacteriol.* **182**:6091–6098.

Friedman, D. B., D. L. Stauff, G. Pishchany, C. W. Whitwell, V. J. Torres, and E. P. Skaar. 2006. *Staphylococcus aureus* redirects central metabolism to increase iron availability. *PLoS Pathog.* **2**:e87.

Fuhlbrigge, R. C., S. M. Fine, E. R. Unanue, and D. D. Chaplin. 1988. Expression of membrane interleukin 1 by fibroblasts transfected with murine pro-interleukin 1 alpha cDNA. *Proc. Natl. Acad. Sci. USA* **85**:5649–5653.

Gambello, M. J., and B. H. Iglewski. 1991. Cloning and characterization of the *Pseudomonas aeruginosa lasR* gene, a transcriptional activator of elastase expression. *J. Bacteriol.* **173**:3000–3009.

Garmendia, J., A. D. Phillips, M. F. Carlier, Y. Chong, S. Schuller, O. Marches, S. Dahan, E. Oswald, R. K. Shaw, S. Knutton, and G. Frankel. 2004. TccP is an enterohaemorrhagic *Escherichia coli* O157:H7 type III effector protein that couples Tir to the actin-cytoskeleton. *Cell. Microbiol.* **6**:1167–1183.

George, E. A., and T. W. Muir. 2007. Molecular mechanisms of agr quorum sensing in virulent staphylococci. *Chembiochem* **8**:847–855.

Goodnow, R. A. 1980. Biology of *Bordetella bronchiseptica*. *Microbiol. Rev.* **44**:722–738.

Gorke, B., and J. Vogel. 2008. Noncoding RNA control of the making and breaking of sugars. *Genes Dev.* **22**:2914–2925.

Gottfert, M. 1993. Regulation and function of rhizobial nodulation genes. *FEMS Microbiol. Rev.* **10**:39–63.

Grant, C. C., M. E. Konkel, W. Cieplak, Jr., and L. S. Tompkins. 1993. Role of flagella in adherence, internalization, and translocation of *Campylobacter jejuni* in nonpolarized and polarized epithelial cell cultures. *Infect. Immun.* **61**:1764–1771.

Gregory, R. A., and H. J. Tracy. 1964. The constitution and properties of two gastrins extracted from hog antral mucosa. *Gut* 5:103–114.

Guo, L., K. B. Lim, J. S. Gunn, B. Bainbridge, R. P. Darveau, M. Hackett, and S. I. Miller. 1997. Regulation of lipid A modifications by *Salmonella typhimurium* virulence genes *phoP-phoQ*. *Science* 276:250–253.

Gupta, S., and R. Chowdhury. 1997. Bile affects production of virulence factors and motility of *Vibrio cholerae*. *Infect. Immun.* 65:1131–1134.

Hahn, P. Y., P. Wang, S. M. Tait, Z. F. Ba, S. S. Reich, and I. H. Chaudry. 1995. Sustained elevation in circulating catecholamine levels during polymicrobial sepsis. *Shock* 4:269–273.

Harmsen, M., L. Yang, S. J. Pamp, and T. Tolker-Nielsen. 2010. An update on Pseudomonas aeruginosa biofilm formation, tolerance, and dispersal. *FEMS Immunol. Med. Microbiol.* 59:253–268.

Hegde, M., T. K. Wood, and A. Jayaraman. 2009. The neuroendocrine hormone norepinephrine increases Pseudomonas aeruginosa PA14 virulence through the las quorum-sensing pathway. *Appl. Microbiol. Biotechnol.* 84:763–776.

Higuchi, T., H. Hayashi, and K. Abe. 1997. Exchange of glutamate and gamma-aminobutyrate in a *Lactobacillus* strain. *J. Bacteriol.* 179:3362–3364.

Hofmann, A. F. 1998. Bile secretion and the enterohepatic circulation of bile acids, p. 937–948. *In* M. Feldman, B. F. Scharschmidt, and M. H. Sleisenger (ed.), *Gastrointestinal and Liver Disease*. The W. B. Saunders Co., Philadelphia, PA.

Horke, S., I. Witte, S. Altenhofer, P. Wilgenbus, M. Goldeck, U. Forstermann, J. Xiao, G. L. Kramer, D. C. Haines, P. K. Chowdhary, R. W. Haley, and J. F. Teiber. 2010. Paraoxonase 2 is down-regulated by the Pseudomonas aeruginosa quorum-sensing signal N-(3-oxododecanoyl)-L-homoserine lactone and attenuates oxidative stress induced by pyocyanin. *Biochem. J.* 426:73–83.

Hsu, S. C., K. R. Johansson, and M. J. Donahue. 1986. The bacterial flora of the intestine of *Ascaris suum* and 5-hydroxytryptamine production. *J. Parasitol.* 72:545–549.

Huang, M. L., G. A. Cangelosi, W. Halperin, and E. W. Nester. 1990. A chromosomal *Agrobacterium tumefaciens* gene required for effective plant signal transduction. *J. Bacteriol.* 172:1814–1822.

Hughes, D. T., M. B. Clarke, K. Yamamoto, D. A. Rasko, and V. Sperandio. 2009. The QseC adrenergic signaling cascade in Enterohemorrhagic E. coli (EHEC). *PLoS Pathog.* 5:e1000553.

Hung, D. T., and J. J. Mekalanos. 2005. Bile acids induce cholera toxin expression in *Vibrio cholerae* in a ToxT-independent manner. *Proc. Natl. Acad. Sci. USA* 102:3028–3033.

Hung, D. T., J. Zhu, D. Sturtevant, and J. J. Mekalanos. 2006. Bile acids stimulate biofilm formation in Vibrio cholerae. *Mol. Microbiol.* 59:193–201.

Jarvis, K. G., J. A. Giron, A. E. Jerse, T. K. McDaniel, M. S. Donnenberg, and J. B. Kaper. 1995. Enteropathogenic *Escherichia coli* contains a putative type III secretion system necessary for the export of proteins involved in attaching and effacing lesion formation. *Proc. Natl. Acad. Sci. USA* 92:7996–8000.

Johansson, C., O. Wisen, B. Kollberg, K. Uvnas-Wallensten, and S. Efendic. 1978. Effects of intragastrically administered somatostatin on basal and pentagastrin stimulated gastric acid secretion in man. *Acta Physiol. Scand.* 104:232–234.

Joseph, I. M., and D. Kirschner. 2004. A model for the study of Helicobacter pylori interaction with human gastric acid secretion. *J. Theor. Biol.* 228:55–80.

Kalra, P. R., J. R. Clague, A. P. Bolger, S. D. Anker, P. A. Poole-Wilson, A. D. Struthers, and A. J. Coats. 2003. Myocardial production of C-type natriuretic peptide in chronic heart failure. *Circulation* 107:571–573.

Kanai, K., E. Kondo, and T. Kurata. 1996. Affinity and response of *Burkholderia pseudomallei* and *Burkholderia cepacia* to insulin. *Southeast Asian J. Trop. Med. Public Health* 27:584–591.

Kaper, J. B., J. P. Nataro, and H. L. Mobley. 2004. Pathogenic *Escherichia coli*. *Nat. Rev. Microbiol.* 2:123–140.

Karlyshev, A. V., E. E. Galyov, O. Smirnov, A. P. Guzayev, V. M. Abramov, and V. P. Zav'yalov. 1992. A new gene of the f1 operon of *Y. pestis* involved in the capsule biogenesis. *FEBS Lett.* 297:77–80.

Kenny, B., and B. B. Finlay. 1995. Protein secretion by enteropathogenic *Escherichia coli* is essential for transducing signals to epithelial cells. *Proc. Natl. Acad. Sci. USA* 92:7991–7995.

Khansari, D. N., A. J. Murgo, and R. E. Faith. 1990. Effects of stress on the immune system. *Immunol. Today* 11:170–175.

Khersonsky, O., and D. S. Tawfik. 2005. Structure-reactivity studies of serum paraoxonase PON1 suggest that its native activity is lactonase. *Biochemistry* 44:6371–6382.

Kostakioti, M., M. Hadjifrangiskou, J. S. Pinkner, and S. J. Hultgren. 2009. QseC-mediated dephosphorylation of QseB is required for expression of genes associated with virulence in uropathogenic Escherichia coli. *Mol. Microbiol.* 73:1020–1031.

Kourie, J. I. 1999. Synthetic mammalian C-type natriuretic peptide forms large cation channels. *FEBS Lett.* 445:57–62.

Kovacs, M., A. Halfmann, I. Fedtke, M. Heintz, A. Peschel, W. Vollmer, R. Hakenbeck, and R. Bruckner. 2006. A functional *dlt* operon, encoding proteins required for incorporation of D-alanine in teichoic acids in gram-positive bacteria, confers resistance to cationic antimicrobial peptides in *Streptococcus pneumoniae*. *J. Bacteriol.* 188:5797–5805.

Krause, A., C. Liepke, M. Meyer, K. Adermann, W. G. Forssmann, and E. Maronde. 2001. Human natriuretic peptides exhibit antimicrobial activity. *Eur. J. Med. Res.* 6:215–218.

Kuroda, M., K. Kuwahara-Arai, and K. Hiramatsu. 2000. Identification of the up- and down-regulated genes in vancomycin-resistant Staphylococcus aureus strains Mu3 and Mu50 by cDNA differential hybridization method. *Biochem. Biophys. Res. Commun.* 269:485–490.

Latifi, A., M. Foglino, K. Tanaka, P. Williams, and A. Lazdunski. 1996. A hierarchical quorum-sensing cascade in Pseudomonas aeruginosa links the transcriptional activators LasR and RhIR (VsmR) to expression of the stationary-phase sigma factor RpoS. *Mol. Microbiol.* 21:1137–1146.

Lau, G. W., D. J. Hassett, H. Ran, and F. Kong. 2004. The role of pyocyanin in Pseudomonas aeruginosa infection. *Trends Mol. Med.* 10:599–606.

Laughlin, R. S., M. W. Musch, C. J. Hollbrook, F. M. Rocha, E. B. Chang, and J. C. Alverdy. 2000. The key role of Pseudomonas aeruginosa PA-I lectin on experimental gut-derived sepsis. *Ann. Surg.* 232:133–142.

Lee, A., J. L. O'Rourke, P. J. Barrington, and T. J. Trust. 1986. Mucus colonization as a determinant of pathogenicity in intestinal infection by *Campylobacter jejuni*: a mouse cecal model. *Infect. Immun.* 51:536–546.

Lee, Y. W., S. Jin, W. S. Sim, and E. W. Nester. 1996. The sensing of plant signal molecules by Agrobacterium: genetic evidence for direct recognition of phenolic inducers by the VirA protein. *Gene* 179:83–88.

Li, L., D. Hooi, S. R. Chhabra, D. Pritchard, and P. E. Shaw. 2004. Bacterial N-acylhomoserine lactone-induced apoptosis in breast carcinoma cells correlated with down-modulation of STAT3. *Oncogene* 23:4894–4902.

Li, M., Y. Lai, A. E. Villaruz, D. J. Cha, D. E. Sturdevant, and M. Otto. 2007. Gram-positive three-component antimicrobial peptide-sensing system. *Proc. Natl. Acad. Sci. USA* 104:9469–9474.

Li, W., M. Lyte, P. P. Freestone, A. Ajmal, J. A. Colmer-Hamood, and A. N. Hamood. 2009. Norepinephrine represses the expression of toxA and the siderophore genes in Pseudomonas aeruginosa. *FEMS Microbiol. Lett.* **299:**100–109.

Lippa, A. M., and M. Goulian. 2009. Feedback inhibition in the PhoQ/PhoP signaling system by a membrane peptide. *PLoS Genet.* **5:**e1000788.

Llama-Palacios, A., E. Lopez-Solanilla, C. Poza-Carrion, F. Garcia-Olmedo, and P. Rodriguez-Palenzuela. 2003. The Erwinia chrysanthemi phoP-phoQ operon plays an important role in growth at low pH, virulence and bacterial survival in plant tissue. *Mol. Microbiol.* **49:**347–357.

Luo, G., D. W. Niesel, R. A. Shaban, E. A. Grimm, and G. R. Klimpel. 1993. Tumor necrosis factor alpha binding to bacteria: evidence for a high-affinity receptor and alteration of bacterial virulence properties. *Infect. Immun.* **61:**830–835.

Lyte, M. 1992. The role of catecholamines in gram-negative sepsis. *Med. Hypotheses* **37:**255–258.

Lyte, M., and S. Ernst. 1992. Catecholamine induced growth of gram negative bacteria. *Life Sci.* **50:**203–212.

Lyte, M., C. D. Frank, and B. T. Green. 1996. Production of an autoinducer of growth by norepinephrine cultured Escherichia coli O157:H7. *FEMS Microbiol. Lett.* **139:**155–159.

Lyte, M., S. G. Nelson, and M. L. Thompson. 1990. Innate and adaptive immune responses in a social conflict paradigm. *Clin. Immunol. Immunopathol.* **57:**137–147.

Maciorkowska, E., A. Panasiuk, K. Kondej-Muszynska, M. Kaczmarski, and A. Kemona. 2006. Mucosal gastrin cells and serum gastrin levels in children with Helicobacter pylori infection. *Adv. Med. Sci.* **51:**137–141.

Martin-Orozco, N., N. Touret, M. L. Zaharik, E. Park, R. Kopelman, S. Miller, B. B. Finlay, P. Gros, and S. Grinstein. 2006. Visualization of vacuolar acidification-induced transcription of genes of pathogens inside macrophages. *Mol. Biol. Cell* **17:**498–510.

Mayer, M. L., J. A. Sheridan, C. J. Blohmke, S. E. Turvey, and R. E. Hancock. 2011. The Pseudomonas aeruginosa autoinducer 3O-C12 homoserine lactone provokes hyperinflammatory responses from cystic fibrosis airway epithelial cells. *PLoS ONE* **6:**e16246.

McDaniel, T. K., K. G. Jarvis, M. S. Donnenberg, and J. B. Kaper. 1995. A genetic locus of enterocyte effacement conserved among diverse enterobacterial pathogens. *Proc. Natl. Acad. Sci. USA* **92:**1664–1668.

McKnight, S. L., B. H. Iglewski, and E. C. Pesci. 2000. The *Pseudomonas* quinolone signal regulates *rhl* quorum sensing in *Pseudomonas aeruginosa. J. Bacteriol.* **182:**2702–2708.

Meehl, M., S. Herbert, F. Gotz, and A. Cheung. 2007. Interaction of the GraRS two-component system with the VraFG ABC transporter to support vancomycin-intermediate resistance in *Staphylococcus aureus. Antimicrob. Agents Chemother.* **51:**2679–2689.

Mekalanos, J. J., E. J. Rubin, and M. K. Waldor. 1997. Cholera: molecular basis for emergence and pathogenesis. *FEMS Immunol. Med. Microbiol.* **18:**241–248.

Mellies, J. L., S. J. Elliott, V. Sperandio, M. S. Donnenberg, and J. B. Kaper. 1999. The Per regulon of enteropathogenic Escherichia coli: identification of a regulatory cascade and a novel transcriptional activator, the locus of enterocyte effacement (LEE)-encoded regulator (Ler). *Mol. Microbiol.* **33:**296–306.

Merighi, M., A. Carroll-Portillo, A. N. Septer, A. Bhatiya, and J. S. Gunn. 2006. Role of Salmonella enterica serovar Typhimurium two-component system PreA/PreB in modulating PmrA-regulated gene transcription. *J. Bacteriol.* **188:**141–149.

Merighi, M., A. N. Septer, A. Carroll-Portillo, A. Bhatiya, S. Porwollik, M. McClelland, and J. S. Gunn. 2009. Genome-wide analysis of the PreA/PreB (QseB/QseC) regulon of Salmonella enterica serovar Typhimurium. *BMC Microbiol.* **9:**42.

Minuk, G. Y. 1986. Gamma-aminobutyric acid (GABA) production by eight common bacterial pathogens. *Scand. J. Infect. Dis.* **18:**465–467.

Moreira, C. G., D. Weinshenker, and V. Sperandio. 2010. QseC mediates *Salmonella enterica* serovar typhimurium virulence in vitro and in vivo. *Infect. Immun.* **78:**914–926.

Morera, S., V. Gueguen-Chaignon, A. Raffoux, and D. Faure. 2008. Cloning, purification, crystallization and preliminary X-ray analysis of a bacterial GABA receptor with a Venus flytrap fold. *Acta Crystallogr. Sect. F Struct. Biol. Cryst. Commun.* **64:**1153–1155.

Morici, L. A., A. J. Carterson, V. E. Wagner, A. Frisk, J. R. Schurr, K. Honer zu Bentrup, D. J. Hassett, B. H. Iglewski, K. Sauer, and M. J. Schurr. 2007. *Pseudomonas aeruginosa* AlgR represses the Rhl quorum-sensing system in a biofilm-specific manner. *J. Bacteriol.* **189:**7752–7764.

Moss, J. E., P. E. Fisher, B. Vick, E. A. Groisman, and A. Zychlinsky. 2000. The regulatory protein PhoP controls susceptibility to the host inflammatory response in Shigella flexneri. *Cell. Microbiol.* **2:**443–452.

Mountfort, D. O., and V. Pybus. 1992. Regulatory influences on the production of gamma-aminobutyric acid by a marine pseudomonad. *Appl. Environ. Microbiol.* **58:**237–242.

Nakka, S., M. Qi, and Y. Zhao. 2010. The Erwinia amylovora PhoPQ system is involved in resistance to antimicrobial peptide and suppresses gene expression of two novel type III secretion systems. *Microbiol. Res.* **165:**665–673.

Neoh, H. M., L. Cui, H. Yuzawa, F. Takeuchi, M. Matsuo, and K. Hiramatsu. 2008. Mutated response regulator *graR* is responsible for phenotypic conversion of *Staphylococcus aureus* from heterogeneous vancomycin-intermediate resistance to vancomycin-intermediate resistance. *Antimicrob. Agents Chemother.* **52:**45–53.

Neudeck, B. L., J. Loeb, and J. Buck. 2003. Activation of the kappa-opioid receptor in Caco-2 cells decreases interleukin-8 secretion. *Eur. J. Pharmacol.* **467:**81–84.

Neudeck, B. L., and J. M. Loeb. 2002. Endomorphin-1 alters interleukin-8 secretion in Caco-2 cells via a receptor mediated process. *Immunol. Lett.* **84:**217–221.

Owens, D. F., and A. R. Kriegstein. 2002. Is there more to GABA than synaptic inhibition? *Nat. Rev. Neurosci.* **3:**715–727.

Oyston, P. C., N. Dorrell, K. Williams, S. R. Li, M. Green, R. W. Titball, and B. W. Wren. 2000. The response regulator PhoP is important for survival under conditions of macrophage-induced stress and virulence in *Yersinia pestis. Infect. Immun.* **68:**3419–3425.

Pal, U., A. M. de Silva, R. R. Montgomery, D. Fish, J. Anguita, J. F. Anderson, Y. Lobet, and E. Fikrig. 2000. Attachment of Borrelia burgdorferi within Ixodes scapularis mediated by outer surface protein A. *J. Clin. Investig.* **106:**561–569.

Pappas, K. M., and S. C. Winans. 2003. A LuxR-type regulator from Agrobacterium tumefaciens elevates Ti plasmid copy number by activating transcription of plasmid replication genes. *Mol. Microbiol.* **48:**1059–1073.

Paris, I., P. Martinez-Alvarado, C. Perez-Pastene, M. N. Vieira, C. Olea-Azar, R. Raisman-Vozari, S. Cardenas, R. Graumann, P. Caviedes, and J. Segura-Aguilar. 2005. Monoamine transporter inhibitors and norepinephrine reduce dopamine-dependent iron toxicity in cells derived from the substantia nigra. *J. Neurochem.* **92:**1021–1032.

Patel, N. J., O. Zaborina, L. Wu, Y. Wang, D. J. Wolfgeher, V. Valuckaite, M. J. Ciancio, J. E. Kohler, O. Shevchenko, S. P. Colgan, E. B. Chang, J. R. Turner, and J. C. Alverdy. 2007. Recognition of intestinal epithelial HIF-1alpha activation by Pseudomonas aeruginosa. *Am. J. Physiol. Gastrointest. Liver Physiol.* **292:**G134–G142.

Pearson, J. P., L. Passador, B. H. Iglewski, and E. P. Greenberg. 1995. A second N-acylhomoserine lactone signal produced by *Pseudomonas aeruginosa. Proc. Natl. Acad. Sci. USA* **92:**1490–1494.

Peschel, A. 2002. How do bacteria resist human antimicrobial peptides? *Trends Microbiol.* **10:**179–186.

Peschel, A., R. W. Jack, M. Otto, L. V. Collins, P. Staubitz, G. Nicholson, H. Kalbacher, W. F. Nieuwenhuizen, G. Jung, A. Tarkowski, K. P. van Kessel, and J. A. van Strijp. 2001. *Staphylococcus aureus* resistance to human defensins and evasion of neutrophil killing via the novel virulence factor MprF is based on modification of membrane lipids with L-lysine. *J. Exp. Med.* **193:**1067–1076.

Pesci, E. C., J. B. Milbank, J. P. Pearson, S. McKnight, A. S. Kende, E. P. Greenberg, and B. H. Iglewski. 1999. Quinolone signaling in the cell-to-cell communication system of *Pseudomonas aeruginosa. Proc. Natl. Acad. Sci. USA* **96:**11229–11234.

Peterson, M. M., J. L. Mack, P. R. Hall, A. A. Alsup, S. M. Alexander, E. K. Sully, Y. S. Sawires, A. L. Cheung, M. Otto, and H. D. Gresham. 2008. Apolipoprotein B is an innate barrier against invasive Staphylococcus aureus infection. *Cell Host Microbe* **4:**555–566.

Peterson, P. K., C. C. Chao, T. Molitor, M. Murtaugh, F. Strgar, and B. M. Sharp. 1991. Stress and pathogenesis of infectious disease. *Rev. Infect. Dis.* **13:**710–720.

Peterson, P. K., T. W. Molitor, and C. C. Chao. 1998. The opioid-cytokine connection. *J. Neuroimmunol.* **83:**63–69.

Piper, K. R., S. Beck von Bodman, and S. K. Farrand. 1993. Conjugation factor of *Agrobacterium tumefaciens* regulates Ti plasmid transfer by autoinduction. *Nature* **362:**448–450.

Piper, K. R., S. Beck Von Bodman, I. Hwang, and S. K. Farrand. 1999. Hierarchical gene regulatory systems arising from fortuitous gene associations: controlling quorum sensing by the opine regulon in Agrobacterium. *Mol. Microbiol.* **32:**1077–1089.

Porat, R., B. D. Clark, S. M. Wolff, and C. A. Dinarello. 1991. Enhancement of growth of virulent strains of *Escherichia coli* by interleukin-1. *Science* **254:**430–432.

Potter, L. R., A. R. Yoder, D. R. Flora, L. K. Antos, and D. M. Dickey. 2009. Natriuretic peptides: their structures, receptors, physiologic functions and therapeutic applications. *Handb. Exp. Pharmacol.* **2009(191):**341–366.

Prost, L. R., M. E. Daley, M. W. Bader, R. E. Klevit, and S. I. Miller. 2008. The PhoQ histidine kinases of Salmonella and Pseudomonas spp. are structurally and functionally different: evidence that pH and antimicrobial peptide sensing contribute to mammalian pathogenesis. *Mol. Microbiol.* **69:**503–519.

Prost, L. R., M. E. Daley, V. Le Sage, M. W. Bader, H. Le Moual, R. E. Klevit, and S. I. Miller. 2007. Activation of the bacterial sensor kinase PhoQ by acidic pH. *Mol. Cell* **26:**165–174.

Provenzano, D., D. A. Schuhmacher, J. L. Barker, and K. E. Klose. 2000. The virulence regulatory protein ToxR mediates enhanced bile resistance in *Vibrio cholerae* and other pathogenic *Vibrio* species. *Infect. Immun.* **68:**1491–1497.

Rashid, M. H., C. Rajanna, D. Zhang, V. Pasquale, L. S. Magder, A. Ali, S. Dumontet, and D. K. Karaolis. 2004. Role of exopolysaccharide, the rugose phenotype and VpsR in the pathogenesis of epidemic Vibrio cholerae. *FEMS Microbiol. Lett.* **230:**105–113.

Rasko, D. A., C. G. Moreira, R. Li de, N. C. Reading, J. M. Ritchie, M. K. Waldor, N. Williams, R. Taussig, S. Wei, M. Roth, D. T. Hughes, J. F. Huntley, M. W. Fina, J. R. Falck, and V. Sperandio. 2008. Targeting QseC signaling and virulence for antibiotic development. *Science* **321:**1078–1080.

Ratledge, C., and L. G. Dover. 2000. Iron metabolism in pathogenic bacteria. *Annu. Rev. Microbiol.* **54:**881–941.

Reading, N. C., D. A. Rasko, A. G. Torres, and V. Sperandio. 2009. The two-component system QseEF and the membrane protein QseG link adrenergic and stress sensing to bacterial pathogenesis. *Proc. Natl. Acad. Sci. USA* **106:**5889–5894.

Reading, N. C., A. G. Torres, M. M. Kendall, D. T. Hughes, K. Yamamoto, and V. Sperandio. 2007. A novel two-component signaling system that activates transcription of an enterohemorrhagic *Escherichia coli* effector involved in remodeling of host actin. *J. Bacteriol.* **189:**2468–2476.

Reichenbach, B., Y. Gopel, and B. Gorke. 2009. Dual control by perfectly overlapping sigma 54- and sigma 70- promoters adjusts small RNA GlmY expression to different environmental signals. *Mol. Microbiol.* **74:**1054–1070.

Ritchie, A. J., A. Jansson, J. Stallberg, P. Nilsson, P. Lysaght, and M. A. Cooley. 2005. The *Pseudomonas aeruginosa* quorum-sensing molecule N-3-(oxododecanoyl)-L-homoserine lactone inhibits T-cell differentiation and cytokine production by a mechanism involving an early step in T-cell activation. *Infect. Immun.* **73:**1648–1655.

Ritchie, A. J., A. O. Yam, K. M. Tanabe, S. A. Rice, and M. A. Cooley. 2003. Modification of in vivo and in vitro T- and B-cell-mediated immune responses by the *Pseudomonas aeruginosa* quorum-sensing molecule N-(3-oxododecanoyl)-L-homoserine lactone. *Infect. Immun.* **71:**4421–4431.

Roberts, M. R. 2007. Does GABA act as a signal in plants? Hints from molecular studies. *Plant Signal. Behav.* **2:**408–409.

Sansonetti, P. J., and C. Egile. 1998. Molecular bases of epithelial cell invasion by Shigella flexneri. *Antonie Van Leeuwenhoek* **74:**191–197.

Scheckelhoff, M. R., S. R. Telford, M. Wesley, and L. T. Hu. 2007. *Borrelia burgdorferi* intercepts host hormonal signals to regulate expression of outer surface protein A. *Proc. Natl. Acad. Sci. USA* **104:**7247–7252.

Schlievert, P. M., L. C. Case, K. A. Nemeth, C. C. Davis, Y. Sun, W. Qin, F. Wang, A. J. Brosnahan, J. A. Mleziva, M. L. Peterson, and B. E. Jones. 2007. Alpha and beta chains of hemoglobin inhibit production of Staphylococcus aureus exotoxins. *Biochemistry* **46:**14349–14358.

Schuhmacher, D. A., and K. E. Klose. 1999. Environmental signals modulate ToxT-dependent virulence factor expression in *Vibrio cholerae. J. Bacteriol.* **181:**1508–1514.

Schuster, M., and E. P. Greenberg. 2006. A network of networks: quorum-sensing gene regulation in Pseudomonas aeruginosa. *Int. J. Med. Microbiol.* **296:**73–81.

Seed, P. C., L. Passador, and B. H. Iglewski. 1995. Activation of the *Pseudomonas aeruginosa lasI* gene by LasR and the *Pseudomonas* autoinducer PAI: an autoinduction regulatory hierarchy. *J. Bacteriol.* **177:**654–659.

Shadel, G. S., J. H. Devine, and T. O. Baldwin. 1990. Control of the lux regulon of Vibrio fischeri. *J. Biolumin. Chemilumin.* **5:**99–106.

Shelp, B. J., A. W. Bown, and M. D. McLean. 1999. Metabolism and functions of gamma-aminobutyric acid. *Trends Plant Sci.* **4:**446–452.

Shiner, E. K., D. Terentyev, A. Bryan, S. Sennoune, R. Martinez-Zaguilan, G. Li, S. Gyorke, S. C. Williams, and K. P. Rumbaugh. 2006. Pseudomonas aeruginosa autoinducer modulates host cell responses through calcium signalling. *Cell. Microbiol.* **8:**1601–1610.

Simpson, A. J., and V. Wuthiekanun. 2000. Interaction of insulin with Burkholderia pseudomallei may be caused by a preservative. *J. Clin. Pathol.* **53:**159–160.

Siraki, A. G., J. Smythies, and P. J. O'Brien. 2000. Superoxide radical scavenging and attenuation of hypoxia-reoxygenation injury by neurotransmitter ferric complexes in isolated rat hepatocytes. *Neurosci. Lett.* **296:**37–40.

Skaar, E. P., M. Humayun, T. Bae, K. L. DeBord, and O. Schneewind. 2004. Iron-source preference of *Staphylococcus aureus* infections. *Science* **305:**1626–1628.

Smith, R. S., E. R. Fedyk, T. A. Springer, N. Mukaida, B. H. Iglewski, and R. P. Phipps. 2001. IL-8 production in human lung fibroblasts and epithelial cells activated by the Pseudomonas autoinducer N-3-oxododecanoyl homoserine lactone is transcriptionally regulated by NF-kappa B and activator protein-2. *J. Immunol.* **167:**366374.

Smith, R. S., S. G. Harris, R. Phipps, and B. Iglewski. 2002. The *Pseudomonas aeruginosa* quorum-sensing molecule N-(3-oxododecanoyl)homoserine lactone contributes to virulence and induces inflammation in vivo. *J. Bacteriol.* **184:**1132–1139.

Smith, R. S., and B. H. Iglewski. 2003. P. aeruginosa quorum-sensing systems and virulence. *Curr. Opin. Microbiol.* **6:**56–60.

Smith, R. S., R. Kelly, B. H. Iglewski, and R. P. Phipps. 2002. The Pseudomonas autoinducer N-(3-oxododecanoyl) homoserine lactone induces cyclooxygenase-2 and prostaglandin E2 production in human lung fibroblasts: implications for inflammation. *J. Immunol.* **169:**2636–2642.

Smith, R. S., M. C. Wolfgang, and S. Lory. 2004. An adenylate cyclase-controlled signaling network regulates *Pseudomonas aeruginosa* virulence in a mouse model of acute pneumonia. *Infect. Immun.* **72:**1677–1684.

Sperandio, V. 2004. Striking a balance: inter-kingdom cell-to-cell signaling, friendship or war? *Trends Immunol.* **25:**505–507.

Sperandio, V., J. L. Mellies, W. Nguyen, S. Shin, and J. B. Kaper. 1999. Quorum sensing controls expression of the type III secretion gene transcription and protein secretion in enterohemorrhagic and enteropathogenic *Escherichia coli. Proc. Natl. Acad. Sci. USA* **96:**15196–15201.

Sperandio, V., A. G. Torres, J. A. Giron, and J. B. Kaper. 2001. Quorum sensing is a global regulatory mechanism in enterohemorrhagic *Escherichia coli* O157:H7. *J. Bacteriol.* **183:**5187–5197.

Sperandio, V., A. G. Torres, B. Jarvis, J. P. Nataro, and J. B. Kaper. 2003. Bacteria-host communication: the language of hormones. *Proc. Natl. Acad. Sci. USA* **100:**8951–8956.

Sperandio, V., A. G. Torres, and J. B. Kaper. 2002. Quorum sensing Escherichia coli regulators B and C (QseBC): a novel two-component regulatory system involved in the regulation of flagella and motility by quorum sensing in E. coli. *Mol. Microbiol.* **43:**809–821.

Stachel, S. E., and P. C. Zambryski. 1986. virA and virG control the plant-induced activation of the T-DNA transfer process of A. tumefaciens. *Cell* **46:**325–333.

Staley, T. E., E. W. Jones, and L. D. Corley. 1969. Attachment and penetration of Escherichia coli into intestinal epithelium of the ileum in newborn pigs. *Am. J. Pathol.* **56:**371–392.

Sternini, C., S. Patierno, I. S. Selmer, and A. Kirchgessner. 2004. The opioid system in the gastrointestinal tract. *Neurogastroenterol. Motil.* **16**(Suppl 2):3–16.

Stoltz, D. A., E. A. Ozer, C. J. Ng, J. M. Yu, S. T. Reddy, A. J. Lusis, N. Bourquard, M. R. Parsek, J. Zabner, and D. M. Shih. 2007. Paraoxonase-2 deficiency enhances Pseudomonas aeruginosa quorum sensing in murine tracheal epithelia. *Am. J. Physiol. Lung Cell. Mol. Physiol.* **292:**L852–L860.

Tateda, K., Y. Ishii, M. Horikawa, T. Matsumoto, S. Miyairi, J. C. Pechere, T. J. Standiford, M. Ishiguro, and K. Yamaguchi. 2003. The *Pseudomonas aeruginosa* autoinducer N-3-oxododecanoyl homoserine lactone accelerates apoptosis in macrophages and neutrophils. *Infect. Immun.* **71:**5785–5793.

Telford, G., D. Wheeler, P. Williams, P. T. Tomkins, P. Appleby, H. Sewell, G. S. Stewart, B. W. Bycroft, and D. I. Pritchard. 1998. The *Pseudomonas aeruginosa* quorum-sensing signal molecule N-(3-oxododecanoyl)-L-homoserine lactone has immunomodulatory activity. *Infect. Immun.* **66:**36–42.

Thedieck, K., T. Hain, W. Mohamed, B. J. Tindall, M. Nimtz, T. Chakraborty, J. Wehland, and L. Jansch. 2006. The MprF protein is required for lysinylation of phospholipids in listerial membranes and confers resistance to cationic antimicrobial peptides (CAMPs) on Listeria monocytogenes. *Mol. Microbiol.* **62:**1325–1339.

Torres, V. J., D. L. Stauff, G. Pishchany, J. S. Bezbradica, L. E. Gordy, J. Iturregui, K. L. Anderson, P. M. Dunman, S. Joyce, and E. P. Skaar. 2007. A Staphylococcus aureus regulatory system that responds to host heme and modulates virulence. *Cell Host Microbe* **1:**109–119.

Tsavkelova, E. A., I. V. Botvinko, V. S. Kudrin, and A. V. Oleskin. 2000. Detection of neurotransmitter amines in microorganisms with the use of high-performance liquid chromatography. *Dokl. Biochem.* **372:**115–117.

Vallejo, R., O. de Leon-Casasola, and R. Benyamin. 2004. Opioid therapy and immunosuppression: a review. *Am. J. Ther.* **11:**354–365.

Van Gennip, M., L. D. Christensen, M. Alhede, R. Phipps, P. O. Jensen, L. Christophersen, S. J. Pamp, C. Moser, P. J. Mikkelsen, A. Y. Koh, T. Tolker-Nielsen, G. B. Pier, N. Hoiby, M. Givskov, and T. Bjarnsholt. 2009. Inactivation of the rhlA gene in Pseudomonas aeruginosa prevents rhamnolipid production, disabling the protection against polymorphonuclear leukocytes. *APMIS* **117:**537–546.

Veron, W., O. Lesouhaitier, X. Pennanec, K. Rehel, P. Leroux, N. Orange, and M. G. Feuilloley. 2007. Natriuretic peptides affect Pseudomonas aeruginosa and specifically modify lipopolysaccharide biosynthesis. *FEBS J.* **274:**5852–5864.

Volmer, J. B., L. F. Thompson, and M. R. Blackburn. 2006. Ecto-5′-nucleotidase (CD73)-mediated adenosine production is tissue protective in a model of bleomycin-induced lung injury. *J. Immunol.* **176:**4449–4458.

Vuckovic, D., M. Abram, M. Bubonja, B. Wraber, and M. Doric. 2006. Host resistance to primary and secondary Campylobacter jejuni infections in C57Bl/6 mice. *Microb. Pathog.* **40:**35–39.

Vuddhakul, V., P. Tharavichitkul, N. Na-Ngam, S. Jitsurong, B. Kunthawa, P. Noimay, A. Binla, and V. Thamlikitkul. 1999. Epidemiology of Burkholderia pseudomallei in Thailand. *Am. J. Trop. Med. Hyg.* **60:**458–461.

Vuopio-Varkila, J., and G. K. Schoolnik. 1991. Localized adherence by enteropathogenic Escherichia coli is an inducible phenotype associated with the expression of new outer membrane proteins. *J. Exp. Med.* **174:**1167–1177.

Wade, D. S., M. W. Calfee, E. R. Rocha, E. A. Ling, E. Engstrom, J. P. Coleman, and E. C. Pesci. 2005. Regulation of *Pseudomonas* quinolone signal synthesis in *Pseudomonas aeruginosa. J. Bacteriol.* **187:**4372–4380.

Watnick, P. I., and R. Kolter. 1999. Steps in the development of a Vibrio cholerae El Tor biofilm. *Mol. Microbiol.* **34:**586–595.

Weiss, D. S., A. Brotcke, T. Henry, J. J. Margolis, K. Chan, and D. M. Monack. 2007. In vivo negative selection screen identifies genes required for *Francisella* virulence. *Proc. Natl. Acad. Sci. USA* **104:**6037–6042.

Wertheim, H. F., M. C. Vos, A. Ott, A. van Belkum, A. Voss, J. A. Kluytmans, P. H. van Keulen, C. M. Vandenbroucke-Grauls, M. H. Meester, and H. A. Verbrugh. 2004. Risk and outcome of nosocomial *Staphylococcus aureus* bacteraemia in nasal carriers versus non-carriers. *Lancet* **364:**703–705.

Woods, D. E., A. L. Jones, and P. J. Hill. 1993. Interaction of insulin with *Pseudomonas pseudomallei. Infect. Immun.* **61:**4045–4050.

Wretlind, B., A. Bjorklind, and O. R. Pavlovskis. 1987. Role of exotoxin A and elastase in the pathogenicity of Pseudomonas aeruginosa strain PAO experimental mouse burn infection. *Microb. Pathog.* **2:**397–404.

Wu, L., O. Estrada, O. Zaborina, M. Bains, L. Shen, J. E. Kohler, N. Patel, M. W. Musch, E. B. Chang, Y. X. Fu, M. A. Jacobs, M. I. Nishimura, R. E. Hancock, J. R. Turner, and J. C. Alverdy. 2005. Recognition of host immune activation by *Pseudomonas aeruginosa. Science* **309:**774–777.

Wu, L., C. Holbrook, O. Zaborina, E. Ploplys, F. Rocha, D. Pelham, E. Chang, M. Musch, and J. Alverdy. 2003. Pseudomonas aeruginosa expresses a lethal virulence determinant, the PA-I lectin/

adhesin, in the intestinal tract of a stressed host: the role of epithelia cell contact and molecules of the quorum sensing signaling system. *Ann. Surg.* **238:**754–764.

Wu, L. R., O. Zaborina, A. Zaborin, E. B. Chang, M. Musch, C. Holbrook, J. R. Turner, and J. C. Alverdy. 2005. Surgical injury and metabolic stress enhance the virulence of the human opportunistic pathogen Pseudomonas aeruginosa. *Surg. Infect. (Larchmt)* **6:**185–195.

Yamamoto, K., K. Hirao, T. Oshima, H. Aiba, R. Utsumi, and A. Ishihama. 2005. Functional characterization in vitro of all two-component signal transduction systems from Escherichia coli. *J. Biol. Chem.* **280:**1448–1456.

Yamashita, K., H. Kaneko, S. Yamamoto, T. Konagaya, K. Kusugami, and T. Mitsuma. 1998. Inhibitory effect of somatostatin on Helicobacter pylori proliferation in vitro. *Gastroenterology* **115:**1123–1130.

Yang, F., L. H. Wang, J. Wang, Y. H. Dong, J. Y. Hu, and L. H. Zhang. 2005. Quorum quenching enzyme activity is widely conserved in the sera of mammalian species. *FEBS Lett.* **579:**3713–3717.

Yang, X. F., U. Pal, S. M. Alani, E. Fikrig, and M. V. Norgard. 2004. Essential role for OspA/B in the life cycle of the Lyme disease spirochete. *J. Exp. Med.* **199:**641–648.

Yarwood, J. M., J. K. McCormick, M. L. Paustian, V. Kapur, and P. M. Schlievert. 2002. Repression of the *Staphylococcus aureus* accessory gene regulator in serum and in vivo. *J. Bacteriol.* **184:**1095–1101.

Yarwood, J. M., J. K. McCormick, and P. M. Schlievert. 2001. Identification of a novel two-component regulatory system that acts in global regulation of virulence factors of *Staphylococcus aureus*. *J. Bacteriol.* **183:**1113–1123.

Yildiz, F. H., N. A. Dolganov, and G. K. Schoolnik. 2001. VpsR, a member of the response regulators of the two-component regulatory systems, is required for expression of *vps* biosynthesis genes and EPS(ETr)-associated phenotypes in *Vibrio cholerae* O1 El Tor. *J. Bacteriol.* **183:**1716–1726.

Zaborina, O., F. Lepine, G. Xiao, V. Valuckaite, Y. Chen, T. Li, M. Ciancio, A. Zaborin, E. O. Petrof, J. R. Turner, L. G. Rahme, E. Chang, and J. C. Alverdy. 2007. Dynorphin activates quorum sensing quinolone signaling in Pseudomonas aeruginosa. *PLoS Pathog.* **3:**e35.

Zav'yalov, V. P., T. V. Chernovskaya, E. V. Navolotskaya, A. V. Karlyshev, S. MacIntyre, A. M. Vasiliev, and V. M. Abramov. 1995. Specific high affinity binding of human interleukin 1 beta by Caf1A usher protein of Yersinia pestis. *FEBS Lett.* **371:**65–68.

Zhang, H. B., L. H. Wang, and L. H. Zhang. 2002. Genetic control of quorum-sensing signal turnover in *Agrobacterium tumefaciens*. *Proc. Natl. Acad. Sci. USA* **99:**4638–4643.

Zhang, L., P. J. Murphy, A. Kerr, and M. E. Tate. 1993. *Agrobacterium* conjugation and gene regulation by N-acyl-L-homoserine lactones. *Nature* **362:**446–448.

Zhou, Y., A. Mohsenin, E. Morschl, H. W. Young, J. G. Molina, W. Ma, C. X. Sun, H. Martinez-Valdez, and M. R. Blackburn. 2009. Enhanced airway inflammation and remodeling in adenosine deaminase-deficient mice lacking the A2B adenosine receptor. *J. Immunol.* **182:**8037–8046.

Zimmermann, S., C. Wagner, W. Muller, G. Brenner-Weiss, F. Hug, B. Prior, U. Obst, and G. M. Hansch. 2006. Induction of neutrophil chemotaxis by the quorum-sensing molecule N-(3-oxododecanoyl)-L-homoserine lactone. *Infect. Immun.* **74:**5687–5692.

Regulation of Bacterial Virulence
Edited by Michael L. Vasil and Andrew J. Darwin
© 2013 ASM Press, Washington, DC   doi:10.1128/9781555818524.ch28

Chapter 28

Regulating the Transition of *Vibrio cholerae* Out of the Host

EMILYKATE MCDONOUGH, EVAN BRADLEY, AND ANDREW CAMILLI

INTRODUCTION

Importance of Waterborne Microbial Pathogens

Waterborne microbial pathogens reside in virtually every aquatic environment on Earth, including fresh, brackish, and marine waters. These pathogens, which include various viruses, bacteria, and protozoa, contribute to significant morbidity and mortality worldwide (Geldreich, 1996). When ingested, most waterborne pathogens cause some type of gastrointestinal disease ranging from mild enteritis to explosive, secretory (watery) diarrhea (Sharma et al., 2003; Skovgaard, 2007). The risk of disease from many of these pathogens is much higher in areas where the infrastructure does not allow for proper sanitation and wastewater treatment. Unfortunately, the impact of disease in the developing world is compounded by the high financial burden of prophylactic and curative treatment plans for many diseases (Geldreich, 1996).

Waterborne pathogens also present a threat in the developed world. In rural areas, well water can be contaminated from sewage runoff, and in urban areas, wastewater treatment plants are less than 100% effective in removing certain waterborne pathogens. For example, *Cryptosporidium* oocysts are able to survive the microbicidal levels of chlorine used in water purification (Sharma et al., 2003; Skovgaard, 2007). Enteric viruses, including rotavirus, are also problematic for treatment facilities, as the physical and chemical filtration systems remove only 50% to 90% of the viruses present in sewage, which is not enough to fully prevent contamination of environmental water sources (Okoh, et al., 2010). Numerous bacterial pathogens, including *Salmonella* spp., *Shigella* spp., and enterohemorrhagic *Escherichia coli*, can also be isolated in the treated effluent of wastewater facilities. Additionally, street level sewage drainage pipes and wastewater treatment plants may be overwhelmed by an abundance of rain or flooding, resulting in contaminated drinking and recreational waters (Geldreich, 1996).

Role of the Aquatic Reservoir in Transmission of Waterborne Pathogens

Marine, brackish, and fresh waters all present different challenges to the survival and persistence of waterborne pathogens, including, but not limited to, macronutrient and micronutrient deprivation, predation, ultraviolet (UV) light, and fluctuations in pH, temperature, dissolved-oxygen concentration, and osmolarity (Table 1) (Alam et al., 2007; Brookes et al., 2004; Freter et al., 1961; Kamal et al., 2007; Lara et al., 2009; Nelson et al., 2008; Schild et al., 2007). In spite of these challenges, many pathogens have evolved to spend some portion of their life cycle in an aquatic environment. There are several benefits of being a waterborne pathogen. As water is essential to life, virtually all hosts of waterborne pathogens come into contact with contaminated water at some frequency. Additionally, water can provide a means to spread pathogens from one region to another, either with the current or via migrating birds, zooplankton, phytoplankton, or insects that pathogens may associate with (Brookes et al., 2004; Geldreich, 1996; Lipp et al., 2002).

Although numerous organisms have been described as waterborne pathogens, the precise role of the aquatic environment in transmission of many of these organisms is unclear. What is known about transmission of waterborne pathogens suggests that a wide variety of strategies for spread and persistence are used by these pathogens. For some, such as the protozoan *Cryptosporidium parvum* and many enteric viruses, the aquatic environment may simply provide a place to quiescently persist until a new host is encountered (Okoh et al., 2010; Sharma et al., 2003; Skovgaard, 2007). In contrast, *Legionella* spp.

EmilyKate McDonough, Evan Bradley, and Andrew Camilli • Department of Molecular Biology & Microbiology and Howard Hughes Medical Institute, Tufts University School of Medicine, Boston, MA 02111.

Table 1. Comparison of physiochemical parameters of typical environments for *V. cholerae*

Parameter	Result for:		
	Rice water stool (RWS)	Urban pond (Dhaka)	Estuary
Salinity (ppt)	3.4 (Nelson et al., 2008)	0.1 (Nelson et al., 2008)	0.1–19 (Lara et al., 2009)
pH	8.3–7 (Freter et al., 1961)	6.6–6.9 (Nelson et al., 2008)	8.0 (Alam et al., 2007)
Temp	37°C	26–28°C (Nelson et al., 2008)	27.2–31.4°C (Lara et al., 2009)
Fixed nitrogen	938 μM NH_4^+ (Nelson et al., 2008)	59.8 μM NO_2^-, 1.8 mM NH_4^+ (Schild et al., 2007)	8–40 μM NO_3^- (Lara et al., 2009)
Phosphate (PO_4^-)	15.2 mM (Nelson et al., 2008)	19 μM (Schild et al., 2007)	3.8 mM (Kamal et al., 2007)
Redox status	Anaerobic[a] (Freter et al., 1961; Merrell et al., 2002)	Relatively high oxygen: 6% dissolved oxygen (Nelson et al., 2008)	Low oxygen: $1 \times 10^{-4}\% \times 10^{-4}\%$ (Kamal et al., 2007)

[a]By direct measurement of redox potential and inference from gene expression data.

and some members of the *Vibrio* genus are natural inhabitants of the aquatic environment, in either free-living or host-associated states. Some of these native inhabitants cause only incidental infection of humans (e.g., *L. pneumophila* and *V. parahaemolyticus*), while others are facultative human pathogens (*V. cholerae*) (Grimes, 1991). Additionally, water contamination has been implicated in the transmission of numerous pathogens, although this may not be their dominant mode of transmission. These include enterohemorrhagic *E. coli*, *Campylobacter jejuni*, *Yersinia enterocolitica*, *Helicobacter pylori*, *Pseudomonas aeruginosa*, and *Aeromonas* spp. (Grimes, 1991; Sharma et al., 2003; Skovgaard, 2007).

V. cholerae as a Model Waterborne Pathogen

Many members of the halophilic *Vibrio* genus have been reported to cause disease in humans. These species are natural members of aquatic environments around the world and thus pose a significant threat to many populations. One member of this genus, *V. cholerae*, is the causative agent of the severe diarrheal disease Asiatic cholera and is a model for waterborne facultative pathogens. Although *Vibrio* species are found within marine or brackish waters, *V. cholerae* is unique in that it can also persist in freshwater environments, a trait that is central to its epidemic potential (Faruque et al., 1998a; Reidl and Klose, 2002).

V. cholerae strains are divided into over 200 serogroups on the basis of their lipopolysaccharide O antigen. While numerous serogroups have contributed to localized outbreaks of cholera, only two serogroups have been reported to cause epidemics: O1 and O139. Within these serogroups, only strains containing genes encoding the major virulence factors, cholera toxin (CT) and the toxin-coregulated pilus (TCP), are considered toxigenic and of epidemic potential (Faruque et al., 1998a, 2011; Reidl and Klose, 2002; Safa et al., 2010). These toxigenic

strains are lysogens of the filamentous bacteriophage CTXΦ, which harbors at least one copy of the CT-encoding genes (*ctxAB*). CTXΦ has been demonstrated to be fully mobilizable and able to lysogenize new strains (Waldor and Mekalanos, 1996). This ability of *ctxAB* to be transferred via CTXΦ infection may contribute to the emergence of new toxigenic *V. cholerae* strains.

The O1 serogroup of *V. cholerae* can be further divided into two biotypes: classical and El Tor. The world is currently experiencing the seventh cholera pandemic since 1817. Despite a lack of data for the first five pandemics, *V. cholerae* O1 of the classical biotype is considered the major biotype responsible for the first six pandemics. However, the seventh pandemic, which began in 1961, has been dominated by *V. cholerae* O1 of the El Tor biotype. It has been suggested, but not conclusively shown, that one evolutionary pressure contributing to the switch from classical to El Tor as the major pathogenic strain may be that El Tor is better able to survive within the aquatic environment (Faruque et al., 1998a, 2011; Reidl and Klose, 2002). In addition to harboring numerous genotypic differences, the classical and El Tor biotypes carry different CTXΦ genomes, annotated as CTXΦ[Cla] and CTXΦ[El], respectively. Among other differences, these two bacteriophages encode different CT; CTXΦ[Cla]-encoded CT induces more severe diarrhea than CTXΦ[El]-encoded CT. Since 1992, the classical biotype has not been isolated from environmental samples and therefore is considered extinct. However, there have been recent reports of altered or atypical El Tor strains that harbor either CTXΦ[Cla] or a hybrid of CTXΦ[Cla] and CTXΦ[El]. It is unclear where the CTXΦ[Cla] was acquired by these atypical El Tor strains (Safa et al., 2010). The combination of El Tor environmental fitness and the high toxicity of the CTXΦ[Cla]-encoded CT may provide a significant survival advantage for these atypical El Tor variants in both the host and the aquatic environment. It will be

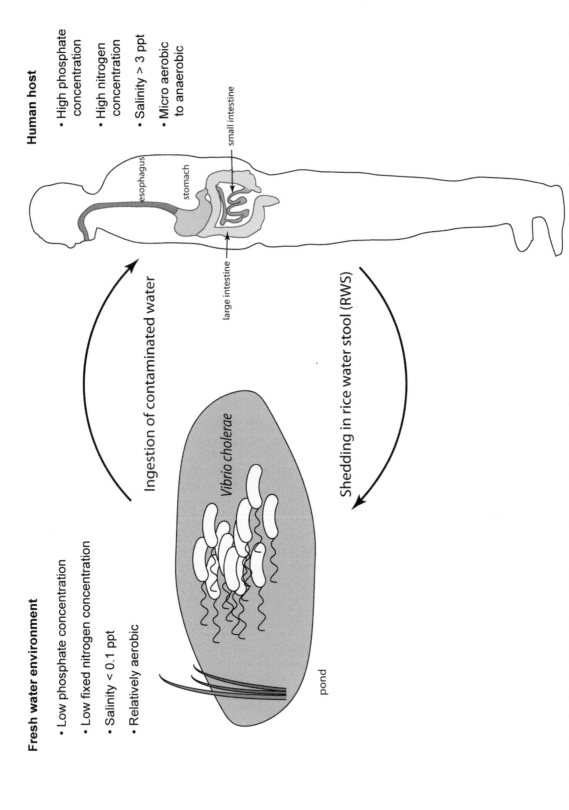

Fresh water environment

- Low phosphate concentration
- Low fixed nitrogen concentration
- Salinity < 0.1 ppt
- Relatively aerobic

Vibrio cholerae

pond

Ingestion of contaminated water

Shedding in rice water stool (RWS)

Human host

- High phosphate concentration
- High nitrogen concentration
- Salinity > 3 ppt
- Micro aerobic to anaerobic

esophagus

stomach

small intestine

large intestine

Figure 1. Simple life cycle of toxigenic *V. cholerae*. *V. cholerae* cells present in contaminated freshwater sources are ingested and cause cholera in humans. The bacteria replicate in the small intestine and are shed back into the environment in rice water stool (RWS). The shed bacteria return to the aquatic environment to initiate another infection cycle. Key chemical differences between these environments are listed and measurements of the typical chemical composition of these environments are presented in Table 1. doi:10.1128/9781555818528.ch24f1

interesting to see if these strains become the dominant epidemic-causing *V. cholerae*.

While *V. cholerae* O1 remains the dominant serogroup causing disease, a new serogroup, *V. cholerae* O139, emerged in India and Bangladesh in 1992 and has been contributing to disease since then (Faruque et al., 1998a, 2011; Faruque and Nair, 2002; Reidl and Klose, 2002). Interestingly, the O139 strain is closely related to O1 El Tor and may have emerged via replacement of the O1-antigen locus with newly transferred O139 O-antigen genes. As with *V. cholerae* O1, some O139 strains are toxigenic and others are not (Faruque and Nair, 2002). Currently, there appears to be an evolutionary battle between toxigenic *V. cholerae* O1 and O139 strains. Since 1992, numerous O139 genetic variants have been isolated, suggesting that this strain is evolving to compete with O1 strains for maintenance in the environmental niche (Faruque et al., 1999, 2000, 2011; Faruque and Nair, 2002). By monitoring the changes in the O139 genome, a clearer picture may arise of why certain strains dominate during an epidemic—information that may prove useful for combating cholera.

A toxigenic *V. cholerae* strain encounters two disparate environments during its life cycle: the human intestinal tract and the aquatic environment (Fig. 1; Table 1). During life in the environment, *V. cholerae* is often found in biofilms associated with phyto- and zooplankton but may also persist in a planktonic state (Lipp et al., 2002). The bacteria enter the human host via ingestion of contaminated water or food and subsequently colonize the small intestine (Peterson, 2002). Eventually, the bacteria are shed back into the environment in secretory diarrhea, termed rice water stool (RWS) due to its appearance. Back in the environment, *V. cholerae* organisms persist or, if conditions permit, multiply and spread via water currents until once again ingested by a host. During outbreaks, the mode of transmission may be dominated by rapid spread of the hyperinfectious form of *V. cholerae* that is shed in the RWS of cholera patients (discussed in more detail below) (Hartley et al., 2006; Merrell et al., 2002). Although direct spread of cholera within households occurs with high frequency (Kendall et al., 2010; Weil et al., 2009), larger outbreaks and epidemics occur via ingestion of contaminated freshwater used for drinking, food preparation, etc.

To mediate the transition from one environment to another, toxigenic *V. cholerae* undergoes adaptive shifts in gene expression throughout the various stages of its life cycle (see, e.g., Childers and Klose, 2007; DiRita, 1992; Merrell et al., 2002; Nielsen et al., 2006, 2010; and Schild, et al., 2007). Properly mediating the transition between host and environment is of particular importance to toxigenic *V.*

cholerae, as it encounters numerous physical and biological stresses during these transitions. Interestingly, there is evidence that *V. cholerae* has evolved to preemptively prepare itself for this transition by turning off virulence genes and turning on environmental survival genes while still in the human host (Schild et al., 2007). This chapter focuses heavily on the genetic changes the bacterium undergoes as it transitions out of the host and into the aquatic environment.

OVERVIEW OF THE VIRULENCE PROGRAM

To better understand the changes in *V. cholerae* gene expression that occur during the transition out of the host, we must first discuss the state of gene expression prior to this transition. This prior, acute stage of infection within the small intestine is when *V. cholerae* multiplies on the epithelium and expresses virulence factors.

Primary Virulence Genes and Their Regulation

For a toxigenic *V. cholerae* strain to colonize the small intestine and cause disease, it requires expression of TCP and CT (Faruque et al., 2011). TCP is a type IV bundle-forming pilus initially expressed upon entry into the host (Lee et al., 1999), and is absolutely essential for *V. cholerae* colonization of the small intestine (Herrington et al., 1988). While the precise mechanism by which TCP contributes to colonization is unknown, the pilus is thought to allow for interactions between bacterial cells rather than bacterium-host interactions (Childers and Klose, 2007). CT, an enterotoxin and a member of the A-B subunit type ADP-ribosylating toxins, is the virulence factor responsible for the profuse secretory diarrhea associated with cholera (Childers and Klose, 2007; Faruque et al., 2011; Herrington et al., 1988; Peterson, 2002).

Some of the *V. cholerae* virulence genes are present in regions of the *V. cholerae* genome that are potentially horizontally transferable (Childers and Klose, 2007; Reidl and Klose, 2002). The *tcpA-F* operon, which encodes TCP, is located within *Vibrio* pathogenicity island 1 (VPI-1), which also contains genes for the accessory colonization factor (Acf) and critical virulence gene regulators (TcpPH and ToxT) (Childers and Klose, 2007). Although VPI-1 was initially shown to be horizontally transferrable (Karaolis et al., 1999), this phenotype was not reproducible (Faruque et al., 2003). Nevertheless, VPI-1 and other virulence gene loci should be subject to low-frequency transfer by natural transformation, a process recently shown to occur in *V. cholerae* during biofilm growth on chitinous surfaces (Blokesch and Schoolnik, 2008; Meibom et al., 2005). As mentioned above, the genes encoding CT (*ctxAB*) are present in the CTXΦ prophage, which is associated with all known pandemic-causing strains (Reidl and Klose, 2002). Horizontal transfer of CTXΦ between strains during coinfection of the small intestine has been observed in experimentally infected animals (Waldor and Mekalanos, 1996) and is likely to occur with some frequency within humans, who can be simultaneously infected with multiple toxigenic *V. cholerae* strains (Kendall et al., 2010). TCP is the receptor for CTXΦ (Childers and Klose, 2007), suggesting that the CTXΦ is most likely acquired by a strain only after acquisition of VPI-1. VPI-1 transfer via natural transformation in the environment, followed by CTXΦ transfer via phage particles during human infection, may contribute to the emergence of new, virulent, toxigenic strains. This hypothesis is supported by the frequent emergence of new toxigenic variants (Faruque et al., 1998b, 1999, 2000, 2011; Faruque and Nair, 2002; Stine et al., 2008).

The regulation of *ctxAB* and *tcpA-F* expression has been studied extensively with respect to the factors that induce and repress transcription, though not exhaustively with respect to the extracellular activating signals. Below is a brief discussion of virulence regulation in *V. cholerae*. For an in-depth discussion of this regulation, see Chapters 12 and 26 of this book.

The major virulence regulator, ToxT, is responsible for inducing the expression of *ctxAB* and *tcpA-F*. The gene encoding ToxT is located within the VPI-1, directly downstream of the *tcpA-F* operon. Expression of *toxT* is under tight regulation by two different membrane-bound transcription factor complexes: TcpP/H (within the VPI-1) and ToxR/S (encoded outside of the VPI-1 and CTXΦ). Although the precise cues which activate signaling through TcpP/H or ToxR/S are yet to be elucidated, it is clear that *toxT* expression is induced through these regulators upon transition into a host and may be affected by changes in bile concentration and/or temperature. In support of this, ToxR/S has been shown to also regulate the expression of *V. cholerae* outer membrane porins in response to bile salts (Childers and Klose, 2007).

Quorum Sensing and the Regulation of Virulence

One pathway that helps orchestrate virulence gene expression is the quorum sensing system, which is the mechanism through which bacteria measure population density. Quorum sensing plays a large role in controlling gene expression and thus the behavior of *V. cholerae* throughout its life cycle. An overview of this system and how it relates to virulence gene expression is shown in Fig. 2. The bacterium has at least two parallel quorum sensing systems, which converge

at the transcription factor HapR, to detect fluctuations in the bacterial population density (Miller et al., 2002). Expression of *hapR* (induced at high cell density) has been associated with repression of virulence (Kovacikova and Skorupski, 2002; Miller et al., 2002; Nielsen et al., 2006; Zhu et al., 2002; Zhu and Mekalanos, 2003). Specifically, HapR represses transcription of *aphA*, whose gene product is a transcription factor required for *tcpPH* expression (Fig. 2) (Kovacikova and Skorupski, 2002).

In the context of a host, HapR appears to orchestrate changes in gene expression necessary for the various stages of the infection. Upon ingestion by a host, *V. cholerae* must pass through the gastric acid barrier in order to colonize the small intestine. It is unclear if *hapR* is expressed in these transitioning bacteria; however, bacteria within a biofilm (where *hapR* is expressed) are more resistant to acid shock and thus may be more likely to reach the small intestine than planktonic bacteria (Zhu and Mekalanos, 2003). After passing through the gastric acid barrier, such HapR-expressing bacteria entering the small intestine become more spread out, diluting the quorum sensing inducer signals (autoinducers) and thus repressing *hapR* expression. Shutting off *hapR* allows *V. cholerae* to express virulence genes, which would otherwise be repressed (Miller et al., 2002; Zhu et al., 2002). Following colonization, *V. cholerae* multiplies, increasing in density until *hapR* is yet again turned on (Nielsen et al., 2006), leading to repression of *ctxAB* and *tcpA-F* (Kovacikova and Skorupski, 2002). This represents our first clue to how the transition out of the host may be regulated. The regulation of *V. cholerae* virulence genes by HapR appears to be evolutionarily selected for, as toxigenic strains have a higher percentage of intact quorum-sensing systems than do nontoxigenic O1 or O139 strains (Wang et al., 2011).

Bistable Expression of Virulence Factors Late in Infection

It was previously shown, using the rabbit ligated ileal loop model of infection, that at 12 hours postinoculation *V. cholerae* makes a coordinated escape from the mucosal epithelial layer and swims into the intestinal lumen (Nielsen et al., 2006). Expanding on this, Nielsen et al. (2010) described a phenomenon in these detached bacteria where the expression of *tcpA*, which encodes the major pilin subunit of TCP, is either low or high; i.e., the population bifurcates, with some bacteria expressing virulence factors and others not doing so. Many of the cells still expressing *tcpA* at the late stage of infection are localized in bacterial clumps, which fits with the idea that TCP may aid in bacterium-bacterium interactions (Childers and Klose, 2007).

This bifurcation late in infection was shown to result from the bistable expression of ToxT. The bistable phenotype is dependent on a positive regulatory feedback loop between ToxT and *tcpA-F* (Fig. 3) (Nielsen et al., 2010). Briefly, ToxT is a known regulator of *tcpA-F* and transcription from this operon can read through into the *toxT* gene, which lies directly downstream of *tcpA-F* (Yu and DiRita, 1999). Protein levels within bacteria are known to vary between cells due to stochastic differences in the efficiency of transcription and translation (Cai et al., 2006). Thus, some *V. cholerae* cells accumulate a large amount of ToxT while others may contain much less. Nielsen et al. (2010) proposed that the

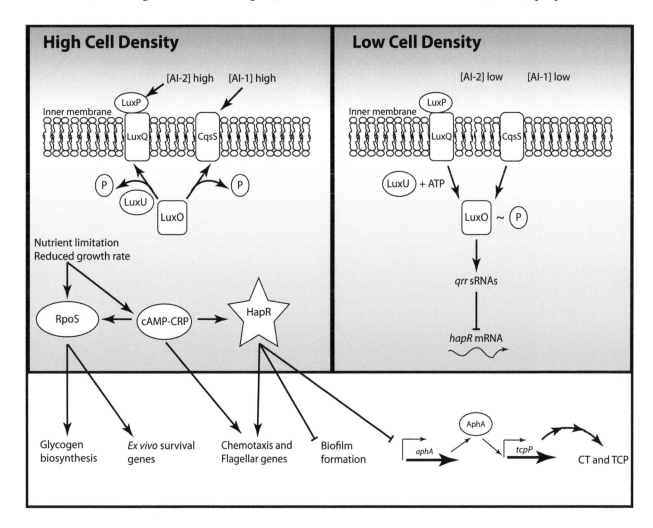

Figure 2. Quorum sensing and starvation conditions coordinate the expression of many *V. cholerae* genes. *V. cholerae* cells are capable of monitoring their population density by production of small molecules known as autoinducers (Waters and Bassler, 2005). The two known autoinducers in *V. cholerae* are labeled as AI-1 and AI-2. At low cell densities, concentrations of these autoinducers are low. Under these conditions, the response regulator LuxO is phosphorylated by the Lux sensor kinases LuxQ and LuxU and the Cqs sensor kinase CqsS, leading to the production of the *qrr* sRNAs (Lenz et al., 2004). The *qrr* sRNAs interfere with *hapR* mRNA translation, leading to reduced levels of this transcription factor. At high cell densities, autoinducers produced by the Lux and Cqs systems are present at high concentrations. Under these conditions LuxU, LuxQ, and CqsS act as phosphatases for LuxO, causing inactivation of the response regulator and lowered *qrr* sRNA levels, which leads to increased levels of HapR. HapR has a number of regulator targets in *V. cholerae*, including activation of chemotaxis and flagellar genes (Nielsen et al., 2006), as well as repression of *aphA* (Lin et al., 2007). AphA is a transcription factor required for *tcpPH* transcription (Skorupski and Taylor, 1999), which encodes a membrane-bound transcription factor that is required to activate ToxT expression (Goss et al., 2010). ToxT is the major transcriptional activator of virulence genes, and it leads to production of cholera toxin (CT) and the toxin-coregulated pilus (TCP) (Childers and Klose, 2007). Stationary-phase signals at high cell density, including slowed growth and nutrient limitation, lead to increased levels of active CRP. In turn, CRP increases activation of HapR and induces activation of RpoS (Hengge-Aronis, 2002; Silva and Benitez, 2004). RpoS contributes to the activation of the chemotaxis and flagellar genes, which are hypothesized to be involved in detachment and escape from the intestinal epithelium at late times of infection (Nielsen et al., 2006). RpoS also regulates genes for glycogen biosynthesis (Bourassa and Camilli, 2009) and genes important for survival in the environment (Schild et al., 2007).
doi:10.1128/9781555818524.ch28f2

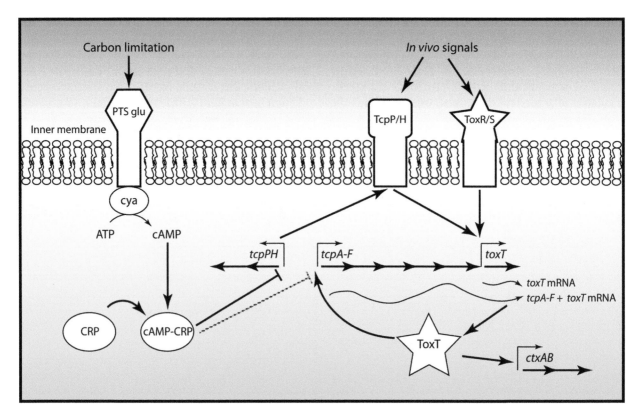

Figure 3. cAMP-CRP repression and ToxT autoregulation create the bistable expression pattern of *toxT*. Upon transition into a host, in vivo signals, possibly including bile salts, temperature, or pH, are sensed through TcpP/H and ToxR/S (Childers and Klose, 2007). These transcription factors activate transcription of the *toxT* gene from its own promoter. Subsequently, ToxT, the major virulence regulator in *V. cholerae*, induces expression from the *tcpA-F* promoter. Read-through transcription from this operon promoter, which is directly upstream of *toxT*, results in an increase in *toxT* mRNA and ToxT protein levels. This generates a situation where *toxT* expression is amplified in a positive feedback loop, which is proposed to stochastically generate bacteria that express high levels of ToxT (Nielsen et al., 2006). Stationary-phase signals, presumably sensed late in infection, are thought to be responsible for extinguishing this regulatory cascade. Carbon limitation sensed though the phosphorylated glucose-specific PTS system (PTS glu) stimulates adenylate cyclase (cya), which increases the pool of cAMP within the cell. Consequently, cAMP can bind and activate the catabolite repressor protein (CRP) (Skorupski and Taylor, 1997). The cAMP-CRP complex directly inhibits the expression of TcpP/H (Kovacikova and Skorupski, 2001). As illustrated in Fig. 2, cAMP-CRP indirectly enhances the expression of HapR (Liang et al., 2007), which also acts to indirectly reduce TcpP/H expression. It is proposed that cAMP-CRP also inhibits expression from the *tcpA-F* promoter by an undetermined mechanism (Liang et al., 2007; Nielsen., 2006). This is hypothesized to dampen the ToxT autoregulatory circuit in a portion of the population (approximately 50%), leading to shutoff of virulence gene expression (Nielsen et al., 2006).
doi:10.1128/9781555818524.ch28f3

positive feedback loop results in amplification of the differences in ToxT levels within the bacteria.

Activation of the catabolite repression protein (CRP) is also necessary for the bistability of ToxT (Nielsen et al., 2010). CRP is turned on in response to carbon limitation, which is presumably encountered in the intestinal lumen late in infection when other high-cell-density cues would be detected (Poncet et al., 2009). Active CRP represses *tcpPH* by competing with AphA at the *tcpPH* promoter (Kovacikova and Skorupski, 2001). This transcriptional repression at the *tcpPH* promoter extinguishes the ToxT autoregulatory loop. Therefore, in response to carbon starvation, CRP appears to play a required role in producing the subpopulation of bacteria late in infection that have turned off expression of ToxT and downstream virulence factors.

The bistable phenotype can be modeled in vitro in a rich broth medium with the addition of bicarbonate, a known inducer of the virulence gene regulatory cascade (Abuaita and Withey, 2009; Nielsen et al., 2010). Using this in vitro system, the bacterial population bifurcates, with respect to *tcpA* expression, upon entry into early stationary phase when carbon becomes limited. Approximately half of the cells repress expression of *tcpA*, whereas the other half continue to express the gene. The stable expression of *tcpA* is maintained for at least 3 hours in vitro after bacterial multiplication has ceased (Nielsen et al., 2010). Therefore, it is possible that a subpopulation of *V. cholerae* cells present in RWS may also be expressing virulence genes. However, contrary to this notion, a study that determined the *V. cholerae* transcriptome in human RWS found that the virulence genes were highly down-regulated (Merrell et al., 2002). Thus, there is a major shift in gene expression between *V. cholerae* bacteria acutely infecting the small intestine and those exiting the host.

The difference between these two studies could be explained by the fact that the bistable switch was observed in the rabbit ligated ileal loop, which is a closed system. Although there is bacterial growth and fluid accumulation in the ligated ileal loop, there is no flow of intestinal contents, nor do the bacteria pass through the large intestine. Thus, it is not clear if the coordination and timing of gene expression in this model is comparable to what occurs in a human host, nor is it evident if the detached bacteria in the lumen of the ligated ileal loop are comparable to bacteria shed in human RWS. Alternatively, the inconsistency between the two studies may be explained by their different approaches. Nielsen et al. (2010) looked at expression of virulence genes in individual cells, which allowed them to discover heterogeneity within the population. Conversely, Merrell et al. (2002) measured gene expression by using microarray analysis. Although microarray data can be quite informative and powerful, these results are an average of all the bacteria in the population. This is a limitation of microarrays and other population-level studies, and it may explain how researchers may have missed a small population of TCP expressors.

LATE STAGE OF INFECTION: PREPARING FOR THE TRANSITION

Despite having turned off the expression of *ctxAB*, *tcpA-F*, and other ToxT-regulated virulence genes, *V. cholerae* cells shed in human RWS have a hyperinfectious phenotype hypothesized to aid in the rapid spread of cholera during outbreaks (Merrell et al., 2002). The discovery of this hyperinfectious form led to the prediction that *V. cholerae* undergoes additional transcriptional

and/or phenotypic changes late in infection other than simply turning off the known virulence genes. Indeed, this and other studies suggest that *V. cholerae* undergoes multiple transcriptional shifts subsequent to the acute stage of infection in the small intestine (Fig. 4) (Larocque et al., 2005; Merrell et al., 2002; Nielsen et al., 2006, 2010; Schild et al., 2007). For example, there is evidence that the bacteria make a coordinated exit from the intestinal surface and out of the host (Nielsen et al., 2006). Additionally, the bacteria appear to turn on genes that are important for life in an aquatic environment prior to exiting the host (Schild et al., 2007).

An RpoS- and HapR-Mediated Transition

In order for *V. cholerae* cells to exit a host, they must detach from the small intestinal epithelium and pass through the rest of the bowels. Using the rabbit ligated ileal loop model, Nielsen et al. (2006) described the *V. cholerae* "mucosal escape response" during which the bacteria detach from the mucus-lined intestinal epithelium and move into the lumen. Activation of the stationary-phase and stress-responsive alternative sigma factor, RpoS, was shown to be necessary for this response. Specifically, RpoS was proposed to induce motility and chemotaxis genes that would allow the bacterium to swim into the intestinal lumen (Nielsen et al., 2006). RpoS was also shown to induce expression of the hemagglutinin (HA) protease, a secreted enzyme that has been proposed to aid in *V. cholerae* detachment from the epithelial mucosal layer (Finkelstein et al., 1992; Nielsen et al., 2006; Yildiz and Schoolnik, 1998). Additionally, RpoS was suggested to play a role in reducing the expression of CT and TCP, possibly through the activation of HapR and the subsequent repression of *aphA* by the quorum-sensing regulator (Nielsen et al., 2006).

The role of RpoS during infection has been controversial. *V. cholerae* cells are likely exposed to stresses that are relevant to alternative sigma factor utilization (Ding et al., 2004; Kovacikova and

Figure 4. Transcriptional changes of *V. cholerae* during infection. During the course of infection, *V. cholerae* is hypothesized to undergo numerous transcriptional changes. After initial attachment, bacteria express high levels of the major virulence regulator, ToxT (Childers and Klose, 2007; Withey and Dirita, 2006). ToxT activates virulence and colonization genes, which allow the bacterium to elaborate the toxin-coregulated pilus (TCP) and cholera toxin (CT) (Childers and Klose, 2007). Additionally, these cells exhibit fast growth (Nielsen et al., 2006). As cell density increases and nutrients become limiting, HapR and cAMP-CRP are hypothesized to inhibit virulence gene expression and begin the process of exit from the small intestine (Kovacikova and Skorupski, 2001; Kovacikova and Skorupski, 2002; Nielsen et al., 2006). While this occurs, RpoS and HapR enhance the expression of factors for chemotaxis and motility, which allow the bacteria to escape the mucosal epithelial layer (Nielsen et al., 2006; Schild et al., 2007). As *V. cholerae* is shed in RWS, chemotaxis becomes repressed and the bacteria enter a hyperinfectious state (Butler et al., 2006; Liang et al., 2007; Merrell et al., 2002). During this transition from the intestinal mucosal layer to RWS, the bacteria regulate genes important for environmental survival (Schild et al., 2007). The regulators important for this transition remain to be determined, although RpoS and PhoB may be key players in this transcriptional shift (Nielsen et al., 2006; Pratt et al., 2009; Schild et al., 2007). doi:10.1128/9781555818524.ch28f4

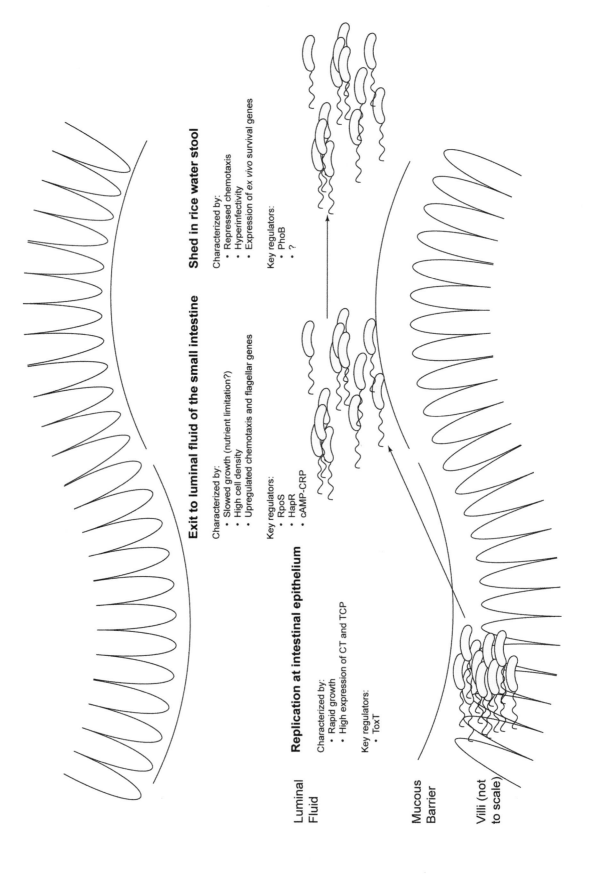

Exit to luminal fluid of the small intestine

Characterized by:
· Slowed growth (nutrient limitation?)
· High cell density
· Upregulated chemotaxis and flagellar genes

Key regulators:
· RpoS
· HapR
· cAMP-CRP

Shed in rice water stool

Characterized by:
· Repressed chemotaxis
· Hyperinfectivity
· Expression of *ex vivo* survival genes

Key regulators:
· PhoB
· ?

Replication at intestinal epithelium

Characterized by:
· Rapid growth
· High expression of CT and TCP

Key regulators:
· ToxT

Luminal
Fluid

Mucous
Barrier

Villi (not
to scale)

Skorupski, 2002), but the specific role of RpoS in bacterial colonization and replication is unclear. Two studies reported that RpoS is dispensable for intestinal colonization (Klose and Mekalanos, 1998; Yildiz and Schoolnik, 1998), while a third reported a subtle but significant role (Merrell et al., 2000). All three of these studies made use of the infant-mouse model of cholera infection, and thus the variation in results may be due to differences in the *V. cholerae* strains used. The strains used in the first two studies harbor natural frameshift mutations in *hapR*, the gene encoding the major quorum-sensing regulator, while the strain used by Merrell et al. (2000) contains an intact *hapR* locus. Signaling of RpoS through HapR was reported to contribute to the *V. cholerae* mucosal escape response via HapR induction of several motility genes (Nielsen et al., 2006). This may help explain why there appear to be differences in the importance of RpoS for virulence depending on the presence of a functional *hapR* locus. Perhaps in the absence of functional HapR signaling, the motility genes important for exit from the epithelium are induced through another pathway.

Since RpoS is a stationary-phase sigma factor, it has been proposed that, late in infection, *V. cholerae* may be phenotypically similar to stationary-phase cells found in broth culture. In vitro, RpoS plays a role in survival of *V. cholerae* during carbon starvation, under hyperosmotic conditions and during oxidative stress (Yildiz and Schoolnik, 1998). It is likely that within a host, the bacterium encounters nutrient deprivation and oxidative stress, which may signal activation of *rpoS*.

In addition to the induction of *rpoS* during the mucosal escape response, there is further evidence that *V. cholerae* exiting a host has entered a stationary-phase-like state. Nielsen et al. (2006) reported that viable counts of *V. cholerae* isolated from a rabbit ligated ileal loop increase exponentially early in infection but begin to level off around 6 to 8 hours postinoculation, a time point that coincides with the early stages of the mucosal escape response. This laboratory also reported that *V. cholerae* found in the lumen at the 8-hour time point has a severe reduction in expression of *rrnB1*, a gene that indicates whether a cell is actively growing (Nielsen et al., 2010). Although these expression data from the ligated ileal loop model may or may not correlate with expression during human infection, a separate study supports the idea of a stationary-phase-like state for *V. cholerae* in its human host. By comparing the transcriptome of *V. cholerae* in vomitus (early infection) to that in RWS (late infection), LaRocque et al. (2005) found that genes involved in DNA repair, protein synthesis, and

energy production were expressed in bacteria present in vomitus but not in the RWS bacteria. The *V. cholerae* cells in vomitus are therefore thought to be derived from an actively growing population of bacteria within the upper small intestine, in contrast to RWS bacteria, which are nongrowing. Additionally, it has long been known that RWS is a medium that does not support the growth of *V. cholerae* (Freter et al., 1961). Taken together, these results suggest that the environment encountered by *V. cholerae* late in infection mimics the conditions of stationary-phase culture growth and that bacteria present in RWS may be phenotypically similar to those from an in vitro stationary-phase culture. This information may prove useful for designing in vitro models of the late stage of *V. cholerae* infection.

V. cholerae Late Genes

A study by Schild et al. (2007) identified 57 *V. cholerae* genes that are expressed specifically at or near the end of the infection cycle in the infant-mouse model of infection. Many of these "late" genes appear to be involved in preparing the bacteria for the shift from the host small intestine into an aquatic environment. A number of these late genes with known or proposed functions are listed in Table 2. During the transition to the aquatic environment, *V. cholerae* typically undergoes numerous physiochemical shocks, including steep reductions in temperature, osmolarity (when shed into fresh water), macronutrient concentration (carbon and energy sources), and micronutrient concentration (namely fixed nitrogen and phosphorus) (Table 1) (Lara et al., 2009; Nelson et al., 2008; Schild et al., 2007). Some of the induced late genes aid in preparing *V. cholerae* for these shocks (a few examples are discussed below). The preinduction of these late genes makes evolutionary sense, as changing gene expression after transitioning to the environment may be too energetically costly or may simply occur too late to prevent killing in the face of the numerous insults the bacterium will endure.

Genes involved in chitin binding and catabolism, as well as genes involved in cyclic diguanylate (c-di-GMP) synthesis, were identified as late genes (Schild et al., 2007). As discussed later, *V. cholerae* can utilize chitin, which is the major component of the exoskeletons of numerous zooplankton, as a sole carbon and nitrogen source (Broza et al., 2008; Huq et al., 1983; Lipp et al., 2002; Watnick and Kolter, 2000). Additionally, in *V. cholerae*, high levels of the bacterial second messenger c-di-GMP are associated with an increase in biofilm formation and repression of virulence and motility genes (Jenal and Malone,

Table 2. *V. cholerae* genes induced late in the infant-mouse model of colonization

Gene/operon	Domains	Associated process	Phenotype
VC0130	GGDEF family protein, EAL domain	c-di-GMP metabolism	Not tested
VC0200-VC0203	Iron(III) ABC transporter, permease protein	Iron uptake	Attenuated early and late in infection
VC0404	MSHA biogenesis protein (*mshN*)	Biofilm formation	Not tested
VC0612	Cellulose degradation product phosphorylase	Chitin utilization	Decreased survival in pond, decreased growth on chitin[a]
VC1593	GGDEF family protein (*acgB*)	c-di-GMP metabolism	Decreased survival in pond and RWS[a]
VC1962	C$_4$-dicarboxylate transport transcriptional regulatory protein (*dctD-1*)	Carbon metabolism	Decreased growth on succinate
VC2369	Sensor histidine kinase FexB (*fexB*)	Oxygen/redox sensor	Decreased survival in pond and RWS[a]
VC2697	GGDEF family protein	c-di-GMP metabolism	Decreased survival in pond and RWS[a]
VCA0601	ABC transporter, permease protein	Unknown	Decreased survival in pond and RWS[a]
VCA0686	Iron(III) ABC transporter, permease protein	Iron uptake	Decreased survival in pond[a]
VCA0774	Glycerol kinase (*gplK*)	Carbon metabolism	Decreased survival in pond, decreased growth on glycerol[a]

[a] In a host-to-pond or host-to-RWS transition assay.

2006; Tischler and Camilli, 2004, 2005). Therefore, *V. cholerae* is proposed to reduce the concentration of c-di-GMP during the early and acute stages of infection, a model that has been experimentally supported (Tamayo et al., 2008; Tischler and Camilli, 2005). However, for the transition from host to aquatic environment, the concentration of c-di-GMP again increases to repress virulence genes and promote biofilm formation and perhaps other processes. The presumed repression of motility by high c-di-GMP levels during this transition is counterintuitive and indeed conflicts with the observation that RWS contains highly motile bacteria, which is the basis for the rapid clinical diagnosis of cholera by dark-field microscopy (Benenson et al., 1964). It is possible that motility is under differential control at this stage or that the *V. cholerae* population in RWS is bifurcated with respect to motility, since the other half of the bacteria are in clumps, as mentioned above. The question of motility aside, the importance of the chitin utilization and c-di-GMP synthesis genes in the aquatic environment is well supported, and it makes sense that *V. cholerae* would have evolved to turn these genes on just prior to exiting the host.

Ten of the late genes found by Schild et al. (2007) are under control of RpoS, although, in contrast, seven of these are repressed by this sigma factor (Nielsen et al., 2006). This suggests that the mucosal escape response may be distinct from the late-gene induction program and that there are two waves of transcriptional changes late in infection: the first corresponds to the mucosal escape response in which RpoS is a dominant regulator, and the second corresponds to the late-gene expression program. Alternatively, differences between the two reports could be due to the use of different infection models; Schild et al. (2007) used the infant-mouse model, in which there is no profuse fluid accumulation, whereas Nielsen et al. (2006) used the rabbit ligated ileal loop model, which exhibits fluid accumulation but is a closed system so the timing and nature of gene expression could be affected.

Current work in our laboratory is investigating how *V. cholerae* mediates the preinduction of aquatic environmental survival genes late in infection in the apparent absence of aquatic environmental inducing signals. The late stage of infection is thought to confront the bacteria with a high population density and nutrient starvation, which is consistent with the hypothesis that the bacteria enter a stationary-phase-like phenotype as they prepare to exit the host (as discussed above). Both nutrient starvation and quorum sensing may play a role in regulating the induction of the mucosal escape response and/or the late gene program. In support of high population density and nutrient starvation acting as important inducing signals, representative genes in all five *V. cholerae* iron acquisition systems are in fact late genes, suggesting that iron becomes growth limiting. Furthermore, Pratt et al. (2009) showed that the late-gene operon *acgAB* is regulated by PhoB/R, a two-component system that is activated by phosphate starvation.

Like in *E. coli*, the PhoB/R two-component system in *V. cholerae* is negatively regulated by the Pst/PhoU system, which functions first and foremost as an ABC transporter of inorganic phosphate. Under high-phosphate conditions, Pst/PhoU represses autophosphorylation of PhoR, thereby repressing the Pho regulon (Lamarche et al., 2008). Null mutations in the Pst/PhoU system of *V. cholerae* result in a constitutively active PhoB (Pratt et al., 2010). Dynamic activity of PhoB/R appears to be important during infection, as deletions in Pst/PhoU or PhoB/R highly attenuate virulence (Lamarche et al., 2008; Merrell et al., 2002; von Kruger et al., 1999). The former is explained by the discovery that PhoB is a repressor of the virulence genes late in infection, mediated through repression of *tcpPH* (Pratt et al., 2010). Further evidence that PhoB/R is active late in infection is that genes in the *V. cholerae* Pho regulon are induced in human RWS (Nelson et al., 2008) despite the fact that RWS contains high levels of phosphate (Table 1). Additionally, there may be overlap between the RpoS and PhoB regulons, since a mutation in *phoB* leads to an increase in the levels of various stress response proteins (von Kruger et al., 2006).

Glycogen Storage in *V. cholerae*

Work from our laboratory has begun to explore the importance of glycogen storage and utilization in *V. cholerae* throughout the various stages of the life cycle. Notably, Bourassa and Camilli (2009) showed that *V. cholerae* cells shed in human RWS contain glycogen granules, suggesting accumulation of the sugar during life in the host. Glycogen is a branched glucose polymer and is used by many bacteria as a carbon and energy reserve. Under nutrient-limiting conditions, bacteria can break down the glycogen stores to provide an internal source of carbon (Wilson, et al., 2010). Therefore, stored glycogen in *V. cholerae* may provide a source of carbon in the aquatic environment.

Glycogen storage in *V. cholerae* is induced under carbon-sufficient but nitrogen- and/or phosphate-limiting conditions in vitro, although the signals during infection are yet to be elucidated (Bourassa and Camilli, 2009; L. Bourassa, unpublished data). Fixed-nitrogen, phosphate, and carbon levels in pond water are typically growth limiting (Table 1) (Nelson et al., 2008; Schild et al., 2007). This, together with the fact that *V. cholerae* cells exiting the human host contain glycogen granules, suggests a potential role for glycogen stores in persistence of *V. cholerae* in the aquatic environment. In support of this, glycogen storage was found to be advantageous to the bacterium for survival in pond water and also in a pond-to-host transmission assay (Bourassa and Camilli, 2009).

As would be expected, *V. cholerae* genes involved in glycogen storage, but not utilization, are induced in human RWS (Merrell et al., 2002; Nelson et al., 2008). Additionally, *glgB*, a gene important for glycogen synthesis (Wilson et al., 2010), is induced during experimental *V. cholerae* infection of human volunteers (Lombardo et al., 2007). When human RWS is incubated in filter-sterilized pond water, glycogen degradation genes are induced, further indicating that glycogen stores provide an internal source of

carbon for the bacterium until an external source is found. Also of note, RpoS appears to be involved in regulating glycogen storage, indicating that part of the mucosal escape response may be to accumulate glycogen (Bourassa and Camilli, 2009).

FATE OF *V. CHOLERAE* IN THE ENVIRONMENT

Although *V. cholerae* is widely accepted to be a natural member of aquatic environments in temperate parts of the world, the fate of the organism upon exiting its human host is a controversial topic within the field. Transmission of *V. cholerae* during outbreaks may occur from human to human within households by contact with RWS, or the bacterium may spend a short time within an aquatic reservoir before infecting another human. Between outbreaks, *V. cholerae* must persist for long periods, possibly lasting from weeks to years, within the aquatic niche until a host ingests it. The phenotypic state of the bacterium during short-term or long-term persistence in the aquatic reservoir is unclear, and it seems that the bacterium may have numerous strategies for survival within this harsh environment.

Dissemination of *V. cholerae*

The aquatic environment is hypothesized to play a key role in the dissemination of *V. cholerae* within and between regions. However, the nature of this role is contentious. During outbreaks the disease is generally transmitted through ingestion of contaminated freshwater, but it has proven difficult to isolate toxigenic strains from these water sources and even more difficult during interepidemic periods (Aulet et al., 2007; Grim et al., 2010; Islam et al., 1992). Furthermore, *V. cholerae* is known to survive optimally in brackish waters with salinities of 0.5% to 3%. This has led to the hypothesis that during seasonal epidemics in areas around the Bay of Bengal (primarily India and Bangladesh), the organism spreads from coastal waters through estuaries and into the freshwater sources (Fig. 5) (Miller et al., 1982). This spread may be due to seasonal weather patterns, such as the monsoon season. Alternatively, spread from coastal to fresh waters could be the result of diseased individuals traveling from one location to another, leading to contamination of new water sources (Lara et al., 2009; Mosley et al., 1968; Sant et al., 1975). In the nonseasonal epidemics that occur elsewhere in the world, such as recent outbreaks in Zimbabwe and Haiti, the origin and pattern of *V. cholerae* spread may be quite different.

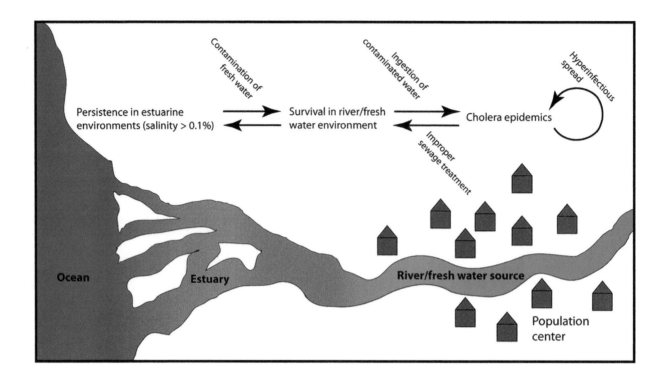

Figure 5. Dissemination and transmission of *V. cholerae* in areas of endemic infection. *V. cholerae* persists in the environment primarily in estuarine waters (salt concentrations greater than 0.1%) (Alam et al., 2007; Lara et al., 2009; Miller et al., 1982, 1985). *V. cholerae* is hypothesized to persist long-term in biofilms associated with chitinous surfaces (Meibom et al., 2004, 2005) or in a viable but nonculturable state (VBNC) (Alam et al., 2007). Weather events and disruption of freshwater ecosystems by human activity (Lara et al., 2009) can lead to contamination of freshwater reservoirs where *V. cholerae* can survive for at least short periods (Bourassa and Camilli, 2009; Nelson et al., 2008). Cholera epidemics begin with ingestion of *V. cholerae* from contaminated water sources. Hyperinfectious spread of freshly shed *V. cholerae* is hypothesized to rapidly amplify the number of infected patients during an outbreak (Hartley et al., 2006; Merrell et al., 2002; Pascual et al., 2006). Improper waste management can lead to continued contamination of drinking-water sources, which prolongs the epidemic.
doi:10.1128/9781555818524.ch28f5

It has long been known that infected people, both symptomatic and asymptomatic, can transport toxigenic *V. cholerae* for long distances, even across oceans (Mosley et al., 1968; Sant et al., 1975). The recent cholera outbreak in Haiti has reiterated the importance of humans as a dissemination vehicle for *V. cholerae*. The strain causing disease in Haiti is more genetically similar to toxigenic, clinical isolates in South Asia than to those causing sporadic disease in South America (Chin, et al., 2011). Although recently isolated environmental strains in South America were not compared with the Haitian strain, it is unlikely that the two spatially distant populations of *V. cholerae* convergently evolved. Similarly, it seems unlikely that water currents or weather events are responsible for transporting strains that originated in South Asia halfway around the globe to Haiti. It is most plausible, in this era of rapid global transportation, that the Haitian strain was delivered to the Caribbean island by one or more asymptomatic people traveling from South Asia. Therefore, while acknowledging the importance of weather events and seasonal patterns

in the dissemination and transmission of *V. cholerae*, it is also important to consider the impact of human activity.

Finding New Carbon Sources

Key to any organism's prolonged survival is the acquisition of nutrients. In the previous section, it was discussed how *V. cholerae* may store glycogen during infection of a host in order to utilize these glycogen stores upon transition into the environment (Bourassa and Camilli, 2009). Although internal glycogen stores may provide carbon to the organism for a short time, they are not an unlimited resource and eventually *V. cholerae* will need to find a new carbon source to survive.

One source of carbon and nitrogen in the aquatic environment is chitin, a polymer of N-acetylglucosamine. Chitin is the major component of the exoskeletons of zooplankton and insects found within and around many water sources (Watnick and Kolter, 2000). It is produced at a rate of 10^{11} metric tons per year, of which 10^9 metric tons are produced by copepods within aquatic reservoirs (Lipp et al., 2002).

All members of the *Vibrio* species that have been tested are chitinolytic, meaning that they have the ability to break down and metabolize chitin (Lipp et al., 2002). In the laboratory, *Vibrio* species can survive in association with a wide variety of chitin sources as the sole supply of carbon and nitrogen, including the exoskeletons of zooplankton and insects found in fresh, brackish, and marine waters (Broza et al., 2008; Huq et al., 1983; Lipp et al., 2002; Watnick and Kolter, 2000). Some *Vibrio* species have been reported to exhibit behavior of both chitin attachment and detachment, suggesting that they are able to remove themselves from a chitinous surface when all or most of the carbon and nitrogen has been consumed or when other signals, such as population density, trigger dispersal (Lipp et al., 2002). In addition, *V. cholerae* exhibits the ability to chemotax toward chitin sources (Meibom et al., 2004). As mentioned earlier, several chitin attachment and metabolism genes are *V. cholerae* late genes. Mutants with mutations of some of these genes showed a defect in their ability to compete with wild-type *V. cholerae* in a host-to-pond transition assay (Table 2) (Schild et al., 2007). Thus, chitin appears to be an important, perhaps the most important, source of carbon, and potentially nitrogen, for *V. cholerae* in the aquatic reservoir.

Biofilm Formation

Chitinous exoskeletons found in the aquatic environment not only provide a source of carbon and nitrogen for *V. cholerae* but also provide a surface on which the organism can form a biofilm (Watnick and Kolter, 2000; Watnick et al., 1999). Biofilms are complex microbial communities that have a three-dimensional structure often consisting of pillars and fluid-filled channels, which are surrounded by bacterially secreted exopolysaccharides and released DNA. These microbial communities attach themselves to a variety of biotic and abiotic surfaces and can be found in aquatic reservoirs, generally associated with zooplankton, insects, or exoskeletons and other detritis. In the laboratory, biofilms are generally composed of one species in order to simplify experiments; however, in the natural environment they may contain a variety of different bacteria. Bacteria within a biofilm may be protected from numerous insults including toxic chemicals, antibiotics, thermal stress, oxidative stresses, UV radiation, bacteriophages, and other predators. This makes the biofilm structure an ideal survival mechanism for many aquatic bacteria that often face these insults (Huq et al., 2008; Watnick and Kolter, 2000).

Formation of biofilms in aquatic environments is considered a developmental process since there are several distinct steps to the formation of a mature biofilm during which the bacteria undergo numerous changes in gene expression (Huq et al., 2008; Watnick and Kolter, 2000). In many bacteria, including *V. cholerae*, quorum sensing plays a major role in controlling biofilm formation (Huq et al., 2008). At low cell density, motile bacteria attach to a surface to form a sparse monolayer. Multiplication on the surface fills in the monolayer. Additional multiplication, coupled with exopolysaccharide production, builds up the mature three-dimensional structure of the biofilm (Moorthy and Watnick, 2004). As the biofilm matures, cell density provides a signal to stop building additional biofilm (Fig. 2). In *V. cholerae*, the quorum-sensing regulator, HapR, represses biofilm formation through two distinct mechanisms. It decreases the level of the biofilm-promoting second messenger c-di-GMP (Waters et al., 2008). It also represses the expression of *vpsT* (Waters et al., 2008), which encodes the positive transcriptional regulator of the *Vibrio* exopolysaccharide synthesis genes (Casper-Lindley and Yildiz, 2004). VpsT is itself regulated posttranslationally by c-di-GMP binding (Krasteva et al., 2010).

Other factors required for *V. cholerae* biofilm formation vary depending on the strain and conditions of the assay (Meibom et al., 2004; Moorthy and Watnick, 2004; Watnick et al., 1999). The mannose-sensitive hemagglutinin (MSHA) pilus, which binds chitin and potentially other surfaces, is the key player in a model of biofilm formation proposed by Moorthy and Watnick (2004). In this model, *V. cholerae* proceeds through three stages en route to a mature biofilm: (i) planktonic, in which free-swimming bacteria express flagella and surface attachment pili; (ii) monolayer, in which a single layer of bacteria transiently attach to a surface via the MSHA pilus; and (iii) the biofilm, a three-dimensional structure supported by an exopolysaccharide. The authors show that the gene expression profiles of the three stages are different, supporting their idea of three distinct phases. They suggest that the MSHA pilus is required for monolayer formation, both mediating attachment to a surface and suppressing the expression of flagellar genes to prevent detachment (Moorthy and Watnick, 2004). The fact that *mshN*, a gene in the MSHA pilus operon, is known to be expressed late in infection (Schild et al., 2007) suggests that the bacteria may begin to elaborate the type IV pilus prior to exiting the host, possibly in preparation for future biofilm formation.

The requirement of the MSHA pilus and chitin in biofilm formation is not absolute, as *V. cholerae* can bypass the MSHA pilus/monolayer step under certain laboratory conditions (Watnick et al., 1999) and biofilms can be formed on surfaces other than chitin (Moorthy and Watnick, 2004; Watnick et al., 1999). However, in nature, monolayer formation on chitinous surfaces, aided by the MSHA pilus, may well be a necessary stage in maturation of a biofilm. One environmental trigger necessary for biofilm formation is the presence of monosaccharides (Kierek and Watnick, 2003), which are normally in low concentrations in aquatic environments (Nelson et al., 2008). The monolayer stage would give *V. cholerae* cells the opportunity to break down some underlying chitin to obtain the monosaccharides necessary for the next stage of biofilm formation.

In addition to providing shelter from a variety of environmental insults, biofilms may contribute to the evolution of bacteria. Rates of bacterial conjugation and exchange of extrachromosomal DNA are accelerated within biofilms (Watnick and Kolter, 2000). It was reported that in *V. cholerae*, in the presence of chitin, HapR can mediate the expression of *comEA*, genes required for bacterial competence (Meibom et al., 2005). HapR is active when the population density is high, suggesting that *V. cholerae* competence is a phenotype of mature biofilms. Exchange of DNA within a biofilm may promote the emergence of new toxigenic strains of *V. cholerae* via horizontal transfer of the O-antigen genes, VPI-1, CTXΦ, or other loci. Indeed, natural transformation may be the only way for VPI-1 to be transferred among strains, as self-transmissibility appears to be lacking (Faruque et al., 2003).

Viable but Nonculturable State

One environmental survival behavior that several enteric bacteria, including *V. cholerae*, have been reported to exhibit is entrance into a viable but nonculturable (VBNC) state (Trevors, 2011) or, more conservatively, an active but nonculturable (ABNC) state (Kell et al., 1998). VBNC *V. cholerae* organisms, which are generally coccoid, are defined as viable cells unable to be cultured on standard media. These cells can be enumerated in the laboratory by epifluorescence microscopy, and their viability (or activity) may in some cases be determined by the addition of yeast extract and nalidixic acid, which induces the metabolically active coccoid cells to elongate (Alam et al., 2007; Aulet et al., 2007; Trevors, 2011). The VBNC state is thought to protect bacteria from environmental insults and allow them to endure long periods of nutrient starvation. In the aquatic environment the VBNC state may protect bacterial cells from fluctuations in pH, salinity, temperature, and nutrient availability. In the laboratory a VBNC state can be

induced by altering many of these variables, although nutrient limitation is most commonly used (Trevors, 2011).

The VBNC phenotype has been found in several pathogenic strains of *V. cholerae*, including *V. cholerae* O1 (classical and El Tor biotypes) and O139 (Alam et al., 2007; Aulet et al., 2007; Baffone et al., 2003; Colwell et al., 1996; Grim et al., 2010; Trevors, 2011). It has been proposed that between cholera outbreaks, toxigenic *V. cholerae* may exist in a quiescent VBNC state in bodies of freshwater. In support of this, VBNC *V. cholerae* strains have been isolated from freshwater sources around the world (Aulet et al., 2007; Broza et al., 2008; Grim et al., 2010). Also, the *V. cholerae* VBNC phenotype can be induced in the laboratory by prolonged incubation in pond water at both 4°C and room temperature (Alam et al., 2007; Colwell, et al., 1996). However, the genetic components that contribute to the VBNC state have yet to be elucidated.

The epidemiological importance of VBNC *V. cholerae* is not yet apparent. The ability of VBNC cells to cause disease depends heavily on their ability to "resuscitate" into fully active, pathogenic bacteria. Although the triggers necessary for resuscitation are largely unknown, the ability of VBNC cells to revert to an infectious form upon passage through a host has been found in several pathogenic bacteria including toxigenic *V. cholerae* (Alam et al., 2007; Baffone et al., 2003; Colwell et al., 1996). One of these studies reported that although *V. cholerae* O1 was resuscitated upon passage through the rabbit ligated ileal loop, there was no appreciable fluid accumulation in the ileal loops, suggesting that the VBNC cells are not as infectious as culturable cells (Alam et al., 2007). Despite these reports, other researchers have argued that the appearance of resuscitated cells upon passage through an animal is merely the result of rare, culturable cells in the inoculum (Bogosian et al., 1998; Kell et al., 1998). Kell et al. (1998) propose the use of strict statistical analysis of resuscitation data, something which is rarely done, taking into account the probability that any culturable cell existed in the inoculum. Until the resuscitation trigger is identified and the genetic mechanisms are described, it will be hard to fully appreciate VBNC *V. cholerae* as epidemiologically important.

Transmission to a New Host

Cholera has been endemic in Southeast Asia throughout recorded history (Barua, 1992). Around the Bay of Bengal there are seasonal outbreaks (Lipp et al., 2002) that appear to originate near coastal waters and travel inland over the course of the outbreak

(Broza et al., 2008). Studies conducted in Bangladesh near the Karnaphuli estuary have shown that weather events and human disruption of the estuary environment can lead to contamination of river water further upstream from where *Vibrio* species are usually detected (Lara et al., 2009). It is possible that these events, which can lead to contamination of freshwater sources for population centers, contribute to the beginning of seasonal outbreaks (Fig. 5). In Bangladesh, the region where most *V. cholerae* ecological research has been conducted, the outbreaks occur in a bimodal pattern—March to June and September to December. This pattern correlates with blooms of zooplankton, insect and alga species with which *V. cholerae* has been reported to associate (Epstein, 1993; Islam, 1992; Lipp et al., 2002). Such blooms may contribute to transmission both by providing a carbon source in the environment on which *V. cholerae* can grow and by generating biofilms, thereby enhancing virulence and concentrating the infectious dose of bacteria.

Biofilm-associated *V. cholerae* may play a significant role in transmission of the bacterium to a human host. A potential role of biofilms in bacterial transmission is not without precedent. Drinking water contaminated with biofilms of *E. coli, Aeromonas,* and *Pseudomonas* species is a potential threat to humans if ingested. Clumps of biofilm found in water have been reported to contain upward of 10^9 cells (Huq et al., 2008). This is a significant number considering that the ID_{50} of in vitro-grown *V. cholerae* is reported to be between 10^8 and 10^{11} bacterial cells (Cash et al., 1974). In addition, a study conducted in Bangladesh found a 50% reduction in cholera cases when folded sari cloth was used to filter drinking water. The folded cloth was able to remove debris larger than ~20 μm, which includes most biofilm-associated *V. cholerae*, but not planktonic cells or smaller clumps of bacteria (Huo et al., 1996). Studies conducted in the lab have shown that *V. cholerae* cells associated with biofilms are more infectious than planktonic cells (Huq et al., 2008; Tamayo et al., 2010). This phenotype is likely due to the physiological state of the bacteria (Tamayo et al., 2010), but it remains a possibility that biofilms on chitinous surfaces may also serve to concentrate *V. cholerae* cells on suspended particles in water, resulting in a higher rate of infection when ingested. These two phenomena are not mutually exclusive, and either or both may be critical to the ability of *V. cholerae* to initiate outbreaks from aquatic reservoirs.

Hyperinfectivity

During an outbreak of cholera, the number of cases in an area can increase explosively. This observation does not fit a model of cholera transmission that relies solely on ingestion of *V. cholerae* of low infectivity (Hartley et al., 2006). Rather, the mode of *V. cholerae* transmission may be dominated by spread of the hyperinfectious form of the organism that is shed in the RWS of cholera patients (Hartley et al., 2006; Merrell et al., 2002). The hyperinfectivity phenotype can be modeled in the laboratory by passaging *V. cholerae* cells through the infant-mouse small intestine (Alam et al., 2005). Hyperinfectious *V. cholerae* cells isolated from either human RWS or infant mice are on average 10- to 100-fold more infectious than in vitro-grown bacteria (Alam et al., 2005; Merrell et al., 2002). In a mathematical model that describes the number of cholera cases that occur during an outbreak, the trend more closely fits the explosive spread of the disease that is described in epidemiological data when the hyperinfectious form of *V. cholerae* is incorporated (Hartley et al., 2006).

Hyperinfectious bacteria are thought to contribute to human-to-human spread of *V. cholerae*. This mode of transmission may be important for household spread of the disease (Harris et al., 2008; Kendall et al., 2010; Weil et al., 2009). Supporting this, it has been shown that individuals who reside within the same household as an index case are at much greater risk of contracting cholera than are individuals living in the same community. Furthermore, the household members who contract cholera typically present with symptoms 2 to 3 days after the index case began shedding (Weil et al., 2009), which is within the range of the incubation period of *V. cholerae* in humans. In addition to household spread via direct contact with RWS, the hyperinfectious bacteria may persist for a short time in water sources before being ingested by a new host. Hyperinfectious *V. cholerae* organisms have been shown to maintain the phenotype for at least 5 hours, but not more than 18 hours, in pond water (Merrell et al., 2002). This suggests a transient nature of hyperinfectivity and helps to explain why cholera outbreaks exhibit a seasonal pattern and are not without end.

The hyperinfectious phenotype goes a long way toward explaining the explosive nature of cholera outbreaks; however, much is still unknown about how the phenotype is induced or maintained in *V. cholerae*. As discussed below, several studies have reported means of artificially inducing hyperinfectivity or have found potential genes involved in the phenotype, but it is unclear if these are artifacts of laboratory manipulation or if they truly mimic the hyperinfectious form found in nature. Additionally, it is unknown if hyperinfectivity is established via one pathway or if there are multiple ways a bacterium can become hyperinfectious.

Acid-tolerized *V. cholerae* has been shown to exhibit a hyperinfectious phenotype in the infant-mouse model of infection (Merrell and Camilli, 1999). The increased infectivity of acid-tolerized bacteria was suggested to result from these bacteria being better able to survive passage through the gastric acid barrier, thus allowing more bacteria to reach and colonize the small intestine. However, a later study ruled out this mechanism by showing that non-acid-tolerized bacteria survive just as well and that, instead, the acid-tolerized bacteria multiply faster once they have colonized the small intestine (Angelichio et al., 2004). Therefore, this suggests that some aspect of the metabolic state of acid-tolerized bacteria may contribute to their hyperinfectivity.

Evidence that suppression of chemotaxis, but not motility, contributes to hyperinfectivity has been presented in several studies from our laboratory (Butler and Camilli, 2004; Butler et al., 2006; Merrell et al., 2002). Chemotaxis genes are repressed in human RWS *V. cholerae* cells compared with in vitro-grown bacteria (Merrell et al., 2002), despite both populations of bacteria being motile. Correspondingly, bacteria isolated from human RWS were shown to be less able to chemotax to amino acid chemoattractants than were in vitro-grown bacteria (Butler et al., 2006). One possible explanation for the increased infectivity of motile but nonchemotactic bacteria is their ability to colonize the entire small intestine, whereas chemotactic strains colonize primarily the distal portion of the organ (Butler and Camilli, 2004). Activation of chemotaxis genes late in infection via RpoS was shown to be important for the mucosal escape response (Nielsen et al., 2006). This suggests that transcriptional changes occur after the mucosal escape response in order to repress the chemotactic phenotype prior to exit from the host. Alternatively, the repression of chemotaxis may occur within a subset of cells exiting the host, similar to the bistable expression of *tcpA* described by Nielsen et al. (2010).

Two Populations of Shed Bacteria

V. cholerae cells exiting a host may end up either encountering a new host or persisting for an extended period in aquatic reservoirs. It is highly unlikely that the organism can predict which of these scenarios it will encounter. In fact, there is evidence that the bacterium may prepare for both situations by expressing genes necessary for environmental survival as well as acquiring a hyperinfectious phenotype.

There are two explanations of how the bacterium can prepare for both distinct situations. One possibility is that every bacterium shed in RWS expresses genes necessary to survive in both situations.

This may be energetically taxing to the bacterium and thus seems unlikely. Alternatively, there could be two subpopulations of *V. cholerae* exiting a host: hyperinfectious (e.g., motile but nonchemotactic) and environmentally ready (e.g., chemotactic and with late genes for chitin utilization expressed).

The bistable switch, described above, serves as a precedent for the latter hypothesis. If RWS *V. cholerae* cells are indeed composed of multiple subpopulations, each specialized for different fates, then investigators using proteomic or RNA-based gene expression analysis, which takes population averages, will have a difficult time revealing the mechanisms of hyperinfectivity or environmental survival. Single-cell analysis, in which the expression of genes that contribute to hyperinfectivity or to environmental survival is assayed, may be the best approach to study this phenomenon.

CONCLUDING REMARKS

Over the course of an infection, waterborne pathogens undergo two major transitions, environment to host and host to environment. During both of these transitions, they experience major physiochemical and nutrient stresses. The study of how waterborne pathogens prepare for, and adapt to, these changes can provide insight into the genes and phenotypes that are important for infection, transmission, and dissemination. Here we focused on the model pathogen *V. cholerae*, whose life cycle has been studied in depth and has revealed fascinating adaptive and evolutionary strategies for moving between host and environment. Determination of where and how *V. cholerae* and other waterborne pathogens are persisting in the environment and the mechanisms by which they cycle between environment and host will allow us to more rationally plan public health interventions with maximum effectiveness. This is especially critical in areas of the world with endemic infection where resources are limited.

REFERENCES

Abuaita, B. H., and J. H. Withey. 2009. Bicarbonate induces *Vibrio cholerae* virulence gene expression by enhancing ToxT activity. *Infect. Immun.* 77:4111–4120.

Alam, A., R. C. Larocque, J. B. Harris, C. Vanderspurt, E. T. Ryan, F. Qadri, and S. B. Calderwood. 2005. Hyperinfectivity of human-passaged *Vibrio cholerae* can be modeled by growth in the infant mouse. *Infect. Immun.* 73:6674–6679.

Alam, M., M. Sultana, G. B. Nair, A. K. Siddique, N. A. Hasan, R. B. Sack, D. A. Sack, K. U. Ahmed, A. Sadique, H. Watanabe, C. J. Grim, A. Huq, and R. R. Colwell. 2007. Viable but non-culturable *Vibrio cholerae* O1 in biofilms in the aquatic environment and their role in cholera transmission. *Proc. Natl. Acad. Sci. USA* **104:**17801–17806.

Angelichio, M. J., D. S. Merrell, and A. Camilli. 2004. Spatiotemporal analysis of acid adaptation-mediated *Vibrio cholerae* hyperinfectivity. *Infect. Immun.* **72:**2405–2407.

Aulet, O., C. Silva, S. G. Fraga, M. Pichel, R. Cangemi, C. Gaudioso, N. Porcel, M. A. Jure, M. C. de Castillo, and N. Binsztein. 2007. Detection of viable and viable nonculturable *Vibrio cholerae* O1 through cultures and immunofluorescence in the Tucuman rivers, Argentina. *Rev. Soc. Bras. Med. Trop.* **40:**385–390.

Baffone, W., B. Citterio, E. Vittoria, A. Casaroli, R. Campana, L. Falzano, and G. Donelli. 2003. Retention of virulence in viable but non-culturable halophilic *Vibrio* spp. *Int. J. Food Microbiol.* **89:**31–39.

Barua, D. 1992. History of cholera, p. 1–36. *In* D. Barua and W. B. Greenough III (ed.), *Cholera.* Plenum Publishing Corp., New York, NY.

Benenson, A. S., M. R. Islam, and W. B. Greenough 3rd. 1964. Rapid identification of *Vibrio cholerae* by darkfield microscopy. *Bull. W. H. O.* **30:**827–831.

Blokesch, M., and G. K. Schoolnik. 2008. The extracellular nuclease Dns and its role in natural transformation of *Vibrio cholerae*. *J. Bacteriol.* **190:**7232–7240.

Bogosian, G., P. J. Morris, and J. P. O'Neil. 1998. A mixed culture recovery method indicates that enteric bacteria do not enter the viable but nonculturable state. *Appl. Environ. Microbiol.* **64:**1736–1742.

Bourassa, L., and A. Camilli. 2009. Glycogen contributes to the environmental persistence and transmission of *Vibrio cholerae*. *Mol. Microbiol.* **72:**124–138.

Brookes, J. D., J. Antenucci, M. Hipsey, M. D. Burch, N. J. Ashbolt, and C. Ferguson. 2004. Fate and transport of pathogens in lakes and reservoirs. *Environ. Int.* **30:**741–759.

Broza, M., H. Gancz, and Y. Kashi. 2008. The association between non-biting midges and *Vibrio cholerae*. *Environ. Microbiol.* **10:**3193–3200.

Butler, S. M., and A. Camilli. 2004. Both chemotaxis and net motility greatly influence the infectivity of *Vibrio cholerae*. *Proc. Natl. Acad. Sci. USA* **101:**5018–5023.

Butler, S. M., E. J. Nelson, N. Chowdhury, S. M. Faruque, S. B. Calderwood, and A. Camilli. 2006. Cholera stool bacteria repress chemotaxis to increase infectivity. *Mol. Microbiol.* **60:**417–426.

Cai, I., N. Friedman, and X. Xie. 2006. Stochastic protein expression in individual cells at the single molecule level. *Nature* **440:**358–362.

Cash, R. A., S. I. Music, J. P. Libonati, M. J. Snyder, R. P. Wenzel, and R. B. Hornick. 1974. Response of man to infection with *Vibrio cholerae*. I. Clinical, serologic, and bacteriologic responses to a known inoculum. *J. Infect. Dis.* **129:**45–52.

Casper-Lindley, C., and F. H. Yildiz. 2004. VpsT is a transcriptional regulator required for expression of *vps* biosynthesis genes and the development of rugose colonial morphology in *Vibrio cholerae* O1 El Tor. *J. Bacteriol.* **186:**1574–1578.

Childers, B. M., and K. E. Klose. 2007. Regulation of virulence in *Vibrio cholerae*: the ToxR regulon. *Future Microbiol.* **2:**335–344.

Chin, C. S., J. Sorenson, J. B. Harris, W. P. Robins, R. C. Charles, R. R. Jean-Charles, J. Bullard, D. R. Webster, A. Kasarskis, P. Peluso, E. E. Paxinos, Y. Yamaichi, S. B. Calderwood, J. J. Mekalanos, E. E. Schadt, and M. K. Waldor. 2011. The origin of the Haitian cholera outbreak strain. *N. Engl. J. Med.* **364:**33–42.

Colwell, R. R., P. Brayton, D. Herrington, B. Tall, A. Huq, and M. M. Levine. 1996. Viable but non-culturable *Vibrio cholerae* O1 revert to a cultivable state in the human intestine. *World J. Microbiol. Biotechnol.* **12:**28–31.

Ding, Y., B. M. Davis, and M. K. Waldor. 2004. Hfq is essential for *Vibrio cholerae* virulence and downregulates sigma expression. *Mol. Microbiol.* **53:**345–354.

DiRita, V. J. 1992. Co-ordinate expression of virulence genes by ToxR in *Vibrio cholerae*. *Mol. Microbiol.* **6:**451–458.

Epstein, P. R. 1993. Algal blooms in the spread and persistence of cholera. *Biosystems* **31:**209–221.

Faruque, S. M., M. J. Albert, and J. J. Mekalanos. 1998a. Epidemiology, genetics, and ecology of toxigenic *Vibrio cholerae*. *Microbiol. Mol. Biol. Rev.* **62:**1301–1314.

Faruque, S. M., Asadulghani, M. N. Saha, A. R. Alim, M. J. Albert, K. M. Islam, and J. J. Mekalanos. 1998b. Analysis of clinical and environmental strains of nontoxigenic *Vibrio cholerae* for susceptibility to CTXPhi: molecular basis for origination of new strains with epidemic potential. *Infect. Immun.* **66:**5819–5825.

Faruque, S. M., A. K. Siddique, M. N. Saha, Asadulghani, M. M. Rahman, K. Zaman, M. J. Albert, D. A. Sack, and R. B. Sack. 1999. Molecular characterization of a new ribotype of *Vibrio cholerae* O139 Bengal associated with an outbreak of cholera in Bangladesh. *J. Clin. Microbiol.* **37:**1313–1318.

Faruque, S. M., M. N. Saha, Asadulghani, D. A. Sack, R. B. Sack, Y. Takeda, and G. B. Nair. 2000. The O139 serogroup of *Vibrio cholerae* comprises diverse clones of epidemic and nonepidemic strains derived from multiple *V. cholerae* O1 or non-O1 progenitors. *J. Infect. Dis.* **182:**1161–1168.

Faruque, S. M., and G. B. Nair. 2002. Molecular ecology of toxigenic *Vibrio cholerae*. *Microbiol. Immunol.* **46:**59–66.

Faruque, S. M., J. Zhu, Asadulghani, M. Kamruzzaman, and J. J. Mekalanos. 2003. Examination of diverse toxin-coregulated pilus-positive *Vibrio cholerae* strains fails to demonstrate evidence for *Vibrio* pathogenicity island phage. *Infect. Immun.* **71:**2993–2999.

Faruque, S. M., G. B. Nair, and Y. Takeda. 2011. Molecular epidemiology of toxigenic *Vibrio cholerae*, p. 115–127. *In* T. Ramamurthy and S. K. Bhattacharya (ed.), *Epidemiological and Molecular Aspects on Cholera*. Springer Science+Business Media, LLC, New York.

Finkelstein, R. A., M. Boesman-Finkelstein, Y. Chang, and C. C. Hase. 1992. *Vibrio cholerae* hemagglutinin/protease, colonial variation, virulence, and detachment. *Infect. Immun.* **60:**472–478.

Freter, R., H. L. Smith, Jr., and F. J. Sweeney, Jr. 1961. An evaluation of intestinal fluids in the pathogenesis of cholera. *J. Infect. Dis.* **109:**35–42.

Geldreich, E. E. 1996. Pathogenic agents in freshwater resources. *Hydrol. Processes* **10:**315–333.

Goss, T. J., C. P. Seaborn, M. D. Gray and E. S. Krukonis. 2010. Identification of the TcpP-binding site in the *toxT* promoter of *Vibrio cholerae* and the role of ToxR in TcpP-mediated activation. *Infect. Immun.* **78:**4122-4133.

Grim, C. J., E. Jaiani, C. A. Whitehouse, N. Janelidze, T. Kokashvili, M. Tediashvili, R. R. Colwell, and A. Huq. 2010. Detection of toxogenic *Vibrio cholerae* O1 in freshwater lakes of the former Soviet Republic of Georgia. *Environ. Microbiol. Rep.* **2:**2–6.

Grimes, D. 1991. Ecology of estuarine bacteria capable of causing human disease: a review. *Estuaries* **14:**345–360.

Harris, J. B., R. C. LaRocque, F. Chowdhury, A. I. Khan, T. Logvinenko, A. S. Faruque, E. T. Ryan, F. Qadri, and S. B. Calderwood. 2008. Susceptibility to *Vibrio cholerae* infection in a cohort of household contacts of patients with cholera in Bangladesh. *PLoS Negl. Trop. Dis.* **2:**e221.

Hartley, D. M., J. G. Morris, and D. L. Smith. 2006. Hyperinfectivity: a critical element in the ability of *V. cholerae* to cause epidemics? *PLoS Med.* **3:**63.

Hengge-Aronis, R. 2002. Signal transduction and regulatory mechanisms involved in control of the sigma(S) (RpoS) subunit of RNA polymerase. *Microbiol. Mol. Biol. Rev.* **66:**373–395.

Herrington, D. A., R. H. Hall, G. Losonsky, J. J. Mekalanos, R. K. Taylor, and M. M. Levine. 1988. Toxin, toxin-coregulated pili, and the *toxR* regulon are essential for *Vibrio cholerae* pathogenesis in humans. *J. Exp. Med.* **168:**1487–1492.

Huo, A., B. Xu, M. A. Chowdhury, M. S. Islam, R. Montilla, and R. R. Colwell. 1996. A simple filtration method to remove plankton-associated *Vibrio cholerae* in raw water supplies in developing countries. *Appl. Environ. Microbiol.* **62:**2508–2512.

Huq, A., E. B. Small, P. A. West, M. I. Huq, R. Rahman, and R. R. Colwell. 1983. Ecological relationships between *Vibrio cholerae* and planktonic crustacean copepods. *Appl. Environ. Microbiol.* **45:**275–283.

Huq, A., C. A. Whitehouse, C. J. Grim, M. Alam, and R. R. Colwell. 2008. Biofilms in water, its role and impact in human disease transmission. *Curr. Opin. Biotechnol.* **19:**244–247.

Islam, M. S., M. J. Alam, and P. K. B. Neogi. 1992. Seasonality and toxigenicity of *Vibrio cholerae* non-O1 isolated from different components of pond ecosystems of Dhaka City, Bangladesh. *World J. Microbiol. Biotechnol.* **8:**160–163.

Jenal, U., and J. Malone. 2006. Mechanisms of cyclic-di-GMP signaling in bacteria. *Annu. Rev. Genet.* **40:**385–407.

Kamal, D., A. N. Khan, M. A. Rahman, and F. Ahamed. 2007. Study on the physico chemical properties of water of Mouri River, Khulna, Bangladesh. *Pak. J. Biol. Sci.* **10:**710–717.

Karaolis, D. K., S. Somara, D. R. Maneval, Jr., J. A. Johnson, and J. B. Kaper. 1999. A bacteriophage encoding a pathogenicity island, a type-IV pilus and a phage receptor in cholera bacteria. *Nature* **399:**375–379.

Kell, D. B., A. S. Kaprelyants, D. H. Weichart, C. R. Harwood, and M. R. Barer. 1998. Viability and activity in readily culturable bacteria: a review and discussion of the practical issues. *Antonie Van Leeuwenhoek* **73:**169–187.

Kendall, E. A., F. Chowdhury, Y. Begum, A. I. Khan, S. Li, J. H. Thierer, J. Bailey, K. Kreisel, C. O. Tacket, R. C. LaRocque, J. B. Harris, E. T. Ryan, F. Qadri, S. B. Calderwood, and O. C. Stine. 2010. Relatedness of *Vibrio cholerae* O1/O139 isolates from patients and their household contacts, determined by multilocus variable-number tandem-repeat analysis. *J. Bacteriol.* **192:**4367–4376.

Kierek, K., and P. I. Watnick. 2003. Environmental determinants of *Vibrio cholerae* biofilm development. *Appl. Environ. Microbiol.* **69:**5079–5088.

Klose, K. E., and J. J. Mekalanos. 1998. Distinct roles of an alternative sigma factor during both free-swimming and colonizing phases of the *Vibrio cholerae* pathogenic cycle. *Mol. Microbiol.* **28:**501–520.

Kovacikova, G., and K. Skorupski. 2001. Overlapping binding sites for the virulence gene regulators AphA, AphB and cAMP-CRP at the *Vibrio cholerae* tcpPH promoter. *Mol. Microbiol.* **41:**393407.

Kovacikova, G., and K. Skorupski. 2002a. Regulation of virulence gene expression in *Vibrio cholerae* by quorum sensing: HapR functions at the *aphA* promoter. *Mol. Microbiol.* **46:**1135–1147.

Kovacikova, G., and K. Skorupski. 2002b. The alternative sigma factor E plays an important role in intestinal survival and virulence in *Vibrio cholerae*. *Infect. Immun.* **70:**5355–5362.

Krasteva, P. V., J. C. Fong, N. J. Shikuma, S. Beyhan, M. V. Navarro, F. H. Yildiz, and H. Sondermann. 2010. *Vibrio cholerae* VpsT regulates matrix production and motility by directly sensing cyclic di-GMP. *Science* **327:**866–868.

Lamarche, M. G., B. L. Wanner, S. Crepin, and J. Harel. 2008. The phosphate regulon and bacterial virulence: a regulatory network connecting phosphate homeostasis and pathogenesis. *FEMS Microbiol. Rev.* **32:**461–473.

Lara, R. J., S. B. Neogi, M. S. Islam, Z. H. Mahmud, S. Yamasaki, and G. B. Nair. 2009. Influence of catastrophic climatic events and human waste on *Vibrio* distribution in the Karnaphuli estuary, Bangladesh. *Ecohealth* **6:**279–286.

Larocque, R. C., J. B. Harris, M. Dziejman, X. Li, A. I. Khan, A. S. Faruque, S. M. Faruque, G. B. Nair, E. T. Ryan, F. Qadri, J. J. Mekalanos, and S. B. Calderwood. 2005. Transcriptional profiling of *Vibrio cholerae* recovered directly from patient specimens during early and late stages of human infection. *Infect. Immun.* **73:**4488–4493.

Lee, S. H., D. L. Hava, M. K. Waldor, and A. Camilli. 1999. Regulation and temporal expression patterns of *Vibrio cholerae* virulence genes during infection. *Cell* **99:**625–634.

Lenz, D. H., K. C. Mok, B. N. Lilley, R. V. Kulkarni, N. S. Wingreen, and B. L. Bassler. 2004. The small RNA chaperone Hfq and multiple small RNAs control quorum sensing in *Vibrio harveyi* and *Vibrio cholerae*. *Cell* **118:**69–82.

Liang, W., A. Pascual-Montano, A. J. Silva, and J. A. Benitez. 2007. The cyclic AMP receptor protein modulates quorum sensing, motility and multiple genes that affect intestinal colonization in *Vibrio cholerae*. *Microbiology* **153:**2964–2975.

Lin, W., G. Kovacikova, and K. Skorupski. 2007. The quorum sensing regulator HapR downregulates the expression of the virulence gene transcription factor AphA in *Vibrio cholerae* by antagonizing Lrp- and VpsR-mediated activation. *Mol. Microbiol.* **64:**953–967.

Lipp, E. K., A. Huq, and R. R. Colwell. 2002. Effects of global climate on infectious disease: the cholera model. *Clin. Microbiol. Rev.* **15:**757–770.

Lombardo, M. J., J. Michalski, H. Martinez-Wilson, C. Morin, T. Hilton, C. G. Osorio, J. P. Nataro, C. O. Tacket, A. Camilli, and J. B. Kaper. 2007. An *in vivo* expression technology screen for *Vibrio cholerae* genes expressed in human volunteers. *Proc. Natl. Acad. Sci. USA* **104:**18229–18234.

Meibom, K. L., X. B. Li, A. T. Nielsen, C. Y. Wu, S. Roseman, and G. K. Schoolnik. 2004. The *Vibrio cholerae* chitin utilization program. *Proc. Natl. Acad. Sci. USA* **101:**2524–2529.

Meibom, K. L., M. Blokesch, N. A. Dolganov, C. Y. Wu, and G. K. Schoolnik. 2005. Chitin induces natural competence in *Vibrio cholerae*. *Science* **310:**1824–1827.

Merrell, D. S., and A. Camilli. 1999. The *cadA* gene of *Vibrio cholerae* is induced during infection and plays a role in acid tolerance. *Mol. Microbiol.* **34:**836–849.

Merrell, D. S., A. D. Tischler, S. H. Lee, and A. Camilli. 2000. *Vibrio cholerae* requires *rpoS* for efficient intestinal colonization. *Infect. Immun.* **68:**6691–6696.

Merrell, D. S., S. M. Butler, F. Qadri, N. A. Dolganov, A. Alam, M. B. Cohen, S. B. Calderwood, G. K. Schoolnik, and A. Camilli. 2002a. Host-induced epidemic spread of the cholera bacterium. *Nature* **417:**642–645.

Merrell, D. S., D. L. Hava, and A. Camilli. 2002b. Identification of novel factors involved in colonization and acid tolerance of *Vibrio cholerae*. *Mol. Microbiol.* **43:**1471–1491.

Miller, C. J., B. S. Drasar, and R. G. Feachem. 1982. Cholera and estuarine salinity in Calcutta and London. *Lancet* **i:**1216–1218.

Miller, C. J., R. G. Feachem, and B. S. Drasar. 1985. Cholera epidemiology in developed and developing countries: new thoughts on transmission, seasonality, and control. *Lancet* **i:**261–262.

Miller, M. B., K. Skorupski, D. H. Lenz, R. K. Taylor, and B. L. Bassler. 2002. Parallel quorum sensing systems converge to regulate virulence in *Vibrio cholerae*. *Cell* **110:**303–314.

Moorthy, S., and P. I. Watnick. 2004. Genetic evidence that the *Vibrio cholerae* monolayer is a distinct stage in biofilm development. *Mol. Microbiol.* **52:**573–587.

Mosley, W. H., S. Ahmad, A. S. Benenson, and A. Ahmed. 1968. The relationship of vibriocidal antibody titre to susceptibility to cholera in family contacts of cholera patients. *Bull. W. H. O.* **38:**777–785.

Nelson, E. J., A. Chowdhury, J. Flynn, S. Schild, L. Bourassa, Y. Shao, R. C. LaRocque, S. B. Calderwood, F. Qadri, and A.

Camilli. 2008. Transmission of *Vibrio cholerae* is antagonized by lytic phage and entry into the aquatic environment. *PLoS Pathog.* 4:e1000187.

Nielsen, A. T., N. A. Dolganov, G. Otto, M. C. Miller, C. Y. Wu, and G. K. Schoolnik. 2006. RpoS controls the *Vibrio cholerae* mucosal escape response. *PLoS Pathog.* 2:e109.

Nielsen, A. T., N. A. Dolganov, T. Rasmussen, G. Otto, M. C. Miller, S. A. Felt, S. Torreilles, and G. K. Schoolnik. 2010. A bistable switch and anatomical site control *Vibrio cholerae* virulence gene expression in the intestine. *PLoS Pathog.* 6:e1001102.

Okoh, A. I., T. Sibanda, and S. S. Gusha. 2010. Inadequately treated wastewater as a source of human enteric viruses in the environment. *Int. J. Environ. Res. Public Health* 7:2620–2637.

Pascual, M., K. Koelle, and A. P. Dobson. 2006. Hyperinfectivity in cholera: a new mechanism for an old epidemiological model? *PLoS Med.* 3:e280.

Peterson, K. M. 2002. Expression of *Vibrio cholerae* virulence genes in response to environmental signals. *Curr. Issues Intest. Microbiol.* 3:29–38.

Poncet, S., E. Milohanic, A. Maze, J. Nait Abdallah, F. Ake, M. Larribe, A. E. Deghmane, M. K. Taha, M. Dozot, X. De Bolle, J. J. Letesson, and J. Deutscher. 2009. Correlations between carbon metabolism and virulence in bacteria. *Contrib. Microbiol.* 16:88–102.

Pratt, J. T., E. McDonough, and A. Camilli. 2009. PhoB regulates motility, biofilms, and cyclic di-GMP in *Vibrio cholerae*. *J. Bacteriol.* 191:6632–6642.

Pratt, J. T., A. M. Ismail, and A. Camilli. 2010. PhoB regulates both environmental and virulence gene expression in *Vibrio cholerae*. *Mol. Microbiol.* 77:1595–1605.

Reidl, J., and K. E. Klose. 2002. *Vibrio cholerae* and cholera: out of the water and into the host. *FEMS Microbiol. Rev.* 26:125–139.

Safa, A., G. B. Nair, and R. Y. Kong. 2010. Evolution of new variants of *Vibrio cholerae* O1. *Trends Microbiol.* 18:46–54.

Sant, M. V., W. N. Gatlewar, and S. K. Bhindey. 1975. Epidemiological studies on cholera in non-endemic regions with special reference to the problem of carrier state during epidemic and non-epidemic period. *Prog. Drug Res.* 19:594–601.

Schild, S., R. Tamayo, E. J. Nelson, F. Qadri, S. B. Calderwood, and A. Camilli. 2007. Genes induced late in infection increase fitness of *Vibrio cholerae* after release into the environment. *Cell Host Microbe* 2:264–277.

Sharma, S., P. Sachdeva, and J. S. Virdi. 2003. Emerging waterborne pathogens. *Appl. Microbiol. Biotechnol.* 61:424–428.

Silva, A. J., and J. A. Benitez. 2004. Transcriptional regulation of *Vibrio cholerae* hemagglutinin/protease by the cyclic AMP receptor protein and RpoS. *J. Bacteriol.* 186:6374–6382.

Skorupski, K., and R. K. Taylor. 1997. Control of the ToxR virulence regulon in *Vibrio cholerae* by environmental stimuli. *Mol. Microbiol.* 25:1003–1009.

Skorupski, K., and R. K. Taylor. 1999. A new level in the *Vibrio cholerae* ToxR virulence cascade: AphA is required for transcriptional activation of the *tcpPH* operon. *Mol. Microbiol.* 31:763–771.

Skovgaard, N. 2007. New trends in emerging pathogens. *Int. J. Food Microbiol.* 120:217–224.

Stine, O. C., M. Alam, L. Tang, G. B. Nair, A. K. Siddique, S. M. Faruque, A. Huq, R. Colwell, R. B. Sack, and J. G. Morris, Jr. 2008. Seasonal cholera from multiple small outbreaks, rural Bangladesh. *Emerg. Infect. Dis.* 14:831–833.

Tamayo, R., S. Schild, J. T. Pratt, and A. Camilli. 2008. Role of cyclic di-GMP during El Tor biotype *Vibrio cholerae* infection:

characterization of the in vivo-induced cyclic di-GMP phosphodiesterase CdpA. *Infect. Immun.* 76:1617–1627.

Tamayo, R., B. Patimalla, and A. Camilli. 2010. Growth in a biofilm induces a hyperinfectious phenotype in *Vibrio cholerae*. *Infect. Immun.* 78:3560–3569.

Tischler, A. D., and A. Camilli. 2004. Cyclic diguanylate (c-di-GMP) regulates *Vibrio cholerae* biofilm formation. *Mol. Microbiol.* 53:857–869.

Tischler, A. D., and A. Camilli. 2005. Cyclic diguanylate regulates *Vibrio cholerae* virulence gene expression. *Infect. Immun.* 73:5873–5882.

Trevors, J. T. 2011. Viable but non-culturable (VBNC) bacteria: gene expression in planktonic and biofilm cells. *J. Microbiol. Methods* 86:266–273.

von Kruger, W. M., S. Humphreys, and J. M. Ketley. 1999. A role for the PhoBR regulatory system homologue in the *Vibrio cholerae* phosphate-limitation response and intestinal colonization. *Microbiology* 145:2463–2475.

von Kruger, W. M., L. M. Lery, M. R. Soares, F. S. de Neves-Manta, C. M. Batista e Silva, A. G. Neves-Ferreira, J. Perales, and P. M. Bisch. 2006. The phosphate-starvation response in *Vibrio cholerae* O1 and *phoB* mutant under proteomic analysis: disclosing functions involved in adaptation, survival and virulence. *Proteomics* 6:1495–1511.

Waldor, M. K., and J. J. Mekalanos. 1996. Lysogenic conversion by a filamentous phage encoding cholera toxin. *Science* 272:1910–1914.

Wang, Y., H. Wang, Z. Cui, H. Chen, Z. Zhong, B. Kan, and J. Zhu. 2011. The prevalence of functional quorum-sensing systems in recently emerged *Vibrio cholerae* toxigenic strains. *Environ. Microbiol. Rep.* 3:218–222.

Waters, C. M., and B. L. Bassler. 2005. Quorum sensing: cell-to-cell communication in bacteria. *Annu. Rev. Cell Dev. Biol.* 21:319–346.

Waters, C. M., W. Lu, J. D. Rabinowitz, and B. L. Bassler. 2008. Quorum sensing controls biofilm formation in *Vibrio cholerae* through modulation of cyclic di-GMP levels and repression of *vpsT*. *J. Bacteriol.* 190:2527–2536.

Watnick, P., and R. Kolter. 2000. Biofilm, city of microbes. *J. Bacteriol.* 182:2675–2679.

Weil, A. A., A. I. Khan, F. Chowdhury, R. C. Larocque, A. S. Faruque, E. T. Ryan, S. B. Calderwood, F. Qadri, and J. B. Harris. 2009. Clinical outcomes in household contacts of patients with cholera in Bangladesh. *Clin. Infect. Dis.* 49:1473–1479.

Wilson, W. A., P. J. Roach, M. Montero, E. Baroja-Fernandez, F. J. Munoz, G. Eydallin, A. M. Viale, and J. Pozueta-Romero. 2010. Regulation of glycogen metabolism in yeast and bacteria. *FEMS Microbiol. Rev.* 34:952–985.

Withey, J. H., and V. J. DiRita. 2006. The toxbox: specific DNA sequence requirements for activation of *Vibrio cholerae* virulence genes by ToxT. *Mol. Microbiol.* 59:1779–1789.

Yildiz, F. H., and G. K. Schoolnik. 1998. Role of *rpoS* in stress survival and virulence of *Vibrio cholerae*. *J. Bacteriol.* 180:773–784.

Yu, R. R., and V. J. DiRita. 1999. Analysis of an autoregulatory loop controlling ToxT, cholera toxin, and toxin-coregulated pilus production in *Vibrio cholerae*. *J. Bacteriol.* 181:2584–2592.

Zhu, J., M. B. Miller, R. E. Vance, M. Dziejman, B. L. Bassler, and J. J. Mekalanos. 2002. Quorum-sensing regulators control virulence gene expression in *Vibrio cholerae*. *Proc. Natl. Acad. Sci. USA* 99:3129–3134.

Zhu, J., and J. J. Mekalanos. 2003. Quorum sensing-dependent biofilms enhance colonization in *Vibrio cholerae*. *Dev. Cell* 5:647–656.

INDEX

Accessory gene regulator. *See* Agr

Acetyl-CoA, and type III secretion system regulation, in
    Pseudomonas aeruginosa, 324
    in *Yersinia enterocolitica*, 324

Acid phosphatases, in *Francisella tularensis*, 413

Acid tolerance response, 95

Actinomyces, iron-dependent virulence regulation in,
    95–96

Actinomyces naeslundii, 95–96

Acyl homoserine lactone receptors, LuxR protein from
    Vibrio fischeri and, 41

Acyl homoserine lactones, 557–559
    encoded by *Pseudomonas aeruginosa*, 548
    in mammalian signaling, 557–559
    Proteobacteria and, 40–41
    quorum sensing-regulated processes of, in
        Pseudomonas aeruginosa, 41
    to regulate virulence factors, 40, 42
    signaling by, 40–42
        in *Pseudomonas aeruginosa*, 43–44
        in *Vibrio fischeri*, 42–43
    synthesis and response of, 40–42

Acyltrehaloses, of *Mycobacterium tuberculosis* outer
    membrane, 467

Adenosine, and enteropathogenic *Escherichia coli*,
    554–555
    and *Pseudomonas aeruginosa*, 555

Aeromonas, 567

Aggregatibacter actinomycetemcomitans, leukotoxin
    produced by, 452
    in outer membrane vesicles, 452

Agr, and pathogenesis of *Staphylococcus aureus*, 68–69
    types of, and Agr interference, 67

Agr locus, expression of, regulators and environmental
    stresses on, 67–68

Agr system, autoinducing peptide of, *Staphylococcus
    aureus* and, 62, 64
    autoinducing peptides of, 62, 64
    interference, types of agr and, 67
    of *Staphylococcus aureus*, 62–63
        effector molecule of, as RNAIII, 63–65
    two component system, regulatory RNA as main
        effector of, 63–65

Agrobacterium tumefaciens, 555
    as model type IV secretion system, 334
    and plasmid R388 type IV secretion system, 348
    and transformed plant cell, chemical signaling
        between, 337
    virulence mechanism of, QS system and, 556

Agrobacterium tumefaciens VirA, activation of, 556

Agrobacterium tumefaciens VirB/VirD4, as effector
    translocator, 336–339

AHQ family, quorum sensing signals and, 326

AlgB, in regulation of alginate production, 177–178

Alginate, 12
    in biofilms, 174
    biosynthesis of, fructose-6-phosphate for, 172–173
    produced by *Pseudomonas aeruginosa*, 171
    production of, AlgR and AlgZ in regulation of, 175–177
        DNA-binding proteins regulating, 178–180
        histidine kinases in, 178
        KinB and AlgB in regulation of, 177–178
        posttranscriptional regulators of, 180–183
        posttranslational regulatory system for, 181–182
        regulation of, 175–178
    regulation of, c-di-GMP in, 183

Alginate biosynthetic genes, transcriptional
    regulation of, 176

Alginate expression, regulators of, and Vfr, links
    between, 28–29

AlgP, in regulation of alginate production, 179

AlgR, in regulation of alginate production, 175–177

AlgZ, in regulation of alginate production, 175–177

Amino acids, type III secretion systems and, 325

Aminoglycoside antibiotics, cell surface disruption by, 446
    outer membrane vesicle formation and, 456

AmrZ, in regulation of alginate production, 179–180

Animals, domestic, *Clostridium perfringens* as pathogen
    in, 282

Anthrax disease, *Bacillus anthracis* and, 262–263
    and iron, 307–313
    iron acquisition during, 307–310

Anthrax toxin, produced by *Bacillus anthracis*, 267–268

Antibiotic resistance, influence of core structural
    modifications on, 221–222

Antibiotics, outer membrane vesicle formation and, 456
    Pseudomonas aeruginosa and, 14–15
    Staphylococcus aureus resistance to, 58
    to treat urinary tract infection, 149

Antimicrobial peptide resistance, and lipopolysaccharide
    modifications, regulation of, 209–238

Antioxidant defenses, regulatory networks
    coordinating, 428–431

Antivirulence compounds, to treat urinary tract
    infections, 149

AphA, as regulatory protein, in *Vibrio cholerae*, 245–246

AphB, and acid survival, in *Helicobacter pylori*, 247
    as regulatory protein, in *Vibrio cholerae*, 246–247

apo-Fur regulation, 114

ApsS histidine sensor kinase, 554

AtxA, as regulator of *Bacillus anthracis*, 265–266

Autoinducers, 319

Autoinducing peptides, of Agr system, 62, 64

Bacilli, iron-dependent virulence regulation in, 90–91

Bacillibactin, 309

Bacillibactin biosynthetic machinery, genes encoding, 311

Bacillus, RNA regulators in, 511–512